# CREATING QUALITY

## CONCEPTS, SYSTEMS, STRATEGIES, AND TOOLS

# McGraw-Hill Series in Industrial Engineering and Management Science

CONSULTING EDITORS

Kenneth E. Case, *Department of Industrial Engineering and Management, Oklahoma State University*
Philip M. Wolfe, *Department of Industrial and Management Systems Engineering, Arizona State University*

**Barnes**: Statistical Analysis for Engineers and Scientists: A Computer-Based Approach
**Bedworth, Henderson, and Wolfe**: Computer-Integrated Design and Manufacturing
**Black**: The Design of the Factory with a Future
**Blank**: Statistical Procedures for Engineering, Management, and Science
**Denton**: Safety Management: Improving Performance
**Dervitsiotis**: Operations Management
**Hicks**: Industrial Engineering and Management: A New Perspective
**Huchingson**: New Horizons for Human Factors in Design
**Juran and Gryna**: Quality Planning and Analysis: From Product Development through Use
**Khoshnevis**: Discrete Systems Simulation
**Kolarik**: Creating Quality: Concepts, Systems, Strategies, and Tools
**Law and Kelton**: Simulation Modeling and Analysis
**Lehrer**: White-Collar Productivity
**Moen, Nolan, and Provost**: Improving Quality through Planned Experimentation
**Niebel, Draper, and Wysk**: Modern Manufacturing Process Engineering
**Polk**: Methods Analysis and Work Measurement
**Riggs and West**: Engineering Economics
**Riggs and West**: Essentials of Engineering Economics
**Taguchi, Elsayed, and Hsiang**: Quality Engineering in Production Systems
**Wu and Coppins**: Linear Programming and Extensions

# CREATING QUALITY

## CONCEPTS, SYSTEMS, STRATEGIES, AND TOOLS

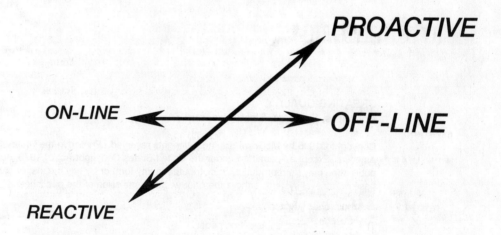

**William J. Kolarik**
Texas Tech University

**McGRAW-HILL, INC.**
New York  St. Louis  San Francisco  Auckland  Bogotá  Caracas  Lisbon
London  Madrid  Mexico City  Milan  Montreal  New Delhi
San Juan  Singapore  Sydney  Tokyo  Toronto

This book was set in Times Roman by Books International.
The editors were Eric M. Munson and James W. Bradley;
the production supervisor was Louise Karam.
The cover was designed by Initial Graphic Services, Inc.
R. R. Donnelley & Sons Company was printer and binder.

CREATING QUALITY
Concepts, Systems, Strategies, and Tools

This book is printed on acid-free paper.


34567890 DOC DOC 909

ISBN 0-07-035217-8


**Library of Congress Cataloging-in-Publication Data**

Kolarik, William J.
   Creating quality: concepts, systems, strategies, and tools / William J. Kolarik.
      p.      cm.
   Includes bibliographical references and index.
   ISBN 0-07-035217-8
   1. Quality control.   2. System analysis.   I. Title.
TS156.K5557      1995
658.2'62—dc20                                              94-31883

# ABOUT
# THE AUTHOR

**WILLIAM J. KOLARIK** is Professor of Industrial Engineering at Texas Tech University and a registered professional engineer. He received his Ph.D. degree with a major in industrial engineering and management from Oklahoma State University. In addition to over 10 years experience in higher education, he has a wide variety of experience in industrial and agricultural production-related endeavors. He has worked through a number of positions ranging from entry level positions to positions of leadership and management, in industry, agriculture, and education.

Professor Kolarik has directed and participated in a number of research projects dealing with both theoretical and applied topics in the quality areas. He has published over 50 technical papers related to his research. Based on his experiences in research, professional practice, and education, he has designed and developed an education-based, project-directed laboratory to supplement student learning and practice in the quality related areas.

The perspective gained from Professor Kolarik's wide variety of experiences and observations in both for-profit and not-for-profit organizations is captured and presented in this comprehensive text on quality.

# CONTENTS

# PREFACE

It has been said that there have been two fundamental shifts in the technical and business worlds in the past few years. The first major shift has been in the way we view and approach the issue of quality relative to customers, products, and processes; the second is the manner in which we use computers.

## Objective

The objective of this book is to address a need for a comprehensive text, focused on problem prevention driven quality engineering and control, from a systems point of view. This writing project was sparked by and grew as a result of my frustration: frustration resulting from not being able to offer my students a well-rounded systems view of the creation of quality in both products and processes.

People (customers), products, and processes are discussed at three levels: the conceptual, the strategic, and the analytical (tools). Quality is viewed as customer satisfaction as a result of both short- and long-term performance in the customer's hands. The text is designed to challenge our readers to think creatively, at a systems level, regarding the issue of product and process quality. We have carefully designed and structured this textbook so that it will be suitable for a wide spectrum of readers, ranging from college students to practicing professionals in the field.

In this textbook, we address the subject of quality from two broad perspectives: (1) the experience of quality and (2) the creation of quality. Our understanding of the experience of quality will ultimately allow us to use our creative powers more effectively and more efficiently, thus providing us with an edge over our competitors. If we have a reasonably good understanding (relative to our competition) of the "experience of quality" our customers need and expect, then we can structure our processes aimed at the "creation of quality" accordingly, and reduce our risk of engineering and business failure.

## Background

The majority of quality assurance concepts and strategies that are taught and practiced today are reactive, aimed at detecting and correcting problems which already exist. Proactive quality assurance is directed at problem prevention. A proactive strategy requires an emphasis on physical cause-effect knowledge, risk analysis, experience, and judgment to justify action. A reactive strategy emphasizes traditional loss accounting and data intensive statistical analyses in order to justify action. A re-active strategy is more comforting for a decision maker, since action can be readily justified, based on past or historic physical observations. A proactive strategy involves a higher level of speculation and risk taking, many times based on cause-effect knowledge rather than historic fact. However, it can lead to accelerated product and process development cycles and avoidance of losses, rather than the management of losses.

## Lesson Plan Overview

Each one of the eight text sections in this book has been carefully designed so that our readers can selectively focus on any one of the eight general topics, or piece the topics together in a manner that is consistent with their instructional goals and ob-jectives. The material is prepared for a target audience of junior or senior level engineering students. However, most of the material is suitable for technology stu-dents at the junior or senior level as well as technical business students. It is also suitable for master's level students in a classroom setting, provided the instructor assigns appropriate outside reading and project work.

The qualitative nature of Sections One, Two, Three, Four, and Eight requires broad thinking skills. The quantitative nature of Sections Five, Six, and Seven requires a reasonable understanding of probability and statistics. For work in the quantitative sections, readers may want to keep a book of basic probability and statistics at their side.

Enough material is contained in this textbook to develop two 3-hour classes on the subject of quality. Hence, in structuring one 3-hour course from this material, instructors must pick and choose their sections according to their specific course purpose. Time estimate ranges for the eight text sections contained in the book are provided below, strictly as a guide. Obviously, the more limited the time on task, the more limited the detail that can be captured from each chapter. With the credit hour compression we are experiencing in contemporary college curriculums, we have made every attempt to compress our sections as much as possible and still pre-sent sufficient information to satisfy both academic demands on college campuses and professional practice demands in the field.

| Section | Lecture hours |
|---|---|
| Section One: Quality—Concept, Philosophy, and Systems | 3–9 |
| Section Two: Experience of Quality | 3–9 |
| Section Three: Creation of Quality—Fundamental Strategic and Tactical Quality Tools | 6–12 |
| Section Four: Creation of Quality—Definition and Design | 6–12 |
| Section Five: Creation of Quality—Statistical Process Control | 9–15 |
| Section Six: Creation of Quality—Designed Experiments | 12–24 |
| Section Seven: Creation of Quality—Reliability Models | 12–21 |
| Section Eight: Creation of Quality—Quality Transformation | 6–15 |

In any treatment of the subject of quality, we recommend that Sections One and Two be used. From the professional practice point of view, Section Three is considered indispensable. Sections Four, Five, Six, and Seven provide instructors with a wide variety of options for presentation to technical students. Section Eight is indispensable from a quality leadership and management point of view.

From the experience of many years of delivering quality assurance related course materials, we have learned that we must provide our students with opportunities for creative, as well as critical thinking. The Review and Discovery Exercises have been designed to encourage critical thinking as well as creative thinking and "doing" through project work. We suggest that any course in the quality area be structured so that about 50 to 75 percent of the evaluation comes from classroom endeavors and the balance comes from creative project assignments. Our discovery level exercises are provided to encourage appropriate and challenging project exercises.

## Acknowledgments

I am deeply indebted to a number of people who, in essence, have made this writing project what it is today. First, and foremost, I am indebted to the leaders of both for-profit and not-for-profit organizations that I have been privileged to work with and observe. I have carefully observed both direct successes and direct failures, and the persistent efforts that have wrought success out of failure. This text has literally grown as the product of observation (the players, the circumstances, the substance, and the results), quality-related literature, and direction taken from my many advisors and reviewers.

I am indebted beyond measure to my students both on-campus and off-campus, undergraduate and graduate. My students have been both my biggest champions as well as my loudest critics in producing, testing, and improving this document. In addition to my traditional on-campus students, three corporate groups of summer on-campus students contributed greatly to this textbook: my AT&T students, my Texas Instruments students, and my E-Systems students. In many instances, my students have contributed ideas, cases, and problems. These contributions are appreciated and recognized with bracketed initials, within the contribution, e.g., [WK] and so on.

I would like to recognize a select group of contributors who have persevered with me throughout this project and added a number of perspectives to the resulting work as well as helped me in the preparation of the work: Kelly Beierschmitt, Babu Chinnam, Robert Fox, Yvonne Kolarik, Jack Pan, Sanjukta Patro, Maheswaran Rajasekharan, and Beverly Wiley. I would also like to recognize my reviewers who have guided me through many revisions and towards the final product you have in your hands: Suraj Alexander, University of Louisville; John English, University of Arkansas; Kailash Kapur, University of Washington; Way Kuo, Texas A&M University; Joseph Pignatiello, Texas A&M University; Steve Vardemann, Iowa State University; and G. S. Wasserman, Wayne State University.

Finally, I would like to express my deepest appreciation to Ken Case and Eric Munson for allowing me the opportunity to develop and publish this material as well as for their support over the course of this project.

*William J. Kolarik*

# CREATING QUALITY

## CONCEPTS, SYSTEMS, STRATEGIES, AND TOOLS

# QUALITY—CONCEPT, PHILOSOPHY, AND SYSTEMS

## VIRTUES FOR LEADERSHIP IN QUALITY
### Vision and Enthusiasm

The purpose of Section One is to introduce our readers to the concept of quality, quality philosophies, and quality systems. In Chapter 1 we present a number of classical definitions pertaining to the concept of quality and then introduce the quality concepts that we will discuss throughout our textbook. We approach quality in this textbook as a two-part concept, centered on customer satisfaction: We distinguish between the experience of quality and the creation of quality. We introduce proactive and reactive strategies and approaches to quality in Chapter 1.

In Chapter 2 we discuss a number of major contributions made by the leaders in the quality movement over the years regarding quality philosophies. These contributions yield a broad perspective in terms of the types of quality philosophies that have molded the quality movement into what it is today. These philosophies are multifaceted and address people (customers), product, and process issues.

In Chapter 3, we introduce five quality paradigms which have emerged over the years as well as the concept of a quality system. We stress the fact that every organization has a quality system, some more effective than others. We develop a set of quality system interfaces and discuss the nature of each interface. We then briefly discuss the Malcolm Baldrige National Quality Award and the ISO 9000 Quality Standards as they relate to quality systems. Integration of the quality system within an organization is stressed in the context of organizational success.

# 1

# CONCEPT

## 1.0 INQUIRY

1. What is quality?
2. Who determines quality?
3. Who creates quality?
4. How are customers satisfied?
5. How do proactive and reactive quality strategies differ?
6. What relationship exists between learning and change?

## 1.1 INTRODUCTION

A survey of over 3000 consumers in the United States, Germany, and Japan was recently commissioned by the American Society for Quality Control (ASQC) and conducted by the Gallup Organization [1]. A summary of tabulated results is shown in Table 1.1. From these results, we see a great deal of commonality in customer thinking regarding quality. We see performance (function, form, and fit), as well as price and reputation, as a driving force.

Market share is a critical factor in business success. Market share (in specialty as well as mass markets) is highly correlated with customer satisfaction and, in turn, business success. In formulating manufacturing strategies, Hill cites quality as the single most important factor in determining market share [2]. In short, *a fundamental understanding of quality is essential to compete effectively in today's international markets*.

Our approach in this textbook is to view quality from a scientific perspective as far as possible. However, we believe that quality assurance is both an art and a science. The quality concepts, systems, strategies, and tools presented and discussed are useful in manufacturing, agricultural production, and service organizations. Throughout this text we stress a systematic approach to quality-related opportunities and problems. We take a unique approach to quality definition and study. In essence, we are proposing a science of quality in the belief that *the study of quality must be aimed at understanding, meeting, and surpassing customer needs and expectations*.

## 1.2 CONCEPT OF QUALITY

The concept of quality with respect to customer satisfaction has been with us since the beginning. However, the formal study of quality is relatively new, dating back to the early part of the twentieth century. Over the years, we have seen a variety of definitions for quality put forth by both practitioners and scholars. These definitions have ranged in length from a few words to comprehensive discussions. *Definitions of*

3

**TABLE 1.1**    CONSUMERS DEFINE QUALITY: SUMMARIZED
RESULTS FROM THE ASQC-GALLUP QUALITY SURVEY

| United States | |
| --- | --- |
| Quality determined by | Buying decision influenced by |
| 1. Well-known name<br>2. Word of mouth<br>3. Past experience<br>4. Performance<br>5. Durability<br>6. Workmanship<br>7. Price<br>8. Manufacturer's reputation | 1. Price<br>2. Quality<br>3. Performance<br>4. Word of mouth<br>5. Well-known name |
| **West Germany** | |
| 1. Price<br>2. Well-known name<br>3. Appearance<br>4. Durability<br>5. Past experience<br>6. Quality itself | 1. Price<br>2. Quality Itself<br>3. Appearance<br>4. Durability<br>5. Well-known name<br>6. Design and style<br>7. Performance |
| **Japan** | |
| 1. Well-known name<br>2. Performance<br>3. Easy to use<br>4. Durability<br>5. Price | 1. Performance<br>2. Price<br>3. Easy to use<br>4. Design and style<br>5. Well-known name |

(Responses mentioned by at least 10% of respondents to open-ended questions.)

Reproduced, with permission, from "ASQC/Gallup Survey," *ON Q— Official Newsletter of the American Society for Quality Control*, vol. VI, no. 9, p. 2, November 1991.

*quality have ranged in scope from narrow definitions such as "meeting engineering specifications on the shop floor" to broad societal-oriented definitions.*

## Classical Definitions of Quality

We have seen a wide variety of definitions proposed for "quality":

Quality is a physical or nonphysical characteristic that constitutes the basic nature of a thing or is one of its distinguishing features, *Webster's New World Dictionary*.

Quality, as applied to the products turned out by industry, means the characteristic or group or combination of characteristics which distinguishes one article from another, or the goods of one manufacturer from those of his competitors, or one grade for product from a certain factory from another grade turned out by the same factory, *Radford* [3].

There are two common aspects of quality, one of these has to do with the consideration of the quality of a thing as an objective reality independent of the existence of man. The other has to do with what we think, feel, or sense as a result of the objective reality—this subjective side of quality is closely linked to value, *Shewhart* [4].

Quality is fitness for use, *Juran* [5].

It is convenient to think of all matters related to quality of manufactured product in terms of these three functions of specification, production, and inspection, *Grant and Leavenworth* [6].

Quality is conformance to requirements (clearly stated), *Crosby* [7].

Quality should be aimed at the needs of the consumer, present and future, *Deming* [8].

Quality is the total composite product and service characteristics of marketing, engineering, manufacture, and maintenance through which the product and service in use will meet the expectations of the customer, *Feigenbaum* [9].

Product quality encompasses those characteristics which the product must possess if it is to be used in the intended manner, *Mizuno* [10].

Quality is the loss (from function variation and harmful effects) a product causes to society after being shipped, other than any losses caused by its intrinsic functions, *Taguchi* [11].

Quality is the totality of features and characteristics of a product or service that bear on its ability to satisfy stated or implied needs, *ISO 9000* [12].

## Classical Definitions of Quality Assurance and Quality Control

"Quality assurance" and "quality control" address the means and techniques of producing quality products. Japanese, U.S., and other international definitions follow:

Quality assurance means to assure quality in a product so that a customer can buy it with confidence and use it for a large period of time with confidence and satisfaction (satisfy the requirements of consumers), *Ishikawa* [13].

Quality assurance: All those planned and systematic actions necessary to provide adequate confidence that a product or service will satisfy given requirements for quality, *ISO 9000* [14].

Quality control is a system of means whereby the qualities of products or services are produced economically to meet the requirements of the purchaser, *Japan Industrial Standards* [*JISZ 8101*].

Quality control is the operational techniques and the activities which sustain a quality of product or service that will satisfy given needs; also the use of such techniques and activities, *American National Standards Institute* [*ANSI ZI.7 1971*].

Total quality control (TQC) is an effective system for integrating the quality-development, quality-maintenance, and quality-improvement efforts of various groups in an organization so as to enable marketing, engineering, production, and service at the most economical levels which allow for full customer satisfaction, *Feigenbaum* [15].

Total quality management (TQM) is both a philosophy and a set of guiding principles that represent the foundation of a continuously improving organization. TQM is the application of quantitative methods and human resources to improve the materials and services supplied to an organization, all the processes within an organization, and the degree to which the needs of customers are met, now and in the future. TQM integrates fundamental management techniques, existing improvement efforts, and technical tools under a disciplined approach focused on continuous improvement, *U.S. Department of Defense* [16].

## Customers, Products, and Processes

*A "customer" is anyone who receives or is affected by the product or process* (Juran [17]). *There are two major classes of customers: (1) external and (2) internal.* We will use "product" in a generic sense to refer to both goods and services in the context of both for-profit and not-for-profit organizations. Along the same lines, the term "process" is used with many different adjectives to refer to a set of activities that are necessary in order to meet and exceed customer needs and expectations.

External customers are people who are affected by the product, for example, the purchaser and ultimate user(s) of a good or service. In a broad sense, external customers also include the general public or society. This societal view is widely held in the Japanese quality philosophy and expressed as "respect for humanity" by Ishikawa [18] and "loss to society" by Taguchi [19].

The term "internal customer" refers to people within the "production organization"; they are the people "up- or down-the-line," in other departments, and stakeholders (owners, investors, and contributors). The internal customer relationship is typically an "upstream-downstream" relationship within a production organization (e.g., between marketing, sales, engineering, production, accounting). We can also become our own customer in a subsequent production step.

The distinction between external and internal customers becomes fuzzy in many cases. For example, people needed in the marketing chain to move the good or service through marketing channels (distributors, wholesalers, retailers, and so on) play the roles of both external and internal customers. The same is true in a supplier-purchaser context where the supplies become a part of the purchaser's product (vendors are an example). Regardless of classification, we desire to satisfy both external and internal customers.

## Science of Quality

The classical definitions of quality are insufficient to approach quality as a science. For our purposes, *we define two fundamental elements of the science of quality*: **(1) *the experience of quality and (2) the creation of quality***.

**Experience of Quality**  The experience of quality results from a product for an external customer or a production process for an internal customer. *The experience of quality results in customer satisfaction or dissatisfaction* and is developed through the customer benefits created by and burdens resulting from our product-process. In other words, *the experience of quality is a function of the fulfillment of human needs and expectations*. The quality experience for external customers is developed in four fundamental dimensions: (1) product performance regarding function, form, and fit, (2) product cost (e.g., initial, operating, maintenance, and disposal costs), (3) product and service timeliness (delivery, product service and repair time, and so on), and (4) customer service (e.g., attention to and respect for customers).

Customers "buy" benefits to make their lives (and the lives of their associates) more productive and pleasant. Benefits must result from products in order to establish customer satisfaction. Customers must believe they can obtain benefits in order to make a purchase. And, furthermore, they must experience significant benefits to become satisfied customers. For example, an automobile may result in transportation, prestige, and so on, but the product by itself is a mass of metal, plastic, and glass. A computer-software system is a mass of plastic, silicon, metals, and so on, which can result in accounting, design, drafting, calculation, and other aids to make our life or business or both more productive. The point is that external customers essentially buy on the basis of benefits, not strictly on the basis of the product.

The experience of quality for internal customers is broadly centered in the production processes. The basis for internal customer satisfaction is the same as for external customers—customer needs and expectations; however, the dimensions are more abstract. These dimensions include job challenge, workplace environment (in both physical and social dimensions) and reward and recognition.

Customers seek both tangible and intangible benefits that will outweigh their "burdens." If we define a benefits-to-burdens ratio, we can say that customers seek a ratio greater than 1.0. Ratios much greater than 1.0 are required to produce enthusiastic customers. We can refer to this ratio as a customer satisfaction index, which can be impacted by changes in either benefits, burdens, or both. We must remember that benefits are measured individually by each customer in different ways, both monetary and otherwise, as are burdens.

All in all, the experience of quality is a complicated phenomenon which is the true determinant of quality. If we fail to recognize the experience of quality as the driving force for individual, organizational, and societal success, we will be unable to focus our efforts to create quality. *If we fail to understand the nature of the experience of quality, we will be unable to systematically create quality.* In other words, we will be forced to rely on trial and error and blind luck in creating quality.

**Creation of Quality**   *We create quality through processes that we develop and maintain.* The creation of quality is accomplished through eight fundamental processes: **(1)** *definition,* **(2)** *design,* **(3)** *development,* **(4)** *production,* **(5)** *delivery,* **(6)** *sales and customer service,* **(7)** *use,* and **(8)** *disposal, which includes recycle.* Each of these general processes has distinctive quality characteristics.

These eight fundamental processes form a sequence of activities that we must approach in a systematic fashion if we are to provide our external and internal customers with a positive quality experience. Figure 1.1 depicts the creation and experience of quality as an interactive sequence or system. This sequence has a profound influence on the internal customer in the early processes. The impact on the external customer develops in the later processes. We must pay attention to each process in the sequence because failure in an earlier process will ultimately impact success in subsequent processes.

## 1.3 RESPONSE TO QUALITY DEMANDS

In this text we address both the experience of quality and the creation of quality from an engineering and business perspective. Our understanding of the quality

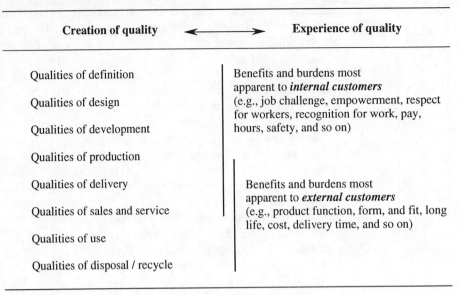

**FIGURE 1.1**   The experience and the creation of quality.

experience will ultimately allow us to use our creative powers more effectively and more efficiently, thus providing us with an edge over our competitors. *If we have a reasonably good understanding (relative to our competition) of the experience of quality our customers need and expect, then we can structure our processes (aimed at the creation of quality) accordingly and reduce our risk of business failure.*

## Proactivity and Reactivity

The majority of quality assurance concepts and strategies that are practiced today are reactive, aimed at detecting and correcting problems which already exist. *Proactive quality assurance is directed at problem prevention*, according to Kolarik and Pan [20]. A proactive strategy requires an emphasis on physical cause-effect knowledge, risk analysis, experience, and judgment to justify action. A reactive strategy emphasizes traditional loss accounting and data intensive statistical inferences to justify action and is more comforting for a decision maker, since action can be readily justified, based on historic physical observations. The proactive strategy involves a higher level of speculation and risk taking and is based on knowledge of cause and effect rather than history. However, it can lead to (1) accelerated product and process development cycles and (2) the avoidance, rather than the management, of losses. Both of these advantages contribute toward a competitive edge in today's fast-paced production environment.

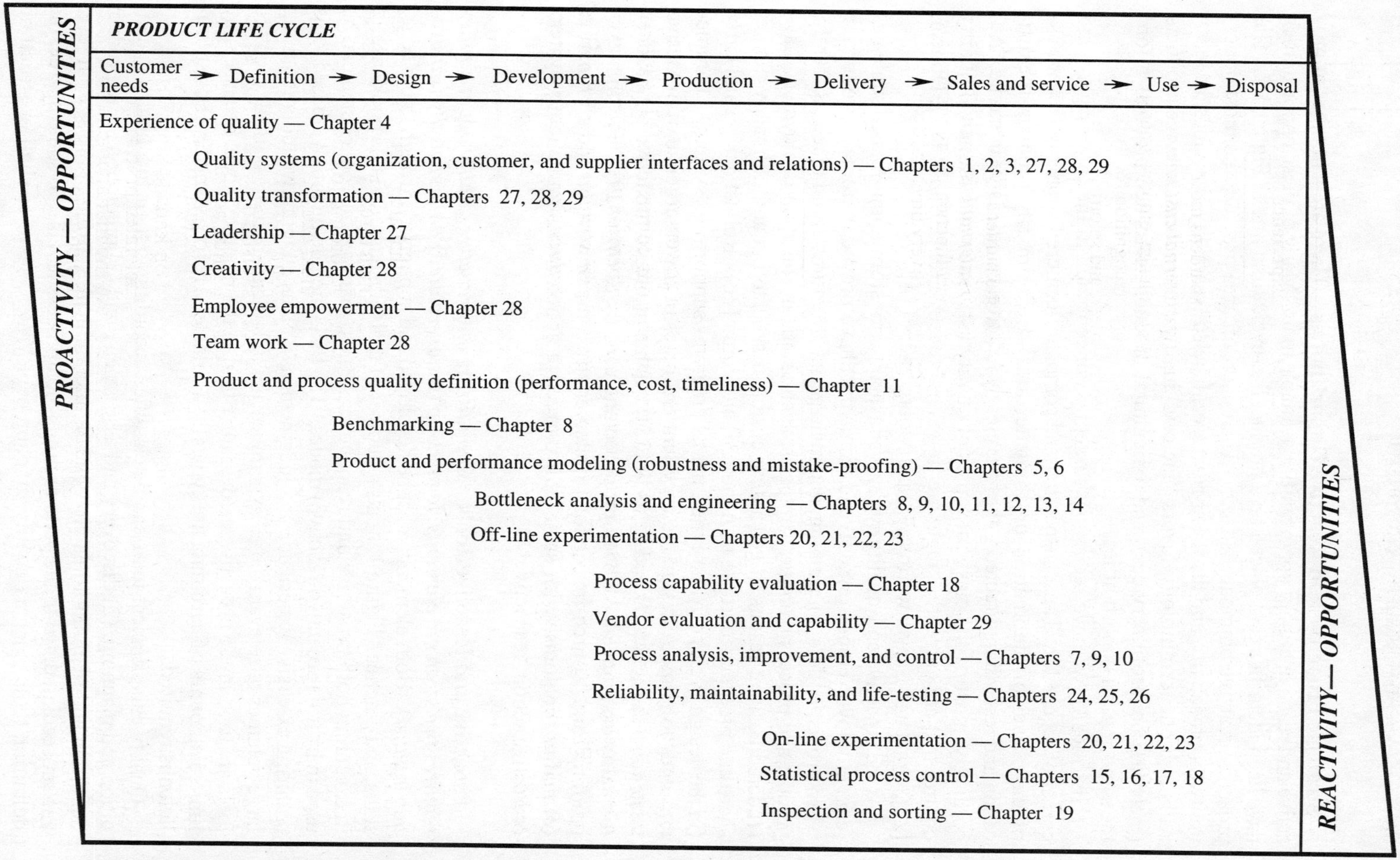
PRODUCT LIFE CYCLE
Customer needs → Definition → Design → Development → Production → Delivery → Sales and service → Use → Disposal
Experience of quality — Chapter 4
Quality systems (organization, customer, and supplier interfaces and relations) — Chapters 1, 2, 3, 27, 28, 29
Quality transformation — Chapters 27, 28, 29
Leadership — Chapter 27
Creativity — Chapter 28
Employee empowerment — Chapter 28
Team work — Chapter 28
Product and process quality definition (performance, cost, timeliness) — Chapter 11
Benchmarking — Chapter 8
Product and performance modeling (robustness and mistake-proofing) — Chapters 5, 6
Bottleneck analysis and engineering — Chapters 8, 9, 10, 11, 12, 13, 14
Off-line experimentation — Chapters 20, 21, 22, 23
Process capability evaluation — Chapter 18
Vendor evaluation and capability — Chapter 29
Process analysis, improvement, and control — Chapters 7, 9, 10
Reliability, maintainability, and life-testing — Chapters 24, 25, 26
On-line experimentation — Chapters 20, 21, 22, 23
Statistical process control — Chapters 15, 16, 17, 18
Inspection and sorting — Chapter 19
PROACTIVITY — OPPORTUNITIES
REACTIVITY— OPPORTUNITIES

**FIGURE 1.2** Proactive and reactive opportunities in the creation of quality.

**Strategic Proactive and Reactive Opportunities** There are a number of proactive and reactive quality opportunities available to us. Many of these opportunities have been identified and placed on a product life-cycle time line in Figure 1.2. The proactive-reactive boundary is fuzzy. The upper-left area provides the most opportunity for proactivity, and the lower-right area, labeled "reactivity," provides the least. Although our terminology favors "factory" hardware products, service products (including nonprofit services), and agricultural production–related products afford the same general opportunities.

Historically, we have seen most strategies concentrated in the lower right-hand corner of Figure 1.2. These inspection-and complaint-centered strategies (regardless of their level of automation or computer aids) are primarily problem-identification and resolution directed—they are clearly reactive, or defensive in nature. Accordingly, most of the classical topics of quality assurance and statistical quality control that are currently taught are in general reactive (e.g., Banks [21], Duncan [22], Grant and Leavenworth [23], and Montgomery [24]).

At the proactive end, quality systems and quality transformation–related strategies (e.g., employee empowerment, and teamwork) have proved extremely productive when proper leadership, training, and managerial structures are in place. Recently, product and process planning strategies, such as quality function deployment, have rendered positive results in generating customer satisfaction and ultimately market share and profits, according to Akao [25] and Juran [26]. Robust design and early (in the design cycle) off-line experimental programs supporting "system," "parameter," and "tolerance" design are excellent examples of the emergence of proactive strategies (Taguchi [27] and Kolarik [28]). Other strategic proactive approaches to quality assurance have been proposed and used by Shingo [29]. His "mistake-proofing" and "source-inspection" strategies stress the prevention and elimination of quality problems, with same-unit corrective action and simultaneous system corrective action, respectively.

**Proactive and Reactive Tools** *Capitalizing on proactive quality opportunities requires two strategic elements: leadership (vision-related) and creativity (innovation-related).* Both elements must be developed and supported through knowledge, skills, and practice. Figure 1.3 lists a number of tools (discussed in this book) useful in executing both proactive and reactive strategies; many of the tools listed can be used in hybrid (proactive-reactive) modes. The tools include qualitative-based tools requiring extensive experience and judgment, as well as quantitative tools requiring large volumes of relevant data. Experience and judgment are more critical when the tools are used in the proactive mode, rather than in the reactive mode, because only limited amounts of hard data are typically available. Hence, some level of speculation is required.

Quality enhancement tools can be classified according to their functions: (1) definition and planning, (2) discovery and assessment, and (3) prediction. The product definition and planning tools are typically hierarchical, starting at the system level and proceeding down to the parts or production process levels. To some degree, the definition tools, such as quality function deployment, serve as shell or structural

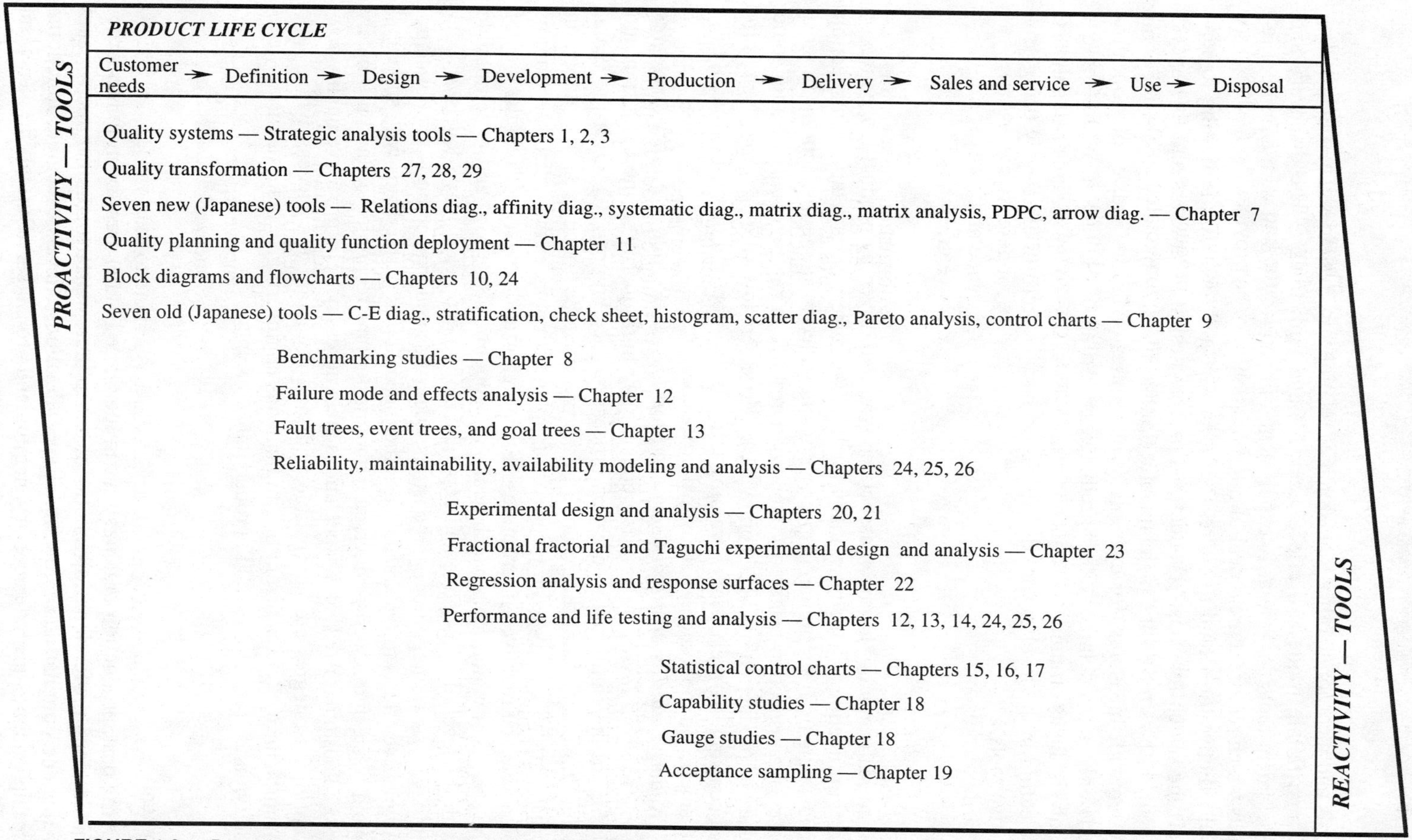

**PRODUCT LIFE CYCLE**

| Customer needs | → | Definition | → | Design | → | Development | → | Production | → | Delivery | → | Sales and service | → | Use | → | Disposal |

Quality systems — Strategic analysis tools — Chapters 1, 2, 3

Quality transformation — Chapters 27, 28, 29

Seven new (Japanese) tools — Relations diag., affinity diag., systematic diag., matrix diag., matrix analysis, PDPC, arrow diag. — Chapter 7

Quality planning and quality function deployment — Chapter 11

Block diagrams and flowcharts — Chapters 10, 24

Seven old (Japanese) tools — C-E diag., stratification, check sheet, histogram, scatter diag., Pareto analysis, control charts — Chapter 9

Benchmarking studies — Chapter 8

Failure mode and effects analysis — Chapter 12

Fault trees, event trees, and goal trees — Chapter 13

Reliability, maintainability, availability modeling and analysis — Chapters 24, 25, 26

Experimental design and analysis — Chapters 20, 21

Fractional fractorial and Taguchi experimental design and analysis — Chapter 23

Regression analysis and response surfaces — Chapter 22

Performance and life testing and analysis — Chapters 12, 13, 14, 24, 25, 26

Statistical control charts — Chapters 15, 16, 17

Capability studies — Chapter 18

Gauge studies — Chapter 18

Acceptance sampling — Chapter 19

PROACTIVITY — TOOLS

**FIGURE 1.3** Proactive and reactive tools for the creation of quality.

11

tools and incorporate other tools (such as relationship, systematic, and matrix diagrams; failure mode and effects analysis; and fault trees). The discovery and assessment tools are primarily experimental in nature and deal with empirical results which require a great deal of interpretation.

Predictive tools tend to address "if then-what if" lines of questioning to support the search for "best" product and process design parameters and tolerances. Some, such as reliability, maintainability, and availability modeling tools, focus on systemwide measures, for example, probabilities of failure and restoration, provide only gross details. Other predictive modeling tools, such as finite element analysis, focus on minute details and require extensive computer aids and support. Selection in the level of tool detail is a primary concern in effectively using proactive tools.

## Practice

Historically, quality has been associated with and driven by product inspection or sorting functions. Currently, many quality assurance efforts retain an inspection focus, ranging from human inspection to automated machine inspection. In most cases, the inspection-centered strategies are reactive in nature (occurring after some phase of production has been completed). Movement toward more proactive efforts is occurring. A great deal of training effort is now focused on on-line quality assessment and tools such as statistical process control.

*Reactive quality strategies serve to limit losses. Proactive quality strategies serve to avoid losses* by assuring consistently high quality in the definition and design of the eight quality-creation processes. (See Figure 1.1.) Over the years, some movement from the reactive to the proactive has occurred. Inevitably, an organization's quality assurance "road map" must consider the entire spectrum of proactive and reactive opportunities. Each organization must focus on the area of the spectrum which makes technical and economic sense to it and is compatible with the level of knowledge and experience of its people. Today's levels of technology, system integration, and customer demands and expectations are not static. Aggressive efforts to understand customer needs, build partnerships with suppliers, avoid waste and rework, understand and master new technologies, and ultimately exceed customer needs and expectations are a necessity. The proactive quality concept and strategies are compatible with and necessary for success with "agile manufacturing," "concurrent or simultaneous engineering," "manufacturing flexibility," and "just-in-time (JIT) production" concepts.

## Professional Isolationism

In the past, production organizations in the United States have been tightly organized along professional or functional disciplines, with the result in many cases being the domination of functional specialization. To some degree, functional specialization is indeed necessary, helpful, and productive. But, when taken to an extreme, functional divisions create "walls" between marketing, design, manufacturing, and other functional groups within the same organization. In practice, these divisions have been labeled by descriptive terms such as "chimneys" and "silos."

*The term "over-the-wall" development refers to product development efforts conducted in a functionally overspecialized environment* (with only minimal inter-functional communications). Figure 1.4 depicts such an extreme arrangement: The product is conceived, designed, manufactured, and sold with little regard for either external or internal customers. Chronic problems with this type of organization have been discussed by well-known consultants, including Crosby [30], Deming [31], Ishikawa [32], Juran [33], and Peters [34].

*The sequential processes involved in the creation of quality must all interact with each other in order for an effective result (quality experience for customers) to be accomplished.* There are many professions and crafts involved in this creative chain. This wide variety of knowledge, experience, and skills must be applied in proactive ways in order to gain a competitive edge in quality. Walls hamper the flow between processes. When this flow is choked, we typically see departmental "empires" competing within an organization, resulting in internal conflicts characterized by reactive countermeasures and "finger pointing" when problems arise. These practices are both inefficient and counterproductive. For example, the classical adversarial relationships between labor and management, engineering and production, supplier and purchaser, and so on, hold back growth in quality improvement.

## Software Project—Over-the-Wall Case [RK]

I was involved as a test engineer at an on-site customer location. My task was to conduct system testing on a large software project that had recently been delivered to my site. Part of the requirements of the contract specified that the contractor supply documentation with the system. The documentation was required to meet certain criteria with respect to formats, but the content specifications were vague. Three organizational entities were involved with the software documentation. The software group wrote specifications describing the software modules. The operations group wrote manuals that described the operation of the software. The test group used both sets of documents during the site integration and testing phase of the program.

The documents developed by the software group and handed off to the testing group described the software design and software algorithms. They were generally not suitable as a user's manual for systems analysts. When a failure occurred, the manuals did not contain information that helped the test engineer isolate the failure. For example, often specific error codes were an indication of a problem in a certain set of modules, but the information linking the two was customarily kept in the developer's notebook only. The software developers wanted to keep their documentation to a minimum. Since they were required contractually to supply a large amount of documentation, they did not often volunteer additional information. I could usually obtain this information by calling them, but occasionally, the developers had left the program by that time, or were busy with other program work.

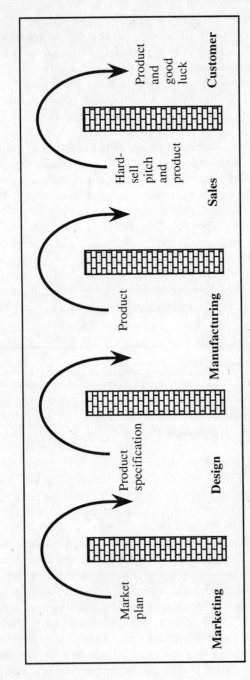

Marketing     Design     Manufacturing     Sales     Customer

Market plan → Product specification → Product → Hard-sell pitch and product → Product and good luck

**FIGURE 1.4**    Classic over-the-wall product development scheme.

The manuals that were developed by the operations group contained information on how to use the system. These manuals assumed that the system was working properly, which it was not (that is why I was there in the first place). The manuals were written for personnel who possessed no technical computer skills and therefore omitted useful technical details.

After several months of testing the system, I managed to collect a volume of notes on the system operation from a useful, technical standpoint. Many of the notes were copies of information from developers' notebooks (not included in the system documents) and notes that I had made when conversing with the developers while solving system problems.

The over-the-wall structure present at the developer's facility created a quality system problem. A more timely flow of information would have produced a functional software product faster and at a lower cost to the customer.

## 1.4 LEARNING AND CHANGE

Dr. Ishikawa maintains that "quality control begins with education and ends in education (study, practice, and participate)," and that "quality control is a discipline that combines knowledge with action." Quality improvement hinges on the acquisition of new knowledge, the mastering of new skills, and the leadership that encourages the activities that make quality improvement happen. Hence, understanding the fundamental nature of the learning process is critical to advancing quality improvement.

The critical point *in the learning process* is that *both knowledge and action are essential to produce positive results.* An incomplete learning process may lead to knowledge without action, action without knowledge, or nothing at all. A complete learning process leads to action supported through knowledge. A brief discussion regarding the components of this knowledge-action combination is provided in Appendix 1A.

## 1.5 THE CADILLAC CASE [36]

Cadillac Motor Car Company is the flagship division of General Motors' North American Automotive Operations. Founded in 1902 by Henry Martin Leland, Cadillac was built on a legacy of superior craftsmanship and unsurpassed quality, which led to its recognition as "the standard of the world." Cadillac's first official quality recognition came in 1908 when it was awarded the world-renowned Dewar Trophy, a prize sponsored annually by the Royal Automobile Club of England to encourage technical progress. Cadillac won for its demonstration of the complete interchangeability of parts. This was the first time an American company had received this prestigious honor. Cadillac won the Dewar Trophy again in 1915 for the first application of the electric self-starter.

Cadillac's quality leadership went unchallenged for decades. Then, the early 1980s ushered in an era of progressively stringent emissions standards and fuel economy requirements. Cadillac responded with new power train components and, ultimately, new exterior designs that did not completely meet customer expectations. By the mid-1980s, Cadillac's prestigious image was in jeopardy. Since 1985, a turnaround has occurred. Cadillac has demonstrated continuous improvement in both quality and customer satisfaction. Cadillac's transformation— and the people, systems, processes, and products responsible for the improvement— earned it the 1990 Malcolm Baldrige National Quality Award, the most prestigious quality award in America.

Three strategies are behind the transformation:

- A cultural change.
- A constant focus on the customer.
- A disciplined approach to planning.

These three strategies support one another and are totally integrated. Together, they reflect Cadillac's total quality process.

Four initiatives are primarily responsible for the greater teamwork and employee involvement at the heart of Cadillac's cultural change: simultaneous engineering, supplier partnerships, the United Auto Workers–General Motors (UAW-GM) Quality Network, and Cadillac's people strategy.

*Simultaneous Engineering*—Defined as "a process in which appropriate disciplines are committed to work interactively to conceive, approve, develop and implement product programs that meet predetermined objectives," simultaneous engineering is the key strategy of Cadillac's product development and improvement plans. The extensive teamwork required for simultaneous engineering demanded that walls between functions come down and cooperation increase. Cadillac's product design and development now begins with integrated knowledge of all key success factors, leading to better decisions and the ability to get to market more quickly with products that are on target.

*Supplier Partnerships*—A close partnership with suppliers is essential to simultaneous engineering. In 1985, Cadillac began redefining its supplier relationships by asking suppliers to take on additional product development responsibilities. This effort, along with the objective to reduce variation, led to a reduction in the supply base, allowing a closer, more focused relationship.

*UAW-GM Quality Network*—Since 1973, General Motors and the United Auto Workers have worked together to improve product and work-life quality. In 1987 corporate management and the UAW recognized that a consistent, joint quality improvement process was needed to increase competitiveness. As a result, the UAW-GM Quality Network was established, comprising joint union and management quality councils at the corporate, group, division, and plant levels. The network's purpose is to focus the entire organization on customer satisfaction as the master plan achieved through people, teamwork, and continuous improvement.

*People Strategy*—Cadillac's people strategy is designed to meet the needs of Cadillac people while achieving the goals of the business plan. It ensures a trusting, empowered workforce able to take responsibility for making things happen.  People strategy includes the development of selection processes to place or reallocate people in concert with the needs of the business, efforts to develop employees and involve them in decision making, and recognition and reward systems that support the behaviors necessary to achieve the business plan.

Cadillac's cultural change broke down the walls between functions and allowed it to focus on the customer—both internal and external—and achieve its master plan.

*Internal Customers*—Cadillac's design for manufacturability strategy provides simultaneous engineering teams with a disciplined approach for considering manufacturing requirements throughout the product development process.  This led to the establishment of the Assembly Line Effectiveness Center, with a simultaneous manufacturing environment used to evaluate the buildability of future models.

*External Customers*—Cadillac's commitment to customer satisfaction begins with gaining a thorough understanding of customer needs and expectations, translating that knowledge into world-class products and services, and then supporting the customer after the purchase by effectively managing the customer relationship.  The Cadillac market assurance process integrates the voice of the customer at every phase of product development.  Customers provide information on their needs and wants and how these are prioritized.  Cadillac uses extensive market research to collect information about its end customers.  Future products and features are tested extensively in vehicle clinics where potential customers rate them against current models from GM and the competition.  The simultaneous engineering teams use this information to maintain the focus on the customer and use the quality function deployment (QFD) tool (which translates the customer's needs and wants into engineering terms and product specifications) to make strategic, customer-driven decisions.

Cadillac's business planning process concentrates on continuously improving the quality of its products and has four objectives:

- To involve every employee in the running of the business.
- To continually reinforce Cadillac's mission and long-term strategic objectives throughout the organization.
- To align short-term business objectives with the long-term goals and action plans developed by every plant and functional staff.
- To institutionalize continuous improvement of products and services.

Each staff and plant develops a quality plan that aligns with the overall Cadillac business plan.

These initiatives have transformed Cadillac.  Since 1986, warranty-related costs have dropped nearly 30 percent.  Productivity has increased by 58 percent.  Lead time for a completely new model has been cut by 40 weeks.  And for the four years 1987–1991, Cadillac was the No. 1 domestic make on the J. D. Power and Associates' Customer Satisfaction Index and Sales Satisfaction Index.

## REVIEW AND DISCOVERY EXERCISES

### Review

**1.1** For the products listed below, give two external customers and two internal customers.
  **a** Aerosol spray paint.
  **b** Pizza.
  **c** Liquor.
  **d** Laundry detergent.
  **e** Automobile.
  **f** College education.

**1.2** Give an example of a situation where one person can switch almost instantaneously from being an internal customer to an external customer. From an external to an internal customer.

**1.3** Extract and list the main points expressed in the classical quality definitions.

**1.4** How is the creation of quality related to the experience of quality?

**1.5** Describe the experience of quality as it relates to a pair of shoes.

**1.6** Describe the creation of quality as it relates to a pair of shoes.

**1.7** How does proactive quality assurance differ from reactive quality assurance?

**1.8** Why is reactive quality assurance so prominent in organizations?

**1.9** How can over-the-wall product scheme development be avoided in an organization?

### Discovery

**1.10** Compare and contrast the Ishikawa and the Taguchi definitions of quality.

**1.11** Why must knowledge and action both be present in effective quality control?

**1.12** Juran's definition of a customer states that "A customer is anyone who receives or is affected by the product or process." Taken literally, this statement can be construed to include "victims" of such events as automobile accidents, child-abuse cases due to alcohol or drug involvement, and so on, as customers. Take a position supporting or refuting Juran's definition of a customer and build a case to defend your position.

**1.13** Studies are sometimes performed to estimate the cost of finding and correcting a specific (individual) quality problem in different stages of the product life cycle. The following break down of cost is typically encountered.

| Life-cycle phase | Cost of correction |
| --- | --- |
| Product definition | $1 |
| Design specification | $10 |
| Prototype | $100 |
| Manufacturing | $1,000 |
| Marketing chain | $10,000 |
| Field use | $100,000 |

Choose a product and match its characteristics with the above break down. Then, briefly explain why the costs tend to increase in order of magnitude as we go through the product life cycle.

**1.14** A concept called "post-sale marketing" is used to describe the process where a technical person is assigned to a customer to stay in touch and help the customer get the most out of the product.
  **a** Develop a list of the potential benefits and burdens associated with such a practice for the customer. For the producer.

**b** List three products for which post-sale marketing would be clearly appropriate.

**c** Is post-sale marketing appropriate for all products (both goods and services)?

## APPENDIX 1A: Learning Theory

Effective change requires learning: new attitudes, new knowledge, new skills, and new behavior. The learning process consists of two distinct domains: the cognitive or knowledge domain [37] and the affective or activity domain [38], described by Bloom et al., and Krathwohl et al., respectively. Table 1.2 summarizes and outlines these two concepts to facilitate our discussion. The cognitive and affective domains are highly interactive and much more complicated than our simplified table implies. The cognitive chain or hierarchy begins with the collection and recognition phase, which can be rote memory work. The next phase moves to comprehension or restatement followed by application or examples. These first three phases represent the classical lecture-test classroom concept. More-advanced cognitive components include analysis (a breakdown of the whole to parts) and synthesis (the reconstruction of the whole from the parts). Most traditional textbooks tend to reach only the analysis phase; synthesis tends to require wide reading exposure and discovery through innovative projects. The evaluation phase is the highest level in the cognitive learning domain. It requires the ability to judge relevance and importance. Evaluation requires professional practice or experience, either actual or simulated.

Mastery of the cognitive domain in any subject is necessary to obtain effective results. However, it is not sufficient. Its activity-enthusiasm-feeling oriented counterpart is the affective domain. The affective domain also has a hierarchical structure. It consists of the receiving or attentiveness phase relative to knowledge at the lowest level. The hierarchy moves to a response (participation in the learning process) phase, the traditional classroom discussion and lecture-test format limit. The valuing or commitment phase marks the threshold level where the knowledge is considered or "felt" to be valuable and hence has potential to be used. The organizational phase leads to a value system within the individual, which in turn leads to the characterization level. Characterization extends the value system to a behavior complex. The latter is necessary if an individual is to take action in a consistent manner.

The structure of this book has been designed to support both the cognitive and affective domains of learning through creative examples and exercises stressing concepts, strategies and tools useful in quality assurance. We should keep these two fundamental, and tightly coupled, domains in mind when we develop and execute training (as well as educational) activities. In short, we must learn and practice simultaneously, in order to produce quality improvements in ourselves, our organizations, and our society.

## REFERENCES

1 "ASQC/Gallup Survey," *On Q—Official Newsletter of the American Society for Quality Control,* vol. VI, no. 9, November 1991, Milwaukee: ASQC.

2 T.Hill, *Manufacturing Strategy,* Homewood, IL: Irwin, 1989.

3 G. S. Radford, *The Control of Quality in Manufacturing,* New York: Ronald Press, 1922.

4 W. A. Shewhart, *Economic Control of Quality Manufactured Product* (originally published by Van Nostrand, 1931), Milwaukee: ASQC, 1980.

5 J. M. Juran, *Juran on Leadership for Quality,* New York: Free Press, 1989.

6 E. L. Grant and R. S. Leavenworth, *Statistical Quality Control,* 6th ed., New York: McGraw-Hill, 1988.

**TABLE 1.2  THE LEARNING PROCESS—OBJECTIVES AND FORMATS**

| Level of learning | Cognitive (knowledge) domain | Functional objectives | Educational format limits | Functional objectives | Affective (activity-enthusiasm) domain |
|---|---|---|---|---|---|
| Low | Recollection and recognition | Rote memory | | Attentiveness | Receiving |
| | Comprehension | Restatement | | | |
| | Application | Examples | Traditional lecture and test limit | Participation | Responding |
| Middle | Analysis | Break down to parts | Traditional textbook limit | Commitment | Valuing |
| | Synthesis | Reconstruction from parts | | Value system | Organization |
| | | | Outside-reading, discovery, and projects limit | | |
| High | Evaluation | Judgment of importance and relevance | Professional practice | Value and behavior complex | Characterization |

**7** P. B. Crosby, *Quality Is Free,* New York: McGraw-Hill, 1979.

**8** W. E. Deming, *Out of the Crisis,* Cambridge, MA: MIT Center for Advanced Engineering Studies, 1986.

**9** A. V. Feigenbaum, *Total Quality Control,* 3d ed., New York: McGraw-Hill, 1983.

**10** S. Mizuno, *Company-Wide Total Quality Control,* Tokyo: Asian Productivity Organization, 1988.

**11** G. Taguchi, *Introduction to Quality Engineering: Designing Quality into Products and Processes,* White Plains, NY: Kraus International, UNIPUB (Asian Productivity Organization), 1986.

**12** "ISO 9000," Geneva, Switzerland: International Organization for Standardization, 1992.

**13** K. Ishikawa, *What Is Total Quality Control? The Japanese Way,* Englewood Cliffs, NJ: Prentice-Hall, 1985.

**14** See reference 12.

**15** See reference 9.

**16** "Total Quality Management Guide," Washington D.C., Department of Defense Office for TQM, DoD 5000.51.G, Final Draft, 1989.

**17** See reference 5.

**18** See reference 13.

**19** See reference 11.

**20** W. J. Kolarik and J. N. Pan, "Proactive Quality: Concept, Strategy, and Tools," *Proceedings—International Industrial Engineering Conference,* pp. 411–420, 1991.

**21** J. Banks, *Principles of Quality Control,* New York: Wiley, 1989.

**22** A. J. Duncan, *Quality Control and Industrial Statistics,* 5th ed., Homewood, IL: Irwin, 1986.

**23** See reference 6.

**24** D. C. Montgomery, *Introduction to Statistical Quality Control,* 2d ed., New York: Wiley, 1991.

**25** Y. Akao (ed.), *Quality Function Deployment,* Cambridge, MA: Productivity Press, 1990.

**26** J. M. Juran, *Juran on Planning for Quality,* New York: Free Press, 1988.

**27** See reference 11.

**28** W. J. Kolarik, "Off-Line Quality Assurance," sec. 11, chap. 3, *Maynard's Industrial Engineering Handbook,* 4th ed., W. Hodson (ed.), New York: McGraw-Hill, 1992.

**29** S. Shingo, *Zero Quality Control: Source Inspection and the Poka-Yoke System,* Cambridge, MA: Productivity Press, 1986.

**30** See reference 7.

**31** See reference 8.

**32** See reference 13.

**33** See reference 5.

**34** T. Peters, *Thriving on Chaos,* New York: Harper & Row, 1988.

**35** See reference 13.

**36** "Cadillac: The Quality Story," (Malcolm Baldrige National Quality Award Winner, 1990), Cadillac Motor Car Company, Detroit, 1990.

**37** B. S. Bloom (ed.), *Taxonomy of Educational Objectives: The Classification of Educational Goals, Handbook I: Cognitive Domain,* New York: Longmans, Green and Company, 1956.

**38** D. R. Krathwohl, B. S. Bloom, and B. B. Masia, *Handbook II: Affective Domain,* New York: David McKay Company, 1956.

# 2

# PHILOSOPHY

## 2.0 INQUIRY

1 What are the critical dimensions of quality assurance?
2 What are the characteristics of contemporary quality philosophies?
3 Who are the quality philosophers that have shaped quality thinking and practices?
4 How has the concept of quality changed and evolved over the years?

## 2.1 INTRODUCTION

This chapter focuses on a variety of quality philosophies as described by the philosophers of contemporary quality practices. Our specific objective in this chapter is to acquaint our readers with a number of broad philosophies which are currently practiced to one degree or another. This overview is critical in the study of quality assurance in order to gain a fundamental perspective of and appreciation for the variety of quality practices that exist. Detailed discussions of quality philosophies, strategies, and analytical tools are developed in later chapters.

*The concept of quality assurance contains three closely coupled dimensions: (1) people, (2) products, and (3) processes.* Efforts to decouple these through rigid functional specialization have led to ineffective and inefficient products and processes as well as dissatisfied customers. Historically, the net result of such specialization has been the over-the-wall concept (Figure 1.4). In practice, this result manifests itself as department clusters focused on internal operations and functional specialties competing with each other for scarce company resources.

*Our systems view of quality holds that every part of the organization must focus on the customers, products, and processes simultaneously.* This view recognizes the interrelatedness of all functional specializations and their obligation to the customer first, rather than to their own narrow disciplinary interests. Ishikawa [1] compares Western professionalism and specialization with that of Japan and concludes that many professionals and professional organizations are more concerned with promoting their own professional interests (e.g., income and influence) than the general good of the organization and society.

A customer-centered philosophy is forged and sustained through leadership at the highest levels. The natural order of rigid, "labeled" divisions throughout contemporary societal structures in the United States, and elsewhere, tends to challenge this fundamental customer-focused strategy. However, the economic and competitive realities of the marketplace tend to support a customer-focused philosophy. In an economic democracy, where customers "vote" with their money (when they buy products or contribute to charitable organizations) and feet (when they join or walk away from organizational membership), a customer focus is an absolute necessity.

## 2.2 CONTEMPORARY QUALITY CONCEPTS AND ISSUES

Product integrity and customer satisfaction resulting from craftsmanship date back to the dawn of recorded history. The hallmarks of craftsmanship are (1) attention to detail through product function, form, and fit, and (2) pride of workmanship through customer satisfaction. The formal study of quality emerged after the industrial revolution and modern production concepts and technologies (interchangeable parts and available factory power). Today, the concept of total quality control has been called a "thought revolution in management" (Ishikawa [2]).

Traditional quality control work was focused on concepts, strategies, and tools to improve the rote inspection and sorting methods typical in manufacturing organizations, as well as on the development of statistical process control and statistical sampling plans, tools which have been used effectively for many years. Today, the concept of quality has been expanded well beyond statistical quality control. Competitive forces, originating in the marketplace, and bold leadership in production and marketing organizations have accelerated the emphasis on total quality control. *Without exception, the "masters" of quality assurance point to leadership and commitment from the top to the bottom as an absolute essential for success.*

*The work of many dedicated people has shaped contemporary quality thinking.* Qualitative as well as quantitative contributions have been critical in the emergence and development of contemporary quality knowledge. A handful of masters have had a profound effect on current quality practices, both reactive and proactive. A limited number of significant contributions are reviewed below. A broader account of quality history can be found in Banks [3].

### Walter A. Shewhart

Shewhart was the pioneer and visionary of modern quality control. Shewhart is most widely recognized for his control chart development and statistical contributions through Bell Laboratories. Indeed, the Shewhart charts (e.g., $X$-bar and $R$ charts) have become fundamental tools of quality control; but, of wider impact, Shewhart published, in 1931, *Economic Control of Quality of Manufactured Product* [4], a landmark book in modern quality control. His book was used by the Japanese after World War II, with the aid of visiting consultants, to help shape modern quality practice in Japan.

Shewhart, using a literal definition of quality (Latin *qualitas*, from *qualis*, meaning "how constituted"), defined two common aspects of quality: (1) "objective quality," which deals with the quality of a thing as an "objective reality" (of the thing) independent of the existence of man and (2) "subjective quality," which deals with the quality of a thing relative to what man thinks, feels, or senses as a result of the "objective reality." Shewhart linked the subjective quality property with value and concluded "it is impossible to think of a thing as having goodness independent of some human want." This definition has been expanded by Ishikawa to include "true" (customer-language based) and "substitute" (technical-language-based) quality characteristics which form the basis for modern quality planning and quality function deployment [5].

It is of great historical interest to point out that *the Shewhart Postulates* (lines of reasoning) and general conclusions published in 1931 *laid the foundation for modern quality theory and practice throughout the industrial world.* His general conclusions are stated below [6]:

> It seems reasonable to believe that there is an objective state of control, making possible the prediction of quality within limits even though the causes of variability are unknown. . . . It has been pointed out that by securing this state of control, we can secure the following advantages:
>
> 1 Reduction in the cost of inspection.
> 2 Reduction in the cost of rejection.
> 3 Attainment of maximum benefits from quantity production.
> 4 Attainment of uniform quality even though the inspection test is destructive.
> 5 Reduction in tolerance limits where quality measurement is indirect.

## W. Edwards Deming

Deming (as well as Joseph M. Juran, discussed in the next section) is widely credited as the master who developed Japan's "road map" to quality. This road map was basic, simple, consisted of readily available technology, and relied on common sense. However, it required a reorientation of priorities toward customer satisfaction, statistical quality control, learning, respect for workers and workmanship, and long-term commitment to true leadership. The Japanese Deming Prize, the premier quality award in the world today, is testimony to the success of Deming's efforts. Although Deming's statistical background permeates his writings on quality and quality measurement, his road map is the central focus.

Shewhart's broad quality concept has lead to the "chain reaction" described by Deming [7]. *The chain reaction,* shown in Figure 2.1, *links quality, productivity, market share, and jobs.* This revolutionary concept ran counter to the belief that quality and productivity were diametric opposites. For example, conventional wisdom (at the time) suggested that quality and productivity were traded off (e.g., we could have one or the other, but not both).

To start and sustain the chain reaction, Deming has proposed a 14-point quality architecture [7]. Many of his lectures and writings also focus on management's responsibility for creating a work environment that is conducive to quality improvement and pride in workmanship. He stresses process stability and system changes (through perceptive managers) as keys to quality improvement. He also stresses the understanding and proper use of quantitative tools, such as statistical process control charts, and experimental design.

The Deming cycle (also referred to as the Shewhart cycle or the PDCA cycle of Plan, Do, Check, Act) is known the world over and serves as the basis for many quality activities and training programs. Its pragmatic nature and systematic PDCA flow are unquestioned in both contemporary Japanese and U.S. quality literature.

*Deming's 14 points serve as the basis of his quality philosophy*. Deming's 14 points are listed below. [Reprinted from *Out of the Crisis* by W. Edwards Deming by permission of MIT and the W. Edwards Deming Institute. Published by MIT,

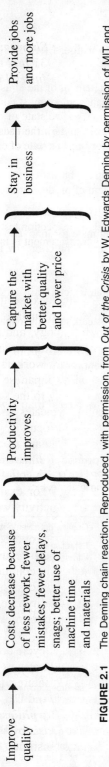

**FIGURE 2.1** The Deming chain reaction. Reproduced, with permission, from *Out of the Crisis* by W. Edwards Deming by permission of MIT and The W. Edwards Deming Institute. Published by MIT, Center for Advanced Engineering Study, Cambridge, MA 02139. Copyright 1986 by W. Edwards Deming.

Center for Advanced Engineering Study, Cambridge, MA 02139. Copyright 1986 (revised 1990) by W. Edwards Deming.]:

1   Create and publish to all employees a statement of the aims and purposes of the company or other organization. The management must demonstrate constantly their commitment to this statement.

2   Learn the new philosophy, top management and everybody.

3   Understand the purpose of inspection, for improvement of processes and reduction of cost.

4   End the practice of awarding business on the basis of price tag alone.

5   Improve constantly and forever the system of production and service.

6   Institute training.

7   Teach and institute leadership.

8   Drive out fear. Create trust. Create a climate for innovation.

9   Optimize toward the aims and purposes of the company the efforts of teams, groups, staff areas.

10  Eliminate exhortations for the workforce.

11a Eliminate numerical quotas for production. Instead learn and institute methods for improvement.

11b Eliminate management by objective. Instead, learn the capabilities of processes, and how to improve them.

12  Remove barriers that rob people of pride of workmanship.

13  Encourage education and self-improvement for everyone.

14  Take action to accomplish the transformation.

Deming considered his 14 points the basis for a road map toward a management philosophy (and many other practitioners agree). His 14 points epitomize a challenge

**FIGURE 2.2**   Juran's trilogy of quality management. Reprinted, with permission of The Free Press, a Division of Simon & Schuster, from J. M. Juran, *Juran on Leadership for Quality*, New York, NY: Free Press, page 22, 1989. Copyright 1989 by Juran Institute, Inc.

| Managing for quality | | |
|---|---|---|
| Quality planning | Quality control | Quality improvement |
| Determine who the customers are | Evaluate actual product performance | Establish the infrastructure |
| Determine the needs of the customers | Compare actual performance to product goals | Identify the improvement projects |
| Develop product features that respond to customers' needs | Act on the difference | Establish project teams |
| Develop processes able to produce the product features | | Provide the teams with resources; training, and motivation to: |
| Transfer the plans to the operating forces | | Diagnose the causes |
| | | Stimulate remedies |
| | | Establish controls to hold the gains |

to leadership in quality assurance. But, the conceptual nature of the 14 points along with the diversity of strategies and tactics available for their planning, execution, and follow-through, preclude a fixed-applications recipe for successfully applying the 14 points. Deming discusses the 14 points in detail [9], as does Scherkenback [10].

### Joseph M. Juran

Juran, like Deming, played a major role in the Japanese quality success story. He primarily introduced the quality management element, whereas Deming introduced the statistical quality-control element to Japanese industry in the 1950s. Juran directs his writings to upper level management, in highly conceptual discussions of both goods and services [11]. The central focus of his work is the Juran trilogy (see Figure 2.2). *The trilogy of (1) quality planning, (2) quality control, and (3) quality improvement serves as the basis for Juran's philosophy for quality assurance.*

Juran advocates a spreadsheet approach to quality planning [12] in an input-process-output format (see Figure 2.3). This approach is very similar to the Japanese quality function deployment approach (discussed later in Chapter 11); it begins with

**FIGURE 2.3**   Juran's quality planning steps. Reprinted, with permission of The Free Press, a Division of Simon & Schuster, from J. M. Juran, *Juran on Planning for Quality*, New York, NY: Free Press, page 15, 1988. Copyright 1988 by Juran Institute, Inc.

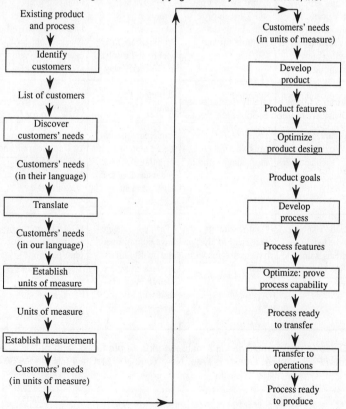

| Measurement Categories | Stage 1: Uncertainty | Stage II: Awakening | Stage III: Enlightenment | Stage IV: Wisdom | Stage V: Certainty |
|---|---|---|---|---|---|
| | **QUALITY MANAGEMENT MATURITY GRID** | | | | |
| | Rater_____ | | | Unit_____ | |

| Measurement Categories | Stage 1: Uncertainty | Stage II: Awakening | Stage III: Enlightenment | Stage IV: Wisdom | Stage V: Certainty |
|---|---|---|---|---|---|
| Management understanding and attitude | No comprehension of quality as a management tool. Tend to blame quality department for "quality problems." | Recognizing that quality management may be of value but not willing to provide money or time to make it all happen. | While going through quality improvement program learn more about quality management; becoming supportive and helpful. | Participating. Understand absolutes of quality management. Recognize their personal role in continuing emphasis. | Consider quality management as essential part of company system. |
| Quality organization status | Quality is hidden in manufacturing or engineering departments. Inspection probably not part of organization. Emphasis on appraisal and sorting. | A stronger quality leader is appointed but main emphasis is still on appraisal and moving the product. Still part of manufacturing or other. | Quality department reports to top management, all appraisal is incorporated and manager has role in management of company. | Quality manager is an officer of company; effective status reporting and preventive action. Involved with consumer affairs and special assignments. | Quality manager on board of directors. Prevention is main concern. Quality is a thought leader. |
| Problem Handling | Problems are fought as they occur; no resolution; inadequate definition; lots of yelling and accusations. | Teams are set up to attack major problems. Long-range solutions are not solicited. | Corrective action communication established. Problems are faced openly and resolved in an orderly way. | Problems are identified early in their development. All functions are open to suggestion and improvement. | Except in the most unusual cases, problems are prevented. |
| Cost of quality as % of sales | Reported: unknown Actual: 20% | Reported: 3% Actual: 18% | Reported: 8% Actual: 12% | Reported: 6.5% Actual: 8% | Reported: 2.5% Actual: 2.5% |
| Quality improvement actions | No organized activities. No understanding of such activities | Trying obvious "motivational" short-range efforts. | Implementation of the 14-step program with thorough understanding and establishment of each step. | Continuing the 14-step program and starting Make Certain. | Quality improvement is a normal and continued activity. |
| Summation of company quality posture | "We don't know why we have problems with quality." | "Is it absolutely necessary to always have problems with quality?" | "Through management commitment and quality improvement we are identifying and resolving our problems." | "Defect prevention is a routine part of our operation." | "We know why we do not have problems with quality." |

**FIGURE 2.4**  Crosby's quality management maturity grid.  Reproduced, with permission from McGraw-Hill, from P. B. Crosby, *Quality is Free*, New York, NY: McGraw-Hill, pages 38 and 39, 1979.

customer identification as either internal or external, and ends with a product and its associated processes ready for full-scale production.

### Philip B. Crosby

Crosby, in his classic book *Quality Is Free,* provides a high level of public visibility for quality issues. The Crosby "Quality Management Maturity Grid," shown in Figure 2.4, traces corporate quality awareness and a quality maturation from a level of uncertainty to one of certainty [13]. His grid addresses quality understanding, organization, problem handling, cost, and improvement.

*Crosby's quality philosophy is characterized by his four absolutes* [14]:

1 *Quality is defined as conformance to requirements*, not goodness or elegance.
2 *The system for causing quality is prevention*, not appraisal.
3 *The performance standard must be zero defects*, not "that's close enough."
4 *The measurement of quality is the price of nonconformance*, not indexes.

Crosby has developed his own quality architecture [15]. This architecture is represented by a 14-step program:

1 Management commitment.
2 Quality improvement team.
3 Quality measurement.
4 Cost-of-quality evaluation.
5 Quality awareness.
6 Corrective action.
7 Ad hoc committee for the zero defects program.
8 Supervisor training.
9 Zero defects day.
10 Goal-setting.
11 Error-cause removal.
12 Recognition.
13 Quality councils.
14 Do it over again.

### Armand V. Feigenbaum

Traditionally (pre-1970s) in the United States, quality assurance was widely associated with establishing and measuring conformance to technical specifications on the shop floor and in inspection departments. The evolution which has occurred in transforming this narrow, reactive view of quality to its current broad companywide approach in the United States can be credited to many people and circumstances. However, Feigenbaum has had a great impact on this transformation through his total quality control concept and strategies. Feigenbaum maintains that the goal of competitive industry, as far as product quality is concerned, can be clearly stated [16]:

> The goal of competitive industry is to provide a product and service into which quality is designed, built, marketed, and maintained at the most economical costs which allow for full customer satisfaction.

Feigenbaum goes on to define total quality control [17]:

> Total quality control is an effective system for integrating the quality-development, quality-maintenance, and quality-improvement efforts of various groups in an organization so as to enable marketing, engineering, production, and service at the most economical levels which allow for full customer satisfaction.

***Feigenbaum stresses a systems approach to quality*** [18] through the definition of a quality system:

> A quality system is the agreed on, company-wide and plant-wide operating work structure, documented in effective, integrated technical and managerial procedures, for guiding the coordinated actions of the work force, the machines, and the information of the company and plant in the best and most practical ways to assure customer quality satisfaction and economical costs of quality.

Feigenbaum expresses total quality control as a horizontal concept stretching across the functional divisions of an organization [19], as shown in Figure 2.5. From this depiction (a clear contrast to an over-the-wall scheme), we see total quality control as a cross-functional concept ranging from marketing to installation and sales.

### Kaoru Ishikawa

The concepts and road maps to quality proposed by masters such as Juran and Deming in the latter part of the 1940s and 1950s were widely available to the world

**FIGURE 2.5**    Feigenbaum's horizontal scope of total quality control. Reproduced, with permission from McGraw-Hill, from A. V. Feigenbaum, *Total Quality Control*, 3d ed., New York, NY: McGraw-Hill, page 83, 1983.

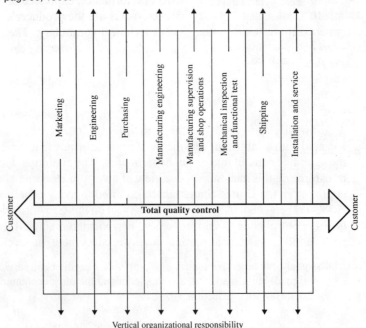

(including the United States).  However, it was the Japanese who seized the opportunity and advanced from a war-devastated economy to their present international quality leadership stature in a period of about 40 to 50 years.

"Made in Japan" in the late 1940s and 1950s implied cheap products that did not perform as well as their counterparts manufactured elsewhere.  However, people like Ishikawa knew that the poor performance of Japanese products did not reflect the virtues of their society: self-discipline, the desire to learn, respect for one another, patience, persistence, and national pride (to name but six).  A pressing economic need, relatively open world markets, and strong quality control leadership resulted in an economic success story that has no equal in history.

Ishikawa provided a great deal of leadership in shaping the Japanese quality movement [20] through his vision and activities associated with the Union of Japanese Scientists and Engineers (JUSE).  By 1967 Japanese quality control could be distinguished from that practiced in the West by six characteristics (Ishikawa [21]):

1 Company-wide quality control; participation by all members of the organization in quality control.
2 Education and training in quality control.
3 Quality control circle activities.
4 Quality control audits (for effectiveness).
5 Utilization of statistical methods.
6 Nationwide quality control promotion (including training) activities.

Ishikawa's impact on quality control practices has been extensive.  *Ishikawa developed the concept of true and of substitute quality characteristics.*  The "true" quality characteristics are the customer's view of product performance, expressed in the customer's vocabulary.  "Substitute" quality characteristics are the producer's view of product performance expressed in the producer's technical vocabulary.  *The degree of match between true and substitute quality characteristics ultimately determines customer satisfaction.*

Ishikawa proposes three steps which are the basis of quality-planning and quality function deployment techniques.

1 Understand true quality characteristics.
2 Determine methods of measuring and testing true quality characteristics.
3 Discover substitute quality characteristics, and have a correct understanding of the relationship between true quality characteristics and substitute quality characteristics.

Ishikawa [22] has been associated with the development and advocacy of universal education in the seven "indispensable" or fundamental tools (of quality control):

1 Cause-effect (Ishikawa) diagram.
2 Stratification.
3 Check sheet.
4 Histogram.

5 Scatter diagram.
6 Pareto chart (vital few, trivial many).
7 Graphs and statistical control charts.

Ishikawa claims that as much as 95 percent of all quality problems within a company can be solved with these tools [23], which are discussed in detail in Chapter 9. **_Ishikawa's concept of total quality control contains six fundamental principles:_**

1 Quality first—not short-term profits first.
2 Consumer orientation—not producer orientation (think from the standpoint of the other party).
3 The next process is your customer—breaking down the barrier of sectionalism.
4 Using facts and data to make presentations—utilization of statistical methods.
5 Respect for humanity as a management philosophy, full participatory management.
6 Cross-functional management (by divisions and functions).

### Genichi Taguchi

Taguchi emphasizes an engineering approach to quality. He stresses producing to target goals or requirements with minimal product performance variation in the customer's environment. Variation is termed noise. Taguchi identifies three distinct types of noise [24]:

1 External noise—variables in the environment or conditions of use that disturb product functions (e.g., temperature, humidity, and dust).
2 Deterioration noise or internal noise—changes that occur as a result of wear or storage.
3 Unit-to-unit noise—differences between individual products that are manufactured to the same specifications.

The objective is to minimize noise through on-line (during production) and off-line (pre- or postproduction) quality activities. The Taguchi concept proposes the use of optimization theory and techniques, along with experimental design, with the ultimate objective of minimizing the loss to society.
**_Taguchi focuses on design for quality by defining three design levels_** [25]:

1 _System design_ (primary)—functional design focused on pertinent technology or architectures.
2 _Parameter design_ (secondary)—a means of both reducing cost and improving performance without removing causes of variation.
3 _Tolerance design_ (tertiary)—a means of reducing variation by controlling causes, but at an increased cost.

Recognition of, and strategies for dealing with, parameter design (whereby performance is increased and costs are decreased by using less expensive materials or processes) is a major contribution. **_The Taguchi approach to both parameter design and tolerance design makes use of cost-performance optimization and experimental design technology._** Taguchi's loss functions (loss to society) and signal-to-noise ratios (SNRs) are critical parts of his optimization procedures.

Essentially, he defines three basic forms of loss functions: (1) smaller is better (e.g., impurity levels, defect counts), (2) bigger is better (e.g., process yields, fuel efficiency), and (3) nominal is best (e.g., outside-inside diameter, humidity level). Taguchi's objective of performance to one of the three target-based functions with minimum variation is universal. Hence, Taguchi's concepts, strategies, and tools have great appeal.

**Shigeo Shingo**

A true "zero defects" level of quality is the ultimate level of conformance to specification. Zero defects (ZD) implies that each and every item built conforms to specification. Shingo maintains that statistical-based quality control is not conducive to zero defects [26]. He states that statistical quality control can lower, but not eliminate, defects. *Shingo proposes the poka-yoke (mistake-proofing) system to totally eliminate defects.*

The mistake-proofing concept is a human- or machine-sensor-based series of 100 percent source inspections, self-checks, or successive checks to detect abnormalities when or as they occur and to correct them on the current unit of production as well as systemwide. Figure 2.6 depicts the classical "long route" and the Shingo cycle, or "short route," dealing with errors and defects. The Shingo Zero Quality Control System consists of four fundamental principles [27]:

1 Use source inspection—the application of control functions at the stages where defects originate.
2 Always use 100 percent source inspections (rather than sampling inspections).
3 Minimize the time to carry out corrective action when abnormalities appear.
4 Set up *poka-yoke* (mistake-proofing) devices, such as sensors and monitors, according to product and process requirements.

We should note at this point that *Shingo's source inspection differs significantly from traditional final product inspection.* Traditional inspection is a "long-route" product-sorting process. *Source inspection is a "short route" production monitoring process which provides real-time feedback and feed-forward information relative to corrective action* (Figure 2.6). Ideally, perfect source inspection will eliminate scrap and rework.

## REVIEW AND DISCOVERY EXERCISES

### Review

**2.1** Why was (is) it so difficult for many people to accept the Deming concept that high quality leads to high productivity (e.g., the Deming chain reaction)?
**2.2** In quality work there is an old saying that "Quality is what the customer says it is." True or false? Support your position.
**2.3** In quality improvement there is an old saying that "You cannot control something that you cannot measure." True or false? Support your position.
**2.4** Explain the difference between Ishikawa's true and substitute quality characteristics.

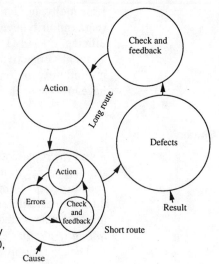

**FIGURE 2.6** Shingo's cycle for managing errors and defects. From *Zero Quality Control: Source Inspection and the Poka-Yoke System*, page 53, by Shigeo Shingo. English translation copyright © 1986 by Productivity Press, Inc., PO Box 13390, Portland, OR 97213-0390, (800) 394-6868. Reprinted by permission.

2.5  List a true quality characteristic for a microcomputer. Then, list three substitute quality characteristics pertaining to the true quality characteristic.

2.6  Choose a product that you use frequently. Next, list a true quality characteristic for your chosen product. Then, list three substitute quality characteristics pertaining to the true quality characteristic.

2.7  If we are using a reactive quality strategy, explain how we can measure the cost of nonconformance.

2.8  If we are using a proactive quality strategy, explain how we can measure the "cost avoidance" results.

2.9  List two examples of smaller is better, bigger is better, and nominal is best quality characteristics.

2.10  In quality work the "management by fact and data" philosophy is very common. Does this philosophy imply that quality control must be approached exclusively through statistics and statistical analysis?

2.11  Is Shingo's 100 percent source inspection relevant for
   a  Production processes involving humans? Explain.
   b  Automated production processes? Explain.
   c  Service-oriented products or processes? Explain.

2.12  What functional areas (e.g., marketing, research and development, engineering, finance) in a production organization should Juran's quality planning process (Figure 2.3) include?

2.13  What advantage do we obtain by designing and developing products and production processes simultaneously?

## Discovery

2.14  Explain the relationship of Figure 2.5 (taken from Feigenbaum's work) to Ishikawa's concept of cross-functional management. You may need to refer to the reference list in order to establish a background for your response.

2.15  Crosby is a strong advocate of measuring cost relative to quality (e.g., the cost of nonconformance). Deming tends to deemphasize measuring costs relative to quality.

Investigate the works of each and comment on the position of each. Whose advice should we take? Why?

**2.16** Many of the quality control techniques and tools that are currently being adopted by U.S. manufacturers and service providers were documented, advocated, and shown to be effective over 50 years ago. Currently, we are seeing a renaissance in quality control activity.

    **a** What factors were responsible for creating the "dark ages" of quality control in the United States (e.g., the 1950s, 1960s, and 1970s)?

    **b** What factors are responsible for the renaissance of quality control in the United States (e.g., the late 1980s and 1990s)?

**2.17** Develop a logical argument for or against Deming's points 11 and 12. Be sure to consult Deming's work, Crosby's work, and Juran's work (as listed in the references) before finalizing your response.

**2.18** Critique Crosby's four quality absolutes in terms of appropriateness regarding the following industries:

    **a** Women's or men's fashions (garments).

    **b** Bolt-making (mechanical fasteners).

    **c** Lawn care service.

**2.19** When cleaning clothes, a laundry is having difficulty with items left in pockets. Develop a mistake-proof analysis, procedure, or device to deal with this problem.

**2.20** Select a process or product and identify a "typical" mistake; then, mistake-proof the process or product.

**2.21** Locate a copy of *Quality Progress* (published by the ASQC) or some other trade journal of interest to you. Look through it and find a quality-related article of interest to you. Read the article and write a brief critical review.

**2.22** Read an original work of one of the quality masters reviewed in this chapter. Then, write a critical book review.

## REFERENCES

**1** K. Ishikawa, *What Is Total Quality Control? The Japanese Way,* Englewood Cliffs, NJ: Prentice-Hall, 1985.

**2** See reference 1.

**3** J. Banks, *Principles of Quality Control,* New York: Wiley, 1989.

**4** W. A. Shewhart, *Economic Control of Quality Manufactured Product* (originally published by Van Nostrand, 1931), Milwaukee: ASQC, 1980.

**5** See reference 1.

**6** See reference 4.

**7** W. E. Deming, *Out of the Crisis,* Cambridge, MA: MIT Center for Advanced Engineering Studies, 1986.

**8** See reference 7.

**9** See reference 7.

**10** W. W. Scherkenbach, *The Deming Route to Quality and Productivity,* Milwaukee: ASQC Press, 1986.

**11** J. M. Juran, *Juran on Leadership for Quality,* New York: Free Press, 1989.

**12** J. M. Juran, *Juran on Planning for Quality,* New York: Free Press, 1988.

**13** P. B. Crosby, *Quality Is Free,* New York: McGraw-Hill, 1979.

**14** P. B. Crosby, *Quality without Tears,* New York: McGraw-Hill, 1984.

**15** See reference 13.

**16** A. V. Feigenbaum, *Total Quality Control,* 3d ed., New York: McGraw-Hill, 1983.

17  See reference 16.
18  See reference 16.
19  See reference 16.
20  See reference 1.
21  See reference 1.
22  K. Ishikawa, *Guide to Quality Control,* Tokyo: Asian Productivity Organization, 1986.
23  See reference 1.
24  G. Taguchi, *Introduction to Quality Engineering: Designing Quality into Products and Processes,* White Plains, NY: Kraus International, UNIPUB (Asian Productivity Organization), 1986.
25  See reference 24.
26  S. Shingo, *Zero Quality Control: Source Inspection and the Poka-Yoke System,* Cambridge, MA: Productivity Press, 1986.
27  See reference 26.

# 3

# PARADIGMS, SYSTEMS, AND SYSTEMS INTEGRATION

## 3.0 INQUIRY

1 What is a quality system?
2 What quality paradigms have emerged over the years?
3 How are quality systems structured?
4 What is the purpose of ISO 9000? ANSI / ASQC Q9000?
5 What role does a quality system play in an organization?

## 3.1 INTRODUCTION

*A quality system,* in its broadest sense, *constitutes a culture* (concepts, beliefs, knowledge, thoughts, skills, and practices) *of people who function as a unit or team to define, design, develop, produce, deliver, sell, service, support, use, and dispose of products that meet customer needs and expectations.* Every organization has a quality system of some kind. Quality systems are constantly changing. This change is driven by dynamic customer needs and expectations, technology, and competition.

## 3.2 QUALITY PARADIGMS

*Five distinct quality paradigms* (ways of thinking about quality and doing quality activities) *are recognizable in organizations today*: **(1)** *custom-craft,* **(2)** *mass production and sorting,* **(3)** *statistical quality control,* **(4)** *total quality management* **(TQM)**, *and* **(5)** *techno-craft*. These different quality paradigms, depicted in Figure 3.1, have emerged over the years due to changes in customer demands, technology, and social systems. Diverse factors, such as the development and availability of interchangeable parts and of power units; advances in transportation, material science, computers, and mathematical science; legislation, and so on, have had a profound effect on the development of quality paradigms.

### Custom-Craft Paradigm

*The custom-craft paradigm focuses on the product and product performance relative to customer demands.* Here, the craftsperson and customer communicate freely. The product is customized to exactly what the customer wants. Figure 3.1*a* illustrates the custom-craft quality paradigm relative to piece-part production. A service industry can be depicted in the same manner, with the box on the bench being a service product, rather than a hard-good product. Examples of this system include the custom building and housing industry, custom clothing, custom firearms, bank loans,

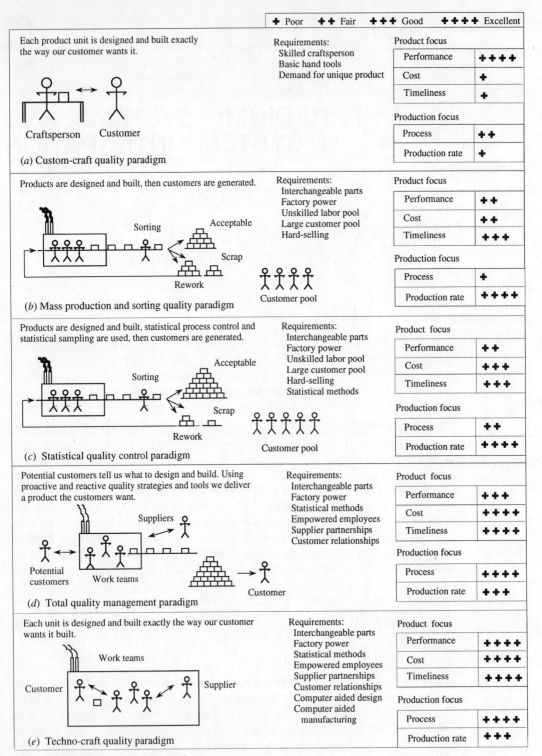

**FIGURE 3.1**    Quality paradigms.

hair styling, and so on.  Customers are usually very excited about product performance, but products are relatively expensive and may require relatively long waiting periods for delivery and service support.

## Mass Production and Sorting Paradigm

*The focus in the mass production and sorting paradigm is on production rates.* Customers are often times gained through mass advertising and hard selling.  The product is defined with the customer in mind, but without direct customer involvement.  Figure 3.1*b* illustrates this quality paradigm.  Here, product performance may be relatively low, cost is usually relatively low, and delivery time is typically short (sales are from stock on hand).  Examples include tract housing and office buildings, hand tools, electronic components, bank counter services, clothes cleaning services, agricultural production, and so on.

   The mass production and sorting quality paradigm emerged with mass production technology (especially mechanization).  To some degree, mass markets and automated inspection and testing technology are driving this paradigm today, whereas in the past, mass markets, factory power, and human inspection drove it.  Service organizations in this paradigm still tend to be rather labor intensive, whereas manufacturing and agricultural production in this paradigm tend to be very mechanized.

## Statistical Quality Control Paradigm

*The statistical quality control paradigm is similar to the mass production and sorting paradigm with the difference that more attention is given to production processes.*  We see, basically, the same products as in the mass production and sorting paradigm.  Here, we see statistical process control and lot-by-lot sampling inspection as a result of the application of statistical techniques.  The result is far less scrap and rework than in the mass production and sorting paradigm.

   We may see product performance and timeliness similar to that obtained with the mass production and sorting paradigm.  Production costs usually will be reduced, sometimes drastically, by using upstream sampling and statistical process control to limit losses due to poor quality.  Since process knowledge is gained, process improvements usually result.  The statistical quality control paradigm is depicted in Figure 3.1*c*.

## Total Quality Management Paradigm

*The total quality management paradigm involves customers and suppliers in addition to mass production and statistical methods.*  A very high level of process focus is a characteristic of this paradigm.  Customer input in product definition and product evaluation are used to increase product performance.  Here, the product lines are very similar to those in the mass production and sorting paradigm.  However, customer involvement contributes to product refinements and production options which tailor products to customer needs.  Customer focus groups, where customers and potential customers provide direct input to organizations, allow products to be developed which will provide higher levels of customer satisfaction. Products such as minivans, performance sedans, graphical computer interfaces, and so on, have resulted from this paradigm. Supplier involvement in the total quality management paradigm helps to reduce waste and improve produceability of products. This paradigm recognizes the importance of the supplier and supplier partnerships. Here, we see higher product performance, lower cost, and faster delivery than in either the mass production and sorting or the statistical quality control paradigm.

The total quality management paradigm is depicted in Figure 3.1*d*. Its distinguishing feature is its focus on employee involvement for the purpose of continuous improvement. In many respects, it has produced very successful results. Its ability to produce large-scale or fundamental gains (breakthroughs) in performance, cost, and timeliness has been questioned since it tends (in practice) to focus on operational improvement in day-to-day processes.

### Techno-craft Paradigm

*The techno-craft paradigm is the sociotechnical counterpart to the custom-craft paradigm*. It is illustrated in Figure 3.1*e*. The techno-craft paradigm is a new frontier in quality that seeks to emulate the custom-craft paradigm in performance, but reduce the cost and the delivery time. The techno-craft paradigm is possible through the proper integration of people, machines, and automation. Headway is being made using the techno-craft paradigm in industries such as apparel and software development. Since this paradigm is new, customers are just now beginning to adjust to its nature and build appropriate expectations. In the near future products such as shoes, firearms, apparel, and other products which must "fit" the user in order to function at their best will emerge at competitive prices.

In the techno-craft paradigm, customers get exactly what they want. A high level of flexibility in product design and process flexibility is necessary to make this paradigm feasible. Computer aided design and computer aided manufacturing technology, along with automated and integrated measuring machines, are rapidly making the techno-craft paradigm a reality.

## 3.3  QUALITY SYSTEM STRUCTURE

*Quality systems are unique to each organization*. We know that every organization has a quality system of some type, some more comprehensive and effective than others. Quality systems are shaped by quality philosophy (e.g., Deming's fourteen points, Juran's trilogy, Ishikawa's six principles, Crosby's four absolutes) and directed through quality strategy. We will describe a broad quality system in terms of fundamental interfaces and identify basic elements of each interface.

### Interfaces

Since quality systems are unique and somewhat dynamic, it is impossible to provide a fixed recipe or best system in detail. However, it is possible to describe the general components of quality systems. ***Three major interfaces constitute the quality system triad: (1) the product interface, (2) the technology interface, and (3) the service interface.*** Figure 3.2 depicts the three major quality system interfaces and nine major interface elements centered on the customer. The nine major elements form expansion triads around each major interface. All components are in one way or another related and tend to interact, making the quality system a very complicated sociotechnical system.

**Product Interface** ***The product interface is made up of product performance, cost, and timeliness elements.*** The performance component refers to both short- and long-term benefits received by customers relative to product function, form, and fit. True performance is ultimately judged by the customer. However, we define and use substitute quality characteristics in order to build the product and measure its performance in technical terms. For example, a customer may demand shoes that fit right (a "true" quality characteristic). The customer will judge his or her shoe fit by wearing the shoes, but at the factory we must use "substitute" characteristics like length, width, and so on, to design, develop, and produce our product.

The cost component refers to both direct and indirect costs of the product to the customer over the product's life cycle. Direct costs include initial, operating, support, and disposal costs. Some of the direct costs are explicitly associated with quality control activities, while other costs are the results of quality as it is defined and built into the product. Early product-process design decisions ultimately drive production costs as well as costs associated with usage, maintenance, and disposal.

**FIGURE 3.2**    Quality system structure and interfaces.

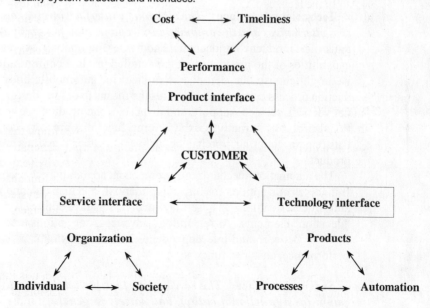

The direct costs explicitly associated with quality can be broken down to internal and external costs. Internal costs can be further broken down to loss prevention, appraisal or inspection costs, and scrap and rework costs. Internal costs occur prior to transfer of ownership (of the product). External costs occur after transfer of ownership and include returns, warranty claims, product-liability claims, and so on. Explicit quality costs are discussed further in Appendix 3A.

In addition to direct costs, indirect costs are encountered. Indirect costs include costs associated with a lack of performance (e.g., costs associated with lost production or inconvenience). Indirect costs also include a loss of customer goodwill or a loss of reputation (e.g., as the U.S. auto industry experienced in the late 1970s and 1980s) as a producer. The cost associated with a loss of reputation is very difficult to measure. But, even though reputation for high-quality products and product service is difficult to quantify explicitly in economic terms, reputation is a major factor in product selection, as we see in Table 1.1. Both the producer and the customer bear the burdens of indirect costs.

The timeliness component is emerging as a critical element in customer satisfaction. A number of timeliness elements can be identified. Design cycle time is the time it takes to define and fully design the product and its processes. Development time is the time required to put the design into production once the design specifications are complete. Production time refers to the time required (actual human and machine time) to build a product unit. Delivery time is the time it takes from the point of a customer order to product delivery and installation in the customer's facility. Service-repair time refers to the time it takes to get a product "up and running," after a problem has been reported. Ultimately, we see a great deal of interaction and dependency between performance, timeliness, and cost elements in the product interface.

**Technology Interface**    *The technology interface involves our product and process alternatives and the automation-mechanization possibilities* we have available to us. The product component here addresses our options for creating the functional capabilities of the product (e.g., the detailed product configuration). For example, we might choose between plastic or steel for automobile body panels. The production process component addresses the means to define, design, develop, produce, deliver, sell, service, support, and dispose of the product. A wide variety of functional process alternatives exist. Some functions are very common to all products (e.g., accounting, purchasing, delivery). Other functions are unique to a given product.

The automation-mechanization component applies to both products and processes. It involves our options for balancing manual, mechanized, and automatic product and process features. These may involve sensing, control, communications and "decision" elements. As technology advances, our options increase. We see many linkages between and interdependencies among products, processes, automation, performance, cost, and timeliness.

**Service Interface**    *The service interface includes the individual, the organization (as a functional entity), and society in general.* Individuals bring needs, expectations, creativity, and leadership potential into the quality system. Individuals are the fundamental building blocks for organizations and society. Individuals are also external and internal customers. Hence, people as individuals and collectively are the driving force in the quality system.

The customer service experience is usually the "sale maker" in a competitive environment when the customer can distinguish very little performance or cost difference in products. In service oriented products (retail sales, food service, banking, and so on) the service element may dominate the customer interfaces. This customer service concept is relevant to internal customers as well as external customers. For example, in the internal customer case, many job positions may pay about the same and/or require the same number of hours each workday. But, the service "on the job" (i.e., the way we are treated and treat our associates) makes a big difference in our experience of quality.

The organization element of the service interface is a collection of individuals with common purpose, willingness (and ability) to contribute toward the purpose, and the ability to communicate with each other. Leadership and management functions must be developed within the organization. Individuals working through the organization ultimately determine purpose as well as success or failure of purpose. This success or failure is ultimately linked to financial solvency (profits in private enterprise) in general.

The society component must be served by organizations, through individuals. Society provides the broad framework (through markets, laws, codes of values and morals, etc.) within which organizations function. For example, legislative issues such as labor laws, tax laws, health and safety requirements, environmental codes, antitrust legislation, and so on, have a profound effect on the organization. In the long run, society influences the beliefs, attitudes, and behavior of the individual. Hence, the individual, organization, and society function in an interactive manner.

In the past, we have seen a quality focus at the individual (external customer satisfaction) level. Now, we are seeing additional emphasis on the organization with respect to workplace environment and teamwork (internal customer satisfaction). Environmental problems associated with product and by-product disposal have broadened the quality focus to include our society in general. We hear the term "social awareness" used to describe quality in a societal context. Hence, in the United States (and elsewhere), we can readily observe significant activity at all three elements in this interface.

---

## Tape Recorder Development—Quality System Differences Case [SM]

A major defense contractor and a manufacturer of commercial tape recorders undertook joint development of a digital data recorder to be used as a computer peripheral. The partnership worked well for both companies since the defense contractor could provide engineers skilled in working with state-of-the-art electronics technology and the recorder manufacturer could provide production expertise and a production facility. As might be expected, the design, development, and production methods typically used by the two companies were quite different.

The defense contractor had traditionally built large one-of-a-kind systems for government customers. Quality control had focused on development since there is really no explicit "production" phase for such systems. The recorder manufacturer had been a commercial producer of tape recorders for several decades. Although they had produced numerous technologically innovative recorder products, a large portion of the innovation was in the area of mechanical transports. In addition, much of their development activity had focused on modifying or upgrading existing designs to produce new models. Quality control in this environment tended to focus on production activities rather than on development.

The program was unique to both companies. It involved a product that was to be mass produced for commercial sale, which was new for the defense contractor, and it required extensive custom hardware development using sophisticated electronic technology, which was unusual for the recorder manufacturer. During production of beta test units the "disconnect" between the companies in terms of quality systems became evident. The defense contractor's engineers had performed board layouts in a way that minimized trace length and cross talk without regard for difficulties of auto-insertion of integrated circuit components. The recorder production facility changed the layouts to facilitate auto-insertion but in the process created "noise" problems on the boards.

In this case the product and technology interfaces in the two companies differed considerably. Eventually, through cooperative efforts, interface differences produced synergistic results.

## 3.4 QUALITY SYSTEM STRATEGIES AND TACTICS

*Wherever a product or process is found there will be an associated quality system.* We know that some form of quality system (sometimes effective, sometimes ineffective) is an integral part of each organization. In other words, it is impossible to operate an organization in a complete quality "vacuum." It is simply impossible for an organization to exist for any period of time with absolutely no product (good or service, tangible or intangible) whatsoever.

Our point here is that *we should proact and consciously define, design, and execute an effective and efficient quality system rather than let one evolve in a reactive manner* (e.g., reacting to quality-related problems, such as product-performance complaints, poor product workmanship, rework, warranty, liability claims, environmental damage). In general, *a proactive quality system strategy must address both the experience of quality and the creation of quality.* Table 3.1 lists a number of critical strategic elements associated with proactive quality systems. These strategic elements involve understanding followed by proaction (also reaction—when we have failed to proact properly). Remaining sections of this text focus on specific issues regarding the experience of quality and the creation of quality.

**TABLE 3.1** QUALITY SYSTEM GOALS AND STRATEGIC ELEMENTS TO ENCOURAGE PROACTIVITY

| Goal | Strategic elements | Comments |
|---|---|---|
| To understand the experience of quality | Human-needs structure and performance<br><br>Product and process performance<br><br>Robust and mistake- proof performance | No amount of engineering and marketing can compensate for ill-defined or unwanted products that do not meet customers' needs and expectations.<br><br>Products and processes with integrity provide customer benefits in a safe and consistent manner under a variety of conditions. |
| To enhance the creation of quality | Leadership processes<br><br>Creative processes<br><br>Product and process definition<br><br>Product and process design<br><br>Product and process improvement | Human resources are the only unique resources available. Through leadership, purpose and direction are set. Through creative endeavors improvements are made.<br><br>A systematic design hierarchy can increase product and process performance, decrease cost, and improve timeliness simultaneously.<br><br>Competition demands both incremental and breakthrough improvement in both products and process results. |

## Intentions and Declarations

In order to design an effective quality system, rather than leave its development to chance, assertive leadership must be exercised. *Leaders must have a clear and consistent understanding of where their organization now stands and a concise and consistent vision of where it will stand in the future.*

Leadership requires that we firmly establish and communicate our intended purpose through declarations and a commitment to action. To do this, we must not only focus on, but also share the vision within the organization. Then, we must position for action through strategies and plans. Finally, we must take action.

A declaration of intentions requires a great deal of forethought, since people have grown skeptical of declarations, slogans, and exhortations. We must follow our declaration with timely action (e.g., training, employee empowerment) if we as individuals, organizations, and a society are to gain from our quality system. An example of the expression of corporate quality intentions, from the AT&T Power Systems Division [1], is shown in Figure 3.3.

Declarations are expressed in both present and future (vision) tense. Present-tense declarations are usually developed on three basic levels, (1) philosophy, (2) mission, and (3) policy. At the highest level, we see a quality philosophy. Usually the philosophy addresses both leadership and creativity relative to external and internal customers. For example, Deming's fourteen points or Ishikawa's six principles (Chapter 2) constitute a generic quality philosophy. In many cases, an organization's quality philosophy is widely publicized within the organization, but not outside the organization. A quality philosophy is inseparable from the "leadership and management philosophies" held within an organization.

The quality mission is verbalized, usually in a broad statement declaring what business or customer needs the organization serves. The quality policy is a blanket statement—usually widely publicized—explicit as to customers, products, service, and satisfaction. It is usually rather brief so it can be remembered readily by both internal and external customers. The policy is often publicly endorsed and signed by the highest operating manager in the organization (e.g., the president or CEO). The AT&T quality intentions (Figure 3.3) are but one example of clarity and brevity while presenting well-organized, effective, and efficient vision, mission, and policy statements.

**Commitment to Action**   *The true test of the effectiveness of a quality system is the manner in which it affects decisions relative to both leadership and management in an organization.* A strong and timely commitment to action is essential to add credibility to declarations. Once again, we see a hierarchy of commitments to action: goals, objectives, targets, and specifications. It is critical to recognize the hierarchical nature of our commitment. We see a general purpose, supported by more detail in purpose as we move through the hierarchy. The specific terms (goal, objective, target, etc.) are arbitrary labels that identify levels of this hierarchy. Figure 3.4 shows Motorola's efforts to commit to action [2], with respect to their quality system. It uses a six sigma theme and clearly lays out a systematic action-oriented strategy backed up with a stated commitment to this theme and to leadership and management efforts to "do" what they "say."

A "goal" is an action-oriented expression of purpose that is typically qualitative. Goals develop from visions. Usually a vision is somewhat fuzzy, while a goal is more definite. For example, a vision in a food-service organization might "see" the organization as a best-in-class food service. Whereas, a goal in the same food-service organization might be stated as "to be recognized as a best-in-class food service" or "to provide the very best food service possible to our customers."

An "objective" is an end that we strive for and is usually narrower and more focused than a goal. For example, we might support our goal in the example above with objectives of "greeting each customer and treating each customer with respect." Here, we see detail and definition added to our goal. In general, high-level goals must "capture" the vision. But, these goals must be broken down and related to organizational activities which can be executed in operations. Objectives help

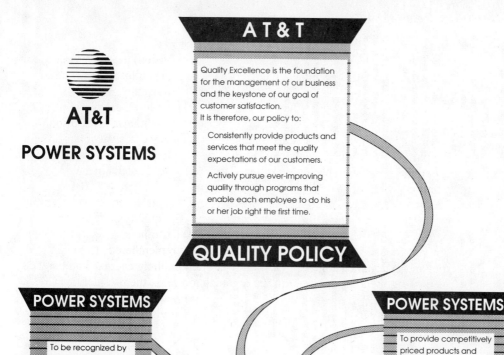

**AT&T POWER SYSTEMS**

**AT&T**

**QUALITY POLICY**

Quality Excellence is the foundation for the management of our business and the keystone of our goal of customer satisfaction.
It is therefore, our policy to:

Consistently provide products and services that meet the quality expectations of our customers.

Actively pursue ever-improving quality through programs that enable each employee to do his or her job right the first time.

**POWER SYSTEMS**

**VISION**

To be recognized by all of our customers around the world as the standard of excellence for power products by 1992.

**POWER SYSTEMS**

**MISSION**

To provide competitively priced products and services by empowering each employee to exceed the customer's expectations for quality, service, and features, while providing a fair return to our investors.

**POWER SYSTEMS THREE IMPERATIVES**

I **CUSTOMER SATISFACTION**

II **EMPLOYEE SATISFACTION**

III **STAKEHOLDER SATISFACTION**

**FIGURE 3.3**    AT&T Power Systems Division corporate quality intentions. Reproduced, with permission, from AT&T Power Systems, "Quality Policy, Vision, and Mission," AT&T Power Systems, Mesquite, TX.

QUALITY
QUALITY
QUALITY
QUALITY
QUALITY
QUALITY

$\sigma$
SIX SIGMA

Ⓜ MOTOROLA INC.

This booklet represents one of our efforts to communicate the quality principles that govern our business and help shape our company's future. Quality at Motorola is company policy aimed essentially at total customer satisfaction.

Take a few moments to read the following statement and see how committed we are to the subject.

*[signature]*
George M. C. Fisher
Chairman of the Board and Chief Executive Officer

*[signature]*
Gary L. Tooker
President and Chief Operating Officer

*[signature]*
Christopher B. Galvin
Senior Executive Vice President and Assistant Chief Operating Officer

## Corporate Quality Policy

Dedication to quality is a way of life at our company, so much so that it goes far beyond rhetorical slogans. Our ongoing program of continued improvement reaches out for change, refinement and even revolution in our pursuit of quality excellence.

It is the objective of Motorola, Inc. to produce and provide products and services of the highest quality. In its activities, Motorola will pursue goals aimed at the achievement of quality excellence. These results will be derived from the dedicated efforts of each employee in conjunction with supportive participation from management at all levels of the corporation.

## Corporate Quality Goal

1. Improve product and services quality ten times by 1989 and at least 100-fold by 1991.

2. Achieve plus or minus SIX SIGMA capability by 1992.

3. Provide total customer satisfaction. Be the best in all products and services as perceived by each customer.

## Why Six Sigma?

Sigma is a statistical unit of measurement that describes the distribution about the mean of any process or procedure. A process or procedure that can achieve plus or minus SIX SIGMA capability can be expected to have a defect rate of no more than a few parts per million, even allowing for some shift in the mean. In statistical terms, this approaches zero defects. Motorola's goal is to achieve this level of quality in everything we do by 1992.

## Implementation of Goal

Every Sector/Group/Staff organization and business unit within the corporation develops its own supportive policies, and the details are oriented toward every phase of its business. Each establishes and maintains regular improvement programs in product quality, reliability and services, with achievements targeted for SIX SIGMA capability by 1992.

The methodology implied in the Corporate Policy Statement is expanded in the following text.

I. Each Sector/Group/Staff organization and business unit will document and install a quality system detailing responsibility for executing its quality goal. Periodic reviews of the system are to be performed without exception to ensure continuing effectiveness.

II. Each Sector/Group/Staff organization and business unit will have a formal process for planning and achieving continuing improvements in quality and reliability of its products and services. Such processes will foster the permanent elimination of the cause of quality problems.

III. Individual Quality Assurance organizations are to perform as the customer's advocate in all areas of the business.

IV. Motorola will maintain a Corporate Quality Council, consisting of senior Quality Managers, for company-wide coordination, promotion and review of the various quality systems and programs to facilitate achievement of these policies.

**FIGURE 3.4** Motorola Corporation Six Sigma commitment to action. Reproduced, with permission, from "Quality, Six Sigma," Motorola Corporation, Schaumburg, IL.

us break down the goals to activity-related "chunks" we can address within our operations.

Objectives may be further broken down into targets and specifications. A "target" is aimed at a specific benchmark, usually quantitative in nature, which addresses a performance, cost, timeliness, and/or customer-service expectation. We might set a target of "greeting a customer within 30 seconds of entering our establishment." An engineering or technical ± (plus or minus) tolerance is termed a "specification" in this discussion. A specification, or requirement, is the tightest commitment to action possible. For our example, with respect to service, we might specify that "each customer is to be enthusiastically greeted by name and with a smile within 30 seconds." As another example, for a cup of coffee we might specify 150°F ± 10°F as the service temperature at the table. In summary, we see visions, we pursue goals, we accomplish objectives, we hit targets, we meet specifications.

**Positioning for Action**     Once our commitment to action is defined, developed, and declared, we must position for action in the most advantageous manner. *Strategies are developed to provide direction and help position our efforts to accomplish our purpose.* Well-formulated strategies provide direction, yet allow flexibility for repositioning or redirection as events unfold. For example, one of our goals might be to increase profits, supported by a strategy of exploiting CAD/CAM technology in our shop to reduce product development and manufacturing time.

*Plans add detail to strategies.* Plans do not have the same level of flexibility as strategies since they lay out action in a step-by-step manner. For example, our plan (with respect to our CAD/CAM strategy) will most likely involve acquiring hardware, software, machine tools, training, and perhaps part redesign for CAM, complete with the "what," "who," "when," and "where" necessary for accomplishing our objective. Here, we pursue our strategies and we carry out our plans.

## 3.5 QUALITY SYSTEM PATTERNS

It is clear that the five quality paradigms parallel advances in technology (e.g., regarding the requirements listed for each in Figure 3.1). Today, we still see each paradigm to one degree or another in our workplaces. *Quality systems vary widely, and most (in practice) manifest characteristics of more than one quality paradigm.* In practice, there is no one best quality system for all organizations. But each organization has a quality system, be it a result of either conscious design and careful nurturing or of haphazard evolution.

At this point, we must realize that *quality systems are indigenous to an organization and reflect philosophies* (both explicitly stated and implicitly held in confidence within the organization and its people) *and shape practices*. We also know that a great deal of interaction between organizations takes place in the form of supplier-purchaser relationships. *Successful interaction demands some degree of compatibility between organizations with respect to their quality systems.* The need for compatibility generally argues for some level of common standards. Since quality is universal, international standards are appealing, as long as they do not discourage improvement.

In the United States, *we see two primary influences shaping the definition, design, and development of quality systems: the ISO 9000 standards and the Baldrige Award.* The ISO 9000 standards series put forth by the International Organization for Standardization (ISO) headquartered in Geneva, Switzerland, is one major influence [3]. The ISO 9000 series is reflected in the ANSI/ASQC Q9000 series of standards in the United States [4]. Another major influence is the Malcolm Baldrige National Quality Award managed by the U.S. Department of Commerce and administered by ASQC [5].

## The ISO 9000 Series

*ISO 9000 provides guidance in quality system structure and operation for both manufacturing and service industries.* The foundation for the ISO 9000 series was laid in the 1980s. Work on the standard was initiated in 1979 and first published as a standard in 1987 [5]. The ISO 9000 series of quality standards is designed for international adoption. The intent is to develop a generic view and structure of quality systems in order to encourage international trade. Certification to standards, through a third-party auditing process, is intended. The ANSI/ASQC Q9000 series of standards are technically equivalent to the ISO 9000–9004 series, but incorporate customary "American" language and spelling. We will use the labels interchangeably.

ISO 9000 defines quality as

The totality of features and characteristics of a product or service that bear on its ability to satisfy stated or implied needs.

In general, *ISO 9000 recognizes four generic product categories* [6]:

1 *Hardware*—Hardware consists of manufactured pieces, parts, and assemblies.
2 *Software*—Software consists of written or recorded information, concepts, transactions, or procedures.
3 *Processed materials*—Processed materials include solids, liquids, gases, or combinations in various forms.
4 *Services*—Services consist of intangible products which may be principal offerings by themselves or part of tangible product offerings (e.g., planning, selling, delivering, training, operating, or servicing).

**ISO 9000 Structure** The ISO 9000 standard is introduced in the Vision 2000 document. It deals with five basic elements: (1) vocabulary, with respect to terms and definitions for international communications (ISO 8402), (2) quality management and quality assurance standards (ISO 9000), (3) specific quality assurance standards and practices (ISO 9001, 9002, 9003, and 9004), (4) guidelines for auditing quality systems (ISO 10011), and (5) quality assurance requirements for measuring equipment (ISO 10012). Figure 3.5 depicts the ISO quality-related documents and standards structure, including the ANSI/ASQC counterparts, available at the time of printing.

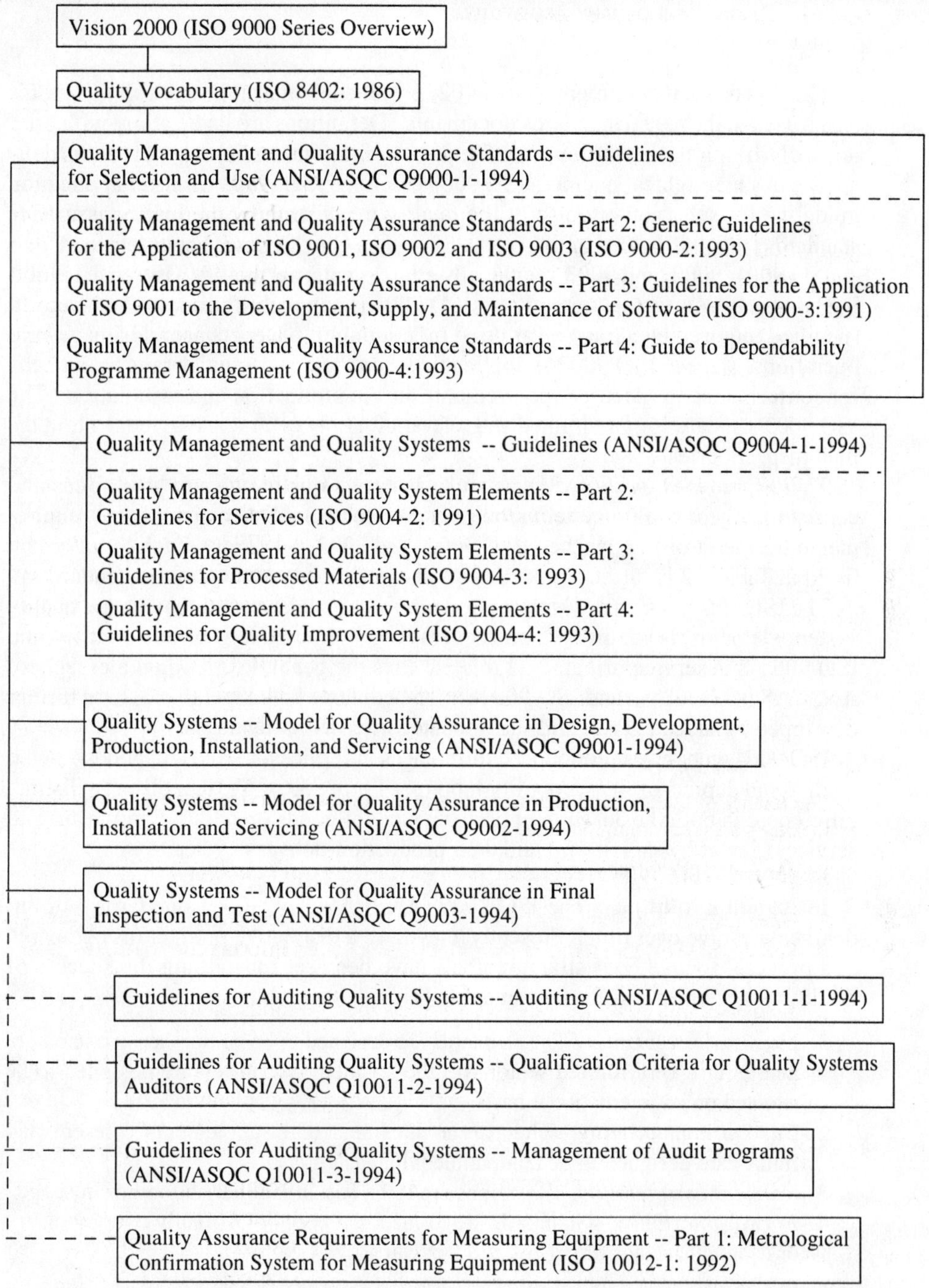

Vision 2000 (ISO 9000 Series Overview)

Quality Vocabulary (ISO 8402: 1986)

Quality Management and Quality Assurance Standards -- Guidelines
for Selection and Use (ANSI/ASQC Q9000-1-1994)

Quality Management and Quality Assurance Standards -- Part 2: Generic Guidelines
for the Application of ISO 9001, ISO 9002 and ISO 9003 (ISO 9000-2:1993)

Quality Management and Quality Assurance Standards -- Part 3: Guidelines for the Application
of ISO 9001 to the Development, Supply, and Maintenance of Software (ISO 9000-3:1991)

Quality Management and Quality Assurance Standards -- Part 4: Guide to Dependability
Programme Management (ISO 9000-4:1993)

Quality Management and Quality Systems -- Guidelines (ANSI/ASQC Q9004-1-1994)

Quality Management and Quality System Elements -- Part 2:
Guidelines for Services (ISO 9004-2: 1991)

Quality Management and Quality System Elements -- Part 3:
Guidelines for Processed Materials (ISO 9004-3: 1993)

Quality Management and Quality System Elements -- Part 4:
Guidelines for Quality Improvement (ISO 9004-4: 1993)

Quality Systems -- Model for Quality Assurance in Design, Development,
Production, Installation, and Servicing (ANSI/ASQC Q9001-1994)

Quality Systems -- Model for Quality Assurance in Production,
Installation and Servicing (ANSI/ASQC Q9002-1994)

Quality Systems -- Model for Quality Assurance in Final
Inspection and Test (ANSI/ASQC Q9003-1994)

Guidelines for Auditing Quality Systems -- Auditing (ANSI/ASQC Q10011-1-1994)

Guidelines for Auditing Quality Systems -- Qualification Criteria for Quality Systems
Auditors (ANSI/ASQC Q10011-2-1994)

Guidelines for Auditing Quality Systems -- Management of Audit Programs
(ANSI/ASQC Q10011-3-1994)

Quality Assurance Requirements for Measuring Equipment -- Part 1: Metrological
Confirmation System for Measuring Equipment (ISO 10012-1: 1992)

**FIGURE 3.5** ISO 9000 and ANSI/ASQC Q9000 standards structure.

The vocabulary document, ISO 8402, establishes international definitions for terms used in the actual standards documents. Definitions are listed along with brief notes clarifying the definitions. Subsequent definitions incorporate previous definitions in a hierarchical manner. ISO 9000 essentially serves as a master application guideline for selection and use of the quality management and quality assurance standards (ISO 9001-9004 and their ANSI/ASQC Q9000-Q9004 counterparts).

ISO 9001, 9002, and 9003 specifically address external quality assurance involving two-party contractual operations. ISO 9001 is applicable when conformance to specified requirements must be assured by a supplier to a purchaser during several operational stages. ISO 9002 is applicable when the contractual arrangement calls for conformance to specified requirements during production and installation. The ISO 9003 standard is more limited and relevant for only contractual agreements at the final inspection-test stage.

These three ISO, ANSI/ASQC standards have a quality system level focus and consist mainly of quality system requirements. Table 3.2 lists major system requirement considerations under the 9001, 9002, and 9003 standards. Each requirement listed in Table 3.2 is further broken down and described in the relevant standard(s).

The ISO 9004 standard, which sets guidelines for quality management and quality system elements, is subdivided into several parts: ISO 9004 is goods oriented and ISO 9004-2 is service oriented. Table 3.3 lists the general functional elements of ANSI/ASQC Q9004-1 and ISO 9004-2. The contents listed in Table 3.3 are further developed by subcategories and sub-subcategories in the standards.

ISO 9004 recognizes pronounced differences in services-related and goods-related quality and depicts each as a quality loop (see Figure 3.6). We can observe distinct differences in both the duration of performance for goods (generally long-term) and services (generally short-term) and their production-delivery cycles.

**ISO 9000 Evolution** The ISO 9000 (ANSI/ASQC Q9000) standards will undoubtedly evolve over time. Presently, they have a distinctive product (as compared to a process) focus. Four strategic goals have been set for judging the success of the ISO 9000 series and future standards [7]:

1 *Universal acceptance*—The standards are designed to be comprehensive in nature and require very limited sector-specific supplements. They are designed to be adopted on an international basis.
2 *Current compatibility*—The parent document and supplements (present and future) are designed to be compatible with each other.
3 *Forward compatibility*—Revisions are to be few in number and minor in scope.
4 *Forward flexibility*—Standards are to meet the needs of virtually every industry or economic sector or generic product category.

### Quality System and Award Criteria

Quality system strategies and tactics are also developed around quality award criteria. The two most prestigious awards, the Deming Prize (Japan) and the Malcolm Baldrige Award (United States), have received a great deal of attention in the United States.

**TABLE 3.2** ANSI/ASQC Q9001-, Q9002-, Q9003-1994 QUALITY SYSTEM REQUIREMENTS

| Requirement | Q9001-1994 | Q9002-1994 | Q9003-1994 |
|---|---|---|---|
| Management responsibility | ✓ | ✓ | ○ |
| Quality system | ✓ | ✓ | ○ |
| Contract review | ✓ | ✓ | ✓ |
| Design control | ✓ | | |
| Document and data control | ✓ | ✓ | ✓ |
| Purchasing | ✓ | ✓ | |
| Purchaser-supplied product | ✓ | ✓ | ✓ |
| Product identification and traceability | ✓ | ✓ | ○ |
| Process control | ✓ | ✓ | |
| Inspection and testing | ✓ | ✓ | ○ |
| Control of inspection, measuring, and test equipment | ✓ | ✓ | ✓ |
| Inspection and test status | ✓ | ✓ | ✓ |
| Control of nonconforming product | ✓ | ✓ | ○ |
| Corrective and preventive action | ✓ | ✓ | ○ |
| Handling, storage, packaging, preservation, and delivery | ✓ | ✓ | ✓ |
| Control of quality records | ✓ | ✓ | ○ |
| Internal quality audits | ✓ | ✓ | ○ |
| Training | ✓ | ✓ | ○ |
| Servicing | ✓ | | |
| Statistical techniques | ✓ | ✓ | ○ |

✓ comprehensive requirement.
○ less comprehensive requirement.

**TABLE 3.3**    ANSI/ASQC Q9004-1-1994 AND ISO 9004-2: 1991 QUALITY MANAGEMENT AND QUALITY SYSTEM ELEMENTS

| ANSI/ASQC Q9004-1-1994 (goods oriented) | ISO 9004-2: 1991 (services oriented) |
|---|---|
| Definitions | Definitions |
| Management responsibility | Characteristics of services |
| Quality system elements |    Service and service-delivery characteristics |
| Financial considerations of quality systems |    Control of service and service-delivery characteristics |
| Quality in marketing | Quality system principles |
| Quality in specification and design |    Key aspects of a quality system |
| Quality in purchasing |    Management responsibility |
| Quality of processes |    Personnel and material resources |
| Control of processes |    Quality system structure |
| Product verification |    Interface with customers |
| Control of inspection, measuring, and test equipment | Quality-system operational elements |
| Control of nonconforming product |    Marketing process |
| Corrective action |    Design process |
| Post-production activities |    Service-delivery process |
| Quality records |    Service performance-analysis and improvement |
| Personnel | |
| Product safety | |
| Use of statistical methods | |

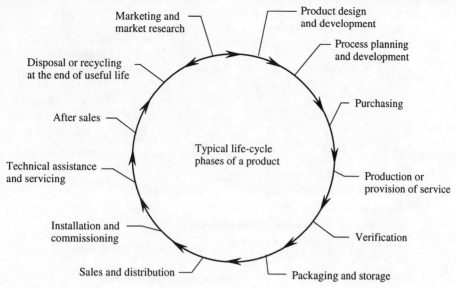

(a) Main activities having an impact on quality. (Reproduced from "ANSI/ASQC Q9004-1-1994," Milwaukee, WI: American Society for Quality Control, 1994.)

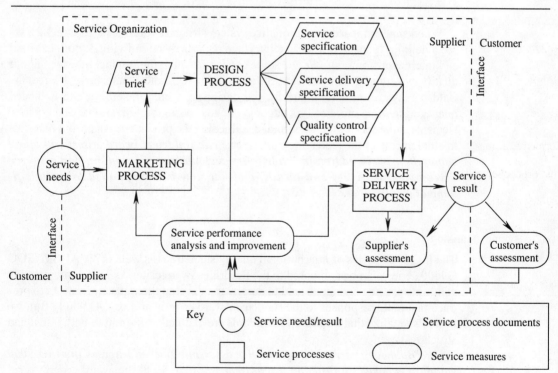

(b) Service quality loop. (Reproduced from "ISO 9004-2: 1991," Geneva, Switzerland, International Qrganization for Standardization, 1992.)

**FIGURE 3.6**    Product and service depictions.

55

The Baldrige Award is described in more detail in our subsequent discussion of total quality management (TQM) in Chapter 29. Here, we will overview the Baldrige Award criteria in order to contrast it with that of the ISO 9000 standard.

The 1995 Baldrige Award examination categories are listed in Table 3.4. An examination of the seven categories indicates that the award criteria are highly results- and performance-based. It is interesting to rank the seven categories by their percent contribution to the total award points [8]:

| | |
|---|---|
| Customer focus and satisfaction | 25.0% |
| Business results | 25.0% |
| Human resource development and management | 14.0% |
| Process management | 14.0% |
| Leadership | 9.0% |
| Information and analysis | 7.5% |
| Strategic quality planning | 5.5% |
| Total | 100.0% |

When superimposed on our quality system structure interfaces (Figure 3.2), we see a definite focus at the service interface regarding the individual (both external and internal customers). We see some concentration at the product interface in our quality system and minimal concentration at the technology interface. Hence, the Baldrige Award, as a template for the design of a comprehensive and balanced quality system, is more or less adequate in addressing organizational and societal elements at the service interface and in addressing the product interface, but marginal in all elements at the technology interface. *The Baldrige Award criteria are quite comprehensive in addressing short-term customer satisfaction, human resource development activities and results, and has now been expanded to include financial results.*

## Quality System Thinking

The ISO 9000 series of international quality standards (as well as the ANSI/ASQC Q9000 series) can be compared to the Baldrige Award criteria. This comparison is valid since some companies tend to design and develop their quality systems to produce favorable responses to the Baldrige Award criteria and/or ISO 9000. But, we must also realize that awards and standards are distinctly different in both intent and structure.

*The Baldrige criteria is performance assessment based whereas the ISO 9000 standard is based on functional requirements.* The Baldrige award criteria is results driven (as we would expect from an award criteria). On the other hand, ISO 9000 addresses structure and requirements with minimal descriptions of, or provisions to elicit results.

**TABLE 3.4** BALDRIGE AWARD EXAMINATION CATEGORIES AND EVALUATION GUIDELINES

| 1995 Examination Categories/Items | Point Values |
|---|---|
| **1.0 Leadership** | **90** |
|    1.1 Senior Executive Leadership . . . . . . . . . . . . . . . . . . . . . . . . . . . . . . . . . . 45 | |
|    1.2 Leadership System and Organization . . . . . . . . . . . . . . . . . . . . . . . . . . 25 | |
|    1.3 Public Responsibility and Corporate Citizenship . . . . . . . . . . . . . . . . . . 20 | |
| **2.0 Information and Analysis** | **75** |
|    2.1 Management of Information and Data . . . . . . . . . . . . . . . . . . . . . . . . . . 20 | |
|    2.2 Competitive Comparisons and Benchmarking . . . . . . . . . . . . . . . . . . . . 15 | |
|    2.3 Analysis and Uses of Company-Level Data . . . . . . . . . . . . . . . . . . . . . . 40 | |
| **3.0 Strategic Quality Planning** | **55** |
|    3.1 Strategic Development . . . . . . . . . . . . . . . . . . . . . . . . . . . . . . . . . . 35 | |
|    3.2 Strategic Deployment . . . . . . . . . . . . . . . . . . . . . . . . . . . . . . . . . . . . . . .20 | |
| **4.0 Human Resource Development and Management** | **140** |
|    4.1 Human Resource Planning and Evaluation . . . . . . . . . . . . . . . . . . . . . . 20 | |
|    4.2 High Performance Work Systems . . . . . . . . . . . . . . . . . . . . . . . . . . . . . 45 | |
|    4.3 Employee Education, Training and Development . . . . . . . . . . . . . . . . . . 50 | |
|    4.4 Employee Well-Being and Satisfaction . . . . . . . . . . . . . . . . . . . . . . . . . .25 | |
| **5.0 Process Management** | **140** |
|    5.1 Design and Introduction of Products and Services . . . . . . . . . . . . . . . . 40 | |
|    5.2 Process Management: Product and Service Production and Delivery . . . . . . . .. . . . . . . . . . . . . . . . . . . . . . . . . . . . . . . . . . . . 40 | |
|    5.3 Process Management: Support Services . . . . . . . . . . . . . . . . . . . . . . . . 30 | |
|    5.4 Management of Supplier Performance . . . . . . . . . . . . . . . . . . . . . . . . . 30 | |
| **6.0 Business Results** | **250** |
|    6.1 Product and Service Quality Results . . . . . . . . . . . . . . . . . . . . . . . . . . . 75 | |
|    6.2 Company Operational and Financial Results . . . . . . . . . . . . . . . . . . . . 130 | |
|    6.3 Supplier Performance Results . . . . . . . . . . . . . . . . . . . . . . . . . . . . . . . 45 | |
| **7.0 Customer Focus and Satisfaction** | **250** |
|    7.1 Customer and Market Knowledge . . . . . . . . . . . . . . . . . . . . . . . . . . . . . 30 | |
|    7.2 Customer Relationship Management . . . . . . . . . . . . . . . . . . . . . . . . . . . 30 | |
|    7.3 Customer Satisfaction Determination . . . . . . . . . . . . . . . . . . . . . . . . . . 30 | |
|    7.4 Customer Satisfaction Results . . . . . . . . . . . . . . . . . . . . . . . . . . . . . . . 100 | |
|    7.5 Customer Satisfaction Comparison . . . . . . . . . . . . . . . . . . . . . . . . . . . . 60 | |
|       TOTAL POINTS | **1000** |

Reproduced from "Malcom Baldrige National Quality Award—1995 Award Criteria," Washington, DC: U.S. Department of Commerce, 1994.

We see *the Baldrige criteria focus squarely on customer satisfaction, relative to marketplace competition.* On the other hand, *ISO 9000 focuses on international trade in a businesslike manner and attempts to expedite quality assurance linkages between suppliers and purchasers.* Both views are relevant and useful.

At this point, we will revisit our broad view of a quality system (Figure 3.2 and Table 3.1). In practice, each interface is unique to an individual organization, just as the customers served or prospective customers targeted will be unique. At the service interface, an organization's employees are unique. Each organization develops its own unique character and develops its own unique relationship with society. The technology interface is also unique in the way each organization selects and uses the sciences and resulting technologies—even though organizations select their technology from a more or less common pool.

Hence, at the detail level, *we see quality system uniqueness as a result of both the leadership and the creativity of people within an organization.* But, on the other hand, *we see similarities in quality systems from the structural and managerial viewpoints.* Diversity at the detail level provides opportunities for uniqueness and to compete successfully. Generality in structure and management allows organizations to form partnerships and interlink their quality systems in order to assure high quality through supplier-purchaser arrangements.

## 3.6  QUALITY SYSTEM INTEGRATION AND IMPLEMENTATION

In Chapter 1 we emphasized that knowledge without action as well as action without knowledge are equally ineffective (but in different ways) in producing quality improvement that leads to significant customer benefits, and benefits in general to individuals, organizations, and society. Our prospects for quality improvement are directly proportional to both the level of our relevant knowledge and of our enthusiasm and energy in our beliefs and actions, respectively.

### Intrasystem Integration

The general applicability (across organizations and industries) of quality system philosophies and standards that we examine in this text argues that quality systems have many common elements (e.g., respect for people, purpose, and communications). We also see a common pool of tools (e.g., discovery, assessment, relationship, control, and prediction based). Yet, we see unique integration arrangements and diverse implementation, driven primarily by the "vision," "style," and "sense of urgency" conveyed by leadership.

At this point, we want to focus our discussion on our choices and opportunities to define, design, and implement a quality system. *We see people-, product-, and process-related issues in the quality system all focused on addressing customer needs and expectations.* These issues must be dealt with and integrated to influence both short- and long-term customer satisfaction. People issues include leadership, management, empowerment, and creativity. Process issues include definition, design, development, production, delivery, sales and service, use, and disposal. Product issues include performance function, form, and fit, as well as timeliness and cost.

*In our workplaces,* we have seen both success and failure coming out of what appears to be (on paper) about the same general quality system recipe. In other words, successful results are not always repeatable from organization to organization. We can generalize and speculate that the *major differences between success and failure are attributable to our effectiveness in seven general areas:*

1  *Understanding the customer's experience of quality.*
2  *Leadership* through a clear, consistent, shared vision of quality and organizational purpose which is based on questioning, experimenting, confirming, acting (i.e., knowledge and action), and which is compatible with available resource bases.
3  *An environment which encourages creativity* directed toward customer benefits such that the individual, organization, and society will all be better off in the future.
4  *Mutual trust* between individuals within and between organizations (e.g., customers, suppliers, purchasers).
5  *Honest and reliable communications.*
6  *Energy and persistence* in mission (purpose) and strategy (direction) in the face of both success and failure (e.g., recognition for success and encouragement to turn failure into success).
7  *Visible achievement of purpose* (e.g., recognizable and meaningful signs of being better off; individuals, organizations, and society).

## Intersystem Integration

A quality system cannot operate in a vacuum. *The quality system must be integrated with other functional systems and provide the vehicle for cross-functional responses to opportunities and challenges.* It is entirely possible to see an organization with a good quality system structure (on paper) fail due to a lack of attention or success in integrating the quality system with other systems in the organization. Typically, this integration will involve seven functional relationships:

1  Marketing and sales.
2  Product-process design.
3  Purchasing and procurement.
4  Finance and accounting.
5  Production and production support.
6  Product support and service.
7  Quality assurance.

Ideally, this *intersystem integration should be somewhat "invisible," in that quality responsibility and authority belong to everyone in the organization.* In other words, quality assurance is indigenous to the other six functions, but with some unique responsibilities for quality specialists or facilitators. *A quality "re-indigenization" process is necessary in many organizations* since we somehow "specialized" quality out when we moved out of the custom-craft quality paradigm and into the mass production and sorting quality paradigm and beyond.

We should note that some organizations never lost sight of the indigenous nature of quality and some have re-indigenized rapidly and effectively. Yet other organizations are still struggling with the re-indigenizing process or are dominated by chronic disbelievers doing little but paying lip service to their quality systems. *Indigenous quality is the vision for our quality system structure.* Converting this vision into reality is challenging, but necessary in our competitive, customer-driven world.

*Empirical experience from the field is now available to support the indigenous quality concept.* AT&T Microelectronics, Power Products, Mesquite, Texas and Marlow Industries, Garland, Texas are both outstanding success stories in quality transformation. Their success in quality transformation is a result of many years of successful strategic planning and action. *The basis for their success is in viewing and integrating their business plan and quality plan together as one* (e.g., they have clearly established that quality can be and is most effective when it is indigenous to the business, rather than a separate business element or function).

The AT&T quality-business function has seven basic facets, as described by Cassidy and Sharma [9]: (1) mission and organization, (2) quality policy deployment, (3) education and training (4) quality improvement method/steps, (5) employee involvement, (6) poka-yoke (mistake-proofing), and (7) just-in-time/focused factory business units. For example, at the outset of quality transformation the plant received about 100 suggestions from employees per year, now they receive more than 5000 per year. They had no poka-yoke (mistake-proofing) devices installed at the outset, now they have over 2000 installed in both manufacturing as well as service areas.

Results to date fall into two distinct categories: recognition for quality by their customers, and business success (Sharma [10] and Sharma and Harrod [11]). Recognition for high quality includes:

- The Deming Prize (AT&T Power Products was the first U.S. manufacturer to win this Japanese prize).
- The Shingo Prize for Excellence in American Manufacturing.
- Pacific Bell Quality Partners Award.
- GTE Supplier Quality Award.
- Hewlett Packard Preferred Supplier Award.
- IBM Product Reliability and Delivery Performance Awards.

Business successes include:

- Average outgoing quality defect reduction of 70%.
- Rework reduction of 70%.
- First pass yield increase from 87% to 95%.
- Inventory reduction by over 50% and investment reduction of 43%.
- Customer order interval reduction of 50%.
- Manufacturing interval reduction of 70%.
- New product introduction interval reduction up to 50%.

Marlow Industries view their business in a holistic manner as a sociotechnical system with many functional needs (e.g., product engineering, process engineering, quality, and so on). However, they have pioneered an organizational structure which has eliminated these specialties as functional entities. *Their approach is one of refocus in business units where employees can capture a vision of the criticality of their relationships to their customers (internal and external) and take action directly to impact quality results.*

The vision of quality at Marlow Industries is that of an indigenous quality function within business units. This vision has been realized and recognized by their winning the Baldrige Award on their first application in 1991. Since then, they have continued their quality transformation in both results and recognition. In 1993 they were designated as one of America's Ten Best Plants by *Industry Week.*

There are, however, many documented cases, of varying degrees of success, of quality improvement efforts more or less isolated from or removed from business improvement effects as well as arguments in the technical literature that the two are somehow separable and should be judged as separate functions within an organization. Currently, the Return on Quality "ROQ" school of thought is gaining ground (Greising [12]). The thought is that quality can be justified using traditional short-term financial analyses. Although these "separability" contentions are continually cultivated by many theoreticians and practitioners, they are vigorously challenged on the grounds that they have been shown to be sub-optimal in theory as well as practice.

---

## Sporting Arms Manufacturing—Indigenous Quality Case [13]

The manufacturing plant of a sporting firearms manufacturer had been in operation for almost a century. At one time, the facility housed more than 10,000 workers, crammed into every available workspace. Like many manufacturers, however, the firm has taken some "hits" during the past few years. In fact, the company almost ceased to exist in the early 1980s and has been in and out of bankruptcy court twice in the past 10 years. Recently, a company from overseas bought the firearms plant with the goal of making the firm a world leader once again.

The current facility is an old six-story building of isolated rooms, long corridors, narrow aisles and low ceilings. Despite the plant's outdated design, the company realized that to "reemerge" it needed to proceed with plans for improving its competitive position, even if the cost associated with maintaining the old building would remain a tremendous burden. It was necessary to develop a new layout and eliminate much of the space the plant had previously used. The company emptied the top two floors, slashed inventory and material handling costs, and reduced lead times by 30 percent.

To help achieve its new quality commitment, the company made capital improvements using newer technology (e.g., purchasing new equipment and refurbishing the old). Newly acquired computer and numerically controlled and coordinate measurement machines provide the company with a much more accurate and precise manufacturing process, as well as enhanced flexibility.

Benchmarking was used to gain insight in process improvement in many areas. For example, benchmarking of an automobile manufacturing facility helped demonstrate the important role of ergonomics. The automobile manufacturer's design principles were incorporated into the various functions and production groups. The firm sought the expertise of a Malcolm Baldrige Award–winner to help with ideas relating to just-in-time production and to gain a better understanding of what makes a world-class manufacturing organization.

Total quality management and teamwork practices were adopted. A minimum of 30 hours of training per year per employee in courses that focus on setup reduction, statistical process control, and various other TQM topics is now provided. The new philosophy is essential to change the work culture that had existed for so many years. In the past, management's perception of what needed to happen on the plant floor was wrong. At one time, the firm's philosophy was to "handcuff" the operators to the machines, tell them they were expected to produce at 100 percent efficiency and make zero defect parts. Management missed the point for two reasons. First, they did not provide a machine that would run all day long to allow the operator to achieve 100 percent efficiency. Second, the equipment was so worn that it could not produce zero defect parts on a consistent basis. Now, workers have the authority to stop the process and the ability to monitor and control quality before a bad part is made.

While significant improvements have been realized at the old plant, the firm plans to combine JIT manufacturing with a totally new facility in order to achieve its long-term goals—to be the "best of the best" in the sporting arms industry.

Plans call for a new plant based on the "focused-factory" concept. Each of the four product lines will be contained within a focused factory—all on a single floor and all positioned in line. Bringing all things necessary to make a product into one area and all workers together in one product line will help employees feel that they are working on the whole product and not just one part. Applying the focused-factory concept will also mean that each of the product lines will be, for the most part, a "profit center." Results expected from the commitment to quality and modern manufacturing technology are lead time reductions from 8 weeks to 10 days, 40 percent reduction in facilities costs, 30 percent savings in energy consumption, 20 percent reduction in material handling costs, and improved throughput capacity.

## REVIEW AND DISCOVERY EXERCISES

### Review

3.1 Provide an example of each of the five quality paradigms regarding a product of your choice:
  a Custom-craft.
  b Mass production and sorting.
  c Statistical quality control.
  d Total quality management.
  e Techno-craft.

3.2 Identify a product that has evolved through the five quality paradigms. Explain what technologies have allowed the paradigm shifts and why the shift actually occurred.

3.3 Write a brief set of quality intentions for an organization of your choice. You may want to refer to Figures 3.3 and 3.4.

3.4 Develop a goal, objective, target, specification chain for the following:
  a An engineering design firm (e.g., building-, production facility-related).
  b A dry cleaning and laundry establishment.

    **c**  A food processing or manufacturing firm (e.g., dairy products, cereal).

    **d**  A foundry casting aluminum parts (e.g., aircraft-airframe brackets, custom wheels).

**3.5**  We make the statement that "every organization has a quality system."  Briefly describe the characteristics of a high-performance and a low-performance quality system with regard to the experience of quality and creation of quality goals and strategic elements shown in Table 3.1.

**3.6**  Identify the proactive quality strategy resource allocation profile and the reactive quality strategy resource allocation profile from the two profiles shown below. Briefly explain the differences and trade-offs.

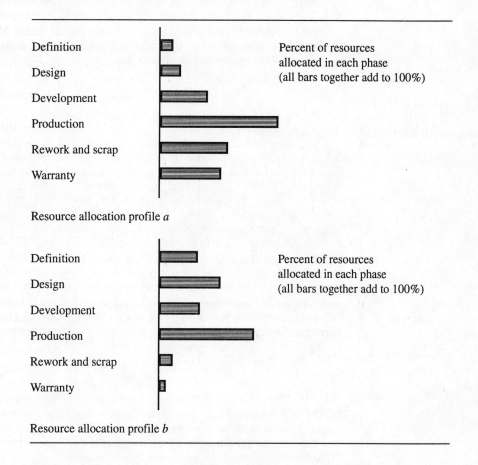

Resource allocation profile *a*

Resource allocation profile *b*

**3.7**  Develop a set of quality goals appropriate for each quality interface (product, technology, and service) for the following organizations. Then, point out the similarities and differences between the two sets of goals you developed.

    **a**  A nonprofit charitable service organization (e.g., United Way, Boy Scouts of America, Red Cross).

    **b**  An automobile manufacturer.

**3.8** Write a set of brief quality intentions for the following businesses. Then, compare and contrast the two sets of declarations. What is the same? What is different?

   **a** A fast-food restaurant (fish and chicken products).

   **b** An exclusive country club restaurant.

**3.9** Write a brief set of quality intentions for the following. Then, compare and contrast the two sets of declarations. What is the same? What is different?

   **a** A manufacturer of $0.25 to $1.00 disposable writing pen products.

   **b** A manufacturer of $100.00 to $500.00 writing pen products.

**3.10** Comment on the Tape Recorder Development Case as to why the quality system differences existed. Was one organization right and the other wrong? Explain.

**3.11** With regard to the Sporting Arms Manufacturing Case (involving indigenous quality) presented in this chapter, research and prepare a brief report on the following concepts. Then, comment as to what role quality plays in these concepts.

   **a** The JIT method of factory production.

   **b** The focused-factory concept.

# APPENDIX 3A: Economics of Quality

Different quality philosophies emphasize quality costs to different extents. For example, Juran compares quality systems to financial systems and stresses the economic impact of quality [14, 15, 16]. Crosby [17] is a strong advocate for economic measures (e.g., his cost of nonconformance). On the other hand, Deming tends to de-emphasize the criticality of concentrating on quality cost measures [18]. Ishikawa includes economics in his fundamentals [19], but does not emphasize economics to the extent of Juran and Crosby.

A sound quality system contributes to customer satisfaction and organization objectives, including financial stability. Significant quality costs are normally overlooked or unrecognized in many organizations simply because most accounting systems are not designed to identify them. The same arguments can be posed toward quality-related revenues associated with quality improvement through product and production process improvements. When we can effectively deal with both the expense and revenue sides of the issue, we can truly develop effective quality systems which play both proactive and reactive roles in furthering strategic financial goals.

## Traditional Quality Cost Thinking

Quality cost objectives have led organizations to develop programs of evaluating quality costs: (1) to quantify the size of the quality program in language that will have a significant impact on upper management, (2) to identify major opportunities for cost reduction, (3) to identify opportunities for reducing customer dissatisfaction and associated threats to product salability, (4) to expand budgetary and cost controls, and (5) to stimulate improvement through publication of quality costs [20]. The strategy for using quality costs is straightforward: (1) identify failure costs and drive them to zero, (2) invest in the right prevention activities to bring about improvement, (3) reduce appraisal (inspection) costs by improving production processes, (4) continuously evaluate and redirect prevention efforts to gain further improvement. This strategy is based on the premise that for each failure there is a root cause, causes are preventable, and prevention is cheaper than quality failures.

## Traditional Quality Cost Management

*Quality costs are defined as* the total of the costs incurred by (1) investing in the ***prevention*** of nonconformance to requirements, (2) ***appraisal*** of a product or service for conformance to requirements, and (3) *failure* to meet requirements [21].

Prevention costs are the costs of all activities specifically designed to prevent nonconformance in deliverable products or service. Appraisal costs are the costs associated with measuring, evaluating or auditing products or services to assure conformance with quality standards and performance requirements. Failure costs are the costs incurred to evaluate and either correct or replace products or services not conforming to requirements or customer/user needs.

Traditional quality cost programs consist of the following steps:

**1** Establish a quality cost measurement system.
**2** Develop a suitable long-range trend analysis.
**3** Establish annual improvement goals for total quality costs.
**4** Develop short-range trend analysis with individual targets that collectively add up to the incremental demands of the annual improvement goal.
**5** Monitor progress against each short-range target and take appropriate corrective action when targets are not being achieved.

The cost element of a quality system should be introduced in a positive manner. It will typically expose a high degree of waste, error, and misdirected expenditures (which are unnecessary in an organization which is well managed for quality). For this reason, it is extremely important that all affected employees, starting with management, be carefully informed and understand that quality cost analysis is a tool for improving the economics of operation and not a means to expose poor management practices and/or sloppy workmanship. A summary of the quality cost elements for each quality cost category is provided in Table 3.5.

## Quality Cost Measurement Bases

The major function of the cost analysis component of a quality system is to help identify opportunities for improvement and then provide a measurement of the resulting improvement over time. In order to expedite this function, it is necessary to provide measurement bases against which quality costs can be compared. Since the volume of business in total or in any particular product or service line will vary with time, real differences (improvements) in the cost of quality can best be measured as a percent of some appropriate base.

For long-range analyses, net sales is the base most often used for presentations to management [22]. For example, the total cost of quality may be 9.0 percent of sales. Other high-level quality cost measures include (1) internal failure costs as a percent of total production costs, (2) external failure costs as an average percent of net sales, (3) procurement appraisal costs as a percent of total purchased material costs, (4) operations appraisal costs as a percent of total production costs, and so on. While these measures may be important from a strategic planning point of view, they are too broad to deal with quality issues on the production floor. Here, we need a short-range base that directly relates the cost of quality to the amount of product produced in each process. We must develop meaningful, timely quality cost measures at the operational level. However, we should realize that there is no perfect base, and each can be somewhat misleading if used alone. Quality metrics (measures) are further discussed in Chapter 29.

**TABLE 3.5**  QUALITY COST ELEMENTS SUMMARY

| 1.0 | Prevention costs | 2.0 | Appraisal costs |
|---|---|---|---|

**1.1 Marketing, customer, user**
1.1.1 Marketing research
1.1.2 Customer-user perception surveys/clinics
1.1.3 Contract/document review

**1.2 Product, service, design development**
1.2.1 Design quality progress reviews
1.2.2 Design support activities
1.2.3 Product design qualification test
1.2.4 Service design qualification
1.2.5 Field trials

**1.3 Purchasing**
1.3.1 Supplier reviews
1.3.2 Supplier rating
1.3.3 Purchase order tech-data reviews
1.3.4 Supplier quality planning

**1.4 Operations (manufacturing or service)**
1.4.1 Operations process validation
1.4.2 Operations quality planning design and development of quality measurement and control equipment
1.4.3 Operations support quality planning
1.4.4 Operator quality education
1.4.5 Operator SPC/process control

**1.5 Quality administration**
1.5.1 Administrative salaries
1.5.2 Administrative expenses
1.5.3 Quality program planning
1.5.4 Quality performance reporting
1.5.5 Quality education
1.5.6 Quality improvement
1.5.7 Quality audits

**1.6 Other prevention**

**2.1 Purchasing appraisal**
2.1.1 Receiving or incoming inspections and tests
2.1.2 Measurement equipment
2.1.3 Qualification of supplier product
2.1.4 Material source inspection and control programs

**2.2 Operations (manufacturing or service) appraisal**
2.2.1 Planned operations inspections, tests, audits
 • Checking labor
 • Product or service quality audits
 • Inspection and test materials
2.2.2 Setup inspections and tests
2.2.3 Special tests (manufacturing)
2.2.4 Process control measurements
2.2.5 Laboratory support
2.2.6 Measurement equipment
 • Depreciation allowances
 • Measurement equipment expenses
 • Maintenance and calibration labor
2.2.7 Outside endorsements and certifications

**2.3 External appraisal**
2.3.1 Field performance evaluation
2.3.2 Special product evaluations
2.3.3 Evaluation of field stock and spare parts

**2.4 Review of test and inspection data**

**2.5 Miscellaneous quality evaluations**

Reprinted, with permission, from J. Campanella (ed.), *Principles of Quality Costs*, 2d ed, Milwaukee, WI: American Society of Quality Control, pages 112 and 113, 1990.

**TABLE 3.5** *(continued)*

| 3.0 | Internal failure costs | 4.0 | External failure |
|-----|------------------------|-----|------------------|
| **3.1** | **Product or service design failure (Internal)** | 4.1 | Complaint investigations and customer or user service |
| 3.1.1 | Design corrective action | | |
| 3.1.2 | Rework due to design changes | 4.2 | Returned goods |
| 3.1.3 | Scrap due to design changes | | |
| 3.1.4 | Production liaison | 4.3 | Retrofit |
| | | 4.3.1 | Recall |
| **3.2** | **Purchasing failure** | | |
| 3.2.1 | Purchased material reject disposition | 4.4 | Warranty claims |
| 3.2.2 | Purchased material replacement | | |
| 3.2.3 | Supplier corrective action | 4.5 | Liability |
| 3.2.4 | Rework of supplier rejects | | |
| 3.2.5 | Uncontrolled material losses | 4.6 | Penalties |
| **3.3** | **Operations (product or service) failure** | 4.7 | Customer/user goodwill |
| 3.3.1 | Material review and corrective action | | |
| | • Disposition | 4.8 | Lost sales |
| | • Troubleshooting or failure analysis | | |
| | • Investigation support | 4.9 | Other external-failure |
| | • Operations corrective action | | |
| 3.3.2 | Operations rework and repair | | |
| | • Rework | | |
| | • Repair | | |
| 3.3.3 | Reinspection/retest | | |
| 3.3.4 | Extra operations | | |
| 3.3.5 | Scrap (operations) | | |
| 3.3.6 | Downgraded end product or service | | |
| 3.3.7 | Internal failure labor losses | | |
| **3.4** | **Other internal failure** | | |

# REFERENCES

1  "Quality Policy, Vision, and Mission," AT&T Power Systems, Mesquite, TX.

2  "Quality, 6 Sigma," Motorola Corporation, Schaumburg, IL.

3  "ISO 9000," Geneva, Switzerland: International Organization for Standardization, 1992.

4  "ANSI/ASQC Q9000-9004-1994," Milwaukee, WI: American Society for Quality Control, 1994.

5  "Malcolm Baldrige National Quality Award—Applications Guidelines," Washington, DC: U.S. Department of Commerce, 1994.

6  See reference 3.

7  See reference 3.

8  See reference 5.

 **9** M. P. Cassidy and S. P. Sharma, "1990's Revolution in Power Systems Manufacturing," Applied Power Electronics Conference, Boston, MA, February 23–27, 1992.

**10** S. P. Sharma, "A Journey Towards Excellence," International Magnetics Conference, Indianapolis, IN, April 25–28, 1994.

**11** S. P. Sharma and W. L. Harrod, "Total Quality Management Process to Achieve Excellence," Applied Power Electronics Conference, San Diego, CA, March 7–11, 1993.

**12** D. Greising, "Quality: How To Make It Pay," Business Week, pp. 54–59, August 8, 1994.

**13** G. A. Ferguson, "U.S. Repeating Arms Sets Sights on World-Class Facility," *Industrial Engineering,* vol. 25, no. 3, pp. 30–32, March 1993.

**14** J. M. Juran, *Juran's Quality Control Handbook,* McGraw-Hill, New York, 1988.

**15** J. M. Juran, *Juran on Leadership for Quality,* New York: Free Press, 1989.

**16** J. M. Juran, *Juran on Planning for Quality,* New York: Free Press, 1988.

**17** P. B. Crosby, *Quality Is Free,* New York: McGraw-Hill, 1979.

**18** W. E. Deming, *Out of the Crisis,* Cambridge, MA: MIT Center for Advanced Engineering Studies, 1986.

**19** K. Ishikawa, *What Is Total Quality Control? The Japanese Way,* Englewood Cliffs, NJ: Prentice-Hall, 1985.

**20** See reference 14.

**21** J. Campanella (ed.), *Principles of Quality Costs,* 2d ed., Milwaukee: American Society of Quality Control, 1990.

**22** See reference 21.

# EXPERIENCE OF QUALITY

## VIRTUES FOR LEADERSHIP IN QUALITY
### Wisdom

The purpose of Section Two is to describe the experience of quality, primarily from the customer's point of view. Chapter 4 develops a human needs-based quality experience model. The model is relevant to the quality experience of both external and internal customers. It is based on consumer behavior concepts, relative to customer satisfaction. Quality chains are developed to describe the transition that people make from external to internal customers as supplier-purchaser linkages form between organizations involved in production.

Chapter 5 examines both true and substitute quality characteristics dealing with physical and human performance. Both short-term and long-term quality characteristics are discussed. Performance in the customer's environment, the field, is stressed, and differences between field and laboratory performance emphasized. The concepts of failure modes and failure mechanisms are presented. Our readers are introduced to both hard and soft failures in the context of both hardware and software.

Chapter 6 focuses on both robust and mistake-proof performance. The Taguchi trilevel design hierarchy concept is presented. The "target" and "goal post" product requirements are introduced as well as the concepts of location and dispersion in performance. We examine the components that define the customer's field experience in the context of robust performance. We discuss the zero defects and six sigma concepts, as well as the concept of source inspection. Finally, we introduce the reader to a strategic mistake-proofing hierarchy useful in the definition and design of both products and processes.

# HUMAN NEEDS AND QUALITY CHAINS

## 4.0 INQUIRY

1 How are human needs structured?
2 How is the experience of quality formulated and expressed?
3 What are the fundamental necessities for a cooperative effort?
4 What is a "chain of quality"?
5 How do customer roles change in quality chains?

## 4.1 INTRODUCTION

*Our focus in this chapter is on the "human" factor in the experience of quality.* We seek to understand the basics of human needs, the experience of quality, cooperative efforts, and the concept of quality chains. Our objective is a fundamental understanding of the individual-organization-society relationship of our service triad (see Figure 3.2). The two primary human performance bottlenecks in the triad are lack of effective leadership in the organizational and societal elements and lack of creativity in the individual element.

## 4.2 HUMAN NEEDS

Humans are very complex—biologically, psychologically, and socially. *Human-needs structures drive human behavior directly and indirectly.* Human needs are defined by Hellriegel et al. [1] as deficiencies that a person experiences at a particular time. Needs are typically viewed as energizers in stimulating human behavior.

Although a number of human needs models have been proposed, we will focus on Maslow's needs hierarchy model. *Maslow [2] has described a very broad hierarchy of human needs that can be ordered into five general categories: (1) physiological, (2) safety and security, (3) social, (4) esteem, and (5) self-actualizational.* This hierarchy is depicted in Figure 4.1.

The first two need categories are fundamental for human survival. *Physiological needs are the most basic and consist of primary life sustaining requirements* such as water, food, shelter, and clothing. *The safety-security needs include self-preservation.* For example, protection from physical elements and dangers are safety- and security-related needs.

*The third level need described by Maslow focuses on social and affiliation needs.* This need to "belong" is very significant because its focus extends beyond the individual to the group. Affiliations may be formed on the basis of personal or pro-

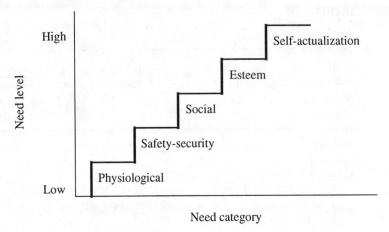

**FIGURE 4.1**   Human needs hierarchy depiction.

fessional characteristics (e.g., joining an engineering association), shared interests, or shared beliefs (e.g., joining a religious organization).

*The esteem level goes beyond the social level in that it relates to both self-esteem and group recognition.* Esteem tends to develop the feelings of self-confidence, prestige, and power in an individual. Here, self-confidence is our belief in our own abilities, knowledge, and skills. Prestige is to some degree a recognition of importance bestowed on an individual by others. It may result from possessions or achievement. Power is a force manifested by compliance from or influence over others and typically results from either personality, position, or situation bases, or a combination of these.

The fifth level is self-actualization. *Self-actualization is the highest order and most intangible need in the Maslow needs structure. It is the pursuit and realization of becoming or being all that you want to or can be* (i.e., fulfilling what is important to you). For example, an artist creates his or her best composition, a machinist produces zero defect parts, an engineer creates unique product features, a work team leader inspires his or her team to process improvement, and so on.

According to the Maslow model, lower level needs are fulfilled before higher level needs are addressed by an individual. In other words, we generally seek to fulfill physiological needs first, then move to the next level of safety and security needs, and so on. In practice, this sequential hierarchy is reasonable, but an individual may also seek to fulfill these needs simultaneously.

Obviously, human needs are much more complex than the Maslow model indicates. But, for our purposes, the model serves to demonstrate the various types of needs individuals seek to fulfill. *Each individual, given his or her genetic makeup, personality, and life experiences, possesses a unique and detailed needs structure that is dynamic over time.*

## 4.3  EXPERIENCE OF QUALITY

*Both external and internal customers develop experiences of quality unique to each and every individual.* We will draw heavily on the consumer behavior literature in order to formulate and support our experience-of-quality concept and individual-quality-experience (IQE) model. In our treatment of the subject, we are seeking a fundamental understanding of customer behavior and will not attempt to speculate on means of modifying customer behavior. The major distinction between our view of the experience of quality and consumer behavior in general is that we are focusing on customer satisfaction (a long-term phenomenon) while a great deal of consumer behavior work focuses on product selection (attitudes of and behavior in purchases and how to influence these attitudes and behaviors).

*Our macro-level IQE model,* depicted in Figure 4.2, *is human needs driven and contains four phases*: **(1) observation, (2) assessment-interpretation, (3) attitude, and (4) behavior. The IQE model is based on three types of processes, (1) sensory-perception, (2) thinking, and (3) decision-initiative.** Our IQE model sheds a unique light on the quality field. It is based on and constructed from many models (and terms) borrowed from the consumer behavior and the organizational behavior fields.

### Human Needs in IQE

*Human needs are the driving force behind the IQE model.* The Maslow hierarchy (Figure 4.1) serves as the focal point of our IQE model. The resultant complex of physiological, safety and security, social, esteem, and self-actualization needs shapes the experience of quality. External customers pursue products that will meet their needs (whether or not they can precisely verbalize their needs); likewise for internal customers. In most cases, these needs seem to be intertwined and confounded together. As we constantly take on the role of both external and internal customers each day, we proceed each day building onto and deleting from our previous day's set of needs.

### IQE Processes and Experience Phases

As we traverse the IQE cycle, we will discuss IQE processes and experience phases together since they act together. These processes and phases tend to overlap and interact to the point that a linear-based model in any pure form will be inaccurate. We will attempt to discuss our model one piece at a time, recognizing that quality related behavior is complex as well as somewhat unpredictable.

**The Observation Phase and the Sensory-Perception Processes**   The observation phase has been placed at the top of the IQE cycle only because it represents a good place to start our description. *The observation phase represents the input stage where customers draw information from their physical and social environments.* For example, when we put on a pair of shoes and become active, we have almost a limitless amount of data available to us regarding our shoes in the context

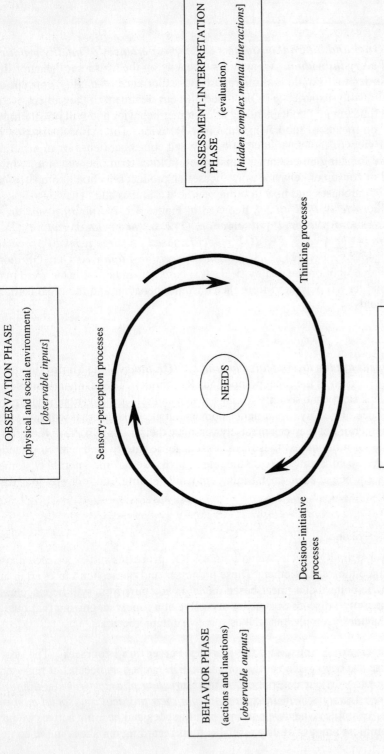

**FIGURE 4.2** Experience of quality concept: Individual Quality Experience, IQE model—processes and phases.

of our physical and social environments (e.g., how they look to us and to others, how they feel on our feet, how they sound when we walk).

*The sensory-perception process involves exposure to the environment.* As a matter of exposure to the physical and social environment, we use our sense organs to gather sensory signals (stimuli) and translate the stimuli into neural impulses that travel through the nervous system to the brain. The senses of sight, hearing, smell, taste, and touch are an individual's only contact with the product (external customers) or the production processes (internal customers).

Sensing the environment is critical, but an individual simply cannot attend to all stimuli present at any one time. An individual's attention serves as a screen, or filter, for recognizing some stimuli as more important than others in the vast array of stimuli present at any one time. Attention to and the selection of the most important stimuli depends on three complex factors, according to Hawkins et al. [3]:

1 The nature of the stimuli present—such as intensity, position in the environment, contrast with other stimuli, format of presentation—and the quantity of stimuli present at any one time.
2 The individual's most pressing needs at the time.
3 The situation as represented by the natural course of other events or endeavors with which the individual is or has been involved (time pressures, past experience, and so on).

"Perception" is defined by Moorehead and Griffin as the set of processes by which the individual becomes aware of and interprets information about the environment [4]. The concept of "perceptual selection" is defined by Hellriegel et al. as a process where people filter out most stimuli so that they can deal with those that are more important to them [5]. Through this filtering process we tend to select stimuli that we believe will help us to address our needs.

**The Assessment-Interpretation and Attitude Phases and the Thinking Processes** *The input to the assessment phase consists of the filtered (perceived) stimuli (the output from the sensory-perception process).* Here, we see the thinking processes taking charge. "Thinking" involves mental representations of our world and ourselves, and the manipulation and further development of these representations and beliefs to aid in addressing some goal (Osherson and Smith [6]).

There is general agreement in the literature (Kinnear and Taylor [7], Moorehead and Griffin [8], Hellriegel et al. [9], Peter and Olson [10]) that *thinking involves two components:* (1) *the affective* and (2) *the cognitive.* These two components were introduced earlier in Appendix 1A and were associated with the learning process.

*The affective component involves feelings, sentiments, moods, emotions, and so on, relative to people, ideas, events, and objects;* it can be manifested by joy, anger, frustration, complacency, preferences (likes-dislikes), and so on. The affective component is believed to be a relatively quick, reactive response which comes about automatically (see Peter and Olson [11]). It produces physical responses that are detectable in the human body (e.g., excitement or anger in voice, varied facial

**TABLE 4.1** TYPES OF AFFECTIVE RESPONSES

| Type of affective response | Level of physiological arousal | Intensity or strength | Examples of positive and negative affects |
|---|---|---|---|
| Emotions | Higher arousal and activation | Stronger | Joy, love<br>Fear, guilt, anger |
| Specific feelings | ↑ | ↑ | Warmth, appreciation<br>Satisfaction<br>Disgust, boredom |
| Moods | ↓ | ↓ | Alert, relaxed, calm<br>Blue, listless |
| Evaluations | Lower arousal and activation | Weaker | Like, good, favorable<br>Dislike, bad, unfavorable |

expression, changes in heart rate). Table 4.1 illustrates the affective component in a consumer behavior context.

*The cognitive component involves mental processes and structures involved with an individual's response to his or her physical and social environments.* Peter and Olson state that the cognitive system makes "sense" of (interprets or understands) the environment [12], developing or forming cognitive representations—subjective meanings that represent personal interpretations. Table 4.2 provides a selected set of cognitive representations in a consumer behavior context.

Cognitive knowledge structures are of two basic types, according to Peter and Olson [13]: (1) schema and (2) script. A schema represents declarative knowledge. It is a network of interrelated meanings regarding an individual's knowledge about a concept or object (e.g., for clothing—size, color, how to wear, where to buy). A script represents procedural knowledge relative to what constitutes appropriate (to the individual) behavior (e.g., for a fast-food establishment—enter, walk to the counter, place an order). These cognitive knowledge structures are continuously undergoing modifications as a result of new information and/or new insights about relationships among existing knowledge structures.

The thinking process is not well understood. General agreement as to the nature of interactions and dominance of the affective-cognitive components is not evident in the literature. In support of the affective component, Smith contends that thinking is much more than information processing—it is action, overt or imagined [14]. That is, doing something and thinking about doing it are inseparable; we cannot do

**TABLE 4.2**    TYPES OF MEANINGS CREATED BY THE COGNITIVE COMPONENT

| | |
|---|---|
| Cognitive representations of physical stimuli:<br>  This sweater is made of lambswool.<br>  This car gets 28 miles per gallon. | Cognitive representation of symbolic meanings:<br>  This car is sexy.<br>  This style of dress is appropriate for older women.<br>  Wearing a Rolex watch means you are successful. |
| Cognitive representation of social stimuli:<br>  The salesperson was helpful.<br>  My friends think Pizza Hut is the best. | Cognitive representations of sensations:<br>  Colors on a box of breakfast cereal.<br>  Sound of a soft drink can being opened and poured.<br>  Sweet taste of chocolate chip cookies.<br>  Smell of your favorite cologne.<br>  Feel of your favorite pair of jeans. |
| Cognitive representations of affective responses:<br>  I love Dove (ice cream) bars.<br>  I feel guilty about not sending a birthday card.<br>  I feel mildly excited and interested in a new store. | Cognitive representations of behaviors:<br>  I drink a lot of Diet Pepsi.<br>  How to pay with a credit card. |

Reprinted, with permission from Richard D. Irwin, from J. P. Peter and J. C. Olson, *Consumer Behavior and Marketing Strategy*, 2d ed., Homewood, IL: Richard D. Irwin, page 47, 1990.

things without thinking, and we cannot think without contemplating doing things. Smith stresses that thinking and learning inevitably involve feelings and that the way in which a person thinks may be primarily determined by emotional or personality considerations rather than intellectual ability.  Smith views thinking in terms of tense: "remembering" is thinking in the past tense, "understanding" is thinking in the present tense, and "learning" is thinking in the future tense.

On the other hand, scholars such as Osherson and Smith [15] view thinking as primarily a cognitive process involving memory, categorization, judgment, and choice.  Here, memory serves as a repository for beliefs and representations, while memory retrieval ability limits our thinking process.  Categorization is the assignment of objects and events to individual experience categories.  Judgment deals with the uncertainty of future events anticipated.  Choice is essentially our means of deciding between options.

Our intuitive experience in quality related work argues that both affective and cognitive components are present in any given thought process.  The IQE model recognizes the existence of both.  For example, we may verbalize a product experience as, "I bought my new sports car because I really love the way it looks and feels and I believe it is dependable since *Consumer Reports* found it to be the most trouble-free automobile in its class."  Furthermore, it is likely that the affective and cognitive components interact in a synergistic manner.  For example, if we like or dislike a product (the affective component), we may use the cognitive component to justify and amplify this feeling.

The thinking process results in the formation of attitudes.  Attitudes are relatively lasting feelings, beliefs, and behavioral intentions directed toward a specific person, group, idea, or object (Hellriegel et al. [16]).  The attitude phase follows the assessment-interpretation phase in the IQE model.  At this point in the IQE cycle, thoughts, feelings, beliefs, and behavioral intentions have emerged from the thinking process and represent measurable outputs (although our ability to measure them may be somewhat crude and unreliable).

**The Behavior Phase and Decision-Initiative Processes** *The inputs to the decision-initiative process are outputs from the thinking processes in the form of thoughts, feelings, beliefs, and behavioral intentions.  In turn, the decision-initiative process produces or results in behavioral actions or inactions on the part of the individual.*  Many possible behaviors can be initiated; i.e., multiple behavioral options exist.  A behavioral intention and a resulting behavior are selected through a decision process.

In the consumer behavior context (regarding product purchases), Peter and Olson [17] describe the decision process in the context of a decision frame.  The decision frame is a cognitive representation of the problem which includes an end goal, a structured set of subgoals, relevant product knowledge, and a set of choice "rules." They state that a consumer's decision does not typically follow the classical linear decision model of problem recognition, a search for alternatives, evaluation, and purchase, and finally, a reevaluation of the product in light of the performance it yields.  They cite three basic reasons for deviation which appear relevant to our IQE model:

1  Consumers may not wait to find or evaluate alternatives.
2  Consumer decision making involves ongoing, complex reciprocal interactions between cognitive and affective processes, behavioral intentions, and the physical and social environment.
3  Most decision alternatives are not isolated, but interrelated in the context of multiple problems and needs and multiple alternative actions and inactions.

Here, "behavior" is defined as what an individual has done or is doing (Kinnear and Taylor [18]).  In other words, behavior refers to observable actions in the quality experience.  For example, an individual telling a friend about their satisfaction with a product is a desirable external-customer behavior.  Expressing job satisfaction in the workplace to a friend is an example of an internal customer's quality experience as manifested through behavior.  Internally (to our customers), thoughts, beliefs, and feelings regarding our products represent the manifestation of the quality experience.  Behavioral intentions may or may not manifest themselves in behaviors, but they are also part of an individual's quality experience.

*In the quality context, the outputs at both the behavior and attitude phases represent an individual's responses to products and processes and constitute customer satisfaction (or dissatisfaction).*  Hence, a quality experience may be manifested in both attitudes and behaviors (see Figure 4.2).

## Sewing Machine—Experience of Quality Case [KD]

My mother has owned a sewing machine for the last 25 years. The first two sewing machines that she purchased were both Dancer machines. Over the years, she had experienced only minor problems with her Dancer products. About 5 years ago, she decided that she wanted a new, computerized sewing machine. She began looking at and trying out the latest models to see exactly what features within a certain price range were available. She decided to buy a Gray sewing machine. The Gray had a computer control that enabled it to do many fancy stitches, button-holes, and various other features.

She was thoroughly pleased with her new purchase for several months—almost a year. Then one day, her machine just quit working. Angrily she packed the machine down to the local fabric store so that it could be fixed. The circuit board had failed. While the Gray machine was being fixed, my mother brought the Dancer machine out of storage and proceeded to sew. After the Gray machine was repaired, she packed away the Dancer and went about sewing clothes for my sister and me. Then, the Gray failed again.

On five separate occasions over the next three years the Gray machine had to be taken in for either a power board or circuit board problem. My mother was tired of having to pack and unpack sewing machines. Each time the Gray had to be taken in to be fixed, she became more displeased with her purchase. She had experienced few problems with the Dancer machines. Finally, she wrote a letter to the president of Gray in hopes of conveying her displeasure in his product and asked that he look into the problem. A few weeks later, she received a letter informing her that a new machine was being sent to her and to return the old machine. However, the new machine that she was actually provided was a different model with only about one-third of the features of the original machine.

The quality experience that my mother went through can, in general, be related to the IQE model. The Observation Phase and the Sensory-Perception Process occurred as my mother was searching for a new sewing machine. Based on her past experience, she had certain features that she expected of her new machine, such as fancy stitching, button hole memory, and so on. When she had evaluated all the machines, she moved into the Assessment-Interpretation Phase. The thinking processes yielded a belief that the Gray machine was the "best" alternative. The Gray machine met her needs and expectations. Then, through the Attitude Phase and into the Behavior Phase, she decided and took action to purchase the new Gray machine.

As she used the machine, the IQE cycle continued. Through observation, assessment, attitudes, and behavior she displayed visible signs of a satisfied customer (for the first several months). Then, after the continual failures of the Gray machine, the IQE cycle continued, but with a different result. She began thinking about her purchase and the quality of what she had purchased. She began to question her purchase and wonder if she had not made a giant mistake. She began to be angry with the company who produced such poor quality (as perceived or judged by her). She developed a negative attitude toward the company. Her negative attitude led her to another decision: to contact the president in hopes of some kind of relief from her problem, her money back or a machine replacement. Even with the company's shipment of a new machine (of lesser physical performance) and the subsequent proper functioning of the lesser machine, her IQE cycle yielded customer dissatisfaction.

## 4.4 COOPERATIVE EFFORT

*The production of goods and services requires cooperative efforts.* Just as brick walls are built one brick at a time, cooperative efforts are built with personal contributions. According to Barnard, *there are three fundamental requirements to establish cooperative efforts* [19]: *(1) common purpose, (2) individual willingness, and (3) interpersonal communication. In addition, cooperative efforts must be effective and efficient over time in order for an organization (cooperative effort) to survive.*

*A common purpose (accepted by all participants) supplies direction and focus.* A hierarchy of purposes must ultimately be defined. This hierarchy starts with a broad vision broken up into goals, subgoals, objectives, and finally, specific targets. Varying levels of abstractness and detail of purpose exist; for example, religious and political purposes tend to be much more abstract than business purposes.

Purpose may be dynamic or it may be relatively static. Nevertheless, a meaningful common purpose must exist at all times and be accepted with enthusiasm in order to generate high performance in a cooperative effort. An individual's acceptance of purpose is related to personal beliefs and values.

Individual willingness is the second necessity. Whereas purpose is accepted by an individual, willingness springs forth from within an individual. *Willingness is the commitment to cooperate, and is energy or action (behavior) related* (i.e., in the affective domain).

Interpersonal communication is the third prerequisite for the formation of a cooperative effort. *Communication links all "cooperators" so that they can function as a unit or team* (i.e., elicit common purpose and willingness in an organization). There are many forms of communication: verbal (words, expressions, numbers), written (words, expressions, numbers), graphical (pictures, drawings), physical expressions (voice tone, body language), and actions (role modeling).

*By definition, organizational effectiveness is the ability of an organization to accomplish, to some degree, its common purpose.* This organizational accomplishment must be recognized by the cooperators. When purposes are totally accomplished, new purposes must take their place to perpetuate the cooperative effort over time; otherwise, an organization will fall apart.

*Organizational efficiency will be defined here simply as a ratio of benefits to burdens.* Generally, if

$$\frac{\text{Benefits obtained by an individual}}{\text{Burdens encountered by an individual}} > 1$$

then an individual will continue to support the cooperative effort. If the ratio is $\gg 1$ (much greater than 1), then a high degree of enthusiasm is generated and efficiency is very high. Otherwise, an individual will fail to contribute in the cooperative effort. With respect to the organization as a whole, individuals may come and go, but the organization must retain a net positive efficiency.

At this point, we should note that benefits and burdens are multidimensional. Usually, we do not receive benefits in like-kind to the burdens we endure. For example, we give our time and ideas (talent) in return for salary (money), interpersonal recognition, and/or intrapersonal feelings of warmth or, perhaps, power.

## 4.5 QUALITY EXPERIENCE CHAINS

*In all organizations, human performance is the critical factor in developing and maintaining a competitive edge.* We must elicit a high level of human performance in both mental and physical activities. These activities range from the visionary and philosophical aspects of quality to the production, delivery, and service aspects of products that our customers want, need, and expect from us. Throughout this process sequence, the experience of quality is inseparable from day-to-day operations on the part of both internal and external customers.

*Internal customers address their individual needs through the IQE processes, while serving the organization through cooperative efforts.* Leadership formulates vision, defines purpose, and builds willingness among members relative to this "service." The managerial function coordinates this service by planning, organizing, and controlling individual efforts and allocating the resources of the organization to elicit effective and efficient cooperative efforts. Next, the "followership" function takes hold of the vision and plans in order to accomplish the purpose, thus moving the organization toward a competitive edge with respect to this service.

*External customers utilize the IQE processes in both individual and cooperative efforts.* The IQE processes are exercised many times each day when we, as external customers, judge the quality of products and services we use in our personal lives (e.g., shampoo, food, automobiles, clothes). The same IQE process is used when we, as internal customers, judge the quality of materials and supplies in the context of our involvement in cooperative efforts.

*We see a chaining sequence of external-internal customer roles involved in all activities.* For example, scrap metal is gathered (a service product) by salvage yard workers (internal customers), sent to a foundry (an external customer), cast into a bulldozer part (by internal customers), sent to a heavy equipment manufacturer (an external customer), where workers (internal customers) build a bulldozer (a hard good) for a contractor (an external customer), who (as an internal customer) uses the bulldozer to level a house site (a service product) for a building contractor (an external customer), who (as an internal customer) builds a house (a hard good product) for a family (external customers), and so on. Figure 4.3 depicts a construction-industry quality chain for the preceding example. The depiction in Figure 4.3 can be extended both upward and downward to include many more organizations.

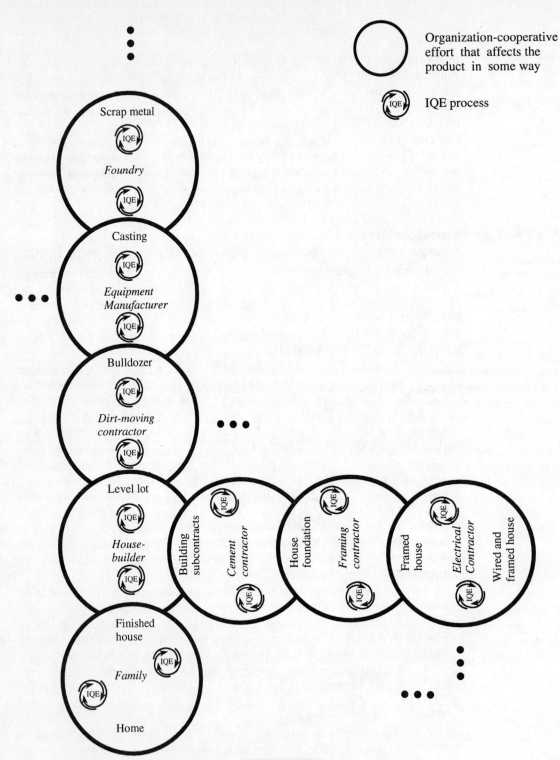

**FIGURE 4.3**  Quality chain example, construction industry.

*What we see in practice are ongoing quality chains involving individuals, or-ganizations, and society in general.* The expectation of net positive (benefits/ burdens > 1) quality experiences at the individual level allow these chains to form. *Satisfaction (of individual needs) and organizational effectiveness and efficiency (the collective satisfaction of the needs of individual contributors) interact to sus-tain quality chains.*

Figure 4.4 depicts a generic quality chain. *We see that products and by-products are passed along the chain, branching as the products split in a "physical" sense or as new customers join the chain as buyers or users.* These chains follow product life cycles from product definition to product disposal or recycle.

*We see external and internal customer roles switching (e.g., from external cus-tomers receiving a product from the preceding link to internal customers sending a product out to the next chain link).* Here, the external-internal customer termi-nology becomes fuzzy, as it depends on whether or not the links are within the same organization (internal) or between organizations (external). We may want to think of the customer roles as "sending" and "receiving" in order to avoid this confusion. Or-ganizations with the strongest chains will, by definition, have the most loyal customer base and therefore can be judged to be high quality-oriented organizations. The same is true for the products within the chains.

## 4.6  REAL-ESTATE TRANSACTIONS — EXPERIENCE OF QUALITY CASE

### Letter

Mr. X
Home Buyer
Anywhere
USA

Mr. Y
Executive Vice President
Anywhere Association of Realtors
Anywhere
USA

Dear Mr. Y:

I have recently made two transactions involving Anywhere realty firms:  (1) a sell through Crown Realtors last fall and (2) a buy through C. Banker this summer.

In both cases I found the services I received to be carried out with questionable ethics (at least judged by my expectations of fair and open business dealings).  I am sure both of these transactions were within the law and perhaps standard operating procedure.  Nevertheless, my level of dissatisfaction in both trans-actions is extremely high.

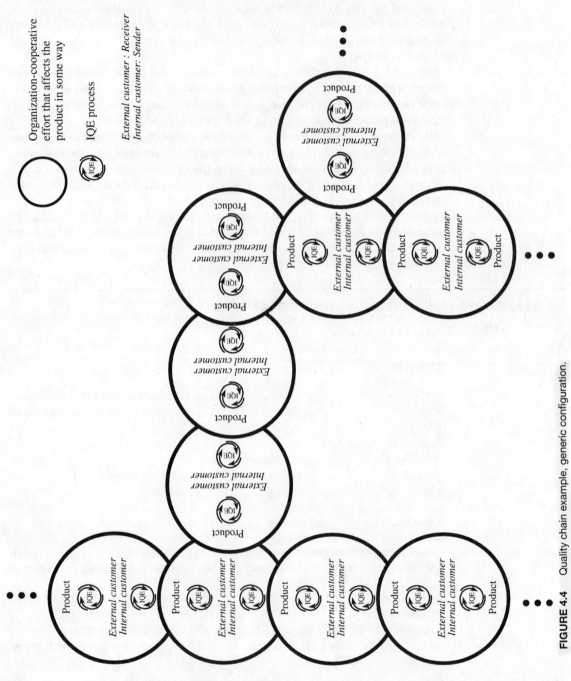

**FIGURE 4.4** Quality chain example, generic configuration.

84

I hope that my experiences were a result of an isolated case of poor management, rather than the level of service provided by Anywhere real-estate firms and advocated or endorsed by your association.

I will briefly describe my experiences and let you be the judge.

*Crown Case:*

Our transactions with Crown up through the buyer's offer (which I accepted) were acceptable and very good at times.  We accepted a VA loan applicant.  During the latter portion of the period before closing, I was informed that everything was going well, and on schedule.  Then, a few days (about 3 to 5 days, as I recall) before closing, I was notified that I was to make repairs.  I thought this notice was very late, but managed to make the repairs in time.

Then, a few days later (about 48 hours prior to closing), after I had moved most of my belongings out of my house, I was told that the applicant did not "lock in" his loan, and I was facing 2 or 3 loan percentage points.  Prior to this, I was led to believe the application had been made and a trivial amount, if any, was to be paid.

After a few communications with Crown, we managed to salvage the deal, as the broker pieced together a last-minute (within 24 hours) deal, which I reluctantly agreed to honor.

*C. Banker Case:*

Our house buying experience with the C. Banker realtors seemed smooth for the first few weeks after our offer was accepted.  Up to about 1 hour before the time of closing everything seemed to be going smoothly.  However, at a last-minute walk-through, prearranged a few days prior to closing by the selling agent (which I appreciated), I discovered that all furniture and appliances (belonging to the seller) were in place, some items were boxed, but the household was intact. This all took place about 30 minutes prior to our closing appointment.

I was expecting a clean or nearly clean house.  I suggested that we postpone our closing until such time that the house was empty.  My agent explained that this was not possible, due to an accommodation to the seller (unknown to us, the buyers, and certainly not in our contract).  I was therefore "bulldozed" into a closing where I was forced to exert my best judgment as to writing a clause to protect myself for liability, a failure by the seller to move out, and/or property damage in moving.  Hence, my closing, which began in the S Title Office at 1 p.m. July 8, was finally completed the next day, about noon.

In addition to the above, we experienced another rather unpleasant surprise in the course of closing.  The appraisal, survey, and termite certificate were not provided for my inspection, despite my request a few weeks prior to closing.  The appraisal was not available at closing, but at my insistence, a copy was faxed to us during closing.  The termite certificate and the survey were pulled out of a stack of papers about 45 minutes into the closing process.  At this time, I was informed that the fence on the west side of the property that I was buying was actually on the neighbor's property and that I could be asked to move the fence.

I expect nothing in return for the two accounts of my real-estate transactions, other than your time in reading this letter.  But, on the other hand, I certainly intend to avoid any further dealings with these two agencies again and now advise my friends to do likewise.

As a dissatisfied customer, I do not appreciate this type of treatment and believe these last-minute pressure tactics, tricks, and surprises are counterproductive.  Based on my first-hand experiences, I have become skeptical of the realty business in Anywhere.  I can state as an absolute fact that these practices are not in the best interest of your industry.

Sincerely,

Mr. X

P.S.  After the rain this morning, I noticed that the roof over the back porch was leaking profusely.  As I recall, I was shown a form where the owner stated there was no leakage on the roof as well as no problems with the plumbing, electrical system, etc.

cc
Crown Realtors
C. Banker Realtors


**Reply**

Mr. X
Home Buyer
Anywhere
USA

Dear Mr. X:

I received your letter today recounting your experience in the sale of your property and your purchase of another and noted your dissatisfaction with the handling of both.

The function of the realtor is that of a communicator and a facilitator while making sure that all parties to a transaction are treated honestly and fairly. The realtor has no authority—he or she performs the roll of a negotiator to do what he or she was hired by the seller to do—to sell and close the property sale while being honest and fair with all parties. It appears from the information supplied in your letter that communication to you might have been slow at times. Certainly this organization does not advocate or endorse any action that is not ethical by any member.

We are very concerned that your transactions weren't as smooth as anticipated and do hope both firms contact you to discuss with you how they can help.

Sincerely,

Mr. Y
Executive Vice President

## REVIEW AND DISCOVERY EXERCISES

### Review

**4.1** Which level(s) in Maslow's hierarchy of needs tends to drive the following:
  **a** Politicians (e.g., mayors, governors, senators).
  **b** Master craftspersons (e.g., tool and die makers, cabinet makers).
  **c** Professionals (e.g., engineers, medical doctors).
  **d** Artisans (e.g., musicians, painters).
  **e** Blue-collar machine operators (e.g., punch press, lathe).
  **f** Poverty victims (e.g., the unemployed, the homeless).
  **g** Teenagers (e.g., those living at home, those living on their own).
**4.2** Reflecting on the Maslow needs model, which level tends to be your driver at the present time? Explain. Do you expect any changes as you progress in your career?
**4.3** Select a social or business organization to which you currently belong:
  **a** Identify the common purpose.
  **b** Describe your willingness to contribute.
  **c** Describe the communication processes within the organization.
  **d** Discuss the organization's level of effectiveness.
  **e** Describe your current level of efficiency (e.g., benefits-to-burdens ratio).
  **f** Describe the organizational efficiency (e.g., aggregate benefits-to-burdens ratio).
  **g** Discuss the prospects for the organization's survival.
**4.4** Does one need purpose, willingness, and communication in a one-person task? Explain.
**4.5** Based on Maslow's hierarchy of needs, what level(s) does a task-focused piece-rate (production-incentive) reward system appeal to? Explain.
**4.6** Based on Maslow's hierarchy of needs, what level(s) does a factory teamwork approach appeal to? Explain.
**4.7** Based on Maslow's hierarchy of needs, what level(s) does "management by fear and intimidation" appeal to? Explain.

**4.8** Based on Maslow's hierarchy of needs, what level(s) does "participative management" appeal to? Explain.

**4.9** Explain how external customers can turn into internal customers in a quality chain.

**4.10** Choose two products you currently use or have used and describe
**a** A good quality experience that you encountered with the first product.
**b** A bad quality experience that you encountered with the second product.

**4.11** With regard to the products you selected for Problem 4.10, explain what it would have taken to
**a** Turn your good quality experience into a bad experience.
**b** Turn your bad quality experience into a good experience.

## Discovery Exercises

**4.12** Develop a quality chain for a product of your choice. Build the chain backward from your product. Then, build the chain forward from your product.

**4.13** Develop a quality chain that starts at the beginning of the product life cycle and continues until the product is disposed of. Develop a recycling loop in your chain. Include by-products (e.g., wastes) in your quality chain.

**4.14** Reread the Real-Estate Transactions Case. After a careful study of the case,
**a** Identify the customers in the case and their relationships to each other.
**b** Review the case from the buyer's, the seller's, and the real-estate firms' perspectives. Determine how each party would define quality in this case.
**c** Taking the perspective of the individual customers, and the major events that transpired as described in the case, discuss the IQE model components (e.g., see if you can identify some of the relevant inputs and outputs by what is documented in the letter).

**4.15** Looking back on your experiences as an external customer, develop a brief case study, similar to the Real-Estate Transactions Case. Relate your quality experience to the IQE model.

**4.16** Looking back on your experiences as an internal customer, develop a brief case study. Use the IQE cycle to describe your experiences. Then, associate your experiences with the principles of cooperative efforts (e.g., describe purpose, willingness, communication, effectiveness, and efficiency perspectives).

## REFERENCES

**1** D. Hellriegel, J. W. Slocum, Jr., and R. W. Woodman, *Organizational Behavior,* 6th ed., St. Paul, MN: West, 1992.

**2** A. H. Maslow, *Motivation and Personality,* 2d ed., New York: Harper and Row, 1970.

**3** D. I. Hawkins, R. J. Best, and K. A. Coney, *Consumer Behavior,* 5th ed., Homewood, IL: Irwin, 1992.

**4** G. Moorehead and R. W. Griffin, *Organizational Behavior,* 2d ed., Boston: Houghton Mifflin, 1989.

**5** See reference 1.

**6** D. N. Osherson and E. E. Smith, *Thinking: An Invitation to Cognitive Science,* vol. 3, Cambridge, MA: MIT Press, 1990.

**7** T. C. Kinnear and J. R. Taylor, *Marketing Research,* 3d ed., New York: McGraw-Hill, 1987.

**8** See reference 4.

   **9** See reference 1.
  **10** J. P. Peter and J. C. Olson, *Consumer Behavior and Marketing Strategy,* 2d ed., Home-
       wood, IL: Richard D. Irwin, 1990.
  **11** See reference 10.
  **12** See reference 10.
  **13** See reference 10.
  **14** F. Smith, *To Think,* New York:  Teachers College Press, 1990.
  **15** See reference 6.
  **16** See reference 1.
  **17** See reference 10.
  **18** See reference 7.
  **19** C. I. Barnard, *The Functions of the Executive,* Cambridge, MA:  Harvard University
       Press, 1938.

# 5

# PRODUCT, PROCESS, AND HUMAN PERFORMANCE

## 5.0 INQUIRY

1 How do the customer's and the producer's views of performance differ?
2 What constitutes product and process integrity?
3 How is product and process performance defined, measured, and modeled?
4 How are failure modes and failure mechanisms related?
5 How do short-term and long-term performance concepts differ?

## 5.1 INTRODUCTION

True product performance is determined in the field, by the customer, through the experience-of-quality processes (Figure 4.2). *Customer satisfaction or dissatisfaction is the final result of field performance (physical performance in function, form, and fit, as well as cost, timeliness, and customer service). Product or process integrity is defined here as the extent to which the product or process performance exceeds the customer's true field performance requirements.* Figure 5.1 illustrates a sequence of product or process physical performance; encompassing both the producer's "factory" world of "substitute" quality characteristics as well as the customer's "field" world of "true" quality characteristics. A high degree of correspondence between factory and field performance helps to assure product integrity.

Customer experience is focused in the field, producer experience is focused in the factory. Here, the terms "field" and "factory" are used loosely and refer to the place and circumstances of use and production, respectively, for both goods and services. *Customers, by definition, determine true performance as a result of their field applications, environments, and methods of operation.* On the other hand, *producers typically determine substitute performance based on their test loads, test environments, and test methods.*

## 5.2 PRODUCT-PROCESS PERFORMANCE AND MEASUREMENT

Physical performance is the resultant of function (works well), form (looks good), and fit (suits the purpose) in action, relative to our customer's requirements and expectations. High or low performance in the long run is the essence of high or low quality, respectively. True performance measures and true quality measures are one in the same. Here, the term "long-run" must be interpreted in the context of requirements. For example, for some products and many services (e.g., food and entertainment) the long-run may be a matter of minutes or hours. While for other products, such as tractors, the long-run may be decades.

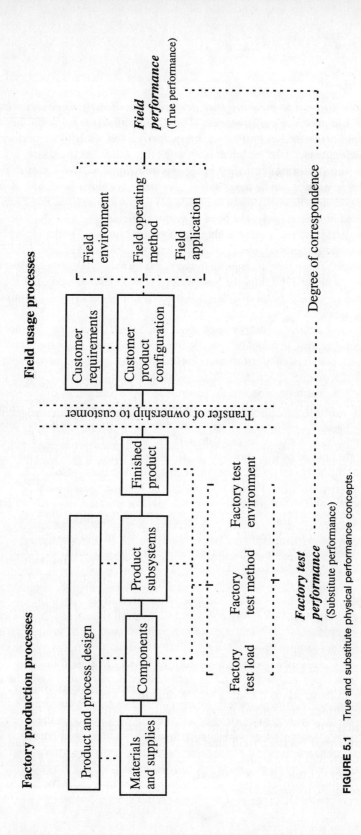

**FIGURE 5.1**   True and substitute physical performance concepts.

## Performance Measurement

*It is difficult to measure true performance directly as each individual customer would measure performance. Therefore, substitute performance measures, both qualitative and quantitative, are typically used.* Substitute performance measures are expressed in a technical language. Regardless of the short- or long-term nature of the performance measures, *two conditions must be met before meaningful performance can be assessed*: **(1)** *performance must be clearly defined in terms of a measurable entity or entities and* **(2)** *sensors or models must exist to measure or predict, respectively, the level of performance.*

Performance measures are of two general types: (1) a variables measure, such as a linear dimension, or (2) an attributes measure, such as acceptable or unacceptable color. These two measurement types are formally defined and discussed in Chapter 9. In either type, high or low performance is established on a relative basis; relative to a fixed point of reference or measure or relative to a competing product or process.

Since each customer may have somewhat unique requirements and expectations, testing product or process integrity becomes a two-part challenge. The first part consists of performance definition and measurement expressed through a performance "profile." The second part consists of overlaying the customer's requirements or expectations (expressed in the same units of measure as the performance profile) on the performance profile. Figure 5.2 depicts a simple performance model for a continuous performance measure (bigger is better) relative to two customers. For example, our true performance characteristics might be the amount of light put out by a flashlight (e.g., the more light, the better). Our substitute measure might be battery voltage level measured over time. Conceptually, as long as the crossover (failure) point is encountered beyond the customer's requirement or expectation of useful service life, the level of product integrity is satisfactory to the customer. If the failure point far exceeds the customer's expectations, we are at a very high or "Wow!" level of integrity. By mapping various levels of customer expectation against the performance profile, different levels of product-process integrity can be observed.

As stated earlier, customers buy products anticipating benefits. For example, we buy an automobile for transportation as well as for intangible self-expression (esteem) needs. We buy a shovel to dig or till the soil so that we can enjoy flowers or vegetables. The automobile is a rather complicated product, while the shovel is a rather simple product. Regardless of product complexity, the customer expects satisfactory field performance both in the short run and the long run.

In a physical sense, product performance develops through a hierarchy beginning with materials and ending with a functional system in the customer's hands. The traditional strength of material substitute performance measures and models depicted in Figure 5.3 are part of such a hierarchy. In the depiction we see specific measures such as yield strength, tensile strength, and breaking strength used as substitutes for a customer's desire for a "strong" product. For example, if our customer demands a strong shovel, we respond by describing the materials, design, and so on, in technical terms.

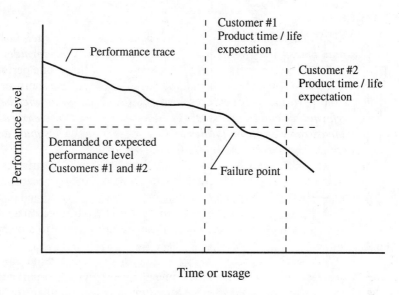

**FIGURE 5.2**    Product-process performance model (bigger is better response).

Field performance is the major factor in customer satisfaction.  A customer may express his or her requirements and expectations as "ease-of-use" and "durability." Engineers must find suitable (meaningful) substitute characteristics compatible with our technical world. ***The definition and discovery of the true performance measures, the corresponding definition of the substitute measures, and understanding the relationship between the two are necessities in quality definition and product design.*** This topic is discussed in detail in Chapter 11.

## Failure Modes and Mechanisms

***Sometimes we learn more about performance by focusing on nonperformance*** (failure rather than success).  An early (preprototype) focus on failure modes and mechanisms helps us to prevent nonperformance.  The failure mode and effects analysis (FMEA) and fault tree analysis (FTA) tools discussed in Chapters 12 and 13, respectively, focus on nonperformance in a proactive manner.

*A failure mode is the consequence or manifestation of a failure mechanism or inadequate performance.  A failure mechanism is the process (physical, psychological, social, or other) responsible for a physical failure, a human error, or other inadequate performance.* Hence, a failure mode is a result and a failure mechanism is a process.  It is typically easier to describe a result than a process.  Results are relatively static, while processes are dynamic (change with time and conditions).  For example, a short circuit is a failure mode, while solder-joint crack propagation (from a small solder defect) due to repeated thermal cycling is a failure mechanism.

Effective proactive quality assurance requires a fundamental understanding of failure modes and mechanisms. ***Failure modes and mechanisms are relevant to***

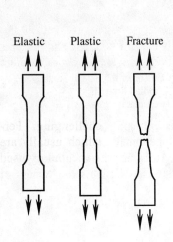

Elastic    Plastic    Fracture

(a) Physical test specimens

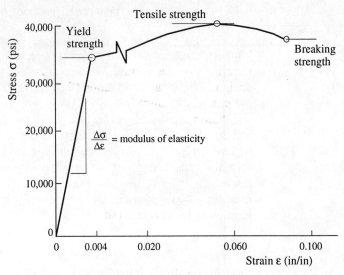

(b) Engineering stress-strain performance curve for an aluminum alloy

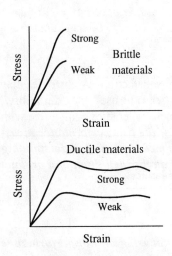

(c) Engineering stress-strain curve patterns

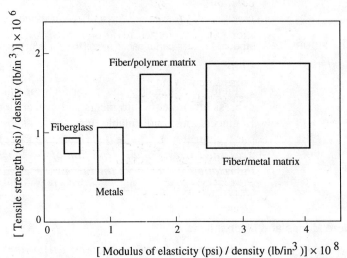

(d) Strength-density ratio measures for selected material groups

Definitions:

*Stress* — applied load on a material.

*Strain* — deformation per unit length of a material.

*Elastic deformation* — deformation of a material which is recovered when a load is released.

*Plastic deformation* — deformation of a material which is not recovered when a load is released.

*Ductility* — measure of a material's ability to be deformed without breaking.

*Yield strength* — stress at which a material moves from elastic to plastic deformation.

*Tensile strength (ultimate strength)* — maximum stress or load encountered in a tensile test.

*Breaking strength* — stress or load observed when a break occurs in a tensile test.

**FIGURE 5.3**    Selected material strength substitute performance measures and models.

***physical products as well as service products.*** Identifying failure modes and determining or describing failure mechanisms are crucial in product and process design. Empirical information (the results of observation, either direct or through the use of specialized sensing devices) is used as the basis for identification, description, and model development. Broad rules of thumb, general qualitative models, and detailed quantitative models are the net results.

The study of failure modes and mechanisms is extremely challenging. Performance and failure are extremely complicated phenomena which usually are not totally understood and not totally predictable. And since the vocabulary used to describe failure modes and mechanisms is not standardized, a wide variety of terminology is encountered.

## Physical Performance

***Product performance is a function of the product configuration's (material-component-subsystem-system) response to exposure from (1) the field environment(s), (2) the operating method(s), and (3) the product application(s).*** Both short- and long-term effects must be considered. Six forms of product exposure are shown, along with selected substitute performance measures, in Table 5.1. This simple display points out the wide variety of exposures a product may encounter. From a quality standpoint, we must identify and deal with exposures ranging from "reasonable" to "worst case."

**TABLE 5.1**    PRODUCT EXPOSURE AND SELECTED SUBSTITUTE PERFORMANCE MEASURES

| Product exposure | Substitute performance measures |
| --- | --- |
| Loading | Forces<br>Moments<br>Vibration<br>Shock |
| Heat, cold | Temperature<br>Thermal emission |
| Atmosphere, surroundings | Elements<br>Concentration<br>Medium (vapor, liquid, solid) |
| Radiation | Wavelength<br>Energy level |
| Age, usage | Time<br>Cycles |
| Usage pattern | Continuous<br>Intermittent<br>Storage, dormant |

In general, a failure mode is a result that limits performance (e.g., a dull knife, a flat tire, flaking house paint, a late arrival, a short circuit). Table 5.2 provides a summary of selected electrical and electronic component failure modes and approximate frequencies as a percent of failures [1]. Here, we see rather broad failure mode descriptions (e.g., short circuit, open circuit, excessive leakage current, and so on).

On the other hand, a failure mechanism is the process that leads to the failure mode. Table 5.3 provides a brief listing of electronic component, subsystem, and system failure mechanisms [2]. In this table, each mechanism has a unique alphanumeric code which can be used in physical failure analysis documentation and database storage and retrieval. In physical failure analysis, it is critical to identify the failure cause and mechanism. For example, we might encounter a failed unit with a short circuit failure mode. We might be able to determine that the failure mechanism was electrical overstress (E1 in Table 5.3). But, we may also discover that during assembly a high-voltage lead wire was switched with a low-voltage lead wire, due to a "fault" of unclear markings. Here, the failure mechanism would be classified as a mismarked (or unclear) package (P1 in Table 5.3).

*To effectively and efficiently improve product-process integrity, we must carefully examine both the mode and the mechanism with the intent of preventing the failure mode in the most cost-effective and timely manner possible.* For example, in the case of our previously discussed electronic component, our quality improvement effort should focus on the marking and assembly processes, rather than the electrical integrity of the design. We might pursue this problem through mistake-proofing (discussed in Chapter 6) in order to eliminate, as far as possible, the wiring error. This observation-analysis-corrective action sequence is primarily reactive. It would be better to eliminate this type of failure mechanism early in the product-process design stage in a proactive manner. For example, we might design a mistake-proof process where the wires could not be physically interchanged.

---

### Memory Boards—Manufacturing Process Failure Case [SS]

An electronics manufacturer building memory boards for an advanced computer system suddenly suffered a number of board shorts. Some shorts were hard shorts and were evident whether power was applied to the board or not. Other shorts could be detected only under certain operating conditions because they caused errors in the data interchange between address lines. The shorts were not confined to any one area of the boards, as is often the case when caused by a fault in the assembly process. Shorts, however, were found to occur only in areas where solder joints had been reworked, but not between the actual reworked solder joints. A short could be eliminated by removing and replacing the solder in the affected plated through holes. Reflowing without solder replacement did not correct the shorted condition.

**TABLE 5.2**  FAILURE MODE DISTRIBUTION OF ELECTRICAL AND ELECTRONIC PARTS

| Part | Failure mode | Approximate % | Part | Failure mode | Approximate % |
|---|---|---|---|---|---|
| Blowers | Winding failure | 35 | Microcircuits | Short high | 33 |
| | Bearing failure | 50 | | Short low | 33 |
| | Slip rings, brushes, and commutators | 5 | | Open | 33 |
| | Other | 10 | Motors, drive, and generators | Winding failures | 20 |
| Capacitors—fixed ceramic dielectric | Short circuit | 50 | | Bearing failures | 20 |
| | Change of value | 40 | | Slip rings, brushes, and commutators | 5 |
| | Open circuit | 5 | | Other | 55 |
| | Other | 5 | Relays—electromechanical | Contact failures | 75 |
| Capacitors—fixed metallized paper or film | Open circuit | 65 | | Open coils | 5 |
| | Short circuit | 30 | | Other | 20 |
| | Other | 5 | Resistors—fixed carbon, and metal film | Open circuits | 80 |
| Capacitors—fixed tantalum electrolytic | Open circuit | 35 | | Change of value | 20 |
| | Short circuit | 35 | Switches, rotary | Intermittent contact | 90 |
| | Excessive leakage current | 10 | | Other | 10 |
| | Decrease in capacitance | 5 | Switches, toggle | Spring breakage (fatigue) | 40 |
| | Other | 15 | | Intermittent contact | 50 |
| Circuit breakers | Mechanical failure of tripping device | 70 | | Other | 10 |
| | Other | 30 | Thermistors | Open circuits | 95 |
| Connectors | Shorts (due to poor seal) | 30 | | Other | 5 |
| | Solder joint failure | 25 | Transformers | Shorted turns | 80 |
| | Degradation of insulation resistance | 20 | | Open circuits | 20 |
| | High contact resistance | 10 | Transistors—silicon | High collector-to-base leakage current | 60 |
| | Miscellaneous mechanical failures | 15 | | Low collector-to-emitter breakdown voltage | 35 |
| Insulators | Mechanical breakage | 50 | | Open circuit | 5 |
| | Deterioration of plastic material | 50 | | | |

Adapted from "Engineering Design Handbook: Design for Reliability," AMCP 706–196, January 1976, AD A027370.

**TABLE 5.3** FAILURE MECHANISM CODES

**M** Metallization (includes polysilicon)
1 Open at oxide step
2 Open at contact
3 Scratched open
4 Smeared causing short
5 Lifted or peeling
6 Corroded or pitted
7 Migration causing open
8 Migration across surface causing short
9 Interlayer short
10 Metal masking fault
11 Under- or overetching
12 Open via (via is window for connecting-layers)
13 Thin-film resistor open
14 Programmed resistor grow back
15 Over alloy

**S** Surface (These codes include descriptive modes which usually are not broken down further)
1 Inversion / channel
2 Foreign matter on surface
3 Beta ($h_{fe}$), gain shift
4 Threshold shift
5 Voltage sensitive
6 Temperature sensitive
7 Marginal electrical
8 Noise

**P** Package
1 Mismarked
2 Index marked wrong
3 External leads or signal damaged
4 Hermeticity lost
5 Die attach
6 Loose particles
7 Package construction
8 Corrosion
9 Solderability
10 Excess seal material
11 Package broken, cracked
12 Contamination of leads
13 Tar in capacitor
14 Mechanism
15 Sticky

**B** Bonding (grouped by interconnect technology)
1 Wire broken
2 Wire corroded
3 Wire mislocated
4 Wire missing
5 Extra wire
6 Wire to scribeline short
7 Wire to lid short
8 Wire to pin short
9 Wire to case short
10 Bond off-center
11 Bond misplaced
12 Package bond lifted
13 Bond broken at heel
14 Bonding pad lifted
15 Bonding tail causing short
16 Intermetallic growth
17 Excessive bonding pressure
18 Beam broken
19 Beam lifted
20 Silicon-beam separation
21 Silicon cracked at beam
22 Beam shorted to silicon
23 Beam open
24 Bump open
25 Bump shorted
26 Chip metal lifted at bump

**C** Crystal
1 Crystal imperfection
2 Cracked chip (die)
3 Chipped chip (die)

**E** External (includes application excessive electrical, mechanical, or thermal stress. This category indicates no device fault)
1 Electrical overstress
2 Electrostatic discharge
3 Wrong application (works OK, but application requires something else)
4 Latch-up (the device may be damaged or not from the experience)
5 Second breakdown
6 Interconnections melted
7 Solder bridge (assembly fault, not device fault)

**U** Unknown (The discussions among types are nebulous but may serve some purpose)
1 Cause unknown
2 Failure confirmed, cause unknown
3 Confirmed specification failure, cause not determined
4 Dc failure, cause unknown
5 Ac failure, cause unknown

**N** No problem found
1 Good device
2 Problem not verified
3 Device retested OK
4 Obsolete device (use when meets old specification but fails new requirements)

**T** Testing (These categories used by device manufacturer to isolate causes of shipping parts never good)
1 Room temperature dc fail
2 Room temperature ac fail
3 High- or low-temp dc fail
4 High- or low-temp ac fail
5 Wrong ROM code
6 Wrong chip revision
7 Wrong device in package
8 Obsolete program
9 Faulty test setup

Adapted, with permission from McGraw-Hill, from D. Burgess, "Physics of Failure," *Handbook of Reliability Engineering and Management*, W. G. Ireson and C. F. Coombs, Jr. (eds.), New York, NY: McGraw-Hill, page 14.11, 1988.

The electrical test methods were reviewed and found to be correct and accurate in identifying operational errors. The shorts were proven to be real, so an investigation into the cause(s) was initiated. Since the shorts only appeared after rework, it was presumed that they were being caused by a process-related phenomenon. Even though the rework materials and processes had been used for years without this type of failure, they were examined. All materials were found to be within material specification limits, and the rework process was being performed within proven and established process parameters.

Since no sole causal factor in the rework process could be found, the investigation traced the assembly process upstream to initial assembly (performed by an external supplier). Materials were found in use in the initial assembly process which could be conductive, and, in residue form, possibly could create shorting between adjacent, noncommon conductors. Cleaning operations designed to remove surface contamination from both the initial assembly and rework operations made no significant change in the shorting condition.

Next, the possibility of subsurface contamination was investigated. Chemical analyses were performed on three sample groups—assembled and reworked boards, assembled and not reworked boards, and unassembled boards—to determine whether any conductive residues were present. The results showed that elevated levels of conductive materials consistent with the chemicals used in the supplier's assembly process were present on all assembled boards, indicating a process residue. Based on available information, a process designed to remove the suspected subsurface contamination was developed. Success with eliminating shorts on affected memory boards using this process provided evidence that the suspected failure mechanism had been correctly diagnosed and reversed.

## Software Performance

Software has become a major part of many systems. Customers are constantly demanding higher performance from software. Typically, these higher performance demands require software to perform more tasks and more complex tasks. For example, the emulation of the human thought-logic process through expert systems and artificial intelligence requires very complex software. Providing interactive capabilities and direct real-time sensor inputs further complicates software definition, design, development, and testing.

Software performance tends to be much more binomial (it operates or it does not operate, complete success or failure) than hardware performance is. It is generally accepted that software does not wear out or degrade through use. Lack of performance in software is attributable to latent defects (sometimes termed "bugs"). Hardware also suffers from defects, both latent and incipient. But, in the hardware case, a defect may not always affect performance in a strictly binomial manner. Rather, we may see degraded performance over time with eventual or abrupt failure (nonperformance).

*Given the nature of software failures, two primary failure modes are widely recognized*: **(1)** *premature termination and* **(2)** *a discrepant result* (Thayer et al. [3]). As in the case of physical hardware, previously discussed, we can classify the associated software failure mechanisms. Table 5.4 contains a list of software defects and software error failure mechanisms pertaining to both premature termination and discrepant results. Since software is defined, designed, and developed by humans, human error is central to the software performance issue.

*Software performance is also subject to function, form, and fit considerations.* The usual discussions of software failure concentrate on function. However, the form of both the input and output is critical to the customer's ability to use the software. The fit of software to specific applications (in the case of purchased software) is also critical.

**TABLE 5.4**    SOFTWARE FAILURE MODES AND MECHANISMS

| Failure modes | Failure mechanisms |
| --- | --- |
| Premature termination | Input-output errors |
| | Logical errors |
| Discrepant results | Computational errors |
| | Data handling errors |
| | Operating system or system support errors |
| | Routine or routine-interface errors |
| | User-interface errors |
| | Database-interface errors |
| | Variable definition errors |
| | Recurrent errors |
| | Documentation errors |
| | Requirement compliance errors |
| | Operator errors |

## Software Performance — Soft Failure Case [DW]

I recently purchased a software package for amortizing a mortgage. I read the description on the label and the software seemed to meet my needs, so I bought it. After I started to use the software, I became dissatisfied with its form. The program was designed so that after you input the data the calculation and solution would go only to a printer—not to the screen. As a result

1  Paper is wasted.
2  The program can't be run without a printer.
3  It is an inconvenient program to use for trial-and-error type scenarios involving a lot of "what ifs" (house price, down payment, mortgage term changes).

The read.me files inform the customer that the "full" version can be purchased for an additional cost. The full version includes all of the features (including presenting results on the screen) that this customer assumed to be standard. Although my software ran without error, I consider the resulting performance a failure.

## 5.3  HUMAN PERFORMANCE

We must recognize that the human resource is the one unique resource available to an organization. ***Physical and mental issues are both critical in human performance.*** Physical issues in the work environment include the surroundings (lighting, temperature, humidity, and so on). The physical environment and workload must be designed to fit the individual, not the other way around. Physical constraints involve the human's ability to sense and respond to his or her surroundings through the five senses. Mental issues involve the process of sensing stimuli in the environment and responding to the stimuli. Information processing is complex and central to mental workload understanding; it includes the reception and conversion of physical stimuli to decision-based responses. Human factors handbooks [4] and texts such as Sanders and McCormick [5] describe physical and mental issues in detail.

A very simple human performance curve is shown in Figure 5.4. This curve implies that there is an optimal level of stress or arousal which maximizes human performance (Kantowitz and Sorkin [6]). While this relationship is well established, capitalizing on it in practice is difficult for a number of reasons. For any given task, physical or mental, it is difficult to identify the level of stress or arousal that results in the optimal performance. Also, different individuals will have different stress or arousal thresholds and limits (e.g., what is optimal for one may be overload or underload for another). This variation is an important consideration, given a heterogeneous workforce (e.g., in terms of age, experience, physical characteristics, and cognitive abilities).

### Human-Machine Differences

*We must use both humans and machines to their best advantage.* Thus, it is important to understand the ways in which human performance differs from the

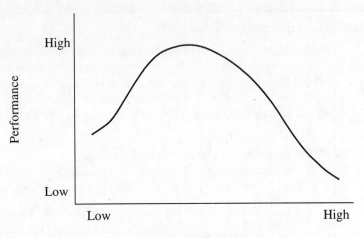

**FIGURE 5.4**    Conceptual human performance curve.

performance of machines.  The human being must not be thought of merely as a living extension of a machine or process.  In considering the differences between human and machine performance, it is possible to assemble a comparison between the two, listing relative capabilities, strengths, and weaknesses.  Table 5.5 has been adapted from Sanders and McCormick [7] and will serve to highlight the relative capabilities.

*We must define and design integrated systems of humans and machines such that each component is performing tasks and is functioning in the roles to which it is best suited.*  There are several key differences in approaching human and machine performance.  Machines and processes may be redesigned and reengineered, using different technologies, design criteria, materials, and so on, in order to achieve the desired performance level.  Human beings, on the other hand, enter the workplace or the field (to use our products) in a largely "as is" physical package and thus cannot successfully be physically engineered for performance.

*The human being cannot and should not be "engineered" to fit a workplace or product; rather, the appropriate strategy should be to design the workplace and tasks or products to fit the human operator in both a physical and a cognitive sense.*  While the human being cannot be engineered, he or she nevertheless brings a highly adaptable and trainable brain that may more than compensate for the lack of reconfigurability.  All in all, as a system, humans and machines should complement each other and capitalize on their respective strengths to ensure that overall system integrity is maintained.

## Human Error

*Human error is defined as an out-of-tolerance action, where the limits of acceptable performance are defined by the system* (Miller and Swain [8]).  Human

**TABLE 5.5   HUMAN-MACHINE PERFORMANCE ADVANTAGES**

| Human advantages | Machine advantages |
|---|---|
| 1 Sense very low levels of certain kind of stimuli (visual, auditory, tactile, olfactory, and taste) in complex situations. | 1 Sense stimuli outside the normal human range of sensitivity. |
| 2 Detect stimuli and recognize objects under widely varying conditions and in "high-noise" backgrounds. | 2 Apply deductive reasoning such as in recognizing stimuli as belonging to a general classification or category. |
| 3 Perceive patterns which may vary from situation to situation in varying conditions. | 3 Monitor for prespecified events, particularly those that are infrequent. |
| 4 Sense and respond to unexpected and low-probability events in the work environment. | 4 Perform precise, routine, and repetitive operations reliably and indefinitely. |
| 5 Reason inductively, generalizing from observations. | 5 Make rapid and consistent responses to input signals. |
| 6 Store (i.e., remember) large amounts of information over long periods of time; retrieve many items of related information. | 6 Store coded information and perform computations quickly and in large quantities. |
| 7 Concentrate on most-important activities when overload conditions require. | 7 Retrieve coded information quickly and accurately when requested. |
| 8 Select alternative modes of operation if certain modes fail. | 8 Perform several programmed activities simultaneously. |
| 9 Make decisions, evaluations, and estimates based on incomplete and unreliable information. | 9 Exert precise amounts of force in a highly controlled and consistent manner. |
| 10 Exercise judgment, flexibility, improvisation, and creativity when properly trained, motivated, and empowered. | 10 Physically resist many environmental stresses. |
| 11 Generally learn from experience. | 11 Maintain efficient operation under conditions of heavy load. |

Adapted, with permission from McGraw-Hill, from M. S. Sanders and E. J. McCormick, *Human Factors in Engineering and Design*, 6th ed., New York, NY: McGraw-Hill, pages 526–527, 1987.

reliability is defined in two general ways: (1) the repeatability of a human activity or (2) the probability of successful human performance of a task or mission, under specified conditions. The latter definition is used throughout this book. Hence, human error represents a failure, while human reliability represents a success.

In the discussion of human performance and human error, several concepts are important. ***Human errors may be divided into five categories*** [9, 10]:

1 *Errors of omission*—Errors of omission occur when part of the task is omitted or skipped.
2 *Errors of commission*—Errors of commission occur when a task is performed incorrectly.
3 *Extraneous acts*—An extraneous act is a task or action that should not have been performed, as it diverted attention from the crucial or major task or action.
4 *Sequential errors*—A sequential error occurs when a task is performed out of sequence.
5 *Timing errors*—A timing error occurs when a task is performed too early or too late or is not completed within the allowed time.

The last three error categories listed above are technically errors of commission, but are important enough in understanding human error to warrant their own categories.

In any consideration of human error, there are a number of rules of thumb that are important. First, the term "error" should not carry an automatic implication of fault. Human error encompasses a wide range of causal factors internal to the human (mental state, fatigue, training) and external to the human (such as interfaces with controls and displays, task design, specified procedures, and so on). The "fault" of many accidents attributed to human error is often a poorly designed human-machine interface. Second, error is usually dependent on the system and situation for its effect (i.e., the severity of its consequences). Finally, humans have the capability to discover and correct their errors given that the task is properly designed and conforms to sound human factors design principles.

## 5.4 LONG-TERM PERFORMANCE

Long-term performance is an important quality characteristic. ***We typically use reliability and maintainability measures to describe long-term performance.*** Reliability is a long-term or extended performance measure related to risk. ***Reliability is defined as the probability that a component, subsystem, or system will perform successfully for a specified amount of time or usage or for a given mission, when operated under specified conditions*** (Lewis [11]). For example, an automobile reliability $R$ of 99 percent for 10,000 mi of city traffic driving, $R(10,000 \text{ mi}) = 0.99$, implies a 99 percent chance of success or a 1 percent chance of failure during 10,000 mi, driven under city driving conditions. If an automobile make or model possesses this 99 percent reliability quality characteristic, it is implied that out of a fleet of 100 cars, 99 would survive the 10,000 mi without failure, on the average.

Traditional reliability definitions and models have been adapted to model human reliability as well as physical reliability, (Dhillon [12] and Swain [13]).

Maintainability is a performance measure related to restoring a system. ***Maintainability is defined as the probability that a failed component, subsystem, or system will be restored to operable condition in a given amount of time or effort when restoration is performed under specified conditions.*** Reliability implies that eventually a successful system will fail, given enough time or usage, while maintainability implies that a failed system will eventually be rejuvenated or restored, given enough time and effort. In Chapters 24, 25, and 26, we will introduce reliability and maintainability modeling concepts. We will also discuss both preventive (before failure) and corrective (after failure) maintenance models.

***Two general types of long-term failures are encountered***: **(1) hard and (2) soft.** A hard failure develops when a stress (environmental-, application-, operation method-related) overwhelms the strength of a component, subsystem, system or configuration, resulting in a "dead" or "crippled" component, subsystem, or system which cannot perform its intended function. A mathematical expression for a hard failure, sometimes called a static failure is

$$\sigma \geq S \tag{5.1}$$

where $\sigma$ represents stress and $S$ represents strength.

A soft failure results in inadequate performance, even though the system is functional (but not to the extent necessary to produce adequate results or to meet customer needs or expectations). The software case previously described is a good example of a soft failure. A mathematical expression for a soft failure can be written as

$$\mathbf{P}_{demanded} \geq \mathbf{P}_{output} \tag{5.2}$$

where $\mathbf{P}$ represents measured performance.

Three general cases of soft failures related to product performance can be identified: (a) bigger is better, (b) smaller is better, and (c) nominal is best. Whenever the performance output does not exceed the demanded performance or expected level (e.g., as in the bigger is better case in Figure 5.2), technically a soft failure is encountered. Hence, a soft failure can occur with a totally functioning, but incapable, deteriorated, or deviated system performance (no hard failures). Whereas, a hard failure will produce an abrupt performance drop in a performance trace (e.g., the bigger is better performance trace in Figure 5.2 would drop off abruptly to 0 or some very low level).

Table 5.6 provides examples of both hard and soft failures. In practice, the distinction between hard and soft failures is fuzzy. Nevertheless, they can be readily distinguished in theory. The point for such a distinction is that we pursue different strategies for preventing hard failures and soft failures. For example, a soft failure may be a form or fit issue, whereas a hard failure is generally a functional performance issue.

**TABLE 5.6**    HARD AND SOFT FAILURE EXAMPLES

| Product | Failure Description | |
| --- | --- | --- |
| | Soft | Hard |
| Cutting tool | Dull edge | Broken edge |
| Automobile | Oil leak | Fractured bearing |
| | Cracked windshield | Shattered windshield |
| Food | Cold meal | Spoiled food |
| | Slow service | Wrong check |

## Consumer Experience Performance

*Customers, before they make a purchase, seek performance-relevant information with respect to the type of product they intend to buy.* Publications such as *Consumer Reports,* as well as special interest publications aimed at enthusiasts and potential purchasers, continually review new products. These reviews are usually developed from two sources: (1) first-hand use and experimentation or (2) user surveys.

Consumer oriented performance evaluations range from simple like-dislike discussions to thorough laboratory tests. Figure 5.5 depicts an automobile assessment for two best-selling family sedans taken from *Consumer Reports* [14]. We see 17 broad categories identified and "trouble-spot" trends tracked over a six year period. For example, the Engine cooling item has improved on the Ford Taurus and regressed on the Honda Accord, according to this report. However, both received the same "average" rating in 1991. Reports such as these along with word-of-mouth consumer experience from friends and associates play a large role in product selection-rejection.

## Common Automotive Failure Modes and Mechanisms—
## Hard-Failure Case [RK]

I have bought EM vehicles all of my life. This experience with a quality issue had momentarily made me rethink my dedication to their products. I bought a used Cross Am while I was in college. One hot summer day a few months after I bought the car, the high-speed position on the blower fan was not working. This caused significant inconvenience because the weather was very hot at the time. I determined that EM used a separate circuit to power the high-speed fan position on their fan motors. A separate relay was used because, presumably, there was too much current in the circuit to directly switch that position. I traced the problem to a connector in the engine compartment that apparently

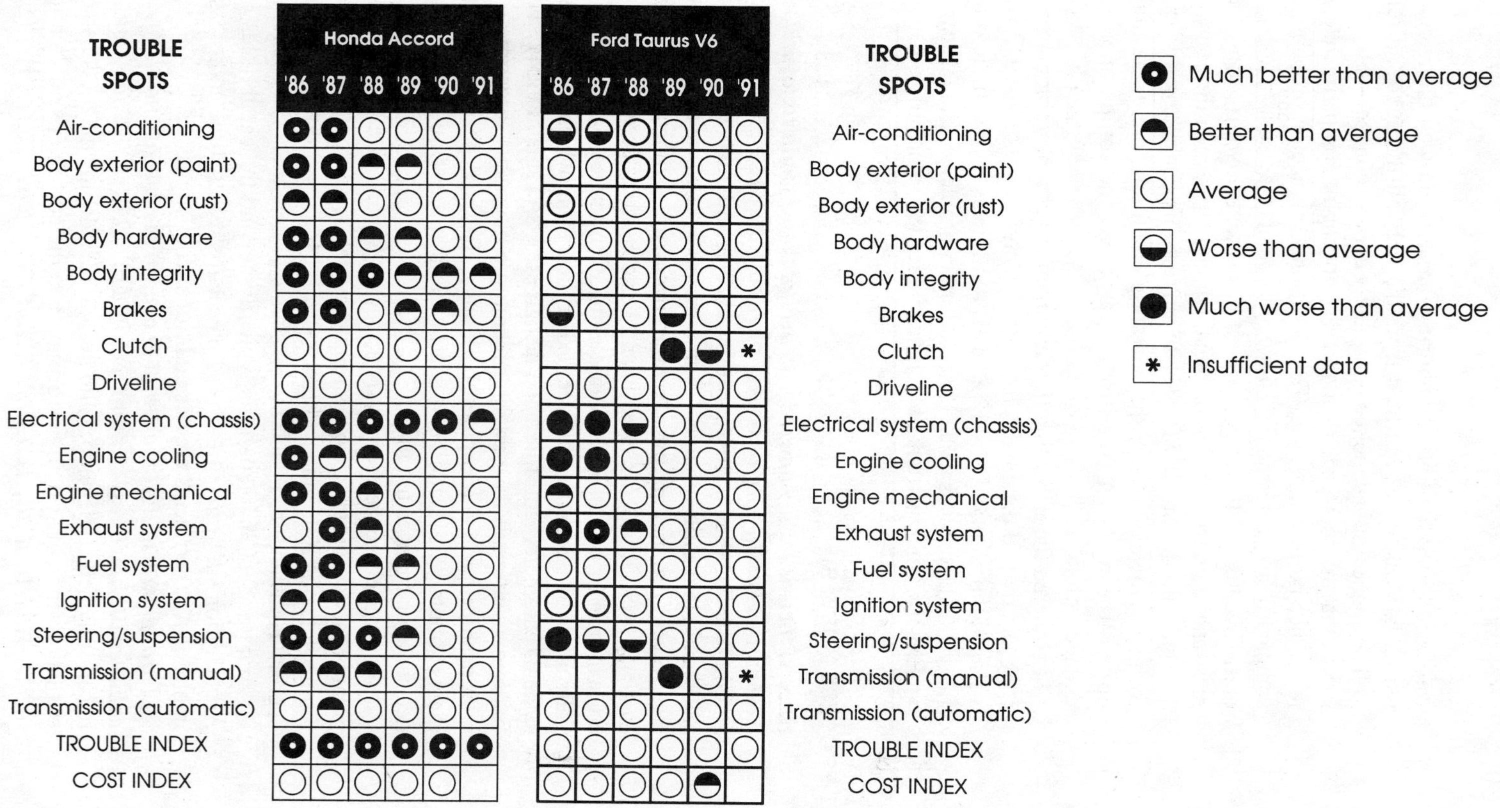

**TROUBLE SPOTS**

Legend:
- ⊙ Much better than average
- ◐ Better than average
- ○ Average
- ◑ Worse than average
- ● Much worse than average
- * Insufficient data

**Honda Accord**

| TROUBLE SPOTS | '86 | '87 | '88 | '89 | '90 | '91 |
|---|---|---|---|---|---|---|
| Air-conditioning | | | | | | |
| Body exterior (paint) | | | | | | |
| Body exterior (rust) | | | | | | |
| Body hardware | | | | | | |
| Body integrity | | | | | | |
| Brakes | | | | | | |
| Clutch | | | | | | |
| Driveline | | | | | | |
| Electrical system (chassis) | | | | | | |
| Engine cooling | | | | | | |
| Engine mechanical | | | | | | |
| Exhaust system | | | | | | |
| Fuel system | | | | | | |
| Ignition system | | | | | | |
| Steering/suspension | | | | | | |
| Transmission (manual) | | | | | | |
| Transmission (automatic) | | | | | | |
| TROUBLE INDEX | | | | | | |
| COST INDEX | | | | | | |

**Ford Taurus V6**

| TROUBLE SPOTS | '86 | '87 | '88 | '89 | '90 | '91 |
|---|---|---|---|---|---|---|
| Air-conditioning | | | | | | |
| Body exterior (paint) | | | | | | |
| Body exterior (rust) | | | | | | |
| Body hardware | | | | | | |
| Body integrity | | | | | | |
| Brakes | | | | | | |
| Clutch | | | | | | * |
| Driveline | | | | | | |
| Electrical system (chassis) | | | | | | |
| Engine cooling | | | | | | |
| Engine mechanical | | | | | | |
| Exhaust system | | | | | | |
| Fuel system | | | | | | |
| Ignition system | | | | | | |
| Steering/suspension | | | | | | |
| Transmission (manual) | | | | | | * |
| Transmission (automatic) | | | | | | |
| TROUBLE INDEX | | | | | | |
| COST INDEX | | | | | | |

**FIGURE 5.5** Consumer oriented performance evaluation. *Source:* "Frequency of Repair Records," Copyright 1992 by Consumers Union of U.S., Inc., Yonkers, NY 10703–1057. Reprinted by permission from *Consumer Reports*, vol. 57, no. 4, pages 251–258, April 1992.

107

was used to facilitate manufacturing, since it allowed the engine compartment wiring harness to be separated from the firewall-mounted fan relay.

The connector had melted completely and the connection was now open. I assumed that the hot weather and the very hot engine compartment created excessive resistance in the connector. This additional connector, probably combined with some corrosion between the connector contacts, acted as a resistor in the fan circuit. Because of the large amount of current flowing through the circuit, the resistance caused considerable heat, which helped melt the connector.

I have no doubts that factory testing of this connector was successfully performed. However, under the operating conditions present (extended use of the circuit to support air conditioning, a hot engine compartment, and so on) the connector failed.

I soldered the wires together and had no further problems with the circuit as long as I owned the car. A few years later, I bought a used Govette, which developed the identical problem. This time I was able to more quickly locate the failure, and fixed it as well. I had no further problems with this car. Later, I bought an EM-built pickup truck, which developed the same problem.

## 5.5 PRODUCT-PROCESS MODELS

*Once sufficient basic knowledge is available, we begin to formulate performance models to explain our observed phenomena.* These models may range from simple rules of thumb, to simple block diagrams, to equations, to sophisticated statistical-stochastic models. Our purpose is to leverage the knowledge available toward product and process improvement. Crude theories and models may eventually (with experimentation and basic cause-effect understanding) lead to the construction of more-sophisticated simulation models. Simulation models essentially allow us a means of indirect experimentation, which may be useful in guiding system, parameter, and tolerance design efforts (Taguchi [15]). For example, the finite element analysis modeling technique allows us to estimate stress-strength relationships through computer modeling. However, physical confirmation experiments should be designed and run to confirm the simulated results before major decisions or changes are made.

Most models represent physical, psychological, and/or social phenomena in a highly abstract form (e.g., our strength model, Figure 5.3, our human needs model, Figure 4.1, our IQE model, Figure 4.2). Model validity is typically a matter of degree. For example, a good model should closely track the phenomena it is modeling, but "perfect" agreement is typically beyond the capabilities of most models. Acceleration (in terms of elevated physical stresses, time compression, and so on) in physical experiments creates additional challenges.

*Through modeling, we attempt to control the cost of knowledge acquisition and enhance the timeliness element required in effective decision making.* The point is to carefully define, develop, validate, and then deploy the model so as to capture the critical essence of system performance in the customer's application and environment. *Model validity is a function of three basic concepts:* **(1)** *relationships,*

(2) *assumptions, and* (3) *calibration.* Subsequent sections focus on detailed models, modeling procedures, and modeling practices, as well as the design and analysis of physical experiments.

## REVIEW AND DISCOVERY EXERCISES

### Review

**5.1** Select a product that you consider to be of high quality.  Describe its
  **a** Function.
  **b** Form.
  **c** Fit.

**5.2** Select a product that you consider to be of poor quality.  Describe its
  **a** Function.
  **b** Form.
  **c** Fit.

**5.3** Explain the difference between a failure mode and a failure mechanism.

**5.4** List three substitute performance measures for a customer's demanded performance of
  **a** A strong garden shovel.
  **b** A powerful computer.
  **c** A fast automobile.
  **d** A good meal.

**5.5** Many times we hear that a product passed all laboratory tests, but did not fulfill customer expectations in the field.  Using the concepts of true and substitute performance measures, explain this discrepancy.

**5.6** One general expression that is used in ergonomics and human factors work is that "a workplace or product should be designed to fit people."  In a physical-work load sense, describe
  **a** The primary considerations a bicycle manufacturer should consider in bicycle design.
  **b** The primary considerations a furniture manufacturer should consider in office furniture (desk and chair) design.

**5.7** Given that a failure occurred in a system made up of humans and machines (e.g., a commercial aircraft and crew), how can we establish the source of the failure?  Remember, the source could be attributed to human error (in operation), design error, or something else.  The objective is not to establish blame, but to locate potential for improvement.

**5.8** With regard to human errors, provide an example of
  **a** An error of omission.
  **b** An error of commission.
  **c** An extraneous act.
  **d** A sequential error.
  **e** A timing error.

**5.9** Provide an example of a hard failure and a soft failure for
  **a** A pair of shoes.
  **b** A flashlight.
  **c** A rifle.
  **d** A writing pen.

### Discovery

**5.10** Develop a performance profile (similar to Figure 5.2) for the following:
   **a**  An automobile.
   **b**  A pair of socks.
   **c**  A movie.
   **d**  Laundry detergent.
   **e**  Hair spray.

**5.11** Failure modes and mechanisms may be developed in a hierarchy as viewed by a sequence of customers until a root cause mechanism specific enough to explain (and prevent) the failure is identified.  See the development below:

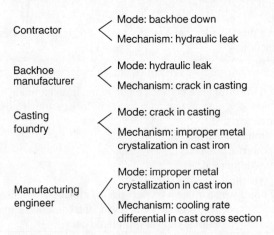

Contractor
   Mode: backhoe down
   Mechanism: hydraulic leak

Backhoe manufacturer
   Mode: hydraulic leak
   Mechanism: crack in casting

Casting foundry
   Mode: crack in casting
   Mechanism: improper metal crystalization in cast iron

Manufacturing engineer
   Mode: improper metal crystallization in cast iron
   Mechanism: cooling rate differential in cast cross section

Based on this example, develop a failure mode and mechanism hierarchy for a product of your choice.  Include at least three levels.

**5.12** Measure the air pressure in a racing bicycle tire (very thin cross-section and high pressure, 120 psi) with a truck tire pressure gauge (which typically measures in the 70 to 150 psi range, but requires a large volume of air).  What result do you expect? What result do you get?  Is this a gauge failure?  Hard or soft?  Explain.

**5.13** In classic cars of the early 1950s, 6-V (volt) electrical systems were common.  In the late 1950s and 1960s manufacturers switched to 12-V systems.  The 6-V and the 12-V automotive bulbs (e.g., tail light bulbs) looked very similar in size and geometric design.  Typically a 6-V or 12-V bulb would fit in either a 6-V or 12-V socket.  A 6-V bulb in a 12-V system would appear very bright for about 1 sec and then burn out.  A 12-V bulb in a 6-V system wouldn't produce light.
   **a**  Was the 6-V bulb in a 12-V socket a soft or a hard failure?  Explain.
   **b**  Was the 12-V bulb in a 6-V socket a soft or a hard failure?  Explain.

## REFERENCES

**1**  N. B. Fuqua, *Reliability Engineering for Electronic Design,* New York:  Marcel Dekker, 1987.
**2**  D. Burgess, "Physics of Failure," *Handbook of Reliability Engineering and Management,* W. G. Ireson and C. F. Coombs, Jr. (eds.), New York:  McGraw-Hill, 1988.

**3** D. J. Thayer, M. Lipow, and E. C. Nelson, *Software Reliability,* Amsterdam, North-Holland Publishing Company, 1978.

**4** G. Salvendy (ed.), *Handbook of Human Factors,* New York:  Wiley, 1987.

**5** M. S. Sanders and E. J. McCormick, *Human Factors in Engineering and Design,* 6th ed., New York:  McGraw-Hill, 1987.

**6** B. H. Kantowitz and R. D. Sorkin, *Human Factors:  Understanding People-System Relationships,* New York:  Wiley, 1983.

**7** See reference 5.

**8** D. P. Miller and A. D. Swain, "Human Error and Human Reliability" (chap. 2.8), *Handbook of Human Factors,* G. Salvendy (ed.), New York:  Wiley, 1987.

**9** A. D. Swain and H. E. Guttman, *Handbook of Human Reliability Analysis with Emphasis on Nuclear Power Plant Applications:  Final Report,* NUREG/CR-1278, SAND80-0200, August, 1983.

**10** See reference 6.

**11** E. E. Lewis *Introduction to Reliability Engineering,* New York:  Wiley, 1987.

**12** B. S. Dhillon, *Human Reliability,* New York: Pergamon Press, 1986.

**13** See reference 9.

**14** "Frequency of Repair Records, 1986–1991," *Consumer Reports,* vol. 57, no. 4, pp. 251–268, April 1992.

**15** G. Taguchi, *Introduction to Quality Engineering: Designing Quality into Products and Processes,* White Plains, NY: Kraus International, UNIPUB (Asian Productivity Organization), 1986.

# 6

# ROBUST AND MISTAKE-PROOF PERFORMANCE

## 6.0 INQUIRY

1 How does design impact quality?
2 What are the two critical parameters in producing or performing to target?
3 What constitutes robust performance?
4 How is quality related to mistake-proofing?
5 What relationship exists between robust performance and mistake-proofing?

## 6.1 INTRODUCTION

Customers buy products in anticipation of benefits, with regard to field performance. They expect high performance in the short-run as well as in the long-run. *Customers expect consistently good performance with little variation, regardless of their application, field environment, product or process configuration, and method of operation.*

Customers expect product and process performance to meet (and hope it will exceed) their present and future needs and expectations even in adverse environments and applications (i.e., beyond the range of reasonable exposure). Customers are virtually relentless in their quest for safe, effective, and efficient products and processes. In the United States many legal entities and resources are positioned to back customer demands for product and process safety and effectiveness. Lack of effectiveness and efficiency in products and processes results in lost work time, environmental damage, wasted energy and materials, late delivery, excessive maintenance, and other costs, inconveniences, and personal aggravations.

## 6.2 PERFORMANCE AND DESIGN

Since performance is designed and built into both products and processes, the manner in which performance is defined is crucial. We begin with the true quality characteristics and translate this "voice" of the customer into substitute quality characteristics, which are expressed in technical language. It is critical that we use systematic and thorough methods to identify true quality characteristics as well as to express our substitute quality characteristics.

### Taguchi Design Levels

Taguchi delineates three levels [1] at which quality is designed into products (these levels also apply to services). The delineation of this systematic design hierarchy is viewed as Taguchi's most significant contribution in the quality field. *The Taguchi trilevel design hierarchy includes (1) the system design level, (2) the parameter design level, and (3) the tolerance design level.*

**System Design Level**    *System design applies to the functional level. At this level, relevant product or process technologies and approaches are identified.* System level issues may include the total system view of the way the customer views and uses the product, or the producer intends to build the product, or both. It may also reach within the product to include technical issues in subsystems, components, materials, production process technologies, assembly strategies, maintainability, and so forth. For example, in order to build a shaft for a golf club, we might select a composite material, rather than steel, in order to reduce shaft weight and provide a variety of "stiffness" options for our customers. This product choice is made at a high level. It then dictates high level process design alternatives which must be selected in order to define a system level product or process technology for our organization (i.e., the process technology to produce composite golf club shafts is distinctly different from that required to produce steel shafts).

**Parameter Design Level**    Parameter design is a secondary design level, within or below the system design level. *Parameter design focuses on determining a "best" level, or target, for the design parameters identified and selected at the system design level.* The point is to meet the performance target with the least expensive materials and processes and to produce a "robust" product (on target and insensitive to variation).

*Parameter design can be thought of as a process of optimizing the functional design with respect to both performance and cost.* Taguchi's approach to parameter design concentrates on designed experiments and specialized signal-to-noise ratio (SNR) measures, discussed in Chapter 23. Whether or not we use Taguchi's SNR measures or other means, a sound and efficient experimental design program is mandatory to elicit an effective parameter design. Systematic experimentation, valid sampling techniques, and valid analyses (involving both location and dispersion measures) are required. For example, at the parameter design level, we determine the best parameter levels (specific materials, temperatures, pressures, additives, and so forth) that will take the least amount of time and create the least expense, in order to gain product-process performance, cost, and timeliness advantages in the marketplace for our golf club shafts.

**Tolerance Design Level**    *Tolerance design is the final step, where the parameter tolerances are set.* The parameter design step sets the best "midvalues," or targets, for the process parameters. *The tolerance design step is a logical extension of parameter design to the point of a complete specification or requirement.* In general, narrow tolerances should be given only to parameters where production variation will create critical performance problems. The point is to recognize that production costs escalate at nonlinear rates, as tolerances are tightened. For example, a tight hole tolerance in the hosel on the golf club head (e.g., $0.3000 \pm 0.0001$ in, or a

mean of 0.3000 with a very small standard deviation) might call for a reaming operation in addition to a drilling operation, whereas the reaming operation may be
eliminated if performance can be assured with a relaxed tolerance, superior tooling,
different materials, or a different processing method.

We use target values and cost-performance calculations to guide tolerancing
when possible.  The ideal case is to produce close to target with a small enough
variation such that we can virtually eliminate sorting inspection.  In some cases this
ideal can be realized, but in many others economic factors still dictate the necessity
for classic tolerance intervals and sorting-inspection of products.

## Targets—Location and Dispersion

*Targets are generally of three types:* **(1)** *smaller is better,* **(2)** *nominal is best, or* **(3)**
*bigger is better.*  Smaller is better refers to production time, operating costs, warranty
costs, defect counts and percentages, impurities, accident statistics, electrical noise,
and so forth.  Nominal is best refers to a specific point target such as a dimension,
chemical content level, such as 2% active ingredient, operating temperature, and so
forth.  Bigger is better refers to process yield, service life, time to failure, and so forth.

*Performance measurement* (*to target*) *involves two critical parameters:* **(1)** *location,* relative to a specific point, such as the center of the bull's eye on a target, and
**(2)** *dispersion,* relative to the center of measurements, such as the arrow pattern
center.  Figure 6.1*a* illustrates possible outcomes relative to four location-dispersion
possibilities in an archery context.  The on location, low dispersion case is the most
desirable case (the highest performance).  It is on target (located in the bull's eye) and
has relatively little dispersion or scatter relative to the center of the arrow pattern.
Hence, we would declare it to have higher performance (relative to the four alternatives shown).

If we were manufacturing general purpose measuring sticks (meter sticks), we
might set a goal, objective, and target such as:

> Goal—to be a "best-in-class" producer of measuring sticks.
> Objective—to produce the most accurate measuring sticks possible (at a modest price).
> Target—to produce measuring sticks with a length of 1 m or 1000 mm.

It is not possible to produce to "exactly" 1 meter, due to product variation resulting
from material and process variations.  Hence, the target can be modified and restated
as a mean (e.g., 1 meter or 1000 mm location measure), with a standard deviation
(e.g. 0.0005 m or 0.5 mm dispersion measure).  The check-sheet histograms in Figure
6.1*b* represent meter stick lengths, measured and classified in intervals of 0.25 mm.
Again, the on-location (1 m or 1000 mm target), low-dispersion case provides the
highest performance of the four alternatives shown.

*One strategy that will help to assure consistent product and service performance
is to develop a production focus on explicit targets, rather than on broad ± specifications* (e.g., lower specification limit = 999 mm, upper specification limit =
1001 mm, or length = 1000 ± 1 mm).  A strategy of placing the production focus on
targets, rather than on broad ± specifications, is superior both from a technical per-

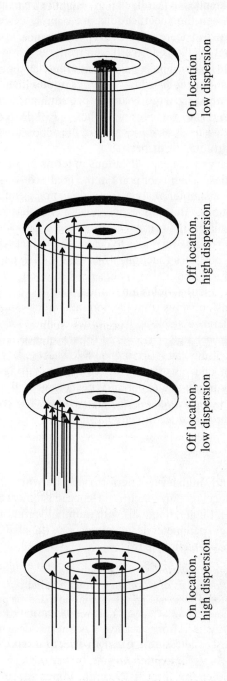

(a) Archery target example, location and dispersion (two-dimensional physical analogy)

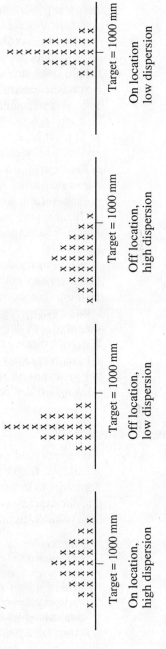

(b) Histogram example for meter stick production (one-dimensional abstraction)

**FIGURE 6.1**    Location and dispersion depiction.

spective as well as a cost perspective (Sullivan [2]).  Relying on inspection and assuming that all items between the specifications are equally effective, not only increases production costs, it also decreases product performance.  Ross [3] refers to ± specifications as the "goal post" syndrome (taken from the game of football regarding extra points and field goals).  Figure 6.2 depicts a target and goal post analogy using smooth product measurement distribution curves, rather than histograms.  In the goal post strategy, we emphasize sorting and inspecting items out of the "tails" beyond the specification limits (i.e., we focus on the tails).  In the target case, we may still use specifications.  But, we focus on the location of the distribution, or center portion, while reducing or controlling the dispersion.

In practice, it is customary to keep specifications in terms of tolerance limits on critical quality characteristics.  The major point in the goal post versus target issue is the strategic focus.  For example, a target focus does not mean that we totally eliminate quality specifications.  But, it does mean we de-emphasize or abandon a product sorting strategy in favor of a process target focused strategy.

Taguchi proposes the use of loss functions to deal with product requirements [4].  Figure 6.3a depicts a goal post loss function, while Figure 6.3b depicts a Taguchi (quadratic) loss function for a nominal is best target value (e.g., 1 m for a meter stick).  This quadratic loss function yields a minimum loss at the target value.  The loss increases as the dimension varies from the nominal target value in a quadratic fashion.  For example, tolerance stack-up in assemblies due to encountering extremes in the ± specification bands can degrade performance as well as create potential rework or "hand-fitting" and "sorting" problems, resulting in higher production costs.  Inspection and rework as well as custom fitting are expensive alternatives to consistently producing to (or very near) target.  *We must focus on the means to hit the target and reduce the dispersion about the target as long as such efforts are economically and technically feasible.*

## 6.3  ROBUST PERFORMANCE

Customers want a consistently high level of performance, notwithstanding operating conditions, from each product unit they purchase.  Hence, customers expect field performance to target (smaller, bigger, nominal) with minimal variation or dispersion.  Phadke describes the Taguchi parameter design philosophy (labeled "robust design" here) as an experimental-based technique [5]:

> *The fundamental principle of robust design is to improve the quality of a product by minimizing the effect of the causes of variation without eliminating the causes.*  This is achieved by optimizing the product and process designs to make the performance minimally sensitive to the various causes of variation, a process called "parameter design."
>
> Robust design uses many ideas from statistical experimental design and adds a new dimension to it by explicitly addressing two major concerns faced by all product and process designers:
>
> a  How to reduce economically the variation of a product's function in the customer's environment.
> b  How to ensure that decisions found optimum during laboratory experiments will prove to be so in manufacturing and in customer environments.

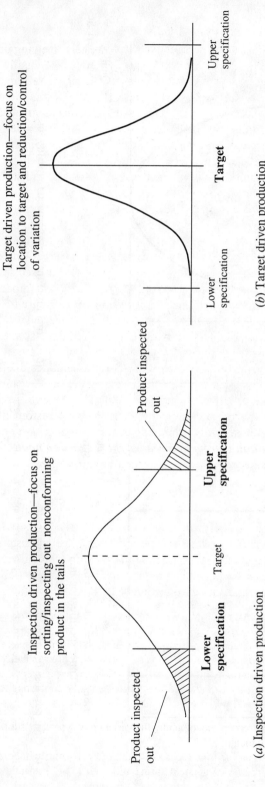

Target driven production—focus on location to target and reduction/control of variation

Upper specification

**Target**

Lower specification

(b) Target driven production

Inspection driven production—focus on sorting/inspecting out nonconforming product in the tails

Product inspected out

**Upper specification**

Target

**Lower specification**

Product inspected out

(a) Inspection driven production

**FIGURE 6.2**   Target versus inspection driven production.

117

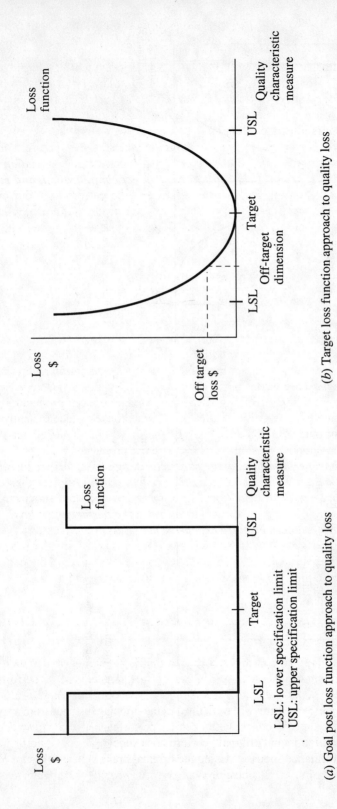

(a) Goal post loss function approach to quality loss

LSL: lower specification limit
USL: upper specification limit

(b) Target loss function approach to quality loss

**FIGURE 6.3** Goal post and target (Taguchi) loss functions, nominal is best case.

In this chapter we will direct our discussion to what we term "robust performance." Here, we will view robust design as one of the tools available to help produce robust performance in our product and production processes. The specific topic of robust design will be discussed in Chapter 23. The robust performance concept is illustrated through a performance variance stack-up model (Figure 6.4). *Product performance variation stack-up,* as shown in Figure 6.4*a, results from four primary factors.* In the case of product performance, *variation or noise in the field environment, field application, product configuration, and method of operation* all stack up to produce variation in the performance our customers (as a group) experience. The *process performance variance stack-up* counterpart, shown in Figure 6.4*b, includes the process environment, the process configuration and conversion equipment, the process materials and supplies, and the method of operation.* The method of operation is depicted in the center of both diagrams, since it is typically developed after the other three inputs are identified. In many cases, operating methods are expected to compensate for variations in environment and application and shortcomings in configuration.

*The term "robust performance" refers to the ability of a product or process to deliver maximum performance with minimal variation. In both the product and process cases, input variation stacks up to produce an amplified level of performance variation.* Here, in Figure 6.4, we can see variation adding up (in the usual statistical sense where we add or subtract independent random variables) or stacking up (in the physical sense) as our customers use our product or process. What we would like to see is a product or process producing a consistently good performance, or output, all the time, under a wide variety of possible inputs, with respect to our population of customers as they use our product.

*Robust performance, at the parameter design level, focuses on reducing stacked variation to assure all of our customers consistent, high performance at a competitive price. The most desirable outcome is to simultaneously reduce product or process performance variation and raise the performance level, without using tighter tolerances or more expensive materials* (which both add to product cost). Hence, we seek both "better" location (bigger is better, smaller is better, or nominal is best) effects and smaller dispersion effects (the spread of the performance distributions in Figure 6.4) in outputs, given variation in our inputs.

## Automotive Brakes—Robust Performance Case

Customers expect that automobile brakes should respond to their need for short stopping distances. Stopping distance is a smaller is better performance measure which benefits the customer by avoiding a collision. Therefore, a substitute performance measure will be defined as the stopping distance from a speed of 55 mph.

Figure 6.5*a* conceptually depicts performance for a conventional brake configuration in an "original" performance subcase and a "modified," or improved, performance subcase. Using the original brake subcase (Figure 6.5*a*) as a reference point, more robust product performance might be obtained by reducing the

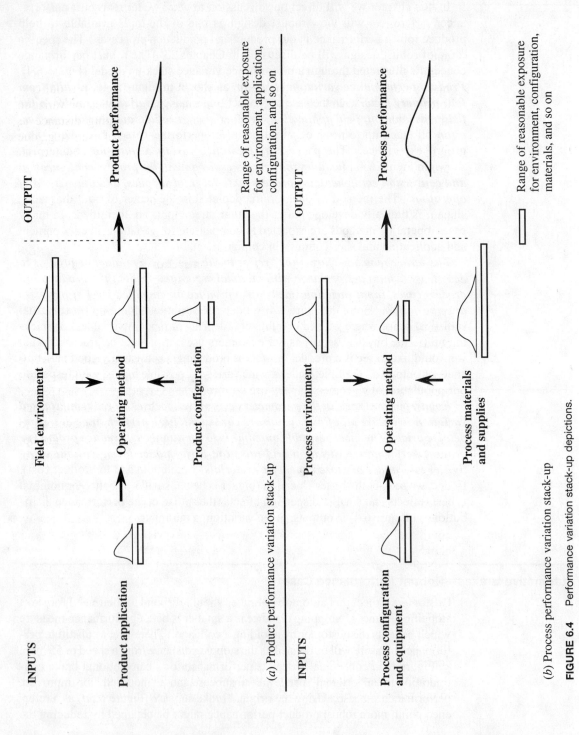

INPUTS

Field environment

Product application

Operating method

Product configuration

OUTPUT

Product performance

Range of reasonable exposure for environment, application, configuration, and so on

(a) Product performance variation stack-up

INPUTS

Process environment

Process configuration and equipment

Operating method

Process materials and supplies

OUTPUT

Process performance

Range of reasonable exposure for environment, configuration, materials, and so on

(b) Process performance variation stack-up

**FIGURE 6.4**  Performance variation stack-up depictions.

variation in any of the four contributing input categories: (1) field environment, (2) field application, (3) product configuration, and (4) method of operation.

With respect to the field environment, we could advise our customers to drive and brake on only warm, dry days.  For an improved product application, we could suggest that customers drive only on paved roads, with moderate loads, and make infrequent stops.  These two alternatives would be effective (if our customers went along with them) in limiting environmental and application variation and producing less performance variance stack-up in stopping distance as well as a shorter average stopping distance for our population of users (e.g., due to the favorable environments and applications we recommended).  Hence, our product's performance would be higher (shorter stopping distances) with less variation.  However, our customers' automobile usage would be quite constrained.  We would limit customer benefits and allow competitors a great opportunity to better serve our customers.

It seems more feasible to address the robustness of our product's performance from both an operating method and a product technology approach.  We might seek to increase braking performance through superior operating methods.  For example, we might establish that a "pedal pumping" method for our brake system is the preferred operating method.  Hence, customer education and training could be pursued to propel customers toward modified (better) operating practices (the modified operating methods curve in Figure 6.5a shows less dispersion than the original curve).

With respect to a product technology approach, three options exist: (1) reduce product unit-to-unit variation through targeting and variance reduction (in our factory) on the brake assembly components, (2) enhance performance and reduce deterioration effects over time and usage through more rigorous engineering specification and design, or (3) innovate changes in fundamental brake technology to produce superior functional performance (through our creative powers).  The options are listed above in ascending order of technical difficulty and development time horizons.

The first technical option, reducing product unit-to-unit variation, typically focuses on the manufacturing processes.  This option has received widespread attention in Japanese versus American product and production process quality comparisons (Sullivan [6]).  This issue is widely viewed as a target versus classical specifications issue as well as a variance reduction issue.

The second technical option is an engineering design change aimed at both short- and long-term brake performance.  For example, automobile brake component characteristics such as brake pad or rotor size and wear, surface friction changes, pressure seal wear or leakage, and so on, impact long-term stopping distance performance.  Our conventional operating method and product configuration improvement activities reduce variation in the modified product configuration in Figure 6.5a.

The net stack-up result (after improvement) appears as reduced variation in our performance stopping distance.  The superior operating method and the

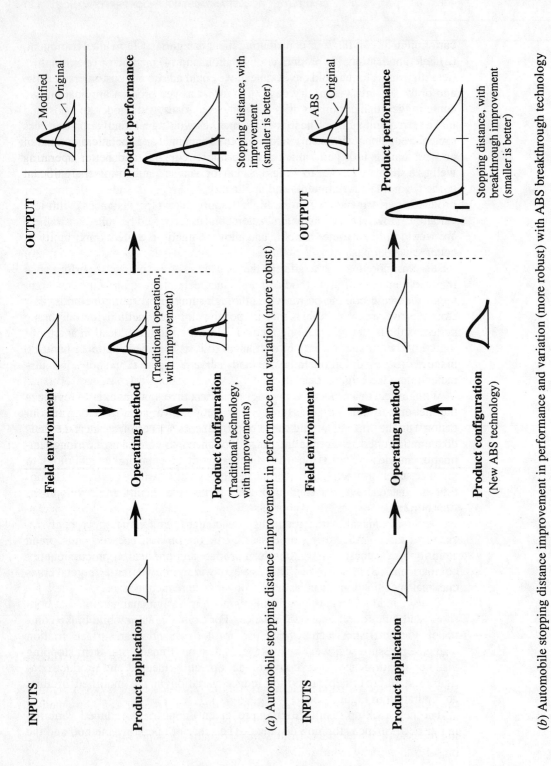

(a) Automobile stopping distance improvement in performance and variation (more robust)

(b) Automobile stopping distance improvement in performance and variation (more robust) with ABS breakthrough technology

**FIGURE 6.5**   Robust peformance variation stack-up depictions (Automotive Brakes Case).

conventional technology improvements shift the product performance distribution to the left and reduce the dispersion. The net result is a more robust performance (less dispersion in stopping distances) and higher performance in shorter average stopping distances (as shown in the modified product performance output curve in Figure 6.5a).

The third technical option involving significant technical innovation deserves special attention. It is more challenging than the preceding technical options of product unit-to-unit variation, limited product redesign, and better operating methods. However, it also holds the most promise for gaining a significant competitive advantage in product performance.

In terms of automotive brakes, the antilock brake system (ABS) falls into this category. Technology has been used to compensate for variation in the field environment, field application, and operating methods. The net result is performance unrivaled by standard brake systems (Figure 6.5a). The comparison of ABS technology to the original brake system in Figure 6.5b shows a significant improvement in product performance level as well as reduced performance variation, even when we see more variation in operating methods (e.g., even a panic stop of slamming on the brakes will result in consistently improved braking performance due to the ABS).

The critical point in technological innovation is to recognize that while variation in the actual field environment, actual field application, and actual field operating methods may remain virtually unchanged or even perhaps increase, performance is significantly improved. The product variation curve now changes to represent the "new" product. The product unit-to-unit and limited product design modification options still apply, and these improvements will produce incrementally greater robust performance. But, they will be unable to match a breakthrough improvement such as the ABS.

---

Process and service examples can be described in a fashion analogous to the automobile brake example. For instance, process techniques involving mechanized, semi-automated, and automated technology have made great improvements in product quality (e.g., improvements in the form of reduced unit-to-unit variation). Service breakthroughs include document delivery and communications (e.g., overnight delivery, fax transmissions, electronic mail, mobile personal communicators).

---

## Umbrella Fabric—Robust Performance Case [SM]

A manufacturer of umbrellas began to export its product to South America under a trade agreement with several South American governments. After an initial surge, sales lagged and the manufacturer undertook customer surveys to determine the reason for the drop in sales. Although the manufacturer had implemented extensive quality testing procedures throughout the production process, the testing focused on the elements from which the umbrellas were intended to provide protection—notably wind and rain.

The new South American market was dominated by large subtropical cities where rainstorms tended to be brief and heavy. The umbrellas performed well during these brief storms. However, a large number of customers used the umbrellas as protection from the sun. Since the markets were subtropical, the levels of ultraviolet radiation to which the umbrellas were subjected were substantially higher than in North American markets. The umbrellas were constructed primarily of a polymer that was extremely lightweight and durable (two qualities determined to be extremely important to customers), but decomposed rapidly when exposed to high levels of ultraviolet radiation. As a result, the manufacturer's umbrellas did not last long in the subtropical sun and had acquired a reputation for poor durability. Based on this information the manufacturer identified a polymer with more robust performance, that was better suited for use in intense sunlight, and employed it in the development of a new product.

## Zero Defects and Six Sigma

The typical customer does not buy products or services in great quantities. Production statistics concerning substitute quality characteristics (the proportion of defective items, defects per unit, and so on) do not greatly interest individual customers. The field performance of the one or two units the customer buys determines his or her product and service experience. For example, as a customer, the performance you get out of the automobile you own influences your opinion of that automobile model much more than factory quality reports or even the performance of your neighbors' similar automobiles. Hence, statements such as "This automobile was thoroughly inspected at the factory," "We have improved our current models, but the older model you own has experienced a number of problems," or "Perhaps you should consider buying our newest model, we have solved the problem you are experiencing" do little to build customer satisfaction.

In the short-run, constant product and process improvement toward defect rates (a substitute quality characteristic) expressed in $x$ defects per thousand or million are reasonable and meaningful in the factory. *Customers now are demanding zero defect (ZD) products. However, their idea of a ZD product is one that meets or exceeds their needs and expectations at all times. Whereas, the factory definition of "zero defects" means that all components, subsystems, and the final product meet all technical specifications (substitute quality characteristics).*

**Zero Defects**   The "zero defect" term was coined by Crosby in 1961 [7]. The original intent of the ZD concept was to position it as a management standard, relative to process expectations. Hence, production systems would seek to meet substitute quality requirements 100 percent of the time. However, the concept was greeted with skepticism in the United States and ultimately was used as an ineffective "motivation" tool, without effective technical and training support. On the other hand, the Japanese, led by Shingo, pursued the "people" and technical issues, used quality training both on and off the job, and were successful in their ZD efforts.

Shingo's "poka-yoke" (mistake-proofing) concept was developed to provide 100 percent product quality (conformance to specifications) [8]. Shingo maintains that production driven by product and process statistics will always have some level of nonconformity. *The mistake-proofing strategy stresses the development of technical and organizational methods to aid in first preventing and then, if necessary, detecting mistakes in real-time, so that corrections can be made on a same-unit basis* (e.g., no nonconforming items will be produced at all). Shingo places a high priority on the development of innovative fabrication and assembly methods, fixtures, and mistake-sensing devices to help assure zero defects.

The Shingo concept of source inspection differs from the classical concept of product sorting inspection [9]. *Source inspection is considered an integral part of the production process, while classical product inspection is considered an independent sorting (acceptable, rework, scrap) process.* Shingo's concept of 100 percent source inspection addresses meeting product or process specifications 100 percent of the time by detecting nonconformance in progress and making appropriate corrections, as soon as possible, so that no nonconforming products are built. Source inspection is of two primary forms: (1) successive checks and (2) self-checks. In successive checks, the next customer (the next operator) examines the previous operator's work, a pass back and correction is made immediately (if necessary). In self-checks, the source inspection and correction is made instantly by the operator (or machine in an automated environment).

*Source inspection demands* (1) *process and product knowledge and* (2) *self-discipline within a quality focus.* Shingo maintains that 100 percent source inspection coupled with process mistake-proofing is capable of producing a ZD product (e.g., one that meets all substitute quality requirements). He cites several examples of its effectiveness—many tied to assembly intensive operations—in Japanese industries. Figure 6.6 depicts a warning class, self-check source inspection ( *poka-yoke*) device. Figure 6.7 depicts an example of a control class, successive source inspection ( *poka-yoke*) device. Both examples demonstrate technical innovation integrated into production processes. The result is virtual defect elimination, yielding a ZD product.

**Six Sigma** *The six sigma concept, developed at Motorola, is process capability focused, rather than product defect (e.g., zero defects) focused* [10]. The six sigma concept in general is characterized by six basic steps: (1) identify your product, (2) identify your customer, (3) identify your needs in producing your product for your customer, (4) define your processes, (5) mistake-proof your process and eliminate waste, and (6) continuously improve your process. In the manufacturing environment, the six sigma steps are more explicit: (1) identify product characteristics that will satisfy your customer, (2) classify the characteristics as to criticality, (3) determine if the classified characteristics are controlled by part

| Inspection method | Setting function | Regulative function | Company name |
|---|---|---|---|
| Source inspection ● | Contact method ● | Control method | Asahi National Lighting Co., Ltd./Gunma Plant |
| Informative inspection (self) | Constant value method | Warning method ● | **Proposed by** |
| Information inspection (successive) | Motion-step method | | Tool plant |

**Theme**

Preventing the omission of tension inspections after L420 units are soldered

**Before improvement**

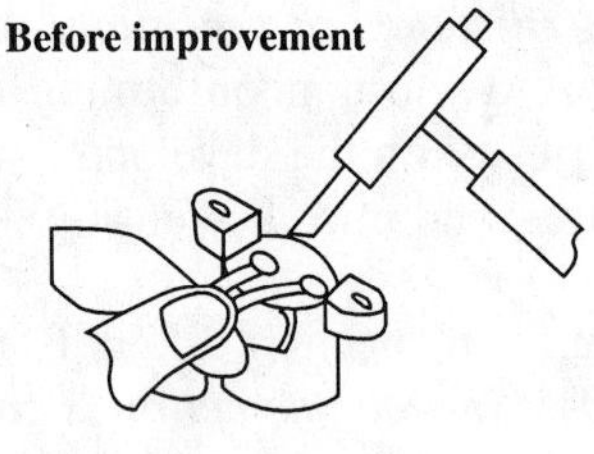

In soldering 504 lead wires to the L420, the lead wires were held by hand while the soldering took place.

After the soldering was over, the lead wires were pulled by hand to verify the state of the solder.

**After improvement**

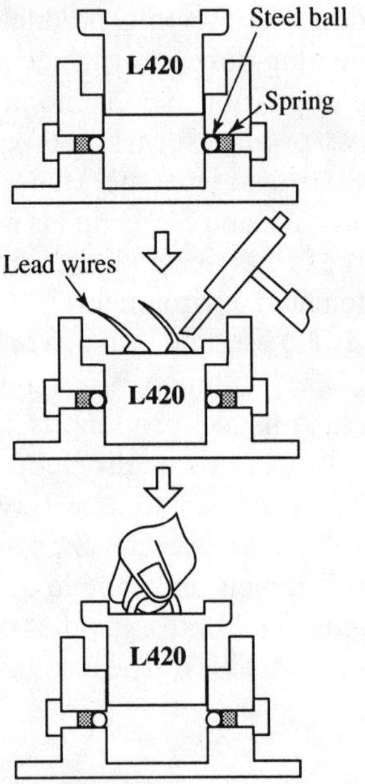

1. Steel balls and springs hold the L420 in place once it is positioned in a jig.

2. 504 lead wires are soldered to the L420's terminals.

3. The lead wires are pulled out of the tension jig. When they are, the resistance of the springs guarantees the tensile force.

| Effects | Post-soldering tension inspections became reliable and soldering defects decreased. | Cost | ¥5000 ($25) |
|---|---|---|---|

**FIGURE 6.6** Warning class, self-check source inspection (*poka-yoke*) example. From *Zero Quality Control: Source Inspection and the Poka-Yoke System*, page 181, by Shigeo Shingo. English translation copyright © 1986 by Productivity Press, Inc., PO Box 13390, Portland, OR 97213-0390, (800) 394-6868. Reprinted by permission.

| Inspection method | Setting function | | Regulative function | | Company name |
|---|---|---|---|---|---|
| Source inspection | Contact method | ● | Control method | ● | Daiho Industries, Ltd. |
| Informative inspection (self) | Constant value method | | Warning method | | **Proposed by** |
| Information inspection (successive) | ● | Motion-step method | | | |

**Theme**

Preventing groove omission

### Before improvement

Groove omission defects would come back to us from customers in the form of returned goods. Each time this happened, either quality groups or line workers would go to the customer and sort the parts out. Since this damaged our relationship of trust with customers, we manufactured the poka-yoke device described below and integrated it into production.

### After improvement

1. In an oil groove cutting process, tool changes and damaged bits would lead to the occasional appearance of products missing grooves.

2. Although a principled approach to this problem would involve mounting a checking device within the process in question, space restrictions made this difficult. For this reason, a device was mounted in the next (punch) process that would catch products without grooves. The device shuts down the press and the previous process and sounds a buzzer to alert workers.

3. This approach made it possible to discover products with misplaced grooves as well as items without grooves.

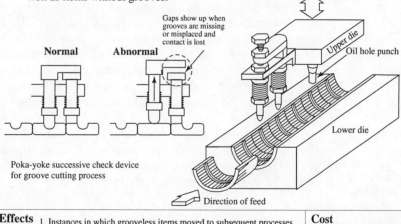

Poka-yoke successive check device for groove cutting process

**Effects** 1. Instances in which grooveless items moved to subsequent processes have been eliminated because missing grooves are discovered and the press is shut down. 2. Grooves cut in the wrong places can be similarly discovered.

**Cost** ¥5000 ($25)

FIGURE 6.7    Control class, successive source inspection (*poka-yoke*) example. From *Zero Quality Control: Source Inspection and the Poka-Yoke System*, page 257, by Shigeo Shingo. English translation copyright © 1986 by Productivity Press, Inc., PO Box 13390, Portland, OR 97213-0390, (800) 394-6868. Reprinted by permission.

and/or process, (4) determine the maximum allowable tolerance for each classified characteristic, (5) determine the process variation for each classified characteristic, and (6) change the design of product, process, or both to achieve a $C_p \geq 2$, a measure of quality discussed in Chapter 18.

The six sigma concept is aligned with both the produce-to-target case and the goal post case previously discussed. It is target focused in that it is attached more to the process than to the product, but it is also goal post focused in that we must have product specifications in order to determine the number of "sigmas" ($\sigma$, a standard deviation measure) we have associated with our process.

In order to determine whether or not a given process is a six sigma process, using a measurable quality characteristic, we must typically have three pieces of information: (1) the process mean, (2) the process standard deviation, and (3) the product specifications. Figure 6.8 depicts the generic six sigma concept. Here, we assume that the quality characteristic measurements are distributed normally; hence, by definition, a six sigma process has a ±6 standard deviation (6$\sigma$, or 6 sigma) distance from the process mean to each product specification limit.

A true six sigma process will yield 2 defects per billion (see Figure 6.8$a$). However, it is usually stated that a six sigma process produces 3.4 or fewer defects (nonconforming to specifications) per 1 million opportunities. This latter statement includes a ±1.5-sigma shift in the mean (see Figure 6.8$b$). Six sigma is relevant even though our production for the life of a product may not actually contain a million or more physical opportunities for a defect, as we use statistical estimation and inference to establish a sigma level, in addition to historical records.

If we base our six sigma assessment on attributes measures (acceptable or unacceptable) where we do not have a continuous measurement (as in Figure 6.8), we must clearly define both a defect and an opportunity. Then, we must "keep score" to develop our measure of defects per million opportunities and proceed in a manner somewhat different from the normal distribution based analysis, shown in Figure 6.8, in order to develop a sigma level for our process.

The six sigma "standard" is used much like a ZD standard. Hence, six sigma processes do not automatically yield satisfied customers for the same reason ZD processes do not automatically yield satisfied customers— both deal with substitute quality characteristics. If we were directly measuring true quality characteristics, then a ZD product would be assured of always creating customer satisfaction and a six sigma product would come very very close.

## Radios—Robustness Case [DS]

I was involved in a venture that was contracted to build radios. The designers "went off" and developed technical specifications from marketing's requirements. A breadboard was built and declared a success—the radio performed wonderfully. Then, production "went off" to build the radios. The first production unit went into test and failed to meet the performance specifications, as did the second one.

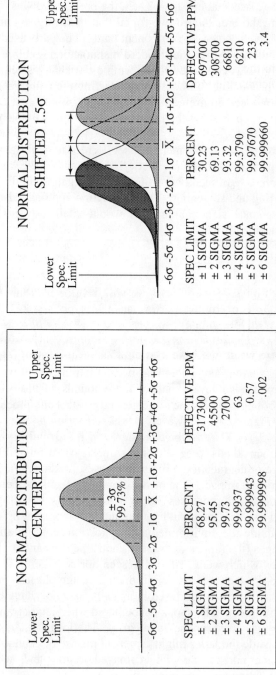

(a) True 6 sigma process

(b) Six sigma process with a ±1.5-sigma shift

**FIGURE 6.8** Six sigma process concept. Reproduced, with permission, from "The Motorola Guide to Statistical Process Control," Phoenix, AZ: Motorola Semiconductor Products Sector, page 5, 1989.

An analysis was performed to determine what was wrong with the manufacturing process. Nothing could be identified; they were following procedures and using standard acceptable parts. Then, the breadboard was examined in detail. In reviewing the breadboard, we found that the designers and developers had hand-tested and selected all component parts. The parts used in the breadboard had performed better than the standard manufacturer's components.

The design was examined to determine the changes required for the units to be more producible (without hand-selecting the components) and still meet their performance requirements. It was determined that extensive redesign would be required and that the radios would not be available until well after the "need" date.

Based on this prognosis, management decided to dedicate personnel to hand-screen the parts used in the production of the radio. The lack of robust design cost the venture significantly in lost time investigating production problems and in parts sorting and screening. This production approach also added risk to the radio maintenance, since maintenance would entail sorting, screening, and testing of replacement parts.

## 6.4 MISTAKE-PROOFING

The concept of mistake-proofing as we will develop it in this textbook is tightly coupled with that of robust performance and poka-yoke. *Mistake-proofing emphasizes safety elements of products and processes in a misuse-survival context, rather than the performance variation emphasis of robustness.* The point is that reasonable ranges or bounds exist for environmental, application, product or process configuration, and operating method parameters. Within these bounds, robust performance is relevant; outside these bounds, mistake-proofing is relevant.

Mistake-proofing contains the same general product and process factors as robust performance (Figure 6.4). However, *mistake-proofing focuses on inputs at unintended levels (beyond the range considered in robust performance).* Hence, the performance focus tends to be personal safety, system reliability or survivability, system damage or degradation, production losses, production inefficiency, and possible environmental impact. The mistake-proofing focus is depicted in Figure 6.9*a* and *b* for products and processes, respectively.

The shaded areas on the input bars (Figure 6.9) represent inputs that are considered to be outside the environmental, application, configuration, and operating parameter ranges (the ranges relevant to robust performance). Environments may be encountered which were neither foreseen nor considered in the environmental envelop in the product-process definition and design phases. For example, solder joint integrity may be compromised in an extremely hot (natural or internal power induced) environment. The shock encountered when a wristwatch is dropped on a concrete floor may compromise its integrity. The former example might represent an oversight, while the latter might represent misuse or product abuse. Nevertheless, the customer may end up with a damaged product, and the resulting degree of customer satisfaction may be compromised.

Misuse and abuse, accidental or otherwise, in product application or process equipment application may compromise personal safety or production efficiency. Misuse and abuse in the field are difficult to eliminate. In addition, product configurations are sometimes altered and process materials are sometimes substituted by customers in the field. For example, safety devices and machine guards may be removed, overload protection such as fuses may be altered, and so on. But, in-

**FIGURE 6.9**    Mistake-proof performance depictions.

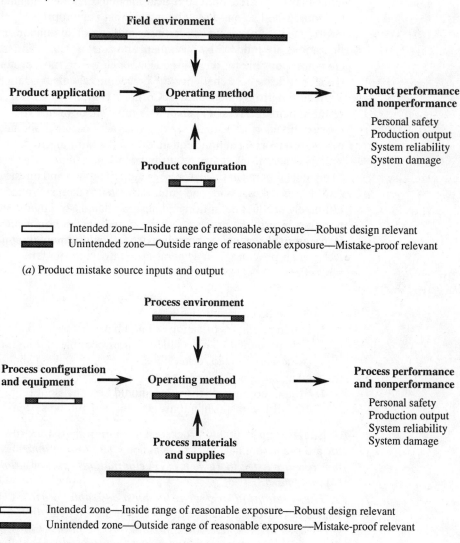

(a) Product mistake source inputs and output

(b) Process mistake source inputs and output

evitably, if we do not look out for our customer's best interests, our competitors will. Hence, "finger pointing" and "blame allocation" are not productive means of dealing with either injured internal or injured external customers.

Operating methods are typically developed after the environment, application, and configuration are identified. Hence, they are often expected to compensate for anticipated shortcomings in the other three inputs. Customers may use product and process operating methods incompatible with recommended methods. Such customer actions may result in personal injury, poor performance and lost production, lower reliability, and/or system damage. For example, machinery startup procedures may be altered, protective gear may be discarded, maintenance lockouts may be ignored, and so on. It can be argued that inappropriate application, unforeseen environments, unexpected operating conditions, and unauthorized equipment configurations are induced by customers and acts of nature and, therefore, producers are not responsible for the consequences. However, most customers and legislative (legal) entities would challenge the preceding argument (Hammer [11]).

According to Sanders and McCormick [12], many mistakes are ultimately the result of human errors: (1) errors of omission, or failure to act; (2) errors of commission, or incorrect action; (3) extraneous acts, or inappropriate action; (4) sequence errors, or incorrect action sequence; and (5) timing errors, or improper timing of action. Other mistakes may be attributable to acts of nature and physical phenomena. Many mistakes result from a combination of human and physical factors.

Many mistakes, errors, and failures charged to human product users and nature are ultimately product definition and design related. In other words, proper mistake-proofing in product-process definition and design could prevent many failures. Hence, *mistake-proofing strategies are critical in proactive quality assurance.*

Norman provides four general system design guidelines to minimize human errors [13]:

1 System state (the current state of a system) feedback (to human operators) should be clear and available.
2 Different classes of actions should have different command sequences.
3 Actions should be reversible as far as possible and high-consequence actions (which could lead to high-consequence mistakes) should be difficult to accomplish.
4 System command structures should be consistent (e.g., operator responses, control levers, pushbuttons, etc.) throughout the system.

Effective product and process mistake-proofing strategies may go a long way toward producing customer satisfaction. *The four primary strategies in mistake-proofing are listed in Table 6.1: (1) elimination, (2) prevention, (3) detection, and (4) loss control.*

*The elimination strategy is the most desirable,* provided an effective alternate technology or design can be discovered or developed. *In this case, the "mistake" is removed or completely designed out of the product or process.* For example, fiber-optic technology virtually eliminates electromagnetic interference or noise; a

**TABLE 6.1   MISTAKE-PROOFING STRATEGIES AND TOOLS**

| Primary tools | Primary strategy and focus | Example |
|---|---|---|
| | **Proactive** | |
| | Elimination | |
| |   Technology (selection, development) | Alternate materials, processes |
| |   Design-out (hazards, threats) | Mechanization, automation |
| Prevention of loss | Prevention | |
|   QFD |   Design-in (redundancy, barriers, safeguards) | Controls, shields, guards |
|   CE |   Methods (operating procedures) | Training, operating displays, labels |
|   FMEA | **Proactive and/or reactive** | |
|   FT | Detection | |
|   ETA |   Warning (predictive, corrective) | Warning displays, gauges, horns, lights, odors |
| |   Fail-safe (shutdown, minor damage) | Mechanical, electrical fuses, emergency shutdown |
| | **Reactive** | |
| | Loss control | |
| |   Containment (system damage) | Personal protection and restraints, "crush" zones, contained releases |
| Management of loss |   Isolation (system loss) | Single-system incident, harmless environmental releases |
| |   Catastrophe management (system and extended loss) | Multiple-system incident, harmful environmental releases |

QFD—quality function deployment, CE—cause-effect analysis, FMEA—failure mode and effects analysis, FTA—fault tree analysis, ETA—event tree analysis.

robot may remove a person from the possibility of a hazardous exposure. Caution must be exercised however, in that new technology may introduce the opportunity for new mistakes.

A wide variety of preventive opportunities may exist. ***Prevention avoids mistakes but does not eliminate mistake possibilities.*** Typically, these opportunities are related to design, retrofit, and operating procedures and methods. For example, machine guards, access barriers, warning labels, product operating instructions and training, process methods training, and so forth, help to prevent mistakes. Medicine and food "tamperproof" processes and packaging help to prevent unauthorized product access. Electrical box lockouts provide mistake-proofing for maintenance activities. Smaller gasoline filler holes and nozzle diameters are used for unleaded versus leaded gasoline, and so on.

Detection relies on both proactive and reactive strategies. ***Detection requires a sensor-based technology whereby critical measures are identified, monitored, and transformed into either advanced warning and corrective action or fail-safe shutdown action.*** The measurement, detection, and action may be of a closed-loop or open-loop (involving a human) configuration. Action may be taken based on predicted problems or problems in progress. Obviously, more opportunity for correction or recovery is afforded by early warnings.

***Loss control is primarily a reactive strategy. It seeks to manage or limit losses.*** Loss control strategies are the lowest form of mistake-proofing. Seat belts and air bags in automobiles, dikes around tanks, levies along rivers, and so on, provide examples of loss control, given that an unfortunate event or act of nature takes place. In some cases, especially in the case of acts of nature, such as floods, loss control may be the only strategy feasible. In many cases, loss control is used as a redundant strategy (in addition to higher order strategies, Table 6.1).

Mistake-proofing strategies must encourage creative efforts to identify potential mistakes resulting from human as well as physical sources. The majority of the tools described in later chapters of this book can be used to support robust performance and mistake-proofing analyses and action. The point is that ***mistake-proofing is effective in preventing product dissatisfaction as well as in yielding product satisfaction.*** Products and processes that are tolerant of mistakes (either accidental or deliberate) will always be favored over those that are not mistake-tolerant, other benefits and costs being about equal.

## REVIEW AND DISCOVERY EXERCISES

### Review

6.1 State a goal, objective, and target sequence for,
   a  A college student regarding his or her major.
   b  A nail manufacturer.
   c  A dog food manufacturer.
   d  An automobile tire manufacturer.

**6.2** Develop a brief system, parameter, and tolerance design statement sequence for the following functions:

    **a** Fastening loose-leaf papers together (2 to 20 sheets).

    **b** Cooking fish in a fast-food restaurant.

    **c** Powering a space satellite.

**6.3** Past design and product improvement efforts in the United States have emphasized the identification and removal of the causes as a means to reduce product variation. Now, robust design techniques attempt to reduce product variation without cause removal. In this chapter, we suggest that engineers increase performance and reduce variation, all without cause removal, and at the same time reduce product cost.

    **a** Explain how good engineering practice can accomplish the above.

    **b** Explain why tolerance design, when used to reduce variation, is generally not capable of reducing product cost.

**6.4** Compare and contrast the ability of parameter design and tolerance design to

    **a** Increase product performance.

    **b** Reduce product cost.

    **c** Accelerate product schedule (timeliness).

**6.5** Do you, as a customer, expect a ZD product? Explain.

**6.6** Will a ZD product bring about customer satisfaction? Explain.

**6.7** Explain the difference between Shingo's concept of 100 percent source inspection and traditional 100 percent product sorting inspection.

**6.8** When can a successive check and a self-check be the same in source inspection?

**6.9** Explain the difference between a warning-class and a control-class source inspection (refer to Figures 6.6 and 6.7).

**6.10** The measurement of performance with respect to a target value includes both location and dispersion parameters. Explain why both parameters are critical.

**6.11** Why is a strategy of producing to target requirements, rather than within ± specification limits, superior?

**6.12** Are tight tolerances desirable? Explain.

**6.13** Explain the similarities and differences between robust performance and product reliability.

**6.14** Compare and contrast the six sigma, ZD, and Taguchi process target concepts as to their focus and ability to deliver customer satisfaction.

**6.15** Why are the detection focused mistake-proofing strategies in Table 6.1 listed as both proactive and reactive?

## Discovery

**6.16** Identify a product of your choice and develop a brief case for robust design similar to the Automotive Brake—Robust Performance Case (see Figures 6.4 and 6.5).

**6.17** Identify a product of your choice and develop a brief mistake-proofing case along the lines shown in Figure 6.9 and Table 6.1.

**6.18** In order to demonstrate the goal post (discrete) and Taguchi (continuous) loss function approaches shown in Figure 6.3, develop an empirical goal post and a Taguchi loss function for your favorite type of shoes by completing the two plots below. Base your estimates on wearing your shoes for an 8-hour day, and walking 2 mi during the day.

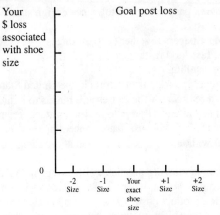

Determine your USL and LSL and associate losses due to ill-fitting shoes subjectively

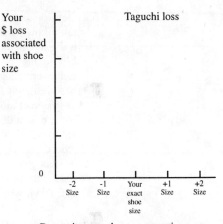

Determine your loss on a continuous scale at 0.5 shoe size increments

## REFERENCES

**1** G. Taguchi, *Introduction to Quality Engineering: Designing Quality into Products and Processes,* White Plains, NY: Kraus International, UNIPUB (Asian Productivity Organization), 1986.

**2** L. P. Sullivan, "Reducing Variation: A New Approach to Quality," *Quality Progress,* pp. 15–17, July 1984.

**3** P. J. Ross, *Taguchi Techniques for Quality Engineering,* New York: McGraw-Hill, 1988.

**4** See reference 1.

**5** M. S. Phadke, *Quality Engineering Using Robust Design,* Englewood Cliffs, NJ: Prentice Hall, 1989.

**6** See reference 2.

**7** P. B. Crosby, *Quality without Tears,* New York: McGraw-Hill, 1984.

**8** S. Shingo, *Zero Quality Control: Source Inspection and the Poka-Yoke System,* Cambridge, MA: Productivity Press, 1986.

**9** See reference 9.

**10** "The Motorola Guide to Statistical Process Control," Phoenix, AZ: Motorola Semiconductor Products Sector, 1989.

**11** W. Hammer, *Product Safety Management and Engineering,* Englewood Cliffs, NJ: Prentice Hall, 1980.

**12** M. S. Sanders and E. J. McCormick, *Human Factors in Engineering and Design,* 6th ed., New York: McGraw-Hill, 1987.

**13** D. A. Norman, "Steps toward a Cognitive Engineering: Design Rules Based on Analysis of Human Error," taken from "Five Papers on Human Machine Interaction," ONR-8205 (AD-A116031, DTIC), 1982.

# THREE

# CREATION OF QUALITY—
# FUNDAMENTAL STRATEGIC AND
# TACTICAL QUALITY TOOLS

## VIRTUES FOR LEADERSHIP IN QUALITY
### Courage and Commitment

The purpose of Section Three is to briefly present a number of strategic and tactical quality improvement tools to our readers. We begin with a presentation of the seven "new" Japanese tools in Chapter 7. These tools are useful in high-level quality planning. They support developmental efforts in formulating strategic quality plans, as well as providing a forum and a format that encourage communication and consensus building. We present the benchmarking tools in Chapter 8 as a means to identify the "best of the best" so that we know what the toughest competitors are capable of. These tools are also useful in assessing the technical capabilities and methods of competitors and competitive organizations that perform the functions we are interested in improving. We introduce the breakthrough thinking process along with benchmarking so that our readers will be able to think in both incremental and breakthrough terms in improving their product or process quality.

In Chapter 9, we introduce our readers to the seven "old" Japanese tools—now widely used by progressive organizations. Most of these tools are simple in concept and in structure; process control charts are the exception. Process improvement is examined in detail in Chapter 10. We introduce our readers to the process improvement cycle and to process flow analysis. All in all, this section contains a number of general and widely applicable quality improvement tools which we can use to address a wide variety of quality challenges, opportunities, and problems.

# THE SEVEN NEW (JAPANESE) TOOLS

## 7.0 INQUIRY

1 What are the requirements for the "new era" in quality?
2 How do the seven "new" tools help us to address strategic planning for quality?
3 How do the seven "new" tools help us to address product planning for quality?
4 How do the seven "new" tools help us to address process planning for quality?
5 How do the seven "new" tools help us to build consensus in quality planning?

## 7.1 INTRODUCTION

In both Japan and the United States (as well as many other countries) tremendous gains have been realized in service and manufacturing organizations using fundamental quality tools. The quality movement is best known for producing incremental improvements in products, processes, and customer service. All told, Japan has led the world in applying fundamental quality improvement tools. Ishikawa maintains that as much as 95 percent of quality-related problems can be solved with the seven fundamental, or "old," tools [1]:

1 Cause-effect diagram.
2 Stratification analysis.
3 Check sheet.
4 Histogram.
5 Scatter diagram.
6 Pareto chart.
7 Control charts.

These seven fundamental tools constitute the nucleus of most quality improvement efforts on the production floor, as they are widely discussed and taught. We will discuss these seven fundamental tools in detail in Chapter 9.

Our purpose in this chapter is to present the seven "new" Japanese tools. These tools are not as widely known and taught as the fundamental tools. However, they offer a good deal more promise in impacting strategic quality planning than the fundamental tools. They are also much more qualitative, open-ended, and difficult to master than the seven old tools.

We have chosen to present the new tools before the old. We believe it is important to address strategic quality issues that provide purpose and direction to our quality efforts before we dive into the detailed operations-level tools. The analogy here is that we need to survey our terrain, determine what kind of road we want to build, where we intend to build it, and how we intend to build it (strategy) before we start clearing trees, moving earth, and paving our road (tactics) to world-class quality.

In other words, the seven new tools serve the leadership function, while the seven old tools serve the management function. We must initiate the leadership function first, if we expect to see really outstanding quality experiences develop for our internal customers and carry through our products to our external customers. (See the quality chain concepts of Chapter 4.)

As early as the late 1960s, the Japanese began to address strategic quality issues. During this quest, they developed and tested many ideas and concepts, several borrowed from other areas of strategic business planning and analysis. Eventually, by the late 1970s and early 1980s, they developed what are now called the seven "new" quality control tools.

*The seven new tools were developed as a means to exploit* what Mizuno and Akao term *the "new era" for quality* [2, 3]:

1 *Relations diagram.*
2 *Affinity diagram.*
3 *Systematic diagram.*
4 *Matrix diagram.*
5 *Matrix data analysis.*
6 *Process decision program chart (PDPC).*
7 *Arrow diagram.*

*This new era is based on two fundamental requirements:* **(1)** *the creation of added value over and above consumer needs and* **(2)** *the prevention, rather than the rectification, of failure in meeting customer needs.* These two requirements are critical in our concept of proactive quality.

*The new tool set was designed to address strategic issues in quality control.* Hence, the tools are broad-based discovery, analysis, and implementation aids which take a systemwide perspective. In essence, they facilitate the discovery and expression of quality-relevant relationships between customer demands, products, and processes. They were designed to stress creative thinking (as opposed to copying strategies) within the framework of the organization. Figure 7.1 provides an overview of the seven new QC tools, relative to three fundamental business objectives: social responsibility, quality (accomplishment) improvement, and human resource development. This figure outlines the Mizuno-Akao new era in quality as it relates to reforming organizational culture in Japan (Mizuno [4]). We can readily observe organizational, societal, and individual elements (e.g., a strong service interface in our quality system triad, Figure 3.2).

We will briefly discuss each tool in this chapter, attempting to retain the Japanese character of each in the descriptions. Readers familiar with the literature in leadership, management, operations research, and statistics will recognize these tools as hybrids, borrowing from many different concepts while focusing tightly on quality-related needs.

## 7.2 RELATIONS DIAGRAM

*The relations diagram serves as a graphical aid in problem cause-effect discovery and description.* It is a useful tool in the problem identification and description

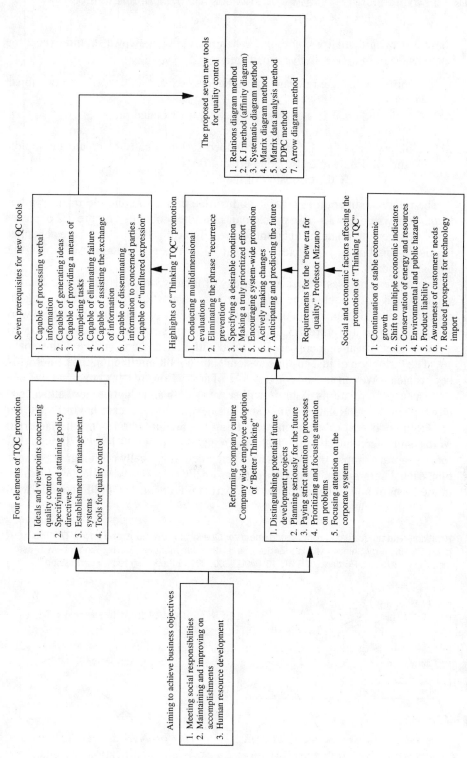

**FIGURE 7.1** The seven "new" Japanese tools positioned in the Mizuno-Akao "new era" for quality. From *Management for Quality Improvement: The 7 New QC Tools*, page 21, edited by Shigeru Mizuno. English translation copyright © 1988 by Productivity Press, Inc., PO Box 13390, Portland, OR 97213-0390, (800) 394-6868. Reprinted by permission.

141

phase of strategic quality planning.  When used effectively, we should expect the relations diagram to help in five broad areas:

1  Identify and isolate relevant causal factors concerning the problem or subject at hand.
2  Clearly and concisely express (verbalize) the isolated factors.
3  Help to place the factors in cause and effect relationship sequences.
4  Link the factors to the problem defined systematically, to create a complete picture of problem, causes, and effects.
5  Help to identify the criticality of the factors.

The relations diagram is flexible and can be drawn in a bubble form or a box form.  It may address a single problem or multiple problems.  Figure 7.2 depicts a simplified, general relations diagram format.  Typically, the problems are contained in double-line bubbles or boxes.  The causal factors are contained in single-line bubbles or boxes.

In complex analyses we see a hierarchy of factors which require more detailed symbols and both double and single arrows.  A portion of a relations diagram concerning the adoption of a Japanese just-in-time (JIT or, in Japanese, *kanban*) work system is shown in Figure 7.3.  Here, we see a box structure interconnected with arrows.  The critical elements are identified with a shaded box outline in this analysis.  The "adoption of just-in-time system" box double-outlined in the center is the key problem.  We can see the convergence of arrows towards this box.

A successful relations diagram will serve as both an analytical tool and a communications tool.  Relations diagrams are developed in a team environment and require a good deal of effort in identifying the factors, expressing the factors in words, and interconnecting the factors in a meaningful way.  Hence, although their format appears simple, the process of development is typically challenging.  In addition, each diagram should not be viewed as totally complete or static.  Modification in the form of additions, deletions, and structural changes should be made when

**FIGURE 7.2**    Relations diagram format. From *Management for Quality Improvement: The 7 New QC Tools*, page 89, edited by Shigeru Mizuno. English translation copyright © 1988 by Productivity Press, Inc., PO Box 13390, Portland, OR 97213-0390, (800) 394-6868. Reprinted by permission.

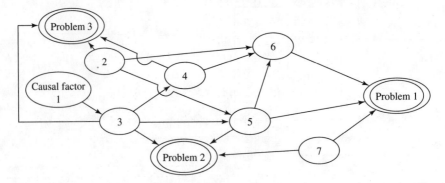

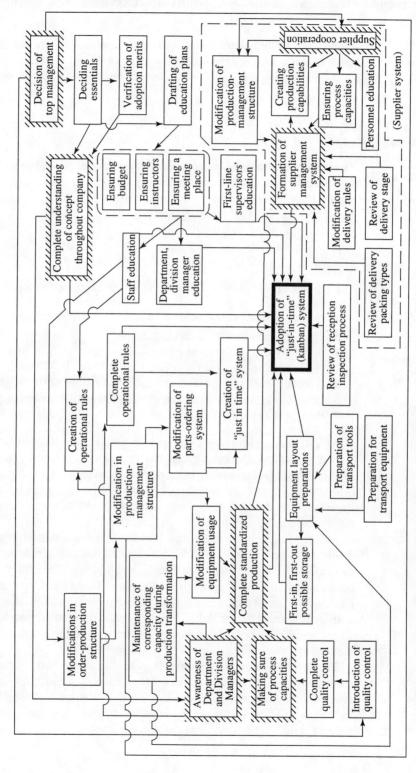

**FIGURE 7.3** Relations diagram illustration. From *Management for Quality Improvement: The 7 New QC Tools*, page 107, edited by Shigeru Mizuno. English translation copyright © 1988 by Productivity Press, Inc., PO Box 13390, Portland, OR 97213-0390, (800) 394-6868. Reprinted by permission.

appropriate.  *The relations diagram is a good tool to use when we want to discover and document high-level thinking, sketch out a high-level strategy, and, at the same time, develop a consensus-building environment to help assure success.*

## 7.3  AFFINITY DIAGRAM

*The affinity diagram is used to collect and organize facts, opinions, and ideas.*  As the name suggests, this tool stimulates creativity in the form of encouraging free expression of both fact and opinion and then seeks to group the information elements with respect to mutual affinity.

Affinity diagram construction requires a form of brainstorming, with the intent of producing a graphic result.  Seven steps are usually executed in the process of developing a formal affinity diagram:

1  Select a theme or purpose which may be expressed as a problem or an opportunity.
2  Collect narrative data.
3  Transfer the narrative data onto cards.
4  Sort the cards into logical groups—typically a wall "post-it" arrangement and "walk-around" review process is used.
5  Label the card groups with respect to the nature of the theme or purpose.
6  Draw the affinity diagram (summarized from the groupings).
7  Present the results.

Affinity diagram construction typically requires an effective group facilitator and a good deal of thought and preparation.  Participants generate cards with facts, opinions, and ideas (sometimes in great numbers) before attempts are made to group and organize the cards.  Criticism is not allowed, but refinements occur as the cards are passed around, shuffled, or exchanged and read out loud by participants.  Developing these cards can be as entertaining as it is informative.

Affinity diagrams are used to address rather complicated strategic themes.  Figure 7.4 depicts a portion of a simplified affinity diagram with a research and development theme.  The basic building blocks are the small boxes.  The content of each small box represents an individual fact, opinion, or idea.  These blocks are then encased with a circle or oval and labeled as a group.  The groups are collected and further encased with a circle and a "super" label, or title.  Then, arrows are drawn to relate the groups.  Thus, we see an organized picture of facts, opinions, and ideas arranged in a hierarchical manner.

*The affinity diagram is an excellent tool to use to start out a cross-functional team effort, where the team has a broad theme.*  The affinity diagramming process puts the team members at ease, while extracting diverse ideas and concerns as well as building a team appreciation for different ideas and concerns on a "level playing field."  *It is excellent for consensus building in associating ideas and*

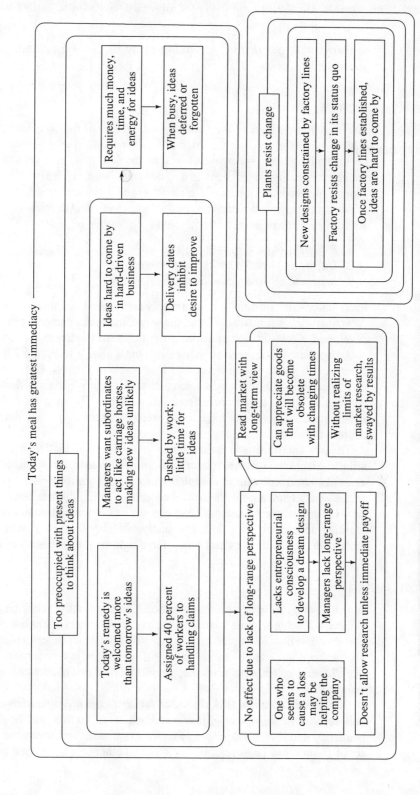

**FIGURE 7.4** Affinity diagram illustration. From *Management for Quality Improvement: The 7 New QC Tools,* page 30, edited by Shigeru Mizuno. English translation copyright © 1988 by Productivity Press, Inc., PO Box 13390, Portland, OR 97213-0390, (800) 394-6868. Reprinted by permission.

*concerns with the theme,* presenting many facets of a complicated problem or opportunity.

## 7.4   SYSTEMATIC DIAGRAM

*The systematic diagram is essentially a logic-based tree diagram which contains a hierarchy of objectives and means to accomplish the objectives.* Development begins with a major objective followed by the deduction of ways and means to meet the objective. Figure 7.5 depicts a systematic diagram construction format, with a generic hierarchy of objectives, questions, and ways and means.

Next, the ways and means are evaluated as to their "practicality." Careful evaluation of each mean leads initially to one of three classifications:

1  *Practical*—marked with the symbol "○."
2  *Uncertain*—marked with the symbol "△."
3  *Impractical*—marked with the symbol "✕."

Each "uncertain" designation is carefully scrutinized and then reclassified as either "practical" or "impractical." Then, the ways and means declared practical are further developed. Classification and scrutiny continue as the tree is constructed.

An application of the systematic diagram is shown in Figure 7.6. Here, we can observe two levels of objectives and four levels of means. Each level is evaluated and associated with implementation activity items. We see a definite refinement in the level of detail as we move from left to right. The number of levels used in a systematic diagram is not fixed, but developed according to the detail necessary to address the objective.

The systematic diagram is a much more focused tool than the relations and affinity diagrams. *It is useful as a planning tool when we have a sequence of objectives that must be accomplished to meet a quality goal.* It systematically traces a path from a goal (first-level objective) to highly detailed objectives necessary for goal accomplishment. We can see our options and choose our path from the diagram, which uses fault tree analysis technology in a positive, rather than fault, context. Fault trees and other tree diagrams are discussed in Chapter 13.

## 7.5   MATRIX DIAGRAM

In the strategic quality planning process, we typically see interrelated factors. Hence, a change in one factor will often affect others. Therefore, we must address trade-offs. We must discover factor relationships critical to our strategic plan as soon as possible, in order to deal with them in a timely fashion. *The matrix diagram offers us a versatile, graphical tool to facilitate both discovery and trade-offs.*

*The matrix diagram is designed to facilitate the identification of relationships between two or more sets of factors.* Each set of factors may be developed independently and then relationships may be examined. Or, in a more usual case, one set of factors may be developed initially and other sets developed subsequently. In

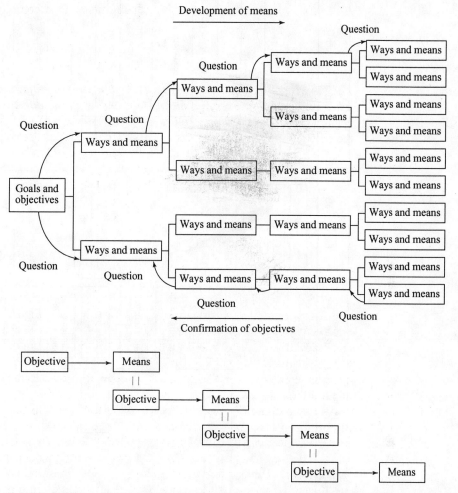

*Note:* In branching, *and/or* can be used when ideas are in the developmental stage. If a distinction between *and* and *or* is necessary, use the logical symbols employed in fault tree analysis (FTA).

**FIGURE 7.5** Systematic diagram construction format. From *Management for Quality Improvement: The 7 New QC Tools*, page 149, edited by Shigeru Mizuno. English translation copyright © 1988 by Productivity Press, Inc., PO Box 13390, Portland, OR 97213-0390, (800) 394-6868. Reprinted by permission.

either case, changes to and reexpression of the factors will take place as the diagram develops.

The matrix diagram is applicable to resource planning and sequencing, product-process definition and design, mistake-proofing and failure prevention, and so on. Figure 7.7 depicts a generic combination of a systematic diagram and a matrix diagram. Here, we see an expression or breakdown of the factors and then relationships developed between their lowest levels.

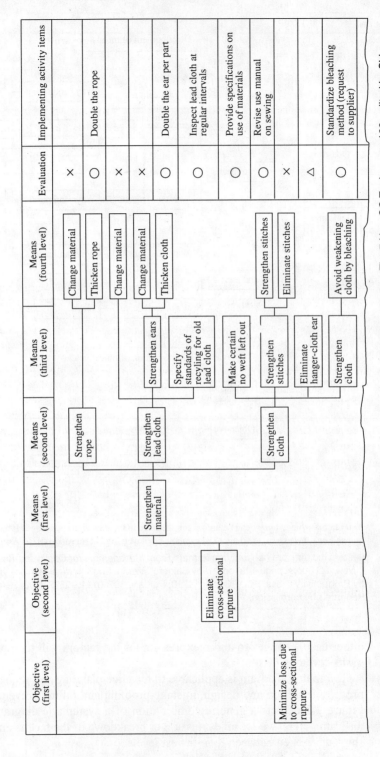

The table in the figure, read with columns left to right:

| Objective (first level) | Objective (second level) | Means (first level) | Means (second level) | Means (third level) | Means (fourth level) | Evaluation | Implementing activity items |
|---|---|---|---|---|---|---|---|
| Minimize loss due to cross-sectional rupture | Eliminate cross-sectional rupture | Strengthen material | Strengthen rope | | Change material | × | |
| | | | | | Thicken rope | ○ | Double the rope |
| | | | Strengthen lead cloth | Strengthen ears | Change material | × | |
| | | | | | Change material | × | Double the ear per part |
| | | | | | Thicken cloth | ○ | |
| | | | | Specify standards of recyling for old lead cloth | | ○ | Inspect lead cloth at regular intervals |
| | | | | Make certain no weft left out | | ○ | Provide specifications on use of materials |
| | | | Strengthen cloth | Strengthen stitches | Strengthen stitches | ○ | Revise use manual on sewing |
| | | | | | Eliminate stitches | × | |
| | | | | Eliminate hanger-cloth ear | | △ | |
| | | | | Strengthen cloth | Avoid weakening cloth by bleaching | ○ | Standardize bleaching method (request to supplier) |

**FIGURE 7.6**  Systematic diagram illustration. From *Management for Quality Improvement: The 7 New QC Tools*, page 162, edited by Shigeru Mizuno. English translation copyright © 1988 by Productivity Press, Inc., PO Box 13390, Portland, OR 97213-0390, (800) 394-6868. Reprinted by permission.

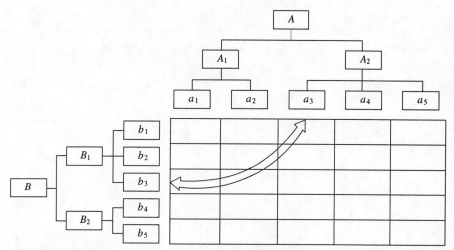

**FIGURE 7.7**   A combined systematic diagram and matrix diagram format. From *Management for Quality Improvement: The 7 New QC Tools*, page 179, edited by Shigeru Mizuno. English translation copyright © 1988 by Productivity Press, Inc., PO Box 13390, Portland, OR 97213-0390, (800) 394-6868.  Reprinted by permission.

The quality function deployment (QFD) method is based on matrix diagrams (Akao [5]).  Figure 7.8, taken from a QFD, illustrates the use of the matrix diagram concept in relating the customer demands, regarding physical performance, timeliness, and cost, to product quality characteristics.  The customer demand factors appear on the vertical axis, while the technical quality characteristics appear on the horizontal axis.  General relationships are indicated in the rectangular body of the matrix.  The triangular relationship matrix at the top serves to document the interrelationships between the quality characteristics.  Hence, this triangular matrix is useful to identify relationships as complementary (positive) or competitive (negative).  The symbols used in Figure 7.8 indicate a subjective level of relationship strength.

The QFD example shown in Figure 7.8 will be explained in detail in Chapter 11. It is shown here so that our readers can gain an appreciation for the matrix diagram. As we will see in Chapter 11, the format is flexible and can support our efforts in product definition.  This strategic planning tool has been credited as a major factor in developing many successful products, including the Ford Taurus and Cadillac Seville automobiles.

## 7.6  MATRIX DATA ANALYSIS

*A matrix data analysis serves to quantify the degree of the relationship that exists between various factors.*  In general, quantification can be attempted in two distinct circumstances:

1  When hard data are unavailable.
2  When hard data are available.

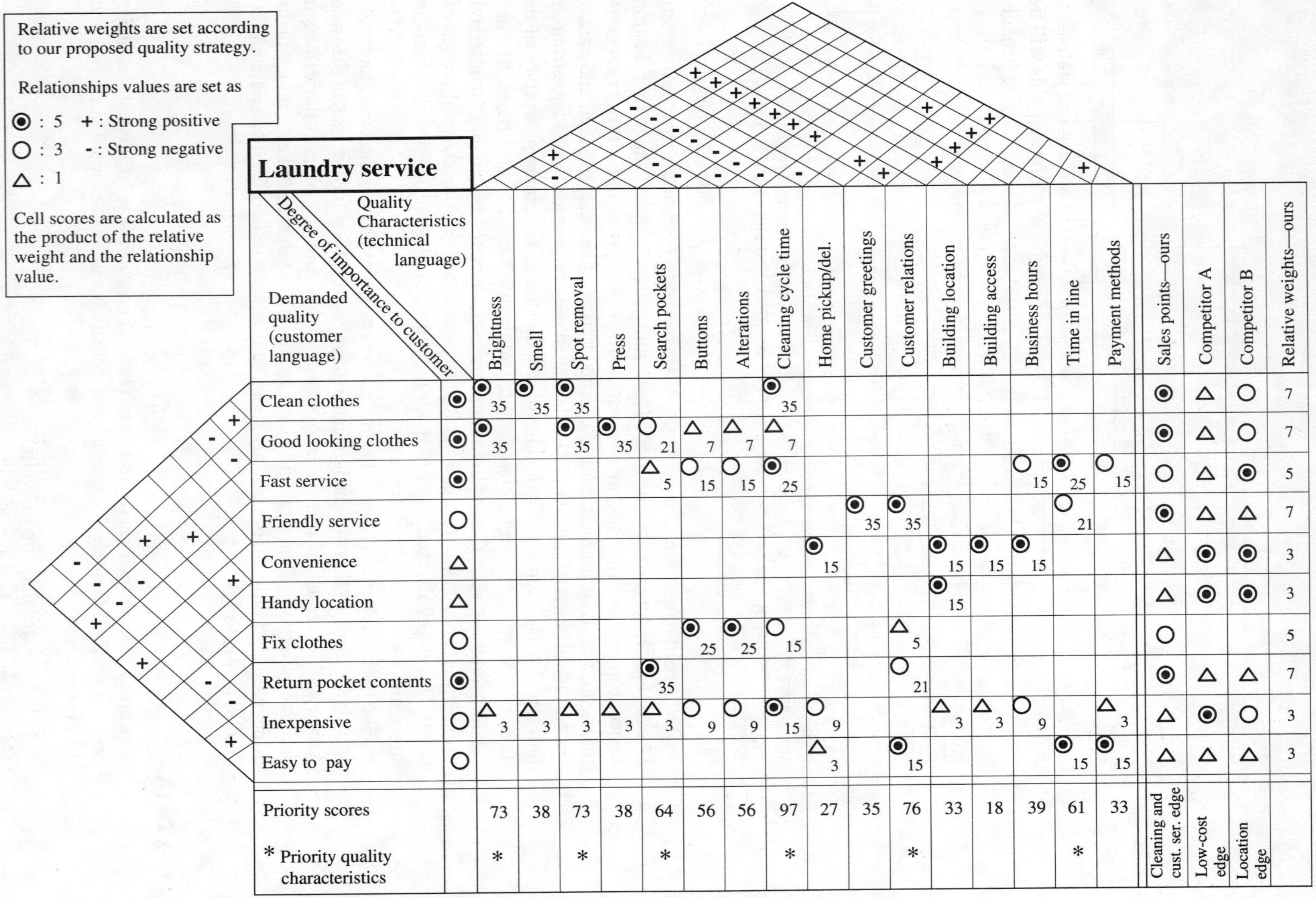

**Laundry service**

Quality Characteristics (technical language)

Degree of importance to customer

Demanded quality (customer language)

| Demanded quality | Importance | Brightness | Smell | Spot removal | Press | Search pockets | Buttons | Alterations | Cleaning cycle time | Home pickup/del. | Customer greetings | Customer relations | Building location | Building access | Business hours | Time in line | Payment methods | Sales points—ours | Competitor A | Competitor B | Relative weights—ours |
|---|---|---|---|---|---|---|---|---|---|---|---|---|---|---|---|---|---|---|---|---|---|
| Clean clothes | ◉ | ◉ 35 | ◉ 35 | ◉ 35 | | | | | ◉ 35 | | | | | | | | | ◉ | △ | ○ | 7 |
| Good looking clothes | ◉ | ◉ 35 | | ◉ 35 | ◉ 35 | ○ 21 | △ 7 | △ 7 | △ 7 | | | | | | | | | ◉ | △ | ○ | 7 |
| Fast service | ◉ | | | | | | ○ 15 | ○ 15 | ◉ 25 | ○ 15 | | | | | ○ 15 | ◉ 25 | | ○ | △ | ◉ | 5 |
| Friendly service | ○ | | | | | | | | | | ◉ 35 | ◉ 35 | | | | ○ 21 | | ◉ | △ | △ | 7 |
| Convenience | △ | | | | | | | | ◉ 15 | ○ 9 | | | ◉ 15 | ◉ 15 | ◉ 15 | ◉ 15 | △ 3 | △ | ◉ | ◉ | 3 |
| Handy location | △ | | | | | | | | ◉ 15 | | | ◉ 15 | ◉ 15 | | | | | △ | ◉ | ◉ | 3 |
| Fix clothes | ○ | | | | | △ 5 | ◉ 25 | ◉ 25 | | | | △ 5 | | | | | | ○ | | | 5 |
| Return pocket contents | ◉ | | | | | ◉ 35 | | | | △ 3 | | ○ 21 | | | | | | ◉ | △ | △ | 7 |
| Inexpensive | ○ | △ 3 | | △ 3 | △ 3 | △ 3 | ○ 9 | ○ 9 | | | | | △ 3 | △ 3 | ○ 9 | | ◉ 15 | △ | ◉ | ○ | 3 |
| Easy to pay | ○ | | △ 3 | | | | | | | | | | | | | | ◉ 15 | △ | △ | △ | 3 |
| **Priority scores** | | 73 | 38 | 73 | 38 | 64 | 56 | 56 | 97 | 27 | 35 | 76 | 33 | 18 | 39 | 61 | 33 | | | | |
| **\* Priority quality characteristics** | | * | | * | | * | | | * | | | * | | | | * | | | | | |

Cleaning and cust. ser. edge (Sales points—ours)
Low-cost edge (Competitor A)
Location edge (Competitor B)

**FIGURE 7.8**  Quality function deployment (QFD) demanded quality deployment chart illustration.

When hard data are not available, subjective weights must be determined and then a means of scoring devised to assess, rank, and prioritize the factors with respect to each other. The numbers shown in Figure 7.8 are one example of how a weighting system can be used to develop a matrix data analysis when hard data are sparse or nonexistent. We will discuss the mechanics of this weighting system in detail in Chapter 11.

If a full range of hard data is available or can be collected, we can use statistical techniques to develop quantitative relationships. Typically, computer aids will be necessary to develop these relationships. Applicable techniques can be broadly divided into two categories: (1) univariate response methods and (2) multivariate response methods. Univariate (single response) methods include linear correlation analysis techniques (Walpole and Myers [6], and SAS [7]) and rank-based correlation analysis techniques (Conover [8], SAS [9]). Multivariate (multiple response) techniques include canonical correlation and factor analysis (Morrison [10], SAS [11]). The mechanics and interpretation of these methods are beyond the scope of our discussion.

## 7.7  PROCESS DECISION PROGRAM CHART (PDPC)

*The process decision program chart (PDPC) method is useful in helping us to evaluate or assess process alternatives so that we might develop a best process in a global (high-level) sense.* PDPC charts vary in their scope and detail. Simple charts may contain a sequence of graphical icons along with brief descriptions and metrics. They are useful for the initial definition and development of processes. Then, we use the more detailed tactical tools such as process flow diagramming (Chapter 10) to further define and refine the process. The PDPC chart may show alternate outcomes, as well as the most desirable outcome.

More complex PDPC charts can be constructed using specialized symbols and brief narrative descriptions. Lines and arrows are used to indicate sequences of activities or events. Specific time lines are typically not included. The PDPC chart shown in Figure 7.9 develops an emergency process sequence resulting in the formulation of countermeasures that will avoid a train derailment. A basic PDPC chart may indicate the need for a more sophisticated and specialized tactical analysis tool such as a process flowchart (Chapter 10), a failure mode and effects chart (Chapter 12), a fault tree diagram (Chapter 13), and so on.

In many cases, the PDPC charts present highly simplified "pictures" of a complicated process. They make excellent communication tools; magazines and newspapers use PDPC-like depictions to present basic details for their readers. *In quality work we use PDPC charts to develop and communicate process strategy.* We will use PDPC charts in our benchmarking example in Chapter 8.

## 7.8  ARROW DIAGRAM

*An arrow diagram is an activity and event sequencing aid.* An arrow diagram contains more detail than a simple Gantt chart (containing activity durations and se-

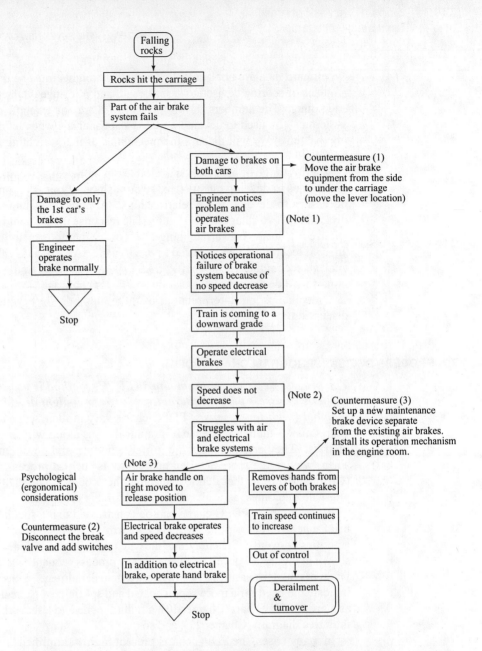

Falling rocks

Rocks hit the carriage

Part of the air brake system fails

Damage to brakes on both cars

Countermeasure (1)
Move the air brake equipment from the side to under the carriage (move the lever location)

Engineer notices problem and operates air brakes

(Note 1)

Damage to only the 1st car's brakes

Engineer operates brake normally

Notices operational failure of brake system because of no speed decrease

Stop

Train is coming to a downward grade

Operate electrical brakes

Speed does not decrease

(Note 2)

Countermeasure (3)
Set up a new maintenance brake device separate from the existing air brakes. Install its operation mechanism in the engine room.

Struggles with air and electrical brake systems

(Note 3)

Psychological (ergonomical) considerations

Air brake handle on right moved to release position

Removes hands from levers of both brakes

Countermeasure (2)
Disconnect the break valve and add switches

Electrical brake operates and speed decreases

Train speed continues to increase

Out of control

In addition to electrical brake, operate hand brake

Stop

Derailment & turnover

Note 1: There are two types of systems to stop trains: the air brake and the electrical brake. There is a separate handle to operate each.
Note 2: There is an order of priority set up for these systems because of peculiar technical reasons. When both handles are operated, the electrical brake does not work because the air brake has priority. In this type of emergency situation, the train cannot be stopped unless a system is developed where the electrical brake automatically works first.
Note 3: As stated in note 2, the air brake has priority when both systems are in operation; however, there is no braking action because of a failure in the air brake system. If the air brake handle were in the release (nonoperating) position, then the electrical brake would be effective for stopping the train.

**FIGURE 7.9**   PDPC illustration for a railroad application. From *Management for Quality Improvement: The 7 New QC Tools*, page 246, edited by Shigeru Mizuno. English translation copyright © 1988 by Productivity Press, Inc., PO Box 13390, Portland, OR 97213-0390, (800) 394-6868. Reprinted by permission.

quence) [12] but less detail than a complete PERT (program evaluation and review technique) or CPM (critical path method) chart (Moder et al. [13]). Figure 7.10 compares the Gantt and arrow charts for a basic residential construction project. Here, we see comparable information on activity duration, but more sequencing detail in the arrow diagram than in the Gantt chart.

Nodes or circles are used in an arrow chart to mark event starts and finishes. Solid arrows are used to indicate an element (activity) that requires time. The dashed lines with arrows are called "dummy" elements and are used to indicate a

**FIGURE 7.10**    Gantt and arrow chart comparison. From *Management for Quality Improvement: The 7 New QC Tools*, page 251, edited by Shigeru Mizuno. English translation copyright © 1988 by Productivity Press, Inc., PO Box 13390, Portland, OR 97213-0390, (800) 394-6868. Reprinted by permission.

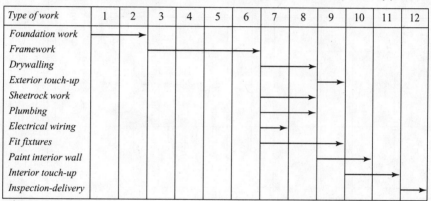

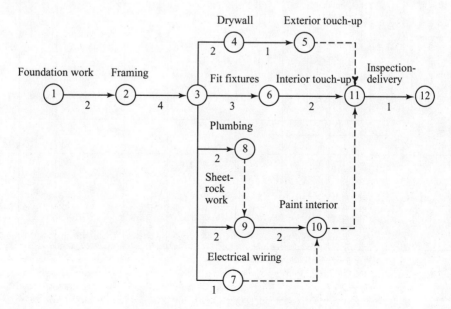

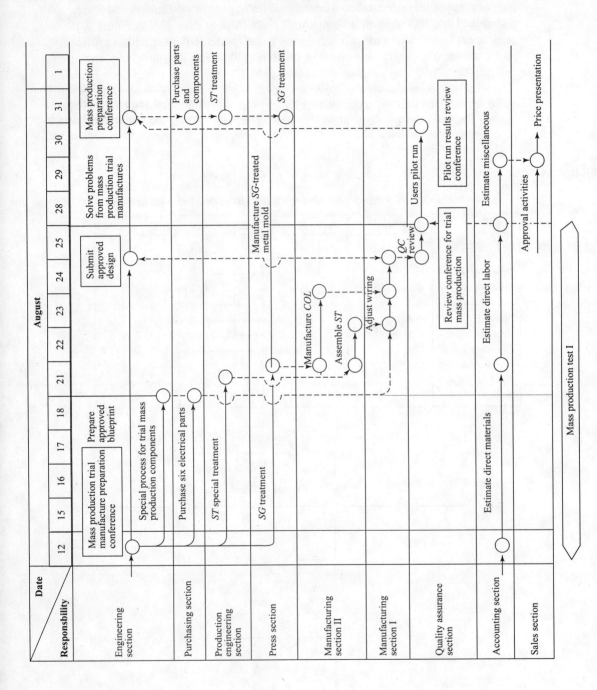

**FIGURE 7.11** Arrow diagram illustration. From *Management for Quality Improvement: The 7 New QC Tools*, page 278, edited by Shigeru Mizuno. English translation copyright © 1988 by Productivity Press, Inc., PO Box 13390, Portland, OR 97213-0390, (800) 394-6868. Reprinted by permission.

sequential relationship between events, therefore time is not a consideration.  The nodes may be depicted as open circles, numbered for identity.

An arrow diagram which deals with a mass production process plan is shown in Figure 7.11.  Here, we see a time scale across the top and a listing of section responsibilities down the left side.  Then, a time sequence of activities is placed in the diagram body.  A series of activity and dummy arrows are used to lay out the flow of the plan.  From this plan depiction, we can see the ordered sequence, responsibility, and time constraints.  Such a plan (when developed on a consensus basis by the sections, groups, or teams involved) is a valuable communications tool as well as an effective planning tool.

Essentially, *the arrow diagram* serves as a tool to delegate responsibility and a time schedule for strategic quality plans.  It *addresses the "who," "where," "when" issues of strategic quality planning.*  In contrast, the preceding six tools primarily address the "what," "how," and "why" issues in strategic quality planning.

## REVIEW AND DISCOVERY EXERCISES

### Review

**7.1**  List the seven new quality tools and briefly describe the purpose of each.

**7.2**  Compare and contrast the characteristics of a strategic quality plan with a tactical quality plan.  Why are the tools for each different?  How are the tools for each different?  (You may want to preview the rest of this section before addressing this question).

**7.3**  Select a product:
   **a**  Briefly describe a strategic quality planning need.
   **b**  Briefly describe a tactical quality planning need.

### Discovery

**7.4**  For a product or process of your choice (a good or a service) develop
   **a**  A relations diagram.
   **b**  An affinity diagram.
   **c**  A systematic diagram.
   **d**  A matrix diagram.
   **e**  A PDPC.
   **f**  An arrow diagram.

## REFERENCES

**1**  K. Ishikawa, *What Is Total Quality Control? The Japanese Way,* Englewood Cliffs, NJ: Prentice-Hall, 1985.

**2**  S Mizuno, *Management for Quality Improvement,* Cambridge, MA: Productivity Press, 1988.

**3**  Y. Akao (ed.), *Quality Function Deployment,* Cambridge, MA: Productivity Press, 1990.

**4**  See reference 2.

**5**  See reference 3.

**6**  R. E. Walpole and R. H. Myers, *Probability and Statistics for Engineers and Scientists,* 5th ed., New York: Macmillan, 1993.

7  *SAS/STAT User's Guide, Volumes 1 and 2,* Version 6, Cary, NC: SAS Institute, 1990.

8  W. J. Conover, *Practical Nonparametric Statistics,* 2d ed., New York: Wiley, 1980.

9  See reference 7.

10  D. F. Morrison, *Multivariate Statistical Methods,* 3d ed., New York: McGraw-Hill, 1990.

11  See reference 7.

12  R. M. Barnes, *Motion and Time Study Design and Measurement of Work,* 7th ed., New York: Wiley, 1980.

13  J. J. Moder, C. R. Phillips, and E. W. Davis, *Project Management with CPM, PERT, and Precedence Diagramming,* 3d ed., New York: Van Nostrand Reinhold, 1983.

# 8

# BENCHMARKING AND STRATEGIC THINKING

## 8.0 INQUIRY

1  What is benchmarking and how can it impact competitive position?
2  How can benchmarking be used to accelerate product and process improvements?
3  How are "best of the best" practices identified?
4  What can be gained by copying and swiping strategies?
5  What characterizes breakthrough thinking?

## 8.1 INTRODUCTION

***Efforts in product and process improvement have traditionally focused on improvements within the same work group, division, or organization.*** Hence, when performance measures such as sales dollars, labor hours per unit, energy per unit of production, fuel economy, and so forth, are trending in desirable directions, all is considered well. ***In highly competitive markets, improvement trends are absolutely necessary but not, in themselves, sufficient for success.*** This is because our competitors may post steeper improvement trends than we do and surpass us in performance. Improvement trend slope is critical. Figure 8.1 illustrates this concept. We see that a competitor (company B) can overtake us (company A), even though we are currently leading and projecting yearly performance improvement.

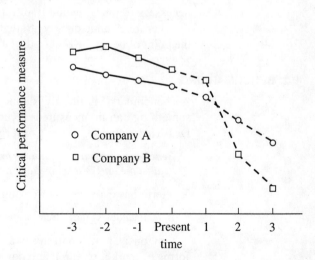

**FIGURE 8.1** Performance trends, smaller is better.

157

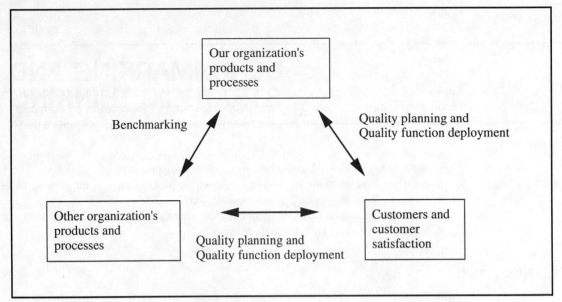

**FIGURE 8.2**    Strategic quality relationships.

*In the improvement process, we must consider competitors, product perform-ance, and customer satisfaction.* Figure 8.2 depicts a strategic quality relationship triad concerning our organization, a competing organization, and our customers. Our purpose in this chapter is to explore the benchmarking linkage between organizations. The quality function planning and deployment linkage will be discussed in Chapter 11. *Benchmarking is the term that has become associated with the practice of studying competitors' and noncompetitors' functional activities in order to enhance or better our own processes and products.* Benchmarking, as traditionally practiced, is short- to medium-range in scope (e.g., 1 to 3 years). Extensions are necessary to push beyond traditional short-range benchmarking in order to piece together long-range quality strategies, plans, and actions that will truly move us to the point of being the best of the best.

## 8.2  BENCHMARKING

A "benchmark" is literally defined as a reference point or a standard by which something can be measured or judged. Benchmarking was formally defined by D. T. Kearns (CEO, Xerox Corporation):

> *Benchmarking is the continuous process of measuring products, services, and practices against the toughest competitors or those companies recognized as industry leaders.*

The term has evolved to a working form, according to Camp [1]:

> Benchmarking is the search for industry best practices that lead to superior performance.

Obviously, benchmarking in a functional sense is not a new concept. Various forms of market research and competitive analysis have focused on customer and competitor behavior and strategies, respectively, for many years.  Most everyone has benchmarked in their personal and professional lives (e.g., in activities such as sports, academics, and work methods).  What is relatively new is the development of benchmarking as a formal and systematic discipline.

Benchmarking, as originally practiced, has been widely accepted and tailored to suit the needs of industry, as Bemowski states [2], and is credited with successes in quality improvement efforts.  ***The benchmarking strategy concentrates on both best practices (performance) and metrics (measurement).***  Figure 8.3 presents a conceptual view of the benchmarking process.

Successful benchmarking focuses first on identifying and understanding best practices; metrics should be considered important, but secondary.  The best practice may be located somewhere in our own organization, in a competitor's organization, or in a functional area of a noncompetitor's organization.  For example, if customer service through rapid order response is the focus, an organization in the fast food industry or mail order industry might be the best of the best and hence a good company to benchmark (even if our business happened to be sheet metal fabrication).  The point is that better practices may be in existence within our organization, within our industry, or outside our industry; we find them, we study them, we learn from them, and we improve our operations.  The best of the best generic functions—such as warehousing, communications, accounting, computer aided engineering, computer aided manufacturing, and so forth, may frequently be outside our own organization and industry.

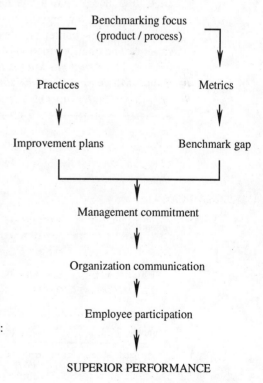

**FIGURE 8.3**   Generic benchmarking process. Adapted, with permission, from R. C. Camp, *Benchmarking*, Milwaukee, WI: Quality Press, American Society for Quality Control, page 5, 1989.

*Camp characterizes benchmarking as consisting of planning, analysis, integration, action, and maturity elements* [3].  These elements are broken down further in Table 8.1.  Benchmarking is viewed as an ongoing feature of corporate culture, hence, an integral part of our competitive master strategy.  Benchmarking is now recognized as a significant factor in quality awards criteria (e.g., the Baldrige Award).

The benchmarking technique, as it is commonly practiced, is tied closely to data collection and analysis.  A sound database is critical for two reasons.  First, we have a human tendency to overrate our own relative position.  In many cases, unless quantitative measures and evidence are established, documented, and accepted as valid, organization members (specifically managers) have difficulty visualizing their true competitive position.

Benchmarking is most effective in proactive or pre-crisis usage as a preventive medicine in an organizational sense.  A "crash" benchmarking exercise may be a "too little, too late" response to a competitive (survival) crisis.  We must recognize the need for benchmarking and commit to action before our organization begins to falter.

Secondly, a good database is imperative in establishing position and tracking progress and improvement through performance measurement.  A number of data sources exist for developing a benchmarking database:

Trade and professional association publications and databases.

Literature and research publications.

Conference presentations and interchanges.

Product literature.

Consultants and experts.

**TABLE 8.1**     STEPS TO SUCCESSFUL BENCHMARKING

| Steps | Activities |
|---|---|
| Planning | Identify product/process<br>Identify comparative organizations<br>Identify data needs |
| Analysis | Collect data<br>Calculate performance gap<br>Project to future performance |
| Integration | Communicate findings<br>Establish functional goals<br>Develop action plans |
| Action | Implement action plans<br>Monitor progress<br>Recalibrate benchmarks<br>Attain leadership position |
| Maturity | Integrate benchmarking into corporate culture |

Adapted, with permission, from R. C. Camp, *Benchmarking*, Milwaukee: Quality Press, American Society for Quality Control, page 17, 1989.

Interviews.

Surveys.

Facility tours and visits.

Historical data on similar products or processes.

Data regarding product and process results are often considered nonproprietary. Hence, our benchmarking data collection, as far as measurements go, is straightforward. However, the processes that produced the results may be proprietary and, hence, more difficult to observe and study. Sometimes, even within the same industry, organizations may form either formal or informal benchmarking partnerships. Typically, the partners do not exchange proprietary information, but a great deal of information (e.g., including methods of operation) can be and is exchanged. Obviously, all partners expect to develop win-win relationships and produce a net gain in the long run. These gains are possible since each partner is typically looking for specific gains in many different areas. Trade associations, professional societies, and so forth, encourage and foster such interchanges.

*Performance gaps represent the difference between our performance level and that of some other organization* (e.g., the best of the best). Performance gaps (good or bad) are typically modeled with Z curves. A Z curve charts the past in terms of the critical metric or metrics. It also projects the future, relative to expected or predicted performance from improvement plans, as shown in Figure 8.4.

The Z curves in Figure 8.4 represent past and expected performance for two companies, where smaller is better. Each curve contains a number of line segments. A vertical drop represents some type of strategic breakthrough (e.g., technology, equipment, etc.) coming on-line. The general slope of the other line segments

**FIGURE 8.4**   Z performance curve, smaller is better.

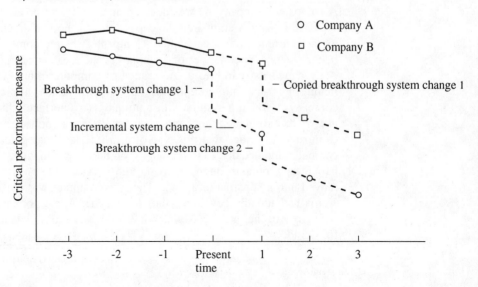

represents a tactical, or incremental, improvement. Both, together, yield total improvement. Hence, company A is widening its performance gap (in a positive sense) and looking better than company B when projected into the future. Since benchmarking is a two-way street, advantages may be reproducible in other companies; hence, company B will copy company A's strategic change and reap the benefits, but at a later date.

## 8.3  COPYING, CREATIVE SWIPING, AND BREAKTHROUGH THINKING

In practice, *benchmarking encourages emulation of successful practices and abandonment of unsuccessful practices across broad areas of operations, relative to both goods and services.* Copying, as such, is not generally a strategy which leads directly to a leadership position. Interindustry copying is one possible exception. In interindustry copying, we may be able to copy a best practice from another industry and gain a competitive advantage in our own industry. But it will probably be only a matter of time until our competitors do the same thing.

*Creative swiping represents an advanced form of copying. Here, creative elements are added in order to tailor outside practices to a workable—and, it is hoped, more effective—form to be used in our own organization* (Peters [4]). Through creative swiping, and depending on the level of creativity realized, gaps on Z curves can be closed, and a leadership position and superior performance gained. In creative swiping, we might find a product or process feature that one organization could not master and perhaps gave up on. If we can find the "handle" on this feature, we may be able to use it to our advantage.

---

### Computer Interface—Creative Swiping Case [DS]

A company (company A) was developing a new desktop computer in the early '80s. Company A compared the capabilities of the market competition against its own current product capability. Company A had to come up with something better for its next computer or lose market share to the competition. Company A expanded its search to a wide range of computer manufacturers, including mainframes and minicomputers. In looking at the minicomputer market, company A found a system with a unique user interface which minimized the use of the command line and involved using a mouse, menus, and icons. Company A decided that this type of user interface was practical for a desktop computer and decided to implement it on the new computer. The interface was hailed as a breakthrough, even though it was not newly invented, but just moved into a new market arena. The new computer was the Apple Macintosh. Xerox had actually developed the concept that Apple popularized. Due to the popularity of the interface style, Windows was, in turn, developed (i.e., copied) for the PC.

---

*Benchmarking is a thinking, planning, and doing process.* The foresight and creativity of people ultimately provide the opportunity for superior performance. Copying and creative swiping both have a role to play in benchmarking. However, neither strategy is capable of producing breakthroughs which can produce huge leaps in performance. As an example, selected major performance breakthroughs are listed in Table 8.2.

*Breakthroughs provide huge potential for performance improvement and added customer satisfaction. However, breakthroughs typically require radically new thought dimensions and lead to new technologies; and as a result create new problems in system integration and management.* Figure 8.5 illustrates the conceptual flow and strategic elements of benchmarking and breakthrough thinking strategies.

Nadler and Hibino [5] provide a pragmatic discussion of breakthrough thinking, based on their work and studies in the United States and Japan in creative problem solving. They advocate a purpose directed approach to problem solving, based on a purpose hierarchy. A purpose hierarchy ranks purpose levels from specific to broad, leading to a sequence of "purpose-of-the-purpose" and "solution-after-next" questions.

Many times people address a purpose so narrow that its breakthrough solution potential is limited, the purpose must be expanded. A simple example of a purpose expansion is provided in Table 8.3. Here we see that a better fan replacement procedure, though it may enhance computer life, is not necessarily the best use of resources. A higher-order purpose might be addressed and some level of breakthrough realized—for example, a fan might not be essential at all.

Many efforts have been made to develop principles or identify common features associated with successful creative efforts. *Nadler and Hibino, have developed seven breakthrough thinking principles relevant to benchmarking* [6], described in Table 8.4. Obviously, these seven principles are not a fixed recipe for perform-

**TABLE 8.2**    PERFORMANCE BREAKTHROUGHS

| Category | Description |
| --- | --- |
| Transportation | Horse, train, automobile |
| Power | Human, animal, steam, internal combustion |
| Lighting | Oil lamp, electric light |
| Refigeration | Ice blocks, refrigeration systems |
| Timekeeping | Sundial, mechanical timepieces, electronic timepieces |
| Recordkeeping | Paper, computer disk |
| Written communications | Messengers, mail, overnight delivery, "fax" |

164

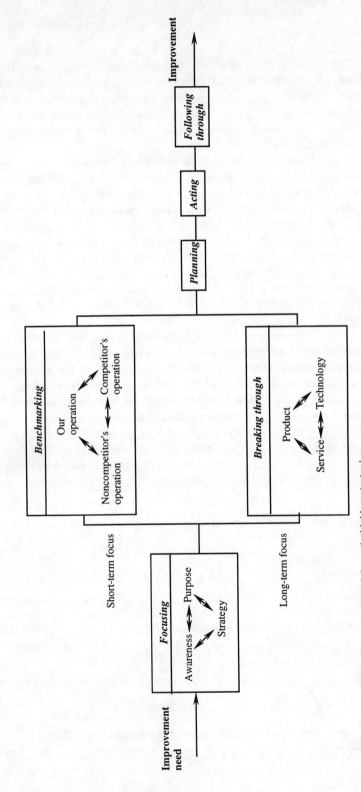

**FIGURE 8.5** Benchmarking and breakthrough thinking strategies.

**TABLE 8.3**    PURPOSE HIERARCHY EXPANSION EXAMPLE

| | |
|---|---|
| **Lower purpose** | Establish replacement procedure for failed fan motor in computer. |
| | Provide working fan in computer. |
| | Provide forced air flow through computer. |
| | Provide forced air flow over PC board. |
| | Keep components below critical temperature. |
| | Control thermal expansion. |
| | Avoid thermal mismatch between materials. |
| | Avoid stress in lead connectors. |
| | Preserve component integrity. |
| | Preserve electrical function of component. |
| | Provide computing circuitry. |
| **Higher purpose** | Provide computing power for customer. |

ance breakthroughs, but they support the strategic element in the benchmarking process and facilitate long-term strategic thinking.

# Document Retrieval—Breakthrough Thinking Case [SM]

A manufacturer of document storage systems was faced with a performance problem that was causing adverse customer reactions. The systems provided archival, retrieval, and display capabilities for documents of different types. Although the systems performed storage and retrieval in a satisfactory manner, the time to display a document from the archive was on the order of minutes, an irritating delay from the point of view of most users.

The manufacturer established target access times on the order of several seconds. The manufacturer experimented with faster archive technologies but achieved only very limited success, with access times remaining in excess of 1 min. Next, faster network technologies were investigated as a means for improving document-file transfer times. Again, some improvements were realized, but an improvement of an order of magnitude was still desired.

Purpose expansion techniques were next employed in an effort to determine the reason for the limited success realized through improvement of image storage and retrieval technologies. The following hierarchy resulted:

| | |
|---|---|
| Lower purposes: | Employ faster archive hardware. |
| | Get documents from archive more rapidly. |
| | Get documents to workstation more rapidly. |
| Higher purposes: | Provide documents to user more rapidly. |

**TABLE 8.4** PRINCIPLES OF BREAKTHROUGH THINKING

The Uniqueness Principle: Each problem is unique and requires a unique solution. The most successful problem solvers do not begin by trying to find out what has worked for someone else; they don't try to clone someone else's solution and impose it on a different situation. The first, principle, then, is that *each problem should be regarded as unique*.

The Purposes Principle: Focusing on purposes helps strip away nonessential aspects of a problem. The second principle calls for being directed by *purposes*. Several studies show that the quality of such solutions is significantly better than the results from conventional approaches.

The Solution-After-Next Principle: Having a target solution in the future gives direction to near-term solutions and infuses them with larger purposes. The third principle states that having an ideal target solution for achieving your purpose can lead to innovative solutions and help guide the development of the actual change you will make.

The Systems Principle: Every problem is part of a larger system of problems, and solving one problem inevitably leads to another. Having a clear framework of what elements and dimensions constitute a solution assures its workability and implementation. Problems don't exist in isolation, each problem is embedded within other problems, and a solution for one needs careful specification in system terms to make it workable in relation to other problems and solutions.

The Limited Information Collection Principle: Excessive data gathering may create an expert in the problem area, but knowing too much about it will probably prevent the discovery of some excellent alternatives. In approaching a problem, a great deal of time and effort can be saved by not collecting a lot of information and by not reviewing all the studies that have already been done. The fifth principle asserts that, at the outset, it is actually better to *limit what you know* about a problem. People, even experts, are better able to cope with incomplete and soft data; successful people often prefer it to hard data.

The People Design Principle: Those who will carry out and use the solution should be intimately and continuously involved in its development. Also, in designing for other people, the solution should include only the critical details in order to allow some flexibility to those who must apply the solution. Outstanding problem solvers are *diverse people who seek many different sources of information* in their problem-solving efforts.

The Betterment Time Line Principle: The only way to preserve the vitality of a solution is to build in and then monitor a program of continual change; the sequence of breakthrough thinking solutions thus becomes a bridge to a better future. The seventh principle refutes the conventional wisdom that you shouldn't fix something if it isn't broken. For a solution to be effective, it has to be *maintained and upgraded continually* toward the target. Even the target needs to be upgraded regularly. You've got to keep improving a situation or thing to prevent it from breaking down due to entropy, the normal wear and tear of events.

It was determined that, since the higher purpose was to improve the timeliness of user access to documents, altering the way in which the documents were provided to the user might substantially reduce access time. After further investigation, local disk caches were added to the viewing stations and software was added to keep the most frequently referenced documents on the local disks. Based on studies of user habits, the archive was restructured so that the indices and front matter of popular documents were kept on a file server, and thus quickly accessible, rather than on archive media.  While the user perused these portions, the remainder of the document could be retrieved from deep archive. Thus, the goal of document access within several seconds was effectively achieved.

## 8.4  TRANSCRIPT SERVICE—CASE STUDY

We are seeking to improve university transcript request-to-receipt response time. Currently, there is a 6- to 8-day receipt-to-response time for a certified copy of a transcript to reach its ultimate destination.  For example, the prospective employer will receive the certified transcript copy about 8 days after the student's request.  We want to improve this smaller is better performance measure.

### Solution

FOCUS

*Awareness*—Need to improve a university transcript service.

*Purpose*—Provide our customers (current and former students) the very fastest service possible (worldwide).

*Strategy*—Offer our customers a number of service levels whereby copies of their transcripts can be sent same day (unofficial copies) and next day (official copies) to other schools, prospective employers, professional certification boards, or the requesting student.

SHORT-TERM FOCUS

*Organizations to benchmark* (conventional benchmarking)—Other universities (transcript services), insurance companies (policies and claims).

*Improvement*—The PDPC in Figure 8.6 indicates potential to substantially reduce in-process office time.  It appears that with better organization the in-office processing time can be cut from about 4 days to 1 day (a 75 percent reduction).  The Z curve in Figure 8.7 shows the improvement in service response time by the sharp drop in the time to customer performance metric. This drop positions our university well with respect to university A.  The improved performance is also in line with insurance claims and policy service. In summary, the short-term conventional thinking improvement alternative would be challenging, and we could produce a 75 percent time saving in in-office processing for our customers.  However, it would still take about 5 days to obtain a transcript.

LONG-TERM FOCUS

*Organizations to benchmark* (strategic benchmarking)—Other universities, insurance companies, phone-in and fax-in mail order services.

*Improvement*—The improved PDPC in Figure 8.8 provides a summarized plan of operations for a radically different transcript service, with an extremely fast performance time response, as well as a wide variety of service choices.  The strategy takes advantage of proven technology, but utilizes

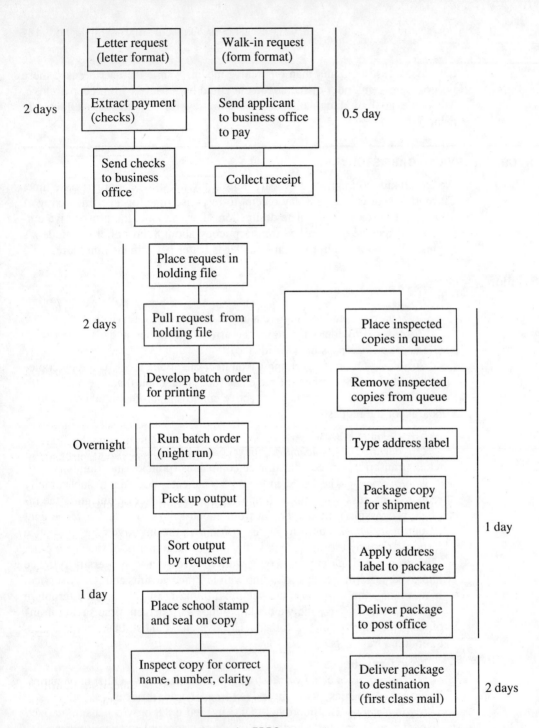

**FIGURE 8.6**    Original transcript service PDPC.

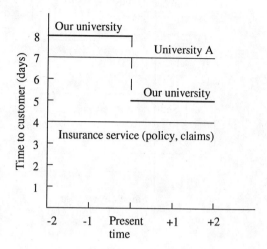

**FIGURE 8.7**   Transcript service Z curve, conventional thinking.

unconventional strategies in the transcript delivery service. Figure 8.9 depicts a plot of projected service times and shows the alternate customer choices, as well as performance comparisons of the alternatives.

PLANNING

*Process planning*—Implementing the PDPC for transcript service improvement (Figure 8.8) will require a great deal of change in the way business is conducted. The plan will provide far faster service, but new bottlenecks may be anticipated.

*Bottlenecks and potential resolutions*

1  Student identification (certification of identity): Use credit card and credit card verification.
2  Payment mode expansion: Modify the university's fee collection processes.
3  Delivery cost increases (for fast deliveries): Charge flat rate fees based on average package weights (foreign and domestic).
4  Phone and fax service: Install additional phone lines; determine a flat-rate fee based on average phone time (foreign and domestic).
5  Training: Develop a comprehensive training package for new procedures and practices.

ACTION

*Training*—Explain the new system and implement a flexible, thorough personnel training program using employee suggestions to detail service procedures and develop critical measurements.

*Execution*—Put the plan and training to use.

FOLLOW-THROUGH

*Fine tuning*—Listen carefully to internal customers (employees) and external customers for suggestions to enhance service quality.

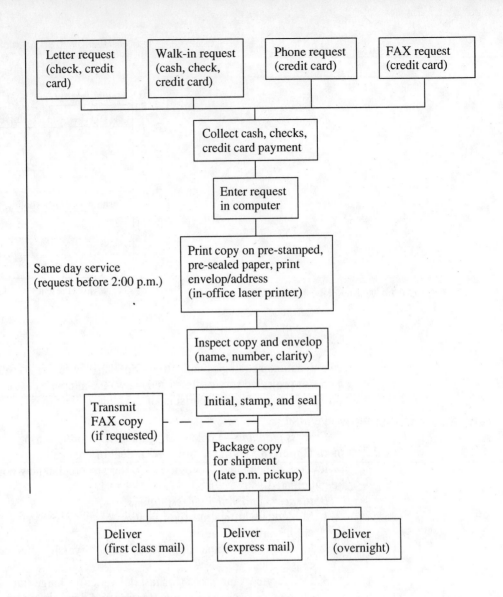

| Service levels | Response time |
|---|---|
| First class | 3 days |
| Express | 2 days |
| Overnight | Next day |
| FAX | Same day |

**FIGURE 8.8** Improved transcript service PDPC.

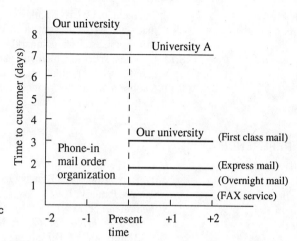

**FIGURE 8.9**    Transcript service Z curve, strategic thinking.

*Measurement*—Provide critical information to employees so that the system will continue to improve and provide customer satisfaction.
*Recognition*—Recognize outstanding performance.

## REVIEW AND DISCOVERY EXERCISES

### Review

**8.1**  Is a benchmarking strategy essentially a copying strategy?  Explain.

**8.2**  Why are metrics an essential part of benchmarking?

**8.3**  Is benchmarking proactive or reactive?  Explain.

**8.4**  What benefits can be expected from a corporate culture that accepts and encourages benchmarking?

**8.5**  Does benchmarking encourage creative thinking?  Explain.

**8.6**  For a Z curve, explain the type of events that might lead to
   **a**  A substantial, nearly vertical, drop or rise in a short time period.
   **b**  A gradual drop or rise on a sustained basis.

**8.7**  Why would two companies agree to form a benchmarking partnership in the same industry?  In different industries?

**8.8**  Is benchmarking a help or a hindrance to breakthrough thinking?  Explain.

**8.9**  Is the "limited information principle" in breakthrough thinking compatible with the "management by fact" principle (slogan) associated with the quality movement?  Explain.

**8.10**  Develop a "purpose-of-the purpose" (refer to Table 8.3) for
   **a**  A nail that can withstand a more forceful blow from a hammer.
   **b**  A tray to catch popsicle drippings while driving an automobile.
   **c**  A table marker to indicate the table where food orders are to be delivered.

### Discovery

**8.11**  One of the problems encountered with the ZD movement in the United States was that it was used as a "motivation" tool (banners or slogans) without a sufficient educational and technical basis. Workers were told, "Work smarter, not harder," and so forth, and that's all. What prevents a Z curve from reflecting wishful thinking, thus allowing a similar misuse of benchmarking?

**8.12** Choose a sport in which you participate and benchmark a leading player or coach. Be sure to include both metrics and practices.

**8.13** Choose a product or process and develop a benchmarking case study which includes both "intra- and inter-industry" benchmarking.

# REFERENCES

**1** R. C. Camp, *Benchmarking,* Milwaukee: Quality Press, ASQC, 1989.

**2** K. Bemowski, "The Benchmarking Bandwagon," *Quality Progress,* vol. 24, no. 1, pp. 19–24, January, 1991.

**3** See reference 1.

**4** T. Peters, *Thriving on Chaos,* New York: Harper & Row, 1987.

**5** G. Nadler and S. Hibino, *Breakthrough Thinking,* Rocklin, CA: Prima Publishing, 1990.

**6** See reference 5.

9

# THE SEVEN BASIC
# JAPANESE TOOLS

## 9.0 INQUIRY

1  What constitutes a cause and effect relationship?
2  What is the stratification principle?
3  How can a check sheet be used to collect data?
4  What is the difference between a histogram and a scatter diagram?
5  What is the Pareto principle?
6  What is a control chart?

## 9.1 INTRODUCTION

Over the past 30 years, the Japanese have studied and practiced what they term "total quality control" (TQC).  One of the leaders in this movement has been Kaoru Ishikawa.  Ishikawa and others repeatedly point to the fact that Japanese industrial workers are among the world's finest in their level of education and quantitative skills.  Consequently, *one of the critical features of the Japanese approach to quality control is its focus on quantitative methods on the factory floor.*

Based on his long experience in Japanese industry, *Ishikawa states that as much as 95 percent of quality related problems in the factory can be solved with seven fundamental quantitative tools,* also termed the seven "old" tools [1]:

1  *Cause-effect diagram.*
2  *Stratification analysis.*
3  *Check sheet.*
4  *Histogram.*
5  *Scatter diagram.*
6  *Pareto analysis.*
7  *Control charts.*

To promote the understanding and usage of these seven fundamental tools, Ishikawa produced a training guide, which is written as a factory floor level handbook [2].  In this chapter, we will briefly describe each tool, using Ishikawa's handbook as our major reference source.  It is important to point out that the Japanese contribution here is primarily in the tailoring, dissemination, and implementation of these tools in their workplaces.  The tools and the underlying technologies can be traced to a variety of sources, many of which are not Japanese.

## 9.2 CAUSE AND EFFECT ANALYSIS

*A cause is a fundamental condition or stimulus of some sort that ultimately creates a result or effect.*  In the analysis process, we may proceed from cause to effect, or conversely from effect to cause. Most analyses will be worked in both directions in order to discover and document causes, effects, and cause-effect linkages. Once the discovery phase is over, we must be careful to address (or treat) causes rather than effects.

*Cause-effect analyses are usually summarized in a cause-effect (CE) diagram. The CE diagram was developed by Ishikawa for the purpose of representing the relationship between an effect and the potential or possible causes influencing it* [3]. The CE diagram, sometimes referred to as a "fish-bone" diagram, is an organized or structured picture with lines and twigs (resembling fish bones) used to stratify and group causes. The effect is typically contained in a box on the right side, while the causes appear on the left side. Figure 9.1 illustrates the general CE-diagram structure.

The CE diagram is a tool which first helps us to discover possible root causes of defects and then helps us understand the failure mechanisms involved, so that we can prevent or eliminate them by proactive-reactive actions. CE diagrams can be developed around a positive effect or functional goal, rather than a negative effect (such as a product defect or undesirable process failure), although this is not widely practiced. *There are three basic types of CE diagrams: (1) cause enumeration, (2) dispersion analysis, and (3) process analysis.*

CE analysis represents a very powerful and versatile, proactive-reactive qualitative tool. Three main benefits of using the CE diagram in the problem-prevention and problem-solving process are listed below:

1 It provides a structure and focus for open discussion of a specific quality concern or challenge.

**FIGURE 9.1** General CE diagram structure.

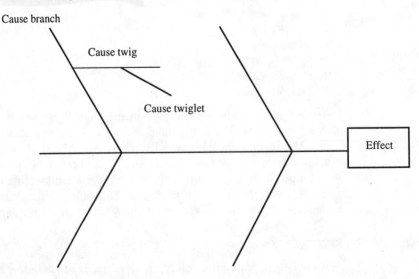

**2** It forces us to discover (sometimes through speculation) many possible causes for a specific effect, making them visible and understandable.

**3** It encourages employee involvement at all levels and promotes better communication within the group or team.

*The benefits obtained from CE analyses ought to outweigh the associated costs and burdens.* As improperly structured CE analysis efforts result in wasted time, procedures should be streamlined to allow maximum participation and effective use of time.

## Cause Enumeration

Cause enumeration analysis is very general. *A cause enumeration CE analysis is a very wide open, free-thinking analysis which may address any important effect.* All possible causes imaginable are freely listed and then placed in major cause categories. The objective is to identify all causes related to the identified effect. The use of a recording device, such as a flip board, is helpful. Wide participation (through teamwork) is encouraged.

Participants are encouraged to think in a very broad fashion. This free-thinking process generates possible causes based on experience, judgment, and speculation, which are recorded on the board as they are identified. As an example, a graphic listing of all the possible causes of a process quality problem regarding welding the hinge bracket and pivot onto an electromechanical relay armature is illustrated in Figure 9.2.

**FIGURE 9.2**   Electromechanical relay cause enumeration analysis CE diagram.

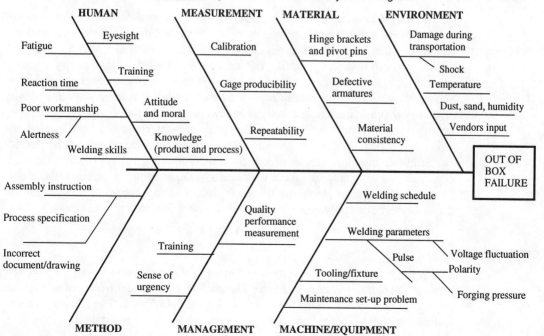

## Dispersion Analysis

*A dispersion CE analysis focuses on product or process variation.*  The typical procedure in dispersion analysis is to systematically ask a sequence of questions starting with, "Why and how does the effect occur?"  The response to the first-level question generally leads to a major branch.  The questioning continues: "Why and how does this branch affect the result?"  This questioning sequence continues through the branches, twigs, and finally to the twiglet level.  Typically, we focus on one major branch at a time.  Once the diagram is complete, the dispersion analysis branch structure looks like the cause enumeration branch structure; however, the development procedures differ significantly.

## Process Analysis

*A process CE analysis is used to investigate potential causes of a specific effect or problem by analyzing each step in the production process flow.*  Each step is labeled and connected by a line from left to right.  The pizza food service CE diagram shown in Figure 9.3 illustrates the process flow approach.  The major causes and their subcauses are drawn as branches and twigs.

The process CE diagram appears noticeably different from the other two CE diagrams.  Its structure is determined by the process flow, and the branches and twigs extend from the process boxes.  Here again, a brainstorming team approach will add perspective to the CE analysis.

## 9.3  STRATIFICATION ANALYSIS

*When we collect data, we should be sure to observe and document the circumstances surrounding our observations* (e.g., time of day, machine number, operator, type of raw material).  If we have collected this information along with our observations, we can divide, or stratify, our data set and look for clues as to the what, when, where, how, who, and why relative to the problem or challenge at hand.

*Stratification is the process of breaking down or sorting a large database so that meaningful subsets, classifications, or summaries can be developed.*  For example, we might sort a quality database by failure categories, by customer classification, by product lines, and so forth, in order to assess performance trends, problems, and so on.  Stratification allows us to effectively and efficiently navigate our way through huge volumes of data, seeking out the clues to quality improvement buried therein.  Later we will demonstrate the stratification principle and its absolute necessity in all data analyses, especially those dealing with cause isolation.

## 9.4  CHECK SHEET

*A check sheet is a simple tool used to record and classify observed data.  We see primarily two types of check sheets: (1) tabular and (2) pictorial.*  Two different types of tabular check sheets are shown in Figure 9.4.  Figure 9.4a illustrates a defective item tally-oriented check sheet.  Here, we see a running tally of five defect

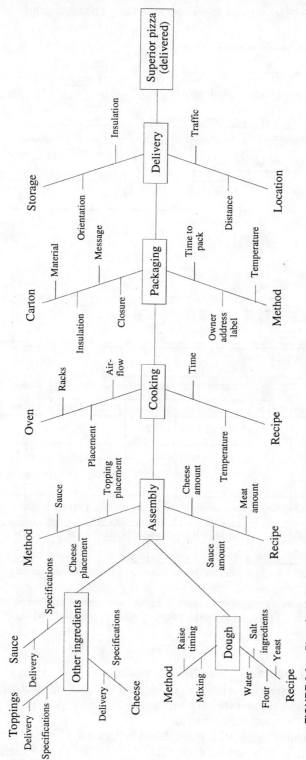

**FIGURE 9.3**  Pizza food service process analysis CE diagram.

**Check Sheet**

| Product: | Date: |
| --- | --- |
| | Factory: |
| Manufacturing stage:   final insp. | Section: |
| | Inspector's |
| Type of defect:   scar, incomplete, mishappen | name: |
| Total no. inspected:   2530 | Lot no.: |
| | Order no.: |
| Remarks:   all items inspected | |

| Type | Check | Sub-total |
| --- | --- | --- |
| Surface scars | ## ## ## ## ## ## // | 32 |
| Cracks | ## ## ## ## /// | 23 |
| Incomplete | ## ## ## ## ## ## ## ## ## /// | 48 |
| Misshappen | //// | 4 |
| Others | ## /// | 8 |
| | Grand total: | 115 |
| Total rejects | ## ## ## ## ## ## ## ## ## ## ## ## ## ## ## ## ## / | 86 |

(*a*) Defect tally check sheet

| Equipment | Worker | Monday | | Tuesday | | Wednesday | | Thursday | | Friday | | Saturday | |
| --- | --- | --- | --- | --- | --- | --- | --- | --- | --- | --- | --- | --- | --- |
| | | am | pm | am | pm | am | pm | am | pm | am | pm | am | pm |
| Machine 1 | A | OOx●O | Ox | OOO | Oxx | OOOxOOxx● | OOOOxxx | OOOOOx●● | Oxx | OOOOOO | O | | xx● |
| | B | Oxx●O | OOOxxO | OOOOOOxx● | OOOxx | OOOOOOxxOx● | OOOO●● | OOOOOxOx | OOOx● | OOxx● | OOOOO | OOx | OOOO●xOx |
| | C | OOx | Ox | OO | ● | OOOOO | OOOOOOx | OO | O● | OOΔ | OO□ | ΔO | O□ |
| Machine 2 | D | OOx | Ox | OO | OOOO●Δ | OOO●Ox | OOOO | O●O | OOΔ | OOΔΔ□ | O●● | □OOx | xxO |

(*b*) Defect cause check sheet (○ : surface scratch; △ : defective finishing; × : blowhole;
● : improper shape; □ : others)

**FIGURE 9.4**   Defect and defect cause check sheets. Adapted, with permission, from K. Ishikawa, *Guide to Quality Control*, Tokyo: Asian Productivity Organization, pages 33 and 36, 1982.

categories and a total count.  Figure 9.4*b* displays a check sheet with five defect categories, represented by special symbols, and stratified by machine, worker, and time of day.  Note, the stratification principle was used to provide a break-out by defect category, day, machine, and so forth.

A graphical check sheet is displayed in Figure 9.5.  Here, we see a pictorial representation of an automotive windshield.  The shaded regions represent the location

Bubble investigation check sheet

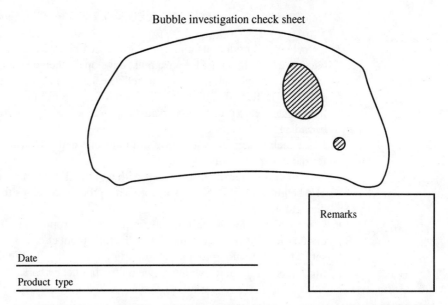

Remarks

Date

Product type

**FIGURE 9.5**   Defect location check sheet for an automotive windshield. Reprinted, with permission, from K. Ishikawa, *Guide to Quality Control,* Tokyo: Asian Productivity Organization, page 34, 1982.

of the highest defect concentrations. We can develop check sheets of this type, where we place a dot or an x where we encounter a defect. Thus, over time, we see areas of high and low defect counts. Sometimes, these charts are called "measles" charts, as the dots or x's resemble the red spot concentrations of the disease.

## 9.5 HISTOGRAM

*The histogram is a graphical data summary tool which allows us to group observed data into cells, or predefined categories, in order to discover data location and dispersion characteristics (without a sophisticated numerical analysis).* The histogram is a very valuable and underrated data analysis tool. Essentially, *we can develop two types of histograms:* **(1)** *a frequency count histogram and* **(2)** *a relative frequency or proportion histogram.* In general, both can be displayed on the same graph using right- and left-hand axes, simultaneously.

We will state the fundamental rules for developing a histogram and illustrate the technique with an example. We need a reasonably large number of observations to form a meaningful histogram—typically, a minimum of 20 to 30. Our procedure follows:

1   Determine the range of the data (i.e., the value of the largest observation minus the smallest observation).

2   Determine the number of cells or division categories desired—typically, from 5 to 15.

3   Calculate cell midpoints and boundaries.  All cells do not have to be of equal width, but usually are.  Here, we must develop a scheme where each observation will fall into only one cell, or "bucket" (i.e., we want to count each observation only once).  If, in defining our cell boundaries, we use just one more significant figure than we used in our observed data, no observation will fall directly on a cell boundary.

4   Place each observation into one and only one cell.  We can use a check sheet format to expedite this task.

5   Display the frequency of each cell with a vertical bar (for a vertical histogram) of a height proportional to its count.  This process develops the frequency count histogram.

6   If we want to develop a relative frequency or proportion histogram, we must divide each cell count (frequency) by the total number of observations in our data set.  This calculation provides a proportion for each cell.  We then graph proportions bars in the same manner as we do the frequency bars.

---

**Example 9.1**

In a factory producing stainless steel meter sticks, we have set up several metal production shearing lines.  One such line is depicted in Figure 9.6.  We have set a target of 1000 mm for our process.  Furthermore, we have set an upper product specification limit (USL) of 1001 mm and a lower product specification limit (LSL) of 999 mm.  Hence, a meter stick outside these specification limits is considered as nonconforming, or defective.  We currently have three metal production shearing lines in operation (lines 1, 2, and 3).  We have collected a total of 150 observations.  We then subtracted 1000 mm from our data to obtain a deviation from target measure.  These results are shown in Table 9.1.

a   Develop a check sheet tally and then a histogram for the observations, disregarding the production shearing line numbers.

b   Using the stratification principle, develop a stratified check sheet and a set of histograms (one for each production line).

c   Comment on your findings.

**FIGURE 9.6**    Simplified meter stick production process (Example 9.1).

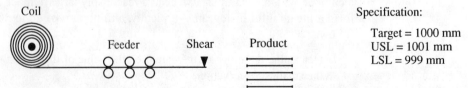

Coil

Feeder          Shear          Product

Specification:

Target = 1000 mm
USL = 1001 mm
LSL = 999 mm

**TABLE 9.1**  METER STICK MEASUREMENTS IN mm
(Actual measurement −1000 mm)

| Production Shear | | |
|:---:|:---:|:---:|
| Line 1 | Line 2 | Line 3 |
| 0.14 | 0.59 | -0.43 |
| -0.09 | 0.28 | -0.39 |
| 0.26 | 0.51 | -0.65 |
| -0.06 | 0.34 | -0.71 |
| 0.32 | 0.50 | -0.57 |
| -0.14 | 0.38 | -0.23 |
| -0.06 | 0.25 | -0.18 |
| 0.74 | 0.48 | -0.33 |
| -0.44 | 0.69 | -0.35 |
| 0.27 | 0.73 | -0.54 |
| 0.49 | 0.46 | -0.60 |
| -0.26 | 0.48 | -0.57 |
| 0.22 | 0.39 | -0.62 |
| -0.06 | 0.51 | -0.75 |
| 0.18 | 0.82 | -0.34 |
| -0.21 | 0.87 | -0.55 |
| -0.08 | 0.28 | -0.42 |
| -0.06 | 0.52 | -0.36 |
| -0.45 | 0.46 | -0.38 |
| 0.11 | 0.29 | -0.25 |
| -0.54 | 0.49 | -0.60 |
| 0.04 | 0.60 | -0.45 |
| 0.32 | 0.27 | -0.58 |
| -0.08 | 0.60 | -0.64 |
| -0.03 | 0.16 | -0.41 |
| 0.45 | 0.47 | -0.77 |
| -0.12 | 0.38 | -0.57 |
| 0.09 | 0.19 | -0.51 |
| 0.20 | 0.84 | -0.73 |
| 0.28 | 0.44 | -0.22 |
| -0.04 | 0.12 | -0.47 |
| -0.52 | 0.46 | -0.73 |
| -0.30 | 0.41 | -0.62 |
| 0.25 | 0.29 | -0.64 |
| -0.05 | 0.51 | -0.69 |
| 0.18 | 0.28 | -0.69 |
| 0.31 | 0.62 | -0.40 |
| 0.11 | 0.38 | -0.62 |
| -0.12 | 0.44 | -0.77 |
| 0.03 | 0.56 | -0.19 |
| 0.28 | 0.69 | -0.68 |
| -0.05 | 0.63 | -0.61 |
| 0.63 | 0.42 | -0.29 |
| 0.20 | 0.22 | -0.09 |
| -0.15 | 0.35 | -0.58 |
| -0.04 | 0.32 | -0.62 |
| 0.46 | 0.50 | -0.45 |
| -0.19 | 0.59 | -0.42 |
| -0.02 | 0.20 | -0.67 |
| -0.25 | 0.54 | -0.39 |

*Note:* All measurements represent a deviation from the 1000 mm target;
(e.g, observation 1 = 1000.14 − 1000 = 0.14 mm).

### Solution

a   We will calculate our range as $0.87 - (-0.77) = 1.64$ mm (See table 9.1).   We will build class intervals of width 0.150, centered at our deviation target point of 0.   This layout will allow a convenient format, relative to our target value and range.   Our check sheet and histogram appear in Figure 9.7.

b   Using the same procedure, we have displayed a stratified check sheet and three histograms (stratified by production line) in Figure 9.8.

c   The power of stratification speaks for itself when we compare Figures 9.7 and 9.8.   We see production shear line 1 more or less on target with what appears to be more dispersion (spread) than the other two lines.   Production shear line 2, on the other hand, centers above target, with what appears to be less dispersion than in line 1.   Production shear line 3 centers below target, with what appears to be less dispersion than in line 1.   We can note that all observations fall between specifications, but line 3 is pushing the LSL, whereas line 2 is pushing the USL.   We might conclude (1) the location should be adjusted in production shear lines 2 and 3, (2) the "secret" of the smaller dispersion in lines 2 and 3 should be uncovered, and (3) the dispersion of line 1 should be reduced.

## 9.6  SCATTER DIAGRAM

*A scatter diagram provides us the opportunity to view a data set in multiple dimensions in order to detect trends, spot best operating regions, explore cause-effect relationships, and so on.   In order to use a scatter diagram, we must observe and record data in observation sets* (e.g., bivariate data pairs such as a process temperature setting and its associated process yield, a tool speed and its associated product surface finish).   Figure 9.9 illustrates graphical correlation (trend) signatures that may appear in a scatter diagram.   This broad view of our data is helpful in assessing our process-product improvement prospects.   The stratification principle can be used to our advantage in a scatter diagram.   We will illustrate through an example.

## Example 9.2

In large cattle finishing feedyards, from a few thousand to maybe 100,000 head of cattle will be kept in semiconfinement (e.g., reasonably small pens).   When it rains, the pens become muddy.   The muddy conditions stress the cattle, and affect their eating habits and rate of weight gain per day.   The goal is to get the animals to gain weight as fast as possible.   One method of counteracting wet weather in the pens is to build what are termed "mounds."   Here, a bulldozer is used to create small hills which might cover as much as 30 to 40 percent of the pen area.   This mound building creates a high, drained place for the cattle to stand and lay (rest).   It also helps the pen to drain more effectively.   However, it requires time and money to build and maintain a system of mounds.

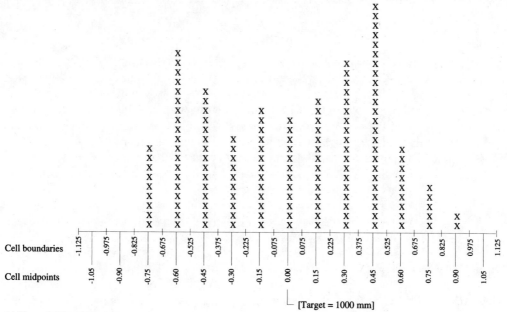

Cell boundaries: -1.125, -0.975, -0.825, -0.675, -0.525, -0.375, -0.225, -0.075, 0.075, 0.225, 0.375, 0.525, 0.675, 0.825, 0.975, 1.125

Cell midpoints: -1.05, -0.90, -0.75, -0.60, -0.45, -0.30, -0.15, 0.00, 0.15, 0.30, 0.45, 0.60, 0.75, 0.90, 1.05

[Target = 1000 mm]

(a) Meter stick length check sheet

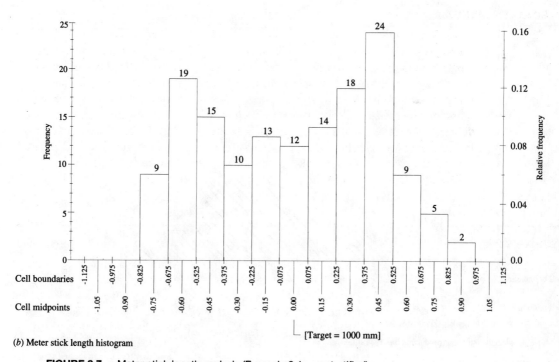

(b) Meter stick length histogram

**FIGURE 9.7**    Meter stick length analysis (Example 9.1a, unstratified).

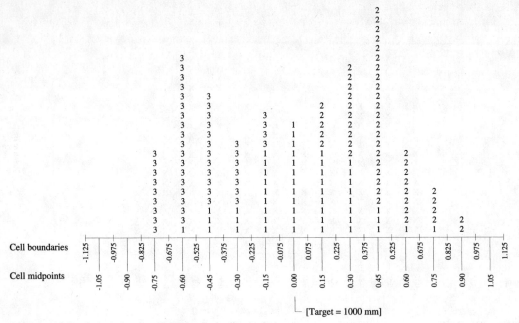

Cell boundaries: -1.125, -0.975, -0.825, -0.675, -0.525, -0.375, -0.225, -0.075, 0.075, 0.225, 0.375, 0.525, 0.675, 0.825, 0.975, 1.125

Cell midpoints: -1.05, -0.90, -0.75, -0.60, -0.45, -0.30, -0.15, 0.00, 0.15, 0.30, 0.45, 0.60, 0.75, 0.90, 1.05

[Target = 1000 mm]

(a) Meter stick length check sheet (stratified by production line number)

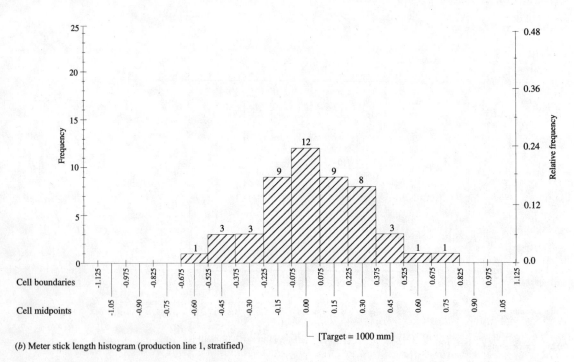

(b) Meter stick length histogram (production line 1, stratified)

**FIGURE 9.8**   Meter stick length analysis (Example 9.1b, stratified).

184

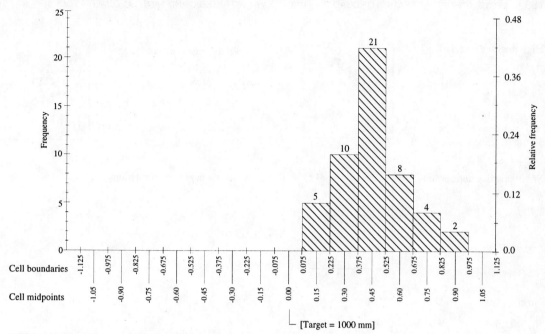

(*c*) Meter stick length histogram (production line 2, stratified)

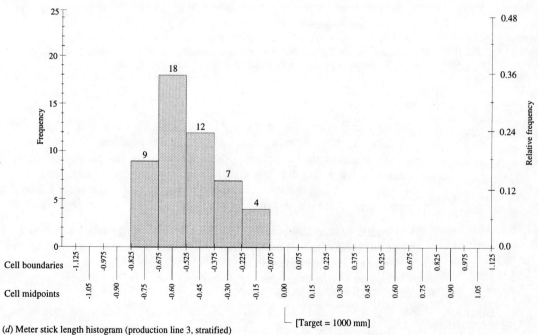

(*d*) Meter stick length histogram (production line 3, stratified)

**FIGURE 9.8** (*continued*).

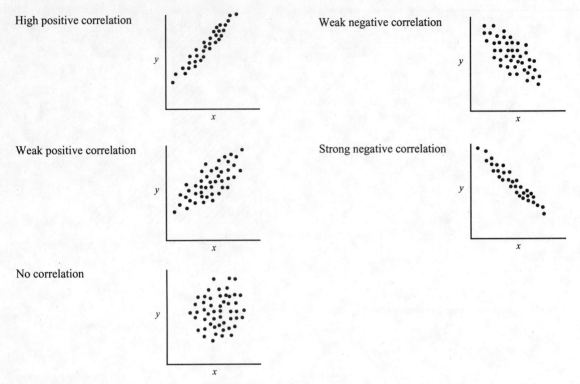

**FIGURE 9.9**   Scatter diagram signatures.  Adapted, with permission, from K. Ishikawa, *Guide to Quality Control*, Tokyo: Asian Productivity Organization, page 91, 1982.

The average daily weight gain (ADG) is a widely used performance measure. ADG is computed when a pen of, say, 30 to 100 head are "finished" and ready for shipment to a packing house.  Table 9.2 contains data pertaining to the ADG calculated for a 100-day feeding cycle and its associated number of recorded "mud days."  Note that one reasonably large rain or snow (say, 1.5 to 2 in of water) may create a string of 3 to 10 mud days.

a  Use the scatter diagram to assess the ADG–mud day relationship as a whole.
b  Stratify the data by pen topology—a mound system and a conventional (no mounds, flat ground) system.
c  Comment on your results.

**Solution**

a  The unstratified scatter diagram is shown in Figure 9.10*a*.
b  The stratified scatter diagram is shown in Figure 9.10*b*.
c  We can clearly see the value of stratification.  Figure 9.10*a* shows a downward trend in ADG as mud days increase.  However, Figure 9.10*b* tells

**TABLE 9.2**   FEEDYARD DATA (EXAMPLE 9.2)

| ADG—mounds | Mud days | AGD—flat | Mud days |
|---|---|---|---|
| 3.12 | 10 | 3.34 | 6 |
| 3.45 | 6 | 4.21 | 3 |
| 2.76 | 16 | 1.67 | 38 |
| 3.89 | 13 | 3.42 | 8 |
| 3.21 | 33 | 1.98 | 23 |
| 2.33 | 38 | 2.34 | 15 |
| 3.24 | 12 | 3.65 | 2 |
| 2.87 | 36 | 3.42 | 18 |
| 4.12 | 2 | 2.23 | 28 |
| 3.56 | 23 | 2.87 | 21 |
| 2.89 | 31 | 2.67 | 19 |
| 3.23 | 25 | 1.87 | 25 |
| 3.87 | 8 | 2.64 | 5 |
| 2.98 | 7 | 3.96 | 9 |
| 2.56 | 19 | 3.70 | 5 |
| 2.87 | 9 | 1.85 | 19 |
| 3.12 | 4 | 3.47 | 11 |
| 3.25 | 11 | 1.26 | 37 |
| 3.35 | 3 | 2.82 | 14 |
| 2.90 | 17 | 3.02 | 6 |
| 3.89 | 5 | 2.75 | 9 |
| 3.76 | 21 | 3.26 | 16 |
| 2.88 | 9 | 1.97 | 29 |
| 3.35 | 8 | 2.73 | 16 |
| 3.76 | 10 | 3.24 | 10 |

another story in that the flat pen technology is associated with a sharp drop in ADG once we get past 10 to 15 mud days. The mound technology appears to be much more robust; ADG does not drop off as fast, even when mud days increase. Here, we can bring in technical arguments of animal stress and so forth to argue a valid cause-effect relationship and justify mound building action as technically useful.

## 9.7  PARETO ANALYSIS

In nineteenth-century Italy, the Italian economist Vilfredo Pareto observed that about 80 percent of the country's wealth was controlled by about 20 percent of the population. This observation lead to what is now known as the Pareto Principle; it is also known as the "80–20" rule. Juran [4] and Juran and Gryna [5] applied this concept to the causes of quality failures. They stated that 20 percent of the causes account for 80 percent of the failures. In general, ***the Pareto principle, applied to quality, suggests that the majority of the quality losses are maldistributed in such a way that a "vital few" quality defects or problems always constitute a high percent of the overall quality losses*** (but not in strictly an 80–20 relationship).

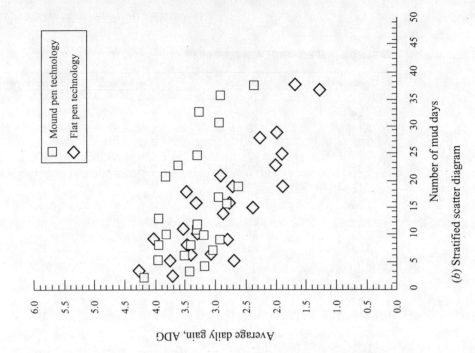

(a) Unstratified scatter diagram

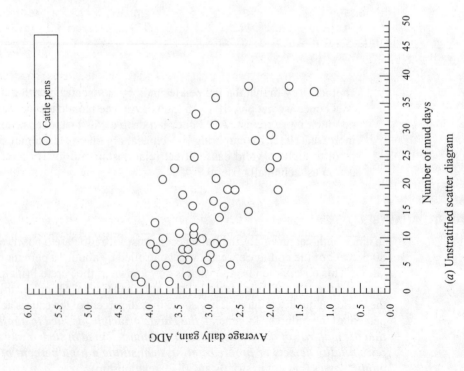

(b) Stratified scatter diagram

**FIGURE 9.10** Cattle feedyard scatter diagrams (Example 9.2).

*The intent of a Pareto analysis is to separate the vital few from the trivial many.* Thus, the Pareto analysis can assist us to identify the most important effects and causes and to stratify the available data *so that we can prioritize our product-process improvement efforts.* In general, *we see two types of Pareto diagrams: (1) result-category diagrams and (2) cause-category diagrams.* Result diagrams focus on the classification and relative importance of observable results (e.g., product defect categories such as scratches, dents). Cause diagrams focus on the classification and relative importance of detected causes (e.g., process related categories such as a dull cutting tool, a misaligned die).

It is very common to see the Pareto diagram used in defect reduction studies. In this case, the result-based diagram refers to defect categories and the cause-based diagram refers to defect causes. The defect-category diagram prioritizes the defects to determine which has the greatest impact on quality. The defect-cause diagram prioritizes the causes of a specific defect to determine which contributes most to the stated quality problem. For instance, a defect-category diagram could be used to identify and highlight the product that has the highest scrap cost. Meanwhile, the defect-cause diagram might rank the order of importance of scrap causes and rejection codes.

*A Pareto analysis is essentially a stratification and ranking process based on available data.* We suggest a three step process in a Pareto analysis:

1  Determine the results or causes to be analyzed and the appropriate unit of measure, typically cost or frequency. Gather quantitative data using appropriate methods, including a check sheet. Figure 9.11*a* provides a sample pizza food service Pareto check sheet. Typically, the analyses will be developed over a given study period (e.g., a day, a week, a month).

2  The Pareto check sheet is used to select and rank the items (results or causes) in descending order of size. Generally, we select from 5 to 10 items and combine the rest in an "others" category. We total the frequency of occurrence for all items and calculate a percent frequency by dividing each individual frequency by the total frequency and multiplying by 100. The cumulative percentages are developed by adding or accumulating item percentages from top to bottom (largest to smallest). The cumulative percent should total 100 percent. Figure 9.11*a* depicts the results of this step.

3  We plot the data from the Pareto table or check sheet summary as a vertical bar graph; the heights of the bars correspond to the frequency count on the left axis. We place the "others" category to the far right on the diagram. Our construction efforts are shown in Figure 9.11*b*. The right axis is used to indicate the cumulative percentage, which should be scaled from 0 to 100 percent and plotted as a line on the diagram. The cumulative percentage line starts at the center of the top of the first bar and is drawn diagonally across, above the bars, to the upper right-hand corner. Figure 9.11*b* illustrates a vertical bar Pareto chart for our pizza food service. When we hit about 80 percent on the cumulative percent axis, we have located the vital few categories, where we typically concentrate our improvement efforts.

| Cause / result | Tally | Total | Percent |
|---|---|---|---|
| Packed upside down | ⵀⵀ | 5 | 2.5 |
| Crust bubble | ⵀⵀ ⵀⵀ ⵀⵀ ⵀⵀ ⵀⵀ ⵀⵀ ⵀⵀ ⵀⵀ II | 42 | 21.0 |
| Burnt edge | ⵀⵀ ⵀⵀ III | 13 | 6.5 |
| Received cold | ⵀⵀ ⵀⵀ ⵀⵀ ⵀⵀ ⵀⵀ ⵀⵀ ⵀⵀ ⵀⵀ ⵀⵀ ⵀⵀ ⵀⵀ II | 57 | 28.5 |
| Received late | ⵀⵀ ⵀⵀ ⵀⵀ ⵀⵀ ⵀⵀ ⵀⵀ II | 32 | 16.0 |
| Wrong address | ⵀⵀ ⵀⵀ | 10 | 5.0 |
| Wrong order | ⵀⵀ ⵀⵀ ⵀⵀ ⵀⵀ | 20 | 10.0 |
| Missing ingredients | ⵀⵀ ⵀⵀ IIII | 14 | 7.0 |
| Other | ⵀⵀ II | 7 | 3.5 |
| | Total | 200 | |

(a) Pareto analysis check sheet

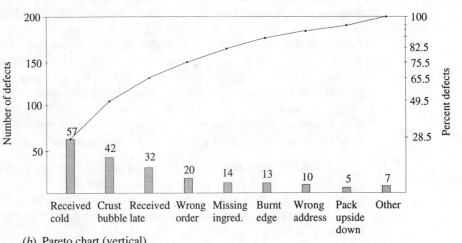

(b) Pareto chart (vertical)

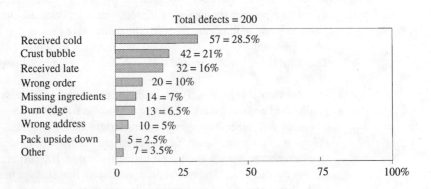

(c) Pareto chart (horizontal)

**FIGURE 9.11**    Pizza food service Pareto analysis and diagram.

The Pareto diagram is a prioritizing tool that helps us isolate and focus on the main problems or concerns. It highlights the vital few results and causes. It helps us to focus on finding effective solutions to the most pressing (costly) problems (e.g., it is easier to reduce a taller bar by half than to reduce a smaller bar, half as tall as the taller bar, to zero). *By comparing Pareto diagrams before and after improvement efforts, the results of improvement can be observed* (i.e., through the change in the bar ordering on the horizontal axis and/or the heights of the vertical bars on the left-hand axis). If the descriptions of results or causes are too long to fit the category area on the vertical bar chart, a horizontal bar chart format can be used (see Figure 9.11c).

## 9.8  CAUSE-EFFECT AND PARETO INTEGRATION

*Pareto analysis and CE analysis are useful in all product life cycle phases: definition, design, development, production, delivery, sales and service, usage, and disposal.* They are systematic, thorough, and based on simple logic. *The Pareto diagram's quantitative structure allows us to stratify and prioritize our information (management by fact). The qualitative structure of the CE diagram allows us to capture experience, engineering judgment, and reasonable technical speculation in our results.*

Pareto diagrams and CE analysis are frequently used together in developing effective proactions or corrective reactions, since each provides different, yet necessary information to support the analysis. Typically, we use a Pareto analysis and the resulting diagram to identify a critical problem or opportunity for quality improvement. Then, we use the CE analysis and diagram to identify causes relative to the effect.

The Pareto diagram helps determine the effect we select for the focus of the CE analysis. The cause-effect diagram then displays all possible causes of a quality problem or challenge and aids in identifying root causes. When the two diagrams are used together, they become an instrument for documenting and communicating quality improvement progress as well as for gaining perspectives on problem solving during the quality improvement process.

## 9.9  INTRODUCTION TO PROCESS CONTROL CHARTS

The seventh and final tool (in the seven fundamental tools) is the statistical process control (SPC) chart. The control charts Ishikawa refers to [6] and the practice of using the charts on the factory floor constitute major elements of a larger set of tools known as "statistical quality control" (SQC) tools. These tools are based on the principles of probability and statistics. Our purpose here is to introduce the control chart concept. In Section V, Chapters 15, 16, 17, 18, and 19, we will discuss control charts and other SQC tools in more detail.

Modern day control charting practices have grown out of the original work performed in the 1920s by Walter A. Shewhart and his peers at the Bell Telephone

Laboratories [7]. *Control charting technology utilizes in-process* (sometimes called "on-line") *sampling techniques to help us monitor a process. The purpose of control charting is to indicate when the process is functioning as intended (at its "best") and when corrective action of some type is necessary,* i.e., to provide early warning of process upsets (shifts or trends). Control charting can be considered proactive or preventive if their warnings allow us to avoid significant amounts of off-specification product.

Control charts are "watchdog" tools that provide us with indications of "in-control" or "out-of-control" status. It is important to note that *an in-control process is considered stable* (Deming [8]). *An out-of-control process is said to be unstable.* Process stability implies the very best operation possible, given the present state of our production system (e.g., incoming materials, machines, operator skills). Improvement in a stable system can occur only through system changes, which are the responsibility of management and empowered employees.

*Instability is created when a special cause, or disturbance, is present.* Special causes shift either the process location or the process dispersion, or both. The shift is judged (in a statistical sense) relative to the process location and dispersion under stable conditions. An unstable or shifted process is not operating up to its potential. Hence, improvement without system level change is possible in unstable processes.

Since process control charts rely on sampling, they are subject to statistical errors. For example, a false alarm (when a process shift is indicated, and in fact, there is none) or a failure to detect a process shift (when a process shift really has occurred and we did not detect the shift from our sample information) may be encountered. Table 9.3 relates these two error classifications to type I and type II statistical errors, respectively. *Once an indication of a process shift is detected, it is up to the operators, engineers, and other technical people to locate the special cause(s) and take corrective action.*

## Attributes and Variables Measures

Every product and production process has many qualities or quality characteristics. *We must choose only the most critical quality characteristics for control charting. Criticality refers to (1) a characteristic related to important product-process function, form, or fit and (2) a strategic position in the production process (e.g., timeliness or cost), so that early warning of nonconforming product is provided.* Table 9.4 lists some typical quality characteristics.

*When a quality characteristic is assessed as being "go" or "no-go" or "acceptable" or "unacceptable" (one of two, or perhaps more, classes) it is termed an attributes quality characteristic, subject to an attributes measure.* A set of such assessments is termed attributes data. *A quality characteristic assessed by means of a measurement (limited only by the resolution of the measuring device) is termed a variables quality characteristic.* We term such a measure a variables measure. A set of such measurements is termed variables data.

**TABLE 9.3** STATISTICAL PROCESS CONTROL INDICATIONS AND ERRORS

Process reality
(god's eye view of the process)

| Sample indication from SPC (sample based view of the process) | In-control (stable) | Out-of-control (unstable) |
|---|---|---|
| In-control (stable) | Correct indication | Failure to detect ( type II error ++ ) |
| Out-of-control (unstable) | False alarm ( type I error * ) | Correct indication |

* Type I error: Declare process out of statistical control (unstable), when the process is really in statistical control (stable).
++ Type II error: Declare process in statistical control (stable), when the process is really out of statistical control (unstable).

**TABLE 9.4**   TYPICAL QUALITY CHARACTERISTICS

|  | Characteristics | Variables measure | Attributes measure |
|---|---|---|---|
| **Product** | | | |
| Ruler | length | centimeters | acceptable, unacceptable |
| Candy | weight | kilograms | acceptable, unacceptable |
| Cloth | surface | flaw width | number of flaws per 100 $m^2$ |
| Paint | color | wave length | acceptable, unacceptable |
| **Process** | | | |
| Machining | speed | meters per second | acceptable, unacceptable |
| Fluid delivery | flow | liters per second | acceptable, unacceptable |
| Soldering | temperature | degrees C | acceptable, unacceptable |
| Forming | pressure | pounds per square inch | acceptable, unacceptable |

Once a quality characteristic is defined, a target set, and specifications stated, we have a benchmark for judging a process or product. For example, a meter stick might be described as follows:

Critical quality characteristic:  overall length

Target:  1000 mm

Specification:  LSL = 999 mm
USL = 1001 mm

Here, each meter stick is perfect if it is exactly 1000 mm long; it is acceptable if 999 mm $\leq$ measured length $\leq$ 1001 mm. Otherwise, it does not conform to specifications.

A simplified production process for our meter stick was shown in Figure 9.6. Assuming we were producing our meter sticks in a reasonably large quantity, the depictions and descriptions in Figure 9.12 illustrate the nature of both attributes and variables data, relative to our meter stick production shearing process—stable process operations in Figures 9.12a and 9.12c, shifts in location or dispersion in Figures 9.12b, 9.12d, and 9.12e.

Binomial probability mass functions are typically used to model quality characteristics when two possible outcomes, i.e., an attributes measure, are considered (see Figure 9.12a and 9.12b). Hence, results are reported as a fraction defective or a fraction acceptable. Figure 9.13 depicts a P chart for our meter stick production process. We see a point that falls above the upper control limit (UCL) at sample 21. Hence, we take corrective action to bring the process back into control. The P and NP control charts are used when we monitor or track processes on a fraction defective and number defective basis, respectively.

Another attributes case involves defects per product unit. Here, the Poisson distribution is used as the basis. In this case, we seek to measure the average number of defects per unit. Hence, we want to detect shifts in our production processes that

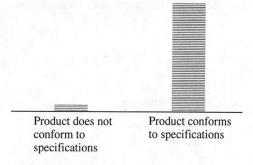

Product does not
conform to
specifications

Product conforms
to specifications

Process is producing a small fraction of
defective or nonconforming meter sticks.
(Product is examined and either meets the
specifications, LSL = 999 mm,
USL = 1001 mm, or product does not meet
specifications).

(*a*) Process operating as it was intended
(stable or in-control)

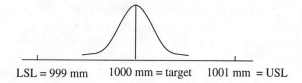

Product does not
conform to
specifications

Product conforms
to specifications

Process is experiencing some sort of
malfunction resulting from a special
cause and hence, is producing
excessive numbers of defective or
nonconforming meter sticks.

(*b*) Process is not operating as it was intended
(unstable or out-of-control)

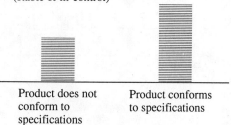

LSL = 999 mm      1000 mm = target      1001 mm  = USL

Process subject to chance variation
due to variation in machines, materials,
methods, environment, and so on.
Result: Product on-target, product
variation about the mean acceptable.

(*c*) Process is operating as it was intended (stable or
in-control)

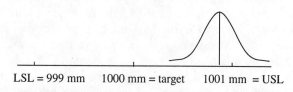

LSL = 999 mm      1000 mm = target      1001 mm  = USL

Process subject to chance variation
plus feeder advance malfunctions (a
special cause).
Result: Product off-target (too long)
with acceptable product variation
about the mean.

(*d*) Process is experiencing a location shift due to feeder
advance malfunction (unstable or out-of-control)

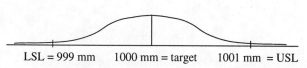

LSL = 999 mm      1000 mm = target      1001 mm  = USL

Process subject to chance variation
plus excessive feeder wear (a special cause).
Result: Product on-target with excessive
product variation about the mean.

(*e*) Process variation increase due to excessive feeder
wear (unstable or out-of-control)

**FIGURE 9.12**     Selected meter stick process operations and measurement possibilities.

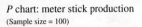

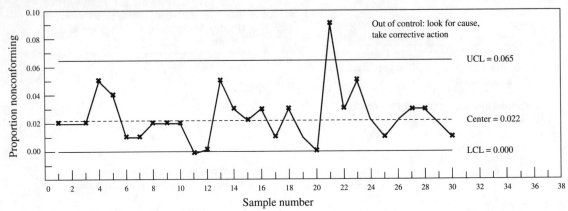

**FIGURE 9.13**   *P* chart illustration.

create excessive defects per unit (lower quality) and correct the process. Or, alternatively, we want to detect shifts to lower defects per unit (higher quality) and to change our process to capture this higher level of performance permanently. The *C* and *U* charts are used when we monitor on a product defect basis. Chart mechanics for attributes measurements are described in detail in Chapter 16.

The meter stick results in Figure 9.12*c*, 9.12*d*, and 9.12*e* portray a single, measurable quality characteristic (length). The normal, or Gaussian, probability mass function is typically used to model quality characteristics when variables measures are used. Since we have a normal model, with two parameters (location and dispersion, e.g., mean and variance, respectively), we will need two charts: one for location (the *X*-bar chart) and one for dispersion (the range, or *R* chart). Illustrative *X*-bar and *R* control charts are shown in Figure 9.14. Here, we see a sample breaking out below the lower control limit (LCL) on the *X*-bar chart (detected in the sample that produced subgroup 10). We take corrective action and get the mean back to the location target. Then, in the range chart, we see subgroup 17 breaking over the UCL, indicating a dispersion-related problem. We take corrective action and get the dispersion back in control. A wide variety of control charts are used for variables measures. These charts are discussed in detail in Chapter 15.

## REVIEW AND DISCOVERY EXERCISES

### Review

**9.1** Compare and contrast the functional applications of the following types of CE diagrams (e.g., what type of applications are best addressed with each, and why?).
   **a** Cause enumeration.
   **b** Dispersion analysis.
   **c** Process analysis.

*X*-bar chart: meter stick length
(Subgroup size = 3)

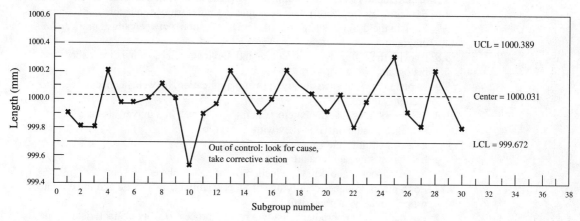

(*a*) *X*-bar chart illustration

*R* chart: meter stick length
(Subgroup size = 3)

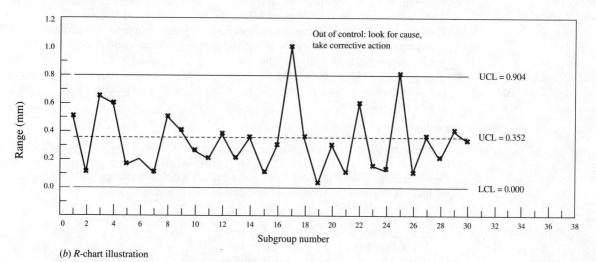

(*b*) *R*-chart illustration

**FIGURE 9.14**    *X*-bar chart and *R* chart illustrations.

**9.2**  What does the term "freethinking" mean with respect to CE analysis?

**9.3**  List the benefits and burdens we should expect to encounter when using CE analysis.

**9.4**  What role does CE analysis play in proactive quality work?

**9.5** Ishikawa is credited with the development of the CE diagram. Is CE analysis and its heavy reliance on speculation consistent with Ishikawa's principle of "using facts and data to make presentations-utilization of statistical methods"? Explain.

**9.6** Explain why the stratification principle is so uniformly applicable in data analysis.

**9.7** Explain the difference, in structure and application, between a histogram and a scatter diagram.

**9.8** Explain the difference, in structure and application, between a check sheet and a histogram.

**9.9** Explain the difference, in structure and application, between a check sheet and a scatter diagram.

**9.10** Do you believe that the Pareto principle holds true in a universal sense? Provide evidence to support your view with respect to
  **a** Sales volume and salespersons.
  **b** Market share and manufacturers.
  **c** Team scores and point production of offensive players in sports.

**9.11** For a given product, identify one product quality characteristic and three related process quality characteristics.

**9.12** For the following products, list two related process quality characteristics (one attributes based and one variables based):
  **a** A plastic (injection molded) toy.
  **b** A motion picture.
  **c** An apple (the edible type).

**9.13** What are the two major parameters in the assessment and analysis of data?

**9.14** Explain the difference between an attributes measure and a variables measure. Which one contains the most information? Explain.

**9.15** How do we decide between (*a*) no measure, (*b*) an attributes measure, or (*c*) a variables measure, regarding the decision to use and develop a process control chart for any given quality characteristic? Explain.

**9.16** Explain the linkage between process control charts and process improvement. For example, given the purpose of each, explain how they can be used in a synergistic fashion.

## Discovery

**9.17** There are about 160 countries in the world at any one time. A final tally of the medals awarded at the 1992 Summer Olympics is provided on the next page. Assuming that any country (out of 160) not listed participated but did not receive metals, devise a method to assess the validity of the Pareto principle and apply it to
  **a** Gold medals.
  **b** Silver medals.
  **c** Bronze medals.
  **d** Total medals.

**9.18** Develop a Pareto chart for the awarding of gold medals at the 1992 Summer Olympics.

**9.19** In a work team structure, we want to violate the Pareto principle in that we do not want approximately 20 percent of the team to do approximately 80 percent of the work. Is it possible to avoid this maldistribution? How?

**9.20** Robust performance is critical in high quality products. How can we use CE analysis in the quest for robust performance? Explain.

**9.21** Mistake-proof performance is critical in high quality products. How can we use CE analysis in the quest for mistake-proof performance?

| | Medals | | | | | Medals | | |
|---|---|---|---|---|---|---|---|---|
| Country | Gold | Silver | Bronze | Total | Country | Gold | Silver | Bronze | Total |
| Unified Team | 45 | 38 | 29 | 112 | Croatia | 0 | 1 | 2 | 3 |
| United States | 37 | 34 | 37 | 108 | Iran | 0 | 1 | 2 | 3 |
| Germany | 33 | 21 | 28 | 82 | Yugoslavia | 0 | 1 | 2 | 3 |
| China | 16 | 22 | 16 | 54 | Greece | 2 | 0 | 0 | 2 |
| Cuba | 14 | 6 | 11 | 31 | Ireland | 1 | 1 | 0 | 2 |
| Hungary | 11 | 12 | 7 | 30 | | | | | |
| South Korea | 12 | 5 | 12 | 29 | Algeria | 1 | 0 | 1 | 2 |
| France | 8 | 5 | 16 | 29 | Estonia | 1 | 0 | 1 | 2 |
| Australia | 7 | 9 | 11 | 27 | Lithuania | 1 | 0 | 1 | 2 |
| Spain | 13 | 7 | 2 | 22 | Austria | 0 | 2 | 0 | 2 |
| | | | | | Namibia | 0 | 2 | 0 | 2 |
| Japan | 3 | 8 | 11 | 22 | South Africa | 0 | 2 | 0 | 2 |
| Britain | 5 | 3 | 12 | 20 | Israel | 0 | 1 | 1 | 2 |
| Italy | 6 | 5 | 8 | 19 | Mongolia | 0 | 0 | 2 | 2 |
| Poland | 3 | 6 | 10 | 19 | Slovenia | 0 | 0 | 2 | 2 |
| Canada | 6 | 5 | 7 | 18 | Switzerland | 1 | 0 | 0 | 1 |
| Romania | 4 | 6 | 8 | 18 | | | | | |
| Bulgaria | 3 | 7 | 6 | 16 | Mexico | 0 | 1 | 0 | 1 |
| Netherlands | 2 | 6 | 7 | 15 | Peru | 0 | 1 | 0 | 1 |
| Sweden | 1 | 7 | 4 | 12 | Taiwan | 0 | 1 | 0 | 1 |
| New Zealand | 1 | 4 | 5 | 10 | Argentina | 0 | 0 | 1 | 1 |
| | | | | | Bahamas | 0 | 0 | 1 | 1 |
| North Korea | 4 | 0 | 5 | 9 | Colombia | 0 | 0 | 1 | 1 |
| Kenya | 2 | 4 | 2 | 8 | Ghana | 0 | 0 | 1 | 1 |
| Czechoslovakia | 4 | 2 | 1 | 7 | Malaysia | 0 | 0 | 1 | 1 |
| Norway | 2 | 4 | 1 | 7 | Pakistan | 0 | 0 | 1 | 1 |
| Turkey | 2 | 2 | 2 | 6 | Philippines | 0 | 0 | 1 | 1 |
| Denmark | 1 | 1 | 4 | 6 | | | | | |
| Indonesia | 2 | 2 | 1 | 5 | Puerto Rico | 0 | 0 | 1 | 1 |
| Finland | 1 | 2 | 2 | 5 | Qatar | 0 | 0 | 1 | 1 |
| Jamaica | 0 | 3 | 1 | 4 | Surinam | 0 | 0 | 1 | 1 |
| Nigeria | 0 | 3 | 1 | 4 | Thailand | 0 | 0 | 1 | 1 |
| | | | | | | | | | |
| Brazil | 2 | 1 | 0 | 3 | | | | | |
| Morocco | 1 | 1 | 1 | 3 | | | | | |
| Ethiopia | 1 | 0 | 2 | 3 | | | | | |
| Latvia | 0 | 2 | 1 | 3 | | | | | |
| Belgium | 0 | 1 | 2 | 3 | | | | | |

**9.22** Choose a product or process.  Identify the critical quality characteristics.  Measure and record relevant data as to one or more of the quality characteristics.

   **a** Develop an appropriate check sheet.

   **b** Develop a histogram(s).

   **c** Develop a scatter diagram.  Remember, you will need bivariate observations here.

**9.23** There are four basic measurement scales:

   **a** *Nominal scale* (the weakest scale)—Here numbers or labels serve only to distinguish between different categories or classes.

   **b** *Ordinal scale*—Here the scale ranks or orders each measurement as "less than," "greater than," or "equal to" other measurements.

    **c** *Interval scale*—Here we see not only an ordering, but a defined interval between measurements (e.g., temperature) where we can assess how far apart two measurements are.  Any 0 point on an interval scale will be arbitrarily defined.

    **d** *Ratio scale* (the strongest scale)—Here we see both order and interval characteristics, as well as a natural zero point whereby the ratio of any two measurements is meaningful.  Distances, lifetimes, weights, costs, profits, and so on, are examples of ratio scales.

Provide an example of each scale relative to a product's quality characteristics. Where does the attributes measure fall?  Where does the variables measure fall?

## REFERENCES

**1** K. Ishikawa, *What Is Total Quality Control? The Japanese Way,* Englewood Cliffs, NJ: Prentice-Hall, 1985.

**2** K. Ishikawa, *Guide to Quality Control,* Tokyo: Asian Productivity Organization, 1982.

**3** See reference 2.

**4** J. M. Juran, *Juran on Leadership for Quality,* New York: Free Press, 1989.

**5** J. M. Juran and F. M. Gryna, *Quality Control Handbook,* 4th ed., New York: McGraw-Hill, 1988.

**6** See reference 2.

**7** W. A. Shewhart, *Economic Control of Quality Manufactured Product* (originally published by Van Nostrand, 1931), Milwaukee: ASQC, 1980.

**8** W. E. Deming, *Out of the Crisis,* Cambridge, MA: MIT Center for Advanced Engineering Studies, 1986.

# PROCESS IMPROVEMENT AND PROCESS FLOW DIAGRAMS

## 10.0 INQUIRY

1 How are production processes defined and described?
2 What is the difference between a product and a process focus?
3 How is it possible to provide high product quality with a process focus?
4 How are process improvement and quality control related?
5 What information is contained in a process flow diagram?

## 10.1 INTRODUCTION

*The production of goods and services can be described as an input-output phenomena. Resources are inputs and products and by-products are outputs.* Figure 10.1 depicts a simplified input-output relationship, in which processes are necessary in converting resources to products and by-products. Hence, *our product quality is a direct result of the coupling of our inputs and our conversion processes.*

*Two general types of strategies exist for creating products that meet or exceed customer needs and expectations: (1) process focused or (2) product focused.* In general, a process focused strategy concentrates on process integrity in a proactive manner. *A process focus is more concerned with how we design and build our products (i.e., how we couple inputs and processes) rather than with what we design and build. A product focus is more concerned with exactly what we design and build rather than how we design and build it.*

**FIGURE 10.1**  Fundamental product-process relationship.

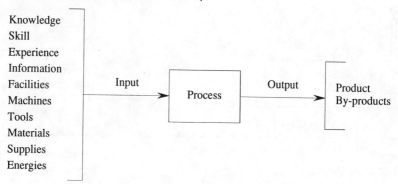

Obviously, we must be concerned with both product and process, formulating strategies for both, although one or the other may dominate at any point in the product life cycle. For example, a product directed strategy will typically dominate during the product definition phase. Both product and process strategies should be strong in the product-process design phase, where technical specifications are developed. When we move into the production phase, we must again address the issues of strategy. Here, our choices are not clear-cut. Figure 10.2 depicts two basic strategies at the production level—Figure 10.2a a dominant process focus, Figure 10.2b a dominant product focus. In either focus, our ultimate goal is customer satisfaction.

We rarely see a strictly process-focused strategy with no regard for product or a strictly product focused strategy with no regard for processes. Our purpose in this chapter is to concentrate on the former and introduce basic process improvement concepts and tools that can be used in any of the eight life cycle process phases (Figure 1.1).

**FIGURE 10.2**    Process and product focused production strategies.

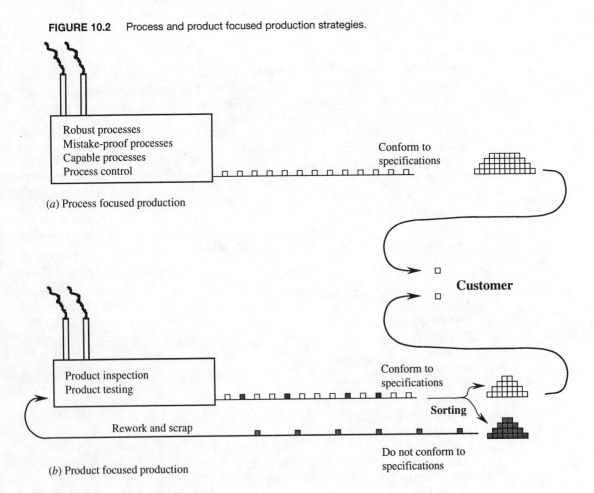

(a) Process focused production

(b) Product focused production

## 10.2 PROCESS IMPROVEMENT CYCLE

***Process improvement thinking has evolved over the years.*** Initially, process improvement consisted of isolated and functionalized applications of "methods engineering," "time study," "flowcharting," "motion study," and other classical industrial engineering tools (Barnes [1]). Today, process improvement consists of integrated applications of the classical tools, as well as statistical methods of process control and off-line experimentation applied at critical process "bottlenecks" or "choke points." This new approach makes up a large part of total quality management (TQM), in the United States, or total quality control (TQC), in Japan. ***Today, we see integrated quality planning, training, analysis, corrective action, and information system networks throughout production facilities (both goods and service related).***

In many cases, the processes necessary to design and build a product are more complicated than the product itself. The inputs and the processes (Figure 10.1) determine, to a large extent, the quality, performance, cost, and delivery time, that our customers experience for any given product. The role of processes in the creation of simple and complex products is critical.

Product performance has always been recognized as a critical component in establishing a competitive edge in a market driven economy. Now more than ever before, process improvement and process quality control are also recognized as critical factors in establishing a competitive edge. Processes transform the inputs, or resources, to the outputs and are the linkage between resources and products. Hence, given adequate inputs, it is the processes that determine the competitive position of the organization.

***We see processes involved in all phases of the product life cycle. That is, processes exist for product definition, design, development, production, delivery, sales and service, use, and disposal.*** The more effective and efficient our conversion processes are, the more competitive our organization will be within its market.

Process improvement, by definition, consists of improving our means of converting resources to products. Figure 10.3 depicts the process quality improvement cycle. ***Our quality improvement cycle begins with questioning current or proposed processes.*** Here, process description is necessary. ***Process description focuses on developing a systematic, thorough, and, of course, accurate picture of our processes.*** Consideration is given to

1 ***Process hierarchy***—the level of description and detail necessary for delineating process components.
2 ***Process function***—the specific nature of each process component.
3 ***Process sequence***—the order or arrangement of functional components.

***The process flowchart (discussed later in this chapter) is the primary tool used for process description.***

***Through observation, we identify critical process and product characteristics. These characteristics must be adequately described so that they can be measured.*** Direct observation, SPC charts, process logs, and product inspection results and

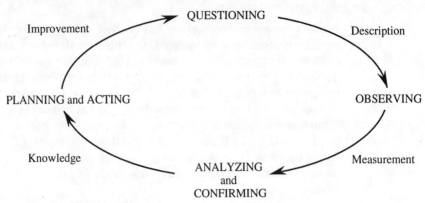

**FIGURE 10.3**    Process quality improvement cycle.

records provide data.  Some level of experimentation may be necessary to yield sufficient data for process improvement efforts to proceed.  Next, *we analyze the process by examining the description and data.*

It is important to confirm our findings in terms of both descriptions and data.  *We then complete our analyses, interpret and confirm our findings, and develop our improvement plan.*  Here, *we use the knowledge gathered to plan and implement process changes and improvements.*  Knowledge supports the planning and acting phase, which ultimately leads to improvement.  Our approach in the improvement cycle is based on the management by fact philosophy.  Here, facts, including empirical observations, form the groundwork for knowledge.

We are continually faced with the question, "When should we commit to action and when should we leave the process alone or take no action?"  The corollary, of course, can be stated as, "If action is necessary, what action is appropriate?"  Action, when appropriate, can take many forms, depending on the exact nature of the inputs and processes.  *The resultant of our actions will manifest itself in change and, ultimately, product improvement when we are successful.*  It is important to be able to measure our degree of success or failure.  The same measures that we used in observing will typically serve this purpose.

We must be prepared to deal with failures in our improvement efforts; we simply will not be successful in all efforts.  But we must take calculated risks, celebrate our successes, and keep working until we eventually turn failures into successes.  Actually, persistence and commitment to purpose are two very underrated virtues in quality transformation.

*The determination of appropriate action requires product-process knowledge and experience, as well as the mental and physical discipline to pursue a course of action.*  Effective action requires leadership and empowered employees who are in touch with their customers, processes, and products.

## 10.3 PROCESS FLOW AND FLOWCHARTS

The essence of outstanding process design is the effective and efficient transformation of input resources to product output. Flowcharting is an excellent tool in describing a process as it relates to this transformation. *Process flowcharting captures both the hierarchical nature and the sequence of a process in a systematic and thorough fashion.*

### Process Flowchart Structure and Development

*A process flowchart provides a graphical or symbolic picture of the components of a physical process, linked to represent the production sequence.* Process flowcharting was originally introduced as a classical industrial engineering tool (Barnes [2]). Over the years, many specialized forms of process flowcharting have evolved, including man-machine charts, gang charts, right-hand–left-hand charts, flow diagrams, process charts, and so on.

Most process flowcharts are composed of generic symbols and simple descriptions. Flowcharts have been used for component fabrication and assembly illustrations, and product-process operation instructions. Other more recent applications include data processing and computer software design and development. The Japanese process decision program chart (PDPC) is yet another example of a process flowchart. Flowcharts are useful whenever hierarchical structure and sequence of activities or events are critical.

Process flowchart symbols vary widely from one field of application to another. Figure 10.4 depicts seven symbols appropriate for quality related analyses. We have adopted the five basic industrial engineering symbols (as described by Barnes [3]) and added two symbols: one for source inspection and one for an SPC charting point.

*Traditionally, we have seen two basic process charting formats: a descriptive format and an analytical format.* A descriptive format starts with inputs and ends with outputs. It is instructive and serves as a guide in implementing the production process. An analytical format provides quantitative detail with regard to process elements, represented by our process flow symbols. We use the analytical format to compare process alternatives. Figure 10.5 illustrates a simple classical analytical format comparing a present and proposed office procedure method; numerical comparison of process elements and distances involved is summarized.

We should note that in the analytical flowchart shown in Figure 10.5, a check box is provided for either present or proposed (improved) method. This flowchart follows a product—in this case, paper, and the process is a classical administrative management activity. Traditional methods improvement procedures suggest following either a product (material) or a person (operator) through the process.

*When flowcharting a process, it is important to proceed in a systematic manner:*

1 Establish and communicate purpose; clearly state the goal and objective(s) to be accomplished in writing.

**Operation.** An operation consists of an activity that changes or transforms an input (material, supply, in-process product, and so forth). An operation may be performed by a person or a machine.

**Transportation.** A transportation consists of the physical movement of an input (material, supply, in-process product, and so forth) when the movement is not an integral part of an operation (e.g., the movement of a part held in a machine necessary to plane a surface is not a transportation, whereas the movement of a part from one machine to another is a transportation).

**Delay.** A delay is created when an input is waiting for the next planned activity. During a delay, no activity is taking place with regard to the input (material, supply, in-process product, and so forth).

**Storage.** A storage is created when a material, supply, in-process product, or finished product is placed somewhere such that authorization must be obtained to remove it from its location (e.g., placed in a stock room or warehouse, etc.).

**Source inspection.** A source inspection is created through a self-check or a successive check of an input at the point of origination of possible defects (e.g., as activities are performed).

**SPC charting point.** Critical points in the production process are identified with a SPC charting point to indicate that a statistical process control chart is positioned to track and help control the production quality of the input.

**Sorting inspection.** A sorting inspection is created when an inspection or test is performed for the purpose of examining the input or final product before it is turned over to the next internal customer or the external customer. At the point of sorting inspection, the input or product is declared as either acceptable or unacceptable as judged against a product or process specification. The sorting inspection point is in essence a formal decision point.

**FIGURE 10.4**    Process flowchart symbols.

2 Define the scope of the flowchart(s) by defining physical and functional boundaries. These boundaries, along with the goals and objectives, focus the flowchart and provide direction and guidance.

3 Define and develop an appropriate process flowchart structure. Start with a broad process view: a high-level block diagram or PDPC developed around a functional or organizational unit structure, whichever is most appropriate. Then, adapt the block diagram or PDPC to the level of detail essential to accomplish the objective.

4 Develop the process flowchart. Capture activity details as they are (for existing processes) or will be (for new process planning). Beware of using formal organizational procedures or documents (operations manuals, etc.) to develop the process flowchart; in many cases, "official" procedure documents do not

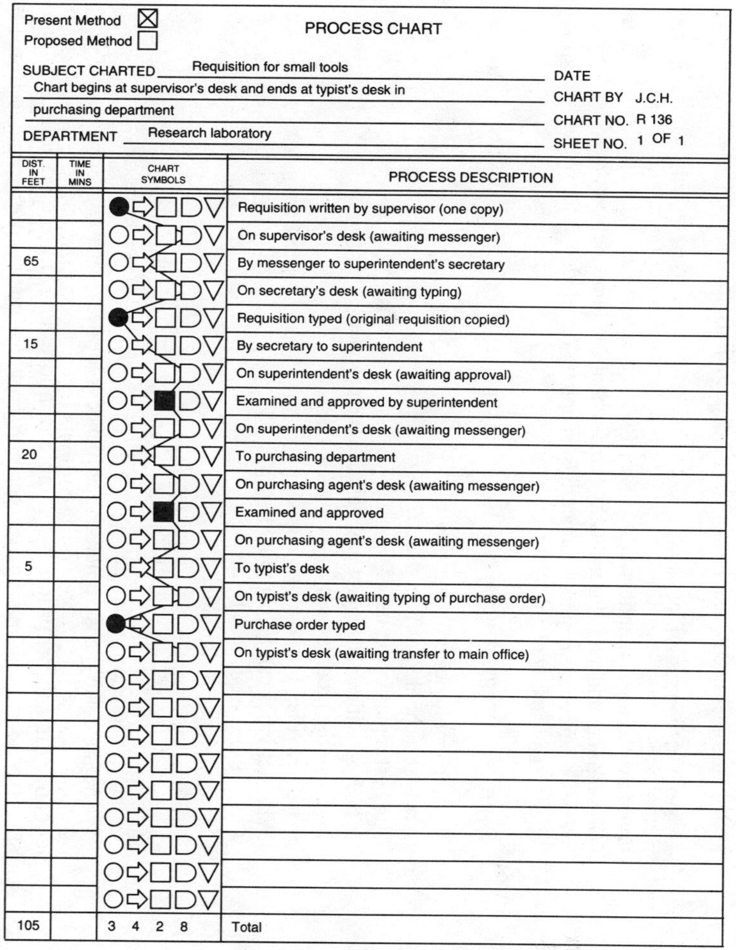

## PROCESS CHART — Present Method

Present Method ☒  Proposed Method ☐

SUBJECT CHARTED: Requisition for small tools
Chart begins at supervisor's desk and ends at typist's desk in purchasing department

DEPARTMENT: Research laboratory

DATE
CHART BY J.C.H.
CHART NO. R 136
SHEET NO. 1 OF 1

| DIST. IN FEET | TIME IN MINS | PROCESS DESCRIPTION |
|---|---|---|
| | | Requisition written by supervisor (one copy) |
| | | On supervisor's desk (awaiting messenger) |
| 65 | | By messenger to superintendent's secretary |
| | | On secretary's desk (awaiting typing) |
| | | Requisition typed (original requisition copied) |
| 15 | | By secretary to superintendent |
| | | On superintendent's desk (awaiting approval) |
| | | Examined and approved by superintendent |
| | | On superintendent's desk (awaiting messenger) |
| 20 | | To purchasing department |
| | | On purchasing agent's desk (awaiting messenger) |
| | | Examined and approved |
| | | On purchasing agent's desk (awaiting messenger) |
| 5 | | To typist's desk |
| | | On typist's desk (awaiting typing of purchase order) |
| | | Purchase order typed |
| | | On typist's desk (awaiting transfer to main office) |
| 105 | Total | 3 4 2 8 |

## PROCESS CHART — Proposed Method

Present Method ☐  Proposed Method ☒

SUBJECT CHARTED: Requisition for small tools
Chart begins at supervisor's desk and ends at purchasing agent's desk

DEPARTMENT: Research laboratory

DATE
CHART BY J.C.H.
CHART NO. R 149
SHEET NO. 1 OF 1

| DIST. IN FEET | TIME IN MINS | PROCESS DESCRIPTION |
|---|---|---|
| | | Purchase order written in triplicate by supervisor |
| | | On supervisor's desk (awaiting messenger) |
| 75 | | By messenger to purchasing agent |
| | | On purchasing agent's desk (awaiting approval) |
| | | Examined and approved by purchasing agent |
| | | On purchasing agent's desk (awaiting transfer to main office) |
| 75 | Total | 1 1 1 3 |

### SUMMARY

| | | PRESENT METHOD | PROPOSED METHOD | DIFFERENCE |
|---|---|---|---|---|
| Operations | ○ | 3 | 1 | 2 |
| Transportations | ⇧ | 4 | 1 | 3 |
| Inspections | □ | 2 | 1 | 1 |
| Delays | D | 8 | 3 | 5 |
| Distance Traveled in Feet | | 105 | 75 | 30 |

Process chart of an office procedure—present method.

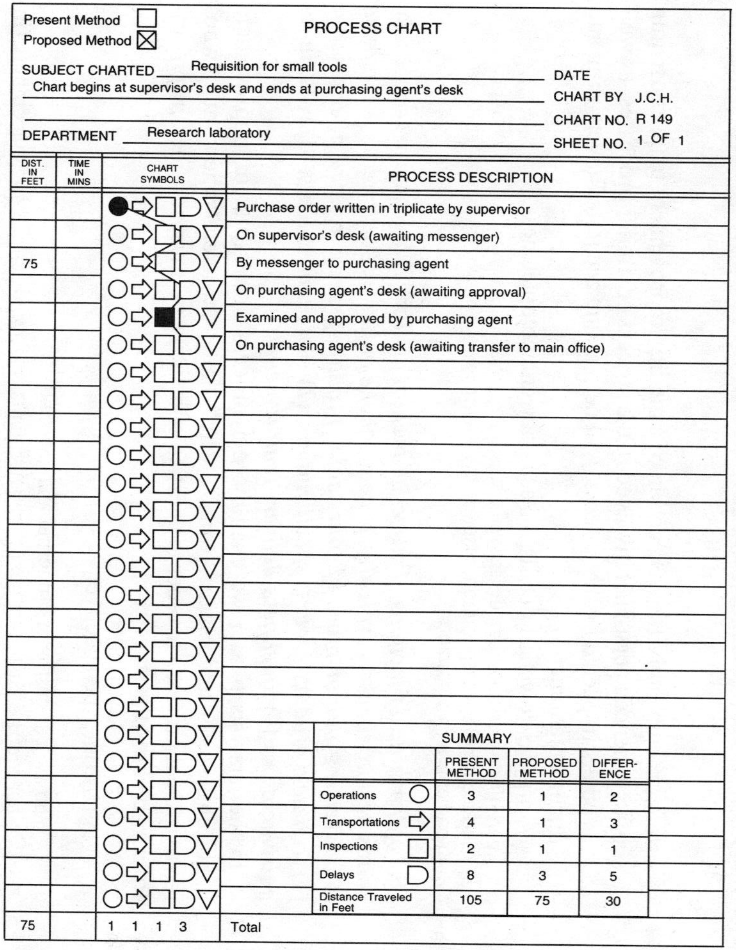

Process chart of an office procedure—proposed method.

**FIGURE 10.5**  Classical present-proposed methods process charts. Adapted, with permission from John Wiley & Sons, Inc., from R. M. Barnes, *Motion and Time Study Design and Measurement of Work*, 7th ed., New York, NY: John Wiley, pages 73 and 75, 1980. Copyright John Wiley & Sons, Inc.

reflect actual activities. Selection of a quality improvement team consisting of experienced personnel involved in the process or similar processes, as well as others impacted by the process, will help assure accurate flowcharting and documentation.

5 Verify the flowchart details for accuracy and completeness.

## Process Flowchart Analysis

*One of the critical functions of process flowchart analysis is to locate process choke points or bottlenecks. Typically, process choke points or bottlenecks are related to addressing one or more of four major product-process related phenomena: (1) technical product-process specifications, (2) cost requirements, (3) volume requirements, or (4) timeliness-delivery requirements.*

In the analysis phase, we examine our process flowcharts for improvement opportunities, and address the "what," "why," "how," "where," "when," and "who" relative to our process. In order to question effectively, we must consider the process from a number of perspectives:

1 A customer perspective:

Who are our customers (internal and external)?
What are the customer needs, requirements and expectations for our outputs (in-process and final)?
What customer needs, requirements, and expectations are not being met?

2 A supplier perspective:

Who are the input suppliers?
Do suppliers understand our quality requirements, volume, delivery?
What resources are required—materials, supplies, knowledge, information, machines, energy, and so on?

3 A process perspective

What technologies are appropriate?
Are best technologies being used?
Are best inputs being used?
Which activities add/do not add value to the product?
How can the process be mistake-proofed?
What activities can be eliminated?
What activities can be combined?
Can activity sequences be changed?
Can activities be simplified?
Are activities sufficient?
Are critical quality points identified for SPC and source inspection?
Is information flow adequate?
Where are quality, cost, timeliness bottlenecks located?

Benchmarking information is valuable in addressing these questions. It helps establish the best-of-the-best processes and provides information relative to many of the above questions.  Process flowcharts not only help us determine and locate process bottlenecks, but locate quality control points where SPC will be most effective or proactive in terms of technical, cost, and timeliness needs.  They also help us address the critical issue of process flexibility.  Our process structure flexibility affects our ability to adjust to different volume levels and different product mixes.

*In our process improvement analysis, we use a five-level hierarchy to guide our creative efforts to enhance process performance.*

1 *Elimination*—We seek to eliminate non–value-added activities.  Sometimes we may not totally eliminate, but replace, the functional essence of the activity with a superior technology.
2 *Combination*—Next, we seek to combine activities in order to extract process improvement.
3 *Change of sequence*—We examine the sequence to see if a reordering will provide improvement.
4 *Simplification*—Here, we examine activities with the expectation of improvement through simplifying the activities themselves.
5 *Addition*—In some cases where processes are clearly ineffective, we may need to add a process step; but additions are our last resort.

We should train ourselves to think in terms of this hierarchy with a breakthrough as well as incremental improvement attitude.  In addition, the robust design and mistake-proofing concepts should guide our thinking in all improvement efforts.  This critical improvement analysis and consideration hierarchy is depicted in  Figure 10.6. *The elimination level is most appropriate for breakthrough improvements, while the simplification and addition levels are more compatible with incremental improvements.*

**FIGURE 10.6**     Process improvement analysis heirarchy.

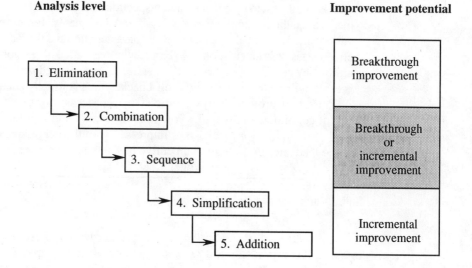

**Example 10.1**

Develop a "gross level of detail" flowchart for the assembly of an electro-mechanical relay.  Indicate the most critical components and inspection points.

**Solution**

The relay is broken out in order of detail from left to right in Figure 10.7.  Here, we are working with a gross level of detail and use boxes rather than the range of flowchart symbols.  Application of the Pareto principle suggests that we can expect the components and assemblies shown by bold outlined boxes to be more critical, in terms of the difficulty in meeting technical specifications, than those in the standard boxes.

**Example 10.2**

Develop a process flow diagram for a cattle feed manufacturing process.

**Solution**

Figure 10.8a depicts a global view of the gross process flow.  We can see the fundamental components and the sequential flow.  The details of the process are shown in Figure 10.8b.  Here, we see a variety of quality control boxes throughout the diagram, as well as a large number of delays and transportation points (essentially non–value-added activities).  We see one component (the molasses additives) where a qualified supplier is used, thus virtually eliminating rejection.  The grain and hay components both require careful scrutiny.  Entire loads (e.g., 30,000 to 40,000 lb of material) may be—and often are—rejected, creating costly supply-receiving problems.

It is of interest to note that source inspections range from quantitative measurements to qualitative, or human sensory-based, judgments.  For example, due to the variety of important alfalfa hay quality characteristics and the lack of timely, cost-effective test instrumentation, we see a qualitative source inspection (in the Ground alfalfa hay column in Figure 10.8b).  Here, an experienced person provides what is essentially a source inspection self-check, which is usually adequate.

This chart provides a systematic start to process improvements and it is a useful communications tool.  Potential customers and regular customers who feed cattle in this yard are always impressed with the aggressive quality attitude in the feedyard. In this cost competitive business a few cents per pound amounts to a significant cost-of-gain edge.

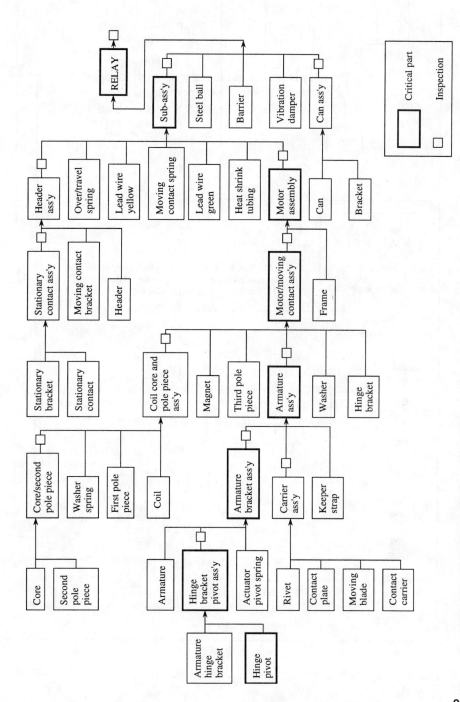

**FIGURE 10.7** Electromechanical relay assembly flowchart.

211

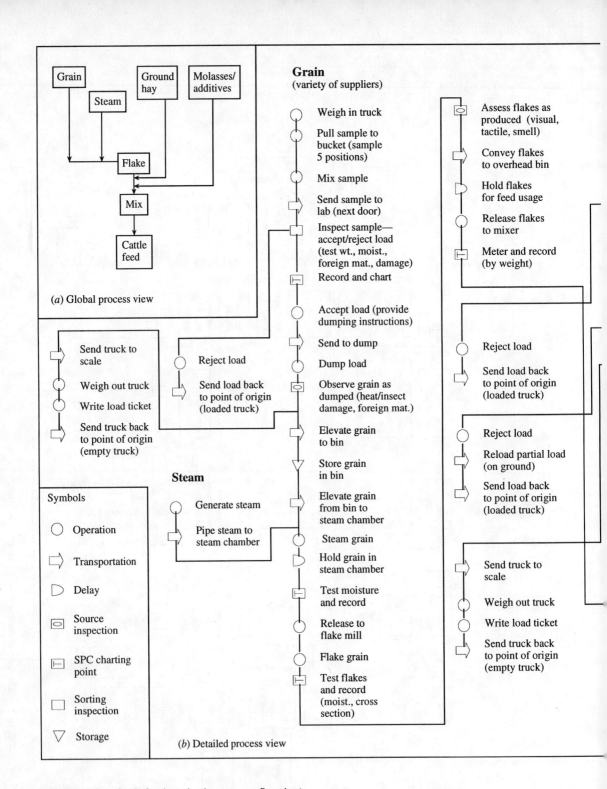

**FIGURE 10.8**  Cattle feed production process flowchart.

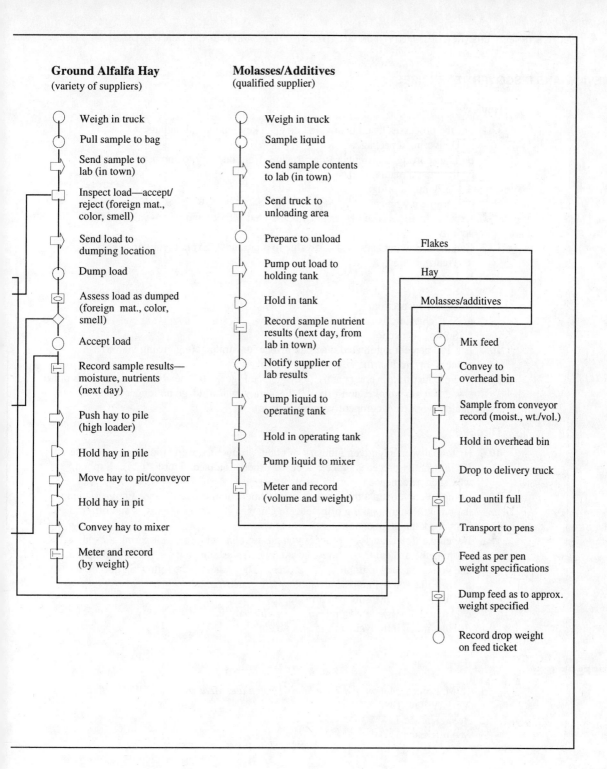

**Ground Alfalfa Hay**
(variety of suppliers)

○ Weigh in truck

○ Pull sample to bag

▷ Send sample to lab (in town)

☐ Inspect load—accept/ reject (foreign mat., color, smell)

▷ Send load to dumping location

○ Dump load

⬭ Assess load as dumped (foreign mat., color, smell)

◇ Accept load

☐ Record sample results— moisture, nutrients (next day)

▷ Push hay to pile (high loader)

▷ Hold hay in pile

▷ Move hay to pit/conveyor

▷ Hold hay in pit

▷ Convey hay to mixer

☐ Meter and record (by weight)

**Molasses/Additives**
(qualified supplier)

○ Weigh in truck

○ Sample liquid

▷ Send sample contents to lab (in town)

▷ Send truck to unloading area

○ Prepare to unload

▷ Pump out load to holding tank

▷ Hold in tank

☐ Record sample nutrient results (next day, from lab in town)

○ Notify supplier of lab results

▷ Pump liquid to operating tank

▷ Hold in operating tank

▷ Pump liquid to mixer

☐ Meter and record (volume and weight)

Flakes

Hay

Molasses/additives

○ Mix feed

▷ Convey to overhead bin

☐ Sample from conveyor record (moist., wt./vol.)

▷ Hold in overhead bin

▷ Drop to delivery truck

⬭ Load until full

▷ Transport to pens

○ Feed as per pen weight specifications

⬭ Dump feed as to approx. weight specified

○ Record drop weight on feed ticket

## REVIEW AND DISCOVERY EXERCISES

### Review

**10.1** For the processes listed below, identify the key inputs and outputs:
   **a** Delivering a package.
   **b** Transporting a passenger (via airplane) from one city to another.
   **c** Running a marathon race.
   **d** Designing a bridge.
   **e** Forging a wrench.

**10.2** Explain the difference between a value-added activity and a non–value-added activity.

**10.3** Can the following activities add value to a product? Explain your answers.
   **a** Transportation.
   **b** Delay.
   **c** Storage.
   **d** Source inspection.
   **e** SPC charting.
   **f** Sorting inspection.

**10.4** Is a 100 percent automated machine inspection strategy (e.g., using vision) a process- or product-focused strategy? Explain.

**10.5** The product improvement activity cycle (see Figure 10.3) tends to contradict the old saw, "If it ain't broke, don't fix it." Explain why this adage no longer holds in quality work (in a competitive marketplace).

### Discovery

**10.6** Historically, in any given industry or plant, product focused production (see Figure 10.2b) usually precedes process focused production (see Figure 10.2a). Explain why such a sequence typically occurs.

**10.7** Compare and contrast the process improvement analysis levels (see Figure 10.6) to the breakthrough thinking principles in Chapter 8. What similarities and differences in concept and application exist?

**10.8** Develop a flowchart for a process you are familiar with (e.g., preparing a meal, washing the dishes at your house or apartment, washing your car). Use your flowchart and your creativity to improve the process. Then, address the following questions:
   **a** How did you define performance in this endeavor?
   **b** How did you measure performance?
   **c** Are timeliness and cost relevant to your analysis?
   **d** How much improvement did you obtain?

## REFERENCES

**1** R. M. Barnes, *Motion and Time Study Design and Measurement of Work,* 7th ed., New York: Wiley, 1980.
**2** See reference 1.
**3** See reference 1.

# CREATION OF QUALITY— DEFINITION AND DESIGN

## VIRTUES FOR LEADERSHIP IN QUALITY
### Self-Discipline

The purpose of Section Four is to discuss a number of quality related tools that can be used proactively to help define quality and design it into both products and processes. Chapter 11 focuses on quality definition through quality planning and quality function deployment (QFD), strategic tools which help form a superstructure within which products and their related processes can be systematically and thoroughly defined and planned. These tools use a matrix diagram format and require a multidisciplinary approach in their development. The use of these tools assures customers a voice in product definition and design and enhances cross-functional communication.

Failure mode and effects analysis (FMEA) is presented in Chapter 12. This tool provides a qualitative means of proactively assessing potential failure modes and their possible effects, with the objective of prevention through proposed counter-measures. FMEA is useful when we desire a broad-spectrum analysis of a product or process with regard to failure or success modes. Chapter 13 extends our discussion of proactive quality tools to include logic tree models. Here, we discuss logic trees in the context of faults, events, and goals. The tree models are useful when we must concentrate on a specific fault, event, or goal. Both the FMEA and the logic tree models are useful in quality planning.

Chapter 14 introduces our readers to design review and value analysis. Design review is an essential tool in proactive quality work. It serves to bring a product or process under close scrutiny by a panel of independent-minded experts, with the objective of producing constructive criticism. Reviewers may be functional experts or they may be potential customers; in either case, new product-process perspectives are gained. The value analysis tool is used to study the functional characteristics of a product or process. Here, functions and their related costs are assessed in order to help provide the function, form, and fit that customers need and expect, at the lowest possible cost. Value analysis can be a very productive tool when used in conjunction with quality planning and QFD.

# QUALITY PLANNING AND QUALITY FUNCTION DEPLOYMENT

## 11.0 INQUIRY

1 How is product and process quality defined?
2 How is quality designed into products and production processes?
3 What is a quality plan?
4 What is quality function deployment (QFD)?
5 How are quality function strategies and plans formulated?

## 11.1 INTRODUCTION

Most organizations profess to having a goal of customer satisfaction. However, such a declaration must be backed up with solid follow-through sufficient to successfully generate and execute sound strategies and functional plans for high quality products and processes. In this chapter, we will explore the customer-organization linkage we introduced in Chapter 8 (see Figure 8.1).

## 11.2 QUALITY PLANNING STRATEGY

In today's highly competitive marketplaces, effective, flexible strategies are necessary to provide the support structure for companywide quality endeavors. *Each product or product line requires unique quality strategies and requires a unique quality plan to be developed through three primary phases*: **(1)** *customer characteristics and demands,* the nature and voice, or demands, of the customer, **(2)** *product and process features,* the technical response of marketing and engineering to the voice of the customer, and **(3)** *the product and process "road map,"* the voice of marketing, engineering, and management, the strategic plan to respond to customer demands. Figure 11.1 depicts these three primary phases. Each phase presents its own unique challenges and adds to the level of detail in a systematic manner. *Due to the nature of the quality plan, interdisciplinary (cross-functional) involvement through teamwork is required. In addition, we must develop effective partnerships with suppliers and potential customers.*

---

### VCR Development—True and Substitute Quality Characteristics Case [SM]

A manufacturer of videocassette recorders (VCRs) conducted a market survey of recent VCR purchasers to determine what qualities they considered most important

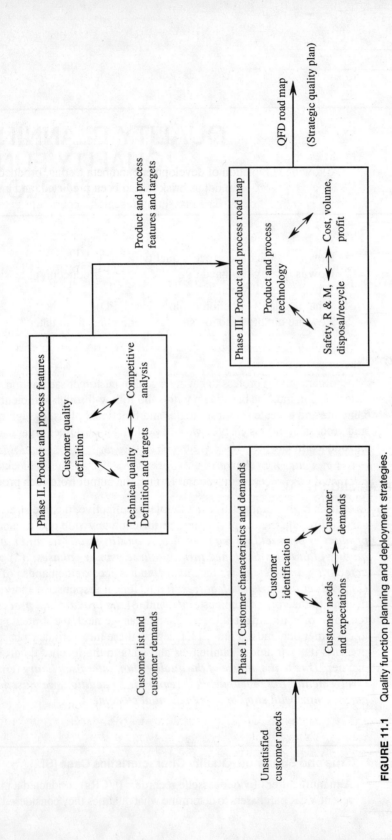

**FIGURE 11.1**  Quality function planning and deployment strategies.

in choosing a particular VCR product. Most customers surveyed indicated that ease of use was the most important criterion. As a result the manufacturer undertook development efforts to increase ease of use for their product. A set of substitute qualities was developed, including such requirements as on-screen programming, multifunction buttons on the remote control, and on-screen help facilities.

Following completion of development and beta testing, production of the new VCR began. Sales were not as brisk as had been predicted, and a second market survey was conducted. The marketing research department determined that a large portion of the market share projected for the new VCR was instead taken by a competitor's product that had targeted the same market.

Rather than providing an extensive set of user interface capabilities, the competitor's product offered a very rudimentary user interface that only required a user to enter a number from the weekly "TV-guide" book. The product then received time, date, and channel information via residential cable. The competitor had translated the request for ease of use into a different set of substitute qualities more closely tuned to the true desires of customers, had secured agreements with "TV-guide"-type book publishers and cable providers, and was positioned to dominate the market.

---

## Process Based Strategy

*Juran has proposed a 10-step process analysis approach to quality planning* [1]. Figure 11.2 depicts the 10-step sequence; *it begins with the identification of customers and ends with the transfer to operations.* Each of the 10 blocks, or processes, is depicted in a process input-output format; the output from each process forms the input to the subsequent process.

Use of a spreadsheet layout to organize the plan is stressed by Juran. Figure 11.3 depicts a spreadsheet layout. We can see the developmental flow from left to right in the layout. We begin with the customer needs expressed in the customer's nontechnical language (e.g., comfortable, fits right, easy to put on) which we translate into our technical language so that we can begin to deal with intraorganizational product definition and design. For each tertiary demand we identify an appropriate unit of measure and a related sensor. Then, we define product features (in a technical sense) that relate to and are capable of satisfying each of the low-level, or tertiary, customer demands. The critical point is to systematically define the product (a good or a service) so as to effectively and efficiently yield customer satisfaction.

Once the product features are identified, this method moves on to plan and optimize the production process (Figure 11.2). The production process related plan usually requires additional spreadsheets (analyses) dealing with process definition and flow. No general process planning format is provided by Juran, as products vary considerably in structure, flow, and complexity.

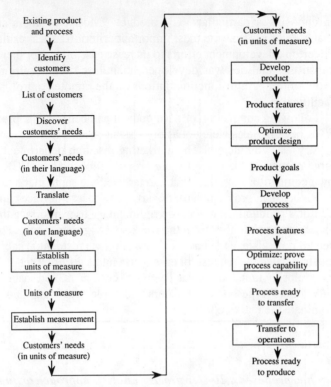

**FIGURE 11.2**   Juran's quality planning steps.  Reprinted, with permission of The Free Press, a Division of Simon & Schuster, from J. M. Juran, *Juran on Planning for Quality*, New York, NY: Free Press, page 15, 1988. Copyright 1988 by Juran Institute, Inc.

## Quality Function Deployment (QFD) Based Strategy

*The Japanese use what they call quality function deployment (QFD) to plan the quality related aspects of products* (Akao [2]).  QFD is more detailed than, but similar to, Juran's quality planning concepts.  Figure 11.4 depicts an eight-element description of a Japanese QFD layout (the triangular protrusions are used for relationships).  Because of the QFD tables' shape, the analyses are sometimes termed "houses of quality."  We will describe methods to build these "houses" in a subsequent section.

In Figure 11.4, elements 1 and 2 represent the same general concept that Juran captures in his quality plan layout, depicted in Figure 11.3.  *The QFDs stress quality targets* (element 6) *and competitive analyses and selling features* (elements 3, 4, and 7), all shown in Figure 11.4.  *They also emphasize product-process alternatives and bottleneck identification and engineering* (elements 5 and 8, Figure 11.4).  "Bottlenecks" are defined as the technical-, cost-, or reliability and safety-related limitations that must be successfully addressed in order to execute the quality plan.  Early identification of these limitations allows timely resolution of potential problems, so that total product-process development cycle time is minimized.  In other words, bottleneck engineering is a form of proactive quality work.

**PRODUCT: HIKING BOOTS**

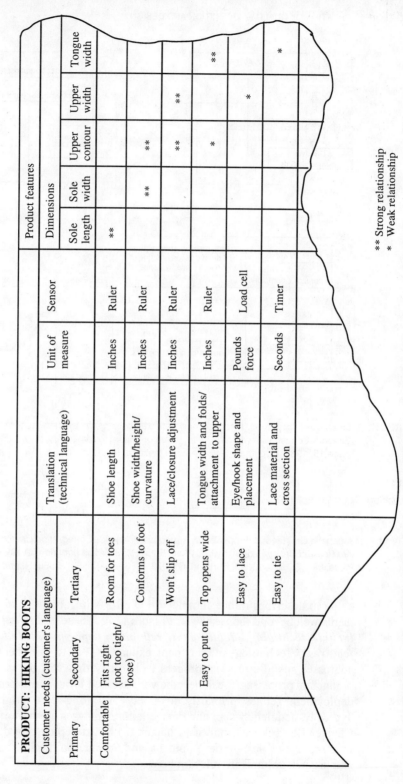

| Customer needs (customer's language) | | | Translation (technical language) | Unit of measure | Sensor | Product features | | | | |
| --- | --- | --- | --- | --- | --- | --- | --- | --- | --- | --- |
| | | | | | | Dimensions | | | | |
| Primary | Secondary | Tertiary | | | | Sole length | Sole width | Upper contour | Upper width | Tongue width |
| Comfortable | Fits right (not too tight/loose) | Room for toes | Shoe length | Inches | Ruler | ** | | | | |
| | | Conforms to foot | Shoe width/height/curvature | Inches | Ruler | | ** | ** | | |
| | | Won't slip off | Lace/closure adjustment | Inches | Ruler | | | ** | ** | ** |
| | Easy to put on | Top opens wide | Tongue width and folds/attachment to upper | Inches | Ruler | | | * | * | |
| | | Easy to lace | Eye/hook shape and placement | Pounds force | Load cell | | | | | |
| | | Easy to tie | Lace material and cross section | Seconds | Timer | | | | | * |

** Strong relationship
 * Weak relationship

**FIGURE 11.3**  Quality plan layout illustration.  Adapted, with permission of The Free Press, a Division of Simon & Schuster, from J. M. Juran, *Juran on Planning for Quality*, New York, NY: Free Press, page 112, 1988.  Copyright 1988 by Juran Institute, Inc.

221

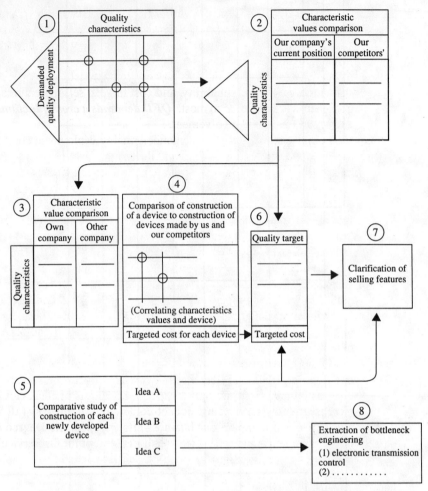

**FIGURE 11.4**    Japanese QFD layout.  From *Quality Function Deployment: Integrating Customer Requirements into Product Design*, edited by Yoji Akao. English translation copyright © 1990 by Productivity Press, Inc., PO Box 13390, Portland, OR 97213-0390, (800) 394-6868. Reprinted by permission.

Akao describes rather complex QFD efforts in Japan [3], many involving elements well beyond those shown in Figure 11.4.  These ***additional elements to QFD include technology deployment, reliability deployment, and cost deployment features.***  Technology deployment expands QFD to the mechanism level and addresses specific product-process technology that will be used in the product-production processes; it can become very detailed for complex products.  Reliability deployment relates primarily to safety and long-term product performance.  Typically, detailed system and device failure analyses are performed (e.g., FMEA, Chapter 12, fault tree analysis, Chapter 13).  Cost deployment deals with the business aspects of design, development, and production; target costs, market prices, and product volume are relevant here.

A detailed QFD resembles a comprehensive technical business plan in many ways.  However, QFD is product-production process specific, rather than company-wide; each QFD layout is generally unique to a product or product line.  Quality function deployment is an emerging science; QFD formats vary, but the components of customer needs; product and process characteristics; competition analysis; performance, safety and reliability, and cost targets; selling features; and engineering bottlenecks are common to most.  *QFD demands a cross-functional team approach* and broad functional involvement.

A simplified, single-level, demanded quality deployment chart for a laundry service is shown in Figure 11.5.  This illustration develops customer demands (in the customer's language) in the left vertical column and then identifies quality characteristics (in technical language) in the horizontal row across the top.  The Degree of Importance to Customer column provides insight as to criticality, to the customer, of the demands.  Relationship symbols are used in the body of the matrix to relate the customer demands and service quality characteristics.  The "hats," or "roofs," identify internal quality demand or quality characteristic relationships.  We will usually encounter a trade-off situation when there are strong negative relationships.  For example, of the quality characteristics (at the top), "Buttons" on the clothing and "Cleaning cycle time" have a negative relationship (i.e., it takes longer to clean and press an item with buttons).  Likewise, of the customer demands (the left side), "Fixing clothes" (e.g., sewing on a button) and "Inexpensive" have a negative relationship.

A number of qualitative symbols are used in the chart.  We can also add quantitative detail by defining relative weights, related to our intended business strategy to win customers.  Here, we will compete on the basis of clean clothes, good-looking clothes, friendly service, and pocket-content return integrity (all 7s on the far right side).  We can compare our business strategy with the Degree of Importance to Customer column and see we are consistent (except for fast service).  We can use our business strategy relative weights along with assigned (by us) relationship weights (5, 3, and 1, in the matrix inset) and calculate a score for each cell:

$$\text{Body cell score}_{ij} = \text{sales point relative weight}_i \times \text{relationship weight}_{ij} \quad (11.1)$$

Here, the $i$ index refers to quality demands (rows) and the $j$ index refers to quality characteristics (columns).  Then, we can sum down each column to obtain its quality characteristic priority score.

$$\text{Quality characteristic priority score}_j = \sum_{\text{all } i} \text{body cell score}_{ij} \quad (11.2)$$

In Figure 11.5 we have arbitrarily marked, with an asterisk "*," quality characteristic scores of 60 points or more, which we consider to be of high priority, given our chosen business strategy and the identified customer quality demands.

Directly below the priorities, a row has been added to address our targets.  These targets pertain to the technical product related characteristics in each column.  Each target helps us to identify the level at which we intend to "build" performance and customer service into our product and production processes.  As we add more detail

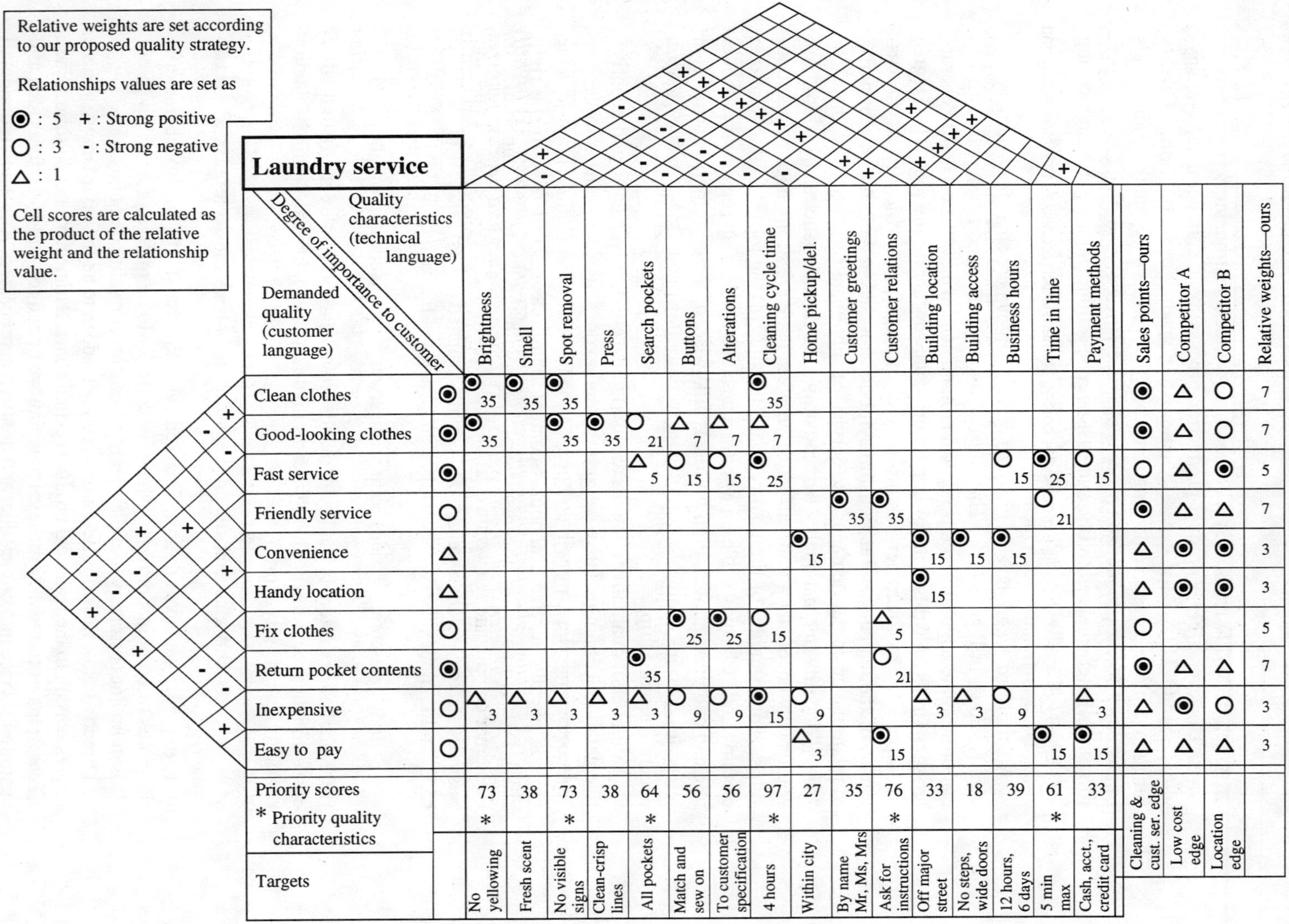

## Laundry service — Demanded quality deployment chart

**Legend (quality strategy notes):**

Relative weights are set according to our proposed quality strategy.

Relationships values are set as
- ◉ : 5
- O : 3
- △ : 1

Roof correlations:
- + : Strong positive
- - : Strong negative

Cell scores are calculated as the product of the relative weight and the relationship value.

The diagram includes a correlation "roof" with scattered + (strong positive) and - (strong negative) marks among the quality characteristics.

### Relationship and cell-score matrix

| Demanded quality (customer language) | Degree of importance | Brightness | Smell | Spot removal | Press | Search pockets | Buttons | Alterations | Cleaning cycle time | Home pickup/del. | Customer greetings | Customer relations | Building location | Building access | Business hours | Time in line | Payment methods |
|---|---|---|---|---|---|---|---|---|---|---|---|---|---|---|---|---|---|
| Clean clothes | ◉ | ◉ 35 | ◉ 35 | ◉ 35 | | | | | ◉ 35 | | | | | | | | |
| Good-looking clothes | ◉ | ◉ 35 | | ◉ 35 | ◉ 35 | | △ 7 | △ 7 | △ 7 | | | | | | | | |
| Fast service | ◉ | | | | | △ 5 | O 15 | O 15 | ◉ 25 | | | | | | O 15 | ◉ 25 | |
| Friendly service | O | | | | | O 21 | | | | | ◉ 35 | ◉ 35 | | | | O 21 | |
| Convenience | △ | | | | | | | | ◉ 15 | ◉ 15 | | | ◉ 15 | ◉ 15 | ◉ 15 | ◉ 15 | △ 3 |
| Handy location | △ | | | | | | | | | | | | ◉ 15 | | | | |
| Fix clothes | O | | | | | | ◉ 25 | ◉ 25 | | | | △ 5 | | | | | |
| Return pocket contents | ◉ | | | | | ◉ 35 | | | | | | O 21 | | | | | |
| Inexpensive | O | △ 3 | △ 3 | △ 3 | △ 3 | △ 3 | O 9 | O 9 | ◉ 15 | O 9 | | ◉ 15 | △ 3 | △ 3 | O 9 | | ◉ 15 |
| Easy to pay | O | | | | | | | | | △ 3 | | | | | | | ◉ 15 |
| **Priority scores** | | 73 | 38 | 73 | 38 | 64 | 56 | 56 | 97 | 27 | 35 | 76 | 33 | 18 | 39 | 61 | 33 |
| **\* Priority quality characteristics** | | * | * | * | | * | | | * | | | * | | | | * | |
| **Targets** | | No yellowing | Fresh scent | No visible signs | Clean-crisp lines | All pockets | Match and sew on | To customer specification | 4 hours | Within city | By name Mr, Ms, Mrs | Ask for instructions | Off major street | No steps, wide doors | 12 hours, 6 days | 5 min max | Cash, acct, credit card |

### Planning matrix

| Demanded quality | Sales points—ours | Competitor A | Competitor B | Relative weights—ours |
|---|---|---|---|---|
| Clean clothes | ◉ | △ | O | 7 |
| Good-looking clothes | ◉ | △ | O | 7 |
| Fast service | O | △ | ◉ | 5 |
| Friendly service | ◉ | △ | △ | 7 |
| Convenience | △ | ◉ | ◉ | 3 |
| Handy location | △ | ◉ | ◉ | 3 |
| Fix clothes | O | | | 5 |
| Return pocket contents | ◉ | △ | △ | 7 |
| Inexpensive | △ | ◉ | O | 3 |
| Easy to pay | △ | △ | △ | 3 |

Competitive edges:
- Sales points—ours: Cleaning & cust. ser. edge
- Competitor A: Low cost edge
- Competitor B: Location edge

**FIGURE 11.5** Demanded quality deployment chart illustration.

224

**TABLE 11.1**     STRATEGIC QUALITY PLAN DEVELOPMENT ELEMENTS

<u>Identification or discovery</u>—The element of identification or discovery is one of perception; it introduces new information and perspectives into the quality plan.

<u>Definition</u>—The definition element is one of refinement and clarification. Its purpose is to transform the "fuzzy" nature of the newly discovered information or perspectives into a clear or refined form which can be communicated in a meaningful manner.

<u>Classification</u>—Classification is a grouping or sorting element which organizes the information and perspectives.

<u>Relationship</u>—The relationship element provides inter- and intraclassification comparisons and contrasts. It establishes interrelationships which must ultimately be recognized in order to structure customer demands, product and process features, sales points, and competitive advantages; it also identifies conflicting or mutually exclusive relationships.

<u>Ranking and priorities</u>—The ranking and priorities element assigns relative importance to relationships in order to support decisions. Priorities are strength indicators which may be expressed qualitatively (e.g., low, medium, high) or quantitatively (e.g., on a number scale).

<u>Decision</u>—The decision element is one of enlightened choice. Making choices is essential in product and process planning for quality.

to our QFD (e.g., through multiple layers of customer demands and technical quality characteristics), we will refine and further define and clarify our targets.

## 11.3  QUALITY FUNCTION PLANNING AND DEPLOYMENT

*Quality function planning and deployment is a three-phase process:* **(1)** *Identify customer characteristics and demands,* **(2)** *define product and process features, and* **(3)** *develop a product and process road map.* This process is depicted in Figure 11.1. *Each phase in quality function planning and deployment requires a hierarchy of open-ended questions and answers.* Six generic elements are involved in each phase (see Table 11.1).

*Quality function planning and deployment leads to the systematic formulation of a quality plan.* This quality function road map, or master plan, should not be viewed as static; rather, flexibility as the "trip" progresses is necessary. Hence, the road map may need to be updated as the plan is executed over the course of weeks, months, or years, because of changes in customer characteristics, competitor activities, technology, and so on. Nevertheless, the road map serves as a reference or basis from which to detour.

### Customer Characteristics and Demands

*Customers for a given product or product line must be identified. Anyone who is impacted in any way by the product or production process is a customer.* Table 11.2 provides an example of an external customer list for a product line of small childproof containers.

**TABLE 11.2**   EXPANDED EXTERNAL CUSTOMER LIST FOR SMALL CHILDPROOF CONTAINERS

| Customers | Functional relationship | Demands |
|---|---|---|
| *Medication users* | *Use and disperse medication* | *Demand priority* |
| Ultimate user | Takes or administers medication | Safe |
| Nurse | Administers medication | Easy to use |
| Doctor | Administers and/or prescribes medication | Long-lived |
| Purchaser (individual or institution) | Makes buying decision, purchases | Inexpensive |
| *Non-medical users* | *Use and dispense contents* | *Demand priority* |
| Ultimate user | Gets contents out, puts contents in | Easy to use |
| Purchaser (individual or company) | Makes buying decision, purchases | Inexpensive |
|  |  | Safe |
|  |  | Long-lived |
| *Manufacturers and distributors* | *Package and distribute packages* | *Demand priority* |
| Drug store | Unpack, repack, label, display | Safe |
| Discount store | Unpack, repack, label, display | Right size |
| Grocery store | Unpack, repack, label, display | Inexpensive |
| Distributor | Pack, unpack, label, distribute | Attractive |
| Manufacturer | Fill, label, pack | Long-lived |
| *Unauthorized customers* | *Keep out of packages* | *Demand priority* |
| Small children | Play, curiosity | Safe |
| Very elderly | Senility, dose mistakes |  |
| Tamperers | Criminal intent |  |

**Customer Identification**   *Customer identity is established by following the product and processes from cradle to grave* (and sometimes beyond, in the case of product disposal) and determining the parties affected.  Next, the parties' role(s) relative to the product and processes must be identified and defined.  The customers' relationships to the product or production process are important and necessary to produce a useful customer list and a customer criticality ranking.  It is important to remember that all customers are important, but some are more critical to organization success than others.  Consideration of criticality is an application of the Pareto principle to the customer list.

The customer list shown in Table 11.2 stratifies the container customers into four general groups: (1) medical users, (2) nonmedical users, (3) manufacturers and distributors, and (4) a special group who should be denied access to the container contents.  Within each group, the external customers are further identified and grouped as to their position in the production-consumer chain.

**Customer Needs, Expectations, and Demands**   Our discussion of the experience of quality in Chapters 4, 5, and 6 is useful here in providing a basis for describing customer demands.  External customer demands represent the needs and expectation of customers relative to product performance and cost, as well as the entire experience of product discovery, acquisition, use, support, and disposal/recycle.  Internal customer demands represent needs and expectations relative to the processes used in product design, development, production, delivery, service, and so forth.  The internal customer's experience may be virtually unrelated to the physical product itself and instead closely related to both direct production processes and existing support service processes throughout the company and its marketing and service affiliates. For example, the people who build airplanes may not fly them or even fly in them, but they have a vested career interest in building them.

*Customer needs and expectations are dynamic and many times difficult if not impossible for the customer to verbalize.* When verbalization is possible, customer language may be difficult to interpret. When clear verbalization is not possible, customer needs and expectations must be discovered in other ways. Means such as past experience in the field, surveys, in-depth interviews, and so on, are useful in establishing needs and expectations. Cases where the buyer and the user are not the same person create special challenges. For example, when parents buy toys for children, the needs and expectations must be assessed for both.

*Fundamental knowledge of market research is helpful in discovering customer demands.* Table 11.3 summarizes three fundamental forms of market research (Kinnear and Taylor [4]). Market research studies generally take one of two forms: (1) cross-sectional or (2) longitudinal. A cross-sectional study design seeks to establish a snapshot in time of relevant market-related phenomena. These studies are usually carried out in a sample survey form of one type or another (e.g., interview, questionnaire). The longitudinal study design usually involves a fixed population sample monitored over time to measure changes in the population as they occur. Relative costs of these two forms of market research depend on the sample size and study duration.

In general, there are four sources of marketing data: (1) respondents, (2) analogous situations, (3) experimentation, and (4) secondary data. The most commonly used data collection method is that of questioning and observing respondents in one form or another. Lines of questioning seek to discover attitudes, perceptions, motivations, knowledge level, intended behavior, and so forth. Analogous situations provide data through case studies of similar situations and computer market-

**TABLE 11.3**    SUMMARY OF MARKET RESEARCH CATEGORIES

| Research category | Research focus |
| --- | --- |
| Exploratory research | Seeks to discover ideas and insights not previously recognized regarding problems and opportunities  (e.g., How could we better serve our customers?  What new features might our customers like?) |
| Conclusive research | Provides information to evaluate alternate courses of action |
| Descriptive research | Characterizes marketing phenomena and frequencies of occurrence (e.g., interview, interrogation, data collection and analysis) |
| Causal research | Studies cause-and-effect relationships regarding marketing variables (e.g., understanding why and how market variables are related) |
| Performance research | Monitors the critical performance measures of the marketing system (e.g., monitoring sales, price, demand trends, economic conditions) |

ing simulations.  Experimentation seeks to design controlled experiments regarding marketing variables to gain marketing causality insights.  Secondary data include available internal and external data not collected specifically for the matter at hand (unlike respondent data).  Secondary data sources include company records, data from previous studies, trade organization publications, and government publications.

*The customer demand list states needs and expectations in customer language (not scientific or engineering language).*  Typically, customer language is vague and lacks clear-cut detail.  For example, customers use words and phrases like "comfortable," "good-looking," "quick and courteous service," "easy to use," "feels right," "inexpensive," and so on.  The right-hand side of Table 11.2 lists basic, first-level customer demands for the childproof container.  At this point of development, the descriptions are very general, such as "safe," "inexpensive," "easy to open," and so forth.  More detail will be developed in the second phase.

Developing a list of customer demands is a challenging, but critical, task.  The customer demands list ultimately drives the quality strategy and planning process by defining the essence of customer satisfaction.  The latter, in turn, defines the ultimate measure of performance and determines the level of quality in the product.  Hence, the customer demands list ultimately serves as a set of master objectives, even though the objectives may be somewhat fuzzy.

## Product and Process Features

*The second phase defines product and process features.*  The input to phase II is the output from phase I (Figure 11.1).  Hence, the demanded quality list is further developed and organized by level of detail and expressed in clearly defined terms.  The customer's language is used to define what constitutes performance to the customer.  Then, *each customer demand is translated into a corresponding technical quality characteristic (expressed in technical language) to establish technical targets for product and process features.*  The quality characteristics establish how performance is to be accomplished.  *Relationships between customer demands and quality characteristics are developed and prioritized.*  Competition, from other products and companies, in satisfying customer demands is also examined in this phase.  The point is to develop product features and establish selling points as well as to begin consideration of technical feasibility and production processes.  Since this phase involves identifying and displaying hierarchical relationships and priorities, tabular tools—such as quad-ruled paper and spreadsheets—can be of great value.  Figure 11.6 shows the development of the product features corresponding to the small childproof containers discussed earlier.  Then, as an example of a service product, product and process features of a university transcript copy service are developed in Figure 11.7.

**Customer Quality Definition**    As stated earlier, a firm grasp of customer needs and expectations is absolutely critical.  The customer demand list from the previous phase is refined and formalized to form a category-by-category, level-by-level hierarchy of product and process related customer needs and expectations.

Typically, each category contains from one to four levels of detail, with two or three levels being most common.  Again, *each entry in the customer demand list is stated in language that customers understand.*  Hence, once developed, the list can be reviewed by customers and potential customers to establish its validity and/or update the customer quality definition hierarchy.  The customer demands are usually laid out in vertical columns on the far-left side of the chart, as shown in Figures 11.5, 11.6, and 11.7.

**Technical Quality Definition and Targets**    *After the hierarchy of customer demands is clearly stated and verified, it must be translated into technical language.* This translation is essential to develop the bridge between true quality characteristics and substitute quality characteristics, as defined by Ishikawa [5].  *The technical language translation makes it possible to define, describe, and finally specify the product so that it can be developed, produced, delivered, sold, serviced, used, and disposed/recycled properly.*  Specification at this level results in establishing the product's critical features and characteristic target values (for further development in the third phase).  These features are laid out in levels across the top of the chart; the customer demands are laid out vertically.  Usually, two or three levels are sufficient, with the lowest level typically setting up a critical quality characteristic target value. The target values are typically rather crude in the chart; however, they add the detail necessary to bring the product definition from the abstraction of words to the concrete reality of product and process engineering.

The point of laying the customer demands hierarchy against the quality characteristic hierarchy and competing products is to establish and qualitatively or quantitatively measure relationships and priorities between the two hierarchies.  The example in Figure 11.5 is laid out in both qualitative and quantitative terms.  The examples in Figures 11.6 and 11.7 use only qualitative measures and symbols to identify relationships.  Akao describes more advanced treatments using quantitative weighting factors [6].

Each demanded quality row should have a strong relationship with at least one quality characteristic column.  Otherwise, the technical definition of the product or process is not complete.  This type of QFD matrix, with the customer demands on the vertical axis and a counterpart hierarchy of technical quality characteristics and competitive products and companies on the horizontal axis, is termed the Demanded Quality Deployment Chart (DQDC) (e.g., the DQDC constitutes a primary element in a detailed or complete QFD).

Development of the DQDC requires a great deal of thought, discussion, and analysis.  Typically, a refinement process is necessary to finalize the DQDC.  Each DQDC is unique with regard to both product and organizational philosophy.  Therefore, layouts and level-of-detail formats, as well as priority schemes, may vary.  The DQDCs shown in Figures 11.5, 11.6, and 11.7 pertain to very simple products.  It is obvious that building DQDCs for complicated products—such as automobiles, airplanes, and so forth—requires extensive work.  In addition, totally new products are treated somewhat differently than upgraded models of existing products.  New

## SMALL-VOLUME CHILDPROOF CONTAINER QFD

QUALITY FUNCTION DEPLOYMENT CHART — QUALITY CHARACTERISTICS DEPLOYMENT / DEMANDED QUALITY DEPLOYMENT — Degree of importance to our customers

First level: Shape—Cylindrical uniform—Dimensions

| First level | Second level | Degree of importance to our customers | Container Small body | | | | Container Medium body | | | | Container Large body | | | | Cap | | | | Gasket | |
|---|---|---|---|---|---|---|---|---|---|---|---|---|---|---|---|---|---|---|---|---|
| Third Level | | | Outside diameter | Wall thickness | Height | Volume | Outside diameter | Wall thickness | Height | Volume | Outside diameter | Wall thickness | Height | Volume | Outside diameter | Inside diameter | Cap thickness | Height | Outside diameter | Gasket thickness |
| Safe | Protect contents | ◎ | | △ | | | | △ | | | | △ | | | | | △ | | | ○ |
| | Childproof | ◎ | | △ | | | | △ | | | | △ | | | | | △ | | | |
| | Tamperproof | ◎ | | | | | | | | | | | | | | | | | | |
| | No contamination from manufacturing processes | ◎ | | | | | | | | | | | | | | | | | | |
| | Contents clearly identified | ◎ | △ | | △ | | △ | | △ | | △ | | △ | | △ | | | | | |
| Easy to use | Easy to grasp | ○ | ○ | | | | ○ | | | | ○ | | | | ○ | | | △ | | |
| | Easy to open and close | ◎ | ○ | | | | ○ | | | | ○ | | | | ◎ | ◎ | | ◎ | | △ |
| | Easy to dispense | ○ | | | △ | | | | △ | | | | △ | | | | | | | |
| | Instructions and logos clearly printed | ◎ | △ | | △ | | △ | | △ | | △ | | △ | | △ | | | | | |
| | Able to see how much of contents left in container | ○ | | | | | | | | | | | | | | | | | | |
| Look attractive | Attractive shape | △ | ○ | | ○ | | ○ | | ○ | | ○ | | ○ | | ○ | | | ○ | | |
| | Color choice | ◎ | | | | | | | | | | | | | | | | | | |
| | Bold store and chain logos | ◎ | ○ | | ○ | | ○ | | ○ | | ○ | | ○ | | ○ | | | | | |
| Right size | Hold right amount of contents | ○ | | | | ◎ | | | | ◎ | | | | ◎ | | | | | | |
| | Fit in hand, purse, pocket | ○ | ○ | | ○ | | ○ | | ○ | | ○ | | ○ | | | | | | | |
| | Won't slip out of hand | ◎ | ○ | | | | ○ | | | | ○ | | | | ○ | | | | | |
| | Will stack on display and cabinet shelves | ○ | ○ | | | | ○ | | | | ○ | | | | ○ | | | | | |
| | Will stack in packing box | ◎ | ○ | | | | ○ | | | | ○ | | | | ○ | | | | | |
| Long life | Will reseal over and over | ◎ | | | | | | | | | | | | | | △ | | | | |
| | Won't break | ◎ | | ◎ | | | | ◎ | | | | ◎ | | | | | ◎ | | | |
| | Won't leak | ○ | | | | | | | | | | | | | | ○ | | | | |
| Inexpensive | Low container cost | △ | | △ | | | | △ | | | | △ | | | | | △ | △ | △ | △ |
| | Easy and quick to fill | ◎ | ○ | ○ | | | ○ | ○ | | | ○ | ○ | | | ○ | | △ | | | |
| | Easy and quick to label | ◎ | ○ | ○ | | | ○ | ○ | | | ○ | ○ | | | ○ | | △ | | | |
| | Easy and quick to serialize | ◎ | | | | | | | | | | | | | ○ | | △ | | | |
| | Easy and quick to open and close in fill lines | ◎ | ○ | | ○ | | ○ | | ○ | | ○ | | ○ | | ○ | ○ | △ | ◎ | | |
| **Target values—our company** | | | 0.85 inch | 0.10 inch | 1.85 inch | 0.61 $in^3$ | 1.10 inch | 0.10 inch | 2.10 inch | 1.34 $in^3$ | 1.35 inch | 0.10 inch | 2.60 inch | 2.70 $in^3$ | 0.45 over body | 0.05 over | 0.10 inch | 0.40 inch | 0.05 under lid | 0.07 inch |
| **Target values—competitor A** | | | | | | | 0.90 | 0.05 | 1.50 | 0.75 | | | | | | | | | | |
| **Target values—competitor B** | | | | | | | | | | | 1.25 | 0.05 | 2.00 | 2.10 | | | | | | |

**FIGURE 11.6** Childproof container demanded quality deployment chart illustration.

| Color | | | | | Package seals | | Opening forces | | | Surface finishes | | | | | Closing and seal mechanism | | | | | | | |
|---|---|---|---|---|---|---|---|---|---|---|---|---|---|---|---|---|---|---|---|---|---|---|
| Other | | Colored opaque | | | | | Pkg. seal | Lid seal | | Body | | Lid | | Gasket | Closing lugs | | Seal mechanics | | | | | |
| White—not opaque | Clear | Red | Orange | Blue | Inter lid seal | Outer package seal | Pull tab | Vertical force | Torque | Inside—outside sides | Inside—outside bottom | Top surface | Outside grip edge | Top and bottom surf. | Lid lug | Body lug | Gasket spring force | Gasket lid snap in force | Cleaning, handling, storage, packing | **Sales points —our product** | **Sales points—competitor A** | **Sales points —competitor B** |
| | | | | | ◎ | | | | | | | | | ◎ | ◎ | ◎ | ◎ | | | ◎ | ○ | ○ |
| △ | | | | | | | | ◎ | ◎ | | | | | | | | | | | ◎ | ○ | ○ |
| | | | | | ◎ | ◎ | | | | | | | | | | | | | | ◎ | ○ | ○ |
| | | | | | | | | | | | | | | | | | | | ◎ | ◎ | ○ | ○ |
| | ◎ | | | | | | | | | ◎ | | ◎ | | | | | | | ◎ | ◎ | ◎ | ◎ |
| | | | | | | | | | | | | | ○ | | | | | | | ○ | ○ | ○ |
| | | | | | | | ◎ | ◎ | ◎ | ◎ | | | ◎ | ○ | ○ | ○ | ◎ | | | ◎ | ○ | ○ |
| | ○ | ○ | ○ | ○ | ○ | | | | | | | | | | | | | | | ○ | ○ | ○ |
| | | | | | | | | | | ○ | | ○ | | | | | | | ◎ | ◎ | ○ | ○ |
| | ◎ | ◎ | ◎ | ◎ | | | | | | | | | | | | | | | | ◎ | △ | △ |
| | | | | | | | | | | ◎ | | ◎ | ◎ | | | | | | | ○ | ○ | ○ |
| ◎ | ◎ | ◎ | ◎ | ◎ | | | | | | | | | | | | | | | | ◎ | △ | △ |
| ◎ | ◎ | ◎ | ◎ | ◎ | | | | | | | | | | | | | | | | ◎ | ○ | ○ |
| | | | | | | | | | | | | | | | | | | | | ○ | ○ | ○ |
| | | | | | | | | | | | | | | | | | | | | ○ | ○ | ○ |
| | | | | | | | | | | △ | | | ◎ | | | | | | | ○ | ○ | △ |
| | | | | | | | | | | | ◎ | △ | | | | | | | | ○ | ○ | △ |
| | | | | | | | | | | | △ | | | | | | | | | ○ | ○ | △ |
| | | | | | ◎ | | | | | | | | | ◎ | ◎ | ◎ | | | | ◎ | ○ | ○ |
| | | | | | | | | | | | | | | | ◎ | ◎ | | | | ◎ | △ | △ |
| | | | | | ◎ | | | | | | | | | ◎ | ◎ | ◎ | ◎ | △ | | ◎ | ○ | ○ |
| △ | △ | △ | △ | | | ○ | ○ | | | | | | | | ◎ | ◎ | △ | | ◎ | △ | ○ | ○ |
| | | | | | | | | | | | | | | | | | | | ○ | ◎ | ○ | ○ |
| | | | | | | | | ○ | ○ | ○ | | ○ | ◎ | | △ | △ | | | | ◎ | ○ | ○ |
| | | | | | | | | | | ○ | | ○ | | | | | | | ◎ | ◎ | ○ | ○ |
| | | | | | ◎ | ○ | | ○ | ○ | △ | | | △ | | △ | △ | △ | | | ◎ | ○ | ○ |
| Lid | Body | Body | Body | Body | Air/water tight | Won't slip | 2 lb | 5 lb | 7 inch lb | Smooth | No burrs | Smooth | No burrs | Smooth | 0.10 inch thick | 0.10 inch thick | 5 lb | 1 lb | Wash, keep clean | | | |
| Lid | | | Body | | | | | Snap | | | | | | | | | | | | | | |
| Lid | | | Body | | | | | Snap | | | | | | | | | | | | | | |

**Symbols**

◎ Critical relationship

○ Important relationship

△ Limited relationship

**QUALITY FUNCTION DEPLOYMENT CHART**

**QUALITY CHARACTERISTICS DEPLOYMENT**

**UNIVERSITY TRANSCRIPT COPY SERVICE**

First level: Transcript information
Second level: Content / Format
Third level (Content): Name, attendance date; Course numbers, terms; Course titles; Title abbreviations; Grades earned; Grade average; Transfer credits; Degree awarded; Legend; Error rate
Third level (Format): Type face; Line spacing; Margins; Term layout; Performance summary

**DEMANDED QUALITY DEPLOYMENT** — Degree of importance to our customers

| First level | Second level | Third level | Importance | Name, attendance date | Course numbers, terms | Course titles | Title abbreviations | Grades earned | Grade average | Transfer credits | Degree awarded | Legend | Error rate | Type face | Line spacing | Margins | Term layout | Performance summary |
|---|---|---|---|---|---|---|---|---|---|---|---|---|---|---|---|---|---|---|
| Easy to order | Minimal paper work for walk-ins and mail-ins | Easy to understand | O | | | | | | | | | | | | | | | |
| | | Easy to fill out | O | | | | | | | | | | | | | | | |
| | | Mail order | ◎ | | | | | | | | | | | | | | | |
| | Phone and FAX service | Phone order | ◎ | | | | | | | | | | | | | | | |
| | | FAX order | ◎ | | | | | | | | | | | | | | | |
| | Easy to pay | Cash on the spot | ◎ | | | | | | | | | | | | | | | |
| | | Credit card payment | ◎ | | | | | | | | | | | | | | | |
| | | Billing service for mail order | O | | | | | | | | | | | | | | | |
| Fast service | Fast office processing | | ◎ | | | | | | | | | | | | | | | |
| | Fast delivery | | ◎ | | | | | | | | | | | | | | | |
| Easy to use | Easy to read | | O | | | | | | | | | | | ◎ | ◎ | ◎ | ◎ | |
| | Easy to understand | | O | ◎ | O | ◎ | ◎ | ◎ | ◎ | O | ◎ | ◎ | | | | | ◎ | |
| Reliable | No mistakes on transcript | | ◎ | | | | | | | | | | ◎ | | | | | |
| | Correct transcript sent | | ◎ | ◎ | | | | | | | | | | | | | | |
| | Transcript sent to correct address | | ◎ | | | | | | | | | | | | | | | |
| | Student authorization for sending | | ◎ | △ | | | | | | | | | | | | | | |
| | Difficult to alter transcript | | O | | | | | | | | | | | O | | | | O |
| | Shipping date traceable | | O | | | | | | | | | | | | | | | |
| Right size to file and use | | | △ | | | | | | | | | | | △ | △ | △ | ◎ | O |
| Professional looking | | | ◎ | △ | △ | △ | | | | | O | | | O | △ | △ | O | O |
| Targets | Targets—our university | | | Full names | Numbers, terms | Full title | N/A | Yes | By term, overall | Numbers and titles | Bold | Top | Zero | Times | Six/inch | One inch | By blocks | Bottom |
| | Targets—university A | | | Full names | Numbers, terms | Abrev. title | Abrev. title | Yes | By term, overall | Numbers only | Yes | Bottom | Unknown | Courier | Eight/inch | One inch | No | No |
| Critical process sequence | Order placed by student | | | | | | | | | | | | | | | | | |
| | Payment made by student | | | | | | | | | | | | | | | | | |
| | Dispatch order | | | | | | | | | | | | | | | | | |
| | Locate transcript | | | | | | | | | | | | | | | | | |
| | Copy transcript | | | | | | | | | | | | | | | | | |
| | Stamp and apply seal to transcript | | | | | | | | | | | | | | | | | |
| | Verify and inspect transcript copy | | | | | | | | | | | | | | | | | |
| | Prepare transcript for delivery | | | | | | | | | | | | | | | | | |
| | Delivery | | | | | | | | | | | | | | | | | |

**FIGURE 11.7** University transcript copy service demanded quality deployment chart illustration.

Quality function deployment matrix.

| | Transcript copies | | | | | | | | | Order information | | | Payment | | | | Shipping options | | | | Shipping integrity | | | Competitive analysis | |
|---|---|---|---|---|---|---|---|---|---|---|---|---|---|---|---|---|---|---|---|---|---|---|---|---|---|
| | File architecture | File access | Copy resolution | Copy output device | Paper color, weight | Seal | Stamp | Inspection | Error rate | Name, address (applicant) | Applicant certification | Copy destination | Cash | Credit card | Billing | Payment error rate | First class mail | Express mail | Next day delivery | FAX | Shipping date trace | Order received to ship time | Address error rate | Selling points —our univ. | Selling points—university A |
| | | | | | | | | | | ◎ | ◎ | ◎ | | | | | | | | | | | | ◎ | △ |
| | | | | | | | | | | ○ | ○ | ○ | | | | | | | | | | | | ◎ | △ |
| | | | | | | | | | | ◎ | ◎ | ◎ | | | | | | | | | ○ | | | ◎ | ○ |
| | | | | | | | | | | ◎ | ◎ | ◎ | | | | | | | | | ○ | | | ◎ | |
| | | | | | | | | | | ◎ | ◎ | ◎ | | | | | | | | | ○ | | | ◎ | |
| | | | | | | | | | | | | | ◎ | | | ◎ | | | | | | | | ○ | |
| | | | | | | | | | | | | | | ◎ | | ◎ | | | | | | | | ◎ | |
| | | | | | | | | | | | | | | | ◎ | ◎ | | | | | | | | △ | |
| | ◎ | △ | | | | | | | | | | | | | | | | | | | ◎ | | | ◎ | △ |
| | ◎ | | | ○ | | △ | △ | ○ | | | | | | | | | ◎ | ◎ | ◎ | ◎ | ◎ | | | ◎ | △ |
| | | | | | | | | | | | | | | | | | | | | | | | | ◎ | ○ |
| | | | | | | | | | | | | | | | | | | | | | | | | ◎ | ○ |
| | | | | | | | | △ | ◎ | | | | | | | | | | | | | | | ◎ | △ |
| | △ | | | | | | | ◎ | ○ | ◎ | ○ | | | | | | | | | | | | ◎ | ◎ | ◎ |
| | | | | | | | | | | | | ◎ | | | | △ | | | | | ○ | ◎ | | ◎ | ◎ |
| | △ | | | | | | | | | ○ | ◎ | | | | | | | | | | | | | ◎ | ◎ |
| | | ◎ | △ | △ | ○ | ◎ | ◎ | | | | | | | | | | | | | | | | | ○ | △ |
| | | | | | | | | | | △ | | △ | | | | △ | | | | | ◎ | △ | △ | ○ | |
| | | | | | | | | | | | | | | | | | | | | | | | | ○ | △ |
| | | | ◎ | ○ | ◎ | ◎ | ◎ | | | | | | | | | | | | | | | | | ◎ | △ |

Value scale (lower / upper):

| Column | Lower | Upper |
|---|---|---|
| File architecture | Batch | On line |
| File access | Terminal | Terminal |
| Copy resolution | Unknown | Sharp |
| Copy output device | Unknown | Laser printer |
| Paper color, weight | White, light | Buff, heavy |
| Seal | None | Embossed |
| Stamp | Black | Red, blue |
| Inspection | Unknown | Clear, complete |
| Error rate | Unknown | Zero |
| Name, address (applicant) | Print | Print |
| Applicant certification | Signed | CC signed |
| Copy destination | Print | Print |
| Cash | Another office | In office |
| Credit card | No | Yes |
| Billing | No | Option |
| Payment error rate | Unknown | Zero |
| First class mail | Yes | Yes |
| Express mail | No | Yes |
| Next day delivery | No | Yes |
| FAX | No | Unofficial copy |
| Shipping date trace | No | Yes |
| Order received to ship time | Three days | Less than one day |
| Address error rate | Unknown | Zero |

Lower matrix:

| | File architecture | File access | Copy resolution | Copy output device | Paper color, weight | Seal | Stamp | Inspection | Error rate | Name, address (applicant) | Applicant certification | Copy destination | Cash | Credit card | Billing | Payment error rate | First class mail | Express mail | Next day delivery | FAX | Shipping date trace | Order received to ship time | Address error rate |
|---|---|---|---|---|---|---|---|---|---|---|---|---|---|---|---|---|---|---|---|---|---|---|---|
| | | | | | | | | | | ◎ | ◎ | ◎ | | | | | | | | | | | |
| | | | | | | | | | | | | | ◎ | ◎ | ◎ | ◎ | | | | | | | |
| | △ | ○ | | | | | | | | | | | | | | | | | | | | | |
| | ◎ | ◎ | | | | | | | | | | | | | | | | | | | | | |
| | | | ◎ | ○ | ◎ | | | | | | | | | | | | | | | | | | |
| | | | | | ◎ | ◎ | ◎ | | | | | | | | | | | | | | | | |
| | | | | | | | | ◎ | ◎ | | | | | | | | | | | | | | |
| | | | | | | | | | | ○ | ○ | ◎ | | | | | | | | | | | |
| | | | | | | | | | | | | | | | | | | | | | | | ◎ |
| | | | | | | | | | | | | | | | | | ◎ | ◎ | ◎ | ◎ | ◎ | | |

products typically will contain more bottlenecks and require more thought as to technical specification, technology selection, reliability, and cost.

**Competitive Analysis** The essence of the DQDC is to define, develop, and describe the critical product features. Competing products and organizations usually exist. Therefore, *fundamental analyses relative to competing products and organizations are essential in formulating a successful quality plan for any product.*

The analysis of a competitive product and organization requires consideration of essential product, marketing, sales, and product support features. The matter of competitive edge or advantage is critical. To gain a competitive edge, we must develop some significant advantage in product field performance, customer service, or both. The determination of our product's selling points, relative to our competition, in the early stages is very important. Otherwise, a great deal of resources may be spent pursuing a fundamentally flawed strategy. For example, if it is clear we cannot be competitive, we should abandon the product.

Simple competitive analyses appear in the far right-hand columns and across the lower rows of the examples in Figures 11.5, 11.6, and 11.7. This arrangement allows competition to be assessed relative to both demanded quality and quality characteristics. The Selling Point column forces early consideration of assuring a competitive edge in customer satisfaction.

*An often overlooked aspect of competitive analysis deals with the analysis of functional alternatives.* For example, a wire paper clip competes with plastic paper clips, staples, spring binders, and so on, in addition to other similar wire paper clips. Therefore, the consideration of competing products in terms of generic function and form are essential for a thorough competitive analysis. The seven principles of breakthrough thinking described in our benchmarking discussion (Table 8.4) apply to quality function planning and deployment.

### Product and Process Road Map

The DQDC (from phase II) defines and develops critical product features and their target values. *The third phase establishes the means to produce and deliver the product features to the customer.* Three major issues are involved in the third phase: (1) product and process technology; (2) product and process related safety, reliability, maintainability, and disposal/recycle; and (3) cost, sales, volume, and profit. *One of the primary functions of this phase is to identify technical, safety, reliability, maintainability, disposal, and cost bottlenecks or limitations so that they can be resolved in a timely fashion.*

Phase III includes the identification of bottlenecks relative to critical product and process issues extending beyond the demanded quality and quality characteristic relationships and priorities in the DQDC. *Product and processes are considered together in this phase, along with critical quality control strategies, critical processes and process points, and the means for assuring customer satisfaction.*

A quality road map is generated and critical bottlenecks in technology, safety and reliability, and cost are identified.

Phase III technology and techniques in QFD are less well developed than are their phase I and phase II counterparts (Akao [7]). Levels of detail, measures, targets, and presentation tend to vary widely in phase III, depending on product nature and product-process complexity, and the level of previous experience (and success). Figures 11.6, 11.8, 11.9, and 11.10 provide a reasonably complete, but simple, quality road map for the childproof container. Figures 11.7 and 11.11 provide a counterpart road map for the university transcript copy service.

**Product and Process Technology**    In the Japanese literature, product and process technology definition is termed "technology deployment" (Akao [8]). *Technology deployment is usually the most difficult and detailed level in QFD.* Product and process technology deployment extends the phase II product definition from one that establishes specific technologies, to one that enables product development and production. *Here, specific product mechanisms are identified. Then, materials and conversion processes are identified.* This identification is done hierarchically and extends downward to a reasonable degree of detail.

At the mechanism level, both product and process detail are required. Relationships and priorities at this level can become quite complicated even with fairly simple products. Nevertheless, the identification of specific bottlenecks expected in the actual product (and its processes) represents a significant contribution—although perhaps bad news. Thus, bottleneck engineering efforts can be planned and executed earlier, rather than later, in the product development cycle. Addressing clearly defined bottlenecks as early as possible is both cost and schedule effective. Plus, if a project "killer" bottleneck is uncovered, resources can be saved by "bailing out" as early as possible.

Product and process technology breakouts usually involve a number of relationship matrices. Typically, a creative series of two-dimensional charts is used. However, some three-dimensional matrices appear in the literature (see Akao [9]). A simple, highly summarized form of technology deployment is displayed in Figures 11.8 and 11.11; here, the products are simple and the technology exists. For more complicated products, which may require technologies not yet well developed, a much more detailed treatment is required.

**Safety, Reliability, Maintainability, and Disposal/Recycle**    The product and process features phase (phase II) develops the critical short-term performance measures and basic targets as well as some long-term performance characteristics of great importance to the customer. *Product-process safety, reliability, maintainability, and disposal/recycle are considered to be long-term performance issues and must be addressed through the definition of appropriate substitute (versus true) measures and targets.* Sometimes, these demands are automatically assumed to be part of the product or process and omitted from customer verbalizations. Nevertheless, external customers expect relatively safe and long-lived

## COST-VOLUME DEPLOYMENT

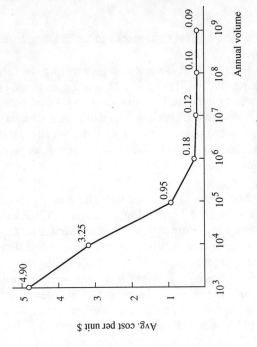

Avg. cost per unit $ (y-axis): 5, 4, 3, 2, 1

Annual volume (x-axis): $10^3$, $10^4$, $10^5$, $10^6$, $10^7$, $10^8$, $10^9$

Data points: 4.90, 3.25, 0.95, 0.18, 0.12, 0.10, 0.09

Projected cost versus volume curve

Critical bottlenecks:

Product volume

Volume for each size offering

Color volume for each size

## PRODUCT TECHNOLOGY DEPLOYMENT

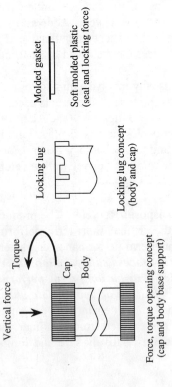

Vertical force

Torque

Cap

Body

Locking lug

Molded gasket

Soft molded plastic
(seal and locking force)

Locking lug concept
(body and cap)

Force, torque opening concept
(cap and body base support)

## PROCESS TECHNOLOGY DEPLOYMENT

| Component | Body | Cap | Gasket |
|---|---|---|---|
| Material | Pellets | Pellets | Sheet |
| Basic process | Injection molding | Injection molding | Stamping (cutting) |
| | Pressure injection | Pressure injection | Pressure forming |

Critical bottlenecks to resolve:

Material specifications (strength, durability, disposal/recycle)

Dimensional specifications (shrinkage, mating parts)

Tooling (cavity, pressure capacity, flashing, mold marks)

Safety (sharp edges out of mold)

Recipe (temperature, pressure, timing)

Ejection (speed, product damage exposure)

SPC points (process parameters, product dimensions)

**FIGURE 11.8** Small childproof container technology and cost deployment illustration.

| Component | Failure modes | Cause of failures | Possible effects | Criticality | Prob. of occurrence | Possible action to reduce failures |
|---|---|---|---|---|---|---|
| Gasket | Loose fit between cap and body | Loss of spring tension in molded plastic, cracking, and fracture in plastic | Leakage—ruins contents | Critical | $1\times10^{-5}$ | Material and dimensional specifications, gasket molding process control, life tests |
| | | | Cap falls off—loss of contents | Critical | $1\times10^{-5}$ | |
| | | | Too easy to open—child opens—child puts contents in mouth | Catastrophic | $1\times10^{-6}$ | |
| Cap | Poor lug contact to body | Lug tolerances, lug deformation, lug fracture | Cap snaps off body—ruins contents, loss of contents | Critical | $1\times10^{-5}$ | Material and dimensional specifications, gasket molding process control, life tests |
| Body | Poor lug contact to cap | Lug tolerances, lug deformation, lug fracture | Loose contents—loss of contents | Critical | $1\times10^{-5}$ | Same as above |
| | Crushed body | Child jumping on container or child hitting container with object | Loose contents—child puts contents in mouth | Catastrophic | $1\times10^{-6}$ | Material and thickness specifications, support rings top and bottom in body, crush tests |

Negligible—loss of function, with no significant effect on system performance.

Marginal—loss of function, with some system performance degradation.

Critical—loss of function, with complete system performance degradation.

Catastrophic—loss of function, with complete system performance degradation and/or possible injuries, fatalities, significant property loss.

**FIGURE 11.9**    Small childproof container reliability deployment failure mode and effects analysis.

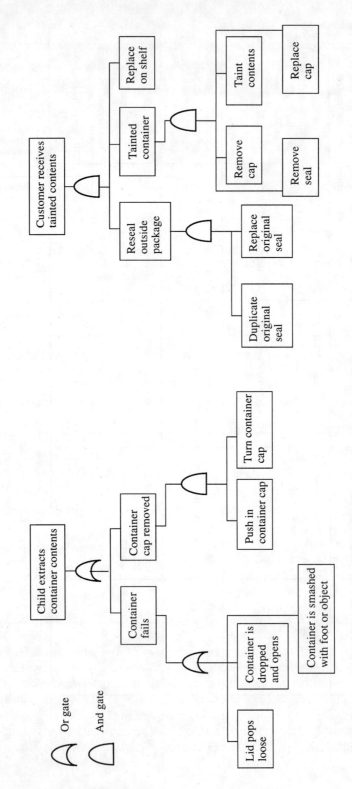

**FIGURE 11.10**  Small childproof container reliability deployment fault tree analysis.

**RELIABILITY DEPLOYMENT—FMEA**

| Component | Failure modes | Cause of failures | Possible effects | Criticality | Possible action to reduce failures |
|---|---|---|---|---|---|
| Request or addressing | Illegible input or error in input | Requester in hurry or misinformed | Copy sent to wrong address, wrong copy sent, no copy sent—loss or delay in applicant's quest for job, school, certification purposes | Catastrophic | Print information, care in checking form, care in addressing, verification |
| Copy | Poor reproduction | Printer malfunction | Difficult to read copy | Marginal | Inspect all copies for clarity of reproduction |
|  | Wrong copy | Retrieval error, duplicate names, request error | Wrong or no copy sent—loss or delay in applicant's quest for job, school, certification purposes | Catastrophic | Require student number, cross check to student number, print requests |

Negligible—loss of function, with no significant effect on system performance.
Marginal—loss of function, with some system performance degradation.
Critical—loss of function, with complete system performance degradation.
Catastrophic—loss of function, with complete system performance degradation and/or possible injuries, fatalities, significant property loss.

| TECHNOLOGY DEPLOYMENT—24 HOUR SERVICE | COST—VOLUME DEPLOYMENT* |
|---|---|

**TECHNOLOGY DEPLOYMENT—24 HOUR SERVICE**

On-line file architecture (protected)

Real-time terminal access (protected)

Laser printer (transcripts and envelopes)

Machine-aided stamp and seal operations

Direct mail, express mail, next day delivery and FAX access

**COST—VOLUME DEPLOYMENT***

| | |
|---|---|
| Labor | $0.50 |
| Paper and copy supplies | 0.45 |
| System time | 0.40 |
| Printer time | 0.15 |
| Stamp and seal time | 0.25 |
| Packaging | 0.20 |
| Subtotal | $1.95 |
| Delivery cost pass-through | |

*Cost per unit,
Volume anticipated—15,000 requests per year

**FIGURE 11.11**   Transcript copy service technology, reliability, and cost deployment illustration.

products, and internal customers expect safe and effective processes and work-places.

Safety in both products and processes is absolutely essential. Many products and processes are subject to safety standards, either mandatory or voluntary. Although meeting applicable standards is obviously necessary, it is not always sufficient in assuring the customer of a safe (low accident probability–low loss potential) product or process. Hence, additional safety measures and targets are necessary. Defining safety measures and setting safety targets is difficult, given the relatively low probability of occurrence–high loss nature of many safety characteristics. Mistake-proofing strategies are appropriate here. Safety analyses are executed using tools such as FMEA and fault tree analysis (FTA) described in Chapters 12 and 13, respectively. It is important to identify safety bottlenecks as early as possible for further study and resolution.

The reliability and maintainability analysis in quality function planning and deployment is approached on a function-by-function or a part-by-part basis, with long-term system level performance as a goal. Reliability and maintainability targets must be clearly defined regarding successful and unsuccessful performance. Appropriate measures (and targets) such as mean time to failure, mean time to repair, parts availability, and so on, are critical guides in product development efforts. Here, too, FMEA and FTA are useful. Some simple applications are depicted in Figures 11.9, 11.10, and 11.11.

**Cost, Volume, and Profit** *The cost, volume, and profit subphase addresses economic-business concerns.* Cost is a critical factor in external customer satisfaction. Profit or economic survival is a critical factor in internal customer satisfaction. Volume is not of primary concern to the customer. However, production costs and profits per unit are tied closely to volume of production as well as to technology (including material, machinery, and tooling). Hence, volume is included in this subphase. Cost measures are reasonably straightforward and are usually defined on a per-product unit basis, assuming some given volume of production. Hence, cost targets may be established working from the top down or from the bottom up. Cost and profit bottlenecks are very common.

Costs are a result, or effect, rather than a cause. Hence, cost is driven by (and controlled through) technical factors such as material specification, process specification, process yield, and labor skill level requirements. Concepts and strategies such as system, parameter, and tolerance design; robust performance; and mistake proofing are effective in delivering performance and benefits to customers at competitive costs. Two simple forms of cost-volume deployment are shown in Figures 11.8 and 11.11 for the childproof container and the transcript copy service, respectively.

## REVIEW AND DISCOVERY EXERCISES

### Review

**11.1** Why might two companies with nearly identical products (in a functional sense) possess different quality strategies?

**11.2** Explain why effective quality plans are cross-functional.

**11.3** We hear about quality professionals (specialists) both within organizations as well as practicing consultants.

    **a** What role does such a quality specialist play in quality planning?

    **b** What role(s) do other functional specialists play in quality planning?

**11.4** Compare and contrast Juran's 10-step quality planning procedure with the Japanese QFD. What is the same? What is different?

**11.5**  Explain the difference between customer language and technical language in quality planning.

**11.6**  Choose a product and provide an example of
  **a**  A technical bottleneck.
  **b**  A cost bottleneck.
  **c**  A reliability bottleneck.

**11.7**  Is bottleneck engineering proactive or reactive? Explain.

**11.8**  Explain the difference between a competitive analysis performed on a product basis and one performed on a functional basis (e.g., metal versus plastic paper clips versus the function of holding sheets of paper together). Provide a brief illustration.

**11.9**  With respect to a first-level customer demand for a shirt that fits right
  **a**  Develop second- and third-level customer demands.
  **b**  Develop a set of quality characteristics, in technical language, to correspond to your third-level customer demands.

**11.10**  Is it possible to use QFD with an over-the-wall organization (see Chapter 1)? Explain.

### Discovery

**11.11**  Choose a relatively simple product (a good or a service) and produce a quality plan using the Juran process method (e.g., Figures 11.2 and 11.3).

**11.12**  Choose a relatively simple product (a good or a service) and produce a quality plan using the Japanese QFD method (e.g., Figures 11.4 through 11.11).

**11.13**  How would you design a QFD team for a good? For a service? What major differences, if any, do you see?

**11.14**  For a product of your choice, demonstrate, by means of role playing, how quality cuts across the following functions in the process of generating a Juran quality plan or a Japanese QFD plan. Assign at least one person to play each role with regard to the five functions below:
  **a**  Marketing.
  **b**  Procurement and purchasing.
  **c**  Product engineering.
  **d**  Process engineering.
  **e**  Manufacturing.
  **f**  Sales and service.

## REFERENCES

**1**  J. M. Juran, *Juran on Planning for Quality*, New York: Free Press, 1988.

**2**  Y. Akao (ed.), *Quality Function Deployment*, Cambridge, MA: Productivity Press, 1990.

**3**  See reference 2.

**4**  T. C. Kinnear, and J. R. Taylor, *Marketing Research*, 3d ed., New York: McGraw-Hill, 1987.

**5**  K. Ishikawa, *What Is Total Quality Control? The Japanese Way*, Englewood Cliffs, NJ: Prentice-Hall, 1985.

**6**  See reference 2.

**7**  See reference 2.

**8**  See reference 2.

**9**  See reference 2.

# 12

# FAILURE MODE AND
# EFFECTS ANALYSIS

## 12.0 INQUIRY

1 When should product failure potential be addressed?
2 What role does engineering judgment play in failure prevention?
3 What differences exist between functional and hardware FMEA?
4 How is failure criticality assessed in proactive quality work?

## 12.1 INTRODUCTION

Products and processes that fail to perform for customers are obviously not going to be considered high quality. ***Proactive quality strategies focus on failure prevention—virtual elimination of the possibility of premature failure—and mistake-proofing, as far as possible.*** Many tools exist for the analysis of performance failure, consequence, and risk (Bell [1] and Dussault [2]). These tools are used in both proactive and reactive modes. One of the most versatile tools, ***failure mode and effects analysis, FMEA, focuses on failure modes, mechanisms, and effects*** [3]. In a proactive application, FMEA can be adapted to focus on failure prevention.

## Radar Performance—FMEA Case [RB]

During the early stages of development of a radar system, our organization conducted an FMEA to identify system failures and their subsequent effects on system operation. Since this system was intended for use in an aircraft, the analysis was extremely important because of flight safety issues.

The FMEA identified two major areas of high criticality, which resulted in substantial system design improvements:

1 The radar transmitter was a pressure-tight enclosure in which air was evacuated and an electrical insulating fluid was introduced. Heaters in the transmitter would vaporize this liquid, providing an insulating gas in the transmitter. This gas would prevent electrical arcing and flashover of high voltage components. The FMEA showed that one possible failure mode of the heater thermostat was a continuous short in the on position, which would cause the heaters to be energized continuously as well as excessive vapor pressure in the transmitter. This failure could eventually lead to vessel rupture. A design change was initiated to provide a pressure sensor for heater cutoff with a secondary pressure relief pop-off valve on the transmitter.

**2** To prevent accidental irradiation of ground personnel, the radar system design included a standby operating mode in which the system was powered up but the transmission of high-power radio frequency energy was inhibited. However, if the operator inadvertently placed the mode switch in the transmit position, the transmitter would be energized. To prevent this, a weight- on-wheels switch was added to the aircraft and was used to disable the transmitter. When the aircraft was on the deck, the switch would detect the aircraft weight on the landing gear and disable the transmitter. In flight, no weight is supported by the landing gear and the transmitter is enabled.

## 12.2 FAILURE MODE AND EFFECTS ANALYSIS

*We usually think of a failure mode as a physical description (result) of a failure, whereas a failure mechanism refers to the process that created the failure.* FMEA is primarily a qualitative tool which can support proactive quality strategies. Successful FMEA requires pertinent knowledge and insight as well as engineering judgment.

*FMEA seeks to identify possible failure modes and mechanisms, the effects or consequences that failure modes may have on performance, methods of detecting the identified failure modes, and possible means for prevention.* In many cases, a criticality assessment of the probability of encountering the failure mode is made. *The net results from effective FMEA work are product and process action plans for elimination, or at least mitigation, of the failure modes.*

Failure mode and effects analysis is absolutely essential in sound design practice, from product and process definition, starting in quality planning and quality function deployment (Akao [4]) and continuing through the development stages. FMEA encourages a systematic cradle-to-grave view of a product or process. To be effective, FMEA must be performed in an iterative fashion as product and process details emerge. FMEA is recognized as a fundamental tool in the reliability engineering field. It is also widely used in maintainability, safety, and survivability analyses. FMEA is emerging as a useful tool in manufacturing process analysis; its broad base makes it effective in both product and process performance improvement.

From a systems development perspective, FMEA encourages

**1** The systematic evaluation of a product or process at specified levels of system complexity (e.g., the system, subsystem, and component levels).
**2** The postulation of single point failures (where a single failure in a component can cause an entire system to fail), the identification of possible failure mechanisms, and the examination of the associated effects, likelihood of occurrence, and preventive measures.
**3** The systematic documentation of potential product or process nonperformance.

Hence, FMEA provides the information on which to base integrity related product and process development decisions.

## 12.3 FMEA DEVELOPMENT

*The classical FMEA is basically a tabulation of system functions or system equipment items, the failure modes for each, and the effects of the failures on the system (people, equipment, and surroundings).*  The failure mode is simply a description of the specific nature or result of the  failure.  The effect is the system response to the failure, or the consequences of that failure.  We will add a proactive element to the classical FMEA with the addition of a preventive measures column or dimension.

*FMEA identifies single failure modes that can cause, or contribute to the cause of, an accident or nonperformance.*  A failure mode effects and criticality analysis (FMECA) is the same as an FMEA, except that the relative ranking (criticality or severity) of each failure mode is included in the analysis.  FMEA is not useful for identifying combinations of failures that can lead to accidents, and generally does not examine operator errors.  The logic tree based analysis tools described in Chapter 13 are capable of dealing with event combinations and operator errors and responses.

### Base Format

*The base format for a proactive FMEA contains eight major columns or entry categories* (see Table 12.1):

1  *Functional or equipment identification*—Here, we must identify specific functions or items to be analyzed.  The identifier should be unique (e.g., a design configuration description or number, system number, part number).  Unique identification is extremely important in tying together existing analyses and providing organization for future analyses.

2  *Functional or equipment purpose*—We must describe the function or equipment in enough detail to communicate its essence relative to potential failure modes.

3  *Failure mode*—We must identify all potential failure modes for the associated function or equipment.  It is extremely important that this column be limited to explicit failure modes.

4  *Failure mechanism*—The failure mechanism, or failure process, for each failure mode must be identified.  Connections between failure mechanisms and the environment, the application, and the operating method are important in assuring product integrity.  For example, a standby system may fail on demand, resulting in failure effects that the standby system was intended to mitigate, or the standby system may fail before it is required.  In the latter case, there may be specific effects that are often overlooked or unforeseen.  For example, when analyzing a fire-protection system, we should consider both a failure of the system to perform in the event of a fire and a system failure when no fire is present, wherein the system "dumps" inadvertently, resulting in the possible loss of other systems or resulting in an accident of greater severity than the event the system was designed to mitigate.

5 *Failure detection*—Failure detection refers to the means of identifying a failure in the process of happening or a failure once it has happened. A sensor, human or otherwise, is involved in detection.

6 *Failure compensation*—Failure compensation, which is not always included, allows us to document those systems which have a certain level of redundancy or backup. In addition, this step can be extremely important to the identification of the failure effects. In cases where some failure compensation is present, it sheds light on the nature of the failure effects and may also be important in identifying other systems to be analyzed.

7 *Failure effects*—Failure effects are the immediate and expected results that the failure will produce relative to people, equipment, or other parts of the system or environment. They represent the consequences of the failure mode and are critical to the FMEA. The effects we are most interested in are those that ultimately result in a high-consequence fault event. Sometimes effects are categorized as "local" (at the point of first failure) and "end" (beyond the point of first failure).

8 *Preventive measures*—Preventive measures can be developed and placed in an FMEA. Here, our attention is focused on preventing the failure mode. Engineering judgment and speculation are required in this column, which constitutes the proactive opportunity for enhancing product or process integrity and developing a mistake-proof product or process.

## Example 12.1

Develop a basic FMEA for a simple no. 2 wooden lead pencil.

### Solution

The pencil usually consists of an eraser, an eraser holder, a wooden dowel, and a pencil lead. Table 12.1 illustrates a simple FMEA for the no. 2 pencil. As we can see from the simple FMEA, the pencil lead and the wooden dowel components must work in order for the system to work, or write. Table 12.1 could be expanded to include varying levels of complexity. For example, the wooden dowel may fail to sharpen because the wood is too hard, or it may sharpen too easily, thus breaking the lead. These situations represent additional failure modes which may be assigned to each component for further analysis across the columns.

The FMEA result is a tabulation of the effects of various equipment failures within a system. As will be discussed later, the criticality of each failure mode can also be included in the FMECA. The failures with high criticality rankings may call for immediate preventive measures.

## 12.4 FMEA FORMAT VARIATIONS

Variations in design complexity and available data will generally dictate the analysis approach to be used. ***There are two primary approaches for performing an***

**TABLE 12.1** PENCIL FMEA ILLUSTRATION

| Functional or equipment identification | Functional or equipment purpose | Failure mode | Failure mechanism | Failure detection | Failure compensation | Failure effects | Preventive measures |
|---|---|---|---|---|---|---|---|
| 1.0 Eraser. | To remove unwanted marks. | Eraser smears pencil marks. | Age hardening of rubber. | Inspection. | Use another eraser. | Reduced operation efficiency or sloppy work. | Reevaluate rubber compound, accelerated aging tests. |
| 2.0 Eraser holder. | To provide support for eraser. | Eraser does not hold to pencil. | Mechanical failure of holder. | Observation, inspection. | Use another eraser. | Reduced operational efficiency. | Assure tight fit of rubber in metal holder. Study more aggressive crimp. |
| 3.0 Wooden dowel. | To provide support and protection for pencil lead. | Lead loose in wooden dowel. | Wood-lead interface tolerance problems. | Inspection. | Use another pencil. | Pencil fails and must be replaced. | Check lead outer diameter, wood inner diameter as well as manufacturing process for best fit combination. |
| 4.0 Pencil lead. | To create marks desired by user. | Lead breaks under pressure. | Physical shear failure of the lead compound. | Inspection. | Use another pencil. | Pencil fails and must be resharpened. | Be sure lead is bonded well. Develop hardening experiment for lead. |

*FMEA.   One is the functional approach, which recognizes that every item is designed to perform functions.* The functions are listed and their failure modes analyzed in the FMEA. *The second approach is the part-level, hardware approach,* which lists individual hardware items and analyzes their possible failure modes. This approach is used more often in conjunction with criticality analysis [5]. *In a service FMEA the term "hardware" would refer to explicit service components, rather than to physical objects.*

For complex systems, a combination of the functional and hardware approaches may be considered. Although the application of these approaches differs, the underlying methodology remains the same. The FMEA may be performed as a functional analysis, a hardware analysis, or a combination analysis and may be initiated at any level of system complexity. Then, it may proceed through decreasing or increasing levels of system complexity to the individual component level or the top system level. Most often, the analysis begins at a high, or system, level; for example, a cooling system analysis is performed and progresses in a graded fashion to an analysis of the individual components of the cooling system—pumps, motors, tanks, and so on. By applying the FMEA in this fashion, limited resources may be focused effectively and efficiently on those subsystems or components whose failure might lead to severe or unacceptable consequences or failure effects. The top-down approach characterizes the functional level analysis. Functional FMEAs are used quite often in industry.

On the other hand, when knowledge of the system is such that the consequences of the overall failure are known to be severe, we may proceed with a bottom-up analysis approach. This method starts with individual components and proceeds up through successive levels of system complexity. The bottom-up approach characterizes the part level analysis methodology.

## Functional Level Analysis

*Functional level analysis is often used when hardware items cannot be uniquely identified or when system complexity requires analysis from the top downward through succeeding complexity levels.* This method of analysis can be applied at any system level and progress in either direction, up or down. This trait is universal with both functional and part level FMEA and allows for a great deal of flexibility in the analysis design.

The initial step in performing a functional level analysis is to determine the level at which the analysis will begin. It is important to consider the resource base available when determining this level. For example, we may have a large number of complex systems to analyze, and only very limited resources to conduct the analysis. In this case, we should perform a very high level analysis, looking at the major subsystems, not at each and every component of the subsystem.

A very coarse analysis may proceed with finer analyses performed later on only those subsystems identified as having severe consequences of failure (the Pareto principle applies here). In this manner, the functional level analysis allows for preliminary screening of subsystems of little importance. This screening process allows for resource expenditure on those subsystems determined to warrant detailed study. Cascading the analysis to finer and finer detail through multiple screenings will eventually lead us to a part level analysis. In addition to the efficient use of resources, this approach provides documentation as to why particular subsystems were not analyzed in detail.

It is very important when performing this type of analysis to always make conservative assumptions in the preliminary stages (e.g., we assume worst cases for all failure effects) to prevent the elimination of subsystems prematurely. If this procedure is executed correctly, we will achieve a complete system analysis, applied in a graded fashion, which is systematic, defensible, and cost efficient.

---

## Example 12.2

Develop a FMEA for a house, with a primary function of providing shelter and comfort in the cold season.

### Solution

In the winter, our house system must provide heat in order for the occupants to be comfortable. Often we find two sources of heat in a house, a fireplace and an electric/gas heating unit. Speaking in broad terms, we may consider three basic functional components. The structure itself which blocks the wind and keeps the occupants sheltered from outside elements. The primary heating unit for the home which provides warmth for the occupants. And finally a fireplace which can also produce limited amounts of heat to provide warmth for the occupants. The last subsystem considered here is a fireplace which has a natural gas fired log starter. Table 12.2 illustrates a functional level FMEA at this basic level of system complexity for our house.

Through a top level analysis such as this one, we can see that subsystem interrelationships can be observed. The house structure, for example, must perform successfully and either the electrical/gas heating unit or the fireplace or both must also perform successfully to yield a successful system performance.

---

### Part Level Analysis

*A part level analysis is used only after hardware items have been identified.* In general, the level of detail in a part level analysis is much more comprehensive than that found in a functional level analysis. The first step in a part level analysis is to develop a hardware list. This list should include detailed information about the individual hardware items, such as

1 Unique hardware identification.
2 Detailed hardware description.

**TABLE 12.2**   HOUSE FMEA ILLUSTRATION

| Functional equipment identification | Functional or equipment purpose | Failure mode | Failure mechanism | Failure detection | Failure compensation | Failure effects | Preventive measures |
|---|---|---|---|---|---|---|---|
| 1.0 House structure. | To isolate the enclosed area from the external elements. | Failure to retain heat in home. | Physical failure caused by poor insulation or faulty structure. | Inspection. | None. | Failure to contain heat from heat sources. Inconvenience to occupants. | Use worst case thermal loss or external temperature calculations to determine insulation value necessary. |
| 2.0 Electrical/gas heating unit (forced air). | To provide a heat source within the house structure. | Electrical failure. | Loss of system power due to storm damage. | Inspection. | Fireplace will provide heat. | Total lack of heat production. | Assess and eliminate potential line interference threats such as tree branches and limbs. |
| 3.0 Fireplace system. | To provide a heat source within the house structure. | Physical failure. | Faulty system construction, or improper installation. | Inspection. | Electrical/gas heating unit will provide heat. | If failure occurs in combination with an electrical heating unit failure, the home fails to provide adequate warmth to occupants. | Provide ample stock of fuel wood in dry condition. |

3  Detailed description of the hardware function.

4  A listing of any interfaces the hardware may have with other unique hardware items as well as a description of the hardware interface(s).

In general, *all hardware component failure events can be described by one of the following:*

1  *Failure on demand*—Certain components must start, change state, or perform a particular function at a specific instant of time. Failure to respond as needed is referred to as failure on demand.

2  *Standby failure*—Some systems or components are normally in standby, but are required to operate on demand. Failure could occur during this nonoperational period, preventing operation when required or causing related systems to failure inadvertently.

3  *Operational failure*—A given system or component may start successfully and operate normally but fail some time in operation. This failure characteristic is referred to as an operational failure.

Once the above considerations are made for all identified hardware items, we can systematically proceed with the FMEA.

---

**Example 12.3**

Develop a detailed FMEA for a residential fireplace. The major components are displayed in Figure 12.1.

**FIGURE 12.1**    Fireplace physical depiction (Example 12.3).

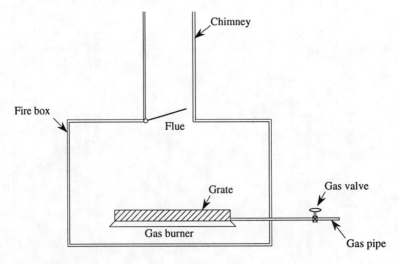

**Solution**

We assume that the fireplace is held primarily in standby, anticipating a credible failure of an electrical or gas heating unit (primarily through an electrical grid failure of some sort). Appropriately, we will cascade the analysis down to an additional level of complexity and look specifically at the system components.

We will first identify the major components of the fireplace subsystem: the chimney, firebox, flue, grate, gas supply, gas valve, and gas burner. These will provide a descriptor for each item. Next, we will assign a unique identification number to each item (e.g., model or part number). Column 1 of Table 12.3 shows this development.

After all system components have been identified and assigned a unique identification number, we will list the intended function of each. For example, the chimney is intended to channel smoke and excess heat out of the firebox. Next, we will identify the component interfaces. Usually, we can divide the interfaces into input, output, and surroundings. Considering the chimney, we must assume that one possible failure mode is that of not allowing smoke to flow out of the firebox. This is just one of many failure modes that may be credible. The chimney may not adequately insulate the effluent heat from the rest of the residential structure, a failure that could eventually lead to a structural fire. Or, the chimney may not release the hot smoke far enough above the structure, eventually leading to a fire on the roof. A failure mechanism is associated with each of these failure modes. In the case of the chimney, not allowing the flow of smoke, one possible failure mechanism would be a physical obstruction in the chimney. Columns 2, 3, 4, and 5 of Table 12.3 show the function, interface, failure mode, and failure mechanism, respectively.

Each of the remaining system components can be developed in a similar fashion. First, we consider its function, and then look at the interfaces, the potential failure modes, and the failure mechanisms. The remaining portion of the analysis considers the mechanisms for failure detection, failure compensation, and the associated effects of the failure. The last column deals with preventive measures.

When considering the failure effects, we most often postulate the worst case effects that are credible. In the case of chimney failure 1.1 (where hot gasses and smoke are not released from the firebox), one possible failure detection mechanism is visual inspection by people in the residence, another is a smoke detector. Regarding failure compensation, the chimney is not a redundant subsystem, and thus there exists no failure compensation.

Obviously, if the chimney cannot perform its primary function, the fireplace subsystem fails. We should carefully consider the potentially high consequence effects. For example, a failure, if undetected, could result in the release of hot gases and smoke in the residence. This effect could lead to personal injury or a house fire. Therefore, we must consider not only the main failure effects but the subsequent or secondary failure effects caused by the main effect. This will be very important when we assign a risk level to each failure effect in a subsequent

**TABLE 12.3  FIREPLACE FMEA**

| Functional or equipment identification | Functional or equipment purpose | Interface | Failure mode | Failure mechanism | Failure detection | Failure compensation | Failure effects | Preventive measures |
|---|---|---|---|---|---|---|---|---|
| 1.0 Chimney. | To allow smoke and excess heat to escape from the firebox. | Input interface with flue and firebox. Interface with structure over length of chimney. Output interface 5 feet over roof. | 1.1 Failure to direct hot gas and smoke for release. | Soot buildup, bird nest. | Inspection and smoke detection. | None. | Possible fire or smoke inhalation harm to residents. | Large opening. Wise fuel wood choice, preventive maintenance. |
| | | | 1.2 Failure to insulate heat from building structure. | Insulation breakdown. | Inspection. | None. | Structural fire, potential harm to residents. | Large design margin on insulation, aging study for insulation. |
| | | | 1.3 Failure to release heated material at an adequate height. | Flow of exhaust over rooftop too close, due to wind. | Inspection. | None. | Fire on roof, potential harm to residents. | Attention to height, top deflector. |
| 2.0 Firebox. | ⋯ | ⋯ | 2. | ⋯ | ⋯ | ⋯ | ⋯ | ⋯ |
| 3.0 Flue. | ⋯ | ⋯ | 3. | ⋯ | ⋯ | ⋯ | ⋯ | ⋯ |
| 4.0 Grate. | ⋯ | ⋯ | 4. | ⋯ | ⋯ | ⋯ | ⋯ | ⋯ |
| 5.0 Gas supply. | ⋯ | ⋯ | 5. | ⋯ | ⋯ | ⋯ | ⋯ | ⋯ |
| 6.0 Gas valve. | To allow or block gas from entering the gas burner. | Input interface with loaded gas pipe, pipe-thread interface. Valve-stem interface with room environment. Output interface with open gas burner. | 6.1 Gas valve fails to prevent gas flow. | Gas leakage around seal (when closed). | None. | If detected gas may be cut off at meter. | Unintentional release of natural gas in residence leading to fire, explosion, or harm to occupants. | Pressure-aging study on valves under worst case condition. Place gas cock for corrective maintenance or replacement. |
| | | | 6.2 Gas valve fails to prevent gas flow. | Seizure of gasket to valve body (when open). | Most likely will be observed by operator. | Detection likely, operator cuts off gas at meter. | Increased exposure to hazard, severe damage or injury unlikely. | Cycle test under worst case condition; look for seizure points. |
| | | | 6.3 Gas valve does not allow gas flow. | Seizure between stem and body materials (when closed). | Inspection or observation by operator. | Other direct means of log starter available. | Inconvenience to operator, delay in operating system. | Cycle test under worst case condition; look for seizure points. |
| 7.0 Gas burner. | ⋯ | ⋯ | 7. | ⋯ | ⋯ | ⋯ | ⋯ | ⋯ |

example. In order to further illustrate FMEA, specifically for a mechanical device, the analysis of the gas valve is expanded in Table 12.3.

## 12.5 CRITICALITY ANALYSIS

*After failure effects have been uniquely defined, we must determine which ones require further attention. This determination is driven by both the consequences associated with the unique failure effects and the likelihood that these failures will occur.*

A logical extension for both the functional level and the parts level FMEA is the criticality analysis. Criticality analysis is a method whereby unique failure effects are ranked relative to one another, thus ranking the criticality of the systems, subsystems, or hardware components analyzed [6]. *Once ranked, the systems, subsystems, or components designated as most severe, in terms of failure effect, are provided analysis resources in a graded fashion.* Those systems determined most critical are given highest priority for detailed analysis (e.g., quantitative analysis), or additional attention in system engineering (e.g., reliability improvement efforts aimed at reducing the likelihood of failure, or the implementation of redundancy aimed at mitigating the consequences of that failure). Then, the Pareto principle is used to determine where process improvements will realize the greatest benefits.

Many schemes for criticality ranking exist. One such scheme is to provide a subjective or qualitative estimate of both the likelihood of the failure and the associated consequences. In order to clearly document this step, it is typical to add another column to the FMEA table. This column may be labeled "criticality," "risk level," or with any other descriptor deemed appropriate for the analysis. Once this additional column is used, the analysis becomes a failure mode and effects criticality analysis (FMECA). We will label this column "risk level" and place it just inside the preventive measures column.

The terms, definitions, or characterizations used for the varying levels of criticality are very important. In industry these terms and definitions vary greatly. Table 12.4 illustrates a criticality ranking system based on risk. The definitions shown in here allow us to discretely classify each failure effect, once these effects are thoroughly understood, at an appropriate level of probability of occurrence and severity of consequence. The probability of occurrence for each failure is determined using the best information available. When information is inadequate, an estimate must be made on the conservative side (i.e., assume the failure will occur more frequently). We determine the consequence definition associated with the failure effect in the same manner; here, should we lack information, we must assume the event will have severe consequences.

Once each failure effect has been classified, the associated systems may be ranked accordingly, with those having the highest consequences, or occurring most frequently, given a high priority ranking, and those having lesser consequences or likelihood of failure, a lower priority. Figure 12.2 illustrates a simple ranking matrix for this process. We can see that each of the sixteen possible consequence likelihood

**TABLE 12.4**   RISK LEVEL DEFINITIONS

| Hazard categories | Consequences to the public, workers, or the environment |
|---|---|
| Category I—catastrophic | May cause death, loss of the facility or operation, or severe impact on the environment |
| Category II—critical | May cause severe injury, severe occupational illness, major damage to a facility or operation, or major impact on the environment |
| Category III—marginal | May cause minor injury, minor occupational illness, or minor impact on the environment |
| Category IV—negligible | Will not result in a significant injury, occupational illness, or a significant impact on the environment |

| Descriptive word | Symbol | Nominal range of frequency per year |
|---|---|---|
| Likely | A | $P(\text{annual occurrence}) \geq 10^{-2}$ |
| Unlikely | B | $10^{-2} > P(\text{annual occurrence}) \geq 10^{-4}$ |
| Extremely unlikely | C | $10^{-4} > P(\text{annual occurrence}) \geq 10^{-6}$ |
| Rare | D | $P(\text{annual occurrence}) < 10^{-6}$ |

combinations falls in exactly one of the three general criticality grade areas: critical, important, or negligible.

It is important to realize that we may develop more than four grades of consequences and more than four classes of likelihood, each with its own definition. Nevertheless, the preceding basic philosophy and application of criticality analysis applies. For example, if the level of information to support the analysis is great, we may be capable of defining more than sixteen consequence-likelihood combinations and more than three grades of criticality, and still be able to distinguish between grades. But, as often is the case in the early design phase, where FMECA is most frequently used, detailed information does not exist.

---

## Example 12.4

Return to the fireplace example, Example 12.3, and develop a criticality analysis (FMECA).

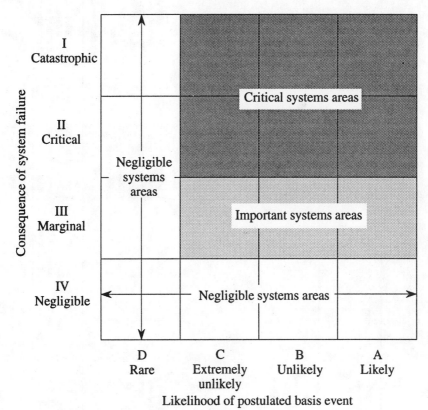

**FIGURE 12.2** Consequence-likelihood grid for criticality ranking.

## Solution

Results are shown in the Risk level column in Table 12.5. In this example, the definitions found in Table 12.4 were used. Likelihood estimates were made in a subjective fashion, on the conservative side. Consequences are worst case consequences associated with each failure effect.

Given the risk level combinations in Table 12.5, the ranking for further analysis and improvement becomes clear once we use the criticality ranking method shown in Figure 12.2. Using the identification numbers, the following ranking was developed and arranged in descending order of priority: 6.1, 1.2, 6.2, 1.1, 1.3, and finally, 6.3. The resources required to address each of these concerns may now be applied in a graded fashion. For example, further analysis and engineering may be required for failure modes 6.1, 1.2, 6.2, 1.1, and 1.3, while no further action may be required for failure mode 6.3. We have not developed the firebox, flue, grate, gas supply, and gas burner in this example.

**TABLE 12.5   FIREPLACE FMECA**

| Functional or equipment identification | Functional or equipment purpose | Interfaces | Failure modes | Failure mechanism | Failure detection | Failure compensation | Failure effects | Risk level | Preventive measures |
|---|---|---|---|---|---|---|---|---|---|
| 1.0 Chimney. | To allow smoke or excess heat to escape from the firebox. | Input interface with flue and firebox. Interface with structure over length of chimney. Output-interface 5 feet over roof. | 1.1 Failure to direct smoke for release. | Soot buildup, bird nest. | Inspection and smoke detection. | None. | Possible fire or smoke inhalation harm to residents. | II-C | Large opening. Wise fuel wood choice, preventive maintenance. |
| | | | 1.2 Failure to insulate heat from building structure. | Insulation breakdown. | Inspection. | None. | Structural fire, potential harm to residents. | I-C | Large design margin on insulation, aging study for insulation. |
| | | | 1.3 Failure to release heated material at an adequate height. | Flow of exhaust over rooftop too close, due to wind. | Inspection. | None. | Fire on roof, potential harm to residents. | II-C | Attention to height, top deflector. |
| 2.0 Firebox. | … | … | 2. | … | … | … | … | … | … |
| 3.0 Flue. | … | … | 3. | … | … | … | … | … | … |
| 4.0 Grate. | … | … | 4. | … | … | … | … | … | … |
| 5.0 Gas supply. | … | … | 5. | … | … | … | … | … | … |
| 6.0 Gas valve. | To allow or block gas from entering the gas burner. | Input interface with loaded gas pipe, pipe-thread interface. Valve-stem interface with room environment. Output-interface with open gas burner. | 6.1 Gas valve fails to prevent gas flow (when closed). | Gas leakage around seal (when closed). | None. | If detected, gas may be cut off at meter. | Unintentional release of natural gas in residence leading to fire, explosion, or harm to occupants. | I-A | Pressure-aging study on valves under worst case condition. Place gas cock for corrective maintenance or replacement. |
| | | | 6.2 Gas valve fails to prevent gas flow. | Seizure of gasket to valve body (when open). | Most likely will be observed by operator. | Detection likely, operator cuts off gas at meter. | Increased exposure to hazard, severe damage or injury unlikely. | II-B | Cycle test under worst case condition, look for seizure points. |
| | | | 6.3 Gas valve does not allow gas flow. | Seizure between stem and body materials (when closed). | Inspection or observation by operator. | Other direct means of log starter available. | Inconvenience to operator, delay in operating system. | IV-C | Cycle test under worst case condition, look for seizure points. |
| 7.0 Gas burner. | … | … | 7. | | | | | … | |

From these simple examples, we can see that FMEA and FMECA, along with the Pareto principle, are two of the most valuable tools we possess in proactive quality work. The flexibility and adaptive nature of FMEA and FMECA allow for broad applications across all phases of system definition, design, manufacture, and operation. Personnel can be trained readily to use FMEA and FMECA. However, sound logic and physical knowledge are required to drive FMEA and FMECA.

## REVIEW AND DISCOVERY EXERCISES

### Review

**12.1** How does FMEA encourage the system analysis perspective?

**12.2** With regard to FMEA approaches, when is it appropriate to use
  **a** A functional level approach?
  **b** A parts level approach?

**12.3** Choose a product (good or service) and develop an FMEA. Use the Table 12.3 format.

**12.4** Expand the FMEA in Problem 12.3, to an FMECA. Use the Table 12.5 format.

### Discovery

**12.5** Robust performance is critical in high quality products. How can we use FMEA analysis in the quest for robust performance? Explain.

**12.6** Mistake-proof performance is critical in high quality products. How can we use FMEA in the quest for mistake-proof performance?

**12.7** CE analysis, Pareto analysis, and FMEA represent three very simple, but powerful tools in proactive quality work. How are they similar? How are they different? How can they be used together?

## REFERENCES

**1** T. E. Bell, "Managing Murphy's Law: Engineering a Minimum Risk System," *IEEE Spectrum,* vol. 26, no. 6, pp. 24–27, 1989.

**2** H. B. Dussault, "The Evolution and Practical Applications of Failure Modes and Effects Analyses," RADC-TR-83-72, Rome Air Development Center, Air Force Systems Command, Griffiss Air Force Base, NY, 1983.

**3** "Procedures for Performing a Failure Mode, Effects and Criticality Analysis," *Military Standard, MIL-STD-1629A,* U. S. Department of Defense, 1980.

**4** Y. Akao (ed.), *Quality Function Deployment,* Cambridge, MA: Productivity Press, 1990.

**5** See reference 3.

**6** See reference 3.

# 13

---

# LOGIC TREE ANALYSIS:
# FAULT, EVENT, AND
# GOAL TREES

---

## 13.0 INQUIRY

1 When are logic tree or hierarchical models most effective?
2 How can logic tree models be used to proactively assess product and process performance?
3 What linkage exists between FMEA and logic tree analyses?
4 How are faults, events, and goals modeled within the logic tree structure?
5 How can logic tree models encourage system improvement?

## 13.1 OVERVIEW

*Logic tree models are hierarchical models which play an important role in performance analysis with respect to safety, reliability, and risk.* Traditionally, these probabilistic risk assessment models have been developed to assess catastrophic accident sequences. With slight modifications, these tools can provide a systematic means to proactively assess product and process performance from a top-down or a bottom-up perspective, thus developing and documenting the hierarchical relationships of subsystems and components (and their functions or malfunctions) within the system. With minor changes, they can also be used effectively to analyze and assess logical steps in system level goal development in conjunction with strategic quality (performance) planning (e.g., the systematic diagram, Chapter 7). Hence, *the logic tree models, as well as FMEA, are indispensable in the quality planning or quality function deployment phase of quality assurance* (Akao [1]).

## 13.2 TREE STRATEGIES AND STRUCTURES

In any system level analysis we seek to describe, understand, and improve our system. *It is often helpful in describing and understanding the system to construct visual aids which depict the logic required to establish and accomplish system performance goals and objectives.* These performance goals and objectives may be measured in terms of safety, reliability, maintainability, or any other meaningful performance measure.

*The primary purpose of the tree structure is to illustrate causal relationships between basic human, hardware, and environmental events.* The tree structure is flexible. Although it offers great versatility in expressing the engineering logic (both qualitatively and quantitatively) inherent in detailed systems, logic tree analysis is not strictly qualitative or quantitative. *The basic logic tree itself is a qualitative characterization of system fault or success sequences.* It lends itself to quantification through Boolean representations.

It is important to realize that the quantification of a logic tree can be costly and should be performed only when the benefits exceed the associated costs. In general, we should quantify those logic trees where success or failure of the top event is very critical or for which the top event cannot be determined, by subjective means, with reasonable confidence, to be either significant or insignificant.

The actual development of the system logic model commences after we have gained a thorough understanding of the system under consideration. FMEA is a useful first-level tool in gaining a basic understanding of product and process performance failures early in system definition and design. Logic tree models are used for more detailed analyses. *FMEA is helpful in establishing the "top" event in logic tree modeling.*

## 13.3 FAULT TREE ANALYSIS

*The purpose of fault tree analysis (FTA) is to identify failure pathways, both physical and human, that could lead to an identified fault event.* FTA is a deductive technique that focuses on one particular fault event and then constructs a logic diagram of all conceivable event sequences (both physical and human) which could lead to the fault event. In effect, the fault tree is a graphic illustration of various combinations of equipment faults and failures and human errors that can result in the predetermined undesirable fault event (the top event). Fault tree analysis is used extensively in the field of reliability, safety, and risk analysis.

As a qualitative tool, FTA is useful because it breaks a given fault event down into basic failures and errors that constitute cause. It also allows us to determine the effect of eliminating, changing, or adding components to a system (e.g., supplying redundant components, independent high-level alarms or shutoffs).

*FTA can be used during definition, design, modification, operation, support, use, or disposal of a system.* It can be especially useful in analyzing new or novel processes for which there is no operating history. The results obtained from FTA include a set of logic diagrams that illustrate how certain combinations of failures and/or errors can result in specific fault events. The results are qualitative, but can be made quantitative if failure rate data or estimates are available for the failure events.

A fault tree does not contain all possible component failure modes or all possible fault events that could cause system failure. Rather, it is tailored to its top event, which corresponds to a specific system failure mode(s) and associated timing constraints. Hence, *the fault tree includes only the fault events and logical interrelationships that contribute to the top event.* Furthermore, the postulated fault events that appear on the fault tree may not be exhaustive. Because of the size of most systems, the tree should include only significant events. It should be noted that the choice of fault events for inclusion is not arbitrary; it is generally guided by detailed procedures; information on system definition, design, and operation; operating histories; input from operating personnel; the available level of detail of basic data; previous analysis results, such as FMEA; and engineering experience (Vesley et al. [2]).

### Fault Tree Analysis Procedure

In general, the construction of a fault tree begins with the consideration of a single failure of a given system. The fault tree is then developed backward using deductive logic until the system failure is described in terms of the failure of its constituent system components. *Development of an FTA consists of the following steps:*

1 *Identify the top event*—Identify the system failure that is to be analyzed and place this event at the top of the tree.

2 *Identify the second-level events*—Proceed to the next lower level of the system, the subsystem level, and identify the subsystem failures that could lead to the top event (system failure).

3 *Develop the tree logic*—Use the AND, OR, or other gate logic structure to show the relationship of subsystem failures that produce the top event. Figure 13.1 depicts commonly used Boolean logic operators for FTA.

4 *Identify lower level events*—Proceed to the next lower system level and determine the logical sequential relationship between this level and the levels above.

5 *Proceed to the desired level of detail*—Repeat steps 3 and 4 until a satisfactory level of detail has been reached.

6 *Quantify the tree logic*—Beginning with component failure rate data, compute the probability of failures described in the fault tree. Follow the logic structure indicated by the AND and OR gates, using the principles of Boolean logic, until the top event probability has been calculated. Step 6 is required only if we desire to provide a quantitative analysis.

---

### Example 13.1

From the fireplace FMEA in Chapter 12, we determined that a gas valve failure could potentially happen and would have severe consequences. For this reason, we will conduct a more detailed analysis to determine the likelihood of this type of failure and explore other potential failure mechanisms that might provoke it.

Develop a fault tree diagram and analysis for the gas valve. The gas valve FMECA appears in Table 12.5; the failure mode of interest is described under entries 6.1 and 6.2.

#### Solution

The first step is to identify the top event on the fault tree: "the gas valve fails to prevent gas flow (when closed)." We must now determine what conditions could lead to this failure state. For example, the handle may inappropriately

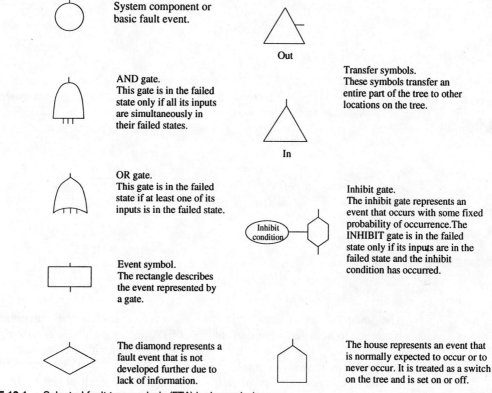

System component or
basic fault event.

AND gate.
This gate is in the failed
state only if all its inputs
are simultaneously in
their failed states.

OR gate.
This gate is in the failed
state if at least one of its
inputs is in the failed state.

Event symbol.
The rectangle describes
the event represented by
a gate.

The diamond represents a
fault event that is not
developed further due to
lack of information.

Out

In

Transfer symbols.
These symbols transfer an
entire part of the tree to other
locations on the tree.

Inhibit gate.
The inhibit gate represents an
event that occurs with some fixed
probability of occurrence.The
INHIBIT gate is in the failed
state only if its inputs are in the
failed state and the inhibit
condition has occurred.

Inhibit
condition

The house represents an event that
is normally expected to occur or to
never occur. It is treated as a switch
on the tree and is set on or off.

**FIGURE 13.1**    Selected fault tree analysis (FTA) logic symbols.

indicate that the valve is in the closed position; or the valve may actually be
closed, as indicated by the handle, however the interior seals may be faulty, thus
allowing continued gas flow; or the valve may not be properly installed to the
gas feed and allows gas to escape at the connection, upstream from the valve.
Note the use of the word "or" in the previous sentence. This word naturally
indicates the use of an OR gate in the fault tree construction. In short, this example assumes that three conditions could lead to the top event. Figure 13.2
shows the top event and the three conditions previously listed as the top two
layers of the FTA diagram.

The fault tree must now be expanded. Ideally, expansion should proceed until
all branches end in a basic event. This expansion is performed using the same
logic as above. For example, for the improper installation state to occur, two
events must also occur, the installation must be faulty, and the installer must
have failed to detect the error when he or she performed the required inspection.
The faulty seal condition is already a basic event and requires no further expansion. The inappropriate valve closure indication could be caused by two possible
events: the valve stem could be broken or the handle indicator could be faulty.

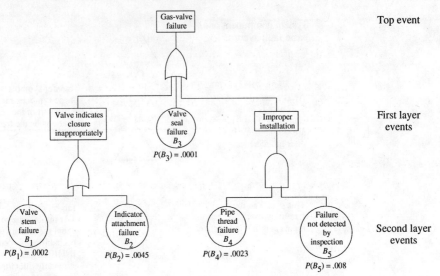

**FIGURE 13.2**   Gas valve fault tree.

The second layer of the diagram in Figure 13.2 depicts these events. It is important to note that a basic event in our FTA (e.g., the valve seal failure) might be a top event—or at least not a basic event—in another FTA, such as one performed by the valve manufacturer.

The next step is to assign failure probabilities to each of the basic events. Then, using the Boolean logic operator we will calculate the probability of the top event. The basic events in the fault tree above have been numbered to assist in understanding the calculations performed (see Figure 13.2):

$$P(\text{top event}) = (B_1 \cup B_2) \cup B_3 \cup (B_4 \cap B_5)$$

where $P(B_1) = .0002$
$P(B_2) = .0045$
$P(B_3) = .0001$
$P(B_4) = .0023$
$P(B_5) = .0080$

Here, the $P(Bs)$ are the probabilities that each of the basic events occur on an annual basis. Thus,

$$P(B_1 \cup B_2) = .0002 + .0045 - .0002(.0045) = .0047$$

$$P(B_4 \cap B_5) = .0023(.0080) = .000019$$

$$P(\text{top event}) = 1 - [(1 - .0047)(1 - .0001)(1 - .000019)]$$

$$= 1 - .99518 = .00482$$

Here, we are assuming independent events. If dependency exists, we must use conditional probabilities and the complexity of the analysis increases. Note that this probability is only slightly less than the subjective probability estimated in the FMEA, Table 12.5 (i.e., less than $1 \times 10^{-2}$, assuming we use the I-A classification). Many times, we find FTA probabilities smaller than FMEA probabilities because, when subjective FMEA probability estimates are provided, they are often on the conservative (worst case) side.

## Faults and Failures

*We must be able to distinguish between the specific term "failure" and the more general term "fault."* This distinction can best be illustrated by an example. If a valve properly closes when a low pressure indication occurs, the valve is said to be in a success state. If, however, the valve fails to properly close under the same circumstance, it is considered to be in a failure state. On the other hand, it is possible that the valve closes at the wrong time because some upstream sensor component functions improperly. This premature closing does not constitute a valve failure; the valve worked as directed. However, the valve's closing at the wrong time may well cause the entire system to enter an unsatisfactory state. Such an occurrence is called a fault. We can state that, in general and regarding any specific component, all failures are faults, but not all faults are failures. Failures are basic abnormal occurrences, whereas faults can be described as "higher order" events.

*In postulating a fault or a failure for inclusion in a fault tree, we must remember that the proper definition of these events includes a specification not only of the undesirable component state but also its timing.* Timing must be incorporated into our thought processes when postulating the top event and all subsequent fault events.

To allow for a quantitative evaluation, failure modes must be clearly defined as they are postulated. Care should be taken to postulate component failure modes so that they are realistic and consistent within the context of the system's operational requirements and environmental factors. In FTA, just as in FMEA, we see three general types of failures: (1) on demand, (2) standby, and (3) operational.

## 13.4 EVENT TREE ANALYSIS

*The purpose of an event tree analysis (ETA) is to identify the sequence of events that follows a given failure or error as it could lead to a loss in system performance. An event tree is a graphical illustration of potential outcomes that can result from a specific equipment failure or human error.* Event tree analysis considers the response of personnel and safety systems in dealing with the failure. The results of an event tree analysis are "accident sequences" or "failure sequences"—a multibranched, chronological set of failures or errors that define an accident or system failure. *ETA is useful in analyzing the effect of safety systems or emergency procedures on accident prevention and mitigation.* It produces useful information when used in parallel with FTA [3].

*A key distinction between FTA and ETA is that in the latter an initiating event is assumed to have occurred,* whereas in FTA this initiating event is usually the event for which the probability of occurrence is determined. This initiating event may be the result of a particular system failure, or it may be caused by some external circumstance such as a natural phenomenon.

ETA can be used during the definition, design, modification, or operation phase of a system. It is particularly useful as a tool for demonstrating the efficiency of accident prevention and mitigation techniques. *Although the ETA is primarily used for safety analysis, it can be quite useful for quality procedure analyses dealing with corrective action procedure design and development.* ETA has great potential to aid in process control when special causes are detected using SPC. SPC methods were introduced in Chapter 9 and are described in detail in Chapters 15, 16, 17, and 18.

ETA produces a series of event trees that illustrates the event sequences effecting a system performance loss following the occurrence of an initiating event. The results are qualitative, but can be quantitative if the event probabilities are known. This quantification is achieved by applying the laws of conditional probabilities.

### Event Tree Analysis Procedure

*An ETA consists of the following steps*:

1 *Identify an initiating event*—The initiating event may be a system failure, equipment failure, human error, or process upset that could have any one of several effects. Actual effects, or results realized, depend on how the system or operator responds to the event.

2 *Identify the response*—Identify which system or operator response is designed to handle the initiating event. This response can include action by subsystems, such as an automatic emergency shutdown triggered by the event, alarms to alert operators, operator actions taken in response to alarms, or even physical barriers to limit the effects of the initiating event. We must identify and list these functions in the chronological order of the designed response. For example, possible responses to "liquid level in a tank is too high and is increasing" might be (1) high-level alarm operates, and (2) operator closes inlet valve. If other affected systems are present, they should also be so listed.

3 *Construct the event tree*—First, we enter the initiating event on the left-hand side of the page. Then, we list the functional responses chronologically across the top of the page. Next, we decide whether or not the success-failure of the function can or does affect the course of the event. If the answer is yes, the event tree is branched to distinguish between success and failure of the function; success always branches upward, failure downward. If the system function has no effect, the tree does not branch, but proceeds to the next system function (to the right).

4 *Describe the event sequences*—The event sequences are a variety of outcomes that could occur following the initiating event. Some of the sequences may represent success (e.g., a return to normal or an orderly shutdown).

Sequences that result in failure should be studied with the objective of improved responses to the event (in order to minimize the probability of failure).

## Example 13.2

Returning to the previous example, we now want to consider the probability that failure will result in the severe consequences postulated in the FMEA, Table 12.5. Use an ETA to assess the probability of each severe consequence identified.

### Solution

We will construct an event tree beginning with the gas flow (see the FTA developed in Example 13.1). The assumed accident sequence for this example is (1) gas flows as a result of a faulty valve, (2) the gas flow rate is not sufficient to cause serious damage, (3) occupants are present, and (4) gas flow is detected and stopped. The event tree shown in Figure 13.3 illustrates the accident sequence defined above, along with the associated probabilities. The simplified sequence shown is just one of many possible accident sequences that could be developed.

The accident event sequence $B_1$, $B_3$, $B_4$, and $B_7$ leads to an outcome event $E_3$ with severe consequences. This sequence will be quantified as follows:

$$P(E_3) = P(B_1)\, P(B_3)\, P(B_4)\, P(B_7)$$
$$= .00482\,(.20)\,(.50)\,(.80)$$
$$= .000386$$

Likewise, the accident event sequence $B_1$, $B_3$, and $B_5$ could potentially lead to an outcome event $E_4$ with severe consequences and can be quantified as

$$P(E_4) = P(B_1)\, P(B_3)\, P(B_5)$$
$$= .00482\,(.20)\,(.50)$$
$$= .000482$$

**FIGURE 13.3**    Gas valve event tree.

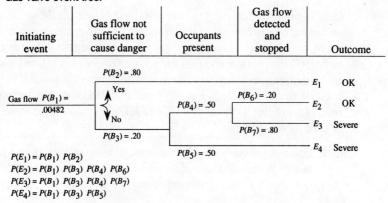

$P(E_1) = P(B_1)\ P(B_2)$
$P(E_2) = P(B_1)\ P(B_3)\ P(B_4)\ P(B_6)$
$P(E_3) = P(B_1)\ P(B_3)\ P(B_4)\ P(B_7)$
$P(E_4) = P(B_1)\ P(B_3)\ P(B_5)$

Each of these events could potentially result in severe consequences. Both events are credible, given the analysis data; therefore, it would seem reasonable to recommend that the valve be reengineered to provide more reliable service. The question now turns to both economics and engineering expertise available, an issue management must ultimately address.

### Event Tree Flexibility

In an event tree, there is considerable latitude in the definition of the event headings; functions, systems, and components can be shown on the same tree. These various event possibilities are listed as headings that represent the functions or systems necessary to mitigate an event's consequences. The end result of each sequence is assumed to be either a successful, or safe, termination of the postulated sequence of events or a system failure state. In developing event trees for a specific system, care must be taken to correctly specify the expected system failure state. Care must also be exercised to ensure that the event headings are consistent with actual system response modes and are precisely related to system success criteria that can be translated to top events for system fault modeling.

The events are placed across the tree either according to the sequence of their occurrence (proceeding from left to right) or some other logical order reflecting operational interdependence. Consequently, the initiating event is always shown first and the total system outcome response is always shown last. The various sequences are represented by the paths of vertical and horizontal lines beneath the event headings. At a horizontal-vertical line junction, the system is successful if the path is upward; the system fails if the path is downward. A column at the far right of the tree identifies the various outcome events resulting from the path sequences.

This information is then used to determine the severity of the event sequence outcome and the level of attention that each component receives. In those areas where the system's state reveals the potential for unacceptable consequences, we should quantify the probability of success and failure at each junction. Then, by the methods of conditional probability the likelihood of the various system states can be estimated.

## 13.5  CAUSE-CONSEQUENCE ANALYSIS

*Cause-consequence analysis can be described as a combination of fault tree analysis and event tree analysis.* It is commonly used to trace an accident from its initiating event, or cause, to its final impact, or consequence. This analysis technique can be helpful as a systematic tool for product or process related quality problems. The cause-consequence diagram illustrates the interrelationships between consequences and causes and serves as a good communication tool. Its purpose, application, and result are similar to those of FTA and ETA. A cause-consequence analysis consists of the following steps:

1 Select an event to be evaluated, e.g., a top event (as in a fault tree) or an initiating event (as in an event tree). Any event that would be of interest in either is suitable for a cause-consequence analysis.

   **2**  Identify system functions that could influence the selected event.  The system functions are similar to those in an event tree (e.g., safety systems, operator actions, procedures).

   **3**  Develop the fault paths resulting from the event and the consequences.  This step is also similar to ETA; however, the graphic representation is different.  Cause-consequence analysis uses a branch-point symbol, while event trees do not.  The branch-point symbol contains the system function description that is normally written in the top row on an event tree.

   **4**  Examine the event selected in step 1 and the system functions, identified in step 2, to determine causes of the event.  This step is similar to an FTA; each failure of a system function is treated as the top event of a fault tree or fault sequence.

   **5**  Determine minimal cut sets for the event (fault or accident) sequence.  A minimal cut set is the smallest set of events which must all occur in order for   the selected (in step 1) event to occur.  This step is analogous to cut set determination for fault trees (discussed later in this chapter).  The fault-sequence fault tree is constructed using all the system function failures, with an AND gate where the fault or accident sequence occurs as the top event.

   **6**  Evaluate the results. Fault sequences can be ranked in order of their severity or importance to overall system performance.  The minimal cut sets for each significant fault sequence can be ranked to determine the most important basic causes, relative to the consequences.

Cause-consequence analysis is not used frequently in industry, as its purpose can be duplicated by the FTA and ETA techniques.  Furthermore, the methods of applying it are highly variable, as well as inconsistent between analyses.

## 13.6  GOAL TREE ANALYSIS

***The purpose of goal tree analysis (GTA) is to identify success pathways, or success trees, that can lead to the fulfillment of a specific goal.***  Like fault tree analysis, GTA is a deductive technique.  However, it focuses not on a specific fault event, but rather on a specific system goal (or system objective).  GTA encourages the construction of a logic diagram of all conceivable event sequences (both physical and human) which could lead to the achievement of the goal event.  ***The goal tree is a graphic illustration of various combinations of goals, subgoals, equipment, or system successes that can result in the success of the top event*** (Roush et al. [4]).  The Japanese systematic diagram described in Chapter 7 is very similar in structure to the GTA.

### Goal Tree Analysis Procedure

The goal tree is developed backward using deductive logic until the system objective or goal is described in terms of the system's components.  ***A GTA generally consists of the following steps***:

   **1**  ***Identify the system goal*** that is to be analyzed and place this event at the top of the tree.

2 Proceed to the next lower level of the system (e.g., subgoal or subsystem level) and *identify the subsystem successes that could lead to the top event*, system success.

3 *Determine the logical relationships between the goal, subgoals, and functions* that are required to produce the top event. Typically, AND logical relationships are found at the higher levels.

4 Proceed to the next lower system level and continue to *identify system success trees or paths* and link them using AND or OR logic gates.

5 *Quantify the goal tree.* (This is an optional step.) Beginning with component success rate data, compute the probability of successes described in the goal tree by following the logic structure indicated by the AND and OR gates. Use the principles of Boolean logic until the probability of the top event has been calculated.

---

## Example 13.3

Apply the GTA tool to the challenge of reducing manufacturing costs. Use "reducing manufacturing costs" (for a given product) as the top event, or goal. Partially expand the tree for this high level goal.

### Solution

Figure 13.4 depicts a partial GTA drawn with logic gates. For a goal as broad as our top event, we can expect a rather large tree. The tree structure allows us to put both our production process and our logical improvement opportunities in perspective.

---

A GTA provides a complete description of a complex system and its operations. It presents the system in a positive light. GTA begins with a precise statement of the system's primary goal and deductively breaks out the immediate subgoals that must be satisfied to achieve the goal above. In turn, functions are identified that impact achievement of the goals and subgoals, thus providing a description of the system in terms of the functional requirements for its operation. The portion of the tree that defines the system goal and subgoals is referred to as the "goal and subgoal portion." The "functional portion" describes the system. Figure 13.4 depicts the goal and functional structures.

Further development of the model identifies the relationships between hardware components and the functions that they support. Understanding these relationships provides a basis for selection of component subsystems by highlighting those performance parameters that are essential for efficient operation. GTA also establishes the relationships between human activities and hardware performance. The lower part of Figure 13.4 is referred to as the "success tree portion."

The top event must be explicitly defined in a single unambiguous statement. It is from this definition that we will identify and relate all the different system

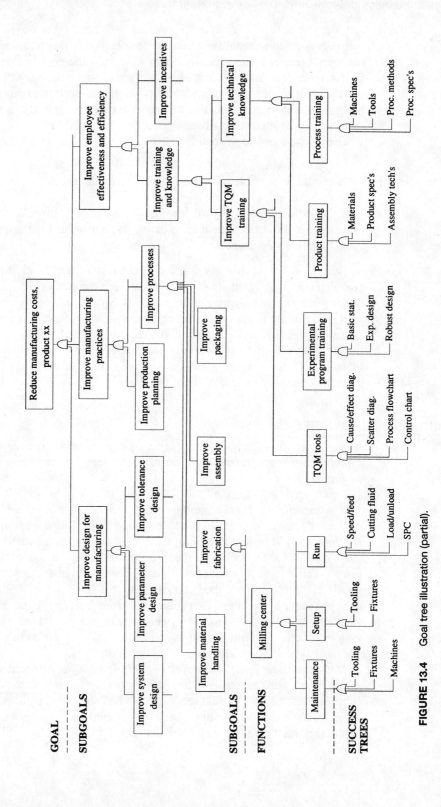

**FIGURE 13.4** Goal tree illustration (partial).

269

subgoals and sub-subgoals which must be achieved to attain the overall goal.  The functional tree is built vertically downward from the goal in levels, wherein we subsequently decompose each identified goal to its necessary and sufficient set of dependent subgoals.  As the vertical development of the tree proceeds, tests must be applied to ensure not only that the tree is accurate and complete, but that it maintains the proper hierarchy between goals and subgoals.

The tests are of two general types:

1  Upon looking upward from any subgoal, i.e., toward the goal or "treetop," it should be possible to define explicitly *why* the specific goal or function must be achieved.
2  Upon looking downward from any subgoal, i.e., toward the bottom of the tree, it should be possible to define explicitly *how* the specific subgoal or function is achieved.

Failure of the tree to pass either of these tests at any level indicates a lack of completeness;  intermediate subgoals have been omitted or the hierarchy of goals or their interdependencies have been carelessly or incompletely defined.

It is important to recognize the boundary between functions and success trees, or paths (Figure 13.4) because this is the point at which the logic structure of the tree changes.  Within the functional tree, all functions are typically connected by logical AND gates, whereas the success paths are typically connected to the functions they serve by logical AND or OR gates.

## 13.7  CUT SETS IN LOGIC TREE ANALYSIS

*A fundamental objective of most logic tree applications is to discover the event combinations most critical to system performance.*  This objective is accomplished by finding the smallest combination of events that, if they all occur, will cause the system to enter a preselected state—e.g., one in which the top event occurs.

*A minimal cut set is the smallest set of primary events, inhibit conditions, undeveloped fault events, or any combination of these, which must all occur in order for the top event to occur.*  A minimal cut set represents the path by which the top event can occur.  For example, the minimal cut set $A_1, A_2$ means that both the primary events $A_1$ and $A_2$ must occur in order for the top event to occur; therefore the set $A_1, A_2$ is an occurrence mode.  If either $A_1$ or $A_2$ does not occur, then the top event does not occur by this mode.  Specifically, in the fault tree, a minimal cut set is the smallest collection of events, all of which are necessary and sufficient, to cause the system failure, even if the other events (not in the minimum cut set) do not occur.  For most systems there are multiple minimum cut sets.

*Determination of the minimal cut sets is significant because they identify which events must be achieved in order for the top event state to happen.*  For very simple system models, the cut sets can often be determined by inspection. For more complex logic trees, algorithms and computer codes have been developed to obtain the minimal cut sets (see Fussel and Arendt [5] and Hansen [6]). MOCUS, one such algorithm (see Fussel et al. [7]), is a computer program that locates minimal cut sets from logic trees.

## 13.8 LARGE SYSTEM ANALYSIS AND LOGIC TREE REDUCTION

*Logic tree analysis techniques have the following common characteristics:*

1 The technique *produces a system model that promotes understanding* of the ways in which the system can fail (or succeed) and the ways in which failure (or success) can be prevented (or achieved).
2 The technique is *applicable to a wide variety of systems.*
3 The technique *provides reasonable assurance of completeness.*
4 The technique *enhances understanding and communication,* regarding system improvement.

Logic trees represent a means for system analysis, documentation, and communication. For complicated systems, they may grow to a rather large size. This growth may evolve over time. Hence, trees may be updated, expanded, modified, and so forth. Trees may be used in a qualitative mode or expanded and used in a quantitative mode. Other analytical tools, such as FMEA, are used in conjunction with the logic tree techniques to support the overall system modeling process. Typically, a highly qualitative FMEA serves as the baseline for the application of logic tree techniques.

All logic tree analyses begin with a clear definition of the performance dimension of interest. Detailed system models are then developed using an appropriate logic tree technique. Minimal cut sets can be identified and their relative importance established; the system models can be evaluated quantitatively (if necessary) to determine the probabilities of minimal cut sets and system failure or success. Sensitivity evaluations can be performed to determine the impact of output changes in the model as a function of input changes. The system model can thus be used to gauge the value of design or procedure improvements on the system performance measures identified.

Since logic trees are frequently lengthy and difficult to evaluate, they are often reduced or reorganized to facilitate their quantification (Hickman [8]). By their very nature, detailed logic trees contain many events that are insignificant in relation to other fault or success events or paths in the tree. It is desirable to include these events in the detailed tree to preserve the rigor and traceability of the analysis. However, in order to evaluate the tree efficiently, it is necessary to group or coalesce these insignificant fault or success events; this also increases ease of handling. This reduction can be accomplished manually or using computer aids. Reduction requires an interpretation of the tree logic and a gathering of the similar inputs under individual logic gates. Often, the original detailed logic tree is considered a worksheet and a reduced or reorganized version is prepared for the evaluation.

The logic tree reduction should not result in the loss of any significant information; rather, it should provide means of focusing on the more important events and eliminating time consuming evaluations of relatively meaningless combinations of insignificant or trivial events.  A detailed tree can be so large that evaluating the complete tree at one time is difficult, even after reduction.  In such a case, the tree is divided into identifiable subtrees that are evaluated separately.  In this approach, a careful search of each subtree is made to ensure that any common elements are identified.  These common elements must be treated consistently in each subtree—not as redundant elements, but as the same elements—for the logic tree to be valid.

In reducing the number of minimal cut sets, we usually select a truncation value. This value is selected such that those cut sets which contribute negligibly to the logic tree are eliminated.  As the truncation process eliminates unnecessary minimal cut sets for the event sequence, it is nonconservative.  If a suitably low truncation value is used, the effect on the total event sequence probability is slight.  Since this process is nonconservative, care must be taken to ensure that an appropriate truncation value is used.  The effect of the truncation value used should be bounded in a probabilistic sense and shown to be insignificant.

## REVIEW AND DISCOVERY EXERCISES

### Review

**13.1** How does FTA differ from FMEA in
    **a** Application objectives?
    **b** Application scope?

**13.2** Is FTA useful in proactive quality work?

**13.3** Can a CE analysis be useful in FTA?  Explain.

**13.4** In Example 13.1 assume we have the following estimates for the basic events:

| Event | Optimistic probability | Pessimistic probability |
|-------|------------------------|--------------------------|
| $B_1$ | .0001 | .005 |
| $B_2$ | .0005 | .005 |
| $B_3$ | .00005 | .001 |
| $B_4$ | .0002 | .01 |
| $B_5$ | .0001 | .05 |

    **a** Calculate the system optimistic (best case) top event occurrence.
    **b** Calculate the system pessimistic (worst case) top event occurrence.

**13.5** What is the difference between FTA and ETA in
    **a** Application objective?
    **b** Application scope?

**13.6** What is the difference between ETA and FMECA in
    **a** Application objective?
    **b** Application scope?

**13.7** How can ETA be useful in quality work concerning a production process?  Explain.

**13.8** Is the ETA compatible with mistake-proofing strategies (e.g., Table 6.1)?  Explain.

**13.9** Is the ETA appropriate for proactive quality work?  Explain.

**13.10** Explain the differences between FTA and GTA.

**Discovery**

**13.11** How can FTA be used in mistake-proofing in order to ultimately provide customer satisfaction?

**13.12** Select a product (good or service).  Identify a critical top event.  Then, develop an FTA.

**13.13** For your FTA in Problem 13.12 develop a set of basic event probabilities and calculate the probability of the top event's occurrence.

**13.14** Develop an ETA for the puncture of an automobile tire by a nail or other sharp object.

**13.15** Develop a set of probabilities for the ETA in Problem 13.14 and calculate the event probabilities.

**13.16** Develop a GTA with the top event being increasing your grade point average this term.

## REFERENCES

**1**  Y. Akao (ed.), *Quality Function Deployment,* Cambridge, MA: Productivity Press, 1990.

**2**  W. E. Vesley et al., *Fault Tree Handbook,* NUREG-0492, U.S. Nuclear Regulatory Commission, Washington, DC, 1983.

**3**  T. E. Bell, "Managing Murphy's Law:  Engineering a Minimum Risk System," *IEEE Spectrum,* vol. 26, no. 6, pp. 24–27, 1989.

**4**  M. C. Roush et al., "Integrated Approach Methodology:  A Handbook for Power Plant Assessment," SAND 87-7138, Sandia National Laboratories, Albuquerque: 1987.

**5**  J. B. Fussel and J. S. Arendt,  "Systems Reliability Engineering Methodology:  A Discussion of the State of the Art," *Nuclear Safety,* vol. 20, no. 5, 1979.

**6**  M. D. Hansen, "Software:  The New Frontier in Safety," *Professional Safety,* vol. 35, no. 10, pp. 20–23, 1990.

**7**  J. B. Fussel et al.,  "MOCUS—A Computer Program to Obtain Minimal Sets from Fault Trees," *Mathematics and Computers,* ANCR-1156, 1974.

**8**  J. W. Hickman et al., *PRA Procedures Guide,* NUREG/CR-2300, U.S. Regulatory Commission, Washington, DC, 1983.

# 14

---

# DESIGN REVIEW AND
# VALUE ANALYSIS

---

**1** How does the design review process enable us to create quality?
**2** Why is an independent design review critical?
**3** How is a value analysis structured?
**4** How is value analysis used in product and process definition?

## 14.1 INTRODUCTION

Our creativity in product-process definition and design is the critical component in creating quality in the product and its associated processes. In Chapter 11, we presented detailed quality planning strategies and tools. Now, we will focus on two additional tools which provide internal control to our creative efforts. We will first introduce the formal design review, which will act as a screen to help identify product and process bottlenecks. Then, we will introduce value analysis as a tool which links customer needs and expectations to the resolution of bottlenecks.

## 14.2 DESIGN REVIEW

*The purpose of a design review is to provide* **(1)** *a systematic and thorough product-process analysis,* **(2)** *a formal record of that analysis, and* **(3)** *feedback to the design team for product-process improvement.* The design review allows for an independent critique of a product and its related processes at appropriate time intervals in the product life cycle. Juran describes a design review as an independent assessment of the product or product plan which provides early warning as to problems that may arise if not addressed with appropriate action [1]. Hence, *the results of a design review are used in a proactive fashion to identify product-process bottlenecks.*

*Formal design reviews should be performed from an independent perspective.* In general, the participants in a design review should be experts in their respective fields and, when possible, hold no vested interest in, nor be directly responsible for, the product or process under review. *The independent perspective is critical in order to elicit unbiased comments, concerns, and recommendations during the review.* Nevertheless, facts, figures, and plans must be presented to the review team. Hence, interaction with those close to the product or process is necessary, even though this interaction is somewhat formal and presentation-response based.

In order to develop an effective design review we require (1) a team of functional experts (the reviewers) and (2) a product-process life cycle plan or program (which is the subject of review). Table 14.1 depicts a design-review matrix which lays out a detailed, generic product life cycle against a set of widely varied functional considerations, or viewpoints, the matrix should represent any and all viewpoints ranging from a customer perspective to a legal perspective.

*Design reviews are typically planned into the design process so that often we see a number of reviews conducted in sequence.* Each design review is focused on one or more aspects of the design or plan as it proceeds across the product life cycle, from needs assessment to disposal considerations. Each design review should yield a report which includes a list of concerns and comments. These items are then fed back to the design and planning team for action and eventual resolution. The action taken and eventual resolution are also entered into the formal record. Thus, a complete and formal documentation of review findings is developed and becomes a permanent part of the quality record.

*Design reviews vary in both formality and structure.* We will describe two extremes in structured design reviews. The first extreme is what we will term an "internal" design review. *In an internal design review four basic concerns are addressed:* (1) *the feasibility of the design with respect to function, form, and fit,* (2) *the produceability of the design with respect to production-process capabilities, sales and service capabilities, and so on,* (3) *cost-volume-profit estimates and the economic feasibility of the product, and* (4) *the technical and timeliness requirements of the product and our ability to meet these requirements.*

The second, or "external," design review extreme focuses on customers and their true quality characteristic demands. This design review extreme may or may not contain highly technical elements. Here, a reasonable representation of the customer base is involved. Customers critically assess the product function, form, fit, cost, timeliness, and so on—elements they feel are important with respect to the design. This extreme is primarily product focused, whereas the first extreme contains elements of both products and processes. *The external or customer review must either* (1) *verify that our product will indeed produce customer satisfaction or* (2) *provide us with information so we can improve our design accordingly.* Hence, sales and service as well as usage and disposal may become major issues in customer focused design reviews.

*The design review process serves as a proactive tool to help us improve our product and processes. Once completed, it also serves as a cradle-to-grave record of our diligence in serving our customers.* An examination of the record should show that all major concerns, or warnings, were adequately investigated and effectively brought to a satisfactory resolution in a timely manner. Otherwise, sloppy and negligent follow-up may eventually serve to encourage product nonperformance, liability accusations, and litigation.

## Radar System—Design Review Case [SR]

We were retained by a customer to serve as a consultant to a contractor that was hired to upgrade a radar system. The contractor was a small group with little

TABLE 14.1  DESIGN REVIEW MATRIX

| Functional consideration (viewpoints) | Life cycle phases | | | | | | | | | |
|---|---|---|---|---|---|---|---|---|---|---|
| | Customer needs | Conceptual definition | Conceptual design | Detailed design | Prototype | Production or manufacturing | Product sales and distribution | Product service and support | Product use | Product disposal |
| Leader-facilitator | ✓ | ✓ | ✓ | ✓ | ✓ | | ✓ | ✓ | ✓ | ✓ |
| Customers | ✓ | ✓ | | | ✓ | | ✓ | ✓ | ✓ | ✓ |
| Marketing | ✓ | ✓ | ✓ | | ✓ | | | | ✓ | ✓ |
| Product design | ✓ | ✓ | ✓ | ✓ | ✓ | ✓ | | ✓ | ✓ | ✓ |
| Production process design | | | ✓ | ✓ | | ✓ | | | | |
| Manufacturing | | | | | ✓ | ✓ | | | | |
| Inspection and testing | | | | ✓ | ✓ | ✓ | | | ✓ | ✓ |
| Distribution and logistics | | ✓ | | | | | ✓ | ✓ | ✓ | ✓ |
| Sales | ✓ | ✓ | ✓ | | ✓ | | ✓ | | | |
| Research and development | ✓ | | ✓ | ✓ | ✓ | | | | | |
| Quality | | ✓ | ✓ | ✓ | ✓ | ✓ | | ✓ | ✓ | ✓ |
| Safety and ergonomics | | ✓ | ✓ | ✓ | ✓ | ✓ | | | ✓ | ✓ |
| Procurement and purchasing | | | | | | | | | | |
| Cost and financial | ✓ | ✓ | | | ✓ | | ✓ | ✓ | ✓ | ✓ |
| Legal | | ✓ | ✓ | ✓ | ✓ | ✓ | ✓ | | ✓ | ✓ |

experience. During our first few meetings with the group members, we attempted to give them advice on the type of tasks that needed to be structured in order to solve technical problems in upgrading the system.

At the first requirements (design) review, the contractor presented schedules and alternative solutions to meet the high-level requirements. They eliminated all quality control related requirements from the schedule and said they could finish the job in two months. We estimated that this process should take about a year and gave them a detailed outline of the tasks that needed to be accomplished.

At the second requirements review, two weeks later, the contractor had failed to develop detailed requirements. They still had basically the same schedule and stated that they had started software coding.

At the third requirements review, two weeks later, they presented very little in the way of a requirements analysis and said that they were almost done coding. The limited analysis presented was full of errors.

At the fourth requirements review, three weeks later, the contractor presented more erroneous analyses and said the coding was done. Their requirements analysis did not come close to showing how they were going to meet their objectives and their code was full of errors. They maintained their original schedule and stated that they would be ready to start flight tests in two weeks. At this point, they had performed no safety analysis or FTA. They planned to execute the safety analysis and FTA after flight had started.

At our request, the customer canceled the contract.

## 14.3 VALUE ANALYSIS

*The purpose of value analysis (VA) is to provide more functional utility (i.e., useful products) to customers at a reduced cost.* The concept of value analysis was developed formally in 1947 at General Electric by L. D. Miles. Over the years, the concept has been practiced under the name of "value engineering" (VE), as well. VA has been widely implemented throughout the industrialized world, including Japan, since the early 1960s. The majority of successful cases reported deal with existing products; however, recently, applications in new product development have been stressed (Fowler [2] and Juran [3]).

We will examine two levels of VA. The high-level VA and the detailed, or detail-level, VA are similar in structure; the major difference is in their purpose. *The high-level VA analysis serves primarily a product-process function identification and screening purpose.* Whereas *the detailed VA provides a creative element in addition to the identification and screening elements.*

### High-Level Value Analysis

Value analysis is described in a variety of ways by different authors. *Juran views VA as a process with four basic inputs*: (1) *customer needs,* (2) *product features,* (3) *cost estimates for product features,* and (4) *information on competing product features and costs* [4] . He describes a matrix format for associating costs and functions

| Departments, operations, assemblies, parts, etc. | Cost | Functions (verb + noun) | | | | | | | | | |
|---|---|---|---|---|---|---|---|---|---|---|---|
| | | Provide landing | Accepts steps | Provide safety | Be durable | Be inter-changeable | Provide rigidity | Be serviceable | Provide identity | Be adjustable | Assembly |
| Comb (4) | £11.58 / 13.14% | 0.58 | 2.90 | 4.05 | 0.23 | 1.16 | | 1.40 | | 1.27 | |
| Safety switch | £13.14 / 14.90% | | | 5.26 | | 4.60 | | 1.31 | | 1.97 | |
| Comb bearer | £28.94 / 32.83% | 11.57 | | | | | 14.47 | | | 2.89 | |
| Tread plate | £ 3.22 / 3.65% | 0.64 | | 0.64 | 1.28 | 0.32 | | 1.53 | 0.32 | | |
| Entry guide | £15.26 / 17.31% | | 5.34 | 3.81 | 1.53 | 1.53 | | | | 1.53 | |
| Logo | £ 1.01 / 1.15% | | | | 0.51 | | | | 0.51 | | |
| Assembly | £15.00 / 17.02% | | | | | | | | | | 15.00 |
| Total cost (£) | 88.15 | 12.79 | 8.24 | 13.76 | 3.54 | 7.61 | 14.47 | 4.23 | 0.82 | 7.66 | 15.00 |
| % of total cost | 100% | 14.51 | 9.33 | 15.61 | 4.01 | 8.63 | 16.41 | 4.80 | 0.93 | 8.69 | 17.02 |

**FIGURE 14.1**   High-level value analysis (VA) format. Reprinted, with permission of The Free Press, a Division of Simon & Schuster, from J. M. Juran, *Juran on Planning for Quality*, New York, NY: Free Press, page 133, 1988. Copyright 1988 by Juran Institute, Inc.

**TABLE 14.2**   EIGHT-STEP DETAILED VALUE ANALYSIS PLAN

| Phase | Step | Comments |
| --- | --- | --- |
| Information and analysis | Preparation | Prepare cost and function data, schedule sessions and participants, collect user-customer data. |
| | Information | Perform function analysis (functions and cost). |
| | Analysis | Define and select specific value analysis targets and projects. |
| Creativity and synthesis | Creativity | Create thoughts and ideas. |
| | Synthesis | Synthesize potential solutions. |
| Evaluation and development | Development | Prove feasibility of potential solutions. |
| | Presentation | Present potential solutions for decision making. |
| | Follow-up | Assure effective implementation. |

*Source:* Adapted, with permission from Van Nostrand Reinhold, from T. C. Fowler, *Value Analysis in Design*, New York: Van Nostrand Reinhold, page 31,1990.

with product and/or processes. The Juran VA format is shown in Figure 14.1; it is compatible with spreadsheet computer aids. This matrix essentially organizes three quality relevant elements: (1) physical product and process components, (2) product features expressed as functions (e.g., using a verb and a noun), and (3) estimated costs allocated by physical product component and by product function.

The Juran approach provides an excellent first-cut analysis format for strategic quality planning. It offers a broad perspective of function and cost issues and associates them with the product-process. This high level of analysis is appropriate for system design level thinking and decision support. It is compatible with quality planning and QFD development. Specifically, it is relevant to cost-volume-profit deployment; it helps in determining and allocating cost estimates to product and process functions. Hence, we can use high-level VA to help identify cost bottlenecks in our quality plan or QFD.

## Detail-Level Value Analysis

In contrast to the high-level VA previously described, *we can develop a detailed VA which not only associates functions and costs, but provides creative resolution to value bottlenecks in product and process design.* Detailed value analysis studies are typically organized by and executed through small work teams—usually around five team members (Fowler [5] and Mudge [6]). *Here, a VA study consists of three phases:* **(1)** *information and analysis,* **(2)** *creativity and synthesis, and* **(3)** *evaluation and development.* Each phase is distinctively different and may be broken up into a number of steps. Table 14.2 outlines an eight-step VA plan described by Fowler [7]. A value analysis study typically stretches over a few months in order to provide ample time to collect function and cost data, provide incubation time for innovative thoughts to develop, and allow for thorough solution synthesis and development.

**Information and Analysis Phase**   *The preparation and information steps take on the characteristics of an exercise in organizing functives, costs, and customer acceptances for analysis.  A "functive" is defined as a two-word term consisting of one verb and one noun* (e.g., inject plastic, filter dust, protect user).  In other words, a functive describes the essence of a function.  The major objective of the analysis step is to identify where the VA team should concentrate its creative effort in order to maximize its impact on the product.  *The net result of the analysis step is the selection of targets (specific functives) where a poor match between function cost and function acceptance is observed.*

Selection of specific VA targets is made by taking a customer perspective.  Essentially, we seek to target functives with relatively low customer acceptance and/or relatively low cost impact.  In other words, we want to improve customer acceptance of our product in the most cost effective manner.  In the case of a new product, we are also searching out basic functives that are overlooked or inadequate and could preclude customer satisfaction altogether.

*Value in the VA is defined as a ratio of worth to cost*:

$$\text{Value} = \frac{\text{worth}}{\text{cost}} \tag{14.1}$$

More specifically, Fowler [8] defines value as

$$\text{Value} = \frac{\text{user's initial impression} + \text{satisfaction in use}}{\text{first cost} + \text{follow-up costs}} \tag{14.2}$$

The VA team's expertise should be representative of six fundamental components: design, operations, costing, field service and marketing, purchasing, team facilitation (Fowler [9]).  *The information and analysis phase consists of a systematic and thorough discovery exercise that identifies both worth and cost elements associated with a product.*  Effective function definition and analysis are the keys to success in this phase.

The function analysis system technique (FAST), as described by Fowler [10], has proven to be an effective tool in both discovery and analysis of functionality.  *The FAST diagram is a structured diagram of "functives."*  A FAST diagram is typically initiated with a display board and a pad of "Post-it" note sheets.  In the FAST technique, we take the perspective of the product user and proceed as follows:

1  Define as many functives as possible.
2  Identify the major functive, sometimes referred to as the "task functive."
3  Identify the basic functives and their supporting functives.
4  Identify the supporting functives and stratify them into four basic groups:
   **a**  Assure dependability.
   **b**  Assure convenience.

**c** Enhance product.

**d** Please senses.

These functives are organized in a tree form (see Figure 14.2). Then, numbers and letters are used to uniquely identify each functive. This tree structure of functives addresses how the product will satisfy customers with respect to product functionality. It is interesting to note that the FAST diagram structure is similar to a combination of the affinity diagram and the systematic diagram (both described in Chapter 7).

Next, *a cost allocation is made against the hierarchy of functives.* This substep establishes cost estimates throughout the hierarchy. For most products this allocation is challenging in that the functives are not totally independent of each other and "splits" must be made (i.e., how much cost is allocated to each function is determined, based on engineering/business judgement).

All splits or cost breakouts must total to the product cost figure. Figure 14.3 illustrates such a breakout in the completed FAST diagram. In this figure, we can see the functives hierarchy and the relative percent of the total cost associated with the basic and support functions. For new products, a cost breakout will consist of rough estimates (as the total product cost figure will be speculative or perhaps a rough cost estimate itself).

The far right-hand column in Figure 14.3 contains function acceptance information. Here, we are dealing with the "worth," or numerator, portion of the value equation. *Function acceptance, or worth, information is usually collected in one of two ways: (1) surveys or (2) focus panels.* The acceptance numbers in Figure 14.3 indicate the strength and the specific like-dislike number from the customer survey or focus panel results.

*Customer Survey*    The customer survey may take the form of interviews or questionnaires. In either case, we must develop a format which will elicit both like and dislike statements with regard to our product. Figure 14.4 shows an example customer survey questionnaire used in either a face-to-face or a telephone interview. The "+" indicates a positive response, the "–" indicates a negative response. The number associated with the "+" or "–" is a strength indicator on a 1 to 10 basis, where 10 is the strongest level (e.g., –10 is the strongest negative and +10 is the strongest positive). The original question represents the planned portion of the interview. The "probe" element in the line of questioning follows up on the original question and extracts detailed information.

*Focus Panel*    The focus panel is a group of perhaps 10 to 15 people, roughly half users or customers and half in-house personnel. Each of these two subgroups plays a distinct role in the panel analysis process. The in-house subgroup must identify prime buyer categories. Prime buyer categories include the identifiable user-buyer-customer players in the marketplace (e.g., for an automobile we could identify the individual owner, the fleet owner, the driver, the passenger, and the mechanic). Then, a unique prime buyer category is selected. The users-customers selected for the focus group should be representative of the selected prime buyer category.

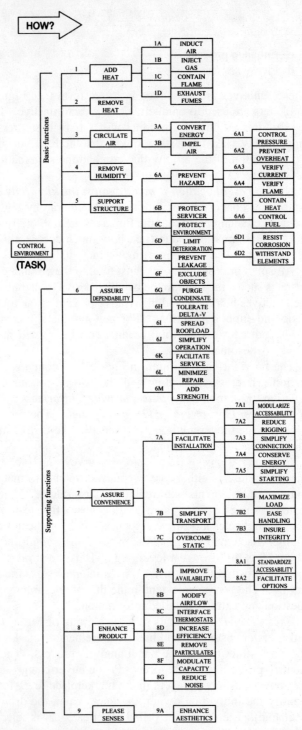

**FIGURE 14.2** Heat pump FAST diagram (partial). Reprinted, with permission from Van Nostrand Reinhold, from T. C. Fowler, *Value Analysis in Design*, New York, NY: Van Nostrand Reinhold, page 86, 1990.

| | | | COST | ACCEPTANCE |
|---|---|---|---|---|
| **TASK FUNCTION** - Control environment | | | 2548.40 [100%] | |
| | | | | |
| **BASIC FUNCTIONS** | | [$711.46 - 28%] | | |
| 1 | Add heat | | 272.64 | |
| | 1A | Induct air | 34.32 | |
| | 1B | Inject gas | 80.02 | |
| | 1C | Contain flame | 127.42 | |
| | 1D | Exhaust fumes | 54.30 | |
| 2 | Remove heat | | 127.42 | |
| 3 | Circulate air | | 84.84 | |
| | 3A | Convert energy | 38.20 | |
| | 3B | Impel air | 46.64 | |
| 4 | Remove humidity | | 47.12 | |
| 5 | Support structure | | 179.44 | |
| | | | | |
| **SUPPORTING FUNCTIONS** | | [$1,836.94 - 72%] | | |
| 6 | Assure dependability | | 686.54 [27%] | |
| | 6A | Prevent hazard | 207.34 | |
| | | 6A1 Control pressure | 28.91 | |
| | | 6A2 Prevent overheat | 32.40 | 6-55 |
| | | 6A3 Verify current | 28.25 | |
| | | 6A4 Verify flame | 26.18 | |
| | | 6A5 Contain heat | 54.10 | |
| | | 6A6 Control fuel | 37.50 | |
| | 6B | Protect services | 41.92 | 8-57 |
| | 6C | Protect environment | 28.20 | 2-39 |
| | 6D | Limit deterioration | 94.22 | |
| | | 6D1 Resist corrosion | 47.80 | |
| | | 6D2 Withstand elements | 46.42 | |
| | 6E | Prevent leakage | 37.80 | 3-21 |
| | 6F | Exclude objects | 28.92 | |
| | 6G | Purge condensate | 14.47 | |
| | 6H | Minimize contamination | 38.17 | 8-43, d6-14 |
| | 6I | Spread roofload | 22.40 | |
| | 6J | Simplify operation | 32.13 | 4-5, 9-7, 4-8, d3-3, d2-29, d3-30 |
| | 6K | Facilitate service | 52.12 | 10-59, 8-63, 8-68, d10-5, d10-9 |
| | 6L | Minimize repair | 24.40 | 8-43, 9-52, 9-67, 8-9, 10-23, 10-29, 9-38, d6-1, d8-8, |
| | 6M | Add strength | 64.25 | 8-17, 10-30, 9-37, d6-20, d5-22, d9-11, d19-17, |
| 7 | Assure convenience | | 481.84 [19%] | \d9-28 |
| | 7A | Facilitate installation | 286.80 | 9-47, 10-48, 1-56, 9-66, 10-1, 10-4 |
| | | 7A1 Modularize accessability | 34.80 | 8-50, 10-62, 9-3, 8-10, 9-15, 9-22, 9-24 |
| | | 7A2 Reduce rigging | 29.25 | 10-49, 8-60, 8-36, d8-6, d8-7, d9-15 |
| | | 7A3 Simplify connection | 117.36 | 9-11, 9-26, 10-16, 6-33, d5-12, d6-16, d6-21, |
| | | 7A4 Conserve energy | 88.20 | \d4-27, d5-32 |
| | | 7A5 Simplify starting | 17.19 | |
| | 7B | Simplify transport | 138.20 | |
| | | 7B1 Maximize load | -0- | 8-64, 10-19, 9-20, 10-32, 10-40, d10-10, d9-13 |
| | | 7B2 Ease handling | 72.56 | 6-13 |
| | | 7B3 Insure integrity | 65.64 | |
| | 7C | Overcome static | 56.84 | |
| 8 | Enhance product | | 562.26 [22%] | |
| | 8A | Improve availability | 76.20 | |
| | | 8A1 Standardize accessability | 56.21 | 10-45 |
| | | 8A2 Facilitate options | 19.99 | 10-46, 5-65, 5-12, 8-41, d8-24, d6-26 |
| | 8B | Modify airflow | 108.30 | 1-18 |
| | 8C | Interface thermostats | 37.30 | |
| | 8D | Increase efficiency | 188.72 | 4-51, 4-54, 10-61, 9-2, 8-6, 10-14, 10-28, 10-35 |
| | 8E | Remove particulates | 30.42 | 2-25, 1-27, d8-4, d10-18, d9-19, d10-23 |
| | 8F | Modulate capacity | 21.20 | |
| | 8G | Reduce noise | 102.12 | 2-31 |
| 9 | Please senses | | 104.30 [4%] | |
| | 9A | Enhance aesthetics | 104.30 | 8-44, 3-34, 4-42 |

**FIGURE 14.3** Completed heat pump FAST diagram, including functives, cost, and customer acceptance. Reprinted, with permission from Van Nostrand Reinhold, from T. C. Fowler, *Value Analysis in Design*, New York, NY: Van Nostrand Reinhold, page 97, 1990.

---

PRODUCT: _Model 8700 8-Ton Gas / Electric Rooftop Heat Pump_

DATE: _August 3, 1993_

MANUFACTURER: _General Corporation_

NAME OF RESPONDENT: _Alex Founder_

LOCATION: _Baltimore, MD_

---

QUALIFY RESPONDENT:

☐ IN-HOUSE _____
(Title and area of responsibility)

OUTSIDER

☒ USER OF PRODUCT BEING ANALYZED ____7____
(months of use)

☒ USER OF COMPETITIVE PRODUCT _Carrier 45 L 8085_
(manufacturer of product)

☐ DEALER/DISTRIBUTOR _____
(Heat Pump product lines)

SPECIFIC JOB AND RELATIONSHIP TO THE HEAT PUMP _Building owner and operator of a Central Laundry processing Building._

---

(1)  Why do you own a General Corporation Heat Pump? (_I like the efficiency._)**+3** Good (_reputation for long life._)**+5**  (probe) (_General is a good name._)**+5**

---

(2)  Overall, how would you rate the effectiveness of this Heat Pump?

☐ Extremely effective          ☐ Fairly effective
☒ Very effective               ☐ Not particularly effective
                               ☐ Not effective at all

(probe) (_Very efficient._)**+6** (_Only problem was_(_a slight leak in piping._)**-2** _Fixed under warranty. (Probe) Problem was_ (_badly soldered joint._)**-2**

---

(3)  Is there anything about the operation of the Heat Pump that annoys you?

YES ☒   NO ☐  (probe) (_Noisy._)**-3** _Vibrates the roof over the office area. Not that bad though._
(Probe) _Maintenance man says it's_ (_Hell to disassemble for repair,_)**-6** _but so far it's_(_so dependable_)**+5** _that it doesn't matter. He also says_(_you need a whole tool kit to repair a General._)**-3** _No two fasteners are alike._

---

(4)  Do you have any complaints about the size or configuration of the Heat Pump?

YES ☒   NO ☐  (probe) (_Well, we did have to put in extra roof beams to hold it._)**-3** _The physical size doesn't matter though. It's on the roof!_

---

**FIGURE 14.4**  Customer survey format for a heat pump. Reprinted, with permission from Van Nostrand Reinhold, from T. C. Fowler, _Value Analysis in Design_, New York, NY: Van Nostrand Reinhold, page 56, 1990.

Typically, a number of the leading competitors (relative to our product) are identified. Inclusion in this set may be determined on the basis of market share in general or specific characteristics regarding the prime buyer category itself (e.g., a best-in-class reputation). The establishment of the prime buyer category and the structure of the competition establish a framework for final selection of the user-customer participants and the identification of product likes and dislikes.

After we select the prime buyer category and assemble the focus group, we identify and verbalize specific likes and dislikes. Once the panel understands the nature of the likes and dislikes, a voting process is initiated. The in-house members are instructed to vote from the perspective of the selected prime buyer, whereas the outside members are instructed to vote or respond from their own frame of reference, as they were selected as typical prime buyers.

The votes are recorded and tallied with a measure of central tendency and variation determined for each like and dislike. Typically, a mode and a range are used for the central tendency and variation measures, respectively, as they are easy to compute and tend to adequately reflect voter response as an aggregate. Figure 14.5 depicts a portion of a focus panel voting summary.

Once the survey and/or panel results are complete, we transfer them to the FAST diagram (see Figure 14.3). The mode (or some other measure of central tendency) is used as a strength indicator. Each like or dislike is matched with a functive on the FAST diagram. Then, the strength and like-dislike numbers are recorded on the FAST diagram. At this point, we have captured and organized functives, cost, and customer acceptance information together.

**Creativity and Synthesis Phase**   *The creativity step of a VA consists of a freewheeling process of encouraging and recording the thoughts of team members with respect to the target functives.* The brainstorming process is used to produce thoughts and ideas in the absence of criticism (Osborn [11]). The purpose is to generate and collect a large volume of thoughts for each functive targeted.

Brainstorming typically produces a wide variety of thoughts, some more developed than others. Each thought originates in an individual's mind. However, it is usually productive to "bounce" thoughts around within the VA team, with the expectation that one thought may stimulate others. In other words, a thought chain may develop within the group and produce a creative series of thoughts with respect to each functive.

In many cases, the previous information and analysis phase resembles a cram session where we load our minds with organized facts, figures, and relationships. Once this information is arranged in a well organized, systematic manner, it leads to the formulation of a specific purpose or purposes relative to a product-process functive; these may be expressed in the form of specific goals, objectives, and targets. In contrast, the creative phase represents a less structured, "off-the-wall" exercise in creative thinking. The creative thinking process and the thoughts it yields are the fragile beginning of the solution, in either an incremental or a breakthrough sense.

*Synthesis refines and scrutinizes the thoughts generated in the creative step.* Essentially, through synthesis, we transform our thoughts into ideas. *We can take basically one of two approaches in synthesis*: (1) *championing or* (2) *critical* . In a championing approach we focus on the virtues of a thought relative to a functive. Hence, we view the thought in an optimistic manner and develop it as an idea in this fashion. The critical approach focuses on identifying the shortcomings of the thoughts; here, we are addressing the same functive, but pessimisticly.

| Description of features (or "likes") | Manufacturer | | | | | | | Users | | | | | Notes | Mode & Range | Function |
|---|---|---|---|---|---|---|---|---|---|---|---|---|---|---|---|
| | 1 | 2 | 3 | 4 | 5 | 6 | 7 | 8 | 9 | 10 | 11 | 12 | | | |
| 43  Less than 1% compressor failure in 5 years | 8 | 8 | 6 | 8 | 5 | 8 | 8 | 8 | 10 | 3 | 4 | 9 | | 8R7 | |
| 44  Low silhouette (less than 3 feet) | 5 | 9 | 8 | 9 | 8 | 8 | 6 | 4 | 6 | 8 | 6 | 8 | (4) | 8R5 | |
| 45  Supply and return sides coincide with Carrier | 6 | 10 | 10 | 8 | 8 | 9 | 10 | 10 | 9 | 8 | 10 | 10 | (4) | 10R4 | |
| 46  Easily field convertible from side output to down | 10 | 10 | 10 | 10 | 10 | 8 | 9 | 9 | 10 | 10 | 10 | 5 | (4) | 10R2 | |
| 47  Single point power, factory installed | 9 | 10 | 9 | 9 | 8 | 10 | 10 | 10 | 10 | 9 | 9 | 9 | | 9R2 | |
| 48  Ability to complete sheet metal work to curb before unit arrives | 9 | 9 | 10 | 10 | 8 | 8 | 9 | 10 | 10 | 8 | 10 | 10 | | 10R2 | |
| 49  Ability to rig without spreader bars | 4 | 8 | 8 | 10 | 9 | 10 | 6 | 10 | 10 | 10 | 3 | 10 | | 10R7 | |
| 50  Parts interchangeable between models | 8 | 9 | 8 | 9 | 8 | 10 | 10 | 9 | 10 | 10 | 8 | 8 | | 8R2 | |
| 51  Thermal expansion valves versus capillary tubes | 8 | 8 | 4 | 4 | 3 | 4 | 4 | 5 | 9 | 6 | 9 | 6 | | 4R6 | |
| 52  Dealer/contractor diagnostic capability | 6 | 9 | - | 6 | 5 | 10 | 9 | 4 | 8 | 9 | 10 | 4 | | 9R6 | |
| 53  Crankcase heater | 9 | 10 | 8 | 8 | 6 | 5 | 6 | 8 | 10 | 5 | 9 | 9 | | 8R5 | |
| 54  Energy management controls available as option | 3 | 4 | 4 | - | 8 | 2 | 5 | 4 | 6 | 9 | 3 | 4 | | 4R7 | |
| 55  5 min. delay on compressor as standard | 6 | 10 | 6 | 8 | 6 | 9 | 6 | 6 | 8 | 8 | 10 | 10 | (2) | 6R4 | |
| 56  Concentric transition within curb | 2 | 3 | - | - | 3 | 1 | - | 0 | 6 | 1 | - | 1 | (4) | 1R5 | |
| 57  Power vent on gas-heated unit | 4 | 6 | 4 | 8 | 5 | 3 | 4 | 9 | 8 | 3 | 9 | 6 | (3) | 8R5 | |

Note (1): Multimodal—rating shown is the MEAN.
Note (2): Multimodal—rating shown is the MODE nearest the MEAN.
Note (3): Data spread is excessive and invalid; rating shown is the MEAN.
Note (4): Not allocable; does not relate to the specific model under study.

**FIGURE 14.5**    Focus group voting summary data sheet for a heat pump. Reprinted, with permission from Van Nostrand Reinhold from T. C. Fowler, *Value Analysis in Design*, New York, NY: Van Nostrand Reinhold, page 64, 1990.

Typically, we exercise both points of view. However, we may select a champion and a critic in two different people or subgroups in order to synthesize an idea thoroughly. Some thoughts will be successfully synthesized and some will not.

**Evaluation and Development Phase**    *Evaluation entails assessing the feasibility of the successfully synthesized ideas.* At this point, we must logically develop functional alternatives from our ideas and assess their potential contributions with respect to our functives. Here, *we must consider three basic types of issues and risks: (1) technical, (2) cost-benefit, and (3) timeliness.*

The technical issues focus on our ability to make the idea work in our organization. The technical risk represents our prospects for successful accomplishment and the likelihood that this accomplishment will mirror or successfully address customer needs, demands, and expectations. In some cases, we may be able to make our idea work directly. In other cases, we may identify bottlenecks that will require implementation of an experimental program to develop the technology and experience necessary to make an idea work. Technically, the more breakthrough thinking based the idea, the more experimentation and development required.

Cost-benefit issues address both the resources required to develop an idea and the benefits likely to accrue as a result of its successful development. Both sides of the issue require estimates, with the benefits usually being the more difficult of the two to estimate.

The timeliness issue is critical in some industries, such as the electronics industry, where product and process technology (as well as customer needs and expectations) change rapidly. In other words, a good technical solution, too late, may be of little value. Or, on the other hand, a good technical solution in a timely fashion may produce significant benefits.

Our ultimate recommendations, which support the decision making process, grow out of the evaluations we perform. Most of our synthesized ideas regarding any given functive will be mutually exclusive (i.e., we must choose or commit to only one). *The evaluation and development phase should set up alternatives for the decision making process.*

## REVIEW AND DISCOVERY EXERCISES

### Review

**14.1** Explain how a design review adds a control element to the quality planning process.

**14.2** Compare and contrast the purpose(s) of a design review with the purpose(s) of a value analysis.

**14.3** Select a product and develop a functive hierarchy for a related FAST diagram.

**14.4** Compare and contrast the high- and detail-level value analysis formats
   **a** In terms of possible applications in the quality planning-QFD (new product) environment.
   **b** In the product and production process improvement (existing product) environment.

### Discovery

**14.5** Select a reasonably simple product (a good or service):
   **a** What is its life cycle?
   **b** What would be an appropriate design review sequence?
   **c** When would you schedule design reviews?
   **d** What general format do you recommend for each design review (e.g., what viewpoints are critical in each design review)?
   **e** How would you get the product-process design facts out in the open for the reviewers to assess?

**14.6** Select a reasonably simple product (a good or service). Develop a high-level VA for the product, use the Juran spreadsheet format (Figure 14.1).

**14.7** Select a reasonably simple product (a good or service). Develop a detail-level VA for the product, use the Fowler format.

## REFERENCES

1  J. M. Juran, *Juran on Planning for Quality,* New York: Free Press, 1988.
2  T. C. Fowler, *Value Analysis in Design,* New York: Van Nostrand Reinhold, 1990.
3  See reference 1.
4  See reference 1.
5  See reference 2.
6  A. E. Mudge, *Value Engineering: A Systematic Approach,* New York: McGraw-Hill, 1971.
7  See reference 2.
8  See reference 2.
9  See reference 2.
10  See reference 2.
11  A. Osborn, *Applied Imagination,* New York: Charles Scribner, 1953.

# CREATION OF QUALITY— STATISTICAL PROCESS CONTROL

## VIRTUES FOR LEADERSHIP IN QUALITY
### Responsibility

Section Five focuses on statistical quality control and features expanded discussions of statistical process control (SPC) and limited discussions pertaining to inspection sampling plans. Our purpose is to introduce our readers to classical statistical quality control and emphasize a process focus, rather than a product focus in production activities. In Chapter 9, we introduced SPC at the conceptual level. In Chapter 15, we briefly review some of this material and again point out the difference between attributes and variables measures. Then, we develop a number of variables control charts, with examples to illustrate both the conceptual application and the statistical mechanics. In Chapter 16, we introduce a number of attributes control charts and emphasize the difference between attributes and variables charting strategies. We follow these discussions up in Chapter 17 with a number of topics necessary to apply and interpret control charts. For example, we discuss sampling schemes, production source charting, runs rules, operating characteristic curves and average run lengths, as well as probability limits on our control charts.

Chapter 18 focuses on process capability and its relation to process stability. Here, we introduce process capability measures as well as assembly tolerance models. We discuss the relationship between traditional product specifications and the Taguchi loss function approach to assessing our production processes. We close Chapter 18 with a gauge study discussion which is applicable to measurement and instrumentation in general. Here, we introduce the reader to gauge study development using the SPC technology. Chapter 19 rounds out the section with a brief introduction to sorting inspection. Here, we discuss acceptance sampling by attributes and introduce our readers to published sampling plans. We also present single sampling plan operating characteristic curves for lot-by-lot acceptance plans based on attributes measures.

# 15

# VARIABLES CONTROL CHARTS

## 15.0 INQUIRY

1  What is the difference between common cause and special cause in production processes?
2  How do control chart limits differ from tolerance limits?
3  How does SPC charting relate to hypothesis testing?
4  What variables control charts see widespread use?
5  How are variables charts constructed?

## 15.1 INTRODUCTION

*Products and production processes are always subject to a certain amount of variation as a result of chance alone.* In other words, there are always inherent chance causes which, as a result of "normal" variation in materials, environments, methods, and so on, are responsible for natural variation in any particular scheme of production. *Variation within a stable pattern of chance causes is inevitable.* If our process begins to operate outside this stable pattern, we want to know about this shift or upset as soon as possible. *Once we have an indication of a shift outside a stable pattern of variation, we must discover the reason for the shift*, i.e., the "special cause," and correct it (Grant and Leavenworth [1]). We want to remove the influence of a special cause if it is adversely affecting product-process quality. If for some reason the special cause influence is improving product-process quality, we want to permanently capture its effect.

*Statistical process control (SPC) charts* were introduced as one of the seven fundamental QC tools in Chapter 9. Control charts, first developed by Shewhart and his colleagues at the Bell Telephone Laboratories (as noted in Chapter 9), *have the ability to indicate the presence of special causes that upset our processes.* SPC charts help us to detect, diagnose, and correct production problems in a timely fashion. The result is substantial improvement in product quality. On the other hand, control charts also tell us when to leave a process alone, thus preventing unnecessary adjustments that tend to increase the variability of the process, rather than reduce it.

In addition to detecting the presence of special causes, control charts help us estimate the natural tolerance of a production process, which in turn permits us to estimate process capability. A natural tolerance is defined as the interval covering $\pm 3\sigma$ ($\pm 3$ sigmas, or standard deviations) from the mean of a measured process quality characteristic (Grant and Leavenworth [2]). Knowing our process capabilities helps us to select the appropriate machines, tools, and so on, to meet engineering specifications consistently in our production operations.

## 15.2  VARIABLES AND ATTRIBUTES

SPC calls for an understanding of the important distinction between variables measures and attributes measures. *When a record is made of an actual, measured quality characteristic, the quality characteristic is said to be "expressed" by variables.* On the other hand, *if a record shows only a summary or classification with regard to any specified set of requirements,* either expressed or implied, *it is said to be a record by "attributes."*

Many requirements are stated as variables. Examples are dimensions, tensile strength, operating temperatures, weight in pounds of the contents of a container, percent yield from a process, life in hours of a product, fuel efficiency, and so on.

Many requirements are also stated in terms of attributes rather than variables. A writing pen writes or it does not. A photograph film is either correctly exposed or it is not. The surface finish of a newly painted automobile presents a satisfactory appearance or it does not. Instructions for taking a particular medicine are either missing from a doctor's written prescription or not. A 1000 square yard piece of carpet has three defects. A shirt has no defects, and so on. In general, the item examined either conforms or does not conform to specifications in one way or another.

*Product and process engineers typically express a quality requirement as a target value, a tolerance interval, or both.* For example, we might state a quality target requirement of 0.5 in for the diameter of a brass bushing product and a tolerance interval of $0.5 \pm 0.01$ in for the same product. The target implies that 0.5 in is the ideal diameter; the tolerance interval implies that a bushing within the interval 0.49 in and 0.51 in, inclusive, is acceptable. We might decide to measure our product on the factory floor as a variable (e.g., we measured the diameter of a specific bushing to be 0.502 in). Or on the other hand, we might have a go–no go gauge that will tell us if a specific bushing falls within the interval or not. If it does, it is acceptable (good). If it doesn't, it is not acceptable (defective).

*We have more information contained in the variables measurement* (e.g., we know more about the bushing diameter) *than in the attributes measurement* (e.g., we know only that the bushing is within the interval, somewhere between 0.49 in and 0.51 in, inclusive). In choosing between variables and attributes measures we must consider specific technical, economic, and timeliness factors. In general, variables sample sizes will be smaller than attributes sample sizes. We will see this fact demonstrated in examples in the next few chapters.

## 15.3  CONTROL CHARTS FOR VARIABLES MEASUREMENTS

Variables data can be summarized and analyzed in a number of ways, including histograms, summary statistics, and so on. However, using these traditional tools to do so is somewhat ineffective and inefficient. We will describe a number of specialized graphical techniques that allow us to monitor both central tendency, or "location," and the amount of variation, or "dispersion," of the production process. When we use a variables measure of assessment, by monitoring these two process characteristics, or parameters, we will be able to assess the stability of a process or product quality characteristic.

*The most commonly used measure of location or central tendency is the "arithmetic mean."* In our discussions, we will develop a number of SPC charts based on the mean. *Two measures of dispersion that are extremely useful in statistical quality control are the "range" and the "standard deviation."* In dealing with variables quality characteristics *it is customary to use separate charts for monitoring location and dispersion measures.* Since our processes operate over time or order of production, we will track location and dispersion over time or order of production and continuously assess process stability.

*In SPC, it is customary to assume that a process is "in control," or stable, until sufficient evidence is discovered to the contrary.* Since the assumption that the process is in control in general implies that both the location parameter and the dispersion or variation parameter are "on target," we can conceptualize a set of statistical hypotheses for each of these parameters. Here, $H_0$ refers to the null hypothesis and $H_1$ refers to the alternate hypothesis.

*Hypothesis for the location parameter:*

$H_0$:   The process population mean is on target

$H_1$:   The process population mean is not on target

*Hypothesis for the dispersion parameter:*

$H_0$:   The process population variation about the mean is on target

$H_1$:   The process population variation about the mean is not on target

In a crude sense, we are testing each of these hypotheses sets each time we plot a point on our variables control charts (e.g., we typically use one control chart to monitor location and another chart (of the set) to monitor dispersion). Hence, we are asking ourselves whether or not to assume that the process is stable (i.e., we do not reject the null hypothesis, $H_0$) with respect to location and likewise for dispersion. *If a special cause is present and the process is upset or distributed, and therefore unstable, we seek to identify the physical cause and remove its influence on the production process* (i.e., we have rejected one or the other or both of the null hypotheses above).

*Variables control charts monitor only one quality characteristic per set of charts and provide three basic benefits: (1) they display the average level of the quality characteristic—e.g., the location or central tendency, (2) they display the variation in the quality characteristic—e.g., the dispersion, and (3) they display the consistency in our production over time or order of production.*

*It is critical to understand that SPC charts are incomplete with respect to total process control in two regards: (1) they do not tell us whether or not we are meeting tolerance specifications consistently, and (2) they neither explicitly identify nor remove special causes.* Once the control charts indicate that a process is in control with respect to both location and dispersion, we may obtain a false feeling of confidence that the product meets specifications. The purpose of process control charts is to address process stability—not to tell us if we are meeting tolerance specifications. Furthermore, process control charts do not identify special causes and they do not tell us how to eliminate effects of special causes. *Control charts basically serve us as statistical based warning devices for process upsets.* It is up to us to develop meaningful process logs to account for and document physical characteristics, actions taken, and results obtained.

## Example 15.1—Meter Stick Case

Before proceeding further with a detailed description of the different types of variables control charts, we will illustrate the use of an $X$-bar chart for monitoring the location parameter and an $R$, or range, chart for monitoring the dispersion parameter. In this example, we will revisit the meter stick production process previously described in Chapter 9. The tolerance limits (the explicit quality requirement applicable to each and every meter stick) for the length of the meter stick were specified as $1000 \pm 1$ mm in order to satisfy customer demands for a household measuring instrument. The characteristic of critical importance is the length of the meter stick. Even though customers desire a meter stick that is exactly 1 m long, we know that we cannot make each meter stick exactly 1 m long due to the nature of our materials, manufacturing processes, and so forth (see Figures 9.6 and 9.12).

The production supervisor in charge of the meter stick production process has never before used any statistical quality control methods. After having attended

**TABLE 15.1**    MEASUREMENTS OF LENGTH (IN MILLIMETERS) FOR METER STICKS (EXAMPLE 15.1)

| Subgroup Number | Measurements of each meter stick (five meter sticks per hour) | | | | | Average X-bar | Range R |
|---|---|---|---|---|---|---|---|
| 1 | 999.90 | 999.80 | 999.60 | 999.70 | 999.50 | 999.70 | 0.40 |
| 2 | 999.40 | 999.40 | 999.70 | 999.50 | 999.30 | 999.46 | 0.40 |
| 3 | 999.30 | 999.30 | 999.50 | 999.30 | 999.50 | 999.38 | 0.20 |
| 4 | 999.50 | 999.60 | 999.60 | 999.50 | 999.80 | 999.60 | 0.30 |
| 5 | 999.50 | 1,000.00 | 1,000.00 | 1,000.00 | 999.80 | 999.86 | 0.50 |
| 6 | 999.50 | 999.50 | 999.40 | 999.60 | 999.60 | 999.52 | 0.20 |
| 7 | 999.60 | 999.60 | 999.90 | 999.50 | 999.40 | 999.60 | 0.50 |
| 8 | 998.80 | 999.60 | 999.80 | 999.80 | 999.80 | 999.56 | 1.00 |
| 9 | 1,000.60 | 999.90 | 999.80 | 998.70 | 999.40 | 999.68 | 1.90 |
| 10 | 999.90 | 999.80 | 999.90 | 1,000.40 | 1,000.40 | 1,000.08 | 0.60 |
| 11 | 999.70 | 1,000.10 | 999.80 | 999.70 | 1,000.10 | 999.88 | 0.40 |
| 12 | 999.90 | 1,000.10 | 1,000.10 | 1,000.20 | 1,000.30 | 1,000.12 | 0.40 |
| 13 | 999.90 | 1,000.30 | 999.80 | 998.90 | 999.60 | 999.70 | 1.40 |
| 14 | 999.90 | 999.80 | 1,000.00 | 999.70 | 999.60 | 999.80 | 0.40 |
| 15 | 999.30 | 1,000.00 | 999.60 | 999.70 | 999.80 | 999.68 | 0.70 |
| 16 | 999.10 | 999.40 | 999.60 | 999.60 | 999.60 | 999.46 | 0.50 |
| 17 | 999.60 | 999.30 | 999.70 | 999.60 | 999.80 | 999.60 | 0.50 |
| 18 | 998.85 | 999.10 | 999.20 | 999.00 | 999.45 | 999.12 | 0.60 |
| 19 | 999.80 | 999.90 | 999.20 | 999.00 | 999.50 | 999.48 | 0.90 |
| 20 | 999.60 | 999.80 | 999.80 | 1,000.20 | 999.90 | 999.86 | 0.60 |
| Totals | ... | ... | ... | ... | ... | 19,993.14 | 12.40 |

a class on the application of variables control charts, he decided to make an experimental application to the meter stick production process in order to become familiar with control charts.

Approximately once every hour the lengths of five meter sticks that had just been produced were measured. For each sample of five meter sticks—we call these samples "subgroups," the average and the range (i.e., the difference between the largest measured value and the smallest measured value in each subgroup) were computed. The figures obtained are shown in Table 15.1. If these measurements had been made without benefit of the supervisor's introduction to the control chart technique, the averages for each subgroup might well have been calculated but probably not the ranges. Figure 15.1 shows two types of charts that, while not control charts, are sometimes made from control chart type information. These charts may be of interest to production personnel, but they do not provide the definite basis for action that control charts supply.

Figure 15.1a shows individual meter stick measurements plotted for each subgroup, as well as the nominal dimension and upper and lower tolerance limits.

**FIGURE 15.1**  Two charts that are *not* control charts but depict control chart information (Example 15.1).

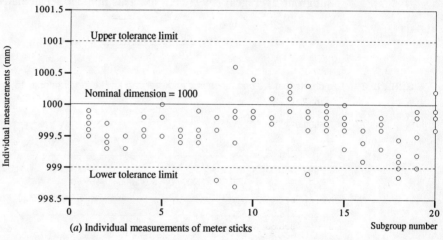

(*a*) Individual measurements of meter sticks

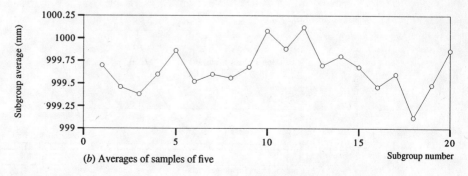

(*b*) Averages of samples of five

All the meter sticks examined met the design specifications, with the exception of four. Figure 15.1b shows the averages of these subgroups (of size 5). The chart in Figure 15.1b shows trends more clearly than the chart shown in Figure 15.1a. However, without imposition of the Shewhart technique limits, which we will subsequently develop, it does not indicate whether the process shows stability in the statistical sense.

It should be noted that because Figure 15.1b shows subgroup averages rather than individual values, it would have been totally misleading and incorrect to indicate the tolerance limits on this chart. It is the individual meter stick, not the subgroup average, that has to meet the tolerances. Averages of a subgroup often fall within tolerance limits even though some of the individual articles in the subgroup are outside the limits. We observed this in samples 8, 9, 13, and 18, in which the averages were between 999 and 1001 mm even though one meter stick in each group measured less that 999 mm. Hence, we can see why a chart for averages that shows tolerance limits is deceptive. Figure 15.2a shows an X-bar control chart for the averages (X-bars). Our readers should note that this is Figure 15.1b with the addition of control limits. Figure 15.2b shows an R control chart, for the ranges.

Each of these control charts is drawn with a solid line to indicate the average value of the statistic that is plotted. The grand average $\overline{\overline{X}}$ (i.e., the average of the subgroup averages) is 999.66 mm. This number is calculated as

$$\overline{\overline{x}} = \frac{\sum\limits_{j=1}^{m} \overline{x}_j}{m} = \frac{19{,}993.14}{20} = 999.66 \text{ mm}$$

**FIGURE 15.2** Two charts that *are* control charts, an X-bar and an R chart (Example 15.1).

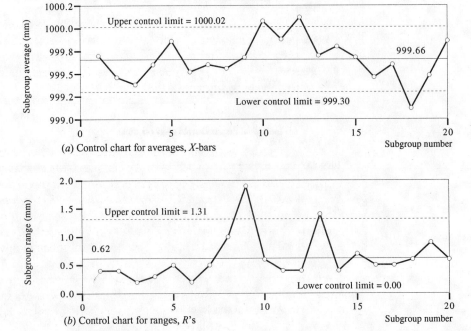

(a) Control chart for averages, X-bars

(b) Control chart for ranges, R's

Likewise, the average of the ranges is 0.62 mm.  This number is calculated as

$$\bar{R} = \frac{\sum\limits_{j=1}^{m} R_j}{m} = \frac{12.40}{20} = 0.62 \text{ mm}$$

The charts show dashed lines marked "upper control limit" and "lower control limit."  The distances of the control limits from the line showing the average value depend on the distribution of the statistics of interest ($\bar{\bar{X}}$ and $\bar{R}$) and the subgroup size.  The methods for calculating control chart limits are explained later in this chapter.  The limits shown in Figure 15.2 are called "trial limits."  Before projecting them into the future (past sample 20) to control future production, they need to be modified.

As the charts now stand, the process shows a lack of control (instability).  Three points (subgroups 10, 12, and 18) are outside the control limits on the $X$-bar chart (for averages).  Two points (subgroups 9 and 13) are outside the control limits on the $R$ chart.  These points indicate the presence of special causes in the manufacturing process; the factors contributing to these upsets in quality should be identified and addressed.  These outlying points in our control chart were found in retrospect, as the control limits were not established until the end of subgroup 20.  But, we must now search for physical causes and remove these influences as best we can, based on our process logs.

The control charts in Figure 15.2 provide evidence that there is a good opportunity to improve our process, without major system changes.  Once the process is stable, only system-level changes will produce further improvements; i.e., when a process is stable, that's all we can ask as managers of our current system, materials, machines, etc.

The dividends from the control chart come in the application of the control limits to future production.  Timely indications of instability, from the charts, aid in hunting down and eliminating the effects of special causes before the process is allowed to operate in an unstable condition for any length of time, as well as helping to prevent recurrence of the process upset.

## 15.4  CONSTRUCTION OF VARIABLES CONTROL CHARTS

A variety of variables charts have been designed over the last seven decades or so.  *In general, variables control charts can be classified into two categories.  The first category includes charts that are applicable for processes where rational subgrouping is possible, with sample or subgroup sizes greater than 1.  The second category includes charts that are applicable for processes where one and only one measurement is available or meaningful at each time process sampling occurs.*  The chart combinations listed in Table 15.2 are the most commonly used chart pairs for monitoring process location and dispersion parameters.

The ± 3-sigma limits are widely used in practice.  Therefore, the control chart mechanics described for constructing the variables control charts in this chapter are typically based on 3-sigma limits—i.e., ± 3 standard deviations associated with the

**TABLE 15.2** VARIABLES CONTROL CHART ALTERNATIVES FOR MONITORING PROCESS LOCATION AND DISPERSION PARAMETERS

Processes that allow rational subgroups with sample size greater than 1

| Parameter | | Advantages | Disadvantages | Comments |
|---|---|---|---|---|
| Location | Dispersion | | | |
| X-bar chart (averages) | Range ($R$) chart or standard ($S$) deviation chart | A good view of the statistical variation of a process. | Complex calculations, slow response, indirect relationship between control limits and tolerance limits. | Select subgroup size, frequency, and number of subgroups used to set and reset control limits carefully. Use runs rules to detect small sustained shifts. $R$ chart works well for $2 \leq n \leq 5$. $S$ chart works well with $n \geq 6$. |

Processes that do not allow rational subgroups with sample size greater than 1

| Parameter | | Advantages | Disadvantages | Comments |
|---|---|---|---|---|
| Location | Dispersion | | | |
| X chart | Moving range ($R_M$) chart | Quick, easy to plot and explain. Compares directly to tolerance limits. | X chart is sensitive to deviations from normality. | Select artificial subgroups size for the moving range chart carefully so as not to inflate within subgroup variability. The interpretations of $X$ and $R_M$ charts are not independent. |
| Exponential weighted moving average (EWMA) chart | Exponential weighted moving deviation (EWMD) chart | Effective in detecting small shifts. | Slow response, complex calculations, indirect relationship between control limits and tolerance limits. | Detection of shifts is based primarily on trend lines. The interpretations of average and deviation charts are not independent. |
| Cumulative sum (CuSum) chart | $R_M$ chart | Fast response to abrupt shifts in mean. Effective in detecting small shifts. | Chart setup is complex, hard to explain. Necessary to specify shifts in advance for chart construction. | The CuSum chart is applicable when the process mean is known to vary and not to be controlled by physical means. The interpretations of CuSum and moving range charts are not independent. |

*Note: n* equals the number of observations or measurements in a subgroup.

chart statistic (e.g., the mean or range). In some cases, it may be beneficial to work with control limits other than 3-sigma limits (e.g., 2.5-sigma limits). In these situations, the usual practice is to (1) calculate 3-sigma limits based on the mechanics presented in this chapter, (2) calculate the 3-sigma distance, UCL value – centerline value, (3) divide by 3 to obtain an estimate for 1 chart-sigma (remember that 1 chart-sigma is equal to 1 standard deviation of the chart-subgroup statistic, e.g., $\hat{\sigma}_{\bar{x}} = \hat{\sigma}/\sqrt{n}$), and (4) calculate the desired UCL and LCL up and down from the centerline in the desired multiple of the chart-sigmas. This applies for all variables control charts discussed in this chapter, other than the cumulated sum (CuSum) control charts and the exponentially weighted moving average (EWMA) and deviation (EWMD) control charts.

## General Control Chart Symbols

$n$: the number of observations or measurements in a subgroup (e.g., subgroup size, or moving subgroup size).

$m$: the number of subgroups in the preliminary data set used to set up a control chart.

$b$: the control limit spread in number of standard deviations (chart-sigmas) of the plotted quantity.

$\mu, \hat{\mu}$ :the population mean, estimated population mean (for the population of individual measurements or items).

$\sigma, \hat{\sigma}$: the population standard deviation, estimated population standard deviation for a population of individual measurements or items (also known as one chart-sigma for the population of individual items for an $X$ chart, where $n = 1$).

$\sigma_{\bar{x}}, \hat{\sigma}_{\bar{x}}$: the population standard deviation, estimated population standard deviation for a population of means (also known as one chart-sigma for the $X$-bar chart, taken from subgroups or samples of size $n$, where $n > 1$).

## R, S, X-bar and X Chart Symbols

$R_j$: a subgroup sample range value for the $j$th subgroup or sample.

$\bar{R}$: the mean of a group of subgroup sample range values.

$LCL_R$, $UCL_R$ : the lower, upper control limits for the $R$ chart.

$s_j$: a subgroup sample standard deviation for the $j$th subgroup.

$\bar{s}$: the mean of a group of subgroup sample standard deviations.

$\sigma_0$: a target or assumed population standard deviation (given rather than calculated).

$LCL_S$, $UCL_S$ : the lower, upper control limits for the $S$ chart.

$x_i$ : the $i$th individual measurement within a subgroup.

$\bar{x}_j$: a subgroup sample mean for the $j$th subgroup or sample.

$\bar{\bar{x}}$: the mean of a group of subgroup sample means.

$\bar{x}_0$: a target or assumed mean value (given rather than calculated).

$LCL_{\bar{x}}$, $UCL_{\bar{x}}$: the lower, upper control limits for the $X$-bar chart.

$LCL_x$, $UCL_x$ : the lower, upper control limits for the $X$ chart.

## Moving Range ($R_M$) Chart Symbols

$R_{Mj}$: a moving subgroup sample range value for the $j$th artificial subgroup or sample.

$\bar{R}_M$: the mean of a group of moving artificial subgroup sample range values.

$LCL_{RM}$, $UCL_{RM}$: the lower, upper control limits for the moving range chart.

## Exponential Weighted Moving Average and Deviation (EWMA and EWMD) Chart Symbols

$r$: a weighting factor (can assume values between 0 and 1).

$W_{j-k}$: weight associated with the observation $x_{j-k}$, where $x_j$ is the most recent observation.

$A_j$: the exponential weighted moving average (EWMA) for observation $j$.

$V_j$: the exponential weighted moving deviation (EWMD) for observation $j$.

$D_j$: the absolute deviation at observation $j$, equal to $|x_j - A_{j-1}|$.

$LCL_A$, $UCL_A$: the lower, upper control limits for the EWMA chart.

$LCL_V$, $UCL_V$: the lower, upper control limits for the EWMD chart.

## Cumulative Sum (CuSum) Chart Symbols

$y_c$: a user-selected CuSum chart scaling factor (vertical over horizontal units), selected so that the CuSum chart horizontal-to-vertical scale and the V-mask angle will produce an attractive and legible chart.

$\theta_c$: a V-mask angle for a CuSum chart.

$D_c$: the process shift magnitude a CuSum chart should detect with very high probability.

$S(j)$: a CuSum value for the $j$th sample.

## Normal Model

*In general, the variables SPC models developed here are based on the normal, or Gaussian, probability mass function.* The normal distribution has two parameters, $\mu$ and $\sigma$, where $\mu$ is the mean and $\sigma$ is the standard deviation, respectively. If we define the random variable $X$ as a measurable quality characteristic with mean $\mu$ and standard deviation $\sigma$, then the probability mass and the cumulative mass functions are

$$f(x; \mu, \sigma) = \frac{1}{\sigma\sqrt{2\pi}} \exp\left[-\frac{(x-\mu)^2}{2\sigma^2}\right] \qquad -\infty < x < \infty$$

and
$$F(x; \mu, \sigma) = \Phi\left[\frac{x-\mu}{\sigma}\right] \qquad -\infty < x < \infty$$

where $\Phi$ represents the standard normal density function cumulated from left to right. A table of the cumulated standard normal distribution, or mass function, appears in Table IX.2, Section IX.

## 15.5 CONSTRUCTION AND INTERPRETATION OF SHEWHART CONTROL CHARTS

*In working with the Shewhart control charts, the information from each sample (a rational subgroup with sample size greater than 1) drawn is judged to determine whether or not it indicates the presence of a special cause disturbance.* Unless the evidence is significant in favor of the occurrence of a special cause disturbance, we declare that the process is "in control" or stable.

*There are two process parameters of general interest in the SPC scenario for variables data: the location and the dispersion.* We are particularly interested in determining any point in time at which these two parameters have changed in magnitude. From a statistical point of view, we are setting up crude hypothesis tests each time we sample and plot subgroup statistics. Hence, type I (false alarm) and type II (failure to detect a process upset) errors are applicable (see Table 9.4). Our use of subgroup sizes $n > 1$ brings the central limit theorem (CLT) to bear on the population of subgroup means, displayed by our $X$-bar chart. (See Appendix 15A for a brief description of the CLT relative to our $X$-bar chart, and see Walpole and Myers [3] for a more detailed discussion of the CLT).

According to the CLT, the distribution of sample means will approach a normal distribution (Gaussian distribution) for reasonably large subgroup sizes, even if the population sampled is not normally distributed. If the population is normally distributed, then the CLT tells us that the distribution of means will also be normal. Grant and Leavenworth maintain that samples of size 3 to 5 offer reasonable protection for $X$-bar charts, even if the underlying population is not normally distributed [4]. Walpole and Myers use a sample size of 30 as a guide to assume normality [5]. In practice, if the population sampled is reasonably normal to begin with, the subgroup size issue tends to become a matter of sampling economics and statistical error probabilities. Dealing with larger subgroups is more expensive and time consuming, but probabilities of false alarms and failures to detect are lower. This topic will be discussed in detail in Chapter 17.

Figure 15.3 graphically depicts the hypothesis testing concept. Figure 15.3*a* depicts a normal population of individual quality characteristic measurements. In Figure 15.3*b*, we see two sample results obtained from the population in Figure 15.3*a*, denoted as $\bar{x}_1$ and $\bar{x}_2$. The critical regions (for $H_0$ rejection) are shaded. They are established on the basis of the sample size $n$, the process standard deviation $\sigma$, and the $\alpha$ "risk" we are willing to accept (type I error probability). Based on the sample result $\bar{x}_1$, we would reject the null hypothesis, while the result $\bar{x}_2$ leads to the conclusion that we cannot reject $H_0$ (e.g., $\bar{x}_1$ fell in the lower critical region while $\bar{x}_2$ did not fall in a critical region).

Figure 15.3*c* shows the two subgroup means $\bar{x}_1, \bar{x}_2$ plotted on an $X$-bar chart. To the right of these two points, we have drawn an "in-control" sequence, or signal (center). An "out-of-control" sequence (far right) is also shown. Under the assumption that the sampling results arise from a constant system of chance causes, the plotted values should evolve randomly over time. The center sequence in Figure 15.3*c* illustrates this case. On the other hand, if the sampling results show evidence of a nonrandom pattern, such as the sequence to the right, a special cause

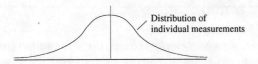

(a) Population of individual items sampled, $N(\mu = \bar{\bar{x}}, \sigma)$

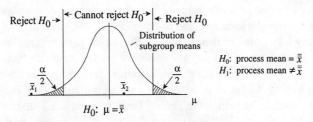

(b) Hypothesis testing formulation, subgroup means, $N(\mu = \bar{\bar{x}}, \sigma_{\bar{x}} = \sigma/\sqrt{n})$

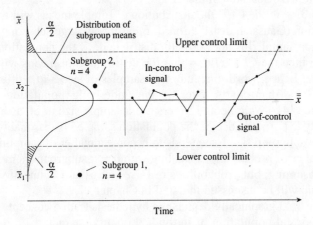

Subgroup 1: reject $H_0$ of a stable, in-control process, based on the subgroup evidence.
Subgroup 2: do not reject $H_0$ of a stable, in-control process, based on the subgroup evidence.

(c) Extended formulation in X-bar chart form

**FIGURE 15.3**   Hypothesis testing formulation extended to establish the Shewhart control chart model.

disturbance may be present and investigation is warranted. From comparing Figures 15.3b and c, we see that the critical regions are defined by the chart control limits.

## X-Bar Control Chart Mechanics

Suppose that we select samples of size *n*, taken at regular and frequent intervals, from a process. ***Each sample indicates how the process is behaving at one "point" in time in terms of the location and dispersion, judged through the subgroup mean*** *and the range or standard deviation of the subgroup.* We seek to determine if the process is stable or unstable. *If the process is subject only to chance causes, the $\bar{x}$'s and the R's or the s's should be randomly distributed within certain probabilistic limits, called "control chart limits," or simply, "control limits."*

The general form to construct the $X$-bar chart control limits, assuming all subgroups are the same size $n$ and that the process mean $\mu$ and the process standard deviation $\sigma$ are known, is

$$\left. \begin{array}{l} \text{UCL}_{\bar{x}} \\ \\ \text{LCL}_{\bar{x}} \end{array} \right\} = \mu \pm b\,\sigma_{\bar{x}} \qquad (15.1)$$

$$\text{Centerline} \quad = \mu$$

It is common practice in the United States to set $b = 3$; in this case, the type I error probability can be found in our standard normal table (Section IX, Table IX.2) and is 0.0026, assuming normality and that we do not use the runs rules (Chapter 17). We typically estimate $\mu$ and $\sigma$ from $\bar{\bar{x}}$ and $\bar{R}$, respectively. Hence, the $\pm$ 3-sigma control limits for our $X$-bar control chart are

$$\left. \begin{array}{l} \text{UCL}_{\bar{x}} \\ \\ \text{LCL}_{\bar{x}} \end{array} \right\} = \bar{\bar{x}} \pm \frac{A\bar{R}}{d_2} = \bar{\bar{x}} \pm \frac{3\bar{R}}{d_2\sqrt{n}} = \bar{\bar{x}} \pm A_2\bar{R} \qquad (15.2)$$

$$\text{Centerline} \quad = \bar{\bar{x}}$$

where

$$\bar{x}_j = \frac{\sum\limits_{i=1}^{n} x_i}{n} \qquad \text{for each subgroup}$$

$$\bar{\bar{x}} = \hat{\mu} = \frac{\sum\limits_{j=1}^{m} \bar{x}_j}{m}$$

$$R_j = (X_{\max} - X_{\min}) \qquad \text{in each subgroup}$$

$$\bar{R} = \frac{\sum\limits_{j=1}^{m} R_j}{m}$$

where $j$ is used as a subgroup index.

For a 3-sigma $X$-bar ($b = 3$) chart where $\mu$ and $\sigma$ are estimated from $\bar{\bar{x}}$ and $\bar{s}$, respectively,

$$\left. \begin{array}{l} \text{UCL}_{\bar{x}} \\ \\ \text{LCL}_{\bar{x}} \end{array} \right\} = \bar{\bar{x}} \pm \frac{3\bar{s}}{c_4\sqrt{n}} = \bar{\bar{x}} \pm A_3\bar{s} \qquad (15.3)$$

$$\text{Centerline} \quad = \bar{\bar{x}}$$

where
$$s_j = \sqrt{\frac{\sum_{i=1}^{n}(x_i - \bar{x}_j)^2}{n-1}} \qquad \text{for each subgroup}$$

$$\bar{s} = \frac{\sum_{j=1}^{m} s_j}{m}$$

The control chart constants $d_2$, $c_4$, $A$, $A_2$ and $A_3$ are as listed in Table IX.10, Section IX. The constants $A$, $A_2$, and $A_3$ were developed for convenience in calculating the 3-sigma limits, whereas the constants $d_2$ and $c_4$ are related to the range and standard deviation distributions, respectively.

Once the X-bar chart is set up using the computed control limits and centerline, we plot the $\bar{x}_j$ values. Next, we connect the points together with a solid line and use the chart to monitor the process.

The fact that all subgroup means $\bar{x}_j$'s fall within the 3-sigma limits on the X-bar chart and exhibit a random pattern indicates a stable process. However, departures from expected process behavior may not always manifest themselves on the control chart immediately, or for that matter, at all. In other words, if the mean has shifted, then the null hypothesis is now false, but the data may not show this. This failure to detect a shift results in a type II statistical error. Although narrower limits (e.g., 2-sigma limits, $b = 2$) would allow easier and faster detection of special causes, such limits would also increase the chance of false alarms, the type I statistical error. That is, when the process is actually in good statistical control but a sample mean $\bar{x}_j$ falls outside the limits, we falsely conclude that the process is out of control and begin looking for a *nonexistent* special cause.

***The selection of the appropriate placement of the UCL and LCL,*** that is, the selection of the type I error probability, ***is an economic (and timeliness) issue. The intent is to design the control limits and subgroup size in such a way as to balance the economic consequences of (1) failing to detect a special cause when it does occur and (2) falsely detecting the presence of a special cause. Practice over the years has led to the use of 3-sigma limits as a good choice for balancing these two undesirable consequences*** (DeVor et al. [6]).

### R and S Control Chart Mechanics

So far, we have confined ourselves to a discussion of the detection of the process mean shifts. However, sporadic disturbances may also cause changes in the dispersion. Another hypothesis testing argument, similar to that of the X-bar chart, can be put forward for the behavior of a series of ranges and standard deviations of the same samples. The sampling distributions of the ranges and the standard deviations provide the bases for the establishment of these control limits.

The R chart is a very popular control chart used to monitor the dispersion associated with a quality characteristic. Its simplicity of construction and maintenance makes it very popular, however, the sample range provides an abstract measure of dispersion. For example, as the sample size increases, we expect the range to increase, but the amount of increase is difficult to visualize.

The range is a good measure of variation for small subgroup sizes (e.g., $2 \le n \le 5$). As $n$ increases, the utility of the range measure as a measure of dispersion falls off, since it only reflects the information in the two extreme points in the subgroup (Montgomery [7]). Once $n > 5$ or so, the standard deviation measure is preferred, since it utilizes all subgroup points. Hence, one rule of thumb is to use $R$ charts for small subgroup sizes and $S$ charts for larger subgroup sizes. The $n = 5$ or $n = 6$ dividing line is arbitrary, but once $n \ge 10$ the $S$ chart is clearly the better chart (from the statistical point of view).

To establish a control chart for the range, we must be a bit more specific about the quality characteristic's underlying distribution. This is so because the distribution of ranges does not exhibit the same robust behavior as that of the distribution of averages (the CLT does not protect us here, as it does in the $X$-bar chart case). If we can assume that the individual measurements are normally distributed, we can establish a relationship between the standard deviation of the ranges $\sigma_R$ and the standard deviation of measurements $\sigma$. We refer our readers to Duncan [8] for details on range distributions.

Statistical theory has established the relationship between the standard deviation of the range and the standard deviation of the (normally distributed) quality characteristic measurement, random variable $X$, as

$$\sigma_R = d_3 \sigma$$

where $d_3$ is a known function of $n$, the subgroup size. Given the relationship between $R$ and $\sigma$, we have

$$\hat{\sigma}_R = \frac{d_3}{d_2} \bar{R}$$

where $d_2$ and $d_3$ are tabulated in Table IX.10, Section IX.

The general form for the $R$ chart, assuming all subgroups are the same size $n$ and the range mean $\mu_R$ is known, is

$$\left.\begin{matrix} \text{UCL}_R \\ \\ \text{LCL}_R \end{matrix}\right\} = \mu_R \pm b\sigma_R \tag{15.4}$$

$$\text{Centerline} = \mu_R$$

For a 3-sigma $R$ chart where $\sigma_R$ is estimated from $\bar{R}$,

$$\text{UCL}_R = \bar{R} + 3\frac{d_3}{d_2}\bar{R} = D_4 \bar{R} \tag{15.5}$$

$$\text{LCL}_R = \bar{R} - 3\frac{d_3}{d_2}\bar{R} = D_3 \bar{R} \tag{15.6}$$

$$\text{Centerline} = \bar{R}$$

where

$$\overline{R} = \frac{\sum\limits_{j=1}^{m} R_j}{m}$$

$$R_j = (x_{max} - x_{min}) \text{ in each subgroup}$$

Here $j$ is used as a subgroup index.

If the LCL calculated is less than 0, no LCL exists (or we set the $LCL_R = 0$). The constants $d_2$, $d_3$, $D_3$, and $D_4$ are as listed in Table IX.10, Section IX.

Once the $R$ chart is set up using the computed control limits and centerline, we plot the $R_j$ values. Next, we connect the points together with a solid line and use the chart to monitor the process.

As explained above, the $S$ chart is preferred as a control chart to monitor dispersion when the sample sizes are relatively large. It tracks the subgroup standard deviations; hence, it is statistically straightforward in its interpretation. As is the case with the $R$ chart, we must assume that the individual measurements are normally distributed in order to justify use of the $S$ chart. Assuming that random variable $X$ is normally distributed, statistical theory has established the relationship between the standard deviation of the subgroup and the standard deviation of the quality characteristic measurement $X$.

For a 3-sigma $S$ chart where $\sigma_s$ is estimated from $\bar{s}$, the control limits are

$$UCL_s = \bar{s} + 3 \; \frac{\bar{s}\sqrt{1 - c_4^2}}{c_4} = B_4 \bar{s} \tag{15.7}$$

$$LCL_s = \bar{s} - 3 \; \frac{\bar{s}\sqrt{1 - c_4^2}}{c_4} = B_3 \bar{s} \tag{15.8}$$

$$\text{Centerline} = \bar{s}$$

If the lower control limit calculated is less than 0, no LCL exists or we set it to 0. The constants $c_4$, $B_3$, and $B_4$ are as listed in Table IX.10, Section IX.

Here,

$$s_j = \sqrt{\frac{\sum\limits_{i=1}^{n} (x_i - \overline{x}_j)^2}{n - 1}}$$

$$\overline{x}_j = \frac{\sum\limits_{i=1}^{n} x_i}{n} \quad \text{for each subgroup}$$

$$\overline{s} = \frac{\sum\limits_{j=1}^{m} s_j}{m}$$

Here $j$ is used as a subgroup index.

Once the $S$ chart is set up using the computed control limits and centerline, we plot the $s_j$ values on the $S$ chart. Next, we connect the points together with a solid line and use the chart to monitor the process.

Since the process standard deviation $\hat{\sigma}$ plays a key role in describing our process dispersion in SPC, a comment is in order.  In statistics, we typically develop estimates of $\mu$ and $\sigma^2$ using

$$\hat{\mu} = \bar{x} = \frac{\sum\limits_{i=1}^{n} x_i}{n}$$

$$\hat{\sigma}^2 = s^2 = \frac{\sum\limits_{i=1}^{n} (x_i - \bar{x})^2}{n-1}$$

Here, $s^2$ can be shown to be an unbiased estimate of $\sigma^2$ (Walpole and Myers [9]) but $s = \sqrt{s^2}$ is biased as an estimate of $\sigma$ and thus, so also is $\bar{s}$.  The $c_4$ factor is used to adjust out this bias.  Scanning the $c_4$ column in Table IX.10, Section IX, we can see that this bias is very small as we approach, say, $n = 30$.  But, for small $n$'s, the usual case in SPC, the bias is large.  Hence, the $c_4$ adjustment in Equation 15.9a is critical.

$$\hat{\sigma} = \frac{\bar{s}}{c_4} \tag{15.9a}$$

We can also estimate $\hat{\sigma}$ from our $R$ chart,

$$\hat{\sigma} = \frac{\bar{R}}{d_2} \tag{15.9b}$$

Here, again, $c_4$ and $d_2$ values are listed in Table IX.10, Section IX.  For more details on how to develop the factors we have previously introduced (e.g., $d_2$, $d_3$, $c_4$) we refer our readers to Duncan [10].

---

## Example 15.2—Meter Stick Case Revisited

We will revisit the meter stick production process discussed in Example 15.1 and construct the 3-sigma trial limits for the $X$-bar and $R$ charts.  What we have are 20 samples ($m = 20$) of size 5 ($n = 5$) meter sticks.  From Table 15.1 it is evident that

$$\bar{\bar{X}} = \frac{19{,}993.14}{20} = 999.66 \text{ mm}$$

$$\bar{R} = \frac{12.40}{20} = 0.62 \text{ mm}$$

From Equation 15.2, the 3-sigma control limits for the $X$-bar chart are

$$\text{UCL}_{\bar{x}} = \bar{\bar{x}} + A_2 \bar{R} = 999.66 + 0.58\,(0.62) = 1000.02 \text{ mm}$$

$$\text{LCL}_{\bar{x}} = \bar{\bar{x}} - A_2 \bar{R} = 999.66 - 0.58\,(0.62) = 999.30 \text{ mm}$$

$$\text{Centerline} = \bar{\bar{x}} = 999.66 \text{ mm}$$

From Equations 15.5 and 15.6, the 3-sigma control limits for the $R$ chart are

$$UCL_R = D_4\overline{R} = 2.11\ (0.62) = 1.31\ \text{mm}$$

$$LCL_R = D_3\overline{R} = 0\ (0.62) = 0\ \text{mm}$$

$$\text{Centerline} = \overline{R} = 0.62\ \text{mm}$$

These control chart limits for the $X$-bar and $R$ charts were shown in Figure 15.2. It was previously mentioned that in order for the limits to be meaningful, the process under consideration must be in statistical control. Figure 15.2 makes it clear that the meter stick production process is not in statistical control. In fact, both the $X$-bar chart and the $R$ chart show points outside the trial control limits.

Since the process appears to be unstable, the variation in the data collected is very likely due to both the variation created by chance causes and that created by special causes. In order to arrive at control chart limits which represent a stable process, the special cause effects must be eliminated. At this point in our example, we will assume these causes were identified and their effects eliminated. Hence, we will eliminate subgroups 10, 12, and 18 on the $X$-bar chart and subgroups 9 and 13 on the $R$ chart and recompute our trial control limits, excluding these subgroups.

We are now left with 15 samples ($m = 15$). The revised grand average, $\overline{\overline{x}} = 999.63$ mm, is calculated as

$$\overline{\overline{x}} = \frac{\sum\limits_{j=1}^{m} \overline{x}_j}{m} = \frac{14{,}994.44}{15} = 999.63\ \text{mm}$$

The average of the ranges is 0.50 mm. This number is calculated as

$$\overline{R} = \frac{\sum\limits_{j=1}^{m} R_j}{m} = \frac{7.50}{15} = 0.50\ \text{mm}$$

The revised control limits for both the $X$-bar chart and the $R$ chart are calculated as follows:

From Equation 15.2, the 3-sigma control limits for the $X$-bar chart are

$$UCL_{\overline{x}} = \overline{\overline{x}} + A_2\overline{R} = 999.63 + 0.58\ (0.50) = 999.92\ \text{mm}$$

$$LCL_{\overline{x}} = \overline{\overline{x}} - A_2\overline{R} = 999.63 - 0.58\ (0.50) = 999.34\ \text{mm}$$

$$\text{Centerline} = \overline{\overline{x}} = 999.63\ \text{mm}$$

From Equations 15.5 and 15.6, the 3-sigma control limits for the $R$ chart are

$$UCL_R = D_4\overline{R} = 2.11\ (0.50) = 1.06\ \text{mm}$$

$$LCL_R = D_3\overline{R} = 0\ (0.50) = 0.00\ \text{mm}$$

$$\text{Centerline} = \overline{R} = 0.50\ \text{mm}$$

These limits, called "revised control limits," are shown in Figure 15.4. All 15 subgroups are now within the control limits on both charts. As such, it is safe to extend the control limits into the future to monitor future production. Now, we may want to estimate $\sigma$, the population standard deviation.

$$\hat{\sigma} = \frac{\bar{R}}{d_2} = \frac{0.50}{2.326} = 0.215 \text{ mm}$$

Should the physical process change, care should be exercised in using these limits as future limits. We want the control charts to reflect our current, stable process; process changes demand that we reevaluate our control charts and assure that they reflect current operations.

## 15.6 SOME CONTROL CHART METHODS FOR INDIVIDUAL MEASUREMENTS

*When we select the X-bar and R, or X-bar and S control chart pairs, the notion of rational subgrouping is central to data collection, chart construction, and interpretation. In some situations the idea of taking several measurements to form a rational sample or subgroup (of a size greater than one) simply does not make sense, because only one measurement is available or meaningful each time a sample is taken.* For example, certain process characteristics (e.g., such as the pH

**FIGURE 15.4**   Revised control charts for meter stick production process (Example 15.2).

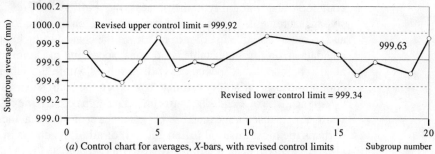

(*a*) Control chart for averages, *X*-bars, with revised control limits      Subgroup number

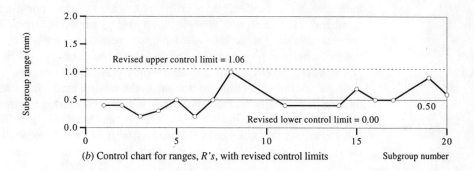

(*b*) Control chart for ranges, *R's*, with revised control limits      Subgroup number

of a chemical compound, temperature of an oven) will not vary in close proximity in time because the medium being sampled is homogeneous.

When we sample a parameter that is homogeneous (over time or order of production) measurement differences within subgroups (measurements taken very close in time) are due more to instrument measurement variation than actual process variation. In other cases, such as batch processing and/or processes with high measurement costs, measurements may become available only infrequently, and special control charts must be used.

In this section, *we introduce several types of control charts which are useful for process control with individual measurements, namely, the X chart, the moving range, $R_M$, chart, the exponentially weighted (EWMA and EWMD) charts, and the cumulative sum (CuSum) charts*. For additional details and advanced treatments of these and other control charts for individual observations the reader may refer to Duncan [11], DeVor et al. [12], Roes et al. [13], Harding et al. [14], Wetherill and Brown [15].

### X Chart and $R_M$ Chart Mechanics

The $X$ chart (sometimes called a "runs chart") and $R_M$ charts are relatively simple to construct. The construction of $X$ and $R_M$ charts is similar to that of $X$-bar and $R$ charts, except that now $x$ is the value of the individual measurement and $R_M$ is the range of a group of $n$ consecutive individual measurements combined "artificially" to form a subgroup of size $n$.

Since the moving range $R_M$ is calculated primarily for the purpose of estimating common cause variability in the process, the artificial samples formed from successive measurements must be small in size and taken as close together in time, space, and so on, as possible in order to minimize the chance of including, in the subgroup, any data arising from unstable conditions, according to DeVor et al. [16]. DeVor et al. also point out that special cause occurrences within the subgroup will inflate the common-cause variability estimate and erode the chart's sensitivity.

Applications of $X$ and $R_M$ control charts receive mixed reviews. Grant and Leavenworth tend to discourage their use [17]. DeVor et al. state that the $X$ and $R_M$ charts are perhaps the most misused (often abused) of all of the charts in common use today [18]. The DeVor et al. reasoning focuses on two points: (1) the charts often have inflated within-subgroup variability, because too much time has transpired from one measurement to another, and (2) these charts are typically constructed with a small number of data points and, hence, subject to large sampling errors. These arguments are generally valid when individual measurements are taken and used in cases where natural subgroup measurements can and should be used instead. But, in other applications, the $X$ and $R_M$ charts may be the only feasible alternatives.

If we assume that random variable $X$, the quality characteristic measurement of interest, is normally distributed, we can set up probabilistic control limits for the $X$ chart that can help us in monitoring the process location parameter. The general form to construct $X$ chart control limits, assuming that $X$ is normally distributed, and the process mean $\mu$ and process standard deviation $\sigma$ are known, is

$$\left. \begin{array}{l} \text{UCL}_x \\ \\ \text{LCL}_x \end{array} \right\} = \mu \pm b\sigma \qquad (15.10)$$

$$\text{Centerline} = \mu$$

For the $X$ chart we use a subgroup size of $n = 1$. For the $R_M$ control chart we should select a small subgroup size ($n = 2$ or $n = 3$) to minimize opportunity for special causes to arise within the artificial samples, or moving subgroups. Larger artificial sample sizes will tend to inflate our estimate of variation since they are collected over a longer time period. We should note here that we are forced to use $n \geq 2$ for the $R_M$ chart, since single points will not, by themselves, allow us to deal with dispersion.

Assuming that the individual observations from our process are at least approximately normally distributed, it is possible to construct the control limits by estimating the process parameters $\mu$ and $\sigma$ from $\bar{x}$ and $\bar{R}_M$, respectively. To do so, it is essential that the number of individual measurements $m$ made for constructing the control limits be at least 25.

We calculate the sample mean as

$$\bar{x} = \hat{\mu} = \frac{\sum\limits_{j=1}^{m} x_j}{m}$$

This sample mean will serve as the centerline for the $X$ chart. We next compute sample moving ranges for the artificial samples of size $n$ starting with $R_{Mn}$, which is the difference between the largest and the smallest value in the first artificial sample $(x_1, \ldots, x_n)$. Each time we obtain a new $x$ value from the process, we repeat this computation for each succeeding moving sample of artificial size $n$ by adding the latest $x$ value and dropping the oldest $x$ value. We calculate an average moving range $\bar{R}_M$ from the $(m - n + 1)$ sample moving ranges,

$$\bar{R}_M = \frac{\sum\limits_{j=n}^{m} R_{Mj}}{m - n + 1}$$

This average will serve as the centerline for the $R_M$ chart.

The 3-sigma control limits for the $X$ chart are

$$\left. \begin{array}{l} \text{UCL}_X \\ \\ \\ \text{LCL}_X \end{array} \right\} = \bar{x} \pm 3\,\frac{\bar{R}_M}{d_2} \qquad (15.11)$$

$$\text{Centerline} = \bar{x}$$

and the 3-sigma control limits for the $R_M$ chart are

$$\text{UCL}_{RM} = D_4 \bar{R}_M \qquad\qquad (15.12)$$

$$\text{LCL}_{RM} = D_3 \bar{R}_M \qquad \text{or 0, whichever is larger} \qquad (15.13)$$

$$\text{Centerline} = \bar{R}_M$$

All constants—$d_2$, $D_3$, and $D_4$—are as listed in Table IX.10, Section IX.

Once the $X$ chart and $R_M$ chart are set up using the computed control limits and centerlines, we plot the $x_j$ values on the $X$ chart and the $R_{Mj}$ values on the $R_M$ chart. Next, for each chart, we connect the points together with a solid line and use the chart to monitor the process.

The $R_M$ chart provides a plotting point for each individual sample taken (once the first $n$ measurements are obtained). For a moving subgroup size of $n$, the effect of any one given sample will be contained in $n$ consecutive plotted points. Hence, we see a relatively slow response, depending on sample frequency, in picking up a shift. In examining $X$ and $R_M$ control charts, given that the artificial subgroups overlap, the interpretation of the $X$ and $R_M$ charts are not independent. Sometimes only the $X$ chart is analyzed for out-of-control signals, while the $R_M$ chart is evaluated primarily to assure proper construction of the $X$ chart.

### Example 15.3—Steel Production Case

Flatt Sheet Metal Company is a major supplier of mild steel sheet metal to a leading automobile manufacturer. Facing tighter automobile quality requirements, the design engineers from the automobile plant have tightened their supplier quality requirements. The sheet metal company is experiencing a situation in which the rolling mills and furnaces require frequent adjustments to maintain the desired tolerances. Each time an adjustment is made, about 1 hour will lapse before the resulting sheets can be evaluated.

To identify potential areas of quality and productivity improvement, it was decided to monitor the production process using $X$ and $R_M$ control charts (here, rational subgrouping with $n > 1$ is not meaningful because intrasheet variation is negligible in this process). A schematic illustration of the flat-rolling process is shown in Figure 15.5. Basically, the process includes feeding cast metal slabs at elevated temperatures iteratively between pressurized cylindrical rollers, which flatten the slabs to sheets. At the end of the process, the sheets are subject to cold rolling, to achieve the desired thickness dimension and surface finish. The produced sheets are sampled for their thickness, uniformity, and roughness using dial comparators and profilometers.

The quality characteristic selected for study in this case is the thickness of the sheet. The product tolerance set by the purchaser is 2.2500 ± 0.0075 mm. The sheet thicknesses (in millimeters) of 32 sheets produced during a 4-hour period are displayed in Table 15.3.

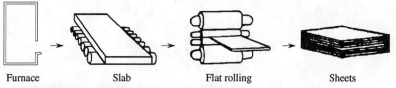

Furnace          Slab          Flat rolling          Sheets

**FIGURE 15.5**    Sheet rolling process depiction (Example 15.3).

To determine the $R_M$ values, an artificial sample size of $n = 2$ was selected. The $R_{Mj}$ values are therefore given by

$$R_{Mj} = |X_j - X_{j-1}| \qquad \text{for } j = 2, ..., m$$

Table 15.3 displays the 31 calculated moving range, $R_{Mj}$, values that were extracted from the data using the artificial subgroup, $n = 2$. From Table 15.3 it is evident that

**TABLE 15.3**    THICKNESS MEASUREMENTS OF 32 SHEETS (EXAMPLE 15.3)

| Roll number | Measurement $x$ | Two-point $R_M$ ($n = 2$) | |
|---|---|---|---|
| | | $R_M$ | Calculations |
| 1 | 2.2331 | | |
| 2 | 2.2403 | 0.0072 | $R_{M2} = |2.2403 - 2.2331| = 0.0072$ |
| 3 | 2.2473 | 0.0070 | $R_{M3} = |2.2473 - 2.2403| = 0.0070$ |
| 4 | 2.2525 | 0.0052 | $R_{M4} = |2.2525 - 2.2473| = 0.0052$ |
| 5 | 2.2497 | 0.0028 | |
| 6 | 2.2511 | 0.0014 | |
| 7 | 2.2521 | 0.0010 | |
| 8 | 2.2507 | 0.0014 | |
| 9 | 2.2342 | 0.0165 | |
| 10 | 2.2395 | 0.0053 | |
| 11 | 2.2510 | 0.0115 | |
| 12 | 2.2459 | 0.0051 | |
| 13 | 2.2518 | 0.0059 | |
| 14 | 2.2547 | 0.0029 | |
| 15 | 2.2550 | 0.0003 | |
| 16 | 2.2502 | 0.0048 | |
| 17 | 2.2341 | 0.0161 | |
| 18 | 2.2423 | 0.0082 | |
| 19 | 2.2500 | 0.0077 | |
| 20 | 2.2487 | 0.0013 | |
| 21 | 2.2488 | 0.0001 | |
| 22 | 2.2522 | 0.0034 | |
| 23 | 2.2548 | 0.0026 | |
| 24 | 2.2477 | 0.0071 | |
| 25 | 2.2316 | 0.0161 | |
| 26 | 2.2453 | 0.0137 | |
| 27 | 2.2511 | 0.0058 | |
| 28 | 2.2445 | 0.0066 | |
| 29 | 2.2531 | 0.0086 | |
| 30 | 2.2511 | 0.0020 | |
| 31 | 2.2523 | 0.0012 | |
| 32 | 2.2509 | 0.0014 | $R_{M32} = |2.2509 - 2.2523| = 0.0014$ |
| Totals | 71.9176 | 0.1802 | |

$$\bar{x} = \frac{71.9176}{32} = 2.2474 \text{ mm}$$

$$\overline{R}_M = \frac{0.1802}{31} = 0.0058 \text{ mm}$$

From Equation 15.11, the 3-sigma control limits for the $X$ chart are

$$\text{UCL}_X = \bar{x} + 3 \frac{\overline{R}_M}{d_2} = 2.2474 + 3 \frac{0.0058}{1.128} = 2.2628 \text{ mm}$$

$$\text{LCL}_X = \bar{x} - 3 \frac{\overline{R}_M}{d_2} = 2.2474 - 3 \frac{0.0058}{1.128} = 2.2320 \text{ mm}$$

Centerline $= \bar{x} = 2.2474$ mm

From Equations 15.12 and 15.13, the 3-sigma control limits for the $R_M$ chart are

$$\text{UCL}_{RM} = D_4 \overline{R}_M = 3.27 \,(0.0058) = 0.0190 \text{ mm}$$

$$\text{LCL}_{RM} = D_3 \overline{R}_M = 0 \,(0.0058) = 0 \text{ mm}$$

$$\text{Centerline} = \overline{R}_M = 0.0058 \text{ mm}$$

**FIGURE 15.6**  $X$ and $R_M$ control charts for the sheet rolling process (Example 15.3).

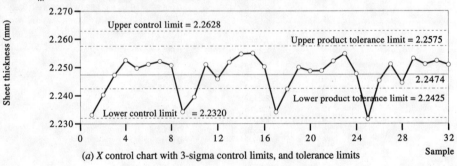

(a) $X$ control chart with 3-sigma control limits, and tolerance limits

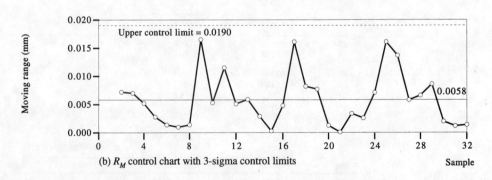

(b) $R_M$ control chart with 3-sigma control limits

These trial control limits for the $X$ and $R_M$ charts, along with tolerance limits for sheet thickness, are shown in Figure 15.6.

In examining the $X$ and $R_M$ control charts in Figure 15.6, we see that only one point, sample 25, is beyond the limits. Also, close observation of the charts shows nonrandom sequences. As was mentioned earlier, the 32 samples shown in the charts represent 4 hours of production. It can be seen from the $X$ control chart that the first two samples from every hour of production (eight samples represent 1 hour) are low in thickness in comparison to the rest of the samples. It can also be seen from Figure 15.6$a$ that these first samples are also outside the tolerance limits, and will end up as scrap or unacceptable sheet rolls. Here, we should note that product tolerance limits can be placed on the $X$ chart, since we are working with subgroups of $n = 1$ sheet.

The production supervisor investigated possible causes for these nonrandom patterns and found an assignable cause. He determined that, because the scrap collected during every production hour goes into the furnace at the beginning of the next hour of production (within 4 or 5 min), the steel obtained during scrap reprocessing is more ductile and weaker in grain strength.

In order to eliminate this cause, the process engineer decided that scrap material will be fed into the furnace at a uniform rate throughout the production

**FIGURE 15.7**   Revised $X$ and $R_M$ control charts for the sheet rolling process after eliminating the special cause (Example 15.3).

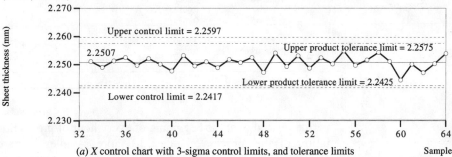

($a$) $X$ control chart with 3-sigma control limits, and tolerance limits          Sample

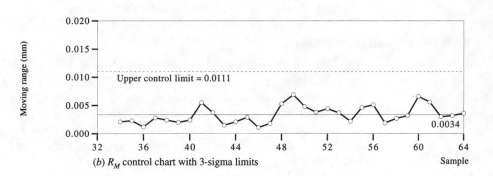

($b$) $R_M$ control chart with 3-sigma limits          Sample

**TABLE 15.4**    THICKNESS MEASUREMENTS OF 32 SHEETS AFTER
ELIMINATING THE SPECIAL CAUSE (EXAMPLE 15.3)

| Roll number | Measurement $x$ | Two-point $R_M$ ($n = 2$) | |
|---|---|---|---|
| | | $R_M$ | Calculations |
| 33 | 2.2511 | | |
| 34 | 2.2490 | 0.0021 | $R_{M34} = \lvert 2.2490 - 2.2511 \rvert = 0.0021$ |
| 35 | 2.2513 | 0.0023 | $R_{M35} = \lvert 2.2513 - 2.2490 \rvert = 0.0023$ |
| 36 | 2.2525 | 0.0012 | $R_{M36} = \lvert 2.2525 - 2.2513 \rvert = 0.0012$ |
| 37 | 2.2497 | 0.0028 | |
| 38 | 2.2521 | 0.0024 | |
| 39 | 2.2501 | 0.0020 | |
| 40 | 2.2477 | 0.0024 | |
| 41 | 2.2532 | 0.0055 | |
| 42 | 2.2495 | 0.0037 | |
| 43 | 2.2510 | 0.0015 | |
| 44 | 2.2489 | 0.0021 | |
| 45 | 2.2518 | 0.0029 | |
| 46 | 2.2507 | 0.0011 | |
| 47 | 2.2525 | 0.0018 | |
| 48 | 2.2472 | 0.0053 | |
| 49 | 2.2541 | 0.0069 | |
| 50 | 2.2493 | 0.0048 | |
| 51 | 2.2531 | 0.0038 | |
| 52 | 2.2487 | 0.0044 | |
| 53 | 2.2524 | 0.0037 | |
| 54 | 2.2502 | 0.0022 | |
| 55 | 2.2548 | 0.0046 | |
| 56 | 2.2497 | 0.0051 | |
| 57 | 2.2516 | 0.0019 | |
| 58 | 2.2543 | 0.0027 | |
| 59 | 2.2511 | 0.0032 | |
| 60 | 2.2445 | 0.0066 | |
| 61 | 2.2501 | 0.0056 | |
| 62 | 2.2471 | 0.0030 | |
| 63 | 2.2503 | 0.0032 | |
| 64 | 2.2539 | 0.0036 | $R_{M64} = \lvert 2.2539 - 2.2503 \rvert = 0.0036$ |
| Totals | 72.0235 | 0.1044 | |

period instead of at the beginning of every hour of production. After implementation of this process change, 32 more samples were collected from the production process (tabulated in Table 15.4). The new calculated control limits for the $X$ and $R_M$ charts, are shown in Figure 15.7, along with the tolerance limits. The new control chart indicates a stable process.

## EWMA and EWMD Control Charts

The $X$ and $R_M$ charts are not extremely effective in detecting shifts in the process mean or process variance, particularly when the shift in either is small. The addition of "runs" rules (discussed in Chapter 17) helps solve this problem to some extent. But, other charts can be used to address this issue as well. In this section, we will introduce the use of exponentially weighted moving averages EWMAs to monitor location and exponentially weighted moving deviations EWMDs for tracking process dispersion.

An EWMA is a moving average of past data where each data point is assigned a weight. These weights decrease in an exponentially decaying fashion from the present into the remote past. Thus, the moving average tends to be a reflection of the more recent process performance, as greater weight is allocated to the most recent data. The amount of decrease in the weights over time is an exponential function of the weighting factor $r$ which can assume values between 0 and 1. The weighting factor associated with observation $x_{j-k}$ is

$$W_{j-k} = r(1-r)^k$$

When a small value of $r$ is used, the moving average at sample point $j$ carries with it a great amount of inertia from the past. Hence, it is relatively insensitive to short-lived changes in the process. For control chart applications where fast response to process shifts is desired, a relatively large weighting factor—say, $r = 0.2$ or $r = 0.5$—may be used [19]. Figure 15.8, based on the weighting relationship above, shows the decaying behavior for some commonly used values of $r$.

In selecting $r$, the following relationship between the weighting factor $r$ and the sample size $n$ for Shewhart control charts showing sample means is often used (Sweet [20]):

$$r = \frac{2}{n+1}$$

We can observe that if $r = 1$, for the EWMA, all the weight is given to the current single observation, which is equivalent to the $X$ chart.

**FIGURE 15.8**   Exponentially decaying weighting depiction $r = 0.2$ and $r = 0.333$.

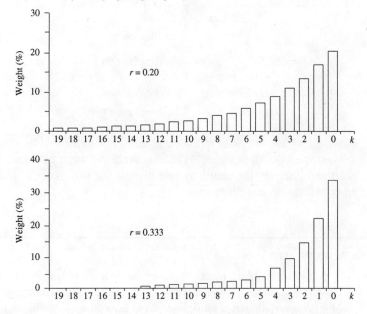

If a shift in the mean occurs, the EWMAs will gradually, depending on $r$, move to the new mean of the process, while the exponentially weighted moving deviations, EWMDs, will remain unchanged. On the other hand, if there is a shift in the process variability, the EWMDs will gradually move to the new level, while the EWMAs still vary about the same process mean. In order to construct the EWMA and EWMD control charts, we first calculate estimates of the process mean and standard deviation by using a sample size of at least 25 to 30 individual measurements,

$$\bar{x} = \hat{\mu} = \frac{\sum_{j=1}^{m} x_j}{m}$$

and

$$s = \hat{\sigma} = \sqrt{\frac{\sum_{j=1}^{m} (x_j - \bar{x})^2}{m - 1}}$$

Then, we compute EWMAs, $A_j$ and EWMDs, $V_j$, as follows:

$$A_j = rx_j + (1 - r)A_{j-1} \qquad \text{where } A_0 = \bar{x} \tag{15.14}$$

$$V_j = rD_j + (1 - r)V_{j-1} \qquad \text{where } D_j = |x_j - A_{j-1}| \text{ and } V_0 = s \tag{15.15}$$

The control limits and the $A_j$ and $V_j$ centerlines are calculated as

$$\left.\begin{array}{l} \text{UCL}_A \\ \text{LCL}_A \end{array}\right\} = \bar{x} \pm A^* s \tag{15.16}$$

$$\text{Centerline} = \bar{x}$$

$$\text{UCL}_V = D_2^* s \tag{15.17}$$

$$\text{LCL}_V = D_1^* s \tag{15.18}$$

$$\text{Centerline} = s d_2^*$$

where all $A^*$, $d_2^*$, $D_1^*$, and $D_2^*$ constants are as listed in Table IX.12, Section IX.

Next, we plot all of the $m$, $A_j$ values on the $A$ chart and the $m$, $V_j$ values on the $V$ chart and interpret the charts to determine if the process is stable in terms of both process location and dispersion.

---

### Example 15.4—Wastewater Treatment Case

A common problem faced by many companies is the treatment of wastewater. Governmental regulation agencies (e.g., the Environmental Protection Agency, or EPA) provide strict guidelines as to how to treat wastewater. The EPA calls

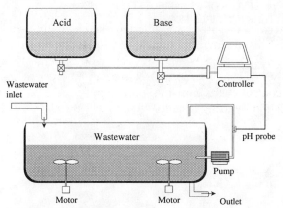

**FIGURE 15.9**    Wastewater neutralization process depiction (Example 15.4).

for neutralization of wastewater before disposal. For example, all solutions can be classified by their pH levels, which can be anywhere between 0 and 14. Solutions with a pH level between 0 and 7 are called acids, while solutions with a pH level close to 0 are referred to as strong acids. Solutions with pH levels between 7 and 14 are termed bases, while those with a level close to 14 are referred to as strong bases. Neutralization is the process of bringing the pH level of the wastewater close to 7; in this case, a pH level within $7 \pm 0.25$ will be declared acceptable.

A schematic illustration of the neutralization process is shown in Figure 15.9. The pH level of the wastewater is monitored through a probe and is fed to a controller. The controller, in turn, is responsible for adding the right amount of acid or base to neutralize the wastewater if the pH level reading from the probe is greater or less than 7, respectively. The paddles in the tank are responsible for mixing the solution and maintaining uniform concentration throughout the tank.

Because we are using a single probe to monitor a large tank, it is possible that the pH level given by the probe is not an accurate representation for the whole tank. We have decided to use control chart information about the pH level of the tank as a controller input in order to avoid unnecessary, frequent neutralization actions (i.e., the addition of acids and bases). The solution-mixing nature of the process makes it impossible to have a rational subgroup with sample size greater than 1. Hence, we will use a sample size of 1 and EWMA and EWMD control charts (built into the process control loop) to monitor the pH level. The probe samples the pH level every 30 sec. The wastewater tank is monitored by EWMA and EWMD charts and must show both charts to be stable before acid or base is added. The pH levels from the last 28 samples are collected and displayed in Table 15.5.

After choosing a weighting factor of $r = 0.333$, in order to gain a fairly rapid response (by emphasizing the newer data), we obtain the following constants from Table IX.12, Section IX:

$$A^* = 1.342 \qquad D_1^* = 0 \qquad D_2^* = 1.780 \qquad d_2^* = 0.874$$

**TABLE 15.5**  pH LEVELS OF 28 SAMPLES FROM THE
WASTEWATER TANK (EXAMPLE 15.4)

| Sample number | pH Level, $x_j$ | Moving average, $A_j$ | Absolute deviation, $D_j$ | Moving deviation, $V_j$ |
|---|---|---|---|---|
| 1 | 7.747 | 7.220 | 0.790 | 0.797 |
| 2 | 5.707 | 6.716 | 1.513 | 1.036 |
| 3 | 6.007 | 6.480 | 0.709 | 0.927 |
| 4 | 7.567 | 6.842 | 1.087 | 0.980 |
| 5 | 6.727 | 6.804 | 0.115 | 0.692 |
| 6 | 7.147 | 6.918 | 0.343 | 0.576 |
| 7 | 7.447 | 7.094 | 0.529 | 0.560 |
| 8 | 7.027 | 7.072 | 0.067 | 0.396 |
| 9 | 6.877 | 7.007 | 0.195 | 0.329 |
| 10 | 6.667 | 6.894 | 0.340 | 0.333 |
| 11 | 7.117 | 6.968 | 0.223 | 0.296 |
| 12 | 5.587 | 6.508 | 1.381 | 0.658 |
| 13 | 7.357 | 6.791 | 0.849 | 0.721 |
| 14 | 8.227 | 7.269 | 1.436 | 0.959 |
| 15 | 8.317 | 7.618 | 1.048 | 0.989 |
| 16 | 6.877 | 7.371 | 0.741 | 0.906 |
| 17 | 6.757 | 7.167 | 0.614 | 0.809 |
| 18 | 7.867 | 7.400 | 0.700 | 0.773 |
| 19 | 6.817 | 7.206 | 0.583 | 0.710 |
| 20 | 6.427 | 6.946 | 0.779 | 0.733 |
| 21 | 6.457 | 6.783 | 0.489 | 0.652 |
| 22 | 7.477 | 7.014 | 0.694 | 0.666 |
| 23 | 8.257 | 7.428 | 1.243 | 0.858 |
| 24 | 6.127 | 6.995 | 1.301 | 1.005 |
| 25 | 6.397 | 6.796 | 0.598 | 0.870 |
| 26 | 7.507 | 7.033 | 0.711 | 0.817 |
| 27 | 7.147 | 7.071 | 0.114 | 0.583 |
| 28 | 5.167 | 6.437 | 1.904 | 1.023 |
| Totals | 194.806 | | | |

The average and standard deviation of the data are calculated as:

$$\bar{x} = \frac{\sum_{j=1}^{m} x_j}{m} = \frac{194.806}{28} = 6.957$$

and

$$s = \sqrt{\frac{\sum_{j=1}^{m} (x_j - \bar{x})^2}{m-1}} = 0.801$$

The EWMAs, $A_j$, and EWMDs, $V_j$, are calculated from Equations 15.14 and 15.15 and are tabulated in Table 15.5.

The 3-sigma control limits for the EWMA control chart are calculated from Equation 15.16 as

$$\text{UCL}_A = \bar{x} + A^* s = 6.957 + 1.342\,(0.801) = 8.032$$

$$\text{LCL}_A = \bar{x} - A^* s = 6.957 - 1.342\,(0.801) = 5.882$$

$$\text{Centerline} = \bar{x} = 6.957$$

The 3-sigma control limits for the exponentially weighted moving deviation control chart are calculated from Equations 15.17 and 15.18 as

$$\text{UCL}_V = D_2^* s = 1.78 \, (0.801) = 1.426$$

$$\text{LCL}_V = D_1^* s = 0 \, (0.801) = 0$$

$$\text{Centerline} = s d_2^* = 0.801 \, (0.874) = 0.700$$

These control limits, along with the tabulated moving averages and deviations of the 28 sample pH values, are shown in Figure 15.10.

Examination of the control charts in Figure 15.10 shows that there are no sample points beyond the control limits and no nonrandom signals, suggesting a stable process. The process is operating with mean pH level at 6.957. Since this pH level is within the tolerance limits of $7 \pm 0.25$, we can safely release the wastewater from the tank. The whole process will be repeated with a new batch of wastewater. The probe will sample the new batch every 30 sec; then, the control-chart data will be tabulated, and the controller will use this information to process the wastewater. Based on the new estimated mean pH level of the tank, an appropriate amount of acid or base will be added to bring the pH level within tolerance limits for disposal. The whole process will be repeated again and again.

**FIGURE 15.10**   EWMA and EWMD control charts for the wastewater neutralization process (Example 15.4).

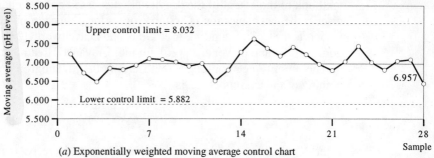

(a) Exponentially weighted moving average control chart

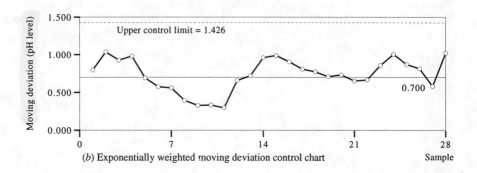

(b) Exponentially weighted moving deviation control chart

## CuSum Control Chart Mechanics

The EWMA and EWMD control charts are constructed in such a way that the last few observations of the process are weighted from the present, backward in time, in an exponentially decreasing manner. A similar type of control chart, the cumulative sum control chart, frequently called the CuSum chart, was designed to identify slight but sustained shifts in a process. The statistic to be accumulated and charted may be an actual observation or the deviation of an observation from a desired target.

One key to understanding the differences between the $X$, CuSum, and exponentially weighted control charts rests in knowing how each technique weights the data obtained from the process. Hunter discusses the weighting functions for the $X$ chart, the CuSum chart, and the EWMA chart [21]. A graphical comparison is shown in Figure 15.11. It is clear that, for the purpose of providing process upset signals, the $X$ chart has no "memory"; That is, it ignores the immediate history. The ordinary CuSum chart, however, gives equal attention to the first data point and the most recent. The EWMA chart gives less and less weight to data as they get older and older. Along these same lines, the $X$-bar, $R$ or $S$ charts focus on the most recent subgroup data. Hence, we see a wide variety of charting alternatives available to us.

CuSum charts work on a "slope" principle, and can be designed for both location and dispersion parameters. If the CuSum chart tool is to be used, we suggest using an $R_M$ chart along with the CuSum location chart (see Table 15.2). The CuSum dispersion charts are awkward to deal with, see Grant and Leavenworth [22]. We view the CuSum chart as a special purpose chart and have not chosen to feature it in our textbook to the extent of the previously discussed control charts. We will briefly discuss the mechanics for a CuSum process location chart in order to introduce our readers to the CuSum concept. (See Grant and Leavenworth [23], Duncan [24], Montgomery [25], and DeVor et al. [26], for more advanced treatments of CuSum charts.)

The point typically plotted, representing the $j$th subgroup ($n \geq 1$) from the beginning of the CuSum chart, is

$$S(j) = \sum_{i=1}^{j} (\bar{x}_i - \bar{x}_0) \tag{15.19}$$

where $i$ is used as an index and $\bar{x}_i = x_i$ if $n = 1$. Here, $\bar{x}_0$ is a target mean value which corresponds to the target (best) level of our quality characteristic.

**FIGURE 15.11** Data weighting for the $X$, CuSum, and EWMA charts. Adapted, with permission from the American Society for Quality Control, from J. S. Hunter, "The Expontially weighted moving average," *Journal of Quality Technology*, vol. 18, no.4, page 205, 1986.

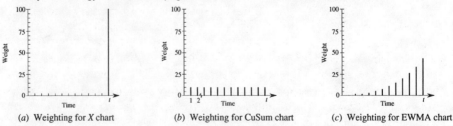

(a) Weighting for $X$ chart     (b) Weighting for CuSum chart     (c) Weighting for EWMA chart

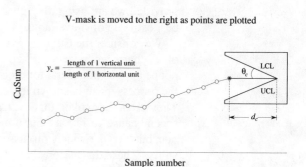

**FIGURE 15.12**   CuSum chart and V-mask construction.

A V-mask is sometimes used to represent the control limits. Every time a statistic is plotted, the V-mask is moved horizontally across the chart to the most recently plotted statistic. If any of the plotted statistics, $S(j)$'s, fall outside the mask limits, the process is declared to be out of statistical control. Since a slope principle is involved, the vertical-to-horizontal scaling is critical in the construction of the V-mask. The general form of the V-mask and its dimensions are illustrated in Figure 15.12.

A feature of the CuSum chart is the necessity of specifying—with virtual certainty—the amount of the shift $D_c$ that must be detected given a normal distribution of the quality characteristic values being measured (Grant and Leavenworth [27]). In setting up the CuSum chart, however, the user specifies the magnitude of $D_c$ independent of the acceptable type I error probability.

Grant and Leavenworth calculate the V-mask parameters $d_c$ and $\theta_c$ as

$$d_c = E_{c,\alpha} \left[ \frac{\hat{\sigma}/\sqrt{n}}{D_c} \right]^2 \qquad (15.20)$$

$$\tan \theta_c = \frac{D_c}{2y_c}$$

$$\theta_c = \tan^{-1} \left( \frac{D_c}{2y_c} \right) \qquad (15.21)$$

where $E_{c,\alpha}$ is a factor that is a function of the acceptable type I error probability [28] and can be read from Table 15.6. Here, $y_c$ is the selected vertical-to-horizontal scaling factor, and

$$\hat{\sigma} = \frac{\bar{R}_M}{d_2}$$

**TABLE 15.6**   SELECTED VALUES OF $E_{c,\alpha}$ FOR USE WITH CuSum CHARTS

| $P$ (type I error) = $\alpha$ | 0.0027 | 0.01 | 0.02 | 0.05 | 0.10 |
|---|---|---|---|---|---|
| $E_{c,\alpha}$ | 13.215 | 10.597 | 9.210 | 7.378 | 5.991 |

*Source:* Reproduced, with permission from McGraw-Hill, from E. L. Grant and R. S. Leavenworth, *Statistical Quality Control*, 6th ed., New York: McGraw-Hill, p. 350, 1988.

provided we are using a $R_M$ chart to monitor the dispersion. Otherwise, we use the best estimate of $\sigma$ available.

We consider the CuSum chart less useful, in most production situations, than the other charts we have described. DeVor et al. suggest that the CuSum chart should not be applied in general for process control, except in some special situations where the process mean is known to vary and not to be controlled by physical means [29]. They go on to explain that the CuSum chart tends to confound common cause and special cause variation. CuSum charts are not widely applied in general SPC practice, but they are widely discussed in the literature and are effective at isolating small shifts. These discussions have lead to new "Shewhart-like" CuSum control charts. These charts are beyond the scope of our SPC treatment in this text; we refer our readers to DeVor et al. [30] for further details.

## REVIEW AND DISCOVERY EXERCISES

### Review

**15.1** Explain the difference between a variables and an attributes measure. Provide two examples.

**15.2** Provide an example of a variables measure that can be converted to an attributes measure.

**15.3** Provide an example of an attributes measure that cannot be converted to a variables measure.

**15.4** What is the purpose of Shewhart control charts? How can these charts be used to improve the quality and productivity of a manufacturing process?

**15.5** How do SPC control charts relate to hypothesis testing?

**15.6** How are control charts for individual measurements different from Shewhart-type control charts?

**15.7** When defining subgroups and selecting samples, what difference does it make if special causes occur within the subgroups or samples or between them? Explain.

**15.8** We have collected a set of data from measurements on the weight of feed additives in pounds of additive per 1000-lb batch of cattle feed. We have data for 30 subgroups, each of size $n = 4$. In summary, $\sum \bar{x}_j = 75$, and $\sum R_j = 2.4$.
   **a** Determine the $X$-bar and $R$ chart parameters for setting up 3-sigma control limits.
   **b** Estimate the process standard deviation.
   **c** Determine $X$-bar and $R$ chart parameters for setting up 2.5-sigma control limits.

**15.9** We have collected a set of data from measurements on the inside diameter, in inches, of machined pulley hubs. We want to develop $X$-bar and $S$ charts. We have data from 25 subgroups each of size $n = 9$. Here, $\sum \bar{x}_j = 50.025$, and $\sum s_j = 0.0255$.
   **a** Determine the $X$-bar and $S$ chart parameters for setting up 3-sigma control limits.
   **b** Estimate the process standard deviation.
   **c** Determine $X$-bar and $S$ chart parameters for setting up 2.5-sigma control limits.

**15.10** A candy bar manufacturer desires to set up a variables control chart for the net weight of its candy bars. It is not clear exactly what charts should be used. A total of 60 candy bar weights (in grams) were obtained and are displayed in the following table.

| Sample number | Candy bar 1 | Candy bar 2 | Sample number | Candy bar 1 | Candy bar 2 |
|---|---|---|---|---|---|
| 1 | 62.671 | 61.740 | 16 | 61.228 | 62.239 |
| 2 | 62.246 | 62.381 | 17 | 62.305 | 61.370 |
| 3 | 61.545 | 59.456 | 18 | 61.286 | 61.483 |
| 4 | 62.233 | 61.409 | 19 | 61.557 | 61.477 |
| 5 | 59.108 | 61.664 | 20 | 62.147 | 61.476 |
| 6 | 61.415 | 60.276 | 21 | 62.333 | 61.675 |
| 7 | 61.872 | 61.979 | 22 | 60.608 | 63.635 |
| 8 | 62.197 | 61.308 | 23 | 60.907 | 60.126 |
| 9 | 60.603 | 61.228 | 24 | 60.890 | 60.231 |
| 10 | 62.007 | 61.305 | 25 | 63.044 | 60.430 |
| 11 | 61.171 | 59.542 | 26 | 60.677 | 61.664 |
| 12 | 62.323 | 61.751 | 27 | 61.284 | 62.112 |
| 13 | 60.707 | 61.921 | 28 | 60.343 | 61.068 |
| 14 | 62.467 | 60.445 | 29 | 61.420 | 61.122 |
| 15 | 59.711 | 60.881 | 30 | 62.146 | 60.837 |

    **a** Set up the 3-sigma $X$-bar and $R$ chart parameters, the centerline, and the control limits ($n = 2$).

    **b** Set up the 3-sigma $X$-bar and $S$ chart parameters, the centerline, and the control limits ($n = 2$).

**15.11** Use the chart parameters you developed in Problem 15.10 (e.g. $X$-bar, $R$ and $X$-bar, $S$ pairs) to define control charts for future candy bar production.

    **a** Calculate the appropriate $\bar{x}$, $R$, and $s$ plotting points from the data below and plot them on your charts.

| Sample number | Candy bar 1 | Candy bar 2 | Sample number | Candy bar 1 | Candy bar 2 |
|---|---|---|---|---|---|
| 31 | 60.381 | 61.841 | 46 | 62.339 | 63.043 |
| 32 | 61.821 | 62.038 | 47 | 70.007 | 64.371 |
| 33 | 61.566 | 63.955 | 48 | 67.858 | 67.673 |
| 34 | 60.399 | 60.908 | 49 | 65.798 | 66.885 |
| 35 | 62.288 | 62.604 | 50 | 65.833 | 67.794 |
| 36 | 60.067 | 61.886 | 51 | 65.712 | 71.446 |
| 37 | 62.660 | 62.403 | 52 | 67.525 | 70.303 |
| 38 | 62.530 | 63.469 | 53 | 68.037 | 72.309 |
| 39 | 62.699 | 62.810 | 54 | 72.449 | 65.958 |
| 40 | 63.676 | 61.905 | 55 | 68.187 | 70.430 |
| 41 | 60.587 | 62.570 | 56 | 67.875 | 64.558 |
| 42 | 62.584 | 63.251 | 57 | 59.381 | 61.288 |
| 43 | 61.937 | 63.574 | 58 | 60.821 | 59.067 |
| 44 | 62.865 | 62.574 | 59 | 60.566 | 60.841 |
| 45 | 63.876 | 63.799 | 60 | 59.399 | 61.038 |

    **b** Comment as to the stability or instability of the candy production process based on your charts.

**15.12** The process logs associated with the data in Problem 15.11 indicate that the process was disturbed three times by assignable causes. The net result was that the process operated at different levels. See if you can locate the shift points. [Actually, the data in Problem 15.11 were simulated from a normal distribution and included samples with $\mu = 61.5$ g, $\sigma = 0.75$ g (samples 31 to 36); $\mu = 62.5$ g, $\sigma = 0.75$ g (samples 37 to 46); $\mu = 68.0$ g, $\sigma = 2.25$ g (samples 47 to 56); $\mu = 60.5$ g, $\sigma = 0.75$ g (samples 57 to 60).]

**15.13** Of the two chart sets ($X$-bar, $R$ and $X$-bar, $S$) developed in Problems 15.10 and 15.11, which chart set would you recommend for the candy bar weight? Why?

**15.14** Given the general $X$-bar and $R$ chart forms below, where the subgroup size was $n = 4$, sketch in the relative position (higher, lower, the same) for all chart parameters if we were to use a subgroup size of $n = 10$. Label the lines after you sketch them in the drawings. You are expected to draw and label a total of four new control limits, and two new centerlines. Explain the reasons for positioning the control limits where you did.

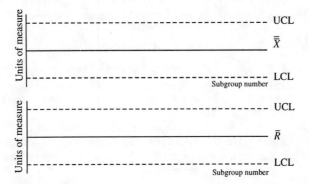

**15.15.** Thirty consecutive samples of size $n = 1$ were collected from a cardboard manufacturing process during an 8-hour shift. This process had not been studied previously by charting, and some problems with poor quality had been noted. The process measurements (cardboard thickness in millimeters) are shown below.

| Sample number | Measurement $x_i$ | Sample number | Measurement $x_i$ |
|---|---|---|---|
| 1 | 2.53 | 16 | 2.53 |
| 2 | 2.49 | 17 | 2.52 |
| 3 | 2.52 | 18 | 2.49 |
| 4 | 2.50 | 19 | 2.52 |
| 5 | 2.51 | 20 | 2.49 |
| 6 | 2.51 | 21 | 2.55 |
| 7 | 2.59 | 22 | 2.47 |
| 8 | 2.58 | 23 | 2.52 |
| 9 | 2.63 | 24 | 2.51 |
| 10 | 2.49 | 25 | 2.50 |
| 11 | 2.52 | 26 | 2.51 |
| 12 | 2.43 | 27 | 2.47 |
| 13 | 2.49 | 28 | 2.51 |
| 14 | 2.51 | 29 | 2.52 |
| 15 | 2.50 | 30 | 2.51 |

    **a** Compute the 3-sigma control limits for the $X$ and $R_M$ charts. Justify the artificial sample size chosen ($n = 2$) for the $R_M$ control chart, interpret the charts, and make comments on any patterns you see.

    **b** Compute the control limits for the EWMA and EWMD charts using the same data; use $r = 0.25$. Then, interpret the charts.

    **c** Comment on the relative appearance of the two sets of charts constructed in parts $a$ and $b$. Are there significant differences? If so, how can these be accounted for or explained?

**15.16** Control charts for $\bar{x}$ and $s$ are maintained on the breaking strength in pounds for deep sea fishing line. The subgroup size is $n = 10$. The values of $\bar{x}$ and $s$ are computed for each subgroup. After collecting 30 subgroups it was calculated that $\sum \bar{x}_j = 4500$ and $\sum s_j = 135$.

    **a** Compute the 3-sigma control limits for the $X$-bar and $S$ charts.

    **b** Estimate $\sigma$, on the assumption that the process is in statistical control.

    **c** Determine $X$-bar and $S$ chart parameters for setting up 2.5-sigma control limits.

**15.17** The first 25 measurements shown in the table below were collected from a chemical process; they represent the concentration of a certain chemical (%) with a product specification of 10%.

| Sample number | Measurement $x_i$ | Sample number | Measurement $x_i$ |
|---|---|---|---|
| 1 | 10.33 | 26 | 9.77 |
| 2 | 10.79 | 27 | 9.87 |
| 3 | 11.23 | 28 | 9.91 |
| 4 | 10.56 | 29 | 10.35 |
| 5 | 9.77 | 30 | 10.69 |
| 6 | 9.55 | 31 | 11.35 |
| 7 | 9.66 | 32 | 12.34 |
| 8 | 9.96 | 33 | 12.45 |
| 9 | 10.23 | 34 | 12.22 |
| 10 | 10.13 | 35 | 11.98 |
| 11 | 10.47 | 36 | 12.04 |
| 12 | 10.65 | 37 | 12.45 |
| 13 | 10.09 | 38 | 11.67 |
| 14 | 10.07 | 39 | 11.23 |
| 15 | 9.85 | 40 | 10.45 |
| 16 | 9.11 | 41 | 10.12 |
| 17 | 9.44 | 42 | 9.88 |
| 18 | 9.77 | 43 | 9.66 |
| 19 | 10.03 | 44 | 9.77 |
| 20 | 10.45 | 45 | 9.88 |
| 21 | 10.44 | 46 | 9.99 |
| 22 | 10.48 | 47 | 10.04 |
| 23 | 10.13 | 48 | 10.08 |
| 24 | 10.04 | 49 | 10.07 |
| 25 | 9.66 | 50 | 10.12 |

    **a** Calculate the estimates of the process mean and standard deviation using the first 25 samples.

**b** Using your estimates in part *a,* construct EWMA and EWMD control charts using *r* values of 0.2 and 0.5. Interpret the two sets of charts.

**c** Plot the "new" data, i.e., samples 26 through 50, on the two sets of control charts developed in part *b.* Interpret each chart set and make comments on any patterns you see. What effects does changing *r* appear to have on these control charts? What factors might influence the selection of an *r* value?

**15.18** Food technologists at a food processing plant are considering the use of a CuSum chart for monitoring the moisture levels taken from an apple dehydration unit which is dehydrating 0.25-in cubes of Granny Smith apples. They intend to maintain a moisture level of between 12 and 16%, by weight. Hourly records of these percentage data are to be charted for process control. The process is designed to maintain an average moisture level at 14%. Experience has shown that the actual moisture level varies, with a standard deviation of 1.25%. Using $\alpha = 0.05$, $y_c = 2.0$, $D_c = 2.0$, construct a CuSum chart.

**15.19** Data have been collected on the apple dehydration line described in Problem 15.18. They are displayed below:

| Hour | Moisture level (%) | Hour | Moisture level (%) |
|------|------|------|------|
| 1 | 12.9 | 13 | 13.1 |
| 2 | 13.2 | 14 | 12.2 |
| 3 | 13.6 | 15 | 15.6 |
| 4 | 13.2 | 16 | 14.5 |
| 5 | 14.3 | 17 | 14.0 |
| 6 | 13.3 | 18 | 13.5 |
| 7 | 14.5 | 19 | 13.4 |
| 8 | 14.7 | 20 | 13.8 |
| 9 | 14.2 | 21 | 13.9 |
| 10 | 15.6 | 22 | 12.8 |
| 11 | 13.4 | 23 | 12.9 |
| 12 | 13.7 | 24 | 13.4 |

**a** Using the chart developed in Problem 15.18, plot the CuSums of the moisture levels over the past 24 hours (shown in the table above).

**b** Interpret the chart to assess how well the moisture levels were controlled during the last 24 hours.

**15.20** Using the data in Problem 15.19, develop an $X$ and $R_M$ chart pair. Use a two-point moving range for the $R_M$ chart. Interpret your results.

## Discovery

**15.21** There are four basic measurement scales:

**a** *Nominal Scale* (the weakest scale)—Here, numbers or labels serve only to distinguish between different categories or classes.

**b** *Ordinal Scale*—Here, the scale ranks or orders each measurement as to less than, greater than, or equal to other measurements.

**c** *Interval Scale*—Here, we see not only an ordering, but a defined interval between measurements (e.g., a temperature measurement) where we can assess how far apart two measurements are. Any zero point on an interval scale will be arbitrarily defined.

    **d** *Ratio Scale* (the strongest scale)—Here, we see both order and interval characteristics but we also see a natural zero point whereby the ratios of any two measurements are meaningful. Distances, lifetimes, weights, costs, profits, and so on are examples of ratio scales.

    Provide an example of each relative to a product's quality characteristic(s). Where does the attributes measure fall? Where does the variables measure fall?

**15.22** Choose a process or product and identify an appropriate quality characteristic that can be measured with a variables measure. Select an appropriate variables control chart pair. Collect a sufficient amount of data, and develop your chart limits. Build a display board which describes your chosen process or product, your selected substitute quality characteristic and how it relates to a true quality characteristic, your measurement procedure, and your chart results.

# APPENDIX 15A: *X*-BAR CHARTS AND THE CENTRAL LIMIT THEOREM

The central limit theorem (CLT) states that a distribution of sample means, with sample size *n,* sampled from any statistical population form, with mean $\mu$ and finite variance $\sigma^2$, will approach the normal distribution as *n* increases. For example,

$$\bar{x} \to N\left(\mu, \sigma_{\bar{x}}^2 = \frac{\sigma^2}{n}\right) \qquad \text{as } n \to \infty$$

This fundamental theorem plays a key role in hypothesis tests involving means (see Walpole and Myers [31]). Since our *X*-bar chart is essentially a crude graphical approximation of a hypothesis test of means, repeated over time, it follows that the CLT will support the *X*-bar chart.

In practice, the CLT supports the use of the normal distribution model for the *X*-bar chart even when the population of individual item measurements being sampled is not normally distributed. Figure 15.13 depicts the CLT concept. Small subgroup sizes, ranging from $n = 3$ to $n = 5$, usually provide reasonably good *X*-bar chart performance (Grant and Leavenworth [32]).

We should note that the previous CLT argument supports only the *X*-bar chart. The *R* and *S* charts are not supported by the CLT, significant deviations from normality in the measurements will distort the effectiveness of both.

# REFERENCES

**1** E. L. Grant and R. S. Leavenworth, *Statistical Quality Control,* 6th ed., New York: McGraw-Hill, 1988.

**2** See reference 1.

**3** R. E. Walpole and R. H. Myers, *Probability and Statistics for Engineers and Scientists,* 5th ed., New York: Macmillan, 1993.

**4** See reference 1.

**5** See reference 3.

**6** R. E. DeVor, T. Chang, and J. W. Sutherland, *Statistical Quality Design and Control: Contemporary Concepts and Methods,* New York: Macmillan, 1992.

**7** D. C. Montgomery, *Introduction to Statistical Quality Control,* 2d ed., New York: Wiley, 1991.

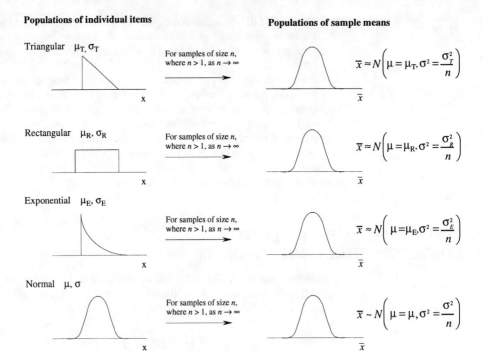

**Populations of individual items**

Triangular   $\mu_T, \sigma_T$

For samples of size $n$,
where $n > 1$, as $n \to \infty$

**Populations of sample means**

$$\overline{x} \approx N\left(\mu = \mu_T, \sigma^2 = \frac{\sigma_T^2}{n}\right)$$

Rectangular   $\mu_R, \sigma_R$

For samples of size $n$,
where $n > 1$, as $n \to \infty$

$$\overline{x} \approx N\left(\mu = \mu_R, \sigma^2 = \frac{\sigma_R^2}{n}\right)$$

Exponential   $\mu_E, \sigma_E$

For samples of size $n$,
where $n > 1$, as $n \to \infty$

$$\overline{x} \approx N\left(\mu = \mu_E, \sigma^2 = \frac{\sigma_E^2}{n}\right)$$

Normal   $\mu, \sigma$

For samples of size $n$,
where $n > 1$, as $n \to \infty$

$$\overline{x} \sim N\left(\mu = \mu, \sigma^2 = \frac{\sigma^2}{n}\right)$$

*Note:*  Above, we are taking many samples of size $n$ to build up the sample distributions.  If $n = 1$, then our populations of "sample" means will be identical to our original populations of individual items.

**FIGURE 15.13**    Graphical depiction of central limit theorem concept.

**8**  A. J. Duncan, *Quality Control and Industrial Statistics,* 5th ed., Homewood, IL: Irwin, 1986.

**9**  See reference 3.

**10**  See reference 8.

**11**  See reference 8.

**12**  See reference 6.

**13**  C. B. K. Roes, J. M. M. R. Does, and Y. Schurink, "Shewhart-Type Control Charts for Individual Measurements," *Journal of Quality Technology,* vol. 25, no. 3, pp. 188–198, 1993.

**14**  A. J. Harding, K. R. Lee, and J. L. Mullins, "The Effect of Instabilities on Estimates of Sigma," *46th Annual Quality Congress Transactions,* ASQC, Milwaukee: pp. 1037–1043.

**15**  G. B. Wetherill and D. W. Brown, *Statistical Process Control: Theory and Practice,* London, England: Chapman & Hall, 1991.

**16**  See reference 6.

**17**  See reference 1.

**18**  See reference 6.

**19**  See reference 6.

**20**  A. L. Sweet, "Control Charts Using Coupled Exponentially Weighted Moving Averages," *Transactions of IIE,* vol. 18, no. 1, pp. 26–33, 1986.

21   S. J. Hunter, "The Exponential Weighted Moving Average," *Journal of Quality Technology,* vol. 18, no. 4, pp. 203–210, 1986.
22   See reference 1.
23   See reference 1.
24   See reference 8.
25   See reference 7.
26   See reference 6.
27   See reference 1.
28   See reference 1.
29   See reference 6.
30   See reference 6.
31   See reference 3.
32   See reference 1.

# 16

# ATTRIBUTES CONTROL CHARTS

## 16.0 INQUIRY

1 How do defects and defectives differ?
2 When do we use an attributes chart? A variables chart?
3 What models are used to develop attributes charts?
4 What is the difference between the *P, NP* chart and the *C, U* chart families?

## 16.1 INTRODUCTION

In our discussion of variables control charts, we pointed out that each chart set can monitor only one quality characteristic. We know that most products have literally hundreds of quality characteristics. We must use variables control charts sparingly and apply them to critical quality characteristics. Grant and Leavenworth state that no dimension should be chosen for variables control charts unless there is an opportunity to save costs—of spoilage and rework, inspection, of excess material—or otherwise effect quality improvements that would in some way more than compensate for the costs of taking the measurements, keeping the charts, and analyzing them [1].

*We can use attributes based process control charts to monitor multiple quality characteristics as well as quality characteristics which are logically defined on a classification scale of measure.* Classification characteristics can include "voids" on a sheet of glass, cracks on an automobile flywheel, surface flaws on sheet metal panels, color inconsistencies on a painted surface, broken potato chips in a packet, paperwork errors, and so on.

*We can design attributes control charts to cover entire product units involving many quality characteristics.* For example, a typical die-cast body for a 35-mm camera, made of copper-aluminum alloy, has hundreds of dimensions. Although any one of these dimensions can be monitored with a variables control chart, we cannot justify hundreds of charts. Hence, we might use an attributes control chart to cover the entire product unit. Then, if we encounter or identify a quality bottleneck by using an attributes chart, we can isolate and study this quality characteristic in detail, using a variables chart. *We sometimes use the attributes chart for broad coverage and "slap" a variables chart at the quality bottleneck point in the production process, once it has been identified.*

Before developing the mechanics of attributes control charts, it is necessary to clearly define a number of important terms. *A "defect" is an individual nonconformity* on a product that causes it to fail to meet a quality specification. *A "zero defect" product refers to a product which meets all technical or engineering speci-*

*fications, and hence has no defects.  An item or article is "defective" if it fails to conform to specifications in some respect.*  Hence, a manufactured item or article that is defective will contain at least one defect or nonconformity.  Our textbook uses the terms "defective item" and "nonconforming item" interchangeably.  Also, the words "defect" and "nonconformity" are used interchangeably.

As was stated in earlier chapters, *zero defect products do not necessarily assure customer satisfaction unless the designers have completely captured the true quality characteristics in the product's substitute quality characteristics, through technical specifications.*  Remember, *we typically chart substitute, rather than true, quality characteristics on the production floor.  As such, we cannot automatically assume that nonconforming products are without value or pose serious risk to our customers (or that conforming products have value and pose no risk).*  It is not uncommon to encounter slightly defective items that still satisfy their functions quite well.  For example, a scratch on a refrigerator door is not going to reduce its functional value in terms of its refrigeration purpose.  But, our customer may object to the scratch because he or she doesn't like the way it looks.  Hence, it is important that we capture true customer needs and expectations in our product features and specifications, and that products be evaluated accordingly.

*The most difficult aspect of quality characterization by attributes is the precise determination of what constitutes the presence of a particular defect* (DeVor et al. [2]).  Since a vast majority of the attribute defects are either visual or cannot be measured precisely because adequate technical measurement tools are not available, some degree of judgment is involved in defect classification.  Variation in human judgment tends to be a problem in attribute characterization.  *It is critical that precise operational definitions of a defect be generated and understood.*  For example, it is necessary to specify exactly what constitutes a crack, crater, fold, inclusion, pit, seam, splatter, and so on, for manufactured components.  It may be necessary to specify a minimum length or depth of a scratch in order to declare it a defect, or the size of a surface blemish, and so on.  In general, providing visual standards, such as photographs or samples of defects, helps in reducing the human variation in attribute characterization.

## 16.2  CONTROL CHARTS FOR ATTRIBUTES QUALITY CHARACTERISTICS

*Attributes quality characteristics are classification oriented.  Typically, we deal with only two classes, good-defective (conforming-nonconforming) product.  Two general families of charts are used: (1) the binomial-based P, NP chart family for monitoring defective-nonconforming individual products or (2) the Poisson-based C, U chart family for monitoring defects per production unit.  Single charts are sufficient to monitor process data when they are expressed in an attributes form.*  The nature of the binomial and Poisson models (specifically their model parameter structure) requires only one control chart.  Table 16.1 summarizes some of the characteristics of each family.

As stated earlier, we must define what constitutes conformance or nonconformance to specifications or requirements for a *P* or an *NP* chart.  For example, *we*

**TABLE 16.1**  CHARACTERISTICS OF ATTRIBUTES CONTROL CHART ALTERNATIVES FOR MONITORING PRODUCTION PROCESSES

| Chart type | Advantages | Disadvantages |
|---|---|---|
| **Charts that monitor information regarding number of defective items** | | |
| *P* | Monitors the proportion of defective items and is based on the binomial model. | A single *P* or *NP* chart can monitor multiple quality characteristics. | Relatively large sample sizes are needed in *P* and *NP* charts when dealing with high process quality levels (a very low proportion defective). |
| *NP* | Monitors the number of defective items per sample and is based on the binomial model. | The statistic monitored with the *P* chart (proportion defective) is easy to interpret (relative to the *NP* chart). | The *NP* chart is calibrated to sample size which makes interpretation more difficult (than interpretation of the *P* chart). |
| **Charts that monitor information regarding number of defects** | | |
| *C* | Monitors the number of defects per inspection unit and is based on the Poisson model. | *C* and *U* charts can be used to monitor multiple types of quality flaws in a product. | Relatively large samples are required in *C* and *U* charts when dealing with high process quality levels (a very low probability of encountering a defect). |
| *U* | Monitors the average number of defects per inspection unit and is based on the Poisson model. | The *U* chart is flexible in dealing with sample sizes that differ from time to time (as compared to the *C* chart). | The *C* and *U* charts are calibrated to defects per inspection unit, which sometimes makes interpretation challenging. |

*Note:* To gain information about each defect category, these charts must be complemented with a process log and Pareto charts.

*can use the P or NP chart as a screening chart* and define more than one inspection point (e.g., length and width dimensions, appearance characteristics) on a product. Hence, *the P and NP charts can monitor many product quality characteristics simultaneously* (versus only one measurable characteristic with a variables chart pair). This feature allows broad coverage possibilities; but it does not provide detailed information. Hence, it has advantages and disadvantages.

Likewise, *a C or U chart may track the number of one or more different types of defects encountered. Our physical product and process knowledge helps us to clearly define what constitutes a defect;* for example, in cloth quality, weaving defects as well as dying defects may be included in the same *U* chart. We must also determine how we will sense or identify and measure each type of defect once it is defined.

*In SPC, it is customary to assume that a process is "in control" or stable until evidence is discovered to the contrary.* Since the assumption that the process is stable implies that the attributes quality measures of interest are on target, we can conceptualize and state a set of statistical hypotheses for each measure. For example, generic hypotheses for the *P* chart are stated below:

$H_0$: The process population proportion defective is as stated.
$H_1$: The process population proportion defective has shifted from the stated value.

Basically, attributes control charts provide a crude, graphical means for assessing the validity of these hypotheses. This assessment leads us to conclude that the process is stable (when we do not reject $H_0$) or that the process is unstable (when we reject $H_0$), based on data collected in our last sample.

*Sample sizes in attributes control charts are typically much larger than sample sizes in variables control charts.* We typically see sample sizes an order of magnitude, or more, larger in attributes charts than we see in variables charts (e.g., a *P* chart might have a sample size of 50, whereas an *X*-bar chart might have a subgroup size of 5). Many authors attempt to compare the costs of attributes charts with those of variables charts. Although costs are very important, one chart type is not necessarily a substitute for another. *We must choose our chart to fit our needs.* Arguing the virtues of one chart type against the other is like arguing whether a hammer is a better tool than a saw. Obviously, it depends on what we want to accomplish. The more we know about our tools and our needs, the wiser we can be in tool selection and application. In general, we can say that an attributes chart paints a landscape picture of our process quality, while a variables chart paints a portrait.

*It is critical to reemphasize that SPC charts (both variables control charts and attributes control charts) are incomplete, in a quality control sense, in that they do not identify special causes, nor do they tell us how to eliminate the effect of a special cause.* Control charts basically serve as statistical based warning devices. They do not help us to account for and document physical characteristics, actions taken, and results obtained. That is the role of process logs and Pareto charts, discussed in the next chapter.

## Example 16.1—Golf Club Shaft Production Case

Before proceeding with a detailed description of attributes control chart mechanics, we will illustrate the use of an attributes control chart, the $P$ chart. In the following example we use the $P$ chart to monitor the pultrusion process used in manufacturing composite components in the sports equipment industry.

The marketing department of a leading sports equipment manufacturer is concerned about losing its product performance edge in marketing golf clubs, as well as losing its market share. A recent survey indicated that customers over the last two years have been unhappy with the performance of their golf clubs. In addition, a significant number of formerly loyal customers have changed brands. Investigations have determined that most of the dissatisfaction is centered on the strength of the club shaft itself. As a response, the company's strategy is to monitor the production process so as to identify new opportunities for quality improvement.

The basic production process involves pultrusion, as described by Kalpakjian [3], where fabric is pulled through a thermosetting polymer bath, and then through a long heated steel die. The shaft is cured as it travels through the die. Figure 16.1 gives a schematic illustration of the process. The materials used are a proprietary thermosetting polymer and a proprietary reinforcing fiber. Experience has shown that the major quality considerations for the process include internal voids, broken strands, gaps between successive layers, and microcracks caused by improper curing.

Groups of ten consecutive shafts were examined every 30 min. Each incidence of each type of defect was noted using nondestructive evaluation equipment. The number of shafts with at least one defect present was noted (i.e., the number of defectives in the group). The data for each 8-hour shift (16 groups of 10 shafts each) were collected into a shift sample of 160 shafts, or, $n = 160$, to provide an overall picture of the process performance. A proportion defective control chart, or $P$ chart, was used to monitor the process. The data in Table 16.2 include the number of defective club shafts in each shift sample of 160. The data were taken over a period of 36 consecutive shifts.

The characteristic of interest here is the process proportion defective. The proportion defective value $p_j$ for each of the shift samples has been calculated

**FIGURE 16.1**    Pultrusion process depiction. Reprinted, with permission, from S. Kalpakjian, *Manufacturing Engineering and Technology*, Second Edition, Reading MA: Addison-Wesley Publishing Company, pp. 564, 1992. Copyright 1992 by Addison-Wesley Publishing Company, Inc.

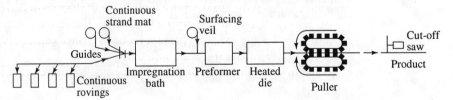

**TABLE 16.2**   PROPORTION DEFECTIVE DATA FOR 36 SHIFT
SAMPLES FROM THE GOLF CLUB SHAFT PULTRUSION
PROCESS (EXAMPLE 16.1)

| Shift number | Number defective | Proportion defective | Shift number | Number defective | Proportion defective |
|---|---|---|---|---|---|
| 1 | 9 | 0.05625 | 19 | 6 | 0.03750 |
| 2 | 6 | 0.03750 | 20 | 12 | 0.07500 |
| 3 | 8 | 0.05000 | 21 | 8 | 0.05000 |
| 4 | 14 | 0.08750 | 22 | 5 | 0.03125 |
| 5 | 7 | 0.04375 | 23 | 9 | 0.05625 |
| 6 | 5 | 0.03125 | 24 | 15 | 0.09375 |
| 7 | 7 | 0.04375 | 25 | 6 | 0.03750 |
| 8 | 9 | 0.05625 | 26 | 8 | 0.05000 |
| 9 | 5 | 0.03125 | 27 | 4 | 0.02500 |
| 10 | 9 | 0.05625 | 28 | 7 | 0.04375 |
| 11 | 1 | 0.00625 | 29 | 2 | 0.01250 |
| 12 | 7 | 0.04375 | 30 | 6 | 0.03750 |
| 13 | 9 | 0.05625 | 31 | 9 | 0.05625 |
| 14 | 14 | 0.08750 | 32 | 11 | 0.06875 |
| 15 | 7 | 0.04375 | 33 | 8 | 0.05000 |
| 16 | 8 | 0.05000 | 34 | 9 | 0.05625 |
| 17 | 4 | 0.02500 | 35 | 7 | 0.04375 |
| 18 | 10 | 0.06250 | 36 | 8 | 0.05000 |
| Totals | 139 | 0.86875 | Totals | 140 | 0.87500 |

and is shown in Table 16.2.  For example, 9 defectives were observed among
160 inspected club shafts in shift 1; hence the proportion defective for the shift
is $p_1 = 9/160 = 0.05625$.

The $p_j$ values for the 36 successive shifts of sample size 160 are plotted in
Figure 16.2.  The plot in Figure 16.2a shows a solid line to indicate the average
proportion defective for the entire data set.  The grand average $\bar{p}$ (i.e., the average
of the 36 proportion defectives) is 0.04844.  This is the sum of the proportion
defectives, 1.74375 (or 0.86875 + 0.87500), divided by the number of shift
samples, 36.

The individual values for the sample proportion defective $p_j$ vary considerably.
But it is too early to determine if the variation about the average proportion defective $\bar{p}$ is due solely to the forces of common cause or if a special cause is
present.  To help us resolve the problem, we develop control limits relative to the
parameter of interest (proportion defective).  At this time, we have not introduced
P chart mechanics, but we have developed our control limits.  Here, we have
only a UCL posted on our chart; in this case we have no LCL.  The methods for
calculating control limits are explained later in this chapter.  The limit shown in
Figure 16.2b is the "trial control limit."  The P chart shows no evidence of process
instability.  Here, all the points plotted are inside the control limit.  Our chart
indicates that the variation present is due to chance causes alone.

Next, we extend the chart into the future; the dividends from control charts
come in the application of the control limits to future production.  If the charts

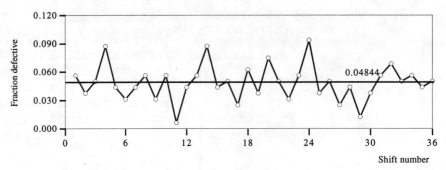

(a) Plot of proportion defective in each shift sample

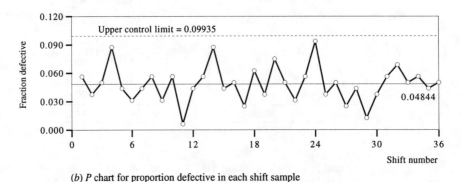

(b) P chart for proportion defective in each shift sample

**FIGURE 16.2**    P chart for the pultrusion process (Example 16.1).

indicate a process upset, we seek assignable causes and bring the process back to a stable condition.

Our control chart here indicates that the process is operating in statistical control (all plotted points are within the control limits), but the estimated process proportion defective value is unsatisfactory, $\bar{p} = 0.04844$. Benchmarking studies suggest that the best-of-the-best manufacturers are operating with less than 1 percent defective, using similar technology. It is clear that the current production process is not acceptable, since almost 5 percent of the golf club shafts produced are defective. The following alternatives are available for managerial consideration:

1  Redesign the existing production process so as to reduce the process proportion defective.
2  Resort to a totally different production process.

We should note here that "pushing" operators for a lower proportion defective will do no good in this case, since the process is already operating in statistical control. Improvement here calls for managerial action with regard to the production system.

## 16.3 CONSTRUCTION OF ATTRIBUTES CONTROL CHARTS

*The control chart mechanics described here for attributes data are applicable when the process in question is considered stable.* If instabilities are present (judged by points that fall beyond the 3-sigma control limits on trial charts), we should pursue engineering work to stabilize the process. *Setting up charts using data from unstable processes will yield misleading chart dimensions and, hence, misleading process control indications.* We must identify special causes, eliminate their effects, and develop revised control chart limits that represent a stable process.

### General Symbols

$n$: the sample size in units of production.

$m$: the number of samples in the preliminary data set used to set up a control chart.

$j$: the sample index.

$b$: the control limit spread in number of standard deviations of the plotted quantity.

### P, NP Chart Symbols

$p$: the population proportion of defective or nonconforming items.

$p_j$: the proportion of defective or nonconforming items in sample $j$.

$\bar{p}$: the mean proportion of defective or nonconforming items in a group of samples.

$\text{LCL}_p$, $\text{UCL}_p$: the lower, upper control limits for the $P$ chart.

$np$: the population mean number of defective or nonconforming units per sample.

$np_j$: number of observed items in sample $j$ that do not meet specifications (defective or nonconforming items).

$n\bar{p}$: the mean number of defective or nonconforming items in a group of samples each of size $n$.

$\text{LCL}_{np}$, $\text{UCL}_{np}$: the lower, upper control limits for the $NP$ chart.

### C, U Chart Symbols

$c$: the population mean number of defects per inspection unit.

$c_j$: the number of defects observed in sample $j$.

$\bar{c}$: the mean number of defects per inspection unit, calculated from a group of samples.

$\text{LCL}_c$, $\text{UCL}_c$: the lower, upper control limits for the $C$ chart.

$u$: the population mean number of defects per inspection unit.

$u_j$: the number of defects observed in sample $j$.

$\bar{u}_j$: the mean number of defects per inspection unit for a single sample $j$.

$\bar{u}$: the mean number of defects per inspection unit of a group of samples.

$\text{LCL}_u$, $\text{UCL}_u$: the lower, upper control limits for the $U$ chart.

## Binomial Model

***The binomial distribution serves as the model for the P, NP chart family.*** The binomial distribution has two parameters, $n$ and $p$, where $p$ is the probability of failure on each trial and $n$ is the number of independent trials (Bernoulli trials). If we define a random variable $X$ as the number of nonconforming or defective items among $n$ produced items with a probability $p$ of failure for each item, then the probability mass and cumulative mass functions for the binomial model are

$$f(x; n, p) = C_x^n \, p^x (1-p)^{n-x}$$

$$= \frac{n!}{x! \, (n-x)!} \, p^x \, (1-p)^{n-x} \qquad x = 0, 1, 2, ..., n$$

and

$$F(x; n, p) = \sum_{i=0}^{x} C_i^n \, p^i (1-p)^{n-i} \qquad x = 0, 1, 2, ..., n$$

Mean: $\mu_x = np$

Variance: $\sigma_x^2 = np \, (1-p)$

## Poisson Model

***The Poisson distribution*** (a limiting case of the binomial distribution, where $n \to \infty$, $p \to 0$, and $\mu = np$ remains constant) ***serves as the basis for the C and U chart family.*** The Poisson distribution has a single parameter, $c$. If we define the random variable $X$ as the number of defects in a specific "area of opportunity" (1 inspection unit), then the probability density and the cumulative density functions are

$$f(x; c) = \frac{e^{-c} c^x}{x!} \qquad x = 0, 1, 2, \cdots$$

$$F(x; c) = \sum_{i=0}^{x} \frac{e^{-c} c^i}{i!} \qquad x = 0, 1, 2, \cdots$$

Mean: $\mu_x = c$

Variance: $\sigma_x^2 = c$

Given the model parameter $c$, the cumulative density function values for different values of $x$ can be obtained from Table IX.6, Section IX.

## P Chart Mechanics

***When we select samples of size n at regular intervals from a process, each sample indicates how the process is behaving at one "point" in time. If the process is sub-***

*ject only to chance causes, the sample proportions defective $p_j$'s should be randomly distributed within certain probabilistic limits.* Occasionally, the process may experience some real change or upset due to a special cause. *The P chart will indicate an upset by exceeding the control chart limits or by showing a non-random pattern.*

As a matter of chance alone, variations in the number of defective items encountered (for a given sample size) are inevitable from sample to sample. The nonconforming proportion in the sample may vary considerably. As long as the nonconforming proportion in the universe remains unchanged, the relative frequencies of nonconforming proportions in the samples may be expected to follow the binomial model. The population or process proportion defective $p$ can be estimated by $\bar{p}$ as long as we have a sufficiently large number of samples ($m \geq 20$). Assuming the binomial model is valid, the sample proportions defective $p_j$'s follow a binomial model with mean $p$ and standard deviation

$$\sqrt{\frac{p(1-p)}{n}}$$

The general form to construct the $P$ chart control limits, assuming all samples are of the same size $n$ and the process proportion defective $p$ is known, is

$$\left.\begin{array}{c} \text{UCL}_p \\[2em] \text{LCL}_p \end{array}\right\} = p \pm b \sqrt{\frac{p(1-p)}{n}} \qquad (16.1)$$

$$\text{Centerline} = p$$

If the lower control limit calculated is less than 0, we say that no lower limit exists or set it equal to 0; a proportion less than 0 is not realistic. Negative numbers are sometimes encountered in the $\text{LCL}_p$ calculations since we are using the traditional $\pm b$-sigma limits (from the centerline) with typically a nonsymmetric distribution (the binomial model is symmetric only when $p = 0.5$).

*It is common practice in the United States to use 3-sigma control chart limits,* where we set $b$ equal to 3. We usually estimate $p$ from $\bar{p}$. As Grant and Leavenworth point out, *it is very important to note that the X-bar chart is the only instance in the application of the Shewhart control chart models for which the distribution of the random variable can be shown to tend toward the normal distribution* [4]. In all other cases (e.g., the $R$ chart, $S$ chart, $P$ chart) it is appropriate to say, for a stable process, that the occurrence of a point falling outside 3-sigma limits at random would be very unlikely. This is not to say that it is impossible to associate probability values with points outside control limits for charts other than the $X$-bar chart. But, such calculations are very tedious and time consuming, and contribute little to the application of the basic decision rule (Grant and Leavenworth [4]). Chapter 17 will discuss this topic in more detail.

The 3-sigma control limits for a $P$ chart where $p$ is estimated from data obtained from a stable process with constant sample size $n$ are

$$\left.\begin{array}{c} \text{UCL}_p \\ \\ \text{LCL}_p \end{array}\right\} = \bar{p} \pm 3 \sqrt{\frac{\bar{p}(1-\bar{p})}{n}} \qquad (16.2)$$

$$\text{Centerline} = \bar{p}$$

Again, if the LCL calculated is less than 0, no lower limit exists. Here,

$$\bar{p} = \frac{\sum\limits_{j=1}^{m} p_j}{m}$$

where $j$ is used as a sample index.

Once the $P$ chart is set up using the computed control limits and centerline, we calculate $p_j$ after each sample is taken and plot the values. Next, we connect the points together with a solid line and use the chart to monitor the process.

We calculate $p_j$ as

$$p_j = \frac{np_j}{n_j} = \frac{\text{number of observed defective items in sample } j}{\text{size of sample } j}$$

## Example 16.2—Golf Club Shaft Production Case Revisited

We will now return to our golf club shaft pultrusion process. In Example 16.1 we examined a pultrusion process for making golf club shafts. At one point in the process the completed golf club shaft was examined for the presence of molding defects. Recall that the data for 36 shift samples of size $n = 160$ are shown in Table 16.2 and the proportion defective values are plotted in Figure 16.2. Given the data in Table 16.2 and the $P$ chart mechanics, the centerline and the 3-sigma control limits for the $P$ chart can be calculated as follows:

$$\bar{p} = \frac{\sum\limits_{j=1}^{m} p_j}{m} = \frac{1.74375}{36} = 0.04844$$

For $\bar{p} = 0.04844$ and $n = 160$,

$$\hat{\sigma}_p = \sqrt{\frac{\bar{p}(1-\bar{p})}{n}} = \sqrt{\frac{(0.04844)(1-0.04844)}{160}} = 0.01697$$

From Equation 16.2, the 3-sigma control limits for the $P$ chart are

$$\text{UCL}_p = \bar{p} + 3\sqrt{\frac{\bar{p}(1-\bar{p})}{n}} = 0.04844 + 3\,(0.01697) = 0.09935$$

$$\text{LCL}_p = \bar{p} - 3\sqrt{\frac{\bar{p}(1-\bar{p})}{n}} = 0.04844 - 3\,(0.01697) < 0$$

Here, $\text{LCL}_p$ is negative; hence, no lower limit exists.

$$\text{Centerline} = \bar{p} = 0.04844$$

The control limit for the $P$ chart is shown in Figure 16.2$b$. We have noted all along that to construct meaningful control chart limits, the process under consideration must be in statistical control. Figure 16.2 shows that there are no signals to indicate the process is unstable. As was mentioned earlier, while the process seems to be stable over the period studied, the proportion defective rate observed was considered too high and process improvement action was taken.

The process improvement team found that microcracks due to improper curing accounted for about 70 percent of the defects. The team decided to redesign the steel die such that curing takes place uniformly and at a faster rate as the product moves through the die. In order to further enhance the process quality, it was decided to implement the following changes:

1 Insulate the die to maintain uniform temperatures.
2 Increase the percentage of catalyst in the polymer so as to increase the stiffness of the club.
3 Modify the puller so as to create a more uniform tension on the product.

After incorporating these changes, the data collection process was resumed.

Table 16.3 provides the proportion defective data for 24 shifts, following the design changes. These data are plotted in Figure 16.3, as a continuation of the original chart (Figure 16.2), using the same control limits. The general appearance of the chart indicates that a significant change has occurred in the production process, and that the control limits must be recomputed.

**TABLE 16.3**    PROPORTION DEFECTIVE DATA FOR 24 SHIFTS AFTER INCORPORATING PULTRUSION PROCESS DESIGN CHANGES (EXAMPLE 16.2)

| Shift number | Number defective | Proportion defective |
|---|---|---|
| 37 | 3 | 0.01875 |
| 38 | 2 | 0.01250 |
| 39 | 5 | 0.03125 |
| 40 | 3 | 0.01875 |
| 41 | 8 | 0.05000 |
| 42 | 4 | 0.02500 |
| 43 | 3 | 0.01875 |
| 44 | 5 | 0.03125 |
| 45 | 6 | 0.03750 |
| 46 | 1 | 0.00625 |
| 47 | 5 | 0.03125 |
| 48 | 3 | 0.01875 |
| 49 | 2 | 0.01250 |
| 50 | 2 | 0.01250 |
| 51 | 4 | 0.02500 |
| 52 | 6 | 0.03750 |
| 53 | 3 | 0.01875 |
| 54 | 8 | 0.05000 |
| 55 | 2 | 0.01250 |
| 56 | 3 | 0.01875 |
| 57 | 1 | 0.00625 |
| 58 | 2 | 0.01250 |
| 59 | 4 | 0.02500 |
| 60 | 3 | 0.01875 |
| Totals | 88 | 0.55000 |

The revised control limits for the $P$ chart are given below:

$$\text{Revised UCL}_p = 0.05841$$

$$\text{Revised LCL}_p < 0 \qquad \text{hence, no lower limit exists}$$

$$\text{Revised centerline} = 0.02292$$

**FIGURE 16.3**    $P$ chart continuation after pultrusion process design changes (Example 16.2).

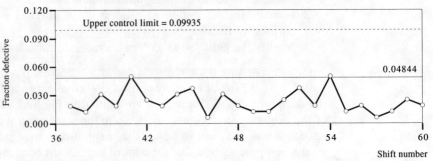

Detailed calculations are not presented, our readers may want to verify the accuracy of these revised limits.

Figure 16.4 presents the revised control limits along with past data and past control limits. It is evident from the plot that the process seems to be in statistical control and does not present any nonrandom patterns. The proportion defective rate was reduced dramatically, to less than half of its original level. We have to remember that, while the control chart showed process stability even during the first 36 shifts of operation, it indicated an unacceptable operating level. The problem faced was not one of special causes, but of system performance itself. We can also see that it was the project team (the users of the control chart) that identified opportunities for improvement—not the control chart.

In some cases, we may want to examine all of our product created during a shift or workday. Or, perhaps we will want to examine a certain percent of our daily production. In these cases, the sample size may vary from one sample to the next; in other words, we are dealing with an $n_j$, rather than a constant sample size of $n$. It was stated earlier that the sample proportion defective follows a binomial model with mean $p$ and standard deviation $\sqrt{p(1-p)/n_j}$ as long as we have a constant sample size. This general form can be modified to incorporate variable sample sizes. The sample proportion defective still follows a binomial model with mean $p$, but now the standard deviation is $\sqrt{p(1-p)/n_j}$. Here, we can see that the control limits will change with each different sample size. Hence, we will see "stair-stepped" control limits on the chart and must judge each sample's calculated $p_j$ on the basis of its specific control limits (i.e., based on its specific $n_j$).

**FIGURE 16.4**   *P* chart for the entire pultrusion process (Example 16.2).

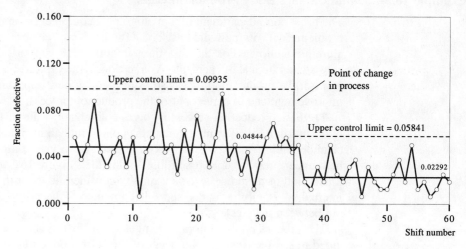

The 3-sigma control limits for a $P$ chart where $p$ is estimated from data obtained from a stable process with variable sample size $n_j$ are

$$\left.\begin{array}{c} \text{UCL}_{p_j} \\[2mm] \text{LCL}_{p_j} \end{array}\right\} = \bar{p} \pm 3 \sqrt{\frac{\bar{p}(1-\bar{p})}{n_j}} \qquad (16.3)$$

$$\text{Centerline} = \bar{p}$$

where

$$\bar{p} = \frac{\sum\limits_{j=1}^{m} np_j}{\sum\limits_{j=1}^{m} n_j}$$

Here $j$ is used as the sample index.

Note, here the centerline is constant, but the control limits change when the sample size changes. Again, if the lower control limit calculated is less than 0, a lower control limit does not exist.

Once the $P$ chart is set up using the computed centerline, we calculate $p_j$ and its control limits after each sample is taken and then, we plot the values. Next, we connect the points with a solid line for $p_j$'s and a dashed line for the control limits. Then, we use the chart to monitor the process. We calculate $p_j$ as

$$p_j = \frac{np_j}{n_j} = \frac{\text{number of observed defective items in sample } j}{\text{sample size } j}$$

---

## Example 16.3—35-mm Camera Body Production Case

The production department of a photographic equipment manufacturer is facing problems with its new die-cast body for a 35-mm camera, made of copper-aluminum alloy. For the last three months, the proportion defective was consistently high at 3.5 percent. As a response, the department's strategy was to monitor the production process so as to identify new opportunities for quality improvement and to better control the production process.

The basic production process involves cold-chamber die casting, as described by Kalpakjian [5]. The molten metal is poured into an injection cylinder or shot chamber and then the metal is forced into the die cavity at high pressure and is held under pressure until it solidifies in the die. Figure 16.5 gives a schematic illustration of the process. Major quality considerations for the process include internal voids, ejector marks, and small amounts of flash (thin material squeezed out between the dies).

The process production rate changes from shift to shift. The engineers decided to sample (inspect) 4 percent of the shift production. Each incidence of

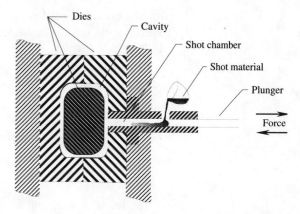

**FIGURE 16.5**   Die-casting process (Example 16.3).

each type of defect was noted, and the values are shown for 20 consecutive shifts in Table 16.4.

Given the data in Table 16.4 and the results above, based on the binomial model, the centerline can be calculated as follows:

$$\bar{p} = \frac{\sum_{j=1}^{20} np_j}{\sum_{j=1}^{20} n_j} = \frac{97}{2735} = 0.0355$$

**TABLE 16.4**   DATA ON NUMBER OF DEFECTS AND PROPORTION DEFECTIVE FOR 20 SHIFTS OF CAMERA BODY PRODUCTION (EXAMPLE 16.3)

| Shift number | Sample size | Number of defects | | | Number defective | Proportion defective |
|---|---|---|---|---|---|---|
| | | Voids | Marks | Flashes | | |
| 1 | 140 | 3 | 2 | 2 | 5 | 0.0357 |
| 2 | 132 | 7 | 3 | 3 | 6 | 0.0455 |
| 3 | 98 | 2 | 0 | 1 | 1 | 0.0102 |
| 4 | 102 | 4 | 4 | 8 | 4 | 0.0392 |
| 5 | 40 | 0 | 0 | 3 | 3 | 0.0750 |
| 6 | 126 | 2 | 3 | 2 | 5 | 0.0397 |
| 7 | 132 | 9 | 5 | 2 | 6 | 0.0455 |
| 8 | 132 | 2 | 1 | 9 | 4 | 0.0303 |
| 9 | 156 | 5 | 4 | 3 | 9 | 0.0577 |
| 10 | 190 | 9 | 3 | 5 | 7 | 0.0368 |
| 11 | 210 | 12 | 6 | 6 | 8 | 0.0381 |
| 12 | 167 | 7 | 4 | 3 | 1 | 0.0060 |
| 13 | 134 | 7 | 5 | 4 | 6 | 0.0448 |
| 14 | 120 | 1 | 2 | 0 | 3 | 0.0250 |
| 15 | 110 | 0 | 0 | 5 | 3 | 0.0273 |
| 16 | 134 | 1 | 1 | 3 | 7 | 0.0522 |
| 17 | 167 | 8 | 3 | 7 | 6 | 0.0359 |
| 18 | 187 | 3 | 4 | 3 | 4 | 0.0214 |
| 19 | 123 | 0 | 0 | 1 | 4 | 0.0325 |
| 20 | 135 | 4 | 3 | 6 | 5 | 0.0370 |
| Totals | 2735 | 86 | 53 | 76 | 97 | |

As was discussed earlier, the control limits will be a function of the sample size. The 3-sigma control limits for shift sample 1 can be calculated from Equation 16.3 as follows:

$$\text{UCL}_{p1} = \bar{p} + 3\sqrt{\frac{\bar{p}(1-\bar{p})}{n_1}} = 0.0355 + 3\sqrt{\frac{0.0355\,(1-0.0355)}{140}} = 0.0824$$

$$\text{LCL}_{p1} = \bar{p} - 3\sqrt{\frac{\bar{p}(1-\bar{p})}{n_1}} = 0.0355 - 3\sqrt{\frac{0.0355\,(1-0.0355)}{140}} < 0$$

Hence, no lower limit exists for sample $j = 1$. In a similar fashion, we can calculate the control limits for the other 19 shift samples. These control limits for the $P$ chart are shown in Figure 16.6.

The control chart shows that there are no signals present that would indicate process instability, suggesting that the calculated control limits are valid. Even though the process seems to be stable over the period studied, the proportion defective was considered too high, and process improvement actions were deemed necessary.

## NP CHART MECHANICS

*Sometimes it is convenient to develop an NP chart—a control chart based on the number of defectives, $np_j$ instead of the fraction defective $p_j$. This is particularly true if the sample size $n$ is constant and the fraction defective $p$ is quite small. The NP chart provides essentially the same information as the P chart. We can show that the number of nonconforming units observed in each sample of size $n$ follows the binomial model with mean parameter $np$ and standard deviation $\sqrt{np\,(1-p)}$ when the process under consideration is in statistical control. For a generic NP chart where we assume that $p$ is known and we have equal sample sizes of size $n$,*

**FIGURE 16.6**   *P* chart for the die-casting process (Example 16.3).

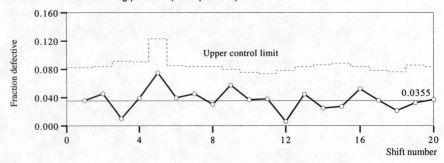

$$\left.\begin{array}{l} \text{UCL}_{np} \\[2ex] \text{LCL}_{np} \end{array}\right\} = np \pm b\sqrt{np\,(1-p)} \qquad (16.4)$$

$$\text{Centerline} = np$$

Just as in the case of the $P$ chart, if the LCL calculated is less than 0, we assume no lower limit exists.

The control limits for a 3-sigma $NP$ chart where $p$ is estimated from the data are

$$\left.\begin{array}{l} \text{UCL}_{np} \\[2ex] \text{LCL}_{np} \end{array}\right\} = n\bar{p} \pm 3\sqrt{n\bar{p}\,(1-\bar{p})} \qquad (16.5)$$

$$\text{Centerline} = n\bar{p}$$

where
$$n\bar{p} = \frac{\displaystyle\sum_{j=1}^{m} np_j}{m}$$

Here, $j$ is used as a sample index, and $m$ should be at least 20.

Once the $NP$ chart is set up using the computed control limits and centerline, we plot the $np_j$ values. Next, we connect the points with a solid line and use the chart to monitor the production process.

---

### Example 16.4—Golf Club Shaft Production Case Revisited

We will again revisit the pultrusion process example of golf club shaft manufacturing. Example 16.1 discusses the details of the case, and Example 16.2 demonstrates the use of the $P$ chart, as well as the strategies used by the quality improvement team to enhance the process quality. After the changes suggested by the team were incorporated in the process, data were collected for 24 shifts (Table 16.3). A $P$ chart for the data was shown in Figure 16.4. Now we shall construct an $NP$ chart for the same 24 shift samples and show that it provides essentially the same information as the $P$ chart. Given the data in Table 16.3, the centerline and the 3-sigma control limits for the $NP$ chart for these 24 shift samples ($m = 24$) of sample size 160 can be determined as follows:

$$n\bar{p} = \frac{\displaystyle\sum_{j=1}^{24} np_j}{24} = \frac{88}{24} = 3.66667$$

Using our previous $\bar{p} = 0.02292$, $n = 160$, and Equation 16.5,

$$\text{UCL}_{np} = n\bar{p} + 3\sqrt{n\bar{p}\,(1-\bar{p})} = 3.66667 + 3(1.89278) = 9.34501$$

$$\text{LCL}_{np} = n\bar{p} - 3\sqrt{n\bar{p}\,(1-\bar{p})} = 3.66667 - 3(1.89278) < 0$$

Hence no lower limit exists.

$$\text{Centerline} = n\bar{p} = 3.66667 \text{ defectives per sample}$$

These control limits for the *NP* chart are shown in Figure 16.7, along with the number of defectives in each of the 24 shift samples.

We can see that the *NP* chart points and the corresponding points (shift samples 37 through 60) on the *P* chart (Figure 16.4) are following the same trend. The only difference is in the vertical scale. We are plotting $np_j$ on the *NP* chart whereas we plot $p_j$ on the *P* chart; $n = 160$ is a constant. The ability to provide indications of process stability-instability is the same for both of these control charts.

## C Chart Mechanics

The *P* and *NP* control charts deal with the notion of a defective product. We must recognize, however, that any one of a number of possible nonconformities will qualify a product as defective. For example, a part with 10 defects, any one of which makes it a defective, is on an equal footing with a part with only 1 defect, in terms of being defective.

*Often, it is of interest to note occurrences of every defect found in our product and chart the number of defects per product unit or sample.* In general, the opportunities for nonconformities or defects may be numerous, even though the chances of a nonconformity occurring in any one location on a product are relatively small.

**FIGURE 16.7**    *NP* control chart, after the process change, for the pultrusion process (Example 16.4).

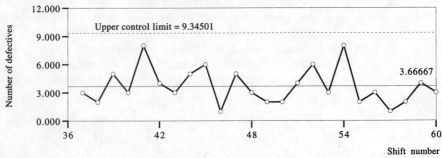

***The Poisson model is appropriate in this case and serves as a basis for the C chart.***
For example, in the case of an automobile, we may find that the opportunity for the
occurrence of a defect is quite large in an entire automobile. However, the probability of occurrence of a defect in any one location on the automobile is typically
very small. ***It is important to note that the area of opportunity for defects to occur
must be constant from sample to sample when we apply the C chart.***

For a generic $C$ chart where we assume that $c$ is known and we have equal
sample sizes of size $n$ inspection units,

$$\left.\begin{array}{c} \text{UCL}_c \\[2mm] \text{LCL}_c \end{array}\right\} = c \pm b\sqrt{c} \qquad\qquad (16.6)$$

$$\text{Centerline} = c$$

Here, an inspection unit must be defined by the chart designer. For example, an
inspection unit could be 1 automobile, 2 automobiles, 1 computer, 3 computers, 100
$m^2$ of cloth, and so on; however, once defined, an inspection unit must remain fixed.
Note that ***each sample in a C chart consists of only one inspection unit.*** In other
words, the area of opportunity for defects to occur is constant from sample to
sample. ***The U chart relaxes this assumption.*** As in the case of any of the other
control charts, if the calculated LCL is less than 0, we assume that the lower limit
does not exist (i.e., we have no LCL).

For a 3-sigma $C$ chart where $c$ is estimated from our data,

$$\left.\begin{array}{c} \text{UCL}_c \\[2mm] \text{LCL}_c \end{array}\right\} = \bar{c} \pm 3\sqrt{\bar{c}} \qquad\qquad (16.7)$$

$$\text{Centerline} = \bar{c}$$

where

$$\bar{c} = \frac{\displaystyle\sum_{j=1}^{m} c_j}{m}$$

Here, $j$ is used as a sample index, and $m$ should be at least equal to 20.

Once the $C$ chart is set up using the computed control limits and centerline, we plot
the $c_j$ values. Next, we connect the points with a solid line and use the chart to monitor the process. Here, $c_j$ is the observed number of defects in the (1) inspection unit.

---

### Example 16.5—Tractor Production Case

A tractor manufacturer desires to use a $C$ chart to monitor his tractor production
process. Two tractors are selected each day for assessment (e.g., 1 inspection
unit is made up of 2 tractors). The data obtained from 30 inspection units (60

**TABLE 16.5**   DATA ON NUMBER OF DEFECTS FOR 30 INSPECTION  UNITS, TRACTOR PRODUCTION PROCESS (EXAMPLE 16.5)

| Date | Sample number | Number of defects |
|---|---|---|
| 1 May | 1 | 2 |
| 2 May | 2 | 14 |
| 3 May | 3 | 1 |
| 4 May | 4 | 2 |
| 5 May | 5 | 2 |
| 8 May | 6 | 2 |
| 9 May | 7 | 11 |
| 10 May | 8 | 5 |
| 11 May | 9 | 4 |
| 12 May | 10 | 1 |
| 15 May | 11 | 5 |
| 16 May | 12 | 6 |
| 17 May | 13 | 2 |
| 18 May | 14 | 5 |
| 19 May | 15 | 4 |
| 22 May | 16 | 7 |
| 23 May | 17 | 13 |
| 24 May | 18 | 0 |
| 25 May | 19 | 4 |
| 26 May | 20 | 6 |
| 29 May | 21 | 8 |
| 30 May | 22 | 12 |
| 31 May | 23 | 1 |
| 1 Jun | 24 | 4 |
| 2 Jun | 25 | 0 |
| 5 Jun | 26 | 2 |
| 6 Jun | 27 | 4 |
| 7 Jun | 28 | 5 |
| 8 Jun | 29 | 0 |
| 9 Jun | 30 | 3 |
| Totals | | 135 |

tractors) are shown in Table 16.5.  Given the data in Table 16.5, the centerline and the 3-sigma control limits for the $C$ chart are calculated below:

$$\text{Centerline} = \bar{c} = \frac{\sum\limits_{j=1}^{30} c_j}{30} = \frac{135}{30} = 4.500 \text{ defects per inspection unit}$$

and from Equation 16.7,

$$\text{UCL}_c = \bar{c} + 3 \sqrt{\bar{c}} = 4.500 + 3 \sqrt{4.500} = 10.864$$

$$\text{LCL}_c = \bar{c} - 3 \sqrt{\bar{c}} = 4.500 - 3 \sqrt{4.500} < 0$$

Hence, no lower limit exists.

These control limits for the $C$ chart, along with the 30 sample points, are shown in Figure 16.8.  Figure 16.8 makes it clear that the tractor production process is not in statistical control.  In fact, there are four sample points outside the trial control limits (samples 2, 7, 17, and 22); hence the calculated limits are not meaningful for future production.

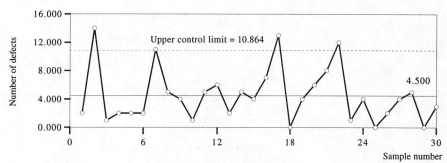

**FIGURE 16.8**    *C* control chart for the tractor production process (Example 16.5).

Since the process seems to be unstable, the current variation in the data collected is likely due to both the variation created by chance causes and that created by special causes. To arrive at control chart limits which represent a stable process, we must find and eliminate the effect of the special causes. At this point in our example, we will assume these causes were identified and their effect eliminated. Hence, we will eliminate subgroups 2, 7, 17, and 22, and we will recompute our trial control limits. Here, we are now left with 26 samples ($m = 26$).

The revised control limits for the tractor production process $C$ chart are calculated as follows. The revised center line is

$$\bar{c} = \frac{\sum_{j=1}^{26} c_j}{26} = \frac{85}{26} = 3.269 \text{ defects per inspection unit}$$

and from Equation 16.7,

$$\text{Revised UCL}_c = \bar{c} + 3\sqrt{\bar{c}} = 3.269 + 3\sqrt{3.269} = 8.693$$

$$\text{Revised LCL}_c = \bar{c} - 3\sqrt{\bar{c}} = 3.269 - 3\sqrt{3.269} < 0$$

Hence, no lower limit exists.

These revised control limits are shown in Figure 16.9. All 26 sample points are now within the control limits. As such, it is safe to extend the control limits into the future to monitor tractor production.

## U Chart Mechanics

When the opportunity space for the occurrence of defects per sample changes from sample to sample we consider it a violation of the constant "area of opportunity" assumption upon which the $C$ chart is based. In this case, we can apply a $U$ chart. *For the U chart, we must define a standard measure of the area of opportunity and*

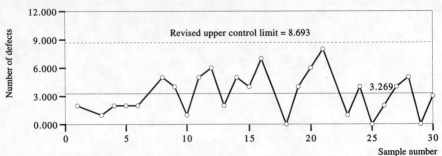

**FIGURE 16.9**    Revised *C* control chart for the tractor production process (Example 16.5).

*chart the average number of defects for this measure.* The symbol $\bar{u}_j$ is used to denote the sample statistic of the average nonconformities per inspection unit $u_j/n_j$, where $u_j$ is the count of nonconformities found in $n_j$ inspection units in sample $j$. In a *U* chart we plot the statistic $\bar{u}_j$, (the average number of nonconformities per inspection unit). The control limits for a generic *U* control chart with known $u$ and variable sample size $n_j$ are

$$\left.\begin{array}{l} \text{UCL}_{uj} \\[4pt] \text{LCL}_{uj} \end{array}\right\} = u \pm b \sqrt{\frac{u}{n_j}} \tag{16.8}$$

$$\text{Centerline} = u$$

This chart has a constant centerline, but the control limits vary as a function of the sample size. Hence, the control chart limits appear stair-stepped.

For a 3-sigma *U* chart where $u$ is estimated from the data, the control limits are

$$\left.\begin{array}{l} \text{UCL}_{uj} \\[4pt] \text{LCL}_{uj} \end{array}\right\} = \bar{u} \pm 3 \sqrt{\frac{\bar{u}}{n_j}} \tag{16.9}$$

$$\text{Centerline} = \bar{u}$$

Again, if the calculated lower limit is less than 0, we assume that we have no LCL.

Here,

$$\bar{u} = \frac{\displaystyle\sum_{j=1}^{m} u_j}{\displaystyle\sum_{j=1}^{m} n_j}$$

Here, $j$ is used as a sample index, and $m$ should be at least equal to 20.

Once the *U* chart is set up using the computed centerline, we calculate the control limits and $\bar{u}_j$ after each sample is taken and plot the control limits and their respective $\bar{u}_j$ values.  Next, we connect the points and use the chart to monitor the process.  We calculate $\bar{u}_j$ as

$$\bar{u}_j = \frac{u_j}{n_j} = \frac{\text{number of observed defects in sample } j}{\text{size of sample } j}$$

As an added note, Grant and Leavenworth point out that the statistic $u$ does not actually follow the Poisson distribution [6], however, the statistic $nu$ does.

## Example 16.6—35-mm Camera Body Production Case Revisited

We will now return to our camera body die-casting process (Example 16.3).  We have been examining a cold-chamber process for making camera bodies using copper-aluminum alloy.  At one point in the process, the completed 35-mm camera bodies are examined for the presence of casting defects.  Recall that the data for 20 shift samples with sizes equal to 4 percent of the shift production are given in Table 16.4, and the proportion defective values are plotted in Figure 16.6.  It was concluded in Example 16.3 that there exists no evidence that the production process is unstable.

Even though the process seemed to be stable over the period studied, the proportion defective was considered too high, and the department concluded that process improvement actions are necessary to compete with the best-of-the-best in the industry.  To gain a further understanding of the production process, and to identify new opportunities for quality improvement, the department decided to maintain a control chart to monitor the number of defects.  The $U$ control chart is appropriate, since the sample sizes are changing from shift to shift.  Table 16.4 presents information regarding the number of defects in each of the three categories for the 20 shift samples collected.  It was decided to treat 40 camera bodies as 1 inspection unit for the $U$ chart.  Average number of defects per inspection unit are calculated for the same 20 shift samples and are shown in Table 16.6.

From the mechanics presented earlier, the 3-sigma control limits for the $U$ chart can be calculated as follows:

$$\text{Centerline} = \bar{u} = \frac{\sum\limits_{j=1}^{20} u_j}{\sum\limits_{j=1}^{20} n_j} = \frac{215}{68.375} = 3.144$$

From Equation 16.9, we have for shift sample $j = 1$,

$$\text{UCL}_{u1} = \bar{u} + 3\sqrt{\frac{\bar{u}}{n_1}} = 3.144 + 3\sqrt{\frac{3.144}{3.500}} = 5.987$$

$$\text{LCL}_{u1} = \bar{u} - 3\sqrt{\frac{\bar{u}}{n_1}} = 3.144 - 3\sqrt{\frac{3.144}{3.500}} = 0.301$$

**TABLE 16.6**  DATA ON AVERAGE NUMBER OF DEFECTS PER INSPECTION UNIT FOR 20 SHIFTS OF CAMERA BODY PRODUCTION (EXAMPLE 16.6)

| Shift number | Sample size | Number of inspection units | Number of defects | | | Total defects | Avg. defects per unit | Control limits | |
|---|---|---|---|---|---|---|---|---|---|
| | | | Voids | Marks | Flashes | | | $UCL_U$ | $LCL_U$ |
| 1 | 140 | 140/40 = 3.500 | 3 | 2 | 2 | 7 | 7/3.5 = 2.0000 | 5.987 | 0.301 |
| 2 | 132 | 132/40 = 3.300 | 7 | 3 | 3 | 13 | 13/3.3 = 3.9394 | 6.072 | 0.216 |
| 3 | 98 | 2.450 | 2 | 0 | 1 | 3 | 1.2245 | 6.542 | 0.000 |
| 4 | 102 | 2.550 | 4 | 4 | 8 | 16 | 6.2745 | 6.475 | 0.000 |
| 5 | 40 | 1.000 | 0 | 0 | 3 | 3 | 3.0000 | 8.463 | 0.000 |
| 6 | 126 | 3.150 | 2 | 3 | 2 | 7 | 2.2222 | 6.141 | 0.147 |
| 7 | 132 | 3.300 | 9 | 5 | 2 | 16 | 4.8485 | 6.072 | 0.216 |
| 8 | 132 | 3.300 | 2 | 1 | 9 | 12 | 3.6364 | 6.072 | 0.216 |
| 9 | 156 | 3.900 | 5 | 4 | 3 | 12 | 3.0769 | 5.838 | 0.450 |
| 10 | 190 | 4.750 | 9 | 3 | 5 | 17 | 3.5789 | 5.585 | 0.703 |
| 11 | 210 | 5.250 | 12 | 6 | 6 | 24 | 4.5714 | 5.466 | 0.822 |
| 12 | 167 | 4.175 | 7 | 4 | 3 | 14 | 3.3533 | 5.747 | 0.541 |
| 13 | 134 | 3.350 | 7 | 5 | 4 | 16 | 4.7761 | 6.050 | 0.238 |
| 14 | 120 | 3.000 | 1 | 2 | 0 | 3 | 1.0000 | 6.215 | 0.073 |
| 15 | 110 | 2.750 | 0 | 0 | 5 | 5 | 1.8182 | 6.352 | 0.000 |
| 16 | 134 | 3.350 | 1 | 1 | 3 | 5 | 1.4925 | 6.050 | 0.238 |
| 17 | 167 | 4.175 | 8 | 3 | 7 | 18 | 4.3114 | 5.747 | 0.541 |
| 18 | 187 | 4.675 | 3 | 4 | 3 | 10 | 2.1390 | 5.604 | 0.684 |
| 19 | 123 | 3.075 | 0 | 0 | 1 | 1 | 0.3252 | 6.177 | 0.111 |
| 20 | 135 | 3.375 | 4 | 3 | 6 | 13 | 3.8519 | 6.040 | 0.248 |
| Totals | 2,735 | 68.375 | 86 | 53 | 76 | 215 | | | |

In a similar fashion we can calculate the control limits for each of the remaining 19 shift samples. The calculated control limits are shown in Table 16.6. These control limits, along with sample points from the 20 shifts, are plotted in Figure 16.10. All 20 samples are within the control limits, and the control chart shows no apparent nonrandom signals. Our chart suggests that the process is in statistical control. But the average number of defects per production unit places the manufacturer in a high-cost situation. The only viable alternative for improving the process quality was to make fundamental changes in the production process. A team of engineers enhanced the quality of the lubricant (parting agent) applied on the die surfaces, and redesigned the die so as to include more taper (draft) to allow easy removal of the casting, and thus avoided ejector-pin marks. The process engineers agreed to increase the pressure at which the molten metal is forced into the die cavity in order to reduce voids. The tooling department accepted the responsibility to improve the die design so as to reduce wear and tear and thereby lower the chances of flashes. Through these actions the production process and final product quality were significantly improved.

**FIGURE 16.10**  *U* chart for the camera die-casting process (Example 16.6).

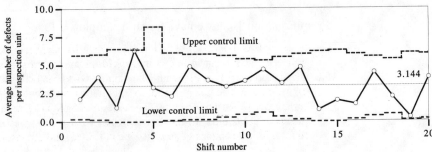

# REVIEW AND DISCOVERY EXERCISES

## Review

**16.1** What situations call for quality characterization with attributes data instead of variables data? Provide two examples.

**16.2** How is it possible to use attributes data to describe quality characteristics that are inherently of a variables measure nature? Explain the possible advantages and disadvantages of doing so. Provide two examples.

**16.3** Why is it that operational definitions of defects are so critical when using attributes measures?

**16.4** Why might it be useful to record all occurrences of all types of defects on a product during an inspection as opposed to simply classifying the product as defective as soon as the first instance of any one of the many possible defects is found?

**16.5** Assume that we are dealing with a large volume manufacturing process. The probability that a single item, drawn from the process, is defective is 0.10. Calculate the exact probability that a sample of size 50 will contain exactly 4 defective items.

**16.6** A total of 30 samples, each of size 75 items, have been collected from a process. A total of 53 defective items were found in the entire 30 samples. Calculate 3-sigma chart parameters for a $P$ chart. For an $NP$ chart.

**16.7** A $P$ chart is to be developed to monitor an eyeglass frame production process. An optical comparator is used to check the dimensions of the frame for conformance with the specifications. Data from 27 shifts of sample size 80 were collected and are shown below:

| Sample number | Nonconforming units found | Sample number | Nonconforming units found |
|---|---|---|---|
| 1 | 4 | 15 | 4 |
| 2 | 2 | 16 | 1 |
| 3 | 3 | 17 | 4 |
| 4 | 4 | 18 | 3 |
| 5 | 7 | 19 | 4 |
| 6 | 4 | 20 | 3 |
| 7 | 4 | 21 | 2 |
| 8 | 2 | 22 | 4 |
| 9 | 1 | 23 | 1 |
| 10 | 7 | 24 | 1 |
| 11 | 0 | 25 | 2 |
| 12 | 8 | 26 | 6 |
| 13 | 0 | 27 | 4 |
| 14 | 2 | | |

**a** Set up the trial control limits for a 3-sigma $P$ chart and plot the data.
**b** Set up the trial control limits for a 3-sigma $NP$ chart and plot the data.
**c** Which chart would you recommend for future use? Explain.

**16.8** The metal case for a cable adapter is composed of two halves made by a combination of extrusion and machining. Then, the two halves are connected with pins. The occurrence of surface cracking can sometimes be a problem in this production process. To monitor the process, it was decided to construct a $P$ chart. During four shifts, 30 samples of size 60 were collected, and the results are as follows:

| Sample number | Nonconforming units found | Sample number | Nonconforming units found |
|:---:|:---:|:---:|:---:|
| 1 | 2 | 16 | 3 |
| 2 | 6 | 17 | 4 |
| 3 | 1 | 18 | 6 |
| 4 | 3 | 19 | 4 |
| 5 | 7 | 20 | 3 |
| 6 | 9 | 21 | 1 |
| 7 | 2 | 22 | 4 |
| 8 | 1 | 23 | 3 |
| 9 | 7 | 24 | 6 |
| 10 | 2 | 25 | 2 |
| 11 | 0 | 26 | 3 |
| 12 | 3 | 27 | 4 |
| 13 | 1 | 28 | 3 |
| 14 | 4 | 29 | 6 |
| 15 | 6 | 30 | 2 |

   **a**  Set up the trial control limits for a 3-sigma $P$ chart and plot the data.

   **b**  Comment as to the stability of the production process.

**16.9**  A $U$ chart is used to control imperfections in the preparation of mats for advertising copy to be used in print media.  The control statistic is number of flaws per 100 $cm^2$ of mat area.  A target value of 1.5 flaws per 100 $cm^2$ is used.  A particular mat 18 cm $\times$ 26 cm was inspected and found to have 16 flaws.  Determine the $\bar{u}_j$ value and the control limits that are appropriate for this entry.

**16.10**  Ceramic boards for classrooms are being produced in $4 \times 8$ foot sheets.  The producer is concerned about surface defects on the sheets, and has decided that a $U$ chart will help her to monitor the production process.  Data have been gathered for 28 days.

| Day number | Sample size | Number of defects | Day number | Sample size | Number of defects |
|:---:|:---:|:---:|:---:|:---:|:---:|
| 1 | 1 | 2 | 15 | 6 | 18 |
| 2 | 5 | 5 | 16 | 3 | 6 |
| 3 | 2 | 3 | 17 | 5 | 8 |
| 4 | 3 | 8 | 18 | 4 | 17 |
| 5 | 2 | 1 | 19 | 7 | 4 |
| 6 | 2 | 4 | 20 | 3 | 6 |
| 7 | 3 | 8 | 21 | 4 | 4 |
| 8 | 1 | 0 | 22 | 2 | 5 |
| 9 | 1 | 2 | 23 | 5 | 2 |
| 10 | 3 | 8 | 24 | 2 | 5 |
| 11 | 4 | 5 | 25 | 2 | 0 |
| 12 | 3 | 7 | 26 | 1 | 15 |
| 13 | 3 | 4 | 27 | 2 | 3 |
| 14 | 4 | 6 | 28 | 3 | 5 |

   **a**  Set up the trial 3-sigma $U$ control chart and plot the data.

   **b**  Comment as to the level of stability shown in the process.

**16.11**  Assume the producer in Problem 16.10 decided to use a $C$ chart.

   **a**  How would this decision affect her sampling strategy?

**b** Assume she chose an inspection unit of 1 board and, out of 25 samples, found 45 defects. What are her 3-sigma $C$ chart parameters?

### Discovery

**16.12** Can one $P$ chart be used to monitor more than one quality characteristic (e.g., three or four quality characteristics)? Explain why or why not. If yes, how is it possible to tell which quality characteristic is creating problems when the $P$ chart indicates that the process is out of control?

**16.13** Given the same process as that described in Problem 16.7, a manager has suggested that a target $p_0 = 1$ percent defective be set and a $P$ chart used. The motivation for this target is to encourage quality improvement. Determine the 3-sigma $P$ chart control limits for this target. How do you think this chart will work on the production floor tomorrow?

**16.14** Choose a process or product, and identify an appropriate quality characteristic that can be measured with an attributes measure. Select an appropriate attributes control chart, collect a sufficient amount of data, and develop your chart limits. Build a display board which describes your chosen process or product, your selected substitute quality characteristics and how they relate to true quality characteristics, your measurement procedure, and your chart results.

## REFERENCES

**1** E. L. Grant and R. S. Leavenworth, *Statistical Quality Control,* 6th ed., New York: McGraw-Hill, 1988.
**2** R. E. DeVor, T. Chang, and J. W. Sutherland, *Statistical Quality Design and Control: Contemporary Concepts and Methods,* New York: Macmillan, 1992.
**3** S. Kalpakjian, *Manufacturing Engineering and Technology,* 2d ed., Reading, MA: Addison-Wesley, 1990.
**4** See reference 1.
**5** See reference 3.
**6** See reference 1.

# 17

# PROCESS SAMPLING, PROCESS STABILITY, AND SPC CHART INTERPRETATION

## 17.0 INQUIRY

1 Why is it important to tailor a sampling scheme to each process?
2 Why are SPC charts "located" at the production process source?
3 What are runs rules?
4 How are process logs and Pareto analysis used in SPC?
5 How do probability limits work in SPC?

## 17.1 INTRODUCTION

SPC charts and SPC chart mechanics have been discussed in Chapters 9, 15, and 16. The concept and mechanics are reasonably straightforward. However, we must discuss chart application and interpretation if we are to gain a fundamental understanding of effective SPC practice.

We may choose to set up SPC charts to monitor processes that are somewhat unstable, as we work to stabilize them. In such cases, we must isolate the cause and eliminate the instability in the process; ultimately, we want to develop a stable process. *We use the statistical nature of our SPC chart and our physical knowledge to detect, locate, and eliminate the effects of special causes.*

Most SPC charts must be fine-tuned to suit specific processes so that relatively small shifts—whether of location, dispersion, or both—can be detected as quickly as possible. The runs rules discussed in this chapter are designed to provide us with a better means of detecting small shifts, instead of always waiting for a point to fall outside control limits. Process logs and Pareto analyses support SPC by relating the abstract data analysis element to the physical process element. *Process logs and Pareto charts* (discussed in Chapter 9) *provide documentation and a record of process experience.*

*SPC chart interpretation requires us to consider the effectiveness and appropriateness of our SPC models.* For example, we may want to assess the probability of both type I errors (false alarms) and type II errors (failures to detect). Operating characteristic (OC) curves provide insight into both type I and type II error probabilities. In practice, we also determine chart effectiveness by assessing average run length (ARL).

## 17.2 SPC CHART SELECTION AND SAMPLING SCHEMES

Once we have described our process in enough detail to understand the "what," "where," "when," "who," "why," and "how" of process operations through flowcharting, we must locate the critical quality process points.  These points, as we have seen, can be marked on our process flowcharts or diagrams (Chapter 10).

Ultimately, we must determine an appropriate SPC chart for the identified or marked process points.  In the previous chapters, we have briefly discussed the attributes and variables chart families.  ***Chart selection must be based on process needs and is usually approached in two steps:***

1  *We choose between an attributes and a variables chart.*
2  *We choose the specific chart type.*

Since chart selection involves quality goals and objectives as well as physical process specifics, ***chart selection will require both engineering and statistical judgment.***  Tables 15.2 and 16.1 provide statistical considerations within the scope of this textbook.  Physical and cost considerations require process familiarity and will differ widely between processes.

## Sample Definition

For the $X$-bar, $R$ control chart sets, we typically see subgroup sizes of 2 to 5 items.  Here, one subgroup constitutes one sample.  In the $X$-bar, $S$ chart pairs we would typically see 6 to 20 items in a subgroup or sample.  The $X$ chart, by definition, has a sample size of 1 unit.  Since one-unit samples cannot provide estimates of dispersion, we generally use a two-point $R_M$ chart along with the $X$ chart.

If we desire to use an attributes chart to monitor the process, we must estimate the level of nonconformance (for the $P$, $NP$ chart family) or the level of nonconformities (for the $C$, $U$ chart family).  We then select a sample size which will produce meaningful statistics for the chart.  For example, low $p$ values require larger sample sizes than high $p$ values.  If we encounter a $p \simeq 0.01$ defective, we must select a sample size large enough to obtain, typically, at least one or two nonconforming items on a frequent basis (e.g., a sample of 50 to 150 items).  For small $p$ values, we can use a rule of thumb of $1/p$ as a starting point in sample size determination.

Assessment of small samples is typically faster and less expensive than that of large samples.  For applications with expensive quality evaluation and/or destructive inspection, large samples are usually infeasible.  Hence, we may consider a variables chart as an alternative to an attributes chart.

When we design $C$ or $U$ control charts, we must also define an inspection unit.  Here again, we like to see samples large enough to be informative (e.g., wherein we can observe at least one or two defects in most samples).  Use of sample sizes so small that we seldom encounter defects makes interpretation difficult.

***Variables based sampling typically requires much smaller sample sizes than attributes based sampling.***  Sample sizes an order of magnitude less than those found in attributes samples, e.g., $P$ and $NP$ charts, are not unusual.  Counterpart statements for $C$ and $U$ charts are not as straightforward, as we must define an inspection unit.  For example, the sample size in a $C$ chart is 1 inspection unit, by definition.  But that unit may include multiple product units or a large area of opportunity for defects (for example, 10 printed circuit boards, 100 yd$^2$ of carpet, and so on).

If we are near zero defect production (i.e., $p$ values are very small) on the order of a few defects or nonconforming items per million or billion, attributes control charts lose their appeal because of extremely large sample size requirements. In general, *we must consider a number of factors when selecting and executing a control chart strategy:*

1  *The benefits our SPC chart will yield*—for example, early detection of quality problems so we can take corrective action in a timely fashion, as well as an enhanced focus on quality.
2  *The cost of sampling and sample evaluation.*
3  *The time it takes to sample and evaluate the sample.*
4  *Our customer's goodwill toward quality conscious suppliers.*

## Sample Timing

*One difficulty in working with control charts is deciding on when to "pull" the production units or product* (especially in the case of continuous flow production) *for assessment.* Basically, this is a sample selection and timing issue. For example, we can choose to sample at regularly scheduled intervals, at randomly determined intervals, or in a judgmental scheme. Figure 17.1 depicts three *major strategies that may be employed in sample timing:*

1  *Use of regular intervals* with grouped spacing (clustering) or with individual unit spacing.
2  *Use of random intervals* with grouped spacing or individual unit spacing.
3  *Use of regular interval sampling with judgment-based modification,* to be implemented when we think we may be getting into trouble.

*Sample timing ultimately comes down to a compromise between* (1) *statistical concerns and* (2) *practical process concerns.* Grant and Leavenworth [1], Deming [2], and others point out that *statistical performance of the X-bar, R chart pair is enhanced with the grouped spacing method.* When we use the grouped spacing method, we generally seek to make each sample (i.e., units within the sample) as homogeneous as possible, thus maximizing our ability to detect heterogeneity between samples (e.g., process changes). However, in some industries where continuous process flow is encountered (chemicals, refining, plastics, etc.), tightly grouped subsamples produce identical material for reasons such as local mixing in the flow stream or tank. Sometimes the differences found are due more to measurement variation than to product variation. In such situations, we can argue that subgroups of size 1 are justified. On the other hand, piece-parts can be readily sampled as consecutive units. Hence, *the sampling intervals developed must be compatible with the physical process, yet retain statistical and cost effectiveness.*

Choosing between regular and random intervals in sampling is difficult. Over the years, we have heard claims that if an operator knows when the process will be sampled, we will get better quality at these sampling times. Perhaps this was true in the past. But now we empower operators to evaluate their own work. Hence, if

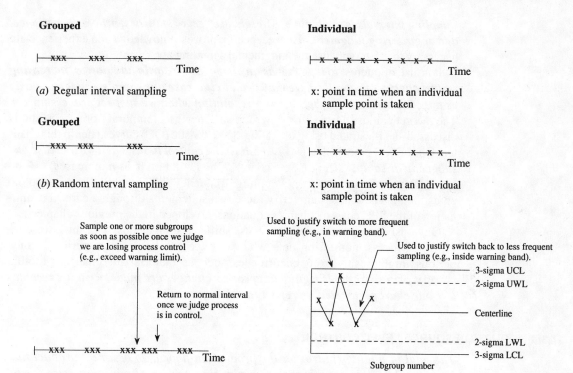

**FIGURE 17.1** Regular, random, and judgmental sampling intervals.

the old argument is still true in any given process, we have deep underlying problems of trust, beyond the technical problems we seek to prevent through SPC. In other cases, we use computer controlled operations, where no readily identifiable human operator exists.

The scheduled time versus random timing argument focuses on systematic process influence issues that often manifest themselves throughout a day, shift, season, or some other natural process influence cycle. Hence, if we think our process may be sensitive to temperature cycles, warm-up cycles, material lot changes, and so on, we should attempt to capture and monitor these effects in our SPC practices. *The objective in effective process control practices is to consistently produce a high-quality product and maintain the documents to support our product's integrity.* We should never lose sight of this objective and we must do what is necessary for its accomplishment.

Judgment based sampling schemes typically rely on operator experience or rules of thumb. Research using average run length (ARL) analysis (described later) supports sampling schemes that increase the sampling frequency when points are encountered near the control limits. One simple judgmental scheme involving 2-sigma warning limits is depicted in Figure 17.1c. *The motivation behind judgmental sampling intervals is to obtain a synergistic effect of statistics, process experience, and engineering judgment.* Hence, specific process knowledge and experience are necessary to effectively use the judgmental sampling approach.

It is also of interest to note that *as an alternative to, or in addition to, increasing our frequency of sampling, we may want to increase our sample size temporarily as a means of obtaining better chart resolution when we suspect process upsets.* The theoretical justification for subgroup size increase, temporary or permanent, is discussed and illustrated using operating characteristic (OC) curves later in this chapter. One critical point is that *we must adjust the control limits when we increase our subgroup or sample size* (i.e., the control limits will pull in as $n$ increases). However, when we increase the sampling frequency our control limits need no adjustment. The reader can verify these two statements through a careful examination of the SPC chart mechanics equations developed in the previous chapters.

Obviously, we want to detect process shifts as soon as possible; thus, we concentrate process monitoring near what we expect to be problem periods. Such periods might result from occurrences such as material lot changes, tooling changes, and so forth. *If significant process changes are made, we must recalculate our chart parameters to reflect the changes.*

## 17.3  PRODUCTION SOURCE LEVEL CONTROL

*The goal in quality control on the production floor is to assure a product that consistently meets specifications, minimizing and/or eliminating rework and scrap.* Control charts measure process stability; capability indices measure process ability to meet specifications. It is possible to develop a control chart for a process, show process stability, and at the same time produce an off-specification, defective, product. The combination of process stability (SPC chart oriented) and product conformance to specification (process capability oriented) is critical. We must remember that the *SPC charts measure the time related process stability performance. The capability indices* (discussed in Chapter 18) *measure the fundamental ability of the process to create a product which meets specifications.*

*The critical point in SPC is to "attach" the chart(s) to a unique production source which can be controlled by an operator or a computer.* Whenever product from two sources is mixed, and product source identity is lost, our ability to impact quality is lost. Figure 17.2*a* and *b* illustrates classical cases of poor SPC practice. Figure 17.2*c* illustrates a case of good SPC practice.

## 17.4  INTERPRETING SPC CHARTS

As we have discussed earlier, SPC charts constitute a sequence of crude hypothesis tests over time, as each point is plotted. The support for hypothesis rejection-acceptance is provided in a graphical format. Examples of appropriate hypotheses have already been shown for both variables and attributes charts (Chapters 15 and 16). *In practice, the null hypothesis of a stable process is rejected in favor of an*

**FIGURE 17.2**   Production process source control depiction.

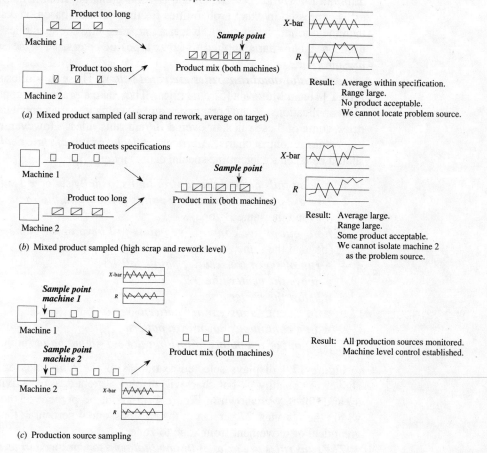

(a)  Mixed product sampled (all scrap and rework, average on target)

(b)  Mixed product sampled (high scrap and rework level)

(c)  Production source sampling

*alternate hypothesis that the process is unstable, out of control, and a special cause is present when any of the following is observed:*

1  ***An extreme point outside of the control limits*** (usually ± 3-sigma limits).
2  ***An unusual pattern.***
3  ***A run or trend*** (upward, downward, or level).

***The most widely used chart control indicators are based on the AT&T runs rules*** [3].  These runs rules are usually applied to a 3-sigma control chart after it has been divided into zones.  Zone A is the innermost zone, which is ± 1 sigma ($b = 1$ in the chart mechanics) from the centerline.  Zone B extends beyond zone A by an additional standard deviation.  Hence, the outer bounds of zone B are at ± 2 sigma ($b = 2$) and the inner bounds of zone B are at ± 1 sigma.  The third zone, zone C, is represented by an outer bound at ± 3 sigma ($b = 3$) and inner bounds at ± 2 sigma ($b = 2$).  Figure 17.3 depicts the zones for an $X$-bar chart.  The probabilities shown are based on a normal distribution; the band probabilities will be different for $R$, $R_M$, $S$, $P$, $NP$, $C$, and $U$ control charts.  In practice, very little interest is expressed in exact probabilities for these charts (based on distributions other than the normal distribution).  It is not uncommon to see runs rules similar to those used on $X$-bar charts applied to other types of charts, with the exception of the CuSum chart.

*Some widely used runs rules are listed below.*  These runs rules are based on the AT&T [4] and Motorola [5] runs rules.  They take a pragmatic approach to control chart applications.  Actual practice in any given facility may entail use of all of these rules, some of these rules, or even different runs rules.  However, rule 1, below, is universal for control charts.  *A process is declared out of statistical control* (action should be taken to determine special cause) *when*

  1 *A point falls beyond the control limits, usually set at ± 3 sigma.*
  2 *Two out of three consecutive points fall beyond the same ± 2-sigma band* (beyond the same B zone).
  3 *Four out of five consecutive points fall beyond the same ± 1-sigma band* (beyond the same A zone).
  4 *A run of seven consecutive points falls*
    a *above the center line.*
    b *below the center line.*
    c *in a continuous upward pattern.*
    d *in a continuous downward pattern.*
  5 *Nine out of ten points fall within the ± 1-sigma band* (both A zones).

Figure 17.4 displays selected examples of the runs rules patterns described above.  A "healthy" $X$-bar chart will display a great deal of activity, or "bounce," up and down.  As shown in Figure 17.3 about 68 percent of the points will fall within the A zones, 27 percent in the B zones, and 4 percent in the C zones, with a great deal of movement from zone to zone.

*The runs rules are rules of thumb that tend to work well in terms of the risk of statistical errors.*  Although the runs rules work well in practice, their theoretical risk probability is not directly calculable when all of the rules are applied simultaneously.  Under certain assumptions, simplified calculations and discussions thereof are pro-

**FIGURE 17.3**   Control chart zones and normal distribution based probabilities for a 3-sigma $X$-bar chart.

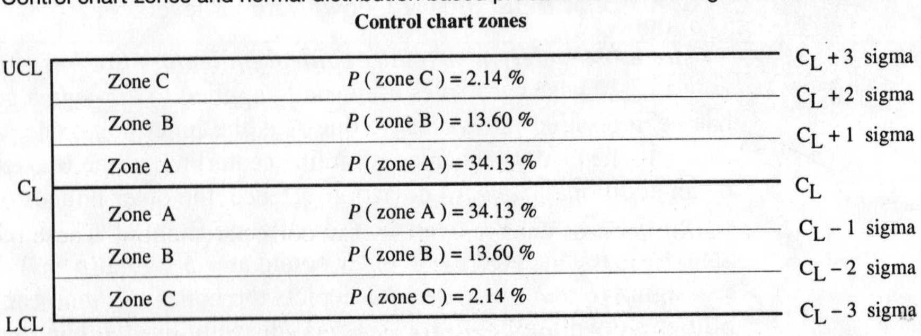

**Control chart zones**

| | | | |
|---|---|---|---|
| UCL | | | $C_L + 3$ sigma |
| | Zone C | $P(\text{zone C}) = 2.14\,\%$ | $C_L + 2$ sigma |
| | Zone B | $P(\text{zone B}) = 13.60\,\%$ | $C_L + 1$ sigma |
| | Zone A | $P(\text{zone A}) = 34.13\,\%$ | |
| $C_L$ | | | $C_L$ |
| | Zone A | $P(\text{zone A}) = 34.13\,\%$ | $C_L - 1$ sigma |
| | Zone B | $P(\text{zone B}) = 13.60\,\%$ | $C_L - 2$ sigma |
| | Zone C | $P(\text{zone C}) = 2.14\,\%$ | $C_L - 3$ sigma |
| LCL | | | |

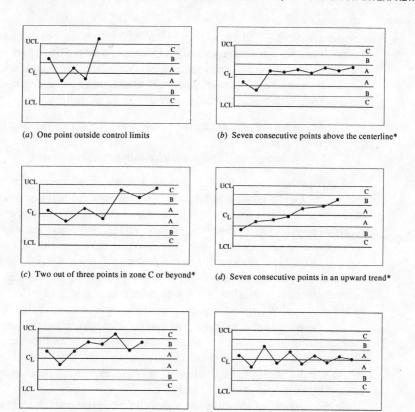

(a) One point outside control limits       (b) Seven consecutive points above the centerline*

(c) Two out of three points in zone C or beyond*       (d) Seven consecutive points in an upward trend*

(e) Four out of five points in zone B or beyond*       (f) Nine out of 10 consecutive points fall within plus or minus one sigma from the centerline*

\* These runs rules are meaningful to only those control charts that assume independence between successive sample points (not applicable to CuSum, EWMA, EWMD, and $R_M$ control charts).

**FIGURE 17.4**    Runs rule based control chart patterns that indicate process instability (a lack of control).

vided [6]. Simulation studies can be developed to estimate type I and type II error probabilities and ARLs for the various chart types. However, little interest in such endeavors has been shown by practitioners. Fundamental OC curves dealing with type I and type II error probabilities and ARLs appear in Appendix 17A.

## 17.5 PROCESS LOGS AND PARETO CHARTS

Process stability indicates the "mastery" of process operations. At the operations level, once process stability is reached, our process performance is as good as it can be. Further improvement, such as location to target and variance reduction, will require process changes, rather than corrective action. These adjustments may involve changes in process inputs, in the process itself, or in both.

Achieving and sustaining process stability is a major accomplishment that deserves recognition. ***Process stability is seldom permanent.*** Process upsets of all imaginable types—and some that are impossible to imagine—are eventually experienced. *Process upsets are usually unpleasant and costly experiences.* However, they do offer a chance to learn more about our inputs, processes, and even our product. *Documentation is essential in order to maximize our learning experience. We must accomplish two major tasks through documentation:*

1 *Add each new process upset* (cause and effect) *remedy and result to our experience base.*
2 *Stratify, categorize, and prioritize our experience base.*

A comprehensive, but brief, process log which addresses the "what," "when," "where," "why," "how," and "who" of critical operations is necessary. The purpose of the process log is to educate and increase awareness. *The process log supplies relevant physical facts regarding cause, effect, corrective action, and results* and must contain both successes and failures in process correction. Successes are easily described, failures are more difficult to describe. Nevertheless, failures offer valuable lessons, even though we may not choose to repeat them. In other cases, the difference between success and failure is not great; sometimes minor adjustments to actions that failed may lead to huge successes, provided we can determine the right adjustment.

*The Pareto chart is an excellent tool for classifying process upset causes.* The rank ordering in the Pareto chart automatically isolates and focuses our attention on the most frequent cause (if ranked by frequency), the most wasteful cause (if ranked by off-specification product volume), the most expensive process upset category (if ranked by cost or downtime), and so on.

A wide variety of effective log styles can be developed. A representative process log for a torque-controlled machining center is shown in Figure 17.5. The point is

**FIGURE 17.5** Process log layout.

| Machining center 2—process log | | | | | | |
|---|---|---|---|---|---|---|
| Date | Time | Corrective action indicator | Special cause | Corrective action taken | Results obtained | Lessons learned |
| 10/8 | 13:15 | High torque requirement at spindle | Dull cutting tool—position 8 | Replace cutting tool—position 8 | Torque back within limits | None |
| 10/10 | 10:45 | High torque requirement/ tool chatter | Chipped cutting edge on carbide insert drill—position 12 | Replace drill—position 12 (carbide insert drill) | Torque back within limits/ quiet operation | None |
| 10/10 | 10:49 | High torque requirement/ tool chatter | Chipped cutting edge on carbide insert drill—position 12 | Replace drill—position 12 (carbide insert drill) , Increase spindle rpm to 3000 from 2500 rpm | Torque back within limits/ quiet operation | Feed problem with the carbide drill, decrease feed/revolution, increasing rpm — result longer tool life with no loss in productivity |

| Cause / result | Tally | Total | Percent |
|---|---|---|---|
| Wrong tool | ꟷꟷ ꟷꟷ ॥ | 12 | 12 |
| Chucking | ꟷꟷ | 5 | 5 |
| Setup | ꟷꟷ ꟷꟷ ॥। | 13 | 13 |
| Feed/speed | ꟷꟷ ॥॥ | 9 | 9 |
| Torque sensor | ꟷꟷ ꟷꟷ ꟷꟷ ꟷꟷ ꟷꟷ ꟷꟷ । | 31 | 31 |
| Workpiece defect | ꟷꟷ ꟷꟷ ꟷꟷ ॥॥ | 19 | 19 |
| Wrong workpiece | ॥ | 2 | 2 |
| Fixture problem | ꟷꟷ । | 6 | 6 |
| Other | ॥। | 3 | 3 |
| | Total | 100 | |

(*a*)  Pareto analysis data sheet with special causes listed

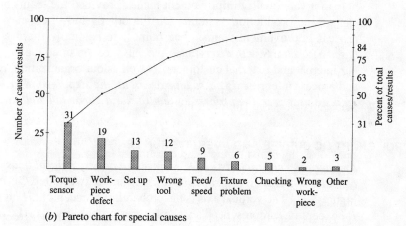

(*b*)  Pareto chart for special causes

**FIGURE 17.6**    Process improvement Pareto analysis for a machining center process.

to match the level of log detail to the level of process sophistication and loss potential present.  For example, complicated processes that are not well understood justify more detail than simple processes that are well understood.  Processes with marginal capability must be logged very carefully.

The Pareto analysis data sheet shown in Figure 17.6*a* basically serves as a stratified process log summary and provides an overview of process problems.  The Pareto chart in Figure 17.6*b* is a graphical representation of special cause occurrences.  It is a useful tool for prioritizing process improvement efforts.  For example, we direct intensive process improvement energies toward the problem classes resulting in the highest losses.  In this case, the torque sensor, workpiece consistency, setup, and tooling warrant further analysis.

## 17.6  TARGET BASED CONTROL CHARTS

We usually develop control chart parameters (UCLs, LCLs, and centerlines) from our actual process data.  The logic behind the use of actual process data is simply to reflect what is happening in our process.  On the other hand, *by using target values* such as $p_0$, $c_0$, $\bar{x}_0$, and $\sigma_0$ *to set up our SPC charts, we reflect what we want to happen in our process—not what is actually happening.* Hence, if we use target values, we should use them carefully. For example, we cannot expect our charts to show stability unless our actual process parameters are extremely close to our targets. If we were to use $p_0 = 0.01$ when our process is running a $\bar{p} = 0.10$ we would clearly have a mismatch. We could not expect to reach our target unless we obtained significant process improvement.

Any charts constructed to targets simply substitute the $p_0$, $c_0$, $\bar{x}_0$, $\sigma_0$ target parameter values into the equations of Chapters 15 and 16 in place of the process generated statistics (e.g., $\bar{p}$, $\bar{c}$, $\bar{\bar{x}}$, $\bar{R}$, and $\bar{s}$). The resulting control charts must be interpreted in the spirit of serious process improvement. Target based charts may help motivate quality improvement teams, provided the teams have the power to change the production system. On the other hand, they may simply frustrate process improvement teams. The point here is simply that *we must use target based SPC charts with discretion.* We must keep in mind the best interests of both our internal and external customers, as well as our organization, regarding product and process integrity. *Process improvement, the step beyond process stability, requires changes in the process system (beyond corrective action).*

## 17.7 CONTROL CHART OC CURVES AND AVERAGE RUN LENGTH

*An operating characteristic (OC) curve can be very helpful in assessing type I and type II statistical error probabilities.* A generic two-tailed OC curve is shown in Figure 17.7. The vertical axis is the probability of accepting $H_0$ as true, all values of $P_a$, except one, represent $\beta$'s (type II error probabilities). The one particular $P_a$, where $H_0$ is true, represents $1 - \alpha$ (i.e., $1 -$ the probability of a type I error). All points other than the point where $H_0$ is true represent cases where some $H_1$ is true. When the parameter value ($p$, $c$, $\mu$) shifts, $1 - P_a$ represents the probability of correctly detecting the shift on the first sample or subgroup taken after the shift occurs.

**FIGURE 17.7** Generic two-tailed OC curve.

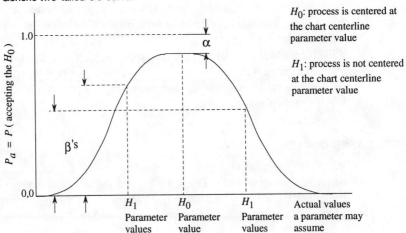

$H_0$: process is centered at the chart centerline parameter value

$H_1$: process is not centered at the chart centerline parameter value

*OC curves can be useful in control chart design by helping to identify the probability of a false alarm, $\alpha$, and the probability of a failure to detect a shift, $\beta$.* Table 17.1 presents generic control chart design strategies (provided that we seek to detect a given process shift) where the shift magnitude is a fixed quantity (e.g., 4 mm, 2 psi, +5 percent defective).

Runs rules significantly confound efforts to develop OC curves. In a general sense, because of the runs rules' complexity (e.g., as in the case of the AT&T rules [7]) and the widespread use of modified AT&T runs rules, OC curve construction has not been a high priority. Large scale computer simulation studies would be required for the development of OC curves for charts used with extensive run rules. Such OC curves could be developed, with acceptable accuracy and precision, but a great deal of time and effort would be required.

*In addition to OC curves, quality control chart performance is measured by ARL. The ARL is the average or expected number of samples or subgroups before the control chart provides an out-of-control signal.* If the null hypothesis is true, the ARL indicates the expected number of correct indications or sample points, before a type I error. In this case a large ARL is desirable.

If, on the other hand, a shift has occurred, the ARL indicates the expected number of "incorrect" indications (failure to detect) before we encounter a "correct" indication of a process shift. Hence, a small ARL is desirable in this case; here, the minimum ARL is 1. In other words, we must encounter at least one sample or subgroup to detect a shift.

## 17.8  PROBABILITY LIMITS

In general, the classical sigma limit control charts (except the $X$-bar and $X$ charts) are awkward for expressing or obtaining exact type I and type II error probabilities even when runs rules are not used. The use of runs rules further complicates this matter. One school of thought suggests we consider using probability limits. In this context, *SPC chart probability limits relate directly to the type I error probability $\alpha$.*

**TABLE 17.1**   GENERIC CONTROL CHART DESIGN STRATEGIES

| Strategy to detect a process shift of a given magnitude | $P$ (false alarm), $\alpha$ | $P$ (failure to detect shift), $\beta$ |
|---|---|---|
| Increase sample size $n$ ; same $b$ (sigma limit) multiple | No change | ↓ |
| Decreasing $b$, in order to pull the control limits in | ↑ | ↓ |
| Use of runs rules in addition to points outside control limits | ↑ | ↓ |

*When we use probability limits, rather than sigma limits, on control charts, we refer to the probability of the distribution which lies beyond or beneath the control limits, rather than to the distance from the centerline out to the control limits.* The probability limit concept is compared with the sigma limit concept in Figure 17.8. It is possible to develop probability limits for any control chart in which the underlying model distribution is quantifiable. $X$-bar and $R$ or $S$ probability limits are frequently encountered. We will briefly discuss probability limits for the $X$-bar, $R$, and $S$ charts. See Oakland [8], Grant and Leavenworth [9], and Montgomery [10] for more details.

$X$-bar chart probability limits are straightforward in that a standard normal distribution is applied once the "tail" areas are associated with the probability levels. For example, a $UCL_{0.975}$ and an $LCL_{0.025}$ would be associated with $b = \pm 1.96$. It is more difficult to deal with the $R$ and $S$ charts; usually, a tabular approach is used. Table IX.11, Section IX, lists UCL and LCL factors for the $X$-bar, $R$ and the $X$-bar, $S$ chart sets.

In the examples below, we will see that the $R$, $S$ chart probability limits are not symmetrically placed about the centerline (as is the case in general for sigma limits). We also notice that $LCL_R$ and $LCL_S$ are not set at 0 for small subgroups sizes, unlike the procedure in previous $\pm b$ sigma examples. Hence, for statistical purposes (regarding type I error probabilities), the probability limits make sense, but in the United States, sigma limits see much wider use owing to their familiarity and the ease in calculation. Of course, for the $X$-bar chart it does not make any difference, we can simply adjust the $b$ value. For convenience, special $A_{2,0.xxx}$ and $A_{3,0.xxx}$ probability limit factors (for use with $X$-bar, $R$ and $X$-bar, $S$ probability limit control chart sets) have been calculated. These values are shown in Table IX.11, Section IX.

## Example 17.1

Using the meter stick production process data from Example 15.2, calculate the probability control limits for the $X$-bar, $R$, and $S$ control charts with type I error

**FIGURE 17.8**   Probability and sigma limit control chart concepts.

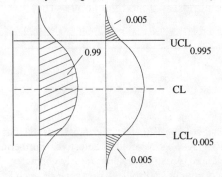

(*a*) Probability limit control chart concept

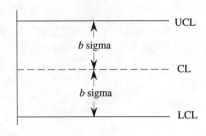

(*b*)  Sigma limit control chart concept

probability equal to .01.  In other words, determine the $UCL_{0.995}$ and $LCL_{0.005}$ for the three control charts.

**Solution**

We can determine from the reduced data set (e.g., after eliminating subgroups 9, 10, 12, 13, and 18) that the subgroup size $n = 5$, the grand average $\bar{\bar{x}} = 999.63$ mm, the mean of subgroup ranges $\bar{R} = 0.50$ mm, and the mean of subgroup sample standard deviations $\bar{s} = 0.206$ mm.

For the $X$-bar control chart (based on $\bar{R}$), the probability limits are

$$UCL_{\bar{x},\,0.995} = \bar{\bar{x}} + A_{2,\,0.995}\,\bar{R} = 999.63 + 0.50(0.50) = 999.88 \text{ mm}$$

$$LCL_{\bar{x},\,0.005} = \bar{\bar{x}} - A_{2,\,0.005}\,\bar{R} = 999.63 - 0.50(0.50) = 999.38 \text{ mm}$$

$$\text{Centerline} = \bar{\bar{x}} = 999.63 \text{ mm}$$

For the $R$ control chart,

$$UCL_{R,\,0.995} = D_{0.995}\,\bar{R} = 2.10\,(0.50) = 1.05 \text{ mm}$$

$$LCL_{R,\,0.005} = D_{0.005}\,\bar{R} = 0.24\,(0.50) = 0.12 \text{ mm}$$

$$\text{Centerline} = \bar{R} = 0.5 \text{ mm}$$

For the $S$ control chart,

$$UCL_{S,\,0.995} = B_{0.995}\,\bar{s} = 1.83(0.206) = 0.377 \text{ mm}$$

$$LCL_{S,\,0.005} = B_{0.005}\,\bar{s} = 0.21(0.206) = 0.043 \text{ mm}$$

$$\text{Centerline} = \bar{s} = 0.206 \text{ mm}$$

All constants are obtained from Table IX.11, Section IX.

*Note:*  Had we calculated the probability limits for the $X$-bar chart based on $\bar{s}$, they would have almost matched the above probability limits based on $\bar{R}$. The reader may want to verify this result.

---

## REVIEW AND DISCOVERY EXERCISES

### Review

**17.1.** We are interested in applying SPC to the production of a printed circuit (PC) board in our factory.  Data collected from inspection shows that out of 1000 boards examined at random, 20 were rejected as defective.  Furthermore, the inspection records show that three types of problems were encountered:

  **a** Solder joints—A total of 250 solder joints exist on each PC board and a total of 35 defective solder joints were found on the 1000 PC boards inspected.

**b** Component placement—A total of 50 components are placed on each PC board. After the PC boards were inspected, a total of 17 components were not in the proper position.

**c** Board circuits—Each board contains 85 circuits within and on top of the board layers. A total of 7 defective circuits were found in the inspected PC boards.

Develop and describe a strategy for an SPC charting program for the PC board discussed above. Then, set up the parameters for the chart or charts that you recommend.

**17.2** The dog food can filling line shown below consists of two filling machines, both feeding onto a single conveyor. Subgroups of size $n = 5$ cans are taken off the conveyor in the position shown. Each hour a subgroup is selected and weighed. $X$-bar and $R$ charts are kept for the fill contents. Each can is labeled to contain 16 oz of wet dog food. If a can is below 16 oz, it is considered defective. If a can contains more than 17 oz, it will not go through the lid sealing process and is considered defective. The operators of the machines take turns sampling the line and plotting the points on the $X$-bar and $R$ charts. Both charts use 3-sigma control limits. Both charts are showing good statistical control.

   The defect rate is running at about 5 percent. About half of the defects are underfills and about half are overfills. Major problems are shaping up both at the lid sealing machine (where three other operators must spoon out a small amount of dog food on overfills in order to get the lids on) and on short weights showing up in the marketplace. Management is motivated. Troubleshoot this process to locate the possible problems. Explain what you would include in your study, how you would go about your study, and what solutions you would recommend. *Note:* The two fill machines are brand new, state-of-the-art machines, capable of producing a standard deviation of 0.1 oz.

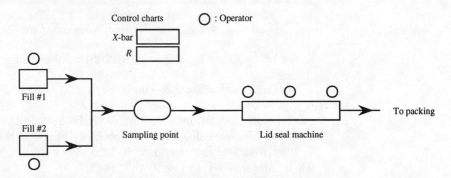

**17.3** Our process has a history of producing about 1.5 percent defective product. One of our engineers has suggested that we use a $P$ chart to monitor the process. Sampling the product destroys the product and is very expensive. Plans call for samples of size $n = 10$ to be used for the chart. Calculate the 3-sigma control limits and comment on the effectiveness of the control chart plan.

**17.4** One of the runs rules for a control chart states that, if seven consecutive points fall above the centerline, we will declare the process to be out of statistical control. Using an $X$-bar chart, calculate the probability of seven consecutive points falling above the centerline given a stable process. Is this probability associated with a type I error or a type II error?

**17.5** Refer back to the revised $X$-bar and $R$ control charts constructed in Example 15.2 (shown in Figure 15.4) and apply our runs rules. Comment on your findings.

**17.6** Refer back to the revised $X$ and $R_M$ control charts of Example 15.3 (shown in Figure 15.7) and apply our runs rules. Comment on your findings.

**17.7** Apply our runs rules to the $X$-bar, $R$ and the $X$-bar, $S$ control chart sets you developed in Problem 15.11 for the candy bars. Comment on your results.

**17.8** Apply our runs rules to the $NP$ control chart of Example 16.4, shown in Figure 16.7. Comment on your specific findings, as well as the appropriateness of using runs rules on the $NP$ or $P$ chart in general.

**17.9** Apply our runs rules to the $C$ chart of Example 16.5, shown in Figure 16.9. Comment on your specific findings, as well as the appropriateness of using runs rules on the $C$ or $U$ chart in general.

**17.10** A $P$ chart, based on a standard value $p_0 = 0.02$, is being designed with 3-sigma limits and a sample size of 50. Compute the control limits for the chart. Comment on the appropriateness of using a $p_0$ value instead of a $\bar{p}$ value. Here, $p_0$ refers to a target value set by management.

**17.11** Using the candy bar data from Problem 15.10, develop the 0.995 and 0.005 probability limits for the $X$-bar, $R$ chart and $X$-bar, $S$ chart pairs.

**17.12** Refer back to the candy bar variables control charting case in Problem 15.10. Now, one of the plant managers suggests that we set a target of 60.1 g for all candy bars (labeled 60 g) and a target of 0.05 g as a candy bar process standard deviation (for the population of individual candy bars). Her contention is that this will constitute world-class candy bar manufacturing and stimulate our production people to "shape up." Provide a quantitative assessment of what our charts would look like as well as what these new targets would do to improve our operations.

**17.13** Refer back to the tractor production process in Example 16.5. Plant management has now decided to follow a ZD program. In a quality council meeting, a target of $c_0 = 0.0$ was suggested as a way to align our SPC efforts with the new ZD program. Calculate our new $C$ control chart parameters. Comment on the new target suggestion and ways in which we may effectively pursue zero defects. If we obtain a ZD tractor production process, will this automatically assure us of customer satisfaction and high market share?

**17.14** In the quality council meeting described in Problem 17.13, a concern was voiced that zero defects seems unrealistic and that we should pursue a six-sigma program instead (3.4 defects per million opportunities). How could we use such a target on a $C$ chart (or $U$ chart)? What would be its probable feasibility and impact relative to the production floor. If we obtain a six-sigma tractor production process, will this automatically assure us of customer satisfaction and high market share?

**17.15** $X$-bar and $S$ control charts are maintained on the breaking strength in pounds for deep sea fishing line. The subgroup size is 10. The values of $\bar{x}$ and $s$ are computed for each subgroup. After 30 subgroups, $\sum \bar{x}_j = 4500$ lb and $\sum s_j = 135$ lb.
  **a** Compute the control limits for the $X$-bar and $S$ charts with .001 and .999 probability limits.
  **b** Estimate $\sigma$, on the assumption that the process is in statistical control.

**17.16** $X$-bar and $R$ control charts are maintained on the weight in ounces of the contents of a box of breakfast cereal (labeled 14 oz). The subgroup size is 4. After 18 subgroups, $\sum \bar{x}_j = 257.8$ oz and $\sum R_j = 27$ oz.
  **a** Compute the 95 percent probability control limits for both charts.
  **b** Estimate $\sigma$, on the assumption that the process is in statistical control.

### Discovery

**17.17** A company that produces nails has retained you as a consultant. The company presently produces six-, ten-, and sixteenpenny common nails. These nails are all produced on the same basic machines, with different wire gauges and dies. Currently, the company is using $P$ charts, one chart for each machine. On a typical day about 35 machines are running. When a nail is inspected, its length, head shape, and point shape are assessed. The fraction defective is currently running at about 0.5 percent, and yesterday all of the machines were in statistical control.

Vendor 1 has proposed a plan where he will install a new vision inspection-sorting system which will 100 percent inspect (i.e., inspect for all three attributes currently being sampled) the nails, and identify and remove nails that do not meet specifications. The resolution of his system is such that it will inspect any of the six-, ten-, and sixteenpenny nails currently in production. The vision inspection-sorting system is claimed to be 100 percent accurate in detecting nonconforming nails. The vendor further states that, since the inspection-sorting system is 100 percent accurate, the nail company will be able to totally discontinue its $P$ chart procedures. He adds that the nail customers will be totally satisfied, since they will be receiving nails that meet all requirements.

Vendor 2 has proposed that the nail company use its variables control chart software. She recommends that, since length is the most important quality characteristic, we should mix all of the sixpenny nails together (from all of the machines set up to produce sixpenny nails) and use only one set of control charts ($X$-bar, $R$) in order to save inspection and charting time. Vendor 2 also recommends that we follow this procedure for the tenpenny and sixteenpenny nails. She also claims that her mixing suggestion will minimize product sampling and will pay for her software in less than six months.

Your client, the nail company, has asked you to study the quality implications of the proposals and make a recommendation on possible courses of action.

*Before addressing the following discovery exercises,*
*you may want to read Appendix 17A.*

**17.18** Assuming that a 3-sigma $X$-bar chart, $n = 4$, is used to monitor candy bar weight, the candy bar process mean is 60.0 g, and the process standard deviation is 0.80 g.

**a** Determine the type I error probability for the $X$-bar chart, assuming one point outside of the control limits indicates an out-of-control process (i.e., ignore the runs rules for this problem).

**b** Determine the type II error probability for the $X$-bar chart, if the process mean shifts to 59 g, assuming one point outside of the control limits indicates an out-of-control process (i.e., ignore the runs rules for this problem).

**17.19** Using the OC curves developed in Figure 17.10 for the 3-sigma $P$ chart:

**a** Determine the type II error probability associated with a shift in the process fraction defective to $p = 0.45$. A sample size of 50 is used.

**b** Determine the probability of detecting this shift within the first two samples taken after the shift.

**17.20** A $C$ chart is used to monitor the number of surface imperfections on sheets of steel. The chart is set up based on $c = 3.0$.

**a** Calculate the 3-sigma control limits for this process.

**b** If the process average shifts to 6.0, determine the probability of not detecting the shift on the first sample taken after the shift occurs.

**17.21** A 3-sigma $P$ chart is used to control brake pad assemblies used in automotive production. The average number of units inspected each shift is 800, and the chart uses a target value $p_0 = 0.015$.

**a** Estimate the probability of a type I error.

**b** If the process proportion defective shifts to a value of 2 percent, estimate the probability of a type II error.

# APPENDIX 17A: OC CURVE CONSTRUCTION AND ARL CALCULATIONS

OC curve construction requires a good deal of effort. The OC curves developed here are highly simplified and do not consider runs rules; only single points beyond the control limits are considered in detecting the shift. OC curves can provide insight as to how control charts work and the design element involved in determining sample size and the number of sigmas, $b$, used for the control chart limits. Figure 17.9 illustrates the general OC curve construction procedure we will use. Notice how the shaded area, $P_a$, changes with the magnitude of the shift. Larger shifts produce smaller $P_a$'s.

**FIGURE 17.9**    Generic OC depiction (no runs rules).

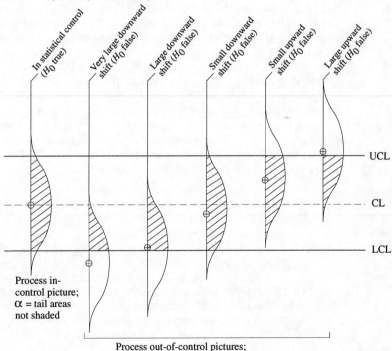

CL :   "Chart" process parameter ( $p_{H_0}$ for a $P$ chart, $c_{H_0}$ for a $C$ chart, $\mu_{H_0}$ for an $X$-bar chart) for a true $H_0$

$\oplus$ :   "Actual" process parameter ( $p$ for a $P$ chart, $c$ for a $C$ chart, $\mu$ for an $X$-bar chart)

⬚ :   $P_a = P$ (accepting the $H_0$ of "in statistical control")

Two OC curves will be constructed to demonstrate the procedure. Modern computer aids—such as most computer-assisted statistical analysis packages and some spreadsheets—allow us to readily calculate "tail area" probabilities and can help us to develop OC curves. Otherwise, we must rely on tabulated values, which many times must be interpolated.

### P Chart OC Curve

The $P$ chart is based on a binomial distribution. Since the binomial is symmetric only when the parameter $p = 0.5$, the OC curve will usually be asymmetric. Hence, we must develop both sides of the OC curve. If the $P$ chart $LCL_p$ is set at 0, then we can develop only the right-hand tail in the OC curve (no left-hand tail exists, as the left-hand tail in this case traces back to $P_a = 1.0$ at $p = 0$).

We will now construct an OC curve for a 3-sigma $P$ chart where $n = 50$, $p = 0.250$, $UCL_p = 0.434$, and $LCL_p = 0.066$. The generic control chart with the shaded $P_a$ values in Figure 17.9 is useful in visualizing the OC curve probability statements that must be constructed in order to estimate the shaded areas. With manual calculations, we typically will use a Poisson approximation for the binomial when $p \leq 0.1$ and a normal approximation for the binomial when $p > 0.1$. Of course, if we had a fast and easy means of evaluating binomial terms directly we would not bother with the approximations. Table 17.2 was developed using manual calculations and the use of a Poisson and a standard normal distribution table (see Tables IX.6 and IX.2, respectively, in Section IX). We will also use the SAS functions [11] to demonstrate computer-aided OC curve development. Our SAS results are shown in Table 17.3. We have used SAS to develop six different OC curves (Table 17.3). Newer versions of spreadsheets such as EXCEL [12] also contain statistical distribution functions.

We have systematically worked from the $p = 0.0$ line to estimate the single-hatched shaded regions (e.g., Figure 17.9) between the $LCL_p$ and the $UCL_p$. The $U$ area (Table 17.2) represents the area under the binomial curve from 0 to the $UCL_p$. The $L$ area represents the area under the binomial curve from 0 to the $LCL_p$. Hence, the difference $U - L$ represents $P_a$ for our OC curve. Three OC curves for 3-sigma $P$ charts are shown in Figure 17.10. Corresponding OC curves for 2-sigma $P$ charts are shown in Figure 17.11. Both the 3-sigma and 2-sigma OC curves are drawn to the same scale. The SAS program output used to obtain the 3-sigma, $n = 50$ column in Table 17.3 is shown in Figure 17.12.

---

### Example 17.2

Use the OC curve in Figure 17.10 and/or the data in Table 17.3, $p = 0.250$, $n = 50$, to answer the following:

    **a** What type I error probability would we expect?
    **b** What type II error probability would we expect if the process shifted to $p = 0.35$?
    **c** What type II error probability would we expect if the process shifted to $p = 0.55$?

#### Solution

    **a** Locating $p = 0.250$ on the horizontal axis and moving up to the OC curve and across to the left we read $P_a = (1 - \alpha) \approx 0.997$. Hence, $\alpha \approx 0.003$ or about a 0.3 percent chance of a false alarm on a single sample. (Actually, we used Table 17.3 to obtain these values, but they can be read—albeit with less accuracy—from Figure 17.10).

**TABLE 17.2**  THREE-SIGMA $P$ CHART OC CURVE DATA FOR $p_{H_0} = 0.25$ AND $n = 50$ (HAND CALCULATED)

Poisson approximation for $p \leq 0.10$

Control limit calculations:

$$UCL_p = p_{H_0} + 3\sqrt{\frac{p_{H_0}(1 - p_{H_0})}{n}}$$

$$= 0.25 + 3\sqrt{\frac{0.25(0.75)}{50}} = 0.434$$

$$LCL_p = p_{H_0} - 3\sqrt{\frac{p_{H_0}(1 - p_{H_0})}{n}}$$

$$= 0.25 - 3\sqrt{\frac{0.25(0.75)}{50}} = 0.066$$

$$p_{H_0} = \text{centerline} = 0.25$$

| Assumed $p$ | $np$ | Area below UCL $P(p < 0.434) = U$ $\approx P(c \leq 21) = U$ | Area below LCL $P(p < 0.066) = L$ $\approx P(c \leq 2) = L$ | Area between UCL and LCL $P_a = U - L$ | $P$ (statistical error \| assumed $p$) |
|---|---|---|---|---|---|
| 0 | 0 | — | — | 0 | 0 |
| 0.02 | 1.0 | $\approx 1.0$ | 0.920 | 0.080 | $\beta = 0.080$ |
| 0.04 | 2.0 | $\approx 1.0$ | 0.677 | 0.323 | $\beta = 0.323$ |
| 0.06 | 3.0 | $\approx 1.0$ | 0.423 | 0.577 | $\beta = 0.577$ |
| 0.08 | 4.0 | $\approx 1.0$ | 0.238 | 0.762 | $\beta = 0.762$ |
| 0.10 | 5.0 | $\approx 1.0$ | 0.125 | 0.975 | $\beta = 0.975$ |

Normal approximation for $p > 0.10$

| Assumed $p$ | $\sigma = \sqrt{\dfrac{p(1-p)}{n}}$ | $Z_U = \dfrac{UCL - p}{\sigma}$ | $Z_L = \dfrac{LCL - p}{\sigma}$ | Area below UCL $P(Z < Z_U) = U$ | Area below LCL $P(Z < Z_L) = L$ | Area between UCL and LCL $P_a = U - L$ | $P$ (statistical error \| assumed $p$) |
|---|---|---|---|---|---|---|---|
| 0.15 | 0.0505 | 5.62 | -1.66 | $\approx 1.0$ | 0.0485 | 0.9515 | $\beta = 0.9515$ |
| 0.25 | 0.0612 | 3.01 | -3.01 | 0.9987 | 0.0013 | 0.9974 | $\alpha = 1 - 0.9974$ |
| 0.35 | 0.0675 | 1.24 | -4.21 | 0.8925 | $\approx 0.0$ | 0.8925 | $\beta = 0.8925$ |
| 0.45 | 0.0704 | -0.23 | -5.45 | 0.4090 | $\approx 0.0$ | 0.4090 | $\beta = 0.4090$ |
| 0.55 | 0.0704 | -1.65 | -6.88 | 0.0495 | $\approx 0.0$ | 0.0495 | $\beta = 0.0495$ |
| 0.65 | 0.0675 | -3.20 | -8.65 | 0.0007 | $\approx 0.0$ | 0.0007 | $\beta = 0.0007$ |

**TABLE 17.3**  $P$ CHART OC CURVE $P_a$ FOR $p = 0.25$ (COMPUTER CALCULATED)

| Assumed | 3-sigma chart control limits ($b = 3$) | | | 2-sigma chart control limits ($b = 2$) | | |
|---|---|---|---|---|---|---|
| | $n = 10$ | $n = 50$ | $n = 100$ | $n = 10$ | $n = 50$ | $n = 100$ |
| $p$ | $P_a = (6 \leq np \leq 0)$ | $P_a = (21 \leq np \leq 3)$ | $P_a = (38 \leq np \leq 12)$ | $P_a = (5 \leq np \leq 0)$ | $P_a = (18 \leq np \leq 6)$ | $P_a = (33 \leq np \leq 16)$ |
| 0.02 | 1.0000 | 0.0784 | 0.0000 | 1.0000 | 0.0005 | 0.0000 |
| 0.04 | 1.0000 | 0.3233 | 0.0007 | 1.0000 | 0.0144 | 0.0000 |
| 0.06 | 1.0000 | 0.5838 | 0.0168 | 1.0000 | 0.0776 | 0.0003 |
| 0.08 | 1.0000 | 0.7740 | 0.1028 | 1.0000 | 0.2081 | 0.0058 |
| 0.10 | 1.0000 | 0.8883 | 0.2970 | 0.9999 | 0.3839 | 0.0399 |
| 0.15 | 0.9999 | 0.9858 | 0.8365 | 0.9986 | 0.7806 | 0.4317 |
| 0.25 | 0.9965 | 0.9973 | 0.9982 | 0.9803 | 0.9642 | 0.9613 |
| 0.35 | 0.9740 | 0.8813 | 0.7699 | 0.9051 | 0.6215 | 0.3803 |
| 0.45 | 0.8980 | 0.3900 | 0.0951 | 0.7384 | 0.1273 | 0.0098 |
| 0.55 | 0.7340 | 0.0444 | 0.0005 | 0.4956 | 0.0053 | 0.0000 |
| 0.65 | 0.4862 | 0.0007 | 0.0000 | 0.2485 | 0.0000 | 0.0000 |
| 0.75 | 0.2241 | 0.0000 | 0.0000 | 0.0781 | 0.0000 | 0.0000 |
| 0.85 | 0.0500 | 0.0000 | 0.0000 | 0.0099 | 0.0000 | 0.0000 |
| 0.95 | 0.0010 | 0.0000 | 0.0000 | 0.0001 | 0.0000 | 0.0000 |

*Source:* SAS Function PROBBNML($p$, $n$, $x$);  binomial tail area estimate, using $np$ integer method.

**b** Again reading the chart, we see $P_a = \beta \simeq 0.88$ or about an 88 percent chance of not detecting on a single sample.

**c** Once again reading the chart, we find $P_a = \beta \simeq 0.04$ or about a 4 percent chance of failing to detect the shift on one sample.

## $P$ Chart OC Curve Generalizations

A conceptual examination of the $P$ chart and its mechanics (Chapter 16) yields the following information as to type I and type II error probabilities.

**FIGURE 17.10**  $P$ chart OC curves, 3-sigma limits.

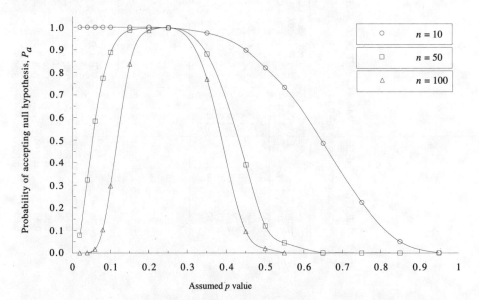

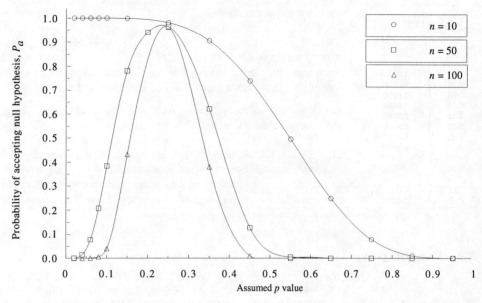

**FIGURE 17.11** *P* chart OC curves, 2-sigma limits.

**FIGURE 17.12** *P* chart OC curve development, SAS code and results, $p_{H_0} = 0.25$, $n = 50$.

```
TITLE 'OC CURVE CALCS, P CHART';
DATA;
FORMAT P 5.2 PU PL PA BEST10.;
P=0.02;    PU=PROBBNML(0.02, 50,21);    PL=PROBBNML(0.02,50,2);    PA=PU-PL;    PUT P PU PL PA;
P=0.04;    PU=PROBBNML(0.04, 50,21);    PL=PROBBNML(0.04,50,2);    PA=PU-PL;    PUT P PU PL PA;
P=0.06;    PU=PROBBNML(0.06, 50,21);    PL=PROBBNML(0.06,50,2);    PA=PU-PL;    PUT P PU PL PA;
P=0.08;    PU=PROBBNML(0.08, 50,21);    PL=PROBBNML(0.08,50,2);    PA=PU-PL;    PUT P PU PL PA;
P=0.10;    PU=PROBBNML(0.10, 50,21);    PL=PROBBNML(0.10,50,2);    PA=PU-PL;    PUT P PU PL PA;
P=0.15;    PU=PROBBNML(0.15, 50,21);    PL=PROBBNML(0.15,50,2);    PA=PU-PL;    PUT P PU PL PA;
P=0.25;    PU=PROBBNML(0.25, 50,21);    PL=PROBBNML(0.25,50,2);    PA=PU-PL;    PUT P PU PL PA;
P=0.35;    PU=PROBBNML(0.35, 50,21);    PL=PROBBNML(0.35,50,2);    PA=PU-PL;    PUT P PU PL PA;
P=0.45;    PU=PROBBNML(0.45, 50,21);    PL=PROBBNML(0.45,50,2);    PA=PU-PL;    PUT P PU PL PA;
P=0.55;    PU=PROBBNML(0.55, 50,21);    PL=PROBBNML(0.55,50,2);    PA=PU-PL;    PUT P PU PL PA;
P=0.65;    PU=PROBBNML(0.65, 50,21);    PL=PROBBNML(0.65,50,2);    PA=PU-PL;    PUT P PU PL PA;
```

| P | PU | PL | PA |
|---|---|---|---|
| 0.02 | 1.000000 | 0.92157225 | 0.07842775 |
| 0.04 | 1.000000 | 0.676714 | 0.323286 |
| 0.06 | 1.000000 | 0.41624647 | 0.58375353 |
| 0.08 | 1.000000 | 0.22597428 | 0.77402572 |
| 0.10 | 1.000000 | 0.11172876 | 0.88827124 |
| 0.15 | 0.99999911 | 0.01418852 | 0.98581059 |
| 0.25 | 0.99738219 | 0.00008709 | 0.9972951 |
| 0.35 | 0.88125959 | 1.69426E-7 | 0.88125942 |
| 0.45 | 0.38996377 | 8.987E-11 | 0.38996377 |
| 0.55 | 0.04437937 | 8.6608E-15 | 0.04437937 |
| 0.65 | 0.00074515 | 6.8988E-20 | 0.00074515 |

1 If $n$ is constant, moving the control limits in (e.g., ± 3 sigma to ± 2 sigma) will increase the type I error probability and decrease the type II error probability. Table 17.3 shows that the $\alpha$ (false alarm) error probability is slightly larger for the 2-sigma chart (e.g., $1 - 0.9642 = 0.0358$ for the 2-sigma chart compared to $1 - 0.9973 = 0.0027$ for the 3-sigma chart, $n = 50$).

2 Increasing the sample size (while holding the control limits at ± 3 sigma) will decrease the type II error probability. If $n$ is decreased, just the opposite will occur.

In our OC curves, we can see the type II (failure to detect) error probability decreasing as the sample size increase draws the OC curve tighter about $p = 0.25$. We can also see that when $n = 10$ the $LCL_p = 0.0000$. Here the $P$ chart has no lower tail (i.e., a sample cannot drop below the $LCL_p$). Hence, the OC curve approaches $P_a = 1.0$ on the lower side. In theory, changing only the sample size will hold the type I error constant (as we will see with the $X$-bar chart OC curves). However, here we are using sigma limits on a discrete, asymmetric distribution (the binomial). Hence, our results in Table 17.3 show some fluctuations from our theoretical expectation.

## X-Bar Chart OC Curves

OC curves can be used to help design $X$-bar and $R$ charts in a fashion similar to our previous discussion for $P$ charts. Our objective, as before, is to quantify our chances of a false alarm and failure to detect. Next, we will develop an OC curve for an $X$-bar chart. $R$ chart OC curves are more difficult to develop (see Duncan for the development of OC curves for an $R$ chart [13]).

Since the $X$-bar chart is based on the normal distribution, its OC curve is straightforward. The $X$-bar chart OC curve will typically be two-tailed, since $LCL_{\bar{x}}$ is usually not backed up against a hard limit (such as the case often is with the 3-sigma $P$ chart or $R$ chart, where we have no LCL). We usually develop the right-hand tail of the $X$-bar chart OC curve and then use the property of symmetry to trace the left-hand tail. This is possible since the normal probability mass function is symmetrical.

We will demonstrate the development of a generic $X$-bar chart OC curve based on process shifts in multiples of $\sigma$ (the standard deviation of a population of individual items). We will assume that the $R$ chart is in control, the process $\sigma$ is known, and the process variation does not shift, even when we shift the process mean. These assumptions allow us to ignore a process dispersion shift in our analysis. We use the standard normal distribution table (see Section IX, Table IX.2) to develop specific probabilities associated with $P_a$, the probability of accepting $H_0$. Here,

$H_0$:  The process mean is as the chart states, $\mu = \mu_0$

$H_1$:  The process mean has shifted, $\mu \neq \mu_0$

where $\mu$ is the true process mean (unknown to us) and $\mu_0$ is the chart centerline. In addition,

$$P(\text{false alarm}) = P(\text{type I error}) = \alpha = P(\text{rejecting } H_0 \mid H_0 \text{ is true})$$

$$P(\text{failure to detect}) = P(\text{type II error}) = \beta = P(\text{not rejecting } H_0 \mid H_1 \text{ is true})$$

Using the standard normal relationship below, and the general approach shown in Figure 17.9, we can reduce the $P_a$ calculation to three factors:

*b:* the sigma level (e.g., 3 sigmas)

*n:* the subgroup size (e.g., 2, 3)

*k:* the shift in the process mean from the *X*-bar chart centerline in multiples of $\sigma$ (e.g., if $k = 0.50$, the process mean shifts by $0.50\ \sigma$)

$$P_a = \Phi\left[\frac{\text{UCL} - (\mu_{\text{centerline}} + k\sigma)}{\sigma\,/\,\sqrt{n}}\right] - \Phi\left[\frac{\text{LCL} - (\mu_{\text{centerline}} + k\sigma)}{\sigma\,/\,\sqrt{n}}\right] \quad (17.1)$$

$$P_a = \Phi\,[b - k\sqrt{n}] - \Phi\,[-b - k\sqrt{n}] \quad (17.2)$$

where $\Phi$ is the tail area under the standard normal curve, with the area accumulated from left to right.

For example, to obtain $P_a$ for a shift of 1.5 $\sigma$ on a 3-sigma *X*-bar chart, with $n = 3$,

$$P_a = \Phi\,[3 - 1.5\sqrt{3}\,] - \Phi\,[-3 - 1.5\sqrt{3}\,] = P(Z \leq 0.402) - P(Z \leq -5.598) \approx 0.6562$$

We have used Equation 17.2 and the SAS functions to obtain the OC curve data shown in Table 17.4. The SAS program output used to generate the 3-sigma, $n = 3$ column is shown in Figure 17.13.

The boxed portions in Table 17.4 emphasize the effectiveness of large subgroup sizes in detecting shifts. The top row, $k = 0.00$, provides $P_a = 1 - \alpha$ values. We can see that the $\beta$ error drops off rapidly as $k$ increases for the larger subgroup sizes (e.g., $n = 5$, $n = 10$, $n = 20$). OC curves are plotted in Figures 17.14 and 17.15 for 3-sigma and 2-sigma *X*-bar charts, respectively. Here again, we can see the dramatic effect of large subgroup sizes on enhancing our probability of shift detection. We should also notice that when we hold the sigma limits constant, the $\alpha$ holds constant. If we want to change $\alpha$, we must change the control limit sigma level. Hence, in practice, if we want to tailor a chart to a given $\alpha$ and $\beta$ we must work with the sigma multiple to get our desired $\alpha$ and then work with the subgroup size to get the desired $\beta$ (at a given shift value, e.g., 1 $\sigma$, 1.5 $\sigma$).

**TABLE 17.4**  *X*-BAR CHART OC CURVE $P_a$ FOR $k\sigma$ SHIFTS IN THE PROCESS MEAN (COMPUTER CALCULATED)

| Shift ($k\sigma$) | 3-sigma chart control limits ($b = 3$) | | | | | 2-sigma chart control limits ($b = 2$) | | | | |
|---|---|---|---|---|---|---|---|---|---|---|
| *k* | *n* = 1 | *n* = 3 | *n* = 5 | *n* = 10 | *n* = 20 | *n* = 1 | *n* = 3 | *n* = 5 | *n* = 10 | *n* = 20 |
| 0.00 | 0.9973 | 0.9973 | 0.9973 | 0.9973 | 0.9973 | 0.9545 | 0.9545 | 0.9545 | 0.9545 | 0.9545 |
| 0.25 | 0.9964 | 0.9946 | 0.9925 | 0.9863 | 0.9701 | 0.9477 | 0.9340 | 0.9200 | 0.8840 | 0.8102 |
| 0.50 | 0.9936 | 0.9835 | 0.9701 | 0.9220 | 0.7776 | 0.9270 | 0.8695 | 0.8102 | 0.6622 | 0.4067 |
| 1.00 | 0.9772 | 0.8976 | 0.7776 | 0.4357 | 0.0705 | 0.8400 | 0.6056 | 0.4067 | 0.1226 | 0.0067 |
| 1.50 | 0.9331 | 0.6562 | 0.3617 | 0.0407 | 0.0001 | 0.6912 | 0.2749 | 0.0879 | 0.0030 | 0.0000 |
| 2.00 | 0.8413 | 0.3213 | 0.0705 | 0.0004 | 0.0000 | 0.5000 | 0.0716 | 0.0067 | 0.0000 | 0.0000 |
| 2.50 | 0.6915 | 0.0918 | 0.0048 | 0.0000 | 0.0000 | 0.3085 | 0.0099 | 0.0002 | 0.0000 | 0.0000 |
| 3.00 | 0.5000 | 0.0140 | 0.0000 | 0.0000 | 0.0000 | 0.1587 | 0.0007 | 0.0000 | 0.0000 | 0.0000 |
| 3.50 | 0.3085 | 0.0011 | 0.0000 | 0.0000 | 0.0000 | 0.0668 | 0.0000 | 0.0000 | 0.0000 | 0.0000 |
| 4.00 | 0.1587 | 0.0000 | 0.0000 | 0.0000 | 0.0000 | 0.0228 | 0.0000 | 0.0000 | 0.0000 | 0.0000 |
| 4.50 | 0.0668 | 0.0000 | 0.0000 | 0.0000 | 0.0000 | 0.0062 | 0.0000 | 0.0000 | 0.0000 | 0.0000 |
| 5.00 | 0.0228 | 0.0000 | 0.0000 | 0.0000 | 0.0000 | 0.0013 | 0.0000 | 0.0000 | 0.0000 | 0.0000 |

*Source:* SAS function PROBNORM(z); normal distribution tail-area estimates.

```
TITLE 'OC CURVE CALCS, X-BAR CHART';
DATA;
FORMAT K 4.2 PU PL PA BEST10.;
K=0.00;     PU=PROBNORM(3.000);     PL=PROBNORM(-3.000);      PA=PU-PL;     PUT K PU PL PA;
K=0.25;     PU=PROBNORM(2.567);     PL=PROBNORM(-3.433);      PA=PU-PL;     PUT K PU PL PA;
K=0.50;     PU=PROBNORM(2.134);     PL=PROBNORM(-3.866);      PA=PU-PL;     PUT K PU PL PA;
K=1.00;     PU=PROBNORM(1.268);     PL=PROBNORM(-4.732);      PA=PU-PL;     PUT K PU PL PA;
K=1.50;     PU=PROBNORM(0.402);     PL=PROBNORM(-5.598);      PA=PU-PL;     PUT K PU PL PA;
K=2.00;     PU=PROBNORM(-0.464);    PL=PROBNORM(-6.464);      PA=PU-PL;     PUT K PU PL PA;
K=2.50;     PU=PROBNORM(-1.330);    PL=PROBNORM(-7.330);      PA=PU-PL;     PUT K PU PL PA;
K=3.00;     PU=PROBNORM(-2.196);    PL=PROBNORM(-8.196);      PA=PU-PL;     PUT K PU PL PA;
K=3.50;     PU=PROBNORM(-3.062);    PL=PROBNORM(-9.062);      PA=PU-PL;     PUT K PU PL PA;
K=4.00;     PU=PROBNORM(-3.928);    PL=PROBNORM(-9.928);      PA=PU-PL;     PUT K PU PL PA;
K=4.50;     PU=PROBNORM(-4.794);    PL=PROBNORM(-10.794);     PA=PU-PL;     PUT K PU PL PA;
K=5.00;     PU=PROBNORM(-5.660);    PL=PROBNORM(-11.660);     PA=PU-PL;     PUT K PU PL PA;
```

| K | PU | PL | PA |
|---|---|---|---|
| 0.00 | 0.9986501 | 0.0013499 | 0.9973002 |
| 0.25 | 0.99487087 | 0.00029847 | 0.9945724 |
| 0.50 | 0.98357861 | 0.00005532 | 0.98352329 |
| 1.00 | 0.89760102 | 1.11159E-6 | 0.89759991 |
| 1.50 | 0.65615799 | 1.05419E-8 | 0.65615798 |
| 2.00 | 0.32132387 | 5.0985E-11 | 0.32132387 |
| 2.50 | 0.09175914 | 1.1508E-13 | 0.09175914 |
| 3.00 | 0.01404597 | 1.2426E-16 | 0.01404597 |
| 3.50 | 0.00109932 | 6.404E-20 | 0.00109932 |
| 4.00 | 0.00004283 | 1.5725E-23 | 0.00004283 |
| 4.50 | 8.17441E-7 | 1.8372E-27 | 8.17441E-7 |
| 5.00 | 7.56865E-9 | 1.0202E-31 | 7.56865E-9 |

**FIGURE 17.13**   *X*-bar chart OC curve development, SAS code and results, 3-sigma, *n* = 3.

**FIGURE 17.14**   *X*-bar chart OC curves, 3-sigma limits.

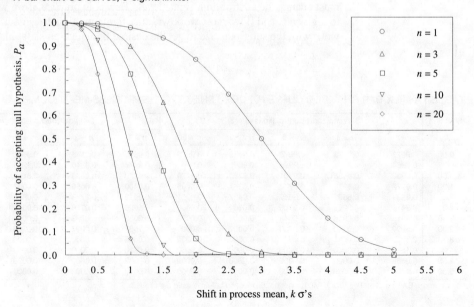

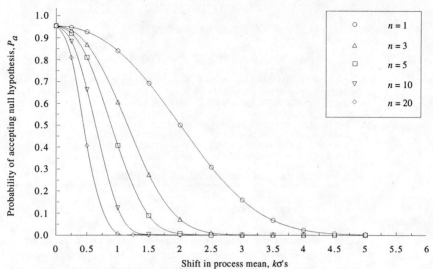

**FIGURE 17.15**     $X$-bar chart OC curves, 2-sigma limits.

---

## Example 17.3

Given a 2-sigma $X$-bar chart and a process where $\bar{\bar{x}} = 150$, $\sigma = 5$, and $n = 3$:

**a** What is the probability that a shift will not be detected after sampling one subgroup (after the shift), if the process mean has shifted to $\mu_1 = 155$?

**b** What is the probability of a false alarm when $\mu_0 = 150$?

**c** How many subgroups would be required in order to have at least a 95 percent chance of detecting a shift from $\bar{\bar{x}} = 150$ to $\bar{\bar{x}} = 155$? (Assume only one point beyond a control limit is used to signal detection.)

### Solution

**a** For a shift to $\mu_1 = 155$,

$$k = \frac{155 - 150}{5} = 1, \; b = 2, \; n = 3$$

$$P_a = \Phi\,[2 - 1\sqrt{3}\,] - \Phi\,[-2 - 1\sqrt{3}\,]$$

$$= P(Z \leq 0.268) - P(Z \leq -3.732) \approx 0.6056 \approx \beta$$

(This answer is also available in Table 17.4, and can be estimated using Figure 17.15.)

**b** When $\mu_0 = 150$,

$$k = \frac{150 - 150}{5} = 0, \; b = 2, \; n = 3$$

$$P_a = \Phi\,[2 - 0] - \Phi\,[-2 - 0]$$

$$= P(Z \le 2) - P(Z \le -2) = 0.9545 = 1 - \alpha$$

$$\alpha = 1 - 0.9545 = 0.0455$$

**c**  From (*a*), $\beta = 0.6056$. Since, $P$(detecting a shift) $= 1 - P$(not detecting a shift),

$P$(not detecting shift after *m* subgroups) $< 1 - 0.95 = 0.05$

Therefore, $(\beta)^m < 0.05$

$$(0.6056)^m = 0.05$$

$$m = \frac{\ln 0.05}{\ln 0.6056} = \frac{-2.996}{-0.5015} = 5.97$$

Hence, we would need at least six subgroups.

## ARL Calculations

Our ARL calculations in this discussion are based on the geometric distribution. The geometric distribution is defined as a sequence or geometric progression of trials, with the success-failure probability held constant from trial to trial (Walpole and Myers [14]). For the random variable *X,* we define the geometric distribution probability mass function as

$$f(x; p) = p(1 - p)^{x-1} \qquad x = 1, 2, 3, \dots \tag{17.3}$$

*p*:  the probability of our control chart signaling an out-of-control process

$(1 - p)$:  the probability of our control chart signaling an in-control process

The expected value, mean, of the geometric distribution is

$$E[X] = \mu = \frac{1}{p} \tag{17.4}$$

If we substitute $1 - P_a$ from our OC curve calculations into Equation 17.4 above for *p,* we can develop the ARL.

$$\text{ARL} = \frac{1}{1 - P_a} \tag{17.5}$$

**TABLE 17.5**    *X*-BAR ARLs FOR $k\sigma$ SHIFTS IN THE PROCESS MEAN (COMPUTER CALCULATED)

| Shift ($k\sigma$) | 3-sigma chart control limits ($b = 3$) | | | | | 2-sigma chart control limits ($b = 2$) | | | | |
|---|---|---|---|---|---|---|---|---|---|---|
| k | $n = 1$ | $n = 3$ | $n = 5$ | $n = 10$ | $n = 20$ | $n = 1$ | $n = 3$ | $n = 5$ | $n = 10$ | $n = 20$ |
| 0.00 | 370.370 | 370.370 | 370.370 | 370.370 | 370.370 | 21.978 | 21.978 | 21.978 | 21.978 | 21.978 |
| 0.25 | 277.778 | 185.185 | 133.333 | 72.993 | 33.445 | 19.120 | 15.152 | 12.500 | 8.621 | 5.269 |
| 0.50 | 156.250 | 60.606 | 33.445 | 12.821 | 4.496 | 13.699 | 7.663 | 5.269 | 2.960 | 1.685 |
| 1.00 | 43.860 | 9.766 | 4.496 | 1.772 | 1.076 | 6.250 | 2.535 | 1.685 | 1.140 | 1.007 |
| 1.50 | 14.948 | 2.909 | 1.567 | 1.042 | 1.000 | 3.238 | 1.379 | 1.096 | 1.003 | 1.000 |
| 2.00 | 6.301 | 1.473 | 1.076 | 1.000 | 1.000 | 2.000 | 1.077 | 1.007 | 1.000 | 1.000 |
| 2.50 | 3.241 | 1.101 | 1.005 | 1.000 | 1.000 | 1.446 | 1.010 | 1.000 | 1.000 | 1.000 |
| 3.00 | 2.000 | 1.014 | 1.000 | 1.000 | 1.000 | 1.189 | 1.001 | 1.000 | 1.000 | 1.000 |
| 3.50 | 1.446 | 1.001 | 1.000 | 1.000 | 1.000 | 1.072 | 1.000 | 1.000 | 1.000 | 1.000 |
| 4.00 | 1.189 | 1.000 | 1.000 | 1.000 | 1.000 | 1.023 | 1.000 | 1.000 | 1.000 | 1.000 |
| 4.50 | 1.072 | 1.000 | 1.000 | 1.000 | 1.000 | 1.006 | 1.000 | 1.000 | 1.000 | 1.000 |
| 5.00 | 1.023 | 1.000 | 1.000 | 1.000 | 1.000 | 1.001 | 1.000 | 1.000 | 1.000 | 1.000 |

ARL = $1/(1 - P_a)$

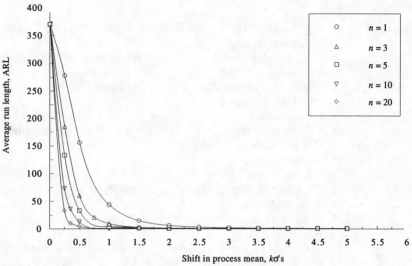

**FIGURE 17.16**    *X*-bar chart ARL curves, 3-sigma limits.

At this point, we can clearly see the ARL–OC curve relationship. We can develop ARL curves from OC curves to provide guidance in assessing the effectiveness of our SPC charts. Table 17.5 provides ARLs for the 2- and 3-sigma *X*-bar chart OC curves developed in the previous section (Table 17.4). The 3- and 2-sigma ARLs are plotted in Figures 17.16 and 17.17, respectively. At this point, we should compare Figure 17.14 with Figure 17.16 and Figure 17.15 with Figure 17.17. We can clearly observe that the ARLs and the OC curve $P_a$'s both provide meaningful performance measures when assessing the sensitivity of SPC charts to process shift detection and to false alarms.

**FIGURE 17.17**    *X*-bar chart ARL curves, 2-sigma limits.

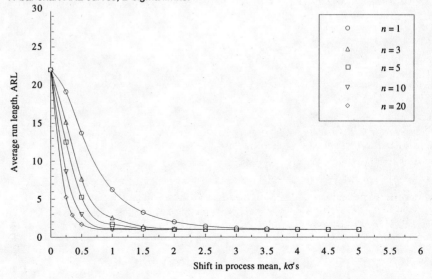

## REFERENCES

1 E. L. Grant and R. S. Leavenworth, *Statistical Quality Control,* 6th ed., New York: McGraw-Hill, 1988.
2 W. E. Deming, *Out of the Crisis,* Cambridge, MA: MIT Center for Advanced Engineering Studies, 1986.
3 AT&T Technologies, *Statistical Quality Control Handbook,* 11th printing, Indianapolis: AT&T Technologies, 1985.
4 See reference 3.
5 J. Hoskins, B. Stuart, and J. Taylor, "Statistical Process Control," Phoenix: Motorola, Inc., BR392/D.
6 See reference 3.
7 See reference 3.
8 J. S. Oakland, *Statistical Process Control,* New York: NY, Wiley, 1986.
9 See reference 1.
10 D. C. Montgomery, *Introduction to Statistical Quality Control,* 2d ed., New York: Wiley, 1991.
11 *SAS Language Reference Version 6,* Cary, NC: SAS Institute, 1990.
12 EXCEL/Function Reference Version 4.0 (Macintosh), Redmond, WA: Microsoft Corporation, 1992.
13 A. J. Duncan, *Quality Control and Industrial Statistics,* 5th ed., Homewood, IL: Irwin, 1986.
14 R. E. Walpole and R. H. Myers, *Probability and Statistics for Engineers and Scientists,* 5th ed., New York: Macmillan, 1993.

# PROCESS CAPABILITY, TOLERANCE, AND MEASUREMENT

## 18.0 INQUIRY

1 What is a capable process?
2 How is process capability measured?
3 How are assembly tolerances modeled?
4 What are quality loss functions and how are they expressed?
5 How does measurement error affect quality control?
6 How are gauge accuracy and gauge precision measured?

## 18.1 INTRODUCTION

As stated earlier, the purpose of SPC charts is to measure and track process stability over time. It is important to know when our production processes are stable and critical to know when they are unstable. We know that a stable process is performing at its best under the given set of process inputs. Unfortunately, *process stability does not assure that our process is meeting product or process specifications.* The purpose of this chapter is to introduce process capability indices, tolerance specifications, loss function models, and gauge measurement studies. A fundamental understanding of these topics is absolutely essential for proper interpretation of quality related data.

## 18.2 PROCESS CAPABILITY

*Process capability is a critical performance measure which addresses process results with respect to product specifications.* Hence, process capability measures are widely used in industry to measure our own ability or a supplier's ability to meet quality specifications. *Two process capability measures or indices are widely used:* (1) *the $C_p$ index,* an inherent or potential measure of capability and (2) *the $C_{pk}$ index,* a realized or actual measure of capability (Kane [1]). Other process capability indices, such as the $C_{pm}$ and $C_{pmk}$, have been developed, discussed, and advocated (Chan et al. [2], and Pearn et al. [3], respectively). An introduction to the $C_{pm}$ and $C_{pmk}$ process capability indices is presented in Appendix 18A.

*In order to evaluate process capability, it is critical to understand that:*

1 *Process specifications pertain to an individual item's quality characteristic(s).*
2 *Capability indices pertain to the population of individual items,* with respect to the quality characteristic of interest, when the process of interest is stable.

**3** *Subgroup based control chart limits pertain to the population of subgroups, and not to the population of individual items,* with respect to the quality characteristic of interest.

Figure 18.1 depicts these critical points, using an $X$-bar chart. The distribution of the plotted statistic, e.g., $\bar{x}$ on the $X$-bar control chart ($n \geq 2$) is "tighter" than the process capability distribution, e.g., $x$. Only in the case of the $X$ chart, where $n = 1$, are they equivalent. It should also be noted that the *product specification limits are set with respect to product design (customer) needs, while the "process spread," as measured by $\hat{\sigma}$, is a function of then process, materials, equipment, tooling, operation methods, and so forth. Capability indices link the product design-related specifications to the process-related results.*

## Potential Capability

*The $C_p$ index measures potential or inherent capability of the production process* (assuming a stable process) and is defined as

$$C_p = \frac{\text{USL} - \text{LSL}}{6\sigma} \tag{18.1}$$

The index is estimated by

$$\hat{C}_p = \frac{\text{USL} - \text{LSL}}{6\hat{\sigma}} \tag{18.2}$$

Here, USL and LSL represent the upper and lower specification limits, respectively. If $C_p = 1$, we declare that the process is potentially capable (in a marginal sense). If $C_p < 1$, we declare that the process is potentially incapable, and if $C_p > 1$, we declare that the process is potentially capable.

**FIGURE 18.1** Global view of the $X$-bar chart, process capability, and specification limits.

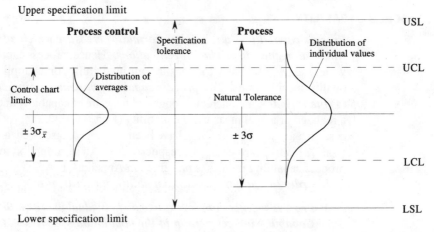

## Actual Capability

***The process $C_{pk}$ index measures realized process capability relative to actual production*** (assuming a stable process), and is defined as

$$C_{pk} = \text{minimum} \left\{ \frac{\mu - \text{LSL}}{3\sigma}, \frac{\text{USL} - \mu}{3\sigma} \right\} \qquad (18.3)$$

The index is estimated by

$$\hat{C}_{pk} = \text{minimum} \left\{ \frac{\hat{\mu} - \text{LSL}}{3\hat{\sigma}}, \frac{\text{USL} - \hat{\mu}}{3\hat{\sigma}} \right\} \qquad (18.4)$$

If $C_{pk} = 1$, we declare that the process is marginally capable. If $C_{pk} < 1$, we declare that the process is incapable, and if $C_{pk} > 1$, we declare that the process is capable.

In some cases, product specifications are set on only one side. For example, purity of a product from a chemical process may be required to be at least 98%, (i.e., LSL = 98%). As another example, we may state that no more than 3 percent cracked grain is allowed in a truckload (i.e., USL = 3 percent). The process capability concept can be applied to such cases by defining the following two measures:

For processes with only an LSL,

$$C_{pL} = \frac{\mu - \text{LSL}}{3\sigma} \qquad (18.5)$$

and for processes with only a USL,

$$C_{pU} = \frac{\text{USL} - \mu}{3\sigma} \qquad (18.6)$$

These two measures are estimated by

$$\hat{C}_{pL} = \frac{\hat{\mu} - \text{LSL}}{3\hat{\sigma}} \qquad (18.7)$$

and $$\hat{C}_{pU} = \frac{\text{USL} - \hat{\mu}}{3\hat{\sigma}} \qquad (18.8)$$

## Interpreting Capability Indices

Potential capability $C_p$ measures the inherent ability of a process to meet specifications, provided the process can be adjusted to target. Hence, it is relevant only to a nominal is best (versus smaller is better or bigger is better) quality characteristic. Since the $C_{pk}$ is developed relative to the process location parameter (the mean) it provides a realized measure of actual production. The $C_{pL}$ and $C_{pU}$ are the single specification counterparts of the $C_{pk}$. Figure 18.2 depicts relationships between the $C_p$ and $C_{pk}$ measures. We should note that high $C_p$'s do not automatically assure us

**FIGURE 18.2**    Relationships between $C_p$ and $C_{pk}$.

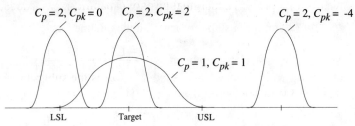

of a high quality product. In practice, the $C_p$ measure is useful since we can typically adjust to target without great expense. However, variance reduction typically is both time consuming and expensive and requires extensive process improvement efforts.

In most cases, we do not know the value of the $\mu$ and $\sigma$ parameters of the production process. We typically use $\hat{\mu}$ and $\hat{\sigma}$ as estimates. These values are usually taken from control chart statistics under stable conditions. Capability analyses based on unstable processes can be expected to produce smaller $C_p$ and $C_{pk}$ indices than those produced in a comparable stable process (instability usually results in greater dispersion estimates).

*Since $\sigma$ and $\mu$ are usually estimated using $\hat{\sigma}$ and $\hat{\mu}$, the results are usually the estimated values $\hat{C}_p$ and $\hat{C}_{pk}$. We expect to see $\hat{C}_p$ or $\hat{C}_{pk}$ greater than 1 in order to support process capability claims* ($C_p$ or $C_{pk}$ = 1). In some cases, we use a rule of thumb that requires a $\hat{C}_p$ or $\hat{C}_{pk}$ of 1.3 (from a sample size of at least 30 to 50 measurements) to ensure clear evidence of process capability.

Capability indices present challenges in statistical inference, as $C_p$ and $C_{pk}$ are themselves random variables, with rather complicated probability mass functions. Kushler and Hurley have published a study which focuses on a number of different approximate methods of establishing lower confidence bounds on $C_{pk}$, assuming that the quality characteristic measure of interest is itself normally distributed [4]. One method discussed results in the approximate $(1 - \alpha)100$ percent lower confidence interval,

$$L_{Cpk,\,\alpha} \approx \hat{C}_{pk} - z_\alpha \sqrt{\frac{1}{9n'} + \frac{\hat{C}_{pk}^{\,2}}{2n' - 2}} \qquad (18.9)$$

where   $\hat{C}_{pk}$ = estimated $C_{pk}$, from Equation 18.4.

$z_\alpha$ = standard normal statistic associated with a right-hand tail area of $\alpha$.

$n'$ = the sample size used to calculate $\hat{C}_{pk}$.

Based on Equation 18.9, we can develop what amounts to an approximate lower confidence limit on $C_{pk}$ and hence, we can structure an approximate hypothesis test for $C_{pk}$. Our test procedure consists of five steps:

*Step 1* Set up the hypotheses and select the level of significance for the approximate hypothesis test:

$$H_0: \ C_{pk} \leq C_{pk0}$$
$$H_1: \ C_{pk} > C_{pk0}$$

Here, we select $C_{pk_0}$ on the basis of our needs (e.g., we would select $C_{pk_0} = 1.0$ if we wanted to represent a marginal level of capability). We choose the desired $\alpha$ level based on our willingness to accept a type I statistical error (i.e., we reject $H_0$, given that is true).

*Step 2* Determine the sample size (e.g., from a control chart or from a separate process capability study). The sample size is the subgroup size multiplied by the number of subgroups (on the control chart) used to estimate $\mu$ and $\sigma$.

*Step 3* Calculate $\hat{C}_{pk}$ using Equation 18.4.

*Step 4* Calculate the $L_{C_{pk}, \alpha}$ using Equation 18.9.

*Step 5* If $C_{pk_0} < L_{C_{pk}, \alpha}$, then reject $H_0$ in favor of the alternate hypothesis at the $\alpha$ level selected, based on the evidence at hand.

---

## Example 18.1

The Motorola Corporation coined and popularized a slogan of the "six-sigma" process. A true six-sigma process, by definition, is obtained when $\mu \pm 6\sigma$ equals the product-specification tolerance interval and when the process mean is centered between the upper and lower product specifications. Assuming that a normal distribution model represents the quality characteristic of interest, answer the following:

**a** What $C_p$ would be associated with six-sigma production?

**b** If the process is centered on its target, what $C_{pk}$ should be realized?

### Solution

**a** Given that the process is a six-sigma process, we know that USL − LSL = $6\sigma$ + $6\sigma = 12\sigma$. Hence,

$$C_p = \frac{USL - LSL}{6\sigma} = \frac{12\sigma}{6\sigma} = 2.0$$

**b** $C_{pk} = \text{minimum} \left\{ \frac{\mu - LSL}{3\sigma}, \frac{USL - \mu}{3\sigma} \right\}$

$$= \text{minimum} \left\{ \frac{6\sigma}{3\sigma}, \frac{6\sigma}{3\sigma} \right\} = 2.0$$

---

## Example 18.2

Using the $X$-bar and $R$ control charts set up in Example 15.2 for the meter stick production process with LSL = 999 mm and USL = 1001 mm, calculate $\hat{C}_p$ and $\hat{C}_{pk}$. Then, test the following hypotheses at the $\alpha = 0.05$ level:

$$H_0: C_{pk} \leq 1.0$$
$$H_1: C_{pk} > 1.0$$

Here, our $H_0$ implies that we assume the process is incapable until and unless we see evidence to the contrary.

### Solution

From Example 15.2 we know that

$$\bar{\bar{x}} = \hat{\mu} = 999.63 \text{ mm}$$

and

$$\hat{\sigma} = \frac{\bar{R}}{d_2} = \frac{0.500}{2.326} = 0.215 \text{ mm}$$

Hence,

$$\hat{C}_p = \frac{\text{USL} - \text{LSL}}{6\hat{\sigma}} = \frac{1001 - 999}{6(0.215)} = 1.550$$

$$\hat{C}_{pk} = \text{minimum} \left\{ \frac{\hat{\mu} - \text{LSL}}{3\hat{\sigma}}, \frac{\text{USL} - \hat{\mu}}{3\hat{\sigma}} \right\}$$

$$= \text{minimum} \left\{ \frac{999.63 - 999}{3(0.215)}, \frac{1001 - 999.63}{3(0.215)} \right\}$$

$$= \text{minimum} \{0.977, 2.124\}$$

$$= 0.977$$

For the hypothesis test, from the revised calculations in Example 15.2 we know that $n = 5$ and $m = 15$; hence, the sample size (the number of measurements used to estimate the parameters) is $nm = 75$. Since $\hat{C}_{pk} = 0.977$ we have no chance of rejecting our $H_0$, but we will follow our steps anyway for illustrative purposes:

$$L_{C_{pk}, 0.05} \simeq 0.977 - 1.645 \sqrt{\frac{1}{9(75)} + \frac{0.977^2}{2(75) - 2}}$$

$$= 0.977 - 1.645 \sqrt{0.00793}$$

$$= 0.831$$

Here,

$$C_{pk0} = 1.0 > L_{C_{pk}, 0.05} \simeq 0.831$$

Therefore, based on the evidence, we cannot reject the null hypothesis that $C_{pk} \leq 1.0$ at the $\alpha = 0.05$ significance level. We conclude that the process is not capable.

## 18.3 TOLERANCE AND INTERFERENCE

*We set product or process requirements on the basis of customer demands. We translate customer demands, expressed in the customer's language, into our technical language of product design and manufacturing. Product-process targets and detailed requirements or specifications are the final result of our product-process definition and design function.*

When we consider quantitative quality characteristics, we set a best or target value and then set up firm specification limits. For example, our meter stick specification could be expressed as $1 \pm 0.001$ m or $1000 \pm 1$ mm. In this case we would consider the target to be 1000 mm, the USL = 1001 mm, and the LSL = 999 mm.

Ultimately, we expect each component to vary in its technical dimension, but we also expect each to fall within specifications. By definition, a component that meets all specifications has zero defects. We should note that *specifications are set relative to customer demands, not process proficiency.* Figure 18.3 illustrates the target, specification, and product quality relationship. We model this relationship relative to our quality characteristic distribution and estimate product nonconformance as a proportion or percentage (see Figure 18.3b).

*When we put components together in a subsystem, or an assembly, we must ensure that the assembly dimension meets its assembly specification.* We also seek to describe the population of assemblies we build from our components, or conversely, to describe component requirements, given an assembly requirement. In general, *we see the assembly dimensional variation growing larger as we add components, or perform additional process operations.* This phenomena is commonly termed "tolerance stack-up."

## Classical Tolerances and Loss Functions

Classical tolerances or specifications have served us well in the past and are still one of the bases for judging quality throughout the world (e.g., quality as conformance to specifications). *An alternate view of classical specifications and tolerances and their conformance-nonconformance dichotomy has been proposed by Taguchi.* Taguchi defines quality in terms of a loss to society [5], as stated in Chapter 1.

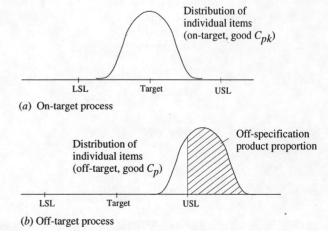

**FIGURE 18.3**    Target, specification, and
product quality relationship.

(a) On-target process

Distribution of
individual items
(on-target, good $C_{pk}$)

LSL    Target    USL

Distribution of
individual items
(off-target, good $C_p$)

Off-specification
product proportion

LSL    Target    USL

(b) Off-target process

Taguchi loss functions view quality as a loss phenomena. Classical specification limits have been termed "goal post" models in contrast to Taguchi's quadratic loss function models (Ross [6]). Figure 18.4 provides a graphical depiction of both the classical goal post and Taguchi quality loss concepts. Figure 18.4*b* shows the classical goal post models alongside the quadratic Taguchi loss function models.

The goal post model gets its name from its graphical shape, depicting a football or soccer goal post (without the top bar), and with respect to a field goal or extra point. Essentially, any ball that passes between the goal posts counts as an equal score, regardless of how close it passed to the center. ***In the classical goal post model, no loss is considered unless the product-process is off specification. In contrast, Taguchi considers loss an increasing function as we move away from the target.*** In the smaller is better model, the zero point is the assumed best target value (e.g., impurity level, rejection rate). The larger is better case assumes some large value as the target.

Taguchi loss functions can be expressed mathematically on both a single-unit and an average-unit basis [7, 8]. Average-unit loss functions deal with units from groups or distributions. The loss functions are provided below, where the random variable $X$ represents the quality characteristic measurement:

**FIGURE 18.4** Comparison of classical tolerance intervals and Taguchi loss function concepts.

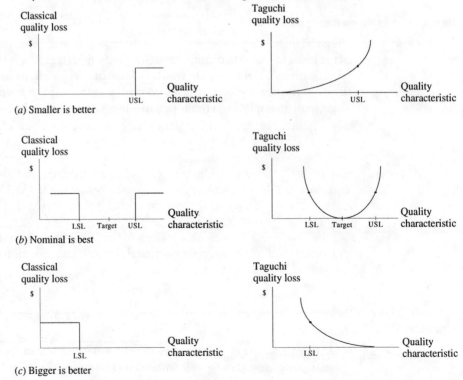

(*a*) Smaller is better

(*b*) Nominal is best

(*c*) Bigger is better

*Smaller is better,* single-unit loss:

$$\text{Loss}_{\text{SIB}} = k(x^2) \tag{18.10}$$

*Smaller is better, average-unit loss function:*

$$L_{\text{SIB}} = k[s^2 + \bar{x}^2] \tag{18.11}$$

*Nominal is best,* single-unit loss:

$$\text{Loss}_{\text{NIB}} = k(x - \mu_0)^2 \tag{18.12}$$

*Nominal is best, average-unit loss function:*

$$L_{\text{NIB}} = k[s^2 + (\bar{x} - \mu_0)^2] \tag{18.13}$$

*Bigger is better,* single-unit loss:

$$\text{Loss}_{\text{BIB}} = k\frac{1}{\bar{x}^2} \tag{18.14}$$

*Bigger is better, average-unit loss function:*

$$L_{\text{BIB}} = k\frac{1}{\bar{x}^2}\left(1 + \frac{3s^2}{\bar{x}^2}\right) \tag{18.15}$$

where $k$ = a constant that fits the loss function to a specific product-process unit.

$\mu_0$ = a target value or best dimension.

$x$ = an actual product unit dimension, as produced.

$\bar{x}$ = the average dimension of a group of product units, as produced.

$s^2$ = the product unit group variance.

and where

$$s^2 = \frac{\sum\limits_{i=1}^{n}(x_i - \bar{x})^2}{n - 1}$$

***By examining the loss functions, especially the nominal is best function, we can see the roles that the location from target and the dispersion play in loss determination.*** This relationship is meaningful and suggests the importance of both location to target and dispersion about the group mean (as does the $C_{pk}$ index, but in a different way). ***In general, it is difficult to fully parameterize and use the Taguchi loss functions on the production floor.*** And furthermore, the specific quadratic nature of the loss function is questionable in an applications context (e.g., the actual loss in the field may not explicitly follow the quadratic function). But if, as a result of studying the loss functions, we better understand and appreciate the critical roles that both parameters play in producing high quality products and preventing quality losses, we are headed in the right direction.

---

**Example 18.3**

In order to demonstrate the use of the Taguchi loss function, let us revisit our meter stick production process with specification limits set at 1000 ± 1 mm.

    **a** If the loss to society for a meter stick of length 1001 mm or 999 mm is \$5, determine the value of $k$ to define parameters for the nominal-is-best situation.

    **b** Given that our process is running at $\bar{x} = 999.63$ mm with $\hat{\sigma} = \bar{R} / d_2 = 0.215$ mm (from Example 15.2), estimate the loss to society per meter stick produced.

    **c** If the process location $\bar{x}$ shifts to 999 mm and process standard deviation $\hat{\sigma}$ shifts to 0.75 mm, estimate the loss to society per meter stick produced.

### Solution

    **a** We will use the single unit form to calculate $k$. We previously stated that the loss due to a meter stick produced with a length of 999 mm is \$5. Hence, using Equation 18.12,

$$\text{Loss}_{\text{NIB}} = \$5 = k(x - \mu_0)^2 = k(999 - 1000)^2$$

$$k = \frac{5}{(1000 - 999)^2} = 5$$

    **b**  We will use the unit average form, Equation 18.13, to calculate the average loss per product unit:

$$
\begin{aligned}
L_{\text{NIB}} &= k[s^2 + (\bar{x} - \mu_0)^2] \\
&= 5[0.215^2 + (999.63 - 1000)^2] = \$0.92 \text{ per unit produced}
\end{aligned}
$$

    **c**     $L_{\text{NIB}} = 5[0.75^2 + (999 - 1000)^2] = \$7.81$ per unit produced

We can clearly see from the above calculations how loss through both product variation and off-target products combine to create the Taguchi quality loss.

---

### Tolerance Stack-up

In order to statistically model tolerance stack-up, we must "convert" tolerances (associated with product function requirements), which are intended for individual items, to statistical models (associated with process function results), which represent populations of individual items. For example, a tolerance may be expressed as $1000 \pm 1$ mm, which implies an LSL = 999 mm, a USL = 1001 mm, and a target value of 1000 mm.

*If we can determine appropriate distributions for quality characteristics, we can use statistical theory to build models for assemblies made up of components or for components resulting from a number of process operations.* For example, if $X$ and $Y$ are defined as random variables, then

$$E(X \pm Y) = E(X) \pm E(Y) = \mu_X \pm \mu_Y \tag{18.16}$$

and

$$\sigma^2_{X \pm Y} = \sigma^2_X + \sigma^2_Y + 2\sigma_{XY} \tag{18.17}$$

where $E(X \pm Y)$ = expected value or the mean, of the sum or difference of $X$ and $Y$.

    $\sigma^2_{X \pm Y}$ = variance of the sum or difference of $X$ and $Y$.

    $\sigma_{XY}$ = covariance of $X$ and $Y$.

If $X$ and $Y$ are statistically independent,

$$\sigma^2_{X \pm Y} = \sigma^2_X + \sigma^2_Y \tag{18.18}$$

We can extend Equations 18.16 and 18.18 to any number of random variables (Walpole and Myers [9]). Furthermore, if $X$ and $Y$ are independent and both distributed normally, the resulting sum or difference of the random variables is distributed normally:

$$\text{If } X \sim N(\mu_X, \sigma^2_X) \quad \text{and} \quad Y \sim N(\mu_Y, \sigma^2_Y)$$

$$\text{Then } (X \pm Y) \sim N(\mu_X \pm \mu_Y, \sigma^2_X + \sigma^2_Y), \text{ respectively} \tag{18.19}$$

where $N(\mu, \sigma^2)$ represents a normal probability mass function with mean $\mu$ and standard deviation $\sigma$.

---

**Example 18.4**

We have purchased a shipment of boards that the manufacturer claims are 8 ft long. After a brief study, we have determined that the board-length population can be modeled with a normal distribution, with a mean of 96 in and a standard deviation of 0.10 in. We are now at our construction site, and we must cut the 8-ft boards to a 7-ft length. Our cutting process capability can be modeled by a normal distribution with a mean of 12 in and a standard deviation of 0.15 in. We will assume a sawblade width of 0 in this example and ignore the cut width.

a  Determine the length characteristic of our 7-ft board population product.
b  If boards less than 83.75 in must be scrapped and boards longer than 84.25 in resawn (reworked), determine the percentage of boards that will be scrapped and the percentage reworked.

**Solution**

a  The original 8-ft length boards $\sim N(96, 0.1^2)$. From the cutting process capability described, we know that the 1-ft cut from the 8-ft board will be distributed as $N(12, 0.15^2)$. From Equation 18.19, we can say that the final 7-ft workpiece is distributed as $N(96 - 12, 0.1^2 + 0.15^2)$. In other words, our 7-ft workpiece will have a mean of 84 in, a standard deviation of $\sqrt{0.1^2 + 0.15^2} = 0.18$ in, and follows a normal distribution.

b  Percentage scrapped:

$$P(X < 83.75) = P\left(Z < \frac{\bar{x} - \mu}{\sigma}\right) = P\left(Z < \frac{83.75 - 84.00}{0.18}\right)$$

$$= P(Z < -1.39)$$

$$= 0.0823 = 8.23\%$$

Percentage reworked:

$$P(X > 84.25) = P\left(Z > \frac{84.25 - 84.00}{0.18}\right)$$

$$= P(Z > 1.39)$$

$$= 0.0823 = 8.23\%$$

## Clearance and Interference

*Some assemblies require mating or matched parts where one must fit with the other.* For example, a common snap-on socket wrench consists of a socket that must fit on the end of a ratchet shaft. A snap-on beverage cap must fit, so that it is watertight, on top of a paper or plastic cup. In both cases, two components are expected to fit and function together. *We may observe interference:* the snap-on component will not fit within or over the other component. For example, our socket will not slide onto the ratchet shaft, or our lid will not push onto our cup. *Or, we may experience excessive clearance, or "play"*—a fit that is too loose. For example, our socket may slide on the ratchet shaft but slip on the shaft when we attempt to tighten a bolt. Or, our lid will set on the cup but leak or fall off when the cup is moved or tipped over.

*In these cases, we must set targets and tolerances so that when we randomly select components, our product assemblies will function properly.* If we model each component with a statistical model, assuming independence, we can determine the clearance mathematically.

If we assume normal distributions for our quality characteristics, we can use a distribution of difference to model interference or clearance. If part A has dimension $X \sim N(\mu_A, \sigma_A^2)$ and part B has dimension $Y \sim N(\mu_B, \sigma_B^2)$ and part A must fit into part B, then the distribution of the clearance or difference between part A and part B is

$$C_{A \text{ into } B} \sim N(\mu_B - \mu_A, \sigma_A^2 + \sigma_B^2) \tag{18.20}$$

and typically, but not always (see Example 18.5, below),

$$P(\text{interference between A and B}) = P(C_{A \text{ into } B} < 0) \tag{18.21}$$

## Example 18.5

A snap-on lid must fit on a cup as shown in Figure 18.5. The metal cup outside diameter follows a normal distribution with a mean of 3.000 in and a standard deviation of 0.005 in. The inside diameter of the plastic lid also follows a normal distribution with a mean of 2.999 in and a standard deviation of 0.006 in. Estimate the percentage of cups that will fail to function properly if a cup-lid clearance of more than 0.003 will leak and an interference of less than −0.010 will not fit (the plastic lid will stretch to some extent).

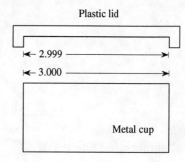

**FIGURE 18.5**    Cup-lid product depiction (Example 18.5).

## Solution

Figure 18.6 depicts the normal cup and lid distributions as well as the clearance-interference distribution. The shaded area to the left on the clearance distribution (Figure 18.6*b*) represents lids that will not snap on. The shaded area to the right represents lids that will leak.

| | |
|---|---|
| Cup outside diameter | $\sim N(3.000, 0.005^2)$ |
| Lid inside diameter | $\sim N(2.999, 0.006^2)$ |
| Hence,    Clearance-interference | $\sim N(2.999 - 3.000, 0.005^2 + 0.006^2)$ |
| | $\sim N(-0.001, 0.0078^2)$ |

**FIGURE 18.6**    Cup-lid distribution depiction (Example 18.5).

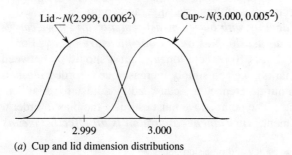

(*a*) Cup and lid dimension distributions

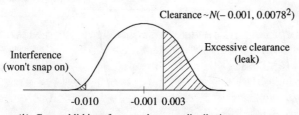

(*b*) Cup and lid interference-clearance distribution

$$P(\text{leak}) = P(\text{clearance} > 0.003)$$

$$= P\left[Z > \frac{0.003 - (-0.001)}{0.0078}\right] = P(Z > 0.51)$$

$$= 0.3050 = 30.5\%$$

$$P(\text{failure to snap on}) = P(\text{clearance} < -0.010)$$

$$= P\left[Z < \frac{-0.010 - (-0.001)}{0.0078}\right] = P(Z < -1.15)$$

$$= 0.125 = 12.5\%$$

## 18.4 GAUGE STUDIES

Whenever we record measurements, we must be aware that they are not perfect. *The two major concerns in measurement technology are* listed below:

1 *Accuracy—the absence of bias in the measurements.*
2 *Precision—the dispersion of the measurements.*

In statistical terms, the estimated value of a parameter should approach the true value of the parameter as the sample size is increased. In practical terms, the average of our measured values should approach the "true dimension" or true value of our product or process, as we make repeated measurements; i.e., it should be accurate. Hence, accuracy is location related, while precision relates to dispersion about a central value, usually the mean of the measurements. Figure 18.7 illustrates these concepts. The measurement process can be one of high precision with low accuracy, high accuracy with low precision, and so on.

*If we are dealing with biased measurements, and if we can determine the bias, then we can adjust the bias out of our measurements.* For example, if we are using a ruler that is 10 percent longer than it should be between tick marks (e.g., centimeter marks), we can simply increase our recorded measurements by the appropriate amount. Hence, we can readily adjust for bias and derive unbiased measurements. The point is, we must estimate the bias in order to make the appropriate adjustment. *Instrument calibration is an effective means to minimize, but*

**FIGURE 18.7** Measurement accuracy and precision depiction.

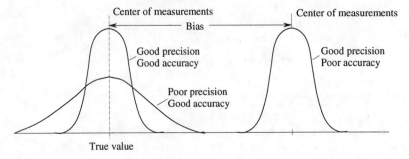

*not totally eliminate, bias.* As long as our bias is small relative to the magnitude of our measurement, we usually tolerate its presence.

*Precision,* on the other hand, *is a more complex issue. Dispersion in measurements has two primary sources: (1) the failure of our gauge or instrument to exactly repeat itself and (2) the failure of an operator or robot to exactly reproduce the measurement technique or method.* We can express the precision of our measurements, assuming independence, as

$$\sigma^2_{\text{measured values}} = \sigma^2_{\text{true dimensions}} + \sigma^2_{\text{gauge repeatability}} + \sigma^2_{\text{operator reproducibility}} \quad (18.22)$$

Sometimes, the term "measurement error" is defined (Grant and Leavenworth [10]) as

$$\sigma^2_{\text{measurement error}} = \sigma^2_{\text{gauge repeatability}} + \sigma^2_{\text{operator reproducibility}} \quad (18.23)$$

*The precision-to-tolerance (P/T) ratio is sometimes used to assess gauge capability:*

$$\text{P/T} = \frac{6\,\hat{\sigma}_{\text{measurement error}}}{\text{USL} - \text{LSL}} \quad (18.24)$$

*If P/T ≤ 0.10, then the gauge or measurement system is typically considered adequate* (Montgomery [11]).

Thus we can clearly see that the data we record on our data sheets or in our hard-disk storage are not the actual dimensions of our products. *Our measured values include variation "inflation" over and above the true variation in the population of product-process dimensions.* We usually assume that the three sources of variation in Equation 18.22 are independent. *If we wish to estimate this variation inflation, we must conduct a controlled instrument or gauge study.* Through a careful procedure of repeatedly measuring the same product dimension, we can extract estimates of gauge repeatability and operator reproducibility.

If either the gauge repeatability or the operator reproducibility is large, in comparison to the true dimensional variance, we must take action to reduce it. Otherwise, our flawed measurement process will limit our ability to assess our product or process. Corrective actions available to us include more precise gauging (better instrumentation), operator training programs, or both.

We may have a very good product (on-target with low dispersion); however, a poor measurement system will obscure this fact. For example, *a large bias in our measurement system or sensor will falsely indicate that we are off-target, while poor measurement precision will falsely indicate that we have excessive variation in our product or process.* The following examples serve to demonstrate how we can use our $X$-bar and $R$ chart technology to estimate $\sigma_{\text{gauge repeatability}}$ and $\sigma_{\text{operator reproducibility}}$.

---

**Example 18.6**

Using $X$-bar and $R$ chart technology, estimate the variation inflation due to gauging for an automatic (no operator) measuring device assigned to our previously described meter stick production process.

   **a** Estimate $\hat{\sigma}_{\text{gauge repeatability}}$.
   **b** Estimate $\hat{\sigma}_{\text{measured values}}$.
   **c** Estimate $\hat{\sigma}_{\text{true dimensions}}$.
   **d** Calculate the P/T ratio.
   **e** Comment on the ability of the automatic measuring machine to produce precise measurements.

## Solution

Here, we assume that $\sigma_{\text{repeatability}}$ is the only gauge related variation component of interest, since we have only one automatic measuring machine. A simple experiment was designed whereby we measured the same meter stick 2 times, and a total of 20 meter sticks were measured. The data, $\bar{x}$, and range values are summarized in Table 18.1. We will use our standard $\hat{\sigma} = \bar{R}/d_2$ calculation on these data.

   **a** $\hat{\sigma}_{\text{gauge repeatability}} = \dfrac{\bar{R}}{d_2} = \dfrac{0.0015}{1.128} = 0.0013$ mm

Here, we use $\bar{R}$ from the data in Table 18.1, $n = 2$.

   **b** $\hat{\sigma}_{\text{measured values}} = \dfrac{\bar{R}}{d_2} = \dfrac{0.50}{2.326} = 0.215$ mm

Here, we use $\bar{R}$ from our revised production control limits from Example 15.2, $n = 5$.

**TABLE 18.1**   AUTOMATIC MEASURING DEVICE METER STICK
GAUGE STUDY DATA (EXAMPLE 18.6)
(All measurements in millimeters.  LSL = 999 mm; USL = 1001 mm)

| Part (meter stick) no. | Automatic measuring device Meas. 1 | Meas. 2 | X-bar | Range |
|---|---|---|---|---|
| 1 | 1000.158 | 1000.157 | 1000.1575 | 0.0010 |
| 2 | 1000.003 | 1000.006 | 1000.0045 | 0.0030 |
| 3 | 999.722 | 999.720 | 999.7210 | 0.0020 |
| 4 | 1000.303 | 1000.305 | 1000.3040 | 0.0020 |
| 5 | 1000.216 | 1000.217 | 1000.2165 | 0.0010 |
| 6 | 999.636 | 999.638 | 999.6370 | 0.0020 |
| 7 | 1000.153 | 1000.153 | 1000.1530 | 0.0000 |
| 8 | 1000.383 | 1000.381 | 1000.3820 | 0.0020 |
| 9 | 999.754 | 999.757 | 999.7555 | 0.0030 |
| 10 | 1000.237 | 1000.238 | 1000.2375 | 0.0010 |
| 11 | 999.968 | 999.968 | 999.9680 | 0.0000 |
| 12 | 1000.073 | 1000.075 | 1000.0740 | 0.0020 |
| 13 | 1000.186 | 1000.184 | 1000.1850 | 0.0020 |
| 14 | 999.948 | 999.945 | 999.9465 | 0.0030 |
| 15 | 999.868 | 999.869 | 999.8685 | 0.0010 |
| 16 | 999.951 | 999.952 | 999.9515 | 0.0010 |
| 17 | 999.900 | 999.903 | 999.9015 | 0.0030 |
| 18 | 1000.387 | 1000.387 | 1000.3870 | 0.0000 |
| 19 | 999.961 | 999.962 | 999.9615 | 0.0010 |
| 20 | 999.813 | 999.812 | 999.8125 | 0.0010 |
| Averages | | | 1000.0312 | 0.0015 |

**c** $\hat{\sigma}_{\text{true dimensions}} = \sqrt{\hat{\sigma}^2_{\text{measured values}} - \hat{\sigma}^2_{\text{gauge repeatability}}}$

$$= \sqrt{(0.215)^2 - (0.0013)^2} = 0.2149 \text{ mm}$$

**d** $\text{P/T} = \dfrac{6\hat{\sigma}_{\text{measurement error}}}{(\text{USL} - \text{LSL})} = \dfrac{6(0.0013)}{(1001 - 999)} = 0.0039 = 0.39\%$

Note, here we assume that $\hat{\sigma}_{\text{measurement error}} = \hat{\sigma}_{\text{gauge repeatability}}$ since we have only one automatic measuring device.

**e** The additional variation added by our measuring machine is extremely small relative to our true product dimension variation. We also see a very small P/T ratio, much less than 10 percent. Hence, we conclude that our automatic measuring device is performing (as far as precision is concerned) very well, relative to the variation inherent in our meter stick production length quality characteristic.

## Example 18.7

A gauge study was run on our meter stick process where measurements were obtained manually by three operators using the same gauge. The data are shown in Table 18.2. Use the $X$-bar and $R$ chart technique to estimate the following measures. Then comment on the ability of our manual measuring process to supply adequate precision.

**a** $\hat{\sigma}_{\text{gauge repeatability}}$
**b** $\hat{\sigma}_{\text{operator reproducibility}}$
**c** The P/T ratio

### Solution

**a** Since we are using the same measuring device with all three operators along with two repeated measurements, $n = 2$,

$$\bar{\bar{R}} = \frac{0.043 + 0.036 + 0.016}{3} = 0.0317 \text{ mm}$$

$$\hat{\sigma}_{\text{gauge repeatability}} = \frac{\bar{\bar{R}}}{d_2} = \frac{0.0317}{1.128} = 0.0281 \text{ mm}$$

**b** Here, we are dealing with three operators. We need a measure of dispersion between operators, $n = 3$,

$$\bar{\bar{X}}_{\text{max}} = \max(1000.040, 999.995, 1000.097) = 1000.097 \text{ mm}$$

$$\bar{\bar{X}}_{\text{min}} = \min(1000.040, 999.995, 1000.097) = 999.995 \text{ mm}$$

$$R_{\bar{\bar{X}}} = 1000.097 - 999.995 = 0.102 \text{ mm}$$

**TABLE 18.2** MANUAL METER STICK MEASURING GAUGE STUDY DATA (EXAMPLE 18.7)

(All measurements in mm. LSL = 999 mm, USL = 1001 mm)

| Part (meter st.) no. | Operator 1 | | | | Operator 2 | | | | Operator 3 | | | |
|---|---|---|---|---|---|---|---|---|---|---|---|---|
| | Meas. 1 | Meas. 2 | X-bar | Range | Meas. 1 | Meas. 2 | X-bar | Range | Meas. 1 | Meas. 2 | X-bar | Range |
| 1 | 1000.16 | 1000.18 | 1000.170 | 0.020 | 1000.08 | 999.99 | 1000.035 | 0,090 | 1000.25 | 1000.28 | 1000.265 | 0.030 |
| 2 | 1000.00 | 999.89 | 999.945 | 0.110 | 999.97 | 999.93 | 999.950 | 0.040 | 1000.11 | 1000.09 | 1000.100 | 0.020 |
| 3 | 999.74 | 999.84 | 999.790 | 0,100 | 999.73 | 999.76 | 999.745 | 0.030 | 999.83 | 999.87 | 999.850 | 0.040 |
| 4 | 1000.29 | 1000.25 | 1000.270 | 0.040 | 1000.28 | 1000.29 | 1000.285 | 0.010 | 1000.33 | 1000.31 | 1000.320 | 0.020 |
| 5 | 1000.21 | 1000.20 | 1000.205 | 0.010 | 1000.12 | 1000.08 | 1000.100 | 0.040 | 1000.27 | 1000.29 | 1000.280 | 0.020 |
| 6 | 999.74 | 999.81 | 999.775 | 0.070 | 999.75 | 999.71 | 999.730 | 0.040 | 999.79 | 999.80 | 999.795 | 0.010 |
| 7 | 1000.16 | 1000.05 | 1000.105 | 0.110 | 1000.03 | 1000.08 | 1000.055 | 0.050 | 1000.21 | 1000.23 | 1000.220 | 0.020 |
| 8 | 1000.38 | 1000.39 | 1000.385 | 0.010 | 1000.33 | 1000.30 | 1000.315 | 0.030 | 1000.44 | 1000.42 | 1000.430 | 0.020 |
| 9 | 999.85 | 999.90 | 999.875 | 0.050 | 999.87 | 999.83 | 999.850 | 0.040 | 999.87 | 999.87 | 999.870 | 0.000 |
| 10 | 1000.24 | 1000.18 | 1000.210 | 0.060 | 1000.21 | 1000.20 | 1000.205 | 0.010 | 1000.21 | 1000.23 | 1000.220 | 0.020 |
| 11 | 999.94 | 999.92 | 999.930 | 0.020 | 999.91 | 999.92 | 999.915 | 0.010 | 999.99 | 999.96 | 999.975 | 0.030 |
| 12 | 1000.07 | 1000.08 | 1000.075 | 0.010 | 1000.08 | 1000.02 | 1000.050 | 0.060 | 1000.12 | 1000.11 | 1000.115 | 0.010 |
| 13 | 1000.19 | 1000.19 | 1000.190 | 0.000 | 1000.14 | 1000.11 | 1000.125 | 0.030 | 1000.23 | 1000.25 | 1000.240 | 0.020 |
| 14 | 999.99 | 1000.02 | 1000.005 | 0.030 | 999.97 | 999.93 | 999.950 | 0.040 | 1000.13 | 1000.11 | 1000.120 | 0.020 |
| 15 | 999.87 | 999.82 | 999.845 | 0.050 | 999.85 | 999.81 | 999.830 | 0.040 | 999.93 | 999.93 | 999.930 | 0.000 |
| 16 | 999.95 | 999.91 | 999.930 | 0.040 | 999.98 | 999.94 | 999.960 | 0.040 | 999.98 | 999.99 | 999.985 | 0.010 |
| 17 | 999.90 | 999.87 | 999.885 | 0.030 | 999.84 | 999.81 | 999.825 | 0.030 | 999.97 | 999.96 | 999.965 | 0.010 |
| 18 | 1000.40 | 1000.47 | 1000.435 | 0.070 | 1000.31 | 1000.34 | 1000.325 | 0.030 | 1000.41 | 1000.41 | 1000.410 | 0.000 |
| 19 | 999.96 | 999.96 | 999.960 | 0.000 | 999.93 | 999.95 | 999.940 | 0.020 | 999.95 | 999.97 | 999.960 | 0.020 |
| 20 | 999.79 | 999.82 | 999.805 | 0.030 | 999.72 | 999.68 | 999.700 | 0.040 | 999.89 | 999.90 | 999.895 | 0.010 |
| Averages | | | 1000.040 | 0.043 | | | 999.995 | 0.036 | | | 1000.097 | 0.016 |

$$\hat{\sigma}_{\text{operator reproducibility}} = \frac{R_{\bar{\bar{X}}}}{d_2} = \frac{0.102}{1.693} = 0.0602 \text{ mm}$$

c    $$\hat{\sigma}_{\text{measurement error}} = \sqrt{0.0281^2 + 0.0602^2} = 0.0664 \text{ mm}$$

$$\text{P/T} = \frac{6(0.0664)}{(1001 - 999)} = 0.1992 = 19.92\%$$

*Comments* In this case, we have a problem with the measurement error (precision) in that the P/T ratio > 0.10. In other words, the measurement error is relatively large in comparison to our length specification interval. Since our estimated operator reproducibility is over twice the size of our estimated gauge repeatability, we would recommend an examination of the consistency of measurement techniques between operators and, possibly, a training program. The gauge repeatability alone would yield P/T = 6(0.0281)/(1001 − 999) = 0.0843. Hence, our gauge instrument appears marginal and, if our operator reproducibility cannot be drastically reduced, may need attention as well.

## REVIEW AND DISCOVERY EXERCISES

### Review

**18.1** How is it possible to have a stable production process and yet not meet specifications? Explain.

**18.2** Explain the difference between the $C_p$ and $C_{pk}$ capability indices. Why do we bother with a $C_p$ index? Explain.

**18.3** Can we use the $C_p$ index on a process or product with a single specification limit (e.g., a USL or an LSL, but not both)? Explain.

**18.4** As we revisit our candy bar manufacturing process (Problem 15.10), we now have an LSL of 58 g. Using the Problem 15.10 data (specifically the $R$ chart and $X$-bar chart data)

    **a** Assess the capability of our process by calculating $C_{pk}$.

    **b** Test the null hypothesis $C_{pk} \leq 1.0$; use an alternative hypothesis of $C_{pk} > 1.0$. Use an $\alpha = 0.10$ level of significance.

**18.5** For the fishing line production process in Problem 15.16, we have a lower product strength specification of 130 lb.

    **a** Is the fishing line process a capable process? Explain.

    **b** Test the null hypothesis that the $C_{pk} \leq 1.0$, use an alternative hypothesis of $C_{pk} > 1.0$. Use an $\alpha = 0.10$ level of significance.

**18.6** In a well known breakfast cereal, there are two scoops of raisins per box. A scoop of raisins contains a mean of 450 raisins with a standard deviation of 10 raisins. We can assume that the number of raisins in a scoop are normally distributed. Statistically describe the distribution of the number of raisins in a box of the cereal.

**18.7** A toy manufacturer is producing small wooden blocks (squares) painted and packaged for sale to toy distributors throughout the world. The target dimension calls for a 2-in cube. Based on production capabilities and actual production, each block's dimensions can be described as on target, $\mu = 2$ in, $\sigma = 0.025$ in. The manufacturer also produces small wooden boxes with ducks painted on the side as packages for the blocks. The dimensions of the wooden boxes (as they are currently produced) are shown below:

Inside depth:     $\mu = 2.1$ in, $\sigma = 0.05$ in

Inside height:    $\mu = 2.1$ in, $\sigma = 0.05$ in

Inside length:    $\mu = 10.25$ in, $\sigma = 0.05$ in

If packaging calls for stacking five blocks in the package, calculate the probability that the blocks will not fit. Assume normal distributions and independent dimensions.

### Discovery

**18.8** In our meter stick production process, we have the following process specifications: LSL = 999 mm, USL = 1001 mm. The process standard deviation is 0.75 mm. We can assume that our dimension can be described by a normal distribution:

    **a** Determine where we should locate our process mean in order to minimize the proportion of defective meter sticks.

    **b** A short meter stick (below the LSL) must be scrapped at a cost of $1.00 per meter stick. A long meter stick (above the USL) can be reworked at a cost of $0.25 per meter stick. Determine where we should locate our process mean in order to minimize our cost of nonconformance. Assume that we can locate only at intervals of 0.1 mm (e.g., 1000.0 mm, 1000.1 mm, 1000.2 mm, 1000.3 mm, 1000.4 mm).

**18.9** The specification on the pressure set point on an air compressor air regulator is $100 \pm 3$ psi. The loss for either a 97-psi or a 103-psi set point is estimated to be

$45.00. The most recent 3-sigma $X$-bar chart for the regulators is running at $\bar{\bar{x}} = 99$ psi, with UCL = 99.75 and LCL = 98.25. A subgroup size of $n = 4$ is used.

**a** Determine the average loss per air compressor, based on the Taguchi loss function concept.

**b** Determine whether the process described above is a capable process. Explain.

**18.10** Our company manufactures screw conveyors. The steel conveyor shaft is supported at each end with a brass bushing. We purchase the shafts and brass bushing stock. We machine the inside diameter of the bushings in our shop. We want to determine where to set our target for the inside bushing diameter in our machine shop. Our shaft diameters are distributed normal ($\mu = 1$ in, $\sigma_{shaft} = 0.001$ in). We must maintain between 0.002 and 0.010 in of clearance (in terms of diameter) for an acceptable fit. We can assume that our machining process results in inside diameters that can be modeled with a normal distribution, with a standard deviation, $\sigma_{bushing} = 0.002$ in.

**a** Where would you center the bushing machining process inside diameter in order to maximize the number of proper fits? Assume we can adjust to an even thousandth of an inch (e.g., 1.000, 1.001, 1.002).

**b** What percent of unacceptable fits will occur at the inside diameter dimension you selected in (a)?

**c** If an inside diameter can be reworked at a cost of $0.50 and one that is oversize must be scrapped at a cost of $2.00, where should we center our process?

**18.11** Select a product and identify a simple measurement on the product. Perform a gauge study similar to the ones in Examples 18.6 and 18.7. Analyze your results and comment on the ability of your measurement process to perform in an adequate fashion. How could your measurement process be improved? Explain.

# APPENDIX 18A: ADDITIONAL CAPABILITY INDICES

Capability indices are used to express the relationship between technical specifications and production abilities on the factory floor. This relationship is important to both suppliers and purchasers. The $C_p$ and $C_{pk}$ process capability indices as defined, discussed, and demonstrated in the text of Chapter 18 see widespread usage. Two other process capability indices have been developed which also link our technical specifications to our ability to produce (meet) them.

The $C_{pm}$ process capability index was independently proposed by Hsiang and Taguchi [12], as well as by Chan et al. [13] (see Rodriguez [14]). The $C_{pm}$ index is expressed as

$$C_{pm} = \frac{USL - LSL}{6\sqrt{(\mu - T)^2 + \sigma^2}} \qquad (18.25)$$

where USL = product USL.

LSL = product LSL.

$T$ = product target (e.g. the best value for the quality characteristic of interest).

$\mu$ = process mean.

$\sigma^2$ = process variance.

The $C_{pm}$ index measures the degree to which the process output is on target. In the special case where $T = \mu$ (by examining Equations 18.1 and 18.25), we can see that $C_{pm} = C_p$. In contrast, the $C_{pk}$, Equation 18.3, measures the degree to which the process output is within the specification limits. One major advantage of the $C_{pm}$ is that it is applicable to an asymmetrical specification interval, where the target $T$ is not in the middle of the interval. The $C_{pm}$ is compatible with the Taguchi loss function concepts [15].

Pearn et al. describe the $C_{pmk}$ as a "third-generation" process capability index (e.g., $C_p$ is the "first generation" and $C_{pk}$ is the "second generation") which is structured to include features of both $C_{pk}$ and $C_{pm}$ [16]. The $C_{pmk}$ index is defined as

$$C_{pmk} = \frac{C_{pk}}{\sqrt{1 + \left(\frac{\mu - T}{\sigma}\right)^2}} \tag{18.26}$$

where the components of the index are as described above in our discussion of the $C_{pm}$ index.

The $C_{pmk}$ index imposes a penalty when the process mean is not on target. For example, if $\mu = T$, then $C_{pmk} = C_{pk}$. But if $\mu \neq T$, then $C_{pmk} < C_{pk}$.

# REFERENCES

1  V. E. Kane, "Process Capability Indices," *Journal of Quality Technology,* vol. 18, no. 1, pp. 41–52, January, 1986.

2  L. K. Chan, S. W. Cheng, and F. A. Spiring, "A New Measure of Process Capability: $C_{pm}$," *Journal of Quality Technology,* vol. 20, no. 3, pp. 162–175, July 1988.

3  W. L. Pearn, S. Kotz, and M. L. Johnson, "Distributional and Inferential Properties of Process Capability Indices," *Journal of Quality Technology,* vol. 24, no. 4, pp. 216–231, October 1992.

4  R. H. Kushler and P. Hurley, "Confidence Bounds for Capability Indices," *Journal of Quality Technology,* vol. 24, no. 4, pp. 188–195, October 1992.

5  G. Taguchi, *Introduction to Quality Engineering: Designing Quality into Products and Processes,* White Plains, NY: Kraus International, UNIPUB (Asian Productivity Organization), 1986.

6  P. J. Ross, *Taguchi Techniques for Quality Engineering,* New York: McGraw-Hill, 1988.

7  See reference 5.

8  See reference 6.

9  R. E. Walpole and R. H. Myers, *Probability and Statistics for Engineers and Scientists,* 5th ed., New York: Macmillan, 1993.

10  E. L. Grant and R. S. Leavenworth, *Statistical Quality Control,* 6th ed., New York: McGraw-Hill, 1988.

11  D. C. Montgomery, *Introduction to Statistical Quality Control,* 2d ed., New York: Wiley, 1991.

12  T. C. Hsiang and G. Taguchi, "A Tutorial on Quality Control and Assurance—The Taguchi Methods," Unpublished presentation given at the Annual Meetings of the American Statistical Association, Las Vegas, Nevada, 1985.

13  See reference 2.

14  R. N. Rodriguez, "Recent Developments in Process Capability Analysis," *Journal of Quality Technology,* vol. 24, no. 4, pp. 176–187, October 1992.

15  See reference 5.

16  See reference 3.

# 19

# PRODUCT ACCEPTANCE AND ACCEPTANCE SAMPLING PLANS

## 19.0 INQUIRY

1 How do process and product focused strategies differ regarding zero defect products?
2 What is consumer's risk? What is producer's risk?
3 What measures are used to assess lot-by-lot sampling plan performance?
4 How are product acceptance sampling plans designed?

## 19.1 INTRODUCTION

Sooner or later we must either deliver our product to our customer, or we must rework or scrap our product. The issue here is our quality strategy. *We can proactively focus on process integrity throughout the production process and generate products that consistently conform to specifications. Or, conversely, we can relax our process discipline and focus on final inspection and prevent nonconforming product from entering the distribution channel. In either case, our customers receive a product that meets technical specifications.* At this point in our discussion, we should again note that a ZD product does meet all technical requirements; however, it may or may not satisfy our customers.

The proactive process focus, advocated throughout this book, emphasizes total quality improvement through education, training, employee empowerment, quality planning, mistake-proofing, capability analysis, SPC, reliability and maintainability deployment, off-line experimentation, and so on. We focus on the processes, which ultimately transform material into products, as well as product definition and design.

On the other hand, a strict, reactive product sorting focus emphasizes product inspection, testing, rework, returned goods adjustments, complaint response, warranty procedures, and so forth. Here, the vanguard methods revolve around inspection and product testing coupled with fast, effective return and rework channels to comfort dissatisfied customers.

*Inspection implies a product examination of some sort, usually passive, with the intent of declaring that the product meets or does not meet specifications.* For example, measuring the length of a meter stick constitutes a passive linear measurement. When this measurement is compared to specifications and a decision as to conformance or nonconformance is made, our inspection is complete.

Testing takes many forms, including burn-in testing and component and device screening. Testing is action oriented, unlike inspection. *Product testing before the product is released to the customer must be viewed as a reactive quality strategy which serves a sorting function.* Hopefully, inspection and test logs are kept, and defects investigated, with the aim of improving both processes and products. Hence, more will be accomplished than merely sorting the product. Nevertheless, *inspection and testing are reactive, after-the-fact stopgap activities.* The focus of this chapter is on designing efficient and effective sample based inspection plans.

*It is important to distinguish between sorting inspection and source inspection. Source inspection,* as discussed in Chapter 6, *is proactive* and directly related to mistake-proofing and process improvement; *sorting inspection is reactive.* In this chapter we are focusing on sorting inspection. Sample based inspection can be described using statistical principles. Census based inspection, by definition, examines all items, that is, 100 percent inspection. All inspection and testing practices involve the technical means to make an assessment. But, 100 percent inspection does not involve statistical inference from samples to populations, it is used to support the direct assessment or disposition of each individual item.

---

## Integrated Circuits—Inspection Case [RB]

Many companies in the United States study the Japanese business environment, and in particular, their quality systems. Sometimes however, it takes a direct business relationship to make the Japanese quality philosophy show through.

Such was the case with the Japanese *MD* program. Our company had a contract to deliver *MD* systems to the Japanese. Because of the success of this contract, a follow-on contract was negotiated whereby in-country maintenance and repair facilities would be constructed so the Japanese could maintain their own *MD* equipment. Because of the uniqueness of and cleanliness requirements for several of the piece parts, the Japanese had to purchase spare parts from our company. One such part was a specialized integrated circuit (IC). We would manufacture the IC and then screen for quality conformance using a sampling method where a small number of ICs were drawn from a production lot and tested. If the samples tested "good," the whole lot was deemed good. At the time, this method was perfectly consistent with our quality policy.

The Japanese customer would receive the shipment of ICs and test every single IC in the shipment, not just a sample. Using this method, they found failures periodically in incoming shipments and would return the complete shipment to us for replacement. We would then pick a sample from the returned shipment and test it. In most cases, all ICs in the sample would test good and the shipment would be returned to the Japanese. In one extreme case, the same shipment was sent back and forth three times, with the sender and receiver each themselves correct.

---

## 19.2  LOT-BY-LOT SAMPLING

*A 100 percent inspection or testing strategy examines each and every product unit.* Assuming a perfect inspection or test (i.e., without human or machine errors), an item is either defect-free or it is not. Here, the entire population is inspected or tested one unit at a time. Our goal is a perfect sorting operation, whereby we are left with two groups of product: product that conforms to specifications and product that does not.

*The alternative to 100 percent assessment is to divide our product into lots of a given number of items and then select a sample from each lot for inspection or test. Based on the sample determination, we either accept or reject the entire lot.* Accepted lots are shipped to our customers and rejected lots are 100 percent inspected, with nonconforming product removed and replaced with conforming product. Sampling plan design provides a statistical challenge in preventing the acceptance of bad lots and the rejection of good. We can develop statistical performance measures relative to sample plan design and the inspection process.

In order to develop meaningful lot-by-lot sampling plans, we will assume that we are dealing with a simple random sampling procedure. Simple random sampling is a method of selecting a sample of $n$ items from a lot of $N$ items such that each one of the $C_n^N$ possible samples is equally likely to be drawn (Cochran [1]).

*Two general forms of lot-by-lot inspection exist. The first, acceptance sampling by attributes, uses attributes data and an accept-reject classification.* It relies on the hypergeometric model for small lot sizes and isolated lots or infrequent lot purchases. The binomial model is used for large lots and more or less regular or frequent lot purchases. The $P$ chart and $P$ chart mechanics are the SPC counterparts to acceptance sampling by attributes; Hence, using attributes sampling plans, we can deal with more than one quality characteristic (e.g. length, width) simultaneously.

*The second form of lot-by-lot inspection sampling is done on a variables data or continuous measurement basis.* Here, the normal distribution is used as a model. The $X$-bar and $R$ or $S$ charts are the SPC counterparts to acceptance sampling by variables. As in the case of variables based SPC, we must restrict our variables sampling plans to a single measurable quality characteristic (i.e., length or width, but not both).

In this chapter we will discuss fundamentals of acceptance sampling by attributes. More advanced treatments of acceptance sampling by attributes as well as acceptance sampling by variables are developed in Duncan [2], Grant and Leavenworth [3], and Montgomery [4]. Military Standard 105 [5] and Military Standard 414 [6] also address attributes and variables based sampling plans, respectively.

## 19.3 ACCEPTANCE SAMPLING BY ATTRIBUTES

*Four basic attributes sampling plans exist: (1) single sampling, (2) double sampling, (3) multiple sampling, and (4) sequential sampling.* Figure 19.1 illustrates the decision mechanics involved in these basic lot-by-lot sampling plans. In single sampling, Figure 19.1a, we take only one sample from a lot and then make a determination as to acceptance or rejection of the entire lot. With double sampling (Figure 19.1b) we take a first sample and make a decision to accept, reject, or take a second sample. The second sample, if taken, always results in either acceptance or rejection of the entire lot. Multiple sampling simply extends the double sampling

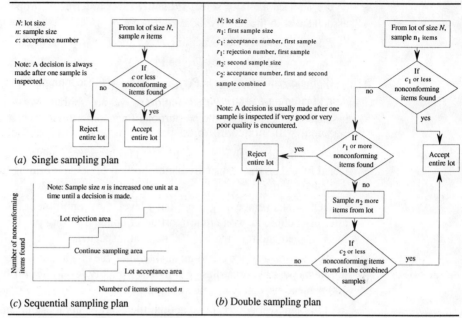

**FIGURE 19.1**    Single, double, and sequential sampling decision rules (attributes sampling plans).

concept beyond a two sample maximum. Sequential sampling (Figure 19.1c) works somewhat differently than the conventional fixed sample size concepts previously discussed. In sequential sampling, we sample one unit at a time until we encounter either enough nonconforming items to reject the lot or few enough nonconforming items to accept the lot.

## Single Sampling Plans

The development of sampling plans presents interesting statistical challenges. In order to develop single sampling plan mechanics, a number of symbols and terms must be defined.

### Symbols and Terms

$N$: lot size.

$n$: sample size.

$c$: single sampling plan acceptance number; if $c$ or fewer nonconforming items are found in a sample of size $n$, the lot is accepted, otherwise it is rejected.

$D$: the number of nonconforming items in a population or lot of size $N$.

$d$: the number of nonconforming items in a sample of size $n$.

$p$: the proportion of nonconforming items in a population.

$p_0$: an acceptable or selected quality level in terms of proportion nonconforming associated with the null hypothesis $H_0$: $p = p_0$.

$P_a$: the probability of accepting $H_0$ in a test of hypothesis context.

$\alpha$: the producer's risk = $P$(type I error) = $P$(rejecting a lot | the lot is "good").

$\beta$: the consumer's risk = $P$(type II error) = $P$(accepting a lot | the lot is "bad").

LTPD: the lot tolerance percent defective, an arbitrary worst quality level selected by the consumer, usually the proportion of nonconforming items in a population associated with $\beta$.

AQL: the acceptable quality level, an arbitrary acceptable or good quality level selected by the producer, usually the proportion of nonconforming items in a population associated with $\alpha$.

AOQ: the average outgoing quality level, the proportion of nonconforming product after inspection, an average proportion of nonconforming product, which includes all lots accepted and rejected but 100 percent inspected.

AOQL: the average outgoing quality limit, the maximum proportion of nonconforming items possible regardless of incoming quality level after lot-by-lot inspection and 100 percent inspection of rejected lots.

ATI: the average total inspection, the average number of items inspected per lot, considering both accepted and rejected lots.

AFI: the average fraction of a lot inspected, considering both accepted and rejected lots.

## OC Curves

*Operating characteristic (OC) curves are helpful in assessing a given sampling plan's statistical performance.* In this context, we will discuss two general types of OC curves: type A and type B. Type A OC curves pertain to isolated or small lots and are modeled with the hypergeometric distribution. Type B OC curves, which pertain to large lots and on-going product acquisition, follow binomial models. The construction of Type B OC curves is discussed in Appendix 19A. We refer our readers to Duncan for a detailed discussion of Type A OC curves [7].

A typical type B single sampling by attributes OC curve is displayed in Figure 19.2*a*. We can see that as the quality level erodes (i.e., $p$ increases) we see smaller $P_a$'s. When $p$ is located to the right of the AQL point $p_0$, the decrease in $P_a$ can be interpreted as a decrease in the probability of accepting lots of poorer quality. When $p$ is located to the left of the AQL point $p_0$, we interpret the $P_a$ to represent the probability of accepting better than expected quality. We notice that the $P_a$ approaches 1.0 as $p$ approaches 0. Hence, for zero defect process quality, lot acceptance is certain.

The small OC curve in Figure 19.2*b* illustrates the perfect OC curve for a given quality level $p_0$ or AQL. Here, we will accept the lot as long as $p \leq p_0$. However, if $p > p_0$ we will reject the lot. This hypothetical, or perfect, OC curve can be ob-

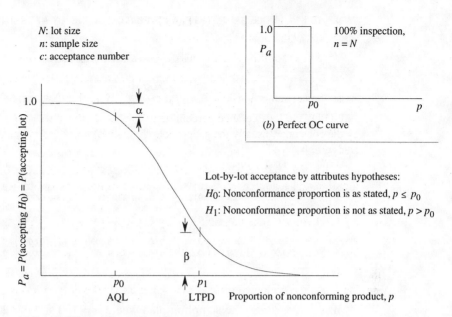

(a) Type B OC curve for $n < N$, attributes sampling

**FIGURE 19.2**    Generic type B OC curve, single sampling plan, by attributes.

tained only by perfect 100 percent inspection, i.e., when $n = N$.  Hence, whenever $n < N,$ we face OC curves such as the one shown in Figure 19.2a.  We must find a sampling plan ($n$ and $c$) that will yield $P_a$'s that are acceptable to both producer and consumer.  ***We want to minimize the probability of rejecting good quality product (low p) and maximize the probability of rejecting bad quality product (high p).***

## Single Sampling Plan Performance

***Given a single sampling plan, an AQL, and an LTPD, we can determine the producer's risk α associated with the AQL and the consumer's risk β associated with the LTPD numerically or from an OC curve.***  Using the proper OC curve for the sampling plan, we can read α and β directly from the graph.  On the other hand, we may decided to use a numerical approach.  We can develop $P_a$ using a Poisson approximation to the binomial (type B OC curve), as long as $np \le 10$.  Otherwise, we may want to use a normal approximation to the binomial to enhance the accuracy of the calculation.  Of course, alternatively, we could use an extensive binomial table or computer aid for the cumulative binomial.  Table 19.1 displays a set of $P_a$'s for a sample size $n = 100$ and acceptance numbers $c = 0$, $c = 2$, and $c = 5$.  Figure 19.3 shows a plot of the corresponding OC curves.

The Poisson table shown in Section IX, Table IX.6, was used to develop the $P_a$'s shown in Table 19.1.  We arbitrarily chose the $p$ values, developed the $np$ values (used as the Poisson mean), and then calculated $P_a = P(d \le c)$ using Table IX.6.

**TABLE 19.1**    SINGLE SAMPLING PLAN TYPE B OC CURVE, BY ATTRIBUTES

| $N$ 2000 | $n$ 100 | Poisson approximation to the binomial | | | | | | |
|---|---|---|---|---|---|---|---|---|
| Proportion nonconforming $p$ | Poisson mean $np$ | $P_a$ $P(d \le c)$ $c = 2$ | AOQ $P_a p$ $c = 2$ | AOQ $P_a p(N - n)/N$ $c = 2$ | ATI $c = 2$ | AFI $c = 2$ | $P_a$ $P(d \le c)$ $c = 0$ | $P_a$ $P(d \le c)$ $c = 5$ |
| 0.000 | 0.000 | 1.000 | 0.0000 | 0.0000 | 100.000 | 0.050 | 1.000 | 1.000 |
| 0.002 | 0.200 | 0.999 | 0.0020 | 0.0019 | 101.900 | 0.051 | 0.819 | 1.000 |
| 0.005 | 0.500 | 0.986 | 0.0049 | 0.0047 | 126.600 | 0.063 | 0.607 | 1.000 |
| 0.010 | 1.000 | 0.920 | 0.0092 | 0.0087 | 252.000 | 0.126 | 0.368 | 0.999 |
| 0.015 | 1.500 | 0.809 | 0.0121 | 0.0115 | 462.900 | 0.231 | 0.223 | 0.996 |
| 0.020 | 2.000 | 0.677 | 0.0135 | 0.0129 | 713.700 | 0.357 | 0.135 | 0.983 |
| 0.030 | 3.000 | 0.423 | 0.0127 | 0.0121 | 1196.300 | 0.598 | 0.050 | 0.916 |
| 0.040 | 4.000 | 0.238 | 0.0095 | 0.0090 | 1547.800 | 0.774 | 0.018 | 0.785 |
| 0.050 | 5.000 | 0.125 | 0.0063 | 0.0059 | 1762.500 | 0.881 | 0.007 | 0.616 |
| 0.060 | 6.000 | 0.062 | 0.0037 | 0.0035 | 1882.200 | 0.941 | 0.002 | 0.446 |
| 0.080 | 8.000 | 0.014 | 0.0011 | 0.0011 | 1973.400 | 0.987 | 0.000 | 0.191 |
| 0.100 | 10.000 | 0.003 | 0.0003 | 0.0003 | 1994.300 | 0.997 | 0.000 | 0.067 |

If we were to use the sampling plan where $n = 100$ and $c = 2$, and if AQL = 1 percent nonconforming product and LTPD = 5 percent nonconforming product, we could easily determine $\alpha$ and $\beta$. The producer's risk $\alpha$ can be read indirectly from Table 19.1 (the $p = 0.01$ row and the $P_a$ column intersection) or estimated from Figure 19.3 as $\alpha = 1 - 0.920 = 8$ percent. Hence, we would expect about 8 percent

**FIGURE 19.3**    Type B OC curve, single sampling plan, by attributes.

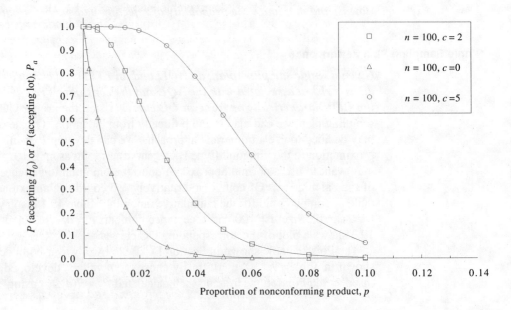

of lots containing 1 percent nonconforming product to be rejected or 92 percent of the same lots to be accepted. The consumer's risk $\beta$ can be determined in a similar manner from Table 19.1 (the $p = 0.05$ row and the $P_a$ column intersection) or from the OC curve in Figure 19.3 as $\beta = 0.125$. We interpret the consumer's risk here as a 12.5 percent chance of accepting lots with 5 percent nonconforming product. Since the AQL and LTPD are selected based on producer and consumer needs and expectations, we could just as well have selected other values, used the same procedure, and obtained the appropriate risk level estimates.

*The AOQ measures the proportion of nonconforming or defective product our customers see.* The AOQ at a given $p$ is developed using OC curve information. Here, we are assuming that the customer receives lots continuously over time. The lots are inspected using a single-sampling plan $(n, c)$. Any nonconforming items found in the sample are replaced with conforming ones. If the lot is accepted, it is shipped to our customer. If the lot is rejected, we 100 percent inspect the remaining units $(N - n)$ and replace all nonconforming units. In this case, our AOQ can be calculated as

$$AOQ = \frac{P_a\, p\, (N - n)}{N} \qquad (19.1)$$

For large lot sizes where we use small sample sizes,

$$AOQ \simeq P_a\, p \qquad (19.2)$$

If we had a choice (at the same price), we would always welcome ZD lots. *There are two ways we can obtain ZD lots. First*, zero defect production processes ($p = 0$ or very near 0) would yield acceptable lots ($P_a \simeq 100\%$). Hence, *a very good AOQ (AOQ = 0) can be obtained directly through ZD production processes. Second, when dealing with very poor production processes* (high $p$, but $p < 1.0$), *a sampling plan can be selected such that it forces rejection of most or all of the lots, hence subjecting the lots to 100 percent inspection, and eventual replacement of the nonconforming items.* This alternative also yields a very good AOQ (low value), but at an economic penalty through inspection, rework, and scrap costs which are ultimately passed on to our customers. Since we can obtain low AOQs in two ways, very good process quality or very bad process quality (with 100 percent inspection or lot by lot acceptance sampling), it is informative to plot our AOQ versus $p$.

This plot yields an interesting AOQ perspective. As an example, we can plot the AOQ versus the $p$ columns in Table 19.1 for a sampling plan with $N = 2000$, $n = 100$, $c = 2$, and obtain the AOQ curve shown in Figure 19.4. *The maximum point on the AOQ curve is called the "average outgoing quality limit" (AOQL). The AOQL is the worst-case AOQ our customer will experience, regardless of our quality level.* For the above sampling plan, our worst-case AOQ would be in the neighborhood of AOQL $\simeq 0.013$, and occurs when $p \simeq 0.02$ (Table 19.1 or Figure 19.4).

Once again, *we can clearly see our process-focused and product-focused quality strategies in action at the far left-hand and right-hand areas of the AOQ*

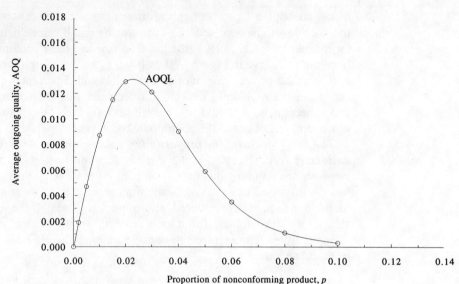

**FIGURE 19.4**    AOQ curve, $N = 2{,}000$, $n = 100$, $c = 2$.

*curve, respectively.* We can ship our customer conforming product (low AOQ) by consistently producing conforming product or by sorting a mixed product produced with relatively sloppy processes. The latter strategy has given way to the former because of economic pressures and our inability to sort product effectively (machine or vision inspection is to some degree an exception).

Two additional inspection performance measures are commonly used to gauge the effectiveness of a lot-by-lot inspection plan. *The average total inspection (ATI) measures the average number of items per lot that are inspected, and the average fraction inspected (AFI) measures the proportion of a lot that is inspected, using the sampling plan:*

$$\text{ATI} = n + (1 - P_a)(N - n) \tag{19.3}$$

$$\text{AFI} = \frac{\text{ATI}}{N} \tag{19.4}$$

We can compare our AOQL with the two extremes; the AOQ $\approx 0.0$ when we have zero defect processes or when we use 100% (perfect) inspection with a high proportion defective. The objective, of course, is to deliver a low AOQ product on time to our customer, and at a competitive cost. Today, *for most products, our preference is to aim for ZD or six-sigma products through a process focus.* However, *we may need to rely on inspection as a stopgap quality control tool until we can master our production processes or incorporate more effective technology.*

## 19.4  PUBLISHED SAMPLING PLANS

Over the years, *a number of sampling plan layouts have been published by both private and governmental organizations.* Duncan [8] and Grant and Leavenworth [9] provide detailed discussions of both attributes and variables sampling plans. For our purposes, we will discuss only two attributes plans: Dodge-Romig and Military Standard 105 (Mil-Std-105).

### Dodge-Romig Plans

*The Dodge-Romig tables focus on minimum ATI under worst-case operations* [10]. They include both single and double sampling plans. Two basic sets of tables were developed: LTPD based tables and AOQL based tables. *The LTPD tables, with minimum ATI, focus on a worst case, considering the probability of accepting bad quality product* (the lower right-hand region of the OC curve). Here, we can select from the Set I—Single Sampling and Set II—Double Sampling tables. These tables are produced for $P_a = \beta = 0.10$ with different LTPD levels (0.5%, 1%, 2%, 3%, 4%, 5%, 7%, and 10%). Table 19.2 is a reproduction of the LTPD = 5%, ß = 0.10 single sampling plan from Set I.

*The AOQL based tables, with minimum ATI focus on a worst-case AOQ shipped to our customer,* i.e., the maximum AOQ level product which typically falls somewhere in the OC curve midrange. The AOQL tables are termed Set III - Single Sampling and Set IV - Double Sampling. These tables are developed for AOQLs 0.1%, 0.25%, 0.5%, 0.75%, 1.0%, 1.5%, 2.0%, 2.5%, 3.0%, 4.0%, 5.0%, 7.0%, and 10.0%. The Set III, AOQL = 2%, single sampling table is reproduced and shown as Table 19.3.

---

### Example 19.1

Determine a single sampling plan with minimum ATI for lots with $N = 500$, given that for an LTPD = 5% the consumer's risk ß should be equal to 0.10, and the process is operating with a nonconformance process average $p = 0.8\%$. Also, develop an OC curve table similar to Table 19.1 for the sampling plan that includes $P_a$, AOQ, ATI, and AFI calculations. Be sure to include $p = 0.008$ in your table. Compare your tabular AOQL with the published AOQL in Table 19.2.

#### Solution

From Table 19.2, the appropriate sampling plan has $n = 100$, $c = 2$, and AOQL = 1.1. Table 19.4 includes $p$, $P_a$, AOQ, ATI, and AFI calculations. Our maximum AOQ = 0.0108 in Table 19.4 represents an approximate AOQL at $p = 2.0\%$, versus the published AOQL = 1.1%.

---

**TABLE 19.2**    EXAMPLE OF DODGE-ROMIG SINGLE SAMPLING LOT TOLERANCE TABLES
(Lot tolerance percent defective = 5.0%; consumer's risk = 0.10)

| | Process average, % | | | | | | | | | | | | | | | | | |
|---|---|---|---|---|---|---|---|---|---|---|---|---|---|---|---|---|---|---|
| | 0–0.05 | | | 0.06–0.50 | | | 0.51–1.00 | | | 1.01–1.50 | | | 1.51–2.00 | | | 2.01–2.50 | | |
| | | | AOQL | | | AOQL | | | AOQL | | | AOQL | | | AOQL | | | AOQL |
| Lot size | n | c | % | n | c | % | n | c | % | n | c | % | n | c | % | n | c | % |
| 1–30 | All | 0 | 0 | All | 0 | 0 | All | 0 | 0 | All | 0 | 0 | All | 0 | 0 | All | 0 | 0 |
| 31–50 | 30 | 0 | 0.49 | 30 | 0 | 0.49 | 30 | 0 | 0.49 | 30 | 0 | 0.49 | 30 | 0 | 0.49 | 30 | 0 | 0.49 |
| 51–100 | 37 | 0 | 0.63 | 37 | 0 | 0.63 | 37 | 0 | 0.63 | 37 | 0 | 0.63 | 37 | 0 | 0.63 | 37 | 0 | 0.63 |
| 101–200 | 40 | 0 | 0.74 | 40 | 0 | 0.74 | 40 | 0 | 0.74 | 40 | 0 | 0.74 | 40 | 0 | 0.74 | 40 | 0 | 0.74 |
| 201–300 | 43 | 0 | 0.74 | 43 | 0 | 0.74 | 70 | 1 | 0.92 | 70 | 1 | 0.92 | 95 | 2 | 0.99 | 95 | 2 | 0.99 |
| 301–400 | 44 | 0 | 0.74 | 44 | 0 | 0.74 | 70 | 1 | 0.99 | 100 | 2 | 1.0 | 120 | 3 | 1.1 | 145 | 4 | 1.1 |
| 401–500 | 45 | 0 | 0.75 | 75 | 1 | 0.95 | 100 | 2 | 1.1 | 100 | 2 | 1.1 | 125 | 3 | 1.2 | 150 | 4 | 1.2 |
| 501–600 | 45 | 0 | 0.76 | 75 | 1 | 0.98 | 100 | 2 | 1.1 | 125 | 3 | 1.2 | 150 | 4 | 1.3 | 175 | 5 | 1.3 |
| 601–800 | 45 | 0 | 0.77 | 75 | 1 | 1.0 | 100 | 2 | 1.2 | 130 | 3 | 1.2 | 175 | 5 | 1.4 | 200 | 6 | 1.4 |
| 801–1000 | 45 | 0 | 0.78 | 75 | 1 | 1.0 | 105 | 2 | 1.2 | 155 | 4 | 1.4 | 180 | 5 | 1.4 | 225 | 7 | 1.5 |
| 1001–2000 | 45 | 0 | 0.80 | 75 | 1 | 1.0 | 130 | 3 | 1.4 | 180 | 5 | 1.6 | 230 | 7 | 1.7 | 280 | 9 | 1.8 |
| 2001–3000 | 75 | 1 | 1.1 | 105 | 2 | 1.3 | 135 | 3 | 1.4 | 210 | 6 | 1.7 | 280 | 9 | 1.9 | 370 | 13 | 2.1 |
| 3001–4000 | 75 | 1 | 1.1 | 105 | 2 | 1.3 | 160 | 4 | 1.5 | 210 | 6 | 1.7 | 305 | 10 | 2.0 | 420 | 15 | 2.2 |
| 4001–5000 | 75 | 1 | 1.1 | 105 | 2 | 1.3 | 160 | 4 | 1.5 | 235 | 7 | 1.8 | 330 | 11 | 2.0 | 440 | 16 | 2.2 |
| 5001–7000 | 75 | 1 | 1.1 | 105 | 2 | 1.3 | 185 | 5 | 1.7 | 260 | 8 | 1.9 | 350 | 12 | 2.2 | 490 | 18 | 2.4 |
| 7001–10,000 | 75 | 1 | 1.1 | 105 | 2 | 1.3 | 185 | 5 | 1.7 | 260 | 8 | 1.9 | 380 | 13 | 2.2 | 535 | 20 | 2.5 |
| 10,001–20,000 | 75 | 1 | 1.1 | 135 | 3 | 1.4 | 210 | 6 | 1.8 | 285 | 9 | 2.0 | 425 | 15 | 2.3 | 610 | 23 | 2.6 |
| 20,001–50,000 | 75 | 1 | 1.1 | 135 | 3 | 1.4 | 235 | 7 | 1.9 | 305 | 10 | 2.1 | 470 | 17 | 2.4 | 700 | 27 | 2.7 |
| 50,001–100,000 | 75 | 1 | 1.1 | 160 | 4 | 1.6 | 235 | 7 | 1.9 | 355 | 12 | 2.2 | 515 | 19 | 2.5 | 770 | 30 | 2.8 |

Source: Reprinted by permission from H. F. Dodge and H. G. Romig, *Sampling Inspection Tables—Single and Double Sampling*, 2d ed., New York, NY: Wiley, page 184, 1959.

**TABLE 19.3**    EXAMPLE OF DODGE-ROMIG SINGLE SAMPLING AOQL TABLES
(Average outgoing quality limit = 2.0%)

| | Process average, % | | | | | | | | | | | | | | | | | |
|---|---|---|---|---|---|---|---|---|---|---|---|---|---|---|---|---|---|---|
| | 0–0.04 | | | 0.05–0.40 | | | 0.41–0.80 | | | 0.81–1.20 | | | 1.21–1.60 | | | 1.61–2.00 | | |
| Lot size | n | c | $100p_{0.10}$ | n | c | $100p_{0.10}$ | n | c | $100p_{0.10}$ | n | c | $100p_{0.10}$ | n | c | $100p_{0.10}$ | n | c | $100p_{0.10}$ |
| 1–15 | All | 0 | — | All | 0 | — | All | 0 | — | All | 0 | — | All | 0 | — | All | 0 | — |
| 16–50 | 14 | 0 | 13.6 | 14 | 0 | 13.6 | 14 | 0 | 13.6 | 14 | 0 | 13.6 | 14 | 0 | 13.6 | 14 | 0 | 13.6 |
| 51–100 | 16 | 0 | 12.4 | 16 | 0 | 12.4 | 16 | 0 | 12.4 | 16 | 0 | 12.4 | 16 | 0 | 12.4 | 16 | 0 | 12.4 |
| 101–200 | 17 | 0 | 12.2 | 17 | 0 | 12.2 | 17 | 0 | 12.2 | 17 | 0 | 12.2 | 35 | 1 | 10.5 | 35 | 1 | 10.5 |
| 201–300 | 17 | 0 | 12.3 | 17 | 0 | 12.3 | 17 | 0 | 12.3 | 37 | 1 | 10.2 | 37 | 1 | 10.2 | 37 | 1 | 10.2 |
| 301–400 | 18 | 0 | 11.8 | 18 | 0 | 11.8 | 38 | 1 | 10.0 | 38 | 1 | 10.0 | 38 | 1 | 10.0 | 60 | 2 | 8.5 |
| 401–500 | 18 | 0 | 11.9 | 18 | 0 | 11.9 | 39 | 1 | 9.8 | 39 | 1 | 9.8 | 60 | 2 | 8.6 | 60 | 2 | 8.6 |
| 501–600 | 18 | 0 | 11.9 | 18 | 0 | 11.9 | 39 | 1 | 9.8 | 39 | 1 | 9.8 | 60 | 2 | 8.6 | 60 | 2 | 8.6 |
| 601–800 | 18 | 0 | 11.9 | 40 | 1 | 9.6 | 40 | 1 | 9.6 | 65 | 2 | 8.0 | 65 | 2 | 8.0 | 85 | 3 | 7.5 |
| 801–1000 | 18 | 0 | 12.0 | 40 | 1 | 9.6 | 40 | 1 | 9.6 | 65 | 2 | 8.1 | 65 | 2 | 8.1 | 90 | 3 | 7.4 |
| 1001–2000 | 18 | 0 | 12.0 | 41 | 1 | 9.4 | 65 | 2 | 8.2 | 65 | 2 | 8.2 | 95 | 3 | 7.0 | 120 | 4 | 6.5 |
| 2001–3000 | 18 | 0 | 12.0 | 41 | 1 | 9.4 | 65 | 2 | 8.2 | 95 | 3 | 7.0 | 120 | 4 | 6.5 | 180 | 6 | 5.8 |
| 3001–4000 | 18 | 0 | 12.0 | 42 | 1 | 9.3 | 65 | 2 | 8.2 | 95 | 3 | 7.0 | 155 | 5 | 6.0 | 210 | 7 | 5.5 |
| 4001–5000 | 18 | 0 | 12.0 | 42 | 1 | 9.3 | 70 | 2 | 7.5 | 125 | 4 | 6.4 | 155 | 5 | 6.0 | 245 | 8 | 5.3 |
| 5001–7000 | 18 | 0 | 12.0 | 42 | 1 | 9.3 | 95 | 3 | 7.0 | 125 | 4 | 6.4 | 185 | 6 | 5.6 | 280 | 9 | 5.1 |
| 7001–10,000 | 42 | 1 | 9.3 | 70 | 2 | 7.5 | 95 | 3 | 7.0 | 155 | 5 | 6.0 | 220 | 7 | 5.4 | 350 | 11 | 4.8 |
| 10,001–20,000 | 42 | 1 | 9.3 | 70 | 2 | 7.6 | 95 | 3 | 7.0 | 190 | 6 | 5.6 | 290 | 9 | 4.9 | 460 | 14 | 4.4 |
| 20,001–50,000 | 42 | 1 | 9.3 | 70 | 2 | 7.6 | 125 | 4 | 6.4 | 220 | 7 | 5.4 | 395 | 12 | 4.5 | 720 | 21 | 3.9 |
| 50,001–100,000 | 42 | 1 | 9.3 | 95 | 3 | 7.0 | 160 | 5 | 5.9 | 290 | 9 | 4.9 | 505 | 15 | 4.2 | 955 | 27 | 3.7 |

Source: Reprinted by permission from H. F. Dodge and H. G. Romig, *Sampling Inspection Tables—Single and Double Sampling*, 2d ed., New York, NY: Wiley, page 201, 1959.

**TABLE 19.4** TYPE B OC CURVE AND SAMPLE PERFORMANCE STATISTICS (EXAMPLE 19.1)

| N | n | c | Poisson approximation | |
|---|---|---|---|---|
| 500 | 100 | 2 | to the binomial | |

| Proportion nonconforming $p$ | Poisson mean $np$ | $P_a$ $P(d \leq c)$ | AOQ $P_a p (N-n)/N$ | ATI | AFI |
|---|---|---|---|---|---|
| 0.000 | 0.000 | 1.000 | 0.0000 | 100.000 | 0.200 |
| 0.002 | 0.200 | 0.999 | 0.0016 | 100.400 | 0.201 |
| 0.005 | 0.500 | 0.986 | 0.0039 | 105.600 | 0.211 |
| 0.008 | 0.800 | 0.953 | 0.0061 | 118.800 | 0.238 |
| 0.010 | 1.000 | 0.920 | 0.0074 | 132.000 | 0.264 |
| 0.015 | 1.500 | 0.809 | 0.0097 | 176.400 | 0.353 |
| 0.020 | 2.000 | 0.677 | 0.0108 | 229.200 | 0.458 |
| 0.030 | 3.000 | 0.423 | 0.0102 | 330.800 | 0.662 |
| 0.040 | 4.000 | 0.238 | 0.0076 | 404.800 | 0.810 |
| 0.050 | 5.000 | 0.125 | 0.0050 | 450.000 | 0.900 |
| 0.060 | 6.000 | 0.062 | 0.0030 | 475.200 | 0.950 |
| 0.080 | 8.000 | 0.014 | 0.0009 | 494.400 | 0.989 |
| 0.100 | 10.000 | 0.003 | 0.0002 | 498.800 | 0.998 |

## Example 19.2

Determine a single sampling plan for an AOQL = 2%, which minimizes ATI, for a lot size $N = 1500$ and a nonconformance process average $p = 2\%$. Develop an OC curve and estimate the LTPD or $p$ associated with consumer's risk $\beta = 0.10$. Then, compare this LTPD with the $100_{p0.10}$ listed in Table 19.3.

### Solution

From Table 19.3, the appropriate sampling plan has $n = 120$, $c = 4$, $100_{p0.10} = 6.5$. Table 19.5 contains the information we developed regarding the sampling plan. The OC curve is shown in Figure 19.5. From Figure 19.5, working with $P_a = 0.10$ back to the OC curve, we can read $p = 6.7\%$, which agrees with the published figure of 6.5% reasonably well. Figure 19.6 contains plots of the AOQ and the AFI for the sampling plan. We can see that AOQL is near 2 percent from Figure 19.6.

## Military Standard 105

*Military Standard 105 (Mil-Std-105) describes an entire sampling scheme.* Mil-Std-105 consists of normal, tightened, and reduced inspection plans linked by a set of rules for switching between the inspection protocols [11]. *This attributes based scheme* (Figure 19.7) *is AQL focused.* It is clear that Mil-Std-105 is designed to encourage high quality through movement towards reduced inspection and to discourage sloppy production through tightened inspection and production shutdown.

**TABLE 19.5**   TYPE B OC CURVE AND SAMPLE PERFORMANCE
STATISTICS (EXAMPLE 19.2)

| N | n | c | Poisson approximation |
|---|---|---|---|
| 1500 | 120 | 4 | to the binomial |

| Proportion nonconforming $p$ | Poisson mean $np$ | $P_a$ $P(d \le c)$ | AOQ $P_a p(N-n)/N$ | ATI | AFI |
|---|---|---|---|---|---|
| 0.000 | 0.000 | 1.000 | 0.0000 | 120.000 | 0.080 |
| 0.002 | 0.240 | 1.000 | 0.0018 | 120.000 | 0.080 |
| 0.005 | 0.600 | 1.000 | 0.0046 | 120.000 | 0.080 |
| 0.010 | 1.200 | 0.992 | 0.0091 | 131.040 | 0.087 |
| 0.015 | 1.800 | 0.964 | 0.0133 | 169.680 | 0.113 |
| 0.020 | 2.400 | 0.904 | 0.0166 | 252.480 | 0.168 |
| 0.030 | 3.600 | 0.706 | 0.0195 | 525.720 | 0.350 |
| 0.040 | 4.800 | 0.476 | 0.0175 | 843.120 | 0.562 |
| 0.050 | 6.000 | 0.285 | 0.0131 | 1106.700 | 0.738 |
| 0.060 | 7.200 | 0.156 | 0.0086 | 1284.720 | 0.856 |
| 0.080 | 9.600 | 0.038 | 0.0028 | 1447.560 | 0.965 |
| 0.100 | 12.000 | 0.008 | 0.0007 | 1488.960 | 0.993 |

**FIGURE 19.5**   Type B OC curve (Example 19.2).

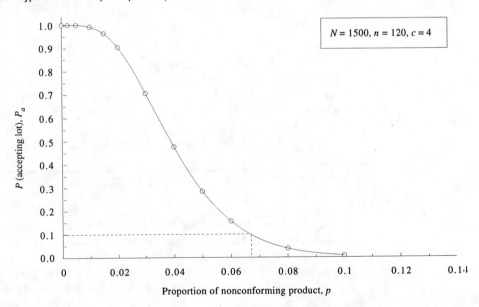

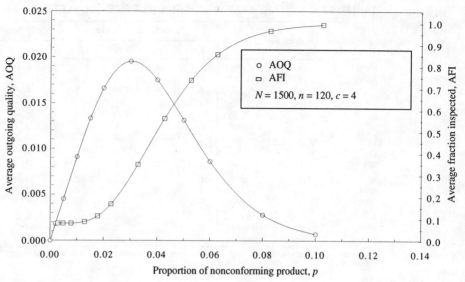

**FIGURE 19.6**    AOQ and AFI curves (Example 19.2).

**FIGURE 19.7**    Simplified Military Standard 105 sampling scheme flowchart.

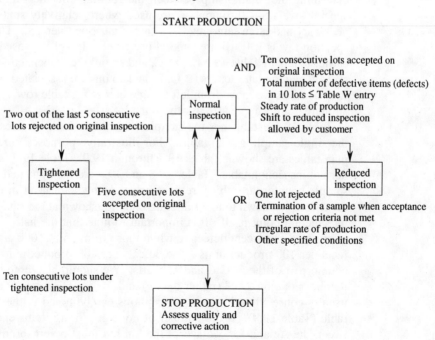

**TABLE 19.6** MILITARY STANDARD 105 SAMPLE SIZE CODE LETTERS

| Lot or batch size | | | Special inspection levels | | | | General inspection levels | | |
|---|---|---|---|---|---|---|---|---|---|
| | | | S-1 | S-2 | S-3 | S-4 | I | II | III |
| 2 | to | 8 | A | A | A | A | A | A | B |
| 9 | to | 15 | A | A | A | A | A | B | C |
| 16 | to | 25 | A | A | B | B | B | C | D |
| 26 | to | 50 | A | B | B | C | C | D | E |
| 51 | to | 90 | B | B | C | C | C | E | F |
| 91 | to | 150 | B | B | C | D | D | F | G |
| 151 | to | 280 | B | C | D | E | E | G | H |
| 281 | to | 500 | B | C | D | E | F | H | J |
| 501 | to | 1200 | C | C | E | F | G | J | K |
| 1201 | to | 3200 | C | D | E | G | H | K | L |
| 3201 | to | 10000 | C | D | F | G | J | L | M |
| 10001 | to | 35000 | C | D | F | H | K | M | N |
| 35001 | to | 150000 | D | E | G | J | L | N | P |
| 150001 | to | 500000 | D | E | G | J | M | P | Q |
| 500001 | and | over | D | E | H | K | N | Q | R |

*Source:* Reproduced from "Sampling Procedures and Tables for Inspection by Attributes, Military Standard 105D," U.S. Department of Defense, Washington, DC: U.S. Government Printing Office, 1963.

***Mil-Std-105 includes single, double, and multiple sampling plans.*** Once we have determined and chosen the most appropriate lot size and inspection level, we can determine the sample size code using Table 19.6. Inspection levels are used to determine the relationship between the lot size and the sample size. The four special levels S-1 through S-4 are used where relatively small sample sizes are necessary and large sampling risks can or must be tolerated. The three general inspection levels I, II, III are general use levels. Level II is used as a default. The earlier letters in Table 19.6 (A, B, C, and so on) are associated with smaller sample sizes, while the later letters (P, Q, R, and so on) are associated with larger samples. In general, as we move from left to right across the table rows, the letters represent an increase in sample size, for the same lot size.

Table 19.6 establishes the sample size code which is necessary to use any of the single sampling plan tables. For illustrative purposes, three single sampling plan tables are shown, Tables 19.7 through 19.9. Table 19.7 is the single sample normal inspection table. Tables 19.8 and 19.9 are the tightened and reduced single sample counterpart tables, respectively. Arrows are used in the tables to direct the user up or down to any sampling plan not shown at the sample size code–AQL intersection point itself. It is important to note that the listed AQLs up to 10 are interpreted as percent defective while those from 15 to 1000 are interpreted as defects per 100 product units. The defects per 100 product units are the inspection counterpart of the SPC $C$ and $U$ charts. We should also note that the acceptance number notation "Ac" (rather than $c$) and the rejection number notation "Re" are usually consecutive integers. Exceptions can be noted in the reduced inspection table (Table 19.9). If the number of nonconforming items encountered falls between the Ac and Re, we accept the lot, but must revert to normal inspection, as depicted in Figure 19.7.

**TABLE 19.7**  MILITARY STANDARD 105 SINGLE SAMPLING PLANS FOR NORMAL INSPECTION (MASTER TABLE)

Each data cell gives the acceptance number (Ac) and rejection number (Re) for the stated acceptable quality level (normal inspection). ↓ = use first sampling plan below arrow; ↑ = use first sampling plan above arrow.

| Sample size code letter | Sample size | 0.010 | 0.015 | 0.025 | 0.040 | 0.065 | 0.10 | 0.15 | 0.25 | 0.40 | 0.65 | 1.0 | 1.5 | 2.5 | 4.0 | 6.5 | 10 | 15 | 25 | 40 | 65 | 100 | 150 | 250 | 400 | 650 | 1000 |
|---|---|---|---|---|---|---|---|---|---|---|---|---|---|---|---|---|---|---|---|---|---|---|---|---|---|---|---|
| | | Ac Re | Ac Re | Ac Re | Ac Re | Ac Re | Ac Re | Ac Re | Ac Re | Ac Re | Ac Re | Ac Re | Ac Re | Ac Re | Ac Re | Ac Re | Ac Re | Ac Re | Ac Re | Ac Re | Ac Re | Ac Re | Ac Re | Ac Re | Ac Re | Ac Re | Ac Re |
| A | 2 | ↓ | ↓ | ↓ | ↓ | ↓ | ↓ | ↓ | ↓ | ↓ | ↓ | ↓ | ↓ | ↓ | ↓ | 0 1 | ↑ | 1 2 | 2 3 | 3 4 | 5 6 | 7 8 | 10 11 | 14 15 | 21 22 | 30 31 | 44 45 |
| B | 3 | ↓ | ↓ | ↓ | ↓ | ↓ | ↓ | ↓ | ↓ | ↓ | ↓ | ↓ | ↓ | ↓ | 0 1 | ↑ | 1 2 | 2 3 | 3 4 | 5 6 | 7 8 | 10 11 | 14 15 | 21 22 | 30 31 | 44 45 | ↑ |
| C | 5 | ↓ | ↓ | ↓ | ↓ | ↓ | ↓ | ↓ | ↓ | ↓ | ↓ | ↓ | ↓ | 0 1 | ↑ | 1 2 | 2 3 | 3 4 | 5 6 | 7 8 | 10 11 | 14 15 | 21 22 | 30 31 | 44 45 | ↑ | ↑ |
| D | 8 | ↓ | ↓ | ↓ | ↓ | ↓ | ↓ | ↓ | ↓ | ↓ | ↓ | ↓ | 0 1 | ↑ | 1 2 | 2 3 | 3 4 | 5 6 | 7 8 | 10 11 | 14 15 | 21 22 | 30 31 | 44 45 | ↑ | ↑ | ↑ |
| E | 13 | ↓ | ↓ | ↓ | ↓ | ↓ | ↓ | ↓ | ↓ | ↓ | ↓ | 0 1 | ↑ | 1 2 | 2 3 | 3 4 | 5 6 | 7 8 | 10 11 | 14 15 | 21 22 | 30 31 | 44 45 | ↑ | ↑ | ↑ | ↑ |
| F | 20 | ↓ | ↓ | ↓ | ↓ | ↓ | ↓ | ↓ | ↓ | ↓ | 0 1 | ↑ | 1 2 | 2 3 | 3 4 | 5 6 | 7 8 | 10 11 | 14 15 | 21 22 | 30 31 | 44 45 | ↑ | ↑ | ↑ | ↑ | ↑ |
| G | 32 | ↓ | ↓ | ↓ | ↓ | ↓ | ↓ | ↓ | ↓ | 0 1 | ↑ | 1 2 | 2 3 | 3 4 | 5 6 | 7 8 | 10 11 | 14 15 | 21 22 | 30 31 | 44 45 | ↑ | ↑ | ↑ | ↑ | ↑ | ↑ |
| H | 50 | ↓ | ↓ | ↓ | ↓ | ↓ | ↓ | ↓ | 0 1 | ↑ | 1 2 | 2 3 | 3 4 | 5 6 | 7 8 | 10 11 | 14 15 | 21 22 | 30 31 | 44 45 | ↑ | ↑ | ↑ | ↑ | ↑ | ↑ | ↑ |
| J | 80 | ↓ | ↓ | ↓ | ↓ | ↓ | ↓ | 0 1 | ↑ | 1 2 | 2 3 | 3 4 | 5 6 | 7 8 | 10 11 | 14 15 | 21 22 | 30 31 | 44 45 | ↑ | ↑ | ↑ | ↑ | ↑ | ↑ | ↑ | ↑ |
| K | 125 | ↓ | ↓ | ↓ | ↓ | ↓ | 0 1 | ↑ | 1 2 | 2 3 | 3 4 | 5 6 | 7 8 | 10 11 | 14 15 | 21 22 | 30 31 | 44 45 | ↑ | ↑ | ↑ | ↑ | ↑ | ↑ | ↑ | ↑ | ↑ |
| L | 200 | ↓ | ↓ | ↓ | ↓ | 0 1 | ↑ | 1 2 | 2 3 | 3 4 | 5 6 | 7 8 | 10 11 | 14 15 | 21 22 | 30 31 | 44 45 | ↑ | ↑ | ↑ | ↑ | ↑ | ↑ | ↑ | ↑ | ↑ | ↑ |
| M | 315 | ↓ | ↓ | ↓ | 0 1 | ↑ | 1 2 | 2 3 | 3 4 | 5 6 | 7 8 | 10 11 | 14 15 | 21 22 | 30 31 | 44 45 | ↑ | ↑ | ↑ | ↑ | ↑ | ↑ | ↑ | ↑ | ↑ | ↑ | ↑ |
| N | 500 | ↓ | ↓ | 0 1 | ↑ | 1 2 | 2 3 | 3 4 | 5 6 | 7 8 | 10 11 | 14 15 | 21 22 | 30 31 | 44 45 | ↑ | ↑ | ↑ | ↑ | ↑ | ↑ | ↑ | ↑ | ↑ | ↑ | ↑ | ↑ |
| P | 800 | ↓ | 0 1 | ↑ | 1 2 | 2 3 | 3 4 | 5 6 | 7 8 | 10 11 | 14 15 | 21 22 | 30 31 | 44 45 | ↑ | ↑ | ↑ | ↑ | ↑ | ↑ | ↑ | ↑ | ↑ | ↑ | ↑ | ↑ | ↑ |
| Q | 1250 | 0 1 | ↑ | 1 2 | 2 3 | 3 4 | 5 6 | 7 8 | 10 11 | 14 15 | 21 22 | 30 31 | 44 45 | ↑ | ↑ | ↑ | ↑ | ↑ | ↑ | ↑ | ↑ | ↑ | ↑ | ↑ | ↑ | ↑ | ↑ |
| R | 2000 | ↑ | ↑ | 1 2 | 2 3 | 3 4 | 5 6 | 7 8 | 10 11 | 14 15 | 21 22 | 30 31 | 44 45 | ↑ | ↑ | ↑ | ↑ | ↑ | ↑ | ↑ | ↑ | ↑ | ↑ | ↑ | ↑ | ↑ | ↑ |

↓ = Use first sampling below arrow. If sample size equals, or exceeds lot or batch size, do 100 percent inspection.

↑ = Use first sampling plan above arrow.

Ac = Acceptance number.

Re = Rejection number.

*Source:* Reproduced from "Sampling Procedures and Tables for Inspection by Attributes, Military Standard 105D," U.S. Department of Defense, Washington, DC: U.S. Government Printing Office, 1963.

**TABLE 19.8**   MILITARY STANDARD 105: SINGLE SAMPLING PLANS FOR TIGHTENED INSPECTION (MASTER TABLE)

Acceptable Quality Levels (tightened inspection). Each cell shows "Ac Re" (Ac = Acceptance number, Re = Rejection number). ↓ = Use first sampling plan below arrow. If sample size equals, or exceeds, lot or batch size, do 100 percent inspection. ↑ = Use first sampling plan above arrow.

| Sample size code letter | Sample size | 0.010 | 0.015 | 0.025 | 0.040 | 0.065 | 0.10 | 0.15 | 0.25 | 0.40 | 0.65 | 1.0 | 1.5 | 2.5 | 4.0 | 6.5 | 10 | 15 | 25 | 40 | 65 | 100 | 150 | 250 | 400 | 650 | 1000 |
|---|---|---|---|---|---|---|---|---|---|---|---|---|---|---|---|---|---|---|---|---|---|---|---|---|---|---|---|
| A | 2 | ↓ | ↓ | ↓ | ↓ | ↓ | ↓ | ↓ | ↓ | ↓ | ↓ | ↓ | ↓ | ↓ | ↓ | ↓ | ↓ | ↓ | 0 1 | 1 2 | 2 3 | 3 4 | 5 6 | 8 9 | 12 13 | 18 19 | 27 28 |
| B | 3 | ↓ | ↓ | ↓ | ↓ | ↓ | ↓ | ↓ | ↓ | ↓ | ↓ | ↓ | ↓ | ↓ | ↓ | ↓ | ↓ | 0 1 | 1 2 | 2 3 | 3 4 | 5 6 | 8 9 | 12 13 | 18 19 | 27 28 | 41 42 |
| C | 5 | ↓ | ↓ | ↓ | ↓ | ↓ | ↓ | ↓ | ↓ | ↓ | ↓ | ↓ | ↓ | ↓ | ↓ | ↓ | 0 1 | 1 2 | 2 3 | 3 4 | 5 6 | 8 9 | 12 13 | 18 19 | 27 28 | 41 42 | ↑ |
| D | 8 | ↓ | ↓ | ↓ | ↓ | ↓ | ↓ | ↓ | ↓ | ↓ | ↓ | ↓ | ↓ | ↓ | ↓ | 0 1 | 1 2 | 2 3 | 3 4 | 5 6 | 8 9 | 12 13 | 18 19 | 27 28 | 41 42 | ↑ | ↑ |
| E | 13 | ↓ | ↓ | ↓ | ↓ | ↓ | ↓ | ↓ | ↓ | ↓ | ↓ | ↓ | ↓ | ↓ | 0 1 | 1 2 | 2 3 | 3 4 | 5 6 | 8 9 | 12 13 | 18 19 | 27 28 | 41 42 | ↑ | ↑ | ↑ |
| F | 20 | ↓ | ↓ | ↓ | ↓ | ↓ | ↓ | ↓ | ↓ | ↓ | ↓ | ↓ | ↓ | 0 1 | 1 2 | 2 3 | 3 4 | 5 6 | 8 9 | 12 13 | 18 19 | 27 28 | 41 42 | ↑ | ↑ | ↑ | ↑ |
| G | 32 | ↓ | ↓ | ↓ | ↓ | ↓ | ↓ | ↓ | ↓ | ↓ | ↓ | ↓ | 0 1 | 1 2 | 2 3 | 3 4 | 5 6 | 8 9 | 12 13 | 18 19 | 27 28 | 41 42 | ↑ | ↑ | ↑ | ↑ | ↑ |
| H | 50 | ↓ | ↓ | ↓ | ↓ | ↓ | ↓ | ↓ | ↓ | ↓ | ↓ | 0 1 | 1 2 | 2 3 | 3 4 | 5 6 | 8 9 | 12 13 | 18 19 | 27 28 | 41 42 | ↑ | ↑ | ↑ | ↑ | ↑ | ↑ |
| J | 80 | ↓ | ↓ | ↓ | ↓ | ↓ | ↓ | ↓ | ↓ | ↓ | 0 1 | 1 2 | 2 3 | 3 4 | 5 6 | 8 9 | 12 13 | 18 19 | 27 28 | 41 42 | ↑ | ↑ | ↑ | ↑ | ↑ | ↑ | ↑ |
| K | 125 | ↓ | ↓ | ↓ | ↓ | ↓ | ↓ | ↓ | ↓ | 0 1 | 1 2 | 2 3 | 3 4 | 5 6 | 8 9 | 12 13 | 18 19 | 27 28 | 41 42 | ↑ | ↑ | ↑ | ↑ | ↑ | ↑ | ↑ | ↑ |
| L | 200 | ↓ | ↓ | ↓ | ↓ | ↓ | ↓ | ↓ | 0 1 | 1 2 | 2 3 | 3 4 | 5 6 | 8 9 | 12 13 | 18 19 | 27 28 | 41 42 | ↑ | ↑ | ↑ | ↑ | ↑ | ↑ | ↑ | ↑ | ↑ |
| M | 315 | ↓ | ↓ | ↓ | ↓ | ↓ | ↓ | 0 1 | 1 2 | 2 3 | 3 4 | 5 6 | 8 9 | 12 13 | 18 19 | 27 28 | 41 42 | ↑ | ↑ | ↑ | ↑ | ↑ | ↑ | ↑ | ↑ | ↑ | ↑ |
| N | 500 | ↓ | ↓ | ↓ | ↓ | ↓ | 0 1 | 1 2 | 2 3 | 3 4 | 5 6 | 8 9 | 12 13 | 18 19 | 27 28 | 41 42 | ↑ | ↑ | ↑ | ↑ | ↑ | ↑ | ↑ | ↑ | ↑ | ↑ | ↑ |
| P | 800 | ↓ | ↓ | ↓ | ↓ | 0 1 | 1 2 | 2 3 | 3 4 | 5 6 | 8 9 | 12 13 | 18 19 | 27 28 | 41 42 | ↑ | ↑ | ↑ | ↑ | ↑ | ↑ | ↑ | ↑ | ↑ | ↑ | ↑ | ↑ |
| Q | 1250 | ↓ | ↓ | ↓ | 0 1 | 1 2 | 2 3 | 3 4 | 5 6 | 8 9 | 12 13 | 18 19 | 27 28 | 41 42 | ↑ | ↑ | ↑ | ↑ | ↑ | ↑ | ↑ | ↑ | ↑ | ↑ | ↑ | ↑ | ↑ |
| R | 2000 | ↓ | ↓ | 0 1 | 1 2 | 2 3 | 3 4 | 5 6 | 8 9 | 12 13 | 18 19 | 27 28 | 41 42 | ↑ | ↑ | ↑ | ↑ | ↑ | ↑ | ↑ | ↑ | ↑ | ↑ | ↑ | ↑ | ↑ | ↑ |
| S | 3150 | ↓ | 0 1 | 1 2 | 2 3 | 3 4 | 5 6 | 8 9 | 12 13 | 18 19 | 27 28 | 41 42 | ↑ | ↑ | ↑ | ↑ | ↑ | ↑ | ↑ | ↑ | ↑ | ↑ | ↑ | ↑ | ↑ | ↑ | ↑ |

↓ = Use first sampling plan below arrow. If sample size equals, or exceeds, lot or batch size, do 100 percent inspection.

↑ = Use first sampling plan above arrow.

Ac = Acceptance number.

Re = Rejection number

Source: Reproduced from "Sampling Procedures and Tables for Inspection by Attributes, Military Standard 105D," U.S. Department of Defense, Washington, DC: U.S. Government Printing Office, 1963.

**TABLE 19.9    MILITARY STANDARD 105 SINGLE SAMPLING PLANS FOR REDUCED INSPECTION (MASTER TABLE)**

| Sample size code letter | Sample size | 0.010 | | 0.015 | | 0.025 | | 0.040 | | 0.065 | | 0.10 | | 0.15 | | 0.25 | | 0.40 | | 0.65 | | 1.0 | | 1.5 | | 2.5 | | 4.0 | | 6.5 | | 10 | | 15 | | 25 | | 40 | | 65 | | 100 | | 150 | | 250 | | 400 | | 650 | | 1000 | |
|---|---|---|---|---|---|---|---|---|---|---|---|---|---|---|---|---|---|---|---|---|---|---|---|---|---|---|---|---|---|---|---|---|---|---|---|---|---|---|---|---|---|---|---|---|---|---|---|---|---|---|---|---|---|---|---|
| | | Ac | Re | Ac | Re | Ac | Re | Ac | Re | Ac | Re | Ac | Re | Ac | Re | Ac | Re | Ac | Re | Ac | Re | Ac | Re | Ac | Re | Ac | Re | Ac | Re | Ac | Re | Ac | Re | Ac | Re | Ac | Re | Ac | Re | Ac | Re | Ac | Re | Ac | Re | Ac | Re | Ac | Re | Ac | Re | Ac | Re |
| A | 2 | ⇩ | | ⇩ | | ⇩ | | ⇩ | | ⇩ | | ⇩ | | ⇩ | | ⇩ | | ⇩ | | ⇩ | | ⇩ | | ⇩ | | ⇩ | | ⇩ | | ⇩ | | ⇩ | | 0 | 1 | 1 | 2 | 2 | 3 | 3 | 4 | 5 | 6 | 7 | 8 | 10 | 11 | 14 | 15 | 21 | 22 | 30 | 31 |
| B | 2 | ⇩ | | ⇩ | | ⇩ | | ⇩ | | ⇩ | | ⇩ | | ⇩ | | ⇩ | | ⇩ | | ⇩ | | ⇩ | | ⇩ | | ⇩ | | ⇩ | | ⇩ | | 0 | 1 | 0 | 2 | 1 | 3 | 2 | 4 | 3 | 5 | 5 | 6 | 7 | 8 | 10 | 11 | 14 | 15 | 21 | 22 | 30 | 31 |
| C | 2 | ⇩ | | ⇩ | | ⇩ | | ⇩ | | ⇩ | | ⇩ | | ⇩ | | ⇩ | | ⇩ | | ⇩ | | ⇩ | | ⇩ | | ⇩ | | ⇩ | | 0 | 1 | 0 | 2 | 1 | 3 | 1 | 4 | 2 | 5 | 3 | 6 | 5 | 8 | 7 | 10 | 10 | 13 | 14 | 17 | 21 | 24 | ⇧ | |
| D | 3 | ⇩ | | ⇩ | | ⇩ | | ⇩ | | ⇩ | | ⇩ | | ⇩ | | ⇩ | | ⇩ | | ⇩ | | ⇩ | | ⇩ | | ⇩ | | 0 | 1 | 0 | 2 | 1 | 3 | 1 | 4 | 2 | 5 | 3 | 6 | 5 | 8 | 7 | 10 | 10 | 13 | 14 | 17 | 21 | 24 | ⇧ | | ⇧ | |
| E | 5 | ⇩ | | ⇩ | | ⇩ | | ⇩ | | ⇩ | | ⇩ | | ⇩ | | ⇩ | | ⇩ | | ⇩ | | ⇩ | | ⇩ | | 0 | 1 | 0 | 2 | 1 | 3 | 1 | 4 | 2 | 5 | 3 | 6 | 5 | 8 | 7 | 10 | 10 | 13 | 14 | 17 | 21 | 24 | ⇧ | | ⇧ | | ⇧ | |
| F | 8 | ⇩ | | ⇩ | | ⇩ | | ⇩ | | ⇩ | | ⇩ | | ⇩ | | ⇩ | | ⇩ | | ⇩ | | ⇩ | | 0 | 1 | 0 | 2 | 1 | 3 | 1 | 4 | 2 | 5 | 3 | 6 | 5 | 8 | 7 | 10 | 10 | 13 | 14 | 17 | 21 | 24 | ⇧ | | ⇧ | | ⇧ | | ⇧ | |
| G | 13 | ⇩ | | ⇩ | | ⇩ | | ⇩ | | ⇩ | | ⇩ | | ⇩ | | ⇩ | | ⇩ | | ⇩ | | 0 | 1 | 0 | 2 | 1 | 3 | 1 | 4 | 2 | 5 | 3 | 6 | 5 | 8 | 7 | 10 | 10 | 13 | 14 | 17 | 21 | 24 | ⇧ | | ⇧ | | ⇧ | | ⇧ | | ⇧ | |
| H | 20 | ⇩ | | ⇩ | | ⇩ | | ⇩ | | ⇩ | | ⇩ | | ⇩ | | ⇩ | | ⇩ | | 0 | 1 | 0 | 2 | 1 | 3 | 1 | 4 | 2 | 5 | 3 | 6 | 5 | 8 | 7 | 10 | 10 | 13 | 14 | 17 | 21 | 24 | ⇧ | | ⇧ | | ⇧ | | ⇧ | | ⇧ | | ⇧ | |
| J | 32 | ⇩ | | ⇩ | | ⇩ | | ⇩ | | ⇩ | | ⇩ | | ⇩ | | ⇩ | | 0 | 1 | 0 | 2 | 1 | 3 | 1 | 4 | 2 | 5 | 3 | 6 | 5 | 8 | 7 | 10 | 10 | 13 | 14 | 17 | 21 | 24 | ⇧ | | ⇧ | | ⇧ | | ⇧ | | ⇧ | | ⇧ | | ⇧ | |
| K | 50 | ⇩ | | ⇩ | | ⇩ | | ⇩ | | ⇩ | | ⇩ | | ⇩ | | 0 | 1 | 0 | 2 | 1 | 3 | 1 | 4 | 2 | 5 | 3 | 6 | 5 | 8 | 7 | 10 | 10 | 13 | 14 | 17 | 21 | 24 | ⇧ | | ⇧ | | ⇧ | | ⇧ | | ⇧ | | ⇧ | | ⇧ | | ⇧ | |
| L | 80 | ⇩ | | ⇩ | | ⇩ | | ⇩ | | ⇩ | | ⇩ | | 0 | 1 | 0 | 2 | 1 | 3 | 1 | 4 | 2 | 5 | 3 | 6 | 5 | 8 | 7 | 10 | 10 | 13 | 14 | 17 | 21 | 24 | ⇧ | | ⇧ | | ⇧ | | ⇧ | | ⇧ | | ⇧ | | ⇧ | | ⇧ | | ⇧ | |
| M | 125 | ⇩ | | ⇩ | | ⇩ | | ⇩ | | ⇩ | | 0 | 1 | 0 | 2 | 1 | 3 | 1 | 4 | 2 | 5 | 3 | 6 | 5 | 8 | 7 | 10 | 10 | 13 | 14 | 17 | 21 | 24 | ⇧ | | ⇧ | | ⇧ | | ⇧ | | ⇧ | | ⇧ | | ⇧ | | ⇧ | | ⇧ | | ⇧ | |
| N | 200 | ⇩ | | ⇩ | | ⇩ | | ⇩ | | 0 | 1 | 0 | 2 | 1 | 3 | 1 | 4 | 2 | 5 | 3 | 6 | 5 | 8 | 7 | 10 | 10 | 13 | 14 | 17 | 21 | 24 | ⇧ | | ⇧ | | ⇧ | | ⇧ | | ⇧ | | ⇧ | | ⇧ | | ⇧ | | ⇧ | | ⇧ | | ⇧ | |
| P | 315 | ⇩ | | ⇩ | | ⇩ | | 0 | 1 | 0 | 2 | 1 | 3 | 1 | 4 | 2 | 5 | 3 | 6 | 5 | 8 | 7 | 10 | 10 | 13 | 14 | 17 | 21 | 24 | ⇧ | | ⇧ | | ⇧ | | ⇧ | | ⇧ | | ⇧ | | ⇧ | | ⇧ | | ⇧ | | ⇧ | | ⇧ | | ⇧ | |
| Q | 500 | ⇩ | | ⇩ | | 0 | 1 | 0 | 2 | 1 | 3 | 1 | 4 | 2 | 5 | 3 | 6 | 5 | 8 | 7 | 10 | 10 | 13 | 14 | 17 | 21 | 24 | ⇧ | | ⇧ | | ⇧ | | ⇧ | | ⇧ | | ⇧ | | ⇧ | | ⇧ | | ⇧ | | ⇧ | | ⇧ | | ⇧ | | ⇧ | |
| R | 800 | ⇩ | | 0 | 1 | 0 | 2 | 1 | 3 | 1 | 4 | 2 | 5 | 3 | 6 | 5 | 8 | 7 | 10 | 10 | 13 | 14 | 17 | 21 | 24 | ⇧ | | ⇧ | | ⇧ | | ⇧ | | ⇧ | | ⇧ | | ⇧ | | ⇧ | | ⇧ | | ⇧ | | ⇧ | | ⇧ | | ⇧ | | ⇧ | |

⇩ = Use first sampling plan below arrow.  If sample size equals, or exceeds, lot or batch size, do 100 percent inspection.

⇧ = Use first sampling plan above arrow.

Ac = Acceptance number.

Re = Rejection number.

† = If the acceptance number has been exceeded, but the rejection number has not been reached, accept the lot, but reinstate normal inspection.

Source: Reproduced from "Sampling Procedures and Tables for Inspection by Attributes, Military Standard 105D," U.S. Department of Defense, Washington, DC: U.S. Government Printing Office, 1963.

## Example 19.3

Determine a single sampling plan with inspection level II, for an AQL = 1.5%, and a lot size $N = 1500$ production units under

a  Normal inspection.
b  Tightened inspection.
c  Reduced inspection.

### Solution

a  From Table 19.6 the code letter for $N = 1500$ is $K$. The appropriate sampling plan (for normal inspection) has $n = 125$, Ac = 5, Re = 6, from Table 19.7.
   *Decision*—If the number of nonconforming items found $\leq 5$, then we accept the lot.  If the number of nonconforming items found $\geq 6$, then we reject the lot.
b  From Table 19.6 the code letter for $N = 1500$ is $K$. The appropriate sampling plan (for tightened inspection) has $n = 125$, Ac = 3, Re = 4, from Table 19.8.
   *Decision*—If the number of nonconforming items $\leq 3$, then we accept the lot. If the number of nonconforming items $\geq 4$, we reject the lot.
c  From Table 19.6 the code letter for $N = 1500$ is $K$. The appropriate sampling plan (for reduced inspection) has $n = 50$, Ac = 2, Re = 5, from Table 19.9.
   *Decision*—If the number of nonconforming items found $\leq 2$, then we accept the lot and continue in reduced inspection.  If the number of nonconforming items found $\geq 5$, then we reject the lot and return to normal inspection.  If the number of nonconforming items found is 3 or 4, then we accept the lot and return to normal inspection, see Figure 19.7.

## Example 19.4

Determine a single sampling plan that operates at inspection level III, with AQL = 0.15%, and a lot size $N = 250$ production units.

### Solution

From Table 19.6 the code letter for $N = 250$ is $H$.
Enter Table 19.7 at code H.  Observe the arrow at the AQL = 0.15% point and follow the arrow down.  We then go back to the sample size column and read $n = 80$, Ac = 0, Re = 1.
*Decision*—If any nonconforming items are found in the sample, we reject the lot.

## REVIEW AND DISCOVERY EXERCISES

### Review

**19.1** For the single sampling plan, by attributes, where $n = 100$, $c = 3$, AQL = 1%, and LTPD = 6%, estimate

   **a** The consumer's risk.
   **b** The producer's risk.
**19.2** Using Table 19.1, develop and plot AOQ and AFI curves, and then determine AOQLs for the following sampling plans:
   **a** $n = 100, c = 0$
   **b** $n = 100, c = 5$
**19.3** Develop type B OC curves for the single sampling plans by attributes, where $N = 1000, n = 50, c = 0, c = 1,$ and $c = 2.$
**19.4** Develop AOQ tables for Problem 19.3 and estimate the associated AOQLs for the three sampling plans.
**19.5** Given the three OC curves in Figure 19.3 (also see Table 19.1),
   **a** Determine the producer's risk $\alpha$ associated with an AQL = 0.5% and the consumer's risk associated with LTPD = 3%.
   **b** Calculate the AOQ for ($N = 2000, n = 100$) for the AQL and LTPD in (a).
**19.6** For an LTPD = 5% and consumer's risk $\beta = 0.10$ find a minimum ATI single sampling plan for a lot size $N = 2500$ items with a process average of 1% defective. What AOQL would we expect for this sampling plan?
**19.7** Find a single sampling plan for an AOQL = 2%, minimum ATI, process average of 0.5% and a lot size $N = 2500$ items. What LTPD would be associated with consumer's risk $\beta = 10\%$ in this plan?
**19.8** Using Mil-Std-105 determine the appropriate single sampling plan (general inspection II) for reduced, normal, and tightened inspection, where AQL = 1% and $N = 4,000$. Explain the decision rules for the sampling plans.
**19.9** For a lot size of $N = 1000$ parts, using Mil-Std-105, general inspection level II, and reduced inspection, determine the appropriate single sampling plan for an AQL = 0.04% defective.
**19.10** In acceptance sampling under Mil-Std-105, single sampling is to be used with inspection level II, AQL = 0.15% and a lot size $N = 15,000$ units. Determine the sampling plans called for under
   **a** Reduced inspection.
   **b** Normal inspection.
   **c** Tightened inspection.

## Discovery

**19.11** Explain the similarities and differences in acceptance sampling by attributes and $P$ charts.
**19.12** How is Mil-Std-105 justified in claiming that AQLs of 15 and above refer to defects per 100 product units, rather than percent defective?
**19.13** How compatible are acceptance sampling by attributes plans to near ZD (very low proportion defective) product quality? Explain.
**19.14** Some companies advertise the fact that 100 percent of their product is inspected and tested. Is this good or bad for you as the customer?
**19.15** Automated inspection through vision systems is highly touted as a high technology approach to quality improvement. Support or attack this position.
**19.16** Is it both possible and likely that sorting inspection based quality control is superior to process based quality control? Explain.
**19.17** Anshutz, a German manufacturer of rifles, includes a paper target shot at 50 m

(10 shots) with each small-bore (.22-caliber) target rifle they sell. They heavily advertise this fact as proof of performance. Is this an inspection-test? Explain.

**19.18** Your company has recently signed a contract with a customer that states that your company will produce and deliver a product, which will be at most 2.0 percent defective, on the average. The contract was signed before your company had built very many of these products, but had built two or three successfully. Only one machine is used in your plant to produce the item and the machine you have is the best in the world.

After a month's production, your $P$ chart has shown a $\bar{p}$ of 4.0 percent, but a gross lack of statistical control (usually on the high side). At the present time, it is improbable that you will be able to establish statistical control for at least one year (even with a high degree of effort) because no one understands the relevant production processes. Shipments to the customer (in lots of $N = 1000$) are to begin next week. The test to determine if an item is good or bad is expensive, but nondestructive.

You are assigned to the product in order to lead in a strategic quality planning effort. Explain your strategy and tactics, in both the short and long run. In other words, how will you resolve the problem? Contract renegotiation is not a feasible alternative.

# APPENDIX 19A: TYPE B OC CURVES

In lot-by-lot acceptance sampling by attributes, we usually assume that our customer buys our product on a regular basis and that we ship relatively large lots. Hence, in the best interests of the customer as well as our own, from a production viewpoint, we develop OC curves based on the binomial model, or "Type B OC curves" (Duncan [12] and Grant and Leavenworth [13]). For convenience in the quantitative development of Type B OC curves, we use a Poisson model as an approximation to the binomial (see Table 19.1). In the lot-by-lot sampling context, the binomial probability mass function is expressed as

$$f_{bin}(d; n, p) = C_d^n p^d (1 - p)^{n - d} \qquad d = 0, 1, 2, ..., n \qquad (19.5)$$

where we use $d$ to represent values of the random variable, which is the number of nonconforming product units.

In the same context, the Poisson (poi) can be expressed as

$$f_{poi}(d; \mu) = \frac{e^{-\mu} (\mu)^d}{d!} \qquad d = 0, 1, 2, ... \qquad (19.6)$$

When $n \to \infty$, $p \to 0$ and $\mu = np$ remain constant,

$$f_{bin}(d; n, p) \to f_{poi}(d; \mu)$$

Hence, for large $n$ and small $p$, we use the Poisson as an approximation to the binomial. To develop the probability of accepting the lot, $H_0$: $p \le p_0$, we use a cumulative Poisson table (Section IX, Table IX.6) and determine

$$P_a = P(d \le c) \qquad (19.7)$$

Tables 19.1, 19.4, and 19.5 were developed using the Poisson approximation. The cumulative Poisson form allows us to obtain $P_a$ directly from the table. We simply locate the $\mu = np$ row, the $c$ column, and read $P_a$ at the row-column intersection in the table. The Type B OC curves shown in Figures 19.3 and 19.5 are smooth curves which treat the proportion of nonconforming product $p$ in a continuous fashion.

## REFERENCES

1 W. G. Cochran, *Sampling Techniques,* 3d ed., New York: Wiley, 1977.

2 A. J. Duncan, *Quality Control and Industrial Statistics,* 5th ed., Homewood, IL: Irwin, 1986.

3 E. L. Grant and R. S. Leavenworth, *Statistical Quality Control,* 6th ed., New York: McGraw-Hill, 1988.

4 D. C. Montgomery, *Introduction to Statistical Quality Control,* 2d ed., New York: Wiley, 1991.

5 "Sampling Procedures and Tables for Inspection by Attributes, Military Standard 105D," U. S. Department of Defense, Washington, DC: U. S. Government Printing Office, 1963.

6 "Sampling Procedures and Tables for Inspection by Variables by Percent Defective, Military Standard 414," U. S. Department of Defense, Washington, DC: U. S. Government Printing Office, 1957.

7 See reference 2.

8 See reference 2.

9 See reference 3.

10 H. F. Dodge and H. G. Romig, *Sampling Inspection Tables—Single and Double Sampling,* 2d ed., New York: Wiley, 1959.

11 See reference 5.

12 See reference 2.

13 See reference 3.

# CREATION OF QUALITY— DESIGNED EXPERIMENTS

## VIRTUES FOR LEADERSHIP IN QUALITY
### Persistence and Patience

The purpose of Section Six is to introduce our readers to the concept of designed experiments and the experimental design protocol. We will first discuss why careful experimental design is critical. In Chapter 20 we provide an introductory discussion of critical concepts, terms, and definitions used in experimental design work. We will examine a number of designs which include single-factor designs with both fixed and random effects factors. In Chapter 21, we discuss multiple-factor experiments as well as factor crossing and nesting. We introduce our reader to regression analysis and response surfaces in Chapter 22. Finally, in Chapter 23 we develop two-level fractional factorial experiments and extend this discussion to the Taguchi signal to noise ratio experiments.

We stress a proactive approach to design so that we can enhance our chances of extracting meaningful information from our experiments. We limit the theoretical derivations to a level compatible with conceptual understanding of the statistical methods we present.

We provide a wide variety of designs and address each with both manual calculations and computer-aided analyses. We stress graphical presentations for the purpose of communicating our experimental findings. Balanced interpretations are provided, where we develop and integrate both graphical and numerical methods.

# 20

# SINGLE-FACTOR EXPERIMENTS

## 20.0 INQUIRY

1  What is a designed experiment?
2  Why should we design our experiments?
3  What constitutes a controllable factor?  What constitutes an uncontrollable factor?
4  What is the difference between statistical significance and practical significance?
5  What is a fixed effect?  What is a random effect?
6  How are single-factor experiments designed?  Analyzed?

## 20.1 INTRODUCTION

***Success in quality improvement requires the acquisition and application of knowledge, relative to our products and processes, in a timely and cost effective manner.***
Figure 20.1 depicts a discovery cycle relative to off-line quality improvement.  This cycle is not only appropriate for both proactive and reactive quality work, it is absolutely essential for off-line bottleneck engineering, where we must resolve product-process limitations regarding physical performance, cost, or timeliness.

The first step, questioning the effectiveness and efficiency of our current product or production process, leads us to examine the "what," "why," "when," "where," "who," and "how" associated with product and process related bottlenecks.  This step yields open-ended questions and general hypotheses about our customer's demands as well as our products and processes.

Once we have developed our lines of questions and our general hypotheses, we structure or design our experiment.  Here, we determine the manner in which we

**FIGURE 20.1**    Off-line quality improvement cycle.

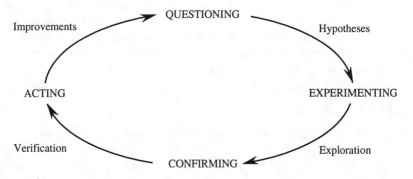

intend to go about collecting data, by observing our customers, product, and/or process. Exploration, the experiment, eventually leads to findings and conclusions.

However, before we draw our conclusions, we must confirm our findings and verify our results. This step is critical to ensure that we can replicate the results our experimenting brought to light. Once we are reasonably sure (we probably never will be absolutely sure) of promising results (which is not always the case) we are ready to draw reasonable conclusions, communicate our results, and take action to create quality improvements.

Deming describes what he calls "enumerative studies" and "analytic studies" [1]. The purpose of an enumerative study is to describe a reasonably well defined population in a static sense. The analytic study views a population in a dynamic state, where we seek to improve or predict what is likely to happen in the future. Analytic studies should provide a "degree of belief" for planning (e.g., to impact decisions and hence address the future).

Our proactive orientation in quality work is compatible with Deming's analytic study concept. Deming points out that most statistical methods (e.g., analysis of variance and tests of hypotheses) are developed for and relevant to enumerative studies. Nevertheless, these methods, along with sound engineering and business judgment and physical knowledge, are useful in analytic or proactive quality related studies.

## 20.2 CAUSE-EFFECT RELATIONSHIPS

*Through experimentation, we seek to discover and confirm cause-effect (CE) relationships in order to help us make product and production process decisions.* Figure 20.2 portrays the input-output nature of a generic process. Our objective is typically to determine the effect that the design and noise variables have on the response variable, or output. *A "design" variable is one that we deem controllable. A "noise," or "nuisance," variable is one that we deem uncontrollable in the field environment.* In a "sterile" laboratory environment, we may well be able to control many of the "uncontrollable" variables; but, in our experimentation we are thinking of product or process operations in the customer's hands, i.e., in the field.

No absolute classification system exists to divide variables into these two categories. We must consider both the physical and the economic feasibility of control. For example, we can place a response in the effect position on a CE diagram, develop a set of causes, or variables, and then identify each as basically controllable or uncontrollable in the field environment. Our competitors across town (or around the world) may classify differently than we do. On occasion, we may change our classification and move a design variable to the noise category or a noise variable to the design category.

## 20.3 DESIGNED EXPERIMENTS

*We design experiments, as opposed to using trial-and-error methods, to gain both effectiveness and efficiency in knowledge acquisition.* We first develop a general

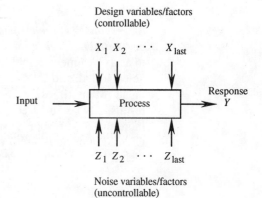

**FIGURE 20.2**    Controllable and uncontrollable factors (variables).

strategy to direct our experimental program and then develop a detailed plan of experimentation. Our experimental strategy constitutes a line, or sequence, of experimentation that will impact our quality improvement goals in the long run. Our experimental plan is a detailed protocol for each experiment.

Experimentation is necessary to gain or verify knowledge about a product or process. It can be performed in a number of ways, ranging from the trial-and-error, "one knob at a time" approach to a carefully planned multifactor experiment. The former has given way to the latter, because of its greater overall effectiveness in isolating main and interaction effects and/or components of variation, efficiency in working with smaller sample sizes, and timeliness in producing results. Hence, *a designed experiment involves a systematic plan of investigation, based on established statistical principles, so that the interpretation of the observations can be defended.* This defense includes concerns of both statistical bias and precision. As previously illustrated in Figure 18.7, bias is related to accuracy (location) and precision is related to variation (dispersion).

## Objectives

*A designed experiment is driven by experimental purpose—formulation of objectives and hypotheses.* In order to clearly state the objective and develop the hypotheses, we must identify the experimental factors and responses.

## Factors

*"Factors" are independent variables* (e.g., surface speed or materials in machining). *"Factor levels" are the physical levels of a factor* (e.g., 100 ft/min, 200 ft/min; steel, aluminum). *Factors may be either quantitative,* such as surface speed, *or qualitative,* such as materials. Sometimes, qualitative factors are termed "class" or "classification" variables. Typically, qualitative variables can take on only a few values (such as steel or aluminum) whereas quantitative variables can take on a great number of values (e.g., any surface speed from 0 to 500 ft/min, pressure from 0 to 400 lb/in$^2$).

## Response

*A "response" is the measured or observed result obtained at a given factor setting or treatment level* (combination of the factor levels involved, sometimes referred to as a "treatment combination"). For example, thrust force in pounds, tool life in holes drilled, and so forth, might be measured responses in a machinability experiment. Many times, multiple responses are recorded in a single experiment. The results may be analyzed one response at a time (the usual case) or, with respect to more than one response, simultaneously in what is termed a multivariate analysis. CE diagrams and Pareto charts are very useful aids in helping us set objectives, identify factors (the causes), and define meaningful responses (the effects).

## Hypotheses

Experimental objectives are typically addressed through stated hypotheses and hypothesis tests. As we noted in using SPC charts, *a null hypothesis $H_0$ is a hypothesis of no difference, while an alternative hypothesis $H_1$ represents some sort of difference or deviation from the null hypothesis. The null hypothesis is typically assumed to be true until enough evidence is produced to justify its rejection.* As we have seen in earlier chapters, $H_0$ formulation is somewhat idealistic, but it allows us a reference point from which to assess our observations.

As an example of a hypothesis structure, we might set up the following set of hypotheses:

$H_0$: There is no difference in the average tool life (the response) with respect to the workpiece materials—steel, copper, and aluminum (treatments or factor levels).

$H_1$: There is a difference in the average tool life with respect to the workpiece materials—steel, copper, and aluminum.

The interpretation and articulation of experimental findings are critical. An interpretation resulting in the statement, "We will accept $H_0$," or "We will not reject $H_0$" is understood as "We do not see enough evidence to reject $H_0$," rather than "We believe $H_0$ to be absolutely true." This point may seem rather trivial at first glance, but it is a critical point. The science of statistics and its usefulness in practice many times suffer from a lack of understanding and gross overinterpretation of results obtained from small samples and experiments of limited scope. We must always state our conclusions as based on the evidence at hand (b.o.e.). When new evidence or more evidence is gathered, we will have to reevaluate our conclusions.

## Test of Hypotheses

*A designed experiment is conducted using controlled physical conditions, along with a random ordering of the treatments (factor levels in a single-factor experiment and factor level combinations in a multifactor experiment).* The design is based on a mathematical or statistical model. The model, along with a set of statistical assumptions, allows the hypotheses of interest to be assessed through formal statistical tests, usually *t*, *F*, or chi-square ($\chi^2$) tests.  For example, with respect to our hypothesis set above, we can statistically assess the evidence provided by a response such as tool life in the form of sample means (locations) or averages pertaining to the treatments—steel, copper, or aluminum workpieces, relative to the dispersion (variation) present in the data.  If the difference in the sample means is relatively large, compared to the dispersion associated with the sample means, evidence supports the rejection of $H_0$, and it is safe to conclude that all the tool lives are not equal.  On the other hand, if the dispersion is large relative to the difference in the sample means, it is not safe to conclude that the tool lives are different.

When drawing conclusions based on samples, we must realize that statistical errors are possible.  Table 20.1*a* summarizes the concept of statistical errors as it relates to statistical significance.  The "absolute truth" is a "god's-eye" view, which we mortals are not privileged to hold.  Hence, we must draw conclusions relative to the sample indication.  Walpole and Myers provide a more detailed, but practical, discussion of statistical hypotheses and inferences [2].

*Results from experiments should be assessed for both statistical and practical significance.*  Table 20.1*b* depicts the relationship between the two.  *Assessing the degree of statistical significance involves evaluating the data collected, relative to the dispersion of the data, with respect to statistical inference models and methods. Assessing the degree of practical significance requires physical understanding and*

**TABLE 20.1**   STATISTICAL AND PRACTICAL SIGNIFICANCE

**a Statistical significance**

| Conclusion from sample | Absolute truth | |
| --- | --- | --- |
| | $H_0$ true | $H_0$ false |
| $H_0$ true | Correct conclusion | Type II error<br>$\beta = P$ (Type II error) |
| $H_0$ false | Type I error<br>$\alpha = P$ (Type I error) | Correct conclusion |

**b Practical significance**

| Statistical significance<br>(statistical dimension) | Practical significance<br>(physical dimension) | |
| --- | --- | --- |
| | Significant difference | Insignificant difference |
| Statistically significant difference ($H_0$ rejected) | Sufficient (and conclusive) physical results presented (decision support present) | Insufficient (but conclusive) physical results presented (decision support present) |
| No statistically significant difference ($H_0$ not rejected) | Promising (but inconclusive) physical results presented (more evidence needed) | Nonpromising (and inconclusive) physical results presented (reevaluate experimental program) |

*engineering or business judgment.* Successes in off-line quality work come about through practical significance, relative to product and process improvements. But, *statistical significance provides support for drawing practical conclusions and making decisions.*

## Treatments

*The terms "treatment" and "treatment combination"* (combinations of factor levels) *are used to describe what is done to the experimental material. The term "experimental unit" (eu) is used to refer to the largest collection of experimental material to which a single independent application of a treatment is made at random.* An experimental unit could be a piece of metal or a group of metal pieces, a volume of a chemical, one animal or a pen of animals, and so forth. An experimental unit is defined in the context of an experiment, considering the experimental objectives, and resources available.

## Balance and Degrees of Freedom

*"Balance" in an experiment implies that the same number of observations are made for each treatment or treatment combination.* Balance is important in order to extract the same amount of information from each treatment level (factor setting). *The term "degrees of freedom" (df) is used to depict the number of* **independent** *pieces of information available to us.* For example, a sample of size $n$ drawn from a population contains $n$ degrees of freedom. In general, when we use information from the sample in computing estimates (i.e., we use a $\bar{y}$ instead of a $\mu$), we lose a degree of freedom (see Walpole and Myers [3]).

## Experimental Error

*"Experimental error" is defined as the measure of variation which exists among responses from experimental units that have been treated alike.* Hence, to obtain a measure of the experimental error directly, we must replicate, or independently observe more than one response, for each setting or treatment combination. *The replication process consists of assigning multiple experimental units to a given treatment combination, treating the experimental units independently, and observing the results, making one observation, or replication, for each eu.* Multiple measurements on the same experimental unit are considered a subsample, not a replication, and can be used to develop a measure of sampling error, which is the variation which exists within an experimental unit. Typically, instrumentation or measurement error can be isolated as a component of sampling error.

## P-Value

A *P*-value is a term used in computerized statistical analysis programs to represent an observed significance level (o.s.l.). *The P-value, or o.s.l., is defined as the*

*smallest statistical significance level at which we can reject the null hypothesis,*
*b.o.e.* In statistical terms, *an o.s.l. is the probability of observing a more extreme*
*test statistic value (t, F, $\chi^2$, etc.) given that $H_0$ is true.* A sample based statistical
analysis and a resulting $P$-value indicate the degree of support, "confidence," for
rejecting $H_0$, in favor of $H_1$, rather than an absolute yes or no answer. In practice,
one rule of thumb of interpretation is to use the following scale for $H_0$ rejection
based on $P$-value support: $P$-values $\leq 0.01$, very strong support (for $H_0$ rejection);
$0.01 < P$-value $\leq 0.05$, strong support; $0.05 < P$-value $\leq 0.10$, some support; and
$P$-value $> 0.10$, little or no support. Thus, the smaller the $P$-value, the stronger the
support for $H_0$ rejection.

Samples (especially small samples) are not perfect indicators of population
characteristics. Larger samples are better (i.e., yield more information) than smaller
samples, but we must still avoid the absolute terms "always" and "never" when
drawing conclusions. Only by examining every member of the population can we
be absolutely sure. However, this alternative is typically not time effective, cost
effective, and/or physically possible.

## Example 20.1

Using a one-sample test of means, demonstrate the statistical significance ($P$-value)
concept and show that we can force statistical significance by increasing our
sample size (when $H_0$ is not absolutely true).

### Solution

We will develop both one-tailed and two-tailed hypotheses and will use a small
sample size, $n = 9$, as well as a large sample size, $n = 81$. Our analytical results are
shown in Figure 20.3. Here, we hold the sample mean $\bar{y} = 11$ and the sample
standard deviation $s = 3$ constant and vary the sample size from $n = 9$ to $n = 81$.
Our $P$-values were developed using our $t$ table (Section IX, Table IX.3) by bracket-
ing the $t_{calc}$ values in the body of the table and noting the $P$-value in the headings.
More detailed discussions of $P$-values appear in Walpole and Myers [4].

We see a pronounced reduction in $P$-values as we increase the sample size,
which shows that we can "force out" a statistical significance accordingly. In the
physical world, we must rely on practical significance to drive our decisions. We
use statistical significance to lend creditability to our analyses in the decision
making process.

## 20.4 EXPERIMENTAL PROTOCOL

We use statistical thinking and methods when we are dealing with samples. Sampling
by its very nature interjects "random variation" into our analyses; hence statistical
methods are called for.

**FIGURE 20.3** *P*-value illustrations, one-sample test of means (Example 20.1).

| | Two-tail hypothesis $H_0: \mu = \mu_0 = 10$ $H_1: \mu \neq 10$ $\mu_0 = 10 \quad \bar{y} = 11 \quad s = 3$ | One-tail hypothesis $H_0: \mu = \mu_0 = 10$ $H_1: \mu > 10$ $\mu_0 = 10 \quad \bar{y} = 11 \quad s = 3$ |
|---|---|---|
| Small sample size $n = 9$ | $s_{\bar{y}} = s/\sqrt{n} = 3/\sqrt{9} = 1$ $t_{calc} = (\bar{y} - \mu_0)/s_{\bar{y}} = (11 - 10)/1 = 1$ *t* distribution (df = 9 − 1 = 8) −1  0  $t_{calc} = 1$ $2(0.15) < P\text{-value} < 2(0.20)*$ $0.30 < P\text{-value} < 0.40$ | $s_{\bar{y}} = s/\sqrt{n} = 3/\sqrt{9} = 1$ $t_{calc} = (\bar{y} - \mu_0)/s_{\bar{y}} = (11 - 10)/1 = 1$ *t* distribution (df = 9 − 1 = 8) 0  $t_{calc} = 1$ $0.15 < P\text{-value} < 0.20*$ |
| Large sample size $n = 81$ | $s_{\bar{y}} = s/\sqrt{n} = 3/\sqrt{81} = 1/3$ $t_{calc} = (\bar{y} - \mu_0)/s_{\bar{y}} = (11 - 10)/(1/3) = 3$ *t* distribution (df = 81 − 1 = 80) 0  $t_{calc} = 3$ $P\text{-value} \ll 2(0.0025)*$ | $s_{\bar{y}} = s/\sqrt{n} = 3/\sqrt{81} = 1/3$ $t_{calc} = (\bar{y} - \mu_0)/s_{\bar{y}} = (11 - 10)/(1/3) = 3$ *t* distribution (df = 81 − 1 = 80) 0  $t_{calc} = 3$ $P\text{-value} \ll 0.0025*$ |

\* Values are taken from the *t* distribution, Section IX, Table IX.3.

We see two general situations in an experimental context: (1) experiments where time- or production-ordered sequence of observations is important in the analysis itself (e.g., in SPC) and (2) experiments where time- or production-ordered sequence is not important in the analysis itself (e.g., typical off-line quality experiments). Our discussion of experimental design (in this chapter and Chapters 21, 22, 23) focuses on the latter, whereas SPC (Chapters 15, 16, and 17) is used to monitor the former.

*We have three classes of statistical tools available to us:* **(1)** *descriptive* (e.g., summary statistics, graphing), **(2)** *inferential* (e.g., formal hypothesis testing), and **(3)** *predictive* (e.g., regression modeling, response surfaces). We typically use a combination of these tools to help us extract and communicate information to support decisions.

Experimental protocol is recognized as a critical factor in experimental design. *We list six fundamental steps involved in a designed experiment* (see Table 20.2). First, we *establish the purpose* and then, we *identify the variables.* Both of these steps are based on practical significance rather than statistical significance, and are driven by the need for knowledge and the nature of the product or process of interest. Next, we *plan the physical ordering of the experiment* relative to a statistical model.

**TABLE 20.2**     FUNDAMENTAL STEPS IN PLANNED EXPERIMENTATION

1  *Establish the purpose*—Clearly define and state the problem at hand. Set goals, objectives, and targets relative to what is needed and expected from each experiment.

2  *Identify the variables*—Clearly delineate response, controllable, and uncontrollable variables. Here we must select or determine our responses, select our factors, and establish factor levels.

3  *Design the experiment*—Our experimental design is based on the way we randomize (organize) our experiment. It is the randomization and the replication structure we use that determines the appropriate experimental model and analysis.

4  *Execute the experiment*—Here, we physically do what we have planned above and record or document our observations carefully. We must be careful to avoid mistakes and biases in our physical setups and execution.

5  *Analyze the results*—Our analysis may consist of descriptive, inferential, and predictive components, based on the theoretical structure of our model and the physical nature as reflected in our observed measurements.

6  *Interpret and communicate the analysis*—Here, we must put our findings in perspective and defend our methodology (from above) in terms of its conformance to good statistical practices as well as its physical relevance.

Fourth, we *execute the experiment* and make our observations. Then, we *analyze the data;* extensive computer aids are available to help us "crunch" the numbers. The final step is to *interpret and communicate the findings* to fulfill our purpose. These steps will be demonstrated through a number of examples. In all cases our interpretation is b.o.e.

Very detailed experimental design protocols have been advocated. Coleman and Montgomery, in their discussion of this issue [5], go so far as to suggest a formal guide-sheet format for designed experiments—a 13-point "predesign" master guide sheet.

Regardless of the sophistication of our experimental protocol, many times we stand to gain knowledge unexpectedly, over and above our original purpose. It is critical that we document our experiment and experimental observations very carefully. In some cases, these notes may prove more valuable than our planned collection of observations. Our notes may lead us to discover a number of unanticipated physical relationships, help us develop the next step in our experimental sequence, and, perhaps, lead us in new directions in our experimental program.

## 20.5 SINGLE-FACTOR EXPERIMENTS

*A single-factor experiment allows for the manipulation of only one factor* (e.g., one $X_i$ from Figure 20.2) *during the course of the experiment.* Ideally, all other factors are controlled or held constant.  Our objective is to isolate the changes in the response, relative to our chosen factor.

*In the simplest single-factor design, the completely randomized design (CRD), initially each treatment has an equal chance of being assigned to any experimental unit.*  In this design, a treatment is defined as a level of the single factor we have selected to study (e.g., a quantitative factor such as temperature, with levels 100°F, 200°F, 300°F, or a qualitative factor, such as workpiece material, with levels brass, steel, aluminum).  Each experimental unit typically produces one response, unless a subsampling procedure is employed.

The single-factor experiment will be developed through a case study which we extend and modify by examples throughout this chapter and Chapters 21, 22, and 23.  This format allows us to point out how a variety of experimental designs are structured and can be used effectively.  Using a common case will help us focus on the design characteristics, relative to our design alternatives, rather than diverting our attention with a number of unrelated physical contexts from one example to another.

## Ultrachewy Chocolate Chip Cookie—Case

We are developing an ultrachewy chocolate chip cookie for what appears to be a large national market (with international possibilities if we use slight modifications in our chocolate ingredients).  Our goal is to develop and distinguish our product by its chewiness (our prime true quality characteristic).  As such, our objective is to develop a high level of chewiness in our product.  We have developed a CE diagram (shown in Figure 20.4) in order to identify and record as many possible design and noise variables as we can.

**FIGURE 20.4**   CE diagram for ultrachewy chocolate chip cookies.

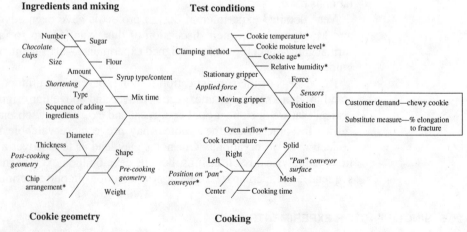

* Noise variables

We may quantify chewiness in two ways: (1) use people to chew our product and provide a quantitative score, or (2) define a substitute quality characteristic and build a measuring-test device. In this case, we think that the actual chewiness of a cookie can be predicted very well by the amount of plastic deformation present when we break the cookie. Hence, for reasons of timeliness and economics, we have opted for the second alternative and defined percent elongation (of a cookie as it is pulled apart) as our substitute quality characteristic. Here, bigger is better in terms of our response. Furthermore, we have developed an automated testing device (shown in Figure 20.5) which clamps the cookie on both edges and pulls it until fracture. Our device outputs the percent elongation up to the fracture point. Each cookie acts as an experimental unit. The test is destructive in nature (tested cookies cannot be packaged for our customers).

---

## Example 20.2

After careful development of a basic cookie recipe, we want to study and determine an appropriate cooking time that will result in a chewy cookie. Develop a single-factor experiment to assess the effect of cooking time on chewiness.

### Solution

We have chosen four cooking time factor levels for our experiment, based on our experience and physical knowledge of cookie making: level 1 = 5 min, level

**FIGURE 20.5** Cookie testing device configuration (plastic elongation).

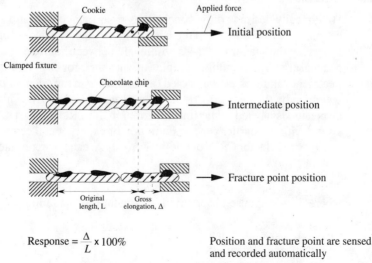

Response = $\dfrac{\Delta}{L} \times 100\%$

Position and fracture point are sensed and recorded automatically

2 = 10 min, level 3 = 15 min, and level 4 = 20 min. We set and control all other cause variables (Figure 20.4) as best we can in our off-line experimental kitchen,—where we have state-of-the-art facilities and well qualified and trained personnel—not on our production line. Hence, we can assume reasonably good control of all cause variables.

For our CRD we develop the experimental layout shown in Figure 20.6a. We randomly selected a cooking time, prepared a small batch of cookie dough, and then baked the batch (see Section IX, Table IX.1, for a table of random numbers). Then, we randomly selected a cookie from the batch, pulled the cookie to fracture, and recorded the response. We repeated this process until we obtained the complete data set for all cooking times, shown in Figure 20.6b. Here, we observed 12 cookies, or experimental units, obtained the 12 elongation percent responses shown in Figure 20.6b, and plotted the scatter plot in Figure 20.6c.

The analysis of the data primarily involves location and dispersion measures (statistics) developed using fundamental statistical tools. The data shown in Figure 20.6b are summarized and arranged to develop treatment totals and treatment means. Then, they are plotted on the scatter plot in Figure 20.6c. Each observed data point $Y_{ij}$ is represented by a + in Figure 20.6c. The $\bar{Y}_{i.}$ symbol refers to the $i$th treatment sample average, and the $\bar{Y}_{..}$ symbol refers to a grand sample average. The subscript dot "." indicates a summation over the subscript that it replaces; for example, $\bar{Y}_{1.}$ is the average of all observations from the first treatment, $i = 1$. Sample variances for each treatment, $s_i^2$, have been calculated in Figure 20.6c. Then, a pooled, weighted, average of the treatment sample variances, $s_{pooled}^2$, was calculated. Here, the degrees of freedom for each treatment, $n - 1 = 2$, was used as the weighting factor. The pooled variance, $s_{pooled}^2$, represents an overall measure of dispersion for all treatments, we assume all within treatment variances are equal (e.g., homogeneity of variance).

## Completely Randomized Design

We typically desire to draw conclusions regarding both location and dispersion. Location is measured by $\bar{Y}_{i.}$ in the CRD (Figure 20.6c). Dispersion can be assessed by observing the scatter in the +'s within and between each treatment. This scatter can be measured by computing sample variances, $s_i^2$, for each treatment. The individual treatment variances can be pooled (a weighted average taken of the sample variances, weighted by the degree of freedom, e.g., $n_i - 1$, for each). This overall measure of dispersion is labeled "experimental error" in the CRD. This statistical process is consistent with our earlier generic definition of experimental error.

The single factor CRD, or one-way ANOVA, can be represented in a model form as

$$Y_{ij} = \mu + \tau_i + \varepsilon_{ij} \tag{20.1}$$

where $Y_{ij}$ = response observation, treatment $i$, observation $j$, $i = 1, 2, \ldots, a$, $j = 1, 2, \ldots, n$

$\mu$ = overall mean

$\tau_i$ = treatment effect

$\varepsilon_{ij}$ = random error associated with observation $ij$

**FIGURE 20.6**   CRD physical layout, data table, and scatter plot (Example 20.2).

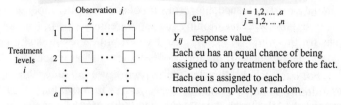

(a)  Generic single-factor completely randomized design (CRD) physical layout

| Treatment levels (cooking time) | Observations | | | $Y_{i.}$ | $\overline{Y}_{i.}$ |
|---|---|---|---|---|---|
| | 1 | 2 | 3 | | |
| 1   (5 min) | 4.0 | 1.0 | 2.0 | 7.00 | 2.333 |
| 2   (10 min) | 9.0 | 7.0 | 10.0 | 26.00 | 8.667 |
| 3   (15 min) | 8.0 | 4.0 | 5.0 | 17.00 | 5.667 |
| 4   (20 min) | 3.0 | 2.0 | 4.0 | 9.00 | 3.000 |
| Response measured in percent elongation | | | | | |

$$Y_{..} = \sum_{i=1}^{a} \sum_{j=1}^{n} Y_{ij}$$

$$Y_{i.} = \sum_{j=1}^{n} Y_{ij}$$

$$\overline{Y}_{i.} = \frac{Y_{i.}}{n}$$

$$\overline{Y}_{..} = \frac{Y_{..}}{an}$$

$Y_{..} = 59.0$

$\overline{Y}_{..} = 4.917$

(b)  Data set for ultrachewy chocolate chip cookies

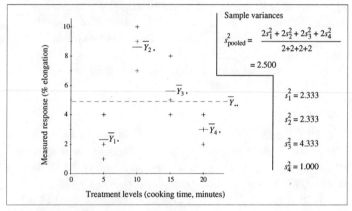

(c)  Scatter plot for ultrachewy chocolate chip cookie data

*The CRD model includes an overall mean as well as a treatment effect and an ex-perimental error.  The error term serves as a "residual" term to capture natural variation between the experimental units as well as variation associated with any uncontrolled variables that may be affecting the responses.*  An expansion, relative to the data, is shown in Equation 20.2.  Notice that the equality is preserved and that the terms on the right-hand side correspond to the model terms in Equation 20.1.  The geometric counterparts of these terms can be identified in Figure 20.6c.  In Figure 20.6c we can see the relationships between the terms in Equation 20.2 with respect to the overall mean, the treatment means, and the individual observations.

$$Y_{ij} = \overline{Y}_{..} + (\overline{Y}_{i.} - \overline{Y}_{..}) + (Y_{ij} - \overline{Y}_{i.}) \tag{20.2}$$

Equation 20.3 can be developed by rearranging, squaring, and summing the terms in Equation 20.2. Hence, the term sum of squares (SS) and, thereafter, the analysis of variance (ANOVA) partitioning concepts are developed. This SS development is the basis for the ANOVA technique. The simplified result is shown in Equation 20.4. The left-hand side of Equation 20.4 represents what we call the "total corrected sum of squares" $SS_{tc}$, where the sum of squares is corrected for the mean. *In the CRD we partition the total corrected sum of squares into two parts, (1) the sum of squares associated with the treatments (between treatments) and (2) the sum of squares associated with experimental error (variation within the treatments).*

$$\sum_{i=1}^{a} \sum_{j=1}^{n} (Y_{ij} - \bar{Y}_{..})^2 = \sum_{i=1}^{a} \sum_{j=1}^{n} [(\bar{Y}_{i.} - \bar{Y}_{..}) + (Y_{ij} - \bar{Y}_{i.})]^2 \tag{20.3}$$

$$\sum_{i=1}^{a} \sum_{j=1}^{n} (Y_{ij} - \bar{Y}_{..})^2 = n \sum_{i=1}^{a} (\bar{Y}_{i.} - \bar{Y}_{..})^2 + \sum_{i=1}^{a} \sum_{j=1}^{n} (Y_{ij} - \bar{Y}_{i.})^2 \tag{20.4}$$

The numerical results in the CRD are typically displayed in ANOVA tables as shown in Table 20.3. Table 20.3a displays a fixed effects model, while Table 20.3b displays a random effects model. These ANOVAs pertain to balanced experiments with $a$ treatment levels and $n$ observations per treatment. The sums of squares shown are the partitioned terms of Equation 20.4, which have been algebraically simplified to expedite hand calculations. The expected mean square terms, EMS, are used to develop appropriate hypothesis tests, discussed later in this chapter. Typically, computer software packages are used in constructing ANOVA tables.

### Fixed Effects in General

*Factors (and their resulting effects) involved in an experiment are classified as either "fixed" or "random." A fixed effects factor is appropriate when we seek to make an inference to a small set of hand-picked (chosen in a non-random manner) factor levels.* For example, if process yield at four temperatures (100°F, 150°F, 200°F, and 250°F) is of interest, then we can use a four-level fixed effects factor (temperature) in the experiment. Statistical inferences, expressed through hypotheses, can be made only to the temperature levels selected. The hypotheses can be stated as

$H_0$: All $\tau_i = 0$ or all $\mu_i$ are equal (all temperatures selected produce the same average response level) $i = 1, 2, ..., a$

$H_1$: At least one $\tau_i \neq 0$ or at least one $\mu_i$ is not equal to the others (at least one temperature selected produces a different average response level than the others) $i = 1, 2, ..., a$

An $F$ statistic (see Figure 20.3a) is appropriate for this test:

$$F_{calc} = \frac{MS_{trt}}{MS_{err}} \qquad df_{numerator} = a - 1, \ df_{denominator} = a(n-1)$$

**TABLE 20.3**   CRD FIXED AND RANDOM EFFECTS MODELS, GENERIC ANOVA, EMS, NOTATIONS, AND ASSUMPTIONS

---

**a  Fixed effects**

Model: $Y_{ij} = \mu + \tau_i + \varepsilon_{ij}$     $i = 1, 2, ..., a$     $j = 1, 2, ..., n$

ANOVA:

| Source | df | SS | MS | F | P | EMS |
|---|---|---|---|---|---|---|
| Total corrected | $an - 1$ | $\sum_{ij} Y_{ij}^2 - CF$ | | | | |
| Treatment (between treatments) | $a - 1$ | $\dfrac{\sum_{i=1}^{a} Y_{i.}^2}{n} - CF$ | $\dfrac{SS_{trt}}{df_{trt}}$ | $\dfrac{MS_{trt}}{MS_{err}}$ | $P_{trt}$ | $\sigma_\varepsilon^2 + n\phi_\tau$ |
| Exp. error (within treatments) | $a(n - 1)$ | $SS_{tc} - SS_{trt}$ | $\dfrac{SS_{err}}{df_{err}}$ | | | $\sigma_\varepsilon^2$ |

where $\phi_\tau = \dfrac{\sum_{i=1}^{a} \tau_i^2}{a - 1}$

**Assumptions and hypotheses**

1 We select each of the $a$ treatments for inclusion in the experiment as we choose (in a nonrandom fashion).

2 $\sum_{i=1}^{a} \tau_i = 0$

3 $\varepsilon_{ij}$ are independent, and distributed normally ($\mu = 0$, $\sigma^2 = \sigma_\varepsilon^2$)

4 $H_0$: All $\tau_i = 0$ (no treatment effect present)
  $H_1$: At least one $\tau_i \neq 0$ (treatment effect present)

---

**b  Random effects**

Model: $Y_{ij} = \mu + \tau_i + \varepsilon_{ij}$     $i = 1, 2, ..., a$     $j = 1, 2, ..., n$

ANOVA

| Source | df | SS | MS | F | P | EMS |
|---|---|---|---|---|---|---|
| Total corrected | $an - 1$ | $\sum_{ij} Y_{ij}^2 - CF$ | | | | |
| Treatment (between treatments) | $a - 1$ | $\dfrac{\sum_{i=1}^{a} Y_{i.}^2}{n} - CF$ | $\dfrac{SS_{trt}}{df_{trt}}$ | $\dfrac{MS_{trt}}{MS_{err}}$ | $P_{trt}$ | $\sigma_\varepsilon^2 + n\sigma_\tau^2$ |
| Exp. error (within treatments) | $a(n - 1)$ | $SS_{tc} - SS_{trt}$ | $\dfrac{SS_{err}}{df_{err}}$ | | | $\sigma_\varepsilon^2$ |

**Assumptions and hypotheses**

1 We randomly select $a$ treatments for inclusion from a population of possible treatments (in a random fashion).

2 $\tau_i \sim$ NID ($\mu = 0$, $\sigma^2 = \sigma_\tau^2$)

3 $\varepsilon_{ij} \sim$ NID ($\mu = 0$, $\sigma^2 = \sigma_\varepsilon^2$)

4 $H_0$: $\sigma_\tau^2 = 0$   (No treatment variation present)
  $H_1$: $\sigma_\tau^2 > 0$   (Treatment variation present)

---

**ANOVA notation**

CF: correction factor $= \dfrac{Y_{..}^2}{an}$

df: degrees of freedom

SS: sum of squares

MS: mean square

F: F test ratio

P-value: observed significance level (o.s.l.)

$SS_{tc}$: total corrected sum of squares
$SS_{trt}$: treatment sum of squares
$SS_{err}$: experimental-error sum of squares

$MS_{trt}$: treatment mean square
$MS_{err}$: experimental-error mean square

$P_{trt}$: P-value associated with ANOVA F test

*Note:* The MS column refers to observed values. The EMS column refers to theoretical expected values. NID: independent and distributed normally.

---

We will not develop the theoretical background for this one-tailed (right-hand tail) $F$ test, but will instead refer the interested reader to Walpole and Myers [6], Montgomery [7], Hicks [8], or Box et al. [9].

A statistically significant result in this $F$ test supports the rejection of $H_0$ and we conclude that sufficient evidence is presented to suggest that the average response is not the same at all treatment levels. Plots of response means are very good at helping to elicit and communicate significant fixed effects, or the lack of fixed effects. Even if the ANOVA indicates no treatment effect, additional observations, more highly controlled conditions in the conduct of the physical experiment, or more precise measurement equipment may eventually lead to the detection of a small effect. On the other hand, there actually may be no effect whatsoever (an unlikely case). This simple illustration points out the criticality of interpretating, wording, and communicating experimental findings and the close relationship of interpretation and the statistical and practical significance concepts (Table 20.1). In general, we seek to discover and understand factor effects on our chosen response, and particularly, factors which heavily influence our response (i.e., we apply the Pareto principle). Large influences are more critical to us than small influences.

The objectives in many off-line quality experiments focus on selecting a best alternative—the level of an ingredient or additive, tool geometry, and so on. In fixed effects analyses, a number of methods exist to compare individual treatment means in pairs or in other combinations: the least significant difference test (LSD), Duncan's multiple range test (DM), the Student-Newman-Keuls test (SNK), Tukey's honestly significant difference test (HSD), and Dunnett's test (DT). These pairwise tests will be discussed in a later section. Scheffe's method and single degree of freedom contrasts can be used to tailor many different types of treatment comparisons as well. (References such as Walpole and Myers, Montgomery, Hicks, and Box et al. describe these contrasting methods in detail [10]).

### Random Effects in General

The second basic statistical model, *the random effects model, is appropriate if we seek to estimate components of variation.* Here, the situation involves a large number of possible factor levels. We would select $a$ of these levels at random. For example, given a process where the cooking temperature fluctuates because of thermostat mechanics, we might want to isolate and measure the response variance caused by cooking temperature fluctuation. Hence, $a$ temperature levels are selected at random (not hand-picked as in a fixed effects case) from possible temperature levels occurring in the process. Then, the experiment is performed and an ANOVA table is developed. Assuming independence between the variance attributable to cooking temperature and other sources of variation, the variance of an observation $Y_{ij}$ is

$$\text{Var}(Y_{ij}) = \sigma_\tau^2 + \sigma_\varepsilon^2 \tag{20.5}$$

Here, $\sigma_\tau^2$ represents the factor related variation, and $\sigma_\varepsilon^2$ represents a variation component comprising the natural variation in our experimental units as well as that resulting from uncontrolled variables (i.e., other than cooking temperature in this case). In Figure 20.3b, the $\sigma_\varepsilon^2$ term is estimated by the mean square error $\text{MS}_{err}$. The mean square treatment term $\text{MS}_{trt}$ estimates $n\sigma_\tau^2 + \sigma_\varepsilon^2$. The hypotheses tested are

$$H_0: \ \sigma_\tau^2 = 0 \ (\text{no temperature variation exists})$$
$$H_1: \ \sigma_\tau^2 > 0 \ (\text{temperature variation exists})$$

The $F$ statistic is calculated as

$$F_{calc} = \frac{\text{MS}_{trt}}{\text{MS}_{err}} \qquad df_{numerator} = a - 1, \ df_{denominator} = a(n-1)$$

A significant result (rejection of $H_0$) implies that we can safely conclude (b.o.e.) that $\sigma_\tau^2 > 0$ (i.e., the cooking temperature variation $\sigma_\tau^2$ is large enough to be detected in the presence of the random error variation $\sigma_\varepsilon^2$). Hence, we solve for an estimate of $\sigma_\tau^2$ numerically.

If we cannot reject $H_0$, then the treatment variation is not detectable (in the presence of the error variation) and is assumed to be 0 or small relative to our estimate of $\sigma_\varepsilon^2$. More evidence may result in a different conclusion. As the sample size increases, our ability to detect even a small treatment variation component increases, provided $H_0$ is not absolutely true.

## Fixed Effects Hypothesis Tests

With fixed effects factors, once we examine the data carefully using graphical techniques, we will usually want to perform a detailed quantitative ANOVA. Many computer aids for both mainframes and microcomputers exist including SAS [11], SPSS [12], SYSTAT [ 13], Minitab [14], Statgraphics [15], JMP [16], and so on. Their use simplifies the analytical task, but does not provide insight into interpretation. In general, with fixed effects factors, we produce analyses and interpretations at two levels: (1) at the experiment, ANOVA table level, and (2) at the treatment or factor level.

**Expected Mean Squares and the Overall $F$ Test**    *At the experiment level, we test for detectable model effects (terms in the model—e.g., $\tau_i$).* The statistical nature of the experimental design model structure permits use of the experiment-wide or ANOVA table tests. *We should think of the ANOVA table as a series of rows, each measuring a certain effect, indicated by each source column label.* For example, with respect to the CRD (Table 20.3a), the Total corrected (for the mean) row measures the failure of each individual experimental unit to respond in exactly the same way. If all $Y_{ij}$'s are exactly the same, then $SS_{tc} = 0$.

We break the Total corrected row into two independent measurements: (1) treatment effect and (2) experimental error. In the case of the Treatment row we are measuring the failure of each treatment mean to be the same (e.g., if the $\bar{Y}_{i.}$'s are all equal, then $SS_{trt} = 0$). The Experimental error row measures the failure of each experimental unit, within each treatment, to respond the same. In this case, if each $Y_{ij}$ (within treatment $i$) is equal to $\bar{Y}_{i.}$ for all $i$, then $SS_{err} = 0$.

*It is critical to associate ANOVA rows with the physical experiment in order to thoroughly understand and interpret the results.* Examination of the relationship of Equations 20.2 through 20.4 with the related ANOVA (Table 20.3a) and plot (Figure 20.6c) help us to "visualize" these statistical measurements in a generic sense. The physical meaning must be extracted from the physical observations.

Table 20.3a displays the CRD model in its fixed effects form along with its expected mean squares. The Total corrected (for the overall mean) row is broken out into two independent measures, one for the between treatment measurement and one for the within treatment measurement. *The expected mean squares (EMS) values represent the expected values for the mean squares.* Essentially, the EMS terms are derived by summing the mean square terms over all possible y values and simplifying. The mean square (MS) terms provide independent estimates of the

EMS terms.  We will not discuss the derivation of the EMS values for the CRD, but instead refer our readers to step-by-step developments in texts such as Walpole and Myers, Montgomery, Hicks, and Box, Hunter, and Hunter [17].

Using the EMS terms, we can construct hypotheses to assess the detection of treatment effect $\tau_i$:

$$H_0: \text{ All } \tau_i = 0$$
$$H_1: \text{ At least one } \tau_i \neq 0$$

If we set up a ratio of $\text{EMS}_{trt}$ to $\text{EMS}_{err}$, we can in essence isolate the $\tau_i$ term. Hence, if the null hypothesis is true, the $\tau_i^2$ terms sum to 0 and we are left with a ratio of $\sigma_\varepsilon^2$ over $\sigma_\varepsilon^2$. In other words, under a true $H_0$, in a perfectly controlled experiment (e.g., we control all of the $X_i$'s and $Z_i$'s) we will obtain two independent estimates of the same value, $\sigma_\varepsilon^2$, and from the one-tailed $F$ test, would expect to see an $F$ ratio value of unity, $F = 1$. In any given experiment, when $H_0$ is true or when the treatment effect $\tau_i$ is small relative to the magnitude of the experimental error $\sigma_\varepsilon^2$, we expect to see $F \simeq 1$ because of the sampling variation present. Hence, in experimental data, if the treatment effects are 0 or relatively small, it is not uncommon to encounter $F < 1$. When this happens, we see no support for rejecting $H_0$, and attribute this unusual $F$ ratio to sampling variation. On the other hand, when large treatment effects are present, we see $F$ ratios where $F \gg 1$.

Because of the model form shown in Equation 20.1, the hypothesis test for the treatment effect above is equivalent to

$$H_0: \mu_1 = \mu_2 = \cdots = \mu_a$$

$$H_1: \text{ At least one } \mu_i \text{ not equal to the others}$$

This result will become more apparent later when we introduce Equations 20.6, 20.7, and 20.8.

We must also note assumption 3 listed in Table 20.3a, where we assume the residuals, $\varepsilon_{ij}$, are normally distributed with $\mu = 0$ and variance $\sigma_\varepsilon^2$. This assumption is necessary to support the validity of the fixed effects treatment hypothesis test. In essence, the graphical analysis and even the ANOVA SS and MS calculations can be made without regard to the normality assumption. In other words, a fixed effects model can be analyzed on a strictly graphical basis. Nevertheless, the ANOVA method adds detail to the analysis, provided the normality assumption can be satisfied.

Here, *we use P-values or traditional $\alpha$ values to assess the probability of a type I error.* We see $\alpha = 0.05$ or $\alpha = 0.01$ typically used as reasonable levels of risk. The smaller the $\alpha$ we choose, the more conservative we are in risk taking, rejecting $H_0$ when it is true. By using and stating the $P$-value, we can provide more information than a simple "do not reject $H_0$" or "reject $H_0$" at a stated $\alpha$ level. Most computer aids provide $P$-values for our convenience, which we may include in our interpretation. The point is, if we have a $P$-value available, we need not refer to an $F$ table for a critical or tabulated $F$ value. But, as is the case with all hypothesis test interpretations, we must state that our conclusions are based on the evidence at hand. Other, or additional, evidence may lead to a different interpretation.

---

**Example 20.3**

The cookie experiment described previously in Example 20.2 was a fixed effects factor model, since we hand-picked the factor levels (cooking times). It yielded the data shown in Figure 20.6b and plotted in Figure 20.6c. Develop an ANOVA table using the technique, associated assumptions, and hypotheses shown in Table 20.3a.

**Solution**

Results and sample calculations are shown in Table 20.4. Our overall $F$ test in the ANOVA produces $F = 9.99$ and $P$-value $= 0.0044$. SAS PROC ANOVA or GLM can be used as a computer aid here [18]. SAS code and selected output for this example are shown in the Computer Aided Analysis Supplement, Solutions Manual. If we rely on manual calculations, using our $F$ table (Section IX, Table IX.4), we would say $P$-value $< 0.01$ (since our $F_{calc} = 9.99$ is greater than the tabulated $F_{0.01,3,8} = 7.59$). For our fixed effects factor of bake time,

$H_0$:  All $\tau_i = 0$ or all $\mu_i$ are equal (no cooking time effect exists or all cooking time percent elongation response means are equal)

$H_1$:  At least one $\tau_i \neq 0$ or not all $\mu_i$ are equal (a cooking time effect exists or at least one cooking time produces a different mean percent elongation response)

Based on our evidence, we reject $H_0$ in favor of $H_1$. This conclusion is also supported qualitatively by our plot (Figure 20.6c).

---

**Prediction and Residual Calculations**

We may want to predict our cooking time treatment mean responses. In this case, we must base our predictions on our model. Here, we can consult our basic CRD model, Equations 20.1 and 20.2, and observe that

$$\hat{\mu} = \bar{Y}_{..} \tag{20.6}$$

$$\hat{\tau}_i = \bar{Y}_{i.} - \bar{Y}_{..} \tag{20.7}$$

and
$$\hat{\mu}_{ij} = \bar{Y}_{..} + (\bar{Y}_{i.} - \bar{Y}_{..}) = \bar{Y}_{i.} \tag{20.8}$$

Since the observation subscript, $j$, is not active on the right-hand side of Equation (20.8), $\hat{\mu}_{ij}$ is interpreted as a predicted treatment mean. Hence, we can calculate our treatment mean predictions for our CRD experiment, Example 20.3, as follows:

**TABLE 20.4**   ULTRACHEWY COOKIE CRD ANOVA RESULTS (EXAMPLE 20.3)

| Source | df | SS | MS | F | P-value |
|---|---|---|---|---|---|
| Total corrected | $12 - 1 = 11$ | 94.917 | | | |
| Treatment (between cooking times) | $4 - 1 = 3$ | 74.917 | 24.972 | 9.99 | 0.0044 |
| Experimental error (within cooking times) | $4(3 - 1) = 8$ | 20.000 | 2.500 | | |

Sample calculations

$SS_{tc}$ = $(4^2 + 1^2 + 2^2 + 9^2 + \cdots + 2^2 + 4^2) - (59^2/12) = 94.917$

$SS_{trt}$ = $(7^2 + 26^2 + 17^2 + 9^2)/3 - (59^2/12) = 74.917$

$SS_{err}$ = $94.917 - 74.917 = 20.000$

$$\hat{\mu}_{i=1} = \overline{Y}_{..} + (\overline{Y}_{1.} - \overline{Y}_{..}) = \overline{Y}_{1.} = 2.333\%$$

$$\hat{\mu}_{i=2} = \overline{Y}_{..} + (\overline{Y}_{2.} - \overline{Y}_{..}) = \overline{Y}_{2.} = 8.667\%$$

$$\hat{\mu}_{i=3} = \overline{Y}_{..} + (\overline{Y}_{3.} - \overline{Y}_{..}) = \overline{Y}_{3.} = 5.667\%$$

$$\hat{\mu}_{i=4} = \overline{Y}_{..} + (\overline{Y}_{4.} - \overline{Y}_{..}) = \overline{Y}_{4.} = 3.000\%$$

By definition, the residual $e_{ij}$ for observation $j$ in treatment $i$ is

$$e_{ij} = Y_{ij} - \hat{\mu}_{ij} \tag{20.9}$$

It then follows for the CRD that

$$\hat{\varepsilon}_{ij} = e_{ij} = Y_{ij} - \hat{\mu}_{ij} = Y_{ij} - \overline{Y}_{i.} \tag{20.10}$$

At this point, we have a good deal of information. We can clearly state that at least one cooking time mean is different from the others, b.o.e., at any reasonable $\alpha$ level, say, $\alpha = 0.05$. However, we are still at a loss to recommend a cooking time for our bigger is better chewiness characteristic.

## Treatment Comparisons (Fixed Effects)

*Pairwise treatment tests allow us to make direct comparisons between treatments, whereas the overall hypothesis test, from the ANOVA table, for fixed effects generally cannot support a decision as to a "best" treatment.* Our plot (Figure 20.6c) suggests that treatment 2 (10-min cooking time) produces the largest percent elongation. Unless we have some way of judging the variation present, we may be fooled by our plot scaling (e.g., we can make the sample treatment means appear closer by reducing the distance between our percent marks and vice versa). So far, we have observed the scatter within the treatments and thus may qualitatively judge the experimental error. However, usually we will want to rely on more quantitative methods.

**Standard Error Interval Arrows for Treatment Means** We will develop a plotting technique based on the standard error bars sometimes used in engineering data displays. We will use our $MS_{err}$ as an estimate of $\sigma_{\hat{\varepsilon}}^2$ and then calculate an interval of $\pm b$ standard errors, which we will depict with arrows. We will next place this interval on our plot, centered about a sample treatment mean, to provide us with a picture of both the estimate of our treatment mean as well as its standard error. We calculate our standard error interval, $\pm b$ s.e.i., as

$$\pm b \text{ s.e.i.} = \pm b \sqrt{\frac{MS_{err}}{n'}} \qquad (20.11)$$

where $n'$ represents the number of observations in the plotted mean.

Technically, we could substitute $b = t_{\alpha/2,\, df_{err}}$ into Equation 20.11 and develop a $(1-\alpha)$ 100 percent confidence interval on the treatment mean, assuming a normal population. We can use our $t$ Table, Table IX.3, to obtain our value for $b$. For example, if we desire a 95 percent confidence interval when $df_{err} = 8$, we would use $b = t_{0.025,8} = 2.306$. We can develop a simpler, but less precise method, using $b = 2$ (e.g., about 95 percent confidence interval for a large sample size). Using $b = 2$ will not provide a precise confidence interval, but it will allow us to picture both our estimated treatment mean and its associated standard error, all on the same plot.

As an illustration, referring back to Example 20.3, a 95 percent confidence interval on a treatment mean is calculated as

$$\pm 2.306 \text{ s.e.i.} = \pm 2.306 \sqrt{\frac{2.500}{3}} = \pm 2.11 \text{ percent elongation}$$

A $\pm 2$ standard error interval, $\pm 2$ s.e.i., is calculated as

$$\pm 2 \text{ s.e.i.} = \pm 2 \sqrt{\frac{2.500}{3}} = \pm 1.83 \text{ percent elongation}$$

The $\pm 2$ s.e.i. arrows are plotted on the highest sample response (the 10 min cooking time treatment), see Figure 20.7, since it is the most promising treatment response (bigger is better). However, the $\pm 2$ s.e.i. interval applies to all of the 4 estimated treatment means plotted in Figure 20.7.

By seeing both the estimated treatment mean and its standard error in the same picture, we obtain a feel for the precision, "goodness," of our estimate. This feel is critical if we are going to base a process or product decision on our experimental results (e.g., recommend a cooking time). The narrower our interval, the more precise our estimate (the more confidence we have in our recommendation).

**Statistical Tests** Now, we will examine five basic quantitative statistical methods, all dealing with pairwise comparisons: (1) the least significant difference (LSD) test, (2) the Student-Newman-Keuls (SNK) range test, (3) the Duncan's Multiple range (DM) test, (4) the Tukey's honestly significant (HSD) difference test, and (5) the Dunnett's test, DT. Here, our pairwise hypotheses are structured as

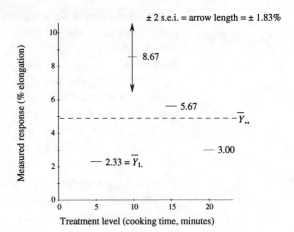

**FIGURE 20.7**    Treatment means plot, chewy chocolate chip cookie data (Example 20.3).

$$H_0: \mu_i = \mu_j \qquad \text{for all } ij \text{ combinations, } i \neq j$$
$$H_1: \mu_i \neq \mu_j$$

Other techniques for developing and assessing contrasts involving more than two treatment means simultaneously can be found in Montgomery, Hicks, and Box, Hunter, and Hunter [19].

Figure 20.8 lays out the five techniques along a type I error probability continuum. Our choice of method generally depends on two factors: (1) the number of treatments we seek to pair and test and (2) the type I error probability classification we seek (a pairwise $\alpha$, an experiment-wise $\alpha$, or a compromise).

***Least Significant Difference (LSD) Test***    The LSD test is developed in the same manner as a two-sample $t$ test of means. Hence, the $\alpha$ significance level applies to each pair of sample means tested. The $\text{LSD}_\alpha$ test statistic is calculated as

$$\text{LSD}_\alpha = t_{\alpha/2,\, df_{err}} \sqrt{\text{MS}_{err}\left(\frac{1}{n_1} + \frac{1}{n_2}\right)} = t_{\alpha/2,\, df_{err}} \sqrt{\frac{2\text{MS}_{err}}{n'}} \qquad (20.12)$$

Here, $n'$ represents the number of observations making up each treatment sample mean (e.g., $n_1 = n_2 = n'$ for a balanced, single-factor CRD experiment). The $t$ is our usual Student's $t$ statistic.

**FIGURE 20.8**    Pairwise test spectrum relative to type I error probability classification.

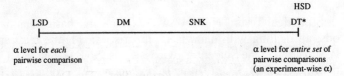

*DT—the Dunnett's test is an experiment wise test for a control (e.g., no active treatment) versus each active treatment.

The $C_2^a$ hypotheses for a full set of comparisons are

$$H_0:\ \mu_i = \mu_j \qquad \text{for all } ij \text{ combinations, } i \neq j$$

$$H_1:\ \mu_i \neq \mu_j$$

where $$C_2^a = \frac{a!}{2!(a-2)!}$$

If $$\text{LSD}_\alpha < |\,\hat{\mu}_i - \hat{\mu}_j\,|$$

then we reject $H_0$ at the $\alpha$ significance level, b.o.e.

If we use the LSD and wish to estimate the overall experiment-wise $\alpha$, we can use the form

$$\alpha_{\text{exp-wise}} \simeq 1 - (1 - \alpha_{\text{pairwise}})^{C_2^a} \tag{20.13}$$

For example, if we use $\text{LSD}_{0.05}$ for a set of five treatments,

$$\alpha_{\text{exp-wise}} \simeq 1 - (0.95)^{C_2^5} = 1 - (0.95)^{10} = 40.1\%$$

Hence, our overall chance of making a type I error is large, considering we are making 10 individual hypothesis tests, with $\alpha = 0.05$ for each of the 10 tests.

***Student-Newman-Keuls (SNK) Range Test***    The SNK test is a range test. Here, we must calculate a series of statistics called "least significant ranges" ($\text{LSR}_S$).

$$\text{LSR}_{S,\alpha,p} = q_\alpha\,(p,\,\text{df}_{\text{err}}) \sqrt{\frac{\text{MS}_{\text{err}}}{n'}} \tag{20.14}$$

where $q_\alpha(p,\,\text{df}_{\text{err}})$ = studentized range statistic (see Section IX, Table IX.7).

$p$ = range index, $p = 2, 3, ..., a'$ where $a'$ = number of treatment levels compared.

$\text{df}_{\text{err}}$ = degrees of freedom of experimental error (from the ANOVA table).

$n'$ = number of observations in each treatment sample mean.

The use of a tabular form helps to keep order in the SNK test. Table 20.5a lays out the LSR calculation sequence, and Table 20.5b provides an organization matrix for the pairwise test of means. Here, if

$$\text{LSR}_{S,\alpha,p} < |\,\hat{\mu}_i - \hat{\mu}_j\,| \qquad i \neq j$$

for the proper value of $p$, as shown in Table 20.5b, then we reject the corresponding $H_0$ at the $\alpha$ significance level, b.o.e., and conclude that $\mu_i \neq \mu_j$ as stated in $H_1$.

***Duncan's Multiple (DM) Range Test***    The DM test is similar in structure to the SNK test. The major difference is that the Duncan's multiple range statistic $r_\alpha(p,\,\text{df}_{\text{err}})$ is used in place of $q_\alpha\,(p,\,\text{df}_{\text{err}})$.

**TABLE 20.5**   LEAST SIGNIFICANT RANGE TABLE LAYOUTS

**a** Least-significant-range (LSR) calculation sequence

| $p$ | 2 | 3 | $\cdots$ | $a'$ |
|---|---|---|---|---|
| $q_\alpha(p, df_{err})$ | $q_\alpha(2, df_{err})$ | $q_\alpha(3, df_{err})$ $\cdots$ | | $q_\alpha(a', df_{err})$ |
| $LSR_{S,\alpha}$ | $LSR_{S,\alpha,2}$ | $LSR_{S,\alpha,3}$ | $\cdots$ | $LSR_{S,\alpha,a'}$ |

**b** Least-significant-range (LSR) comparison layout

| Treatment means | Largest mean | Next-to-largest mean | $\cdots$ | Next-to-smallest mean | Smallest mean |
|---|---|---|---|---|---|
| Smallest mean | Compare difference with LSR for $p = a'$. | Compare difference with LSR for $p = a' - 1$. | $\cdots$ | Compare difference with LSR for $p = 2$. | |
| Next-to-smallest mean | Compare difference with LSR for $p = a' - 1$. | Compare difference with LSR for $p = a' - 2$ | | | |
| $\vdots$ | $\vdots$ | $\vdots$ | | | |
| Next-to-largest mean | Compare difference with LSR for $p = 2$ | | | | |
| Largest mean | | | | | |

$$LSR_{D,\,\alpha,\,p} = r_\alpha\,(p,\,df_{err})\,\sqrt{\frac{MS_{err}}{n'}} \qquad (20.15)$$

where $p$, $df_{err}$, and $n'$ are defined as in the SNK test.

The Duncan's multiple range statistic $r_\alpha\,(p,\,df_{err})$ appears in Section IX, Table IX.8. We use the same tabular layout and comparison rules as in the SNK range test (Table 20.5). Hence, if

$$LSR_{D,\,\alpha,\,p} < |\,\hat{\mu}_i - \hat{\mu}_j\,| \qquad i \neq j$$

for the proper $p$, as in the SNK case, then we reject $H_0$ at the $\alpha$ significance level, b.o.e., and conclude that $\mu_i \neq \mu_j$, as stated in $H_1$.

***Tukey's Honestly Significant Difference* (HSD) *Test***   The HSD test is the experiment-wise $\alpha$ counterpart to the pairwise $LSD_\alpha$ test. In the HSD we use

$$HSD_\alpha = q_\alpha\,(p = a',\,df_{err})\,\sqrt{\frac{MS_{err}}{n'}} \qquad (20.16)$$

The $\text{HSD}_\alpha$ is simply the largest $\text{LSR}_{S,\,\alpha,\,p}$, where $q_\alpha$ $(p = a',\,\text{df}_{\text{err}})$, $p = a'$, and $n'$ are defined as in the SNK test (see Section IX, Table IX.7).  If

$$\text{HSD}_\alpha < |\,\hat{\mu}_i - \hat{\mu}_j\,| \qquad i \neq j$$

then, we reject $H_0$ at the $\alpha$ significance level, b.o.e., and conclude that $\mu_i \neq \mu_j$, as stated in $H_1$.  We should also note the following relationships:

$$\text{LSD}_\alpha = \text{LSR}_{S,\,\alpha,\,p=2} \tag{20.17}$$

$$\text{HSD}_\alpha = \text{LSR}_{S,\,\alpha,\,p=a'} \tag{20.18}$$

Hence, the LSD, SNK, and HSD tests form a consistent continuum regarding pairwise to experiment-wise type I error probabilities.

***Dunnett's Test (DT)***    The DT is used where we seek to compare a "control" response, or a special treatment where nothing was actually done to the experimental units, to treated responses.  A control establishes a baseline from which to judge the responses of the actual treatments.  For example, controls are widely used in experiments to evaluate drug, fertilizer, and additive effectiveness.

The DT provides an experiment-wise $\alpha$ for a comparison of the control, labeled treatment $a'$, with each actual treatment on a pairwise basis.  The DT can be developed for one- or two-tailed alternate hypotheses.

Here $\text{DT}_\alpha$ is calculated as

$$\text{DT}_\alpha = d_\alpha\,(a' - 1,\,\text{df}_{\text{err}})\,\sqrt{\text{MS}_{\text{err}}\left(\frac{1}{n_i} + \frac{1}{n_{a'}}\right)} \tag{20.19}$$

where $d_\alpha\,(a' - 1,\,\text{df}_{\text{err}})$ = Dunnett's statistic one- or two-tailed (see Section IX, Table IX.9).

$a' - 1$ = the number of true treatments (not including the control).
$\text{df}_{\text{err}}$ = degrees of freedom error (from the ANOVA table).
$n_i$ = number of observations in the sample mean of treatment $i$.
$n_{a'}$ = number of observations in the sample mean of the control.

For the two-tailed $H_1$, if

$$\text{DT}_\alpha < |\,\hat{\mu}_{a'} - \hat{\mu}_i\,| \qquad i = 1, 2, \ldots, a' - 1$$

then reject $H_0$ at the $\alpha$ significance level, b.o.e.
For a one-tailed $H_1$, if

$$\text{DT}_\alpha < \hat{\mu}_{a'} - \hat{\mu}_i \quad \text{and} \quad H_1\colon \;\mu_{a'} > \mu_i \qquad i = 1, 2, \ldots, a' - 1$$

or if $\quad \text{DT}_\alpha < \hat{\mu}_i - \hat{\mu}_{a'} \quad \text{and} \quad H_1\colon \;\mu_{a'} < \mu_i$

then reject $H_0$ at the $\alpha$ significance level, b.o.e.  Here, $\hat{\mu}_{a'}$ represents the sample mean of the control group response.

**Example 20.4**

Using the data (Figure 20.6*b*) from Example 20.1 and the resulting ANOVA (Table 20.4), develop pairwise treatment mean comparisons using the LSD, SNK, DM, and HSD methods and a 0.05 significance level. (Use Tables IX.3, IX.7, and IX.8, as needed.) Then, comment on the results.

**Solution**

The LSD, SNK, DM, and HSD calculations are developed below. Table 20.6 contains the layout for the four comparisons. The * marks the significant differences detected.

*LSD:*

$$LSD_{0.05} = t_{0.025,8} \sqrt{\frac{2MS_{err}}{n'}} = 2.306 \sqrt{\frac{2(2.500)}{3}} = 2.977$$

*SNK:*

$$LSR_{S,0.05,p} = q_{0.05}(p, 8) \sqrt{\frac{MS_{err}}{n'}} = q_{0.05}(p, 8) \sqrt{\frac{2.500}{3}}$$

| $p$ | 2 | 3 | 4 |
|---|---|---|---|
| $q_{0.05}(p,8)$ | 3.26 | 4.04 | 4.53 |
| $LSR_{S, 0.05, p}$ | 2.976 | 3.688 | 4.135 |

*DM:*

$$LSR_{D, 0.05, p} = r_{0.05}(p, 8) \sqrt{\frac{MS_{err}}{n'}} = r_{0.05}(p, 8) \sqrt{\frac{2.500}{3}}$$

| $p$ | 2 | 3 | 4 |
|---|---|---|---|
| $r_{0.05}(p, 8)$ | 3.26 | 3.39 | 3.47 |
| $LSR_{D, 0.05, p}$ | 2.976 | 3.095 | 3.168 |

*HSD:*

$$HSD_{0.05} = q(4, 8) \sqrt{\frac{MS_{err}}{n'}} = 4.53 \sqrt{\frac{2.500}{3}} = 4.135$$

We have already noticed that our LSD, SNK, DM, and HSD critical values calculated above vary (e.g., $LSD_\alpha \leq HSD_\alpha$). A close examination of Table 20.6

**TABLE 20.6**    PAIRWISE TREATMENT MEANS COMPARISONS FOR THE CHEWY COOKIE EXPERIMENT (EXAMPLE 20.4)

**a** LSD test

| Trt | Trt \ Avg / Avg | 2 | 3 | 4 | 1 |
|---|---|---|---|---|---|
|  |  | 8.667 | 5.667 | 3.000 | 2.333 |
| 1 | 2.333 | 6.334* <br> LSD = 2.977 | 3.334* <br> LSD = 2.977 | 0.667 <br> LSD = 2.977 |  |
| 4 | 3.000 | 5.667* <br> LSD = 2.977 | 2.667 <br> LSD = 2.977 |  |  |
| 3 | 5.667 | 3.000* <br> LSD = 2.977 |  |  |  |
| 2 | 8.667 | * Significant at α = 0.05 b.o.e. |  |  |  |

**b** SNK test

| Trt | Trt \ Avg / Avg | 2 | 3 | 4 | 1 |
|---|---|---|---|---|---|
|  |  | 8.667 | 5.667 | 3.000 | 2.333 |
| 1 | 2.333 | 6.334* <br> LSR = 4.135 | 3.334 <br> LSR = 3.688 | 0.667 <br> LSR = 2.976 |  |
| 4 | 3.000 | 5.667* <br> LSR = 3.688 | 2.667 <br> LSR = 2.976 |  |  |
| 3 | 5.667 | 3.000* <br> LSR = 2.976 |  |  |  |
| 2 | 8.667 | * Significant at α = 0.05 b.o.e. |  |  |  |

**c** DM test

| Trt | Trt \ Avg / Avg | 2 | 3 | 4 | 1 |
|---|---|---|---|---|---|
|  |  | 8.667 | 5.667 | 3.000 | 2.333 |
| 1 | 2.333 | 6.334* <br> LSR = 3.168 | 3.334* <br> LSR = 3.095 | 0.667 <br> LSR = 2.976 |  |
| 4 | 3.000 | 5.667* <br> LSR = 3.095 | 2.667 <br> LSR = 2.976 |  |  |
| 3 | 5.667 | 3.000* <br> LSR = 2.976 |  |  |  |
| 2 | 8.667 | * Significant at α = 0.05 b.o.e. |  |  |  |

**d** HSD test

| Trt | Trt \ Avg / Avg | 2 | 3 | 4 | 1 |
|---|---|---|---|---|---|
|  |  | 8.667 | 5.667 | 3.000 | 2.333 |
| 1 | 2.333 | 6.334* <br> HSD = 4.135 | 3.334 <br> HSD = 4.135 | 0.667 <br> HSD = 4.135 |  |
| 4 | 3.000 | 5.667* <br> HSD = 4.135 | 2.667 <br> HSD = 4.135 |  |  |
| 3 | 5.667 | 3.000 <br> HSD = 4.135 |  |  |  |
| 2 | 8.667 | * Significant at α = 0.05 b.o.e. |  |  |  |

shows that, b.o.e., treatment 1 (trt 1) and treatment 3 (trt 3) differ significantly in the LSD, but not in the HSD. Here, we are seeing a demonstration of the pairwise α and experiment-wise α concept shown in Figure 20.8.

## Fixed Effects Extensions

In Example 20.4, we were dealing with a bigger is better response and performing pairwise difference tests. To assess a best treatment setting, we must essentially deal with one-tailed alternative hypotheses. The LSD test can be readily converted for one-tailed comparisons. We will leave this conversion and the development of a nominal is best–response type comparison test as Discovery exercises.

## Random Effects Hypothesis Tests

As stated earlier, the random effects model is appropriate when we want to study the variation associated with a factor. Table 20.3b lists the model, ANOVA, EMS, and assumptions for the CRD random effects model. The overall $F$ test is based on an EMS ratio. For the Treatment row, the EMS is

$$EMS_{trt} = \sigma_\varepsilon^2 + n\sigma_\tau^2 \qquad\qquad (20.20)$$

and the EMS for the Experimental error row is

$$EMS_{err} = \sigma_\varepsilon^2 \qquad\qquad (20.21)$$

The ratio of $MS_{trt}$ over $MS_{err}$ forms the basis for the overall $F$ test. Here, we are in essence assessing our ability to measure the variation associated with the treatment factor relative to the experimental error present. Our hypotheses are

$$H_0: \ \sigma_\tau^2 = 0 \ \text{(no treatment factor variation present)}$$

$$H_1: \ \sigma_\tau^2 > 0 \ \text{(treatment factor variation present)}$$

Here, relatively large $F$ values (obviously greater than 1) support rejection of $H_0$. If we reject $H_0$, we usually proceed to estimate $\sigma_\tau^2$:

$$\hat{\sigma}_\tau^2 = \frac{MS_{trt} - MS_{err}}{n} \qquad\qquad (20.22)$$

If we do not reject $H_0$, we need not attempt to estimate $\sigma_\tau^2$, because, in the extreme case, when $F < 1$, we would obtain $\hat{\sigma}_\tau^2 < 0$, which makes no sense at all. Hence, if the $F$ test is not significant, we typically state that we cannot distinguish $\sigma_\tau^2$ from 0 in the presence of the experimental error $\sigma_\varepsilon^2$ at the $\alpha$ significance level, b.o.e. In other words, if $H_0$ is not absolutely true, and we fail to reject $H_0$, our experimental error is large enough to obscure our ability to detect $\sigma_\tau^2$. Hence, if a true $H_0$ does not seem physically feasible, and we desire to estimate $\sigma_\tau^2$, we must attempt to obtain more precise estimates by increasing our sample size, tightening our control, being more careful in our physical laboratory procedures, and in breaking out (from the experimental error) other sources of variation (e.g., by blocking).

---

## Example 20.5

In the case of our chewy chocolate chip cookies our CE diagram (Figure 20.4) indicates that the cookie's position on the baking pan may affect the elongation percent response. Design an experiment to study the random effect that cookie placement on the baking pan plays in determining the chewiness of our cookie.

### Solution

We have designed an experiment to measure the baking pan position variation $\sigma_\tau^2$ relative to our cookie percent elongation response. First, we will fix the cooking time at 10 min. Next, we randomly select four general positions, or zones, on our baking pan: right side, left side, right center, and left center. Next, we

| Treatment (baking-pan position) | Observation | | | $Y_{i.}$ |
|---|---|---|---|---|
| | 1 | 2 | 3 | |
| 1 (right) | 4.0 | 5.0 | 3.0 | 12.0 |
| 2 (right center) | 8.0 | 10.0 | 11.0 | 29.0 |
| 3 (left center) | 7.0 | 9.0 | 6.0 | 22.0 |
| 4 (left) | 6.0 | 5.0 | 7.0 | 18.0 |

Response measured in percent elongation.   $Y_{..} = 81.0$

**TABLE 20.7** RANDOM EFFECTS DATA FOR PRODUCT VARIATION DUE TO COOKIE POSITION ON BAKING PAN (EXAMPLE 20.5)

bake a large batch of cookies. Finally, we randomly sample three cookies from each position. Here, we have used a rather crude division of our baking pan position, analogous to a discrete distribution with four pan locations, which we will sample.

Using our testing device, we pull each cookie apart and record the percent elongation. Our data table is shown in Table 20.7. Using the ANOVA method (Table 20.3*b*) we developed the ANOVA shown in Table 20.8. The hypotheses associated with the overall $F$ test are

$H_0$:  $\sigma_\tau^2 = 0$ (no variation due to cookie placement on the baking pan exists; i.e., response is not influenced by cookie position)

$H_1$:  $\sigma_\tau^2 > 0$ (variation due to cookie placement on the baking pan exists; i.e., response is influenced by cookie position)

The results yielded $F_{calc} = 10.18$ with $P$-value $= 0.0042$. Here our $P$-value was extracted from our computer aided analysis (see the Computer Aided Analysis

**TABLE 20.8** RANDOM EFFECTS ANOVA TABLE FOR PRODUCT VARIATION DUE TO COOKIE POSITION ON BAKING PAN (EXAMPLE 20.5)

| Source | df | SS | MS | F | P-value |
|---|---|---|---|---|---|
| Total corrected | 12 − 1 = 11 | 64.250 | | | |
| Treatment (between positions) | 4 − 1 = 3 | 50.917 | 16.972 | 10.18 | 0.0042 |
| Exp. error (within positions) | 4(3 − 1) = 8 | 13.333 | 1.667 | | |

Sample calculations

$SS_{tc}$ = $(4^2 + 5^2 + 3^2 + 8^2 + \cdots + 5^2 + 7^2) - (81^2/12) = 64.250$

$SS_{trt}$ = $(12^2 + 29^2 + 22^2 + 18^2)/3 - (81^2/12) = 50.917$

$SS_{err}$ = $64.250 - 50.917 = 13.333$

Supplement, Solutions Manual). Or, using our manual $F$ table method (see Section IX, Table IX.4), we could state that $P < 0.01$.

The evidence clearly supports the rejection of $H_0$. Hence, we will estimate both $\sigma_{\varepsilon}^2$ and $\sigma_{\tau}^2$. Since

$$\text{EMS}_{\text{err}} = \sigma_{\varepsilon}^2$$

$$\text{MS}_{\text{err}} = \hat{\sigma}_{\varepsilon}^2 = 1.667$$

Since

$$\text{EMS}_{\text{trt}} = \sigma_{\varepsilon}^2 + n\sigma_{\tau}^2$$

$$\hat{\sigma}_{\tau}^2 = \frac{\text{MS}_{\text{trt}} - \text{MS}_{\text{err}}}{n} = \frac{16.972 - 1.667}{3} = 5.102$$

In order to interpret our results, we can state that the variance associated with the population of elongation percentages $Y_{ij}$ is

$$\text{Var}\,(Y_{ij}) = \sigma_{\varepsilon}^2 + \sigma_{\tau}^2$$

and its estimate is

$$\widehat{\text{Var}}\,(Y_{ij}) = 1.667 + 5.102 = 6.769$$

In other words, we are looking at an elongation average estimate of $\bar{Y}_{..} = 6.75\%$ and a standard deviation estimate $(6.769)^{1/2} = 2.60\%$.

We can also state that our evidence suggests that roughly $5.102/6.769 = 0.754$, or about 75 percent, of the variation encountered is attributable to cookie position on the baking pan. Our interpretation is that our product's chewiness variation is heavily influenced by the position the cookie occupies on the baking pan, for the recipe used, baked for 10 min., b.o.e.. If this product variation is judged to be excessive, a thorough "variance reduction" project concentrating on the cooking pan and cooking process will be in order.

### Random Effects Extensions

*Random effects factor experiments are useful in estimating components of product-process variation attributable to factors such as machines, operators, tooling, measurement and instruments, and so on.* Exercises in this chapter as well as in Chapter 21 address this application.

## 20.6 BLOCKING

*Blocking can be very effective in helping to isolate treatment or factor effects when the experimental units are somewhat heterogeneous with respect to the response of interest.* Blocking is a stratification process where experimental units are grouped together to form blocks. Blocking will be ineffective when the experimental units are either homogeneous or highly heterogeneous and thus do not lend themselves to stratification.

*Blocking variables are sometimes called "nuisance" or "noise" variables, in that we want to isolate their effect (which is not usually of interest) as far as possible from the experimental error.* This isolation allows us to detect smaller treatment differences, in the case of fixed effects, and smaller treatment variances, in the case of random effects. Table 20.9 lists some typical blocking situations and variables.

The physical requirements in blocking create some difficulties, since the experimental units within each block should be more alike (with respect to the anticipated response) than between different blocks. For example, in an experiment involving hand assembly methods, we might find it beneficial to block on operators (as individuals) in order to remove operator-to-operator variation from the experimental error mean square. This manner of blocking may help us isolate assembly method effects. Here, each operator, one block, must be observed using each operating method.

In another case, for example, where we may want to test for fertilizer effects in crops, we might block on soil characteristics of the land. Here, we might have sandy, sandy loam, and clay based soil blocks, where each fertilizer treatment must be used in each soil type. Physically, we must locate suitable plots of land, blocks, and work within each.

**TABLE 20.9   FACTOR BLOCKING DESCRIPTIONS**

| Objective | Factor of interest | Blocking variable | Comment |
|---|---|---|---|
| Determine fastest assembly method. | Assembly methods (fixed effect) | Human operators (random effect) | Each operator acts as a block since we think that a good deal of variation may exist from person to person relative to our chosen response. |
| Determine best drying temperature for product quality. | Selected temperatures (fixed effect) | Time of day, humidity (fixed or random effect) | The humidity of ambient air may effect the process, blocks may be morning, afternoon, and night. |
| Increase effectiveness of fuel additive to reduce pollution. | Selected fuel additives (fixed effect) | Automobile age (fixed or random effect) | Age may reflect technology in engine performance (e.g., 0–1, 2–3, 4–5 years). |
|  |  | Automobile wear or usage (fixed or random effect) | Wear or usage may reflect ability of engine to respond (e.g., 0–20 k, 21–40 k, 41–60 k, 61–80 k, 81–100 k mi). |
| Reduce variation in a product dimension. | Selected product units (random effect) | Actual measurement instruments used on factory floor (random effect) | We expect that different measuring instruments may produce variation in the measurements of the dimensions. |

Experiments blocked in one direction (stratified by one nuisance variable) are simply termed blocked experiments. Experiments blocked in two directions (stratified by two nuisance variables simultaneously) are termed Latin square (LS) designs. In a Latin square design, two restrictions on randomization are involved, typically called "row" and "column" effects. The Latin square restriction, that the number of treatments be equal to the number of columns (blocks) in one direction and also equal to the number of rows (blocks) in the second direction, places rather harsh restrictions on the physical layout of the experiment. Experiments blocked in three directions are referred to as Graeco-Latin squares. Figure 20.9 depicts the blocking concept in terms of the total corrected sum of squares ($SS_{tc}$). Blocking is compatible with both single- and multiple-factor experiments.

### Randomized Complete Block Design

*In the randomized complete block (RCB) we introduce one restriction on randomization where the experimental units are stratified by blocks and then randomized within each block:* Each treatment must appear once and only once in each block. Hence, we group the experimental units into blocks and then assign treatments randomly to the units within each block, one block at a time. We must have at least as many experimental units within each block as we have treatments. Figure 20.10 provides a depiction of the physical layout and data table layout for a RCB design.

The RCB model and sum of squares expansion are

*RCB Model:*

$$Y_{ij} = \mu + \tau_i + \beta_j + \varepsilon_{ij} \tag{20.23}$$

where $Y_{ij}$ = response observation
$\mu$ = overall mean
$\tau_i$ = treatment effect; $i = 1, 2, \ldots, a$
$\beta_j$ = block effect; $j = 1, 2, \ldots, n$
$\varepsilon_{ij}$ = random error

**FIGURE 20.9**  Blocking concept in terms of sum of squares.

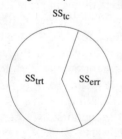

(a)  No blocking—completely randomized design

(b)  Blocking on one nuisance or noise variable–randomized complete block design

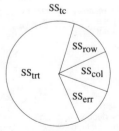

(c)  Blocking on two nuisance or noise variables–Latin square design

**FIGURE 20.10** RCB physical layout and data table.

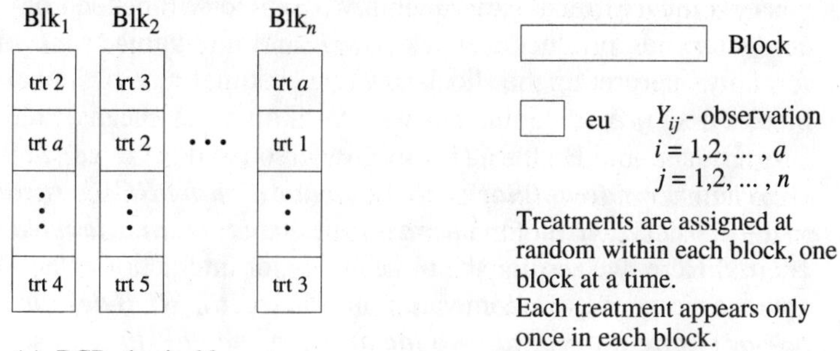


(*a*) RCB physical layout

| | | $Blk_j$ | | | | Treatment statistics | |
|---|---|---|---|---|---|---|---|
| | | 1 | 2 | ⋯ | $n$ | $Y_{i.}$ | $\overline{Y}_{i.}$ |
| | 1 | $Y_{11}$ | $Y_{12}$ | ⋯ | $Y_{1n}$ | $Y_{1.}$ | $\overline{Y}_{1.}$ |
| | 2 | $Y_{21}$ | $Y_{22}$ | ⋯ | $Y_{2n}$ | $Y_{2.}$ | $\overline{Y}_{2.}$ |
| $Trt_i$ | ⋮ | ⋮ | ⋮ | | ⋮ | ⋮ | ⋮ |
| | $a$ | $Y_{a1}$ | $Y_{a2}$ | ⋯ | $Y_{an}$ | $Y_{a.}$ | $\overline{Y}_{a.}$ |
| Block statistics | | $Y_{.1}$ | $Y_{.2}$ | ⋯ | $Y_{.n}$ | $Y_{..}$ | $\overline{Y}_{..}$ |

(*b*) RCB data layout

*RCB Sum of Squares:*

$$\sum_{i=1}^{a}\sum_{j=1}^{n}(Y_{ij}-\overline{Y}_{..})^2 = n\sum_{i=1}^{a}(\overline{Y}_{i.}-\overline{Y}_{..})^2 + a\sum_{j=1}^{n}(\overline{Y}_{.j}-\overline{Y}_{..})^2 + \sum_{i=1}^{a}\sum_{j=1}^{n}(Y_{ij}-\overline{Y}_{i.}-\overline{Y}_{j.}+\overline{Y}_{..})^2 \tag{20.24}$$

We can see similarities to the CRD sum of squares model, Equation 20.4. However, in the RCB an additional term breaks out the block effect.

An ANOVA table breakout for the RCB is shown in Table 20.10. Treatments may be developed from either fixed or random effects factors, as in the case of the CRD. Assumptions associated with and interpretation of the Treatment row are similar to the CRD case (e.g., the hypothesis tests and *P*-value). In the fixed effects treatment case, also, we may proceed to pairwise comparisons. The distinction of the RCB here is that we develop the Block row to "explain" some of the sum of squares that would ordinarily end up in the Experimental error row in a CRD ANOVA.

We may interpret the Block row in order to satisfy our curiosity as to our "payoff" from blocking. Here, we typically ratio the $MS_{blk}$ over the $MS_{err}$. A large ratio value indicates effectiveness, a small ratio value ineffectiveness. This ratio represents an $F$ test in the Blocks row, which is not considered to be theoretically sound since our experimental units are stratified into blocks in a nonrandom fashion, whereas the treatments are randomly assigned within each block. Although most computer aids produce a block row $F$ and a $P$-value, we typically refrain from formally interpreting the Block row (e.g., estimating a block variation for a random block effect or performing pairwise tests on block means) (see Montgomery [20], Box, Hunter, and Hunter [21], and Anderson and McLean [22]).

*Because we do not have true replication in the RCB structure, the block variable by treatment factor interaction is used as (and labeled as) "experimental error."* Here, we assume the block by factor interaction effect is negligible and we are measuring only random variation. Hence, *the RCB design is not a good substitute or shortcut for a full two factor experimental study* (discussed in Chapter 21, along with the interaction concept).

## Example 20.6

At this point in our chocolate chip cookie case we want to design a third experiment where we consider cooking time and the nuisance variable cookie position on the baking pan simultaneously. We want to "pull" the position effect out of our experimental error term so that we can isolate the cooking time effect more precisely.

## Solution

We will develop a RCB using the cookie position on the baking pan as our blocking factor. Here, we will consider three positions: (1) right side, (2) center, (3) left side. We will again use four cooking time treatments: (1) 5 min, (2) 10 min,

**TABLE 20.10**  RCB ANOVA TABLE BREAKOUT

| Source | df | SS | MS | F |
|--------|-----|-----|-----|-----|
| Total corrected | $an - 1$ | $\sum_{ij} Y_{ij}^2 - CF$ | | |
| Blocks | $n - 1$ | $\dfrac{\sum_{j=1}^{n} Y_{.j}^2}{a} - CF$ | $SS_{blk}/df_{blk}$ | $MS_{blk}/MS_{err}$* |
| Treatments | $a - 1$ | $\dfrac{\sum_{i=1}^{a} Y_{i.}^2}{n} - CF$ | $SS_{trt}/df_{trt}$ | $MS_{trt}/MS_{err}$ |
| Exp. error | $(a - 1)(n - 1)$ | $SS_{tc} - SS_{blk} - SS_{trt}$ | $SS_{err}/df_{err}$ | |

Here $CF = \dfrac{Y_{..}^2}{an}$

* The validity of this $F$ value is questionable due to the blocking procedure (the experimental units were stratified into blocks in a nonrandom fashion, whereas the treatments were assigned to experimental units within each block in a random fashion). However, this $F$ test does provide a general indication of blocking effectiveness.

**FIGURE 20.11**    Physical layout and ordering of RCB cookie experiment (EXAMPLE 20.6).

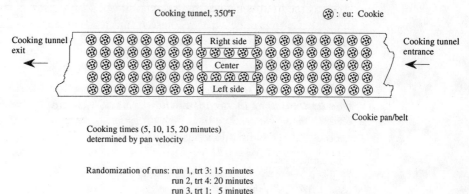

Cooking tunnel, 350°F                          ⊛ : eu: Cookie

Cooking tunnel exit ←

Cooking tunnel entrance ←

Right side

Center

Left side

Cookie pan/belt

Cooking times (5, 10, 15, 20 minutes)
determined by pan velocity

Randomization of runs: run 1, trt 3: 15 minutes
run 2, trt 4: 20 minutes
run 3, trt 1:  5 minutes
run 4, trt 2: 10 minutes

(3) 15 min, and (4) 20 min. We will need 3(4) = 12 eus, or cookies for analysis. We will mix a batch of cookie dough, choose a pan position, choose a cooking time, at random, bake our cookies, sample one cookie from that position at random, and record the elongation percent response. We will repeat this process until we have gathered our 12 observations. Figure 20.11 depicts our physical process. Table 20.11 contains the data collected and Table 20.12 develops the ANOVA. The treatment sample means and $LSD_{0.05}$ bars are plotted in Figure 20.12. SAS results are shown in the Computer-Aided Analysis Supplement, Solutions Manual.

The results show a significant cooking time effect ($F = 8.83$, $P$-value = 0.013). Our blocking seems to have helped, as we see a reasonable reduction in $SS_{err}$ and $P$-value = 0.041 in the Block row. We have calculated $LSD_{0.05} = 2.92$ and displayed it graphically as an interval centered at the "best" sample treatment mean response (10 min cooking time). Here, the lower $LSD_{0.05}$ bar extends below the

**TABLE 20.11**    EXPERIMENTAL RESULTS FOR THE RCB (EXAMPLE 20.6)

| Treatment (cooking time) | Blocks (Cookie position on pan) | | | Treatment totals $Y_{i.}$ | Treatment means $\bar{Y}_{i.}$ |
|---|---|---|---|---|---|
| | 1 (right) | 2 (center) | 3 (left) | | |
| 1 (5 min) | 1.0 | 4.0 | 2.0 | 7.0 | 2.333 |
| 2 (10 min) | 7.0 | 11.0 | 6.0 | 24.0 | 8.000 |
| 3 (15 min) | 7.0 | 8.0 | 3.0 | 18.0 | 6.000 |
| 4 (20 min) | 2.0 | 5.0 | 4.0 | 11.0 | 3.667 |
| Block totals $Y_{.j}$ | 17.0 | 28.0 | 15.0 | $Y_{..} = 60.0$ | $\bar{Y}_{..} = 5.000$ |

Response measured in percent elongation.

**TABLE 20.12**    ANOVA FOR THE RCB EXPERIMENT (EXAMPLE 20.6)

| Source | df | SS | MS | F | P-value[**] |
|---|---|---|---|---|---|
| Total corrected | 12 – 1 = 11 | 94.000 | | | |
| Block (cookie position on pan) | 3 – 1 = 2 | 24.500 | 12.250 | 5.73* | 0.0406 |
| Treatment (cooking time) | 4 – 1 = 3 | 56.667 | 18.889 | 8.83 | 0.0128 |
| Exp. error | (4 – 1)(3 – 1) = 6 | 12.833 | 2.139 | | |

Sample calculations

$SS_{tc}$ = $(1^2 + 4^2 + 2^2 + 7^2 + \cdots + 5^2 + 4^2) - (60^2/12)$ = 94.000

$SS_{blk}$ = $(17^2 + 28^2 + 15^2)/4 - (60^2/12)$ = 24.500

$SS_{trt}$ = $(7^2 + 24^2 + 18^2 + 11^2)/3 - (60^2/12)$ = 56.667

$SS_{err}$ = $SS_{tc} - SS_{blk} - SS_{trt}$ = 94.000 – 24.500 – 56.667 = 12.833

\* Validity of F test is questionable.
\*\* Obtained from Computer-Aided Analysis Supplement, Solutions Manual.

sample mean of the 15-min cooking time, indicating no statistically significant difference. However, there is a statistically significant difference between the 10-min cooking time response and the 5- and 20-min cooking times, as the lower $LSD_{0.05}$ bar does not reach their sample mean values.

## Prediction and Residual Calculations

Based on our RCB model, we can predict our mean cooking time treatment responses. It follows from Equations 20.23 and 20.24 that

$$\hat{\mu} = \overline{Y}_{..} \tag{20.25}$$

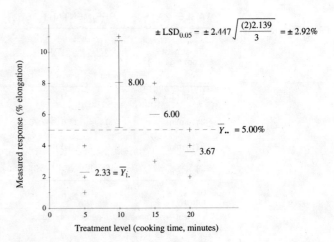

$$\pm LSD_{0.05} = \pm 2.447 \sqrt{\frac{(2)2.139}{3}} = \pm 2.92\%$$

**FIGURE  20.12**    Scatter plot for RCB (Example 20.6).

$$\hat{\tau}_i = \bar{Y}_{i.} - \bar{Y}_{..} \tag{20.26}$$

$$\hat{\beta}_j = \bar{Y}_{.j} - \bar{Y}_{..} \tag{20.27}$$

and therefore the predicted mean for treatment $i$ and block $j$ is

$$\hat{\mu}_{ij} = \bar{Y}_{..} + (\bar{Y}_{i.} - \bar{Y}_{..}) + (\bar{Y}_{.j} - \bar{Y}_{..}) \tag{20.28}$$

Finally, the residual for treatment $i$ and block $j$ is

$$\hat{\varepsilon}_{ij} = e_{ij} = Y_{ij} - \hat{\mu}_{ij} \tag{20.29}$$

For example, using data taken from Example 20.6, Table 20.11,

$$\hat{\mu}_{i=1\,j=1} = 5.000 + (2.333 - 5.000) + (17.0/4 - 5.000) = 5.000 + (-2.667) + (-0.75) = 1.583$$

and
$$e_{11} = Y_{11} - \hat{\mu}_{11} = 1.0 - 1.583 = -0.583$$

## 20.7 MODEL ADEQUACY

We typically make a number of assumptions regarding our residual $\varepsilon_{ij}$ terms. ***In order to support our hypothesis testing, we assume our residuals to be independent and normally distributed,*** $\mu = 0$, $\sigma^2 = \sigma_\varepsilon^2$. We can assess our assumption in two general ways: (1) with a probability plot or graph or (2) with a statistical goodness-of-fit test. Without computer aids, the first alternative is usually more feasible; with computer aids, both alternatives may be reasonable.

***During the course of an experiment, we also assume that no systematic change, or "bias" takes place.*** In residuals order sequence analysis, we record the order of experimentation (as the experiment is performed) and plot our residuals against our order sequence. If we see a systematic pattern (e.g., during the physical execution of the experiment, residuals have been mostly negative, but switch to mostly positive at some point), we have evidence that something changed during the course of the experiment and our results may be misleading.

***We should also assess the homogeneity of variance assumption, used to develop our experimental error.*** A residuals-predicted values plot helps assess the validity of this assumption (e.g., we are assuming that the variances, within each of the treatments, are equal and can be "pooled" to form the experimental error). This pooling effect, in general, is responsible for the "power" of the ANOVA. This plot should display no obvious patterns.

Computer aids can free us of the drudgery involved in residual analysis. Most computer aids, such as SAS, have provisions to deal with normality, order, and homogeneity of variance analyses [23].

## Example 20.7

Use the data and results of Examples 20.2 and 20.3 to
  **a**  Develop a normal probability plot of residuals and residual-order and residual-predicted value plots.
  **b**  Produce a normal test and a residual-order plot with SAS.

### Solution

**a**  We must develop our $\hat{\mu}_{ij}$ and $e_{ij}$ values as well as our plotting positions, $\hat{F}(t)$'s. We will use normal probability plotting paper (Section IX, Table IX.14) and the traditional mean rank plotting form,

$$\hat{F}(t) = \frac{k - 0.5}{an} \tag{20.30}$$

where $k$ = rank number
  $an$ = number of residuals, each observation produces one residual

Here, the plotting positions, $\hat{F}(t)$'s represent estimated cumulative mass points, relative to the ranks (e.g., 1 through 12) of the residuals.

Table 20.13a lists the ordering of the experiment (not mentioned in Example 20.2) and the observed values $Y_{ij}$'s. Since the CRD was used, we then use Equations 20.8, 20.9, and 20.10 to develop our $\hat{\mu}_{ij}$ and $e_{ij}$ values. In Table 20.13b we use the mean rank method, as shown in Equation 20.30, to develop our $\hat{F}(t)$ values, and assign our $e_{ij}$'s to the mean plotting positions in ascending order (by rank).

**TABLE 20.13**    OBSERVED AND PREDICTED ANALYSES OF RESIDUALS AND PLOTTING POSITIONS (EXAMPLE 20.7)

**a  Observed data**

| Observation | Treatment | $Y_{ij}$ | $\hat{\mu}_{ij} = \overline{Y}_{i.}$ | $e_{ij} = Y_{ij} - \hat{\mu}_{ij}$ |
|:---:|:---:|:---:|:---:|:---:|
| 1 | 2 (10 min) | 9 | 8.67 | 0.33 |
| 2 | 1 (5 min) | 4 | 2.33 | 1.67 |
| 3 | 3 (15 min) | 8 | 5.67 | 2.33 |
| 4 | 1 (5 min) | 1 | 2.33 | -1.33 |
| 5 | 3 (15 min) | 4 | 5.67 | -1.67 |
| 6 | 4 (20 min) | 3 | 3.00 | 0.00 |
| 7 | 2 (10 min) | 7 | 8.67 | -1.67 |
| 8 | 2 (10 min) | 10 | 8.67 | 1.33 |
| 9 | 1 (5 min) | 2 | 2.33 | -0.33 |
| 10 | 4 (20 min) | 2 | 3.00 | -1.00 |
| 11 | 3 (15 min) | 5 | 5.67 | -0.67 |
| 12 | 4 (20 min) | 4 | 3.00 | 1.00 |

**b  Ranked Residuals (in ascending order)**

| Rank $k$ | Mean plotting position $\hat{F}(t) = (k - 0.5)/an$  percentiles | $e_{ij}$ |
|:---:|:---:|:---:|
| 1 | 0.5/12 = 0.042  4.2% | -1.67 |
| 2 | 1.5/12 = 0.125  12.5% | -1.67 |
| 3 | 2.5/12 = 0.201  20.1% | -1.33 |
| 4 | 3.5/12 = 0.292  29.2% | -1.00 |
| 5 | 4.5/12 = 0.375  37.5% | -0.67 |
| 6 | 5.5/12 = 0.458  45.8% | -0.33 |
| 7 | 6.5/12 = 0.542  54.2% | 0.00 |
| 8 | 7.5/12 = 0.625  62.5% | 0.33 |
| 9 | 8.5/12 = 0.708  70.8% | 1.00 |
| 10 | 9.5/12 = 0.792  79.2% | 1.33 |
| 11 | 10.5/12 = 0.875  87.5% | 1.67 |
| 12 | 11.5/12 = 0.958  95.8% | 2.33 |

We have developed a normal probability plot (Figure 20.13). Here, we see a reasonable pattern of our plotted points about a straight line. The horizontal scale is linear on our normal plotting paper, while the vertical scale is developed relative to the cumulative probability of a normal distribution. Due to the scaling of the axes, plotted points on normal plotting paper will fall roughly on a straight line, if the points represent a sample from a normal population. Once we plot our $e_{ij}$, $\hat{F}(t)$ coordinates, we sketch in a "model" line. If the points fall near the line, we deem the normal assumption as more or less reasonable and we can defend the validity of the $F$ test in our ANOVA. If, on the other hand, we are skeptical of the normality plot, we have a dilemma. We can try either a transformation or nonparametric methods. We refer our readers to Montgomery [24] and Conover [25] for more-detailed discussions on transformations and nonparametric methods, respectively.

Our residual-order and residual-predicted value plots are shown in Figure 20.14. Here, we simply plot our residuals across time on the horizontal axis (simi-

**FIGURE 20.13**    Residual plot, normal probability plotting paper (Example 20.7).

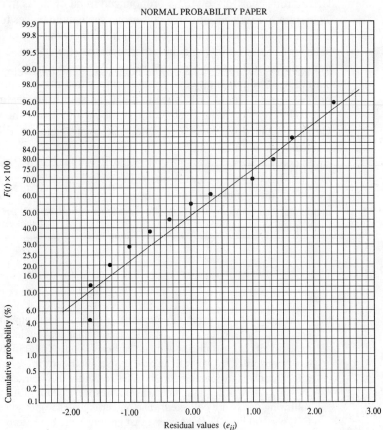

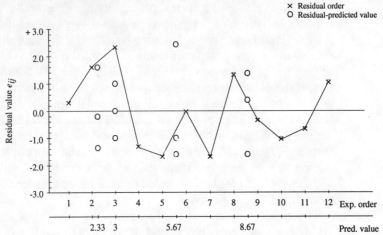

**FIGURE 20.14**    Residual plots, residual-order and residual-predicted value (Example 20.7).

lar to a runs chart). We see no pronounced trend or shifts in our $e_{ij}$'s across time. Therefore, we will assume that no serious systematic biases took place during the course of our experiment. If, on the other hand, we see trends or biases, our results may be misleading and the experiment may have to be rerun. We also plotted our residuals against the predicted values $\hat{\mu}_{ij}$, shown in Figure 20.14. Here, the assumption of homogeneity of variance (across treatments) appears valid, as we do not see a great deal of difference in the dispersion from treatment to treatment.

**b** The SAS code and selected output for our assessment are shown in the Computer Aided Analysis Supplement, Solutions Manual. PROC UNIVARIATE provides us with a wealth of information. We focus on the goodness-of-fit, Shaprio-Wilk test, SAS [26], statistic $W = 0.947$ with $P$-value = 0.556. Here, our null hypothesis assumes normality. With such a large $P$-value, we do not reject $H_0$ and conclude that our normality assumption is valid. We also see a crude normal probability plot, the SAS equivalent to our plot on normal probability paper.

## REVIEW AND DISCOVERY EXERCISES

### Review

20.1 Corrosion and the prevention of corrosion in natural gas pipelines are major concerns. In order to address these issues, a chemical engineer conducted an experiment to determine how much carbon dioxide ($CO_2$) in a natural gas stream can be absorbed by water in an acid gas scrubbing tower. She was interested in the effect of tower temperature on the $CO_2$ solubility in water, at a specific tower pressure. Five different tower temperatures were chosen for the test. The following data were collected. The bigger is better response recorded is in pounds $CO_2$ per 100 lb of $H_2O$. Assume that all runs were made in random order; the sequence is indicated by the numbers in parentheses. [CT]

| Tower temperature, °C | Replication | | | | |
|---|---|---|---|---|---|
| | 1 | 2 | 3 | 4 | 5 |
| 40 | 3.9 (2) | 3.1 (6) | 2.6 (9) | 3.5 (14) | 2.3 (18) |
| 60 | 2.5 (5) | 1.7 (10) | 2.8 (15) | 2.0 (19) | 2.3 (22) |
| 80 | 1.4 (1) | 1.9 (4) | 2.3 (11) | 1.7 (21) | 2.1 (24) |
| 100 | 1.6 (3) | 1.5 (12) | 2.0 (13) | 1.2 (17) | 1.8 (20) |
| 120 | 1.3 (7) | 1.6 (8) | 1.1 (16) | 1.5 (23) | 1.8 (25) |

**a** Estimate the overall mean and the treatment means.  Develop a response plot.

**b** Using the ANOVA, $P$-values, and $\alpha = 0.05$, determine whether there is a difference in $CO_2$ solubility due to tower temperature.  Add s.e.i. arrows to your response plot in part *a*.

**c** Develop LSD, DM, SNK, and HSD pairwise test tables for the data.

**d** Assuming that a tower temperature of 100°C is currently used, make recommendations to the plant manager.  Explain your recommendations.

**e** Provide an analysis of residuals to assess the ANOVA assumptions.

**20.2** Three engineers each analyzed the same set of experimental data.  They also performed pairwise comparisons on the four treatment means (treatments *A, B, C, D*).  Their comparison results were as follows:

| | Treatments | | | |
|---|---|---|---|---|
| Scientist 1: | A | B | C | D |
| Scientist 2: | A | B | C | D |
| Scientist 3: | A | B | C | D |

Here, the underbars indicate treatments that are not significantly different from each other (e.g., if  treatment *A* and treatment *B* are both over the same bar, they were not judged to be significantly different).  One of the engineers used the LSD method, one the HSD method, and the other the SNK method.  Is it possible to determine which comparison method was used by each engineer?  If so, make the determination, if not, explain why you cannot make the determination.

**20.3** Potato growers are paid a premium for both size and uniformity of product.  A grower wants to study the average weight and the weight variation in the population of potatoes.  He also wants to estimate the weight variation within plants and between different plants.  He uses a random effects model CRD experiment and samples five potato plants, selected at random.  From each plant, he chooses three potatoes at random.  The data appear below, where the response is the weight of each potato in pounds.

| Plant | Observation | | |
|---|---|---|---|
| | 1 | 2 | 3 |
| 1 | 0.75 | 0.83 | 0.64 |
| 2 | 0.86 | 0.91 | 0.54 |
| 3 | 0.73 | 0.61 | 0.57 |
| 4 | 0.93 | 0.98 | 1.15 |
| 5 | 1.17 | 1.23 | 0.96 |

    a  Identify the experimental unit in the experiment described and write out the model statement.

    b  Develop an ANOVA for the potato data and include the $P$-value.

    c  Estimate the within plant and the between plant variance.

    d  Estimate the weight of a typical potato. Estimate the standard deviation of the population of potato weights.

    e  Estimate the percent of the population variance attributable to within plant variation, and that due to between plant variation.

**20.4**  In an energy intensive food dehydration process an experiment was designed to assess the dehydrating effects of three different slicing-dicing geometries for apples. Treatment 1 used a rectangular geometry, treatment 2 used a cubic geometry, and treatment 3 used a spiral geometry. In every other respect, the apples were prepared for dehydration in the same manner and dehydrated for 2 hr. The resulting percent (by weight) moisture levels were measured and recorded. The smaller is better response data are shown below. The numbers in parentheses indicate the experimental order.

| Treatments | Replications | | | |
|---|---|---|---|---|
| | 1 | 2 | 3 | 4 |
| 1 | 22 (2) | 28 (6) | 18 (7) | 33 (12) |
| 2 | 26 (3) | 34 (5) | 27 (9) | 44 (10) |
| 3 | 13 (1) | 17 (4) | 21 (8) | 24 (11) |

    a  Identify the experimental unit and develop a response plot.

    b  Develop the ANOVA table, including a $P$-value, for the CRD experiment and add s.e.i. arrows to your plot in part *a*.

    c  Calculate the LSD at the $\alpha = 0.05$ significance level and estimate the experiment-wise type I error probability. Place $LSD_{0.05}$ bars on your plot.

    d  Interpret the experimental results above. Be sure to point out any clear "winner" among the treatments in the experiment.

    e  Develop an analysis of the residuals.

**20.5**  After some discussion with the technical people who designed and performed the experiment described in Problem 20.4, additional facts were discovered. There was a good deal of concern that apple varieties might dehydrate differently, thus creating a considerable amount of variation that might obscure the ability to detect slicing geometry effects. Hence, a new experiment was designed, and four apple varieties that were commonly dehydrated were chosen as blocks. An RCB was run (rather than the CRD); the data are shown below, along with the runs sequence in parentheses:

| Treatments | Blocks (apple varieties) | | | |
|---|---|---|---|---|
| | 1 | 2 | 3 | 4 |
| 1 | 22 (2) | 28 (6) | 18 (7) | 33 (12) |
| 2 | 26 (3) | 34 (5) | 27 (9) | 44 (10) |
| 3 | 13 (1) | 17 (4) | 21 (8) | 24 (11) |

**a** Develop the ANOVA table, including a *P*-value, for the RCB experiment and prepare a response plot with s.e.i. arrows.

**b** Calculate the LSD at the $\alpha = 0.05$ significance level and estimate the experiment-wise type I error probability. Place $LSD_{0.05}$ bars on your plot.

**c** Interpret the experimental results above in parts (*a*) and (*b*). Be sure to point out any clear "winner" among the treatments in the experiment.

**d** Develop an analysis of the residuals.

**e** If you worked Problem 20.4 as well as this problem, comment on the differences in the two analyses and interpretations.

**20.6** In a meter stick production facility, the length of meter sticks is measured with an automatic measuring machine (there is only one device). We want to estimate the variation associated with the population of recorded measurements (i.e., the "labeled" variation) of our product. We also want to estimate the proportion of variation attributable to the product and that attributable to the measuring machine. We have designed a CRD experiment where we have taken 20 meter sticks out of our production output at random and carefully measured each meter stick 2 times (i.e., two independent measurements). The data in millimeters appear below:

| Meter stick | Observation 1 | 2 |
|---|---|---|
| 1 | 1000.158 | 1000.157 |
| 2 | 1000.003 | 1000.006 |
| 3 | 999.722 | 999.720 |
| 4 | 1000.303 | 1000.305 |
| 5 | 1000.216 | 1000.217 |
| 6 | 999.636 | 999.638 |
| 7 | 1000.153 | 1000.153 |
| 8 | 1000.383 | 1000.381 |
| 9 | 999.754 | 999.757 |
| 10 | 1000.237 | 1000.238 |
| 11 | 999.968 | 999.968 |
| 12 | 1000.073 | 1000.075 |
| 13 | 1000.186 | 1000.184 |
| 14 | 999.948 | 999.945 |
| 15 | 999.868 | 999.869 |
| 16 | 999.951 | 999.952 |
| 17 | 999.900 | 999.903 |
| 18 | 1000.387 | 1000.387 |
| 19 | 999.961 | 999.962 |
| 20 | 999.813 | 999.812 |

**a** Develop a random effects ANOVA table.

**b** Estimate the measuring machine variation component.

**c** Estimate the product variation component.

**d** Estimate the labeled variation (associated with the population of meter sticks) on the outgoing product.

**e** Estimate the percent of the labeled variation due to the measuring machine.

20.7  If you studied the material in Chapter 18—specifically, Example 18.6, compare your results in Problem 20.6 to the results in Example 18.6. Comment on the two different approaches to conducting gauge studies.

20.8  The Night Hauler Trucking Company is considering the use of a new diesel fuel additive that is claimed to effectively capture more usable energy from diesel fuel by enhancing the combustion process. The manufacturer of the additive recommends using 150 ml of additive per 10 gal of diesel fuel. Since the Night Haulers are not capable of developing detailed measurements in combustion engineering, they have decided to use a substitute quality performance measure, miles per gallon of diesel fuel, in their field operations. In order to test the effectiveness of the additive a work team of Night Haulers has devised an RCB experiment. They have selected three light trucks and tested the mileage of each truck with 0, 50, 100, and 150 ml of additive per 10 gal of diesel. The data collected are shown below [JM]:

| Additive level, ml/10 gal | Truck | | |
|:---:|:---:|:---:|:---:|
| | 1 | 2 | 3 |
| 0 | 13.5 | 14.6 | 15.6 |
| 50 | 17.8 | 18.8 | 18.2 |
| 100 | 16.8 | 20.6 | 22.8 |
| 150 | 18.9 | 17.7 | 23.9 |

  a  Develop a response plot and the RCB ANOVA table, including a $P$-value, for the experimental results. Add s.e.i. arrows to your plot.
  b  Use a one-tailed Dunnett's test, $\alpha = 0.05$, to determine whether the additive is effective.
  c  Using an LSD criteria, determine whether you can justify the manufacturer's claim that 150 ml/10 gal is the best usage rate. Recommend a rate.
  d  Determine how effective the Night Haulers were in blocking on the trucks as a nuisance variable.
  e  Critique the experimental protocol used by the Night Haulers.

20.9  Assume the experimenter failed to record the truck numbers in Problem 20.8. Now, we don't know which response resulted from which truck, but we do know which treatment each response resulted from.
  a  Analyze the data from Problem 20.8 using the CRD model.
  b  Using a one-tailed Dunnett's test, $\alpha = 0.05$, determine whether the additive is effective.
  c  Using an LSD criteria, determine whether you can justify the manufacturer's claim that 150 ml/10 gal is the best usage rate. Recommend a rate.
  d  Comment on the differences between your results in Problem 20.8 and the CRD model results of this problem.

20.10  Return to Example 20.6 and develop a full set of the following pairwise tests. Then interpret your results relative to the $LSD_{0.05}$ bars used in the example.
  a  LSD, $\alpha = 0.05$
  b  DM, $\alpha = 0.05$
  c  SNK, $\alpha = 0.05$
  d  HSD, $\alpha = 0.05$

**20.11** A nutritionist wants to express the importance of eating high-protein, low-fat foods. In order to enhance her presentation, she decides to discuss body composition measurements. She decides to use three methods to illustrate the variety of ways to measure body fat (all are substitute measures for actual body fat composition). Specifically, she needs to know if the methods provide consistent responses (i.e., if any one method gives an unusually high or low body fat count). She chose three methods: circumference measuring, skin fold measuring, and hydrostatic weighing. Then, she designed an RCB experiment to address her purpose. The data from her experiment are shown below [LD]:

| Measurement method | Subject | | | | |
|---|---|---|---|---|---|
| | 1 | 2 | 3 | 4 | 5 |
| Circumference | 17.67 | 8.78 | 23.30 | 33.11 | 21.12 |
| Skin fold | 19.40 | 11.80 | 25.00 | 23.80 | 18.96 |
| Hydrostatic weighing | 11.28 | 16.06 | 23.50 | 22.80 | 22.60 |

    **a** Assuming an RCB design was used, determine whether there is a difference in the percent body fat obtained using the different methods. Use $\alpha = 0.05$ as a significance level for your interpretation.

    **b** Calculate the values of $\hat{\sigma}_{\varepsilon}^{2}$, $e_{23}$, $\hat{\mu}$, and $\hat{\tau}_{2}$.

**20.12** Referring back to Problem 20.11, assume the nutritionist failed to record the subject number and only recorded the observations for each of the three body fat measurements. Further assume that the CRD design is appropriate.

    **a** Determine whether there is a difference in the percent body fat obtained using the different methods. Use $\alpha = 0.05$ as a significance level for your interpretation.

    **b** Calculate the values of $\hat{\sigma}_{\varepsilon}^{2}$, $e_{23}$, $\hat{\mu}$, and $\hat{\tau}_{2}$.

    **c** Compare your analysis and results to those obtained in Problem 20.11.

**20.13** Samples of groundwater were taken from five different toxic waste dump sites by each of three different agencies: (1) the EPA, (2) the company that owned the site, and (3) an engineering lab. Each sample was analyzed for the presence of a certain contaminant by whatever laboratory method was customarily used by the agency collecting the sample, with the following results:

| Agency | Site | | | | |
|---|---|---|---|---|---|
| | A | B | C | D | E |
| 1 | 23.8 | 7.6 | 15.4 | 30.6 | 4.2 |
| 2 | 19.2 | 6.8 | 13.2 | 22.5 | 3.9 |
| 3 | 20.9 | 5.9 | 14.0 | 27.1 | 3.0 |

    **a** Develop an ANOVA table for the RCB experiment, including the *P*-value, where the agencies act as the treatments.

    **b** Determine whether there is reason to believe that the agencies are not consistent with one another in their measurements.

    **c** Determine if the blocking paid off in a statistical sense in this experiment. Briefly explain your position.

    **d** Calculate the residual value associated with the agency 3, site C observation.

**20.14**  AAA Tours owns a fleet of buses, built by different European and U.S. manufacturers. All of the buses at AAA Tours are used interchangeably for over-the-road and local tours. Maintenance personnel have encountered a high level of brake replacement demands. They have suggested that the company perform an independent evaluation of brake pad materials. A decision was made to design an RCB experiment in order to remove bus effects from the experimental error. Four brake pad types were of interest. Each bus constituted a block, and each type of brake pad constituted a treatment. Brake pad materials were assigned at random to the different axles on each bus, one bus at a time. All brakes were adjusted to manufacturer's specifications, and the buses were operated for 50,000 mi to measure brake pad wear. The amount of wear in inches is shown in the data table below [DM]:

| Brake pad type (treatments) | Buses (blocks) | | | |
|:---:|:---:|:---:|:---:|:---:|
| | 1 | 2 | 3 | 4 |
| 1 | 0.18 | 0.21 | 0.15 | 0.26 |
| 2 | 0.28 | 0.31 | 0.35 | 0.32 |
| 3 | 0.37 | 0.35 | 0.42 | 0.46 |
| 4 | 0.25 | 0.29 | 0.27 | 0.23 |

a  Develop an appropriate graphical analysis.
b  Develop an appropriate ANOVA table. Add s.e.i. arrows to your graph from part *a*. Interpret your results.
c  Perform pairwise treatment comparisons using the LSD and SNK methods, with $\alpha = 0.05$. Interpret your results.

## Discovery

**20.15**  A roller chain manufacturer is investigating the effect of four specific press operating speeds on the pitch of link plates. The press speeds in surface feet per minute chosen are 90 sfm, 120 sfm, 140 sfm, and 160 sfm. For this experiment, four different presses, each of a different capacity, are chosen. The observations recorded are deviations, in thousandths of an inch, from the nominal pitch, or target value (e.g., zero is best) [MM]:

| Operating speed (sfm) | Press type | | | |
|:---:|:---:|:---:|:---:|:---:|
| | 1 | 2 | 3 | 4 |
| 90 | 0.35 | 0.15 | −0.46 | −0.53 |
| 120 | 0.54 | −0.25 | −0.11 | −0.05 |
| 140 | 0.98 | 0.68 | 0.65 | 0.14 |
| 160 | 0.97 | 0.70 | 0.85 | 0.25 |

a  Develop a set of hypotheses for the fixed effects model.
b  Develop a plot and an ANOVA from the data collected.
c  Discuss and devise an experimental analysis protocol for a nominal is best experimental objective.
d  Recommend a speed.

**20.16** Develop a one-tailed LSD pairwise test. Then, using the smaller is better LSD you developed and an $\alpha = 0.05$ significance level, revisit Problem 20.5 and assess the apple geometry treatments for a clear winner. How do the results of this one-tailed LSD comparison differ from those originally obtained in Problem 20.5? Explain.

**20.17** As the co-owner of a custom laundry at the corner of 50th and Avenue Q, you have been approached by three vendors each claiming their spot remover will outclean all others (guaranteed). Furthermore, each vendor has multicolored charts and brochures, including test results, claiming that third party tests at a leading university "prove" ($\alpha = 0.05$) that their spot removers are more effective than those of their competitors. The products offered to you are as follows:

Vendor 1—Spotfree, tested at the Reputable University

Vendor 2—Spotless, tested at the University of Knowledge

Vendor 3—Spotaway, tested at Correct University

At present you are using a product, called "Nospot," furnished by a fourth vendor. It so happens that all of the three calling vendors' published test results include Spotfree, Spotless, Spotaway, and Nospot.

**a** Determine whether such a situation, where each vendor has third party, statistical proof that its product is the best, is possible. Explain.

**b** Your co-owner (who is statistically illiterate) is very puzzled as to which spot remover to use. Describe an experiment you would design in order to decide for yourself which spot remover you should purchase. There is very little cost difference between the four spot removers.

**20.18** In our factory, we are producing precision brass bushings with an outside diameter target value of 0.750 in. Our customer expects a product on target with a standard deviation of no more than 0.001 in. We use one automated turning center and one instrument to measure our bushings. We know that the labeled variation of our shipped product is a result of both the true variation of our product and that resulting from our measurement process. Design an experiment that will allow us to estimate the proportion of the variation due to the machining process and that due to the measuring process. Lay out and describe your recommended design and model; then point out how we might determine whether both the outside diameter and the product standard deviation are on target.

**20.19** We have called you in as a second consultant to help plan a follow-on (subsequent) experiment to that described in Problem 20.13. Now, we believe the intraagency variation in the physical analysis methods may contribute to the differences in the site analyses at each site. Explain how we can modify our next experiment so that we can *also* measure the variation within agency methods. Remember, we want to be efficient in our experiment—we do not have a great deal of money to spend.

**20.20** Select a product or process. Design, execute, and interpret a single-factor (either fixed or random effect) experiment regarding some facet of the product or process you selected. Follow the six steps involved in experimental protocol (Table 20.2). Construct a display board to communicate your experimental protocol and conclusions. After completion of your experiment, critique your design's effectiveness in fulfilling your objective.

## REFERENCES

1 W. E. Deming, *Out of the Crisis,* Cambridge, MA: MIT Center for Advanced Engineering Studies, 1986.

2 R. E. Walpole and R. H. Myers, *Probability and Statistics for Engineers and Scientists,* 5th ed., New York: Macmillan, 1993.

3 See reference 2.

4 See reference 2.

5 D. E. Coleman and D. G. Montgomery, "A Systematic Approach to Planning for a Designed Industrial Experiment," *Technometrics,* vol. 35, no. 1, 1993.

6 See reference 2.

7 D. C. Montgomery, *Design and Analysis of Experiments,* 3d ed., New York: Wiley, 1991.

8 C. R. Hicks, *Fundamental Concepts in the Design of Experiments,* 3d ed., New York: Holt, Rinehart, and Winston, 1982.

9 G. E. P. Box, W. G. Hunter, and J. S. Hunter, *Statistics for Experimenters,* New York: Wiley, 1978.

10 See references 2, 7, 8, and 9.

11 SAS Institute, Inc., SAS Circle, Box 8000, Cary, NC 27512–8000.

12 SPSS Inc., Suite 3000, 444 North Michigan Avenue, Chicago, IL 60611.

13 SYSTAT, Inc. 1800 Sherman Avenue, Evanston, IL 60201-3793.

14 Minitab, Inc. 3081 Enterprise Drive, State College, PA 16801.

15 STSC, Inc. 2115 East Jefferson Street, Rockville, MD 20852.

16 See reference 11.

17 See references 2, 7, 8, and 9.

18 *SAS/STAT User's Guide, Volumes 1 and 2,* Version 6, Cary, NC: SAS Institute, 1990.

19 See references 7, 8, and 9.

20 See reference 7.

21 See reference 9.

22 V. L. Anderson and R. A. McLean, *Design of Experiments: A Realistic Approach,* New York: Dekker, 1974.

23 *SAS Procedures Guide*, Version 6, Cary, NC: SAS Institute, 1990.

24 See reference 7.

25 W. J. Conover, *Practical Nonparametric Statistics,* 2d ed., New York: Wiley, 1980.

26 See reference 23.

# MULTIPLE-FACTOR EXPERIMENTS

## 21.0 INQUIRY

1. How can we study multiple factors simultaneously?
2. What do simple effects, main effects, and interaction effects measure?
3. How can EMSs be used to structure hypothesis tests?
4. What is the difference between factors that are crossed and factors that are nested in experiments?
5. How can pseudo experimental error terms be developed when no replications are present?

## 21.1 INTRODUCTION

In off-line quality there are typically many variables that serve to establish a process recipe or dictate product performance. *Multiple factors (two or more) can be designed into experiments with what is termed a "factorial arrangement of treatments," or FAT. In a complete or full factorial arrangement, all possible treatment combinations (of factor levels) are observed and their responses recorded.* Both graphical and numerical analyses are appropriate.

In a FAT, two, three, or more levels of the selected factors will be of interest. When two levels of each factor are studied, the arrangement is termed a $2^f$ FAT, where $f$ represents the number of factors. A $2^f$ FAT will require $2^f$ factor level treatment combinations. When three levels are involved, the arrangement is termed a "$3^f$," and so on. We may also design factorial experiments with a different number of levels of each factor (e.g., a four-factor experiment with $2 \times 3 \times 2 \times 4$ respective levels of the four factors will require 48 factor level combinations). The number of treatment combinations grows large very quickly with the addition of factors and/or factor levels. In most cases, experiments using a complete factorial with more than three factors and at least three levels soon become unwieldy. Fractional factorials, discussed in Chapter 23, are one way to solve this problem by reducing the size of an experiment.

## 21.2 EXPERIMENTS INVOLVING TWO FACTORS

*When we design two-factor experiments, we seek to learn about both factors and their possible interrelationship (interaction) with respect to the response, simultaneously.* The single-factor experiment fundamentals apply to multiple-factor experiments in general; however, additional principles are involved. The concept of a treatment now becomes a matter of factor level combinations. For example, if factor $A$ is cooking temperature with $a$ levels, and factor $B$ is cooking time with $b$ levels, we would see $ab$ treatment combinations of these two factors.

With multiple factors, we do not usually develop a Treatment row in our ANOVA. We break out the $A$ and $B$ main effects and $AB$ interaction so that we can measure all three. This breakout will assist us in our process setup and product development work. Typically, we structure our process by setting two "knobs," $A$ and $B$, each representing a factor. That is, we are breaking a complicated process with two factors down to its constituent parts for analysis.

## Simple, Main, and Interaction Effects

Many times we seek to discover cause-effect relationships between factors and responses and hence develop better process recipes. To accomplish this objective, we must exercise great care in the experimental execution and resulting analysis and interpretation. This task calls for a thorough and systematic approach. *We can develop three levels of effects in FATs: (1) simple effects, (2) main effects, and (3) interaction effects.* The main and interaction effects are typically graphed, and then analyzed in ANOVA form. We will use simple effects here only to demonstrate the analysis process in both a numerical and a graphical format.

*Simple effects are elementary level-to-level changes in the average response when we move from the lower level to a higher level of a given factor, at a fixed level of another factor.* In a $2^2$ experiment (two levels of factor $A$, and two levels of factor $B$) we can develop four simple effects. We will let $Y_{ijk}$ represent an observed response

where $i = 1, 2, ..., a$ levels of factor $A$

$j = 1, 2, ..., b$ levels of factor $B$

$k = 1, 2, ..., n$ replications within the factor $A$ and factor $B$ combinations

Then,

$$\text{Average simple effect of } A \text{ at low } B = \text{SA}_{\text{low } B} = \bar{Y}_{21.} - \bar{Y}_{11.} \qquad (21.1)$$

$$\text{Average simple effect of } A \text{ at high } B = \text{SA}_{\text{high } B} = \bar{Y}_{22.} - \bar{Y}_{12.} \qquad (21.2)$$

$$\text{Average simple effect of } B \text{ at low } A = \text{SB}_{\text{low } A} = \bar{Y}_{12.} - \bar{Y}_{11.} \qquad (21.3)$$

$$\text{Average simple effect of } B \text{ at high } A = \text{SB}_{\text{high } A} = \bar{Y}_{22.} - \bar{Y}_{21.} \qquad (21.4)$$

Simple effects can be observed on interaction graphs, but are not developed in the ANOVA as row breakouts.

We use simple effects to develop main effects and interaction effects. *Main effects are simply the average of the simple effects for the appropriate factors.* In a $2^2$ experiment,

$$\text{Average main effect of } A = A_{\text{avg}} = \tfrac{1}{2} (\text{SA}_{\text{low } B} + \text{SA}_{\text{high } B}) \qquad (21.5)$$

$$\text{Average main effect of } B = B_{\text{avg}} = \tfrac{1}{2} (\text{SB}_{\text{low } A} + \text{SB}_{\text{high } A}) \qquad (21.6)$$

The interaction effect of $A$ and $B$ can be calculated in two ways. Here *AB interaction is the average of the difference in simple effects taken from higher levels to lower levels* and can be calculated as

$$\text{Average } AB \text{ interaction effect} = AB_{\text{avg}} = \tfrac{1}{2}\,(\text{SA}_{\text{high } B} - \text{SA}_{\text{low } B}) \qquad (21.7)$$

or

$$\text{Average } AB \text{ interaction effect} = AB_{\text{avg}} = \tfrac{1}{2}\,(\text{SB}_{\text{high } A} - \text{SB}_{\text{low } A}) \qquad (21.8)$$

## Example 21.1

We have designed a $2 \times 2$, or $2^2$, factorial arrangement of treatments in a completely randomized design (FAT-CRD), experiment to investigate the effects of syrup content and cooking temperature on the chewiness of our chocolate chip cookies. Factor $A$ represents syrup content measured in pounds per batch and factor $B$ represents cooking temperature. Our response is measured in percent elongation, as before. Both factors $A$ and $B$ are fixed effects factors. Figure 21.1$a$ shows the experimental, or physical, boundary of this experiment. In essence, we have four syrup–cooking temperature treatment combinations ($a_1b_1$, 40 lb and 300°F; $a_2b_1$, 50 lb and 300°F; $a_1b_2$, 40 lb and 350°F; and $a_2b_2$, 50 lb and 350°F). The experiment was designed with three replications. We selected one of the four combinations and mixed dough, formed cookies, baked the cookies, tested a randomly selected cookie, and recorded the response. We repeated the process until we obtained our 12 observations (shown in Figure 21.1$b$). Develop a set of plots and calculations illustrating simple, main, and interaction effects.

FIGURE 21.1    Physical boundary and data for a $2^2$ FAT-CRD experiment (Example 21.1).

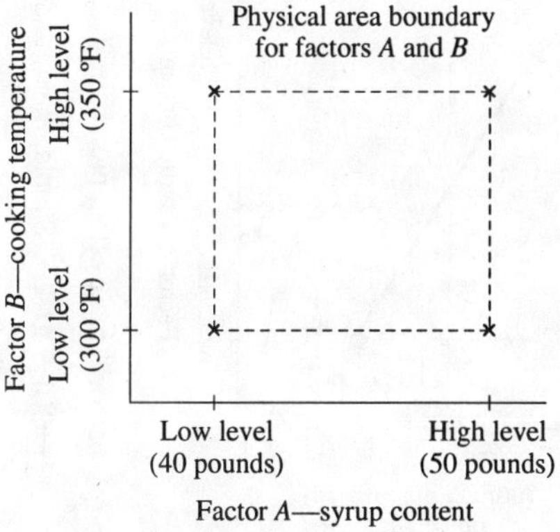

Response: $Y_{ijk}$ (percent elongation)

Factor $A$:   $i = 1,2$
Factor $B$:   $j = 1,2$
Replicates: $k = 1,2,3$

| Factors | | Factor B | | A totals | A means |
|---|---|---|---|---|---|
| | | Low level | High level | | |
| Factor A | Low level | 2  $Y_{11\bullet} = 12$<br>4  $\overline{Y}_{11\bullet} = 4$<br>6 | 4  $Y_{12\bullet} = 18$<br>6  $\overline{Y}_{12\bullet} = 6$<br>8 | $Y_{1\bullet\bullet} = 30$ | $\overline{Y}_{1\bullet\bullet} = 5$ |
| | High level | 6  $Y_{21\bullet} = 24$<br>8  $\overline{Y}_{21\bullet} = 8$<br>10 | 14  $Y_{22\bullet} = 48$<br>16  $\overline{Y}_{22\bullet} = 16$<br>18 | $Y_{2\bullet\bullet} = 72$ | $\overline{Y}_{2\bullet\bullet} = 12$ |
| B totals | | $Y_{\bullet 1\bullet} = 36$ | $Y_{\bullet 2\bullet} = 66$ | $Y_{\bullet\bullet\bullet} = 102$ | |
| B means | | $\overline{Y}_{\bullet 1\bullet} = 6$ | $\overline{Y}_{\bullet 2\bullet} = 11$ | | $\overline{Y}_{\bullet\bullet\bullet} = 8.5$ |

($a$) $2^2$ physical boundary of factors $A$ and $B$        ($b$) Data layout and summary statistics

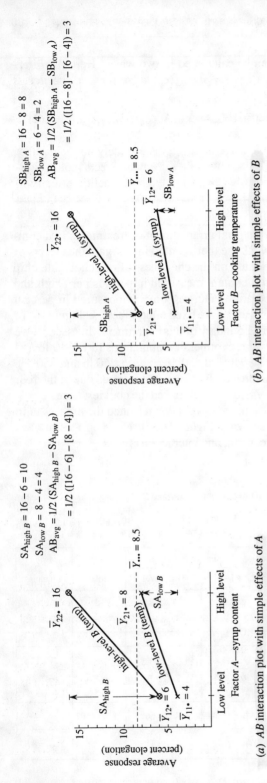

$SA_{high\ B} = 16 - 6 = 10$
$SA_{low\ B} = 8 - 4 = 4$
$AB_{avg} = 1/2\ (SA_{high\ B} - SA_{low\ B})$
$\qquad = 1/2\ ([16 - 6] - [8 - 4]) = 3$

(a) *AB* interaction plot with simple effects of *A*

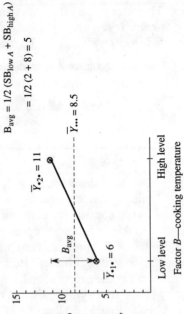

$SB_{high\ A} = 16 - 8 = 8$
$SB_{low\ A} = 6 - 4 = 2$
$AB_{avg} = 1/2\ (SB_{high\ A} - SB_{low\ A})$
$\qquad = 1/2\ ([16 - 8] - [6 - 4]) = 3$

(b) *AB* interaction plot with simple effects of *B*

$A_{avg} = 1/2\ (SA_{low\ B} + SA_{high\ B})$
$\qquad = 1/2\ (4 + 10) = 7$

(c) Factor *A* main effect plot

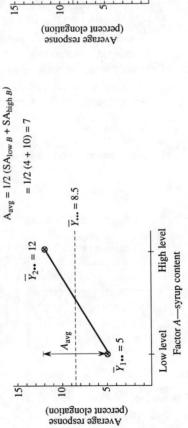

$B_{avg} = 1/2\ (SB_{low\ A} + SB_{high\ A})$
$\qquad = 1/2\ (2 + 8) = 5$

(d) Factor *B* main effect plot

**FIGURE 21.2**  Simple, main, and interaction effects for a $2^2$ FAT-CRD experiment (Example 21.1).

### Solution

We have developed a full set of graphs from the data table. The average simple effects are displayed and calculated in Figure 21.2a and b. The average main effects appear in Figure 21.2c and d. Average interaction terms have been developed using both Equations 21.7 and 21.8 (Figure 21.2a and b).

At this point, we should note that observations with absolutely no main effect will produce "flat," or level, graphical signatures on the main effects plots. Interaction graphs where the observations have absolutely no interaction effect will produce "parallel" factor level lines. In our case (Figure 21.2) we see both main effect and interaction signatures in the plots.

### Analysis of Variance (ANOVA)

Using the ANOVA technique, we can take our analysis beyond the graphical display in Figure 21.2. *In the ANOVA table we can break out rows to quantitatively measure both main effects and interactions.* Our design of interest here is a two-factor FAT-CRD. The FAT is developed as combinations of the factor levels, while the CRD is developed by the randomization process where each eu has an equal chance of being assigned to any one of the factor level treatment combinations. The statistical model is

$$Y_{ijk} = \mu + A_i + B_j + AB_{ij} + \varepsilon_{k(ij)} \tag{21.9}$$

where $Y_{ijk}$ = a response observation

$\mu$ = overall mean

$A_i$ = factor A main effect, $i = 1, 2, ..., a$

$B_j$ = factor B main effect, $j = 1, 2, ..., b$

$AB_{ij}$ = factor A, factor B interaction effect

$\varepsilon_{k(ij)}$ = random error, $k = 1, 2, ..., n$

Taken together the $A$, $B$, and $AB$ effects constitute what we called a "treatment effect" in our single factor experiment. The $\varepsilon_{k(ij)}$ term represents the experimental error developed from the replicates (we assume $n$ replicates) of the experimental units, which have been treated alike within each treatment combination. The notation is read "$k$ within $ij$ factor level combinations."

Using the dot notation introduced in Chapter 20, and the FAT-CRD model form of Equation 21.9, we can express the $SS_{tc}$ equation for our two factor experiment as

$$\sum_{i=1}^{a} \sum_{j=1}^{b} \sum_{k=1}^{n} (Y_{ijk} - \bar{Y}_{...})^2 = \sum_{i=1}^{a} \sum_{j=1}^{b} \sum_{k=1}^{n} [(\bar{Y}_{i..} - \bar{Y}_{...})$$
$$+ (\bar{Y}_{.j.} - \bar{Y}_{...}) + (\bar{Y}_{ij.} - \bar{Y}_{i..} - \bar{Y}_{.j.} + \bar{Y}_{...}) + (Y_{ijk} - \bar{Y}_{ij.})]^2 \tag{21.10}$$

**TABLE 21.1  GENERIC TWO-FACTOR FAT-CRD ANOVA TABLE WITH EMS FORMS**

| Source | df | SS | MS | F* | EMS** Fixed | EMS** Random |
|---|---|---|---|---|---|---|
| Total corrected | $abn - 1$ | $\sum_{ijk} Y_{ijk}^2 - CF$ | | | | |
| A | $a - 1$ | $\dfrac{\sum_i Y_{i..}^2}{bn} - CF$ | $SS_A/df_A$ | $MS_A/MS_{err}$ | $\sigma_\varepsilon^2 + bn\phi_A$ | $\sigma_\varepsilon^2 + n\sigma_{AB}^2 + bn\sigma_A^2$ |
| B | $b - 1$ | $\dfrac{\sum_j Y_{.j.}^2}{an} - CF$ | $SS_B/df_B$ | $MS_B/MS_{err}$ | $\sigma_\varepsilon^2 + an\phi_B$ | $\sigma_\varepsilon^2 + n\sigma_{AB}^2 + an\sigma_B^2$ |
| AB | $(a - 1)(b - 1)$ | $\dfrac{\sum_{ij} Y_{ij.}^2}{n} - CF - SS_A - SS_B$ | $SS_{AB}/df_{AB}$ | $MS_{AB}/MS_{err}$ | $\sigma_\varepsilon^2 + n\phi_{AB}$ | $\sigma_\varepsilon^2 + n\sigma_{AB}^2$ |
| Exp. error | $ab(n - 1)$ | $SS_{tc} - SS_A - SS_B - SS_{AB}$ | $SS_{err}/df_{err}$ | | $\sigma_\varepsilon^2$ | $\sigma_\varepsilon^2$ |

where $\quad CF = \dfrac{Y_{...}^2}{abn}$

and where $\quad \phi_A = \dfrac{\sum A_i^2}{(a - 1)} \qquad \phi_B = \dfrac{\sum B_j^2}{(b - 1)} \qquad \phi_{AB} = \dfrac{\sum AB_{ij}^2}{(a - 1)(b - 1)}$

\* *F* tests are structured for fixed effects factors.

\*\* See Figure 20.3 for fixed and random effects model assumptions and notation format.

Expanding and simplifying Equation 21.10, we have

$$\sum_{i=1}^{a}\sum_{j=1}^{b}\sum_{k=1}^{n}(Y_{ijk}-\bar{Y}_{...})^2 = bn\sum_{i=1}^{a}(\bar{Y}_{i..}-\bar{Y}_{...})^2 + an\sum_{j=1}^{b}(\bar{Y}_{.j.}-\bar{Y}_{...})^2$$

$$+ n\sum_{i=1}^{a}\sum_{j=1}^{b}(\bar{Y}_{ij.}-\bar{Y}_{i..}-\bar{Y}_{.j.}+\bar{Y}_{...})^2 + \sum_{i=1}^{a}\sum_{j=1}^{b}\sum_{k=1}^{n}(Y_{ijk}-\bar{Y}_{ij.})^2 \qquad (21.11)$$

In Equation 21.11, we have established $SS_{tc}$ on the left-hand side, and partitioned it into $SS_A$, $SS_B$, $SS_{AB}$, and $SS_{err}$, respectively. As an extension to our EMS discussion in Chapter 20, Table 21.1 displays the simplified ANOVA row calculations, and the EMSs of both fixed and random effects factors.

---

## Example 21.2

Using the fixed effects experiment data in Figure 21.1b, develop an ANOVA table, perform F tests, and interpret the results. Develop s.e.i. arrows for both main effect and interaction plots. Then, perform LSD tests as appropriate using $\alpha = 0.05$.

### Solution

In our solution, factor A represents syrup content and factor B represents cooking temperature. The ANOVA table is displayed in Table 21.2, along with sample calculations. We have applied the generic formulas of Table 21.1. Since both factors (A and B) are fixed, our F ratios are straightforward, based on the Fixed EMS column of Table 21.1. The SAS code [1, 2] and selected output for this example are shown in the Computer Aided Analysis Supplement, Solutions Manual. In our interpretation, we will examine main effects first. In general, it does not matter which order is used, but we must consider all effects.

*Model:*

$$Y_{ijk} = \mu + A_i + B_j + AB_{ij} + \varepsilon_{k(ij)}$$

*Main Effect A:*

$H_0$:   All $A_i = 0$ or all $\mu_{A_i}$ are equal

$H_1$:   At least one $A_i \neq 0$ or at least one $\mu_{A_i}$ is not equal to the others

The term $\mu_{A_i}$ represents a syrup mean.

In this case, $F_{calc} = 36.75$ and P-value $= 0.0003$; thus P-value $\ll 0.01$, judging from our F tables, since $F_{0.01, 1, 8} = 11.26$. Hence, we see overwhelming evidence to support the rejection of $H_0$ and conclude that the average elongation (averaged over cooking temperature and our replications) is not the same at all syrup content levels. Figure 21.2c presents this story graphically and should be a part of our results report.

TABLE 21.2    ANOVA TABLE FOR $2^2$ COOKIE EXPERIMENT (EXAMPLE 21.2)

| Source | df | SS | MS | F | P-value* |
|---|---|---|---|---|---|
| Total corrected | 12 – 1  = 11 | 281.000 | | | |
| A (syrup main effect) | 2 – 1  = 1 | 147.000 | 147.000 | 36.75 | 0.0003 |
| B (cooking temp. main effect) | 2 – 1  = 1 | 75.000 | 75.000 | 18.75 | 0.0025 |
| AB (syrup * cooking temp interaction) | (2 – 1)(2 – 1)  = 1 | 27.000 | 27.000 | 6.75 | 0.0317 |
| Experimental error | 2(2)(3 – 1)  = 8 | 32.000 | 4.000 | | |

Sample calculations

$SS_{tc} = (2^2 + 4^2 + 6^2 + \ldots + 16^2 + 18^2) - (102^2/12) = 281.000$

$SS_A = (30^2 + 72^2)/6 - (102^2/12) = 147.000$

$SS_B = (36^2 + 66^2)/6 - (102^2/12) = 75.000$

$SS_{AB} = (12^2 + 24^2 + 18^2 + 48^2)/3 - (102^2/12) - 147.0 - 75.0 = 27.000$

$SS_{err} = SS_{tc} - SS_A - SS_B - SS_{AB} = 281.000 - 147.000 - 75.000 - 27.000 = 32.000$

* P-values were obtained from SAS computer output, see Computer Aided Analysis Supplement, Solutions Manual.

*Main Effect B:*

$H_0$:   All $B_j = 0$ or all $\mu_{B_j}$ are equal

$H_1$:   At least one $B_j \neq 0$ or at least one $\mu_{B_j}$ is not equal to the others

The term $\mu_{B_j}$ represents a cooking temperature mean.

In this case, we see $F_{calc} = 18.75$ which yields P-value = 0.0025. Using our F tables we could say that the P-value $\ll$ 0.01, judging $F_{calc}$ against $F_{0.01,1,8} = 11.26$. We see overwhelming evidence to support the rejection of $H_0$. We therefore conclude that the average elongation here is not the same at all levels of cooking temperature. Figure 21.2d depicts this result and should be included as part of our findings report.

*Interaction AB:*

$H_0$:   All $AB_{ij} = 0$ or all $\mu_{AB_{ij}}$ are equal

$H_1$:   At least one $AB_{ij} \neq 0$ or at least one $\mu_{AB_{ij}}$ is not equal to the others

The term $\mu_{AB_{ij}}$ represents a syrup–cooking temperature combination mean.

Here, we observe $F_{calc} = 6.75$ which has an associated $P$-value = 0.0317; we would say $0.025 < P$-value $< 0.05$, based on our $F$ tables, since $F_{0.05,1,8} = 5.32$ and $F_{0.025,1,8} = 7.57$ (Section IX, Table IX.4). We see strong evidence here to reject $H_0$, given the $\alpha = 0.05$ significance level. On the other hand, if we used $\alpha = 0.01$, we would not reject $H_0$.

We will go along with the $\alpha = 0.05$ criterion and interpret as follows:

**1** We will reject $H_0$ (that all $AB_{ij} = 0$) and conclude that significant interaction has been detected, b.o.e.

**2** Now, we must interpret factor $A$ and factor $B$ together in assessing the elongation response.

**3** We must focus on our interaction plots and their lack of parallelism (Figure 21.2$a$ and $b$) and comment on the average responses (averaged over replicates only). For example, in Figure 21.2$b$ when we move from low $B$ to high $B$ at low $A$, we see an average of 2 percent additional elongation in the response. Likewise, when we move from low $B$ to high $B$ at high $A$, we see an average of 8 percent additional elongation in the response.

Now, we will consider the case had we used $\alpha = 0.01$ as our criterion for judging $AB$ interaction:

**1** We would not reject $H_0$ and would conclude that the interaction detected in the sample is not significant (when compared to our experimental error), b.o.e.

**2** Now, we interpret factors $A$ and $B$ independently in assessing our elongation response.

**3** We use the same main effects hypotheses and $P$-values as above. For factor $A$ (syrup) we would use Figure 21.2$c$ as our prime graphical support and say that the average (averaged over factor $B$ and replicates) increases by 7 percent elongation (from 5 percent to 12 percent) as we move from low $A$ (40 lb) to high $A$ (50 lb). Likewise, for factor $B$ (cooking temperature) we would use Figure 21.2$d$ and point out that the average increases by 5 percent (from 6 percent to 11 percent) as we move from low $B$ (300°F) to high $B$ (350°F). Here, we have discussed each factor "independent" of the other factor. The implication in our physical world (when we have no significant interaction) is that we can set each factor at its "best" level without considering the other factor's level.

In this $\alpha = 0.01$ case, we see that we are simply interpreting the main effects and ignoring the signs of interaction we actually see in Figures 21.2$a$ and $b$ as well as in our $AB$ row in the ANOVA . Hence we need to be careful to state that our conclusions (that we cannot detect a significant interaction) is drawn for $\alpha = 0.01$, b.o.e. The word "significant" is critical here. What we really mean is that the interaction actually measured is relatively small compared to the experimental error. Other experimenters using a larger sample size and/or a more carefully designed (e.g., blocked) experiment might produce a statistically significant interaction at the $\alpha = 0.01$ level.

In order to develop the $\pm 2$ s.e.i. arrows for the main effect plots, we notice that each main effect sample mean contains six observations. Hence, for main effects $A$ and $B$,

$$\pm 2 \text{ s.e.i.} = \pm 2 \sqrt{\frac{MS_{err}}{n'}} = \pm 2 \sqrt{\frac{4.000}{6}} = \pm 1.63\%$$

Remember, $n'$ equals the number of observations in the sample mean of interest here. For the interaction plots, each sample mean contains three observations; hence, $\pm 2$ s.e.i. arrows are:

$$\pm 2 \text{ s.e.i.} = \pm 2 \sqrt{\frac{MS_{err}}{n'}} = \pm 2 \sqrt{\frac{4.000}{3}} = \pm 2.31\%$$

Figure 21.3 depicts both interaction and main effect plots with the s.e.i. arrows.

The $LSD_{\alpha=0.05}$ analysis is shown in Table 21.3. Table 21.3a and Table 21.3b develop the main effects $A$ and $B$ LSD analysis, respectively, where

$$LSD_{0.05} = t_{0.025,\,8} \sqrt{\frac{2\,MS_{err}}{n'}} = 2.306 \sqrt{\frac{2(4.000)}{6}} = 2.663$$

The $AB$ interaction LSD analysis is developed in Table 21.3c, where

$$LSD_{0.05} = t_{0.025,\,8} \sqrt{\frac{2\,MS_{err}}{n'}} = 2.306 \sqrt{\frac{2(4.000)}{3}} = 3.766$$

The LSD development points to a number of significant differences, which were marked with the asterisk in Table 21.3, and supports what we see in the plots. Namely, treatment combination $a_2b_2$ (high $A$, high $B$), or the 50-lb syrup and 350°F cooking temperature, appears to produce a significantly different elongation response than the other treatment combinations, b.o.e. Hence, we expect to obtain a different (higher) degree of chewiness at this factor combination than at the others. We must note at this time that the other pairwise tests, see Chapter 20, are also applicable in factorial experiments.

## 21.3 EXPERIMENTS INVOLVING THREE OR MORE FACTORS

The typical form of a three-factor FAT-CRD model is

$$Y_{ijkl} = \mu + A_i + B_j + C_k + AB_{ij} + AC_{ik} + BC_{jk} + ABC_{ijk} + \varepsilon_{l(ijk)} \qquad (21.12)$$

where $i = 1, 2, ..., a$

$j = 1, 2, ..., b$

$k = 1, 2, ..., c$

$l = 1, 2, ..., n$

Here, each $Y_{ijkl}$ represents an observation. *The $A_i$, $B_j$, and $C_k$ terms are main effect terms.* For instance, *the A main effect measures the failure of the average response of the factor A levels to be the same, when averaged over all levels of factor B and factor C as well as the replications.* The main effect of $B$ and $C$ are interpreted similarly. *The $AB_{ij}$, $AC_{ik}$, and $BC_{jk}$ terms are referred to as two-factor interaction terms.* For example, *the AB interaction term measures the failure of A*

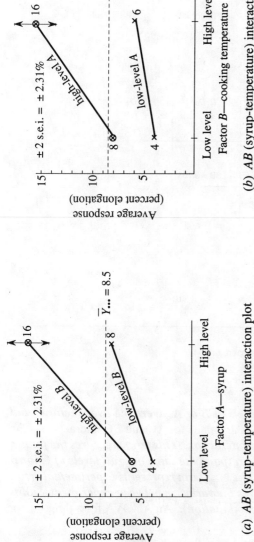

(a) AB (syrup-temperature) interaction plot

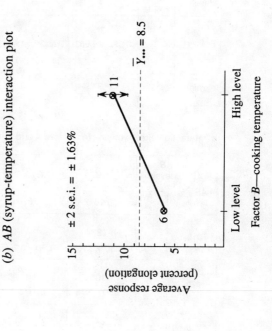

(b) AB (syrup-temperature) interaction plot

(c) Factor A main effect plot

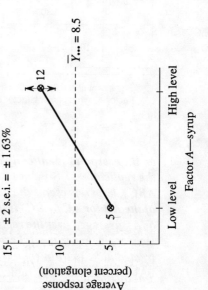

(d) Factor B main effect plot

**FIGURE 21.3** Experimental results plots for a $2^2$ FAT-CRD experiment (Example 21.2).

**TABLE 21.3    LSD COMPARISONS FOR THE PAIRWISE TREATMENT COMBINATIONS OF FACTORS A AND B (EXAMPLE 21.2)**

**a** Main effect means for factor A, LSD

| A levels / A levels (Avg/Avg) | 2 (high) | 1 (low) |
|---|---|---|
| | 12 | 5 |
| 1 (low) — 5 | 7* LSD = 2.663 | |
| 2 (high) — 12 | | |

**b** Main effect means for factor B, LSD

| B Levels / B Levels (Avg/Avg) | 2 (high) | 1 (low) |
|---|---|---|
| | 11 | 6 |
| 1 (low) — 6 | 5* LSD = 2.663 | |
| 2 (high) — 11 | | |

**c** Interaction effects means for factors A and B, LSD

| AB levels / AB levels (Avg/Avg) | $a_2 b_2$ | $a_2 b_1$ | $a_1 b_2$ | $a_1 b_1$ |
|---|---|---|---|---|
| | 16 | 8 | 6 | 4 |
| $a_1 b_1$ — 4 | 12* LSD = 3.766 | 4* LSD = 3.766 | 2 LSD = 3.766 | |
| $a_1 b_2$ — 6 | 10* LSD = 3.766 | 2 LSD = 3.766 | | |
| $a_2 b_1$ — 8 | 8* LSD = 3.766 | | | |
| $a_2 b_2$ — 16 | | | | |

\* Significant at $\alpha = 0.05$ level, b.o.e.
Factor A, syrup content; factor B, cooking temperature.

or B to respond identically over all levels of B or A, averaged over factor C and the replications. Similar interpretations are made for AC and BC interaction. The $ABC_{ijk}$ term represents the three-factor interaction. This term measures the failure of the AB (or AC or BC) interaction to respond the same over all levels of C (or B or A), averaged over the replications. The $\varepsilon_{l(ijk)}$ term represents experimental error. Here again, experimental error measures the variation between experimental units treated alike (within each treatment combination). An ANOVA for a three-factor FAT-CRD is shown in Table 21.4.

**TABLE 21.4    GENERIC THREE-FACTOR FAT-CRD ANOVA TABLE**

| Source | df | SS | MS | $F^*$ | P-value |
|---|---|---|---|---|---|
| Total corrected | $abcn - 1$ | $\sum_{ijkl} Y_{ijkl}^2 - CF$ | | | |
| A | $a - 1$ | $\dfrac{\sum_i Y_{i...}^2}{bcn} - CF$ | $SS_A/df_A$ | $MS_A/MS_{err}$ | $P_A$ |
| B | $b - 1$ | $\dfrac{\sum_j Y_{.j..}^2}{acn} - CF$ | $SS_B/df_B$ | $MS_B/MS_{err}$ | $P_B$ |
| C | $c - 1$ | $\dfrac{\sum_k Y_{..k.}^2}{abn} - CF$ | $SS_C/df_C$ | $MS_C/MS_{err}$ | $P_C$ |
| AB | $(a-1)(b-1)$ | $\dfrac{\sum_{ij} Y_{ij..}^2}{cn} - SS_A - SS_B - CF$ | $SS_{AB}/df_{AB}$ | $MS_{AB}/MS_{err}$ | $P_{AB}$ |
| AC | $(a-1)(c-1)$ | $\dfrac{\sum_{ik} Y_{i.k.}^2}{bn} - SS_A - SS_C - CF$ | $SS_{AC}/df_{AC}$ | $MS_{AC}/MS_{err}$ | $P_{AC}$ |
| BC | $(b-1)(c-1)$ | $\dfrac{\sum_{jk} Y_{.jk.}^2}{an} - SS_B - SS_C - CF$ | $SS_{BC}/df_{BC}$ | $MS_{BC}/MS_{err}$ | $P_{BC}$ |
| ABC | $(a-1)(b-1)(c-1)$ | $\dfrac{\sum_{ijk} Y_{ijk.}^2}{n} - SS_A - SS_B - SS_C$ $- SS_{AB} - SS_{AC} - SS_{BC} - CF$ | $SS_{ABC}/df_{ABC}$ | $MS_{ABC}/MS_{err}$ | $P_{ABC}$ |
| Experimental error | $abc(n-1)$ | $SS_{tc} - SS_A - SS_B - SS_C$ $- SS_{AB} - SS_{AC} - SS_{BC} - SS_{ABC}$ | $SS_{err}/df_{err}$ | | |

$$CF = \frac{Y_{....}^2}{abcn}$$

*Note:* EMS terms for a three-factor FAT-CRD with fixed effects factors are shown in Table 21.12. EMS terms for a three-factor FAT-CRD with random effects factors are shown in Table 21.13.

* $F$ tests are structured for fixed effects factors $A$, $B$, and $C$.

When we compare the two-factor ANOVA (Table 21.1) with the three-factor ANOVA, we can see patterns that will assist in our df and SS calculations. The df forms are straightforward. For the SS form for main effect terms, we start with factor level totals. We then square each total, sum the totals, and divide by the number of observations in each. Finally, we subtract the correction factor. A similar procedure is used for the interaction terms. Hence, after some reflection, we could write the ANOVA for a four-factor case by inspection.

Just as in the previous cases, ***when significant interaction is not present, the factors can be interpreted independently of one another.*** When interaction is present, the factors cannot be interpreted independently of one another; they must be interpreted simultaneously. In other words, ***when significant interaction is present, we describe the observed response relative to the levels of the interacting factors. Plots are very helpful tools to illustrate main effects and interactions in multifactor interpretation, i.e., involving fixed effects factors.***

*If multiple replications are not made for each treatment combination, the error term cannot be estimated directly.* In such cases the higher order interactions (e.g., three factors or more) may be pooled (combined) and used to estimate the experimental error term. Here, essentially, we pool the mean square values, weighted by the degrees of freedom.

Factors in our experiments may be either fixed or random effects related. The former are used to assess the response relative to a hand–selected group of factor levels, just as we did in our single-factor experiments. Random effects factors are useful to estimate components of variation. For example, if we were assessing product variation in a manufacturing process where machines, operators, and materials are involved, a three-factor random effects model could be used. Here we could determine the proportion of product variation created by each factor; Then, we could address the most critical elements through a well structured quality (variance reduction) project.

In both random and fixed effects models, we use computer aids to develop our ANOVAs. For fixed effects factors, we typically use computer aids to develop our graphs as well. We select appropriate information from the ANOVA output, develop or piece together graphs and other analyses (e.g., pairwise tests on the main effect and interaction terms, as in Examples 21.1 and 21.2).

---

## Example 21.3

Based on our chocolate chip cookie manufacturing process, we want to develop an experiment that can provide information regarding three factors relative to chewiness: cooking temperature, cooking time, and syrup content. We want to measure all three main effects, all two-factor interactions, the three-factor interaction, and the experimental error. Furthermore, we want to broaden our experiment to include three levels of each factor. Develop an experiment which can accommodate our needs and analyze the results.

### Solution

We will develop a $3^3$ FAT-CRD with two replicates. The FAT design will be necessary to measure our interactions. The replicates will be necessary to capture (estimate) our experimental error. Here, we will expand our previous $2^2$ FAT-CRD example. Our factors are

Factor $A$:   Cooking temperature—300°F, 350°F, and 400°F

Factor $B$:   Cooking time—5 min, 10 min, 15 min

Factor $C$:   Syrup content per batch:—40 lb, 50 lb, 60 lb

We have selected these factor levels based on our physical knowledge of the process. Our experimental boundaries are depicted in Figure 21.4. All factors here are fixed effect factors in that we have hand-picked each factor level and

FIGURE 21.4    Physical boundaries for a $3^3$ FAT-CRD experiment (Example 21.3).

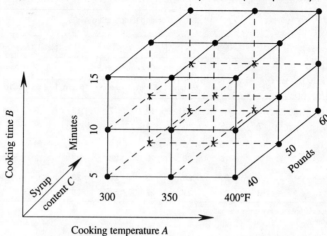

wish to make inferences specifically to only those levels. We randomized our experiment (as in a CRD) and recorded our $3 \times 3 \times 3 \times 2 = 54$ observations. The data appear in Table 21.5. Now, we will develop an ANOVA and a set of plots with s.e.i. arrows, as well as both LSD and HSD comparisons for our main effects, with $\alpha = 0.05$.

The data shown in Table 21.5 were stratified, and $AB$, $AC$, and $BC$ tables developed in Table 21.6. These tables were then used to develop the sum of squares shown in the ANOVA in Table 21.7. The ANOVA was developed using both hand calculations and SAS. The $P$-values shown are taken from the SAS output. The SAS code and selected output are shown in the Computer Aided Analysis Supplement, Solutions Manual.

A full set of main effect and two-factor interaction plots is presented in Figure 21.5. The $\pm 2$ s.e.i. arrows are calculated and plotted for the entire set, as follows:

*Main Effects:*

$$\pm 2 \text{ s.e.i.} = \pm 2 \sqrt{\frac{\text{MS}_{\text{err}}}{n'}} = \pm 2 \sqrt{\frac{14.13}{18}} = \pm 1.77$$

*Two-Factor Interactions:*

$$\pm 2 \text{ s.e.i.} = \pm 2 \sqrt{\frac{\text{MS}_{\text{err}}}{n'}} = \pm 2 \sqrt{\frac{14.13}{6}} = \pm 3.07$$

Based on our ANOVA indications, and the plots, we conclude that our $AB$ interaction and main effects $A$, $B$, and $C$ are all significant at the $\alpha = 0.05$ level, b.o.e. (e.g., we would reject their respective $H_0$'s). Furthermore, factor $C$ (syrup content) and interaction $AB$ (cooking temperature–cooking time) tend to show

TABLE 21.5    DATA LAYOUT FOR $3^3$ FAT-CRD (EXAMPLE 21.3)

| Syrup content per batch C | Cooking temperature A | | | | | | | | |
|---|---|---|---|---|---|---|---|---|---|
| | 300°F | | | 350°F | | | 400°F | | |
| | Cooking time B | | | | | | | | |
| | 5 min | 10 min | 15 min | 5 min | 10 min | 15 min | 5 min | 10 min | 15 min |
| 40 lb | 3 | 7 | 6 | 10 | 7 | 13 | 6 | 13 | 4 |
| | 7 | 9 | 13 | 4 | 14 | 19 | 15 | 5 | 1 |
| 50 lb | 17 | 20 | 23 | 18 | 32 | 21 | 16 | 21 | 11 |
| | 12 | 16 | 28 | 20 | 28 | 24 | 19 | 16 | 19 |
| 60 lb | 19 | 25 | 29 | 23 | 26 | 23 | 23 | 30 | 13 |
| | 13 | 23 | 31 | 27 | 21 | 28 | 19 | 24 | 21 |

Response units are in percent elongation.

up as very strong effects, ($P$-values of 0.0001 and 0.0009, respectively). We see some weak indications of a three-factor interaction ($P$-value = 0.0984). The other effects, $AC$ and $BC$ appear insignificant. Obviously, we are drawing these conclusions relative to our experimental error. As we made two replications for each treatment combination, the 27 df in the error row gives us a reasonably good estimate of $\sigma_\varepsilon^2$ (e.g., $MS_{err} = \hat{\sigma}_\varepsilon^2$).

We used our $MS_{err} = 14.13$ in our ±2 s.e.i. arrow calculations. Now, we will develop our $LSD_{0.05}$ and $HSD_{0.05}$ comparison values using our $MS_{err}$. For the pairwise main effects comparisons,

$$LSD_{0.05} = t_{0.025,\, 27} \sqrt{\frac{2\, MS_{err}}{n'}} = 2.052 \sqrt{\frac{2(14.13)}{18}} = 2.571$$

$$HSD_{0.05} = q_{0.05}\, (3,\, 27) \sqrt{\frac{MS_{err}}{n'}} \simeq 3.53 \sqrt{\frac{14.13}{18}} = 3.128$$

For the pairwise two-factor interaction comparisons,

$$LSD_{0.05} = 2.052 \sqrt{\frac{2\, MS_{err}}{n'}} = 2.052 \sqrt{\frac{2(14.13)}{6}} = 4.453$$

$$HSD_{0.05} = q_{0.05}(9,\, 27) \sqrt{\frac{MS_{err}}{n'}} \simeq 4.81 \sqrt{\frac{14.13}{6}} = 7.381$$

In calculating $HSD_{0.05}$, we have used $q_{0.05}\,(3, 24) = 3.53$ and $q_{0.05}\,(9, 24) = 4.81$, since our table (Section IX, Table IX.7) does not contain $df_{err} = 27$. Using the $df_{err} = 24$ value will tend to overstate our $HSD_{0.05}$ slightly and produce a conservative test.

The resulting LSD and HSD data for main effects $A$, $B$, and $C$ are shown in Table 21.8. The pairwise comparisons denoted with a * or + are statistically

## TABLE 21.6   AB, AC, AND BC TABLES (EXAMPLE 21.3)

**a** *AB* table

Factor B (cooking time)

| Factor A (cooking temperature) | 1 (5 min) | 2 (10 min) | 3 (15 min) | $Y_{i..}$ | $\bar{Y}_{i..}$ |
|---|---|---|---|---|---|
| 1 (300°F) | Total 71 / Obs. 6 / Mean 11.83 | 100 / 6 / 16.66 | 130 / 6 / 21.66 | 301 / 18 | 16.7222 |
| 2 (350°F) | 102 / 6 / 17.0 | 128 / 6 / 21.33 | 128 / 6 / 21.33 | 358 / 18 | 19.8888 |
| 3 (400°F) | 98 / 6 / 16.33 | 109 / 6 / 18.16 | 69 / 6 / 11.5 | 276 / 18 | 15.3333 |
| $Y_{.j.}$ | 271 / 18 | 337 / 18 | 327 / 18 | $Y_{....} = 935.00$ | Obs. 54 |
| $\bar{Y}_{.j.}$ | 15.0555 | 18.7222 | 18.1667 | $\bar{Y}_{....} = 17.3148$ | |

**b** *AC* table

Factor C (syrup content)

| Factor A (cooking temperature) | 1 (40 lb) | 2 (50 lb) | 3 (60 lb) | $Y_{i..}$ | $\bar{Y}_{i..}$ |
|---|---|---|---|---|---|
| 1 (300°F) | Total 45 / Obs. 6 / Mean 7.5 | 116 / 6 / 19.33 | 140 / 6 / 23.33 | 301 / 18 | 16.7222 |
| 2 (350°F) | 67 / 6 / 11.16 | 143 / 6 / 23.83 | 148 / 6 / 24.66 | 358 / 18 | 19.8888 |
| 3 (400°F) | 44 / 6 / 7.33 | 102 / 6 / 17.00 | 130 / 6 / 21.66 | 276 / 18 | 15.3333 |
| $Y_{..k}$ | 156 / 18 | 361 / 18 | 418 / 18 | $Y_{....} = 935.00$ | Obs. 54 |
| $\bar{Y}_{..k}$ | 8.6666 | 20.0555 | 23.2222 | $\bar{Y}_{....} = 17.3148$ | |

**c** *BC* table

Factor C (syrup content)

| Factor B (cooking time) | 1 (40 lb) | 2 (50 lb) | 3 (60 lb) | $Y_{.j.}$ | $\bar{Y}_{.j.}$ |
|---|---|---|---|---|---|
| 1 (5 min) | Total 45 / Obs. 6 / Mean 7.50 | 102 / 6 / 17.00 | 124 / 6 / 20.66 | 271 / 18 | 15.0555 |
| 2 (10 min) | 55 / 6 / 9.16 | 133 / 6 / 22.16 | 149 / 6 / 24.83 | 337 / 18 | 18.7222 |
| 3 (15 min) | 56 / 6 / 9.33 | 126 / 6 / 21.00 | 145 / 6 / 24.16 | 327 / 18 | 18.1666 |
| $Y_{..k}$ | 156 / 18 | 361 / 18 | 418 / 18 | $Y_{....} = 935.00$ | Obs. 54 |
| $\bar{Y}_{..k}$ | 8.6666 | 20.0555 | 23.2222 | $\bar{Y}_{....} = 17.3148$ | |

**TABLE 21.7**   ANOVA TABLE FOR $3^3$ FAT-CRD (EXAMPLE 21.3)

| Source | df | SS | MS | $F^*$ | P-value |
|---|---|---|---|---|---|
| Total corrected | 54 − 1 = 53 | 3463.65 | | | |
| A | 3 − 1 = 2 | 196.26 | 98.13 | 6.94 | 0.0037 |
| B | 3 − 1 = 2 | 140.59 | 70.30 | 4.98 | 0.0145 |
| C | 3 − 1 = 2 | 2109.59 | 1054.80 | 74.65 | 0.0001 |
| AB | (3 − 1)(3 − 1) = 4 | 366.96 | 91.74 | 6.49 | 0.0009 |
| AC | (3 − 1)(3 − 1) = 4 | 31.96 | 7.99 | 0.57 | 0.6897 |
| BC | (3 − 1)(3 − 1) = 4 | 19.96 | 4.99 | 0.35 | 0.8395 |
| ABC | (3 − 1)(3 − 1)(3 − 1) = 8 | 216.81 | 27.10 | 1.92 | 0.0984 |
| Exp. error | (3)(3)(3)(2 − 1) = 27 | 381.50 | 14.13 | | |

Sample calculations

$$SS_{tc} = (3^2 + 7^2 + \cdots + 13^2 + 21^2) - (935^2/54) = 3463.65$$

$$SS_A = (301^2 + 358^2 + 276^2)/18 - (935^2/54) = 196.26$$

$$SS_B = (271^2 + 337^2 + 327^2)/18 - (935^2/54) = 140.59$$

$$SS_C = (156^2 + 361^2 + 418^2)/18 - (935^2/54) = 2109.59$$

$$SS_{AB} = (71^2 + 100^2 + \cdots + 69^2)/6 - 196.26 - 140.59 - (935^2/54) = 366.96$$

$$SS_{AC} = (45^2 + 116^2 + \cdots + 130^2)/6 - 196.26 - 2109.59 - (935^2/54) = 31.96$$

$$SS_{BC} = (45^2 + 102^2 + \cdots + 145^2)/6 - 140.59 - 2109.59 - (935^2/54) = 19.96$$

$$SS_{ABC} = (10^2 + 29^2 + \cdots + 30^2 + 34^2)/2 - 196.26 - 140.59 - 2109.59 - 366.96 - 31.96$$
$$-19.96 - (935^2/54) = 216.81$$

$$SS_{err} = SS_{tc} - SS_A - SS_B - SS_C - SS_{AB} - SS_{AC} - SS_{BC} - SS_{ABC} = 3463.65 - 196.26$$
$$-140.59 - 2109.59 - 366.96 - 31.96 - 19.96 - 216.81 = 381.50$$

*Factors A, B, and C are fixed effects factors.

significant at the $\alpha = 0.05$ level, b.o.e.  The differences in the LSD and HSD results are due to the pairwise $\alpha$, in the former, and the experiment-wise $\alpha$ in the latter.  When we write our conclusions, we must be careful to state both our significance level (e.g., $\alpha = 0.05$) and our testing criterion (e.g., LSD or HSD), otherwise our interpretation may be confusing or misleading.

## 21.4  CROSSED AND NESTED FACTORS

So far, we have designed experiments where we can measure factor-by-factor interaction (e.g., two factors, *AB*; three factors, *ABC*; etc.).  *It is possible to measure interaction only when factors are "crossed" with each other.  Two factors A and B are said to be crossed if every level of A occurs with every level of B.*  When we deal with quantitative factors, meeting the crossing condition is usually straightforward (e.g., as with our cookie factors of Example 21.3).  *If,* on the other hand, *our physical requirements or resource base will not allow us to cross our factors, we cannot*

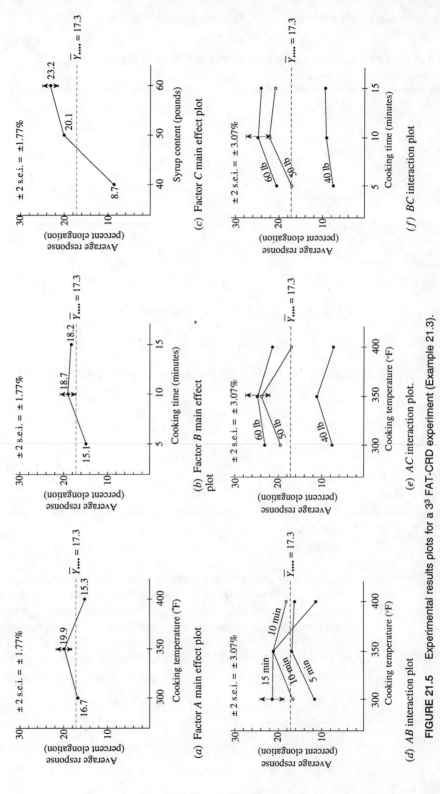

**FIGURE 21.5** Experimental results plots for a $3^3$ FAT-CRD experiment (Example 21.3).

(a) Factor A main effect plot

(b) Factor B main effect plot

(c) Factor C main effect plot

(d) AB interaction plot

(e) AC interaction plot.

(f) BC interaction plot

501

**TABLE 21.8    LSD AND HSD PAIRWISE COMPARISONS (EXAMPLE 21.3)**

**a** Main effect means for factor A, cooking temperature

| A levels ╲ | | 2 (350 °F) | 1 (300 °F) | 3 (400 °F) |
|---|---|---|---|---|
| A levels | Avg ╲ Avg | 19.889 | 16.772 | 15.333 |
| 3 (400 °F) | 15.333 | 4.556*+ <br> LSD = 2.571 <br> HSD ≅ 3.128 | 1.439 <br> LSD = 2.571 <br> HSD ≅ 3.128 | |
| 1 (300 °F) | 16.772 | 3.117* <br> LSD = 2.571 <br> HSD ≅ 3.128 | | |
| 2 (350 °F) | 19.889 | | | |

**b** Main effect means for factor B, cooking time

| B levels ╲ | | 2 (10 min) | 3 (15 min) | 1 (5 min) |
|---|---|---|---|---|
| B levels | Avg ╲ Avg | 18.722 | 18.167 | 15.056 |
| 1 (5 min) | 15.056 | 3.666*+ <br> LSD = 2.571 <br> HSD ≅ 3.128 | 3.111* <br> LSD = 2.571 <br> HSD ≅ 3.128 | |
| 3 (15 min) | 18.167 | 0.555 <br> LSD = 2.571 <br> HSD ≅ 3.128 | | |
| 2 (10 min) | 18.722 | | | |

**c** Main effect means for factor C, syrup content

| C levels ╲ | | 3 (60 lb) | 2 (50 lb) | 1 (40 lb) |
|---|---|---|---|---|
| C levels | Avg ╲ Avg | 23.222 | 20.056 | 8.667 |
| 1 (40 lb) | 8.667 | 14.555*+ <br> LSD = 2.571 <br> HSD ≅ 3.128 | 11.389*+ <br> LSD = 2.571 <br> HSD ≅ 3.128 | |
| 2 (50 lb) | 20.056 | 3.166*+ <br> LSD = 2.571 <br> HSD ≅ 3.128 | | |
| 3 (60 lb) | 23.222 | | | |

\* Significant $LSD_{0.05}$, b.o.e.

\+ Significant $HSD_{0.05}$, b.o.e.

*measure interaction, (i. e. we cannot break out meaningful interaction terms). In this case, we have what are termed "nested" factors.* We may nest factors (such as operators, batches, tools, and so on) in many situations. As an experimental design rule of thumb, nesting may be necessary when qualitative factor levels are involved. Figure 21.6 depicts both crossed and nested configurations. The crossed configuration, Figure 21.*a*, has already been discussed. In the nested or hierarchical model (Figure 21.6*b*),

$$Y_{ijk} = \mu + A_i + B_{j(i)} + \varepsilon_{k(ij)} \qquad (21.13)$$

We read $B_{j(i)}$ as "*B* nested within *A*." We read $\varepsilon_{k(ij)}$ as "replications nested within *AB* factor level combinations" (i.e., the subscripts and parentheses indicate our nesting structure).

*We can extract more information from crossed factors than from nested factors.*  Nevertheless, we usually nest factors either to accommodate physical restrictions or to allow greater convenience in experimentation.  If we consider the operators in Figure 21.6*b*, we may nest the operators within work methods in order to expedite the execution of the experiment.

The $\varepsilon_{k(ij)}$ term has been nested in our previous models (e.g., our two-factor FAT-CRD).  We nest due to a physical restriction.  In this case (and in all of our previous crossed factor cases), we allocate a distinct or different experimental unit to each treatment combination (i.e., 1 eu for each treatment combination replication).  This practice is obviously a nesting device (e.g., we use *abn* different experimental units during the course of an experiment when we have *a* levels of factor *A*, *b* levels of factor *B*, and *n* independent replicates within each *AB* combination).

When designing nested-crossed experiments we must carefully determine the nature—fixed or random—of each factor.  Then, the factor relationships—nested or crossed—must be determined.  Finally, we develop our model statement.  The nature of the subscript structure reflects the crossing-nesting nature of our model (e.g., as shown in Equation 21.13).  To generate our ANOVA, we develop our df and SS terms.  Typically, we will use a computer aid such as SAS for our analysis, but if we do not use a computer aid, the df and SS terms can be developed by combining the multifactor "pseudo crossed" term calculations (see Figure 21.6*a* and *b*).

---

## Example 21.4

In our cookie factory, we must form the raw cookie "blanks" from the dough for placement on our cooking pans before baking.  This process consists of extruding the cookie dough through a tube and a die.  The raw cookie blanks are then cut from the extrusion.  We are interested in producing cookie blanks of uniform weight and will explore the effects of three factors on this process response: temperature of the raw dough, extrusion pressure, and die-to-die variation.  Develop, execute, and analyze an experiment to study the resulting cookie blank weight phenomenon.

### Solution

We will define and develop our factors as follows:

*Factor T*:  The temperature of raw cookie dough in the extrusion tube

Here, temperature is somewhat difficult to control precisely as dough still warm from mixing as well as cooled dough from storage may be used.  Factor *T* will be developed as a random effects factor with four levels, randomly selected from the distribution of dough temperatures likely—50°F, 65°F, 70°F, and 75°F.

*Factor P*:  The extrusion pressure

Factor *P* will be developed as a fixed effects factor with two levels—75 psi and 100 psi.

*Factor D*:  The extrusion die factor

ANOVA: Crossed factors

| Source | df | SS |
|---|---|---|
| Total corrected | $abn - 1$ | $\sum_{ijk} Y_{ijk}^2 - CF$ |
| A | $a - 1$ | $\left(\sum_i Y_{i..}^2 \big/ bn\right) - CF$ |
| B | $b - 1$ | $\left(\sum_j Y_{.j.}^2 \big/ an\right) - CF$ |
| AB* | $(a-1)(b-1)$ | $\left(\sum_{ij} Y_{ij.}^2 \big/ n\right) - CF - SS_A - SS_B$ |
| Exp. error | $ab(n-1)$ | $SS_{tc} - SS_A - SS_B - SS_{AB}$ |

* AB (e.g., the failure of each operator to respond the same with each work method) can be measured here since identical operators (levels of B) appear with each work method (levels of A).

ANOVA: Nested factors

| Source | df | SS |
|---|---|---|
| Total corrected | $abn - 1$ | $\sum_{ijk} Y_{ijk}^2 - CF$ |
| A | $a - 1$ | $\left(\sum_i Y_{i..}^2 \big/ bn\right) - CF$ |
| B(A)* (B within A) | $a(b-1)$ | $\sum_i \left(\sum_j Y_{ij.}^2 \big/ n - Y_{i..}^2 \big/ bn\right)$ |
| Exp. error | $ab(n-1)$ | $SS_{tc} - SS_A - SS_{B(A)}$ |

* $B_{j(i)}$ or B(A) in the nested design consists statistically of the B and AB rows combined from the crossed analysis breakout (e.g., $df_{B(A)} = df_B + df_{AB}$; $SS_{B(A)} = SS_B + SS_{AB}$). Here, we can not extract AB interaction, but we can breakout $SS_{B(A)}$ for each A level and develop each A level's contribution to $SS_{B(A)}$. This breakout may provide insight as to the variability within the A levels.

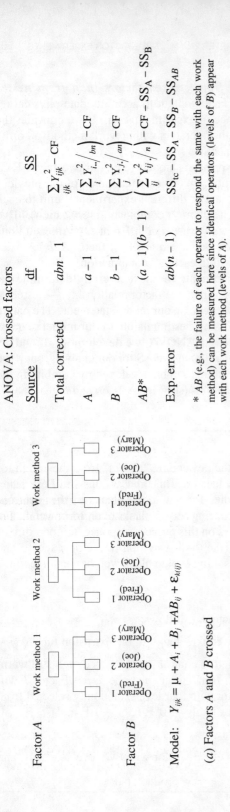

Model: $Y_{ijk} = \mu + A_i + B_j + AB_{ij} + \varepsilon_{k(ij)}$

(a) Factors A and B crossed

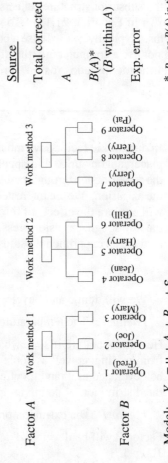

Model: $Y_{ijk} = \mu + A_i + B_{j(i)} + \varepsilon_{k(ij)}$

(b) Factor B nested in factor A

FIGURE 21.6 Nested and crossed factor design comparison.

We purchase dies from an outside vendor and we want to assess the uniformity of our die inventory. Each die is built to our (identical) specifications, but we know some die-to-die variation exists. We therefore are setting up the die factor as a random effects factor. We will randomly select eight different dies (with identical part numbers), which we then randomly assign throughout the temperature-pressure combinations. We will take two replications at each of the treatment combinations so that we can estimate the experimental error.

We have crossed factors $T$ and $P$. However, since we are using eight different dies (not just one die) we have failed to cross the die factor $D$ within the $T$ by $P$ combinations. Hence, our model will be

$$Y_{ijkl} = \mu + T_i + P_j + TP_{ij} + D_{k(ij)} + \varepsilon_{l(ijk)} \qquad (21.14)$$

We have used the CRD for our three-factor experiment. The data are displayed in Table 21.9. Stratified data summaries are shown in Table 21.10. We have developed our ANOVA for the nested-crossed design; it appears in Table 21.11. The SAS code and output appear in the Computer Aided Analysis Supplement, Solutions Manual.

Many computer aids allow us to either cross or nest our factors directly in the analysis. For example, we have used the $D(T*P)$ term in our SAS model statement to nest $D$ within $T$ by $P$ combinations. Hence, here, $T$ and $P$ are crossed, while $D$ is nested within the $T*P$ combinations.

If our computer aid has no provisions for nesting our factors, we can develop a "crossed" analysis and then piece together nested rows. For example, the second part of the SAS code is developed to calculate the $D$, $DT$, $DP$, and $DTP$ terms shown in the sample calculations in Table 21.11. We can piece together the nested terms by adding the proper "crossed" term calculations from our output. For example,

$$\mathrm{df}_{D(TP)} = \mathrm{df}_D + \mathrm{df}_{DT} + \mathrm{df}_{DP} + \mathrm{df}_{DTP}$$

and $$\mathrm{SS}_{D(TP)} = \mathrm{SS}_D + \mathrm{SS}_{DT} + \mathrm{SS}_{DP} + \mathrm{SS}_{DTP}$$

Most computer aids also allow us to structure specific hypothesis tests in our analyses. For example, in SAS, we have used TEST statements to develop specific hypothesis tests. The specific $F$-test structure used in this example will be developed and discussed in Example 21.5.

We see a highly significant dough temperature effect ($P$-value = 0.0027) which suggests that dough temperature will have to be controlled in order to produce uniform cookie blanks, b.o.e. We also see a reasonably significant pressure effect ($P$-value = 0.0433) which suggests we will also need to control our extrusion pressure. The $TP$ interaction is not significant ($P$-value = 0.1203). The die effect appears insignificant ($P$-value = 0.1930) suggesting that our die inventory will not create excessive variation in our raw product, b.o.e.

TABLE 21.9 NESTED AND CROSSED DATA FOR COOKIE BLANK EXTRUSION (EXAMPLE 21.4)

| Temperature T | | | | | | | | | | | | | | | |
|---|---|---|---|---|---|---|---|---|---|---|---|---|---|---|---|
| 50°F | | | | 65°F | | | | 70°F | | | | 75°F | | | |
| Pressure P | | | | | | | | | | | | | | | |
| 75 psi | | 100 psi | | 75 psi | | 100 psi | | 75 psi | | 100 psi | | 75 psi | | 100 psi | |
| Die D | | | | | | | | | | | | | | | |
| 4 | 6 | 8 | 5 | 1 | 5 | 3 | 7 | 2 | 8 | 4 | 7 | 6 | 4 | 1 | 3 |
| 35 | 34 | 39 | 40 | 41 | 38 | 48 | 47 | 36 | 39 | 38 | 44 | 38 | 44 | 49 | 51 |
| 32 | 38 | 36 | 37 | 37 | 42 | 50 | 43 | 40 | 41 | 42 | 50 | 39 | 41 | 54 | 58 |

Response: Raw cookie blank weight in grams.

## 21.5 MIXED MODELS, FIXED-RANDOM EFFECTS

We have previously discussed both fixed effects factors and random effects factors. Previously (with the exception of Example 21.4), our $F$ tests were structured by placing $MS_{err}$ in the denominator and the MS associated with the effect we seek to assess in the numerator. This procedure is appropriate for fixed effects models, but not in general for random effects models and models with both fixed and random effects, "mixed models."

***Our F test ratios are developed from EMS terms***, by isolating the term of interest in our model (e.g., expressed in our hypotheses as $\tau_i$, $A_i$, $AB_{ij}$). ***Generally, the F ratio based hypothesis test is used to isolate a specific main effect or interaction effect.*** The theoretical derivation of EMS terms are quite complicated mathematically (Hicks [3]). But when all factors are fixed or when all factors are random, we can write out the EMS for each model term by inspection. We will consider the three-factor FAT in a balanced CRD:

$$Y_{ijkl} = \mu + A_i + B_j + C_k + AB_{ij} + AC_{ik} + BC_{jk} + ABC_{ijk} + \varepsilon_{l(ijk)} \qquad (21.15)$$

We can develop the fixed effects case (for $A$, $B$, and $C$) as shown in Table 21.12. We can write the terms out in either full or abbreviated notation.

In the full notation scheme, each EMS entry contains the experimental error term. Each main effect and interaction entry has an additional term, made up of a denominator and a numerator. The denominator is simply the degrees of freedom for its row. The numerator is made up of the product of the number of replications and of the factor levels not included in the denominator, all multiplied by the sum of the squared "hypothesis" effect. In the abbreviated notation, we simply use $\phi$ to represent the sum of the factor effects terms divided by the df term.

The random effects EMS development is shown in Table 21.13. Here, we can see terms accumulating in the EMS as we move up the Source column. We observe the random effects expressed in the form of variation with common terms (e.g., for source $C_k$, we include the $ABC$, $BC$, $AC$, and $C$ components) representing factor levels other than those in the source entry. In other words, we look for subsets of our subscript in the terms we include in our EMS entries.

**TABLE 21.10 STRATIFIED DATA TABLES FOR ANOVA CALCULATIONS (EXAMPLE 21.4)**

**a** *TP* table

Factor *P* (pressure)

| Factor *T* (temperature) | 1 (75 psi) | 2 (100 psi) | $Y_{i..}$ | $\overline{Y}_{i..}$ |
|---|---|---|---|---|
| 1 (50°F) | Total 139 / Obs. 4 / Mean 34.75 | 152 / 4 / 38.0 | 291 / 8 | 36.375 |
| 2 (65°F) | 158 / 4 / 39.5 | 188 / 4 / 47.0 | 346 / 8 | 43.250 |
| 3 (70°F) | 156 / 4 / 39.0 | 174 / 4 / 43.5 | 330 / 8 | 41.250 |
| 4 (75°F) | 162 / 4 / 40.5 | 212 / 4 / 53.0 | 374 / 8 | 46.750 |
| $Y_{.j.}$ | 615 / 16 | 726 / 16 | $Y_{....}$ = 1341 | Obs. 32 |
| $\overline{Y}_{.j.}$ | 38.4375 | 45.375 | $\overline{Y}_{....}$ = 41.90625 | |

**b** *TD* table

Factor *D* (die)

| Factor *T* (temperature) | 1 | 2 | $Y_{i..}$ | $\overline{Y}_{i..}$ |
|---|---|---|---|---|
| 1 (50°F) | Total 142 / Obs. 4 / Mean 35.5 | 149 / 4 / 37.25 | 291 / 8 | 36.375 |
| 2 (65°F) | 176 / 4 / 44.0 | 170 / 4 / 42.5 | 346 / 8 | 43.250 |
| 3 (70°F) | 156 / 4 / 39.0 | 174 / 4 / 43.5 | 330 / 8 | 41.250 |
| 4 (75°F) | 180 / 4 / 45.0 | 194 / 4 / 48.5 | 374 / 8 | 46.750 |
| $Y_{..k}$ | 654 / 16 | 687 / 16 | $Y_{....}$ = 1341 | Obs. 32 |
| $\overline{Y}_{..k}$ | 40.875 | 42.9375 | $\overline{Y}_{....}$ = 41.90625 | |

**c** *PD* table

Factor *D* (die)

| Factor *P* (pressure) | 1 | 2 | $Y_{.j.}$ | $\overline{Y}_{.j.}$ |
|---|---|---|---|---|
| 1 (75 psi) | Total 298 / Obs. 8 / Mean 37.25 | 317 / 8 / 39.625 | 615 / 16 | 38.4375 |
| 2 (100 psi) | 356 / 8 / 44.5 | 370 / 8 / 46.25 | 726 / 16 | 45.375 |
| $Y_{..k}$ | 654 / 16 | 687 / 16 | $Y_{....}$ = 1341 | Obs. 32 |
| $\overline{Y}_{..k}$ | 40.875 | 42.9375 | $\overline{Y}_{....}$ = 41.90625 | |

**TABLE 21.11** ANOVA TABLE FOR THREE FACTOR CROSSED AND NESTED CRD (EXAMPLE 21.4)

| Source | df | SS | MS | $F^*$ | | P-value |
|---|---|---|---|---|---|---|
| Total corrected | $32 - 1 = 31$ | 1164.72 | | | | |
| $T$ | $4 - 1 = 3$ | 450.34 | 150.11 | $MS_T / MS_{D(TP)}$ | $= 11.74$ | 0.0027 |
| $P$ | $2 - 1 = 1$ | 385.03 | 385.03 | $MS_P / MS_{TP}$ | $= 11.37$ | 0.0433 |
| $TP$ | $(4-1)(2-1) = 3$ | 101.59 | 33.86 | $MS_{TP} / MS_{D(TP)}$ | $= 2.65$ | 0.1203 |
| $D(TP)$ | $4(2)(2-1) = 8$ | 102.25 | 12.78 | $MS_{D(TP)} / MS_{err}$ | $= 1.63$ | 0.1930 |
| Exp. error | $4(2)(2)(2-1) = 16$ | 125.50 | 7.84 | | | |

Sample calculations

$SS_{tc} = (35^2 + 32^2 + \cdots + 51^2 + 58^2) - (1341^2/32) = 1164.72$

$SS_T = (291^2 + 346^2 + 330^2 + 374^2)/8 - (1341^2/32) = 450.34$

$SS_P = (615^2 + 726^2)/16 - (1341^2/32) = 385.03$

$SS_{TP} = (139^2 + 152^2 + \cdots + 212^2)/4 - (1341^2/32) = 101.59$

$SS_{D(TP)} = SS_D + SS_{DT} + SS_{DP} + SS_{DTP} = 102.25$

$SS_{err} = SS_{tc} - SS_T - SS_P - SS_{TP} - SS_{D(TP)} = 1164.72 - 450.34 - 385.03 - 101.59 - 102.25 = 125.50$

$SS_D = (654^2 + 687^2)/16 - (1341^2/32) = 34.031$

$SS_{DT} = (142^2 + 149^2 + \cdots + 194^2)/4 - 34.031 - 450.344 - (1341^2/32) = 41.594$

$SS_{DP} = (298^2 + 317^2 + \cdots + 370^2)/8 - 34.031 - 385.031 - (1341^2/32) = 0.781$

$SS_{DTP} = (67^2 + 72^2 + \cdots + 103^2 + 109^2)/2 - 34.031 - 450.344 - 385.031 - 41.594 - 0.781 - 101.594 - (1341^2/32) = 25.844$

* The F ratios shown here are developed from the EMS algorithm in Example 21.5 (shown in Table 21.14).

Sometimes, no exact $F$ test exists, as in Table 21.13 for the main effects rows (we cannot isolate $\sigma_C^2$, $\sigma_B^2$, or $\sigma_A^2$). In these cases, we may develop pseudo $F$ tests using linear combinations of available EMS terms. For main effect $C$, Table 21.13, our pseudo term, $F'$ is

$$F' = \frac{MS_C}{MS_{AC} + MS_{BC} - MS_{ABC}}$$

Its corresponding EMS ratio, EMS' ratio, is

$$\text{EMS' ratio} = \frac{\sigma_\varepsilon^2 + n\sigma_{ABC}^2 + bn\sigma_{AC}^2 + an\sigma_{BC}^2 + abn\sigma_C^2}{\sigma_\varepsilon^2 + n\sigma_{ABC}^2 + bn\sigma_{AC}^2 + \sigma_\varepsilon^2 + n\sigma_{ABC}^2 + an\sigma_{BC}^2 - \sigma_\varepsilon^2 - n\sigma_{ABC}^2}$$

When we simplify, we can isolate the $\sigma_C^2$ term

$$\text{EMS' ratio} = \frac{\sigma_\varepsilon^2 + n\sigma_{ABC}^2 + bn\sigma_{AC}^2 + an\sigma_{BC}^2 + abn\sigma_C^2}{\sigma_\varepsilon^2 + n\sigma_{ABC}^2 + bn\sigma_{AC}^2 + an\sigma_{BC}^2}$$

We can use the ANOVA table to obtain our degrees of freedom in the numerator. The df value for the denominator is calculated using squared terms from the $F'$ denominator, as shown below:

$$\text{df}_{\text{denominator}} = \frac{(MS_{AC} + MS_{BC} - MS_{ABC})^2}{(MS_{AC}^2/\text{df}_{AC} + MS_{BC}^2/\text{df}_{BC} - MS_{ABC}^2/\text{df}_{ABC})}$$

**TABLE 21.12    THREE-FACTOR FIXED-EFFECTS EMS FORM**

| Source | df | EMS(full notation) | EMS (abbreviated notation) | F ratio |
|---|---|---|---|---|
| $A_i$ | $a-1$ | $\sigma_\varepsilon^2 + \dfrac{bcn \sum_i A_i^2}{a-1}$ | $\sigma_\varepsilon^2 + bcn\,\phi_A$ | $MS_A/MS_{err}$ |
| $B_j$ | $b-1$ | $\sigma_\varepsilon^2 + \dfrac{acn \sum_j B_j^2}{b-1}$ | $\sigma_\varepsilon^2 + acn\,\phi_B$ | $MS_B/MS_{err}$ |
| $C_k$ | $c-1$ | $\sigma_\varepsilon^2 + \dfrac{abn \sum_k C_k^2}{c-1}$ | $\sigma_\varepsilon^2 + abn\,\phi_C$ | $MS_C/MS_{err}$ |
| $AB_{ij}$ | $(a-1)(b-1)$ | $\sigma_\varepsilon^2 + \dfrac{cn \sum_{ij} AB_{ij}^2}{(a-1)(b-1)}$ | $\sigma_\varepsilon^2 + cn\,\phi_{AB}$ | $MS_{AB}/MS_{err}$ |
| $AC_{ik}$ | $(a-1)(c-1)$ | $\sigma_\varepsilon^2 + \dfrac{bn \sum_{ik} AC_{ik}^2}{(a-1)(c-1)}$ | $\sigma_\varepsilon^2 + bn\,\phi_{AC}$ | $MS_{AC}/MS_{err}$ |
| $BC_{jk}$ | $(b-1)(c-1)$ | $\sigma_\varepsilon^2 + \dfrac{an \sum_{jk} BC_{jk}^2}{(b-1)(c-1)}$ | $\sigma_\varepsilon^2 + an\,\phi_{BC}$ | $MS_{BC}/MS_{err}$ |
| $ABC_{ijk}$ | $(a-1)(b-1)(c-1)$ | $\sigma_\varepsilon^2 + \dfrac{n \sum_{ijk} ABC_{ijk}^2}{(a-1)(b-1)(c-1)}$ | $\sigma_\varepsilon^2 + n\,\phi_{ABC}$ | $MS_{ABC}/MS_{err}$ |
| $\varepsilon_{l(ijk)}$ | $abc(n-1)$ | $\sigma_\varepsilon^2$ | $\sigma_\varepsilon^2$ | |

What we are doing is setting up our $F'$ test denominator as a linear combination of available EMS terms, in order to isolate the term of interest. This procedure is attributed to Satterthwaite [4] and is advocated by Hicks [5] and Montgomery [6]. It provides an approximate $F$ test and is applicable to both random effects or mixed models (a totally fixed effects model will not require the approximation). Typically, the denominator df will not be an integer, thus we usually interpolate an $F$ value between those shown in our $F$ tables (section IX, Table IX.4). In off-line quality practice, approximate $F$ tests can be used to guide and support our recommendations.

*Mixed effects models present difficulties in developing EMS terms.* One method commonly used to develop the mixed model EMS is based on an algorithm developed by Bennett and Franklin [7] and advocated by Hicks [8] and Montgomery [9]. This algorithm is applicable for balanced, single-factor CRD and RCB designs and for balanced multiple-factor crossed and nested designs. We will illustrate the procedure with an example.

TABLE 21.13   THREE-FACTOR RANDOM EFFECTS EMS FORM

| Source | df | EMS | F-ratio |
|--------|-----|-----|---------|
| $A_i$ | $a-1$ | $\sigma_\varepsilon^2 + n\sigma_{ABC}^2 + cn\sigma_{AB}^2 + bn\sigma_{AC}^2 + bcn\,\sigma_A^2$ | * |
| $B_j$ | $b-1$ | $\sigma_\varepsilon^2 + n\sigma_{ABC}^2 + cn\sigma_{AB}^2 + an\sigma_{BC}^2 + acn\,\sigma_B^2$ | * |
| $C_k$ | $c-1$ | $\sigma_\varepsilon^2 + n\sigma_{ABC}^2 + bn\sigma_{AC}^2 + an\sigma_{BC}^2 + abn\,\sigma_C^2$ | * |
| $AB_{ij}$ | $(a-1)(b-1)$ | $\sigma_\varepsilon^2 + n\sigma_{ABC}^2 + cn\,\sigma_{AB}^2$ | $MS_{AB}/MS_{ABC}$ |
| $AC_{ik}$ | $(a-1)(c-1)$ | $\sigma_\varepsilon^2 + n\sigma_{ABC}^2 + bn\sigma_{AC}^2$ | $MS_{AC}/MS_{ABC}$ |
| $BC_{jk}$ | $(b-1)(c-1)$ | $\sigma_\varepsilon^2 + n\sigma_{ABC}^2 + an\,\sigma_{BC}^2$ | $MS_{BC}/MS_{ABC}$ |
| $ABC_{ijk}$ | $(a-1)(b-1)(c-1)$ | $\sigma_\varepsilon^2 + n\sigma_{ABC}^2$ | $MS_{ABC}/MS_{err}$ |
| $\varepsilon_{l(ijk)}$ | $abc(n-1)$ | $\sigma_\varepsilon^2$ | |

* Exact $F$ tests using available EMS terms directly cannot be developed to isolate $\sigma_A^2$, $\sigma_B^2$, or $\sigma_C^2$.

## Example 21.5

Develop an EMS table for a proposed experiment where $A$ and $B$ are crossed and $C$ is nested in $AB$ combinations. For this example, we will assume $B$ is a fixed effects factor and $A$ and $C$ are random effects factors. The error term is always random (i.e., experimental units are nested in treatment combinations).

### Solution

*Rule 1*   Express the model in a detailed form, making sure to capture crossed and nested factors in the subscript notation:

$$Y_{ijkl} = \mu + A_i + B_j + AB_{ij} + C_{k(ij)} + \varepsilon_{1(ijk)}$$

We can recognize this model from Example 21.4, where

$$A = \text{Factor } T, \text{ temperatures} \quad i = 1, 2, 3, 4$$
$$B = \text{Factor } P, \text{ pressures} \quad j = 1, 2$$
$$C = \text{Factor } D, \text{ dies} \quad k = 1, 2$$

*Rule 2*   Develop the EMS table structure. Place the model terms out in row format and lay the subscripts and levels out as column headings. It is also a good idea to include the df column so that we can assess our ability to measure effects before we commit to physical experimentation.

| Source | df | R a i | F b j | R c k | R n l | EMS |
|--------|-----|-----|-----|-----|-----|-----|
| $A_i$ | $a-1$ | | | | | |
| $B_j$ | $b-1$ | | | | | |
| $AB_{ij}$ | $(a-1)(b-1)$ | | | | | |
| $C_{k(ij)}$ | $ab(c-1)$ | | | | | |
| $\varepsilon_{l(ijk)}$ | $abc(n-1)$ | | | | | |

Here $F$ indicates a fixed effect and $R$ a random effect. Subscripts inside parentheses are termed "dead" subscripts and those without parentheses are termed "live" subscripts.

*Rule 3*   Fill in the body of the EMS table.

**a** Starting with the bottom row (and moving upward), place a 1 in the column corresponding to each dead subscript:

| Source | df | $R$ $a$ $i$ | $F$ $b$ $j$ | $R$ $c$ $k$ | $R$ $n$ $l$ | EMS |
|---|---|---|---|---|---|---|
| $A_i$ | $a-1$ | | | | | |
| $B_j$ | $b-1$ | | | | | |
| $AB_{ij}$ | $(a-1)(b-1)$ | | | | | |
| $C_{k(ij)}$ | $ab(c-1)$ | 1 | 1 | | | |
| $\varepsilon_{l(ijk)}$ | $abc(n-1)$ | 1 | 1 | 1 | | |

**b** Place a 0 in the fixed effect columns corresponding to each live subscript.

**c** Place a 1 in the random effect columns corresponding to each live subscript:

| Source | df | $R$ $a$ $i$ | $F$ $b$ $j$ | $R$ $c$ $k$ | $R$ $n$ $l$ | EMS |
|---|---|---|---|---|---|---|
| $A_i$ | $a-1$ | 1 | | | | |
| $B_j$ | $b-1$ | | 0 | | | |
| $AB_{ij}$ | $(a-1)(b-1)$ | 1 | 0 | | | |
| $C_{k(ij)}$ | $ab(c-1)$ | 1 | 1 | 1 | | |
| $\varepsilon_{l(ijk)}$ | $abc(n-1)$ | 1 | 1 | 1 | 1 | |

**d** Fill in the remaining open cells, by column, with the level number symbols in each column heading.

| Source | df | $R$ $a$ $i$ | $F$ $b$ $j$ | $R$ $c$ $k$ | $R$ $n$ $l$ | EMS |
|---|---|---|---|---|---|---|
| $A_i$ | $a-1$ | 1 | $b$ | $c$ | $n$ | |
| $B_j$ | $b-1$ | $a$ | 0 | $c$ | $n$ | |
| $AB_{ij}$ | $(a-1)(b-1)$ | 1 | 0 | $c$ | $n$ | |
| $C_{k(ij)}$ | $ab(c-1)$ | 1 | 1 | 1 | $n$ | |
| $\varepsilon_{l(ijk)}$ | $abc(n-1)$ | 1 | 1 | 1 | 1 | |

*Rule 4*   Develop the EMS column. Start at the bottom row and work upward. Cover the live subscript's column(s) for each row as you work on it. Then, multiply across the uncovered columns, one row at a time, to develop the individual terms in the entries. A term may be added for each row which contains a subset of the row subscripts being addressed (disregard the difference between live and dead subscripts). For example, for the $A_i$ row's EMS, we cover the $i$ column. The $A_i$ row EMS may contain terms associated with all other rows that contain the $i$ subscript (e.g., $AB_{ij}$, $C_{k(ij)}$, and $\varepsilon_{l(ijk)}$). The final results of our algorithm and a sample calculation are shown in Table 21.14.

Two observations can be made regarding the EMS development:

**1** We have used the abbreviated notation $\phi$ (see Table 21.12) for fixed effects.
**2** When a fixed effects factor and a random effects factor interact, they will yield a random effects interaction term.

In Table 21.14, a column has been included to identify the $F$ test numerators and denominators suggested by the EMS structure. In this case no pseudo $F$ tests are required. Our previous EMS development is relevant to the mixed effects experiment of Example 21.4. The ANOVA $F$ tests shown in Table 21.11 were structured by using this EMS development technique (e.g., Table 21.14). These same $F$ tests were coded in SAS (see the Computer Aided Analysis Supplement, Solutions Manual) in the form of "TEST  H = numerator term E = denominator term" statements.

---

## 21.6  FULL FACTORIALS WITHOUT REPLICATION

Up to this point, we have discussed experimental designs with replication. Recall that we define experimental error as the failure of experimental units, when treated alike, to respond alike. Replication allows us to obtain a direct estimate of our experimental error. *If we do not replicate, we cannot obtain a direct estimate of our experimental error $\sigma_\varepsilon^2$. In the case where we have no replication, we sometimes pool the mean squares associated with the higher order interaction terms in our ANOVA in order to obtain a "pooled" estimate of experimental error. An alterna-*

**TABLE 21.14    GENERIC EMS ALGORITHM RESULTS (EXAMPLE 21.5)**

| Source | df | R a i | F b j | R c k | R n l | EMS | F test ratios |
|---|---|---|---|---|---|---|---|
| $A_i$ | $a - 1$ | 1 | $b$ | $c$ | $n$ | $\sigma_\varepsilon^2 + n\sigma_C^2 + bcn\,\sigma_A^2$ | Num |
| $B_j$ | $b - 1$ | $a$ | 0 | $c$ | $n$ | $\sigma_\varepsilon^2 + n\sigma_C^2 + cn\,\sigma_{AB}^2 + acn\,\phi_B$ | Num |
| $AB_{ij}$ | $(a - 1)(b - 1)$ | 1 | 0 | $c$ | $n$ | $\sigma_\varepsilon^2 + n\sigma_C^2 + cn\,\sigma_{AB}^2$ | Num  Den |
| $C_{k(ij)}$ | $ab(c - 1)$ | 1 | 1 | 1 | $n$ | $\sigma_\varepsilon^2 + n\sigma_C^2$ | Num  Den   Den |
| $\varepsilon_{l(ijk)}$ | $abc(n - 1)$ | 1 | 1 | 1 | 1 | $\sigma_\varepsilon^2$ | Den |

Example calculation for $A_i$ row EMS
   Cover the $i$ column.
   The four terms that contain $i$ are identified:  $\varepsilon_{l(ijk)}$, $C_{k(ij)}$, $AB_{ij}$, $A_i$.
   For the $\varepsilon_{l(ijk)}$ row possibility: $1(1)(1)\,\sigma_\varepsilon^2$.
   For the $C_{k(ij)}$ row possibility: $1(1)(n)\sigma_C^2$.
   For the $AB_{ij}$ row possibility: $0(c)(n)\,\sigma_{AB}^2 = 0$.
   For the $A_i$ row possibility: $b(c)(n)\sigma_A^2$ .

---

Num: Numerator
Den: Denominator

*tive to pooling* (as well as an aid to determining what to pool), *in the case where we have a $2^f$ with no replication, consists of plotting the ranked main and interaction effect values on normal probability paper.*

### Error Pooling

In error pooling we will rely on (1) the Pareto principle and (2) the EMS structure. First, we will assume in any list of factors we may develop initially, some factors will influence the response much more than others; but we do not know at the outset which ones have the most influence. These few factors are the ones we seek to identify. Second, we also know that experimental error is a component of an EMS term (whether or not we can actually measure it). Hence, if we can develop a convincing argument (physically and statistically) that some factors and/or interactions are of negligible impact on the response, relative to other factors and interactions, we can pool the negligible ones and use the pooled term as a pseudo experimental error.

Fabricating a pseudo experimental error term (pseudo $MS_{err}$) poses problems. Different approaches have been studied and advocated in the literature (Ross [10], Berk and Picard [11], and Lenth [12]). For example, Ross discusses the "pooling up" and "pooling down" processes and concludes that pooling up is a rather conservative approach to pooling mean squares, that tends to prevent elimination of active factor effects by "mistake" that is, our pooled-up estimate will have a tendency to be smaller than the true, but unknown, experimental error.

At this point, we are caught between the proverbial rock and a hard place. We simply do not have an available measurement of experimental error, but we must develop one in order to help assess our factor effect measurements. In Example 21.6, we will demonstrate two general pooling options available to us: (1) we can use judgment to select the effects we will pool, or (2) we can develop a sequence of $F$ tests starting with the smallest MS, then taking the next-to-the-smallest MS, and so on, continuing to pool up the ANOVA until we find a significant difference.

### Example 21.6

Develop an experiment to study cooking temperature effect, syrup content effect, and variation due to cookie position on the baking pan. Use only 16 observations in the experiment. Develop the EMS for the experiment and analyze the results. Use pooling techniques to develop a pooled experimental error term.

### Solution

We will develop a $2 \times 2 \times 4$ FAT-CRD with one observation per treatment combination. Factor $A$, syrup content, will be a fixed effects factor with levels of 40 lb and 50 lb per batch. Factor $B$, cooking temperature, will be a fixed effects factor, with levels of 300°F and 350°F. For factor $C$, cookie position, we will randomly select four cookie positions on the baking pan (right side, left side, right center, and left center); hence it is a random effects factor. We should note that this experiment is somewhat akin to a hybrid combining the designs in Examples 21.1 and 20.5 in that we will study two fixed effect factors and one random effects factor in the same experiment.

Our model can be written as

$$Y_{ijkl} = \mu + A_i + B_j + C_k + AB_{ij} + AC_{ik} + BC_{jk} + ABC_{ijk} + \varepsilon_{l(ijk)}$$

where $i = 1, 2$
$\quad j = 1, 2$
$\quad k = 1, 2, 3, 4$
$\quad l = 1$, no replications are involved

Our data table is shown in Table 21.15, our EMS terms in Table 21.16, and our ANOVA in Table 21.17.

Since our design did not allow us to directly estimate the experimental error (as we did not include replications), we will pool a pseudo error term from the smaller MS terms, using the first pooling technique. Our pooled MS is simply a weighted average (weighted by the degrees of freedom) of the mean squares thrown into the pool. In this case, we will subjectively select and pool $MS_{ABC}$, $MS_{BC}$, $MS_{AC}$, and $MS_{AB}$ since they are all relatively small:

$$MS_{\text{pooled err}} = \frac{df_{ABC}(MS_{ABC}) + df_{BC}(MS_{BC}) + df_{AC}(MS_{AC}) + df_{AB}(MS_{AB})}{df_{ABC} + df_{BC} + df_{AC} + df_{AB}}$$

$$= \frac{3(0.56) + 3(1.90) + 3(4.73) + 1(7.56)}{3 + 3 + 3 + 1} = 2.91$$

Our reasoning for this pooling is as follows: (1) $\sigma_\varepsilon^2$ is a part of the EMS for $ABC$, $BC$, $AC$, and $AB$; (2) $MS_{ABC}$, $MS_{BC}$, $MS_{AC}$, $MS_{AB}$ are all relatively small; and (3) $ABC$, $BC$, $AC$, and $AB$ effects in a physical sense are unlikely or negligible. Obviously, if our reasoning is flawed, our estimate will be flawed.

TABLE 21.15    DATA FROM $2 \times 2 \times 4$ FAT-CRD (EXAMPLE 21.6)

| | Syrup A | | | |
| --- | --- | --- | --- | --- |
| | 40 lb | | 50 lb | |
| | Temperature B | | | |
| Positions C | 300°F | 350°F | 300°F | 350°F |
| Right side | 2 | 7 | 5 | 12 |
| Right center | 5 | 8 | 11 | 18 |
| Left center | 6 | 7 | 12 | 17 |
| Left side | 3 | 6 | 7 | 11 |

Response: Percent elongation.

TABLE 21.16    EMS TERMS FOR $2 \times 2 \times 4$ FAT-CRD (EXAMPLE 21.6)

| Source | df | F<br>a = 2<br>i | F<br>b = 2<br>j | R<br>c = 4<br>k | R<br>n = 1<br>l | EMS |
|--------|-----|-----|-----|-----|-----|-----|
| $A_i$ | 1 | 0 | b | c | n | $\sigma_\varepsilon^2 + bn\,\sigma_{AC}^2 + bcn\,\phi_A$ |
| $B_j$ | 1 | a | 0 | c | n | $\sigma_\varepsilon^2 + an\,\sigma_{BC}^2 + acn\,\phi_B$ |
| $C_k$ | 3 | a | b | 1 | n | $\sigma_\varepsilon^2 + abn\,\sigma_C^2$ |
| $AB_{ij}$ | 1 | 0 | 0 | c | n | $\sigma_\varepsilon^2 + n\,\sigma_{ABC}^2 + cn\,\phi_{AB}$ |
| $AC_{ik}$ | 3 | 0 | b | 1 | n | $\sigma_\varepsilon^2 + bn\,\sigma_{AC}^2$ |
| $BC_{jk}$ | 3 | a | 0 | 1 | n | $\sigma_\varepsilon^2 + an\,\sigma_{BC}^2$ |
| $ABC_{ijk}$ | 3 | 0 | 0 | 1 | n | $\sigma_\varepsilon^2 + n\,\sigma_{ABC}^2$ |
| $\varepsilon_{l(ijk)}$ | 0 | 1 | 1 | 1 | 1 | $\sigma_\varepsilon^2$    no direct estimate, $n = 1$ |

Now we will demonstrate the second technique, pooling up. First, we select our two smallest mean squares and test. We will use $\alpha = 0.05$.

$$F_{\text{pooled err 1}} = \frac{\text{MS}_{\text{next to smallest}}}{\text{MS}_{\text{smallest}}} = \frac{\text{MS}_{BC}}{\text{MS}_{ABC}} = \frac{1.90}{0.56} = 3.39$$

$$F_{0.05,3,3} = 9.28$$

Since, $F_{\text{pooled err 1}} < F_{0.05,3,3}$, $F_{\text{pooled err 1}}$ is not in the critical region, therefore we will pool $\text{MS}_{BC}$ and $\text{MS}_{ABC}$:

$$\text{MS}_{\text{pooled err 1}} = \frac{\text{df}_{ABC}\,(\text{MS}_{ABC}) + \text{df}_{BC}\,(\text{MS}_{BC})}{\text{df}_{ABC} + \text{df}_{BC}} = \frac{3(0.56) + 3(1.90)}{3 + 3} = 1.23$$

Now, we move to the next comparison:

$$F_{\text{pooled err 2}} = \frac{\text{MS}_{AC}}{\text{MS}_{\text{pooled err 1}}} = \frac{4.73}{1.23} = 3.85$$

$$F_{0.05,3,6} = 4.76$$

Here $F_{\text{pooled err 2}}$ is not in the critical region; hence we pool $\text{MS}_{\text{pooled err 1}}$ and $\text{MS}_{AC}$:

$$\text{MS}_{\text{pooled err 2}} = \frac{\text{df}_{\text{pooled err 1}}\,(\text{MS}_{\text{pooled err 1}}) + \text{df}_{AC}\,(\text{MS}_{AC})}{\text{df}_{\text{pooled err 1}} + \text{df}_{AC}} = \frac{6(1.23) + 3(4.73)}{6 + 3} = 2.40$$

Now, we proceed to the next pooling test:

$$F_{\text{pooled err 3}} = \frac{\text{MS}_{AB}}{\text{MS}_{\text{pooled err 2}}} = \frac{7.56}{2.40} = 3.15$$

$$F_{0.05,1,9} = 5.12$$

TABLE 21.17   ANOVA FOR 2 × 2 × 4 FAT-CRD (EXAMPLE 21.6)

| Source | df | SS | MS | F | P-value |
|---|---|---|---|---|---|
| Total corrected | 15 | 315.94 | | | |
| A (Syrup content) | 1 | 150.06 | 150.06 | · | · |
| B (Cooking temperature) | 1 | 76.56 | 76.56 | · | · |
| C (Pan position) | 3 | 60.19 | 20.06 | · | · |
| AB | 1 | 7.56 | 7.56 | · | · |
| AC | 3 | 14.19 | 4.73 | · | · |
| BC | 3 | 5.69 | 1.90 | · | · |
| ABC | 3 | 1.69 | 0.56 | · | · |
| Experimental error* | · | · | · | · | · |

*No direct estimate available due to a lack of replications.
Note: We are unable to calculate the experimental error row, *F*, and *P*-values due to the lack of an error term.  See the EMSs in Table 21.16.  See the SAS output, Computer Aided Analysis Supplement, Solutions Manual.

Here $F_{\text{pooled err 3}}$ is not in the critical region; and we pool $MS_{\text{pooled err 2}}$ and $MS_{AB}$:

$$MS_{\text{pooled err 3}} = \frac{df_{\text{pooled err 2}}\,(MS_{\text{pooled err 2}}) + df_{AB}\,(MS_{AB})}{df_{\text{pooled err 2}} + df_{AB}} = \frac{9(2.40) + 1(7.56)}{9 + 1} = 2.91$$

Now, we perform the next pooling test:

$$F_{\text{pooled err 4}} = \frac{MS_C}{MS_{\text{pooled err 3}}} = \frac{20.06}{2.91} = 6.89$$

$$F_{0.05,3,10} = 3.71$$

Here, we have encountered a significant difference, based on our $\alpha = 0.05$ significance level, and we stop our pooling process with

$$MS_{\text{pooled err}} = MS_{\text{pooled err 3}} = 2.91$$

Both our subjective pooling and pooling up processes led to the same result, primarily because of the nature of our problem.  Agreement between the subjective and quantitative methods is not assured.  Table 21.18 displays our resulting ANOVA after this pooling process.  Our *P*-values were taken from the SAS output in the Computer Aided Analysis Supplement, Solutions Manual.

At this point, we will proceed to test our remaining MS terms using our EMS table and following the line of reasoning developed for $MS_{\text{pooled err}}$.  Here, we assume that the EMS components pooled are not significant and ignore them in our EMS ratios (when we structure our *F* tests).  We used our SAS computer aid to make the tests, but we could also have used hand calculations and *F* table look-ups, Table IX.4.

An examination of Table 21.18 indicates that both main effects *A* and *B* are significant (i.e., we see syrup content and cooking temperature main effects in our observations).  Based on our evidence, main effect *C* is also significant;

TABLE 21.18    POOLED ERROR ANOVA FOR $2 \times 2 \times 4$ FAT-CRD (EXAMPLE 21.6)

| Source | df | SS | MS | $F^*$ | $P$-value |
|---|---|---|---|---|---|
| Total corrected | 15 | 315.94 | | | |
| A (Syrup content) | 1 | 150.06 | 150.06 | 51.5 | 0.0001 |
| B (Cooking temperature) | 1 | 76.56 | 76.56 | 26.3 | 0.0004 |
| C (Pan position) | 3 | 60.19 | 20.06 | 6.9 | 0.0085 |
| Pooled experimental error | 10 | 29.13 | 2.91 | | |
| AB | 1 | 7.56 | 7.56 | | |
| AC | 3 | 14.19 | 4.73 | | |
| BC | 3 | 5.69 | 1.90 | | |
| ABC | 3 | 1.69 | 0.56 | | |

* $F$ tests are based on MS pooled err.

hence, we conclude that the cookie position on the baking pan is influencing the elongation response variation.

We may want to estimate the pan position component of variation. From our EMS table, and our ANOVA results,

$$MS_{\text{pooled err}} = \hat{\sigma}_{\varepsilon}^2 = 2.91$$

$$MS_C = \hat{\sigma}_{\varepsilon}^2 + abn\, \hat{\sigma}_C^2 = 20.06$$

Hence,

$$\hat{\sigma}_C^2 = \frac{MS_C - \hat{\sigma}_{\varepsilon}^2}{abn} = \frac{20.06 - 2.91}{4} = 4.29$$

Here, as in Example 20.5, we can assess the variation impact:

$$\hat{V}ar\,(Y_{ijkl}) = \hat{\sigma}_{\varepsilon}^2 + \hat{\sigma}_C^2 = 2.91 + 4.29 = 7.20$$

This impact can be expressed as a ratio, as in Example 20.5,

$$\frac{4.29}{7.20} = 0.596$$

Here, our analysis and estimates indicate about 60 percent of the variation is due to cookie position on the baking pan, b.o.e., assuming that our error pooling procedure is reasonable in this experimental analysis.

## Normal Plots in $2^f$ Factorial Arrangements of Treatments

Box, Hunter, and Hunter [13] as well as Montgomery [14], advocate using normal plots of effects, rather than the error pooling method for assessing significance of effects when no actual experimental error term is available. This procedure works very well in $2^f$ FATs with no replication and $2^f$ fractional factorial experiments (discussed in Chapter 23). It consists of four steps: (1) developing effect columns, (2) calculating the effect values, (3) plotting the effect values on normal probability paper, and (4) identifying outlying points on the plot and associating them with significant effects.

We previously discussed the main and interaction effect concepts. We will now extend that discussion to $2^f$ factorials with no replication, where we cannot obtain an estimate of error directly. We will consider a FAT-CRD model such as

$$Y_{ijkl} = \mu + A_i + B_j + C_k + AB_{ij} + AC_{ik} + BC_{jk} + ABC_{ijk} + \varepsilon_{l(ijk)}$$

and the following sets of hypotheses:

$$H_0: \quad \text{All } A_i = 0$$
$$H_1: \quad \text{At least one } A_i \neq 0$$
$$H_0: \quad \text{All } B_j = 0$$
$$H_1: \quad \text{At least one } B_j \neq 0$$
$$\vdots$$
$$H_0: \quad \text{All } ABC_{ijk} = 0$$
$$H_1: \quad \text{At least one } ABC_{ijk} \neq 0$$

In the case where all $H_0$'s are true, a normal plot of the effects will show a pattern that falls near a straight line on our normal probability paper (see Section IX, Table IX.14). In the case where only some of the $H_0$'s are true, effects that fall away from the model line are termed "outliers" and interpreted as significant effects. We seek to identify the outlying effects. The farther out (more extreme) the outlier, the more evidence we see to support the rejection of that effect's $H_0$. Our normal plot is constructed using the general procedure we developed in Chapter 20 for our residual terms. We will develop the normal effects plotting procedure through an example.

---

**Example 21.7**

Develop a normal effects plot for a $2^3$ FAT-CRD with no replication. The data collected are shown below.

| | | | |
|---|---|---|---|
| (1) = 0.55 | $a = -0.37$ | $b = 1.05$ | $c = -0.25$ |
| $ab = 1.23$ | $ac = -0.57$ | $bc = -1.27$ | $abc = 0.16$ |

Here, we are using a new treatment factor combination notation where:

| Treatment combination | Factor levels | | |
|---|:---:|:---:|:---:|
| | **A** | **B** | **C** |
| (1) | − | − | − |
| a | + | − | − |
| b | − | + | − |
| c | − | − | + |
| ab | + | + | − |
| ac | + | − | + |
| bc | − | + | + |
| abc | + | + | + |

Here, the – and + notation indicate the low and high levels of each factor, respectively.

**Solution**

We will develop this procedure for the $2^3$ FAT-CRD; however, $2^f$ FAT–CRD (single replication) effects in general can be developed using the same four steps:

*Step 1* Lay out the main effects in a column heading and the treatment factor combinations in rows using + and − signs (+ for the high factor levels, − for the low factor levels).

| Observation | Treatment combination | Factor columns (main effects) | | | | | | |
|---|---|---|---|---|---|---|---|---|
| | | A | B | C | | | | |
| 1 | (1) | − | − | − | | | | |
| 2 | a | + | − | − | | | | |
| 3 | b | − | + | − | | | | |
| 4 | c | − | − | + | | | | |
| 5 | ab | + | + | − | | | | |
| 6 | ac | + | − | + | | | | |
| 7 | bc | − | + | + | | | | |
| 8 | abc | + | + | + | | | | |
| | | | | | | | | |

*Step 2* Add the interaction terms as column headings and multiply through, row by row, to obtain the sign for each interaction column [e.g., the row (1), *AB* column intersection yields (–)(–) or +, the row *ac*, *ABC* column intersection yields (+)(–)(+), or –]. Notice that when we finish we have half +'s and half –'s in each column.

| Observation | Treatment combination | Factor columns (main effects and interactions) | | | | | | |
|---|---|---|---|---|---|---|---|---|
| | | A | B | C | AB | AC | BC | ABC |
| 1 | (1) | − | − | − | + | + | + | − |
| 2 | a | + | − | − | − | − | + | + |
| 3 | b | − | + | − | − | + | − | + |
| 4 | c | − | − | + | + | − | − | + |
| 5 | ab | + | + | − | + | − | − | − |
| 6 | ac | + | − | + | − | + | − | − |
| 7 | bc | − | + | + | − | − | + | − |
| 8 | abc | + | + | + | + | + | + | + |
| | | | | | | | | |

*Step 3*   Record the response values and then add all the + responses and all the − responses down each column. Once we have the column total, we divide it by the number of + signs (or the number of − signs, as it is the same) that appear in the column. The result at the bottom will be the effect value we will plot. Our results are displayed in Table 21.19.

*Step 4*   Rank the effect values in ascending order and develop their plotting positions $\hat{F}(y)$ using

$$\hat{F}(y) = \frac{\text{rank number} - 0.5}{\text{number of effects}} \qquad (21.16)$$

For example, for rank 1, $\hat{F}(y) = (1 - 0.5)/7$; for rank 2, $\hat{F}(y) = (2 - 0.5)/7$; and so on for our $2^3$ case. Then, plot the effect value, $\hat{F}(y)$ coordinates on normal probability paper and draw in the model line. Our rankings and plotting positions are shown in Table 21.20. Our normal plot is shown in Figure 21.7.

Drawing the model line and then selecting outliers from the plot is a subjective process. Judgment is required (e.g., model line placement is based on favoring clusters of non-significant effect points, not passing as near all points as possible). Different line placement may lead to different conclusions. Once we identify the effects using the normal plot, we may proceed to pool the terms associated with a failure to reject the null hypotheses (i.e., those terms near the model line). Hence, in practice, we can use the normal plotting method as a guide for our pooling (e.g., rather than pooling up). Then, we can use our pooled error term to develop s.e.i. arrows or pairwise comparisons for our fixed effects factors and to solve for components of variation for our random effects factors.

It is of interest to note that the eight responses in this example were randomly sampled from a standard normal distribution ($\mu = 0$, $\sigma^2 = 1$). If we had taken large samples (many replications) in a full factorial, we would expect to see all treatment combination means near 0, with a common variance near 1. Hence, we would

**TABLE 21.19**   FACTOR VALUE TABULATION FOR $2^3$ FAT-CRD WITHOUT REPLICATION (EXAMPLE 21.7)

| Observation | Treatment combination | A | B | C | AB | AC | BC | ABC | Response $Y_{ijk}$ |
|---|---|---|---|---|---|---|---|---|---|
| | | Factor columns (main effects and interactions) | | | | | | | |
| 1 | (1) | − (0.55) | − | − | + | + | + | − | 0.55 |
| 2 | a | + (−0.37) | − | − | − | − | + | + | −0.37 |
| 3 | b | − (1.05) | + | − | − | + | − | + | 1.05 |
| 4 | c | − (−0.25) | − | + | + | − | − | + | −0.25 |
| 5 | ab | + (1.23) | + | − | + | − | − | − | 1.23 |
| 6 | ac | + (−0.57) | − | + | − | + | − | − | −0.57 |
| 7 | bc | − (−1.27) | + | + | − | − | + | − | −1.27 |
| 8 | abc | + (0.16) | + | + | + | + | + | + | 0.16 |
| Effect sums (column × response total) | | 0.37 | 1.81 | −4.39 | 2.85 | 1.85 | −2.39 | 0.65 | |
| Effect value (sums divided by 4) | | 0.093 | 0.453 | −1.098 | 0.713 | 0.463 | −0.598 | 0.163 | $\bar{Y}... = 0.066$ |

Sample calculation—A:
  −0.55 − 0.37 − 1.05 + 0.25 + 1.23 − 0.57 + 1.27 + 0.16 = 0.37
  0.37 / 4 = 0.093

TABLE 21.20   RANKED EFFECTS AND THEIR MEAN PLOTTING
POSITIONS FOR $2^3$ FAT-CRD WITHOUT REPLICATION (EXAMPLE 21.7)

| Effect rank $k$ | Ranked effect value | $\hat{F}(y) =$ $(k - 0.5)$/number of effects |
|---|---|---|
| 1 | $C = -1.098$ | $(1 - 0.5)/7 = 7.1\%$ |
| 2 | $BC = -0.598$ | $1.5/7 = 21.4\%$ |
| 3 | $A = 0.093$ | $2.5/7 = 35.7\%$ |
| 4 | $ABC = 0.163$ | $3.5/7 = 50.0\%$ |
| 5 | $B = 0.453$ | $4.5/7 = 64.3\%$ |
| 6 | $AC = 0.463$ | $5.5/7 = 78.6\%$ |
| 7 | $AB = 0.713$ | $6.5/7 = 92.9\%$ |

FIGURE 21.7   Effects plot, normal probability plotting paper (Example 21.7).

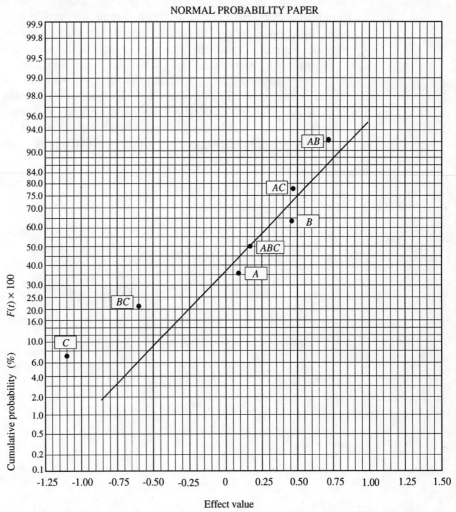

NORMAL PROBABILITY PAPER

reject none of the $H_0$'s. We would expect that all of our effects would line up neatly on the normal plot, about half on each side of our model line and about half on each side of 0. But, in our case, we see a good deal of scatter about the normal model line (e.g., effects $C$ and $BC$ appear to be outliers). This brief discussion re-inforces our warning regarding the danger of basing critical decisions on the results of small samples. Our off-line quality improvement cycle (Figure 20.1) contains a verification component as a safeguard for this very reason.

## 21.7  MIXTURE EXPERIMENTS

*A general assumption in most experimental designs is that the levels of one factor or variable are completely independent of the levels of all others.* For example, tem-perature levels are typically independent of cooking time. A number of quality related experiments may not meet this criterion. For example, if chemical composi-tion is important, reducing or increasing the proportion of one ingredient will affect the proportion of another ingredient; hence, a two-solution mixture of 20 to 40 per-cent of chemical A and 80 to 60 percent of chemical B requires special treatment.

*A number of specialized experimental designs have been developed to address mixture related studies.* Typically, the layout of these designs are portrayed on variable scales from 0 to 100 percent; a triangular representation for three variables is common. Figure 21.8 depicts three- and four-factor mixture compounds. A point somewhere on or within the geometric figure represents a unique blend or mixture of the components. Diamond [15] and Cornell [16] provide discussions regarding useful mixture study strategies.

FIGURE 21.8    Mixture experiment configurations.

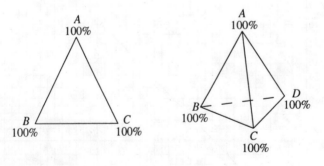

## REVIEW AND DISCOVERY EXERCISES

### Review

**21.1** Given the $2^2$ FAT interaction plot axes below, sketch the following effects (i.e., connect four points with two lines in an appropriate manner):

**a** Significant $AB$ interaction, significant $A$ main effect, significant $B$ main effect

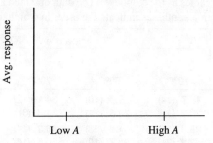

**b** No $AB$ interaction, significant $A$ main effect, significant $B$ main effect

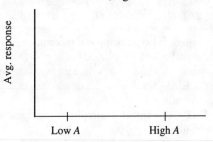

**c** No $AB$ interaction, significant $A$ main effect, no $B$ main effect

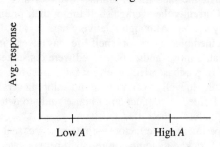

**d** Significant $AB$ interaction, no $A$ main effect, no $B$ main effect

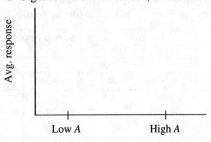

**21.2** A wastewater treatment plant utilizes hydrogen peroxide at various temperatures to destroy organic contaminants in the incoming water. The manager of the facility wants to determine whether the treatment is influenced by operating temperatures and peroxide flow rates. He selects three peroxide flow rates and four temperatures. He then designs and conducts a FAT-CRD experiment (fixed effects) on contaminated water drawn from a lagoon. In each case, he measures the percentage removal of organics as the response. The data obtained are shown below. The numbers in parentheses indicate the order in which data were obtained. [MR]

| Hydrogen peroxide flow rate | Temperature | | | |
|---|---|---|---|---|
| | 10°C | 50°C | 70°C | 90°C |
| 50 ml/hr | 17.7 (14) | 26.1 (10) | 34.1 (12) | 33.2 (6) |
| | 23.3 (1) | 28.3 (24) | 31.3 (3) | 35.8 (22) |
| 100 ml/hr | 29.8 (15) | 31.0 (23) | 46.3 (11) | 53.1 (4) |
| | 23.4 (21) | 28.7 (19) | 50.6 (16) | 56.5 (8) |
| 150 ml/hr | 29.3 (13) | 52.1 (17) | 60.3 (20) | 61.2 (5) |
| | 27.7 (9) | 54.3 (7) | 57.2 (2) | 58.8 (18) |

**a** Assuming a bigger is better response, recommend an operating setting. Use both graphical and ANOVA methods to support your conclusions (i.e., develop the ANOVA and draw the appropriate plots using s.e.i. arrows).

**b** Determine the EMSs for this experimental design and verify that your $F$ tests were properly structured.

**c** Assess the FAT-CRD model assumptions (e.g., normality of residuals and order of experimentation).

**21.3** Fantasy Candies manufacturers pralines and several other types of candies. They want to evaluate the shelf life of pralines manufactured and packaged under various conditions. Currently the average shelf life of their pralines is 23 days; after this time rancidity sets in. Although relatively harmless in small amounts, rancid food is practically inedible. The short shelf life was never a major concern of Fantasy Candies, because most candies purchased were consumed on the spot or within two to three days of purchase. However, recent market research has indicated that the company would do well to offer order and shipping services for the candies. Fantasy has decided to conduct an in-house study to determine the best methods to prolong the product's shelf life.

In this experiment, three factors—container size, seal strength of the container, and packaging temperature—will be evaluated. These factors are treated as fixed effects. The response is shelf life in days for candies selected at random from the processing line. A data table for the results is shown below:

| Container size | Packing temperature | | | | | |
|---|---|---|---|---|---|---|
| | 5°C | | | -5°C | | |
| | Seal strength | | | Seal strength | | |
| | Light | Medium | Strong | Light | Medium | Strong |
| 8 oz | 16 | 18 | 29 | 18 | 23 | 34 |
| | 15 | 22 | 30 | 20 | 26 | 33 |
| 16 oz | 16 | 21 | 28 | 20 | 27 | 33 |
| | 18 | 23 | 32 | 22 | 28 | 31 |
| 32 oz | 15 | 20 | 28 | 19 | 28 | 32 |
| | 18 | 24 | 30 | 23 | 30 | 31 |

It was difficult to obtain the responses, in that a number of containers in each treatment combination were manufactured and set on the shelf under identical storage conditions. Then, after five days, one of each container was selected at random and was opened each day until a rancid candy was encountered. A local "taste expert" was brought in each day to execute the taste test. [DM]

a For this bigger is better response, develop an appropriate set of plots.

b Develop an ANOVA and add s.e.i. arrows to the plots.

c Interpret the results.

d Since a true quality characteristic, taste, was used, the experiment was very large and expensive. Suggest how we might develop a substitute quality characteristic and reduce our costs.

**21.4** An employee was constantly using tools requiring high grip strength, and he was beginning to complain of muscle soreness in his forearms. After observing him at work and finding that hand position was not the problem, it was assumed that the force required to run the tools was causing muscle fatigue. A grip strength test was performed on this employee. A maximum grip strength measurement was taken immediately prior to a fatigue test. The response was measured in pound-force. Two fatigue tests were performed, 50 percent and 100 percent of maximum force, followed by another maximum grip strength measurement. The purpose was to measure a subsequent drop in strength. The procedure was repeated four times. The data collected are given below. Assume fixed effects for all factors. [LD]

| Fatigue test level | Testing protocol | | | |
|---|---|---|---|---|
| | Pretest | | Posttest | |
| 50% max | 23.5 | 24.0 | 9.6 | 10.4 |
| | 19.2 | 24.8 | 14.4 | 12.0 |
| 100% max | 17.0 | 17.6 | 12.3 | 8.8 |
| | 19.2 | 18.4 | 8.0 | 10.4 |

a Develop appropriate plots and an appropriate ANOVA.

b Use the HSD method, $\alpha = 0.01$, to test for differences.

c Assess the assumption of normality of residuals.

d Interpret your analyses, comment on the design.

**21.5** An engineer in charge of a short pipeline conducted a FAT-CRD experiment to determine the energy required to deliver light crude oil from one end of the pipeline to the other. He used three pump speeds and varied the temperature of the oil. The responses recorded are in energy units per barrel of crude oil. His data are presented below. [JM]

| Pump speed | Temperature | | | |
|---|---|---|---|---|
| | 50°F | 70°F | 90°F | 110°F |
| 1050 rpm | 0.86 | 0.90 | 0.93 | 0.88 |
| | 0.87 | 0.88 | 0.85 | 0.92 |
| 1200 rpm | 0.68 | 0.85 | 0.81 | 0.89 |
| | 0.78 | 0.76 | 0.77 | 0.88 |
| 1800 rpm | 0.86 | 0.85 | 0.89 | 0.95 |
| | 0.79 | 0.85 | 0.91 | 0.90 |

**a** What pump speed uses the least energy to deliver a barrel of oil?

**b** At what oil temperature is the least energy consumed?

**c** Do the pump speed and oil temperature interact?

**21.6** In the last few years many entrepreneurs have entered the field of aquaculture. Aquaculture is defined as the regulation and cultivation of water plants and animals for human consumption. In order to learn more about the aquaculture practices best suited to raising oysters in captivity, a three-factor experiment was designed. The three factors considered for this experiment were bed density, salinity, and water flow position.

| _Response Variable_ | _Measurement technique_ |
|---|---|
| Average weight of oyster meat | Laboratory scale (ounces) |
| _Factors Under Study_ | _Levels_ |
| F (water flow position) | 1: near water inlet |
| | 2: middle between positions 1 and 3 |
| | 3: middle between positions 2 and 4 |
| | 4: near water outlet |
| D (bed density) | 600, 300 spat per tray—a spat is a |
| | young oyster(s) |
| S (salinity) | 10 ppt, 15 ppt |

This experiment was conducted for 12 months in order to gather the necessary data. Four different troughs were prepared to simulate the possible conditions under which oysters could be grown. The troughs were loaded with trays (each trough contained 4 trays). A total of 8 trays were loaded with 600 spat (young oysters) per tray, and 8 trays were loaded with 300 spat per tray. Spat were assigned at random to the trays. The trays were laid in the troughs. Bay water was pumped over the trays at a controlled rate. The troughs were numbered and arranged 1 through 4, with number 1 being located at the water inlet point and number 4 near the outlet. At the end of the experiment oysters were selected at random from each tray, within each trough. The oyster meat was carefully removed and weighed (in oz). Two oysters were removed from each tray and designated by the corresponding tray and trough numbers. The weight of the oyster meat and the relationship to the various combination of growing conditions were the focus of this experiment. The factors and response data are shown below:

| Water flow position | Low density population (300 spat) | | High density population (600 spat) | |
|---|---|---|---|---|
| | Salinity | | Salinity | |
| | 10 ppt | 15 ppt | 10 ppt | 15 ppt |
| 1 | 2.6 | 2.9 | 2.8 | 3.0 |
| | 2.5 | 2.7 | 2.9 | 2.9 |
| 2 | 2.5 | 2.6 | 2.8 | 2.8 |
| | 2.4 | 2.5 | 2.7 | 2.8 |
| 3 | 2.3 | 2.5 | 2.4 | 2.4 |
| | 2.3 | 2.3 | 2.2 | 2.2 |
| 4 | 1.5 | 2.1 | 1.6 | 1.8 |
| | 1.6 | 2.0 | 1.4 | 1.6 |

a Develop plots for the main effects and appropriate two-factor interactions.
b Develop the ANOVA and add s.e.i. arrows to your plots.
c Develop an $\alpha = 0.01$ HSD analysis for the significant effects.
d Using the bigger is better criteria, make recommendations based on the experimental results.

**21.7** Refer to Problem 20.6. A competing plant uses a different measurement process to measure its meter sticks—a manual measuring device operated by machine attendants. An experiment was designed to estimate the variation associated with the population of recorded product measurements (labeled variation). The purpose was to estimate actual product, operator, and measuring device variation components. A total of 20 meter sticks and 3 operators were selected at random. The same measuring device was used by all attendants to measure each meter stick two times. It was carefully calibrated before the experiment. The data collected are shown below:

| Meter stick | Operator 1 | | Operator 2 | | Operator 3 | |
|---|---|---|---|---|---|---|
| | Observation 1 | Observation 2 | Observation 1 | Observation 2 | Observation 1 | Observation 2 |
| 1 | 1000.16 | 1000.18 | 1000.08 | 999.99 | 1000.25 | 1000.28 |
| 2 | 1000.00 | 999.89 | 999.97 | 999.93 | 1000.11 | 1000.09 |
| 3 | 999.74 | 999.84 | 999.73 | 999.76 | 999.83 | 999.87 |
| 4 | 1000.29 | 1000.25 | 1000.28 | 1000.29 | 1000.33 | 1000.31 |
| 5 | 1000.21 | 1000.20 | 1000.12 | 1000.08 | 1000.27 | 1000.29 |
| 6 | 999.74 | 999.81 | 999.75 | 999.71 | 999.79 | 999.80 |
| 7 | 1000.16 | 1000.05 | 1000.03 | 1000.08 | 1000.21 | 1000.23 |
| 8 | 1000.38 | 1000.39 | 1000.33 | 1000.30 | 1000.44 | 1000.42 |
| 9 | 999.85 | 999.90 | 999.87 | 999.83 | 999.87 | 999.87 |
| 10 | 1000.24 | 1000.18 | 1000.21 | 1000.20 | 1000.21 | 1000.23 |
| 11 | 999.94 | 999.92 | 999.91 | 999.92 | 999.99 | 999.96 |
| 12 | 1000.07 | 1000.08 | 1000.08 | 1000.02 | 1000.12 | 1000.11 |
| 13 | 1000.19 | 1000.19 | 1000.14 | 1000.11 | 1000.23 | 1000.25 |
| 14 | 999.99 | 1000.02 | 999.97 | 999.93 | 1000.13 | 1000.11 |
| 15 | 999.87 | 999.82 | 999.85 | 999.81 | 999.93 | 999.93 |
| 16 | 999.95 | 999.91 | 999.98 | 999.94 | 999.98 | 999.99 |
| 17 | 999.90 | 999.87 | 999.84 | 999.81 | 999.97 | 999.96 |
| 18 | 1000.40 | 1000.47 | 1000.31 | 1000.34 | 1000.41 | 1000.41 |
| 19 | 999.96 | 999.96 | 999.93 | 999.95 | 999.95 | 999.97 |
| 20 | 999.79 | 999.82 | 999.72 | 999.68 | 999.89 | 999.90 |

a Analyze the data using the ANOVA procedure as a FAT-CRD with two random effects factors—meter sticks and operators.
b Develop the EMSs for this random effects model.
c Estimate the components of variation associated with the meter sticks, the operators, and the measuring machine.
d Estimate the product's labeled variation and comment on the results.

**21.8** If you studied the material in Chapter 18—specifically, Example 18.7, compare your results in Problem 21.7 to the results in Example 18.7. Comment on the two different approaches to conducting gauge studies.

**21.9** Given the model equation

$$Y_{ijkl} = \mu + A_i + B_{j(i)} + C_{k(ij)} + \varepsilon_{l(ijk)}$$

assume that there are $a$ levels of $A$, $b$ levels of $B$, and $c$ levels of $C$ and $n$ replications at each treatment combination. Here, $A$ is a fixed effects factor, $B$ is a fixed effects factor, and $C$ is a random effects factor.

  **a** Use our EMS generation method to develop the EMSs for this model.

  **b** Write out general formulas to calculate the df and the SS associated with each term in the model.

  **c** Write out hypotheses corresponding to the ANOVA rows.

**21.10** Given the four-factor CRD, nested-crossed model below, where factors $B$ and $C$ are fixed effects factors and factors $A$ and $D$ are random effects factors, develop an appropriate EMS table and indicate appropriate $F$ tests and hypotheses statements for an ANOVA.

$$Y_{ijklm} = \mu + A_{i(j)} + B_j + C_k + D_{l(k)} + BC_{jk} + \varepsilon_{m(ijkl)}$$

where $i = 1, 2, ..., a$
$j = 1, 2, ..., b$
$k = 1, 2, ..., c$
$l = 1, 2, ..., d$
$m = 1, 2, ..., n$

**21.11** Revisit Example 21.7 and develop an ANOVA. Then, use the pooling up process, $\alpha = 0.05$, to develop a pooled error. Comment on your results, considering the nature of the data used.

**21.12** AA Inc. manufactures fiberglass parts and wishes to conduct an in-house experiment to measure the effect of four factors on the gel time of polyester-fiberglass resin. Gel time is the time interval when fiberglass resin is workable. According to the manufacturer of the polyester resin, gel time may be affected by age of the resin, ambient humidity, ambient temperature, and the amount of catalyst used. The engineering department of AA Inc. designed and conducted a FAT-CRD experiment to assess these effects. No replications were run in the experiment. The data collected appear below, where the response is measured in minutes [DM]:

| | | Temperature C | | | |
| | | 60°F | | 80°F | |
| | | Catalyst D | | | |
| Age A | Humidity B | 1% | 2% | 1% | 2% |
|---|---|---|---|---|---|
| 0–6 months | 30% | 50 | 12 | 22 | 6 |
| | 70% | 75 | 20 | 38 | 15 |
| 6–12 months | 30% | 65 | 18 | 35 | 12 |
| | 70% | 90 | 25 | 60 | 36 |

  **a** Develop the ANOVA for these experimental data.

  **b** Use the normal plotting technique to determine which effects to pool in order to develop a pooled experimental error. Then, calculate the pooled error.

  **c** Develop an appropriate set of plots. Include 95 percent confidence s.e.i. arrows on your plots.

    **d** Develop LSD bars for the significant effects, $\alpha = 0.05$.

    **e** Interpret your results and recommend a "best" factor combination; you may assume either a bigger or smaller is better response, depending on the size of the part and time needed to lay the part up (i.e., for small parts, smaller is better; for large parts, bigger is better).

## Discovery

**21.13**  For the Problem 21.12 ANOVA results:

    **a** How would we approach the analysis if we are faced with a nominal is best response?

    **b** Assume a nominal is best response time of 60 minutes. Use the s.e.i. arrow technique to recommend a "best" factor combination.

**21.14**  A nail factory is currently operating with a large number of machines, each set up and attended by operators. Wire is fed into the machines from large coils, then compressed to form the nail head and trimmed to form the point. Finally, the nail is ejected onto a conveyor and moved to a packing area. Each operator attends seven machines and is responsible for setting up the machines as well as monitoring them as they produce nails. Each operator has a pair of handheld dial calipers which he or she uses to measure the length of the nails when setting up a machine and when monitoring production. The engineering specifications for nail length are set by a national standard, expressed in both inches and millimeters.

    Nail production has become very competitive and international competition has developed. In our nail plant there is a great deal of concern that our product length is not as uniform as that of one of our international competitors. However, no formal study of this concern has been attempted. The only information recorded to date is the average length of the outgoing product taken from all machines, as the product is mixed on the conveyor before packing.

    A work team of operators has suggested that we improve our product length uniformity. So far, four variables have been identified as potential causes of the problem: the raw material (wire) hardness, operator training and procedures in machine setup, machine differences in the wire feed mechanisms and the impact force of the forming dies, and measurement instruments.

    Confronted with this situation, develop an appropriate experimental protocol to help resolve these concerns. Develop and present your response in a systematic and thorough fashion.

**21.15**  Gasoline is often moved from a refinery in an isolated location to a mass marketing area, such as a large city, by pipeline. The octane rating, or number, is used to classify gasoline in terms of its ability to combust properly in automobile engines. Presently, we see posted minimum octane ratings between 86 and 90 at most gasoline stations. Hence, octane ratings are used as one characteristic of gasoline quality.

    A purchaser of gasoline specifies a minimum octane rating of 88 for a 250,000-gal purchase to be transported by pipeline from a West Texas refinery to Kansas City (about 600 mi). The gasoline is tested (sampled once) at the refinery, found to be 88.25 octane, and put into the pipeline. When the gasoline arrives at Kansas City, one sample is pulled and tested by the purchaser, and it is found to be 87.50. The purchaser requires the supplier to provide additives that will bring the gasoline up to the minimum specification of 88 octane. The supplier loads a truck with the expensive additive and sends it to Kansas City to the purchaser. The additive is added to the gasoline, mixed, and tested one time with the result being 88.40 octane, and the supplier is satisfied.

As a consultant you have been retained by the supplier. The supplier tells you that this is a rather common occurrence and he cannot understand why it happens. He tells you that he has the latest octane testing equipment, with a carefully planned calibration program and an excellent laboratory staff. He also tells you that the customer has older equipment, but maintains it carefully.

As the supplier's consultant, develop a strategy and protocol for helping to solve this problem. Describe your strategy and protocol clearly.

**21.16** At a party, two intellectuals were discussing the size of individual potato chips, relative to package size and different potato chip brands. Two major concerns were aired: (1) whether or not small (lunch-box size) bags of potato chips contain smaller potato chips than the larger bags (family-size bags) and (2) whether or not chips vary in size from one manufacturer to another manufacturer (e.g., Manufacturer L's chips are bigger than Manufacturer G's). Design one experiment to address these concerns. Describe your experimental design and protocol in detail.

**21.17** Select a product or process. Design, execute, and interpret a multifactor experiment regarding some facet of the product or process you selected. Follow the six steps involved in experimental protocol (discussed in Table 20.2). Construct a display board which will summarize your experiment from beginning to end. After completion of your experiment, critique the effectiveness of its design to help fulfill your experimental objective.

## REFERENCES

1 *SAS Procedures Guide,* Version 6, Cary, NC: SAS Institute, 1990.

2 *SAS/STAT User's Guide,* Volumes 1 and 2, Version 6, Cary, NC: SAS Institute, 1990.

3 C. R. Hicks, *Fundamental Concepts in the Design of Experiments,* 3d ed., New York: Holt, Rinehart, and Winston, 1982.

4 F. E. Satterthwaite, "An Approximate Distribution of Estimates of Variance Components," *Biometrics Bulletin*, vol. 2, pp. 110–112, 1946.

5 See reference 3.

6 D. C. Montgomery, *Design and Analysis of Experiments,* 3d ed., New York: Wiley, 1991.

7 C. A. Bennett and N. L. Franklin, *Statistical Analysis in Chemistry and the Chemical Industry,* New York: Wiley, 1954.

8 See reference 3.

9 See reference 6.

10 P. J. Ross, *Taguchi Techniques for Quality Engineering,* New York: McGraw-Hill, 1988.

11 K. N. Berk and R. R. Picard, "Significance Tests for Saturated Orthogonal Arrays," *Journal of Quality Technology*, vol. 20, pp. 79–89, 1991.

12 R. V. Lenth, "Quick and Easy Analysis of Experiments," *Technometrics,* vol. 31, pp. 469–473, 1989.

**13** G. E. P. Box, W. G. Hunter, and J. S. Hunter, *Statistics for Experimenters,* New York: Wiley, 1978.

**14** See reference 6.

**15** W. Diamond, *Practical Experiment Designs,* Belmont, CA: Lifetime Learning, 1981.

**16** J. A. Cornell, *Experiments with Mixtures,* 2d ed., New York: Wiley, 1990.

# 22

# REGRESSION AND RESPONSE SURFACES

## 22.0 INQUIRY

1 How are regression models developed from experimental observations?
2 What is the least squares estimation method?
3 How are regression models simplified?
4 What is a response surface?
5 How are response surfaces developed?
6 How are response surfaces used to improve process recipes?

## 22.1 INTRODUCTION

In general, *graphical and ANOVA analyses help to isolate design and noise variables which are driving product or process performance.* When we are working with quantitative factors in a fixed effects model, we may want to go beyond identifying critical factors. *Regression and response surface models allow us to predict responses at factor levels other than those originally included in our experiments* (e.g., between our chosen factor levels). We will develop a model of the form

$$\hat{\mu} = f(\text{design and noise variables}) \tag{22.1}$$

This general model will help us improve our process recipe by providing a mathematical representation of our response as a function of the variables that are driving it. This predictor will allow us to search out better recipe settings faster than we could otherwise.

## 22.2 REGRESSION—LEAST SQUARES ESTIMATION

Once we have collected our data, we are in a position to hypothesize and fit a model. *We typically hypothesize some form of a general linear model and use the least squares technique to "fit" our model.* In the least squares estimation technique we develop our predictor $\hat{\mu}$ by minimizing the sum of the squares of deviation between the observed values and the predicted values (e.g., from the observed $Y_i$'s to our predicted $\hat{\mu}_i$'s, where $i$ represents the observation number).

### Least Squares Estimators

*The least squares estimation technique is associated with a general linear model of the form*

$$Y = \beta_0 + \beta_1 X_1 + \beta_2 X_2 + \cdots + \beta_m X_m + \varepsilon \tag{22.2}$$

where    $Y$ = a dependent random variable or response variable
$X_1, X_2, ..., X_m$ = independent or input model variables
$\beta_0, \beta_1, ..., \beta_m$ = model parameters
$\varepsilon$ = an error or residual term

In general,

$$Y_i = b_0 + b_1 X_{1i} + b_2 X_{2i} + \cdots + b_m X_{mi} + e_i \qquad i = 1, 2, ..., n \tag{22.3}$$

which is also expressed as

$$\hat{\mu}_i = b_0 + b_1 X_{1i} + b_2 X_{2i} + ... + b_m X_{mi} \qquad i = 1, 2, ..., n \tag{22.4}$$

Here,    $Y_i$ = $i$th response observation
$\hat{\mu}_i$ = $i$th predicted response (the predicted mean of a population of $Y$ values, indexed by a given level of $X$)
$b_0, b_1, ..., b_m$ = estimated model parameters, estimated from the data
(e.g., $\hat{\beta}_0 = b_0, ..., \hat{\beta}_m = b_m$)
$e_i$ = specific error (or residual) value that represents the difference between an observed $Y_i$ and its predicted value $\hat{\mu}_i$
$n$ = number of observations

***We use least squares estimators to minimize the sum of the squared deviations:***

$$\sum_{i=1}^{n} e_i^2 = \sum_{i=1}^{n} (Y_i - \hat{\mu}_i)^2 = \sum_{i=1}^{n} [Y_i - (b_0 + b_1 X_{1i} + \cdots + b_m X_{mi})]^2 \tag{22.5}$$

The sum of squared deviations shown above is minimized by setting the partial derivatives with respect to the parameters—$b_0, b_1, ..., b_m$—equal to 0 and solving. The result is the general set of classical normal equations (see Walpole and Myers [1], Draper and Smith [2], Box and Draper [3] and Neter, Wasserman, and Kutner [4]):

$$\sum_{i=1}^{n} Y_i = nb_0 + b_1 \sum_{i=1}^{n} X_{1i} + b_2 \sum_{i=1}^{n} X_{2i} + \cdots + b_m \sum_{i=1}^{n} X_{mi} \tag{22.6}$$

$$\sum_{i=1}^{n} X_{1i} Y_i = b_0 \sum_{i=1}^{n} X_{1i} + b_1 \sum_{i=1}^{n} X_{1i}^2 + b_2 \sum_{i=1}^{n} X_{1i} X_{2i} + \cdots + b_m \sum_{i=1}^{n} X_{1i} X_{mi} \tag{22.7}$$

$$\sum_{i=1}^{n} X_{2i} Y_i = b_0 \sum_{i=1}^{n} X_{2i} + b_1 \sum_{i=1}^{n} X_{2i} X_{1i} + b_2 \sum_{i=1}^{n} X_{2i}^2 + \cdots + b_m \sum_{i=1}^{n} X_{2i} X_{mi} \tag{22.8}$$

$$\cdot$$
$$\cdot$$
$$\cdot$$

$$\sum_{i=1}^{n} X_{mi} Y_i = b_0 \sum_{i=1}^{n} X_{mi} + b_1 \sum_{i=1}^{n} X_{mi} X_{1i} + b_2 \sum_{i=1}^{n} X_{mi} X_{2i} + \cdots + b_m \sum_{i=1}^{n} X_{mi}^2 \tag{22.9}$$

*There are a number of fundamental assumptions associated with this general linear regression model.*

1  *The Xs and Ys are observed in sets,* (e.g., pairs or bivariate random samples in simple linear regression).
2  *The X values are measured with little or no error.*
3  *Each X value indexes a distribution of Y values,* and these distributions have a common variance $\sigma_\varepsilon^2$.
4  *The residuals are independent and distributed normally* with $\mu = 0$, $\sigma^2 = \sigma_\varepsilon^2$.

The predicted responses, $\hat{\mu}_i$'s, represent the estimated means of the distribution of $Y$ values mentioned in assumption 3. Assumption 4 is necessary to develop our hypothesis tests (e.g., our $t$ or $F$ tests, discussed later). In this chapter, we will use the basic graphical techniques we developed in Chapter 20 to assess normality, independence, and constant variation. (Our discussions and presentation of graphical tools for assessing the validity of these assumptions are limited; however, texts by authors such as Walpole and Myers, Draper and Smith, Box and Draper, and Neter, Wasserman, and Kutner provide thorough treatments and examples of advanced graphical as well as statistical tests [5]).

A generic simple linear regression model (a special case of Equation 22.2 with only one independent variable, $X$) is of the form

$$Y = \beta_0 + \beta_1 X_1 + \varepsilon \tag{22.10}$$

A generic fitted predictor for the model is of the form

$$\hat{\mu} = b_0 + b_1 X_1 \tag{22.11}$$

Here, $Y$ = dependent response variable
$\hat{\mu}$ = predicted response (estimated mean value)
$\beta_0$ = intercept parameter
$b_0$ = estimated value for $\beta_0$
$\beta_1$ = slope parameter
$b_1$ = estimated value for $\beta_1$
$X_1$ = independent variable
$\varepsilon$ = error term used to compensate for imperfect fits.

Figure 22.1 illustrates the regression concept for the simple linear regression case. Here, we can see the distributions of $Y$ values indexed by each $X_1$ value.

---

**Example 22.1**

Given a simple linear regression model, $Y = \beta_0 + \beta_1 X_1 + \varepsilon$, develop the least squares parameter estimates for a bivariate $(X_{1i}, Y_i)$, sample.

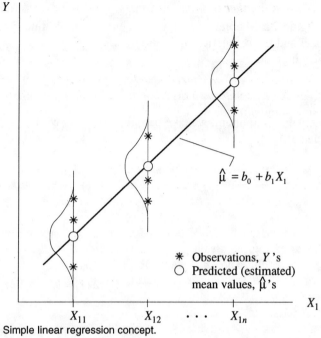

**FIGURE 22.1**    Simple linear regression concept.

**FIGURE 22.2**    Simple linear regression and the least squares concept.

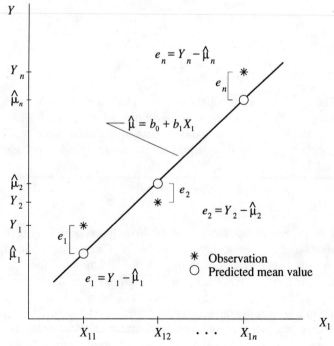

### Solution

Figure 22.2 depicts the plot of a generic simple linear regression predictor model and three generic observations.

By inspection,

$$\sum_{i=1}^{n} e_i^2 = \sum_{i=1}^{n} [Y_i - (b_0 + b_1 X_{1i})]^2$$

To obtain a minimum, we take partial derivatives with respect to $b_0$ and $b_1$, and equate them to 0.

$$\frac{\partial \sum_{i=1}^{n} (Y_i - b_0 - b_1 X_{1i})^2}{\partial b_0} = \sum_{i=1}^{n} [2 (Y_i - b_0 - b_1 X_{1i})(-1)]$$

$$= -2 \sum_{i=1}^{n} Y_i + 2nb_0 + 2b_1 \sum_{i=1}^{n} X_{1i} = 0$$

$$\frac{\partial \sum_{i=1}^{n} (Y_i - b_0 - b_1 X_{1i})^2}{\partial b_1} = \sum_{i=1}^{n} [2 (Y_i - b_0 - b_1 X_{1i})(-X_{1i})]$$

$$= -2 \sum_{i=1}^{n} X_{1i} Y_i + 2b_0 \sum_{i=1}^{n} X_{1i} + 2b_1 \sum_{i=1}^{n} X_{1i}^2 = 0$$

Rearranging,

$$\sum_{i=1}^{n} Y_i = nb_0 + b_1 \sum_{i=1}^{n} X_{1i} \tag{22.12}$$

$$\sum_{i=1}^{n} X_{1i} Y_i = b_0 \sum_{i=1}^{n} X_{1i} + b_1 \sum_{i=1}^{n} X_{1i}^2 \tag{22.13}$$

These equations are solved simultaneously for estimators of $\beta_0$ and $\beta_1$. Once the $X_{1i}$'s and $Y_i$'s are substituted, the estimates for $\beta_0$ (i.e., $b_0$) and $\beta_1$ (i.e., $b_1$) are obtained. It should be noted that these two normal equations are special cases of the classical normal equations shown earlier (the left portion of Equations 22.6 and 22.7).

---

### Regression ANOVAs

*Least squares estimators are used heavily in engineering work in regression analysis and response surface modeling.* Cross product and quadratic or higher order terms are introduced through transformations in the $X$'s. The least squares concept is also the basis for graphical curve fitting using plotting paper (e.g., the normal plot of residuals we developed in Chapter 20). Usually we "eyeball" the regression line on the graph paper.

The regression approach to model fitting and data analysis is general in that it applies to both designed and "undesigned" experiments, whether balanced or unbalanced, containing both quantitative and qualitative factors. In our subsequent regression discussion, we will assume that we have data available from a designed, balanced experiment (such as those discussed in Chapters 20 and 21). For details on applying the regression approach in general, see Draper and Smith, Box and Draper, and Neter, Wasserman, and Kutner, [6].

*Least squares predictor models and their corresponding regression ANOVAs can be developed from the results of our designed experiments.* If we have replications in our experiment, we can extract the experimental error term (see Chapters 20 and 21) and develop what is called a "pure error" term. Figure 22.3 depicts a simple linear regression predictor model which is compatible with our single-factor CRD experimental results and notation. A group of four replications is shown at $X_{11}$. Here $X_{1i}$ represents the *i*th level of our treatment factor $X_1$. In order to present a regression discussion consistent with our previous experimental notation, we will reintroduce the $Y_{ij}$ response for treatment factor $X_{1i}$ (at treatment level *i*, $i = 1, 2, ..., a$) used in Chapter 20. *We now have a total of n responses from our experiment.*

Labeling the largest observed value at $X_{11}$ as $Y_{ij}$, we can develop the regression ANOVA terms. Expanding,

$$Y_{ij} = \bar{Y}_{..} + (\hat{\mu}_{ij} - \bar{Y}_{..}) + (\bar{Y}_{i.} - \hat{\mu}_{ij}) + (Y_{ij} - \bar{Y}_{i.}) \qquad (22.14)$$

At this point our readers may want to review Equation 20.2 and compare it to Equation 22.14. Now, by rearranging, summing, and squaring both sides, we obtain

$$\sum_{ij} (Y_{ij} - \bar{Y}_{..})^2 = \sum_{ij} (\hat{\mu}_{ij} - \bar{Y}_{..})^2 + \sum_{ij} (\hat{\mu}_{ij} - \bar{Y}_{i.})^2 + \sum_{ij} (Y_{ij} - \bar{Y}_{i.})^2 \qquad (22.15)$$

Or, in other words, we can represent the terms above as

$$SS_{tc} = SS_{reg} + SS_{lof} + SS_{perr} \qquad (22.16)$$

**FIGURE 22.3**   Sum of squares components, relative to a single observation.

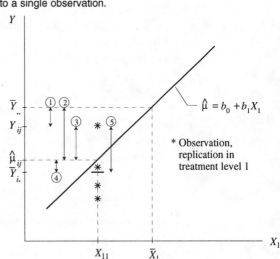

Sum of squares expansion for point $Y_{ij}$:

① $(Y_{ij} - \bar{Y}_{..})$ is $SS_{tc}$ related

② $(\hat{\mu}_{ij} - \bar{Y}_{..})$ is $SS_{reg}$ related

③ $(Y_{ij} - \hat{\mu}_{ij})$ is $SS_{rerr}$ related

④ $(\hat{\mu}_{ij} - \bar{Y}_{i.})$ is $SS_{lof}$ related

⑤ $(Y_{ij} - \bar{Y}_{i.})$ is $SS_{perr}$ related

$\hat{\mu} = b_0 + b_1 X_1$

* Observation, replication in treatment level 1

An alternate form (which is used when no replication is present) in our notation is

$$\sum_{ij} (Y_{ij} - \overline{Y}_{..})^2 = \sum_{ij} (\hat{\mu}_{ij} - \overline{Y}_{..})^2 + \sum_{ij} (Y_{ij} - \hat{\mu}_{ij})^2 \tag{22.17}$$

$$SS_{tc} = SS_{reg} + SS_{rerr} \tag{22.18}$$

where $SS_{tc}$ = total corrected sum of squares
$SS_{reg}$ = regression sum of squares
$SS_{rerr}$ = residual error sum of squares
$SS_{lof}$ = lack of fit sum of squares
$SS_{perr}$ = pure error sum of squares

A generic regression ANOVA is displayed in Table 22.1. Here, we see source entries for each of the terms identified above. We typically use computer aids, such as SAS to develop these ANOVA tables [7, 8], but with enough effort, they can be developed with hand calculations, as indicated in Table 22.1.

Typically, regression ANOVA computer aids (such as PROC REG in SAS) do not develop the $SS_{perr}$ or $SS_{lof}$. Instead, they usually display a Total corrected row (labeled "C TOTAL"), a Regression row (labeled "MODEL"), and a Residual error row (labeled "ERROR"). By examining the last terms in Equations 22.15 and 20.4, we can see that the $SS_{perr}$ and the $SS_{err}$ (experimental error from the experimental ANOVA) are identical for our balanced designs (e.g., the sum of squares within each treatment). Hence, if our computer aid is incapable of breaking out pure error and lack-of-fit terms, we can piece together these terms from the experimental ANOVA and the regression ANOVA.

**TABLE 22.1   GENERIC REGRESSION ANOVA FORMAT FOR BALANCED SINGLE-FACTOR EXPERIMENTS**

| Source | df | SS | MS | F | P-value |
|---|---|---|---|---|---|
| Total corrected (TOTAL row in SAS PROC REG) | $n-1$ | $\sum_{ij} (Y_{ij} - \overline{Y}_{..})^2 = SS_{tc}$ | | | |
| Regression (MODEL row in SAS PROC REG) | $m$ | $\sum_{ij} (\hat{\mu}_{ij} - \overline{Y}_{..})^2 = SS_{reg}$ | $SS_{reg}/df_{reg}$ | $MS_{reg}/MS_{rerr}$ | $P_{reg}$ |
| Residual error (ERROR row in SAS PROC REG) | $n-m-1$ | $\sum_{ij} (Y_{ij} - \hat{\mu}_{ij})^2 = SS_{rerr}$ | $SS_{rerr}/df_{rerr}$ | | |
| Lack of fit | $a-m-1$ | $\sum_{ij} (\hat{\mu}_{ij} - \overline{Y}_{i.})^2 = SS_{lof}$ | $SS_{lof}/df_{lof}$ | $MS_{lof}/MS_{perr}$ | $P_{lof}$ |
| Pure error * (ERROR row in SAS PROC GLM or PROC ANOVA) | $n-a$ | $\sum_{ij} (Y_{ij} - \overline{Y}_{i.})^2 = SS_{perr}$ | $SS_{perr}/df_{perr}$ | | |

Where $m$ represents the number of independent or input model variables and $a$ represents the number of treatment levels.

* Pure error is identical to experimental error in a balanced CRD.

When we use computer aids, our focus shifts from fitting calculations to model structure and interpretations. ***We formulate model structures that make sense from both the statistical and physical points of view—i.e., the "best" model. Best here refers to the simplest model (i.e., having the fewest terms) which will adequately describe the data.***

It is of interest to note that we may not be able to find a suitable model in all cases. When our experimental error is relatively large, modeling becomes very challenging. If this "noise," or dispersion within treatments (e.g., from a failure of experimental units treated alike to respond the same) becomes too great, we will not be able to find a suitable model under any circumstances, regardless of how brilliant we are (e.g., we may not have captured the driving variable(s) in our experiment).

## Response Surface Structure

***Response surface models can be of assistance to us in two primary ways: (1) in the design of a follow-on experiment and (2) in the optimization of a process response.*** For example, we typically use our model to predict response values at process settings we did not observe in our experiment. In the early stages of discovery, this knowledge aids in configuring our follow-on experiment. In later stages, where we have more confidence in our model, we may use it to establish optimal process settings under different operating conditions (e.g., additional levels of our design and noise variables).

We will present an introduction to response surface modeling. A more detailed treatment is available in Box and Draper [9]. A "full" first-order model in two factors, factor $A$ (e.g., $X_1$) and factor $B$ (e.g., $X_2$) is

$$Y = b_0 + b_1 (X_1) + b_2 (X_2) + e \tag{22.19}$$

In general, we can establish first-order models from two level (e.g., $2^f$) experiments. A full second-order response surface using the same two factors, $A$ and $B$ is

$$Y = b_0 + b_1(X_1) + b_2(X_2) + b_3(X_1X_2) + b_4(X_1)^2 + b_5(X_2)^2 + e \tag{22.20}$$

In general, a second-order model requires more than two factor levels (e.g., a $3^f$ FAT would work, but we will also discuss a more efficient design in this chapter).

We will take a rather fundamental approach in our response surface modeling in this chapter. Much more sophisticated approaches are available. We will start with a full model and then eliminate terms that do not explain a significant amount of the total corrected sum of squares. In practice, we may improvise somewhat by including or excluding a borderline statistically significant term when doing so allows a more reasonable physical defense of the model. We must justify our model on a physical and a statistical basis, otherwise we risk advocating what is termed a spurious model: a good statistical model that fits to the data available but with little, if any, ability to explain the physical process.

**Model Simplification**

Our approach here will be to develop our response surface model in its full form first. Then, we will attempt to eliminate terms in the following order:

1   Second-order terms and two-factor interaction terms.
2   First-order terms, provided all associated interaction terms and second-order terms have also been eliminated (i.e., we will not eliminate a first-order term when the factor remains in an interaction term or its second-order term remains in the model).

We will also assess the adequacy of our model, assuming a pure error (the experimental error from the experimental data) is available for our analysis.

In order to simplify a full model, we use what are called "partial" sums of squares, generated by most computer aids. These terms develop the sum of squares corresponding to the situation where each regression model term is fit last (after all other terms have been fit). By examining the fitting of each term last, we can assess its impact in explaining part of the $SS_{tc}$. For example, we can examine the residual error terms resulting from the full model and reduced models (each missing one term). Once we see the impact of each term fit last, we develop either $F$ or $t$ tests (they are equivalent when we have 1 numerator df), Draper and Smith [10]. Then, we set a threshold significance level (say, $\alpha = 0.15$) and eliminate the term with the highest $P$-value (as long as this value exceeds our threshold value). In other words, we attempt to eliminate the term which explains the least amount of the $SS_{tc}$ first. We then rerun the model without this term (provided we can justify its elimination) and repeat the process until all terms left are deemed significant. We typically keep the intercept term $b_0$, regardless of its significance. If we eliminate the intercept term, we force the predictor through the origin.

*We assess the significance of our regression model terms quantitatively using t or F tests.* Here, we will assess each term individually. We set up our hypotheses as

$H_0$: $\beta_i = 0$   (the $i$th term in our regression model does not help in explaining the response)

$H_1$: $\beta_i \neq 0$   (the $i$th term in our regression model does help in explaining the response)

The calculated $t$ value is

$$t_{calc} = \frac{b_i - 0}{\sqrt{MS_{rerr}\ c_{ii}}} = \frac{b_i - 0}{s_{bi}} \tag{22.21}$$

where $b_i$ = least squares estimate of $\beta_i$     $i = 0, 1, 2, ..., m$
   $c_{ii}$ = diagonal element of the matrix, $(X^T X)^{-1}$, see Walpole and Myers and Draper and Smith [11]
$MS_{rerr}$ = mean square residual from the model which includes the $i$th term
   $s_{bi}$ = standard error associated with the least squares estimate of $\beta_i$

If

$$| t_{calc} | > t_{\alpha/2, \, df_{rerr}}$$

then we reject $H_0$: $\beta_i = 0$ and conclude that the $i$th term should remain in the model, based on the evidence at hand, b.o.e.

If we develop an $F$ test format,

$$F_{calc} = \frac{SS(\beta_i \mid \beta_0, \beta_1, ...., \beta_{i-1}, \beta_{i+1}, ..., \beta_m)}{MS_{rerr}} \qquad (22.22)$$

where $SS(\beta_i \mid \beta_0, \beta_1, ..., \beta_{i-1}, \beta_{i+1}, ..., \beta_m)$ is the sum of squares explained by adding the $i$th term after all $m$ terms (all terms but the $i$th term) are already in the model

$$SS\,(\beta_i \mid \beta_0, \beta_1, ..., \beta_{i-1}, \beta_{i+1}, ..., \beta_m) = SS\,(\beta_0, \beta_1, ..., \beta_m) - SS\,(\beta_0, \beta_1, ..., \beta_{i-1}, \beta_{i+1}, ..., \beta_m) \qquad (22.23)$$

Here, $MS_{rerr}$ is the mean square residual from the model which includes the $i$th parameter, term $\beta_i$.

For this $F$ test,

$$df_{numerator} = 1 \quad \text{and} \quad df_{denominator} = df_{rerr}$$

If

$$F_{calc} > F_{\alpha, 1, \, df_{rerr}}$$

then we reject the $H_0$: $\beta_i = 0$ and conclude that the $i$th term should remain in the model, b.o.e. In this case, the $t$ and $F$ tests above are equivalent tests (see Walpole and Myers [12] or Draper and Smith [13]). Since, it can be shown that

$$t^2_{\alpha/2, \upsilon} = F_{\alpha, 1, \upsilon} \qquad (22.24)$$

Obviously, these calculations are difficult to develop manually. But computer aids such as SAS PROC REG provide regression ANOVA rows specifically developed for our convenience in these analyses. Some computer aids use the $t$ test and some, the $F$ test, when assessing the significance of individual terms in regression models. The $F$ test allows us to test $H_0$: $\beta_i = 0$, as opposed to $H_1$: $\beta_i \neq 0$. The $t$ test is a more general test in that we can test for values of $\beta_i = \beta_{i0}$ other than 0 (e.g., we can test $\beta_{i0} = 0.5$, and so forth, in place of $\beta_{i0} = 0$ in Equation 22.21). In most cases, we will be using computer aids and their associated $P$-values (regardless of which test was used) to decide which model terms to include or exclude.

The process we have just described, where we start with a full model and simplify out terms, is a form of the "backward elimination" method. There are many other techniques used to develop regression models with respect to selecting or rejecting terms.  Draper and Smith provide a more detailed discussion of other regression development techniques [14].

## Model Fit and Adequacy

The coefficient of multiple determination measure $R^2$ is associated with a regression fit. *The $R^2$ value is the proportion of the $SS_{tc}$ "explained" by the regression model.*  It is calculated as

$$R^2 = \frac{SS_{model}}{SS_{corrected\ total}} = \frac{SS_{reg}}{SS_{tc}} \tag{22.25}$$

Hence, $0 \le R^2 \le 1$, with $R^2 = 1$ representing the "perfect" model fit.  The $R^2$ measure is useful, but can be misleading if used alone in assessing model fit.  For $n$ data points, with no replication, an $R^2 = 1$ can be obtained by employing $n$ properly selected coefficients in the model, including $\beta_0$, since a model can then be chosen which fits the data exactly (Draper and Smith [15]).

*If we use a designed experiment with replication, we can calculate the maximum $R^2$ possible* as

$$R^2_{max} = \frac{SS_{tc} - SS_{perr}}{SS_{tc}} = \frac{SS_{tc} - SS_{err}}{SS_{tc}} \tag{22.26}$$

where $SS_{err}$ represents the sum of squares experimental error in a designed experiment (see Chapter 20).  *No matter what we do in the model, we cannot explain the sum of squares due to the pure error.*

We sometimes develop what is called an adjusted $R^2$ (Draper and Smith [16]):

$$R^2_{adj} = 1 - \left( \frac{n-1}{n-m+1} \right)(1 - R^2) \tag{22.27}$$

where $R^2$ = usual coefficient of multiple determination
$\quad\quad n$ = number of observations in the data set
$\quad\quad m$ = number of parameters (excluding $b_0$) in the model being fit.

We can see that $R^2_{adj}$ considers both the number of observations we have and the number of model terms we are using.  As we add (delete) model terms, $R^2$ will either remain constant or increase (decrease).  The $R^2_{adj}$, on the other hand, may or may not increase (decrease) when we add (delete) terms.  We will follow tradition and focus primarily on $R^2$ rather than $R^2_{adj}$.  Most computer aids provide both measures.

In the case of designed experiments with replications, a pure error term can be developed.  *If the pure error term is available, we can test for a lack of fit associated with the regression model.*  We develop an $F$ test where

$H_0$:   The model form is adequate
$H_1$:   The model form is inadequate

$$F = \frac{(SS_{rerr} - SS_{perr})/(df_{rerr} - df_{perr})}{SS_{perr}/df_{perr}} = \frac{MS_{lof}}{MS_{perr}} = \frac{MS_{lof}}{MS_{err}} \qquad (22.28)$$

If we reject $H_0$ at the $\alpha$ level of significance, we see strong evidence to suggest that our predicted values do not correspond to the treatment means. This correspondence is measured by the lack-of-fit row in the ANOVA, see Table 22.1. Its significance is then assessed relative to the pure error, see Equation 22.28. In Figure 22.3, item 4 corresponds to lack of fit, and item 5 corresponds to pure error.

*The pure error term measures the variation within treatments (a function of only the replicates within each treatment). The pure error row in the ANOVA will not change when we add or delete terms in our regression model.* On the other hand, *the lack-of-fit term will change because it depends on the terms we include in our regression model* (i.e., it is developed from the residual error, which depends on the model selected, see Table 22.1).

---

**Example 22.2**

Based on the results of the single-factor CRD data and response plot from Example 20.2, Figure 20.6b and c, develop a predictor model for the percent elongation response as a function of cooking time in minutes for our chocolate chip cookies.

**Solution**

First, we will develop a simple linear regression model of the form

$$\hat{\mu} = b_0 + b_1 X_1$$

We will assess its fit and examine a residual plot. Then, if necessary, we will develop a second-order model of the form

$$\hat{\mu} = b_0 + b_1 X_1 + b_2 X_1^2$$

An examination of the response plot, Figure 20.6c, indicates that a second-order model will probably be necessary (e.g., the treatment means appear in a rainbow shape), but we will proceed step by step and develop and test both models.

Restating our data,

| Replication | Treatment level (cooking time) | | | |
| --- | --- | --- | --- | --- |
| | 5 min | 10 min | 15 min | 20 min |
| 1 | 4 % | 9 % | 8 % | 3 % |
| 2 | 1 % | 7 % | 4 % | 2 % |
| 3 | 2 % | 10 % | 5 % | 4 % |

Developing the terms necessary for our set of classical normal equations,

| $i$ | $j$ | $X_{1ij}$ | $Y_{ij}$ |
|-----|-----|-----------|----------|
| 1 | 1 | 5 | 4 |
| 1 | 2 | 5 | 1 |
| 1 | 3 | 5 | 2 |
| 2 | 1 | 10 | 9 |
| 2 | 2 | 10 | 7 |
| 2 | 3 | 10 | 10 |
| 3 | 1 | 15 | 8 |
| 3 | 2 | 15 | 4 |
| 3 | 3 | 15 | 5 |
| 4 | 1 | 20 | 3 |
| 4 | 2 | 20 | 2 |
| 4 | 3 | 20 | 4 |

$$\sum_{ij} X_{1ij} = 150$$

$$\sum_{ij} X_{1ij}^2 = 2250$$

$$\sum_{ij} X_{1ij} Y_{ij} = 730$$

$$\sum_{ij} Y_{ij} = Y.. = 59$$

$$n = 12$$

Reexpressing the first normal equation (from Equation 22.6 or 22.12),

$$\sum Y_{ij} = nb_0 + b_1 \sum X_{1ij}$$

$$59 = 12b_0 + 150b_1$$

For the second normal equation (from Equation 22.7 or 22.13),

$$\sum X_{1ij} Y_{ij} = b_0 \sum X_{1ij} + b_1 \sum X_{1ij}^2$$

$$730 = 150b_0 + 2250b_1$$

Solving the two equations for $b_0$ and $b_1$, we obtain

$$b_0 = 5.167$$
$$b_1 = -0.020$$

Restating our simple linear regression predictor,

$$\hat{\mu} = 5.167 - 0.020T$$

where $T$ is cooking time in minutes; $T = X_1$.

Developing our residual terms, where

$$e_{ij} = Y_{ij} - \hat{\mu}_{ij}$$

| $T = X_1$ | $Y_{ij}$ | $\hat{\mu}_{ij}$ | $e_{ij}$ |
|---|---|---|---|
| 5 | 4 | 5.067 | −1.067 |
| 5 | 1 | 5.067 | −4.067 |
| 5 | 2 | 5.067 | −3.067 |
| 10 | 9 | 4.967 | 4.033 |
| 10 | 7 | 4.967 | 2.033 |
| 10 | 10 | 4.967 | 5.033 |
| 15 | 8 | 4.867 | 3.133 |
| 15 | 4 | 4.867 | −0.867 |
| 15 | 5 | 4.867 | 0.133 |
| 20 | 3 | 4.767 | −1.767 |
| 20 | 2 | 4.767 | −2.767 |
| 20 | 4 | 4.767 | −0.767 |

Our residuals are plotted against our $X_1$'s in Figure 22.4. Here, we see a "rainbow" pattern around the zero line. A rainbow or inverted rainbow is an indication that we should add a second-order term to our linear model. Hence, we must introduce a second independent variable, called $X_2$, which we will define as "$X_1$ squared" [i.e., $X_{2ij} = X_{1ij} (X_{1ij})$]. In effect, $X_2$ is a transformation of our original cooking time variable that constitutes a second-order term.

**FIGURE 22.4**    Residuals plot for first-order regression model (Example 22.2).

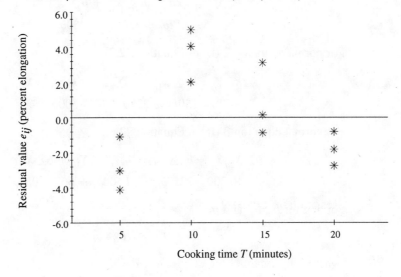

| $X_{1ij}$ | $X_{2ij} = X_{1ij}(X_{1ij})$ | $Y_{ij}$ |
|---|---|---|
| 5 | 25 | 4 |
| 5 | 25 | 1 |
| 5 | 25 | 2 |
| 10 | 100 | 9 |
| 10 | 100 | 7 |
| 10 | 100 | 10 |
| 15 | 225 | 8 |
| 15 | 225 | 4 |
| 15 | 225 | 5 |
| 20 | 400 | 3 |
| 20 | 400 | 2 |
| 20 | 400 | 4 |

$$\sum X_{1ij} = 150$$

$$\sum X_{1ij}^2 = 2250$$

$$\sum X_{2ij} = 2250$$

$$\sum X_{2ij}^2 = 663{,}750$$

$$\sum X_{1ij}X_{2ij} = \sum X_{2ij}X_{1ij} = 37{,}500$$

$$\sum Y_{ij} = 59$$

$$\sum X_{1ij}Y_{ij} = 730$$

$$\sum X_{2ij}Y_{ij} = 10{,}200$$

$$n = 12$$

Reformulating normal equation 1 (from Equation 22.6),

$$\sum Y_{ij} = nb_0 + b_1\sum X_{1ij} + b_2\sum X_{2ij}$$
$$59 = 12b_0 + 150b_1 + 2250b_2$$

For normal equation 2 (from Equation 22.7),

$$\sum X_{1ij}Y_{ij} = b_0\sum X_{1ij} + b_1\sum X_{1ij}^2 + b_2\sum X_{1ij}X_{2ij}$$
$$730 = 150b_0 + 2250b_1 + 37{,}500b_2$$

For normal equation 3 (from Equation 22.8),

$$\sum X_{2ij}Y_{ij} = b_0\sum X_{2ij} + b_1\sum X_{2ij}X_{1ij} + b\sum X_{2ij}^2$$
$$10{,}200 = 2250b_0 + 37{,}500b_1 + 663{,}750b_2$$

Solving for $b_0$, $b_1$, and $b_2$,

$$b_0 = -6.083$$

$$b_1 = 2.230$$

$$b_2 = -0.090$$

Restating our second-order regression predictor,

$$\hat{\mu} = -6.083 + 2.230\, T - 0.090\, T^2$$

where $T$ = cooking time in min, $T = X_1$
$T^2$ = square of cooking time in min, $T^2 = X_2 = X_1\,(X_1)$

Developing our residual terms,

| $T = X_1$ | $Y_{ij}$ | $\hat{\mu}_{ij}$ | $e_{ij}$ |
|---|---|---|---|
| 5 | 4 | 2.817 | 1.183 |
| 5 | 1 | 2.817 | −1.817 |
| 5 | 2 | 2.817 | −0.817 |
| 10 | 9 | 7.217 | 1.783 |
| 10 | 7 | 7.217 | −0.217 |
| 10 | 10 | 7.217 | 2.783 |
| 15 | 8 | 7.117 | 0.883 |
| 15 | 4 | 7.117 | −3.117 |
| 15 | 5 | 7.117 | −2.217 |
| 20 | 3 | 2.517 | 0.483 |
| 20 | 2 | 2.517 | −0.517 |
| 20 | 4 | 2.517 | 1.483 |

We have plotted our residuals against our $X_1$'s in Figure 22.5. If we see a random pattern of dispersion around our zero line in a plot of residuals, we typically refrain from adding higher order terms. In Figure 22.5, we see at least some statistical evidence of a "hill and valley" pattern in the residuals, which indicates that we should add a third-order term to our model and assess the results. We have left this as a discovery exercise.

FIGURE 22.5    Residuals plot for second-order regression model (Example 22.2).

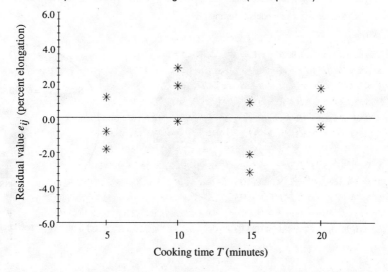

Both the first-order and second-order predictors are superimposed on a scatter plot of the data in Figure 22.6*a*. Here, we see that the second-order model appears to be much more attractive in that it "explains" the data better. We can also see the dispersion within each treatment level, relative to the treatment means and the predictors. The dispersion looks reasonably large; hence, we may not be able to develop as good a model as we had hoped. Figures 22.6*b* and *c* show pie charts where the SS terms (Equations 22.15 and 22.17) are developed. In these pie charts, we can see the proportion of $SS_{tc}$ explained by our models (e.g., the proportional sizes of the pie sections), as well as the residual error, pure error, and lack-of-fit sum of squares components. The numerical values for the sum of squares shown are discussed in Example 22.2A. The sums of squares shown in Figure 22.6 were computer generated.

**FIGURE 22.6**   Scatter plot for first- and second-order predictors and sum of squares (Example 22.2).

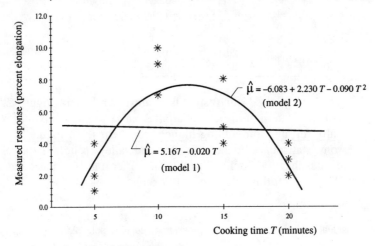

(*a*) Scatter plot with models superimposed

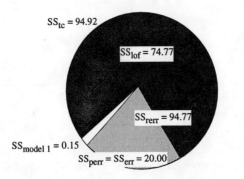

(*b*) Sum of squares breakout for model 1

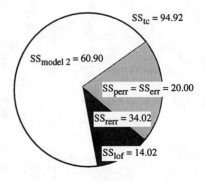

(*c*) Sum of squares breakout for model 2

### Example 22.2A (Computer Aided Solution)

Revisit Example 22.2 and develop solutions with a computer aided regression analysis package.

### Solution

Up to this point, we have used our normal equations and hand calculations to develop our two predictors. We will now pursue the development of regression ANOVAs and our hypothesis tests for establishing the model parameter significance levels with the data of Example 22.2. Authors such as Walpole and Myers [17], Draper and Smith [18], and others develop matrix notation to expedite hand analysis. We will use SAS PROC REG to avoid the laborious task of hand calculations. SAS [19] and other computer aids use similar matrix calculations and are very convenient, powerful tools for regression analysis.

Computer aids will output results arranged like those shown in Table 22.1, with the addition of parameter rows corresponding to each parameter fit. The parameter rows usually include a df value, a parameter estimate, a standard error for the parameter estimate, a $t_{calc}$ value ($H_0$: $\beta_i = 0$; $H_1$: $\beta_i \neq 0$), and an associated $P$-value. When we perform the hypothesis tests on our parameters, and our pure error analysis for lack of fit, we are calling into action our previously stated fourth assumption—$\varepsilon \sim N(\mu = 0, \sigma^2 = \sigma_\varepsilon^2)$. The validity of this assumption can be assessed using a normal plotting procedure for residuals, see Chapter 20.

Summarized regression ANOVA results appear in Table 22.2a. The SAS PROC REG code and output for our example are shown in the Computer Aided Analysis Supplement, Solutions Manual. Here, the first-order model is a poor fit, $R^2 = 0.0016$, b.o.e. We can construct hypothesis tests to assess the contributions of our parameters as follows:

$H_0$:  $\beta_0 = 0$    (the intercept parameter $\beta_0$ is not helping us in our predictor model)
$H_1$:  $\beta_0 \neq 0$    (the intercept parameter $\beta_0$ is helping us in our predictor model)

Here, we see $b_0 = 5.167$ with a $t_{calc} = (b_0 - 0)/s_{b0} = 5.16667/2.1768 = 2.374$ and a $P$-value $= 0.0390$. We will conclude that the intercept term is helping to explain our response. In general, we will tend to ignore this test (for the reason explained before, we do not wish to force the model through the origin) and hence we leave $b_0$ in all of our models. Next, we develop

$H_0$:  $\beta_1 = 0$:    (the parameter $\beta_1$ is not helping us predict the response)
$H_1$:  $\beta_1 \neq 0$:    (the parameter $\beta_1$ is helping us predict the response)

Here, we see $b_1 = -0.020$ with a $t_{calc} = -0.126$ and a $P$-value $= 0.9024$. We cannot reject $H_0$ here, b.o.e.

At this point we are reasonably sure that we do not have a good predictor, but our ANOVA does not provide clues as to what we should do next—try a new model or give up. Our original scatter plot (Figure 20.6c) tends to indicate that a second order model should be attempted. Our previous hand calculated

**TABLE 22.2    SUMMARIZED COMPUTER AIDED ANALYSIS ANOVA TABLES, SAS PROC REG (EXAMPLES 22.2A)**

*(a)* First-order (simple linear) regression model

### REGRESSION WITH SINGLE-FACTOR CRD

Model: MODEL1
Dependent Variable: ELON    ELONGATION PERCENT

#### Analysis of Variance

| Source | DF | Sum of Squares | Mean Square | F Value | Prob>F |
|--------|----|----|----|----|----|
| Model | 1 | 0.15000 | 0.15000 | 0.016 | 0.9024 |
| Error | 10 | 94.76667 | 9.47667 | | |
| C total | 11 | 94.91667 | | | |

| | | | | |
|--|--|--|--|--|
| Root MSE | 3.07842 | R-square | 0.0016 | |
| Dep Mean | 4.91667 | Adj R-sq | -0.0983 | |
| C.V. | 62.61192 | | | |

#### Parameter Estimates

| Variable | DF | Parameter Estimate | Standard Error | T for H0: Parameter=0 | Prob > |T| | Variable Label |
|----------|----|----|----|----|----|----|
| INTERCEP | 1 | 5.166667 | 2.17677131 | 2.374 | 0.0390 | Intercept |
| T | 1 | -0.020000 | 0.15896890 | -0.126 | 0.9024 | COOKING TIME |

*(b)* Second-order (curvilinear) regression model

### REGRESSION WITH SINGLE-FACTOR CRD

Model: MODEL1
Dependent Variable: ELON    ELONGATION PERCENT

#### Analysis of Variance

| Source | DF | Sum of Squares | Mean Square | F Value | Prob>F |
|--------|----|----|----|----|----|
| Model | 2 | 60.90000 | 30.45000 | 8.056 | 0.0099 |
| Error | 9 | 34.01667 | 3.77963 | | |
| C total | 11 | 94.91667 | | | |

| | | | | |
|--|--|--|--|--|
| Root MSE | 1.94413 | R-square | 0.6416 | |
| Dep Mean | 4.91667 | Adj R-sq | 0.5620 | |
| C.V. | 39.54157 | | | |

#### Parameter Estimates

| Variable | DF | Parameter Estimate | Standard Error | T for H0: Parameter=0 | Prob > |T| | Variable Label |
|----------|----|----|----|----|----|----|
| INTERCEP | 1 | -6.083333 | 3.12474690 | -1.947 | 0.0834 | Intercept |
| T | 1 | 2.230000 | 0.57012994 | 3.911 | 0.0036 | COOKING TIME |
| TSQ | 1 | -0.090000 | 0.02244884 | -4.009 | 0.0031 | T SQUARE TRANSFORMATION VAR |

residual versus predicted value results provide quantitative support for a second-order model (Figure 22.4). A plot of the residuals from our computer aided solution would look exactly like the plot in Figure 22.4. Selected residual and predicted plots generated with SAS are displayed in the Computer Aided Analysis Supplement, Solutions Manual. Hence, we are justfied in building a second-order model.

Since we are using data collected in a balanced, replicated experiment, we can develop a pure error analysis to test for lack of fit in our first order model. This test is provided mostly for illustrative purposes, as our previous discussions and plots leave little doubt that our first-order model will fail this lack-of-fit test.

$H_0$: The linear model proposed is an adequate response predictor
$H_1$: The linear model proposed is an inadequate response predictor

Here, we will have to piece together information from two sources: (1) our regression output and (2) our ANOVA from Example 20.3 (Table 20.4). Now, from the ANOVA,

$$SS_{perr} = SS_{err} = 20.000$$

$$MS_{perr} = MS_{err} = 2.500$$

$$df_{perr} = df_{err} = 4(3 - 1) = 8$$

And from our first-order SAS output (see Table 22.2a),

$$SS_{rerr} = 94.767$$
$$df_{rerr} = 10$$

Hence,

$$SS_{lof} = SS_{rerr} - SS_{perr} = SS_{rerr} - SS_{err} = 94.767 - 20.000 = 74.767$$

$$df_{lof} = df_{rerr} - df_{err} = 10 - 8 = 2$$

$$MS_{lof} = \frac{SS_{lof}}{df_{lof}} = \frac{74.767}{2} = 37.384$$

Our result is

$$F_{calc} = \frac{MS_{lof}}{MS_{perr}} = \frac{37.384}{2.500} = 14.95$$

Since,

$$F_{0.01,2,8} = 8.65 \qquad \text{and} \qquad F_{calc} = 14.95$$

we can state that $P$-value $< 0.01$ and, hence, conclude that our first-order model is inadequate, b.o.e.

In order to get our computer aid, SAS in this case, to develop our second-order model, we must include additional code lines:

$$TSQ = T * T;$$

is added to develop our transformed variable $X_2$ for $T^2$ and

$$MODEL\ ELON = T\ TSQ\ /\ P;$$

is added to build our second-order model in $X_1$ and $X_2$ for $T$ and $T^2$, respectively, and to obtain a set of predicted values.

A summarized second-order ANOVA is shown in Table 22.2b (the code and output are shown in the Computer Aided Analysis Supplement, Solutions Manual). From the results of our model we see that $R^2$ has increased to 0.6416. We see that both our $T$ and $TSQ$ terms are highly significant ($P$-values of 0.0036 and 0.0031, respectively).

We will next develop our second-order model's pure error analysis for the lack-of-fit test. Our generic hypotheses are as previously stated. Here, our pure error terms remain the same as above, since our data set remains the same:

$$SS_{perr} = SS_{err} = 20.000$$

$$MS_{perr} = MS_{err} = 2.500$$

$$df_{perr} = df_{err} = 8$$

Now, from our second-order model ANOVA in Table 22.2b,

$$SS_{rerr} = 34.017$$

$$df_{rerr} = 9$$

Hence,

$$SS_{lof} = SS_{rerr} - SS_{perr} = SS_{rerr} - SS_{err} = 34.017 - 20.000 = 14.017$$

$$df_{lof} = df_{rerr} - df_{err} = 9 - 8 = 1$$

$$MS_{lof} = \frac{SS_{lof}}{df_{lof}} = \frac{14.017}{1} = 14.017$$

and

$$F_{calc} = \frac{MS_{lof}}{MS_{perr}} = \frac{14.017}{2.500} = 5.61$$

Since,

$$F_{0.01,1,8} = 11.26 \qquad F_{0.05,1,8} = 5.32 \qquad \text{and} \qquad F_{calc} = 5.61$$

we can state that $0.01 \ll P$-value $< 0.05$. We see reasonable evidence to believe that our more extensive model,

$$\hat{\mu} = -6.083 + 2.230T - 0.090T^2$$

will perform satisfactorily between the 5 and 20 min cooking time levels.  For now, we will consider our second-order model as marginally adequate;  Problem 22.9 will explore a third-order model and its fit to these same data.

In addition to using residual plots to help us structure our models, we use them for assessing our normality assumption.  This assessment is done using the procedure developed in Chapter 20.  We have left the residual analysis for the assessment of the normality as an exercise.

Our previous statement, in which we limit the range of $T$ to the range of the experimental treatments, is critical.  *We want to be sure that our predictor models are adequate and that they are used within a relevant range.*  For example, someone using our model for a cooking time of 17 min, would obtain

$$\hat{\mu}_{T=17} = -6.083 + 2.230(17) - 0.090(17^2)$$

$$= -6.083 + 37.91 - 26.01$$

$$= 5.82\% \text{ elongation}$$

but for a cooking time of 30 min, would obtain

$$\hat{\mu}_{T=30} = -6.083 + 2.230(30) - 0.090(30^2)$$

$$= -6.083 + 66.900 - 81.000$$

$$= -20.18\% \text{ elongation}$$

This latter case is physically nonsensical.  Hence, our previous caution.

---

Our examples have dealt with only a single-factor experiment.  For multifactor experiments, it is likely that we will want to develop a response surface.  In this case, we will have a number of design or noise variables.  We can use the general normal equations, Equations 22.6 through 22.9, to develop our predictor model.  However, the hand solution method is rather challenging.  Typically, we use a computer aid, such as SAS, to expedite our calculations and develop a regression ANOVA at the same time.  We will illustrate the multifactor response surface development process with another example.

---

**Example 22.3**

We can develop a full, second-order response surface from a three-level factorial experiment.  We will start with the $3^3$ factorial experiment in Example 21.3 and develop a response surface for the percent elongation response.  The experimental layout and boundaries are shown in Figure 21.4.  The data are shown in Table 21.5.

## Solution

We have three factors at three levels, each corresponding to an independent variable in our normal equations:

$A$: $X_1$:     Cooking temperature—300°F, 350°F, and 400°F

$B$: $X_2$:     Cooking time—5 min, 10 min, 15 min

$C$: $X_3$:     Syrup content—40 lb, 50 lb, 60 lb

We will use our SAS computer aid to fit a full second-order model. Then, we will examine the $t$ test $P$-values and eliminate terms. In order to fit the interaction (cross product terms) and the second-order terms, we will introduce the transformed independent variables:

$$AB: X_4 = A*B \text{ (e.g., } 300(5), ..., 400(15))$$
$$AC: X_5 = A*C \text{ (e.g., } 300(40), ..., 400(60))$$
$$BC: X_6 = B*C \text{ (e.g., } 5(40), ..., 15(60))$$
$$ASQ: X_7 = A*A \text{ (e.g., } 300(300), ..., 400(400))$$
$$BSQ: X_8 = B*B \text{ (e.g., } 5(5), ..., 15(15))$$
$$CSQ: X_9 = C*C \text{ (e.g., } 40(40), ..., 60(60))$$

Hence, we are faced with solving 10 normal equations (for $b_0, b_1, ..., b_9$). At this point, our computer aid will be appreciated. Our SAS code and output are shown in the Computer Aided Analysis Supplement, Solutions Manual. In order to expedite the progress, we have placed multiple SAS model statements in our code.

The summarized results of the full model analysis are shown in Table 22.3$a$. The fitted full model is

$$\hat{\mu} = -371.310 + 1.251(A) + 6.717(B) + 5.018(C) - 0.0147(AB) - 0.00075(AC) + 0.00833(BC) - 0.00154(ASQ) - 0.0844(BSQ) - 0.0411(CSQ)$$

with $R^2 = 0.802$. In the same output, in the column to the right of Parameter estimate (the $b_i$'s), we see the standard errors associated with the $b_i$'s. Hence, using Equation 22.21, we can verify the $t_{calc}$ value in our outputs.

From the regression ANOVA, we can see that the $AC$ term has the highest $P$-value (0.6443). We remove the $AC$ term from our SAS model and rerun. We continue the simplification process following the rules we set forth earlier in this chapter. Here, we will not eliminate the intercept term $b_0$ or eliminate a main effect term unless we eliminate its second-order and interaction term.

After we fit and examine a number of models for relating our response to our input variables, we will finally recommend the following responce surface model (see Table 22.3$b$):

$$\hat{\mu} = -355.315 + 1.214(A) + 5.444(B) + 4.839(C) - 0.0147(AB) - 0.00154(ASQ) - 0.0411(CSQ)$$

with $R^2 = 0.784$.

**TABLE 22.3**    SUMMARIZED COMPUTER AIDED ANALYSIS ANOVA TABLES, SAS PROC REG (EXAMPLE 22.3)

(a) Full second-order response surface model

Model: MODEL1
Dependent Variable: ELON    ELONGATION PERCENT

### Analysis of Variance

| Source | DF | Sum of Squares | Mean Square | F Value | Prob>F |
|---|---|---|---|---|---|
| Model | 9 | 2776.65278 | 308.51698 | 19.760 | 0.0001 |
| Error | 44 | 686.99537 | 15.61353 | | |
| C total | 53 | 3463.64815 | | | |

| | | | | |
|---|---|---|---|---|
| Root MSE | 3.95140 | R-square | 0.8017 | |
| Dep Mean | 17.31481 | Adj R-sq | 0.7611 | |
| C.V. | 22.82090 | | | |

### Parameter Estimates

| Variable | DF | Parameter Estimate | Standard Error | T for H0: Parameter=0 | Prob > ITI | Variable Label |
|---|---|---|---|---|---|---|
| INTERCEP | 1 | -371.310185 | 69.62811002 | -5.333 | 0.0001 | Intercept |
| A | 1 | 1.251389 | 0.33125281 | 3.778 | 0.0005 | COOKING TEMPERATURE |
| B | 1 | 6.716667 | 1.66605488 | 4.031 | 0.0002 | COOKING TIME |
| C | 1 | 5.018056 | 1.28462582 | 3.906 | 0.0003 | SYRUP CONTENT |
| AB | 1 | -0.014667 | 0.00322630 | -4.546 | 0.0001 | |
| AC | 1 | -0.000750 | 0.00161315 | -0.465 | 0.6443 | |
| BC | 1 | 0.008333 | 0.01613151 | 0.517 | 0.6080 | |
| ASQ | 1 | -0.001544 | 0.00045627 | -3.385 | 0.0015 | |
| BSQ | 1 | -0.084444 | 0.04562679 | -1.851 | 0.0709 | |
| CSQ | 1 | -0.041111 | 0.01140670 | -3.604 | 0.0008 | |

(b) Final response surface model

Model: MODEL4
Dependent Variable: ELON    ELONGATION PERCENT

### Analysis of Variance

| Source | DF | Sum of Squares | Mean Square | F Value | Prob>F |
|---|---|---|---|---|---|
| Model | 6 | 2715.62963 | 452.60494 | 28.438 | 0.0001 |
| Error | 47 | 748.01852 | 15.91529 | | |
| C total | 53 | 3463.64815 | | | |

| | | | | |
|---|---|---|---|---|
| Root MSE | 3.98940 | R-square | 0.7840 | |
| Dep Mean | 17.31481 | Adj R-sq | 0.7565 | |
| C.V. | 23.04037 | | | |

### Parameter estimates

| Variable | DF | Parameter Estimate | Standard Error | T for H0: Parameter=0 | Prob > ITI | Variable Label |
|---|---|---|---|---|---|---|
| INTERCEP | 1 | -355.314815 | 63.62686232 | -5.584 | 0.0001 | Intercept |
| A | 1 | 1.213889 | 0.32437283 | 3.742 | 0.0005 | COOKING TEMPERATURE |
| B | 1 | 5.444444 | 1.14779447 | 4.743 | 0.0001 | COOKING TIME |
| C | 1 | 4.838889 | 1.15355749 | 4.195 | 0.0001 | SYRUP CONTENT |
| AB | 1 | -0.014667 | 0.00325733 | -4.503 | 0.0001 | |
| ASQ | 1 | -0.001544 | 0.00046066 | -3.353 | 0.0016 | |
| CSQ | 1 | -0.041111 | 0.01151640 | -3.570 | 0.0008 | |

Our model development is driven by a need to predict our physical response, given a set of values for our independent variables (e.g., $A$, cooking temperature between 300 and 400°F; $B$, cooking time between 5 and 15 min; $C$, syrup content between 40 and 60 lb). We can perform a pure error analysis to test for lack of fit for this final model, using both our regression ANOVA (Table 22.3$b$) and our experimental ANOVA (Table 21.7):

$$SS_{perr} = SS_{err} = 381.50$$

$$df_{perr} = df_{err} = 27$$

$$SS_{lof} = SS_{rerr} - SS_{perr} = 748.02 - 381.50 = 366.52$$

$$df_{lof} = df_{rerr} - df_{perr} = 47 - 27 = 20$$

$$F_{calc} = \frac{MS_{lof}}{MS_{perr}} = \frac{366.52/20}{381.50/27} = \frac{18.326}{14.130} = 1.30$$

From our $F$ table (Section IX, Table IX.4), $F_{0.05,20,27} \cong 2.05$; hence, this $F$ test results in $P$-value $> 0.05$. We will not reject $H_0$ of adequate fit and we conclude that we have an adequate model, b.o.e.

---

## 22.3 RESPONSE SURFACE DESIGNS

*In many cases, our objective in off-line quality control is to develop an optimal process recipe where our process settings are described by quantitative variables.* To accomplish this objective we must usually run a sequence of experiments. *It is unlikely that we can capture a best recipe setting in only one experiment.*

Many strategies exist for meeting the challenge of response optimization. We can use any experimental design we have discussed so far. The point is to find the optimum in a cost effective and timely fashion; hence, relatively small, efficient, multifactor experiments are usually desirable. *One strategy used very often is first to develop a relatively broad based initial experiment that includes the variables we think are driving the response. Next, we run the experiment and fit a first-order response surface from the results. We then use the response surface model to guide us in developing our second experiment* (thus moving in closer to the optimal setting or recipe). This systematic practice tends to work very well when we are dealing with quantitative factors. (The alternative is a more or less haphazard trial-and-error search over our feasible region or study grid).

*Our sequence of experiments may include second-order response surfaces if a first-order response surface is judged to be inadequate.* Nevertheless, *our objective remains to close in on our best operating setting as quickly and economically as possible.* We will briefly discuss and demonstrate two designs which are useful in response surface work: (1) a $2^f$ FAT-CRD with a center point and (2) a central composite design.

## $2^f$ with a Center Point

*The $2^f$ design can be used to develop first-order response surfaces.* If replicates are run, an estimate of pure error can be calculated and we can assess model adequacy. In this case, the Experimental error row from the $2^f$ experimental ANOVA is identical to, and will serve as, a Pure error row for our regression ANOVA and allow us to assess lack of fit.

*Using a $2^f$ design as our base, we can expand our plan to include a center point setting,* which we arbitrarily call the "zero" point or "origin," for convenience. This point would be, by definition, our best guess at a process setting before the experiment. All other settings can be considered as step-out points to be investigated. Hence, we can define the "step-out" as the planned expansion distance from the center point, in each direction (relative to each factor).

We will take at least one observation at each step-out point and take $n_c \geq 2$ observations at the origin. Using this design we can assess nonlinearity, but generally cannot develop a significant full second-order regression model. *For the $2^f$ with a center point (with no replications on the corners or step-outs) we develop the pure error out of the multiple center point observations:*

$$SS_{perr} = SS_{err} = \sum_{i=1}^{n_c} (Y_{ci} - \bar{Y}_c)^2 \tag{22.29}$$

and

$$df_{err} = n_c - 1 \tag{22.30}$$

hence

$$MS_{perr} = MS_{err} = \frac{SS_{err}}{df_{err}} \tag{22.31}$$

where $Y_{ci}$ = $i$th center point response
  $\bar{Y}_c$ = mean of the $n_c$ center point responses
  $n_c$ = number of observations at the design center point

A single degree of freedom curvature row can be added to our ANOVA (Montgomery [20]). We develop the curvature row sum of squares as

$$SS_{curve} = \frac{n_f n_c (\bar{Y}_f - \bar{Y}_c)^2}{n_f + n_c} \tag{22.32}$$

with

$$df_{curve} = 1 \tag{22.33}$$

where $Y_f$ = average of factorial (corner or step-out) responses
  $n_f$ = total number of factorial (corner) observations

*The center point addition to a $2^f$ design allows us to assess the linearity-nonlinearity characteristics in a two level design (otherwise we would require three or more levels of the factors, e.g., a $3^f$ or possibly a central composite design). It also allows us the option of obtaining a pure error analysis (to assess model adequacy) by adding replicates only at the center of the design space.*

## Example 22.4

At this point in our cookie product experimentation, we want to examine the effects of cooking temperature $T$ and syrup content $S$ on the chewiness response, measured by percent elongation. We are proposing to employ a response surface technique using a sequence of designed experiments.

### Solution

As we stated earlier, it is unlikely that an optimal recipe can be developed with one experiment. Typically, we will use a sequence of experiments in order to close in on our best recipe. For our first experiment, we will develop a $2^2$ FAT-CRD with a center point. This design choice is reasonable here, since we are in the early stages of process analysis and wish to limit our first experiment in terms of cost and time. We will take one observation at each corner and three observations in the center. Figure 22.7 depicts our experimental boundary as well as our seven observations. Here, our strategy is to design and execute a simple experiment (which, in this case, is not compatible with development of a full second-order model). Then, we will use the results to guide us in setting up a more extensive second experiment capable of producing a full second-order model.

Due to our factor level selections (i.e., our center point structure), we can create two scales: (1) the actual scale of measure, pounds and degrees Fahrenheit, and (2) a coded scale of measure— $-1, 0, +1$. Both scales are shown in Figure 22.7.

An experimental ANOVA is shown in Table 22.4. Our curvature row's $P$-value $> 0.05$ indicates that a first-order model will be reasonable. Here, we would not reject the null hypotheses that our second-order parameters, $b_i$'s, for the terms $T^2$ and $S^2$ together are equal to 0. We see the $ST$ interaction as not significant, while both $S$ and $T$ linear effects are significant.

FIGURE 22.7    Physical boundaries and response data, $2^2$ with a center point design (Example 22.4).

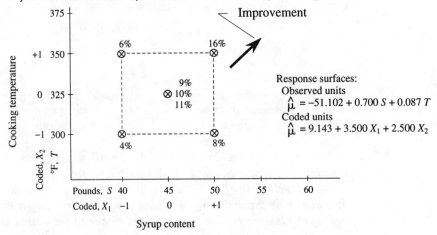

Response surfaces:
Observed units
$\hat{\mu} = -51.102 + 0.700\,S + 0.087\,T$
Coded units
$\hat{\mu} = 9.143 + 3.500\,X_1 + 2.500\,X_2$

We will now develop our first-order predictor for both the observed scale and the coded scale. Our SAS code and selected output appear in the Computer Aided Analysis Supplement, Solutions Manual. Our first-order response surface models appear in Figure 22.7. We can see that our response will increase as we move upward and to the right (due to the positive slope parameters in both response surfaces). Hence, our next experiment should be structured somewhere in the direction of the "improvement" arrow.

## Central Composite Design

The $3^f$ designs allow us to develop second-order response surface models. But the $3^f$ models have two primary disadvantages:

1 They are usually inefficient in that they require a great deal of experimentation, especially if we have a relatively large number of factors and want to develop a pure error analysis.
2 They are not rotatable. (When the variance of a predicted response $\hat{\mu}$ at some selected set of $X$'s is a function of only the distance from the center of the design space, and is not a function of direction from the center, the form is said to be "rotatable.")

**TABLE 22.4   EXPERIMENTAL ANOVA FOR $2^2$ WITH A CENTER POINT (EXAMPLE 22.4)**

| Source | df | SS | MS | F | P-value |
|---|---|---|---|---|---|
| Total corrected | $7 - 1 = 6$ | 88.857 | | | |
| $S$ | $2 - 1 = 1$ | 49.000 | 49.000 | 49.00 | $0.01 < P\text{-value} < 0.05$ |
| $T$ | $2 - 1 = 1$ | 25.000 | 25.000 | 25.00 | $0.01 < P\text{-value} < 0.05$ |
| $ST$ | $(2 - 1)(2 - 1) = 1$ | 9.000 | 9.000 | 9.00 | $P\text{-value} > 0.05$ |
| Curvature | 1 | 3.857 | 3.857 | 3.86 | $P\text{-value} \gg 0.05$ |
| Exp. error | 2 | 2.000 | 1.000 | | |

Sample calculations

$\text{SS}_{tc} = (6^2 + 4^2 + 8^2 + 16^2 + 9^2 + 10^2 + 11^2) - (64^2/7) = 88.857$

* $\text{SS}_S = [(6 + 4)^2 + (16 + 8)^2]/2 - (34^2/4) = 49.000$

* $\text{SS}_T = [(4 + 8)^2 + (6 + 16)^2]/2 - (34^2/4) = 25.000$

* $\text{SS}_{ST} = (6^2 + 4^2 + 8^2 + 16^2)/1 - (34^2/4) - 49.000 - 25.000 = 9.000$

$\overline{Y}_F = (6 + 4 + 8 + 16)/4 = 8.500$

$\overline{Y}_C = (9 + 10 + 11)/3 = 10.000$

$\text{SS}_{cur} = 4(3)(8.5 - 10)^2/(4 + 3) = 3.857$

$\text{SS}_{err} = (9 - 10)^2 + (10 - 10)^2 + (11 - 10)^2 = 2.000$

*Based only on the four corner responses.

*The central composite designs are widely used for second-order response surface modeling both because of their statistical properties and the practical appeal of their expanded coverage around a center point.* Figure 22.8 depicts both a two-factor and a three-factor central composite design layout. By observation, we can see that both layouts can be built up from the $2^f$ design or the $2^f$ with a center point design. In other words, we might perform a $2^f$ with a center point and then add the axial points, if necessary, after an initial analysis. We also see broad coverage of the design space surrounding the center point and extending beyond the $2^f$ borders on the axes. The $\alpha$ symbol here represents a distance and has no relationship to a type I error probability. Here, we use the $\alpha$ symbol only because it is commonly encountered in the central composite design literature.

**FIGURE 22.8**    Generic two- and three-factor central composite design layouts.

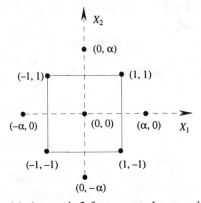

(*a*) A generic 2-factor central composite design layout.

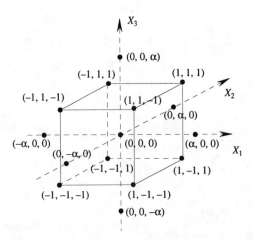

(*b*) A generic 3-factor central composite design layout

**TABLE 22.5    STRUCTURAL REQUIREMENTS FOR THE UNIFORM-PRECISION CENTRAL COMPOSITE DESIGN**

|  | Full factorials | | | | | One-half fractional factorials | | | |
|---|---|---|---|---|---|---|---|---|---|
| Number of factors | 2 | 3 | 4 | 5 | 6 | 5 | 6 | 7 | 8 |
| Number of vertice points* | 4 | 8 | 16 | 32 | 64 | 16 | 32 | 64 | 128 |
| Number of axial points* | 4 | 6 | 8 | 10 | 12 | 10 | 12 | 14 | 16 |
| Number of center observations | 5 | 6 | 7 | 10 | 15 | 6 | 9 | 14 | 20 |
| Total observations required | 13 | 20 | 31 | 52 | 91 | 32 | 53 | 92 | 164 |
| $\alpha$ value | 1.414 | 1.682 | 2.000 | 2.378 | 2.828 | 2.000 | 2.378 | 2.828 | 3.364 |

*Only one observation is taken at each vertice point and at each axial point.
*Source:* Adapted, with permission, from G.E.P. Box and J.S. Hunter, "Multifactor Experimental Designs for Exploring Response Surfaces," *The Annals of Mathematical Statistics*, vol. 28, page 227, 1957.

Different classes of the central composite design exist. Here, we will discuss only the uniform-precision and the orthogonal central composite designs; both are rotatable and allow a pure error analysis (see Montgomery [21]). The reader is referred to Cochran and Cox [22] and Box and Draper [23] for further discussions. Tables 22.5 and 22.6 provide the structural requirements for the uniform-precision and the orthogonal central composite design, respectively. Corresponding one-half fractional factorial ($\frac{1}{2}$ FFE) central composite designs are also listed. The FFE design will be discussed in the next chapter. We will develop a uniform-precision central composite design in Example 22.5. The orthogonal central composite design can be developed in a similar fashion. The only difference here is in the number of center points used.

**TABLE 22.6    STRUCTURAL REQUIREMENTS FOR THE ORTHOGONAL CENTRAL COMPOSITE DESIGN**

|  | Full factorials | | | | | One-half fractional factorials | | | |
|---|---|---|---|---|---|---|---|---|---|
| Number of factors | 2 | 3 | 4 | 5 | 6 | 5 | 6 | 7 | 8 |
| Number of vertice points* | 4 | 8 | 16 | 32 | 64 | 16 | 32 | 64 | 128 |
| Number of axial points* | 4 | 6 | 8 | 10 | 12 | 10 | 12 | 14 | 16 |
| Number of center observations | 8 | 9 | 12 | 17 | 24 | 10 | 15 | 22 | 33 |
| Total observations required | 16 | 23 | 36 | 59 | 100 | 36 | 59 | 100 | 177 |
| $\alpha$ value | 1.414 | 1.682 | 2.000 | 2.378 | 2.828 | 2.000 | 2.378 | 2.828 | 3.364 |

*Only one observation is taken at each vertice point and at each axial point.
*Source:* Adapted, with permission, from G.E.P. Box and J.S. Hunter, "Multifactor Experimental Designs for Exploring Response Surfaces," *The Annals of Mathematical Statistics*, vol. 28, page 227, 1957.

## Example 22.5

Based on the results of the linear response surfaces developed in Example 22.4, design a follow-on experiment to generate a second-order response surface.

### Solution

We will develop a central composite design of the uniform-precision class and then fit a full second-order model. This design is built to the two-factor design specifications in Table 22.5. We will make 13 observations in all.

Figure 22.9 depicts our experimental boundaries as well as our observed data. We have moved our center up and over to the right to coordinates ($S = 52$ lb, $T = 350°F$). We then set our corners at ($\pm5$, $\pm25$) from the center. Our coded coordinates retain their original values, but have been shifted. Our previous experimental boundaries appear in the background.

The amount of shift used is a matter of judgment. Once we decide how much to shift in one dimension (here, we selected a shift of 7 lb in syrup, or a center at 52 lb), the coded response surface helps us determine our shift in the other dimension. Here our coded values are

$$X_1 = \frac{S - 45}{5} \quad \text{and} \quad X_2 = \frac{T - 325}{25}$$

A seven-unit shift in $S$ (to 52 lb) results in 7/5, or a 1.4-unit shift ($\Delta X_1$) in $X_1$. Since

$$\hat{\mu} = 9.143 + 3.500X_1 + 2.500X_2$$

FIGURE 22.9    Physical boundaries and response data, two-factor central composite (uniform-precision) design (Example 22.5).

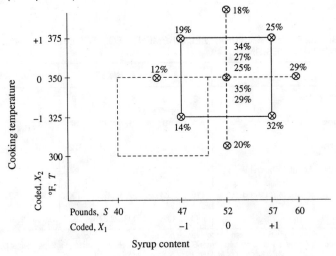

and
$$\frac{\Delta X_1}{b_1} = \frac{\Delta X_2}{b_2}$$

will follow a path of steepest ascent,

$$\Delta X_2 = \frac{\Delta X_1\, b_2}{b_1} = \frac{1.4\,(2.5)}{3.5} = 1.0$$

Hence our new $T$ center is

$$T = \Delta X_2\,(25) + 325 = 1\,(25) + 325 = 350$$

Therefore, our new center is at (52 lb, 350°F). We could also use our best judgment in locating the center point in this second experiment, rather than use the method of steepest ascent. Nevertheless, we would still follow the direction of our arrow (Figure 22.7).

We can see the four corners, or vertex points, and the four axial points in Figure 22.9. We use the $\alpha = 1.414$ value from Table 22.5 to construct the axial points. Our 13 experimental coordinates and responses are summarized below:

| S | T | $X_1$ | $X_2$ | Observed response |
|---|---|---|---|---|
| 47 | 325 | −1 | −1 | 14% |
| 47 | 375 | −1 | 1 | 19% |
| 57 | 325 | 1 | −1 | 32% |
| 57 | 375 | 1 | 1 | 25% |
| 52 | 314.65 | 0 | −1.414 | 20% |
| 52 | 385.35 | 0 | 1.414 | 18% |
| 44.93 | 350 | −1.414 | 0 | 12% |
| 59.07 | 350 | 1.414 | 0 | 29% |
| 52 | 350 | 0 | 0 | 34% |
| 52 | 350 | 0 | 0 | 27% |
| 52 | 350 | 0 | 0 | 25% |
| 52 | 350 | 0 | 0 | 35% |
| 52 | 350 | 0 | 0 | 29% |

Here,

$$X_1 = \frac{S - 52}{5} \quad \text{and} \quad X_2 = \frac{T - 350}{25}$$

We will use our SAS computer aid to develop our full second-order model. The SAS code and selected results are shown in the Computer Aided Analysis Supplement, Solutions Manual. Our full fitted response surface in the observed units for $S$ (pounds) and $T$ (degrees Fahrenheit) is

$$\hat{\mu} = -1843.590 + 26.502(S) + 6.614(T) - 0.024(ST) - 0.163(S^2) - 0.0077(T^2)$$

Our model is generally applicable over the ranges of $44.93 \leq S \leq 59.07$ and $314.65 \leq T \leq 385.35$ (the range included in our experiment). In terms of our coded model, again from our computer-aided analysis, our full fitted response surface in the $X_1$ and $X_2$ units is

$$\hat{\mu} = 29.999 + 6.006(X_1) - 0.604(X_2) - 3.000(X_1^2) - 4.063(X_1 X_2) - 4.813(X_2^2)$$

Our model here, in the coded scale, is applicable over the ranges of $-1.414 \leq X_1 \leq 1.414$ and $-1.414 \leq X_2 \leq 1.414$.

---

In Example 22.5, after we examine the SAS output for the two forms of response surface fitting, we see rather large discrepancies; the significance of the model terms tends to vary depending on the form we use. The literature is divided on whether or not to center and code our independent variables or to fit our response surfaces in the units of measure. In general, we prefer the units of measure approach. Typically, communication with our customers is enhanced when we can express our results in the units of measure found in the field. The subsequent section has been developed to provide more perspective on this matter.

## 22.4  OBSERVED VERSUS CENTERED DATA IN RESPONSE SURFACES

*The results from response surface experiments can be analyzed in two basic ways: (1) in the observed units on the independent variables and (2) in coded units on the independent variables.* In Examples 22.4 and 22.5 we have produced response surface regression models using both observed units and centered, or coded, units. In general, our models predict the same response (cookie elongation), but we must code our independent variables ($X$ values) when we want to make a prediction with the centered model. *Both practical and theoretical arguments can be posed as to whether or not we should code our data.*

From the practical side, if we build our response surface with observed units it is much easier to use (since we simply substitute our desired independent variable values directly into the prediction model). Also on the practical side, if we center our data by defining a new origin (as we did in our two examples), we can readily and visually compare the magnitudes of the model coefficients (the $b$'s). Hence, we can obtain a rough feel for the strength of each term intuitively.

On the theoretical side, if two or more of our independent variables are highly correlated, we face what is termed "multicollinearity," an ill-conditioning in the model. Centering the data with respect to the independent variables will allow us, using our computer aid, to avoid this problem to some extent.

In practice, we see both methods in wide use. We must recognize that, given the nature of the calculations involved, our solution will not match exactly (as we have seen in our example). This brief introductory treatment of data centering is presented here as a matter of interest. Draper and Smith [24] and Neter, Wasserman, and Kutner [25] provide a more thorough discussion of this issue.

## REVIEW AND DISCOVERY EXERCISES

### Review

**22.1** We will revisit Problem 20.1. Corrosion and the prevention of corrosion in natural gas pipelines are major concerns. In order to address these issues, a chemical engineer conducted an experiment to determine how much carbon dioxide ($CO_2$) in a natural gas stream can be absorbed by water in an acid gas scrubbing tower. She was interested in the effect of tower temperature on the $CO_2$ solubility in water at a specific tower pressure. Five different tower temperatures were chosen for the test. The following data were collected. The bigger is better response recorded is in pounds $CO_2$ per 100 lb of $H_2O$. Assume that all runs were made in random order. [CT]

| Tower temperature, °C | Replication | | | | |
|---|---|---|---|---|---|
| | 1 | 2 | 3 | 4 | 5 |
| 40 | 3.9 | 3.1 | 2.6 | 3.5 | 2.3 |
| 60 | 2.5 | 1.7 | 2.8 | 2.0 | 2.3 |
| 80 | 1.4 | 1.9 | 2.3 | 1.7 | 2.1 |
| 100 | 1.6 | 1.5 | 2.0 | 1.2 | 1.8 |
| 120 | 1.3 | 1.6 | 1.1 | 1.5 | 1.8 |

**a** Estimate the overall mean and the treatment means. Develop a response plot.
**b** Develop a linear response surface model that will predict the absorption response as a function of the tower temperature. Predict the average response at a tower temperature of 50°C, 110°C, and 0°C. Comment on your predictions.
**c** Calculate the residual terms from the linear model; then construct and interpret your residual plot.
**d** Execute a lack-of-fit test for your linear model.
**e** Develop a second-order response surface model that will predict the absorption response as a function of the tower temperature. Predict the average response at a tower temperature of 50°C, 110°C, and 0°C. Comment on your predictions.
**f** Calculate the residual terms from the second-order model; then construct and interpret a residual plot.
**g** Execute a lack-of-fit test for your second-order model.
**h** Calculate $R_{max}^2$ for this data set. Recommend one of the two models fitted above and explain your decision.

**22.2** An engineer is interested in the effect of temperature on the vapor pressure of ethylene glycol. She designed a CRD experiment using temperature as the treatment factor. The following experimental data were obtained in random order, response units are mmHg [MR]:

| Temperature, °C | Replications | | | | |
|---|---|---|---|---|---|
| 100 | 14.8 | 16.2 | 16.4 | 16.6 | 17.3 |
| 125 | 53.2 | 53.5 | 56.9 | 54.4 | 55.1 |
| 150 | 149.3 | 150.0 | 155.7 | 154.8 | 151.6 |
| 175 | 379.2 | 371.3 | 368.5 | 380.0 | 378.9 |
| 200 | 836.8 | 853.0 | 847.3 | 847.4 | 846.9 |

a Estimate the overall mean and the treatment means. Develop a response plot.

b Develop a linear response surface that will predict the vapor pressure response as a function of the ethylene glycol solution temperature. Predict the average response at a temperature of 160°C, 199°C, and 50°C. Comment on your predictions.

c Calculate the residual terms from the linear model; then construct and interpret your residual plot.

d Execute a lack-of-fit test for your linear model.

e Develop a second-order response surface that will predict the vapor pressure response as a function of the ethylene glycol solution temperature. Predict the average response at a temperature of 160°C, 199°C, and 50°C. Comment on your predictions.

f Calculate the residual terms from the second-order model; then construct and interpret a residual plot.

g Execute a lack-of-fit test for your second-order model.

h Calculate $R^2_{max}$ for this data set. Recommend one of the two models fitted above and explain your decision.

**22.3** We will revisit Problem 21.2. A wastewater treatment plant utilizes hydrogen peroxide at various temperatures to destroy organic contaminants in the incoming water. The manager of the facility wants to determine whether the treatment is influenced by operating temperatures and peroxide flow rates. He selects three peroxide flow rates and four temperatures. He then designs and conducts a FAT-CRD experiment (fixed effects) on contaminated water drawn from a lagoon. In each case, he measures the percentage removal of organics as the response. The data obtained are shown below. [MR]

| Hydrogen peroxide flow rate | Temperature, °C | | | |
|---|---|---|---|---|
| | 10 | 50 | 70 | 90 |
| 50 ml/hr | 17.7 | 26.1 | 34.1 | 33.2 |
| | 23.3 | 28.3 | 31.3 | 35.8 |
| 100 ml/hr | 29.8 | 31.0 | 46.3 | 53.1 |
| | 23.4 | 28.7 | 50.6 | 56.5 |
| 150 ml/hr | 29.3 | 52.1 | 60.3 | 61.2 |
| | 27.7 | 54.3 | 57.2 | 58.8 |

a Estimate the overall mean and the treatment means. Develop a response plot; stratify your plot by temperature.

b Develop a full linear response surface that will predict the percent removal of organics response as a function of the hydrogen peroxide flow rate and the temperature.

c Calculate the residual terms from the linear model; then construct and interpret residual plots.

d Execute a lack-of-fit test for your linear model.

e Develop a full second-order response surface that will predict the percent removal of organics response as a function of the hydrogen peroxide flow rate and the temperature. Then, simplify your full model. Using the simplified model, predict the average percent removal at a temperature of 80°C and a hydrogen peroxide flow rate of 120 ml/hr; at 100°C and 200 ml/hr flow. Comment on your results.

**f** Calculate the residual terms from the simplified model; then construct and interpret residual plots.

**g** Execute a lack-of-fit test for your simplified model.

**h** Calculate $R^2_{max}$ for this data set. Recommend a model and explain your decision.

**22.4** We will revisit Problem 21.5. An engineer in charge of a short pipeline conducted a FAT-CRD experiment to determine the energy required to deliver light crude oil from one end of the pipeline to the other. He used three pump speeds and varied the temperature of the oil. The responses recorded are in energy units per barrel of crude oil. His data are presented below. [JM]

| Pump speed | Temperature, °F | | | |
|---|---|---|---|---|
| | 50 | 70 | 90 | 110 |
| 1050 rpm | 0.86 | 0.90 | 0.93 | 0.88 |
| | 0.87 | 0.88 | 0.85 | 0.92 |
| 1200 rpm | 0.68 | 0.85 | 0.81 | 0.89 |
| | 0.78 | 0.76 | 0.77 | 0.88 |
| 1800 rpm | 0.86 | 0.85 | 0.89 | 0.95 |
| | 0.79 | 0.85 | 0.91 | 0.90 |

**a** Estimate the overall mean and the treatment means. Develop a response plot; stratify your plot by temperature.

**b** Develop a full linear response surface model that will predict the energy required per barrel response as a function of the pump rpm and the crude oil temperature.

**c** Calculate the residual terms from the linear model; then construct and interpret residual plots.

**d** Execute a lack-of-fit test for your linear model.

**e** Develop a full second-order response surface model that will predict the energy required per barrel response as a function of the pump rpm and the crude oil temperature. Then, simplify your full model. Using the simplified model, predict the average energy required per barrel of crude oil at a temperature of 80°F and a pump speed of 1000 rpm; at 75°F and 500 rpm. Comment on your results.

**f** Calculate the residual terms from the simplified model; then construct and interpret residual plots.

**g** Execute a lack-of-fit test for your simplified model.

**h** Calculate $R^2_{max}$ for this data set.

**22.5** We will revisit Problem 21.13. AA Inc. manufactures fiberglass parts and wishes to conduct an in-house experiment to measure the effect of four factors on the gel time of polyester-fiberglass resin. Gel time is the time interval when fiberglass resin is workable. According to the manufacturer of the polyester resin, gel time may be affected by age of the resin, ambient humidity, ambient temperature, and the amount of catalyst used. The engineering department of AA Inc. designed and conducted a FAT-CRD experiment to assess these effects No replications were run in the experiment. The data collected appear below, where the response is measured in minutes [DM]:

| | | Temperature C | | | |
|---|---|---|---|---|---|
| | | 60°F | | 80°F | |
| | | Catalyst D | | | |
| Age A | Humidity B | 1% | 2% | 1% | 2% |
| 3 months | 30% | 50 | 12 | 22 | 6 |
| | 70% | 75 | 20 | 38 | 15 |
| 9 months | 30% | 65 | 18 | 35 | 12 |
| | 70% | 90 | 25 | 60 | 36 |

**a** Develop a full first-order response surface that will predict the gel time response as a function of the age, humidity percent, temperature, and catalyst percent. Predict the average gel time at an age of six months, a humidity of 40 percent, a temperature of 70°F, and a 1.5 percent catalyst level using your full first-order model. Then, simplify your full model.

**b** Recommend a model. Explain.

**c** At this point, again predict the average gel time at an age of six months, a humidity of 40 percent, a temperature of 70°F, and a 1.5 percent catalyst level using your simplified model. Comment on your results.

**22.6** In the past nearly all sheet metal parts for the external surfaces of an airplane have been assembled to the framework with rivet joints. Research has been done on the application of cohesive connections. In order to attain maximum holding force (bigger is better), tensile strength tests have been performed, where at each experimental setup the tensile breaking stress, measured in kilopounds (klb) was recorded. All experimental units were small, same-size partial sections of the airplane's external surfaces. Three factors were assumed to influence the quality and strength of the cohesive connection:

Factor A: Hardening temperature (degrees Celsius)

Factor B: Press time of parts (hours)

Factor C: Relative humidity of the environment (percent)

All experiments consist of two phases: (1) preparing the experimental units (bonding the parts together at the selected humidity, curing the parts at selected temperature and time) and (2) performing the tensile stress test. A $3^3$ FAT-CRD experiment was designed. For each of the 27 treatment combinations, 2 replications were taken, a total of 54 observations in all. The data obtained are shown below [HO]:

| | Temperature A | | | | | | | | |
|---|---|---|---|---|---|---|---|---|---|
| | 22 °C | | | 40 °C | | | 65 °C | | |
| | Press time B | | | Press time B | | | Press time B | | |
| Relative Humidity | 20 hr | 40 hr | 60 hr | 20 hr | 40 hr | 60 hr | 20 hr | 40 hr | 60 hr |
| 25% | 4 | 8 | 7 | 9 | 14 | 13 | 7 | 12 | 5 |
| | 8 | 10 | 15 | 5 | 11 | 18 | 16 | 9 | 7 |
| 50% | 17 | 26 | 28 | 18 | 25 | 32 | 17 | 23 | 17 |
| | 21 | 21 | 29 | 19 | 23 | 28 | 25 | 19 | 14 |
| 75% | 12 | 16 | 20 | 13 | 19 | 29 | 11 | 21 | 11 |
| | 15 | 21 | 24 | 19 | 23 | 25 | 19 | 17 | 7 |

   **a** Fit an appropriate response surface to the experimental results.
   **b** Develop a lack-of-fit test for your model.
   **c** Comment on the location of the best operating range in the fitted response surface.
   **d** Comment on the efficiency and effectiveness of the experimental design used.

**22.7** A $2^2$ FAT-CRD experiment with a center point was designed to assess an apple dehydration process. The goal is to dry the fruit to a target level of 10% moisture by weight in a 2-hr drying period. Two factors were studied: temperature and cross sectional (cube) dimensions. The experiment consisted of preparing Granny Smith apples according to the cross sectional dimensions and drying them in a laboratory oven under simulated bulk volume production conditions. The results of the experiment are shown below:

| Observed | Temperature, °F | U temperature | Cross section, in | U cross section | Response, % moisture |
|---|---|---|---|---|---|
| 1 | 200 | -1 | 0.250 | -1 | 10 |
| 2 | 200 | -1 | 0.500 | 1 | 20 |
| 3 | 300 | 1 | 0.250 | -1 | 5 |
| 4 | 300 | 1 | 0.500 | 1 | 8 |
| 5 | 250 | 0 | 0.375 | 0 | 7 |
| 6 | 250 | 0 | 0.375 | 0 | 9 |
| 7 | 250 | 0 | 0.375 | 0 | 10 |
| 8 | 250 | 0 | 0.375 | 0 | 8 |

   **a** Determine whether a linear model is reasonable (develop an ANOVA table which includes Main effects, Interaction, Curvature, and Experimental error rows).
   **b** Using the above data, fit a linear model in the units of measure and then in the coded units.
   **c** Assume that we choose to use the model

   $$\hat{\mu} = b_0 + b_1 \text{ temp} + b_2 \text{ cross section}$$

   Develop a pure error (lack-of-fit) analysis.
   **d** Using the model in part *c*, recommend a location for a follow-on experiment. Explain.

**22.8** The Gear Corporation manufactures injected molded gears, or "toothed wheels." Gear has established a good reputation for quality over the past several years. However, recently, Gear has been having trouble with a new product line of small gears made exclusively for one of its automotive customers. This product line's scrap rate varied from 3 to 10 percent while the other product lines' scrap percentages were running between 2 to 4 percent. To solve this scrap-variation problem, Gear organized a team of engineers, technicians, and operators to determine what was affecting the process and to minimize scrap.

   After a brainstorming session, the team concluded that the major factors affecting the process were

   Factor *A*: Cure time (seconds)
   Factor *B*: Injection pressure (klb)
   Factor *C*. Mold temperature (degrees Fahrenheit)

After reviewing records of past settings for these factors, the team members decided to perform a three-factor central composite design experiment and use a response surface approach to optimizing their product's scrap rate. They chose this experimental design, since only 20 trials were needed—several less than the other designs which had been discussed. The data collected are shown below.

| Trial | Cure time, sec | Cure time, coded units | Injection pressure, (klb) | Injection pressure, coded units | Mold temperature, °F | Mold temperature, coded units | Scrap, % |
|-------|------|--------|------|--------|------|--------|------|
| 1  | 1   | -1     | 10   | -1     | 70   | -1     | 8.6 |
| 2  | 3   | 1      | 10   | -1     | 70   | -1     | 8.0 |
| 3  | 3   | 1      | 10   | -1     | 90   | 1      | 7.5 |
| 4  | 1   | -1     | 10   | -1     | 90   | 1      | 8.6 |
| 5  | 1   | -1     | 30   | 1      | 70   | -1     | 9.2 |
| 6  | 3   | 1      | 30   | 1      | 70   | -1     | 8.2 |
| 7  | 3   | 1      | 30   | 1      | 90   | 1      | 7.9 |
| 8  | 1   | -1     | 30   | 1      | 90   | 1      | 8.5 |
| 9  | 2   | 0      | 20   | 0      | 80   | 0      | 2.0 |
| 10 | 2   | 0      | 20   | 0      | 80   | 0      | 2.8 |
| 11 | 2   | 0      | 20   | 0      | 80   | 0      | 2.7 |
| 12 | 2   | 0      | 20   | 0      | 80   | 0      | 3.0 |
| 13 | 2   | 0      | 20   | 0      | 80   | 0      | 2.6 |
| 14 | 2   | 0      | 20   | 0      | 80   | 0      | 2.8 |
| 15 | 0.3 | -1.682 | 20   | 0      | 80   | 0      | 9.2 |
| 16 | 3.7 | 1.682  | 20   | 0      | 80   | 0      | 8.3 |
| 17 | 2   | 0      | 4.2  | -1.682 | 80   | 0      | 5.2 |
| 18 | 2   | 0      | 36.8 | 1.682  | 80   | 0      | 6.9 |
| 19 | 2   | 0      | 20   | 0      | 64.2 | -1.682 | 8.5 |
| 20 | 2   | 0      | 20   | 0      | 96.8 | 1.682  | 7.5 |

a  Starting with a full second-order model, determine the best response surface to model the scrap.

b  Test for lack of fit.

c  Comment on the normality assumptions regarding the residuals of your model choice.

d  Briefly critique the design chosen for this problem.

### Discovery

22.9  Revisit Example 22.2.

a  Add a third-order term to the regression model; in this case, fit a model of the form

$$\hat{\mu} = b_0 + b_1\, T + b_2\, T^2 + b_3\, T^3$$

b  Overlay a plot of the new model on Figure 22.6a.

c  Develop an analysis of residuals for the second-order model (i.e., develop a normal plot for the residuals already calculated in Example 22.2 for the second-order model).

d  For the third-order model developed, calculate the residuals and assess the model form as well as the normality assumption.

e  Comment on the fits. Recommend a model.

**22.10** The Sure-Stick Adhesive Corporation manufactures several product lines of commercial and industrial adhesives for the automotive industry. A new product line of high shear strength industrial adhesives, made exclusively for one of its automotive customers, is being developed.

In trials at the automotive customer's facilities, problems in the sheer strength of the new adhesive have surfaced. In fact, the shear strength varied from 600 to1000 psi while the customer's specification required its strength to be at least 800 psi. Sure-Stick Corporation guarantees that its minimum shear strength will be at least 900 psi. The automotive customer, who has been a loyal customer of Sure-Stick for years, has requested that Sure-Stick correct this problem. If Sure-Stick cannot fix this problem, the customer will consider other adhesive suppliers.

In an attempt to keep one of their most profitable accounts and maintain their quality reputation, Sure-Stick developed an experimental program. First, Sure-Stick organized a team of engineers, technicians, and operators from both their own company as well as from their customer's facilities. This team's objective was to maximize the shear strength of the adhesive so as to regain customer satisfaction. After a brainstorming session, this group identified the factors which were most likely affecting the process. They concluded the major factors affecting the process were

Factor *A*: Volume of additive in the material mixed within their factory

Factor *B*: Application pressure (psi) used by their customer

Factor *C*: Application temperature used by their customer

After reviewing records of past settings for these factors, they decided to perform a three-factor central composite design experiment and develop a response surface to identify the best combination of these factors. The data collected are shown below [RM]:

| Trial | Additive, ml | Additive, coded units | Application pressure, psi | Application pressure, coded units | Application temperature, °F | Application temperature, coded units | Shear strength, x 100 psi |
|---|---|---|---|---|---|---|---|
| 1 | 20 | -1 | 50 | -1 | 100 | -1 | 6.6 |
| 2 | 40 | 1 | 50 | -1 | 100 | -1 | 8.0 |
| 3 | 40 | 1 | 50 | -1 | 200 | 1 | 7.5 |
| 4 | 20 | -1 | 50 | -1 | 200 | 1 | 7.0 |
| 5 | 20 | -1 | 100 | 1 | 100 | -1 | 7.2 |
| 6 | 40 | 1 | 100 | 1 | 100 | -1 | 9.0 |
| 7 | 40 | 1 | 100 | 1 | 200 | 1 | 8.3 |
| 8 | 20 | -1 | 100 | 1 | 200 | 1 | 6.0 |
| 9 | 30 | 0 | 100 | 0 | 150 | 0 | 10.0 |
| 10 | 30 | 0 | 75 | 0 | 150 | 0 | 11.4 |
| 11 | 30 | 0 | 75 | 0 | 150 | 0 | 11.6 |
| 12 | 30 | 0 | 75 | 0 | 150 | 0 | 9.7 |
| 13 | 30 | 0 | 75 | 0 | 150 | 0 | 10.2 |
| 14 | 30 | 0 | 75 | 0 | 150 | 0 | 9.8 |
| 15 | 13.18 | -1.682 | 75 | 0 | 150 | 0 | 10.0 |
| 16 | 46.82 | 1.682 | 75 | 0 | 150 | 0 | 6.0 |
| 17 | 30 | 0 | 33.0 | -1.682 | 150 | 0 | 6.8 |
| 18 | 30 | 0 | 117.0 | 1.682 | 150 | 0 | 6.3 |
| 19 | 30 | 0 | 75 | 0 | 66.9 | -1.682 | 6.5 |
| 20 | 30 | 0 | 75 | 0 | 234.1 | 1.682 | 8.1 |

    **a** Start with a full second-order model and simplify it until you have the best model.

    **b** Test for lack of fit.

    **c** Comment on the normality assumptions relative to the residuals of your best model.

    **d** Briefly critique the design chosen for this problem.

    **e** Comment on the likelihood of Sure-Stick's meeting the customer's 800 psi specification. Sure-Stick's 900 psi guarantee.

**22.11** Select a product or process. Design, execute, and interpret a sequence of multifactor experiments (all quantitative factors) regarding some facet of the product or process. Follow the six steps involved in experimental protocol (discussed in Table 20.2). Use a response surface approach and develop predictor equations for the response. Construct a display board, presenting your experimental protocol, results, and conclusions. After completion of your experiment, critique your experimental design as to its effectiveness in meeting your experimental objective.

# REFERENCES

**1** R. E. Walpole and R. H. Myers, *Probability and Statistics for Engineers and Scientists*, 5th ed., New York: Macmillan, 1993.

**2** N. R. Draper and H. Smith, *Applied Regression Analysis*, 2d ed., New York: Wiley, 1981.

**3** G. E. P. Box and N. R. Draper, *Empirical Model-Building and Response Surfaces*, New York: Wiley, 1987.

**4** J. Neter, W. Wasserman, and M. H. Kutner, *Applied Linear Statistical Models*, 3d ed., Homewood, IL: Irwin, 1990.

**5** See references 1, 2, 3, and 4.

**6** See references 2, 3, and 4.

**7** *SAS Procedures Guide*, Version 6, Cary, NC: SAS Institute, 1990.

**8** *SAS/STAT User's Guide, Volumes 1 and 2*, Version 6, Cary, NC: SAS Institute, 1990.

**9** See reference 3.

**10** See reference 2.

**11** See references 1 and 2.

**12** See reference 1.

**13** See reference 2.

**14** See reference 2.

**15** See reference 2.

**16** See reference 2.

**17** See reference 1.

**18** See reference 2.

**19** See references 7 and 8.

**20** D. C. Montgomery, *Design and Analysis of Experiments*, 3d ed., New York: Wiley, 1991.

**21** See reference 20.

**22** W. G. Cochran and G. M. Cox, *Experimental Designs*, 2d ed., New York, New York: Wiley, 1957.

**23** See reference 3.

**24** See reference 2.

**25** See reference 4.

# 23

# BLOCKED, SPLIT-PLOT, FRACTIONAL FACTORIAL, AND TAGUCHI SNR EXPERIMENTS

## 23.0 INQUIRY

1 How is it possible to block out nuisance variables in FAT experiments?
2 What is a split-plot experiment?
3 How can we increase the number of factors and reduce the number of observations simultaneously in our experiments?
4 What is resolution in a designed experiment?
5 How can experiments be designed to help develop robust products and processes?
6 What are the differences between classical and Taguchi approaches to experimental design?

## 23.1 INTRODUCTION

*The use of multiple factors in designed experiments provides a systematic means of developing both effective and efficient off-line quality investigations to address product or process bottlenecks.* In this chapter, we will extend our previous discussions to include blocking on a nuisance variable, as well as the use of a split-plot experiment to relax the FAT randomization requirements. We will also develop two-level fractional factorials and introduce the Taguchi signal-to-noise ratio (SNR) transformations and approach to robust design, as well as orthogonal arrays (OAs) at two and three factor levels.

Blocking out a nuisance variable is a logical extension of the FAT-CRD that we developed in Chapter 21. The blocking concept is also fundamental in developing fractional factorial experiments (FFEs). Use of the FFEs and FFE based Taguchi SNR methods requires a sound knowledge of basic experimental design in order to develop proper interpretations. Blind use of these techniques may lead to serious misunderstandings. We will discuss the underlying assumptions and pitfalls of these powerful experimental design tools so that we can properly choose and apply these methods and then interpret our results.

## 23.2 FACTORIAL ARRANGEMENT OF TREATMENTS IN A RANDOMIZED COMPLETE BLOCK (FAT-RCB) DESIGN

So far we have discussed the FAT in a CRD. *We can use the FAT in a RCB design when we desire to break out the effect of a noise or nuisance variable in our*

*ANOVA.  First, we must clearly identify the blocks* (as previously discussed in our single-factor RCB design discussion in Chapter 20). *Then, we randomize each of our treatment combinations within each block, one block at a time.* Using our previous notation schemes, our model for a blocked two-factor FAT is

$$Y_{ijkl} = \mu + \beta_k + A_i + B_j + AB_{ij} + \varepsilon_{l(ijk)} \tag{23.1}$$

where $Y_{ijkl}$ = response observation
$\mu$ = overall mean
$\beta_k$ = block effect, $k = 1, 2, ..., n$
$A_i$ = factor A main effect, $i = 1, 2, ..., a$
$B_j$ = factor B main effect, $j = 1, 2, ..., b$
$AB_{ij}$ = factor A, factor B interaction effect
$\varepsilon_{l(ijh)}$ = random error, $l = 1$

In this case, the $l$ subscript is placed in the model in order to accommodate our EMS algorithm, $l = 1$, by definition for the RCB design.

Here, we see the treatment factor level combinations acting as an expanded form of our treatment term ($\tau_i$ as in our single-factor RCB discussion). Figure 23.1 depicts the relevant blocks, treatment combinations, and experimental units for this design. Table 23.1 develops the FAT-RCB ANOVA table, including the EMS rows for two fixed effects factors $A$ and $B$, and a random block effect. We can see by examination of the model, Equation 23.1, as well as in the ANOVA table, that the experimental error is developed from the experimental units nested in the blocks (i.e., each experimental unit is a unique piece of experimental material). Hence, we do not have the ability in this design to measure the experimental error directly. What we essentially do is pool all of the block by factor interaction terms to form an Experimental error row. *To justify this design, we typically assume the block variable does not interact with the factor variables. Otherwise, we should elevate the nuisance variable to factor status and develop a three-factor experiment with replication.*

**FIGURE 23.1**    FAT-RCB randomization scheme and experimental layout.

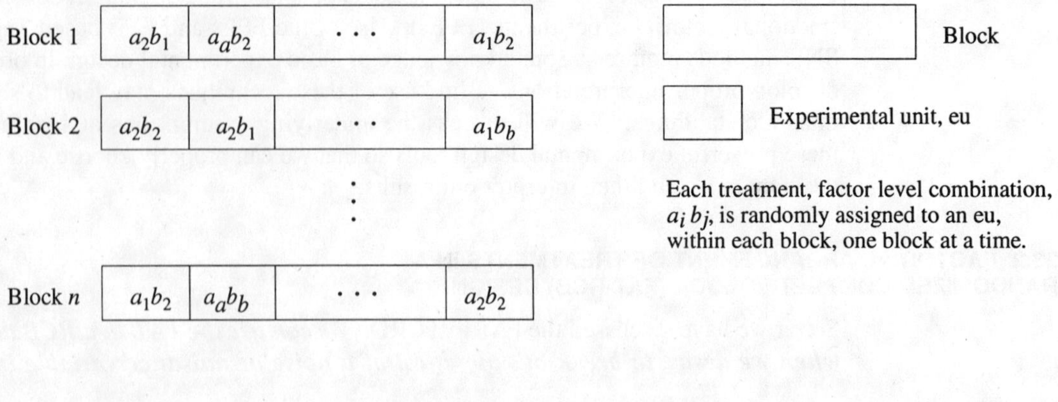

TABLE 23.1  GENERIC TWO-FACTOR FAT-RCB ANOVA TABLE AND EMS DEVELOPMENT

| Source | df | SS | MS | F | EMS** $\begin{matrix} F & F & R & R \\ a & b & n & 1 \\ i & j & k & l \end{matrix}$ | EMS |
|---|---|---|---|---|---|---|
| Total corrected | $abn - 1$ | $\sum_{ijk} Y_{ijk}^2 - CF$ | | | | |
| $Block_k$ | $n - 1$ | $\dfrac{\sum_k Y_{..k}^2}{ab} - CF$ | $SS_{blk}/df_{blk}$ | $MS_{blk}/MS_{err}^*$ | $a\ b\ 1\ 1$ | $\sigma_\varepsilon^2 + ab\sigma_{blk}^2$ |
| $A_i$ | $a - 1$ | $\dfrac{\sum_i Y_{i..}^2}{bn} - CF$ | $SS_A/df_A$ | $MS_A/MS_{err}$ | $0\ b\ n\ 1$ | $\sigma_\varepsilon^2 + bn\phi_A$ |
| $B_j$ | $b - 1$ | $\dfrac{\sum_j Y_{.j.}^2}{an} - CF$ | $SS_B/df_B$ | $MS_B/MS_{err}$ | $a\ 0\ n\ 1$ | $\sigma_\varepsilon^2 + an\phi_B$ |
| $AB_{ij}$ | $(a-1)(b-1)$ | $\dfrac{\sum_{ij} Y_{ij.}^2}{n} - CF - SS_A - SS_B$ | $SS_{AB}/df_{AB}$ | $MS_{AB}/MS_{err}$ | $0\ 0\ n\ 1$ | $\sigma_\varepsilon^2 + n\phi_{AB}$ |
| Exp. error$_{(ijk)}$ | $(ab-1)(n-1)$ | $SS_{tc} - SS_{blk} - SS_A - SS_B - SS_{AB}$ | $SS_{err}/df_{err}$ | | $1\ 1\ 1\ 1$ | $\sigma_\varepsilon^2$ |

Here  $CF = \dfrac{Y_{...}^2}{abn}$

* The validity of this F test is questionable due to the blocking procedure, see Chapter 20.
** Block is a random effect; A, B are fixed effects.

## Example 23.1

Back in our cookie factory, we want to study the effects of two factors—syrup content and cooking temperature—on the chewiness of our chocolate chip product. Also, based on past experience, we suspect that the placement of cookies on our pan or conveyor (through our cooking tunnel) has a significant effect on product quality, in that cookies near the edges sometimes appear to be over- or undercooked compared to those in the center of the pan. We think this is due to air flow along the pan edge, but we are not sure. Develop an experiment capable of studying both factors and effectively dealing with the cookie position phenomenon.

## Solution

Here, we will try an approach somewhat different from the one we used in Example 21.6. We will use a two-factor FAT in an RCB design, $\alpha = 0.05$. We will develop our two factors as

Factor $S$:   Syrup content with three fixed levels—50 lb, 55 lb, and 60 lb.

Factor $T$:   Cooking temperature with two fixed levels—350°F and 375°F.

We will block on pan position

BLK:   Cookie position on the baking pan with three random levels—right side, middle, and left side.

Our experimental data and ANOVA are shown in Table 23.2 and Table 23.3, respectively. We should note that each of the six treatment combinations appears once in each block. The ANOVA and EMS in Table 23.1 are applicable to our design. The SAS code and selected output for this example are shown in the Computer Aided Analysis Supplement, Solutions Manual.

TABLE 23.2    EXPERIMENTAL DATA FOR TWO-FACTOR FAT-RCB (EXAMPLE 23.1)

| Cookie position on pan Blk | Syrup content S | | | | | |
|---|---|---|---|---|---|---|
| | 50 lb | | 55 lb | | 60 lb | |
| | Cooking temperature T | | | | | |
| | 350°F | 375°F | 350°F | 375°F | 350°F | 375°F |
| Right side | 29 | 26 | 38 | 31 | 32 | 28 |
| Middle | 37 | 31 | 41 | 40 | 35 | 26 |
| Left side | 26 | 29 | 36 | 38 | 31 | 27 |

Response units are in percent elongation.

TABLE 23.3    ANOVA TABLE FOR TWO-FACTOR FAT-RCB (EXAMPLE 23.1)

| Source | df | SS | MS | F | P-value[**] |
|---|---|---|---|---|---|
| Total corrected | 3(2)(3) − 1 = 17 | 435.611 | | | |
| Blk | 3 − 1 = 2 | 67.444 | 33.722 | 4.33* | 0.0442 |
| S | 3 − 1 = 2 | 230.111 | 115.056 | 14.77 | 0.0010 |
| T | 2 − 1 = 1 | 46.722 | 46.722 | 6.00 | 0.0343 |
| ST | (3 − 1)(2 − 1) = 2 | 13.444 | 6.722 | 0.86 | 0.4511 |
| Experimental error | [3(2) − 1](3 − 1) = 10 | 77.889 | 7.789 | | |

Sample calculations

$SS_{tc} = (29^2 + 37^2 + \cdots + 26^2 + 27^2) - (581^2/18) = 435.611$

$SS_{Blk} = (184^2 + 210^2 + 187^2)/6 - (581^2/18) = 67.444$

$SS_S = (178^2 + 224^2 + 179^2)/6 - (581^2/18) = 230.111$

$SS_T = (305^2 + 276^2)/9 - (581^2/18) = 46.722$

$SS_{ST} = (92^2 + 86^2 + 115^2 + 109^2 + 98^2 + 81^2)/3 - (581^2/18) - 230.111 - 46.722 = 13.444$

$SS_{err} = SS_{tc} - SS_{blk} - SS_S - SS_T - SS_{ST} = 77.889$

* F test in Blk row is not theoretically sound; use only as a general guide.

** P-values were obtained from SAS PROC GLM.

In the ANOVA, we see significant syrup $S$ and temperature $T$ main effects (P-value = 0.0010 and P-value = 0.0343, respectively). Our interaction effect is not statistically significant, b.o.e. Figure 23.2 develops main effect and interaction plots with ± 2 s.e.i. arrows. We see by examining the ANOVA that the blocking has pulled out a substantial sum of squares ($SS_{Blk} = 67.444$, or about 15 percent of $SS_{tc}$). Hence, our evidence indicates that blocking is effective. The $MS_{err} = 7.789$ seems rather large, based on our past experience. Checking with the testing lab, we have found that the cookie temperature at the time of testing was not controlled. This oversight may be responsible for some of the $SS_{err}$. We should be more careful next time.

*Once we complete our ANOVA, we should assess the validity of our assumptions regarding the homogeneity of variance across the treatments as well as the normality of our residuals.* We should also record the order of our observations *and assess the randomness of the residuals with respect to the sequence of experimentation.* Computer aids will usually have provisions to assess many of these assumptions, or we can resort to the manual calculations and plotting techniques we introduced in Chapter 20.

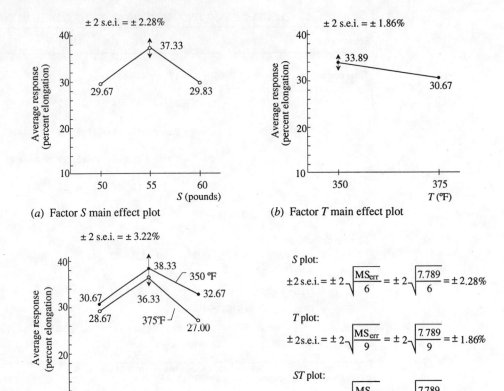

**FIGURE 23.2**    FAT-RCB main effects and interaction plots (Example 23.1).

## 23.3 SPLIT-PLOT DESIGN

When we encounter one factor which is difficult or impossible to randomize with the other factors, a split-plot design may be useful. *A split-plot, or split-unit, experiment is essentially one designed to use a multistage randomization process.* In essence, *it is an experiment within an experiment in the context of our experimental units, which are of two types—a whole-plot experimental unit and a split-plot experimental unit. The split-plot experiment is very useful when we have a factor which is physically difficult, time consuming, or costly to change from one level to the next.*

In a FAT-CRD or FAT-RCB we must randomly choose our factor level combinations every time we prepare to make an observation. For example, in a two-factor FAT, we randomly select one combination (e.g., low *A*, high *B*), assign it to our experimental unit, treat our experimental unit, and make an observation. Then, we repeat the process until we have made all of our observations. On the other hand, in the split-plot design, we select one level of factor $A$ at random and then run all of our factor $B$ levels at random within the selected $A$ level. Next, we select another $A$ level and repeat the process. From the physical standpoint, we have to change our $A$ level only $a$ times. This method saves a great deal of setup time and expense. This multi-level randomization concept distinguishes the split plot from the FAT-CRD. In the split plot, both factors are crossed, as in the FAT-CRD, but the randomization requirements are less rigorous.

*In a split plot, we will be able to measure both main effects and interactions.* But we must modify our analysis procedures somewhat. Our model looks about the same as before, except that *we divide the effects into two groups: (1) whole-plot terms and (2) split-plot terms.* Our SS calculations are straightforward and can be accomplished using our previous techniques. We will illustrate the split-plot mechanics by example.

---

## Example 23.2

In our cookie factory, we want to design an experiment in order to study two factors, cooking temperature and cooking time, as well as their interaction. We wish to avoid as many temperature change setups as possible. Futhermore, we only have a limited amount of time each day to perform our experiment. We typically produce a new batch of dough each day and the possible day-to-day variation in the batches of dough is also of concern here.

### Solution

We can satisfy all of our desires by developing a blocked, split-plot experiment where we place cooking temperature in our main plot. We will lose precision in our main-plot analysis in order to gain the benefit of reduced setup on the main-plot factor. In this example, we will stress a computer aided solution, but also provide hand calculations. We will define our terms here in a more descriptive, contextual, manner as

BAT: A batch of dough, made up each day; batches serve as blocks in our experiment

TEMP: Cooking temperature with three levels—300°F, 350°F, 400°F

TIME: Cooking time with three levels—5 min, 10 min, 15 min

Our blocked, split-plot model is written as

$$
\begin{aligned}
Y_{ijkl} = {} & \mu \\
& + \text{BAT}_i + \text{TEMP}_j + \text{BAT*TEMP}_{ij} \\
& \text{(whole plot)} \\
& + \text{TIME}_k + \text{TEMP*TIME}_{jk} + \text{BAT*TIME}_{ik} + \text{BAT*TEMP*TIME}_{ijk} \\
& \text{(split plot)} \\
& + \varepsilon_{l(ijk)}
\end{aligned}
$$

$$(23.2)$$

Figure 23.3 depicts our randomization process. We will set the oven for 400°F (on day 1, batch 1) and start cooking; for 15 min, we will pull out a cookie at random, for 5 min we will pull out another cookie, and for 10 min we will pull out a third cookie. We will test our cookies and record the results. We next randomly select a temperature level (300°F), set up and repeat our procedure, still using batch 1. We continue for four batches (days). Here, temperature acts as our whole-plot factor and cooking time as our split-plot factor. We can think of the oven full of cookies, baked at a certain temperature, as a whole-plot experimental unit. In the split plot, we consider an individual cookie, baked at a certain temperature for a certain amount of time, as a split-plot experimental unit.

Our data are shown in Table 23.4. Our analysis is straightforward. We proceed just as in a crossed-factor FAT experiment. We have used SAS to produce our SS rows. Our SAS code and output are shown in the Computer Aided Analysis Supplement, Solutions Manual. We have run two model versions in order to obtain our complete ANOVA and *P*-value report.

Our ANOVA (Table 23.5) was developed by piecing together the ANOVA from our computer aid, SAS in this case (hand calculations are also shown), and includes our whole plot and split plot. We should note that *our whole-plot error estimate has less precision* (df = 6) *than our split-plot error estimate* (df = 18). *In the split-plot design we use the whole-plot error to test terms* and calculate

**FIGURE 23.3**   Split-plot design randomization (Example 23.2).

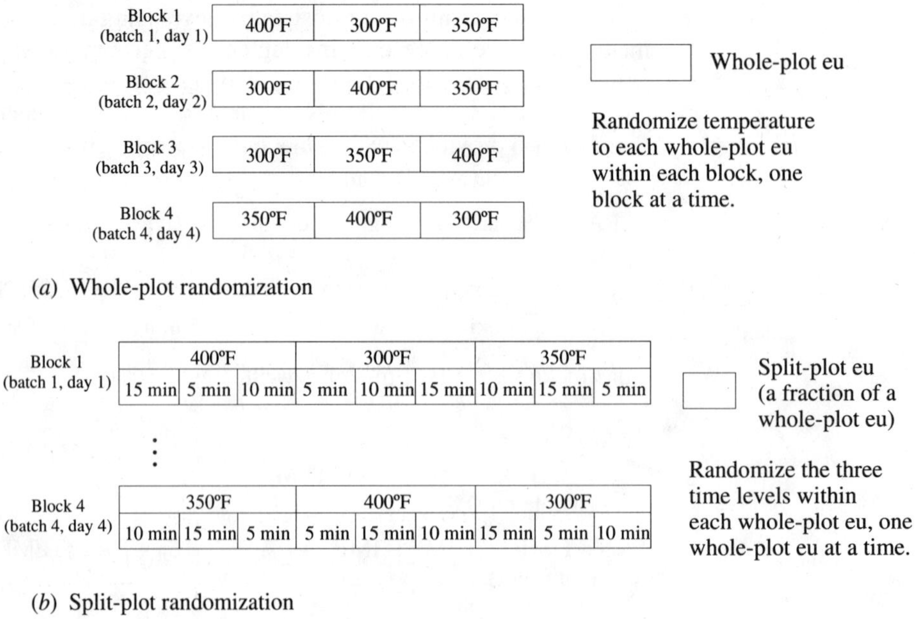

(*a*)  Whole-plot randomization

(*b*)  Split-plot randomization

TABLE 23.4    SPLIT-PLOT COOKIE EXPERIMENT DATA (EXAMPLE 23.2).

| Batch (block) | Cooking temperature | | | | | | | | |
|---|---|---|---|---|---|---|---|---|---|
| | 300°F | | | 350°F | | | 400°F | | |
| | Cooking time | | | | | | | | |
| | 5 min | 10 min | 15 min | 5 min | 10 min | 15 min | 5 min | 10 min | 15 min |
| 1 | 16 | 21 | 25 | 19 | 25 | 26 | 23 | 22 | 18 |
| 2 | 19 | 17 | 28 | 21 | 31 | 28 | 19 | 17 | 17 |
| 3 | 15 | 18 | 24 | 18 | 29 | 25 | 20 | 20 | 14 |
| 4 | 17 | 22 | 23 | 18 | 28 | 21 | 19 | 21 | 15 |

Response units are in percent elongation.

s.e.i. arrows and pairwise test statistics *associated with the whole-plot factor.* Likewise, *we use the split-plot error to test terms* and calculate s.e.i. arrows and pairwise test statistics *associated with the split-plot factor*.

Based on the evidence we have collected through our experiment, the temperature and time main effects are both highly significant, as is our temperature-time interaction. Even though we do not have a clear test on our batches (blocks), we can see that our $SS_{BAT} = 17.64$ is small. This result gives us comfort in that our dough batches seem to be homogeneous from day to day, b.o.e. Our readers are urged to study the hypothesis test structure used in Table 23.5 carefully and notice that we worked within the main and split plots with our main- and split-plot errors, respectively.

Now we will plot our results and develop our $\pm 2$ s.e.i. arrows for both main-plot and split-plot analyses. For the main effect, cooking temperature, from the whole plot,

$$\pm 2 \text{ s.e.i.} = \pm 2 \sqrt{\frac{MS_{err\ wp}}{12}} = \pm 2 \sqrt{\frac{7.185}{12}} = \pm 1.55\%$$

For the main effect cooking time, from the split plot

$$\pm 2 \text{ s.e.i.} = \pm 2 \sqrt{\frac{MS_{err\ sp}}{12}} = \pm 2 \sqrt{\frac{3.583}{12}} = \pm 1.09\%$$

For our temperature-time interaction plot,

$$\pm 2 \text{ s.e.i.} = \pm 2 \sqrt{\frac{MS_{err\ sp}}{4}} = \pm 2 \sqrt{\frac{3.583}{4}} = \pm 1.89\%$$

TABLE 23.5    SPLIT-PLOT ANOVA TABLE (EXAMPLE 23.2)

| Source | df | SS | MS | F | P-value |
|---|---|---|---|---|---|
| Total corrected | $4(3)(3) - 1 = 35$ | 666.750 | | | |
| BAT (block) | $4 - 1 = 3$ | 17.639 | | | |
| TEMP | $3 - 1 = 2$ | 178.667 | 89.339 | $89.33/7.19 = 12.43$ | 0.0073 |
| Whole-plot error (BAT * TEMP) | $(4 - 1)(3 - 1) = 6$ | 43.111 | 7.185 | | |
| TIME | $3 - 1 = 2$ | 107.167 | 53.583 | $53.58/3.58 = 14.95$ | 0.0001 |
| TEMP * TIME | $(3 - 1)(3 - 1) = 4$ | 255.667 | 63.917 | $63.92/3.58 = 17.84$ | 0.0001 |
| Split-plot error (BAT*TIME and BAT*TEMP*TIME pooled together) | 18<br>$(4 - 1)(3 - 1) = 6$<br>$(4 - 1)(3 - 1)(3 - 1) = 12$ | 64.500<br>35.944<br>28.556 | 3.583 | | |

Sample calculations

$$SS_{tc} = (16^2 + 21^2 + \cdots + 21^2 + 15^2) - (759^2/36) = 666.750$$
$$SS_{BAT} = (195^2 + 197^2 + 183^2 + 184^2)/9 - (759^2/36) = 17.639$$
$$SS_{TEMP} = (245^2 + 289^2 + 225^2)/12 - (759^2/36) = 178.667$$
$$SS_{err\ wp} = SS_{BAT*TEMP} = (62^2 + 70^2 + \cdots + 55^2)/3 - 17.639 - 178.667 - (759^2/36) = 43.111$$

$$SS_{TIME} = (224^2 + 271^2 + 264^2)/12 - (759^2/36) = 107.167$$
$$SS_{TEMP*TIME} = (67^2 + 78^2 + \cdots + 64^2)/4 - 178.667 - 107.167 - (759^2/36) = 255.667$$
$$SS_{err\ sp} = 35.944 + 28.556 = 64.500$$

$$SS_{BAT*TIME} = (58^2 + 68^2 + \cdots + 59^2)/3 - 17.639 - 107.167 - (759^2/36) = 35.944$$
$$SS_{BAT*TEMP*TIME} = SS_{tc} - SS_{BAT} - SS_{TEMP} - SS_{BAT*TEMP} - SS_{TIME} - SS_{TEMP*TIME} - SS_{BAT*TIME} = 28.556$$

Our plots and s.e.i. arrows are shown in Figure 23.4. We can use a pairwise test (e.g., LSD) to establish statistical significance between treatment means (i.e., we would use $MS_{err\ wp}$ for whole-plot comparisons and $MS_{err\ sp}$ for split-plot comparisons).

Split-plot designs are popular in both industrial and agricultural applications. One strategy described by Hicks involves setting up blocks, as we did in Example 23.2, and running our experiment one day at a time in a modular fashion, then stopping our experiment after we obtain the level of precision we seek [1]. For example, we might detect significant effects after three days. Steel and Torrie discuss split plots in the CRD, RCB, and Latin square designs [2].

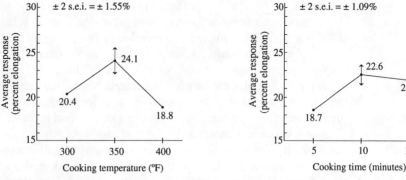

(a) Cooking temperature whole-plot main effect      (b) Cooking time split-plot main effect

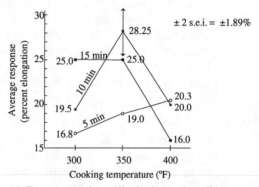

(c) Temperature-time split-plot interaction effect

**FIGURE 23.4**   Split-plot experimental plots (Example 23.2).

## 23.4 CONFOUNDING AND FRACTIONAL FACTORIALS

Since there are typically a number of factors affecting even simple product or process performance, a FAT is a very effective tool. *As we noted in Chaper 21, the amount of experimentation required in a full factorial* (all possible treatment combinations) *outstrips the available resources after about three or four factors at, say, three or four levels.* For example, four factors at three levels require $3^4 = 81$ observations for each replication of the experiment. In the case of four factors and four levels, $4^4 = 256$ observations would be required for one replication. The point is that for initial screening experiments, where we seek to determine which of a number of variables are critical in affecting a given response, a full FAT is usually not feasible. Even the $2^f$ experiments where $f = 5$ or $f = 6$ require a great deal of experimentation. *One alternative to a full factorial is to develop a fractional factorial experiment, FFE, where only a fraction* (e.g., one half, one fourth,) *of a full FAT is run*.

The pursuit of an FFE design is in general sensible only when we are reasonably sure that some main effects and/or factor interactions will not be active in our physical world. For example, the Pareto principle argues that for any set of possible effects, only a small proportion will be of great significance, relative to the others. *An FFE is a useful tool to screen a number of factors and identify the most critical factors for further study, possibly through a more detailed follow-up experiment.*

We typically confound selected effects with blocks, lay out our blocks, and then choose only one block to run, hence the term "fractional factorial." Confounding an interaction effect with blocks results in a total loss of our ability to measure that particular interaction effect. Confounding a main effect is clearly counterproductive. Aliasing, which is determined by the confounding scheme used, occurs as a result of confounding. Here, we see two or more effects with equivalent statistical values (e.g., sum of squares) due to the reduction in observations taken. *We must confound wisely in order to alias "important" effects—those we are interested in measuring—with "unimportant" ones—those we are not particularly interested in measuring.*

Our discussion will focus on only the $2^f$ class of fractional factorials. The two-level, FFEs have been shown to be very effective when used properly, (see Box, Hunter, and Hunter [3]). Two-level FFEs are widely advocated because of their statistical properties and their ease of use, when compared to three-level FFEs, (Hunter [4, 5]). Hunter provides extensive arguments to discourage the use of $3^f$ FFEs, especially when two-factor interaction is possible.

## Confounding

In order to discuss the concept of FFEs, we will first develop a method where we can split a two-level full FAT-CRD (with only one replication) into a number of blocks, each containing a fraction (e.g., $\frac{1}{2}, \frac{1}{4}, \frac{1}{8}$) of the treatment factor level combinations. We must begin by choosing a confounding scheme. Here, we will choose a noncritcal interaction effect for confounding.

To develop a $\frac{1}{2}$ FFE, we must confound one effect (e.g., $L = ABCDE$ in a $\frac{1}{2}$ FFE of a $2^5$ FAT). The $L$ notation is refered to as a "design generator." To develop a $\frac{1}{4}$ FFE, we will have to choose two effects, which will determine a third confounded effect (e.g., we choose $L_1 = ABE$ and $L_2 = CDE$ and then, $L_1$ and $L_2$ combine as $L_1 * L_2 = ABE \cdot CDE = ABCDE^2 = ABCD$; we drop the "squared" terms for the $2^f$ case). In other words, this particular $\frac{1}{4}$ FFE would confound $ABE$, $CDE$, and $ABCD$ with blocks: four blocks will yield 3 df and the three confounded effects, at two levels for each factor also yield 3 df. Hence, we choose our $L$s so as to accommodate the confounding structure we desire. *When we confound an effect* as mentioned above, *the effect becomes hopelessly mixed (we cannot break it out separately) with the block effect.*

---

## Example 23.3

Given a $2^5$ FAT with factors $A$, $B$, $C$, $D$, and $E$,

**a** Develop the two possible blocks for $\frac{1}{2}$ FFEs with $L = ABCDE$.

**b** Then, develop the four possible blocks for $\frac{1}{4}$ FFEs with $L_1 = ABE$, $L_2 = CDE$, and $L_1 * L_2 = ABCD$.

## Solution

**a** Here, we will use the factor level–treatment combination notation introduced in Example 21.7, where

| | | |
|---|---|---|
| $(1) \rightarrow$ | $a_1\, b_1\, c_1\, d_1\, e_1 \rightarrow$ | low A, low B, low C, low D, low E |
| $a \rightarrow$ | $a_2\, b_1\, c_1\, d_1\, e_1 \rightarrow$ | high A, low B, low C, low D, low E |
| $b \rightarrow$ | $a_1\, b_2\, c_1\, d_1\, e_1 \rightarrow$ | low A, high B, low C, low D, low E |
| . | | |
| . | | |
| . | | |
| $abc \rightarrow$ | $a_2\, b_2\, c_2\, d_1\, e_1 \rightarrow$ | high A, high B, high C, low D, low E |
| . | | |
| . | | |
| . | | |
| $abcde \rightarrow$ | $a_2\, b_2\, c_2\, d_2\, e_2 \rightarrow$ | high A, high B, high C, high D, high E |

The worksheet in Table 23.6 contains the experimental factors $A$ through $E$ and their levels for each treatment combination. The $-$ notation indicates the low factor level while the $+$ indicates the high factor level. To obtain the sign for a confounded effect, we simply multiply through the treatment rows (row by row), using only the factors included in the confounded effect. For example, in row (1) for the one-half FFE, $L = ABCDE$,

$$ABCDE \rightarrow (-)(-)(-)(-)(-) \rightarrow (-)$$

For the $abc$ row, with $L = ABCDE$,

$$ABCDE \rightarrow (+)(+)(+)(-)(-) \rightarrow (+)$$

We continue in this fashion until we complete all 32 rows (i.e., each row represents a treatment factor level combination in the $2^5$ full FAT).

Block assignments are made from the $L$ function signs. For the $\frac{1}{2}$ FFE, we assign the $+$'s to block 1 and the $-$'s to block 2 (both blocks are equivalent $\frac{1}{2}$ FFEs with $L = ABCDE$, so the block number given is arbitrary). The $+$'s and $-$'s divide the 32 treatment combination settings into two groups, or blocks. Figure 23.5$a$ develops the block layouts from this confounding scheme. We see all 32 treatment combinations, but we have divided the 32 combinations into two 16-block groups. Our strategy will be to select one group of 16 treatment combinations and run only that group in our $\frac{1}{2}$ FFE experiment.

**b** In order to develop the $\frac{1}{4}$ FFE, $L_1 = ABE$, $L_2 = CDE$ and $L_1 * L_2 = ABCD$, we use the same 32 combinations and follow the procedure outlined above. Here we use the 4 right-hand columns in the worksheet. We multiply through the signs of the factors making up the three confounded columns $ABE$, $CDE$, and $ABCD$ to develop the four equivalent $\frac{1}{4}$ FFEs. Then, we assign all of the $+ + +$ combinations to block 1, the $- - +$ combinations to block 2, the $+ - -$ combinations to block 3, and the $- + -$ combinations, arbitrarily, to block 4.

**TABLE 23.6**   $\frac{1}{2}$ FFE (L = ABCDE) AND $\frac{1}{4}$ FFE ($L_1$ = ABE, $L_2$ = CDE) WORKSHEET

| Treatment combination | Main Effects Factors | | | | | $\frac{1}{2}$ FFE L = ABCDE | | $\frac{1}{4}$ FFE $L_1$ = ABE $L_2$ = CDE $L_1 \cdot L_2$ = ABCD | | | |
|---|---|---|---|---|---|---|---|---|---|---|---|
| | A | B | C | D | E | ABCDE | Block assignment | ABE | CDE | ABCD | Block assignment |
| (1) | − | − | − | − | − | − | 2 | − | − | + | 2 |
| a | + | − | − | − | − | + | 1 | + | − | − | 3 |
| b | − | + | − | − | − | + | 1 | + | − | − | 3 |
| c | − | − | + | − | − | + | 1 | − | + | − | 4 |
| d | − | − | − | + | − | + | 1 | − | + | − | 4 |
| e | − | − | − | − | + | + | 1 | + | + | + | 1 |
| ab | + | + | − | − | − | − | 2 | − | − | + | 2 |
| ac | + | − | + | − | − | − | 2 | + | + | + | 1 |
| ad | + | − | − | + | − | − | 2 | + | + | + | 1 |
| ae | + | − | − | − | + | − | 2 | − | + | − | 4 |
| bc | − | + | + | − | − | − | 2 | + | + | + | 1 |
| bd | − | + | − | + | − | − | 2 | + | + | + | 1 |
| be | − | + | − | − | + | − | 2 | − | + | − | 4 |
| cd | − | − | + | + | − | − | 2 | − | − | + | 2 |
| ce | − | − | + | − | + | − | 2 | + | − | − | 3 |
| de | − | − | − | + | + | − | 2 | + | − | − | 3 |
| abc | + | + | + | − | − | + | 1 | − | + | − | 4 |
| abd | + | + | − | + | − | + | 1 | − | + | − | 4 |
| abe | + | + | − | − | + | + | 1 | + | + | + | 1 |
| acd | + | − | + | + | − | + | 1 | + | − | − | 3 |
| ace | + | − | + | − | + | + | 1 | − | − | + | 2 |
| ade | + | − | − | + | + | + | 1 | − | − | + | 2 |
| bcd | − | + | + | + | − | + | 1 | + | − | − | 3 |
| bce | − | + | + | − | + | + | 1 | − | − | + | 2 |
| bde | − | + | − | + | + | + | 1 | − | − | + | 2 |
| cde | − | − | + | + | + | + | 1 | + | + | + | 1 |
| abcd | + | + | + | + | − | − | 2 | − | − | + | 2 |
| abce | + | + | + | − | + | − | 2 | + | − | − | 3 |
| abde | + | + | − | + | + | − | 2 | + | − | − | 3 |
| acde | + | − | + | + | + | − | 2 | − | + | − | 4 |
| bcde | − | + | + | + | + | − | 2 | − | + | − | 4 |
| abcde | + | + | + | + | + | + | 1 | + | + | + | 1 |

For example, in row (1) for the $\frac{1}{4}$ FFE,

| | | A | B | C | D | E | | |
|---|---|---|---|---|---|---|---|---|
| ABE | → | (−) | (−) | | | (−) | → | (−) |
| CDE | → | | | (−) | (−) | (−) | → | (−) |
| ABCD | → | (−) | (−) | (−) | (−) | | → | (+) |

Here, we assign this − − + outcome to block 2.  In row *abc*,

| | | A | B | C | D | E | | |
|---|---|---|---|---|---|---|---|---|
| ABE | → | (+) | (+) | | | (−) | → | (−) |
| CDE | → | | | (+) | (−) | (−) | → | (+) |
| ABCD | → | (+) | (+) | (+) | (−) | | → | (−) |

| Treatment combinations in block 1 | Treatment combinations in block 2 |
|---|---|
| a | (1) |
| b | ab |
| c | ac |
| d | ad |
| e | ae |
| abc | bc |
| abd | bd |
| abe | be |
| acd | cd |
| ace | ce |
| ade | de |
| bcd | abcd |
| bce | abce |
| bde | abde |
| cde | acde |
| abcde | bcde |

*Note:* Do not get treatment combinations and factors/factor effects confused. Lowercase letters are treatment combinations in the $2^5$ factorial. Upper case letters represent factor effects. For example, *abcde* represents the high *A*, high *B*, high *C*, high *D*, high *E* factor combination, while *ABCDE* represents the 5-factor interaction term or measurement.

(a) $\frac{1}{2}$ FFE, $L = ABCDE$ confounded with blocks

| Treatment combinations in block 1 | Treatment combinations in block 2 | Treatment combinations in block 3 | Treatment combinations in block 4 |
|---|---|---|---|
| e | (1) | a | c |
| ac | ab | b | d |
| ad | cd | ce | ae |
| bc | ace | de | be |
| bd | ade | acd | abc |
| abe | bce | bcd | abd |
| cde | bde | abce | acde |
| abcde | abcd | abde | bcde |

(b) $\frac{1}{4}$ FFE, $L_1 = ABE$, $L_2 = CDE$, $L_1*L_2 = ABCD$ confounded with blocks

**FIGURE 23.5**   Block assignments for $\frac{1}{2}$ and $\frac{1}{4}$ FFE experiments (Example 23.3).

Here, we assign this $- + -$ outcome to block 4. Actually, we can examine any two *L* columns, two at a time (e.g., $+ +, + -, - +, - -$), and obtain the identical block assignment structures.

Figure 23.5b depicts the four block layouts we obtained from this confounding scheme. For our $\frac{1}{4}$ FFE, we would select one block and run the eight treatment combinations. Regardless of which of the four blocks we select to run, we have completely given up our ability to measure the *ABE*, *CDE*, and *ABCD* interactions.

If we wanted to develop a $\frac{1}{8}$ FFE, we would need three *L* functions ($L_1, L_2, L_3$) which, taken singly, in pairs, and all together, would yield seven confounded effects and eight block combination designations (e.g., $+ + +, + + -, ..., - - -$).

We would choose one of the eight blocks and run our experiment. Hence, the procedures are algorithmic in nature and allow us to generate other fractions.

Figure 23.6 illustrates the FFE concept through our cookie example; here we have five factors at two levels, a $2^5$ FAT-CRD (Figure 23.6a). Careful observation will show that block 2 of the $\frac{1}{2}$ FFE is represented by the white squares and block 1 by the black squares (Figure 23.6b). Likewise, block 2 of the $\frac{1}{4}$ FFE is represented by the black squares (Figure 23.6c). The eight observation settings for the five factors are shown in Figure 23.6d; here, observation 1 is low A, low B, low E, low C, and low D and observation 7 is high A, high B, low E, low C, and low D. The word "observation" is used here in a general sense; hence the observations are numbered 1 through 8. We typically randomize the order (e.g., we might run the experiment physically by choosing a random number 1 through 8, and observing that treatment combination, and then another random number, and so on, until we have completed the eight listed combinations). For an abbreviated table of random numbers see Section IX, Table IX.1. The careful reader will note that we have placed the E column between the B and C columns in Figure 23.6d. As long as we switch entire columns (+'s, −'s, and headings), rearranging the array format makes no difference in the experimental design. We will see later, when we dis-

**FIGURE 23.6**   Layouts for $\frac{1}{2}$ and $\frac{1}{4}$ FFEs (Example 23.3).

| Factor | Example | Level 1 | Level 2 |
|---|---|---|---|
| A | Cooking temp. | Low (−) | High (+) |
| B | Syrup content | Low (−) | High (+) |
| C | Cooking time | Short (−) | Long (+) |
| D | Cooking pan type | Solid (−) | Mesh (+) |
| E | Shortening type | Corn (−) | Coconut (+) |

(a) $2^5$ cookie plastic deformation (chewiness) example

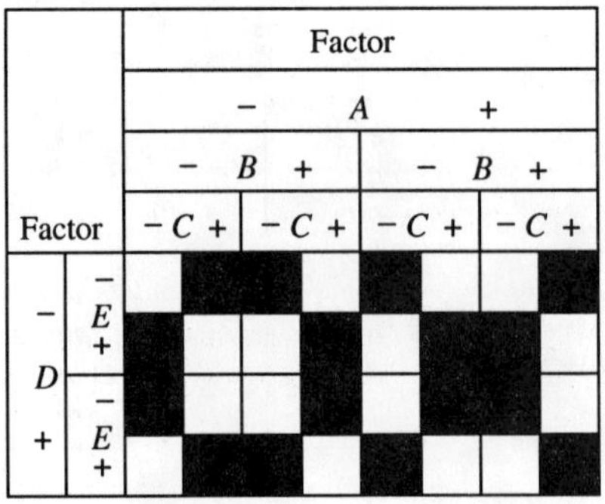

(b) $\frac{1}{2}$ FFE layout

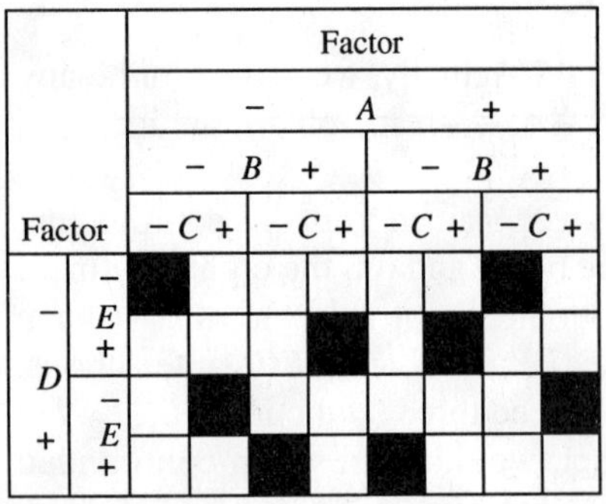

(c) $\frac{1}{4}$ FFE layout

| ANOVA aliases | ACDE BCD BE | BCDE ACD AE | ABCDE CD AB | ABCE ABD DE | ABDE ABC CE | BCE ADE BD | BDE ACE AD |
|---|---|---|---|---|---|---|---|
| Factors | A | B | E | C | D | AC | BC |
| Obs. 1 | − | − | − | − | − | | |
| 2 | − | − | + | + | + | | |
| 3 | − | + | + | − | + | | |
| 4 | − | + | + | + | − | | |
| 5 | + | − | + | − | + | | |
| 6 | + | − | + | + | − | | |
| 7 | + | + | − | − | − | | |
| 8 | + | + | − | + | + | | |

(d) $\frac{1}{4}$ FFE treatment combinations and aliases

cuss orthogonal arrays, what happens when we switch only factor headings. More detailed discussions regarding confounding in blocks may be found in Hicks [6], Box, Hunter, and Hunter [7], and Montgomery [8].

## Aliases

The interpretation of a full factorial is straightforward. However, *in FFEs, we run only a fraction of a full factorial.* Hence, *we cannot break out and uniquely measure all main and interaction effects.* The columns shown in Figure 23.6*d* represent ANOVA rows and their aliases for our $\frac{1}{4}$ FFE. The four factor effects in each column heading (e.g., in the first column $A$, $BE$, $BCD$, $ACDE$) are aliases. The $A$ effect, the $BE$ effect, the $BCD$ effect, and the $ACDE$ effect ANOVA sum of squares result in the same numerical value, since only a carefully selected fraction of the treatment combinations (the black squares in Figure 23.6*c*) are run. Therefore, in order to attribute the $A$ row in this particular design's ANOVA to the $A$ main effect measurement, we must assume that the $BE$, $BCD$, and $ACDE$ effects are 0, or not statistically significant. Otherwise, we are at a loss to interpret the results. In a $2^f$ FFE, we can develop our aliases directly. We will demonstrate this procedure through an example.

## Example 23.4

Given the block layouts for the $\frac{1}{2}$ FFE and the $\frac{1}{4}$ FFE of the $2^5$ experiment in Example 23.3, develop the alias structure.

## Soution

In order to generate the alias structure, we will systematically proceed from the main effects to the two-factor interaction. In the $2^f$ case, we drop the squared terms as we multiply each through the confounded effects. For the $\frac{1}{2}$ FFE we will have 16 observations and 15 df:

| $L = ABCDE$ | | Aliases | df |
|---|---|---|---|
| $A(ABCDE)$ | $= A^2BCDE$ | $A, BCDE$ | 1 |
| $B(ABCDE)$ | $= AB^2CDE$ | $B, ACDE$ | 1 |
| $C(ABCDE)$ | $= ABC^2DE$ | $C, ABDE$ | 1 |
| $D(ABCDE)$ | $= ABCD^2E$ | $D, ABCE$ | 1 |
| $E(ABCDE)$ | $= ABCDE^2$ | $E, ABCD$ | 1 |
| $AB(ABCDE)$ | $= A^2B^2CDE$ | $AB, CDE$ | 1 |
| $AC(ABCDE)$ | $= A^2BC^2DE$ | $AC, BDE$ | 1 |
| $AD(ABCDE)$ | $= A^2BCD^2E$ | $AD, BCE$ | 1 |
| $AE(ABCDE)$ | $= A^2BCDE^2$ | $AE, BCD$ | 1 |
| $BC(ABCDE)$ | $= AB^2C^2DE$ | $BC, ADE$ | 1 |
| $BD(ABCDE)$ | $= AB^2CD^2E$ | $BD, ACE$ | 1 |
| $BE(ABCDE)$ | $= AB^2CDE^2$ | $BE, ACD$ | 1 |
| $CD(ABCDE)$ | $= ABC^2D^2E$ | $CD, ABE$ | 1 |
| $CE(ABCDE)$ | $= ABC^2DE^2$ | $CE, ABD$ | 1 |
| $DE(ABCDE)$ | $= ABCD^2E^2$ | $DE, ABC$ | 1 |

For the $\frac{1}{4}$ FFE we will have eight observations and 7df:

| $L_1 = ABE; L_2 = CDE; L_1{}^*L_2 = ABCD$ | Aliases | df |
|---|---|---|
| $A \quad = A^2BE = ACDE = A^2BCD$ | $A, BE, ACDE, BCD$ | 1 |
| $B \quad = AB^2E = BCDE = AB^2CD$ | $B, AE, BCDE, ACD$ | 1 |
| $C \quad = ABEC = C^2DE = ABC^2D$ | $C, ABCE, DE, ABD$ | 1 |
| $D \quad = ABDE = CD^2E = ABCD^2$ | $D, ABDE, CE, ABC$ | 1 |
| $E \quad = ABE^2 = CDE^2 = ABCDE$ | $E, AB, CD, ABCDE$ | 1 |
| $AC = A^2BCE = AC^2DE = A^2BC^2D$ | $AC, BCE, ADE, BD$ | 1 |
| $BC = AB^2CE = BC^2DE = AB^2C^2D$ | $BC, ACE, BDE, AD$ | 1 |

A few comments are in order:

1  The confounded effects—*ABE*, *CDE*, and *ABCD*—do not appear in the alias structure; they are lost when confounded with the blocks.
2  Once we select the effects we wish to confound, the alias structure is set. If we do not like the alias structure, we can change it only by changing the fraction (e.g., going from one fourth up to one half) or by changing the confounding effects and starting over (e.g., next time selecting $L_1 = ABD$, $L_2 = ACE$, and $L_1{}^* L_2 = BCDE$).
3  If we choose an effect (the left-hand side in our layout above) that is already in an alias row, we will merely duplicate an existing row (e.g., if we were to use *AB* instead of *AC* for row 6, in the $\frac{1}{4}$ FFE case above, we would merely duplicate row 5). As we complete our alias structure, we typically check preceeding rows to prevent choosing an effect already in the generated rows (e.g., *AC* was picked for row 6 since it did not appear above, likewise for *BC*).

## Resolution

*FFEs are classified numerically by their resolution. Resolution refers to the amount of information that can be obtained from an experiment.* Resolution numbers indicate the degree of resolution possible and, for a given FFE, can be determined relative to the technical confounding-alias structure. In general, an experimental design is said to be of resolution *N* if no *n*-factor effect is aliased with another effect containing less than $N - n$ factors (Box, Hunter, and Hunter [9]). Typically, Roman numerals are used as resolution numbers. For our purpose, resolution III, IV, and V classifications are of major interest.

In a *resolution III design* we see

1  No main effects aliased with other main effects.
2  Main effects aliased with two-factor interactions.
3  Two-factor interactions aliased with other two-factor interactions.

In a *resolution IV design* we see

1  No main effects aliased with other main effects.
2  No main effects aliased with two-factor interactions.
3  Two-factor interactions aliased with other two-factor interactions.

In a *resolution V design* we see

**1** No main effects aliased with other main effects.
**2** No main effects aliased with two-factor interactions.
**3** No two-factor interactions aliased with other two-factor interactions.

From our previous development (Example 23.4), we can classify the $\frac{1}{2}$ FFE of the $2^5$ FAT, where $L = ABCDE$, as a resolution V. The $\frac{1}{4}$ FFE of the $2^5$ FAT, where $L_1 = ABE$, $L_2 = CDE$, and $L_1*L_2 = ABCD$, is classified as a resolution III. *In general, if we suspect that we will see significant two-factor interactions, we should use a resolution V design (or, of course, a full factorial).* The resolution IV and III designs both present biased model estimates in the presence of two-factor interaction (see Hunter for details [10]).

It is important to note that we must take care to develop FFEs of the highest resolution possible. We do this by carefully selecting the confounding scheme. Table 23.7, adapted from Montgomery [11], summarizes a number of $2^f$ FFE possibilities which result in the highest possible resolution. We should point out that, although the tabled confounding schemes result in the highest possible resolution, they are not necessarily the only such schemes available. For example, the $\frac{1}{4}$ FFE with $L_1 = ABE$, $L_2 = CDE$, and $L_1*L_2 = ABCD$ design in Examples 23.3 and 23.4 produced a resolution III, $\frac{1}{4}$ FFE from a $2^5$, whereas the $2^5$ design of a resolution III, $\frac{1}{4}$ FFE in Table 23.7 calls for $L_1 = ABD$, $L_2 = ACE$, and $L_1*L_2 = BCDE$.

**Plots**

As in all previous designs, plots are indispensable in FFE analysis and communications of FFE findings. *We can develop plots from fractional factorial experimental results in exactly the same way we do in full factorial designs.* Plots for main effects and interactions are developed from factor level stratification of the data and arranged in the usual manner. However, we cannot expect to plot all interaction effects because we do not have observations for all treatment factor combinations. In the low-resolution FFEs (such as the resolution III, above), we can usually develop main effect and two-factor interaction plots, but three-factor interaction plots are typically infeasible. We will develop a series of plots in Example 23.5 and discuss the features of plot interpretation that are unique to FFEs.

**Example 23.5**

For the $\frac{1}{4}$ FFE designed in Example 23.3 and displayed in Figure 23.6c, the data listed in Table 23.8 were collected. Stratify and summarize the data, produce a sequence of main effects and two-factor interaction means. Then, develop the main effect plots and the $AB$, $AC$, and $BC$ two-factor interaction plots.

**TABLE 23.7  SELECTED $2^f$ FFEs WITH HIGHEST POSSIBLE RESOLUTION**

| Complete factorial | | Fractional factorial | | | Design specifications | |
|---|---|---|---|---|---|---|
| Notation $2^f$ | Number of runs | Notation for FFE | Fraction of FFE | Number of runs | Resolution of FFE | Design generators $L$ |
| $2^3$ | 8 | $2_{III}^{3-1}$ | $\frac{1}{2}$ | $2^{3-1}=4$ | III | $L = ABC$ |
| $2^4$ | 16 | $2_{IV}^{4-1}$ | $\frac{1}{2}$ | $2^{4-1}=8$ | IV | $L = ABCD$ |
| $2^5$ | 32 | $2_{V}^{5-1}$ | $\frac{1}{2}$ | $2^{5-1}=16$ | V | $L = ABCDE$ |
| $2^5$ | 32 | $2_{III}^{5-2}$ | $\frac{1}{4}$ | 8 | III | $L_1 = ABD$ |
| | | | | | | $L_2 = ACE$ |
| $2^6$ | 64 | $2_{VI}^{6-1}$ | $\frac{1}{2}$ | 32 | VI | $L = ABCDEF$ |
| $2^6$ | 64 | $2_{IV}^{6-2}$ | $\frac{1}{4}$ | 16 | IV | $L_1 = ABCE$ |
| | | | | | | $L_2 = BCDF$ |
| $2^6$ | 64 | $2_{III}^{6-3}$ | $\frac{1}{8}$ | 8 | III | $L_1 = ABD$ |
| | | | | | | $L_2 = ACE$ |
| | | | | | | $L_3 = BCF$ |
| $2^7$ | 128 | $2_{VII}^{7-1}$ | $\frac{1}{2}$ | 64 | VII | $L = ABCDEFG$ |
| $2^7$ | 128 | $2_{IV}^{7-2}$ | $\frac{1}{4}$ | 32 | IV | $L_1 = ABCDF$ |
| | | | | | | $L_2 = ABDEG$ |
| $2^7$ | 128 | $2_{IV}^{7-3}$ | $\frac{1}{8}$ | 16 | IV | $L_1 = ABCE$ |
| | | | | | | $L_2 = BCDF$ |
| | | | | | | $L_3 = ACDG$ |
| $2^7$ | 128 | $2_{III}^{7-4}$ | $\frac{1}{16}$ | 8 | III | $L_1 = ABD$ |
| | | | | | | $L_2 = ACE$ |
| | | | | | | $L_3 = BCF$ |
| | | | | | | $L_4 = ABCG$ |
| $2^8$ | 256 | $2_{V}^{8-2}$ | $\frac{1}{4}$ | 64 | V | $L_1 = ABCDG$ |
| | | | | | | $L_2 = ABEFH$ |
| $2^8$ | 256 | $2_{V}^{8-3}$ | $\frac{1}{8}$ | 32 | V | $L_1 = ABCF$ |
| | | | | | | $L_2 = ABDG$ |
| | | | | | | $L_3 = BCDEH$ |
| $2^8$ | 256 | $2_{IV}^{8-4}$ | $\frac{1}{16}$ | 16 | IV | $L_1 = BCDE$ |
| | | | | | | $L_2 = ACDF$ |
| | | | | | | $L_3 = ABCG$ |
| | | | | | | $L_4 = ABDH$ |
| $2^9$ | 512 | $2_{VI}^{9-2}$ | $\frac{1}{4}$ | 128 | VI | $L_1 = ACDFGH$ |
| | | | | | | $L_2 = BCEFGJ$ |
| $2^9$ | 512 | $2_{IV}^{9-3}$ | $\frac{1}{8}$ | 64 | IV | $L_1 = ABCDG$ |
| | | | | | | $L_2 = ACEFH$ |
| | | | | | | $L_3 = CDEFJ$ |
| $2^9$ | 512 | $2_{IV}^{9-4}$ | $\frac{1}{16}$ | 32 | IV | $L_1 = BCDEF$ |
| | | | | | | $L_2 = ACDEG$ |
| | | | | | | $L_3 = ABDE$ |
| | | | | | | $L_4 = ABCEJ$ |
| $2^9$ | 512 | $2_{III}^{9-5}$ | $\frac{1}{32}$ | 16 | III | $L_1 = ABCE$ |
| | | | | | | $L_2 = BCDF$ |
| | | | | | | $L_3 = ACDG$ |
| | | | | | | $L_4 = ABDH$ |
| | | | | | | $L_5 = ABCDJ$ |

*Source:* Adapted, with permission from John Wiley and Sons, Inc., from D.C. Montgomery, *Design and Analysis of Experiments*, 3d ed., New York: John Wiley, pages 358 and 359, 1991. Copyright John Wiley and Sons, Inc.

**TABLE 23.8   RESULTS FROM THE $\frac{1}{4}$ FFE, $2^5$ COOKIE EXPERIMENT (EXAMPLE 23.5)**

| Exp. trial | Main effects factors | | | | | Treatment combination | Response (% elongation) |
|---|---|---|---|---|---|---|---|
| | A | B | E | C | D | | |
| 1 | − | − | − | − | − | (1) | 5 |
| 2 | − | − | − | + | + | cd | 10 |
| 3 | − | + | + | − | + | bde | 31 |
| 4 | − | + | + | + | − | bce | 35 |
| 5 | + | − | + | − | + | ade | 7 |
| 6 | + | − | + | + | − | ace | 13 |
| 7 | + | + | − | − | − | ab | 12 |
| 8 | + | + | − | + | + | abcd | 24 |

## Solution

The stratified data layouts appear in Table 23.9. The plots appear in Figure 23.7. A superficial analysis of the main effect plots suggests that main effects $A$, $B$, $C$, and $E$ are active, whereas main effect $D$ is relatively inactive. The $AB$ interaction plot seems to suggest a possible interaction. The $AC$ and $BC$ plots suggest very limited interaction.

**TABLE 23.9   MAIN EFFECTS AND $AB$, $AC$, AND $BC$ INTERACTION TABLES (EXAMPLE 23.5)**

**a  Main effects data summary**

| Factor | Level | Total | Mean | Number of observations | Notation |
|---|---|---|---|---|---|
| A | − | 81 | 20.25 | 4 | $Y_{i...}$ |
| | + | 56 | 14.00 | 4 | |
| B | − | 35 | 8.75 | 4 | $Y_{.j..}$ |
| | + | 102 | 25.50 | 4 | |
| C | − | 55 | 13.75 | 4 | $Y_{..k.}$ |
| | + | 82 | 20.50 | 4 | |
| D | − | 65 | 16.25 | 4 | $Y_{...l}$ |
| | + | 72 | 18.00 | 4 | |
| E | − | 51 | 12.75 | 4 | $Y_{....m}$ |
| | + | 86 | 21.50 | 4 | |
| | $Y_{.....} = 137$ | | $\bar{Y}_{.....} = 17.125\%$ | | |

A:  Cooking temperature, $i = 1, 2$
B:  Syrup content, $j = 1, 2$
C:  Cooking time, $k = 1, 2$
D:  Cooking pan type, $l = 1, 2$
E:  Shortening type, $m = 1, 2$
Responses are in percent elongation

**b  $AB$, $AC$, $BC$ data summaries, respectively**

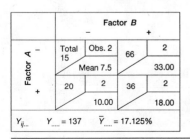

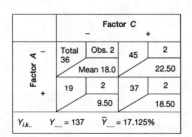

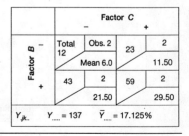

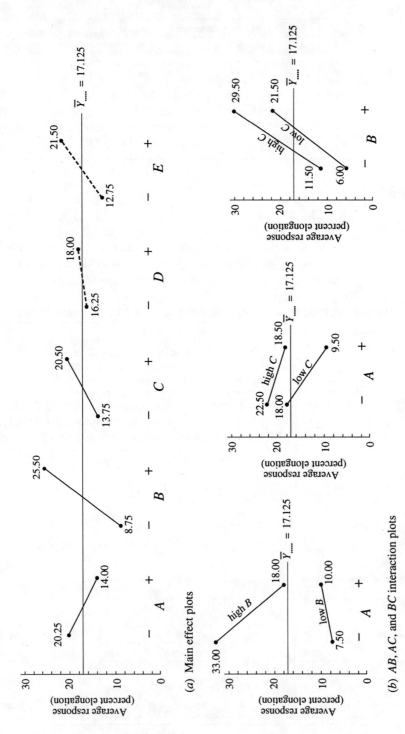

(a) Main effect plots

(b) AB, AC, and BC interaction plots

**FIGURE 23.7** Main effects and *AB*, *AC*, and *BC* interaction plots from $\frac{1}{4}$ FFE (Example 23.5).

594

On the surface, the observations seem very neat and clear. We developed a $\frac{1}{4}$ FFE, ran only 8 of the 32 treatment combinations and came out with a nice set of graphs. However, we must carefully examine our alias structure. For example, we see $E$ aliased with $AB$. We cannot distinguish the $E$ and $AB$ effects from each other. At this point, we must resort to a physical argument as to syrup-temperature interaction or shortening effects. In reality, we might have detected one or the other, or both. We simply cannot tell or distinguish which one, based solely on our experimental results. Hence, our analysis must be reconciled with the alias structure. Anything less may be misleading, unless we decide to extend our experiment, (e.g., perhaps by running the other 24 treatment combinations to complete the full factorial).

## Sum of Squares

If we use our previous procedures, we can calculate the sums of squares relative to our *aliases* and see that they *are identical in value and indistinguishable statistically.* Hence, *the statistical gridlock caused by alias confusion must be broken using physical arguments and/or additional experimentation, such that we can isolate and measure the extent of these aliased effects.* In Example 23.6, using limited hand calculations and our computer aid, SAS PROC ANOVA, we will demonstrate the alias sums of squares and the dilemma they present.

## Example 23.6

Using the data from Example 23.5, shown in Tables 23.8 and 23.9,

**a** Develop the $A$, $B$, $C$, $D$, $E$, $AC$, $BC$, and $AB$ sums of squares by hand.

**b** Use the SAS computer aid to develop a full set of sums of squares for the seven effects $A$, $B$, $C$, $D$, $E$, $AC$, and $BC$.

**c** Finally, rerun the data to develop a "collapsed" full FAT in $A$ and $B$ with two replications (i.e., ignore the $C$, $D$, and $E$ effects).

### Solution

**a** From Tables 23.8 and 23.9 and our previous FAT ANOVA forms,

$$SS_{tc} = (5^2 + 10^2 + 31^2 + 35^2 + 7^2 + 13^2 + 12^2 + 24^2) - (137^2/8) = 902.875$$

$$SS_A = (81^2 + 56^2)/4 - (137^2/8) = 78.125$$

$$SS_B = (35^2 + 102^2)/4 - (137^2/8) = 561.125$$

$$SS_C = (55^2 + 82^2)/4 - (137^2/8) = 91.125$$

$$SS_D = (65^2 + 72^2)/4 - (137^2/8) = 6.125$$

$$SS_E = (51^2 + 86^2)/4 - (137^2/8) = 153.125$$

$$SS_{AC} = (36^2 + 19^2 + 45^2 + 37^2)/2 - (137^2/8) - SS_A - SS_C = 10.125$$

$$SS_{BC} = (12^2 + 43^2 + 23^2 + 59^2)/2 - (137^2/8) - SS_B - SS_C = 3.125$$

$$SS_{AB} = (15^2 + 20^2 + 66^2 + 36^2)/2 - (137^2/8) - SS_A - SS_B = 153.125$$

Here, we can see that $SS_E = SS_{AB}$ as a result of the alias structure. More rigorous derivations dealing with factor effect aliases are shown in Hicks; Box, Hunter, and Hunter; and Montgomery [12].

**b** Table 23.10$a$ displays a summary of our computer-aided ANOVA results. The SAS code and the resulting output are shown in the Computer Aided Analysis Supplement, Solutions Manual.

**c** If we were to run this $\frac{1}{4}$ FFE and only factors $A$ and $B$ were considered significant, we could collapse our $\frac{1}{4}$ FFE in five factors (eight observations) into a two factor FAT with two replications. This collapse is not warranted by our resulting plots and analysis. Nevertheless, a summarized computer aided ANOVA is shown

**TABLE 23.10** SUMMARIZED COMPUTER AIDED ANOVA TABLES, SAS PROC ANOVA (EXAMPLE 23.6)

**a** Five-factor $\frac{1}{4}$ FFE

Dependent Variable: RESP ELONGATION PERCENT

| Source | DF | Sum of Squares | Mean Square | F Value | Pr > F |
|---|---|---|---|---|---|
| Model | 7 | 902.87500000 | 128.98214286 | . | . |
| Error | 0 | . | . | | |
| Corrected Total | 7 | 902.87500000 | | | |

| | R-Square | C.V. | Root MSE | RESP Mean |
|---|---|---|---|---|
| | 1.000000 | 0 | 0 | 17.12500000 |

| Source | DF | Anova SS | Mean Square | F Value | Pr > F |
|---|---|---|---|---|---|
| A | 1 | 78.12500000 | 78.12500000 | . | . |
| B | 1 | 561.12500000 | 561.12500000 | . | . |
| C | 1 | 91.12500000 | 91.12500000 | . | . |
| D | 1 | 6.12500000 | 6.12500000 | . | . |
| E | 1 | 153.12500000 | 153.12500000 | . | . |
| A*C | 1 | 10.12500000 | 10.12500000 | . | . |
| B*C | 1 | 3.12500000 | 3.12500000 | . | . |

**b** Five-factor $\frac{1}{4}$ FFE reduced to factors $A$ and $B$ only

Dependent Variable: RESP ELONGATION PERCENT

| Source | DF | Sum of Squares | Mean Square | F Value | Pr>F |
|---|---|---|---|---|---|
| Model | 3 | 792.37500000 | 264.12500000 | 9.56 | 0.0269 |
| Error | 4 | 110.50000000 | 27.62500000 | | |
| Corrected Total | 7 | 902.87500000 | | | |

| | R-Square | C.V. | Root MSE | RESP Mean |
|---|---|---|---|---|
| | 0.877613 | 30.69167 | 5.25594901 | 17.12500000 |

| Source | DF | Anova SS | Mean Square | F Value | Pr > F |
|---|---|---|---|---|---|
| A | 1 | 78.12500000 | 78.12500000 | 2.83 | 0.1679 |
| B | 1 | 561.12500000 | 561.12500000 | 20.31 | 0.0108 |
| A*B | 1 | 153.12500000 | 153.12500000 | 5.54 | 0.0781 |

in Table 23.10$b$. We should notice that $SS_{tc}$, $SS_A$, $SS_B$, and $SS_{AB} = SS_E$ from before remain the same. Now, with only factors $A$ and $B$, we would have an actual estimate of experimental error (e.g., the $C$, $D$, $AC$, and $BC$ rows are essentially pooled).

## Error Pooling and Factor Effects Plotting

*The error pooling and factor effects plotting techniques* introduced in Chapter 21 *are directly applicable in FFEs.* We will demonstrate both techniques through an example format.

### Example 23.7

For the $\frac{1}{4}$ FFE results shown in Table 23.8 (Example 23.5),

**a** Use the ANOVA (Table 23.10) from Example 23.6 and the plots in Figure 23.7 to subjectively develop a pseudo experimental error term. Then pool up to develop an experimental error term; use $\alpha = 0.05$.

**b** Develop a set of main and interaction effects and construct a normal effects plot.

### Solution

**a** From an examination of the mean squares in the ANOVA in Table 23.10$a$ and, to some degree, of the plots, we can identify three general groups of effects: (1) effect $B$, very pronounced; (2) effects $A$, $C$, and $E$, substantial; and (3) $D$, $AC$, and $BC$, rather small in comparison to the others. Based on these groups, it appears reasonable (subjectively) to pool the third group to form a pseudo error term:

$$MS_{pooled\ err} = \frac{df_D\ (MS_D) + df_{AC}\ (MS_{AC}) + df_{BC}\ (MS_{BC})}{df_D + df_{AC} + df_{BC}}$$

$$= \frac{6.125 + 10.125 + 3.125}{3}$$

$$MS_{pooled\ err} = 6.458$$

Just as in Chapter 21, we can see that the pooled error term is simply a weighted average of the mean squares, weighted by the degrees of freedom of each mean square.

When we pool up, we select our two smallest mean squares, and test.

$$F_{pooled\ err\ 1} = \frac{MS_{next\ to\ smallest}}{MS_{smallest}} = \frac{MS_{BC}}{MS_D} = \frac{6.125}{3.125} = 1.96$$

From Table IX.4,

$$F_{0.05,1,1} = 161.4$$

Here, $F_{\text{pooled err 1}}$ is not in the critical region, therefore we will pool $MS_{BC}$ and $MS_D$:

$$MS_{\text{pooled err 1}} = \frac{df_D (MS_D) + df_{BC} (MS_{BC})}{df_D + df_{BC}} = \frac{3.125 + 6.125}{2} = 4.625$$

Now, we move to the next comparison:

$$F_{\text{pooled err 2}} = \frac{MS_{AC}}{MS_{\text{pooled err 1}}} = \frac{10.125}{4.625} = 2.189$$

$$F_{0.05,1,2} = 18.51$$

Here $F_{\text{pooled err 2}}$ is not in the critical region, hence we pool $MS_{\text{pooled err 1}}$ and $MS_{AC}$:

$$MS_{\text{pooled err 2}} = \frac{df_{\text{pooled err 1}} (MS_{\text{pooled err 1}}) + df_{AC} (MS_{AC})}{df_{\text{pooled err 1}} + df_{AC}}$$

$$= \frac{2(4.625) + 10.125}{2 + 1}$$

$$= 6.458$$

Now, we proceed to the next pooling test:

$$F_{\text{pooled err 3}} = \frac{MS_A}{MS_{\text{pooled err 2}}} = \frac{78.125}{6.458} = 12.10$$

$$F_{0.05,1,3} = 10.13$$

Here, we have encountered a significant test, based on our $\alpha = 0.05$ significance level, and we stop our pooling process with

$$MS_{\text{pooled err}} = MS_{\text{pooled err 2}} = 6.458$$

Table 23.11 displays our resulting ANOVA table after this pooling process. Our P-values were taken from the SAS output in the Computer Aided Analysis Supplement, Solutions Manual.

TABLE 23.11   POOLED ANOVA FOR $\frac{1}{4}$ FFE $2^5$ EXPERIMENT (EXAMPLE 23.7)

| Source | df | SS | MS | F | P-value* |
|---|---|---|---|---|---|
| Total corrected | 7 | 902.875 | | | |
| A | 1 | 78.125 | 78.125 | 12.10 | 0.0401 |
| B | 1 | 561.125 | 561.125 | 86.88 | 0.0026 |
| C | 1 | 91.125 | 91.125 | 14.11 | 0.0330 |
| E | 1 | 153.125 | 153.125 | 23.71 | 0.0165 |
| Error pooled | 3 | 19.375 | 6.458 | | |
| AC | 1 | 10.125 | | | |
| D | 1 | 6.125 | | | |
| BC | 1 | 3.125 | | | |

*Taken from SAS output, Computer Aided Analysis Supplement.

**TABLE 23.12** FACTOR EFFECTS VALUE TABULATION FOR $\frac{1}{4}$ FFE $2^5$ EXPERIMENT (EXAMPLE 23.7)

| Exp. trial | Treatment combination | Main effects | | | | | Interaction effects | | Response |
|---|---|---|---|---|---|---|---|---|---|
| | | A | B | E | C | D | AC | BC | |
| 1 | (1) | − (5) | − | − | − | − | + | + | 5 |
| 2 | cd | − (10) | − | − | + | + | − | − | 10 |
| 3 | bde | − (31) | + | + | − | + | + | − | 31 |
| 4 | bce | − (35) | + | + | + | − | − | + | 35 |
| 5 | ade | + (7) | − | + | − | + | − | + | 7 |
| 6 | ace | + (13) | − | + | + | − | + | − | 13 |
| 7 | ab | + (12) | + | − | − | − | − | − | 12 |
| 8 | abcd | + (24) | + | − | + | + | + | + | 24 |
| Effect sums (column x response total) | | −25 | 67 | 35 | 27 | 7 | 9 | 5 | |
| Effect value (sums divided by 4) | | −6.25 | 16.75 | 8.75 | 6.75 | 1.75 | 2.25 | 1.25 | $\overline{Y}..... = 17.125$ |

**b** Using the same general procedure as we used in Chapter 21, we will now develop our alternative analysis and prepare a normal plot of the effects. Table 23.12 contains our factor effects results. Again, we obtained our effect values by multiplying each response by the + or − in the effect column and summing down each column. Then, we divide by the number of + signs, 4 in this case. Table 23.13 contains our plotting position calculations for each ranked effect value.

The results are plotted in Figure 23.8. This type of plot provides a challenge in that we do not have a large number of points and those that we have are somewhat scattered. Our model line appears to pass near E, C, D, AC, and BC (or maybe right through AC, D, BC). Hence, we see an indication that effects B and, to some extent, A are outliers and can be declared significant. Here, just as in our pooling exercise, our skills of interpretation are tested.

## Standard Error Interval Arrows

In order to enhance our plots, we will once again use our ± b s.e.i. arrows. We can use $b = 2$ or substitute $b = t_{\alpha/2, \, \text{df pooled err}}$ to obtain a $(1 − \alpha)$ 100 percent confidence interval on the mean. For example,

$$\pm b \text{ s.e.i.} = \pm b \sqrt{\frac{MS_{\text{pooled err}}}{n'}} \qquad (23.3)$$

where $n'$ is the number of observations in the plotted mean.

| Effect Rank $k$ | Ranked effect value | $\hat{F}(y) = (k − 0.5) /$ number of effects |
|---|---|---|
| 1 | A = −6.25 | (1−0.5)/7 = 7.1% |
| 2 | BC = 1.25 | 1.5/7 = 21.4% |
| 3 | D = 1.75 | 2.5/7 = 35.7% |
| 4 | AC = 2.25 | 3.5/7 = 50.0% |
| 5 | C = 6.75 | 4.5/7 = 64.3% |
| 6 | E = 8.75 | 5.5/7 = 78.6% |
| 7 | B = 16.75 | 6.5/7 = 92.9% |

**TABLE 23.13** RANKED EFFECTS AND THEIR MEAN PLOTTING POSITIONS (EXAMPLE 23.7)

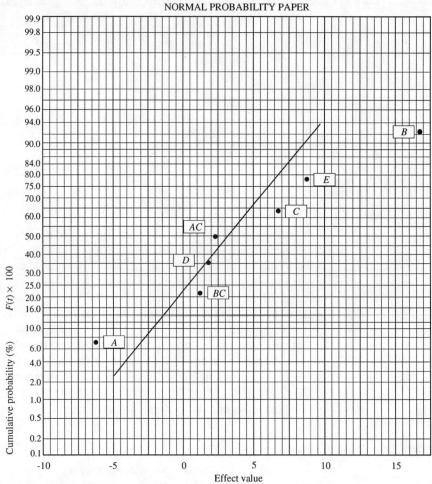

**FIGURE 23.8**     Effects plot, normal probability plotting paper (Example 23.7).

---

## Example 23.8

Enhance the plots in Figure 23.7 with s.e.i. arrows to reflect dispersion so that we can develop a more sophisticated graphical analysis.

### Solution

We will develop 95 percent confidence interval arrows ($b = t_{0.025,3} = 3.182$) for the main effects and the interaction effects displayed in Figure 23.7. We will use our $MS_{pooled\ err}$ from Example 23.7 ($MS_{pooled\ err} = 6.458$) and place the appropriate arrows on our plots.

*Main effect plots*:

$$\pm 3.182 \text{ s.e.i.} = \pm 3.182 \sqrt{\frac{MS_{\text{pooled err}}}{n'}} = \pm 3.182 \sqrt{\frac{6.458}{4}} = \pm 4.04\%$$

*Two-factor interaction plots*:

$$\pm 3.182 \text{ s.e.i.} = \pm 3.182 \sqrt{\frac{6.458}{2}} = \pm 5.72\%$$

The results are plotted in Figure 23.9.

We should note that the addition of the arrows is a major enhancement of the plots in Figure 23.7. In Figure 23.7, we can judge relative size of effects, but not experimental error. This point is critical because, although we can see what appear to be large effects, a large experimental error may lead us to falsely interpret noise disturbances as actual factor effects. Product-process decisions made on the basis of such analyses, with little or no regard to variation as measured by experimental error, will create confusion and frustration at the design stage or on the production floor. In other words the results we see in our experimental plots will not, in general, be consistent in actual operations.

**FIGURE 23.9** Main effects and *AB*, *AC*, *BC* interaction plots with s.e.i. arrows (Example 23.8).

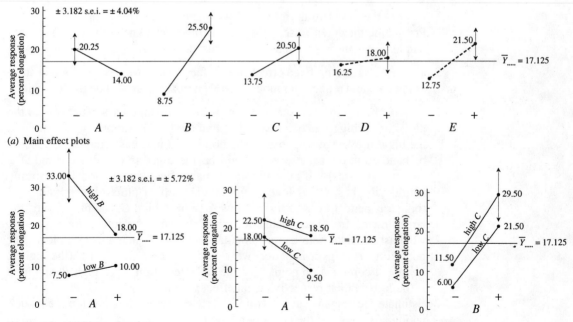

(a) Main effect plots

(b) *AB*, *AC*, and *BC* interaction plots

### Response Estimates

*We can develop estimates for factor level responses we do not examine in the FFE using a response model.* Our plots and ANOVA help us identify the significant effects. Sometimes, we deal only with main effects and two-factor interactions in our response model. A response model containing all main effects and two-factor interaction terms for a $2^5$ FAT-CRD is

$$Y_{ijklm} = \mu + A_i + B_j + C_k + D_l + E_m + AB_{ij} + AC_{ik} + \cdots + DE_{lm} + \varepsilon \qquad (23.4)$$

An examination of the evidence in our plots and ANOVA enables us to identify the terms which significantly affect the response. The results from Example 23.7 suggest one possibility.

*Possibility I— main effects model in A, B, C, and E:*

$$Y_{ijklm} = \mu + A_i + B_j + C_k + E_m + \varepsilon \qquad (23.5)$$

Strictly for illustrative purposes (our data clearly do not warrant this possibility), we will consider a second possibility,

*Possibility II—main effects model with the addition of AC interaction:*

$$Y_{ijklm} = \mu + A_i + B_j + C_k + E_m + AC_{ik} + \varepsilon \qquad (23.6)$$

At this point, we typically want to

1 Identify the best setting from those included in the experiment.
2 Predict the mean response at the best setting or at some other setting, for example, if we are in production, at the current setting.

Since we are assuming fixed effects for all factors and we have a plot that summarizes our data and reflects our model (see Figure 23.9), we will do the following:

1 Pick the best settings (usually bigger is better or smaller is better) off of the plot; e.g., if bigger is better we must find the factor combinations that yield the biggest average response. (The nominal is best case cannot, in general, be handled in the same way and will be addressed in our Review and Discovery Exercises). If all we have are main effects in the model, we simply identify the best factor level from the main effect plots. For possibility I (bigger is better) we would select low $A$, high $B$, high $C$, and high $E$ (i.e., the largest means from each main effect plot). If two-factor interaction is present, we must consider both factors simultaneously and use interaction plots. For possibility II (bigger is better) we would select low $A$, high $C$ (the highest treatment combination point from the $AC$ interaction plot), high $B$, and high $E$ (both from their respective main effect plots).
2 Estimate the response at the setting selected. Here we redevelop the model, taking advantage of the equality relationship of the model effect (e.g., $A_i$) to

the effect's mean, less the overall mean (e.g., $\overline{Y}_{1...} - \overline{Y}_{....}$). We obtain our numbers from the plots in Figure 23.9. For our possibility I example,

$$\hat{\mu}_{best} = \hat{\mu}_{A_1 B_2 C_2 E_2} = \hat{\mu} + \hat{A}_1 + \hat{B}_2 + \hat{C}_2 + \hat{E}_2$$

$$= \overline{Y}_{....} + (\overline{Y}_{1...} - \overline{Y}_{....}) + (\overline{Y}_{.2..} - \overline{Y}_{....}) + (\overline{Y}_{..2.} - \overline{Y}_{....}) + (\overline{Y}_{...2} - \overline{Y}_{....})$$

$$= 17.125 + (20.25 - 17.125) + (25.50 - 17.125) + (20.50 - 17.125)$$
$$+ (21.50 - 17.125)$$

$$= 36.38$$

For possibility II, we must include the *AC* term, since the two pertinent main effects by themselves will not reflect the interaction activity:

$$\hat{\mu}_{best} = \hat{\mu}_{AC_{12} B_2 E_2} = \hat{\mu} + \hat{A}_1 + \hat{B}_2 + \hat{C}_2 + \hat{E}_2 + \hat{AC}_{12}$$

$$\hat{\mu}_{AC_{12} B_2 E_2} = \overline{Y}_{....} + (\overline{Y}_{1...} - \overline{Y}_{....}) + (\overline{Y}_{.2..} - \overline{Y}_{....}) + (\overline{Y}_{..2.} - \overline{Y}_{....}) + (\overline{Y}_{...2} - \overline{Y}_{....})$$
$$+ (\overline{Y}_{1.2.} - \overline{Y}_{1...} - \overline{Y}_{..2.} + \overline{Y}_{....})$$

$$= 17.125 + (20.25 - 17.125) + (25.50 - 17.125)$$
$$+ (20.50 - 17.125) + (21.50 - 17.125) + (22.50 - 20.25$$
$$- 20.50 + 17.125)$$

$$= 17.125 + 3.125 + 8.375 + 3.375 + 4.375 - 1.125$$

$$= 35.25$$

It is critical to observe three facts concerning the preceding analysis:

**1** By definition, in FFEs we do not examine all treatment combinations. However, in this case, we did examine low *A*, high *B*, high *C*, high *E* in observation 4 (at low *D*).

**2** We can use our model across all treatment combinations (both observed and unobserved).

**3** We can see that our model does not include any effect attributable to factor *D*. Factor *D* was not significant, b.o.e., we will choose either low *D* or high *D*, whichever is more economical.

It is likely that we will be developing response estimates for treatment combinations that we did not examine (e.g., in a $\frac{1}{4}$ FFE, there is only a 1 in 4 chance of a given factor combination being selected). ***Unless we have a great deal of confidence in our model, we should make model based decisions with reservation.*** For example, the results from most FFEs, especially those of resolution III or IV, will be helpful in devising follow-on experiments, but will probably be insufficient to support major decisions especially where factor interaction is an issue.

Our discussion of main effects and interaction in Chapter 21 is relevant here. We must remember that main effects measure average responses (averaged over all other factors). With no physical or statistical indication regarding interaction, declaring a factor as having little or no influence on a response solely on the basis of a main effect analysis is irresponsible. In short, from a physical point of view, prematurely and erroneously eliminating a key factor or interaction, in our opinion, is a greater sin than retaining a reasonably inactive factor in our model and including it in more detailed future experiments.

The previous examples developed two-level FFEs. The orthogonal arrays (OAs) discussed later in this chapter offer another alternative to structuring FFEs. We can also structure central composite designs based on two-level FFEs, as shown in Tables 22.5 and 22.6. Additional design guidance is available in a number of sources, including Cochran and Cox [13],and Diamond [14], as well as Montgomery, and Box, Hunter, and Hunter [15].

### Decisions Based on FFE Results

***Working with low resolution FFEs and very small sample sizes always presents the danger of obtaining imprecise (unusually large or small) estimates.*** For example, we may declare some effects significant when they are not or, conversely, declare some effects not significant when they are. Example 21.7, with the eight standard normal response values in the $2^3$ complete factorial (all sampled from the same population $\sim N$, where ($\mu = 0$, and $\sigma^2 = 1$) demonstrates the element of risk we assume when we rely on small sample sizes. The use of fractional factorials tends to increase this risk.

We must realize that any conclusions that we draw from FFEs may be misleading, or inadequate, regardless of the sophistication of our statistical methods. Small sample sizes just do not yield precise estimates. ***FFEs can be effective when used as screening devices in order to produce knowledge that will help us set up a more complete confirmation or follow-up experiment*** (e.g., a higher resolution FFE or a full FAT experiment with fewer variables and perhaps more levels) ***before we totally commit to major product-process changes.*** And as always, we should seek a sensibility check in terms of practical significance before such changes are implemented.

### 23.5  ROBUST PRODUCTS AND PROCESSES

***Products which generate performances that are robust, on target and tolerant to variation, resulting from varying environments or applications, will typically provide a higher level of customer satisfaction than less robust competing products.*** The actual degree of product robustness is determined by customers in the field, but is heavily influenced by product-process design decisions.

### Taguchi Loss Functions and Experimental Methods

***The Taguchi loss functions and signal-to-noise ratios (SNRs)*** [16], ***represent a pragmatic (and controversial) integration of classical statistics with the economic realities of manufacturing***, according to Nair et al.[17]. These tools are generally applied to off-line applications, mainly at the parameter design and tolerance design stages. ***The Taguchi approach to off-line quality improvement is driven by a quadratic loss function concept, whereby a "loss to society" is modeled and optimized.***

***Taguchi methods have received both accolades and criticism.*** Pignatiello and Ramberg published a list of the top 10 triumphs and tragedies associated with Taguchi methods [18], shown in Table 23.14. Although a comprehensive review and discussion of the Taguchi methods is well beyond our scope in this textbook, we will introduce some of the more basic methods in this section so that our readers will understand their structure and intended purposes. The reader seeking more in-depth knowledge is referred to Kacker [19], Box [20], Phadke [21], and Ross [22], in addition to Hunter, Nair, Pignatiello and Ramberg, and Taguchi [23].

**Parameter Design Experiments** ***In general, parameter design is useful to improve quality without removing or controlling the cause of variation.*** In other words, parameter design aims to make the product or process robust to noise or tol-

TABLE 23.14    TAGUCHI TRIUMPHS AND TRAGEDIES

Triumphs

  1  Won the attention of a new audience

  2  Expanded the role of quality beyond that of control

  3  Formulated a complete methodology for quality improvement

  4  Focused attention on the cost associated with variability

  5  Demonstrated that experimentation produces results

  6  Established new directions for quality-engineering research

  7  Attracted a significant level of attention for education in quality engineering

  8  Popularized the concept of robust product design

  9  Pioneered the simultaneous study of both the mean and variability

 10  Simplified tolerance analysis through designed experiments

Tragedies

  1  Spawned a cult of extremists that accept only his teachings

  2  Experienced a backlash of criticism from Western statisticians

  3  Introduced misleading signal-to-noise statistics

  4  Neglected to explain the assumptions underlying his methodology

  5  Ignored modern graphical, data-analytic approaches

  6  Recommended potentially misleading three-level orthogonal arrays

  7  Failed to advocate randomization

  8  Discouraged the adaptive, sequential approach to experimentation

  9  Maintained a dogmatic position on the importance of interactions

 10  Advocated the invalid accumulation of minute analyses

*Source:* Reproduced, with permission, from J. J. Pignatiello and J. S. Ramberg, "Top Ten Triumphs and Tragedies of Genichi Taguchi," *Quality Engineering*, vol. 4, no. 2, pp. 221–225, 1991–1992.

erant of variation resulting from applications and environments, as well as product nonuniformities. For example, we seek to determine the least expensive materials, methods, and so forth, while maintaining adequate performance. In other cases, performance might be increased and costs lowered simultaneously when innovative thinking and technology are applied (as discussed in Chapter 6).

*The Taguchi school of thought brings two conceptual additions to classical experimental design and experimental data analysis:* **(1)** *the systematic introduction of response variation due to noise variables and* **(2)** *The use of the SNR transformation to optimize or minimize, the loss to society, which is expressed by a quadratic loss function* (discussed in Chapter 18). In general, the SNR is built up from quality characteristic performance responses, as well as economic considerations or costs due to both location and variation deviations in the performance response. *The SNR method of parameter design proposed by Taguchi entails an FFE approach to experimental design using orthogonal arrays (OAs) and three types of specialized response transformations—nominal is best, bigger is better, or smaller is better* [24].

Both of the Taguchi innovations above are controversial. Regarding addition 1, injecting variation into an experiment runs counter to the concept of designed experiments, where tight control is generally advocated to shrink experimental error and thus isolate the smallest factor effects possible. Nevertheless, by injecting variation, Taguchi seeks to test or experiment under conditions expected in the field environment.

This technique has both advantages and disadvantages. Statistically, it has disadvantages in that variation tends to collect in our Experimental error row and enlarge our $MS_{err}$ or $MS_{pooled\ err}$ term. This enlargement in turn prevents us from isolating effects in an efficient manner (i.e., we can always increase our sample size, but at increased cost). On the positive side, we observe our product or process in a pseudo field environment, rather than a sterile laboratory environment, and if we are perceptive, we may well detect shortcomings in our product or process that we can address proactively, to some extent. Hence, we may be encouraged to seek and find improvements we otherwise might not have pursued. In other words, we may thus recognize and root out bottlenecks in our product-process designs faster than our competitors, and produce more robust (variation tolerant) products for our customers.

Addition 2, the use of SNRs, is very controversial. The logic used to develop a loss function and associate it with a quality characteristic in a quantitative fashion is a major contribution. However, the specific nature of the loss function (quadratic) and other inherent assumptions, as described by Phadke [25], tend to limit the direct applicability of these techniques.

*Taguchi introduces three categories of SNRs, wherein each η transformation is expressed in decibel (dB) units, rather than the observed units.*

*Smaller is better:*

$$\eta = -10 \log_{10} \left( \frac{1}{r} \sum_{i=1}^{r} Y_i^2 \right) \tag{23.7}$$

*Bigger is better:*

$$\eta = -10 \log_{10} \left( \frac{1}{r} \sum_{i=1}^{r} \frac{1}{Y_i^2} \right) \tag{23.8}$$

*Nominal is best:*

$$\eta = 10 \log_{10} \left( \frac{\bar{Y}^2}{s^2} \right) \tag{23.9}$$

and

$$\eta = -10 \log_{10} \left[ \frac{1}{r-1} \sum_{i=1}^{r} (Y_i - \bar{Y})^2 \right] \tag{23.10}$$

where  $\eta$ = SNR transformation, measured in dB; $\eta$ is always judged by a bigger
is better criterion
   $r$ = number of outer array observation combinations used for each inner
array combination
   $Y_i$ = $i$th observed response at a given inner array combination
   $\bar{Y}$ = average of observed responses for each inner array combination
   $s^2$ = sample variance of observed responses for each inner array combination

and where    $\bar{Y} = \frac{1}{r} \sum_{i=1}^{r} Y_i$    and    $s^2 = \frac{1}{r-1} \sum_{i=1}^{r} (Y_i - \bar{Y})^2$

*By maximizing the SNRs, the quadratic loss functions,* as discussed in Chapter 18,
*are minimized, without the need to develop a loss constant k* (see Taguchi [26] and
Phadke [27] for a detailed development).

   *Inadequacies in these transformations,* especially the mean-to-variance ratio in
Equation 23.9, *have been pointed out by many authors,* including Hunter [28], as
well as Montgomery, Kacker, and Box [29]. On the other hand, the SNR transformations have among their advocates Taguchi, Phadke, and Ross [30]. In general,
*the critics point out that by optimizing the SNR, we may not be optimizing the
product or process in a physical sense.* This criticism is very apparent in Equation
23.9 for the nominal is best SNR. Here, we see both a location and a dispersion
measure. Optimization of the SNR might lead to suboptimal values of the constituent parts. Many authorities advocate the Taguchi methods as universally
efficient and effective. This claim has been discredited in the published statistical
literature (see Pignatiello and Ramberg for a general discussion [31]).

   *The usual procedure in a Taguchi robust design experiment is to utilize OAs for
the inner design factors and the outer noise factors.* These arrays are typically displayed along with linear graphs and triangular tables, which aid in designing the
experiment (see Section IX, Table IX.18, for selected OAs). These experimental
designs or layouts are sometimes referred to as "matrix experiments." The plan
consists of assigning factors to array columns to maintain the desired degree of resolution and minimize the number of trials. The linear graphs are useful in making
column assignments and the triangular tables are useful in determining aliases (discussed earlier).

   Example 23.9 will serve to illustrate the robust design procedure. Here, we will
also incorporate the SNR, but we must stress that such a study typically includes two

parts: (1) the experimental layout or design and (2) the analysis. We should clearly differentiate the two components of the robust design procedure and understand that we may use other analysis techniques (rather than a strict SNR transformation).

**Analysis and Interpretation**   *Analyses of the SNR experiments generally use main effect plots and two-factor interaction plots. ANOVA tables, where we pool an experimental error, are also used. We use the transformed response in decibel units throughout the analysis. Then, at the end,* once we have made a "best-setting" response estimate in decibels from our model, *we attempt to "reverse" the transformation and express the estimated response in the original observation units.*

## Example 23.9

Example 23.9 is an extension of the previous cookie chewiness examples (Examples 23.5 through 23.8). We are using five design factors and two noise factors, as depicted in Table 23.15. We want to perform a robust design study on the chewiness of our chocolate chip cookies. The design factors, over which we can exercise control, are listed in Table 23.15*a*; the noise factors, over which we have less control, are listed in Table 23.15*b*. Develop the robust design experiment and analyze the results.

### Solution

We will use the $L_8$ and $L_4$ orthogonal arrays, shown in Table 23.16. The five design factors $A$ through $E$ are assigned to $L_8$ columns 1, 2, 4, 7, and 3, respectively, as shown in Table 23.17*a*. Columns 5 and 6 are not assigned main effects. Note that the $L_8$ array in Table 23.17*a* and the $\frac{1}{4}$ FFE of a $2^5$ FAT developed earlier (Figure 23.6*d*) are identical with respect to the assignment of factor effect rows and aliases (e.g., the + and − assignments in the observation rows are identical), only the column position of factor $D$ differs. The alias structure is included

TABLE 23.15    DESIGN AND NOISE FACTOR DESCRIPTIONS (EXAMPLE 23.9)

**a** Design factors at two levels each

| Factor | Description | Level 1 (−) | Level 2 (+) |
|--------|-------------|-------------|-------------|
| A | Cooking temperature | Low | High |
| B | Syrup content | Low | High |
| C | Cooking time | Short | Long |
| D | Cooking pan | Solid | Mesh |
| E | Shortening type | Corn | Coconut |

**b** Noise factors at two levels each

| Factor | Description | Level 1 (−) | Level 2 (+) |
|--------|-------------|-------------|-------------|
| $Z_1$ | Cookie position | Side | Middle |
| $Z_2$ | Temperature at test | Low | High |

in both of the displays. Hence, in our previous FFE exercises we actually developed the $L_8$ orthogonal array. Or, in other words, the OA tables can be viewed as generic FFE layouts in which we assign our factors to their columns.

The outer array used in the example is an $L_4$ with two noise factors at two levels each, Table 23.17b shows our assignments. Columns 1 through 3 represent $Z_1$ and $Z_2$ main effects and the $Z_1Z_2$ interaction effects, respectively. Four experimental trials (set by the outer array rows) are required for each combination (one row) of the design array. Altogether, 32 observations (indicated by the y's in Table 23.17c) must be collected. Next, the SNRs, or the $\eta$'s, are computed for each row of four observations. The SNR transformation selected depends on the type of response—bigger is better, smaller is better, or nominal is best (see Equations 23.7 through 23.10).

We executed our experiment as described in Table 23.17. We systematically introduced the noise variables—cookie position on pan and cookie temperature at test—in order to emulate our process field environment. Our response is bigger is better; therefore, we will use the SNR transformation of Equation 23.8:

$$\eta = -10 \log_{10}\left(\frac{1}{4}\sum_{i=1}^{4}\frac{1}{Y_i^2}\right)$$

We should note that, *in using the outer array, we are injecting the anticipated field variation into our experiment, but we are not able to measure its effect directly.* If we want to measure the effect of these noise variables, we must develop a full FAT or FFE in which they are factors. Here, we merely want to develop a recipe by selecting only the design factor levels which will be tolerant of the noise variables and produce a chewy cookie.

**TABLE 23.16**   $L_8$ AND $L_4$ ORTHOGONAL ARRAY STRUCTURES (EXAMPLE 23.9)

**a** $L_8$ ($2^7$) orthogonal array

| Exp. trial | Column | | | | | | |
|---|---|---|---|---|---|---|---|
| | 1 | 2 | 3 | 4 | 5 | 6 | 7 |
| 1 | − | − | − | − | − | − | − |
| 2 | − | − | − | + | + | + | + |
| 3 | − | + | + | − | − | + | + |
| 4 | − | + | + | + | + | − | − |
| 5 | + | − | + | − | + | − | + |
| 6 | + | − | + | + | − | + | − |
| 7 | + | + | − | − | + | + | − |
| 8 | + | + | − | + | − | − | + |

**b** $L_4$ ($2^3$) orthogonal array

| Exp. trial | Column | | |
|---|---|---|---|
| | 1 | 2 | 3 |
| 1 | − | − | − |
| 2 | − | + | + |
| 3 | + | − | + |
| 4 | + | + | − |

− represents level 1.
+ represents level 2.

**TABLE 23.17   EXPERIMENTAL PROTOCOL IN A ROBUST DESIGN EXPERIMENT (EXAMPLE 23.9)**

**a** Inner, $L_8$ design array

| ANOVA aliases | ACDE BCD BE | BCDE ACD AE | ABCDE CD AB | ABCE ABD DE | BCE ADE BD | BDE ACE AD | ABDE ABC CE |
|---|---|---|---|---|---|---|---|
| Factor columns | A 1 | B 2 | E 3 | C 4 | AC 5 | BC 6 | D 7 |
| 1 | − | − | − | − | | | − |
| 2 | − | − | − | + | | | + |
| 3 | − | + | + | − | | | + |
| 4 | − | + | + | + | | | − |
| 5 | + | − | + | − | | | + |
| 6 | + | − | + | + | | | − |
| 7 | + | + | − | − | | | − |
| 8 | + | + | − | + | | | + |

(Design factor combinations, rows 1–8)

**b** Outer, $L_4$ noise array

| Noise factor comb. | $Z_1$ 1 | $Z_2$ 2 | $Z_1Z_2$ 3 |
|---|---|---|---|
| Obs.   1 | − | − | |
| 2 | − | + | |
| 3 | + | − | |
| 4 | + | + | |

**c** Data and SNR format

| Design factor comb. | Observed data — Noise factor trial 1 2 3 4 | | | | Calculated SNRs* |
|---|---|---|---|---|---|
| 1 | y | y | y | y | η |
| 2 | y | y | y | y | η |
| 3 | y | y | y | y | η |
| 4 | y | y | y | y | η |
| 5 | y | y | y | y | η |
| 6 | y | y | y | y | η |
| 7 | y | y | y | y | η |
| 8 | y | y | y | y | η |

*We select one of the eight observations from the inner design array at random. We set up the experiment according to the design factor levels described by the observation row. Next, we select, at random, one of the four observation rows in the outer noise array and complete our experimental setup. We execute this combination of inner and outer factor levels and obtain one observation y. We then select another outer array row at the same inner array combination of factors and obtain another y. We repeat until all four outer array factor combinations are completed. Finally, we repeat all of the above steps seven more times to obtain all 32 individual observations listed. Then, we compute the SNRs, η, one row at a time.

Our data and calculated SNRs are shown in Table 23.18. Summary statistics for main effect plots are shown in Table 23.19a; the AC, BC and AB interaction plot means are shown in Table 23.19b. We note that E and AB are aliases due to our $L_8$ structure and our column-factor assignments. Our alias structure for the $L_8$ OA was developed using the corresponding interaction table, which we will discuss in Example 23.10. We used SAS and pooling up techniques to develop our ANOVA [32,33], which was pieced together from the SAS output and is shown in Table 23.20. The SAS program and ANOVA tables are shown in the Computer Aided Analysis Supplement, Solutions Manual. In pooling up, we used BC, AC, E, and D and an $\alpha = 0.05$ significance level.

TABLE 23.18    DATA AND SNRs FOR CHEWY COOKIE ROBUST
DESIGN STUDY (EXAMPLE 23.9)

| Design factor comb. | Observed data from noise factor combinations (percent elongation) | | | | Calculated SNR (decibels, dB) |
|---|---|---|---|---|---|
| | 1 | 2 | 3 | 4 | Bigger is better |
| 1 | 2 | 6 | 4 | 8 | 10.507 |
| 2 | 7 | 12 | 6 | 15 | 18.270 |
| 3 | 23 | 35 | 27 | 39 | 29.267 |
| 4 | 29 | 41 | 34 | 36 | 30.680 |
| 5 | 3 | 12 | 2 | 11 | 10.265 |
| 6 | 2 | 23 | 8 | 19 | 11.702 |
| 7 | 3 | 18 | 5 | 22 | 14.082 |
| 8 | 16 | 28 | 21 | 31 | 26.732 |

Sample calculation (trial 1)

$$\eta_1 = -10 \log_{10}\left[ \frac{1}{4} \left( \frac{1}{2^2} + \frac{1}{6^2} + \frac{1}{4^2} + \frac{1}{8^2} \right) \right] = 10.507 \text{ dB}$$

Using our resulting $MS_{\text{pooled err}} = 15.88$, we can develop 95 percent confidence interval arrows, $b = t_{0.025,4} = 2.776$, in decibels for our main effect plots:

$$\pm 2.776 \text{ s.e.i.} = \pm 2.776 \sqrt{\frac{MS_{\text{pooled err}}}{n'}} = \pm 2.776 \sqrt{\frac{15.88}{4}} = \pm 5.53 \text{ dB}$$

Also for our two-factor interaction plots

$$\pm 2.776 \text{ s.e.i.} = \pm 2.776 \sqrt{\frac{15.88}{2}} = \pm 7.82 \text{ dB}$$

The main effect plots are developed in Figure 23.10.  For illustrative purposes, $AC$, $BC$, and $AB$ interaction plots are also shown.  At this point, we should note that the ANOVA can also be developed using our FFE hand calculation methods and decibel units.

From an examination of our plots and ANOVA, factor $B$ (syrup content) appears to be very significant.  Factors $A$ and $C$ (cooking temperature and cooking time, respectively) appear to be somewhat significant.  Factors $D$ and $E$ (cooking pan type and shortening type) were judged insignificant and then pooled into our error term.  Here again, we are faced with aliases in a resolution III design, so interpretation will be a challenge.

At this point, we can develop a main effects model with factors $A$, $B$, and $C$ and predict our best response (i.e., we will use low $A$, high $B$, high $C$ from the main effect plots and assume no significant interaction exists):

$$\eta = \mu' + A_i' + B_j' + C_k' + \varepsilon' \tag{23.11}$$

# TABLE 23.19   MAIN EFFECTS AND INTERACTION DATA SUMMARIES (EXAMPLE 23.9)

## a Main effects data summary

A:  Cooking temperature, $i = 1, 2$
B:  Syrup content, $j = 1, 2$
C:  Cooking time, $k = 1, 2$
D:  Cooking pan type, $l = 1, 2$
E:  Shortening type, $m = 1, 2$
Response in dBs

| Factor | Level | Total | Mean | Nbr. obs | Notation |
|---|---|---|---|---|---|
| A | – | 88.723 | 22.181 | 4 | $\eta_{i....}$ |
|   | + | 62.781 | 15.695 | 4 | |
| B | – | 50.745 | 12.686 | 4 | $\eta_{.j...}$ |
|   | + | 100.760 | 25.190 | 4 | |
| C | – | 64.121 | 16.030 | 4 | $\eta_{..k..}$ |
|   | + | 87.383 | 21.846 | 4 | |
| D | – | 66.971 | 16.743 | 4 | $\eta_{...l.}$ |
|   | + | 84.533 | 21.133 | 4 | |
| E | – | 69.591 | 17.398 | 4 | $\eta_{....m}$ |
|   | + | 81.914 | 20.478 | 4 | |

$\eta_{.....} = 151.504$   $\bar{\eta}_{.....} = 18.938$ dB

## b AB, AC, and BC data summaries

**AB**

| Factor A | Factor B – | Factor B + |
|---|---|---|
| – | Total 28.78, Obs. 2, Mean 14.39 | 59.95, 2, 29.97 |
| + | 21.97, 2, 10.98 | 40.81, 2, 20.41 |

$\eta_{ij...}$   $\eta_{.....} = 151.504$   $\bar{\eta}_{.....} = 18.938$ dB

**AC**

| Factor A | Factor C – | Factor C + |
|---|---|---|
| – | Total 39.77, Obs. 2, Mean 19.89 | 48.95, 2, 24.48 |
| + | 24.35, 2, 12.17 | 38.43, 2, 19.22 |

$\eta_{i.k.}$   $\eta_{.....} = 151.504$   $\bar{\eta}_{.....} = 18.938$ dB

**BC**

| Factor B | Factor C – | Factor C + |
|---|---|---|
| – | 20.77, 2, 10.39 | 29.97, 2, 14.99 |
| + | 43.35, 2, 21.67 | 57.41, 2, 28.71 |

$\eta_{.jk.}$   $\eta_{.....} = 151.504$   $\bar{\eta}_{.....} = 18.938$ dB

TABLE 23.20   POOLED ANOVA FOR $L_8$ INNER ARRAY DESIGN
AND $L_4$ OUTER ARRAY DESIGN, TAGUCHI EXPERIMENT
(EXAMPLE 23.9)

| Source | df | SS | MS | F | P-value* |
|---|---|---|---|---|---|
| Total corrected | 7 | 527.962 | | | |
| A | 1 | 84.126 | 84.126 | 5.30 | 0.0828 |
| B | 1 | 312.689 | 312.689 | 19.70 | 0.0114 |
| C | 1 | 67.643 | 67.643 | 4.26 | 0.1079 |
| Error pooled | 4 | 63.504 | 15.876 | | |
| D | 1 | 38.553 | | | |
| E | 1 | 18.981 | | | |
| AC | 1 | 3.015 | | | |
| BC | 1 | 2.955 | | | |

*Taken from SAS output, Computer Aided Analysis Supplement, Solutions Manual.

Here, our "prime" notation is introduced to indicate the decibel domain. Using our main effects model and Figure 23.10, we develop our estimates as we did in our FFE (Equations 23.5 and 23.6):

$$\hat{\mu}_{\eta\,best} = \hat{\mu}_{\eta\,A_1'B_2'C_2'} = \bar{\eta}_{.....} + (\bar{\eta}_{1....} - \bar{\eta}_{.....}) + (\bar{\eta}_{.2..} - \bar{\eta}_{.....}) + (\bar{\eta}_{..2.} - \bar{\eta}_{.....})$$

$$= 18.94 + (22.18 - 18.94) + (25.19 - 18.94) + (21.85 - 18.94)$$

$$= 31.34 \text{ dB}$$

At this point, we can develop the reverse transform of Equation 23.8 and convert to our original units of measure (percent elongation):

$$\hat{\mu}_{best} = \hat{\mu}_{A_1B_2C_2} = \left[ \frac{1}{\text{antilog}_{10}\left( -\frac{\hat{\mu}_{\eta A_1'B_2'C_2'}}{10} \right)} \right]^{1/2}$$

$$= \left[ \frac{1}{\text{antilog}_{10}\left( -\frac{31.34}{10} \right)} \right]^{1/2} = 36.90\%$$

We should note that, in general, a reverse transform is not always possible.

## Confirmation Experiments

*After the Taguchi experiment is run and analyzed, some type of confirmation experiment is recommended.* One possibility typically recommend by Taguchi SNR experiment advocates (Taguchi, Phadke, and Ross [34]) is to perform a limited series of replications at the best identified combination of design factors. In our case, we would set up a confirmation experiment under the noise array levels. For example, we would set our process at low $A$, high $B$, and high $C$, with $D$ and $E$ set to the most economical levels (since $D$ and $E$ were not considered significant) and obtain a few observations. Then, we would convert the results into decibels and compare these results to those predicted by our model. If the confirmation results match the model results, we conclude the model is basically valid and proceed with recipe changes. On the other hand, if there is a disparity or inadequate agreement, we must redevelop our experimental protocol.

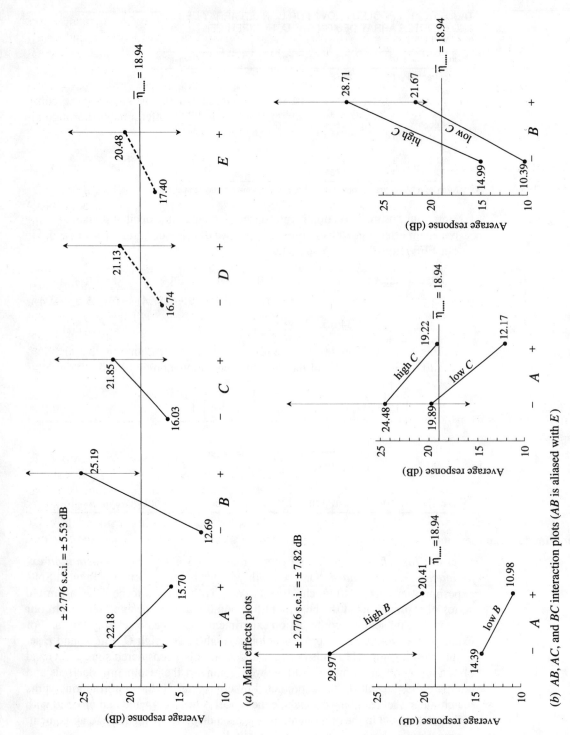

(a) Main effects plots

(b) *AB*, *AC*, and *BC* interaction plots (*AB* is aliased with *E*)

**FIGURE 23.10**   Main effects and *AB*, *AC*, *BC* interaction plots (Example 23.9).

The Taguchi SNR ratio experimental confirmation process seems attractive in that it allows us to limit our experimentation; thus it is a very economical and timely protocol. However, it is somewhat open-ended in that a subjective "match" in decibel units is assessed. Also, the assumption that one of the factor level combinations considered (in the experiment) will be a best or optimal process setting seems somewhat reckless, given that quantitative factors are included. The sequential response surface methods introduced in Examples 22.4 and 22.5 offer a more thorough alternative when quantitative factors are involved.

## 23.6 ORTHOGONAL ARRAY TABLES

As we have seen in Example 23.9, the OA tables form the basis for the Taguchi inner and outer arrays. OAs predate the Taguchi work. They may be used in both classical and Taguchi design experiments. In this section, we will demonstrate both two- and three-level experimental layouts using the OA linear graphs and the OA interaction tables included in Section IX, Table IX.18.

*Once we understand the nature of the OAs, we can use them as inner or outer robust design arrays, or as the basis for classical FFEs.* In general, *we may assign one factor to each column in an OA. If all columns are assigned, we refer to the OA or experiment as "saturated."* We should note that in a two-level OA the "1" level is equivalent to the "−" level and the "2" level is equivalent to the "+" level in our previous $2^f$ FFE discussion. In the three-level OAs the "1" refers to the low factor level, the "2" to the middle factor level, and the "3" to the high or third factor level.

---

## Example 23.10

The $L_8$ OA can be used to develop a full $2^3$ FAT experiment with factors $A$, $B$, and $C$. The $L_8$ OA demands eight observations and yields 7 df. Each column contains 1 df. The $L_8$ OA can be assigned a minimum of two factors at two levels each, to yield a full FAT with two replications. It can be assigned a maximum of seven factors (one to each column) to yield a saturated resolution III FFE. Once we make our assignments and collect our data, we can develop our ANOVA using a computer aid or by hand calculations.

**a** For a $2^3$ FAT, develop the column assignments using the $L_8$ linear graph.

**b** Then, justify the column assignments using the $L_8$ interaction table.

### Solution

For our $2^3$ FAT, since we have eight observations in the $L_8$ OA, we should be able to assign our factors to the OA columns in such a way as to obtain a full FAT with one replication. But not all column assignments will yield this full FAT.

**a** Our $L_8$ OA, one of its linear graphs, and its interaction table appear in Figure 23.11. We assign factors to the columns (Figure 23.11a) using the linear graph (Figure 23.11b). If our factors are assigned to the nodes, then our interactions will appear in the column numbers listed on the branches. Here, we assign factor $A$ to node 1 (column 1), $B$ to node 2 (column 2), and $C$ to node 4 (column 4), as shown in Figure 23.11$a$ and $b$. Hence, $AB$ interaction will appear in column 3, $AC$ in column 5, and $BC$ in column 6; column 7 is isolated, or open.

**b** The linear graphs can be built from the interaction tables, which typically contain more information than any one linear graph. Our seven $L_8$ OA columns are arranged in a matrix form in the interaction table. When we assign our factors to columns, we can list them above and to the left of the column entries and then use the relationships in the body of the table to develop our interactions. In Figure 23.11$c$, we have assigned $A$ to column 1, $B$ to column 2, and $C$ to column 4. Now, in Figure 23.11$d$, we have set up our interactions by moving to the table body, reading the intersection number, and recording the interaction against the column number. Hence, columns 3, 5, 6, and 7 contain the $AB$, $AC$, $BC$, and $ABC$ interactions, respectively.

---

*Note*: If we include our 5 factors $A$, $B$, $C$, $D$, and $E$ as in Table 23.17$a$ we can use the above procedure to generate alias information. With persistence and patience, we generated the aliases shown in Table 23.17$a$ using the $L_8$ OA interaction table and this column-row intersection technique.

**FIGURE 23.11**    $L_8$ OA factor assignments for a $2^3$ experiment (Example 23.10).

| Expt. No. | A 1 | B 2 | 3 | C 4 | 5 | 6 | 7 |
|---|---|---|---|---|---|---|---|
| 1 | 1 | 1 | 1 | 1 | 1 | 1 | 1 |
| 2 | 1 | 1 | 1 | 2 | 2 | 2 | 2 |
| 3 | 1 | 2 | 2 | 1 | 1 | 2 | 2 |
| 4 | 1 | 2 | 2 | 2 | 2 | 1 | 1 |
| 5 | 2 | 1 | 2 | 1 | 2 | 1 | 2 |
| 6 | 2 | 1 | 2 | 2 | 1 | 2 | 1 |
| 7 | 2 | 2 | 1 | 1 | 2 | 2 | 1 |
| 8 | 2 | 2 | 1 | 2 | 1 | 1 | 2 |

($a$) $L_8$ OA with factor assignments
($1 \rightarrow -$ level, $2 \rightarrow +$ level)

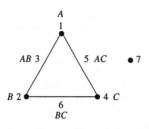

($b$) $L_8$ OA linear graph with factor assignments

| Column | A 1 | B 2 | 3 | C 4 | 5 | 6 | 7 |
|---|---|---|---|---|---|---|---|
| A 1 | (1) | 3 | 2 | 5 | 4 | 7 | 6 |
| B 2 | | (2) | 1 | 6 | 7 | 4 | 5 |
| 3 | | | (3) | 7 | 6 | 5 | 4 |
| C 4 | | | | (4) | 1 | 2 | 3 |
| 5 | | | | | (5) | 3 | 2 |
| 6 | | | | | | (6) | 1 |
| 7 | | | | | | | (7) |

($c$) $L_8$ OA interaction table with factor assignments listed

| Column | A 1 | B 2 | AB 3 | C 4 | AC 5 | BC 6 | ABC 7 |
|---|---|---|---|---|---|---|---|
| A 1 | (1) | 3 | 2 | 5 | 4 | 7 | 6 |
| B 2 | | (2) | 1 | 6 | 7 | 4 | 5 |
| AB 3 | | | (3) | 7 | 6 | 5 | 4 |
| C 4 | | | | (4) | 1 | 2 | 3 |
| AC 5 | | | | | (5) | 3 | 2 |
| BC 6 | | | | | | (6) | 1 |
| ABC 7 | | | | | | | (7) |

($d$) $L_8$ OA interaction table completed

## Example 23.11

The $L_9$ OA can be used to design three-level experiments. The $L_9$ OA demands nine observations and contains four columns. Each column has 2 df associated with it; hence, we have a total of 8 df. In a three-level experiment each main effect has two df and each two-factor interaction has four df. Thus, one column can support a main effect, but two columns are required to support a two-factor interaction effect. After we develop our column assignments and collect our data, our conventional ANOVA calculation procedures and computer aids can be used (just as in our two-level FFE examples).

**a** Use the $L_9$ and its linear graph to develop a $3^2$ full FAT with factors $A$ and $B$.
**b** Use the $L_9$ to develop a $\frac{1}{3}$ FFE of a $3^3$ FAT.

## Solution

We can use the same general procedure for the three-level OAs as we did for the two-level OAs. However, we must remember that here two columns (4 df) are necessary for two-factor interactions.

**a** Using the $L_9$ OA shown in Figure 23.12a, and our linear graph in Figure 23.12b, we have assigned factor $A$ to node 1 (column 1, Figure 23.12a) and factor $B$ to node 2 (column 2, Figure 23.12a). We then would set up our treatment combinations as indicated in our OA rows (e.g., for observation 1, we set $A$ at level 1 and $B$ at level 1; for observation 7, we set $A$ at level 3 and $B$ at level 1).

**b** Here, we have three factors at three levels, which would require $3^3 = 27$ observations for a full FAT. But, we will take only nine observations (thus running a $\frac{1}{3}$ FFE). Figure 23.12c depicts our column assignments; we will assign factor $A$ to column 1, factor $B$ to column 2, and factor $C$ to column 3. The best result we

**FIGURE 23.12**    $L_9$ OA factor assignments for $3^2$ and $3^3$ experiments (Example 23.11).

| Expt. No. | A 1 | B 2 | AB 3 | AB 4 |
|---|---|---|---|---|
| 1 | 1 | 1 | 1 | 1 |
| 2 | 1 | 2 | 2 | 2 |
| 3 | 1 | 3 | 3 | 3 |
| 4 | 2 | 1 | 2 | 3 |
| 5 | 2 | 2 | 3 | 1 |
| 6 | 2 | 3 | 1 | 2 |
| 7 | 3 | 1 | 3 | 2 |
| 8 | 3 | 2 | 1 | 3 |
| 9 | 3 | 3 | 2 | 1 |

(a) $L_9$ OA with $3^2$ factor assignments (1 → level 1, 2 → level 2, 3 → level 3)

$A$ 1 ●————————● 2 $B$
              3,4 $AB$

(b) $L_9$ OA linear graph with $3^2$ factor assignments

| Expt. No. | A 1 | B 2 | C 3 | 4 |
|---|---|---|---|---|
| 1 | 1 | 1 | 1 | 1 |
| 2 | 1 | 2 | 2 | 2 |
| 3 | 1 | 3 | 3 | 3 |
| 4 | 2 | 1 | 2 | 3 |
| 5 | 2 | 2 | 3 | 1 |
| 6 | 2 | 3 | 1 | 2 |
| 7 | 3 | 1 | 3 | 2 |
| 8 | 3 | 2 | 1 | 3 |
| 9 | 3 | 3 | 2 | 1 |

(c) $L_9$ OA linear graph with $3^3$ factor assignments

can obtain here is to extract effects *A*, *B*, and *C*, realizing that interactions will be "hopelessly" aliased with these main effects. Because, we only use three columns for main effects, we have 2 df (from the remaining column) to develop as a pseudo experimental error—assuming a main effects model is appropriate.

---

### Example 23.12

Use the $L_{16}$ OA to develop a resolution V, $\frac{1}{2}$ FFE of a $2^5$ FAT.

### Solution

The $L_{16}$ OA can hold up to fifteen factors at two levels before saturation. A saturated OA yields a resolution III. Here, we want to include five factors and maintain a resolution V, if possible. Figure 23.13a and b contain the $L_{16}$ OA and one of its linear graphs, respectively. For resolution V, we must not see any main effects or two-factor interactions aliased. Hence, we will have five distinct columns for main effects and $C_2^5 = 10$ distinct columns devoted to two-factor interactions. The linear graph in Figure 23.13b meets this condition. Therefore, the factors A through E are assigned to columns 1, 2, 4, 8, and 15, respectively.

FIGURE 23.13    $L_{16}$ OA factor assignments for a $2^5$, resolution V, experiment (Example 23.12).

| Expt. No. | A 1 | B 2 | 3 | C 4 | 5 | 6 | 7 | D 8 | 9 | 10 | 11 | 12 | 13 | 14 | E 15 |
|---|---|---|---|---|---|---|---|---|---|---|---|---|---|---|---|
| 1 | 1 | 1 | 1 | 1 | 1 | 1 | 1 | 1 | 1 | 1 | 1 | 1 | 1 | 1 | 1 |
| 2 | 1 | 1 | 1 | 1 | 1 | 1 | 1 | 2 | 2 | 2 | 2 | 2 | 2 | 2 | 2 |
| 3 | 1 | 1 | 1 | 2 | 2 | 2 | 2 | 1 | 1 | 1 | 1 | 2 | 2 | 2 | 2 |
| 4 | 1 | 1 | 1 | 2 | 2 | 2 | 2 | 2 | 2 | 2 | 2 | 1 | 1 | 1 | 1 |
| 5 | 1 | 2 | 2 | 1 | 1 | 2 | 2 | 1 | 1 | 2 | 2 | 1 | 1 | 2 | 2 |
| 6 | 1 | 2 | 2 | 1 | 1 | 2 | 2 | 2 | 2 | 1 | 1 | 2 | 2 | 1 | 1 |
| 7 | 1 | 2 | 2 | 2 | 2 | 1 | 1 | 1 | 1 | 2 | 2 | 2 | 2 | 1 | 1 |
| 8 | 1 | 2 | 2 | 2 | 2 | 1 | 1 | 2 | 2 | 1 | 1 | 1 | 1 | 2 | 2 |
| 9 | 2 | 1 | 2 | 1 | 2 | 1 | 2 | 1 | 2 | 1 | 2 | 1 | 2 | 1 | 2 |
| 10 | 2 | 1 | 2 | 1 | 2 | 1 | 2 | 2 | 1 | 2 | 1 | 2 | 1 | 2 | 1 |
| 11 | 2 | 1 | 2 | 2 | 1 | 2 | 1 | 1 | 2 | 1 | 2 | 2 | 1 | 2 | 1 |
| 12 | 2 | 1 | 2 | 2 | 1 | 2 | 1 | 2 | 1 | 2 | 1 | 1 | 2 | 1 | 2 |
| 13 | 2 | 2 | 1 | 1 | 2 | 2 | 1 | 1 | 2 | 2 | 1 | 1 | 2 | 2 | 1 |
| 14 | 2 | 2 | 1 | 1 | 2 | 2 | 1 | 2 | 1 | 1 | 2 | 2 | 1 | 1 | 2 |
| 15 | 2 | 2 | 1 | 2 | 1 | 1 | 2 | 1 | 2 | 2 | 1 | 2 | 1 | 1 | 2 |
| 16 | 2 | 2 | 1 | 2 | 1 | 1 | 2 | 2 | 1 | 1 | 2 | 1 | 2 | 2 | 1 |

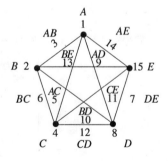

(*a*) $L_{16}$ OA with $2^5$ factor assignments

(*b*) $L_{16}$ OA linear graph with $2^5$ factor assignments

**Example 23.13**

Use the $L_{27}$ OA to develop a three-level FFE that includes four factors—$A$, $B$, $C$, and $D$—and captures all main effects, as well as $AB$, $AC$, and $BC$ interactions free of main effects. How many degrees of freedom would we have left over to form a pseudo error term?

**Solution**

Here, we will use a linear graph and also illustrate the $L_{27}$ interaction table in making our column assignments. The $L_{27}$ OA, one of its linear graphs, and the interaction table appear in Figure 23.14$a$, $b$, and $c$, respectively. Based on the linear graph, we will assign factors $A$, $B$, $C$, and $D$ to nodes 1, 2, 5, and 9, respectively (Figure 23.14$b$). Then, from Figure 23.14$c$, columns 3 and 4 will together contain the $AB$ interaction; columns 6 and 7, the $AC$; and columns 8 and 11, the $BC$. Columns 10, 12, and 13 are open, leaving three columns at 2 df per column or 6 df for a pseudo error term.

*Note*: Factor D could just as well be assigned to columns 10, 12, or 13 and Figure 23.14$b$ could have been used to determine the interaction columns.

## REVIEW AND DISCOVERY EXERCISES

### Review

**23.1** A plastics research and development firm has developed a new plastic material in its laboratory and is now refining the product. It is specifically interested in two factors—the plastic manufacturing process recipe and the temperature to which the material is heated during the molding process. Three recipes and four temperature levels are to be studied. The measured response is product breaking strength tension (psi). The firm decided to split (block) the experiment among three of its largest customer's plants, in order to assess the product through the customers' eyes. The data obtained from the FAT-RCB experiment are shown below [ME]:

| Customer | Process recipe | Molding temperature | | | |
| --- | --- | --- | --- | --- | --- |
| | | 1 | 2 | 3 | 4 |
| | 1 | 1000 | 1050 | 1090 | 960 |
| 1 | 2 | 1350 | 1300 | 1350 | 1210 |
| | 3 | 920 | 940 | 980 | 850 |
| | 1 | 1000 | 1140 | 900 | 820 |
| 2 | 2 | 1150 | 1200 | 1250 | 1160 |
| | 3 | 870 | 900 | 940 | 820 |
| | 1 | 1080 | 1130 | 1150 | 1020 |
| 3 | 2 | 1300 | 1330 | 1350 | 1250 |
| | 3 | 950 | 1000 | 1020 | 900 |

Response unit: psi.

| Expt. No. | A 1 | B 2 | 3 | 4 | C 5 | 6 | 7 | 8 | D 9 | 10 | 11 | 12 | 13 |
|---|---|---|---|---|---|---|---|---|---|---|---|---|---|
| 1 | 1 | 1 | 1 | 1 | 1 | 1 | 1 | 1 | 1 | 1 | 1 | 1 | 1 |
| 2 | 1 | 1 | 1 | 1 | 2 | 2 | 2 | 2 | 2 | 2 | 2 | 2 | 2 |
| 3 | 1 | 1 | 1 | 1 | 3 | 3 | 3 | 3 | 3 | 3 | 3 | 3 | 3 |
| 4 | 1 | 2 | 2 | 2 | 1 | 1 | 1 | 2 | 2 | 2 | 3 | 3 | 3 |
| 5 | 1 | 2 | 2 | 2 | 2 | 2 | 2 | 3 | 3 | 3 | 1 | 1 | 1 |
| 6 | 1 | 2 | 2 | 2 | 3 | 3 | 3 | 1 | 1 | 1 | 2 | 2 | 2 |
| 7 | 1 | 3 | 3 | 3 | 1 | 1 | 1 | 3 | 3 | 3 | 2 | 2 | 2 |
| 8 | 1 | 3 | 3 | 3 | 2 | 2 | 2 | 1 | 1 | 1 | 3 | 3 | 3 |
| 9 | 1 | 3 | 3 | 3 | 3 | 3 | 3 | 2 | 2 | 2 | 1 | 1 | 1 |
| 10 | 2 | 1 | 2 | 3 | 1 | 2 | 3 | 1 | 2 | 3 | 1 | 2 | 3 |
| 11 | 2 | 1 | 2 | 3 | 2 | 3 | 1 | 2 | 3 | 1 | 2 | 3 | 1 |
| 12 | 2 | 1 | 2 | 3 | 3 | 1 | 2 | 3 | 1 | 2 | 3 | 1 | 2 |
| 13 | 2 | 2 | 3 | 1 | 1 | 2 | 3 | 2 | 3 | 1 | 3 | 1 | 2 |
| 14 | 2 | 2 | 3 | 1 | 2 | 3 | 1 | 3 | 1 | 2 | 1 | 2 | 3 |
| 15 | 2 | 2 | 3 | 1 | 3 | 1 | 2 | 1 | 2 | 3 | 2 | 3 | 1 |
| 16 | 2 | 3 | 1 | 2 | 1 | 2 | 3 | 3 | 1 | 2 | 2 | 3 | 1 |
| 17 | 2 | 3 | 1 | 2 | 2 | 3 | 1 | 1 | 2 | 3 | 3 | 1 | 2 |
| 18 | 2 | 3 | 1 | 2 | 3 | 1 | 2 | 2 | 3 | 1 | 1 | 2 | 3 |
| 19 | 3 | 1 | 3 | 2 | 1 | 3 | 2 | 1 | 3 | 2 | 1 | 3 | 2 |
| 20 | 3 | 1 | 3 | 2 | 2 | 1 | 3 | 2 | 1 | 3 | 2 | 1 | 3 |
| 21 | 3 | 1 | 3 | 2 | 3 | 2 | 1 | 3 | 2 | 1 | 3 | 2 | 1 |
| 22 | 3 | 2 | 1 | 3 | 1 | 3 | 2 | 2 | 1 | 3 | 3 | 2 | 1 |
| 23 | 3 | 2 | 1 | 3 | 2 | 1 | 3 | 3 | 2 | 1 | 1 | 3 | 2 |
| 24 | 3 | 2 | 1 | 3 | 3 | 2 | 1 | 1 | 3 | 2 | 2 | 1 | 3 |
| 25 | 3 | 3 | 2 | 1 | 1 | 3 | 2 | 3 | 2 | 1 | 2 | 1 | 3 |
| 26 | 3 | 3 | 2 | 1 | 2 | 1 | 3 | 1 | 3 | 2 | 3 | 2 | 1 |
| 27 | 3 | 3 | 2 | 1 | 3 | 2 | 1 | 2 | 1 | 3 | 1 | 3 | 2 |

(a) $L_{27}$ OA with $3^4$ factor assignments

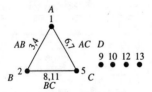

(b) $L_{27}$ OA linear graph with $3^4$ factor assignments

**FIGURE 23.14**   $L_{27}$ OA factor assignments for a $3^4$ experiment (Example 23.13).

| Column | A 1 | B 2 | AB 3 | AB 4 | C 5 | AC 6 | AC 7 | BC 8 | D 9 | 10 | BC 11 | 12 | 13 |
|---|---|---|---|---|---|---|---|---|---|---|---|---|---|
| A 1 | (1) | 3 4 | 2 4 | 2 3 | 6 7 | 5 7 | 5 6 | 9 10 | 8 10 | 8 9 | 12 13 | 11 13 | 11 12 |
| B 2 | | (2) | 1 4 | 1 3 | 8 11 | 9 12 | 10 13 | 5 11 | 6 12 | 7 13 | 5 8 | 6 9 | 7 10 |
| AB 3 | | | (3) | 1 2 | 9 13 | 10 11 | 8 12 | 7 12 | 5 13 | 6 11 | 6 10 | 7 8 | 5 9 |
| AB 4 | | | | (4) | 10 12 | 8 13 | 9 11 | 6 13 | 7 11 | 5 12 | 7 9 | 5 10 | 6 8 |
| C 5 | | | | | (5) | 1 7 | 1 6 | 2 11 | 3 13 | 4 12 | 2 8 | 4 10 | 3 9 |
| AC 6 | | | | | | (6) | 1 5 | 4 13 | 2 12 | 3 11 | 3 10 | 2 9 | 4 8 |
| AC 7 | | | | | | | (7) | 3 12 | 4 11 | 2 13 | 4 9 | 3 8 | 2 10 |
| BC 8 | | | | | | | | (8) | 1 10 | 1 9 | 2 5 | 3 7 | 4 6 |
| D 9 | | | | | | | | | (9) | 1 8 | 4 7 | 2 6 | 3 5 |
| 10 | | | | | | | | | | (10) | 3 6 | 4 5 | 2 7 |
| BC 11 | | | | | | | | | | | (11) | 1 13 | 1 12 |
| 12 | | | | | | | | | | | | (12) | 1 11 |
| 13 | | | | | | | | | | | | | (13) |

(c) $L_{27}$ OA interaction table with AB, AC, BC interaction completed

**FIGURE 23.14**   $L_{27}$ OA factor assignments for a $3^4$ experiment (Example 23.13).

    **a** Develop a graphical analysis of the results.
    **b** Develop an appropriate ANOVA table. Add $LSD_{0.05}$ bars to you plots.
    **c** Interpret the experimental results.
    **d** Critique the design of this experiment (bigger is better).
**23.2** A hay equipment manufacturer is redesigning a forage pickup mechanism for use on a new line of hay balers. The pickup mechanism has been developed in three different design configurations. One of the project objectives is to provide a design that will be robust when used under various field and crop conditions. Three

commonly baled crops—alfalfa, native grass, and sorghum—were selected in a nonrandom fashion for the test. Field conditions of light, medium, and heavy windrows are considered random and vary both within and between crops. Hence, they were set in a random fashion within each field. The data are collected and shown in the table below. The response is measured in percent material left in the field (a smaller is better measure).

| | Blocks | | | | | | | | |
|---|---|---|---|---|---|---|---|---|---|
| | Light | | | Medium | | | Heavy | | |
| | | | | Design A | | | | | |
| Crop B | 1 | 2 | 3 | 1 | 2 | 3 | 1 | 2 | 3 |
| Alfalfa | 4.5 | 7.5 | 9.0 | 2.5 | 5.5 | 3.5 | 2.5 | 3.5 | 2.0 |
| Native grass | 7.5 | 10.0 | 12.5 | 6.0 | 7.5 | 5.0 | 3.5 | 4.5 | 4.0 |
| Sorghum | 3.0 | 5.0 | 4.5 | 1.0 | 2.0 | 2.5 | 1.5 | 2.5 | 1.0 |

a  Develop an ANOVA table for this experiment.
b  Develop the EMS rows for the ANOVA table and perform the appropriate hypothesis tests.
c  Develop a set of plots and include s.e.i. arrows.
d  Recommend a pickup design configuration. Use the HSD test method in your analysis and develop appropriate $HSD_{0.05}$ bars in your plots.

23.3  Four brands of paper towels were selected for a customer-demand test for strength under different moisture conditions. Controlling the moisture variable was determined to be difficult and time consuming. To avoid possible bottlenecks, a split-plot design was selected. The moisture levels (fixed effects) served as the main plots, and the four brands (a fixed effects factor) were randomized as split plots within the humidity main plots. The experiment was designed to include three replications, one each day. A total of 36 experimental units were cut and prepared. Strength was measured in pounds force. The data layout is shown below.

| | | Humidity % | | |
|---|---|---|---|---|
| Replications | Brand | 20 | 60 | 100 (wet) |
| 1 | 1 | 3.2 | 2.9 | 0.5 |
| | 2 | 2.9 | 2.7 | 1.8 |
| | 3 | 3.3 | 2.6 | 1.2 |
| | 4 | 4.1 | 3.1 | 0.9 |
| 2 | 1 | 3.9 | 2.6 | 0.7 |
| | 2 | 3.4 | 3.0 | 1.6 |
| | 3 | 3.1 | 2.8 | 0.9 |
| | 4 | 4.3 | 2.8 | 1.2 |
| 3 | 1 | 3.6 | 2.3 | 0.9 |
| | 2 | 2.8 | 2.6 | 1.9 |
| | 3 | 3.2 | 2.9 | 0.7 |
| | 4 | 3.9 | 2.7 | 0.6 |

a  Develop an appropriate ANOVA table.
b  Develop an appropriate set of main effect and interaction plots, including s.e.i. arrows.
c  Recommend a brand for low humidity use and one for high humidity use. Use an LSD technique in your assessment.

**23.4** The supervisor in charge of an electrolytic grinding process in a machine shop is confronted with a product surface finish problem. She wants to rectify the problem by trying to remove the adverse effects produced by the various factors instead of eliminating the causes of variation themselves. She has identified five of the most crucial factors influencing the surface finish (measured as $\mu$ in RMS) and has decided to conduct an experiment to estimate their effects on the surface finish (a smaller is better response). She decides on two levels for each of the factors, as shown below:

| Factor | Description | Level 1 | Level 2 |
|---|---|---|---|
| A | Workpiece feed rate | 250 mm/min | 750 mm/min |
| B | Electrolyte flow rate | 314 cm³/min | 1570 cm³/min |
| C | Amperage per square inch | 100 A/cm² | 200 A/cm² |
| D | Wheel rpm | 1000 | 1500 |
| E | Effective electrode diameter | Full diameter | 30% diameter |

The supervisor is interested in studying both the individual factor effects and the two-factor interactions. [GN]

**a** Using the FFE approach, suggest an experimental design that will yield a resolution V experiment. Then determine the factor combinations that will be run, as well as the alias structure.

**b** Using OAs, suggest a suitable experiment design strategy that will yield a resolution V experiment and then describe your recommended experiment.

**23.5** AA Inc. manufactures fiberglass parts for boat hulls and other products and wishes to conduct an in-house experiment to measure the effect of four factors on the gel time of polyester-fiberglass resin. Gel time is the time interval when fiberglass resin is workable. According to the manufacturer, gel time may be affected by age of the resin, ambient humidity, ambient temperature, and the amount of catalyst used. The engineering department of AA Inc. desires to conduct an experiment to assess these possible effects.

Initially, a $2^4$ factorial design was selected for this experiment. Due to budget limitations, the engineering department opted to conduct a $2_{IV}^{4-1}$, or $\frac{1}{2}$ FFE, experiment instead. The two levels selected for each factor are shown in the data table below. The $\frac{1}{2}$ FFE was designed with $L = ABCD$ and performed, but the alias structure was ignored in the initial experiment design process. Assume a bigger is better response is desirable for AA's boat hull manufacturing operations.

| | | Temperature C | | | |
|---|---|---|---|---|---|
| | | 60°F | | 80°F | |
| | | Catalyst D | | | |
| Age A | Humidity B | 1% | 2% | 1% | 2% |
| 0–6 months | 30% | 50 | | | 6 |
| | 70% | | 20 | 38 | |
| 6–12 months | 30% | | 18 | 35 | |
| | 70% | 90 | | | 36 |

**a** Develop the block layout and alias structure to verify that the $\frac{1}{2}$ FFE assignments were correct and to expose the aliases. Then, identify the resolution number that applies.

**b** Develop a set of main effect plots and an AB interaction plot.

    **c** Use the pooling up process to develop an $MS_{pooled\ err}$ term ($\alpha = 0.05$).

    **d** Place appropriate s.e.i. arrows on your plots.

    **e** Develop a table of effects and then develop a corresponding normal plot of effects.

    **f** Interpret your results. Compare the results obtained from the pooling up method and the normal plotting method.

    **g** Develop a suitable predictor, select the best factor setting, and estimate $\hat{\mu}_{best}$.

**23.6** Refer to Problem 23.5. After the initial data were collected, the alias structure was examined for the $\frac{1}{2}$ FFE. The engineering department at AA felt that the $\frac{1}{2}$ FFE design might not meet its needs and might be rather inconclusive. The team then requested permission to conduct a full factorial experiment for additional evaluation. They went back into the laboratory and ran the other half of the FFE. The complete data table is shown below (16 observations). The starred observations constitute the first half run, the nonstarred observations are the data collected in the process of completing the full FAT. Assume a bigger is better response is desirable for AA's boat hull manufacturing operations.

| | | Temperature C | | | |
|---|---|---|---|---|---|
| | | 60°F | | 80°F | |
| | | Catalyst D | | | |
| Age A | Humidity B | 1% | 2% | 1% | 2% |
| 0–6 months | 30% | 50* | 12 | 22 | 6* |
| | 70% | 75 | 20* | 38* | 15 |
| 6–12 months | 30% | 65 | 18* | 35* | 12 |
| | 70% | 90* | 25 | 60 | 36* |

    **a** Repeat parts *a* through *g* from Problem 23.5 for the data here.

    **b** Describe how the interpretation of the full FAT differs from that of the $\frac{1}{2}$ FFE in Problem 23.5.

**23.7** A company conducted laboratory research to study the effectiveness of air stripping combined with chemical oxidation to clean their process wastewater contaminated with dissolved organics. It was hoped that the volatile organics could be removed by air stripping, and the nonvolatile organics simultaneously oxidized in solution. A catalyst platinum wire, 0.152 mm diameter, 52-mesh, was chosen, as it would provide a solid surface for the oxidation reaction. Since an integrated treatment approach such as this one was novel, neither prior experience nor documentation was available. The firm decided to design an experiment to study the significance of the identified variables. The experimental factors and their levels are shown below:

| | Levels | |
|---|---|---|
| Factor | 1 | 2 |
| Temperature A | 25°C | 100°C |
| Air flow B | 0 l/min | 2 l/min |
| Ammonium nitrate C | 20 ml/hr | 50 ml/hr |
| Treatment time D | 60 min | 120 min |
| Platinum catalyst E | No catalyst | Catalyst |

Since it was quite expensive and time consuming to perform each experimental run, an FFE was proposed. Each experiment consisted of treating 450 ml of wastewater. A condenser was used to minimize water vapor losses due to evaporation. The chemical oxygen demand (COD) was measured before and after treatment. The experimental response was the percent removal of organics (bigger is better), which was estimated using the COD values before and after treatment. The data obtained are shown below. The numbers in parentheses indicate the order in which data were collected (determined at random).

$$(1) = 3.02 \ (8) \qquad cde = 10.11 \ (4) \qquad bde = 7.76 \ (7) \qquad bc = 0.14 \ (5)$$

$$ad = 15.26 \ (2) \qquad ace = 27.22 \ (3) \qquad abe = 63.96 \ (6) \quad abcd = 43.87 \ (1)$$

**a** Verify that the design generators used were $L_1 = ADE$ and $L_2 = BCE$.
**b** Develop the alias structure for this design. Determine the resolution of the design.
**c** Develop a graphical analysis.
**e** Analyze the data using the normal probability plotting (of effects) method.
**d** Generate an appropriate ANOVA table, add $LSD_{0.05}$ bars to your graphs.
**e** Analyze the residuals.
**f** Interpret the analyses.
**g** Using the following predictor model, estimate the best mean response:

$$\hat{\mu}_{best} = \hat{\mu} + \hat{A}_i + \hat{B}_j + \hat{E}_m + \hat{AB}_{ij}$$

**h** Assuming that only factors A, B, and E are active (i.e., ignore factors C and D) in the process, generate an appropriate $2^3$ ANOVA table.

**23.8** A manufacturer is confronted with a problem of variation in strength (measured in pounds per square inch, psi) at the thin-wall molded section of a distributor cap. He seeks to address the problem by trying to remove the adverse effects produced by the various factors instead of eliminating the causes of variation. The supervisor has isolated six of these factors and has decided to conduct an experiment to estimate their effects on strength, using two levels for each, as shown below [GN]:

| Factor | Description | Level 1 | Level 2 |
|--------|-------------|---------|---------|
| A | Mold type | Old design | New design |
| B | Compound dosage | D1 | D2 |
| C | Filling pressure | 500 lb/in² | 1200 lb/in² |
| D | Rate of mold fill | $Ram_{speed1}$ | $Ram_{speed2}$ |
| E | Initial mold temperature | No preheat | Preheat to 300°F |
| F | Material temperature before injection | 350°F | 500°F |

The supervisor is interested in studying only the main effects in the limited time he has available. Using OAs, suggest a suitable experimental design strategy. Determine the resolution of your design.

**Discovery**

**23.9** A manufacturing company has the opportunity to bid on a contract that includes a requirement to perform precision drilling using an excimer laser. The energy for the drilling operation must be maintained between 1875 and 1975 millijoules (mJ) This range is within the capability of the company's laser but is also very near the

2000-mJ damage threshold for the collimating optics and beam-steering lenses. The company's laser has never been used at these levels before. The laser manufacturer has estimated that as few as 10 occurrences of energy over the 2000-mJ threshold may catastrophically damage the optics and has identified four factors that affect the damage threshold. They also helped to establish threshold levels for each factor in order to prevent damage to the laser, while providing the required energy level for the experiment. These factors are $A$, the pressure inside the lasing cavity (from 2 to 5 atmospheres); $B$, the current used to excite the excimer (from 150 mA to 600 mA); $C$, the wavelength of the laser (from 139 to 350 nanometers, nm); and $D$, the percentage of inert buffer gas mixed with the active excimer gas within the cavity (from 90 to 99 percent). Management has asked the manufacturing engineers to determine if the laser drilling operation is feasible using the company's laser.

A $2^4$ full factorial experiment was developed. This experiment was expected to be very demanding physically, time consuming, and costly. In addition, the time allowed for experimentation in the proposal schedule was unexpectedly reduced. At this point, the engineers have decided to use a $2^{4-1}$ fractional factorial design ($\frac{1}{2}$ FFE of a $2^4$) that would reduce the amount of experimental runs from sixteen to eight. The experimental runs for the fractional factorial (using $L = ABCD$) include the factor combinations, levels, and responses shown below [DK]:

| Observation | Factors and factor-level settings | | | | Response, |
|:---:|:---:|:---:|:---:|:---:|:---:|
| | $A$ | $B$ | $C$ | $D$ | mJ |
| 1 | low | low | low | low | 2128 |
| 2 | high | high | low | low | 1952 |
| 3 | high | low | high | low | 2200 |
| 4 | high | low | low | high | 1700 |
| 5 | low | high | high | low | 1328 |
| 6 | low | high | low | high | 1520 |
| 7 | low | low | high | high | 1820 |
| 8 | high | high | high | high | 906 |

**a** Verify the block and factor settings of the experimental protocol using the confounding scheme. Develop the alias structure, and determine the resolution.

**b** Develop an appropriate set of plots.

**c** Develop and interpret a normal probability plot analysis for the effects.

**d** Develop an ANOVA analysis and plot the appropriate s.e.i. arrows on the plots. Then interpret your results, using your s.e.i. arrows and the nominal response level.

**e** Develop a suitable predictor, select the best factor setting, and estimate $\hat{\mu}_{best}$.

**f** Critique the design protocol in terms of its meeting the nominal is best experimental objectives.

**23.10** AQUACULTURE CASE STUDY

*Background*

In the last few years, many entrepreneurs have entered the field of aquaculture. Aquaculture is defined as the regulation and cultivation of water plants and animals for human consumption. One of the most visible uses of aquaculture has been catfish farming. This form of aquaculture has provided a market for fresh catfish to supermarkets, restaurants, and avid fishermen.

In recent years, aquaculture research has also been applied to raising shrimp and oysters in a controlled environment. This trend was generated as a result of increased demand for shellfish products, and a decrease in quality of marketable shellfish due to contamination of habitat. For example, the contamination of oyster beds in many areas has occurred as a result of frequent flooding of oyster beds with fresh water and other contaminants.

For these reasons, entrepreneurs are looking at different methods of oyster farming. Some of the factors important to the propagation, growth, and survival of oysters are bottom type, water circulation, salinity, temperature, food, sedimentation, pollution, competition, disease, and predation.

*Purpose*

This case study will address the impact of four variables—bed density, salinity, temperature, and water flow—that can be controlled under commercial conditions.

---

*Experimental Variables*

| Response variable | Measurement technique |
|---|---|
| Average weight of oyster meat | Laboratory scale (ounces) |

| Factors under study | Levels |
|---|---|
| A  Bed density | 300, 600 spat (young oysters) per tray |
| B  Salinity | 10 ppt, 15 ppt |
| C  Temperature | 15°C , 20°C |
| D  Water flow | 4, 8 l/min |
| E  Tray position in water flow | 1 (inlet), 2 (near inlet), 3 (near outlet), 4 (outlet) |

| Background variables | Method of control |
|---|---|
| Food | Grown at location |
| Percent oxygen | Aeration at water outlet |
| Other chemicals in seawater | Not measured at this time |

---

*Experimentation*

The experiment was conducted for 12 months in order to gather the necessary data  Sixteen different troughs were prepared in order to simulate all possible conditions under which oysters could be grown. Eight troughs contained 600 spat per tray, and eight troughs contained 300 spat per tray. Spat were assigned at random to 64 trays. Four trays were laid back to back in a trough. Bay water was pumped over the trays at a controlled rate. The trays were numbered and arranged 1 through 4 in each trough, with tray 1 being located at the water inlet point and tray 4 at the outlet point. Salinity was controlled by the addition of artificial sea salt, and concentrations were measured using a temperature compensated refractometer. At the end of the experiment, oysters were selected at random from each tray, within each trough. The oysters were then measured umbo to bill, and the oyster meat was carefully removed and weighed. Two oysters were removed from each tray and identified by the corresponding tray and trough numbers. The weight of the oyster meat and the relationship to the various combinations of growing conditions were the focus of this experiment. The factors and response data are shown below [DM]:

*Experimental Data*

| | Water flow (4 l/min) | | | |
| | Temperature 1 (15°C) | | | |
| | Low density population (300 spat) | | High density population (600 spat) | |
| Water flow position | Salinity (10 ppt) | Salinity (15 ppt) | Salinity (10 ppt) | Salinity (15 ppt) |
|---|---|---|---|---|
| 1 | 2.6 | 2.9 | 2.8 | 3.0 |
|   | 2.5 | 2.7 | 2.9 | 2.9 |
| 2 | 2.5 | 2.6 | 2.8 | 2.8 |
|   | 2.4 | 2.5 | 2.7 | 2.8 |
| 3 | 2.3 | 2.5 | 2.4 | 2.4 |
|   | 2.3 | 2.3 | 2.2 | 2.2 |
| 4 | 1.5 | 2.1 | 1.6 | 1.8 |
|   | 1.6 | 2.0 | 1.4 | 1.6 |

| Water flow position | Temperature 2 (20°C) | | | |
|---|---|---|---|---|
| 1 | 2.8 | 2.9 | 3.0 | 3.1 |
|   | 2.7 | 2.8 | 2.9 | 2.8 |
| 2 | 2.7 | 2.8 | 2.9 | 2.6 |
|   | 2.7 | 2.6 | 2.8 | 2.7 |
| 3 | 2.5 | 2.5 | 2.5 | 2.4 |
|   | 2.2 | 2.5 | 2.3 | 2.3 |
| 4 | 2.0 | 2.2 | 1.9 | 1.9 |
|   | 1.8 | 2.1 | 1.9 | 1.8 |

| | Water flow (8 l/min) | | | |
| | Temperature 1 (15°C) | | | |
| | Low density population (300 spat) | | High density population (600 spat) | |
| Water flow position | Salinity (10 ppt) | Salinity (15 ppt) | Salinity (10 ppt) | Salinity (15 ppt) |
|---|---|---|---|---|
| 1 | 3.6 | 3.8 | 4.0 | 4.2 |
|   | 3.5 | 3.8 | 3.9 | 4.1 |
| 2 | 3.3 | 3.6 | 3.9 | 4.0 |
|   | 3.1 | 3.5 | 3.7 | 3.9 |
| 3 | 2.8 | 3.5 | 3.7 | 3.9 |
|   | 2.6 | 3.3 | 3.6 | 3.7 |
| 4 | 2.2 | 2.7 | 3.2 | 3.5 |
|   | 2.1 | 2.6 | 3.1 | 3.4 |

| Water flow position | Temperature 2 (20°C) | | | |
|---|---|---|---|---|
| 1 | 3.6 | 3.8 | 4.1 | 4.3 |
|   | 3.7 | 3.9 | 4.0 | 4.1 |
| 2 | 3.5 | 3.7 | 3.9 | 4.0 |
|   | 3.4 | 3.6 | 3.8 | 3.9 |
| 3 | 3.0 | 3.2 | 3.6 | 3.9 |
|   | 2.8 | 3.0 | 3.5 | 3.8 |
| 4 | 2.4 | 2.5 | 3.5 | 3.6 |
|   | 2.2 | 2.6 | 3.3 | 3.4 |

**a** Analyze the five factor experimental data as a $2^4 \times 4$ FAT-RCB with two replications.

**b** Analyze the data as a bigger is better Taguchi SNR transformation in a full factorial, using tray position (i.e., water flow position; tray 1 = upstream, ..., tray 4 = downstream) as a noise variable. Here, you will use a $2^4$ FAT-CRD as the design array and a bigger is better SNR transformation built up from the eight oyster meat weight observations in a given water flow, temperature, population density, salinity combination.

**c** Compare and contrast your findings.

**23.11** EMISSIONS CASE STUDY

*Background*

Increasing demand from the general public for cleaner air emissions has led to tighter EPA standard emission requirements for acid gases, including sulfur dioxide ($SO_2$). To reduce the emission level of $SO_2$ in the stack gas of a processing plant, robust performance of the absorption unit which affects the sulfur removal process is necessary.

*Process*

In the acid gas absorption process an $SO_2$-rich gas is fed to the bottom of an absorption column where it is contacted with an $SO_2$-lean absorbent solution. $SO_2$ is absorbed into the solution and exits from the bottom of the column. The vent gas, containing a lesser amount of $SO_2$, is released from the top of the absorption column into the atmosphere.

*Purpose*

A processing plant has decided to study absorption unit performance. The objective of this experimental study was to choose the combination of the design factors, under selected noise conditions, that would yield the best response. A Taguchi experimental design was used to assess robust performance of the $SO_2$-absorption process. The responses were collected: (1) the percent $SO_2$ absorption (e.g., percent $SO_2$ removed, a larger is better response), (2) the $SO_2$-emission level (ppmv) in the vent gas, a smaller is better response.

*Detailed experiment*

A CE diagram identifying factors which may affect the absorption process responses was developed. The current state of knowledge indicated that absorption column operating pressure was not significant. Therefore, a common operating pressure was set at 1 atmosphere; hence, no additional coating or material reinforcement on the absorption column was necessary. In addition, relatively low capital requirements and an explosion-proof process resulted. The ratio of base ($Na^+$) and acid ($PO_4^{2-}$) added to the absorbent solution was set at 1.5. This ratio was consistent with most $SO_2$-removal processes. The common number of beds in the absorption column was set at two, for economic reasons.

The design factors chosen for the detailed experiment are listed below. The noise factor was the percent $SO_2$ in the feed gas.

| Design factor | Level 1 | Level 2 |
|---|---|---|
| $A$: Temperature, °F | 113 | 165 |
| $B$: Acid $PO_4^{2-}$, M | 0.5 | 1.0 |
| $C$: Solution flow rate, gal/min | 0.2 | 2.0 |
| $D$: Feed gas flow rate, scfm | 1.0 | 10.0 |
| $E$: Base $Na^+$, M | 0.75 | 1.50 |

| Noise factor | Level 1 | Level 2 | Level 3 |
|---|---|---|---|
| $SO_2$ in gas, % | 0.2 | 1.0 | 1.8 |

An $L_8$ orthogonal array was chosen for the design factors. The array and column assignments are shown below.

| | Column Assignments | | | | | | |
|---|---|---|---|---|---|---|---|
| | 1 | 2 | 3 | 4 | 5 | 6 | 7 |
| Exp. trial | A | B | AB | C | E | BC | D |
| 1 | 1 | 1 | 1 | 1 | 1 | | 1 |
| 2 | 1 | 1 | | 2 | 2 | | 2 |
| 3 | 1 | 2 | | 1 | 1 | | 2 |
| 4 | 1 | 2 | | 2 | 2 | | 1 |
| 5 | 2 | 1 | | 1 | 2 | | 2 |
| 6 | 2 | 1 | | 2 | 1 | | 1 |
| 7 | 2 | 2 | | 1 | 2 | | 1 |
| 8 | 2 | 2 | | 2 | 1 | | 2 |

The experimenter's log has been summarized and displayed in the form of data tables. Separate tables were created for each of the two responses—percent absorption and $SO_2$ emission (ppmv) [ST]:

| Exp. trial | Experimental treatments summary | | | | |
|---|---|---|---|---|---|
| | temperature, °F | Acid $PO_4^{2-}$, M | Solution flow rate, gal/min | Base $Na^+$, M | Gas flow rate, scfm |
| 1 | 113 | 0.5 | 0.2 | 0.75 | 1.0 |
| 2 | 113 | 0.5 | 2.0 | 1.50 | 10.0 |
| 3 | 113 | 1.0 | 0.2 | 0.75 | 10.0 |
| 4 | 113 | 1.0 | 2.0 | 1.50 | 1.0 |
| 5 | 165 | 0.5 | 0.2 | 1.50 | 10.0 |
| 6 | 165 | 0.5 | 2.0 | 0.75 | 1.0 |
| 7 | 165 | 1.0 | 0.2 | 1.50 | 1.0 |
| 8 | 165 | 1.0 | 2.0 | 0.75 | 10.0 |

| Exp. trial | % $SO_2$ in gas | | |
|---|---|---|---|
| | 0.2 | 1.0 | 1.8 |
| 1 | 62.86 | 64.33 | 63.78 |
| 2 | 62.87 | 65.39 | 65.47 |
| 3 | 7.89 | 7.93 | 8.60 |
| 4 | 94.75 | 99.90 | 99.94 |
| 5 | 12.73 | 14.10 | 15.07 |
| 6 | 99.33 | 98.67 | 98.55 |
| 7 | 46.23 | 43.99 | 45.56 |
| 8 | 39.71 | 44.15 | 44.28 |

Response—absorption percent

| Exp. trial | % $SO_2$ in Gas | | |
|---|---|---|---|
| | 0.2 | 1.0 | 1.8 |
| 1 | 37.2 | 180.8 | 877.4 |
| 2 | 37.2 | 175.5 | 319.0 |
| 3 | 90.4 | 451.8 | 806.7 |
| 4 | 5.3 | 0.5 | 0.5 |
| 5 | 79.8 | 393.5 | 701.9 |
| 6 | 0.5 | 5.3 | 10.6 |
| 7 | 42.5 | 223.3 | 393.5 |
| 8 | 47.9 | 223.3 | 404.1 |

Response — $SO_2$ emission (ppmv)

**a** Analyze the experimental data and develop graphical displays as appropriate. Interpret your graphs and analyses.

**b** Develop suitable predictor models and estimate the response in both decibels and original units at the best operating combination of the design variables.

**c** Comment on the adequacy or inadequacy of the Taguchi based procedure and models developed in this experiment.

**23.12** Select a product or process, then design, execute, and interpret an FFE experiment using either a classical approach or a Taguchi design with noise variables. Follow the six experimental protocol steps (Table 20.2). After completion of your experiment, critique your design's effectiveness in meeting your experimental objectives. Develop a display board documenting your experiment and results.

# REFERENCES

**1** C. R. Hicks, *Fundamental Concepts in the Design of Experiments*, 3d ed., New York: Holt, Rinehart, and Winston, 1982.

**2** R. G. D. Steel and J. H. Torrie, *Principles and Procedures of Statistics*, New York: McGraw-Hill, 1960.

**3** G. E. P. Box, W. G. Hunter, and J. S. Hunter, *Statistics for Experimenters*, New York: Wiley, 1978.

**4** J. S. Hunter, "Statistical Design Applied to Product Design," *Journal of Quality Technology*, vol. 17, no. 4, pp. 210–221, October 1985.

**5** J. S. Hunter, "Let's All Beware the Latin Square," *Quality Engineering*, vol. 1, no. 4, pp. 453–465, 1989.

**6** See reference 1.

**7** See reference 3.

**8** D. C. Montgomery, *Design and Analysis of Experiments*, 3d ed., New York: Wiley, 1991.

**9** See reference 3.

**10** See references 4 and 5.

**11** See reference 8.

**12** See references 1, 3, and 8.

**13** W. G. Cochran and G. M. Cox, *Experimental Designs*, 2d ed., New York: Wiley, 1957.

**14**  W. Diamond, *Practical Experiment Designs*, Belmont, CA: Lifetime Learning, 1981.

**15**  See references 8 and 3.

**16**  G. Taguchi. *Introduction to Quality Engineering: Designing Quality into Products and Processes*, White Plains, NY: Kraus International, UNIPUB (Asian Productivity Organization), 1986.

**17**  V. N. Nair et al. "Taguchi's Parameter Design: A Panel Discussion," *Technometrics*, vol. 34, no. 2, pp. 127–161, 1992.

**18**  J. J. Pignatiello and J. S. Ramberg, "Top Ten Triumphs and Tragedies of Genichi Taguchi," *Quality Engineering*, vol. 4, no. 2, pp. 221–225, 1991–1992.

**19**  R. N. Kackar, "Off-line Quality Control, Parameter Design, and the Taguchi Method," *Journal of Quality Technology*, vol. 17, no. 4, pp. 176–188, 1985.

**20**  G. E. P. Box, "Discussion, Off-line Quality Control, Parameter Design, and the Taguchi Method," *Journal of Quality Technology*, vol. 17, no. 4, pp. 189–190, 1985.

**21**  M. S. Phadke, *Quality Engineering Using Robust Design*, Englewood Cliffs, NJ: Prentice Hall, 1989.

**22**  P. J. Ross, *Taguchi Techniques for Quality Engineering*, New York: McGraw-Hill, 1988.

**23**  See references 4, 5, 17, 18, and 16.

**24**  See reference 16.

**25**  See reference 21.

**26**  See reference 16.

**27**  See reference 21.

**28**  J. S. Hunter, "Signal to Noise Ratio Debated," *Quality Progress*, pp. 7–9, May, 1987.

**29**  See references 8, 19, 20.

**30**  See references 16, 21, and 22.

**31**  See reference 18.

**32**  *SAS/STAT User's Guide*, Volumes 1 and 2, Version 6, Cary, NC: SAS Institute, 1990.

**33**  *SAS Procedures Guide*, Version 6, Cary, NC: SAS Institute, 1990.

**34**  See references 16, 21, and 22.

# SEVEN

# CREATION OF QUALITY— RELIABILITY MODELS

## VIRTUES FOR LEADERSHIP IN QUALITY
### Faith and Compassion

In Section Seven, we present a number of quantitative reliability and maintainability models to complement our qualitative discussions of long-term performance (see Chapters 5, 6, 12, and 13). Our purpose in this section is to introduce our readers to both mission and time or usage based reliability models and to the concepts of maintainability and availability. In Chapter 24, we develop block diagram construction methods and mission based reliability models. We present a number of strategies that aid in assuring long-term, sustained, performance. We also briefly discuss the stress-strength concepts and modeling techniques.

In Chapter 25, we discuss the concept of time based reliability and maintainability models, as well as the concept of availability. We present the concept of modeling redundancy as well as the concept of combining both reliability and maintainability characteristics to assess system availability. We discuss both analytical and simulation modeling formats.

We close out the section in Chapter 26 with a treatment of both nonparametric and parametric reliability model building in the context of designed experiments. Graphical model fitting is introduced in the form of probability plotting. We include a basic discussion of covariate reliability models where such factors as environmental stresses, product configurations, and so on, are represented by covariates. Throughout, we stress the use of computer aids to model our empirical data.

# 24

# RELIABILITY CONFIGURATIONS AND STRESS-STRENGTH MODELS

## 24.0 INQUIRY

1 What is reliability? Maintainability?
2 How can product-process configurations affect sustained performance?
3 How is mission based reliability defined and modeled?
4 What is reliability allocation?
5 How do stress-strength relationships impact reliability?

## 24.1 INTRODUCTION

Both reliability and maintainability are critical quality characteristics. The former deals with sustained, failure-free product-process performance that meets or exceeds our customer's needs and expectations; the latter deals with critical long-term product-process field performance in terms of failure prevention, as well as restoration. *Reliability is defined as the probability that a system, subsystem, or component will perform successfully for a specified amount of time or usage or for a given mission, when operated under specified conditions. Maintainability is defined as the probability that a failed system, subsystem, or component will be restored to operable condition in a given amount of time or effort when restoration is performed under specified conditions.*

Reliability and maintainability are critical product-process quality characteristics. *Reliability-maintainability engineering seeks to improve both product and process performance in the long run, and when successful, results in improved customer satisfaction.* Hence, reliability-maintainability is a measurable, predictable characteristic while reliability-maintainability engineering is a means to impact the measure. *Two major challenges are present in reliability engineering, analyses, and predictions*:

1 *Customer demands*—variation in product and process expectations
2 *Product and process performance*—variation as a result of
   a configurations
   b applications and materials
   c environments
   d operating methods

These two challenges are encountered, to one degree or another, in reliability engineering for all products, services, and production processes. Through careful product and process definitions, robust design, and mistake-proofing, we can favorably influence their impact.

A proactive view of quality requires that we consider reliability as it relates to expected performance, rather that on a historic basis. It is beneficial to treat the reliability characteristics of products and processes which can change with time as probabilistic rather than deterministic. We will use applied probability theory in building our reliability models, and hence will be able to deal with both unconditional and conditional analyses. For example, we might buy an automobile and make the following mission related statements regarding true and substitute quality characteristics expressed in terms of conditional and unconditional reliability:

*Unconditional reliability*: The probability of driving 50,000 mi without a transmission failure is 0.75.

*Conditional reliability*: The probability of driving 50,000 mi without a transmission failure given that we have already successfully completed 40,000 mi of the 50,000 is 0.916.

*Conditional reliability*: The probability of driving an additional 1000 mi without a transmission failure given that grinding noises are now coming from our transmission is 0.01.

*We deliver sustained performance to our customers through a four-level hierarchical strategy, which involves both reliability and maintainability considerations*:

1 *Sustained performance at the component level throughout the system*—In this "pure" reliability strategy we rely on the component itself. Reliability engineering includes analysis of the failure modes and mechanisms and improvements with respect to the materials, processes, methods of manufacture, and so on to enhance component performance. Here, we basically increase the "design margin" (i.e., the ratio of the component's capacity relative to the loads to which the component is subjected).

2 *Sustained performance through redundant configurations*—Here, we use standby and redundant systems where components may fail, but the overall system performance functions are not interrupted. Reliability engineering here deals with redundant configurations and switching mechanisms (physical or otherwise). In systems where the cost factor is dominant, components that increase the reliability most per dollar will see greater redundancy.

3 *Sustained performance through effective preventive maintenance policies and practices (before failure)*—Here, reliability-maintainability engineering focuses on locating preventive maintenance "points," determining the frequency and extent of preventive maintenance, and developing maintenance training.

4 *Sustained performance through effective corrective maintenance policies and practices (after failure)*—Reliability-maintainability engineering in this

strategy deals with the development of effective diagnosis and repair procedures, repair training, spare parts inventories, and so on.

The failure mode and effects analysis (FMEA), fault tree analysis (FTA), and event tree analysis (ETA), tools discussed in Chapters 12 and 13 can be used to proactively affect long-term performance through a reliability focus at all of these four levels. However, now we want to expand our discussion to include configuration, mission, and time or usage based models with both qualitative and quantitative elements.

## 24.2  PHYSICAL AND RELIABILITY CONFIGURATIONS

The architecture of modern systems typically includes physical, human, and/or software elements. *A "system" can be defined as an orderly arrangement of components that interact among themselves and with external components, other systems, and humans to perform some intended function.* If we work from the top down, the system can be defined at a functional level. Then, the functional definition can be brought down to a system configuration level and, finally, to the component level. *Reliability modeling and analyses may be performed at the functional level, as well as at the system, subsystem, and/or component levels, in order to study system reliability characteristics.*

### Block Diagrams

*Block diagrams are useful to represent system configurations.* Each block represents either a functional characteristic, a subsystem, or component, and has an input and an output. The box itself represents either a functional or component "mechanism" to convert an input to an output. Block diagrams are fundamental and valuable tools in product-process definition, design, analysis, and improvement.

In building block diagrams, we first describe the physical system at the functional level to the desired degree. Then, we construct additional block diagrams to describe the reliability characteristics of the critical elements of the system. *Five basic configurations*, shown in Figure 24.1, *represent the fundamental building blocks in developing system design alternatives*.

The *series configuration* (shown in Figure 24.1*a*) is very common and is denoted as *n/n* or *n* out of *n*, where *n* represents the number of components (physical/functional) in the system. In other words, all of the *n* components must be successful for the system to perform successfully in a series configuration. For example, a basic automobile drive system can be described by the simple series configuration of subsystems shown in Figure 24.2. Both a functional and a hardware oriented system are laid out. Each subsystem can be further "broken out" by increasing levels of detail. Here we can see that a "single-point" failure (a failure isolated in one subsystem or block) of any of the subsystems or blocks in a series configuration will result in an overall system failure. There is no redundancy in a series system.

| Configuration | Comments |
|---|---|

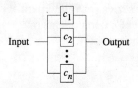

*(a)* Series configuration (*n/n* configuration)

All components must function for system success.

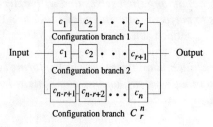

*(b)* Parallel configuration (1/*n* configuration)

At least one component must function for system success.

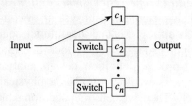

*(c)* *r/n* configuration

At least $r$ out of a total of $n$ components must function for system success, where ($r \leq n$). The series is an $n$ out of $n$ configuration and the parallel is a 1 out of $n$ configuration.

$$C_r^n = \frac{n!}{r!\,(n-r)!}$$

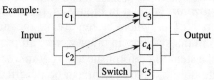

*(d)* Standby configuration

Standby configurations allow for reserve components which are placed in action upon failure of the main component. The standby units may or may not be identical to the main unit.

Example:

Input — Output

*(e)* General directed configuration

The general configuration contains components arranged in any manner so long as one input and one output exists.

**FIGURE 24.1**    Basic physical and reliability configurations.

The *parallel configuration* (shown in Figure 24.1b) denoted as $1/n$ or 1 out of $n$, is extremely tolerant of single-point component failures. Here, the configuration is such that if at least one component survives, the overall system performance is sustained. For example, an airplane with two engines, designed to fly successfully with only one engine, represents a parallel configuration.

The *r out of n (r/n) configuration* (shown is Figure 24.1c) is a general category which technically includes the series ($n/n$) and parallel ($1/n$) configurations as extreme cases. Here, at least $r$ out of $n$ components ($n \geq r$) must function successfully to sustain a successful overall system performance. For example, a four-engine aircraft that can fly successfully on any two of the four engines represents a two-out-of-four ($2/4$) $r/n$ configuration. Here, as in all cases, we must initially define what constitutes a successful performance at the component (physical-functional) level in technical terms for our configurations to be meaningful as product or process system alternatives.

The *standby configuration* (shown in Figure 24.1d) is similar to the parallel configuration with the exception that only one unit is in service at any given time. In a physical context, some means of switching from the primary, or first, unit to the standby unit must exist. The standby unit or units do not have to be identical to the primary unit. For example, an electrical power user might develop the standby system shown in Figure 24.3, where the primary power unit is the local utility grid and the standby unit is an internal combustion motor-generator set.

Figure 24.1e depicts a *general directed configuration,* which must include one input and one output, but otherwise is limited only by our imagination and the basic laws of nature. This configuration must be depicted with a directional arrow (or arrows) to identify the input-to-output paths, hence the designation "directed." Examples of general configurations are found in electronic packaging, computer networks, manufacturing fabrication and assembly channels within a plant, and so on. A brief manufacturing fabrication example for a sheet metal part family is shown in Figure 24.4.

Block diagrams are useful in both physical and reliability analyses. Usually, we start with a block diagram sequence which represents a physical system. Then, we develop a reliability configuration to represent the reliability characteristics of the system. Once the system alternative has been identified, we can identify system maintainability or support points. As a rule, the physical system block diagram and the reliability configuration are similar. However, exceptions exist, so care must be exercised when moving from a physical block diagram to a reliability block diagram. Figure 24.5 depicts one of these exceptions; in this case, a series arrangement of electric circuit breakers in an electric line in the physical system (shown in Figure 24.5a) generates a parallel reliability configuration that provides a means of interrupting current flow (shown in Figure 24.5b).

## 24.3 MISSION BASED RELIABILITY MODELS

Physical system configuration alternatives are proposed in the product-process definition phase of the product-process life cycle. We can develop reliability block

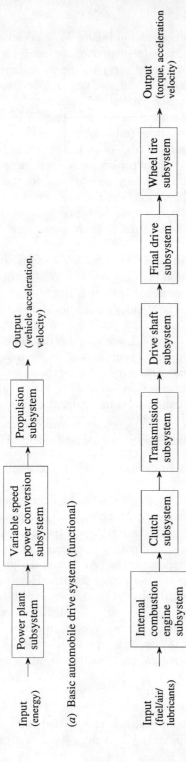

Input
(energy)  →  Power plant
subsystem  →  Variable speed
power conversion
subsystem  →  Propulsion
subsystem  →  Output
(vehicle acceleration,
velocity)

(*a*) Basic automobile drive system (functional)

Input
(*fuel/air/
lubricants*)  →  Internal
combustion
engine
subsystem  →  Clutch
subsystem  →  Transmission
subsystem  →  Drive shaft
subsystem  →  Final drive
subsystem  →  Wheel tire
subsystem  →  Output
(torque, acceleration
velocity)

(*b*) Basic automobile drive system (hardware)

**FIGURE 24.2**  Example for a series configuration system.

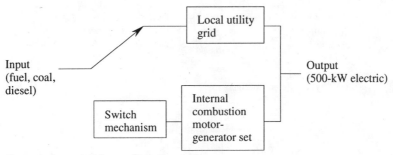

**FIGURE 24.3**    Example for a standby configuration system.

diagrams as our next step. Hence, we can study high-level product-process reliability alternatives in a proactive manner at this early stage of the life cycle.

*In order to study system configuration reliability in a meaningful way, on a mission level basis, we must follow three primary steps:*

1  *Define an acceptable level of physical performance* in enough detail so that it is meaningful to our customers.
2  *Define the mission* (e.g., the goal, objective, target) in enough detail so as to reflect our customers' field environments, applications, and operating methods.
3  *Develop reasonable reliability estimates* for the system blocks, where the former constitute a mission success probability. Here, system reliability is the probability that system performance will exceed the customer's demanded physical performance (defined in step 1), considering the customer's field environment, application, and operating method, (defined in step 2).

## Mission Success and Failure Measures

*The usual mission reliability case deals with one of two outcomes*: **(1)** *success or* **(2)** *failure.* We may use engineering judgment (expert opinion) and develop an *a priori* estimate without hard experimental evidence. Or, on the other hand,

**FIGURE 24.4**    Example for a general directed configuration system.

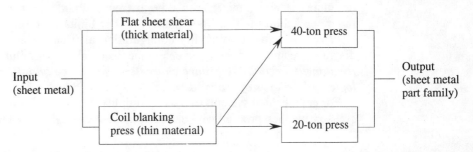

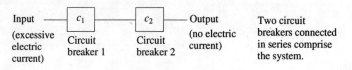

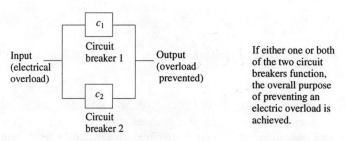

(a)  Physical system configuration

(b)  System reliability configuration

FIGURE 24.5    Physical and reliability configurations for an electric line with electrical circuit breakers installed.

we might base our estimate on results obtained from either laboratory or field experience with similar technology or actual performance tests.  Typically, a combination of engineering judgment and past experience with similar technology will be used.  We may work  with either the failure probability $p$ or the success probability $q = 1 - p$.  If we work with $p$, we develop a configuration failure probability; if we work with $q$,  we develop a configuration success probability or reliability.  In either case, we know that

$$P(\text{mission reliability}) = 1 - P(\text{mission failure}) \qquad (24.1)$$

If *a priori* failure probability estimates are used in a mission based system failure model, we may want to establish a pessimistic estimate (worst case), a most likely estimate, and an optimistic estimate (best case).  By using the worst and best case estimates, we can develop crude upper and lower bounds on our system's mission failure probability estimate.  If hard data are used, we may also want to calculate crude optimistic and pessimistic bounds for either configuration reliability or failure.  Here, we might use the limits from a confidence interval to serve as our failure probability limits.  Binomial confidence limits are discussed in Chapter 26.

Once we define a reliability configuration alternative and develop its block diagram, we are in a position to determine the reliability or failure probability. *The fundamental reliability-failure characteristics can be quantified using Boolean logic and the laws of probability*.

For complicated systems, we try to simplify the system by breaking it into subsystems and then proceed with our analysis.  For example, we search the system for series, parallel, *r/n,* standby, and generic directed subsystems, analyze these sub-systems, and, finally put the developed subsystem analyses together to obtain the system reliability or failure probabilities. Assumptions dealing with independence of component failures and identical-nonidentical components will simplify our calculations.

### Evaluation of Series, Parallel, and *r/n* Configurations

For successful performance or system success in a series configuration, all components must function; in a parallel system configuration, at least one must function; in an *r/n* configuration, at least *r* out of *n* $(r \leq n)$ components must function. Block diagrams and generic reliability models for each of these configurations are shown in Figure 24.6.

As previously stated, the series and parallel configurations are actually special cases of the more general *r/n* configuration. A series system is an *n/n* configuration, while the parallel system is a *1/n* configuration. The series system is termed a "worst case system", the parallel, a "best case system." Here, the worst or best case identification refers to possible *r/n* configurations of the *n* components.

***Independent, Identical Components (IIC)***   The IIC case is the simplest to quantify. Here, independent implies that one component's failure is not influenced by another component's failure or some common cause condition. The generic models in Figure 24.6 are relatively easy to develop for the IIC case. Given the two possible outcomes (success or failure) for each component and the IIC assumption, the binomial model is appropriate. We let *q* represent the probability of mission success (the reliability of the component) and *p* represent the probability of component failure during the mission $(q = 1 - p)$.

*Series Configuration:*

$$R_{sys} = C_n^n \, q^n p^0 = q^n = (1 - p)^n \tag{24.2}$$

*Parallel Configuration:*

$$R_{sys} = 1 - C_n^n \, q^0 p^n = 1 - p^n \tag{24.3}$$

*r/n Configuration:*

$$R_{sys} = \sum_{i=r}^{n} C_i^n \, q^i \, p^{n-i} \tag{24.4}$$

where $R_{sys}$ represents overall system reliability, and $C_i^n = \dfrac{n!}{i! \, (n-i)!}$ is the combination of *n* things taken *i* at a time.

***Independent, Nonidentical Components (INC)***   If the components are not identical in their failure characteristics, then the quantification is more difficult. The generic models in Figure 24.6 are also applicable for INC components. However, considerably more effort is required here than in the IIC case. Each component's

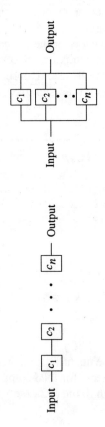

$R_{sys} = P(c_1 \cap c_2 \cap \cdots \cap c_n)$

$R_{sys} = 1 - P(\tilde{c}_1 \cup \tilde{c}_2 \cup \cdots \cup \tilde{c}_n)$

where $c_i$: success of component $i$

$\tilde{c}_i$: failure of component $i$

(a) Series configuration

$R_{sys} = P(c_1 \cup c_2 \cup \cdots \cup c_n)$

$R_{sys} = 1 - P(\tilde{c}_1 \cap \tilde{c}_2 \cap \cdots \cap \tilde{c}_n)$

(b) Parallel configuration

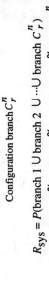

$R_{sys} = P(\text{branch 1} \cup \text{branch 2} \cup \cdots \cup \text{branch } C_r^n)$

$R_{sys} = 1 - P(\text{branch 1} \cap \text{branch 2} \cap \cdots \cap \text{branch } C_r^n)$

$C_r^n = \dfrac{n!}{r!\,(n-r)!}$

(c) $r$ out of $n$ configuration

**FIGURE 24.6**  Reliability models for series, parallel, and $r/n$ configurations.

reliability $R(c_i)$, which is also denoted by $P(c_i)$, is the probability of component success throughout the mission. Unreliability (i.e., the failure probability) is denoted by $P(\tilde{c}_i)$.

*Series Configuration:*

$$R_{sys} = \prod_{i=1}^{n} P(c_i) \tag{24.5}$$

*Parallel Configuration:*

$$R_{sys} = 1 - \prod_{i=1}^{n} P(\tilde{c}_i) \tag{24.6}$$

*r/n Configuration:*

$$R_{sys} = P(B_1 \cup B_2 \cup \cdots \cup B_{C_r^n}) \tag{24.7}$$

It can be shown for the *r/n* configuration that

$$R_{sys} = P(B_1) + P(B_2) + \cdots + P(B_{C_r^n}) - [P(B_1 \cap B_2) + P(B_1 \cap B_3) + \cdots + P(B_{C_{r-1}^n} \cap B_{C_r^n})]$$
$$+ [P(B_1 \cap B_2 \cap B_3) + \cdots + P(B_{C_{r-2}^n} \cap B_{C_{r-1}^n} \cap B_{C_r^n})] - [\ldots] + [\ldots]$$

where $B_i$ = configuration branch $i$, $i = 1, 2, \ldots, C_r^n$.

We must work through the above terms carefully to reflect the effects of the common elements in each of the $C_r^n$ branches. In other words, we must recall that multiple branches ($B_i$'s) include common components; hence, branch reliabilities are not independent.

---

**Example 24.1**

For a three-component system, develop a general expression for
**a** A series INC system reliability and a series IIC system reliability.
**b** A parallel INC system reliability and a parallel IIC system reliability.
**c** A two out of three INC system reliability and a two out of three IIC system reliability.

**Solution**

**a** INC:   $R_{sys} = P(c_1)\, P(c_2)\, P(c_3)$
   IIC:   $R_{sys} = q^3$

**b** INC:   $R_{sys} = 1 - [P(\tilde{c}_1)\, P(\tilde{c}_2)\, P(\tilde{c}_3)]$
   IIC:   $R_{sys} = 1 - p^3 = 1 - (1 - q)^3$

**c** INC:   $R_{sys} = P(B_1 \cup B_2 \cup B_3) = P(B_1) + P(B_2) + P(B_3) - P(B_1 \cap B_2)$
$\qquad\qquad - P(B_1 \cap B_3) - P(B_2 \cap B_3) + P(B_1 \cap B_2 \cap B_3)$

where $B_1$ = branch including components $c_1$ and $c_2$
$B_2$ = branch including components $c_1$ and $c_3$
$B_3$ = branch including components $c_2$ and $c_3$

Then,

$$R_{sys} = P(c_1 \cap c_2) + P(c_1 \cap c_3) + P(c_2 \cap c_3) - P(c_1 \cap c_2 \cap c_3) - P(c_1 \cap c_2 \cap c_3)$$
$$- P(c_1 \cap c_2 \cap c_3) + P(c_1 \cap c_2 \cap c_3)$$

$$R_{sys} = P(c_1 \cap c_2) + P(c_1 \cap c_3) + P(c_2 \cap c_3) - 2P(c_1 \cap c_2 \cap c_3)$$

$$R_{sys} = P(c_1)\,P(c_2) + P(c_1)\,P(c_3) + P(c_2)\,P(c_3) - 2P(c_1)\,P(c_2)\,P(c_3)$$

IIC:    $R_{sys} = q^2 + q^2 + q^2 - 2q^3 = 3q^2 - 2q^3$

---

## Reduction of Systems Combining Series, Parallel, and *r/n* Configurations

***In some cases, systems are made up of many components arranged in combinations of series, parallel, and/or r/n clusters.*** Two simple series-parallel combinations are shown in Figure 24.7. In such a system, we can reduce the series-parallel combinations to determine a system reliability. In general, hybrid series-parallel-*r/n* models are analyzed by systematically grouping the components into subsystems by configuration. Then, each subsystem is analyzed (reduced) until the system reliability is obtained.

## Evaluation of Standby Configurations

Standby configurations vary widely in their physical design. Therefore, we expect their reliability configurations to vary as well. In general, a standby system may consist of components such as those found in the other (series, parallel, and *r/n*) configurations. The uniqueness of the standby configuration is expressed by two prerequisites:

**FIGURE 24.7**    Simple series, parallel combinations.

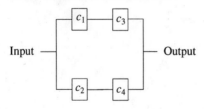

IIC: $R_{sys} = 1 - (1 - q^2)^2$

(*a*) Series configuration in parallel

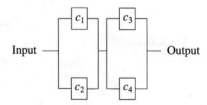

IIC: $R_{sys} = [1 - (1 - q)^2]^2$

(*b*) Parallel configuration in series

**1** A switching mechanism must be present to switch from the primary component to the standby component.

**2** Conditional probabilities must be used to capture the nature of the standby component's reliability.

It is difficult to draw a highly descriptive reliability block diagram for a standby system. Figure 24.8 depicts two possible configurations and basic reliability model statements. The first system (Figure 24.8$a$) is a classical "perfect switch" or "switchless" configuration; the second (Figure 24.8$b$) is a complete series switch configuration, where each switch block corresponds to the switching sequence that must take place in order to use the standby components. For example, $s_{1-2}$ is the switchover from $c_1$ to $c_2$; $s_{2-3}$ is the switchover from $c_2$ to $c_3$.

## Evaluation of General Directed Configurations

*Configurations that cannot be grouped in series, parallel, r/n, or standby subsystems are termed general directed configurations.* Figure 24.9$a$ depicts a simple directed configuration. *A number of analysis methods exist for the solution of general directed configurations.* Four methods will be developed in this section: (1) event counting, (2) tie set, (3) cut set, and (4) key component or decomposition. Each of these methods will be discussed and illustrated for a simple configuration where component failures are independent.

## Event Counting Method

*The event counting method consists of enumerating all system component success-failure possibilities.* Once all possibilities have been identified, probabilities are calculated for each possible system outcome, relative to each possible component outcome. Here, a 0 represents a failure result, and a 1 represents a success.

**FIGURE 24.8**   General standby configuration models.

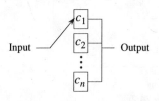

$$R_{sys} = P[c_1 \cup (c_2 \mid \tilde{c}_1) \cup \cdots \cup (c_n \mid \tilde{c}_{n-1} \cap \cdots \cap \tilde{c}_1)]$$

$$R_{sys} = P[c_1 \cup (s_{1-2} \cap (c_2 \mid \tilde{c}_1)) \cup (s_{2-3} \cap (c_3 \mid \tilde{c}_1 \cap \tilde{c}_2)) \cup$$
$$\cdots \cup (s_{(n-1)-n} \cap (c_n \mid \tilde{c}_1 \cap \tilde{c}_2 \cap \cdots \cap \tilde{c}_{n-1}))]$$

($a$) Standby configuration (perfect switches)          ($b$) Standby configuration with switches

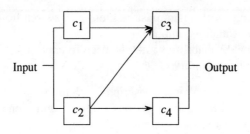

(a)  Reliability block diagram for a four-component
     directed configuration

| Component results | | | | System | Event |
|---|---|---|---|---|---|
| $c_1$ | $c_2$ | $c_3$ | $c_4$ | result | reliability |
| 0 | 0 | 0 | 0 | 0 | $(1-q)^4$ |
| 0 | 0 | 0 | 1 | 0 | $(1-q)^3 q$ |
| 0 | 0 | 1 | 0 | 0 | $(1-q)^3 q$ |
| 0 | 1 | 0 | 0 | 0 | $(1-q)^3 q$ |
| 1 | 0 | 0 | 0 | 0 | $(1-q)^3 q$ |
| 0 | 0 | 1 | 1 | 0 | $(1-q)^2 q^2$ |
| 0 | 1 | 0 | 1 | 1 | $(1-q)^2 q^2$ |
| 1 | 0 | 0 | 1 | 0 | $(1-q)^2 q^2$ |
| 0 | 1 | 1 | 0 | 1 | $(1-q)^2 q^2$ |
| 1 | 0 | 1 | 0 | 1 | $(1-q)^2 q^2$ |
| 1 | 1 | 0 | 0 | 0 | $(1-q)^2 q^2$ |
| 0 | 1 | 1 | 1 | 1 | $(1-q) q^3$ |
| 1 | 0 | 1 | 1 | 1 | $(1-q) q^3$ |
| 1 | 1 | 0 | 1 | 1 | $(1-q) q^3$ |
| 1 | 1 | 1 | 0 | 1 | $(1-q) q^3$ |
| 1 | 1 | 1 | 1 | 1 | $q^4$ |

0: failure result,  1: success result

$$R_{sys} = 3 (1 - q)^2 q^2 + 4 (1 - q) q^3 + q^4 = 3q^2 - 2q^3$$

(b)  Component events and system results

**FIGURE 24.9**    Event counting reliability model (Example 24.2).

Because the two possible outcomes are mutually exclusive, all system success
possibilities can be added together to develop the system reliability.  This method,
while thorough, becomes lengthy for all but the smallest configurations.

**Example 24.2**

Develop a complete, mutually exclusive, set of component success-failure states for the IIC directed system shown in Figure 24.9a.  Then, develop a system reliability expression.

**Solution**

Figure 24.9b shows the event counting solution for this four-component system configuration.  As the solution is based on two outcomes for each identical component, we see $2^4$, or 16, possible outcomes.  Here, $p$ represents the probability of component failure and $(1 - p) = q$ represents the component reliability.

**Tie Set Method**

*The tie set method is based on success paths in the system.*  A "tie set" is defined as a group of components, or branches, which form a connection between input and output on a block diagram, when traversed in the arrow direction.  Furthermore, a "minimum" tie set is a tie set containing a minimum number of components or elements (see Figure 24.10a).  If one component in a minimum tie set fails, and all components outside of the minimum tie set fail, then the system will fail.  If we let $T_i$ represent minimum tie set $i$, where $i = 1, 2, ..., m$, then

$$R_{sys} = P(T_1 \cup T_2 \cup \cdots \cup T_m) = 1 - P(\tilde{T}_1 \cap \tilde{T}_2 \cap \cdots \cap \tilde{T}_m) \qquad (24.8)$$

Here, the minimum tie sets are arranged in a parallel configuration, with the elements in individual tie sets arranged in a series configuration.  Hence, Equation 24.8 can be solved using Equation 24.7 and the techniques illustrated in Example 24.1, part c.

**Cut Set Method**

*The cut set method is related to system interrupt sets.*  A "cut set" is defined as a set of branches or components which interrupt all connections between input and output when removed from a block diagram.  A "minimum" cut set is a cut set which contains a minimum number of elements or components (see Figure 24.10b).  If the components in a minimum cut set fail, when all other components in the system succeed, the system will fail.  If we let $C_i$ represent minimum cut set $i$, where $i = 1, 2, \cdots, m$, then

$$R_{sys} = P(C_1 \cap C_2 \cap \cdots \cap C_m) = 1 - P(\tilde{C}_1 \cup \tilde{C}_2 \cup \cdots \cup \tilde{C}_m) \qquad (24.9)$$

Here, the minimum cut sets are arranged in series, with individual components in each cut set arranged in parallel.

## Key Component (Decomposition) Method

*The key component method is based on identifying a key component and removing it from the configuration.* This method breaks or decomposes the analysis down to two smaller, mutually exclusive configuration analyses (see Figure 24.10c). One of the two analyses assumes that the selected component is certain to succeed. The other analysis assumes that the selected component is certain to fail. The key component method can be repeated to further break the smaller configurations down and further reduce the complexity of a complicated system configuration. When using the key component, or decomposition, method, it is sometimes difficult to choose the key component. A rule of thumb is to choose a component located in "heavy traffic." For example, a component which has many branches entering and/or exiting it.

---

## Example 24.3

Calculate the system reliability for the configuration shown in Figure 24.9a using

a  The tie set method
b  The cut set method
c  The key component method

Show that all three of the results are identical to the event counting solution for IIC components.

### Solution

**a** *Tie Set Method:*
The minimum tie sets are

$$T_1 = \{c_1, c_3\} \qquad T_2 = \{c_2, c_4\} \qquad T_3 = \{c_2, c_3\}$$

Arranging the minimum tie sets as series configurations, all in parallel, as shown in Figure 24.10a, we can solve for the INC system reliability,

$$R_{sys} = P(T_1 \cup T_2 \cup T_3)$$
$$= P(T_1) + P(T_2) + P(T_3) - P(T_1 \cap T_2) - P(T_2 \cap T_3) - P(T_1 \cap T_3) + P(T_1 \cap T_2 \cap T_3)$$

$$R_{sys} = P(c_1 \cap c_3) + P(c_2 \cap c_4) + P(c_2 \cap c_3) - P(c_1 \cap c_2 \cap c_3 \cap c_4) - P(c_2 \cap c_3 \cap c_4)$$
$$\quad - P(c_1 \cap c_2 \cap c_3) + P(c_1 \cap c_2 \cap c_3 \cap c_4)$$

$$= P(c_1 \cap c_3) + P(c_2 \cap c_4) + P(c_2 \cap c_3) - P(c_2 \cap c_3 \cap c_4) - P(c_1 \cap c_2 \cap c_3)$$

For IIC failures,

$$R_{sys} = 3q^2 - 2q^3$$

**b** *Cut Set Method:*
The minimum cut sets are

$$C_1 = \{c_1, c_2\} \qquad C_2 = \{c_3, c_4\} \qquad C_3 = \{c_2, c_3\}$$

Arranging the minimum cut sets as parallel configurations, all in series, as shown in Figure 24.10*b*, we will solve the INC system reliability:

$$\begin{aligned}
R_{sys} &= 1 - P(\tilde{C}_1 \cup \tilde{C}_2 \cup \tilde{C}_3) \\
&= 1 - [P(\tilde{C}_1) + P(\tilde{C}_2) + P(\tilde{C}_3) - P(\tilde{C}_1 \cap \tilde{C}_2) - P(\tilde{C}_2 \cap \tilde{C}_3) - P(\tilde{C}_1 \cap P\tilde{C}_3) \\
&\quad + P(\tilde{C}_1 \cap \tilde{C}_2 \cap \tilde{C}_3)]
\end{aligned}$$

$$\begin{aligned}
R_{sys} &= 1 - [P(\tilde{c}_1 \cap \tilde{c}_2) + P(\tilde{c}_3 \cap \tilde{c}_4) + P(\tilde{c}_2 \cap \tilde{c}_3) - P(\tilde{c}_1 \cap \tilde{c}_2 \cap \tilde{c}_3 \cap \tilde{c}_4) \\
&\quad - P(\tilde{c}_2 \cap \tilde{c}_3 \cap \tilde{c}_4) - P(\tilde{c}_1 \cap \tilde{c}_2 \cap \tilde{c}_3) + P(\tilde{c}_1 \cap \tilde{c}_2 \cap \tilde{c}_3 \cap \tilde{c}_4)] \\
&= 1 - [P(\tilde{c}_1 \cap \tilde{c}_2) + P(\tilde{c}_3 \cap \tilde{c}_4) + P(\tilde{c}_2 \cap \tilde{c}_3) - P(\tilde{c}_2 \cap \tilde{c}_3 \cap \tilde{c}_4) \\
&\quad - P(\tilde{c}_1 \cap \tilde{c}_2 \cap \tilde{c}_3)]
\end{aligned}$$

For IIC failures,

$$R_{sys} = 1 - [3(1 - q)^2 - 2(1 - q)^3] = 3q^2 - 2q^3$$

**c** *Key Component Method:*
Choosing component $c_2$ as the key component in the original configuration, we can break the four component arrangement down to the two simplified configurations shown in Figure 24.10*c*. Then, we can develop a conditional INC solution relative to the key component's success $c_2$ and failure $\tilde{c}_2$ labeled as $R_A$ and $R_B$, respectively:

$$R_A = P(c_2) [1 - P(\tilde{c}_3 \cap \tilde{c}_4)]$$
$$R_B = P(\tilde{c}_2) P(c_1 \cap c_3)$$
$$R_{sys} = R_A + R_B$$

For IIC failures,

$$R_{sys} = q[1 - (1 - q)^2] + (1 - q)(q^2) = 3q^2 - 2q^3$$

So far, we have discussed systems where component failures are independent. However, in reality, this is not always the case. Sometimes an event can cause two or more components to fail in the same manner simultaneously, the failures are thus called "common-mode" failures (Sundararajan [1]).

The causes for common-mode failures can be mechanical, electrical, chemical, and environmental factors such as common electrical connections, common maintenance, excessive humidity. Even in redundant systems, common-mode failures may

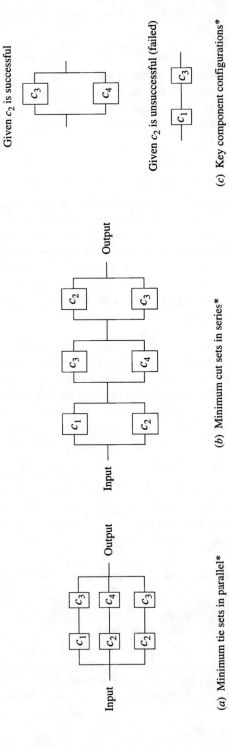

Given $c_2$ is successful

Given $c_2$ is unsuccessful (failed)

(c) Key component configurations*

(a) Minimum tie sets in parallel*

(b) Minimum cut sets in series*

* Original reliability configuration block diagram appears in Figure 24.9a.

**FIGURE 24.10** Tie set, cut set, and key component methods reliability solution diagrams (Example 24.3).

develop as a result of common environmental factors such as radiation, moisture, and vibration. Common-mode failure modeling is difficult since dependencies must be considered.

Many component failure dependencies or interactions—as well as systems with independent failures—may be modeled effectively as Markov processes, provided that the failure can be approximated as time independent, according to Lewis [2]. Markovian methods are popular for examination of two kinds of failure interactions—shared-load systems and standby systems. Markovian analysis is beyond the scope of this textbook; interested readers may refer to Kapur and Lamberson [3] and Lewis [4].

## 24.4  RELIABILITY ALLOCATION AND RELIABILITY GROWTH

*Reliability allocation is a process whereby we set a reliability target for the system, then use this target to develop reliability targets for each subsystem and, finally, for each component. Reliability growth refers to the reliability improvement we see as components, subsystems, and systems are developed, refined, and improved.* Hence, we use reliability allocation to set reliability goals, objectives, and targets, and reliability growth to develop action plans and metrics to accomplish and verify our product-process reliability results.

### Reliability Allocation

As stated earlier, reliability allocation is addressed early in the product-process design phase; hence, it offers a proactive method of establishing reliability targets. Reliability allocation is a complicated process when complex systems, containing human, software, and hardware elements are involved. Major factors in product reliability allocation include (1) customer demands and expectations, (2) field environment, (3) product application, (4) product configuration, and (5) operating method.

A good deal of engineering judgment is involved in reliability allocation. In many cases, customer-driven reliability needs and expectations call for technical capability well beyond current levels. For example, our customers may desire much longer product life than our current technology will allow. Or, our customers may desire much longer warranty periods, with no increase in product cost, and so on.

We usually begin the reliability allocation process by setting reliability-maintainability goals for the entire system in the product-process definition phase (see Chapter 11). First, we obtain a general idea of our customer's needs and expectations regarding extended product performance. We must then translate this demanded quality into substitute quality characteristics (i.e., mission reliability). Next, we work down, level by level, through our defined system and establish mission reliability targets compatible with our system level target. Example 24.4 provides a simple illustration for a series system.

**Example 24.4**

Given a series system with five components, and a system mission reliability target of at least 0.99, determine target component reliability levels such that all component targets are equal.

**Solution**

*Series:*

$$R_{sys} = q^n$$

Therefore,

$$q = \sqrt[n]{R_{sys}} = \sqrt[5]{0.99} = 0.997992$$

Hence, each component's mission reliability must equal or exceed 0.997992.

Working through configuration models both forward and backward provides insights as to component and system level reliability requirements. We should explore different configurations regarding component level reliability and redundancy before we commit to a specific system configuration. A wide variety of reliability allocation tools and approaches exist, ranging from simple trial-and-error methods, to analytical solution methods involving operations research–mathematical programming methods and sophisticated computer simulation methods (see Dai and Wang [5]).

## Reliability Growth

Often the reliability allocation process yields targets which may be beyond our current capabilities. Hence, we are "betting" that our ability to improve the reliability of our products and processes will exceed any "reliability gap" encountered in the product-process definition and early design phases. The benchmarking concept discussed in Chapter 8 is relevant here. We can think of the $Z$ curve (in a bigger is better sense) used in benchmarking as representing the reliability improvement we need to accomplish in order to meet our reliability target. To succeed, we must formulate and execute both incremental and breakthrough action plans.

It is crucial that quality and reliability targets be met before we begin full scale production. Otherwise, customer dissatisfaction and financial problems will eventually result from excessive process scrap and rework, as well as returns, warranty claims, and loss of goodwill. Figure 24.11 depicts the reliability growth concept. We can see that proactive means (i.e., preproduction efforts) to increase our reliability are critical. If we cannot meet or exceed our reliability and maintainability targets proactively, we are forced into a protracted reactive game plan. Reactive plans for reliability growth carry huge cost burdens internally and externally (e.g., engineering changes, returns, warranty, and product liability). Our only chance to succeed in a reactive mode of operation is if our competitors are worse than we are in internal and external quality losses before and after we transfer our product to the customer, respectively.

In such a case, our product's selling price may be able to absorb these losses and still yield a profit.

Many mathematical models to track and predict reliability growth have grown out of defense and weapons programs. Some, such as the Duane model (see Lewis [6]), provide interesting projections and are sometimes used to justify recommendations for applying more or fewer resources to accelerate or retard reliability growth, respectively. Most reliability growth models are speculative—based more or less on past incremental improvements that are expected to continue. We suggest, instead, an emphasis on quality-reliability improvement efforts which can be brought to focus on product-process bottlenecks, and their resolution, often as a result of breakthroughs in product and process technology.

## 24.5 STRESS-STRENGTH MODELS

***Stress-strength models provide the mechanisms to develop physical based reliability analyses.*** Stress-strength models also form the basis for stress-cycle modeling. The finite element analysis (FEA) can be extended to provide stress and strength based reliability predictions. Stress-strength relationships, modeled in either a deterministic or a probabilistic fashion, are fundamental tools in proactive quality-reliability efforts.

We will begin our model development by defining two random variables, $L$ and $C$, to represent stress or load and strength or capacity, respectively. Reliability in stress-strength models is defined as

*Reliability:*

$$R = P(\text{success}) = P(L < C) \tag{24.10}$$

*Failure:*

$$P(\text{failure}) = P(L \geq C) \tag{24.11}$$

where $L \geq 0$ and $C \geq 0$.

Classical stress-strength reliability modeling treats both stress and strength as time independent. This treatment leads to the concepts of the safety factor *sf* and safety margin *sm*. Since $l$ and $c$ (some given values of $L$ and $C$, respectively) are expressed in the same units, the safety factor is a unitless number defined as

$$sf = \frac{c}{l} \tag{24.12}$$

The margin of safety carries the same units as $l$ and $c$, and is defined as

$$sm = c - l \tag{24.13}$$

In this classical approach $l$ and $c$ are treated as deterministic quantities. Obviously, however, in reality both load and capacity are probabilistic, random variables.

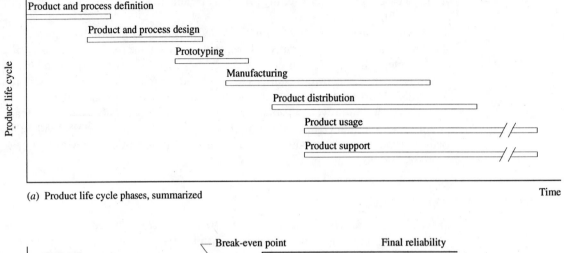

(a) Product life cycle phases, summarized

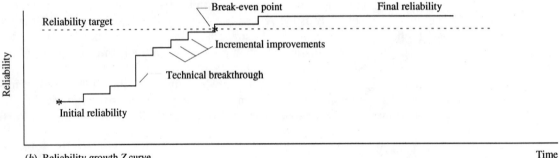

(b) Reliability growth Z curve

**FIGURE 24.11** Reliability growth concept integrated with the benchmarking concept.

A number of possible deterministic and probabilistic treatments of load and capacity are depicted in Figure 24.12. In all four cases, load and capacity are assumed to be independent of time.

### Case I—Both L and C Deterministic

Since both load and capacity are deterministic, case I (Figure 24.12a), the reliability expression can be developed by inspection. Using the definitions of reliability and failure (Equations 24.10 and 24.11, respectively), we can compare the $l$ and $c$ values and declare the device to be successful or reliable if $l < c$. On the other hand, if $l \geq c$ we declare the system unreliable. Case I is compatible with the classical safety factor and safety margin measures.

Case I is also used in deterministic FEA models to assess success or failure, using Equations 24.10 and 24.11. In FEA we might develop a stress $l$ and compare it to a capacity $c$; or we might model a load deflection $l$ and compare it to a capacity deflection $c$.

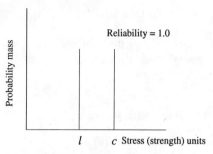

(a) Case I, both L and C are deterministic

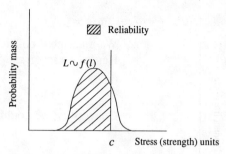

(b) Case II, L probabilistic, C deterministic

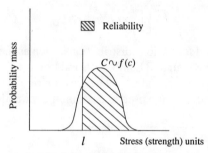

(c) Case III, L deterministic, C probabilistic

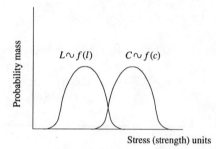

(d) Case IV, both L and C are probabilistic

**FIGURE 24.12**    Deterministic and probabilistic treatments of load L and capacity C.

### Cases II and III—*L* Probabilistic, *C* Deterministic and *C* Probabilistic, *L* Deterministic

Cases II and III can be considered together, due to their mathematical similarity. Using our definition of reliability, and by inspection of Figure 24.12b and c, we can develop reliability models for both cases.

For Case II, the reliability expression is

$$R(c) = P(L < c) = \int_0^c f(l)\, dl \tag{24.14}$$

For Case III, the reliability expression is

$$R(l) = P(l < C) = \int_l^\infty f(c)\, dc = 1 - \int_0^l f(c)\, dc \tag{24.15}$$

Again, these forms are applicable to FEA, provided our model is capable of dealing with a probabilistic stress (or strength).

### Case IV—Both *L* and *C* Probabilistic

Case IV (Figure 24.12d) presents a challenge, since both load and capacity are probabilistic. In order to develop this treatment, we can begin with Case II, where

$$R(c) = \int_0^c f(l)\, dl \tag{24.16}$$

We must now consider the capacity as probabilistic, $f(c)$, ranging from 0 to $\infty$. Hence, we integrate, "averaging" over all possible $c$ values:

$$R = \int_0^\infty \left[ \int_0^c f(l)\, dl \right] f(c)\, dc \tag{24.17}$$

Conversely, we could start with Case III, where

$$R(l) = \int_l^\infty f(c)\, dc \tag{24.18}$$

Again, consider the load as probabilistic, $f(l)$, ranging from 0 to $\infty$. Then, integrating over all possible $l$ values,

$$R = \int_0^\infty \left[ \int_l^\infty f(c)\, dc \right] f(l)\, dl \tag{24.19}$$

Results have been developed for various distributions, such as the exponential, Weibull, normal, and so on (see Kapur and Lamberson and Lewis [7]). (Kapur and Lamberson also describe a flexible graphical model where empirical data, rather than parametric probability mass functions, are used to develop the reliability expression [8]).

---

## Example 24.5

Redevelop Case IV as a distribution of difference between load and capacity. Use independent normal distributions to model both load $L$ and capacity $C$.

### Solution

Since, by definition,

$$\text{Reliability} = P(L < C) \tag{24.20}$$

we can also state that

$$\text{Reliability} = P[(C - L) > 0] \tag{24.21}$$

Hence, if we can determine the form of the distribution of the difference between capacity and load, we can then express reliability as

$$\text{Reliability} = \int_0^\infty f(\phi)\, d\phi \tag{24.22}$$

where $f(\phi)$ represents the distribution of difference.

For normally distributed load and capacity models, the difference of the two normal distributions is itself normally distributed, with mean

$$\mu_d = \mu_{c-l} = \mu_c - \mu_l \tag{24.23}$$

and variance (assuming independence between capacity and load)

$$\sigma_d^2 = \sigma_{c-l}^2 = \sigma_c^2 + \sigma_l^2 \tag{24.24}$$

Figure 24.13 depicts the difference concept with normal distributions. For independent normal distributions, the reliability can be expressed as

$$R = P\left(Z > \frac{0 - \mu_{c-l}}{\sigma_{c-l}}\right) = P\left[Z > \frac{0 - (\mu_c - \mu_l)}{\sqrt{\sigma_l^2 + \sigma_c^2}}\right]$$    (24.25)

where $Z$ represents a standard normal variable.

---

Probabilistic design concepts, such as those described by Lewis [9] and by Kapur and Lamberson [10], use Cases II, III, and IV extensively. This approach to design views reliability as the probabilistic result of stress-strength interference (e.g., Equations 24.17 and 24.19).

## Finite Element Models

*The finite elements analysis (FEA) method has become a powerful tool for the numerical solution of a wide range of engineering problems involving stress-strength relationships* (Grandin [11]). It essentially breaks a large, difficult problem into a number of smaller problems, which can be more readily solved, and then pieces the parts back together as a whole. Mechanical integrity of products and processes in the presence of defects, such as cracks, can be modeled with finite elements. Hence, finite element models are extensively used to study the design of engineering components and systems in a proactive manner.

There are a number of finite element analysis codes (Brebbia [12]). Personal computer applications are emerging which increase FEA availability to small organizations (Champion [13]). Applications range from deformation and stress analyses of automotive, aircraft, building, bridge, and electronic circuit board structures to field analyses of heat flux, fluid flow, magnetic flux, and other flow problems. With the refinement of computer aided design, CAD, systems, most finite element packages provide graphical user interfaces and displays, allowing the user to interpret the results by sight, without going through huge data files. Some FEA codes have an animation feature to play-replay the analysis results.

**FIGURE 24.13**    Capacity and load difference concept, assuming normality (Example 24.5).

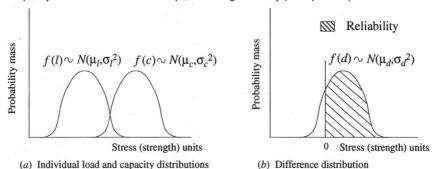

(*a*) Individual load and capacity distributions          (*b*) Difference distribution

*Very sophisticated and detailed physical stress-strength reliability-failure analysis tools are emerging* (Dai and Wang [14]). Many perform basic reliability analyses in a deterministic fashion (e.g., as we discussed in Case I). Others involve probabilistic relationships such as those described in Case IV. FEA models such as NESSUS, developed by NASA and Southwest Research Institute [15], treat both stress and strength as probabilistic.

## REVIEW AND DISCOVERY EXERCISES

### Review

**24.1** An aircraft landing gear has a probability of $1 \times 10^{-5}$ per landing of being damaged from excessive impact. What is the probability that the landing gear will survive a 10,000-landing design life without damage?

**24.2** Given the reliability block diagrams and component reliabilities shown below, determine which design configuration will yield the highest system reliability.

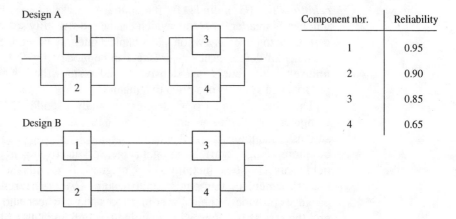

| Component nbr. | Reliability |
|---|---|
| 1 | 0.95 |
| 2 | 0.90 |
| 3 | 0.85 |
| 4 | 0.65 |

**24.3** Assuming that all components are IIC with reliability $q$, calculate the reliability of the configuration shown below using each of the following methods [BH]:
  **a** Series-parallel reduction
  **b** Event enumeration
  **c** Cut set
  **d** Tie set

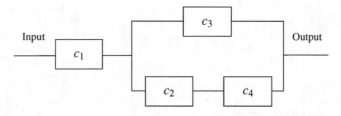

**24.4** An airplane has four identical engines. It can successfully complete a mission if any three of the four engines function successfully.

     **a** Sketch a branch reliability diagram for the airplane engine system.

     **b** Assuming the airplane engines to be IIC, develop an expression for the system reliability.

     **c** Determine the engine reliability necessary to yield a mission reliability of at least 99.9 percent for the system of airplane engines.

**24.5** Assuming that all components of reliability configurations 1 through 3 below are IIC with reliability $q$,

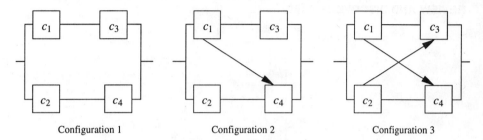

Configuration 1            Configuration 2            Configuration 3

     **a** Calculate the configuration reliabilities using the event enumeration method.

     **b** Calculate the configuration reliabilities using the tie set method.

     **c** Calculate the configuration reliabilities using the cut set method.

     **d** Calculate the configuration reliabilities using the key component method.

     **e** Construct a single plot of system reliability as a function of component reliability; include all configurations. Describe any benefits in reliability we see as a result of the crossover branches.

**24.6** A student attending school has two used cars that are six and nine years old. The reliability of a successful engine start is 0.79 and 0.71, respectively, for the two old cars. The student has the option of trading both cars for a two-year-old car with a reliability of 0.93. [RC]

     **a** Based on starting reliability, would it be wise to buy the two-year-old car? Provide reliability block diagrams for the two alternatives.

     **b** A trip to school would require two successful engine starts and two successful one-way drives. Given a successful start, the probability of a successful one-way trip to school with any one of the two old cars is 0.95, versus 0.999 for the two-year-old car, would it be wise to buy the two-year-old car? Provide block diagrams for the alternatives, and base your decision on reliability considerations.

**24.7** Given the reliability block diagram below, and the reliabilities listed, determine the system reliability.

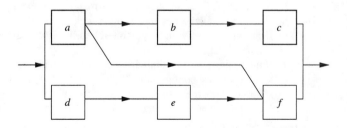

$$R_a = 0.90 \qquad R_d = 0.95$$
$$R_b = 0.95 \qquad R_e = 0.95$$
$$R_c = 0.95 \qquad R_f = 0.90$$

**24.8** The reliability block diagram and estimated individual subsystem reliabilities for a spot-welding process in an automobile assembly line are shown below [JG]:

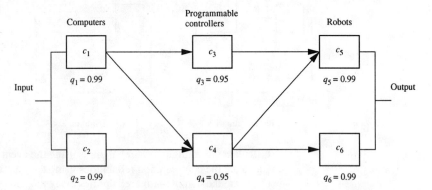

The criticality of this process calls for an overall system reliability of 0.99. Assuming all components fail independently, will the process reliability meet such a requirement?

**24.9** A hospital buys power from the local utility grid and maintains a diesel motor-generator set as a backup power source. For a given one-year period, the reliability of the utility system providing power is 0.99. For the same one-year period, the conditional probability of motor-generator operation given the failure of the grid is 0.95. The reliability of the automatic switching device between the grid and the motor-generator set is 0.98. Determine the reliability of the standby system for the one-year period.

**24.10** A design engineer in an electronics company is faced with a reliability-cost decision regarding an electronic assembly that can be designed as a parallel sub-system. The mission reliability requirement for the device is at least 0.98. There are three components available to the engineer. Component A has 0.90 reliability, a space requirement of $0.5 \times 0.5 \times 0.5$ in³, and a price of $38. Component B has 0.85 reliability, with a space requirement of $0.7 \times 0.7 \times 0.7$ in³, and a price of $20. Component C has 0.80 reliability, and a space requirement of $1 \times 1 \times 1$ in³, and a price of $15. The components can be mixed. For example, we can place A, B, and/or C in parallel, or A and A in parallel, and so forth. [TY]

   **a** If minimum cost with acceptable reliability is the selection criterion, what is the best configuration?

   **b** If the design space is limited to $1.5 \times 1.5 \times 1.5$ in³ and cost is not a consideration, what is the best configuration?

   **c** If space is not a limitation, and the configuration must yield a reliability of at least 0.99, what arrangement would you recommend and how much would it cost?

**24.11** Assume that a system consists of seven components that have the same functional and physical nature but different reliability characteristics. In other words, any component can be placed in any position in the reliability block diagram below. For example, component 1 may be placed in position B, and so on, but a given component can be used only once. The reliability of each of the seven components is shown below. Determine the position in which each component should be placed for highest system reliability. [MC]

| | Component | | | | | | |
|---|---|---|---|---|---|---|---|
| | 1 | 2 | 3 | 4 | 5 | 6 | 7 |
| Reliability | 0.99 | 0.85 | 0.99 | 0.85 | 0.7 | 0.5 | 0.5 |

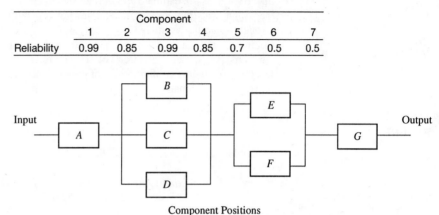

Component Positions

**24.12** A safety system on an aircraft has three subsystems: a sensor, a processor, and a servomechanism. The sensor senses low oxygen levels in the passenger compartment and sends the signal to a processor, which then processes the signal and prompts a servomechanism to drop oxygen masks for passengers in the aircraft. In order to produce a successful performance, all three subsystems must function in a successful manner. Reliability estimates are shown below for each subsystem. [RC]

| | Sensor | Processor | Servomechanism |
|---|---|---|---|
| Reliability | 0.60 | 0.8 | 0.98 |

**a** Assuming a series system of single components, and using the subsystem reliabilities given above, calculate the system reliability.

**b** A reliability engineer at the aircraft company has designed a redundant system shown below to increase the system reliability. Assuming identical and independent processors and servomechanisms, calculate the system reliability.

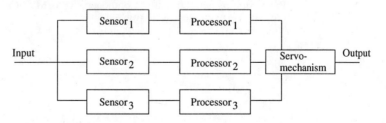

**c** You are asked to design this safety system with, at most, three sensors, two processors, and one servomechanism; recommend a design configuration to maximize reliability. Calculate the system reliability. State any considerations concerning common-mode failures.

### Discovery

**24.13** Identify a system of your choice.

    **a** Develop a functional block diagram for the system.

    **b** Develop a reliability block diagram for your selected system.

    **c** Specify a mission, estimate appropriate success or failure probabilities for the components in your system, and develop an appropriate system level mission reliability estimate.

    **d** Comment on possible ways to improve the system's reliability. Consider our four-level strategic hierarchy of (1) pure reliability at the component level, (2) redundant configurations of components in the system, (3) preventive maintenance, and (4) corrective maintenance.

## REFERENCES

**1** C. R. Sundararajan, *Guide to Reliability Engineering,* New York: Van Nostrand Reinhold, 1990.

**2** E. E. Lewis, *Introduction to Reliability Engineering,* New York: Wiley, 1987.

**3** K. C. Kapur and L. R. Lamberson, *Reliability in Engineering Design,* New York: Wiley, 1977.

**4** See reference 2.

**5** S. H. Dai and M. O. Wang, *Reliability Analysis in Engineering Applications*, New York: Van Nostrand Reinhold, 1992.

**6** See reference 2.

**7** See references 3 and 2.

**8** See reference 3.

**9** See reference 2.

**10** See reference 3.

**11** H. Grandin, *Fundamentals of the Finite Element Method,* Prospect Heights, IL: Waveland Press, 1991.

**12** C. A. Brebbia (ed.), *Finite Element Systems,* 3d ed., Berlin, W. Germany: Springer-Verlag, 1985.

**13** E. R. Champion, *Finite Element Analysis in Manufacturing Engineering,* New York: McGraw-Hill, 1992.

**14** See reference 5.

**15** T. A. Cruse et al., "Probabilistic Structural Analysis Methods for Select Space Propulsion System Structural Components (PSAM)," *Computers & Structures,* vol. 29, no. 5, pp. 891–901, 1988.

# RELIABILITY, MAINTAINABILITY, AND AVAILABILITY

## 25.0 INQUIRY

1 How are time based reliability and maintainability characteristics defined?
2 How are reliability and maintainability characteristics measured?
3 How are time dependent reliability and maintainability models developed?
4 How can preventive and corrective maintenance policies be modeled?
5 How are reliability and maintainability characteristics modeled simultaneously?

## 25.1 INTRODUCTION

In addition to delivering customer demanded performance in the short-term, we must ensure that our product or process will continue to deliver long-term, sustained, performance. *Reliability is a critical quality characteristic which measures sustained product-process performance; maintainability,* also a critical quality characteristic, *measures sustained product-process support. Reliability and maintainability characteristics must be defined, designed, and manufactured into each product that we deliver to our customer.*

Reliability and maintainability needs and expectations are inherent in customer demands, and must be translated into long-term substitute quality characteristics, as stated earlier. For example, a footwear customer may express the demand for long-term foot comfort. This demand must be translated through substitute quality characteristics such as dimensions, material specifications, process specifications, wear rates, and so on, to address the concept of quality and its mission of superior customer satisfaction. When customers use footwear, they declare it to be comfortable or uncomfortable. Assuming that a declaration of "very comfortable" is obtained, we must ensure that the comfort is sustained in the long-run (days, months, years). The essence of this sustained performance is reliability. Sustained performance may need support, if the comfort level begins to fail. The essence of this support is maintainability. For example, our shoes may be designed to readily accept new innersoles, new outer soles, polish, and so forth, to extend their field performance.

## 25.2 TIME OR USAGE BASED PERFORMANCE MEASURES

*Reliability measures are used to predict the probability that a given unit will perform satisfactorily over a specified time or usage horizon. For example, $R(t)$*

*represents the probability of successful operation from time or usage points 0 to t.*
Here, time 0 represents the instant a component is placed in service, and time $t$, an arbitrarily selected "instant" in its life (in the future). We could use this measure to predict the reliability of a computer in delivering satisfactory performance for a period of three years. Our prediction might be expressed as $R(3 \text{ yr}) = 0.98$; i.e., we are forecasting a 98 percent chance of failure-free performance for our computer over the three-year usage period. The same measure can be used to represent the proportion of a population expected to survive beyond some time—say, $t$. For example, if 50 computers are to be placed in service at time 0, then at time 3 yr, we would expect that $50 (0.98) = 1$ computer would fail to perform in a satisfactory manner.

*The maintainability measure M(t) is used to predict the probability that a non-performing unit will be restored to a satisfactory level of performance within a given time frame (0 to t).* For example, $M(1.5 \text{ hr}) = 0.99$ indicates a 99 percent chance of restoring a failed unit within 1.5 hr of repair time. Hence, the quantitative reliability and maintainability interpretations are analogous. We will discover that reliability and maintainability models also have many identical developmental characteristics.

## Reliability Development

The mechanics of probability mass functions (pmf's) or probability density functions (pdf's), usually denoted by $f(x)$, and their cumulative forms (cmf or cdf, respectively), denoted by $F(x)$, are well defined and developed in basic probability and statistics texts (see Walpole and Myers [1]). *Based on the pmf and cmf concepts, we will define a random variable T that represents time or usage, $t \geq 0$, since negative life or usage ($t < 0$) is unrealistic in the physical world.* By definition of the cmf,

$$F(t) = P(T \leq t) = \int_0^t f(\zeta)\, d\zeta \qquad t \geq 0 \tag{25.1}$$

and *by the definition of reliability,*

$$R(t) = P(T > t) = 1 - F(t) = 1 - \int_0^t f(\zeta)\, d\zeta = \int_t^\infty f(\zeta)\, d\zeta \qquad t \geq 0 \tag{25.2}$$

Here, $R(0) = 1$ and $\lim_{t \to \infty} R(t) = 0$. We interpret these statements to mean that the item is sure to perform successfully at $t = 0$ and sure to fail, given sufficient time, as $t$ increases.

By definition, we know that the pmf measures the rate of change in the failure related cmf, $F(t)$, and that $R(t) = 1 - F(t)$. Hence,

$$f(t) = \frac{d}{dt} F(t) = -\frac{d}{dt} R(t) \tag{25.3}$$

*We can develop an instantaneous failure rate $\lambda(t)$ known as the failure "hazard" rate.* By defining a small increment of time $\Delta t$, we can relate the probability of failure in $\Delta t$, given survival up to the beginning of the interval, to the instantaneous failure rate:

$$\lambda(t)\,\Delta t = P(T < t + \Delta t \mid T > t) \tag{25.4}$$

From the definition of conditional probability, if $A$ and $B$ are defined as events, we know that

$$P(A \mid B) = \frac{P(A \cap B)}{P(B)} \tag{25.5}$$

From Equations 25.4 and 25.5, it follows that

$$\lambda(t)\,\Delta t = P(T < t + \Delta t \mid T > t) = \frac{P[(T < t + \Delta t) \cap (T > t)]}{P(T > t)}$$

From the definition of pmf, we know that

$$P\,[(T < t + \Delta t) \cap (T > t)] = f(t)\,\Delta t$$

Since $P(T > t) = R(t)$, we have

$$\lambda(t) = \frac{f(t)}{R(t)} \tag{25.6}$$

Now, it follows from Equation 25.3 that

$$\lambda(t) = -\frac{1}{R(t)}\,\frac{d}{dt}\,R(t)$$

which can also be stated as

$$\lambda(t)\,dt = -\frac{dR(t)}{R(t)}$$

Integrating both sides, we obtain

$$\int_0^t \lambda(\zeta)\,d\zeta = -\ln\,[R(t)]$$

and thus
$$R(t) = \exp\left[-\int_0^t \lambda(\zeta)\,d\zeta\right] \tag{25.7}$$

Then, from Equation 25.6, we obtain

$$f(t) = \lambda(t)\,\exp\left[-\int_0^t \lambda(\zeta)\,d\zeta\right] \tag{25.8}$$

where $\lambda(t) \geq 0$ and $\int_0^t \lambda(\zeta)\,d\zeta \to \infty$ as $t \to \infty$.

Two additional, critical reliability measures are helpful in reliability analysis. *The mean time to failure (MTTF) is the expected value of T, E(T):*

$$\text{MTTF} = E(T) = \int_0^\infty t\, f(t)\, dt \tag{25.9}$$

Now, substituting Equation 25.3 and integrating by parts, we obtain

$$\text{MTTF} = -\int_0^\infty t\, \frac{dR(t)}{dt} = -tR(t)\, \Big|_0^\infty + \int_0^\infty R(t)\, dt$$

Since the term $tR(t) = 0$ at $t = 0$ and $tR(t) \to 0$ as $t \to \infty$,

$$\text{MTTF} = \int_0^\infty R(t)\, dt \tag{25.10}$$

At this point, we should note that MTTF is interpreted as mean time to first failure. We use the term "mean time between failures" MTBF in cases where we have repair capabilities. In general, MTTF is straightforward to deal with mathematically, while MTBF is more complicated to deal with in a predictive sense. Both measures are straightforward when we use them in a historic sense, i.e., when we calculate from operating history.

The second useful measure is the failure time quantile $t_p$, where $0 < p < 1$. *The failure time quantile represents the point in time or usage where the cumulative failure probability is equal to p.* Hence, $t_p$ must satisfy the following relationship:

$$F(t_p) = p = \int_0^{t_p} f(\zeta)\, d\zeta \tag{25.11}$$

In general, the failure time quantile (or percentile) measure is very useful for expressing failure times relative to early failures. Failure time quantiles are widely used in warranty analysis.

Our critical reliability measures are displayed in Figure 25.1a. *The decreasing failure rate (DFR), constant failure rate (CFR), and increasing failure rate (IFR), are depicted graphically to form, together, the classic "bathtub" curve.* We associate the negative sloped DFR region with early failures. These failures, or "infant mortalities," are typically attributed to latent or incipient defects. Infant mortalities in hardware devices are usually dealt with by specifying a period of time during which the device undergoes "burn-in" or "wear-in." During this time, loading and use are controlled in such a way that weaknesses are likely to be detected and repaired without failure, or so that any resulting failures will not cause inordinate harm or financial loss (Lewis [2]). The CFR phase, which produces a straight line signature, is associated with "random failures." Here, an environmental or application-driven overload, rather than an inherent defect in the device or system, typically leads to failure. The IFR region is usually termed a "wear-out" region. Here we see a positive slope which might be attributable to mechanical wear, corrosion, or some other fundamental product degradation process. Variation in product configurations and production processes, customer applications, field environments, and operating methods significantly affect the overall nature of the bathtub curve.

Generic pmf, cmf, and reliability functional relationships are depicted in Figure 25.1a. Specific models such as the exponential, Weibull, and so on, are discussed and illustrated in a later section. The $t_p$, or quantile failure time, measure is superimposed on the pmf, cmf, and reliability curves. As we can see, the $t_p$ location depends on the value of $p$ we select ($0 < p < 1$) and the functional shape and location of the distribution.

At this point, we should point out the graphical significance of the functional relationship shown in Equation 25.7:

$$R(t) = \exp\left[-\int_0^t \lambda(\xi)\, d\xi\right]$$
$$= \exp\left[-(\text{cumulated area under the } \lambda(t) \text{ curve, } 0 \text{ to } t)\right] \qquad (25.12)$$

An understanding of this relationship will help us to visualize the effects of DFR, CFR, and IFR. It is also useful if we seek to develop $R(t)$ in a discrete or numeric fashion, or to form a piecewise linear failure hazard curve $\lambda(t)$. For example, we could define a set of piecewise linear functions to form a bathtub—a DFR, CFR, IFR sequence. Then, we could solve our $R(t)$ relationship by applying Equation 25.12 across the pieces, from left to right, from 0 to $t$.

---

**Example 25.1**

An engineer has approximated the pmf for failures of an electronic transistor in a harsh environment as

$$f(t) = 2e^{-2t}$$

where $t$ represents the life of the transistor in years. Determine the reliability of the transistor at the end of two years and its MTTF. What can we say about the transistor's instantaneous failure rate $\lambda(t)$?

**Solution**

From Equation 25.2, we know that

$$R(t) = 1 - \int_0^t f(\zeta)\, d\zeta$$

$$R(t = 2) = 1 - \int_0^2 2 \exp(-2\zeta)\, d\zeta = 1 - [-\exp(-2\zeta)]_0^2 = 1 - [-\exp(-4) + 1]$$

$$= 0.01832$$

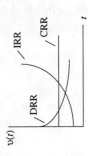
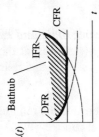
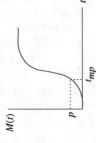

## Reliability (performance → nonperformance state)

Instantaneous failure rate

$\lambda(t)$

IFR: increasing failure rate
CFR: constant failure rate
DFR: decreasing failure rate

Failure probability density function

$f(t)$

IFR models: lognormal, Weibull
CFR models: exponential, Weibull
DFR models: lognormal, Weibull

Failure cumulative density function

$F(t)$

Reliability

$$R(t) = 1 - F(t) = 1 - \int_0^t f(\zeta)\,d\zeta$$

$$R(t) = \exp\left[-\int_0^t \lambda(\zeta)\,d\zeta\right]$$

(a) Reliability relationship summary

## Maintainability (nonperformance → performance state)

Instantanecus repair rate

$\upsilon(t)$

IRR: increasing repair rate
CRR: constant repair rate
DRR: decreasing repair rate

Repair probability density function

$m(t)$

IRR models: lognormal, Weibull
CRR models: exponential, Weibull
DRR models: lognormal, Weibull

Maintainability

$M(t)$

$$M(t) = \int_0^t m(\zeta)\,d\zeta$$

$$M(t) = 1 - \exp\left[-\int_0^t \upsilon(\zeta)\,d\zeta\right]$$

(b) Maintainability relationship summary

$$R(t_2|t_1) = R(t_2)/R(t_1) = \exp\left[-\int_{t_1}^{t_2}\lambda(\zeta)\,d\zeta\right]$$

$$F(t_2|t_1) = 1 - R(t_2|t_1) = 1 - \exp\left[-\int_{t_1}^{t_2}\lambda(\zeta)\,d\zeta\right]$$

$$M(t_2|t_1) = 1 - \exp\left[-\int_{t_1}^{t_2}\upsilon(\zeta)\,d\zeta\right]$$

$$t_2 > t_1$$

(c) Conditional reliability, failure, and maintainability relationships

**FIGURE 25.1** Fundamental reliability and maintainability relationships.

670

From Equation 25.9, we know that

$$\text{MTTF} = \int_0^\infty t f(t)\, dt$$

Thus,

$$\text{MTTF} = \int_0^\infty t\, 2 \exp(-2t)\, dt$$

Integrating by parts, we get

$$\text{MTTF} = -t \exp(-2t) \Big|_0^\infty + \int_0^\infty \exp(-2t)\, dt = 0 + \frac{\exp(-2t)}{-2} \Big|_0^\infty = 0.5 \text{ yr}$$

From Equation 25.6, we know that

$$\lambda(t) = \frac{f(t)}{R(t)}$$

$$\lambda(t) = \frac{f(t)}{1 - \int_0^t f(\zeta)\, d\zeta} = \frac{2\exp(-2t)}{1 - \int_0^t 2\exp(-2\zeta)\, d\zeta} = \frac{2\exp(-2t)}{\exp(-2t)} = 2 \text{ failures per yr}$$

Since $\lambda(t) = 2$, a constant, we can say that the electronic transistor has a constant failure rate, or CFR.

## Maintainability Development

*We can establish maintainability relationships using the same logic and mathematics that we used for the reliability development.* Let $m(t)$ represent the pmf associated with component restoration time. From our definition of maintainability, we have

$$M(t) = \int_0^t m(\zeta)\, d\zeta \qquad t \geq 0 \tag{25.13}$$

where $M(0) = 0$ and $\lim_{t \to \infty} M(t) = 1$.

If $\upsilon(t)$ represents an instantaneous repair rate [the maintainability counterpart of $\lambda(t)$, the instantaneous failure rate], we have

$$\upsilon(t)\, \Delta t = P(T < t + \Delta t \mid T > t) \tag{25.14}$$

We can then express maintainability as

$$M(t) = 1 - \exp\left[-\int_0^t \upsilon(\zeta)\, d\zeta\right] \qquad t \geq 0 \tag{25.15}$$

where $\upsilon(t) \geq 0$ and $\int_0^t \upsilon(\zeta)\, d\zeta \to \infty$ as $t \to \infty$.

We can also show that

$$\upsilon(t) = \frac{m(t)}{1 - M(t)} \tag{25.16}$$

If *MTTR denotes the expected value of the time to repair a system*, then

$$\text{MTTR} = E[T] = \int_0^\infty t\, m(t)\, dt \tag{25.17}$$

Figure 25.1*b* summarizes these basic maintainability relationships and illustrates their graphical counterparts. *We interpret the instantaneous maintainability repair rates (IRR, CRR, and DRR) in the same manner as we interpret their failure counterparts*.

Thus, if we want to develop a repair time quantile $t_{mp}$, where $0 < p < 1$, then $t_{mp}$ must satisfy the following relationship:

$$M(t_{mp}) = p = \int_0^{t_{mp}} m(\zeta)\, d\zeta \tag{25.18}$$

*The $t_{mp}$ value is the direct counterpart to $t_p$ in failure-reliability models*. The major distinction is that *maintainability focuses on transforming a "down" system to an "up" state, while reliability focuses on an "up" system state transforming to a "down" or failure state*.

## Conditional Reliability and Maintainability Development

The measures of reliability $R(t)$ and maintainability $M(t)$ address reliability or maintainability over the entire interval from 0 to $t$. Many times we wish to make conditional predictions. For example, $R(t_2 \mid t_1)$ *refers to the conditional probability of survival to some future time $t_2$ given survival to present time $t_1$, where $t_2 > t_1$. The conditional measure* (see Figure 25.2) *is useful when we want to assess the reliability for a unit already in service that is still functioning successfully, or when we want to predict the reliability over a defined interval—say, $t_2 - t_1$.*

In developing a quantitative model for conditional reliability, we can observe that

$$
\begin{aligned}
F(t_2 \mid t_1) &= P(t_1 < T \le t_2 \mid T > t_1) \\
&= \frac{P[(t_1 < T \le t_2) \cap (T > t_1)]}{P(T > t_1)} \\
&= \frac{F(t_2) - F(t_1)}{R(t_1)} \qquad t_2 > t_1
\end{aligned}
\tag{25.19}
$$

Now,

$$
\begin{aligned}
R(t_2 \mid t_1) &= 1 - F(t_2 \mid t_1) \\
&= 1 - \frac{F(t_2) - F(t_1)}{R(t_1)} \\
&= \frac{R(t_1) - [F(t_2) - F(t_1)]}{R(t_1)} \qquad t_2 > t_1
\end{aligned}
$$

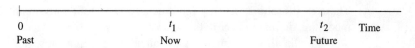

0                $t_1$               $t_2$    Time

Past            Now            Future

(*a*) Component active time line (unit was placed in service at $t = 0$)

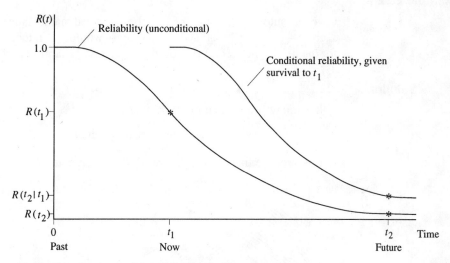

(*b*) Conditional and unconditional reliability models

**FIGURE 25.2**    Conditional reliability concept.

Substitution of $R(t_1) - R(t_2)$ for $F(t_2) - F(t_1)$, since $R(t) = 1 - F(t)$, yields

$$R(t_2 \mid t_1) = \frac{R(t_1) - [R(t_1) - R(t_2)]}{R(t_1)} = \frac{R(t_2)}{R(t_1)} \qquad t_2 > t_1 \qquad (25.20)$$

Using Equation 25.7 in the above form, we obtain

$$R(t_2 \mid t_1) = \exp\left[-\int_0^{t_2} \lambda(\zeta)\, d\zeta\right] \exp\left[+\int_0^{t_1} \lambda(\zeta)\, d\zeta\right]$$

$$= \exp\left[-\int_{t_1}^{t_2} \lambda(\zeta)\, d\zeta\right] \qquad t_2 > t_1 \qquad (25.21)$$

Conditional failure and maintainability relationships can be developed in a manner similar to that used for the conditional reliability forms (Equations 25.20 and 25.21), as follows:

$$F(t_2 \mid t_1) = 1 - R(t_2 \mid t_1)$$

$$= 1 - \frac{R(t_2)}{R(t_1)}$$

$$= 1 - \exp\left[-\int_{t_1}^{t_2} \lambda(\zeta)\, d\zeta\right] \qquad t_2 > t_1 \qquad (25.22)$$

and $$M(t_2 \mid t_1) = 1 - \exp\left[-\int_{t_1}^{t_2} \upsilon(\zeta)\, d\zeta\right] \qquad t_2 > t_1 \qquad (25.23)$$

These relationships are useful in time-advanced reliability-maintainability simulation modeling, discussed later in this chapter. Figure 25.1c provides a summary of the conditional reliability, failure, and maintainability relationships.

## Example 25.2

Given an automobile whose $\lambda(t)$ can be described mathematically for time periods $t_1$ and $t_2$—say, $t_1 = 1$ year and $t_2 = 2$ years—would you rather own the automobile for the first or the second year of life, based on a reliability performance measure?

### Solution

We know from Equation 25.12 that

$$R(t) = \exp\ [-(\text{cumulated area under the } \lambda(t) \text{ curve, 0 to } t)]$$

Let us suppose that the $\lambda(t)$ curve for the given automobile is as shown below.

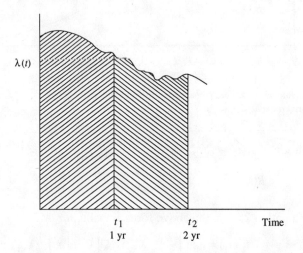

Now, the probability that the automobile will deliver satisfactory performance during the first year of its life may be expressed as

$$R_1(t_1 = 1 \text{ yr}) = \exp\ [-(\text{cumulated area under the } \lambda(t) \text{ curve, 0 to } t_1 = 1 \text{ yr})]$$

Given that the automobile has survived the first year, $P(t > 1) = 1.0$, the probability that it would deliver satisfactory performance during the second year of its life is

$$R_2(t_2 = 2 \text{ yr} \mid t > 1 \text{ yr})$$

From Equation 25.21, we obtain

$$R_2(t_2 = 2 \text{ yr} \mid t > 1 \text{ yr}) = \exp\left[-\int_{t_1}^{t_2} \lambda(\zeta)\,d\zeta\right]$$

$$= \exp\ [-(\text{cumulated area under the } \lambda(t) \text{ curve, } t_1 = 1 \text{ year to } t_2 = 2 \text{ yr})]$$

Thus, from $R_1$ ($t_1 = 1$) and $R_2$ ($t_2 = 2 \mid t > 1$), clearly it would be best to own the automobile for the year which has less cumulated area under the $\lambda(t)$ curve. Therefore, if $\int_0^{t_1} \lambda(\zeta)\, d\zeta > \int_{t_1}^{t_2} \lambda(\zeta)\, d\zeta$, then we would choose to own the automobile for the second year of its life; if $\int_0^{t_1} \lambda(\zeta)\, d\zeta < \int_{t_1}^{t_2} \lambda(\zeta)\, d\zeta$ the first year. If $\int_0^{t_1} \lambda(\zeta)\, d\zeta = \int_{t_1}^{t_2} \lambda(\zeta)\, d\zeta$, there would be no difference in performances, thus we would be indifferent. That is, we can say that, for a DFR, we would opt for second year ownership; for an IFR, first year ownership; and, for a CFR, either year.

## 25.3 TIME AND USAGE BASED RELIABILITY AND MAINTAINABILITY MODELS

As stated earlier, reliability and maintainability performance is improved through our creative product-process efforts early in the product life cycle. In other words, what we do in the definition and design phases profoundly affects our final results. The models we will present in this section are useful in adding measurement and prediction detail to the four-level hierarchical reliability-maintainability strategy.

In the preceding sections we have developed reliability and maintainability in a generic manner for time or usage based models. Now we will enhance these relationships to better predict system reliability and maintainability characteristics. *A number of parametric distributions have been found to be useful for modeling time dependent reliability and maintainability characteristics.* Here, we will present only a few of the more widely used models; many others are developed and discussed in the literature (see Nelson [3], Kececioglu [4], and Lawless [5]). In addition, we can also use sequences of rather simple models, such as piecewise linear hazard models and, using our generic forms (Equations 25.2 and 25.7, and Equations 25.13 and 25.15), develop custom tailored models for reliability and maintainability, respectively.

Regardless of the model selected or developed, *it is critical that we justify our model in terms of our product or process. Model justification takes two distinct forms: physical and statistical.* A physical justification starts with the identification of the physical nature of the product or process. Then, models are developed which abstractly describe this physical nature.

Physical models vary widely in both complexity and form. For example, our block diagrams (Chapter 24) are simple forms of physical models. Dai and Wang describe a number of complex physical models [6]. If we are dealing with a product in its useful-life phase (the flat bottom of the bathtub curve), the exponential model may be defensible. Here, we might ignore initial DFR and subsequent IFR portions of the failure hazard curve on the basis that they are not relevant to our analysis (e.g., the product was "burned-in" used before delivery to eliminate the DFR region and will be retired before the IFR region becomes relevant). The Weibull model is a smallest extreme value model and, hence, is sometimes called a "weakest link model." Depending on the stress-strength aspect of system failure, a weakest link model may be defensible.

Statistical justification is developed by matching data characteristics (primarily location and shape) to model characteristics. In other words, we are fitting a model to the data, with limited emphasis on physically justifying the model. Various goodness-of-fit tests (such as the chi-square goodness-of-fit test, the Kolmogorov-Smirnov tests, and many other statistical tests) are used to justify statistical models (see Conover [7]). Details regarding fitting a model to data will be developed in Chapter 26, where we deal with reliability experiments. For now, we will discuss the basic models and their ability to add detail to our reliability and maintainability analyses with respect to the reliability configurations introduced in Chapter 24.

## Exponential Model

Three forms of the exponential model are useful in engineering work: (1) the one-parameter, (2) the two-parameter, and (3) the covariate. In some cases, the covariate exponential model is referred to as a "proportional hazard" model, (Landers and Kolarik [8]) or as a "regression" model (Lawless [9]). Graphical representations of the hazard, time to failure pmf, and reliability for a one-parameter exponential model are shown in Figure 25.3.

**One-Parameter Exponential Model** *The one-parameter exponential model is a widely used failure and repair model. Its most distinguishing characteristic is the constant (time independent) hazard rate* $\lambda(t) = \lambda$. It is also referred to as a "random" failure model because of its constant hazard rate. The conditional probability of failure $F(t_2 \mid t_1)$ and the conditional reliability $R(t_2 \mid t_1)$ are independent of time in service and depend only on the difference $t_2 - t_1$. For example, if the hazard rate holds constant, $R(1000 \text{ hr} \mid 500 \text{ hr}) = R(2000 \text{ hr} \mid 1500 \text{ hr}) \doteq R(50,000 \text{ hr} \mid 45,500 \text{ hr})$, and so on. This characteristic is sometimes referred to as the "forgetfulness" or "memoryless" property, in that we may ignore the past in projecting or predicting future performance, as long as our component is presently operational. The reliability, pmf, and hazard rate forms are

$$R(t) = \exp(-\lambda t) = \exp\left(-\frac{t}{\theta}\right) \qquad t \geq 0 \qquad (25.24a)$$

$$f(t) = \lambda \exp(-\lambda t) \qquad t \geq 0 \qquad (25.24b)$$

$$\lambda(t) = \lambda \qquad (25.24c)$$

**FIGURE 25.3**    Graphical representation of hazard rate, pmf, and reliability for the one-parameter exponential model.

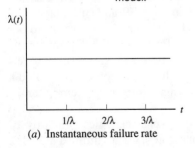

(a)  Instantaneous failure rate

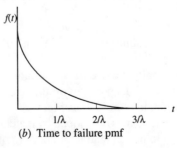

(b)  Time to failure pmf

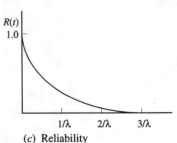

(c)  Reliability

The mean and the variance parameters are

$$\mu = \frac{1}{\lambda} = \theta \qquad (25.24d)$$

$$\sigma^2 = \frac{1}{\lambda^2} \qquad (25.24e)$$

where $\lambda$ = the failure hazard rate
$\theta$ = the MTTF = $1/\lambda$

**Two-Parameter Exponential Model**   The two-parameter exponential model is useful when virtually perfect reliability is encountered for a time period on the order of $t_0$, and thereafter, we observe component failures, which can be described by the exponential model.  It is compatible with studies for high reliability products or products that have been screened or survived a burn-in to remove weak components.  The reliability, pmf, and hazard rates are expressed as

$$R(t) = \exp\left[-\lambda(t - t_0)\right] \qquad t > t_0 > 0 \qquad (25.25a)$$

$$f(t) = \lambda \exp\left[-\lambda(t - t_0)\right] \qquad t > t_0 > 0 \qquad (25.25b)$$

$$\lambda(t) = \begin{cases} 0 & 0 \le t \le t_0 \\ \\ \lambda & t > t_0 \end{cases} \qquad (25.25c)$$

**Covariate Exponential Model**   A covariate is a variable that is measured along with a response (e.g., we might observe a failure time, a response, and also record covariables associated with the field environment, application, configuration, or operating method).  By capturing covariates in our reliability models, we can use one general model to represent a variety of conditions, rather than a separate model for each specific condition.  Covariate models lend themselves to designed reliability experimentation, which is discussed later in Chapter 26.  The covariate exponential form is

$$R(t;\, \theta) = \exp\left(\frac{-t}{\theta}\right) \qquad t \ge 0 \qquad (25.26)$$

where $\theta = \exp(\mathbf{Z}\, \boldsymbol{\beta})$
$\mathbf{Z} = (Z_1, Z_2, ..., Z_m)$
$\boldsymbol{\beta}^T = (\beta_1, \beta_2, ..., \beta_m)$

Typically, $Z_1 = 1$.

$Z_2, Z_3, ..., Z_m$ = continuous covariates, such as temperature, or classification covariates, such as design configuration.

$\beta_1$ is a regression intercept coefficient.

$\beta_2, \beta_3, ..., \beta_m$ = regression coefficients associated with the $Z_2, Z_3, ..., Z_m$ covariates, respectively.

The boldface notation indicates a vector, and the $T$ superscript notation indicates a transposed vector.

**Exponential Model Comments**   *The exponential model is the most widely used failure and repair model in the reliability and maintainability field.* For example, *electronics failure modeling and analysis are dominated by applications of the exponential model.* Military Standard 217 (Mil-Std-217), "Reliability Prediction of Electronic Equipment," is based on exponential failure models [10]. The "π" factors used in Mil-Std-217 to adjust base failure rates for environment, temperature, quality level, and so forth, have been shown to be equivalent to the covariate form of the exponential model (see Landers and Kolarik [11]). The parts-count method described later in this chapter is also based on the exponential model.

Mil-Std-781, "Reliability Design Qualification and Production Acceptance Tests: Exponential Distribution," is based on the exponential failure distribution [12]. This test design standard uses the exponential failure model as a base and develops stress test guidelines for temperature, vibration, and moisture related stresses, focusing on MTTF.

## Weibull Model

*The Weibull model is widely used in the reliability field because of its relationship to a weakest-link physical process, smallest extreme value statistical property.* The Weibull hazard function is of a basic power function form (see Equation 25.27c), and hence, can take many different graphical shapes (e.g., DFR, CFR, or IFR). Three general Weibull forms are useful: (1) the two-parameter, (2) the three-parameter, and (3) the covariate.

**Two-Parameter Weibull Model**   The two-parameter Weibull model contains a scale parameter or characteristic life $\theta$ and a shape parameter $\delta$. Graphical representations of the hazard, time to failure pmf, and reliability models for a two-parameter Weibull model are shown in Figure 25.4. The hazard plot (Figure 25.4a) clearly indicates that *the Weibull can accommodate the IFR, DFR, and CFR forms, but not all three simultaneously.* For example, it cannot be used to model the complete bathtub curve.

**FIGURE 25.4**   Graphical representation of hazard rate, pmf, and reliability for the two-parameter Weibull model.

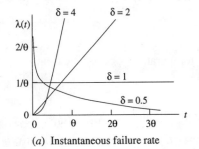

(*a*) Instantaneous failure rate

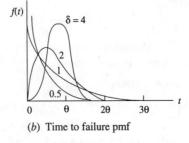

(*b*) Time to failure pmf

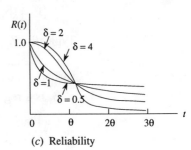

(*c*) Reliability

The forms for reliablility, pmf, and hazard rate are

$$R(t) = \exp\left[-\left(\frac{t}{\theta}\right)^{\delta}\right] \qquad\qquad t \geq 0 \qquad\qquad (25.27a)$$

$$f(t) = \frac{\delta}{\theta}\left(\frac{t}{\theta}\right)^{\delta-1}\exp\left[-\left(\frac{t}{\theta}\right)^{\delta}\right] \qquad t \geq 0 \qquad\qquad (25.27b)$$

$$\lambda(t) = \frac{\delta}{\theta}\left(\frac{t}{\theta}\right)^{\delta-1} \qquad\qquad t \geq 0 \qquad\qquad (25.27c)$$

The mean and the variance parameters are

$$\mu = \theta\,\Gamma\left(1 + \frac{1}{\delta}\right) \qquad\qquad (25.27d)$$

$$\sigma^2 = \theta^2\left\{\Gamma\left(1 + \frac{2}{\delta}\right) - \left[\Gamma\left(1 + \frac{1}{\delta}\right)\right]^2\right\} \qquad\qquad (25.27e)$$

where $\theta$ = scale parameter or characteristic life
$\qquad \delta$ = shape parameter ($\delta = 1$ yields the exponential model)

The complete gamma function $\Gamma(\omega)$ is given by

$$\Gamma(\omega) = \int_0^{\infty} \zeta^{\omega-1}\,\exp\{-\zeta\}\,d\zeta \qquad\qquad (25.27f)$$

The *CRC Mathematical Tables* provide tabularized values for the gamma function [13]. We will demonstrate use of this model in Example 25.3.

**Three-Parameter Weibull Model** The three-parameter Weibull model is useful under the same general circumstances as the two-parameter exponential model. And, like the latter, it is suitable for applications involving high reliability or screened products.

$$R(t) = \exp\left[-\left(\frac{t - t_0}{\theta}\right)^{\delta}\right] \qquad t > t_0 > 0 \qquad\qquad (25.28a)$$

$$f(t) = \frac{\delta}{\theta}\left(\frac{t-t_0}{\theta}\right)^{\delta-1}\exp\left[-\left(\frac{t-t_0}{\theta}\right)^{\delta}\right] \qquad t > t_0 > 0 \qquad\qquad (25.28b)$$

$$\lambda(t) = \begin{cases} 0 & 0 \leq t \leq t_0 \\[2mm] \dfrac{\delta}{\theta}\left(\dfrac{t-t_0}{\theta}\right)^{\delta-1} & t > t_0 \end{cases} \qquad\qquad (25.28c)$$

**Covariate Weibull Model** The covariate Weibull model resembles the exponential covariate in that covariates can be captured in the characteristic-life parameter. The covariate Weibull model, in common with its exponential counterpart, lends itself to designed reliability experimentation.  Additionally, however it is well suited to Arrhenius and power-law stress-acceleration modeling (discussed in Chapter 26).  The form here is

$$R(t; \theta, \delta) = \exp\left[-\left(\frac{t}{\theta}\right)^{\delta}\right] \qquad t \geq 0 \tag{25.29}$$

where $\theta = \exp(\mathbf{Z}, \boldsymbol{\beta})$
$\mathbf{Z} = [Z_1, Z_2, ..., Z_m]$
$\boldsymbol{\beta}^T = [\beta_1, \beta_2, ..., \beta_m]$

Here, $\mathbf{Z}$ and $\boldsymbol{\beta}$ are developed as in the exponential covariate model case.

**Weibull Model Comments**   The Weibull model is a very flexible failure-reliability model.  Figure 25.4 depicts only a few of the many different shapes it may take.  Typically, the ability of a model to conform to many different shapes adds to the model's utility.  In the Weibull case, when $\delta = 1$, the result is an exponential model.  Hence, *the exponential model is a special case of the Weibull model.*
*The Weibull model is widely used for systems, subsystems, and components subject to wear-out and degradation.*  Here, we would expect to see $\delta > 1$, which is the IFR form.  As described earlier, the relationship of the Weibull to a smallest extreme value model justifies, to some extent, its widespread usage in electromechanical system failure-reliability modeling (see Nelson [14]).  In general, for any given device where there are a number of potential failure points "competing" with each other to induce a device failure, the Weibull model may be applicable.

---

**Example 25.3**

A spark plug manufacturer believes that its spark plugs have an IFR characterized by a two-parameter Weibull distribution with $\theta = 4$ yr and $\delta = 2$.
**a** What is the $p = 0.25$ failure time quantile, $t_{p=0.25}$, for the spark plug?
**b** What is the MTTF for the spark plug?

**Solution**

**a** From Equation 25.27a, we know that

$$R(t) = \exp\left[-\left(\frac{t}{\theta}\right)^{\delta}\right] \qquad t \geq 0$$

hence, solving for $t$,

$$t = \theta \left\{ \ln\left[\frac{1}{R(t)}\right] \right\}^{1/\delta}$$

It was given that $p = 0.25$, therefore,

$$R(t_{p=0.25}) = 0.75$$

and
$$t_{p=0.25} = 4 \left[ \ln\left(\frac{1}{0.75}\right)\right]^{1/2} = 2.145 \text{ yr}$$

**b** From Equation 25.27d, the MTTF for the spark plug is $\mu = \theta\Gamma\left(1+\frac{1}{\delta}\right)$; thus

$$\text{MTTF} = 4\Gamma\left(\frac{3}{2}\right) = 4 \, (0.88623) = 3.545 \text{ yr}$$

The *CRC Mathematical Tables* were used to evaluate the gamma function in this case [15].

## Normal Model

The normal, or Gaussian, model is not widely used in reliability work, due to its limited flexibility. However, it will be discussed briefly in order to develop the lognormal model. The normal model has two parameters, which are directly associated with the mean $\mu$ and the variance $\sigma^2$. Figure 25.5 depicts the hazard, pmf, and reliability model forms for the normal model, which are stated below:

$$R(t) = 1 - \Phi\left(\frac{t-\mu}{\sigma}\right) \qquad t \geq 0 \qquad (25.30a)$$

$$f(t) = \frac{1}{\sqrt{2\pi}\,\sigma} \exp\left[-\frac{(t-\mu)^2}{2\sigma^2}\right] \qquad t \geq 0 \qquad (25.30b)$$

$$\lambda(t) = \frac{1}{\sqrt{2\pi}\,\sigma} \exp\left[-\frac{1}{2}\frac{(t-\mu)^2}{\sigma^2}\right]\left[1 - \Phi\left(\frac{t-\mu}{\sigma}\right)\right]^{-1} \qquad t \geq 0 \qquad (25.30c)$$

The normal pmf is symmetrical, whereas many failure processes tend to produce right-skewed pmf signatures.

**FIGURE 25.5**   Graphical representation of hazard rate, pmf, and reliability for the normal model.

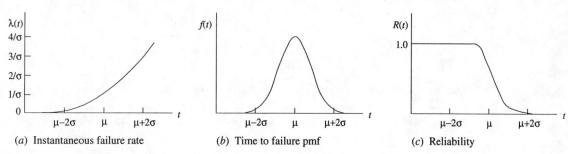

(a) Instantaneous failure rate     (b) Time to failure pmf     (c) Reliability

## Lognormal Model

The lognormal model has been used frequently as a reliability model. Essentially, the lognormal model uses the logarithmically transformed observed failure times (e.g., $X = \ln T$, where $T$ is a random variable representing lifetime or usage). This transformation produces a model which is commonly described by the lognormal median parameter $t_{0.50}$ and the lognormal shape parameter $\sigma_{\ln t}$. The analogy to the normal model is clear. When we define $X = \ln T$, if $X$ follows a normal model, then $T$ follows a lognormal model. Here, it is customary to develop a reverse transformation and obtain the mean $\mu$, in the observed lifetime units, and the variance $\sigma^2$, in the observed lifetime units squared (Lewis [16]). *The lognormal model is very flexible and, like the Weibull Model, can take on IFR, DFR, and CFR characteristics.* Figure 25.6 depicts a variety of lognormal hazard, pmf, and reliability shapes.

$$R(t) = 1 - \Phi \left[ \frac{\ln (t / t_{0.50})}{\sigma_{\ln t}} \right] = 1 - \Phi \left( \frac{\ln t - \mu_{\ln t}}{\sigma_{\ln t}} \right) \qquad t \geq 0 \quad (25.31a)$$

$$f(t) = \frac{1}{\sqrt{2\pi}\, \sigma_{\ln t}\, t} \; \exp - \left[ \frac{1}{2} \left( \frac{\ln t - \mu_{\ln t}}{\sigma_{\ln t}} \right)^2 \right] \qquad t \geq 0 \quad (25.31b)$$

$$\lambda(t) = \frac{f(t)}{R(t)} \qquad\qquad (25.31c)$$

The mean and the variance parameters in the observed life units are

$$\mu = E[T] = \exp \left[ \mu_{\ln t} + \left( \frac{\sigma_{\ln t}^2}{2} \right) \right] \qquad (25.31d)$$

$$\sigma^2 = \text{var}\,[T] = \exp\,(2\mu_{\ln t} + \sigma_{\ln t}^2)\,[\exp\,(\sigma_{\ln t}^2) - 1] \qquad (25.31e)$$

where $t_{0.50}$ = lognormal median (50% failure point in the observed time units)

$\mu_{\ln t} = \ln t_{0.50}$; mean of normal distribution of $X = \ln T$

$\sigma_{\ln t}$ = lognormal shape parameter; standard deviation of normal distribution of $X = \ln T$

**FIGURE 25.6**    Graphical representation of hazard rate, pmf, and reliability for the lognormal model.

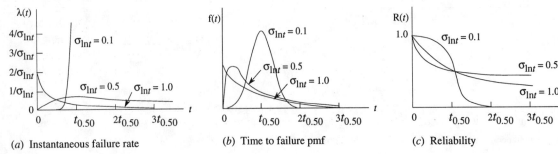

(a) Instantaneous failure rate    (b) Time to failure pmf    (c) Reliability

Practitioners and researchers are more comfortable working with the normal model; therefore, we typically work with data using the normal model with parameters $\mu_{\ln t}$ and $\sigma^2_{\ln t}$ after we transform each observed failure time by taking its natural log.

**Covariate Lognormal Model**    We can develop a covariate form for the lognormal model using our basic notation from above. The covariate lognormal model allows us to describe the mean $\mu_{\ln t}$ as a function of environment, application, configuration, and so on. The covariate lognormal model will be further developed in the context of failure data analysis in Chapter 26.

$$R(t; \mu_{\ln t}, \sigma_{\ln t}) = 1 - \Phi \frac{\ln t - \mu_{\ln t}}{\sigma_{\ln t}} \qquad t \geq 0 \qquad (25.32)$$

where $\mu_{\ln t} = \mathbf{Z}\, \boldsymbol{\beta}$
$\mathbf{Z} = [Z_1, Z_2, ..., Z_m]$
$\boldsymbol{\beta}^T = [\beta_1, \beta_2, ..., \beta_m]$

Here, $\mathbf{Z}$ and $\boldsymbol{\beta}$ are developed as described in the exponential covariate model discussion.

**Lognormal Model Comments**    The lognormal model is used pragmatically in reliability work. As Lawless [17] and Nelson [18] point out, the lognormal's failure hazard rate $\lambda(t)$ is not consistent across time. For example, it starts at 0, increases up to a certain time, or change point, and then decreases beyond that time. (Some of this behavior can be seen in Figure 25.6a.) Typically, as we are interested in an interval of time or usage prior to this change point, it is not a major concern. *The lognormal model is used for applications where physical degradation is present.* Tobias and Trindade describe the lognormal model in the context of a multiplicative growth model which is useful in modeling processes where failure is brought about by a gradual degradation process, located in many sites, over time [19]. For example, it finds usage in modeling chemical based failure processes such as corrosion as well as failure processes involving crack propagation. It has also found application in molecular diffusion or migration in semiconductor devices.

---

**Example 25.4**

The life of an electrical relay was observed to be dependent upon the temperature of its surroundings and fits a lognormal model. It was estimated that $\mu_{\ln t} = 7.2 - 0.02C$, where $C$ is the temperature of the surroundings in degrees Celsius. The original scale of measure $t$ is in hours. Given the lognormal shape parameter $\sigma_{\ln t} = 0.7$,

   **a** Compare the relay's MTTF's when operating at 20°C and at 150°C.
   **b** Calculate its reliability at the end of 250 hr if operated at 100°C.

**Solution**

**a** Operating at 20 °C,

$$\mu_{\ln t} = 7.2 - 0.02(20) = 6.800$$

From Equation 25.31*d*, we know that

$$\text{MTTF} = \mu = \exp\left[(\mu_{\ln t} + (\sigma^2_{\ln t}/2)\right] = \exp\left[6.800 + (0.7^2/2)\right] = 1147.1 \text{ hr}$$

Operating at 150°C,

$$\mu_{\ln t} = 7.2 - 0.02(150) = 4.200$$

$$\text{MTTF} = \mu = \exp\left[4.200 + (0.7^2/2)\right] = 85.2 \text{ hr}$$

**b** From Equation 25.32, we know that

$$R(t) = 1 - \Phi\left(\frac{\ln t - \mu_{\ln t}}{\sigma_{\ln t}}\right)$$

When operating at 100°C,

$$\mu_{\ln t} = 7.2 - 0.02(100) = 5.200$$

$$R(250 \text{ hr}) = 1 - \Phi\left[\frac{\ln 250 - 5.200}{0.7}\right]$$

Now, using our standard normal table, Table IX.2,

$$R(250) = 1 - \Phi(0.46) = 1 - P(Z \le 0.46) = 1 - 0.6772 = 0.323$$

If operated at 100°C, the relay reliability for 250 hr of operation is estimated to be 0.323.

---

## 25.4 RELIABILITY-MAINTAINABILITY MODEL APPLICATIONS

The reliability and maintainability models previously developed can be applied in many different ways. Obviously, model applications are driven by system analysis goals and objectives. *We may use the time and usage based models directly in block diagrams in place of a priori success-failure estimates.* This approach yields a time and usage based system block diagram model. Either of two general approaches can be taken in solving a time and usage based system block diagram.

1 Model each block individually and solve for an $R(t)$ block point estimate, and then use the standard Boolean deterministic probability solution method described in Chapter 24. Here, the Boolean operations are performed on point estimates developed from the time based models.

2 Embed the $R(t)$ models directly in the general Boolean statements and solve for an overall system expression as a function of time-usage $t$. Here, the Boolean operations take place at the $R(t)$ model level.

The second approach yields the most general result, but requires the most effort. The second approach is widely used for series and standby configurations, where component failure characteristics are described with one-parameter exponential models. The more complicated models, such as the Weibull and lognormal models, are extremely difficult (if not impossible) to deal with in forming an overall system expression. In these cases, we would most likely develop a point estimate from our time and usage based model and work through our block diagrams.

## Parts-Count Model

*The parts-count model views a system as a series arrangement of independent components, regardless of the actual configuration.* Each component's reliability is represented by an exponential model. Hence, an $n$ component system is modeled as

$$R_{sys}(t) = \prod_{i=1}^{n} R_i(t) = \prod_{i=1}^{n} e^{-\lambda_i t} = \exp\left(-t \sum_{i=l}^{n} \lambda_i\right) \tag{25.33}$$

We know from our series system discussions in Chapter 24 that the parts-count model will be a worst-case model. We also know that the failure hazard for an exponential model is constant (time independent). In the above equation, we can see that the failure hazards are additive and that we assume that all $n$ components operate for time $t$. Hence, each individual part makes its contribution to the overall system failure hazard rate. We can also observe that the overall system reliability $R(t)$ takes an exponential model form, with a system hazard rate of

$$\lambda' = \sum_{i=1}^{n} \lambda_i \tag{25.34}$$

Hence, the system MTTF (from Equation 25.24d) is

$$\theta' = \frac{1}{\lambda'} = \frac{1}{\displaystyle\sum_{i=1}^{n} \lambda_i} \tag{25.35}$$

In practice, we may obtain our $\lambda$ values from a database, such as *Military Handbook 217* [20]. They may be expressed as failures per $10^6$ hr or some other appropriate time unit along with adjustment "$\pi$" factors (see Fuqua [21]).

---

## Example 25.5

A preamplifier used in a radar system is made up of interconnected components whose failure rates (under normal operating conditions) and respective quantities are given below. Using the parts-count method, calculate

**a** The assembly failure rate.
**b** The reliability for a six-day mission.

| Component | Function | Number of components | Failure rate per $10^6$ hr | Total failure rate per $10^6$ hr |
|---|---|---|---|---|
| Resistor type A | Voltage divider | 4 | 1.50 | 6.00 |
| Capacitor type A | Decoupling | 4 | 0.22 | 0.88 |
| Diode | Voltage divider | 3 | 1.00 | 3.00 |
| Transistor | Amplifier | 3 | 3.00 | 9.00 |
| Transformer | Coupling | 2 | 0.30 | 0.60 |
| Resistor type B | Bias | 2 | 0.005 | 0.01 |
| Capacitor type B | Bypass | 4 | 0.48 | 1.92 |
| Total | | 22 | | 21.41 |

## Solution

**a**  We have calculated the total failure rate for each component type (shown in column 5, above).  For a nonredundant system the assembly failure rate is the sum of these numbers, or, as indicated, $\lambda' = 21.41 \times 10^{-6}$/hr.

**b**  The reliability for a six-day mission is calculated from $R = e^{-\lambda' t}$:

$$R(144 \text{ hr}) = \exp\left[(-21.41 \times 10^{-6} (6)(24)\right]$$

$$= 0.9969$$

## Standby Models

As discussed earlier, reliability models for standby systems (depicted in Figure 24.8) can be developed using *a priori* probabilities.  Using these same basic configurations, we can also apply time or usage based reliability models.  *In general, we may pursue one of the following two strategies in developing probabilistic standby models*:  (1) *an analytical approach or* (2) *a simulation approach*.  Simulation techniques are discussed in a later section of this chapter.

Analytical approaches require us to apply basic probability logic and tools as well as advanced mathematical tools such as Markovian state models.  The mathematical demands of these models rapidly become burdensome.  For our purposes, we will discuss a few relatively simple standby configurations, with the assumption that exponential failure models are applicable to all system components.  More detailed developments can be found in Lewis [22] and Kapur and Lamberson [23].

We will first develop a time based reliability model for a two-component standby system (shown in Figure 25.7a).  Here, we have two success possibilities (Figure 25.7b and 25.7c).  Assuming that both components are identical in their failure characteristics, a failure-proof switch exists, and the standby component cannot fail while held in reserve,

$$R_{sys}(t) = P\{(T_1 > t) \cup [(T_1 \le t) \cap (T_2 > t - T_1)]\} \tag{25.36}$$

where $T_1$ = random variable representing the lifetime of primary component, $c_1$

$T_2$ = random variable representing the lifetime of standby component, $c_2$

Applying time based models for each component,

$$R_{sys}(t) = R_{c_1}(t) + \int_0^t f_{c_1}(t_1)\, R_{c_2}(t - t_1)\, dt_1 \tag{25.37}$$

Now, assuming identical exponential failure models for both systems,

$$R_{sys}(t) = e^{-\lambda t} + \int_0^t \lambda e^{-\lambda t_1}\, e^{-\lambda(t - t_1)}\, dt_1$$

$$= e^{-\lambda t}(1 + \lambda t) \qquad t \geq 0 \tag{25.38}$$

Equation 25.36 can be extended to include $n$ components. For an $n$ component standby system with identical components, perfect switching, no failure in the standby condition, and exponential failure models, we have

$$R_{sys}(t) = e^{-\lambda t} \sum_{i=0}^{n-1} \frac{(\lambda t)^i}{i!} \qquad t \geq 0 \tag{25.39}$$

If a switching success (switch reliability) is expressed as $q_s$, then, for a two-component standby system,

$$R_{sys}(t) = R_{c_1}(t) + q_s \int_0^t f_{c_1}(t_1)\, R_{c_2}(t - t_1)\, dt_1 \tag{25.40}$$

Now, assuming identical exponential failure characteristics for both systems,

$$R_{sys}(t) = e^{-\lambda t} + q_s\, \lambda t\, e^{-\lambda t} \qquad t \geq 0 \tag{25.41}$$

Furthermore, if we assume that switch failure can be modeled with an exponential failure model with parameter $\lambda_s$, then, for a two-component standby system,

$$R_{sys}(t) = P\{(T_1 > t) \cup [(T_1 \leq t) \cap (T_s > T_1) \cap (T_2 > t - T_1)]\} \tag{25.42}$$

where $T_s$ is the switch failure time.

FIGURE 25.7   Two-component standby system description.

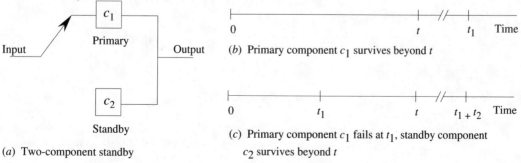

(a) Two-component standby

(b) Primary component $c_1$ survives beyond $t$

(c) Primary component $c_1$ fails at $t_1$, standby component $c_2$ survives beyond $t$

With the assumption of exponential models and simplifying, we obtain

$$R_{sys}(t) = e^{-\lambda t}\left[1 + \frac{\lambda}{\lambda_s}\left(1 - e^{-\lambda_s t}\right)\right] \qquad t \geq 0 \qquad (25.43)$$

In the case where we have a two-component standby system, as above, but with a standby component that may fail while in the standby or dormant mode, a more complicated analysis is required. Lewis uses a Markovian state model to develop a general system reliability form, assuming exponential failure models for both non-identical components and a perfect switch [24]. His result is

$$R_{sys}(t) = e^{-\lambda_1 t} + \frac{\lambda_1}{\lambda_1 + \lambda_2^+ - \lambda_2}\left(e^{-\lambda_2 t} - e^{-(\lambda_1 + \lambda_2^+)t}\right) \qquad (25.44)$$

where $\lambda_1$ = primary component's instantaneous failure rate.

$\lambda_2$ = standby component's instantaneous failure rate when in the operation mode.

$\lambda_2^+$ = standby component's instantaneous failure rate when in the standby or dormant mode.

---

## Example 25.6

An overhead projector bulb-circuit in a classroom has a CFR of $\lambda_1 = 1.63 \times 10^{-2}$ failures per hr. A standby bulb-circuit in the same projector has a CFR of $\lambda_2 = 4.32 \times 10^{-2}$ failures per hr. In the event that the first overhead projector bulb fails, the operator flips a switch to start the standby bulb-circuit. The switch has a reliability of $q_s = 0.95$. Calculate the reliability of the standby system for a 3-hr classroom presentation.

### Solution

Let $c_1$ denote the first bulb and $c_2$ the standby. From Equation 25.40, we have

$$R_{sys}(t) = R_{c_1}(t) + q_s \int_0^t f_{c_1}(t_1) R_{c_2}(t - t_1)\, dt_1 \qquad (25.45)$$

$$= e^{-\lambda_1 t} + q_s \int_0^t \lambda_1 e^{-\lambda_1 t_1} e^{-\lambda_2 (t - t_1)}\, dt_1$$

$$= e^{-\lambda_1 t} + q_s \lambda_1 e^{-\lambda_2 t} \int_0^t e^{(\lambda_2 - \lambda_1)t_1}\, dt_1$$

$$= e^{-\lambda_1 t} + q_s \lambda_1 e^{-\lambda_2 t} \left[\frac{e^{(\lambda_2 - \lambda_1)t_1}}{\lambda_2 - \lambda_1}\right]_0^t$$

$$= e^{-\lambda_1 t} + q_s \left(\frac{\lambda_1}{\lambda_2 - \lambda_1}\right) e^{-\lambda_2 t} \left(e^{(\lambda_2 - \lambda_1)t} - 1\right)$$

Thus,

$$R_{sys} = e^{-\lambda_1 t} + q_s \left(\frac{\lambda_1}{\lambda_2 - \lambda_1}\right)(e^{-\lambda_1 t} - e^{-\lambda_2 t}) \qquad (25.46)$$

Accordingly, for a three-hour classroom presentation, the reliability of the system is

$$R_{sys} (t = 3 \text{ hr}) = e^{-[(1.63 \times 10^{-2})(3)]}$$

$$+ (0.95) \left(\frac{1.63 \times 10^{-2}}{4.32 \times 10^{-2} - 1.63 \times 10^{-2}}\right)(e^{-[(1.63 \times 10^{-2})(3)]} - e^{-[(4.32 \times 10^{-2})(3)]})$$

$$= 0.99478$$

Note that the failure characteristics of $c_1$ and $c_2$ are different in this case, hence we could not use Equation 25.41 directly.

## 25.5 AVAILABILITY MODELS

*Availability models are useful for studying both preventive and corrective maintenance strategies.* They combine reliability and maintainability characteristics simultaneously. *Availability A(t) is defined as the probability that a system, subsystem, or component is available for use at some given time, considering given conditions of operation and restoration.* In a special case, where no restoration, corrective or preventive maintenance, is possible, availability is identical to reliability [$A(t) \to 0$ as $t \to \infty$]. Whereas, typically, the presence of restoration provisions leads to a cyclical or steady-state value where $A(t) \neq 0$ as $t \to \infty$. A variety of $A(t)$ curves are shown in Figure 25.8. We should note at this point that the cyclical signatures of $A(t)$ are a result of the assumption or condition that a preventive or corrective maintenance act will always produce an operational system upon re-start (immediately after the maintenance act is completed).

The conceptual $A(t)$ curves in Figure 25.8 illustrate the performance advantage that restoration provisions offer; we can see that the availability measure of performance clearly is enhanced by both preventive and corrective maintenance. By definition, availability measures a ratio of available state time to available and unavailable state time. Sometimes this ratio is expressed as up time to "total" time, where total time includes up time and downtime. Figure 25.9 shows a graphical representation of this relationship. *The availability measure by itself is not capable of providing detailed information as to the duration of individual operation and restoration intervals* (shown in Figure 25.9).

### Preventive Maintenance

In the case of "as-good-as-new" preventive maintenance, provided at some time interval $t_{pm}$, a system will demonstrate the availability characteristics shown in Figure 25.8b. In this case, for the first cycle, $A(t) = R(t)$, $0 \leq t \leq t_{pm}$. Thereafter we observe an identical periodic cycle of $A(t)$, beginning immediately after $t = t_{pm}$ and

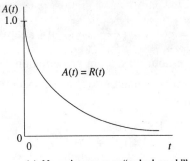

(*a*)  No maintenance or "as-bad-as-old"
preventive maintenance

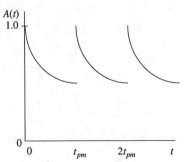

(*b*)  Preventive maintenance "as-good-as-new" only

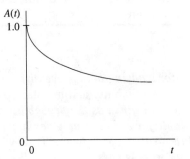

(*c*)  Corrective (repair) maintenance only

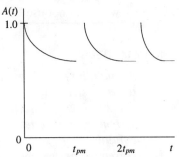

(*d*)  Preventive maintenance "almost-as-good-as-new"
and corrective (repair) maintenance

**FIGURE 25.8**   Generic availability curves.

**FIGURE 25.9**   Empirical availability performance measure.

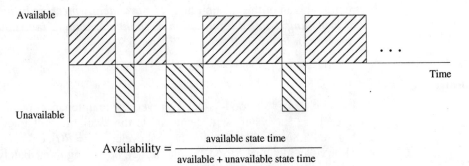

$$\text{Availability} = \frac{\text{available state time}}{\text{available} + \text{unavailable state time}}$$

again at each of the following multiples of $t_{pm}$ (provided preventive maintenance restores the system to as-good-as-new condition). The other restoration extreme, "as-bad-as-old," produces no change in the system; hence the original reliability curve (Figure 25.8$a$) applies. Here, we assume the maintenance is instantaneous and the system will not fail on re-start. From this comparison, we can see that as-bad-as-old preventive maintenance is pointless, as it requires resources but provides no improvement in $A(t)$:

$$A(t) = R(t) \qquad t \geq 0 \tag{25.47}$$

In the case of as-good-as-new preventive maintenance (Figure 25.8$b$), where we see repeated availability cycles, we can calculate the cyclical availability using the reliability measure, for the first cycle and repeating cycles thereafter, $n = 0, 1, 2, \ldots$ :

$$A(t) = \exp\left[-\int_0^{(t-nt_{pm})} \lambda(\zeta)\, d\zeta\right] \qquad nt_{pm} \leq t < (n+1)t_{pm} \tag{25.48}$$

Figure 25.10 shows a generic hazard curve, a corresponding reliability curve, and the resulting availability curve for this case. We can observe that the hazard curve here is made up of identical hazard "cycles." Hence, we can use Equation 25.48 to model availability and develop a reliability expression:

$$R(t) = [R(t_{pm})]^n\, R(t - nt_{pm}) \qquad nt_{pm} \leq t < (n+1)\, t_{pm} \tag{25.49}$$

Here, $n$ represents the number of preventive maintenance cycles we have completed as a function of time or usage, $n = 0, 1, 2, \ldots$

The as-good-as-new case is of great interest because it can be directly related to a "replacement" analysis. For example, the depiction in Figure 25.10 would correspond to system replacements at $t_{pm}$, $2t_{pm}$, and so on. Hence, we can use the previous equations to study replacement reliability–availability alternatives.

We can also model the situation where each preventive maintenance operation at $t_{pm}$, $2t_{pm}$, $3t_{pm}$, and so on, may be imperfect. Here, we assume that a chance $p$ exists of encountering a preventive maintenance operation that is totally faulty (albeit unintentionally) (e.g., the maintenance results in a dead system). Under this assumption,

$$R(t) = [R(t_{pm})]^n\, (1-p)^n\, R(t - nt_{pm}) \qquad nt_{pm} \leq t < (n+1)\, t_{pm} \tag{25.50}$$

When preventive maintenance produces a system somewhere between the as-good-as-new and as-bad-as-old cases (almost-as-good-as-new, see Figure 25.8$d$), a determination of the exact hazard curve characteristics from one preventive maintenance cycle to the next must be made. Kolarik and Case have suggested that this relationship can be associated with the complexity of the system and the extent of the preventive maintenance operation [25]. In any event, sound engineering judgment is required to establish this relationship. Again, we assume the maintenance is instantaneous and the system will not fail on re-start. Equations 25.48 and 25.49 can be modified and expressed as:

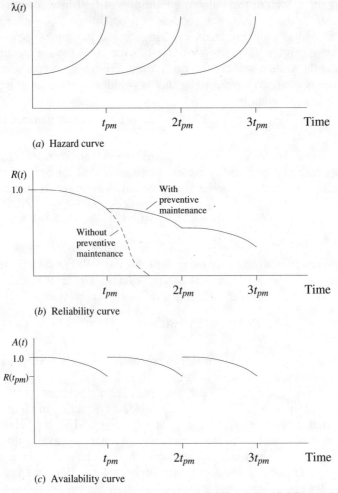

(a)  Hazard curve

(b)  Reliability curve

(c)  Availability curve

**FIGURE 25.10**     As-good-as-new preventive maintenance reliability and availability models.

$$R(t) = \left[ \prod_{i=1}^{n+1} R_{i-1}(t_{pm}) \right] R_{n+1}(t - nt_{pm}) \tag{25.51}$$

where $nt_{pm} \le t < (n+1)\, t_{pm}$, $R_0(t_{pm}) = 1$, $n = 0, 1, 2, \ldots$.

Accordingly, the $A(t)$ cycle must correspond to a sequence of hazard curves. The result is a series of unique $A(t)$ segments where

$$A(t) = \exp\left[ -\int_0^{(t - nt_{pm})} \lambda_{n+1}(\zeta)\, d(\zeta) \right] \qquad nt_{pm} \le t\, (n+1)\, t_{pm} \tag{25.52}$$

## Example 25.7

Assume that automotive brakes are designed for 12 years of operation. There are two significant contributions to the failure rate. The first, due to wear of the brake linings, is described by a Weibull distribution with $\theta = 16$ years and $\delta = 1.7$. The second, which includes all other causes, is a CFR of $\lambda_1 = 3.688 \times 10^{-3}$ failures per yr.

**a** What is the availability if no preventive maintenance is performed over the 12-year design life?

**b** If the reliability of the 12-year design life is to be at least 0.8, through periodic replacement of the brake linings, how often must the linings be replaced?

## Solution

Let $t_d = 12$ represent the design life.

**a** The system availability will be equal to the system reliability:

$$A(t_d) = R(t_d) = R_1(t_d) \, R_2(t_d)$$

Using Equation 25.24a to deal with the CFR portion,

$$R_1(t_d) = \exp(-\lambda_1 t_d) = \exp[(-3.688 \times 10^{-3})(12)] = 0.95671$$

Similarly, using Equation 25.27a for the Weibull (IFR) portion,

$$R_2(t_d) = \exp\left[-\left(\frac{t_d}{\theta}\right)^\delta\right] = \exp\left[-\left(\frac{12}{16}\right)^{1.7}\right] = 0.54161$$

Thus,

$$A(t_d) = R(t_d) = 0.95671\,(0.54161) = 0.51816$$

**b** Suppose that we divide the design life into $n$ equal intervals; the time interval $t_{pm}$ at which preventive maintenance is carried out is then $t_{pm} = t_d/n$. Correspondingly, $t_d = nt_{pm}$. For brake lining replacement at time interval $t_{pm}$ based on Equation 25.49, we can write

$$R(t) = [R(t_{pm})]^n$$

For the criterion to be met, we must have

$$R(t_d = 12) = [R_1(t_{pm}) \, R_2(t_{pm})]^n \geq 0.8$$

where $t_{pm} = t_d/n$.

Therefore, we must determine a value of $n$ such that the above condition is met, using a crude trial and error method:

| $n$ | 1 | 2 | 3 | 4 | 5 | 6 |
|---|---|---|---|---|---|---|
| $t_{pm} = t_d / n$ | 12 | 6 | 4 | 3 | 2.4 | 2 |
| $R(t_d)$ | 0.518 | 0.656 | 0.720 | 0.758 | 0.784 | 0.803 |

Thus the criterion is met for $n = 6$, and the time interval for brake lining replacement is

$$t_{pm} = \frac{t_d}{6} = \frac{12}{6} = 2 \text{ yr}$$

## Corrective Maintenance

Corrective maintenance is more difficult to model than preventive maintenance. Again, the availability measure is of interest. Figure 25.8c depicts a solely corrective maintenance, or repair, case. It is clear that $A(t)$ is initially falling with time and then approaches a steady-state value, which is dependent on the relative failure and repair characteristics. In this case, three primary measures of availability are typically of interest, according to Lewis [26]: (1) point availability as a function of time or usage, defined earlier, $A(t)$, (2) steady-state availability $A_\infty$, and (3) average, or mission, availability $A_{avg}$. The $A_\infty$ and $A_{avg}$ measures are defined in terms of $A(t)$:

$$A_\infty = \lim_{t \to \infty} A(t) \tag{25.53}$$

$$A_{avg}(t_0) = \frac{1}{t_0} \int_0^{t_0} A(t) \, dt \qquad t_0 = \text{mission time} \tag{25.54}$$

These three availability measures hold regardless of the form of the reliability model and corrective maintenance (repair) model. Most analytical treatments of repairable systems are based on exponential (constant hazard) failure-repair models. Various analytical approaches and solution methods have been advocated. We will not attempt to redevelop the detailed solutions which are readily available in Shooman [27], as well as in texts by Lewis, Kapur and Lamberson, and others [28]. Here, we will restate availability measures for a single-component system. Cases involving nonexponential failure-repair models, as well as multiple-component analyses, can be analyzed through simulation techniques, discussed later.

**Single Component With Repair** Assuming an exponential failure model where $\lambda(t) = \lambda$ and an exponential repair model where $\upsilon(t) = \upsilon$, the $A(t)$, $A_\infty$, and $A_{avg}$ relationships can be developed as follows:

$$A(t + \Delta t) = A(t) - \lambda \Delta t \, A(t) + \upsilon \Delta t \, [1 - A(t)] \tag{25.55}$$

that is, $\qquad \dfrac{A(t + \Delta t) - A(t)}{\Delta t} = -(\lambda + \upsilon) \, A(t) + \upsilon$

Hence,

$$\frac{dA(t)}{dt} = -(\lambda + \upsilon) \, A(t) + \upsilon \tag{25.56}$$

and
$$A(t) = \frac{\upsilon}{\lambda + \upsilon} + \frac{\lambda}{\lambda + \upsilon} \exp\left[-(\lambda + \upsilon)t\right] \tag{25.57}$$

$$A_\infty = \frac{\upsilon}{\lambda + \upsilon} = \frac{1/\lambda}{1/\lambda + 1/\upsilon} = \frac{\text{MTTF}}{\text{MTTF} + \text{MTTR}} \tag{25.58}$$

$$A_{\text{avg}}(t_0) = \frac{\upsilon}{\lambda + \upsilon} + \frac{\lambda}{t_0(\lambda + \upsilon)^2}\left\{1 - \exp\left[-(\lambda + \upsilon)t_0\right]\right\} \tag{25.59}$$

More complicated single-component cases involving testing are presented in Lewis [29]. The text includes a number of models regarding test time and repair rate assumptions.

## 25.6 RELIABILITY AND MAINTAINABILITY SIMULATION MODELS

*Simulation analysis offers a flexible method for assessing reliability and maintainability characteristics of complex systems. Two general forms of simulation models are appropriate for reliability and maintainability applications: (1) event-advanced and (2) time-advanced.* Each has its own advantages and disadvantages. The event-advanced model is relatively fast and is compatible with event-advanced simulation tools such as GPSS/H [30], SLAMII [31], SIMAN [32], and so forth, as well as generic Monte Carlo methods [33]. The time-advanced model requires more execution time and programming expertise than does the event-advanced model. However, it offers more modeling flexibility, in return.

*Simulation is essentially an abstract time or usage sequence that includes failure, corrective maintenance, and preventive maintenance activities and events.* The simulation study method is a logical approach which represents critical activities and events in an abstract sense, over time, based on the probabilistic nature of the actual activities and events. We use the fundamental component models (hazard functions or pmf's) along with the system configuration diagram and success/failure logic to create an availability simulation.

### Event-Advanced Model

The event-advanced model is driven by a time based schedule of failure, repair (corrective maintenance), and preventive maintenance events. In such simulations, we generate this time based sequence for each component. Essentially, we carry out a pseudo-random sampling process: Samples are taken from the basic component failure and repair models, thus simulating component-level performance. The component results are then superimposed on the system configuration to determine the system failure-repair or operation characteristics.

Figure 25.11 provides a simple graphical example of an event-advanced model for a two-component parallel system. A block diagram for the system is shown in Figure 25.11a. For each component we must develop failure and repair characteristics, depicted as curves in Figure 25.11b. We then sample from the failure functions to obtain the times to failure and from the repair functions to obtain the times to repair. The resulting event times are portrayed in Figure 25.11c for each component. Next, based on our system configuration (in this case, parallel) we determine the system status, (also shown in Figure 25.11c). This process continues inside our computer until we have collected sufficient data. Then, we develop a report on component and system status to meet our objectives.

## Time-Advanced Model

The time-advanced model develops the conditional probability of failure or repair as time is incremented. This method requires more computer time than is required in the event-advanced method. However, it has the flexibility needed to modify the conditional probabilities; this feature allows us to represent different applications and/or environments over the life of the system being simulated.

Figure 25.12 shows a simple graphical example of a time-advanced model for a two-component parallel system. Here, in each operation and repair increment $\Delta t$, the conditional probability of failure or repair is generated according to the conditional relationships of Equations 25.60 and 25.61, respectively, (also shown

**FIGURE 25.11** Simplified event-advanced reliability and maintainability simulation.

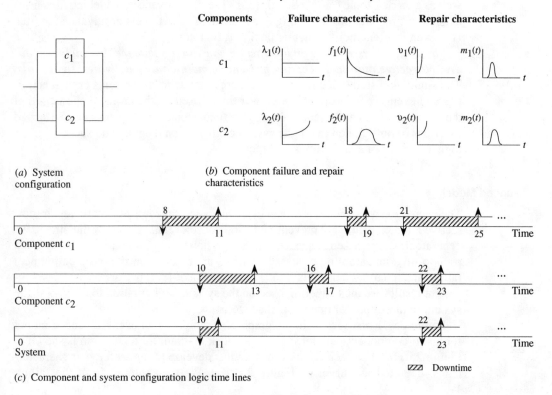

(a) System configuration

(b) Component failure and repair characteristics

(c) Component and system configuration logic time lines

⧅ Downtime

in Figure 25.12c). Then, in each time step, another pseudo-random variable between 0 and 1 is generated and compared to the conditional probability. If the pseudo-random variable is less than the conditional probability of failure-repair, a failure-repair completion event is triggered and a switch from an "up" to a "down" state (or vice versa) is made. Otherwise, time is again incremented in the same up-down mode, and the process is repeated. The conditional probability forms are

$$F(t + \Delta t_i \mid t) = 1 - \exp\left[-\int_t^{t+\Delta t_i} \lambda(\zeta)\, d\zeta\right] \tag{25.60}$$

$$M(t + \Delta t_j \mid t) = 1 - \exp\left[-\int_t^{t+\Delta t_j} \upsilon(\zeta)\, d\zeta\right] \tag{25.61}$$

Once the component failure-repair events have been developed, the system reliability and maintainability characteristics can be assessed. Again, as in the case of the event-advanced simulation method, component and system level analyses can be made. Figure 25.12d depicts the up-down status of each component, as well as the system status, and thus, the performance. As system availability is simply the ratio

**FIGURE 25.12**    Simplified time-advanced reliability and maintainability simulation.

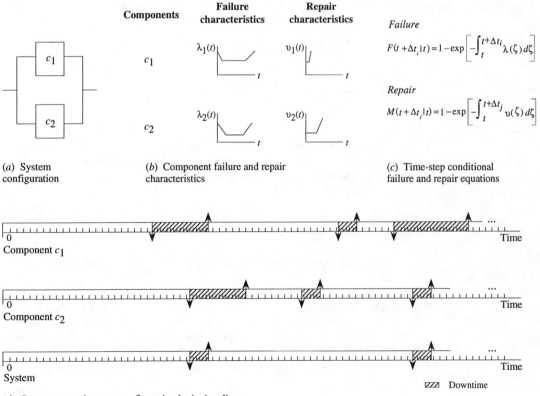

(a) System configuration

(b) Component failure and repair characteristics

(c) Time-step conditional failure and repair equations

(d) Component and system configuration logic time lines

of up time to up time plus downtime (see Figure 25.9). In simulation, we view availability from a "historic" perspective (e.g., our ratio of up time to total time serves as our availability estimate).

## REVIEW AND DISCOVERY EXERCISES

### Review

**25.1** Study the reliability curves below. Then, determine which ones are theoretically feasible and which ones are not. Finally, for each theoretically feasible curve, sketch a compatible hazard curve on the axes below its reliability curve.

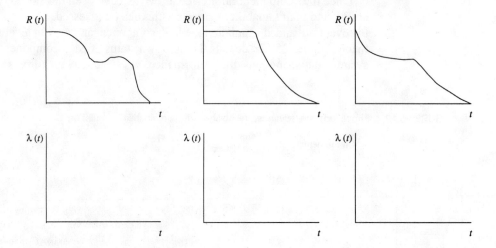

**25.2** The reliability configuration for a six-component electronic device is shown below. It is reasonable to assume that the components exhibit a constant hazard rate. Each component is identical and failures are independent, with $\lambda = 0.02$/yr. [PB]

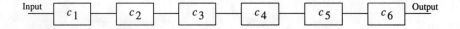

**a** Calculate the probability that the system will operate failure-free for 5 years. For 20 years.

**b** Calculate the MTTF for the configuration.

**25.3** A particular thermocouple, when used in a harsh environment, has a failure rate of $\lambda = 0.008$/hr. How many thermocouples must be placed in parallel if the system is to run for 100 hr with a system failure probability of no more than 0.05? Assume that all failures are independent.

**25.4** Given the failure hazard curve shown on the next page, predict the reliability at time $t = 100$.

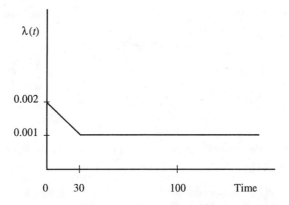

**25.5** A pneumatic hammer has been tested and the following data for a piecewise linear bathtub failure hazard curve have been established (time $t$ is measured in hours):

$$\lambda(t) = 0.005 - 0.0001(t) \qquad\qquad 0 \leq t \leq 47$$

$$\lambda(t) = 0.0003 \qquad\qquad 47 < t \leq 146$$

$$\lambda(t) = -0.007 + 0.00005(t) \qquad\qquad 146 < t$$

**a** Draw the bathtub curve.
**b** Estimate $R(t = 10 \text{ hr})$.
**c** Estimate $R(t = 100 \text{ hr})$.
**d** Estimate $R(t = 200 \text{ hr})$.

**25.6** The fatigue life of a machine member that is a critical part of a holding device in an automated machining process is modeled with a lognormal distribution; the estimated parameters are $\sigma_{\ln t} = 0.43$ and $t_{0.50} = 3.6 \times 10^7$ cycles. [PP]

**a** Estimate $t_{0.10}$, a usage period for the holding device that will yield a reliability of 0.9.
**b** Determine whether a wear-in period (i.e,. the mechanical counterpart to an electronics burn-in period) would be appropriate. Explain your reasoning.

**25.7** A NASA Goddard Space Flight Center report was published which maintained, based on communications satellite data, that a Weibull reliability model with a scale parameter of 5.2 yr and a shape parameter of 1.9 was justified.

**a** Estimate the satellite reliability for a service life of 7 years.
**b** Considering the Goddard model to be appropriate, develop a conditional reliability model for a communications satellite that was launched and put into service 2 years ago and is presently fully operational.
**c** Using the relationship developed above, predict the reliability or probability of the satellite's survival for the next 5 years (assume that it is working properly at the present time).

**25.8** Given the reliability block diagrams and failure hazard rates shown below, determine which design configuration will yield the highest system reliability for $t = 100$ hr of operation.

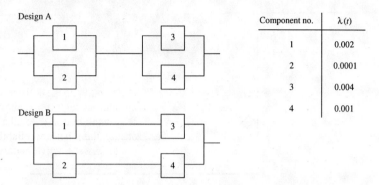

| Component no. | $\lambda(t)$ |
|:---:|:---:|
| 1 | 0.002 |
| 2 | 0.0001 |
| 3 | 0.004 |
| 4 | 0.001 |

**25.9** A PC board contains the following parts and solder joints with the failure rates (per $10^6$ hr) shown below. Using the parts-count method, estimate the PC board's reliability for 10,000 hr of operation.

| Component | Quantity | Failure rate / $10^6$ hr |
|:---|:---:|:---:|
| Capacitors | 5 | 0.025 |
| Resistors | 10 | 0.020 |
| ROMs | 2 | 0.050 |
| Board | 1 | 1.600 |
| Solder joints | 40 | 0.030 |

**25.10** A college professor is leaving home for the summer, but would like to have a light burning at all times in her house to discourage burglars. She decides to build a device that will hold two light bulbs. Design option 1 will switch the current to the second bulb if the first bulb fails. However, a 20 percent chance of failure in making the bulb switch is anticipated. Design option 2 is less technical and will simply allow both bulbs to burn simultaneously. The box in which the light bulbs are packaged says, "average life 1000 hr, exponentially distributed, bulbs will not deteriorate in storage." The professor will be gone 90 days. Which design alternative would you recommend? [NG]

**25.11** An electric power line is to be designed with circuit breakers in the line to interrupt the flow of current if an overload occurs. A sketch of the line is shown below. Assume the circuit breakers operate and fail independently. Operation is considered successful when at least one circuit breaker manages to interrupt the current flow. If the line is to have a reliability of at least 99.9 percent for a 5-year period and each circuit breaker has a hazard function of $\lambda = 0.02 + 0.008t$, how many circuit breakers should be placed in the line?

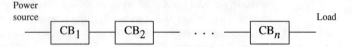

**25.12** A heart pacemaker has a failure hazard rate of $\lambda(t) = 0.072 + 0.01t$, where $t$ is measured in years. The failure hazard rate reflects both random failures and time-dependent failures due to wear and age effects. Regular preventive maintenance may erase the wear and age effects and restore the pacemaker to as-good-as-new condition. [BF]

    **a**  Suppose that the patient and cardiologist want to schedule preventive maintenance on the basis of the time period from installation ($t = 0$) to the point at which reliability reaches 0.95.  Determine this time interval.

    **b**  Calculate the reliability of the maintained system for a four-year period using the preventative maintenance interval determined above.

**25.13**  Heart pacemakers are powered by small button-type alkaline batteries.  Patient as well as cardiologist concern for maximum battery reliability has led to a design where two standby batteries supplement the active battery.  If the charge in the active battery is unable to power the pacemaker, the system automatically switches to the first standby and, if and when it runs down, to the second standby.  [BF]

    **a**  If the failure hazard rate of each battery is 0.004 failures per yr, what is the reliability of the standby system for a four-year period (assuming a perfect switch)?

    **b**  Contrast what is meant by "spare" as opposed to "standby" in maintenance and repair terms.  In the case of the pacemaker, why might a more expensive standby arrangement be preferable to an economical spare arrangement.

**25.14**  Six months ago Fred purchased a used car which has not performed as well as he expected.  The following table contains the times, in days, when the car experienced a failure $t_f$ and the times at which the car was repaired $t_r$ and road-ready once again, over the 6-month period. [PB]

| Failure $i$ | Failure $t_f$ | Repair completion time $t_r$ | Failure $i$ | Failure $t_f$ | Repair completion time $t_r$ |
|---|---|---|---|---|---|
| 1 | 10 | 12 | 6 | 120 | 122 |
| 2 | 30 | 34 | 7 | 140 | 143 |
| 3 | 65 | 70 | 8 | 151 | 152 |
| 4 | 85 | 87 | 9 | 160 | 163 |
| 5 | 90 | 91 | 10 | 177 | 181 |

    **a**  Determine the 6-month availability from the data.

    **b**  Estimate the MTTF and MTTR, assuming exponential failure and reliability models.

    **c**  Based on the data, estimate $A(t = 60$ days).

    **d**  Based on the data, estimate $A_\infty$.

**25.15**  Studies indicate that, on the average, a television set is used 5.7 hr per day.  A manufacturer has the option of specifying picture tube A with an MTTF of 10,000 hr at a cost of $150 or picture tube B with an MTTF of 20,000 hr at a cost of $200.  The MTTFs are believed accurate for typical usage conditions for the average television set.  The cost to replace a picture tube under warranty in the field is estimated to be $300. Failure in the off state and on-off switching are not considered critical in the analysis.  Assuming an exponential distribution is relevant for modeling the reliability of a picture tube, what tube should be specified?  Base your analysis on the one-year warranty period and cost considerations.

**25.16**  We want to develop and analyze the reliability of a battery-powered electronic game.  We believe an exponential model will be adequate.  The battery set has a

40-hr MTTF in our game device. On a long trip, the game is to be operating continuously for 18 hr. [DS]

**a** Determine the battery set reliability for the trip.

**b** Assuming a battery regains no strength when turned off, find the MTTF necessary for one battery set to have a reliability of 0.925 for 30 hr of operation.

### Discovery

**25.17** An airline company calculates that, on the average, one of its planes is in service 15 hr per day and 7 days a week. The company estimates the landing lights are on 60 min per 15 hr of in-service time. The lights are assumed to have a CFR of 0.02 failures per hr while on; 0.001 failures per hr when off, but in service; and 0.0005 failures per hr when the aircraft is out of service. [DS]

**a** Estimate the MTTF of the landing lights.

**b** Assuming that all the landing lights will be replaced the instant the system reliability goes below 0.7, calculate the mean time between replacements.

**c** Determine the point in time at which the probability of failure would equal the reliability.

**25.18** An air traffic controller's human reliability can be modeled with a hazard rate of $\lambda(f) = 4 \times 10^{-7} f$, where $f$ is the number of flights he has directed since the beginning of his shift. (The controller becomes fatigued as the day wears on). The controller's actions are being monitored by a computer system which can identify 99.9 percent of his errors in time for the controller to avert an accident. If the airport must shut down when the reliability of the controller falls below 0.9999, how many flights can a controller handle in the event that a complete computer failure occurs as his shift begins? [DC]

**25.19** A cooling unit is used for temperature control in a clean room. The unit is operated by means of computer software interfaced with a sensor. The physical system configuration is shown in the following figure:

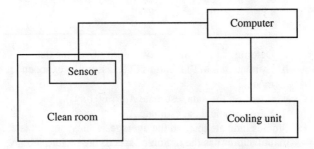

The failure rates of the components and demand data are as follows:

$\lambda_{sensor} = 2.4 \times 10^{-4}$ failure per day

$\lambda_{computer} = 1.2 \times 10^{-5}$ failure per day

$\lambda_{cooling\ unit\ on} = 7.2 \times 10^{-4}$ failure per hr

$\lambda_{cooling\ unit\ off} = 5.0 \times 10^{-5}$ failure per hr

Probability of failure on demand of cooling unit = $5 \times 10^{-6}$

Number of demands per day = 10

The cooling unit has a 15 percent duty cycle (i.e., it is in operation 15 percent of the time). Estimate the system reliability for a 6-month period. [PP]

**25.20** It is becoming fairly common for equipment development contracts to state reliability, maintainability and/or availability measures of merit for equipment (e.g., MTBF of 5000 operating hours, mean time between adjustments of 100 operating hours, MTTR of 5 hr, availability of 99.9 percent). Given the above practice, comment on the technical soundness of this approach. Then, describe how we can establish that equipment will meet the performance measures in the field (i.e., for the customer).

**25.21** Electric power is needed at a remote beachside resort. A number of proposals for motor-generator sets have been developed. The failure characteristics of the type of motor-generator being considered can be modeled by a Weibull distribution with scale parameter $\theta = 2$ years and a shape parameter $\delta = 2.5$. This set can be leased for $125 per month, with expenses of $75 per month for fuel and other supplies. [RC]

    **a** For a single motor-generator unit, estimate the 6-month, 1-year, 3-year, and 5-year reliabilities.

    **b** For two motor generators connected in parallel, estimate the 6-month, 1-year, 3-year, and five-year reliabilities.

    **c** The power company will run power lines to the resort at a cost to the resort owners of $15,000 and will charge a flat monthly rate of $50. The local power-grid reliability can be modeled by an exponential distribution with MTTF of 5 years. Estimate the 6-month, 1-year, 3-year, and 5-year reliabilities.

    **d** Based on reliability and cost considerations, develop a set of recommendations. Plot your results.

**25.22** Identify a product or process of interest to you. Research its technical nature from both a physical as well as a reliability perspective. Then, develop a reliability study which includes

    **a** A description of the system and its components.

    **b** A study objective.

    **c** Block diagrams appropriate for the system.

    **d** A reasonable set of data.

    **e** A systematic analysis and interpretation.

**25.23** Refer to Problem 25.22. Carry out the same tasks in terms of maintainability (i.e., conduct a maintainability study).

**25.24** Refer to Problem 25.22. Carry out the same tasks in terms of availability (i.e., conduct an availability study).

## REFERENCES

**1** R. E. Walpole and R. H. Myers, *Probability and Statistics for Engineers and Scientists*, 5th ed., New York: Macmillan, 1993.

**2** E. E. Lewis, *Introduction to Reliability Engineering*, New York: Wiley, 1987.

**3** W. Nelson, *Applied Life Data Analysis*, New York: Wiley, 1982.

**4** D. Kececioglu, *Reliability Engineering Handbook*, vols. 1 and 2, Englewood Cliffs, NJ: Prentice Hall, 1991.

**5** J. F. Lawless, *Statistical Models and Methods for Lifetime Data*, New York: Wiley, 1982.

**6** S. H. Dai and M. O. Wang, *Reliability Analysis in Engineering Applications*, New York: Van Nostrand Reinhold, 1992.

7 W. J. Conover, *Practical Nonparametric Statistics*, 2d ed., New York: Wiley, 1980.

8 T. L. Landers and W. J. Kolarik, "Proportional Hazards Models and Mil-Hdbk-217," *Microelectronics and Reliability*, vol. 26, no. 4, pp. 763–771, 1986.

9 See reference 5.

10 "Reliability Prediction of Electronic Equipment," *Military Handbook 217*, U. S. Department of Defense, Washington, DC: U. S. Government Printing Office, 1986.

11 See reference 8.

12 "Reliability Design Qualification and Production Acceptance Tests Exponential Distribution," *Military Standard 781*, U. S. Department of Defense, Washington, DC: U. S. Government Printing Office, 1977.

13 *CRC Standard Mathematical Tables*, 18th ed., Cleveland: The Chemical Rubber Company, 1990.

14 See reference 3.

15 See reference 13.

16 See reference 2.

17 See reference 5.

18 See reference 3.

19 P. A. Tobias and D. C. Trindade, *Applied Reliability*, New York: Van Nostrand Reinhold, 1986.

20 See reference 10.

21 Norman B. Fuqua, *Reliability Engineering for Electronic Design*, New York: Marcel Dekker, 1987.

22 See reference 2.

23 K. C. Kapur and L. R. Lamberson, *Reliability in Engineering Design*, New York: Wiley, 1977.

24 See reference 2.

25 W. J. Kolarik and K. E. Case, "Availability Modeling Techniques for Restoration Adjustments and Multiple Working Conditions," *IEEE Transactions on Reliability*, vol. 29, no. 4, p. 324, 1980.

26 See reference 2.

27 M. L. Shooman, *Probabilistic Reliability: An Engineering Approach*, New York: McGraw-Hill, 1968.

28 See references 2 and 23.

29 See reference 2.

30 T. J. Schriber, *An Introduction to Simulation Using GPSS/H*, New York: Wiley, 1991.

31 A. A. B. Pritsker, *Introduction to Simulation and SLAMII*, 3d ed., New York: Wiley (Halsted Press), 1986.

32 C. D. Pegden, R. E. Shannon, and R. P. Sadowski, *Introduction to Simulation Using SIMAN*, New York: McGraw-Hill, 1990.

33 A. M. Law and W. D. Kelton, *Simulation Modeling and Analysis*, 2d ed., New York: McGraw-Hill, 1991.

# 26

# RELIABILITY AND MAINTAINABILITY EXPERIMENTS

## 26.0 INQUIRY

1  How do reliability experiments differ from other experiments?
2  How do nonparametric and parametric reliability models differ?
3  How are trial or mission based experiments developed and analyzed?
4  How are life-test experiments developed and analyzed?
5  How can plotting techniques be used to fit reliability models?
6  How can covariates be used to expand the scope of reliability experiments?

## 26.1 INTRODUCTION

*Our previous discussion of designed or planned experiments*, in Chapters 20 through 23, *can be extended to incorporate reliability experiments, commonly called "life testing," where experimental units (eu's) are essentially run until (1) the mission is completed or (2) the experimental unit fails.* Reliability experiments present additional challenges to us in that we must deal with both failure modes and failure mechanisms.

Essentially, we will follow the experimental design protocol. However, the classical experimental design models and assumptions do not, in general, work well given the nature of failure phenomena and reliability objectives. In most cases, our objective is to model the reliability or maintainability characteristics of a device, system, or process. This objective differs from the classical experimental objective, where we typically focus on a mean response value, while considering the variation present (e.g., as in the fixed effects ANOVA). These two major differences require us to develop and apply rather specialized data analysis tools for life testing.

All of our discussions and examples in this chapter address reliability experiments. However, maintainability experiments can be designed, conducted, and analyzed in the same manner. Maintainability is a probabilistic failed-state-to-operational-state change phenomenon, whereas reliability is a probabilistic operational-state-to-failed-state change phenomenon. Failure and repair phenomena tend to produce distributions which are skewed to the right; symmetric distributions, such as the normal distribution, lack the flexibility to deal with our responses. Hence, we rely on nonsymmetric distributions, such as the binomial, exponential, Weibull, and lognormal models (introduced in Chapter 25).

## 26.2  MISSION SUCCESS AND FAILURE TESTING

*The binomial distribution serves as a useful reliability model when we desire to assess mission success prospects.*  In order to justify use of a binomial model, we must satisfy the following four conditions:

1  The experiment consists of $n$ repeated trials.
2  Each trial can end in only one of two outcomes (here, success or failure).
3  The probability of failure $p$ remains constant throughout the experiment. Here, the probability of success is $q = 1 - p$, as stated in Chapter 24.
4  The repeated trials are conducted and respond in an independent manner.

*The model contains two parameters—the probability of failure p and the number of trials or missions n.*  If we define the random variable $Y$ as the number of failures, the pmf represents a "failure" distribution.  However, if we define the random variable $Y$ as the number of successes, as we do in Equation 26.1, we can develop a "success" or reliability distribution:

$$f(y;n,q) = P(Y = y) = C_y^n \, q^y \, p^{n-y} \qquad y = 0, 1, 2, ..., n \qquad (26.1)$$

where
$$C_y^n = \frac{n!}{y! \, (n - y)!}$$

### Trial or Mission Based Data

Equation 26.1 represents the binomial reliability model in a theoretical sense.  For our testing, we may desire to fit a set of empirical mission success and failure data to the binomial model.  Such data may be collected from a mission based life test or from field experience with our product.

Provided our data include the test size $n$, the number of failures observed $(n - y)$, and the number of successes observed $y$, we can estimate the probability of failure of a single mission as

$$\hat{p} = \frac{n - y}{n} \qquad (26.2)$$

or the probability of success, or reliability, of a single mission as

$$\hat{R} = \hat{q} = 1 - \hat{p} = \frac{y}{n} \qquad (26.3)$$

In each case, we assume we do not know the true value of $p$ or $R$ (that is why we are conducting the experiment in the first place).  It stands to reason that a $\hat{p}$ or $\hat{R}$ estimated from a small $n$ will not be as precise an estimate as a $\hat{p}$ or $\hat{R}$ from a large $n$.  In practice, smaller experiments cost less than larger experiments, but provide less precise estimates.

*If we present only the calculated $\hat{p}$ or $\hat{R}$ and do not account for the sample size n, serious misunderstandings in interpretations and decisions may result.*  For example, we might execute a trial of $n = 2$, where both units successfully fulfill a mission. Here, $\hat{p} = 0/2 = 0$ implies a perfect product for this mission; however, the implication of perfect reliability is misleading here. If we observe a trial where $\hat{p} = 0$ and $n = 10,000$, our evidence of high reliability is much stronger.

*Confidence intervals (**both one-sided and two-sided**) for the mission failure probability or reliability represent a quantitative means to measure the precision of the estimate.* Since mission failure probabilities are estimated with the binomial model, we can take advantage of statistical developments for binomial distributions (see Kececioglu [1] and Lipson and Sheth [2]). To obtain the $(1 - \alpha)$ 100 percent, two-sided, upper and lower, binomial confidence limits $R_{U2}$ and $R_{L2}$, respectively, the following equations must be solved for $R_{U2}$ and $R_{L2}$:

$$\frac{\alpha}{2} = \sum_{i=n-y}^{n} C_i^n (1 - R_{U2})^i R_{U2}^{n-i} \tag{26.4a}$$

$$\frac{\alpha}{2} = \sum_{i=0}^{n-y} C_i^n (1 - R_{L2})^i R_{L2}^{n-i} \tag{26.4b}$$

Using a transformation, $R_{U2}$ and $R_{L2}$ can be calculated from Equations 26.4a and 26.4b as

$$R_{U2} = \frac{1}{1 + \dfrac{n-y}{y+1} F_{1-\alpha/2;2(n-y);2y+2}} \tag{26.5}$$

and

$$R_{L2} = \frac{1}{1 + \dfrac{n-y+1}{y} F_{\alpha/2;\,2(n-y)+2;2y}} \tag{26.6}$$

where $F_{1-\alpha/2}$ and $F_{\alpha/2}$ refer to the $F$ statistics tabulated in Table IX.4. To obtain the $(1 - \alpha)$ 100 percent one-sided, upper and lower binomial reliability confidence limits, $R_{U1}$ and $R_{L1}$, respectively, the following equations must be solved:

$$\alpha = \sum_{i=n-y}^{n} C_{n-y}^{n} (1 - R_{U1})^i R_{U1}^{n-i} \tag{26.7a}$$

and

$$\alpha = \sum_{i=0}^{n-y} C_i^n (1 - R_{L1})^i R_{L1}^{n-i} \tag{26.7b}$$

Again using a transformation, we can calculate the one-sided confidence limits as

$$R_{U1} = 1 + \frac{1}{\dfrac{n-y}{y+1} F_{1-\alpha;2(n-y);2y+2}} \tag{26.8}$$

and

$$R_{L1} = 1 + \frac{1}{\dfrac{n-y+1}{y} F_{\alpha;2(n-y)+2;2y}} \tag{26.9}$$

Binomial confidence intervals, or bands, for mission success-failure have been extensively graphed. Figures 26.1 through 26.3 depict two-sided 90 percent, 95 percent, and 99 percent binomial confidence intervals, respectively, for a number of sample sizes $n$.

Here, the case $y = n$, where we observe only successful outcomes in a field trial or experiment, represents a best case. We can use Equation 26.7b to estimate an optimistic or best-case, test-unit size and result (no failures) to meet a statistical requirement or to provide a quantitative demonstration of reliability. From Equation 26.7b, we have

$$\alpha = \sum_{i=0}^{0} C_i^n (1 - R_{L1})^i R_{L1}^{n-i} = R_{L1}^n$$

and hence,

$$n = \frac{\ln \alpha}{\ln R_{L1}} \tag{26.10}$$

---

## Example 26.1

An advanced automotive battery is under development for use in both hot and cold environments. A severe (worst-case) cold-cranking test is devised with a mission of delivering 15 min of continuous, cranking power at a temperature of –50°F. Test results show that of 15 batteries tested, 12 were successful in fulfilling the mission and 3 failed.

FIGURE 26.1   Two-sided, 90 percent, binomial confidence bands for mission success or failure probabilities. *Source*: Reproduced, with permission of McGraw-Hill, from W. J. Dixon and F. J. Massey, Jr., *Introduction to Statistical Analysis*, New York, NY: McGraw-Hill, page 321, 1951.

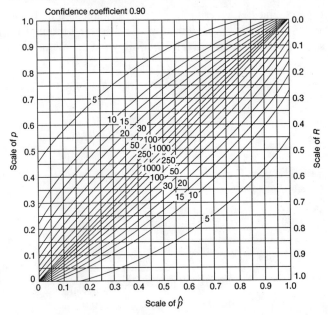

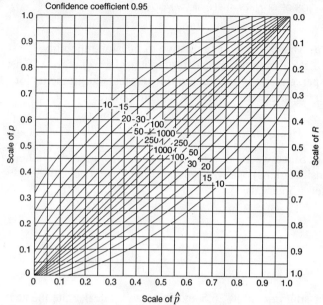

**FIGURE 26.2** Two-sided, 95 percent, binomal confidence bands for mission success or failure probabilities. *Source*: Reproduced from C. J. Clopper and E. S. Pearson, "The Use of Confidence or Fiducial Limits Illustrated in the Case of the Binominal," *Biometrika*, vol. 26, page 410, 1934. Reproduced with the permission of the Biometrika Trustees.

**FIGURE 26.3** Two-sided, 99 percent, binomal confidence bands for mission success or failure probabilities. *Source*: Reproduced from C. J. Clopper and E. S. Pearson, "The Use of Confidence or Fiducial Limits Illustrated in the Case of the Binominal," *Biometrika*, vol. 26, page 410, 1934. Reproduced with the permission of the Biometrika Trustees.

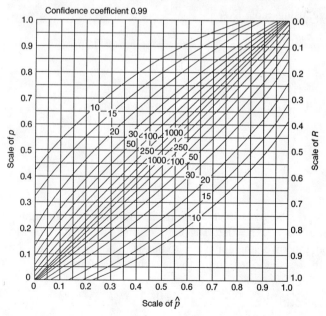

**a**  Analyze these data relative to mission reliability.

**b**  Assuming we wanted to support a claim of at least 99 percent reliability, with 90 percent confidence, determine our test requirements for a best case (smallest number of batteries on test, with zero failures).

**Solution**

**a**  Since we think a binomial model is appropriate here (based on physical considerations), we will estimate $\hat{R}$ and $\hat{p}$ and develop 90 percent two-sided confidence intervals:

$$\hat{p} = \frac{3}{15} = 0.200$$

$$\hat{R} = \frac{12}{15} = 0.800$$

Using Equations 26.5 and 26.6, we can calculate the analytical limits as

$$R_{U2} = \frac{1}{1 + \dfrac{15 - 12}{12 + 1} F_{0.95;2(15-12);2(12)+2}}$$

$$= \frac{1}{1 + \dfrac{3}{13} \, 0.261} = 0.943$$

Here, using Table IX.4,

$$F_{0.95,6,26} = \frac{1}{F_{0.05,26,6}} = \frac{1}{3.83} = 0.261$$

$$R_{L2} = \frac{1}{1 + \dfrac{15 - 12 + 1}{12} F_{0.05;\, 2(15-12)+2;\, 2(12)}}$$

$$= \frac{1}{1 + \dfrac{4}{12} \, 2.36} = 0.560$$

It is possible to estimate these limits graphically from Figure 26.1 as follows. We will use $\hat{p} = 3/15 = 0.200$ as our coordinate on the horizontal axis and project up to both the lower and upper $n = 15$ bands. Then, we will read our $p$ confidence limits to the left and our $R$ limits to the right. These limits are:

$$p_{U2} \approx 0.46 \qquad p_{L2} \approx 0.07 \qquad R_{U2} \approx 0.93 \qquad R_{L2} \approx 0.54$$

**b** In order to determine the minimum test size to support our claim, we will use Equation 26.10.  Here,

$$\alpha = 1 - 0.90 = 0.10 \qquad \text{and} \qquad R_{L1} = 0.99$$

hence the size is

$$n = \frac{\ln \alpha}{\ln R_{L1}} = \frac{\ln 0.10}{\ln 0.99} = 229.1 \text{ batteries}$$

Therefore, we would have to test at least 230 batteries with absolutely no failures in order to support our statement of at least 99 percent reliability with 90 percent confidence.

## 26.3  LIFE TESTING

*Life testing can be thought of as a specially designed experiment where the primary objective is to observe and study the long-term performance, failure modes, and failure mechanisms associated with a product or process.*  In addition to product improvement knowledge, life testing provides information useful for predicting product lifetimes, and thus enhances our warranty and maintenance planning abilities.  Hence, long-term performance or quality issues drive life testing.

The objective of this section is to develop and demonstrate effective procedures for analyzing failure data, whether life-test results or field results.  In either case, we assume that the data do not contain serious biases resulting from poor judgmental manipulation or instrumentation problems.

In practice, we have to use engineering judgment to establish the integrity of our data, relative to its representation of the particular population of devices.  If we encounter true outliers in our data, we must decide to include/exclude each datum on two bases:  (1) a physical argument that the datum is or is not relevant or (2) a statistical argument that the datum is or is not so extreme (e.g., well over $\pm$ 3 or 4 standard deviations) that its elimination can be defended.  Obviously the physical argument is the strongest argument in reliability engineering.

*Life data are typically collected in two forms*:  **(1)** *individual-failure data and* **(2)** *class-interval data.*  Individual-failure data contain a measured failure time for each failure.  Hence, if 10 failures occur, 10 corresponding failure times are recorded.  In working with class intervals, the number of failures in each class (i.e., "bucket" of time) are known, but exact failure times cannot be determined.  Hence, class-interval data provide less information than individual-failure data.  However, in general, collection of failure data based on class intervals is the easier of the two methods as it does not require detecting and logging of actual failure times.

*Two unique characteristics of life testing which create challenges in analyses are censoring and acceleration.  Censoring involves the withdrawal of a test unit from the test for* **(1)** *a reason other than failure or* **(2)** *failure from a failure mode other than the specific failure mode of concern.  Acceleration refers to testing a device at a stress level or usage frequency higher than the nominal-use stress* *level or usage frequency, respectively, expected in the customer's environment or application.* The point in accelerated testing is to shorten the test duration, in order to gain a timeliness advantage in assessing long-term performance.

## Censoring Schemes

As discussed previously, failure must be precisely defined in a technical or engineering sense. In some cases, it may be necessary to remove certain units from test before the defined failure occurs. In other cases, the test may have to be truncated, or stopped, before all units fail. In general, *we can classify censoring practices into three categories: (1) type I, or time censoring, (2) type II, or failure censoring, and (3) multiple censoring.* Figure 26.4 illustrates the three censoring schemes.

*In type I censoring, experimental units are run over a fixed time period* such that an individual experimental unit's lifetime will be known exactly only if it is less than that time period. This censoring scheme is convenient because the duration of the test can be specified in the test planning stage. *In Type II censoring, the experiment is terminated after some predetermined number of failures have taken place.* Thus, in a test consisting of $n$ experimental units the test is terminated after $r$ experimental units failure. Censoring schemes are frequently used to save time and monetary resources, since in many cases, it could take a long time for all the experimental units to fail.

*Multiple censoring represents a general case where units are removed during a test.* This scheme is needed when: (1) a failure mode not under study caused the failure or (2) a test unit is no longer available for testing. A special case of multiple censoring involves the stratification of a set of failure data by failure mode; hence, an analysis may be performed by failure mode group. In this case, the failure mechanisms and modes must act or be independent of each other to produce mean-

**FIGURE 24.4**    Common sensoring schemes in engineering life tests.

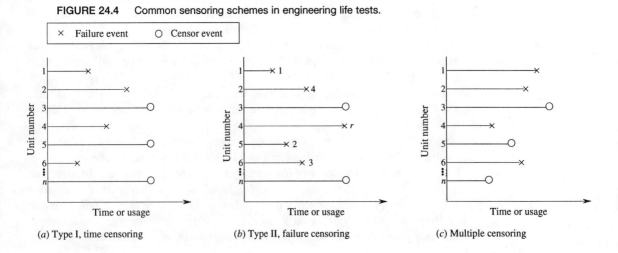

(a) Type I, time censoring          (b) Type II, failure censoring          (c) Multiple censoring

ingful analyses in the strata. This phenomenon is sometimes referred to as "competing" failure modes.

A major point of interest in analyzing censored experiments is to be able to use all data, regardless of a failure or a censoring result, in the analysis. Otherwise, we would have to "throw away" data by discarding the censored unit's performance. In addition, we assume that a censored item's life distribution is the same as that of a noncensored item. Otherwise, censoring might be used to remove units from test immediately before failure, thereby presenting a biased picture of failure characteristics.

## 26.4 NONPARAMETRIC ANALYSIS

*Sometimes it is useful to develop empirical reliability analyses without assuming the applicability of any specific failure-reliability model* (e.g., exponential, Weibull, lognormal, etc.). *Such methods are termed "nonparametric."* Although many nonparametric methods exist, only two widely used methods will be presented here: (1) life-table and (2) product-limit.

### Life-Table Method

*The life-table method is used to analyze class-interval data in order to develop reliability related functions of product-process performance.* The calculations can be made by hand, or computer aids such as SAS PROC LIFETEST [3] can be used. The SAS procedure allows for plots as well as tabulated results. It is beyond the scope of this book to deal with the analytical development of the generic relationships necessary to estimate reliability, pmf, and hazard functions using the class-interval data. Analytical terms and relationships adapted from Lawless [4] (and compatible with the SAS computer aid) are provided below, and can be used to build nonparametric models for reliability characteristics. All of the relationships below are based on time or usage intervals defined in the context of the data collection scheme (e.g., $j$ is an interval index and $t_j$ is some time or usage point within interval $j$).

### Symbols

$I_j$ :  $j$th time or usage interval

$D_j$ :  number of failed units observed in interval $j$

$W_j$ :  number of censored units in interval $j$

$n_j$ :  number of units in service at the beginning of interval $j$

$\hat{p}_j$:  estimated probability of failure in interval $j$

$\hat{s}_{\text{int}\,j}$:  estimated standard error associated with $\hat{p}_j$

$\hat{R}(t_j)$:  estimated reliability (survival) for interval $j$

$\hat{F}(t_j)$:  estimated cumulative failure probability for interval $j$

$\hat{s}_{\text{rel}}(t_j)$:  estimated standard error associated with $\hat{R}(t_j)$

$\hat{f}(t_j)$: estimated probability mass at the midpoint of interval $j$

$\hat{s}_{pmf}(t_j)$: estimated standard error associated with $f(t_j)$

$\hat{\lambda}(t_j)$: estimated failure hazard at the midpoint of interval $j$

$\hat{s}_{haz}(t_j)$: estimated standard error associated with $\lambda(t_j)$

## Relationships

$$\hat{p}_j = \frac{D_j}{n_j - \frac{1}{2}W_j} \qquad j = 1, 2, \ldots \tag{26.11}$$

$$\hat{s}_{int\,j} = \left[\frac{\hat{p}_j(1 - \hat{p}_j)}{n_j - \frac{1}{2}W_j}\right]^{1/2} \qquad j = 1, 2, \ldots \tag{26.12}$$

$$\hat{R}(t_{j+1}) = (1 - \hat{p}_j)\hat{R}(t_j) \qquad R(t_1) = 1,\ j = 1, 2, \ldots \tag{26.13}$$

$$\hat{F}(t_j) = 1 - \hat{R}(t_j) \qquad j = 1, 2, \ldots \tag{26.14}$$

$$\hat{s}_{rel}(t_{j+1}) = \left[\hat{R}^2(t_{j+1}) \sum_{i=1}^{j} \frac{D_i}{\left(n_i - \frac{1}{2}W_i\right)\left(n_i - \frac{1}{2}W_i - D_i\right)}\right]^{1/2} \hat{s}(t_1) = 0,\ j = 1, 2, \ldots \tag{26.15}$$

$$\hat{f}(t_j) = \frac{\hat{R}(t_j)\,\hat{p}_j}{t_j - t_{j-1}} \qquad t_0 = 0, j = 1, 2, \ldots \tag{26.16}$$

$$\hat{s}_{pmf}(t_j) = \left(\hat{f}^2(t_j)\left\{\left[\sum_{i=1}^{j-1} \frac{D_i}{(n_i - \frac{1}{2}W_i)(n_i - \frac{1}{2}W_i - D_i)}\right] + \frac{1 - \hat{p}_j}{(n_j - \frac{1}{2}W_j)\,\hat{p}_j}\right\}\right)^{1/2} j = 1, 2, \ldots \tag{26.17}$$

$$\hat{\lambda}(t_j) = \frac{2\hat{p}_j}{(t_j - t_{j-1})[1 + (1 - \hat{p}_j)]} \qquad t_0 = 0,\ j = 1, 2, \ldots \tag{26.18}$$

$$\hat{s}_{haz}(t_j) = \left(\frac{\hat{\lambda}^2(t_j)\,\{1 - [(t_j - t_{j-1})\,\hat{\lambda}(t_j)/2]^2\}}{(n_j - \frac{1}{2}W_j)\hat{p}_j}\right)^{1/2} \qquad t_0 = 0,\ j = 1, 2, \ldots \tag{26.19}$$

The estimated standard errors associated with reliability, pmf, and hazard estimates can be used to construct "sigma" bands around the estimates. As discussed earlier, these crude error bands help to assess our confidence in the estimates. Many analysts, in general, set up bands based on large sample size approximations, using the standard normal statistic Z; see Table IX.2:

$$\text{Upper, lower bands} = \text{estimated value} \pm Z \text{(estimated standard error)} \tag{26.20}$$

Here, for example, we could use ± 2 (estimated standard error) to represent an approximate two-sided, 95 percent confidence interval, provided our sample size was reasonably large. In other words, we can use a technique similar to our ± s.e.i. arrow method (Chapters 20 through 23).

## Example 26.2

A total of 100 light bulbs were put on test. Failure and censoring results were recorded on an interval basis and are shown below. Develop a life-table analysis by hand calculations and then use SAS PROC LIFETEST to develop a life table.

| Time interval, hr | Number of failed units | Number of censored units |
|---|---|---|
| 0 – 300 | 5 | 0 |
| 300+ – 500 | 4 | 2 |
| 500+ – 600 | 7 | 5 |
| 600+ – 700 | 9 | 0 |
| 700+ – 800 | 15 | 0 |
| 800+ – 950 | 22 | 2 |
| 950+ – 1000 | 12 | 17 |

### Solution

Using the preceding relationships, Equations 26.11 through 26.19, we have produced the life table shown in Table 26.1. The summarized SAS PROC LIFETEST output appears in Figure 26.5. The actual SAS code and abbreviated output are shown in the Computer Aided Analysis Supplement, Solutions Manual. We can see that the hand calculated solution and SAS solutions match; however, the latter is more extensive in that it includes all standard errors.

## Product-Limit Method

*The product-limit method is used to develop nonparametric reliability or failure functions and curves using individual-failure data.* In this case, we need exact

**TABLE 26.1** ANALYSIS OF LIGHT BULB INTERVAL DATA, LIFE-TABLE METHOD (EXAMPLE 26.2)

| Interval $j$ | Failure Time $t_j$ | Number failed $D_j$ | Number censored $W_j$ | Number in service $n_j$ | $\hat{p}_j$ | $\hat{s}_{int j}$ | $\hat{R}(t_j)$ | $\hat{F}(t_j)$ | $\hat{s}_{rel}(t_j)$ | $\hat{f}(t_j)$ | $\hat{\lambda}(t_j)$ |
|---|---|---|---|---|---|---|---|---|---|---|---|
| 1 | 300 | 5 | 0 | 100 | 0.0500 | 0.0218 | 1.0000 | 0.0000 | 0.0000 | 0.000167 | 0.000171 |
| 2 | 500 | 4 | 2 | 95 | 0.0426 | 0.0208 | 0.9500 | 0.0500 | 0.0218 | 0.000202 | 0.000217 |
| 3 | 600 | 7 | 5 | 89 | 0.0809 | 0.0293 | 0.9096 | 0.0904 | 0.0288 | 0.000736 | 0.000843 |
| 4 | 700 | 9 | 0 | 77 | 0.1169 | 0.0366 | 0.8360 | 0.1640 | 0.0375 | 0.000977 | 0.001241 |
| 5 | 800 | 15 | 0 | 68 | 0.2206 | 0.0503 | 0.7383 | 0.2617 | 0.0451 | 0.001630 | 0.002479 |
| 6 | 950 | 22 | 2 | 53 | 0.4231 | 0.0685 | 0.5754 | 0.4246 | 0.0511 | 0.001620 | 0.003577 |
| 7 | 1000 | 12 | 17 | 29 | 0.5854 | 0.1088 | 0.3320 | 0.6680 | 0.0492 | 0.003890 | 0.016552 |

Sample calculations

Interval 1 (estimates)

$p_1 = 5/100 = 0.0500$

$R(t_1) = 1$

$f(t_1) = 1.0(0.05)/(300 - 0) = 0.000167$

$s_{int1} = [0.05(1 - 0.05)/100]^{1/2} = 0.0218$

$s_{rel}(t_1) = 0$

$\lambda(t_1) = 2(0.05)/(300(1 + 1 - 0.05)) = 0.000171$

Interval 2 (estimates)

$p_2 = 4/(95 - 1) = 0.0426$

$R(t_2) = (1 - 0.05)(1.00) = 0.9500$

$f(t_2) = 0.95(0.0426)/200 = 0.000202$

$s_{int2} = [0.0426(1 - 0.426)/(95-2/2)]^{1/2} = 0.0208$

$s_{rel}(t_2) = \{0.95^2 [5/100(95)]\}^{1/2} = 0.0218$

$\lambda(t_2) = 2(0.0426)/(200(1 + 1 - 0.0426)) = 0.000217$

THE LIFETEST PROCEDURE

Life Table Survival Estimates

| Interval (Lower, Upper) | | Number Failed | Number Censored | Effective Sample Size | Conditional Probability of Failure | Conditional Probability Standard Error | Survival | Failure | Survival Standard Error | Median Residual Lifetime | Median Standard Error |
|---|---|---|---|---|---|---|---|---|---|---|---|
| 0 | 300 | 5 | 0 | 100.0 | 0.0500 | 0.0218 | 1.0000 | 0 | 0 | 846.5 | 30.8083 |
| 300 | 500 | 4 | 2 | 94.0 | 0.0426 | 0.0208 | 0.9500 | 0.0500 | 0.0218 | 561.9 | 30.1875 |
| 500 | 600 | 7 | 5 | 86.5 | 0.0809 | 0.0293 | 0.9096 | 0.0904 | 0.0288 | 374.3 | 30.1299 |
| 600 | 700 | 9 | 0 | 77.0 | 0.1169 | 0.0366 | 0.8360 | 0.1640 | 0.0375 | 297.0 | 29.3502 |
| 700 | 800 | 15 | 0 | 68.0 | 0.2206 | 0.0503 | 0.7383 | 0.2617 | 0.0451 | 227.1 | 27.5817 |
| 800 | 950 | 22 | 2 | 52.0 | 0.4231 | 0.0685 | 0.5754 | 0.4246 | 0.0511 | • | • |
| 950 | 1000 | 12 | 17 | 20.5 | 0.5854 | 0.1088 | 0.3320 | 0.6680 | 0.0492 | • | • |

Evaluated at the Midpoint of the Interval

| Interval (Lower, Upper) | | PDF | PDF Standard Error | Hazard | Hazard Standard Error |
|---|---|---|---|---|---|
| 0 | 300 | 0.000167 | 0.000073 | 0.000171 | 0.000076 |
| 300 | 500 | 0.000202 | 0.000099 | 0.000217 | 0.000109 |
| 500 | 600 | 0.000736 | 0.000268 | 0.000843 | 0.000318 |
| 600 | 700 | 0.000977 | 0.000309 | 0.001241 | 0.000413 |
| 700 | 800 | 0.00163 | 0.000384 | 0.002479 | 0.000635 |
| 800 | 950 | 0.00162 | 0.0003 | 0.003577 | 0.000735 |
| 950 | 1000 | 0.00389 | 0.000924 | 0.016552 | 0.00435 |

**FIGURE 26.5**    Life-table method computer output summary, light bulb data, PROC SAS LIFETEST (Example 26.2).

failure or censor times for each unit placed in the test. Again, the calculations can be made by hand, or computer aids such as SAS PROC LIFETEST can be used [5]. We will not address the analytical development of the generic relationships necessary to estimate reliability functions using the individual failure data. However, summarized analytical relationships adapted from Lawless [6] and Kaplan and Meier [7] and compatible with SAS PROC LIFETEST, are provided below, and can be used to develop nonparametric reliability models.

## Symbols

$t_j$:  observed failure or censor time for unit $j$

$n_j$:  number of units remaining in service after the $j$th failure or censor event

$D_j$:  number of units failing at $t_j$ ($D_j > 1$ if ties occur)

$C_j$:  number of units censored at $t_j$ ($C_j > 1$ if ties occur)

$\varphi$:  the failure or censor exponent indicator variable

$\hat{R}(t_j)$:  estimated reliability at $t_j$

$\hat{F}(t_j)$:  estimated cumulative failure probability at $t_j$

$\hat{s}_{rel}(t_j)$:  estimated standard error associated with $\hat{R}(t_j)$

**Relationships**

$$\hat{R}(t_{j+1}) = \hat{R}(t_j) \left(\frac{n_j - D_j}{n_j}\right)^{\varphi} \qquad\qquad R(t_0) = 1, \ j = 0, 1, 2, \ldots \quad (26.21)$$

$$\varphi = \begin{cases} 1 \text{ for a failure at } t_j \\ 0 \text{ for a censor at } t_j \end{cases} \qquad j = 1, 2, \ldots \qquad\qquad (26.22)$$

$$\hat{F}(t_j) = 1 - \hat{R}(t_j) \qquad\qquad j = 0, 1, 2, \ldots \qquad\qquad (26.23)$$

$$\hat{s}_{rel}(t_{j+1}) = \left[\hat{R}^2(t_{j+1}) \sum_{i=1}^{j} \left(\frac{D_j}{n_j(n_j - D_j)}\right)\right]^{1/2} \qquad \hat{s}_{rel}(t_0) = 0, \ j = 0, 1, 2, \ldots \quad (26.24)$$

we will illustrate the product-limit method by example.

---

**Example 26.3**

Large high-voltage capacitors have been configured to provide bursts of power for electromagnetic launch devices (called "rail guns"). High reliability is essential. We have developed an automated testing device which will charge and discharge our capacitors and count the cycles. We put 15 capacitors on test. Due to time constraints, we stop the test after 150,000 cycles, or 150 kilocycles. Our results are shown below, rounded to the nearest 100 cycles:

| Failure time, kilocycles | Censor time, kilocycles |
|---|---|
| 9.4 | |
| 10.8 | |
| | 15.4 |
| 16.1 | |
| 20.8 | |
| 40.2 | |
| | 48.0 |
| 51.1 | |
| 60.4 | |
| 97.4 | |
| 129.2 | |
| 143.7 | |
| | 150.0 |
| | 150.0 |
| | 150.0 |

Use the product-limit method of analysis to develop the nonparametric reliability estimates for these data. Develop both a hand solution and an SAS PROC LIFETEST solution.

**Solution**

The hand solution, based on Equations 26.21 through 26.24, is shown in Table 26.2. A summarized SAS output is shown in Figure 26.6. The complete SAS code and an abbreviated output listing appear in the Computer Aided Analysis Supplement, Solutions Manual.

**TABLE 26.2** ANALYSIS OF THE MULTIPLY CENSORED CAPACITOR FAILURE DATA, PRODUCT-LIMIT METHOD (EXAMPLE 26.3)

| Interval $j$ | Failure time $t_i$ | Censor time $t_j$ | Indicator $\varphi$ | Number failing $D_j$ | Number in service $n_j$ | $\hat{R}(t_j)$ | $\hat{F}(t_j)$ | $\hat{s}(t_j)$ |
|---|---|---|---|---|---|---|---|---|
| 0 | 0 | | | 0 | 15 | 1 | 0 | 0 |
| 1 | 9.4 | | 1 | 1 | 14 | 0.9333 | 0.0667 | 0.0644 |
| 2 | 10.8 | | 1 | 1 | 13 | 0.8667 | 0.0133 | 0.0878 |
| 3 | | 15.4 | 0 | 0 | 12 | — | — | — |
| 4 | 16.1 | | 1 | 1 | 11 | 0.7944 | 0.2056 | 0.1061 |
| 5 | 20.8 | | 1 | 1 | 10 | 0.7222 | 0.2778 | 0.1185 |
| 6 | 40.2 | | 1 | 1 | 9 | 0.6500 | 0.3500 | 0.1268 |
| 7 | | 48.0 | 0 | 0 | 8 | — | — | — |
| 8 | 51.1 | | 1 | 1 | 7 | 0.5688 | 0.4313 | 0.1345 |
| 9 | 60.4 | | 1 | 1 | 6 | 0.4875 | 0.5125 | 0.1376 |
| 10 | 97.4 | | 1 | 1 | 5 | 0.4062 | 0.5938 | 0.1366 |
| 11 | 129.2 | | 1 | 1 | 4 | 0.3250 | 0.6750 | 0.1312 |
| 12 | 143.7 | | 1 | 1 | 3 | 0.2437 | 0.7563 | 0.1210 |
| 13 | | 150.0 | 0 | 0 | 2 | — | — | — |
| 14 | | 150.0 | 0 | 0 | 1 | — | — | — |
| 15 | | 150.0 | 0 | 0 | 0 | — | — | — |

Sample calculations

Interval 1

$\hat{R}(t_1) = [(15 - 1)/15]^1 = 0.9333$

$\hat{s}(t_1) = [0.9333^2 \{1/[15(15 - 1)]\}]^{1/2} = 0.0644$

Interval 2

$\hat{R}(t_2) = 0.9333[(14 - 1)/14]^1 = 0.8667$

$\hat{s}(t_2) = (0.8667^2 [(1/210) + \{1/[14(14 - 1)]\}])^{1/2} = 0.0878$

Interval 3

$\hat{R}(t_3) = —$

$\hat{s}(t_3) = —$

Interval 4

$\hat{R}(t_4) = 0.8667[(12 - 1)/12]^1 = 0.7944$

$\hat{s}(t_4) = (0.7944^2 \{[1/210] + [1/182] + [06]$
$+ [1/(12(12-1))]\})^{1/2} = 0.1061$

## 26.5 PARAMETRIC PLOTTING TECHNIQUES

*The nonparametric life-table and product-limit methods,* introduced above, *are useful to assess measured failure characteristics free of model distribution assumptions. However, in many cases, a parametric distribution assumption (e.g., exponential, Weibull, normal, or lognormal) may be justified.* We will now follow up our introduction of these models (Chapter 25) by demonstrating how we fit, or parameterize, them.

Fitting a model from empirical data is not a simple matter. We emphasize graphical as well as numerical methods, including software aids to estimate the model parameters. For our discussion, *we will utilize three basic forms of parameter estimation:* (1) *the method of moments,* see Appendix 26A; (2) *the graphical, linear rectification, method* (an empirical application of least squares estimation, see Chapter 22); *and* (3) *the maximum-likelihood method,* see Appendix 26A.

Our approach is engineering oriented. We focus on relating the physical characteristics of the product or process to the model form, rather than taking a traditional statistical approach which concentrates exclusively on mathematical fit. In other

THE LIFETEST PROCEDURE

Product-Limit Survival Estimates

| KCYCLE | Survival | Failure | Survival Standard Error | Number Failed | Number Left |
|---|---|---|---|---|---|
| 0.000 | 1.0000 | 0 | 0 | 0 | 15 |
| 9.400 | 0.9333 | 0.0667 | 0.0644 | 1 | 14 |
| 10.800 | 0.8667 | 0.1333 | 0.0878 | 2 | 13 |
| 15.400* | • | • | • | 2 | 12 |
| 16.100 | 0.7944 | 0.2056 | 0.1061 | 3 | 11 |
| 20.800 | 0.7222 | 0.2778 | 0.1185 | 4 | 10 |
| 40.200 | 0.6500 | 0.3500 | 0.1268 | 5 | 9 |
| 48.000* | • | • | • | 5 | 8 |
| 51.100 | 0.5688 | 0.4313 | 0.1345 | 6 | 7 |
| 60.400 | 0.4875 | 0.5125 | 0.1376 | 7 | 6 |
| 97.400 | 0.4062 | 0.5938 | 0.1366 | 8 | 5 |
| 129.200 | 0.3250 | 0.6750 | 0.1312 | 9 | 4 |
| 143.700 | 0.2437 | 0.7563 | 0.1210 | 10 | 3 |
| 150.000* | • | • | • | 10 | 2 |
| 150.000* | • | • | • | 10 | 1 |
| 150.000* | • | • | • | 10 | 0 |

• Censored Observation

| Quantities | 75% | 143.700 | Mean | 81.088 |
|---|---|---|---|---|
| | 50% | 60.400 | Standard Error | 15.522 |
| | 25% | 20.800 | | |

NOTE: The last observation was censored so the estimate of the mean is biased.

Summary of the Number of Censored and Uncensored Values

| Total | Failed | Censored | %Censored |
|---|---|---|---|
| 15 | 10 | 5 | 33.3333 |

FIGURE 26.6    Product-limit method computer output summary, capacitor data, SAS PROC LIFETEST (Example 26.3).

words, to meet our objectives a good model should make sense in the physical, as well as statistical, context.

## Rank Distribution Plotting Positions

*Linear rectification or probability plotting techniques constitute a crude, but relatively simple, form of statistical analysis.* They provide a good deal of support in model assessment and justification. Our strategy is to develop nonparametric plotting positions, based on the rank distribution, and then develop the parametric plots on probability paper. Authors such as Nelson extend plotting methods to include hazard plots, as well as probability plots [8].

The theory of order statistics (Nelson [9]), or rank distributions (Kapur and Lamberson [10]), forms the basis for the plotting positions we will discuss. Given an ordered random sample of failure times $t_1$, $t_2$, $t_3$, ..., $t_n$, where $t_1 \leq t_2 \leq \cdots \leq t_n$, we must associate a cumulative probability mass point with each observation in order to plot the failure observations in a meaningful fashion. ***We seek to use the plot to estimate the model parameters and to assess the model fit in a graphical sense.*** In general, our horizontal axis coordinates represent the time-usage scale and hence are straightforward. Our vertical coordinates require more consideration, as they represent cumulative probability, or $\hat{F}(t)$, plotting positions. The estimates from the life-table and product-limit methods can be laid out in the same fashion, but axis scaling is not an issue.

Traditionally, mean or median ranks of the corresponding failure times are used for model fitting purposes. Then, a straight line is fit—usually by sight, rather than by a least squares method (see Chapter 22)—to the plotted data. Plotting paper is generally scaled such that a perfect model will plot as a straight line. Failure of the plotted points to form a straight line is attributed to either lack of fit (model incompatibility) or sampling error (random error). We ultimately judge the fit as adequate or inadequate, based on the graph. However, statistical goodness-of-fit tests, such as the K-S test, are applicable (see Conover [11] and Kapur and Lamberson [12]). Once a model is established graphically, crude, but useful, nonparametric confidence bands may be plotted.

The $\hat{F}(t)$ plotting positions for the mean rank and the median rank approximations are shown below (Kapur and Lamberson [13]):

$$\hat{F}(t)_{\text{mean rank}} = \frac{j}{n+1} \qquad j = 1, 2, \ldots n \qquad (26.25)$$

$$\hat{F}(t)_{\text{median rank}} \cong \frac{j - 0.3}{n + 0.4} \qquad j = 1, 2, \ldots n \qquad (26.26)$$

Median ranks are usually preferred over mean ranks since many life tests produce nonsymmetric life distributions, and the median is a better measure of central tendency for skewed data. Typically, 5 percent and 95 percent failure confidence bands are plotted. The rank table (Section IX, Table IX.13) supports median, 5 percent, and 95 percent plotting positions for relatively small samples. In general, failure bands can be computed as follows, according to Kapur and Lamberson [14]:

$$w_\alpha = \frac{j/(n-j-1)}{F_{1-\alpha;2(n-j+1);2j} + [j/(n-j+1)]} \qquad \alpha \geq 0.5 \qquad (26.27a)$$

$$w_\alpha = \frac{[j/(n-j+1)]\, F_{\alpha;2j;2(n-j+1)}}{1 + [j/(n-j+1)]\, F_{\alpha;2j;2(n-j+1)}} \qquad \alpha < 0.5 \qquad (26.27b)$$

$F$ statistics are available in Section IX, Table IX.4. Since failure probability plotting methods are commonly used, failure bands (rather than reliability bands) result. In order to interpret the results in terms of reliability, we use the relationship $R(t) = 1 - F(t)$.

**Example 26.4**

A life test was designed to test ten 3-V rated dc motors to failure.  Each motor was run at 3 V until it failed.  The resulting failure times (in hours of run time) were 153.49, 119.32, 75.34, 141.53, 215.92, 135.55, 189.43, 171.64, 140.98, and 256.71.  Develop the median, 5 percent, and 95 percent rank plotting positions for our motor test results.

**Solution**

We will first rank the failure times and then use our rank tables in Section IX, Table IX.13, to build the failure bands, where $n = 10$.  The results are shown in Table 26.3.

**TABLE 26.3    FAILURE TIMES AND RANKS FOR THE COMPLETE DC-MOTOR FAILURE DATA (EXAMPLE 26.4)**

| Failure time, hr | Failure | Order $j$ | Median rank | 5% rank | 95% rank |
|---|---|---|---|---|---|
| 75.34 | 1 | 1 | 6.70 | 0.51 | 25.89 |
| 119.32 | 2 | 2 | 16.23 | 3.68 | 39.42 |
| 135.55 | 3 | 3 | 25.86 | 8.73 | 50.69 |
| 140.98 | 4 | 4 | 35.51 | 15.00 | 60.66 |
| 141.53 | 5 | 5 | 45.17 | 22.24 | 69.65 |
| 153.49 | 6 | 6 | 54.83 | 30.35 | 77.76 |
| 171.64 | 7 | 7 | 64.49 | 39.34 | 85.00 |
| 189.43 | 8 | 8 | 74.14 | 49.31 | 91.27 |
| 215.92 | 9 | 9 | 83.77 | 60.58 | 96.32 |
| 256.71 | 10 | 10 | 93.30 | 74.11 | 99.49 |

*Censoring presents special problems in dealing with parametric analysis.*  For simple time or failure censored data, it is customary to use the failure positions up to, but not including, the censor times, essentially dropping off or ignoring the censored items, since they would indicate "life" beyond the last recorded failure time.  However, in the case of multiple censoring, units are removed during test.  In this case, average order numbers can be calculated and then linear interpolation used to develop plotting positions for noninteger values.  Successive average order numbers are calculated by adding an increment $I$, as shown below (Kapur and Lamberson [15]):

$$\text{Subsequent average order number} = \text{previous order number} + I \quad (26.28)$$

$$\text{where} \qquad I = \frac{(n + 1) - \text{previous order number}}{1 + \text{number of items following the censored item(s)}} \quad (26.29)$$

**Example 26.5**

Using the capacitor data from Example 26.3, develop a complete set of order numbers and ranks for the failed capacitors.

TABLE 26.4   FAILURE CYCLES AND RANKS FOR THE MULTIPLY CENSORED, HIGH-VOLTAGE CAPACITOR FAILURE DATA, $n = 15$ (EXAMPLE 26.5)

| Failure time $t_j$, kilocycles | Censor time $t_j$, kilocycles | Order $j$ | Median rank | 5% rank | 95% rank |
|---|---|---|---|---|---|
| 9.4 | | 1 | 4.52 | 0.34 | 18.10 |
| 10.8 | | 2 | 10.94 | 2.42 | 27.94 |
| | 15.4 | — | — | — | — |
| 16.1 | | 3.08 = 2+(16 − 2)/(1+(15-3)) | 17.95* | 6.00 | 36.95 |
| 20.8 | | 4.16 = 3.08 +1.08 | 24.98 | 10.39 | 45.11 |
| 40.2 | | 5.24 = 4.16 +1.08 | 32.02 | 15.35 | 52.68 |
| | 48.0 | — | — | — | — |
| 51.1 | | 6.44 = 5.24+(16 − 5.24)/(1 + 8) | 39.83 | 21.41 | 60.52 |
| 60.4 | | 7.64 = 6.44 +1.20 | 47.65 | 27.97 | 67.86 |
| 97.4 | | 8.84 = 7.64 +1.20 | 55.47 | 35.00 | 74.73 |
| 129.2 | | 10.04 = 8.84 +1.20 | 63.29 | 42.52 | 81.11 |
| 143.7 | | 11.24 = 10.04 +1.20 | 71.11 | 50.63 | 86.91 |
| | 150.0 | — | — | — | — |
| | 150.0 | — | — | — | — |
| | 150.0 | — | — | — | — |

*Noninteger ranks are linearly interpolated [e.g., 17.95 = 17.432 + 0.08(23.939 − 17.432)].

## Solution

Average order numbers were developed using Equations 26.28 and 26.29. Then, median, 5 percent, and 95 percent ranks were extracted and interpolated using Table IX.13, Section IX. Results are shown in Table 26.4.

## Plotting Papers

*Once the nonparametric plotting data are prepared, the next step is to choose and construct a parametric plotting paper and develop the parameter estimates.* Our plotting discussion will deal exclusively with the exponential, Weibull, normal, and lognormal models as discussed in Chapter 25.

*In order to develop plotting paper, we must find suitable transformations* to produce a model of the form

$$\hat{Y} = b_0 + b_1 X \tag{26.30}$$

This form is identical to a simple linear regression model. (However, if we plot median positions, $\hat{Y}$ represents a median estimator.) Here, we would "eyeball" the model and assess the fit by sight. Blank plotting paper appears in Section IX, Tables IX.14 through IX.17, and is intended for academic practice, since copying has distorted the scales somewhat. More accurate paper is available from Team [16].

## Exponential Models

*The single-parameter exponential model is relatively easy to fit. The single parameter can be estimated graphically or mathematically.* The basis for a graphical fit is the cumulative mass function, $F(t)$, from Equation 25.24a, and reformulated here as

$$F(t) = 1 - R(t) = 1 - \exp(-\lambda t) = 1 - \exp(-t/\theta)$$

Using a logarithmic transformation and algebraic manipulations, we have

$$\ln\left[\frac{1}{1 - F(t)}\right] = \lambda t = \frac{1}{\theta}t$$

The linearized form is

$$t = \theta\left\{\ln\left[\frac{1}{1 - F(t)}\right]\right\} \tag{26.31}$$

Comparing this result with the general linear form in Equation 26.30, we see that here $b_0 = 0$, i.e., the model line will pass through the origin (0, 0).

Now, for $t = \theta$,

$$F(\theta) = 1 - \exp(-1) = 0.632 \tag{26.32}$$

After calculating our $\hat{F}(t)$ values and plotting $\hat{F}(t)$ versus time or usage failure points, we can read $\hat{\theta}$ directly from the graph as the time or usage coordinate associated with $F(t) = 0.632$. It should be noted that we typically draw the model line by sight.

Because of the nature of the exponential distribution, we can calculate $\hat{\theta}$ directly as

$$\hat{\theta} = \frac{\text{total operating time from all units}}{\text{total number of failures observed}} \tag{26.33a}$$

and, from Equation 25.24d,

$$\hat{\lambda} = \frac{1}{\hat{\theta}} \tag{26.33b}$$

The graphical estimate of $\theta$, available from the plot, is a crude approximation. The calculated $\hat{\theta}$ provides a better statistical estimate (at this point, interested readers may want to refer to Examples 26.14 and 26.15 in Appendix 26A). However, the graph allows us to visually assess the model fit. Here, a straight-line plot indicates a CFR model; an inverted-rainbow pattern in the plotted points, an IFR with time or usage; and a rainbow shape, a DFR. A pattern of plotted points falling on a straight line which intersects the time-usage axis at a point $t_0$, without passing through the origin suggests the necessity for a two-parameter exponential model. A two-parameter exponential model can be fit using the one-parameter technique and subtracting the constant $t_0$ from each observed failure time or usage measurement, after which the model line will pass through the origin.

---

## Example 26.6

Using our dc-motor and capacitor data sets from Tables 26.3 and 26.4, respectively,

    **a** Fit each to the exponential model using exponential plotting paper.
    **b** Fit each to the exponential model using exponential model calculations.

**c** Estimate the failure hazard function $\lambda(t)$ for our population of capacitors.

**d** Predict the reliability of a capacitor for 50 kilocycles.

## Solution

**a** Exponential plots for the dc motors and the capacitors are shown in Figures 26.7 and 26.8, respectively. In both cases, we plotted our failure time, median rank $\hat{F}(t)$ value coordinates on the exponential plotting paper. Notice that our model line must pass through the origin point (0, 0). For the motors, $\hat{\theta} = 137$ hr as indicated in Figure 26.7. Here, our data do not appear to fit well (i.e., the inverted rainbow in the points indicates an IFR rather than a CFR model). For the capacitors, $\hat{\theta} = 111$ kilocycles as indicated in Figure 26.8. Here, the exponential model seems reasonable (i.e., the points fall reasonably close to the line, an indication of an adequate CFR model).

**FIGURE 26.7** DC-motor exponential model probability plot, complete data (Example 26.6).

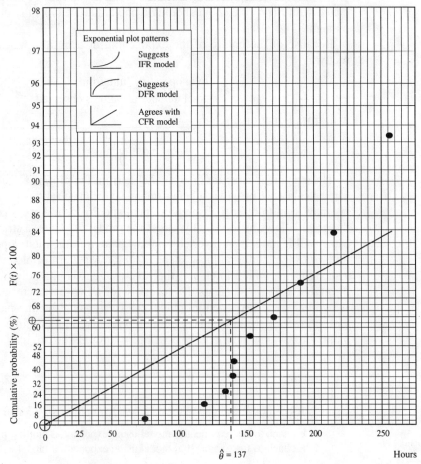

EXPONENTIAL PROBABILITY PAPER

EXPONENTIAL PROBABILITY PAPER

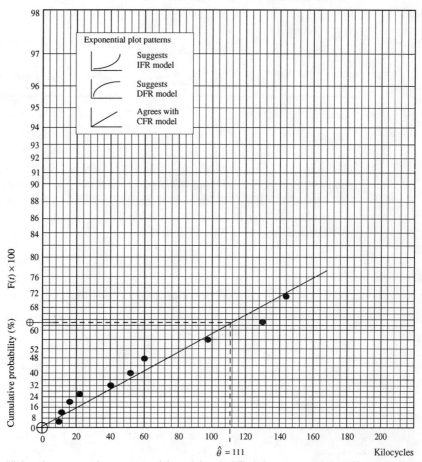

**FIGURE 26.8**   High-voltage capacitor exponential model probability plot, censored data (Example 26.6).

**b**  For the complete (uncensored) motor data, using Equation 26.33a, we can obtain

$$\hat{\theta} = \frac{75.34 + 119.32 + \cdots + 256.71}{10} = \frac{1599.91}{10} = 159.99 \text{ hr}$$

Here, we do not have an indicator of fit as we do using the plot.

For the censored capacitor data we must include all operating times (from both failed and censored units):

$$\hat{\theta} = \frac{9.4 + 10.8 + 15.4 + 16.1 + 20.8 + \cdots + 150}{10} = \frac{1092.5}{10} = 109.25 \text{ kilocycles}$$

**c**  Using Equation 26.33b, the estimated failure hazard function is

$$\hat{\lambda} = \frac{1}{\hat{\theta}} = \frac{1}{109.25} = 0.00915 \text{ failure per kilocycle}$$

**d** Our capacitor reliability estimate for the exponential model is

$$\hat{R}(t) = \exp{(-t/\hat{\theta})}$$

Substituting $t = 50$ kilocycles,

$$\hat{R}(50) = \exp{(-50/109.25)} = 0.633$$

---

## Weibull Models

*Unlike the exponential model, the two-parameter Weibull model has no simple mathematical estimator. We typically rely on either graphical estimates of the parameters θ and δ or maximum-likelihood estimates provided by computer aids* such as SAS PROC LIFEREG (described later). Weibull plotting paper can be developed using a double logarithmic transformation and algebraic manipulations.

We know from Equation 25.27a that

$$F(t) = 1 - R(t) = 1 - \exp{[-(t/\theta)^{\delta}]}$$

Using algebraic manipulation and taking a logarithm transformation, we obtain

$$\ln\left[\frac{1}{1 - F(t)}\right] = \left(\frac{t}{\theta}\right)^{\delta}$$

and

$$\ln\left\{\ln\left[\frac{1}{1 - F(t)}\right]\right\} = \delta \ln t - \delta \ln \theta$$

The linearized form is

$$\ln t = \frac{1}{\delta} \ln\left\{\ln\left[\frac{1}{1 - F(t)}\right]\right\} + \ln \theta \qquad (26.34)$$

The Weibull slope parameter $\delta$ can be estimated from two points on the model line since:

$$\delta = \frac{\ln\{-\ln[1 - F(t_2)] / -\ln[1 - F(t_1)]\}}{\ln (t_2/t_1)} \qquad t_2 > t_1 \qquad (26.35)$$

Now, for $t = \theta$,

$$F(\theta) = 1 - \exp{[-(1)^{\delta}]} = 1 - e^{-1} = 0.632 \qquad (26.36)$$

After calculating the plotting positions $\hat{F}(t)$, and plotting $\hat{F}(t)$ versus the failure points, we can read $\hat{\theta}$ from the graph directly as the time or usage coordinate associated with $\hat{F}(t) = 0.632$. If we have Weibull plotting paper with a slope parameter scale, the Weibull slope $\hat{\delta}$ can be determined from the graph by constructing a line parallel to the model line through the marked point and reading $\hat{\delta}$ from the scale. Otherwise, we must use Equation 26.35. A three-parameter Weibull model is fit using the two-parameter technique and subtracting a constant $t_0$ from each observed failure time or usage measurement.

### Example 26.7

Develop a Weibull plot for the dc electric motor data of Table 26.3. Then, estimate the Weibull model parameters. Finally, estimate $R(75 \text{ hr})$.

### Solution

Our Weibull plot is shown in Figure 26.9. Based on our plot, it appears that our model is reasonable as we assess the fit of our median plotting positions to the model line. Our parameters are estimated graphically as

$$\hat{\theta} = 185 \text{ hr} \qquad \text{and} \qquad \hat{\delta} = 3.3$$

**FIGURE 26.9**    DC-motor Weibull model probability plot, complete failure data (Example 26.7).

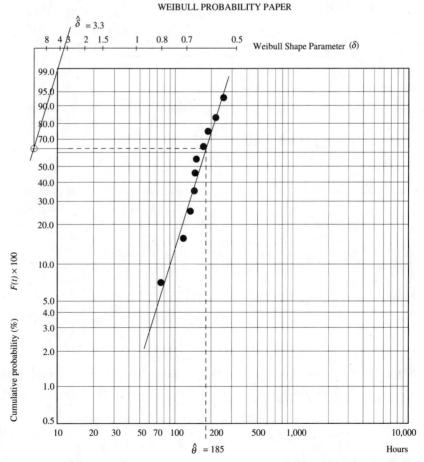

Hence,

$$\hat{R}(t) = \exp\ [-(t/\ \hat{\theta})^{\hat{\delta}}] = \exp\ [-(t/185)^{3.3}]$$

and

$$\hat{R}(75) = \exp\ [-(75/185)^{3.3}] = 0.950$$

---

## Example 26.8

Develop a Weibull plot for the capacitor data of Table 26.4. Estimate the Weibull parameters $\theta$ and $\delta$, and compare the fitted Weibull with the fitted exponential of Example 26.6.

### Solution

Our Weibull plot is shown in Figure 26.10. Our model appears reasonable, as the plotted points fall basically around the model line. The parameters are estimated graphically as

$$\hat{\theta} = 105\ \text{hr} \qquad \text{and} \qquad \hat{\delta} = 1.1$$

Here, we can see that our Weibull has $\hat{\delta}$ near 1. Since the exponential model is a special case of the Weibull ($\delta = 1$), we can use Weibull paper to fit an exponential model if we fix $\delta = 1$ and draw our model line accordingly.

---

## Normal Models

***Normal probability plotting paper can be used to develop estimates of the mean $\mu$ and standard deviation $\sigma$.*** The standard normal transformation is used to simplify the plot. From Equation 25.30a we can state that

$$F(t) = \Phi\left(\frac{t - \mu}{\sigma}\right)$$

where $\Phi(z)$ represents a cumulative standard normal distribution, cumulated from left to right up to $z$, and $Z_{F(t)} = \Phi^{-1}[F(t)]$ represents the $Z$ value associated with a cumulated probability of $F(t)$. A standard normal distribution table appears in Section IX, Table IX.2.

Using the inverse standard normal transformation,

$$\Phi^{-1}\ [F(t)] = \frac{1}{\sigma}\ t - \frac{\mu}{\sigma}$$

Hence, a linear relation can be developed between $t$ and $Z_{F(t)}$:

$$t = \mu + \sigma\ \Phi^{-1}\ [F(t)] = \mu + \sigma\ Z_{F(t)} \tag{26.37}$$

The estimate of $\mu$, $\hat{\mu}$, is found as the point $t_{0.50}$ on the horizontal axis corresponding to $F(t) = 0.50$ or $Z_{0.50} = 0$. The standard deviation is developed from the bell-shaped characteristics of the normal distribution. It is calculated as

$$\hat{\sigma} = \hat{t}_{0.84} - \hat{t}_{0.50} \tag{26.38}$$

since $t_{0.50} - t_{0.84}$ represents the distance from $\mu$ to $\mu + 1\sigma$ for the normal model.

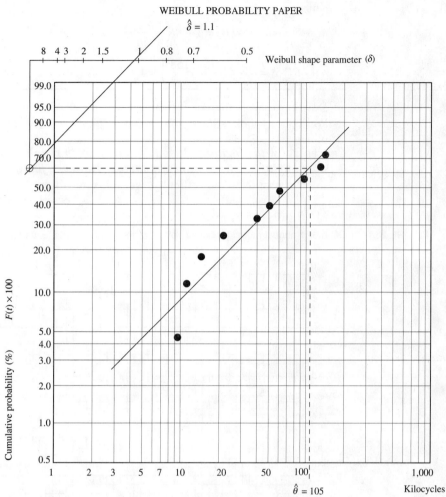

**FIGURE 26.10**    High-voltage capacitor Weibull model probability plot, censored data (Example 26.6).

---

## Example 26.9

Fit the dc electric motor data, Table 26.3, to a normal model.  Estimate $\mu$ and $\sigma$. Then, estimate $R$(75 hr).

### Solution

Our normal plot is shown in Figure 26.11.  We will use the median point $F(t) = 0.50$ and move across and down to establish $\hat{\mu}$ as

$$\hat{\mu} = 158 \text{ hr}$$

Next, we will use Equation 26.38 to estimate $\hat{\sigma}$:

$$\hat{\sigma} = \hat{t}_{0.84} - \hat{t}_{0.50} = 214 - 158 = 56 \text{ hr}$$

Hence,

$$\hat{R}(t) = 1 - \hat{F}(t) = 1 - \Phi\left(\frac{t - 158}{56}\right)$$

$$\hat{R}(75) = 1 - \Phi\left(\frac{75 - 158}{56}\right) = 1 - P(Z \le -1.48) = 1 - 0.0694 = 0.931$$

## Lognormal Models

*Conceptually, the lognormal model is reasonably simple (e.g., a normal model with a straightforward ln transformation). However, fitting it can be confusing.* In order to minimize the difficulty, we will refer to our notation in Chapter 25.

FIGURE 26.11     DC-motor normal model probability plot, complete failure data (Example 26.7).

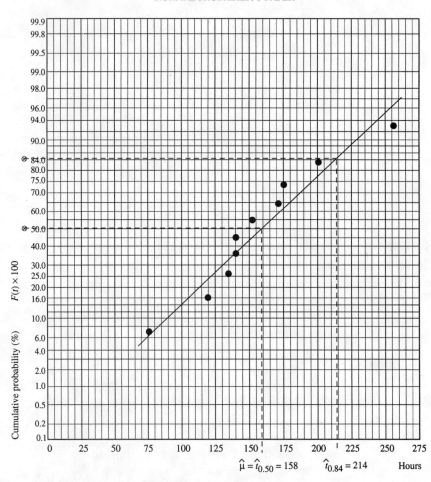

NORMAL PROBABILITY PAPER

$\hat{\mu} = \hat{t}_{0.50} = 158$     $\hat{t}_{0.84} = 214$     Hours

From Equation 25.31a we can write

$$F(t) = \Phi \left[ \frac{\ln t - \mu_{\ln t}}{\sigma_{\ln t}} \right]$$

That is,

$$\Phi^{-1} [F(t)] = \frac{\ln t}{\sigma_{\ln t}} - \frac{\mu_{\ln t}}{\sigma_{\ln t}}$$

Hence, a linear relationship can be developed between $t$ and $F(t)$:

$$\ln t = \mu_{\ln t} + \sigma_{\ln t} \, \Phi^{-1} [F(t)] = \mu_{\ln t} + \sigma_{\ln t} \, Z_{F(t)} \qquad (26.39)$$

where $\Phi^{-1} [F(t)]$ represents the inverse of the standard normal cmf associated with $F(t)$ and is denoted as $Z_{F(t)}$. For example, when $F(t) = 0.50$, $Z_{0.50} = 0$. **_Here, we must deal with the distribution of ln t, which is normal._** Hence, on our lognormal plotting paper, the time or usage scale is logarithmic, not linear (as on normal probability plotting paper). The cumulative probability scale is associated with the inverse standard normal, and can accommodate $F(t)$ values directly. Furthermore, the cumulative probability, $F(t)$, scales on the normal and lognormal paper are identical.

Using lognormal plotting paper, we see that $\exp (\hat{\mu}_{\ln t}) = \hat{t}_{0.50}$ is the time or usage point associated with $F(t) = 0.50$. The lognormal shape parameter is estimated as

$$\hat{\sigma}_{\ln t} = \ln \hat{t}_{0.84} - \ln \hat{t}_{0.50} = \ln \left( \frac{\hat{t}_{0.84}}{\hat{t}_{0.50}} \right) \qquad (26.40a)$$

Then, in terms of the observed time or usage units (e.g., the original $t$ scale), from Equations 25.31d and 25.31e,

$$\hat{\mu} = \exp \left( \hat{\mu}_{\ln t} + \frac{\hat{\sigma}^2_{\ln t}}{2} \right) \qquad (26.40b)$$

and

$$\hat{\sigma}^2 = \exp (2\hat{\mu}_{\ln t} + \hat{\sigma}^2_{\ln t}) \, [\exp (\sigma^2_{\ln t}) - 1] \qquad (26.40c)$$

---

## Example 26.10

Fit the dc electric motor data, Table 26.3, to a lognormal model. Estimate the model parameters in both the $\ln t$ domain and the $t$ domain. Estimate $R(75 \text{ hr})$. Finally, comment on the lognormal fit.

### Solution

The data from Table 26.3 are plotted on lognormal plotting paper (Figure 26.12). From the plot,

$$\hat{\mu}_{\ln t} = \ln \hat{t}_{0.50} = \ln 150 = 5.011$$

and
$$\hat{\sigma}_{\ln t} = \ln\left(\frac{\hat{t}_{0.84}}{\hat{t}_{0.50}}\right) = \ln\left(\frac{210}{150}\right) = 0.336$$

Moving from the $\ln t$ domain to the $t$ (hour) domain,

$$\hat{\mu} = \exp\left(\hat{\mu}_{\ln t} + \frac{\hat{\sigma}_{\ln t}^2}{2}\right) = \exp\left[5.011 + \left(\frac{0.336^2}{2}\right)\right] = 158.8 \text{ hr}$$

$$\hat{\sigma}^2 = \exp\left(2\hat{\mu}_{\ln t} + \hat{\sigma}_{\ln t}^2\right)\left[\exp\left(\hat{\sigma}_{\ln t}^2\right) - 1\right]$$
$$= \exp\left[2(5.011) + 0.336^2\right]\left[\exp\left(0.336^2\right) - 1\right] = 3012.68$$

and   $\hat{\sigma} = 54.89$ hr

$$\hat{R}(75) = 1 - \Phi\left(\frac{\ln t - \hat{\mu}_{\ln t}}{\hat{\sigma}_{\ln t}}\right) = 1 - \Phi\left(\frac{\ln 75 - 5.011}{0.336}\right) = 1 - P(Z \le -2.06) = 0.980$$

**FIGURE 26.12**   DC-motor lognormal model probability plot, complete failure data (Example 26.10).

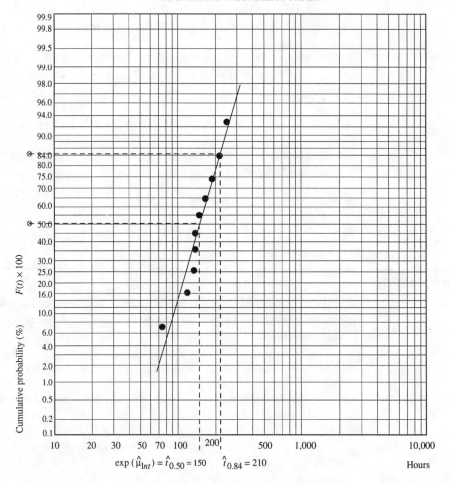

LOGNORMAL PROBABILITY PAPER

$F(t) \times 100$

Cumulative probability (%)

$\exp\left(\hat{\mu}_{lnt}\right) = \hat{t}_{0.50} = 150$      $\hat{t}_{0.84} = 210$

Hours

In Figure 26.12, we see a reasonably good lognormal fit, as we also observed in the Weibull fitting case (Figure 26.9). Graphically, with this motor data, we just cannot distinguish which fit is better.

## Confidence Intervals

*The four models we have presented provide reliability estimates, $\hat{R}(t)$, for us in the form of a median value. It is helpful to develop a confidence interval to obtain a feel for the estimate's precision or integrity; the narrower our confidence interval, the more faith we have in the estimated reliability. Nonparametric confidence intervals,* based on the rank distribution, previously discussed, *can be placed on our parametric plots*. Care must be taken to place them about the model "line" rather than the points plotted. We will demonstrate the construction technique through an example.

---

## Example 26.11

Develop the 5 percent and 95 percent confidence bands for the Weibull model fit in Example 26.7 for the dc electric motors.

### Solution

We will start with the Weibull model plotted in Figure 26.9, and then plot the 95 percent and 5 percent confidence limits developed earlier in Table 26.3. Care must be exercised in the plotting process to ensure that the confidence bands are located vertically above and below the fitted median line. For example, we project each plotted point over (horizontally) to the model line and then locate the upper and lower confidence interval points directly above and below, respectively. Here, we are constructing the confidence bands about the model line, not the plotted points. The plotted confidence limits are shown in Figure 26.13. Traditionally, we draw smooth arcs to join the points, in each confidence band. These connections help us to visualize confidence intervals at points other that the failure times.

---

We should point out that, in general, the 5 percent and 95 percent (as well as the median) rank points can be plotted on any of our four plotting papers since they are nonparametric (i.e., no specific distribution model is assumed; the points are based only on the failure ranks or order).

Having developed our model and the corresponding confidence intervals, for the Weibull case, Figure 26.13, we can make a number of statements regarding the reliability characteristics of our dc electric motors. For example, we can state that we are 90 percent confident that 50 percent to 92 percent of our population will fail by 200 hr. In this case, we locate the point 200 hr on the time scale, project it onto the confidence bands, and then trace a line from these intersected points across to the $F(t)$ scale. We end up with the points 92 percent and 50 percent, which represent the 90 percent, two-sided, confidence limits at time $t = 200$ hr (i.e., the two-sided interval is developed from the two bands).

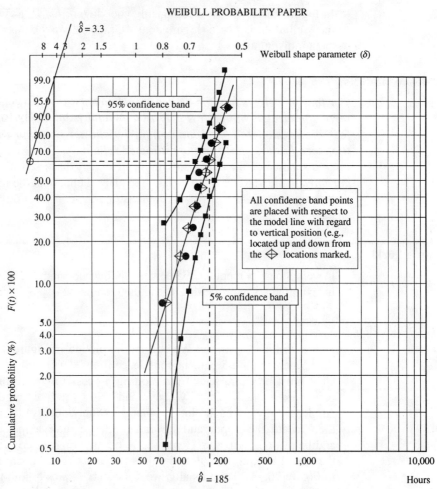

**FIGURE 26.13**   DC-motor Weibull model probability plot with 5 percent and 95 percent confidence bands, complete failure data (Example 26.11).

On the other hand, we could focus on the $F(t)$ scale and select a value of, say, 30 percent. Then, we could proceed across to the confidence bands and back down to the time scale. Here, we could say that we are 90 percent confident that our 30 percent failure point (in the population of dc motors) will be from 86 to 170 hr. We can also use the fact that $R(t) = 1 - F(t)$ and say that we are 90 percent confident that the 70 percent reliability point will be from 86 to 170 hr.

The above statements are two-sided confidence interval oriented. We can use single confidence bands to make one-sided statements. For example, we could select $F(t) = 80\%$ [or $R(t) = 1 - F(t) = 20\%$] on the $F(t)$ scale, project across to the 95 percent confidence band and down to the $t$ scale. We could state that we are 95 percent confident that 20 percent of our population will survive beyond 165 hr. As another example, we are 95 percent confident that less than 37 percent of the population will fail by 100 hr.

## Exponential Model Confidence Intervals

*We can also construct confidence intervals numerically.* To demonstrate this procedure, we will develop the confidence interval numerically for the exponential model. Following the line of development used by Kapur and Lamberson [17], we have

$$\frac{2T}{\chi^2_{\alpha/2;2r}} \le \theta \le \frac{2T}{\chi^2_{1-\alpha/2;2r}} \tag{26.41}$$

where $T$ = total time on test for all items (those that failed and those that did not)
$\quad\quad r$ = number of failures in the test
$\quad\quad \chi^2$ = chi-square statistic (Section IX, Table IX.5)

Since MTTF = $\theta$ for the exponential model, Equation 26.41 represents a two-sided confidence interval on the MTTF. We will apply this interval directly to our exponential reliability model in order to set a two-sided confidence interval on $R(t)$. Since from Equation 25.24a,

$$R(t) = \exp(-\lambda t) = \exp(-t/\theta)$$

we state that the confidence interval is

$$\exp(-t/L) \le R(t) \le \exp(-t/U) \tag{26.42}$$

where $U$ = the upper confidence limit on $\theta$ (the right side of Equation 26.41)
$\quad\quad L$ = the lower confidence limit on $\theta$ (the left side of Equation 26.41)

We can also estimate a failure time $t_p$ quantile for the exponential model as

$$\hat{R}(t_p) = \exp(-t_p/\hat{\theta}) = 1 - p$$

Using a logarithmic transformation and algebraic manipulations, we obtain

$$\hat{t}_p = \hat{\theta} \ln \frac{1}{1-p} \tag{26.43}$$

---

## Example 26.12

Use the capacitor life test data, Table 26.4, and the model results from Example 26.6 to

**a** Develop the 90 percent two-sided confidence interval on the capacitor MTTF.

**b** Develop the 90 percent confidence interval on the reliability of a capacitor for 50 kilocycles.

**c** Estimate the number of kilocycles by which we would expect 15 percent of the capacitors to fail (i.e., $t_{0.15}$).

### Solution

**a** Using Equation 26.41,

$$\frac{2T}{\chi^2_{\alpha/2;2r}} \leq \theta \leq \frac{2T}{\chi^2_{1-\alpha/2;2r}}$$

where $T$ = 1092.5 kilocycles
$r$ = 10 failures
$\chi^2_{0.05;20}$ = 31.410
$\chi^2_{0.95;20}$ = 10.851

We develop our confidence interval as

$$\frac{2(1092.5)}{31.410} \leq \theta \leq \frac{2(1092.5)}{10.851}$$

$$69.56 \text{ kilocycles} \leq \theta \leq 201.36 \text{ kilocycles}$$

We also recall, from Example 26.6,

$$\hat{\theta} = \frac{1092.5}{10} = 109.25 \text{ kilocycles}$$

**b** Using Equation 26.42, with $L$ = 69.56 and $U$ = 201.36, from above,

$$\exp(-t/L) \leq R(t) \leq \exp(-t/U)$$

$$\exp(-50/69.56) \leq R(50) \leq \exp(-50/201.36)$$

$$0.487 \leq R(50) \leq 0.780$$

**c** Using Equation 26.43,

$$\hat{t}_p = \hat{\theta} \ln\left(\frac{1}{1-p}\right)$$

$$\hat{t}_{0.15} = 109.25 \ln\left(\frac{1}{1-0.15}\right) = 17.76 \text{ kilocycles}$$

## 26.6  COMPUTER-AIDED MODEL FITTING

It is possible to computerize the graphical reliability modeling methods previously discussed and use a least squares estimation (fitting) method.  However, *a fitting method called "maximum-likelihood estimation" (MLE) is usually preferred when computer aids are available.* MLE usually requires rather involved mathematical manipulations and solutions, hence, hand solutions to most MLE problems are infeasible.  We should note that when we use the MLE method, we do not require plotting positions as we do in the least squares based graphical, "linear rectification" method. Maximum-likelihood estimation and estimators are introduced in Appendix 26A. The SAS procedure PROC LIFEREG, which supports MLE solutions, will be demonstrated in this section to develop exponential, Weibull, and lognormal model parameters. PROC LIFEREG (which is based on a Newton-Raphson algorithm) is a very powerful MLE fitting tool for reliability, or survival, study purposes and is capable of fitting many survival models. Specifically, *we will use SAS to fit the following models*: **(1)** *the one-parameter exponential,* **(2)** *the covariate exponential,* **(3)** *the two-parameter Weibull,* **(4)** *the covariate Weibull,* **(5)** *the lognormal,* *and* **(6)** *the covariate lognormal.*

We have previously discussed the basic nature of all six of these models in Chapter 25. At this point, we will concentrate on applying PROC LIFEREG, rather than presenting a theoretical treatment of the models themselves. The reader is referred to Lawless [18], Cox and Oakes [19], and Kalbfleisch and Prentice [20], as well as the SAS Institute [21] for further details regarding model development and the accompanying theory.

---

### Example 26.13

Using our dc-motor failure data, Table 26.3, develop MLE estimates for the exponential, Weibull, and lognormal models. Estimate $R(t = 200$ hr). Compare your results with those developed for the same data set in our previous examples using graphical techniques (Examples 26.6, 26.7, and 26.10).

### Solution

Our summarized SAS PROC LIFEREG output appears in Figure 26.14. The SAS code and an abbreviated output appear in the Computer Aided Analysis Supplement, Solutions Manual. We have defined three SAS variables:

HOUR:  observed time to failure

CONFIG:  design configuration (here only one configuration was studied; if we had two or more, we would assign 1s, 2s, etc, and sequences of indication variables.)

CENSOR:  censor indicator variable (here no items were censored; a value of 1 indicates a failure in our code)

This example hardly challenges our PROC LIFEREG tool in that we are not using an active covariate. However, our output, Figure 26.14, is in a covariate model form, so we must uncode the printed results accordingly. Here, we are using the covariate notation introduced in Chapter 25 in conjunction with our SAS output.

*Covariate Exponential Model Summary:*

With the covariate exponential model we have one parameter to deal with:

$$\theta = \exp(Z\,\beta)$$

MLE METHOD COMPLETE FAILURE DATA

Lifereg Procedure

| Variable | DF | Estimate | Std Err | ChiSquare | Pr>Chi | Label/Value |
|----------|-----|----------|---------|-----------|--------|-------------|
| INTERCPT | 1 | 5.07511686 | 0.316228 | 257.5683 | 0.0001 | Intercept |
| CONFIG | 0 | | | 0 | 0.0001 | CONFIGURATION INDICATOR COVARIATE |
| | 0 | 0 | 0 | • | • | 1 |
| SCALE | 0 | 1 | 0 | | | Extreme value scale parameter |

Lagrange Multiplier ChiSquare for Scale     • Pr>Chi is • •

Log Likelihood for EXPONENT -10.4909759

(a) Exponential model

MLE METHOD COMPLETE FAILURE DATA

Lifereg Procedure

| Variable | DF | Estimate | Std Err | ChiSquare | Pr>Chi | Label/Value |
|----------|-----|----------|---------|-----------|--------|-------------|
| INTERCPT | 1 | 5.17975671 | 0.094029 | 3034.578 | 0.0001 | Intercept |
| CONFIG | 0 | | | 0 | 0.0001 | CONFIGURATION INDICATOR COVARIATE |
| | 0 | 0 | 0 | • | | 1 |
| SCALE | 1 | 0.28118101 | 0.067468 | | | Extreme value scale parameter |

Log Likelihood for WEIBULL -2.779973836

(b) Weibull model

MLE METHOD COMPLETE FAILURE DATA

Lifereg Procedure

| Variable | DF | Estimate | Std Err | ChiSquare | Pr>Chi | Label/Value |
|----------|-----|----------|---------|-----------|--------|-------------|
| INTERCPT | 1 | 5.02601997 | 0.101796 | 2437.744 | 0.0001 | Intercept |
| CONFIG | 0 | | | 0 | 0.0001 | CONFIGURATION INDICATOR COVARIATE |
| | 0 | 0 | 0 | • | | 1 |
| SCALE | 1 | 0.32190678 | 0.071981 | | | Normal scale parameter |

Log Likelihood for LNORMAL -2.854453296

(c) Lognormal model

**FIGURE 26.14**    MLE computer output summary, dc-motor failure data, exponential, Weibull, lognormal models, SAS PROC LIFEREG (Example 26.13).

Here, $\qquad Z_1 = 1 \quad$ and $\quad \hat{\beta}_1 = 5.07512$

$$\hat{\theta} = \exp (Z_1 \hat{\beta}_1) = \exp [1(5.07512)] = 159.99 \text{ hr}$$

Therefore,

$$\hat{R}(t) = \exp [-(t/\hat{\theta})] = \exp [-(t/159.99)]$$

Our measure of merit here is

$$\text{Loglikelihood value} = -10.491$$

In the MLE technique, see Appendix 26A, we are maximizing a loglikelihood function, hence the larger the loglikelihood value, the better our model fit.

Using our fitted exponential model,

$$\hat{R}(200) = \exp\,[-(200/159.99)] = 0.286$$

*Covariate Weibull Model Summary:*

With the covariate Weibull model, we have two parameters to consider:

$\delta$:  Weibull shape parameter
$\theta$:  Weibull scale parameter or characteristic life = $\exp\,(\mathbf{Z}\,\boldsymbol{\beta})$

For the Weibull, we use the reciprocal of the SAS scale estimate:

$$\hat{\delta} = \frac{1}{\text{scale estimate}} = \frac{1}{0.28118} = 3.56$$

and $\qquad\qquad Z_1 = 1 \qquad \hat{\beta}_1 = 5.17976$

Hence,

$$\hat{\theta} = \exp\,(Z_1\,\hat{\beta}_1) = \exp\,[1(5.17976)] = 177.64 \text{ hr}$$

Therefore,

$$\hat{R}(t) = \exp\,[-(t/\hat{\theta})^{\hat{\delta}}] = \exp\,[-(t/177.64)^{3.56}]$$

Our measure of merit here is

$$\text{Loglikelihood value} = -2.780$$

Applying our fitted Weibull model,

$$\hat{R}(200) = \exp\,[-(200/177.64)^{3.56}] = 0.218$$

*Covariate Lognormal Model Summary:*

The covariate lognormal model includes two parameters,

$\sigma_{\ln t}$:  lognormal shape parameter
$\mu_{\ln t}$:  lognormal mean = $\mathbf{Z}\,\boldsymbol{\beta}$

Here,

$$\hat{\sigma}_{\ln t} = 0.32191$$

$$Z_1 = 1 \qquad \hat{\beta}_1 = 5.02602$$

Hence,

$$\hat{\mu}_{\ln t} = Z_1\,\hat{\beta}_1 = 5.02602$$

Therefore,

$$\hat{R}(t) = 1 - \Phi\left(\frac{\ln t - \hat{\mu}_{\ln t}}{\hat{\sigma}_{\ln t}}\right) = 1 - \Phi\left(\frac{\ln t - 5.0260}{0.32191}\right)$$

Our measure of merit here is

$$\text{Loglikelihood value} = -2.854$$

Applying our fitted lognormal model,

$$\hat{R}(200) = 1 - \Phi \left( \frac{\ln 200 - 5.0260}{0.32191} \right) = 1 - P(Z \le 0.85)$$

$$= 1 - 0.8023 = 0.198$$

In addition, converting back to our original response units, hours,

$$\hat{\mu} = \exp \left( \hat{\mu}_{\ln t} + \frac{\hat{\sigma}_{\ln t}^2}{2} \right) = \exp \left[ 5.026 + \left( \frac{0.322^2}{2} \right) \right] = 160.4 \text{ hr}$$

$$\hat{\sigma} = \{ \exp (2\hat{\mu}_{\ln t} + \hat{\sigma}_{\ln t}^2)[\exp ( \hat{\sigma}_{\ln t}^2) - 1] \}^{1/2} = \{ \exp [2(5.026) + 0.322^2]$$

$$\times [\exp (0.322^2) - 1] \}^{1/2} = 53.03 \text{ hr}$$

Our comparisons are summarized, with comments, in Table 26.5. Our results here, using our computer aid, are consistent with those we obtained using our graphical method. We should note that the loglikelihood values can be compared, in a relative sense, to assess the best-fitting model of those examined. The MLE is a bigger is better criterion; the bigger the loglikelihood, the better the model fit. The Weibull and lognormal models fit much better than the exponential in this example.

When we use reasonably small sample sizes, as is usually the case, model choice may not be clear-cut from a statistical point of view. For example, the

**TABLE 26.5**  PLOTTING VERSUS MLE COMPARISON OF MODELS AND MODEL FIT FOR DC MOTORS (EXAMPLE 26.13)

| Model | Plot fit | MLE fit | Comments |
|---|---|---|---|
| Exponential | $\hat{\theta} = 137$ hr<br>Poor model fit, IFR indicated | $\hat{\theta} = 160.0$ hr<br>Loglikelihood $= -10.491$ | Our plot indicates a poor fit (i.e., the exponential model is not appropriate). We have a reasonably small loglikelihood, when judged against the Weibull and lognormal loglikelihoods. |
| Weibull | $\hat{\delta} = 3.3$<br>$\hat{\theta} = 185$ hr<br>Generally a good fit | $\hat{\delta} = 3.56$<br>$\hat{\theta} = 177.6$ hr<br>Loglikelihood $= -2.780$ | We see a reasonably good fit on the Weibull plot. Our plot estimates and MLE estimates agree in general. Our loglikelihood has improved over the exponential model, indicating a better fit. |
| Lognormal | $\hat{\mu}_{\ln t} = 5.011$<br>$\hat{\sigma}_{\ln t} = 0.336$<br>$\hat{\mu} = 158.8$ hr<br>$\hat{\sigma} = 54.89$ hr<br>Generally a good fit | $\hat{\mu}_{\ln t} = 5.026$<br>$\hat{\sigma}_{\ln t} = 0.322$<br>$\hat{\mu} = 160.4$ hr<br>$\hat{\sigma} = 53.03$ hr<br>Loglikelihood $= -2.854$ | We see a reasonably good fit on our plot. Here again, we see plot and MLE estimates in reasonably good agreement. We have a major dilemma in that both the Weibull and the lognormal produce similar fits (graphically and in terms of loglikelihood) with such a small sample—10 motor failures—we cannot resolve this dilemma statistically. We might base our arguments on physical (failure mode-mechanism) aspects. We actually observed a brush wearout failure mode in the physical test. |

Weibull plot and the lognormal plot may both look reasonable. Our loglikelihood statistic from the MLE method will help to distinguish the two, but often these will be rather close together. At this point, we may want to resort to a physical argument (see the comments on the Weibull and lognormal in Chapter 25) in order to select a model. In any case, we must be cautious when making decisions based on small sample sizes.

## Accelerated Testing and Covariate Models

*Accelerated testing makes use of one or more stress variables, applied at levels or usage frequencies elevated above the design or nominal stress conditions. The basic objective is to obtain failures and estimates of reliability parameters in a more timely fashion.* Accelerated testing is especially useful when components have very long lives at nominal stress levels. For example, it might take a time frame of years to observe a few failures in the case of stored, dormant, or high-reliability components. In such cases, accelerated testing may be our only viable alternative to assess long-term performance. *Accelerated testing should be*

1 *Based on knowledge of the failure physics,* or at least a reasonable engineering judgment based on evidence available.
2 *Empirically verified* through exploratory tests.

*It is critical that the elevated levels of stress yield the same failure mode and mechanism as does the nominal level of stress.* Various acceleration models have been developed, such as the power rule (or inverse power law), Arrhenius, Eyring, generalized Eyring, step stress, and so on (see Nelson [22], Kececioglu [23], Trindade and Tobias [24]).

**Covariates as Explanatory Variables** *An "explanatory" variable, or covariate, is defined as an independent variable upon which the time to failure depends. Covariates can be used in reliability experiments much as factor levels are used in traditional experiments.* They are useful in accounting for

1 Nonhomogeneity of the test sample (e.g., test units representing different design revisions or different manufacturers).
2 Different treatments (e.g., different levels of environmental stresses applied to different test articles).

Multiple stress levels (nominal and elevated) such as applied voltage and temperature are examples of covariates in a designed reliability experiment. Our objective is to relate, mathematically, the values of a covariate to a parameter of the life distribution, such as $\theta$, the Weibull characteristic life.

Rarely is the failure physics completely understood, i.e., to permit straightforward development of a physical acceleration model. However, knowledgeable and experienced scientists and engineers can usually identify major explanatory variables. Covariate reliability experiments present both physical and statistical challenges in design, execution, analysis, and interpretation.

*There are several classifications of covariates*:

1 *Fixed.* A fixed covariate has a value measured in advance of the study (test) and is fixed for the duration of the study. Most classical life tests fix all covariates, such as temperature, voltage, load, and so forth. Hence, inferences are made only to the fixed levels (e.g., Example 26.13).

2 *Time-dependent.* A time-dependent covariate changes in value as a function of time. Such covariates are very complicated to model and are outside the scope of our discussion.

3 *Continuous.* Continuous covariates have values along a continuous scale, such as applied electrical stress measured in volts.

4 *Discrete.* Discrete covariates have discrete values, or levels, which essentially stratify the population. For example, discrete covariates include units produced by different manufacturers, units subject to different maintenance and support policies, units of different design levels (i.e., before and after some design change implementation), or units either tested or field deployed in different "test beds," such as devices installed in stationary land-based or mobile environments.

Our treatment in this chapter will include fixed, continuous, and discrete covariates.

**Treatment of Covariates** *There are many ways to incorporate covariates into a reliability analysis.* The most common approach defines the failure hazard as a function of the covariates:

$$\lambda(t) = \lambda(t; \mathbf{Z} \, \boldsymbol{\beta}) \qquad (26.44)$$

where $\mathbf{Z}$ = row vector of covariates
$\boldsymbol{\beta}$ = column vector of regression coefficients

We will estimate the $\boldsymbol{\beta}$ vector by the MLE method using our computer aid SAS PROC LIFEREG.

The general multiplicative hazard function is

$$\lambda(t; \mathbf{Z} \, \boldsymbol{\beta}) = \lambda_0(t)g(\mathbf{Z} \, \boldsymbol{\beta}) \qquad (26.45)$$

where $\lambda_0(t)$ is the baseline hazard function and $g(\mathbf{Z} \, \boldsymbol{\beta})$ is a function relating the covariates. The $g(\cdot)$ can take various forms, but must assure that the $\lambda(\cdot)$ function is nonnegative for all values of the time-to-failure random variable $T$. A function which has been found most useful, according to Lawless [25], is

$$g(\mathbf{Z} \, \boldsymbol{\beta}) = \exp(\mathbf{Z} \, \boldsymbol{\beta}) \qquad (26.46)$$

In Equation 26.46, the first covariate in the $\mathbf{Z}$ vector is generally defined as $Z_1 = 1$ such that when all other covariates are set to 0, the following case arises:

$$g(\mathbf{Z} \, \boldsymbol{\beta}) = \exp(\beta_1)$$

We saw this general form in Example 26.13.

Advanced treatments of these models appear in Lawless, Cox and Oakes, and Kalbfleisch and Prentice [26]. The $\lambda_0(t)$ function can be taken as either a parametric or a nonparametric hazard form.

## Acceleration Factors and the Covariate Models

The covariate models introduced in Chapter 25 have been extensively developed in the literature, but not widely applied in practice. What we will briefly describe as the "covariate exponential" model, Lawless terms simply an "exponential regression" model [27]. Likewise, what we term a "covariate Weibull" model is also known as a "Weibull regression" model, and our "covariate lognormal," a "lognormal regression" model. All of the covariate models we discuss contain one parameter which is expressed in an exponential-regression form.

*The covariate exponential and Weibull reliability models allow for variables representing environment, application, and configuration differences to be captured.* The covariate Weibull model also lends itself to designed reliability experimentation including Arrhenius and power law stress acceleration. Both the covariate exponential and Weibull are proportional hazard models in that the ratio of any two parameterized hazard functions (i.e., defined by sets of covariates) will equal a constant value. Therefore, when we develop plots for our treatments, stratified by our covariates/covariate levels, using approaches such as the product-limit method or our plotting positions, we will be looking for parallel model lines. Hence, we will plot each treatment with our $\ln t$ versus $\ln[-\ln R(t)]$ or the relationships of Equation 26.34 and look for evidence of linear and parallel model lines to support or justify our given model assumptions (remember that the exponential is a special case of the Weibull when $\delta = 1$).

*In traditional reliability practice, we seek an acceleration model relationship where we obtain linear acceleration* (Tobias and Trindade [28]). In this case, *we can define an acceleration factor, AF, where every time to failure, $t_p$, and every failure time quantile, $F(t_p)$, are related across stress levels described by our covariates:*

$$t_{p1} = AF_{1:2} t_{p2}$$

$$F_1(t_p) = F_2\left(\frac{t_p}{AF_{1:2}}\right)$$

where $t_{p1}$ and $t_{p2}$ represent corresponding failure times for stress levels 1 and 2, respectively, where $\text{stress}_2 > \text{stress}_1$. $F_1(t_p)$ and $F_2(t_p)$ represent corresponding failure-time quantiles (or percentiles) for stress levels 1 and 2, respectively, $\text{stress}_2 > \text{stress}_1$.

For example, if $AF_{3:6} = 20$ for a 6-V stress as compared to a 3-V stress, we can say that a 10-hr device life at 6 V corresponds to a $10(20) = 200$-hr device life at 3 V. On the other hand, we could say that for a population of devices with a $t_{0.05} = 4$ hr at 6 V, then our corresponding $t_{0.05}$ for a population of devices at 3 V would be $4(20) = 80$

hr. In addition, we could also say that $F_1(80) = F_2(\frac{80}{20})$. Weibull based acceleration models are capable of demonstrating this linear acceleration.

*The Weibull covariate model meets the linear acceleration conditions.* Our Weibull acceleration factor for stress$_1$ and stress$_2$ is

$$R_2(t) = \exp[-(t/\theta_2)^\delta] \tag{26.47}$$

$$R_1(AF_{1:2}t) = \exp[-(AF_{1:2}\,t/\theta_1)^\delta] \tag{26.48}$$

Now, since we can use $R(t)$ just as well as $F(t)$,

$$R_2(t) = R_1(AF_{1:2}t)$$

and

$$\exp[-(t/\theta_2)]^\delta = \exp[-(AF_{1:2}\,t/\theta_1)^\delta]$$

Removing the $\delta$, since both are assumed to have the same Weibull slope $\delta$, and making a logarithmic transformation on each side, we have

$$\left(\frac{t}{\theta_2}\right) = \left(\frac{AF_{1:2}t}{\theta_1}\right)$$

Solving for $AF_{1:2}$, we get

$$AF_{1:2} = \frac{\theta_1}{\theta_2} \tag{26.49}$$

Now, in the Weibull or exponential covariate case,

$$\theta = \exp(\mathbf{Z}\,\boldsymbol{\beta}) \tag{26.50}$$

and

$$AF_{1:2} = \exp(\mathbf{Z}_1\boldsymbol{\beta} - \mathbf{Z}_2\boldsymbol{\beta}) \tag{26.51}$$

*The lognormal reliability model may take a covariate form where we describe the lognormal mean* $\mu_{\ln t}$ *as a function of environment, application, configuration differences, and so on.* Here, our lognormal shape parameter $\sigma_{\ln t}$ is held constant across all covariates. SAS PROC LIFEREG allows us to fit the covariate lognormal in the same manner as the covariate Weibull. The covariate lognormal can be a useful model, but it does not exhibit the acceleration properties that we have seen in the covariate Weibull or exponential. PROC LIFETEST will not develop a lognormal plot; hence, our plot verification must be pursued outside of SAS. SAS PROC LIFEREG will print a logliklihood value which may be compared to those from other models. However, this relative comparison will not serve as a justification of a "good fit" for any of the models being compared in an absolute sense.

## Covariate Model Hypothesis Tests

*Our computer aids,* SAS PROC LIFETEST and PROC LIFEREG, *develop a number of test statistics for hypothesis testing. These hypothesis tests provide sta-*

***tistical guidance to us in our interpretation activities.*** Our graphical displays also help us to assess the effects of our selected covariates on device reliability.

**Tests of Homogeneity**   SAS PROC LIFETEST performs a series of tests for homogeneity between strata, or data sets, with regard to our covariates. The tests basically determine whether the differences among the strata, e.g., defined by a covariate, are statistically significant. The statistical hypotheses are

$H_0$:   All $R_i(t)$ are equal [e.g., $R_1(t) = R_2(t) = \cdots = R_c(t)$]     $i = 1, 2, ..., c$

$H_1$:   At least one $R_i(t)$ not equal to the others

where we have $c$ strata, e.g., $c$ levels of the covariate. There are three nonparametric tests commonly used, and available in SAS LIFETEST:

**1** Logrank
**2** Wilcoxon
**3** Likelihood ratio [labeled –2 log (LR) in SAS PROC LIFETEST]

We will briefly discuss these tests below. The theoretical background may be found in SAS and in Lawless [29].

**Logrank and Wilcoxon Tests**   The same formula is used to calculate the logrank and the Wilcoxon test statistics, but with a difference in weighting factors. The test statistics employ the chi-square distribution as a limiting (asymptotic, or large-sample) probability distribution. The chi-square is widely tabulated in statistics texts (see Section IX, Table IX.5).

The logrank test places greater weight on larger survival times, while the Wilcoxon test places more weight on early survival times (SAS [30]). Consequently, the logrank test is more sensitive in detecting significant differences among strata at the higher values of time $t$. The Wilcoxon is more useful in detecting differences among the reliability functions at the lower values of $t$. Ideally, both tests will provide results in basic agreement.

Our test-of-homogeneity hypotheses for both the logrank and Wilcoxon tests are as stated above, where we are considering the nonparametric $\hat{R}_i(t)$ data we develop from the life-table or product-limit methods. SAS estimates a $P$-value, which, in this case, is the probability of observing a larger chi-square value, assuming that the null hypothesis is true. Just as in our discussion of classical designed experiments, the $P$-value is sometimes referred to as the "observed level of significance." Hence, small $P$-values indicate support for rejecting $H_0$.

**Likelihood-Ratio Test**   The likelihood-ratio test in PROC LIFETEST assumes an exponential distribution for all strata. This amounts to assuming a Weibull distribution, with shape parameter $\delta = 1$ for all strata. This assumption is graphically interpreted as follows: the parametric Weibull plots are parallel, with a Weibull slope $\delta = 1$. Under this assumption, the experimental hypotheses are equality of the

Weibull scale parameters across strata:

$$H_0: \text{ All } \theta_i \text{ are equal (e.g., } \theta_1 = \theta_2 = \cdots = \theta_c) \qquad i = 1, 2, ..., c$$

$$H_1: \text{ At least one } \theta_i \text{ is not equal to the others}$$

where $c$ represents the number of strata (covariate levels). The test statistic is calculated and treated as a chi-square random variable, with $c - 1$ df. $P$-values are estimated to help interpret the results.

**Other PROC LIFETEST Tests**   Typically, there are multiple covariates defined for an individual item. The general procedure is to

1  Stratify the data based on one covariate.
2  Calculate the product-limit estimates for each strata.
3  Test the equality of reliability distribution (homogeneity across strata).
4  Test for association of other covariates.

SAS calculates four tests of association between a covariate and the response (the time-to-failure random variable). These tests are generalizations of the Logrank and Wilcoxon tests for homogeneity. The statistics are computed for each covariate and treated as chi-square variates, with 1 df. The tests are as follows:

1  Univariate logrank
2  Univariate Wilcoxon
3  Stepwise logrank
4  Stepwise Wilcoxon

In the stepwise tests, a candidate covariate is selected by a method of steepest ascent, then a marginal test statistic is calculated, conditioned upon the covariates already selected. When only one covariate is considered, the univariate and stepwise statistics are equal.

**Chi-Square Test on Regression Coefficients**   PROC LIFEREG provides a chi-square test of the hypotheses

$$H_0: \quad \beta_j = 0$$

$$H_1: \quad \beta_j \neq 0$$

for all $j$, where $j$ represents the covariate strata index. Statistics are calculated and treated as chi-square deviates with 1 df. Here, again, SAS estimates $P$-values, which we interpret in the usual manner. For example, small $P$-values provide support for the rejection of $H_0$, b.o.e.

---

## 26.7  ON-OFF CYCLING CASE STUDY

A life-test experiment was designed and performed to study the on-off cycling characteristics of small dc motors (Teng and Kolarik [31]). The general procedure called for the motors to run until they ceased operation; then the failure times resulting from on-off switching, as well as continuous operation, under multiple stresses were recorded. All of the motors were identical-part-number, permanent-magnet motors (rated 1.5 to 3.0 V dc). Dc power supplies were used, and the motors were started using computer-controlled relays.

### Test Plans

Seven test plans (summarized in Table 26.6a) were developed for this experiment. Every plan tested at least six motors until all of them failed. Voltages of 3 and 6 V dc were selected to represent nominal and accelerated stress levels, respectively. Load currents were selected to represent typical unloaded and loaded conditions for the motors.

**TABLE 26.6**   TEST PLAN AND DATA SUMMARY, DC-MOTORS
DATA (ON-OFF CYCLING CASE)

**a** Test plan summary

| Plan | Dc voltage, V | Load, A | Operation type |
|------|------|------|------|
| A | 3 | 0.20 | on-off |
| B | 3 | 0.48 | on-off |
| C | 6 | 0.30 | on-off |
| D | 6 | 0.65 | on-off |
| E | 3 | 0.48 | continuous run |
| F | 6 | 0.65 | continuous run |
| G | 3 | 0.08 | continuous run |

**b** Failure time data for the seven operating policies

| Test plan | Lifetime, hr | Test plan | Lifetime, hr | Test plan | Lifetime, hr |
|------|------|------|------|------|------|
| A | 71.79 | D | 4.27 | F | 6.57 |
| A | 85.24 | D | 5.32 | F | 8.25 |
| A | 96.01 | D | 5.76 | F | 8.50 |
| A | 100.68 | D | 6.68 | F | 8.87 |
| A | 104.21 | D | 7.62 | F | 8.96 |
| A | 112.65 | D | 8.66 | F | 9.74 |
| A | 114.84 | D | 8.78 | F | 9.98 |
| A | 115.03 | D | 9.19 | F | 11.58 |
| A | 128.66 | D | 10.33 | F | 12.09 |
| B | 51.99 | D | 10.78 | F | 13.89 |
| B | 74.19 | E | 84.57 | G | 152.74 |
| B | 74.69 | E | 115.79 | G | 176.32 |
| B | 91.05 | E | 151.20 | G | 195.35 |
| B | 91.56 | E | 154.85 | G | 213.32 |
| B | 100.11 | E | 164.55 | G | 214.07 |
| B | 112.99 | E | 186.24 | G | 228.26 |
| C | 9.03 | | | G | 242.18 |
| C | 10.62 | | | G | 264.56 |
| C | 12.38 | | | | |
| C | 13.92 | | | | |
| C | 15.48 | | | | |
| C | 18.82 | | | | |

The failure time data are summarized in Table 26.6b. The voltage, load, and operation mode (on-off or continuous) were considered stress factors. For on-off operation, off time was ignored in order to analyze only the "active" phase of both operation modes. A single failure mode was observed: brush-commutator wear-out.

### Analysis

The SAS procedures LIFETEST (product-limit method) and LIFEREG were used to analyze the failure data. The code and selected output appear in the Computer Aided Analysis Supplement, Solutions Manual. An output summary is shown in Figure 26.15. The covariates studied in this case are load, voltage, and operating mode. Although this case contains no censored motors, these same procedures could also have accommodated a censored test. Initially, a graphical method was used to analyze the data; they were plotted both within and across strata, which were defined by the covariate combinations. The "regression" variables in the experiment were

$Z_1 = 1$

$Z_2 = $ dc voltage, V

$Z_3 = $ operation type, on-off or continuous; 1 or 0, respectively

$Z_4 = $ amperes load, A

Figure 26.15 contains a plot of $\ln t$ versus $\ln(-\ln R)$. Given the nature of the $\ln t$-by-$\ln(-\ln R)$ transformation used, linearity on this plot supports the use of a Weibull failure model. Furthermore, parallel and linear strata support the use of the covariate Weibull (proportional-hazards) model. Table 26.7a summarizes the PROC LIFETEST analysis, which establishes that the strata yield meaningful results and should be retained for further evaluation and modeling. Table 26.7a also displays the three overall (strata) tests for equality. We will ignore the third, likelihood ratio, test. From the very small $P$-value (logrank and Wilcoxon tests), there is strong evidence to reject the null hypothesis that data should be merged into one large group, rather than considered in separate strata. The conclusion at this point is that, overall, the data should be stratified. The PROC LIFETEST results support the usage of the seven distinct strata.

The next step was to fit the data to parametric failure models. PROC LIFEREG was used to fit the data to both the covariate exponential and covariate Weibull models, using MLE (see Figure 26.15). The results for the Weibull model are shown in Table 26.7b. Visual examination of the loglikelihood values obtained ($-2.23$ for the Weibull and $-57.67$ for the exponential) suggest that the Weibull model is a much better fit than the more restricted ($\delta = 1$) exponential model, e.g., its loglikelihood has a much larger value. The Weibull distribution was chosen for model development.

The fitted Weibull model was of the form

$$\hat{\delta} = 4.50$$

$$\hat{\boldsymbol{\beta}}^T = (7.667, -0.724, -0.385, -1.245)$$

**TABLE 26.7**   ANALYSIS RESULTS, DC-MOTORS DATA (ON-OFF CASE)

**a** Overall tests of equality over strata for the seven test plans
(SAS PROC LIFETEST)

| Test | Chi-square | df | Approximate P-value |
|------|-----------|-----|---------------------|
| Logrank | 102.29 | 6 | 0.0001 |
| Wilcoxon | 79.48 | 6 | 0.0001 |
| −2 log(LR) | 78.22 | 6 | 0.0001 |

**b** Maximum likelihood parameter estimates, covariate Weibull model
(SAS PROC LIFEREG)

| Parameter estimate | Description | Standard error |
|--------------------|-------------|----------------|
| $\hat{\delta} = 1/0.2222 = 4.50$ | Common shape parameter | — |
| $\hat{\beta}_1 = 7.667$ | Intercept coefficient | 0.096 |
| $\hat{\beta}_2 = -0.724$ | Voltage coefficient | 0.024 |
| $\hat{\beta}_3 = -0.385$ | Operation-type coefficient | 0.067 |
| $\hat{\beta}_4 = -1.245$ | Load coefficient | 0.169 |

$\mathbf{Z} = (1;$ voltage, V dc; operation type = 1 for on-off or = 0 for continuous operation; load, A)

For example, for Plan A,

$$\hat{\theta}_A = \exp(\mathbf{Z}_A\hat{\boldsymbol{\beta}}) = \exp[7.667 - 0.724(3\ \text{V}) - 0.385(1) - 1.245(0.20\ \text{A})]$$

$$= \exp(4.861) = 129.15\ \text{hr}$$

and $\quad R_A(t; \hat{\theta}_A, \hat{\delta}) = \exp[-(t/\hat{\theta}_A)^{\hat{\delta}}] = \exp[-(t/129.15)^{4.50}]$

For Plan F,

$$\hat{\theta}_F = \exp(\mathbf{Z}_F\hat{\boldsymbol{\beta}}) = \exp[7.667 - 0.724(6\ \text{V}) - 0.385(0) - 1.245(0.65\ \text{A})]$$

$$= \exp(2.514) = 12.35\ \text{hr}$$

and $\quad \hat{R}_F(t; \hat{\theta}_F, \hat{\delta}) = \exp[-(t/\hat{\theta}_F)^{\hat{\delta}}] = \exp[-(t/12.35)^{4.50}]$

A voltage-load interaction was also considered, but is omitted in this model.

**Acceleration Factors**

Based on the experimental data and fitted Weibull model, we can use Equation 26.51 to calculate a number of stress acceleration factors. We will first estimate

The LIFETEST Procedure

Product-Limit Survival Estimates
TRT = 1

| FTIME | Survival | Failure | Survival Standard Error | Number Failed | Number Left |
|---|---|---|---|---|---|
| 0.000 | 1.0000 | 0 | 0 | 0 | 9 |
| 71.790 | 0.8889 | 0.1111 | 0.1048 | 1 | 8 |
| 85.240 | 0.7778 | 0.2222 | 0.1386 | 2 | 7 |
| 96.010 | 0.6667 | 0.3333 | 0.1571 | 3 | 6 |
| 100.680 | 0.5556 | 0.4444 | 0.1656 | 4 | 5 |
| 104.210 | 0.4444 | 0.5556 | 0.1656 | 5 | 4 |
| 112.650 | 0.3333 | 0.6667 | 0.1571 | 6 | 3 |
| 114.840 | 0.2222 | 0.7778 | 0.1386 | 7 | 2 |
| 115.030 | 0.1111 | 0.8889 | 0.1048 | 8 | 1 |
| 128.660 | 0 | 1.0000 | 0 | 9 | 0 |

| Quantiles | 75% | 114.840 | Mean | 103.234 |
|---|---|---|---|---|
| | 50% | 104.210 | Standard Error | 5.751 |
| | 25% | 96.010 | | |

Product-Limit Survival Estimates
TRT = 7

| FTIME | Survival | Failure | Survival Standard Error | Number Failed | Number Left |
|---|---|---|---|---|---|
| 0.000 | 1.0000 | 0 | 0 | 0 | 8 |
| 152.740 | 0.8750 | 0.1250 | 0.1169 | 1 | 7 |
| 176.320 | 0.7500 | 0.2500 | 0.1531 | 2 | 6 |
| 195.350 | 0.6250 | 0.3750 | 0.1712 | 3 | 5 |
| 213.320 | 0.5000 | 0.5000 | 0.1768 | 4 | 4 |
| 214.070 | 0.3750 | 0.6250 | 0.1712 | 5 | 3 |
| 228.260 | 0.2500 | 0.7500 | 0.1531 | 6 | 2 |
| 242.180 | 0.1250 | 0.8750 | 0.1169 | 7 | 1 |
| 264.560 | 0 | 1.0000 | 0 | 8 | 0 |

| Quantiles | 75% | 235.220 | Mean | 210.850 |
|---|---|---|---|---|
| | 50% | 213.695 | Standard Error | 12.674 |
| | 25% | 185.835 | | |

Summary of the Number of Censored and Uncensored Values

| TRT | Total | Failed | Censored | %Censored |
|---|---|---|---|---|
| 1 | 9 | 9 | 0 | 0.0000 |
| 2 | 7 | 7 | 0 | 0.0000 |
| 3 | 6 | 6 | 0 | 0.0000 |
| 4 | 10 | 10 | 0 | 0.0000 |
| 5 | 6 | 6 | 0 | 0.0000 |
| 6 | 10 | 10 | 0 | 0.0000 |
| 7 | 8 | 8 | 0 | 0.0000 |
| Total | 56 | 56 | 0 | 0.0000 |

FIGURE 26.15    MLE computer output summary, dc-motor on-off cycling experiment, covariate Weibull model SAS PROC LIFETEST and LIFEREG (On-Off Cycling Case).

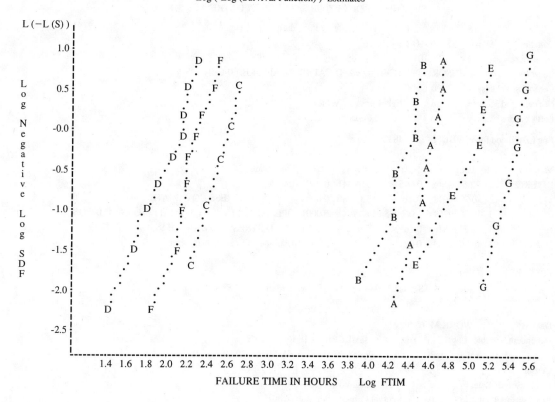

CASE STUDY -- DC MOTORS AT MILTIPLE STRESSES
The LIFETEST Procedure
Log (-Log (Survival Function) ) Estimates

FAILURE TIME IN HOURS     Log  FTIM

CASE STUDY  --  DC MOTORS AT MULTIPLE STRESSES

The LIFETEST Procedure

Testing Homogeneity of Survival Curves over Strata
Time Variable FTIME

Test of Equality over Strata

| Test | Chi-Square | DF | Pr > Chi-Square |
|------|------------|-----|-----------------|
| Logrank | 102.2871 | 6 | 0.0001 |
| Wilcoxon | 79.4814 | 6 | 0.0001 |
| -2Log (LR) | 78.2201 | 6 | 0.0001 |

**FIGURE 26.15**    (*continued*)

CASE STUDY — DC MOTORS AT MULTIPLE STRESSES

Lifereg Procedure

Data Set            =WORK.MOTOR
Dependent Variable=Log (FTIME)          FAILURE TIME IN HOURS
Censoring Variable=CENSOR          CENSOR INDICATOR  1-FAILURE,  0-CENSOR
Censoring Value(s)=        0
Noncensored Values=      56        Right Censored Values=        0
Left Censored Values=      0        Interval Censored Values=        0

Log Likelihood for WEIBULL-2.225483063

| Variable | DF | Estimate | Std Err | ChiSquare | Pr>Chi | Label/Value |
|----------|----|----------|---------|-----------|--------|-------------|
| INTERCPT | 1 | 7.66737868 | 0.096489 | 6314.441 | 0.0001 | Intercept |
| V | 1 | -0.7240642 | 0.024497 | 873.5986 | 0.0001 | VOLTAGE STRESS COVARIATE |
| OP | 1 | -0.3849869 | 0.067225 | 32.79641 | 0.0001 | OPERATION TYPE COVARIATE 1-ON/OFF, 0-CON |
| L | 1 | -1.2446846 | 0.169392 | 53.99218 | 0.0001 | LOAD STRESS COVARIATE IN AMPS |
| SCALE | 1 | 0.22218111 | 0.022579 | | | Extreme value scale parameter |

CASE STUDY — DC MOTORS AT MULTIPLE STRESSES

Lifereg Procedure

Data Set            =WORK.MOTOR
Dependent Variable=Log(FTIME)          FAILURE TIME IN HOURS
Censoring Variable=CENSOR          CENSOR INDICATOR  1-FAILURE,  0-CENSOR
Censoring Value(s)=        0
Noncensored Values=      56        Right Censored Values=        0
Left Censored Values=      0        Interval Censored Values=        0

Log Likelihood for EXPONENT -57.67043193

| Variable | DF | Estimate | Std Err | ChiSquare | Pr>Chi | Label/Value |
|----------|----|----------|---------|-----------|--------|-------------|
| INTERCPT | 1 | 7.62452371 | 0.426084 | 320.2095 | 0.0001 | Intercept |
| V | 1 | -0.7303728 | 0.11486 | 40.43469 | 0.0001 | VOLTAGE STRESS COVARIATE |
| OP | 1 | -0.3890585 | 0.274258 | 2.012379 | 0.1560 | OPERATION TYPE COVARIATE 1-ON/OFF, 0-CON |
| L | 1 | -1.2879579 | 0.806185 | 2.552312 | 0.1101 | LOAD STRESS COVARIATE IN AMPS |
| SCALE | 0 | 1 | 0 | | | Extreme value scale parameter |

**FIGURE 26.15**    (continued)

the scale parameter of the Weibull distribution for the seven operating policies, A through G (indicated by the subscript):

$$\exp(\mathbf{Z}_A\hat{\boldsymbol{\beta}}) = \exp(4.861)$$

$$\exp(\mathbf{Z}_B\hat{\boldsymbol{\beta}}) = \exp(4.512)$$

$$\exp(\mathbf{Z}_C\hat{\boldsymbol{\beta}}) = \exp(2.565)$$

$$\exp (\mathbf{Z}_D \hat{\boldsymbol{\beta}}) = \exp (2.129)$$

$$\exp (\mathbf{Z}_E \hat{\boldsymbol{\beta}}) = \exp (4.898)$$

$$\exp (\mathbf{Z}_F \hat{\boldsymbol{\beta}}) = \exp (2.514)$$

$$\exp (\mathbf{Z}_G \hat{\boldsymbol{\beta}}) = \exp (5.395)$$

Next, using policy E as the nominal level, the acceleration factors are estimated:

Plan E vs. A: $\hat{AF}_{E:A} = \exp (4.898 - 4.861) = 1.038$

Plan E vs. B: $\hat{AF}_{E:B} = \exp (4.898 - 4.512) = 1.471$

Plan E vs. C: $\hat{AF}_{E:C} = \exp (4.898 - 2.565) = 10.309$

Plan E vs. D: $\hat{AF}_{E:D} = \exp (4.898 - 2.129) = 15.943$

Plan E vs. F: $\hat{AF}_{E:F} = \exp (4.898 - 2.514) = 10.848$

Plan E vs. G: $\hat{AF}_{E:G} = \exp (4.898 - 5.395) = 0.608$

---

## REVIEW AND DISCOVERY EXERCISES

### Review

**26.1** A success-failure test was conducted for four-wheel drive vehicles operated in an extremely harsh environment. Out of 30 vehicles tested, a total of 6 vehicles failed to complete the mission successfully. Estimate the reliability for this vehicle configuration on this particular mission and determine the 95 percent reliability confidence limits.

**26.2** Two new robotic manipulator designs have been developed for performing a specific task—grasping cylindrical parts from a parts bin. A test was developed whereby each manipulator would be "asked" to grasp 50 parts. The test results were as follows: Design 1 grasped 45 parts successfully, design 2 grasped 35 parts successfully.

    **a** Develop reliability and failure probability estimates for each design.

    **b** Develop 90 percent confidence intervals for both reliability and failure probability estimates for each design.

    **c** Can one design be recommended over the other? Use the 90 percent confidence intervals of part *b* in your analysis and conclusions.

**26.3** Our company has contracted to produce a small arc-jet engine-thruster assembly for course corrections on a deep-space, unmanned space probe. The specification states that the thruster assembly is to have at least a 99 percent reliability, with 97.5 percent confidence, for a 5-year mission. During this mission, it is anticipated that the thruster will be fired 2000 times and will not be suitable for reuse. Briefly outline a reliability assurance plan to address the specification.

**26.4** The reliability of window glass subjected to a given load is time-dependent. The longer a window glass panel is subjected to a uniform load, the weaker it gets, and failure occurs. A sample of 80 window glass panels were tested. Data collected relative to the loaded time intervals and the number of failures observed are shown below. Estimate the reliability, the pmf, cmf, and the failure-hazard rate. Indicate whether the failure-hazard rate appears to be increasing or decreasing. Use the life-table method of analysis. [PB]

| Time interval, min | Number of failures observed |
|---|---|
| $0 < t \leq 1$ | 2 |
| $1 < t \leq 2$ | 4 |
| $2 < t \leq 3$ | 8 |
| $3 < t \leq 4$ | 14 |
| $4 < t \leq 5$ | 21 |
| $5 < t \leq 6$ | 31 |

**26.5** A life-test experiment was performed on high-density capacitors. The capacitors were subjected to charge-discharge cycling in a simulated space environment. The life data are presented below in thousands of cycles:

5.3    9.1+    12.2    17.4    21.3+    24.5    32.4    38.5    40+    40+

The + indicates a censored capacitor. The censoring at 9.1 and 21.3 kilocycles resulted from a malfunction in the charging circuit which destroyed the capacitors, but the fault was not attributed to the capacitors. Using the product-limit method, make a nonparametric plot of the capacitor reliability.

**26.6** A prototype strawberry harvesting machine was tested for 200 hr. At the end of the test, 23 failures were observed. Once a failure was encountered, a repair operation was initiated and the failed machine was brought back to operating condition. The following table gives the system failure times $t_f$ and the repair completion times $t_r$ over the 200 hr test [MC]:

| $t_f$ | $t_r$ | $t_f$ | $t_r$ |
|---|---|---|---|
| 2.8 | 7.9 | 115.4 | 116.1 |
| 11.7 | 11.9 | 126.3 | 129.2 |
| 43.2 | 43.9 | 132.4 | 133.0 |
| 53.4 | 55.6 | 139.9 | 140.2 |
| 56.2 | 56.6 | 145.1 | 146.6 |
| 64.2 | 64.3 | 155.8 | 157.9 |
| 77.8 | 77.9 | 166.6 | 169.4 |
| 78.7 | 79.3 | 181.6 | 182.2 |
| 86.2 | 86.6 | 182.3 | 183.5 |
| 90.8 | 91.1 | 188.6 | 188.7 |
| 96.0 | 96.8 | 198.8 | 199.7 |
| 109.9 | 111.8 | | |

**a** Calculate the average availability over the time interval $0 \leq t \leq 200$ hr from the data.

**b** Assuming that the failure times and repair times are exponentially distributed, estimate the MTTF and MTTR.

c Estimate the 90 percent confidence interval on the MTTF and MTTR. Assume the failure and repair times to be exponentially distributed.

**26.7** The following data contain elevator failure times and elevator restoration times in working days over a year period. A comprehensive service contract for the elevator costs $10,000 per year, while each service call costs $200 plus $100 per hour, with a 3-hour minimum. Assume that a working day consists of 10 hr, and address the following parts on an hourly basis [DS]:

| Failure-repair | Failure time | Repair time |
| --- | --- | --- |
| 1 | 10.8 | 11.3 |
| 2 | 42.1 | 43.0 |
| 3 | 70.1 | 70.8 |
| 4 | 100.0 | 100.9 |
| 5 | 120.3 | 120.5 |
| 6 | 180.0 | 181.2 |
| 7 | 230.8 | 231.1 |
| 8 | 261.7 | 262.3 |
| 9 | 289.3 | 290.0 |
| 10 | 310.0 | 310.9 |
| 11 | 330.3 | 330.7 |
| 12 | 359.1 | 360.8 |

a Estimate the availability over a 1-year period.

b Assuming that the failure and repair times are exponentially distributed, estimate the MTTF and the MTTR.

c Determine whether the company should buy a service contract or simply pay for the repairs as they occur.

**26.8** The times to failure for four compressors are 240, 550, 630, and 1080 days.

a Fit the data to a Weibull model and plot the 5 percent and 95 percent confidence bands. Clearly state the values of your parameter estimates. Then, determine a two-sided 90 percent reliability confidence interval for 500 days.

b Fit the data to a lognormal model and plot 5 percent and 95 percent confidence bands. Clearly state the values of your parameter estimates. Then determine a two-sided 90 percent reliability confidence interval for 500 days.

c Which model is the best. Explain your choice.

**26.9** A one-day experiment was run on the time to repair the Grasshopper lawn mower in a factory shop, in order to determine a warranty prediction model. Assuming that an exponential repair model is to be fit, use the repair times given below to determine the MTTR for the Grasshopper lawn mower and predict the probability of repairing a mower in 1.5 hr:

1.25 hr    2.60 hr    5.75 hr    6.25 hr    8.00 hr*    8.00 hr*

*Note: The Grasshopper lawn mower was not repaired by the end of the day.

**26.10** Four long-life light bulbs were life tested. During the experiment, one bulb was removed due to a failure in the test-sensing equipment. The result was a multiple-censored experiment. The times to failure, and censor time, in continuous days of usage, are shown below. The censored unit's time on test is marked with a +.

45.0    75.5    95.0+    150.5

Using a graphical approach, fit the data to either an exponential or a Weibull model, whichever is most appropriate. Support your decision.

**26.11** A large airline is assessing bids to supply landing lights. Two potential suppliers have submitted the most competitive contract bids. The two proposals are essentially identical in terms of cost. However, the suppliers' bids are also being assessed on the basis of product reliability. The following data, in hours, were obtained from life-tests conducted by an independent testing agency [DS]:

| Company A | Company B |
|-----------|-----------|
| 980 | 850 |
| 1050 | 900+ |
| 1100+ | 900 |
| 1290 | 1000 |
| 1320 | 1020 |
| 1410 | 1050 |
| 1530+ | 1065+ |
| 1650 | 1070 |
| 1820 | 1080 |
| 1880 | 1120 |
| + indicates a censored unit. | |

a Develop a nonparametric reliability plot for each supplier.
b Produce an exponential reliability model fit for each supplier.
c Produce a Weibull reliability model fit for each supplier.
d Produce a normal reliability model fit for each supplier.
e Produce a lognormal reliability model fit for each supplier.
f Determine which model produces the best fit, and comment on the fits you observed.
g Recommend a supplier. Explain your decision.

**26.12** A gear manufacturer has developed two new heat-treatment processes to improve the life of a worm gear. Two samples of 10 gears each were randomly selected for each of the two heat treatment methods. The number of cycles to failure (in thousands) are shown below.

| Heat treatment 1 | Heat treatment 2 |
|------------------|------------------|
| 4.50 | 12.25 |
| 6.50 | 16.00 |
| 8.00 | 17.00 |
| 9.00+ | 19.00 |
| 9.50 | 21.00 |
| 11.20+ | 21.50+ |
| 12.00 | 22.00 |
| 13.00 | 25.00 |
| 15.00 | 25.50+ |
| 16.50 | 26.00 |
| + indicates a censored gear. | |

The manufacturer is interested in comparing the reliability effects of the two heat-treatment processes. [RC]

a Develop Weibull reliability-predictor equations for each treatment using the plotting method.

**b** Develop and plot 5 and 95 percent nonparametric confidence bands for both heat treatments.

**c** Use your results to determine the best heat treatment process, considering reliability characteristics.

**26.13** In many advertisements, battery manufacturers state that their batteries have the longest life. In order to support these statements, an experiment was designed. Three different brands (Pan, Ever, and Dura) were chosen. Six batteries of each brand were placed on test. A failure criteria of 1.1 V dc was defined (each battery was a 1.5-V rated battery). Equal resistive loads were placed on each battery. The failure data for the three different battery brands are shown below [MC]:

| Brand name | Failure time, min | | |
|---|---|---|---|
| Pan | 420.12, | 412.52, | 474.01 |
| | 351.14, | 325.24, | 385.16 |
| Ever | 521.65, | 468.30, | 543.15 |
| | 501.23, | 525.19, | 552.13 |
| Dura | 525.33, | 508.45, | 602.13 |
| | 543.86, | 521.43, | 583.68 |

**a** Develop nonparametric reliability models for each battery brand. You will probably want to use PROC LIFETEST here.

**b** Develop appropriate reliability models for each brand separately. You will probably want to use PROC LIFEREG here.

**c** Taking the brands two at a time (e.g., Pan vs. Ever; Pan vs. Dura; and Ever vs. Dura) develop covariate reliability model analyses. You will probably want to use PROC LIFEREG here.

**d** Determine which battery has the longest life.

**e** Develop a covariate reliability model analysis for the three brands, using each brand as a classification covariate. This analysis will call for PROC LIFEREG.

## Discovery

**26.14** The inverse power law has been applied in analyses of aging for multicomponent systems, as well as in testing of fatigue in metals and dielectric breakdown in capacitors. The inverse power law model below is compatible with the covariate Weibull model, with the shape parameter $\delta$ independent of stress and the scale parameter $\theta$ (the Weibull characteristic life) expressed as an inverse power function of a positive stress $V$:

$$\theta(V) = \frac{1}{aV^b}$$

where $V$ = positive stress covariate (e.g., voltage)

$a, b$ = positive coefficients, or constants

Typically, the inverse power law as applied to two stress levels, $stress_1$ and $stress_2$, is expressed as

$$\theta_1 = \theta_2 \left( \frac{V_2}{V_1} \right)^b$$

where stress level 2 is greater than stress level 1.  The acceleration factor typically stated in the inverse power law model is

$$AF_{1:2} = \left( \frac{V_2}{V_1} \right)^b$$

Develop a covariate Weibull model form for the inverse power law (i.e., transform $\theta(V)$ to a regression form with logs).

**26.15** An accelerated experiment was developed to test small capacitors designed to operate in the 20-V dc range.  The response was defined as the number of kilocycles (charge-discharge cycles) to failure.  Voltage levels of 20 V dc and 30 V dc were used.  The experiment was censored at 25 kilocycles.  The data are displayed below:

| Voltage | Life response (kilocycles) | | |
|---|---|---|---|
| 20 V dc | 12.8 | 17.5 | 19.2 |
| | 22.6 | 25.0* | 25.0* |
| | 25.0* | 25.0* | 25.0* |
| | 25.0* | 25.0* | 25.0* |
| 30 V dc | 4.5 | 5.1 | 6.2 |
| | 6.3 | 7.1 | 7.8 |
| | 8.5 | 9.3 | 10.5 |
| | 11.0 | 12.7 | 15.0 |

*Censored capacitors

**a** Using the inverse power law as the basis for a reliability model, analyze the data collected.  Use SAS PROC LIFETEST and PROC LIFEREG in combination.  Then, predict $R(t = 12$ kilocycles; $V = 25$ V dc).

**b** Estimate the acceleration factor $AF_{20:25}$ for the results you obtained above.

CAPACITOR CASE STUDY—Problems 26.16 through 26.19
An extensive life test experiment was designed to assess the reliability of high-voltage capacitors.  These capacitors were designed for a space environment where the temperature was expected to vary widely due to orbital periods of light and darkness.  Voltage requirements were expected to be in the 2-kV range during normal use.  In general, the capacitor design called for a life of charge-discharge cycles on the order of 100 to 200 kilocycles with high reliability.  In order to ensure a comprehensive life test, a FAT-CRD experiment (see Chapter 21) was developed whereby both temperature and voltage would be used as acceleration factors.  The following temperatures and voltage levels were selected:

Temperature: −20°C, 25°C, and 50°C

Voltage: 2 kV, 3 kV, and 4 kV

A total of 54 capacitors were randomly assigned to the nine treatment combinations.  The response was defined as the number of kilocycles (charge-discharge) to failure. A censoring point of 600 kilocycles was selected to expedite the completion of the experiment. The data collected are summarized in the table below.

**Accelerated capacitor experiment data**

| Temperature, °C | Voltage, kV | | |
|---|---|---|---|
| | 2 | 3 | 4 |
| −20 | 600.000* | 11.151 | 0.575 |
| | 600.000* | 4.363 | 0.865 |
| | 600.000* | 20.128 | 1.817 |
| | 600.000* | 2.860 | 0.230 |
| | 293.771 | 39.738 | 4.823 |
| | 525.320 | 2.559 | 3.875 |
| 25 | 600.000* | 18.763 | 0.175 |
| | 115.631 | 1.566 | 0.102 |
| | 600.000* | 5.266 | 0.770 |
| | 600.000* | 2.163 | 0.300 |
| | 524.625 | 12.345 | 0.513 |
| | 600.000* | 10.932 | 0.048 |
| 50 | 428.212 | 1.032 | 0.040 |
| | 600.000* | 4.090 | 0.030 |
| | 249.828 | 5.177 | 0.033 |
| | 499.627 | 4.656 | 0.169 |
| | 585.878 | 2.214 | 0.049 |
| | 600.000* | 0.064 | 0.142 |

*Indicates censored capacitors.

**26.16** Using the capacitor data from the 50°C, 3 kV treatment combination:
  **a** Develop a nonparametric plot using the product-limit method for reliability.
  **b** Develop a Weibull probability plot with 5 percent and 95 percent nonparametric confidence bands, and estimate the reliability of the capacitor to survive 2 kilocycles.
  **c** Develop a lognormal probability plot with 5 percent and 95 percent nonparametric confidence bands, and estimate the reliability of the capacitor to survive 2 kilocycles.
  **d** Determine which of the analyses above you would recommend to model the life of this capacitor.

**26.17** Using the capacitor data from the 50°C, 2 kV treatment combination:
  **a** Develop a nonparametric plot using the product-limit method for reliability.
  **b** Develop a Weibull probability plot with 5 percent and 95 percent nonparametric confidence bands, and estimate the reliability of the capacitor to survive 200 kilocycles.
  **c** Develop a lognormal probability plot with 5 percent and 95 percent nonparametric confidence bands, and estimate the reliability of the capacitor to survive 200 kilocycles.
  **d** Determine which of the analyses above you would recommend to model the life of this capacitor.

**26.18** Using the capacitor data from the 50°C, 2 kV treatment combination as the nominal operating environment and that from the 50°C, 3 kV treatment combination as an accelerated operating environment, develop a covariate Weibull accelerated model analysis by performing the following tasks:

**a** Develop nonparametric reliability plots.

**b** Develop and parameterize an appropriate accelerated model.

**c** Estimate the acceleration factor.

**26.19** Using all nine treatment combinations, conduct the following covariate model analysis:

**a** Develop nonparametric tables and plots for the reliability estimates.

**b** Develop a covariate exponential model. Estimate the reliability of the capacitor to survive 300 kilocycles at a temperature of 22°C and a voltage of 2.5 kV.

**c** Develop a covariate Weibull model. Estimate the reliability of the capacitor to survive 300 kilocycles at a temperature of 22°C and a voltage of 2.5 kV.

**d** Develop a covariate lognormal model. Estimate the reliability of the capacitor to survive 300 kilocycles at a temperature of 22°C and a voltage of 2.5 kV.

**e** Recommend a model and explain your choice.

**26.20** Design and perform a reliability experiment relative to a product or process of your choice. Include one or two covariates at two or three levels. Develop a report on your experiment that includes the following:

**a** the experimental objective.

**b** the experimental design.

**c** the experimental protocol.

**d** the data taken.

**e** an appropriate analysis.

**f** an interpretation and conclusion.

# Appendix 26A:  Estimation and Estimators

## Estimation

To be useful, the reliability models previously described must be tailored to specific applications. For example, the $\lambda$ or $\theta$ parameter in an exponential model must be estimated for the model to be useful quantitatively. The process of developing the parameter estimate $\hat{\lambda}$ or $\hat{\theta}$ is termed "estimation" in statistics. The parameter estimates (graphical and numerical) for the binomial, exponential, Weibull, normal, and lognormal distributions shown earlier are based on estimation theory principles. Estimating the single parameter in the exponential model or the binomial parameter $p$ in the binomial model is a straightforward process; it becomes more complicated in the case of models having two or more parameters.

Estimation is a rather involved branch of statistical theory. Only a brief summary will be presented here. If we define $X$ to be a random variable, we can define the estimator $W$ as a new random variable. Here, $W$ and $X$ are related by some formula, or functional relationship. When we substitute our data values—say, $x_1, x_2, x_3, ..., x_n$—into $W$, we obtain a value $w$ which we call an "estimate." For example, a point estimator of the mean (a parameter that describes the population of $X$) is

$$W = \hat{\mu}_X = \overline{X} = \frac{1}{n} \sum_{i=1}^{n} X_i \qquad (26.52)$$

To develop a point estimate, we will plug in a specific set of data:

If $x_1 = 2$, $x_2 = 5$, $x_3 = 6$, $x_4 = 4$, $x_5 = 3$, then

$$w = \hat{\mu}_x = \bar{x} = \frac{20}{5} = 4$$

As another example, we can define a different estimator, for the variance of $X$, $\hat{\sigma}_X^2$ as

$$\hat{\sigma}_X^2 = S^2 = \frac{\sum_{i=1}^{n} (X_i - \bar{X})^2}{n-1} \qquad (26.53)$$

Then, to develop a point estimate, we plug in our specific data set: If $x_1 = 2$, $x_2 = 5$, $x_3 = 6$, $x_4 = 4$, $x_5 = 3$, then,

$$\hat{\sigma}_x^2 = s^2 = \frac{(2-4)^2 + (5-4)^2 + (6-4)^2 + (4-4)^2 + (3-4)^2}{5-1} = \frac{10}{4} = 2.5$$

From the above examples, we note that a new data sample will yield a new estimate, although the estimator will not change. Hence, the estimator has a unique distribution. Statisticians have developed many estimators over the years, some better than others. We use a number of criteria to judge the goodness of our estimator. Some desirable properties of estimators are listed below.

**Unbiased Estimator**  A point estimator, $\hat{\Theta}$, of some parameter $\theta$ is said to be unbiased if $E[\hat{\Theta}] = \mu_{\hat{\Theta}} = \theta$ (see Figure 26.16 for a depiction). In order to extend our example above, we may decide to estimate the mean of $X$, $\mu_X = E[X]$, with the statistic $\bar{X}$. Hence, to show that $\bar{x}$ is an unbiased estimate of $\mu_X$, we must show that

$$E[\bar{X}] = E[X] = \mu_X$$

$$E[\bar{X}] = E\left(\frac{1}{n} \sum_{i=1}^{n} X_i\right) = \frac{1}{n} E\left(\sum_{i=1}^{n} X_i\right)$$

$$= \frac{1}{n} (E[X_1] + E[X_2] + \cdots + E[X_n])$$

Assuming that $X_1, X_2, ..., X_n$ are independent random variables, where

$$E[X_1] = E[X_2] = \cdots = E[X_n]$$

we have

$$E[\bar{X}] = \frac{1}{n} (n \, E[X]) = E[X] = \mu_X$$

For practical purposes, unbiased estimates are preferred, all other measures being equal.

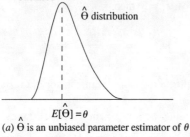

**FIGURE 26.16**    Concept of bias in parameter estimators.

(a) $\hat{\Theta}$ is an unbiased parameter estimator of $\theta$

(b) $\hat{\Theta}$ is a biased estimator of $\theta$

**Consistent Estimator**    An estimator $\hat{\Theta}$ is said to be consistent if the probability that $\hat{\Theta}$ differs from $\theta$ by more than an arbitrary constant $e$ approaches 0 as the sample size $n$ approaches infinity.  Hence, for large sample sizes, we would expect very little error.  In practice, we tend to put more faith in estimates from larger samples, than from smaller samples, as stated earlier.

**Relative Efficiency of Two Estimators**    Where two or more estimators exist, relative efficiency is measured by the ratio of their variances.  The most efficient estimator is the one with the smallest variance.  For practical purposes, we prefer estimates with minimal variation.

**Best Unbiased Estimator**    The best unbiased estimator is that estimator among all possible unbiased estimators which has the smallest variance.

## Estimators

We will now proceed to briefly summarize two of three fundamental classes of estimators useful in engineering:  (1) moment estimators and (2) maximum-likelihood estimators.  A third class, least squares estimators, was discussed in Chapter 22.  The type of estimator we use is important as it affects our ability to estimate a model parameter.  For example, should we wish to estimate $\lambda$ for an exponential model or $\theta$ and $\delta$ for a Weibull model, we would choose the most appropriate method.

**Moment Estimators**    Using the method of moments, we develop a theoretical moment, from our chosen model, for each parameter we wish to estimate (e.g., two moments for two parameters).  The $r$th theoretical moment about the origin is defined as:

$$\mu_r' = E[X^r] = \sum_{all\ x} x^r f(x) \qquad \text{if } X \text{ is discrete}$$

$$= \int_{-\infty}^{\infty} x^r f(x)\, dx \qquad \text{if } X \text{ is continuous} \qquad (26.54)$$

For example, $\mu = \mu_1'$ is the first moment about the origin, and $\sigma^2 = E[(X - \mu)^2]$ is the second moment about the mean.  Next, we develop the empirical moment counterparts.  The $r$th empirical moment about the origin is defined as

$$m_r = \frac{1}{n} \sum_{i=1}^{n} x_i^r \qquad (26.55a)$$

and the $r$th moment about the mean is

$$m_{rm} = \frac{1}{n} \sum_{i=1}^{n} (x_i - \bar{x})^r \qquad (26.55b)$$

Finally, we equate the corresponding theoretical and empirical moments, and solve for the estimators.

---

**Example 26.14**

Using the method of moments, determine an estimator for the exponential model parameter $\lambda$.

**Solution**

From Equation 25.24b, we know that the pmf for the exponential model is

$$f(x) = \lambda \exp(-\lambda x)$$

The first theoretical moment about the origin is

$$\mu_1' = E[X] = \int_0^\infty x \lambda \exp(-\lambda x) \, dx = \frac{1}{\lambda}$$

The first empirical moment about the origin is

$$m_1 = \frac{1}{n} \sum_{i=1}^{n} x_i$$

Equating the two, we have

$$\frac{1}{\hat{\lambda}} = \frac{1}{n} \sum_{i=1}^{n} x_i \quad \text{or} \quad \hat{\theta} = \frac{1}{\hat{\lambda}} = \frac{\sum_{i=1}^{n} x_i}{n}$$

---

**Maximum-Likelihood Estimators**    MLEs are based on maximizing a likelihood or loglikelihood function. The likelihood function is typically defined as a "joint density" function. Given a random sample $x_1, x_2, ..., x_n$, we assume each $x_i$ represents a value for the random variables $X_1, X_2, ..., X_n$, respectively. Furthermore, each random variable has a pmf $f(x_i; \boldsymbol{\theta})$, where $\boldsymbol{\theta}$ represents a parameter set for the pmf, $\boldsymbol{\theta} = \theta_i, i = 1, 2, ...$ Assuming all random variables to be independent, the likelihood function can be written as the product of the pmf's.

$$L = f(x_1, x_2, ..., x_n; \boldsymbol{\theta}) = f(x_1; \boldsymbol{\theta}) \; f(x_2; \boldsymbol{\theta}) \cdots f(x_n; \boldsymbol{\theta}) \qquad (26.56)$$

Now, since $f(x_i; \boldsymbol{\theta}) \geq 0$, by definition, and $f(x_i; \boldsymbol{\theta}) \Delta x = P(X_i = x_i)$, we can calculate $\boldsymbol{\theta}_{\text{MLE}}$, which is the parameter set estimate that maximizes the likelihood function. The maximum is developed by setting $\partial L / \partial \theta_i = 0$ for all $i$ and solving the equations simultaneously.

Figure 26.17 illustrates the MLE method by superimposing three cases or sets of pmf's, each with different parameter values. It should be noted that, within each case, or set, the pmf's parameter values are constant. The MLE parameter values maximize the likelihood function by maximizing the product of all pmf values (see Figure 26.17c).

Optimizing the likelihood function may be difficult because of its product nature. Therefore, typically a loglikelihood $\Lambda$ function is developed and maximized, provided a one-to-one mapping of the loglikelihood and likelihood functions exists.

---

**Example 26.15**

Develop the MLE for the exponential model parameter.

**Solution**

In general, for the exponential pmf,

$$L = \prod_{i=1}^{n} f(x_i; \lambda) = \prod_{i=1}^{n} [\lambda \exp(-\lambda x_i)]$$

In order to simplify the mathematics,

$$\Lambda = \log L = \log \left[ \prod_{i=1}^{n} f(x_i; \lambda) \right] = \sum_{i=1}^{n} \log[f(x_i; \lambda)]$$

Taking the partial derivative with respect to $\lambda$ on both sides, we have

$$\frac{\partial \Lambda}{\partial \lambda} = \frac{\partial}{\partial \lambda} \left\{ \sum_{i=1}^{n} \log[f(x_i; \lambda)] \right\} = \sum_{i=1}^{n} \left\{ \frac{\partial}{\partial \lambda} \log[f(x_i; \lambda)] \right\}$$

Applying the general relation above to the exponential pmf,

$$\frac{\partial \Lambda}{\partial \lambda} = \sum_{i=1}^{n} \frac{\partial}{\partial \lambda} \left\{ \log[\lambda \exp(-\lambda x_i)] \right\} = \sum_{i=1}^{n} \frac{\partial}{\partial \lambda} (\log \lambda - \lambda x_i)$$

$$= \sum_{i=1}^{n} \left( \frac{1}{\lambda} - x_i \right) = \frac{n}{\lambda} - \sum_{i=1}^{n} x_i$$

Now, to obtain our MLE, we equate the partial derivative to 0. We obtain

$$\frac{\partial \Lambda}{\partial \lambda} = 0 = \frac{n}{\lambda} - \sum_{i=1}^{n} x_i$$

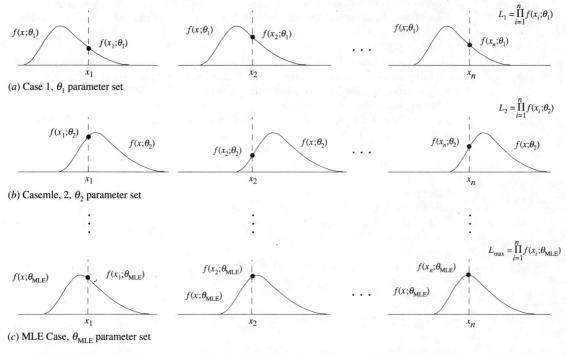

$$L_1 = \prod_{i=1}^{n} f(x_i;\theta_1)$$

(a) Case 1, $\theta_1$ parameter set

$$L_2 = \prod_{i=1}^{n} f(x_i;\theta_2)$$

(b) Casemle, 2, $\theta_2$ parameter set

$$L_{max} = \prod_{i=1}^{n} f(x_i;\theta_{MLE})$$

(c) MLE Case, $\theta_{MLE}$ parameter set

**FIGURE 26.17**    Maximum-likelihood estimation MLE concept.

Solving for $\lambda$, we obtain

$$\hat{\lambda}_{MLE} = \frac{n}{\sum_{i=1}^{n} x_i} \qquad \text{or} \qquad \hat{\theta}_{MLE} = \frac{\sum_{i=1}^{n} x_i}{n}$$

The previous example involved a single-parameter distribution. More complicated cases where two or more parameters are to be estimated present more difficult solutions. In many such cases, iterative search procedures are used to satisfy the partial differential equations and maximize the likelihood or loglikelihood function. Shooman [32] and Lawless [33] provide further discussion on MLE's, their properties, and solution techniques, relative to reliability.

## REFERENCES

**1** D. Keceioglu, "Lecture Notes for the 11th Annual Reliability Testing Institute," 11th Annual Reliability Testing Institute, Tucson, AZ, 1985.

**2** C. Lipson and N. J. Sheth, *Statistical Design of Analysis of Experiments,* New York: McGraw-Hill, 1973.

**3** *SAS/STAT User's Guide, Volumes 1 and 2,* Version 6, Cary, NC: SAS Institute, 1990.

**4** J. F. Lawless, *Statistical Models and Methods for Lifetime Data,* New York: Wiley, 1982.

**5** See reference 3.

**6** See reference 4.

**7** E. L. Kaplan and P. Meier, "Nonparametric Estimation from Incomplete Observations," *Journal of the American Statistical Association,* vol. 53, pp. 457–481, 1958.

**8** W. Nelson, *Applied Life Data Analysis,* New York: Wiley, 1982.

**9** See reference 8.

**10** K. C. Kapur and L. R. Lamberson, *Reliability in Engineering Design,* New York: Wiley, 1977.

**11** W. J. Conover, *Practical Nonparametric Statistics,* 2d ed., New York: Wiley, 1980.

**12** See reference 10.

**13** See reference 10.

**14** See reference 10.

**15** See reference 10.

**16** Team Graph Papers, Box 25, Tamworth, NH 03886.

**17** See reference 10.

**18** See reference 4.

**19** D. R. Cox and D. Oakes, *Analysis of Survival Data,* London: Chapman and Hall, 1984.

**20** J. D. Kalbfleisch and R. L. Prentice, *The Statistical Analysis of Failure Time Data,* New York: Wiley, 1980.

**21** See reference 3.

**22** W. Nelson, *Accelerated Testing, Statistical Models, Test Plans, and Data Analyses,* New York: Wiley, 1990.

**23** See reference 1.

**24** P. A. Tobias and D. C. Trindade, *Applied Reliability,* New York: Van Nostrand Reinhold, 1986.

**25** See reference 4.

**26** See references 4, 19, and 20.

**27** See reference 4.

**28** See reference 24.

**29** See references 3 and 4.

**30** See reference 3.

**31** N. H. Teng and W. J. Kolarik, "On/Off Cycling under Multiple Stresses," *IEEE Transaction on Reliability,* vol. 38, no. 4, pp. 494–498, 1989.

**32** M. L. Shooman, *Probabilistic Reliability: An Engineering Approach,* New York: McGraw-Hill, 1968.

**33** See reference 4.

# CREATION OF QUALITY— QUALITY TRANSFORMATION

## VIRTUES FOR LEADERSHIP IN QUALITY
### Trustworthiness and Honesty

The purpose of Section Eight is to explore the nature of quality transformations within organizations. We reintroduce our readers (See Chapters 1 and 2) to the quality transformation concept and discuss it in more detail, focusing on several bottlenecks in transforming to higher levels of quality within an organization. Chapter 27 presents the philosophy of total quality from both a people and a production perspective. A discussion of the leadership concept and how it differs from the management concept is provided. Then, classical leadership and management models are presented. Every attempt is made to stress the fact that leadership, not management, makes the difference in quality transformation. The leadership element is the most decisive factor in quality transformation and ultimately responsible for either its success or its failure.

Employee empowerment and communications issues are addressed in Chapter 28. Here, we discuss the teamwork concept and the empowerment issue together, since teamwork requires employee empowerment to work successfully in quality transformation. Then, we discuss the critical importance of communications. We describe quality transformation as it evolves from a conventional to an empowered environment. Finally, we focus on the issue of creativity. Effectively tapping the creative powers of people in the workplace is usually a bottleneck in quality improvement. Hence, we discuss creativity and environments that encourage creativity.

Chapter 29 concludes our treatment of quality transformation. First, we focus on field experiences in total quality management, TQM, implementation. Then, we discuss TQM themes and structures relative to the Baldrige Award and the ISO 9000 Quality Standards. We briefly introduce our readers to supplier certification, as well as internal process certification. Quality audits are discussed, along with quality-business metrics. Finally, we conclude the textbook with a brief discussion of the diversity and commonality found in quality transformations.

# 27

# PHILOSOPHY AND LEADERSHIP

## 27.0 INQUIRY

1  Why is quality indigenous to organizations, products, and processes?
2  What constitutes a total quality philosophy?
3  How are the major total quality philosophies alike?  Different?
4  How do leadership and management differ?
5  How do leaders lead?

## 27.1 INTRODUCTION

At this point, our readers may benefit by reviewing Chapters 1, 2, and 3.  In Chapter 1 we provided a technical introduction to quality and listed a number of definitions of quality as well as quality control related concepts (e.g., total quality control, TQC, total quality management, TQM, and so on).  In Chapter 2, we listed a number of quality masters and summarized their major contributions.  In Chapter 3, we introduced our readers to a global view of quality systems and to the critical point that *quality is indigenous to organizations, products, and processes.*

*An indigenous view of quality is somewhat difficult to grasp when we are bombarded with jargon-like terms such as "quality program," "quality management," "quality culture," and even TQC and TQM.*  At best, these terms represent complex, abstract concepts which are not clearly defined (e.g., their definitions contain other abstract terms); at worse, they are completely misleading.  Taken literally, these terms suggest that quality is a separable part of an organization, a product, or a process.  This separability concept can be shown to be false by observation (e.g., *we see many world-class organizations in terms of the customer satisfaction that they generate, both for-profit and not-for-profit, who do not have formal quality "programs" as such*).  Their internal culture is such that they go about organization, product, and process improvement with a vision of higher customer satisfaction without a quality "department."  *Their leaders are customer, organization, product, and process focused (e.g., focused on an outstanding business), not quality program focused.*

On the other hand, over the last 75 years or so, there have been a number of specialists and managers that have lead people to believe that quality can be specialized, separated, or isolated from the organization, products, and processes, and that a small group of individuals can be somehow accountable or responsible for quality.  Many of these groups now are essentially attempting to actively reverse this thinking (and practice) through a quality re-indigenization process. Some will be successful and some will not; currently the rate of successful transformation to the TQM paradigm (see Chapter 3) is about 33 percent, Brown et al. [1].

*The re-indigenization of quality is a very extensive socio-technical undertaking for organizations where quality has been "specialized" out of, or never recognized as an integral part of, all the everyday business activities. TQM in the United States and TQC in Japan* (see Chapter 1 for explicit definitions) *are "labels" applied to coordinated efforts to address re-indigenization of quality.* In order to conform to the contemporary literature, we will use the TQM label in our discussions at times, but with some reservation that the label has been overworked by proponents, and when literally taken is misleading, as our readers will quickly pick up as we discuss the bottlenecks in quality transformations. We prefer, and use, the term quality transformation, QT, since it is more descriptive of what we are doing, moving to ever higher levels of customer satisfaction and business success in the long run.

Our discussions in Chapters 27, 28, and 29 will include contemporary TQM concepts and strategies, but will also probe well beyond TQM as it is practiced today. We will introduce relevant concepts and strategies that affect the entire organization. We will focus on capturing the true nature of indigenous quality—*quality is a critical and integrated part of the business, not something apart from the business.*

## 27.2  TOTAL QUALITY OVERVIEW

*The major functional components of a typical TQM structure* are depicted in Figure 27.1. The three components are **(1)** *a customer focus*, **(2)** *continuous improvement*, *and* **(3)** *total involvement.* We see consideration for both external and internal customer satisfaction involved in the customer focus. Continuous improvement is

**FIGURE 27.1**    TQM functional components.

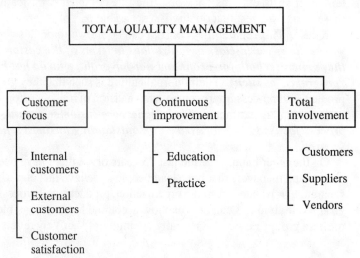

addressed through constant education and practice in the workplace. The theme of many TQM initiatives is "total involvement," where customers, suppliers, and vendors become "partners" in relevant product-process issues. Employee empowerment is a feature of all three of these TQM functions.

The practice of TQM has evolved from the works of the quality masters such as Deming [2], Ishikawa [3], Juran [4, 5], Crosby [6, 7], and others, even though they do not necessarily condone the term TQM. For example, Deming and Juran, in their work, are careful to distance themselves from the TQM label; e.g., Juran [8] maintains that TQM, as such, lacks specific definition with respect to implementation. TQM practice has also borrowed from modern management principles such as "participative management," discussed later in this chapter. *Sound TQM practice includes employee empowerment, teamwork, and quality planning coupled with problem-solving tools and techniques, performance measurement, and systematic analyses of both products and processes. The purpose of TQM is to develop a "workplace" environment (comprising the organization and its suppliers and vendors) that encourages and challenges people to learn, cooperate, and perform to their full potential in pursuit of external and internal customer satisfaction.*

## Total Quality Practice in Electronics Manufacturing — Work-Cell Case [SS]

An engineering design and manufacturing company had a microcircuit facility for producing thick- and thin-film hybrids and complex microwave assemblies. One of the primary microwave products was a log amplifier. The work cell members were trained in using a structured teamwork approach. The members of the work cell were challenged to increase their production rate and were empowered (within boundaries) to make changes as necessary for continuous improvement.

Within four months, the production rate had increased by over 380 percent. The almost-immediate improvements were achieved by corrective action and enhancement initiatives defined entirely by the team. For example, the team requested that the supplier of a particularly troublesome part be brought to the microcircuit facility and shown the problems the assemblers were experiencing. This initiative resulted in the manufacturer making a very low-cost change to the part. The result was a significant reduction in assembly and rework time. When asked why they had not identified this problem earlier, the team members said it had been mentioned to a midlevel supervisor who told them that he wa engineering had considered the problem and that the current design was th possible. The seven fundamental tools (Chapter 9) were used for problem fication and focus.

In addition to the throughput increase, labor requirements decreased cent immediately after the team was formed and empowered. Much of tion resulted from removing mandatory in-process inspection steps. T wanted and assumed more responsibility for their work. The team a 67 percent reduction in hybrid cycle time simply by laying out a d flow, identifying queue times, and reducing them as much as pos

The above accomplishments directly contributed to higher profits, but if you asked the team members they would say that the best result was an improvement in their quality of life. They felt better about themselves and their jobs. The net result was that both the individual and the organization saw a significant increase in benefits.

## 27.3  TOTAL QUALITY PHILOSOPHIES

*Ishikawa states that "TQC is a thought revolution in management"* [9].  Deming clearly points out the need for a transformation in Western management practices with regard to both quality and productivity [10]. *Deming states that "the transformation can only be accomplished by man, not by hardware (computers, gadgets, automation, new machinery)."*  The Ishikawa and Deming positions represent a major shift in both manufacturing and service related organizational thinking.  They both point to the critical role of people, with regard to both leadership and creativity, in reshaping our organizations, e.g., moving to the TQM paradigm described in Chapter 3.

Over the past few years, we have seen the fruits of this thinking emerge in the form of "total quality" philosophies.  Table 27.1 summarizes the broad total quality philosophies put forth by four of the quality experts.  These philosophical declarations are quite similar in some respects and quite different in other respects.  Our purpose here is to glean both similarities and differences as well as to expand on the common elements in order to discover the fundamental characteristics of effective quality philosophies.

*Quality philosophies*, as shown in Table 27.1, *can be divided into two groups on the basis of their orientation*: (1) *people-focused or* (2) *production-focused.*  The Deming and Ishikawa quality philosophies [11] tend to be more people-focused, while the Juran and Crosby philosophies [12,13] tend to be more production-focused.  Regardless of orientation, customer satisfaction, business success, and societal benefits are the ends sought.

### used Philosophies

*People-focused quality philosophies tend to be very broad in scope and less well-defined and sequenced than their production-focused counterparts.*  The Deming and Ishikawa theories are excellent examples of people-focused philosophies. Observation shows that each point in these belief systems represents a subphilosophy in itself.  The points are not generally sequential and do not form a clear step-by-step recipe for success.

The Ishikawa philosophy lays out six broad rules, frequently stated as "do–don't" chotomies (see Table 27.1).  The first two points provide overall guidance and di- tion.  The third, "The next process is your customer," speaks to the entire kforce in order to develop and establish a single organizational purpose and eness that everyone must do their part in creating a quality product.  The fourth stresses the use of facts and data as evidence and justification for quality im-

TABLE 27.1 WIDELY USED AND PUBLICIZED TOTAL QUALITY PHILOSOPHIES

| Deming | Ishikawa | Crosby | Juran |
|---|---|---|---|
| 1 Create and publish to all employees a statement of the aims and purposes of the company or other organization. The management must demonstrate constantly their commitment to this statement. | 1 Quality first — not short-term profits first. | Four quality absolutes | 1 Quality planning: Determine who the customers are. Determine the needs of the customers. Develop product features that respond to customer's needs. Develop processes able to produce the product features. Transfer the plans to the operating forces. |
| 2 Learn the new philosophy, top management and everybody. | 2 Consumer orientation — not producer orientation (think from the standpoint of the other party). | 1 Quality is defined as conformance to requirements, not goodness or elegance. | 2 Quality control: Evaluate actual product performance. Compare actual performance to product goals. Act on the differences. |
| 3 Understand the purpose of inspection, for improvement of processes and reduction of cost. | 3 The next process is your customer — breaking down the barrier of sectionalism. | 2 The system for causing quality is prevention, not appraisal. | 3 Quality improvement: Establish the infrastructure. Identify the improvement projects. Establish project teams. Provide the teams with resources; training, and motivation to Diagnose the causes. Stimulate remedies. Establish controls to hold the gains. |
| 4 End the practice of awarding business on the basis of price tag alone. | 4 Using facts and data to make presentations — utilization of statistical methods. | 3 The performance standard must be zero defects, not "that's close enough." | |
| 5 Improve constantly and forever the system of production and service. | 5 Respect for humanity as a management philosophy — full participatory management. | 4 The measurement of quality is the price of nonconformance, not indexes. | |
| 6 Institute training. | 6 Cross-functional management (by divisions and functions). | Fourteen steps | |
| 7 Teach and institute leadership. | | 1 Management commitment. | |
| 8 Drive out fear. Create trust. Create a climate for innovation. | | 2 Quality improvement team. 3 Quality measurement. 4 Cost of quality evaluation. | |
| 9 Optimize toward the aims and purposes of the company the efforts of teams, groups, staff areas. | | 5 Quality awareness. 6 Corrective action. 7 Ad hoc committee for the zero defects program. | |
| 10 Eliminate exhortations for the workforce. | | 8 Supervisor training. 9 Zero defects day. 10 Goal setting. | |
| 11a Eliminate numerical quotas for production. Instead, learn and institute methods for improvement. | | 11 Error-cause removal. 12 Recognition. 13 Quality councils. 14 Do it over again. | |
| 11b Eliminate Management by Objective. Instead, learn the capabilities of processes, and how to improve them. | | | |
| 12 Remove barriers that rob people of pride of workmanship. | | | |
| 13 Encourage education and self-improvement for everyone. | | | |
| 14 Take action to accomplish the transformation. | | | |

773

provement. The fifth point—the broadest—addresses the elements of leadership and creativity in order to elicit full utilization of human resources. The sixth point stresses the synergistic use of functional specialization and communication to impact quality improvement.

The Deming philosophy is squarely directed at top management and many traditional beliefs, attitudes, and practices thereof [14]. Its 14 points specifically address fundamental changes in traditional business practices in the United States. Specifically, points 3, 4, 8, 10, 11, and 12 challenge a variety of widely held business philosophies and practices. The Deming philosophy represents the most radical of the four outlined in Table 27.1. Roughly half of Deming's points pose significant challenges to traditional management practices.

## Production-Focused Philosophies

*The production-focused philosophies stress quality improvement projects and measurement of progress,* with the expectation of relatively short-term (one to three years), measurable impacts. *These philosophies tend to be fairly well defined and systematic with regard to their integration into the workplace.*

The Crosby philosophy starts with the four "absolutes," quality beliefs which constitute a very businesslike, production centered approach to quality definition and measurement. The 14 steps provide an activity sequence, centered on a ZD performance standard, through which the philosophy may be implemented. This very "shop-floor" centered methodology is capable of producing and tracking (through the cost of nonconformance) quality improvement in a relatively short time.

The Juran philosophy, as expressed through the Juran Trilogy, is very business oriented. It is the quality counterpart to financial planning and control systems found in most organizations. Juran compares the quality planning phase to the financial budgeting and business planning functions; the quality control and improvement phases to the cost reduction and profit improvement functions.

Juran views quality in a process sense, especially in the quality planning phase. Each process has an input and an output. This approach encourages a systematic and thorough examination of customer needs, product features, and production processes (see Chapters 2 and 11). The quality control phase approaches quality in a very classical sense. Here, product-process performance measures of quality characteristics are compared to a standard. Then, action is taken to deal with any differences. Hence, accurate and precise measurement is necessary. The quality improvement phase is viewed on a project basis and carried out by teams.

## People-Oriented versus Production-Oriented
## Quality Philosophies

Deming's system is more detailed than Ishikawa's. However, in both cases, the philosophical points do not provide a stepwise sequence or plan as is the case in the Crosby and Juran philosophies. Deming and Ishikawa demand radical new mind-sets, whereas the Juran and Crosby approaches are more compatible with traditional business practices.

*Many companies develop contingency-based philosophies that borrow heavily from both people- and production-focused quality philosophies.* Regardless of the primary focus, people and their ability to define and create quality in products and processes is the major issue in quality transformation.

Effective quality philosophies encourage leadership vision regarding purpose. In most cases, a "substantial transformation" (in Deming's words) or "revolution" (in Ishikawa's words) is necessary to release both the leadership and the creative potential of people in the workplace. The prerequisite for quality transformation is the definition and delineation of purpose, which is a function of leadership vision. Belief in and commitment to purpose must then follow. Beliefs must be visibly transformed to actions. This transformation depends on two elements: (1) a passion for creating the quality experience for both external and internal customers and (2) the dogged persistence and patience necessary to make the quality philosophy work.

Peters states that quality efforts fail for one of the following reasons [15]: (1) they have system without passion (for quality) or (2) they have passion without a system. Peters discounts the difference between the philosophies of the leading quality masters and suggests that "religious" implementation of a quality system is more important than which system is ultimately chosen.

*Sound quality philosophies contain a number of common focal points*:

1 *Definition and delineation of purpose* (in the context of the organization, e.g., a quality-business–based purpose).
2 *Commitment* (to a common purpose).
3 *Leadership* (through words, action, and role models).
4 *Employee empowerment* (and creative license).
5 *Organizationwide teamwork and communications*.
6 *Education and training* (to prepare people to lead and create).
7 *Customer and supplier relationships* (partnerships to define, produce, and deliver customer satisfaction).
8 *Metrics and measurement* (to gauge progress).
9 *Recognition* (for both efforts and results in the creation of quality).

## 27.4 LEADERSHIP

Leadership or, more bluntly stated, a lack of leadership is usually the biggest bottleneck in quality transformation. We see the word highlighted in most quality philolophies (e.g., the Deming points) but we see very little, other than check lists to help us understand the concept in quality related writings. Hence we will focus on leadership in this section.

*Reitz defines leadership as the process of influencing a group toward the achievement of a goal (purpose)* [16]. This basic definition is subject to considerable expansion and elaboration by various leadership theories and approaches.

Whatever approach is used, virtually all agree that leadership does not occur in isolation. Chemers states that leadership is a social phenomenon, not divorced from a broader situational context [17]. For the most part, a given group of followers will reflect the social, economic, and cultural characteristics of the overall society from which they spring.

In thinking about leadership, the image that first comes to mind is that of an elected politician, official, executive, or someone recognized as a "boss," one who possesses authority granted in some way by an organization or through some recognized process (e.g., a formal or legal confirmation). In actuality, leadership may be defined in many cases as a process, authority, and influence which can exist without formal or legal confirmation, according to Shannon [18].

Some leadership studies have addressed the phenomena of "emergent leaders," or the rise of informal (although effective and influential) leaders in casual or largely unstructured situations. Since a leader is anyone attempting to influence the behavior of one or more others in order to attain some goal, it is very common for subordinates to act as leaders in their dealings with either superiors or peers on occasion (e.g., an employee on the shop floor may lead a supervisor or manager to change a process or machine to improve a process and product).

***Leadership must be distinguished from management.*** It is very typical, even in the literature on the subject, to equate one with the other. For example it is not unusual for a discussion to start out using the term "leader" and replace it with "manager" as the discourse proceeds. In noting the differences between the concepts, "manager" usually carries the connotation of legalistic or formal legitimacy in some type of organization. "Leader," on the other hand, does not always conform to a formal or legalistic context. The two terms are neither mutually inclusive nor exclusive. Table 27.2 presents an interesting comparison and contrast of leaders and managers (see Johnson [19]). In summary, leaders begin with vision, while managers begin with orders.

Good managers may be poor leaders and outstanding leaders may be poor managers. In general, ***leadership deals with vision and the subsequent defining and delineating of purpose as well as cultivating belief in purpose and motivating people to act accordingly. Management entails providing, metering, and monitoring the resources necessary to sustain the action sparked by leadership.*** These two coupled functions work together in synergistic harmony in effective and efficient organizations. It is very important to keep in mind that although the above distinction may be very fine, and fuzzy at times, the processes and mechanisms involved in each function are very different.

The manager exercises power and influence based on his or her position in the organization—power to give rewards (e.g., promotions and pay raises) and power to punish (e.g., firing and demotion). A leader, on the other hand, may have no formal power base at all, but nevertheless exercises considerable influence over some constituency who perceive the leader as capable of facilitating the attainment of some goal (purpose). The power base of such a leader is much harder to qualify and quantify and often involves the perceived personal characteristics and abilities of that leader.

TABLE 27.2   LEADERSHIP-MANAGEMENT COMPARISON

| Leader | Manager |
|---|---|
| Leaders gain power through their actions and personal relations. | Managers have positional power on which to rely. |
| Leaders are found throughout an organization. | Managers are found in the organization's higher echelons. |
| Leaders have followers who desire to be on the team. | Managers have subordinates who have been assigned to them. |
| Leaders depend on people for success. | Managers depend on the system for success. |
| Leaders provide vision in terms of, "The real benefit to you is...." | Managers use the "This is your job ..." approach. |
| Leaders have self-conceived goals to better the organization. | Managers attempt to meet the goals provided by the organization. |
| Leaders strive to change the organization to best meet needs as they perceive them. | Managers work to maintain the organization's status quo. |
| Leaders often view rules and procedures as bureaucratic red tape. | Managers view rules and procedures as necessary controls to provide order. |
| Leaders work for results. | Managers follow orders. |
| Leaders work through their people. | Managers work with charts and computer printouts. |

*Source:* Reproduced, with permission, from R. S. Johnson, *TQM—Leadership for the Quality Transformation,* Milwaukee, WI: Quality Press, American Society for Quality Control, page 60, 1993.

### Emergent Leadership

The emergent leader with little or no formal or legalistic power should never be underestimated. It is a fact that emergent leaders, as such, have had a significant impact on world history. A contemporary example is Lech Walesa, who rose from obscurity as an electrician in the Lenin shipyard in Gdansk, Poland, to lead one of the twentieth century's most crucial political-reform movements. Repercussions of his leadership contributed to the changing of the political face of eastern Europe. Walesa's fortunes in a decade's time ran the gamut from political prisoner to president of a democratic Poland.

### Leadership Theories

The history of approaches to and thinking on leadership is generally regarded as having moved through several phases [20, 21, 22]. Dividing the history of the study of leadership into distinct periods may be misleading in that it implies that the approach of a given period entirely supersedes that of the preceding period. In general, earlier approaches were found not to explain leadership in its entirety, and while superseded to a degree, they are not, and ought not to be, entirely discarded.

Broadly speaking, ***there are three historical phases of leadership theory***:

1 ***Trait period.*** The trait period dates back to early human civilization, even though it was not formally recognized and investigated empirically until early in this century. It remained strong until the 1940s.

**2** *Behavioral period.* The behavioral period dates roughly from the late 1940s into the 1960s.

**3** *Situational-contingency period.* The situational-contingency period dates from the 1960s to the present.

We must remember that *although the trait and behavioral theories have given way to the situational-contingency theory, they remain influential to some degree today.*

**Leader-Trait Theories**    *The trait approach emphasizes the examination of leader characteristics in an attempt to identify a set of universal characteristics which allow leaders to be effective in all situations* [23]. Circumstances or characteristics which traditionally conferred leadership status in ancient societies included royal birth, ability to withstand some physical test or initiation rite, military prowess, or religious talent.

*The trait theories of leadership generally assume that leadership is a personal trait, like intelligence, physical features, and so on, which some have and others do not.* For much of the twentieth century, research was directed toward attempting to identify traits or characteristics that leaders have in common, often using written examinations of some kind [24]. A variety of mental tests were devised to ostensibly distinguish those individuals with leadership potential. The implication was that such traits in and of themselves are indicators of leadership ability or success in leadership endeavors. Hence, using this school of thought, organizations need only seek out individuals with those traits and characteristics and acquire their services. The traits themselves might be physical features, personality characteristics, or a combination thereof; in the past, six traits of leaders mentioned most often include [25, 26]:

**1** Leaders are taller than average (more physically imposing).

**2** Leaders are more physically attractive or are better-looking.

**3** Leaders are more intelligent.

**4** Leaders are more self-confident.

**5** Leaders are more outgoing, extroverted.

**6** Leaders possess superior health, vitality, and endurance.

Modern political candidates continue to be very sensitive to public perceptions of physical appearance and image. Physical appearance continues to be highly regarded in many, if not most, organizational contexts. However, reviews of the research have failed to demonstrate a consistent and definite relationship between leadership ability and physical traits, personality characteristics, or any combinations of the two [27].

The trait approach to leadership may fail to distinguish between traits of those in leadership positions versus traits of truly effective leaders. Reitz points out that having a certain characteristic, such as extroversion, may make an individual's election to certain positions more likely, although it has little or no connection to leadership ability or effectiveness [28]. Also, traits which are desirable for leaders in some circumstances may not be so in other circumstances.

While leadership approaches based on individual leader traits alone have faded, there remains an often powerful popular focus on the personality characteristics of leaders. As a recent example, Torrey, in an editorial entitled "What Defines a Quality Leader?" discussed and summarized a consultant's research report which provided a list of common attributes of successful total quality leaders [29] including the following:

**1** Are clearly focused on goals from an early age.
**2** Exhibit strong work ethic from an early age.
**3** Are strong-willed.
**4** Have a strong customer orientation.
**5** Shun bureaucracy.
**6** Handle change and uncertainty well.

The discussion provided by Torrey is superficial and makes no mention of how such attributes were assessed or scored (most being considerably subjective) and makes absolutely no concession to differing circumstances or constituencies. Torrey strongly implies that the listed qualities transcend circumstance and situation, a position largely discredited by research. In fact, Suriesheim et al. state that the search for universal personal characteristics of effective leaders is doomed to failure [30]. Nevertheless, *there remains a strong popular societal and organizational mythology around leadership traits.*

**Behavioral Theories of Leadership**    *The behavioral approach turned away from leader traits and instead focused on the nature of the behavior exhibited by an individual acting as a leader.* In this approach, leadership is defined as the conduct of an individual when directing the activities of a group toward a shared goal (Adams and Yoder [31]). Studies in the late 1930s by leader behaviorists such as Kurt Lewin defined styles of leadership and investigated the effects of each on group performance and satisfaction. Since then, *three classical behavioral leadership styles have been identified and studied* [32, 33].

**1** *Authoritarian.* Authoritarian leaders make all or almost all decisions regarding group activities such as policy determination, job techniques, planning, and job assignments.
**2** *Democratic.* Democratic leaders delegate some decision making to the group, may suggest choices, provide information for the group to make decisions, and provide performance evaluation against objective standards.
**3** *Laissez-faire.* The laissez-faire leader in effect abdicates his or her role; work organization, vision, policy making, and planning are left completely to individuals in the group.

These behavior types may be said to form a spectrum. Research has reported that, in most instances, group members prefer the democratic style of leader behavior, to one degree or another. However, the particulars of the situation and circumstances influence both group performance and satisfaction. In some instances, group members may prefer to leave all of the decision making to the leader.

In this case, authoritarian leadership behavior may exact greater performance from groups, particularly in terms of simple tasks and with unskilled group members [34].

*The behavioral approach identifies two categories of leadership conduct:* **(1)** *showing concern for "task accomplishment" and* **(2)** *showing consideration for "subordinate satisfaction."* Leaders of the first category assign tasks, emphasize deadlines, expect subordinates to follow rules, and stress competition. Concern for the accomplishment of a task prompts a leader to structure the work for subordinates; hence the term "initiating structure" is commonly used for this category. Considerate leaders, the second category, show concern for workers, are open to their input and suggestions, and act as though the group members are their peers. *Alternative terms for these two behaviors are "task-oriented" and "employee-oriented,"* first defined and described by Katz et al. based on observations in the insurance industry [35].

*One shortcoming of the leadership-behavior approach is its "one-way street," basic model, with the leader doing the influencing and the subordinates responding to that influence.* As Reitz points out, in the behavioral approach there is no questioning of how the leader arrives at his or her style or of what situations or circumstances (e.g., the behavior of the subordinates) influence or shape his or her behavior [36].

**Situational Theories of Leadership** *The situational, or contingency, approach examines the interrelationships among leaders and subordinates in various situations.* The main theme of this approach is that different situations and circumstances may require different leader behavior and that an effective leader will need to change behavior responsively. *Here, the leadership process is viewed as dynamic, with the nature of the subordinates (and other factors as well) playing a role in determining leader behavior.* It has been stated that the trait and behavioral leader schools of thought have regarded subordinates as passive elements. This statement is not totally true. However, what the subordinates do, think, and feel in any depth has been generally excluded from consideration.

In actuality, subordinates are only one of the forces external to the leader that act to create or influence the situation, as Reitz states in his definition of this theory. *Reitz lists seven potential sources of influence on leadership behavior* [37]:

1 *Subordinates.* The subordinate effect is the most obvious, perhaps the most discussed and perhaps still the most elusive effect in terms of understanding its impact on leadership. Differences in background, age, gender, education, skill, and maturity all may influence the leader-subordinate relationship. The degree to which a leader delegates responsibility may have much to do with his or her perception of subordinates.

2 *The task.* The nature of the task, whether it is tightly or loosely structured, simple or complex, and so on, may have much to do with how a leader will behave. Leaders who are very good at leading or directing some types of tasks or activities may be very poor at directing others.

3  *Organizational policy and climate.* Organizational climate, as determined by both written rules and de facto practices, may determine, in large part, the behavior that leaders and subordinates display towards each other. For example, traditional organizations with roots in the industrial revolution may enforce a strict distinction between hourly workers and management, while more-contemporary organizations dispense with many of the symbols and "perks" of executive office and privilege.

4  *Superiors.* In most cases, the leader is in fact a subordinate. Superiors may influence a leader's behavior through the same mechanisms that the leader uses to influence his or her subordinates. For example, the subordinate leader may consciously or unconsciously imitate the leadership style of his or her superior. Indeed, the leadership training to which a superior exposes his or her managers may greatly influence their leadership styles and behavior in the future.

5  *Peers.* Usually, the leader has similarly situated colleagues and, most likely, competitors among his or her peers for advancement and promotion possibilities. Some organizations encourage a more cooperative environment and long evaluation periods among peers. Others thrive on a high degree of "cut-throat" competition among rising managers with short-term evaluations and fast-track promotion policies. The leader, in addition to the influence of superiors, must also deal with peer pressures.

6  *The leader's own characteristics.* The sum of the leader's upbringing, attitudes, basic perceptions about people and work, and self-image may play a major role in determining leader style. The leadership styles mentioned earlier are the net result of genetics and experiences.

7  *Subordinates' responses.* In addition to the preexistent characteristics of the subordinates, how the subordinates respond to the leader and his or her style will generally influence the leader to some degree, in a process which can be termed "reciprocal causation." A leader will often do "what seems to work" in getting the appropriate responses from subordinates, be it showing consideration, initiating structure, or both. Given that subordinates are unlikely to be uniform in maturity, skills, and attitudes, it is not unusual for leaders to show at least some degree of adjustment in their style according to the needs of the work situation and follower (e.g., providing closer supervision for the new, inexperienced, or marginal individual and allowing more latitude for the experienced, or "trusted" ones).

**Hybrid Theories**  *In some cases we see leadership discussed in a context that utilizes two, or all three, of the previously discussed theories.* We will term these combined approaches "hybrid" theories. For example, military leadership approaches tend to be hybrids to one degree or another. (We should note, however, that the goals, objectives, and task content of military organizations tend to differ from those of our civilian production organizations. Our focus in this text is clearly not on military leadership and we are not advocating wholesale use of the military leadership principles. But study of this approach does provide insight for our brief treatment of leadership in general.)

The armed services have focused on the leadership issue for many years. There has also been a good deal of evolution in military leadership thinking over the years, which effectively mirrors the progress of leadership research findings. The U. S. Army has published a list of 19 important leadership characteristics identified in military studies [38]. We would agree with the Army classificaition of the list as a "trait list," on the surface. However, careful examination reveals that the 19 items listed and the manner in which they are dealt with take on both behavior and contingency theory qualities (see Table 27.3). We will henceforth refer to these developmental leadership traits as the "Army points of leadership," in accordance with their use in study, training, and practice.

The Army has also published 11 principles of leadership (shown in Table 27.4) [39]. It is interesting to compare both the 19 points and the 11 principles to the quality philosophies of Deming and Ishikawa. We see the concepts of responsibility, decisiveness, role models, the welfare of the group, training, teamwork, and individual capabilities in each.

## Leader Power

Theories of leadership and leadership style only partially explain how a leader leads. The types of power that a leader can wield, what roles the leader may play, and how effectiveness of a leader can be described are also important and pertinent factors.

*Much of leadership has to do with the exercise of power and authority.* Power in a behavioral sense can mean a number of different things. *Shannon* [40] *and Adams and Yoder* [41] *identify different types of power*:

1 *Reciprocal.* Reciprocal power results from an exchange relationship, a *quid pro quo* (one thing in return for another) situation where mutual interdependency and interest exist.
2 *Expert.* Power and influence by virtue of knowledge or ability that others need is termed expert power.
3 *Legitimate.* Legitimate power comes from formal authority or a recognized position or standing in an organization.
4 *Reward.* Reward power emanates from the control over the dispensing of some sort of recompense or recognition.
5 *Penalty or coercive.* Penalty or coercive power emanates from the control over the dispensing of punishments and/or the ability to force others to one's will.
6 *Referent.* Referent power arises from the leader's possession of attributes which followers are strongly attracted to or identify with.

A leader may possess several kinds of power, and power may be wielded by agents or individuals that are not necessarily leaders. For example, a police officer exercises several powers:

1 Legitimate power as a designated agent of a governmental body charged with law enforcement.
2 Penalty-coercive power through the power to write tickets, make arrests, and so on.
3 Some degree of expert power in knowledge of the law and of handling emergency situations.

TABLE 27.3    U.S. ARMY LEADERSHIP POINTS

**1** "Integrity" is utter sincerity, honesty, and candor.  A leader of integrity strictly adheres to his code of professional beliefs and values.

**2** "Maturity" refers to the sense of responsibility a person has developed.  A mature leader does not make impulsive decisions based on childlike emotional desires or feelings.

**3** "Will" is the perseverance to accomplish a goal, regardless of seemingly insurmountable obstacles.

**4.** "Self-discipline" is forcing yourself to do your duty — what you ought to do — regardless of how tired or unwilling you may be.

**5** "Flexibility" is the capability to make timely and appropriate changes in thinking, plans, or methods when you see, or when others convince you, that there is a better way.

**6** "Confidence" is the assurance that you and your soldiers will be successful in whatever you do.  Confidence shows in your bearing, in the look in your eye, in the tone of your voice, in your enthusiasm, in what you say, and in what you do.

**7** "Endurance" includes mental, spiritual, and physical stamina.  It is the ability to keep going mentally, spiritually, and physically over prolonged periods of stress.

**8** "Decisiveness" is the ability to use sound judgment and make a good decision at the right time.  Timing is a critically important element of decisiveness.

**9** "Coolness under stress" is a confident calmness in looks and behavior.

**10** "Initiative" is the ability to take actions that you believe will accomplish unit goals without waiting for orders or supervision.

**11** "Justice" is the fair treatment of all people regardless of race, religion, color, sex, age, or national origin.  As a just leader, you must give rewards according to merit and performance.

**12** "Self-improvement" is shown by reading, studying, seeking challenging assignments, and working to strengthen beliefs, values, ethics, character, knowledge, and skills.

**13** "Assertiveness" is taking charge when necessary, making your ideas known, helping to define the problem, and getting others to do the right thing to solve the problem.

**14** "Empathy" or "compassion" is being sensitive to the feelings, values, interests, and well-being of others.

**15** A "sense of humor" is shown by not taking yourself too seriously and by contributing to the laughter and morale of the people around you.  A sense of humor eases tension; combats fear and depression; and enhances communication, trust, and respect.

**16** "Creativity" is demonstrated by thinking of new and better goals, ideas, programs, and solutions to problems.

**17** "Bearing" is shown by posture, overall appearance, and manner of physical movement.

**18** "Humility" is admitting weaknesses or imperfections in your character, knowledge, and skills.  It is acknowledging mistakes and taking the appropriate action to correct those mistakes.

**19** "Tact" is a sensitive perception of people — their values, their feelings, and their views — which allows positive interaction.  We know, however, that a leader's character — that combination of his personality traits — can be the determining force of victory or defeat.

*Source:* Adapted from "Military Leadership," FM-22-100, Washington, DC: Headquarters, U. S. Army, 1983.

## Leader Performance

Effective leaders have risen and fallen over the course of history;  biographical accounts are plentiful.  ***We can develop five general observations regarding high-performance leaders,*** based on biographies.

**TABLE 27.4** U.S. ARMY PRINCIPLES OF LEADERSHIP

1 Know yourself and seek self-improvement.
2 Be technically and tactically proficient.
3 Seek responsibility and take responsibility for your actions.
4 Make sound and timely decisions.
5 Set the example.
6 Know your soldiers and look out for their well-being.
7 Keep your soldiers informed.
8 Develop a sense of responsibility in your subordinates.
9 Ensure that the task is understood, supervised, and accomplished.
10 Train your soldiers as a team.
11 Employ your unit in accordance with its capabilities.

*Source:* Adapted from "Military Leadership," FM-22-100,
Washington, DC: Headquarters, U. S. Army, 1983.

1 *Each is fixed, and can fix followers, on particular visions and goals (purpose), i.e., followers follow-pursue the vision rather than the leader, the vision "out-lives" and transcends the leader.*

2 *Each leader possesses an intricate and sound understanding of human nature in general and of the needs, wants, and desires of his or her followers in particular.* The result is a charismatic leader-follower relationship.

3 *Each systematically controls, manipulates, or eliminates constraints on leadership through various means* — whether through diplomatic, political, ruthless, or benevolent means, or a combination thereof. Each concentrates power in his or her position.

4 *Each skillfully assembles talented and dedicated subordinates* to assist in leading and to "root out" the disloyal.

5 *Each rises to a challenge or circumstance which requires a tremendous leadership energy level.*

Judging by the contemporary criteria, a good leader is capable of changing his or her behavior and style to match the situation and circumstances. There is no "one best way" to lead. There are a number of leader abilities that are valued. As pointed out by Shannon, the leader must win the confidence of followers and inspire their enthusiasm and commitment [42]. Consistency, dependability, flexibility, and predictability are also valued in leaders, who must be able to interact and deal effectively with the various constituencies and constraints in performing his or her tasks. This interaction is essential for effective leadership.

*Most accounts of success in quality transformations are traceable to strong and dedicated leadership by pathfinders who guided their organizations from the classical mode of operations to a TQM paradigm. In general, as Deming points out, leadership is the bottleneck in transforming to higher levels of quality* [43].

## 27.5 MANAGEMENT-LEADERSHIP-MOTIVATION MODELS

In this section, we will review TQM-paradigm relevant management-leadership-motivation (MLM) behavior concepts, which date back roughly to the late 1800s. *The urgency of need and goal-objective-target structure at hand in a cooperative effort play a large role in determining the MLM model used in any given appli-*

*cation.* For example, an emergency or time-pressure situation involving life or death (e.g., accidents, hostage negotiations) and one with lower time pressure (e.g., an engineering design or industrial production-assembly task) usually require different leadership and management approaches. As another example, interaction with a group of highly skilled craftspersons and with a group of unskilled, untrained persons also call for different MLMs. "Best," in terms of MLM for performance, is a relative term—relative to organizational goals and objectives, situations, and available resources, as well as to personal needs, beliefs, and attitudes.

*We see two distinct schools of thought regarding MLM practices*: **(1)** *task-focused,* scientific management, *and* **(2)** *people-focused,* participative management (human relations, quality of worklife). The total quality, TQ, movements (e.g., TQM and TQC), due to their pragmatic evolution from within the workplace, manifest characteristics of both, and in general, research findings and observations made by both camps are relevant to quality transformation.

Effective total quality practice deals with both external and internal customer satisfaction simultaneously in a systemwide sense. Both the task-focused and people-focused MLM models have limited their scope, in general, to the confines of the workplace and do not explicitly include the external customers in their treatments.

## Scientific-Management Concept

Credit for the scientific-management concept is usually given to Frederick W. Taylor [44]. *Taylor viewed the workplace as a "system" and formulated four fundamental principles of scientific management* in the late 1800s and early 1900s (see Polk [45]):

1 *Develop a science for each element of work* (to replace the old rule-of-thumb methods).
2 *Select, train, teach, and develop the worker.*
3 *Cooperate with the worker to insure that the work is properly done.*
4 *Divide equally the work and responsibility between management and the workers, each doing what it is best fitted to do.*

Contemporary scientific-management theory developed (broadly) around the Taylor principles, which promote specialization in terms of the design of work and the workplace, as well as in the execution of the work itself. The result of these principles has been narrow job definition justified by scientific time and motion analyses, learning curves, and so forth.

*The specific focus in scientific management is on the task and on the self-interest of the individual in economic terms. This concept stresses efficiency.* The factory piece-rate system (where individuals are paid for the number of pieces produced) was one of the early products of scientific management. Implementation of Taylor's principles have yielded significant results; in some cases, production costs were cut in half and wages were doubled, simultaneously. However, scientific management is generally criticized for disregarding the social needs of people.

Scientific management has grown to include broad-based optimization and operations-research techniques such as linear programming and operations simulation. The strict quantitative focus of scientific management is most productive in situations where social structure is not a major factor. For example, primarily economic or schedule-driven problems can be successfully addressed with scientific-management models and tools. These tools are also useful in highly automated systems where the human element has been reduced on the production floor.

## Human-Relations Concept

*The human-relations concept focuses on interpersonal relations within an organization,* and is almost diametrically opposite that of scientific management. Elton Mayo is credited as the founder of the human relations school of thought [46], which grew out of the research program at the Western Electric Company in Hawthorne, Illinois, in the 1920s.

The original objectives of the Hawthorne research were to determine the most productive levels of physical work-environment variables (e.g., illumination) in the workplace in order to increase production. A number of experiments involving physical working conditions were developed, and group production measured. It was found that significant production gains were made across the board, not as a result of physical conditions, but because of the increased attention directed to the operators during the course of the experiment. The experimental findings led to the conclusion that personal attention and interpersonal relationships are significant factors in organizational performance. To this day, we sometimes refer to a "Hawthorne effect" when we see improved worker attitudes and production as a result of enhanced interpersonal worker-manager relationships.

## Participative Management

In Germany in the 1920s, Kurt Lewin, trained in psychology, was studying Taylor's work and published a paper dealing with "the humanization of the Taylor system." Lewin's work is described as an extension of Taylor's work that included the human or social side of people [47]. Lewin proposed the combination of scientific thinking and democratic values in the workplaces.

Lewin pointed out that all problems (including technical and economic problems) have social consequences that include human perceptions, self-worth, motivation, and commitment. *Lewin is credited with the development of the fundamental principle of participative management* [48]:

> *People are more likely to modify their behavior when they participate in problem analysis and solution and more likely to carry out decisions they help make (rather than decisions made by someone else).*

## Quality of Worklife

*During the 1950s, Eric Trist and Fred Emery advocated the elimination of authoritarian management behavior from the workplace.* Taylor had essentially attempted to accomplish this to some extent by introducing "specialists" and, for the most part, failed. Trist and Emery took a different approach by experimenting with teams of workers as a means of replacing the classical foreman and authoritarian management configuration.

The idea was to use staff experts to transfer knowledge to the team members and thus equip them to operate effectively and efficiently. *Management was placed in the system as a resource for, rather than directors of, worker teams.* Success was encountered in coal mining (Britain), textiles (India), and automobile production (Sweden) [49]. *This experience with worker teams has led to the "quality of work-life" movement.*

## Contemporary Management-Leadership Models

*The result of the task focus of the scientific-management concept is greater accommodation of the lower human need levels* (physiological and safety-security, refer to Maslow's hierarchy of needs, Figure 4.1) through its economic emphasis. *The human-relations concept, with its focus on interpersonal relationships, tends to accommodate the middle-order needs* (social and, to some degree, esteem). Both schools of thought have relevant points. The major issue here is that the individual, the organization, and society must all be clear-cut "winners" (and perceive themselves as such) in order to provide a stable and prosperous economy and healthy social environment in the long run.

*Lewin* observed that scientific management raised output, cut costs, increased wages, and reduced working hours, but had very limited "life value" (for workers). He *maintained that work should not limit personal potential, but develop it.* By 1920, Lewin was the first to identify and understand what would later be called "job satisfaction" (see Weisbord [50]). *Lewin developed his "force-field analysis," which stated that you move a problem by either increasing the intensity of the driving forces behind it or by reducing the restraining forces that resist the driving forces.* His philosophy suggested that reducing the restraining forces in the workplace was in general better than increasing the driving forces.

Relying primarily on Lewin's work, *Douglas McGregor produced his theory X-theory Y attitude model* in the 1950s [51]. *Later, Chris Argyris produced his A-B behavior pattern model* [52] *to complement McGregor's work.* Table 27.5 describes both of these models. *Theory X takes the view that work is inherently distasteful to people, while theory Y takes the opposing view—that work is as natural as play to people.*

The theory X attitude and pattern A behavior and the theory Y attitude and pattern B behavior form consistent combinations; whereas theory X and pattern B, and theory Y and pattern A are both inconsistent, but still possible. *QT literally demands the theory Y, pattern B combination.* Hence, some people must undergo a certain degree of attitude and behavior change to work in a TQ environment. Some people adjust well, and some never do. A chronic lack of adjustment may eventually lead to employment termination, either voluntary or involuntary.

The 1950s and 1960s proved to be very fruitful for MLM research. A contemporary of McGregor and Argyris, *Frederick Herzberg, developed his motivation-hygiene factors model* based on personal interviews with engineers and accountants [53, 54]. Table 27.6 lists motivation and hygiene factors. In general, *hygiene factors must be in place to effect a productive work environment (to avoid dissatisfaction), i.e., a lack of effective hygiene factors will be de-motivating. But they*

**TABLE 27.5  LEADERSHIP ATTITUDES AND BEHAVIORS SUMMARY**

| Theory X | Theory Y | Pattern A | Pattern B |
|---|---|---|---|
| 1 The typical person inherently dislikes work and will avoid work if possible. | 1 The typical person likes work and sees work as natural as play or rest. | 1 Rigid supervision of work. | 1 Relaxed supervision of work. |
| 2 The typical person must be coerced to and directed in work. | 2 The typical person will exercise self-direction and self-control in work. | 2 Co-workers are not trusted. | 2 Co-workers are trusted. |
| 3 The typical person prefers to avoid responsibility for work. | 3 The typical person will seek and accept responsibility in work. | 3 High structured work environment. | 3 Loosely structured work environment. |
| 4 The typical person is incapable of producing creative solutions to problems in work. | 4 The typical person is capable of producing creative solutions to problems in work. | 4 Little concern for development of coworkers. | 4 High concern for development of coworkers. |

Source: Adapted from D. McGregor, The Human Side of Enterprise, New York: McGraw Hill, 1960, and from C. Argyris, Management and Organizational Development: The Path from XA to YB, New York, McGraw Hill, 1971.

**TABLE 27.6**   MOTIVATION-HYGIENE FACTORS

| Motivation factors (Job related "satisfier" factors) | Hygiene factors (Work-environment related "dissatisfier" factors) |
|---|---|
| Achievement | Policies and administration |
| Recognition | Supervision |
| Work itself | Salary |
| Responsibility | Interpersonal relations |
| Advancement | Working conditions |

*Source:* Adapted from F. Herzberg, M. Mausner, and B. Snyderman, *The Motivation to Work*, New York: Wiley, 1959.

*do not provide significant motivation (satisfaction) in themselves. Herzberg associates the motivation factors* (in the list) *with the power to produce significant motivation in the workplace.*

*Rensis Likert*, in his work at the University of Michigan's Institute for Social Research, *emphasizes the importance of moving from theory X toward theory Y in management structures, expectations, and operations* [55, 56]. *A four-system model or organizational-transformation sequence has been developed* (see Table 27.7) [57, 58]. Likert and his colleagues have developed extensive, detailed question-response instruments to measure organizational position-progress in system transformation.

In this same time period, *Blake and Mouton developed the Managerial Grid Model* [59], *which views leadership style as a function of two major concerns*: **(1)** *concern for people, relationships, and* **(2)** *for production, tasks*. The grid is broken into four quadrants, with five identified positions (shown in Figure 27.2). These five positions are labeled in terms of grid coordinates, reflecting various approaches. It is of interest to note that the Blake and Mouton "9, 9 management position" description is necessary for effective quality transformation.

The trend over the past few years has been toward situational leadership models. *The situational-leadership concept considers the situation and the maturity of the people involved as the determinant of leadership style. Paul Hersey and Kenneth Blanchard have proposed the situational-leadership model* shown in Figure 27.3 [60]. Although developed in a more pragmatic context, this model is compatible with our previous discussions on leadership theories. This model consists of a scheme of four quadrants, labeled "telling," "selling," "participating," and "delegating." These four quadrants are associated with human ability and willingness, presented as "job readiness" and "follower(s) readiness," respectively. It is interesting to note that "ability" and "willingness" as defined in the Hersey and Blanchard model can be directly related to the cognitive and affective domains respectively (Figure 1.5) in learning theory.

Relatively new research in the MLM field has centered on what is termed the "female," or "web," management model, according to Helgesen [61]. *Helgesen describes the web model* through observations of successful female managers in their day-to-day operations. *Here, the typical pyramid or organizational chart concept of an organizational hierarchy is replaced with a set of concentric rings.* The leader is located in the center, as depicted in Figure 27.4; the surrounding circles represent "levels" of personnel within the organization.

## TABLE 27.7  LIKERT'S SYSTEMS 1, 2, 3, AND 4 SUMMARY

### System 1 — An exploitative, authoritarian model (autocratic).

*Description:* System hoards control and direction at the very top of the organization; decisions are made and orders issued from the top. There is some downward communication; a great deal of effort may be expended in an attempt to communicate accurately. These communications are received with hesitancy and suspicion by subordinates. There is very little upward communication and literally no lateral communication. The organization relies on fear, the need for money, and a desire for status or power.

*Result:* Decisions which are made at the top of the organization are based upon partial, often inaccurate, information. Data are often distorted and falsified by subordinates in an attempt at self-protection. There is no built-in effort to develop the natural motivation which involvement produces. Mistrust, hostility, and dissatisfaction are present in the organization. There is little motivation among subordinates to fulfill the organization's goals and only those at the very top of the organization feel the responsibility for the organization's success.

### System 2 — A benevolent authoritarian model (improves somewhat on system 1)

*Description:* Policy is decided at the top, specific implementation decisions may be delegated to somewhat lower levels. Orders are issued, but some opportunity may be provided for individual subordinates to comment on the orders. There is a great deal of downward communication—viewed, ordinarily, with mixed feelings by subordinates. What little upward communication exists is often distorted and filtered. Practically no lateral communication exists. The organization relies more heavily upon the need for money, desire for status, and other allied individual ego motives.

*Result:* Untapped motives exist, as they do in system 1, and, in some measure, cancel out those motivational forces relied on by the organization. Attitudes are sometimes hostile, sometimes favorable, but there is ordinarily a substantial degree of dissatisfaction present in the organization. Managers usually feel responsible for the organization's well-being. Real control usually is less than what is presumed to exist by the management personnel.

### System 3 — A consultative model (improves on system 2)

*Description:* Only the broad policy is determined at the top, and more specific decisions are made at the lower levels. Practically no reliance is placed on fear or coercion as motivational forces. Most major motive sources are inherent in the individual; the need for money, ego motives, the desire for new experiences are more generally in use. Those motivational forces inherent in group processes are used very little.

*Result:* Goals are set or orders issued after discussion with subordinates. Subordinates' attitudes are usually favorable and there is little hostility. Most persons in the organization feel a responsibility for its welfare, and a low level of resistance to the organization's directives is evident. An informal organization may, to some degree, still exist. To the extent it does, this informal organization dilutes the real control available to the formal organization.

### System 4 — Participative group model (the most democratic on the system 1–4 continuum)

*Description:* Decision by consensus is the rule. Decisions are made face-to-face by work groups. The decisions are coordinated by persons called "linking pins" who hold overlapping membership in groups (e.g., a superior in one group is a subordinate in another). The decisions made, or coordinated by each of these groups, are appropriate for that group's hierarchical level. The higher its position in the organization, the more the group concerns itself with organization-wide matters.

*Result:* Decisions are made by group participation; they are fully accepted both overtly and covertly. Information flows freely upward, downward, and laterally, and there exist practically no forces to distort or filter that information. System 4 taps all of the major positive motives. No use is made of fear or coercion, and, as a result, group attitudes are quite favorable.

*Source:* Adapted, with permission, from D. G. Bowers, *Systems of Organization: Management of the Human Resource*, Ann Arbor, MI: University of Michigan Press, 1976 and from R. Likert and J. G. Likert, *New Ways of Managing Conflict*, New York, NY: McGraw-Hill, 1976.

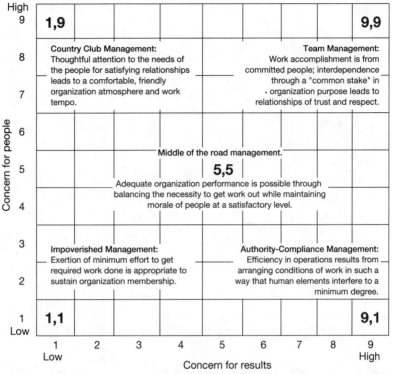

**FIGURE 27.2** Leadership grid model. Reproduced, with permission, from R. R. Blake and A. A. McCanse, *Leadership Dilemmas—Grid Solutions,* Houston, TX: Gulf Publishing Company, page 29, 1991. Copyright 1991 by Scientific Methods, Inc. Reproduced by permission of the owners.

**FIGURE 27.3** Situational-leadership model of style and maturity. Adapted with permission from P. Hersey and K. Blanchard, *Management of Organizational Behavior*, 6th ed., Englewood Cliffs, NJ: Prentice-Hall, page 207, 1993. © 1993 by Center for Leadership Studies, Inc. Escondido, CA. All rights reserved. Reproduced by permission of the owners.

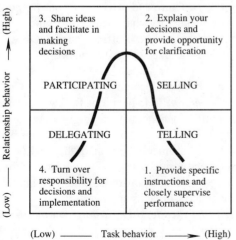

Readiness indices—Match the readiness idex numbers with the quadrant positions for successful combinations of leadership style and readiness requirements.

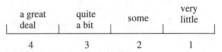

Job readiness—ability—degree of knowledge and skills

Follower's readiness—willingness— degree of confidence and commitment

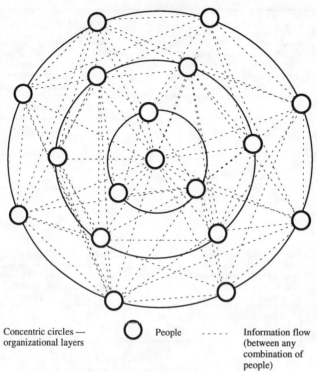

Concentric circles —          People          - - - - -    Information flow
organizational layers                                      (between any
                                                           combination of
                                                           people)

FIGURE 27.4    Web leadership-management model.

*In the web model, emphasis is placed on sharing information, rather than a top-down directional flow of information, as is typical in a classical pyramid or authoritarian model.* Assignments are rotated within rings, and "promotions" essentially occur between rings. Helgesen states that the information sharing is typical of female managers, but not of male managers. We present the web model as an alternative MLM to be assessed on its structure and merits. For our purposes, we want to scrutinize the web model with respect to what it can contribute to QT, rather than enter into an argument of gender-based MLM effectiveness. Observations in the workplace provide ample evidence that both men and women can be effective leaders and managers.

Principle-centered leadership, PCL, has been advocated by Covey [62]. *Covey postulates PCL on the basis of the "whole person."* He positions PCL with respect to preceding MLMs and states that the scientific-management paradigm views people primarily as "stomachs" (economic beings). He states that the human-relations paradigm acknowledges that people are not only stomachs, but also "hearts" (social beings). Here, people are viewed as socioeconomic beings. Furthermore, Covey characterizes the human-resource paradigm as one of fairness, kindness, and efficiency (e.g., people have "minds" in addition to stomachs and hearts).

*PCL is positioned (by Covey) as the fourth leadership paradigm; one of fairness, kindness, efficiency, and effectiveness.* PCL considers and works with the "whole person." Here, Covey characterizes PCL as adding (and developing) the

| Level | Principle | Paradigm | Comments |
|-------|-----------|----------|----------|
| Personal | Trustworthiness | Self | The personal level focuses on individual character and competence. |
| Interpersonal | Trust | People | The interpersonal level focuses on relationships between people and their effectiveness in organizations. |
| Managerial | Empowerment | Style | Managerial style facilitates the empowerment of people. |
| | | Skill | Managerial initiatives encourage learning with respect to team building, delegation, communication, negotiation, self-management, and so forth. |
| Organizational | Alignment | Shared vision and principles | At the organizational level, a common mission (vision) is developed in a cooperative sense. It has three components: strategy, structure, and systems All three are aligned towards the mission. |
| (*a*) Principles | | (*b*) Paradigm | |

**FIGURE 27.5**    Description of Covey's Principle-Centered Leadership concept.

"spiritual" dimension in addition to the economic, social, and psychological dimensions. The spiritual dimension is related to an individual's establishment of "meaning" or a sense of doing something that matters (to themselves and others) that will bring people to their "highest selves." Reviewing the quality system interfaces, Figure 3.2, we can relate this concept to the service interface, specifically at the society element.

*The primary basis for the PCL model is trust,* the four levels and key principles are pictured in Figure 27.5*a* [63]. These four levels center on the trustworthiness of the individual and move out through interpersonal, managerial, and organizational levels. Obviously, without trustworthiness on the part of each individual in the organization, the PCL paradigm and its application will not function. Hence, Covey and his associates focus on this virtue and cultivate and develop it in their training. The "big picture" of the PCL model is shown in Figure 27.5*b*. Here, we see the four levels and the key principles integrated into a socio-logical paradigm divided or stratified by the four levels, one through four, top to bottom.

In our discussion of PCL, we could consider PCL as addressing the selfactualization level of Maslow's hierarchy of needs (see Figure 4.1). Hence, we can clearly see that over the years, *the architects of MLMs have stressed each and every level of Maslow's hierarchy and produced models that progressively moved up the hierarchy.* The initial models started with the basic necessities and then progressed or grew up the hierarchy, as each model typically lacked the ability to completely describe or capture the human needs that were (or could) drive human performance. We also see added complexity in the growth of these MLM models (e.g., the Covey model is extremely complex, driven by all five basic human needs, more or less simultaneously, whereas the scientific management model is relatively simple, as it is driven by the physiological and safety-security needs).

*Accurate and precise measurement of MLM effectiveness and the variables that drive MLM effectiveness is a difficult task.* Difficulties stem from both the dynamic nature of people and the complexity of the phenomena being measured. Covey and his associates, as well as Likert and his associates, produce "instruments" that can be used to measure, or at least provide some indication of the influence of the key elements and variables in their models. Teleometrics offers similar instruments and services [64]. These instruments typically ask the respondent to answer a number of basic questions or ask them to respond to a number of basic statements. Analyses result in summarized response scores and respondent profiles. These responses are then used to mark and monitor progress at critical points in time (e.g., the beginning, during training, after training). By assessing the responses, both on an individual basis and on a group basis, and comparing them to "norms" and other benchmarks or targets, progress is measured over time.

The thinking which is reflected in the workplace MLM models we have described has evolved over the course of about 100 years as researchers have gained access to more information. Empirical observations and interpretations in both laboratory and workplace settings have increased our knowledge and broadened our perspectives of human performance on the job. However, the variety of human needs, knowledge, and abilities coupled with the variety of products, processes, time and cost pressures, and so on, that we see in the field preclude advocacy of one universal and undisputed "best" model.

In summary, leader effectiveness seems to be most often discussed in the literature in terms of what style of leadership is most successful in positively impacting task performance and/or employee satisfaction. The consensus seems to be that, in much of the research, groups in nonstressful situations perform better under more-democratic leadership while groups in stressful situations perform better under more-authoritarian leadership [65]. Many of the studies conducted in this area involve rather stylized, short-term task situations (e.g., Lewin's 1939 study of ten-year-olds performing hobby tasks and subjected to different leadership styles). Many studies are abstractions of real-world situations; real-world situations are bound to be more complex, dynamic, and of a considerably greater duration. As the situational approach advocates, different leadership styles may apply in different situations and with different individuals. The PCL approach goes even further in involving the "whole person." *Hersey and Blanchard* [66] *cite studies that seem to indicate that high-productivity work teams tend to have supervisors who allow subordinates freedom to do the job (consideration aspects), while relating clear objectives and goals (task and initiating-structure aspects).*

## Night Vision System — Over-the-Wall Case [TW]

A marketing representative came up with an idea for a new airborne night-vision system (NVS). The idea was, in fact, a good one. The problem was the schedule to which we were committed, because the difficulty of the task had been grossly underestimated. The goal was to take an existing ground-based system and, in words burned indelibly in my mind, "slap it in a can."

The "can" turned out to be a 360° servodriven turret which had to go into a "too-small" space. Slapping things into the can sounded easy; in practice, it turned out to be an exercise in frustration. The requirements called for 360°-rotation capabilities. A slip-ring design was used. The slip rings had to be custom-made and proved to be a major bottleneck. The servosystem experienced numerous delays for both electrical and mechanical reasons. Also, the mechanical engineers—the makers of the "can"—had a difficult time working with the space constraints and getting the heat out.

However, the flight tests went well, and the customers who had contracted to buy the NVS liked what they saw. After we had made several systems and put them in the field, management canceled the program. Even after a difficult technical success, no one felt good about the project and the final result.

Since the project was considered a minor modification, we had been provided only limited resources. At the design level, there was ample evidence that upper management was not committed to the project. This case illustrates a major disconnect between those who came up with the idea and those who had to make it work. Not only did marketing throw the problem over the wall to us, but our management didn't seem to see or to care what we were doing behind our wall.

---

## REVIEW AND DISCOVERY EXERCISES

### Review

**27.1**  Compare and contrast the Deming and Crosby quality philosophies (Table 27.1).

**27.2**  Is Feigenbaum's definition of TQC (refer back to Chapter 1) compatible with environmental concerns (pollution, scarce resources, etc.)? Explain.

**27.3**  Compare and contrast the Deming quality philosophy with the Likert systems (Tables 27.1 and 27.7).

**27.4**  What does Deming mean by his statement that "The transformation can only be accomplished by man, not by hardware (computers, gadgets, automation, new machinery)?" Is he correct? Explain.

**27.5**  Examine Table 27.1 and develop a set of characteristics common to people- and production-focused TQM philosophies.

**27.6**  Deming's 14 points, with a few exceptions, challenge many long-held practices in U.S. business, yet are well received by businesspeople as long-term goals. How is it that they are not simply dismissed as "good ideas that won't work in our company?"

**27.7**  Explain Peter's statement that quality efforts fail for one of the two following reasons: (1) they have a system without passion or (2) they have passion without a system.

**27.8**    Name three famous emergent leaders who have shaped the course of events in the United States: one political leader, one labor leader, and one business leader. Then briefly explain their contributions.

**27.9**    It is common to see subordinates acting as leaders when attempting to influence both peers and superiors. Provide two examples, one each of peer influence and of superior influence.

**27.10**    Explain why we see so many MLM models described in the literature. Which one is best? Explain.

## Discovery

**27.11**    One classical argument with regard to TQ philosophy is that it takes a special type of worker, the "willing worker," to practice TQ successfully. The argument goes on to state that only a select few individuals in the United States are willing workers; therefore, TQ cannot be implemented in all organizations, given the lack of such employees. These "others" must resort to classical theory X, pattern A leadership styles. Support or refute the argument.

**27.12**    After examining the five basic MLM models (scientific management, human relations, participative management, situational leadership, and principle-centered leadership), explain why they evolved in this precise chronological order. Was it an accident or not?

**27.13**    Some people having field experience with TQ practice praise Crosby's philosophy as being compatible with business; others reject it as too simplistic, stating that a given company has "outgrown" it. Explain why we see such mixed reviews.

**27.14**    Locate a political advertisement or poster with a candidate's picture on it. Examine the portrait for evidence of leadership-trait characteristics. Describe what you see.

**27.15**    Observe a strong leader in action directly or indirectly (through television, a magazine, or a newspaper account). Extract and document as many elements of leadership regarding situational leadership theory as possible relative to
   **a** Subordinates.
   **b** Task.
   **c** Organizational policy and climate.
   **d** Superiors.
   **e** Peers.
   **f** Leader characteristics.
   **g** Subordinate responses.

**27.16**    Discuss the appropriateness of the U.S. Army's 19 points of leadership (Table 27.3) when applied to:
   **a** A manufacturing organization in the United States.
   **b** A service organization in the United States.

**27.17**    Discuss, in detail, the appropriateness of the U.S. Army's 11 principles of leadership (Table 27.4) for nonmilitary leadership.

**27.18**    Assume the objective in an organization is to produce a high-quality product at a competitive cost, on schedule, for a customer. Describe the advantages and disadvantages the following types of leadership models offer:
   **a** Task based.
   **b** Human-relations based.
   **c** Situational-contingency based.

**27.19**    Do you tend to be task, human-relations, or situation based in your attitudes toward leadership? How can you tell?

**27.20** Referring to the night-vision case, describe how leadership and management were exercised. Comment on how leadership and management could have been improved.

**27.21** Select one of the MLM models and concepts (e.g., Taylor's principles, theory X, theory Y, the web model, PCL, and so on); research and discuss its source and substance. Provide your own critique of the model.

## REFERENCES

**1** M. G. Brown, D. E. Hitchcock, and M. L. Willard, *Why TQM Fails and What To Do About It*, New York: Irwin, 1994.

**2** W. E. Deming, *Out of the Crisis,* Cambridge, MA; MIT Center for Advanced Engineering Studies, 1986. (Deming's 14 Points, copyright 1986, revised 1990, by W. Edwards Deming).

**3** K. Ishikawa, *What is Total Quality Control? The Japanese Way,* Englewood Cliffs, NJ: Prentice-Hall, 1985.

**4** J. M. Juran, *Juran on Leadership for Quality,* New York: Free Press, 1989.

**5** J. M. Juran, *Juran on Planning for Quality,* New York: Free Press, 1988.

**6** P. B. Crosby, *Quality is Free,* New York: McGraw-Hill, 1979.

**7** P. B. Crosby, *Quality without Tears,* New York: McGraw Hill, 1984.

**8** J. M. Juran, *Juran on Quality by Design,* New York: Free Press, 1992.

**9** See reference 3.

**10** See reference 2.

**11** See references 2 and 3.

**12** See references 4 and 5.

**13** See references 6 and 7.

**14** See reference 2.

**15** T. Peters, *Thriving on Chaos,* New York: Harper and Row, 1987.

**16** H. J. Reitz, *Behavior in Organizations,* The Irwin Series in Management and the Behavior Sciences, Homewood, IL: Richard D. Irwin, Inc., 1977.

**17** M. H. Chemers, "The Social, Organizational, and Cultural Context of Effective Leadership," in *Leadership: Multidisciplinary Perspectives,* B. Kellerman (ed.), Englewood Cliffs, NJ: Prentice Hall, Inc., 1984.

**18** R. E. Shannon, *Engineering Management,* New York: John Wiley and Sons, 1980.

**19** R. S. Johnson, *TQM—Leadership for the Quality Transformation,* Milwaukee: ASQC Quality Press, 1993.

**20** See references 16 and 17.

**21** C. A. Suriesheim, J. M. Tolliver, and O. C. Behling. "Leadership: Some Organizational and Managerial Implications," in *Perspectives in Leadership Effectiveness,* P. Hersey and J. Stinson (eds.), Center for Leadership Studies, Ohio University Press, 1980.

**22** J. Adams and J. D. Yoder. *Effective Leadership for Women and Men,* Norwood, NJ: Abley Publishing Corporation, 1985.

**23** See reference 21.

**24** See reference 16.

**25** See references 18 and 22.

**26** C. A. Gibb, "Leadership," in *The Handbook of Social Psychology,* 2d ed., Volume 4, G. Lindzey and E. Aronson (eds.), Reading, PA: Addison-Wesley Publishing Company, 1969.

**27** See references 16, 21, 22, and 26.

**28** See reference 16.

**29** E. E. Torrey, "What Defines a Quality Leader?" *Industrial Engineering,* vol. 24, no. 4, April, pg. 4, 1992.

**30** See reference 21.

**31** See reference 22.

**32** See reference 16.

**33** P. Hersey and K. Blanchard, *Management of Organizational Behavior,* 6th ed., Englewood Cliffs, NJ: Prentice-Hall, 1993.

**34** See reference 16.

**35** D. Katz, N. Maccoby, and N. C. Morse. *Productivity, Supervision and Morale in an Office Situation,* Survey Research Center, Institute for Social Research, University of Michigan, Ann Arbor, MI, 1950.

**36** See reference 16.

**37** See reference 16.

**38** "Military Leadership," FM-22-100, Washington, DC: Headquarters, U.S. Army, 1983.

**39** See reference 38.

**40** See reference 18.

**41** See reference 22.

**42** See reference 18.

**43** See reference 2.

**44** F. W. Taylor, *The Principles of Scientific Management,* New York: Harper and Row, 1915.

**45** E. J. Polk, *Methods Analysis and Work Measurement,* New York: McGraw-Hill, 1984.

**46** E. Mayo, *The Human Problems of Industrial Civilization,* New York: Macmillan, 1933.

**47** M. R. Weisbord, *Productive Workplaces, Organizing and Managing for Dignity, Meaning, and Community,* San Francisco: Josey Bass, 1991.

**48** See reference 47.

**49** See reference 47.

**50** See reference 47.

**51** D. McGregor, *The Human Side of Enterprise,* New York: McGraw-Hill, 1960.

**52** C. Argyris, *Management and Organizational Development: The Path from XA to YB,* New York: McGraw-Hill, 1971.

**53** F. Herzberg, M. Mausner, and B. Snyderman, *The Motivation to Work,* New York: Wiley, 1959.

**54** F. Herzberg, *Work and the Nature of Man,* New York: World Publishing, 1966.

**55** R. Likert, *The Human Organization,* New York: McGraw-Hill, 1967.

**56** R. Likert, *New Patterns of Management,* New York: McGraw-Hill, 1961.

**57** D. G. Bowers, *Systems of Organization: Management of the Human Resource,* Ann Arbor, MI: University of Michigan Press, 1976.

**58** R. Likert and J. G. Likert, *New Ways of Managing Conflict,* New York: McGraw-Hill, 1976.

**59** R. R. Blake and J. S. Mouton, *The Managerial Grid,* Houston, TX: Gulf Publishing, 1964.

**60** See reference 33.

**61** S. Helgesen, *The Female Advantage,* New York: Doubleday Currency, 1990.

**62** S. R. Covey, *Principle Centered Leadership,* New York: Simon and Schuster, Fireside, 1992.

**63** See reference 62.

**64** Teleometrics International, "Learning System Catalog," Woodlands, TX: Teleometrics International, 1994.

**65** See reference 16.

**66** See reference 33.

# 28

# EMPOWERMENT AND CREATIVITY

## 28.0 INQUIRY

1 What is empowerment?
2 Why is empowerment necessary for effective total quality practices?
3 What is creativity?
4 How can total quality practices encourage creativity?

## 28.1 INTRODUCTION

*In order to more effectively create quality, our entire workforce must be empowered to make changes in our products and processes; management must provide all employees with the knowledge, opportunity, and encouragement to improve both products and processes* (and thereby improve our customer's experience of quality). A team structure within an environment which encourages creativity allows empowerment to work in quality transformation. In this chapter, we will address empowerment and teamwork, as well as creativity issues.

## 28.2 EMPLOYEE EMPOWERMENT AND TEAMWORK

*Empowerment provides each employee an opportunity (e.g., both authority and responsibility) to be creative and make changes regarding our products and the life-cycle processes (those that define, design, develop, produce, deliver, sell and service, use, and dispose/recycle our product and by-products).* On the surface, empowerment seems simple and straightforward. We might confuse empowerment with simply giving each employee permission to incorporate changes into products-processes associated with his or her job. In both theory and practice, empowerment is much more complex.

From a theoretical standpoint, *Covey lists six conditions of, or for, empowerment* [1]: *(1) character, (2) skills, (3) win-win agreement, (4) self-supervision, (5) helpful structures, and (6) accountability.* These six conditions are displayed in Figure 28.1. Examination of the model indicates that it is a "trust" based model. Character, as used in the model, refers to what a person is, while skills refer to what a person can do. The win-win agreement condition is the most intricate condition. It establishes, or in many cases reestablishes, the boundaries within which activities go on in an organization. Covey describes it as a psychological or social "contract" which includes consideration of results, guidelines for activities, resources, performance standards, and organizational and personal consequences. The self-supervision condition requires a high level of personal maturity. The help-

| Element | Comments |
|---------|----------|
| Character | Refers to what a person is (i.e., personal integrity as expressed through the virtues listed at the beginning of each section in our textbook): <br> Vision and enthusiasm <br> Wisdom <br> Courage and commitment <br> Self-discipline <br> Responsibility <br> Persistence and patience <br> Faith and compassion <br> Trustworthiness and honesty |
| Skills | Refers to what a person can do (i.e., personal knowledge of and proficiency in job related activities). |
| Win-win agreement | Refers to a social contract which delineates results (desired outcomes), guidelines (policies and procedures), resources (human, machine, financial), accountability (performance standards amd methods of evaluation), and consequences (organizational and personal impact). |
| Self-supervision | Refers to self-initiation and self-control with respect to the win-win agreement. |
| Structures | Refers to the organizational format and functional activities with respect to executing the win-win agreement. |
| Accountability | Refers to the establishment and acceptance of personal responsibility for affecting and producing results. |

**FIGURE 28.1**   Description of Covey's six conditions of empowerment.

ful structure and systems condition is necessary in order to assure alignment between the win-win agreement and the organizational structure or environment (e.g., strategic planning, information flow, selection of personnel, placement, training, and so forth). The sixth and final condition, accountability, is necessary in order to foster personal responsibility.

The Covey empowerment model is an extension of his principle-centered leadership, PCL, paradigm, see Chapter 27. It is built around what Covey terms "abundance mentality" (e.g., making the "pie" bigger so that you can have a larger piece, rather than taking "pie" away from your "competitor"). ***Covey's PCL and empowerment*** *models are primarily directed to the individual, and are very idealistic in their demands on both the individual,* as far as trustworthiness and "whole person involvement," *and the organization,* as far as fostering a favorable environment for the win-win agreement (as compared to other MLM and empowerment models).

*In practice, empowerment demands a good deal of change in most organizations.* We seek improvement resulting from this change. But, *we must remember that not all change results in improvement and that a change in one part of our product or production system may impact other parts* (in either a positive or negative manner). We will experience both successes and failures, which must be anticipated and effectively dealt with at all production system levels. We seek action supported by knowledge in order to assure our success. Thus, we must examine empowerment and a number of related critical dimensions as they impact empowerment: (1) teamwork and communication, (2) the evolution of empowerment, (3) the bounds of empowerment, and (4) education and training.

## Teamwork

*One cornerstone of TQ is a team-based structure.* Wellins et al. compare and contrast traditional and team, or employee group, based organization structure and operations regarding a number of elements [2], see Table 28.1. As noted in Chapter 27, traditionally, organizations have been structured hierarchically, or as rigid pyramids. Team-based organizations are much "flatter" in structure, with only a few layers of management. Jobs are designed differently and are "whole-process" structured, rather than narrowly focused. Leadership and management functions change from a directive to a facilitative style. Information flow is more open and information must be shared. For example, we might see organizations approaching the web model depicted in Figure 27.4 emerge. Rewards and recognition shift from an individual basis to a team or group basis. And, just like athletic teams, a strong esprit de corps (team pride), as well as trust and confidence in "our team," is necessary for effective teamwork.

**TABLE 28.1**    KEY DIFFERENCES BETWEEN TRADITIONAL ORGANIZATIONS AND EMPOWERMENT-TEAM ORGANIZATIONS

| Element | Traditional organization | Self-directed teams |
|---|---|---|
| Organizational structure | Layered/individual | Flat/team |
| Job design | Narrow single task | Whole process/ multiple tasks |
| Management role | Direct/control | Coach/facilitate |
| Leadership | Top down | Shared with team |
| Information flow | Controlled/limited | Open/shared |
| Rewards | Individual/seniority | Team-based/ skills-based |
| Job process | Managers plan, control, and improve processes | Teams plan, control, and improve processes |

*Source:* Reproduced, with permission, from R. S. Wellins, W. C. Byham, and J. M. Wilson, *Empowered Teams,* San Francisco, CA: Jossey-Bass, page 6, 1991.

*Team (employee) empowerment is crucial to effective teamwork.* Empowerment requires a parceling out of functions that have traditionally been reserved for management and functional specialists. In one sense, we can think of empowerment as a redistribution of both authority and responsibility throughout an organization. *Sound team structures and effective communications are necessary to gain synergistic effects from employee empowerment.* Otherwise, the change that results from empowerment may be misdirected, due to a lack of coordination. Coordination is critical since change in a product or process at any given point or level will generally impact other products and processes at other points and levels.

*Three basic types of teams exist within the TQ structure*: **(1)** *quality council,* **(2)** *work-unit, and* **(3)** *cross-functional.* The quality council team is typically made up of high-level functional leaders or managers (with participation by others also) and has a high-level coordinating function. *The council is responsible for establishing and sustaining commitment, direction, and energy for organizationwide quality improvement.*

Crosby [3] and others stress the importance of quality councils and the role of quality professionals at the council level. Responsibility for the creation of quality within an organization goes well beyond the one or, at most, handful of quality professionals on the council. It extends throughout upper-level management and the entire organization. The council must initiate quality commitment and spark a quality passion throughout the entire organization. *All top-level leaders or managers must actively demonstrate both commitment and passion, in both words and deeds, thus acting as positive role models for the entire organization.* They must provide the energy necessary to initiate quality transformation.

*Work-unit teams (also called self-directed work teams) represent the means of accomplishing day-to-day production.* Wellins et al. define *a "self-directed" work team as an intact group of employees responsible for a whole process or process segment that delivers a product or service to an internal or external customer.* An effective work-unit team structure essentially redefines traditional worker and first-line supervision roles, as well as peripheral roles such as maintenance, production scheduling, and so on (depending on the specific nature of the work group). Team structures for work units vary widely; but in most cases, the team is responsible for its processes (e.g., meeting technical specifications, meeting schedule deadlines, solving basic production-related problems, and interfacing to some degree with suppliers and customers outside the plant). *First-line supervisors and functional experts take on the roles of "coaches" and facilitators, rather than bosses in the traditional sense.*

*Cross-functional teams are special teams put together to address specific situations (either problems, concerns, or opportunities) that call for knowledge and experience from different fields.* There are few rules that guide team selection. Generally, the team members include individuals (1) that are affected by the problem/concern or the situation, (2) that possess knowledge and experience relative to the situation, and (3) that will be involved in carrying out the solution (through action) relative to the situation.

*The dynamic nature of cross-functional teams poses two distinct advantages:* **(1)** *use of the consensus approach to solving problems and addressing concerns* *and opportunities helps assure that the solutions will work, and* (2) *team members gain diverse perspectives from those outside their own area of influence and expertise. These teams are constantly being formed and disbanded as problems and concerns are successfully dealt with.*

Drawing on both research and experience, *Weisbord provides four basic conditions that must be satisfied for successful team building* [4]:

1 *Interdependence.* The team is working on important problems in which each person has a stake. In other words, teamwork is central to future success, not an expression of ideology or some misplaced "ought-to" speculation.
2 *Effective leadership.* The "leader" wants so strongly to improve group performance that he or she will take risks.
3 *Joint decision.* All members agree to participate.
4 *"Equal" influence.* Each person has a chance to influence the agenda.

Teams are small groups of people working toward a common purpose. *Empowered TQ work teams must become proficient in both the problem-solving and decision-making processes.* Problem solving requires group members to think individually and to focus their thoughts collectively. *Two general thinking models tend to be used in small group problem solving*, according to Beebe and Masterson [5]: (1) *reflective, or conventional, and* (2) *brainstorming* (Table 28.2, *a* and *b*, respectively). The reflective process (Table 28.2, *a*) closely follows the classical, logic-based scientific method. The brainstorming process (Table 28.2, *b*) concentrates on generating and capturing a multitude of thoughts, without a structured, disciplined, logical format.

*Team members typically play three types of roles*: (1) *group task*, (2) *group maintenance, and* (3) *individual.* Table 28.3 describes a number of roles people in these categories might play. These roles are dynamic and varied in their impact on group success and failure. One person will typically play many roles in the course of effective group problem solving and decision making.

**TABLE 28.2** SMALL GROUP THINKING AND PROBLEM SOLVING PROCESSES

---

**a Reflective thinking**
1. Identify and define the problem.
2. Analyze the problem.
3. Suggest possible solutions.
4. Suggest the best solution(s).
5. Test and implement the solution(s).

**b Brainstorming**
1. Select a problem that needs creative solutions.
2. Tell group members to withhold judgments and evaluations.
3. Tell the group to generate as many solutions as possible.
4. Tell the group it's OK to piggyback off someone else's idea.
5. Have someone record the ideas generated.
6. Evaluate the ideas when the time allotted for brainstorming has elapsed.

---

*Source:* From *Communicating in Small Groups: Principles and Practices*, 3d ed., by S. A. Beebe and J. T. Masterson, pages 182 and 185. Copyright © 1990, 1986, 1982, Scott Foresman and Company. Reprinted by permission of Harper Collins College Publishers, Glenview, IL.

## TABLE 28.3    TEAM MEMBER ROLES IN SMALL GROUPS

### Group task

*Initiator-contributor*—Proposes new ideas or approaches to group problem solving; a person who occupies this role may suggest a different approach to procedure or organizing the problem-solving task.

*Information seeker*—Asks for clarification of suggestions; an information seeker also asks for facts or other information that may help the group deal with the issues at hand.

*Opinion seeker*—Asks for a clarification of the values and opinions expressed by other group members.

*Information giver*—Provides facts, examples, statistics, and other evidence that pertain to the problem the group is attempting to solve.

*Opinion giver*—Offers beliefs or opinions about the ideas under discussion.

*Elaborator*—Provides examples based upon his or her experience or the experience of others that helps to show how an idea or suggestion would work if the group accepted a particular course of action.

*Coordinator*—Tries to clarify and note relationships among the ideas and suggestions that have been provided by others.

*Orienter*—Attempts to summarize what has occurred and tries to keep the group focused on the task at hand.

*Evaluator-critic*—makes an effort to judge the evidence and conclusions that the group suggests.

*Energizer*—Tries to spur the group to action and attempts to motivate and stimulate the group to greater production.

*Procedural technician*—Helps the group achieve its goal by performing tasks such as distributing papers, rearranging the seating, or running errands for the group.

*Recorder*—Writes down suggestions and ideas of others; makes a record of the group's progress.

### Group building and maintenance

*Encourager*—Offers praise, understanding, and acceptance of others' ideas and suggestions.

*Harmonizer*—Mediates disagreements among group members.

*Compromiser*—Attempts to resolve conflicts by trying to find an acceptable solution to disagreements among group members.

*Gatekeeper and expediter*—Encourages less-talkative group members to participate and tries to limit lengthy contributions of other group members.

*Standard setter*—Helps to set standards and goals for the group.

*Group observer*—Keeps records of the group's process and uses the information that is gathered to evaluate the group's procedures.

*Follower*—Basically goes along with the suggestions and ideas of other group members; serves as an audience in group discussions and decision making.

### Individual

*Team player*—Facilitates teamwork by yielding immediate self-interests to group purpose.

*Aggressor*—Destroys or deflates the status of other group members; may try to take credit for someone else's contribution.

*Blocker*—Is generally negative, stubborn, and disagreeable without apparent reason.

*Recognition-seeker*—Seeks the spotlight by boasting and reporting on his or her personal achievements.

*Self-confessor*—Uses the group as an audience to report personal feelings, insights, and observations.

*Playboy*—Lacks involvement in the group's process; lack of interest may result in cynicism, nonchalance, or other behaviors that indicate lack of enthusiasm for the group.

*Dominator*—Makes an effort to assert authority by manipulating group members or attempting to take over the entire group; may use flattery or assertive behavior to dominate the discussion.

*Help-seeker*—Tries to evoke a sympathetic response from others; often expresses insecurity or feelings of low self-worth.

*Special-interest pleader*—Speaks for a special group or organization that best fits his or her own biases to serve an individual need.

*Source:* From *Communicating in Small Groups: Principles and Practices*, 3d ed., by S. A. Beebe and J. T. Masterson, pages 64–66. Copyright © 1990, 1986, 1982, Scott Foresman and Company. Reprinted by permission of Harper Collins College Publishers, Glenview, IL.

*One of the attributes of effective teams is a result of their nature—that of consensus building in the decision-making process. Consensus occurs when all group members agree on a certain course of action relative to the problem at hand.* Concensus is very different from group democracy and a majority voting process. *A consensus agreement ties all group or team members to a common purpose, as well as a common course of action.*

*In the consensus-building process, two undesirable results are sometimes encountered: groupthink and group conflict. The term "groupthink" is used to refer to the situation wherein a group strives to avoid or minimize conflict, and reach a consensus without critically testing, analyzing, and evaluating thoughts relative to the problem at hand* (see Beebe and Masterson [6]). Groupthink typically results when one or a few group members dominate the group and, consequently, when a large portion of the group become passive.

Since *some level of conflict is inevitable in team functions,* a good deal of research has been directed at conflict resolution. *Beebe and Masterson* [7] *classify group conflict into three categories: (1) pseudo conflict, (2) simple conflict, and (3) ego conflict.* "Pseudo conflict" results from misunderstandings with regard to a given member's perception of the problem. Hence, it can be resolved readily, if approached carefully. "Simple conflict" is a real conflict that is problem related. "Ego conflict" is person or member based. Emotions play a large part in ego conflict, making it by far the most difficult type of conflict to resolve. Pseudo and simple conflict may lead to ego conflict if handled improperly. Table 28.4 summarizes these three types of conflict, as well as suggested countermeasures.

Weisbord suggests that teams focus both on their task and their teamwork process in order to grow and be successful in the workplace [8]. Task effectiveness is measured by the degree of problem resolution or opportunity captured. Weisbord proposes three success criteria to measure team effectiveness:

1 The team resolves important dilemmas, often ones on which little progress was made before.
2 People emerge more confident of their ability to influence the future.
3 Members learn the extent to which output is linked to their own candor, responsibility for themselves, and willingness to cooperate with others.

---

## PCB Measles — Corrective-Action Team Case [SS]

Many electronic devices, machines, and systems use printed circuit (PC) boards as the mounting platform for electronic parts. A PC board is an epoxy-fiberglass composite which, like a sponge, absorbs moisture. Because it expands in response to heat, moisture within the epoxy-fiberglass matrix increases the tendency for the PC board to delaminate and "measle," or separate between plies or fibers. Separation within the fiberglass weave, especially when occurring close to the board surface, appears as discrete white spots or crosses visible through the laminate. Though measling has been shown to cause no functional

TABLE 28.4    TYPES OF SMALL GROUP CONFLICT

| Conflict type | Pseudo conflict | Simple conflict | Ego conflict |
|---|---|---|---|
| Source of conflict | Misunderstanding individuals' perceptions of the problem. | Individual disagreement over which course of action to pursue. | Defense of ego—Individual believes he or she is being attacked personally. |
| Suggestions for managing conflict | **1** Ask for clarification of perceptions.<br><br>**2** Establish a supportive rather than a defensive climate.<br><br>**3** Employ active listening: Stop—Look—Listen—Question—Paraphrase content—Paraphrase feelings | **1** Listen and clarify perceptions.<br><br>**2** Make sure issues are clear to all group members.<br><br>**3** Use a problem-solving approach to manage differences of opinion.<br><br>**4** Keep discussion focused on the issues.<br><br>**5** Use facts rather than opinions for evidence.<br><br>**6** Look for alternatives or compromise positions.<br><br>**7** Make the conflict a group concern, rather than an individual concern.<br><br>**8** Determine which conflicts are the most important to resolve.<br><br>**9** If possible, postpone the decision while additional research can be conducted. This delay also helps relieve tensions. | **1** Let members express their concerns but do not permit personal attacks.<br><br>**2** Employ active listening.<br><br>**3** Call for a cooling-off period.<br><br>**4** Try to keep discussion focused on issues (simple conflict).<br><br>**5** Encourage parties to be descriptive, rather than evaluative and judgmental.<br><br>**6** Use a problem-solving approach to manage differences of opinion.<br><br>**7** Speak slowly and calmly.<br><br>**8** Develop rules or procedures that create a relationship which allows for the personality difference. |

*Source:* From *Communicating in Small Groups: Principles and Practices*, 3d ed., by S. A. Beebe and J. T. Masterson, page 214. Copyright © 1990, 1986, 1982, Scott Foresman and Company. Reprinted by permission of Harper Collins College Publishers, Glenview, IL.

problems, both short- and long-term, it is aesthetically unappealing and contributes to the perception of a poor product quality.

The typical approach to reducing measling is to make minor adjustments to process parameters after measles are found, repair the condition, and go on. Here, by implication, some level of the measling defect remains and is "acceptable," and only when this level is exceeded is reactive remedial action required.

On the other hand, a proactive (cross-functional) strategy to eliminate measling was used by one electronics manufacturer, and a striking change occurred. Research and experimentation into new materials and techniques were performed. The materials and production processes used in assembly were reviewed and evaluated. Many changes were made: the epoxy-laminate material was changed to one more resistant to measling; regular vacuum baking cycles were introduced to the assembly process; rework techniques were modified to reduce thermal stress levels; storage environments were made more inert; and manufacturing personnel were empowered to evaluate alternative methods and materials, thus acquiring "ownership" in the process and material changes. As a result, the incidence of measles was virtually eliminated. Product quality was improved (as measured by customer satisfaction), and repair and material disposition costs were reduced.

---

## Communications

*Communication is one of the three basic requirements of cooperative effort*; common purpose and willingness are the other two. *Burgoon et al.* [9] *define human communication as a dynamic and ongoing process whereby people create shared meaning through the sending and receiving of messages in commonly understood "codes." Interpersonal communication can be divided into three major types*: (1) *written*, (2) *spoken or verbal, and* (3) *unwritten or nonverbal*.

**Written and Verbal Communications**    Written and verbal communications, either formal or informal, constitute a literal form of communication. Because of the contractual nature of business, *the written word is typically considered more forceful than the spoken word* (e.g., "put that in writing please"), even when (at face value) the meaning is the same. In quality transformations we hear and see (in writing) mission statements, vision statements, policy statements, and so on. Both internal and external customers look carefully at who is saying, writing, and signing these statements—as well as at their literal content. Figures 3.3 and 3.4 illustrated two examples of quality-oriented statements. We see brevity as an important attribute of effective statements, which, in some cases, may be reduced to slogans (e.g., Motorola's "six sigma"). *The spoken and written word is essential in communicating quality transformation intent and indispensable in supporting its practice*.

**Nonwritten-Nonverbal Communications**    In interpersonal (face-to-face) communications in a social setting (consisting of both verbal and nonverbal channels, where nonverbal excludes written communication in our discussion) *research findings estimate that nearly two-thirds of meaning is derived from nonverbal cues,* according to Burgoon et al. [10]. Beebe and Masterson cite three reasons for the importance of nonverbal variables in small group communications [11].

1  People in groups spend more time communicating nonverbally than verbally, since only one member speaks at a time while the others watch the speaker and their group peers.
2  People tend to believe nonverbal cues more than verbal communications. (That is, "Actions speak louder than words.")
3  People communicate emotions and feelings primarily by nonverbal cues (e.g., voice tone and facial expression carry a large share of emotional content).

*Nonverbal communication is received through the five human senses* with sight, hearing, and touch being the most important.  Seven major areas of nonverbal communication have received a good deal of attention and appear to play critical roles in communication processes, depending on the situation at hand (Burgoon et al. [12]).

*Kinesics* (body language)—body movement, facial activity, and gaze or eye behavior.
*Vocalics*—vocal activity such as tone, pitch, and intensity.
*Physical appearance*—dress and adornment, as well as physical attributes (e.g., clothing, stature, grooming).
*Haptics*—touch and the use of touching (e.g., shaking hands).
*Proxemics*—use of space and spatial arrangements (e.g., workplace or office layout).
*Chronemics*—timing and the use of time cues (e.g., looking at one's watch during a conversation, arriving late at a meeting).
*Artifacts*—environmental surroundings and objects or possessions displayed or used in the proximity of the communication activity or by the communicator (e.g., workplace or office decor, personal and company vehicles, meeting rooms).

*Research findings indicate that expectations and interpretation of nonverbal communication vary between individuals, cultures, and genders.*  This variation leads to complex intra- and interrelationships of these seven areas, which in many cases are not well understood.  With the advent of the global organization and increased emphasis on international markets, this becomes an important issue.  Our reliance on nonverbal cues may create unintended problems when we communicate with people from another culture, having a different nonverbal "code."

**Total Effects of Communication**    *When taken together, the written, spoken, and nonverbal communication channels may produce one of three general effects*:

1  *Reinforcement and emphasis*.  The message is strengthened (e.g., saying "good morning" and smiling).
2  *Complement*.  The message is completed (e.g., introducing yourself and shaking hands as a gesture of friendship).
3  *Contradiction*.  The communication channels are inconsistent (e.g., saying you feel great with a grimace of pain on your face).

We typically associate words and actions.  Hence, *we must align our written, spoken, and nonverbal communications so that they are interpreted as comple-*

*mentary rather than contradictory.* We must use all channels available to express, reinforce, and emphasize quality philosophy, vision, concepts, and practice. In addition, we must be sincere in our intent and actions. *Deception and insincerity are readily detectable.*

All in all, based on our knowledge of nonverbal communications, we can expect the success or failure of QT efforts to hinge on the success or failure of nonverbal communications. The reasoning for this is relatively simple. Virtually every organization claims to have an effective quality system and some type of TQ-oriented program. Hence, customers cannot distinguish effectiveness at the written or spoken level. Internal as well as external customers are watching and listening for signs of action or inaction at the strategic level and at the day-to-day tactical level of operations. Our customers and people we want to become our customers are constantly asking themselves "Are these folks serious about quality or is it a new slogan?" They will be assessing our behavior through the spectrum of communications we exhibit, looking for evidence that suggests (to them) we are serious. They will evaluate the evidence relative to their own personal belief or value systems and draw their own conclusions. We, and our quality system structure, will be assessed at our product, service, and technology interfaces (Figure 3.2). This assessment may be immediate or it may take a while, but today's customers—both internal and external—will form attitudes and manifest behavior with respect to our efforts (see Figure 4.2), according to their individual quality experiences.

**Networks of Communication**    Networks of communications are important enough to merit special consideration. *Communication networks are typically thought of as reflecting the vertical and horizontal transfer of information.* Discussions of communications and networks applied to organizations are presented in the behavioral and human factors literature (Reitz [13] and Kantowitz and Sorkin [14]). Here, we will briefly examine a few fundamental types of communication networks (shown in Figure 28.2) and their significance in team effectiveness, regarding quality transformations.

FIGURE 28.2    Fundamental communication patterns.

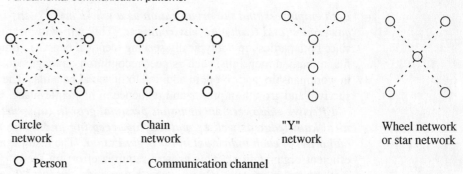

| Circle network | Chain network | "Y" network | Wheel network or star network |

O  Person        ------  Communication channel

The major difference between these five-person networks is the degree to which they are centralized. The circle network is the least centralized, and the wheel or star network, the most. The chain and Y networks fall in between. The circle network requires participation by all in sharing information to perform a task, while the wheel network is effectively dominated by the central element. We can see similarities to the MLM models discussed in the previous chapter (e.g., circles in our web model and the wheel associated with the authoritarian leadership style).

Comparative studies in task performance utilizing these networks have shown that, for simple tasks involving information collection, the wheel pattern is faster, although it results in lower satisfaction scores among its users (compared to the circle network). For more complex tasks, the circle network is generally superior in performance and allows free communication for cross-checking and error correction. The more participative circle network, while slower and requiring more communication interactions, is more adaptable, flexible, and generally superior in performance and user satisfaction levels to the other four networks for many tasks. *Over time, use of the circle network fosters involvement and commitment on the part of the participants, while the star network configuration may create resentment and hostility.* These findings have important implications for organizational change—specifically, as it relates to group or team communications.

Beebe and Masterson have introduced a graphical means (see Figure 28.3) of identifying communication network patterns and documenting communications flow in a small group [15]. They develop a bubble for each member. Then, they observe the directional flow of the communication and indicate the first transaction with an arrow from the addresser to the addressee. Each subsequent transaction is marked with a crossbar. Directional arrows lead from one person to another. Communications from one person to the entire group are represented by an arrow pointed outward from the communicator. Figure 28.3 depicts a simple communications flow pattern. More detailed systems can chronologically number the communications and document their content, thus yielding a summarized set of interactions. These analyses indicate how groups are really communicating versus how they are supposed to be communicating.

### Evolution of Empowerment

*Both employees and the organization as a whole must "win" in an effective empowerment and teamwork environment.* This win-win arrangement is possible since both parties are not typically seeking identical benefits. Employees seek benefits in many dimensions, such as peer recognition, management recognition, pride in workmanship, and so on, in addition to financial benefits. The organization seeks survival and growth in profits and influence in its marketplace.

*Effective empowerment demands personal growth (in terms of trust and technical knowledge as well as maturity in accepting and using responsibility and authority) of each individual in the organization.* The result is a more creative and efficient organization. This growth requires effort on the employee's part, both workers and managers. *Due to growth demands, successful employee empower-*

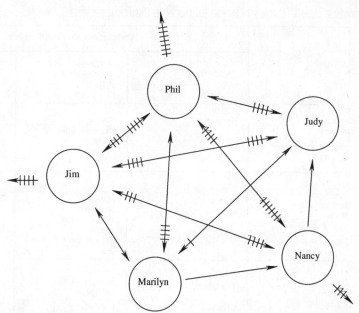

FIGURE 28.3    Communication pattern interaction diagram. From *Communicating in Small Groups*: Principles and Practices, 3d ed., by S. A. Beebe and J. T. Masterson, page 264. Copyright © 1990, 1986, 1982, Scott Foresman and Company. Reprinted by permission of Harper Collins College Publishers, Glenview, Il.

*ment is an evolutionary process and cannot be accomplished instantly.* People simply need time to learn and to adjust to their new roles and culture. Some people must grow more than others, and different people grow at faster or slower rates. Certain individuals cannot or refuse to grow, which may result in an irreconcilable incompatibility between a worker or manager and the quality transformation. Sometimes the only remedy is either a voluntary or an involuntary separation.

Table 28.5 shows three stages of evolution which lead from the traditional workplace environment to employee involvement and finally to empowerment. (See Table 28.1 for key differences between traditional and empowered organizations.) *Six critical elements are developed relative to the evolutionary stages of empowerment: (1) organizational structure, (2) improvement efforts, (3) decision power, (4) recognition, (5) management focus, and (6) organizational focus. Each element must evolve in order for empowerment to become effective.* Hence, even in a relatively participative environment (versus an authoritarian environment) the organization culture must evolve over time.

Many of the employee empowerment and quality-related teamwork concepts, strategies, and structures owe a debt to Japanese quality circle literature [16] and closely parallel the philosophies and structures of the human-relations focused MLM movement (discussed in the previous chapter). For example, the transition from Likert's system 1 to system 4 (Table 27.7) closely resembles the empowerment evolution described in Table 28.5. Hence, we are using the MLM knowledge

TABLE 28.5    EVOLUTION OF EMPLOYEE EMPOWERMENT

| | Workplace environment stages | | |
| --- | --- | --- | --- |
| | Traditional | Employee involvement | Employee empowerment |
| Organizational structure | Hierarchical. | Hierarchical and work-unit teams. | Customer focused quality council, work-unit, and cross-functional teams. |
| Improvement efforts | Suggestion program. Work measurement. | Teams work on problems identified and selected by management. | Teams consistently analyze and improve processes; problems are identified by teams. |
| Decision power | Top down. Chain of command | Recommend changes. Limited, team-based decision making. | Teams make process-related decisions. Teams provide input into strategic decisions. |
| Recognition | Individuals paid for suggestions. Individual work performance reviews and bonuses. | Team recognition for recommendations. Companywide gainsharing. | Team recognition for solutions. Team-based gainsharing. |
| Management focus | Supervising. | Coaching. | Facilitating. |
| Organizational focus | Skill, speciality, professional discipline. | Department, product, production processes. | Customer, product, production processes. |

base, currently available, to help us pragmatically evolve toward employee empowerment in quality transformations.

## Bounds on Empowerment

It is important to transform the concepts of employee empowerment and TQ teams into action. To do this, authority and responsibility must be redistributed. All employees—both workers and managers—must understand and accept this redistribution.

*New boundaries develop regarding authority and responsibility in the product life-cycle processes (to define, design, develop, build, deliver, sell and service, use, and dispose/recycle products).* The new boundaries must be identified and communicated to those involved for empowerment to function effectively. *These boundaries redefine the old authority and responsibility structure and establish the limitations for new team authority and responsibility.* A new set of expectations for teams and team members must emerge.

This restructuring of the workplace requires careful planning and communication. Otherwise, excessive and serious misunderstandings and coordination problems are likely to develop and literally wreck the quality transformation, resulting in losses rather than gains to employees and their organization.

## Quality Education and Training

*Both workers and managers must grow in preparation for quality transformation, otherwise it will be ineffective and slow to take hold or it will result in a total failure. Education and training are recognized as the essential components of TQC and TQM.* Most leading Japanese companies have rigorous training programs. All employees in a company ranging from top management to line operators are taught to use the fundamental TQC tools and techniques. Each new employee undergoes weeks of intensive training on company procedures and operations, plus several days of on-the-job technical training. Ishikawa states that in Japan, "Quality control begins with education and ends in education" [17]. Systematic and continuous TQC education and training are the key contributing factors in the success of Japanese industries.

Many U.S. companies are recognizing and associating quality transformation success with the importance of quality education and training. Many have developed or are developing companywide education and training programs. According to findings published by the Goal/QPC Company [18], commitment, total involvement, and workplace integration are the three key features of a successful education and training program.

*Top-down commitment is essential in order to provide resources and time for training all employees.* Top management must recognize that quality education and training are means to companywide, continuous improvement and thus must be included in their strategic plans. Education and training emphasize the total involvement of employees at all levels, from top executives to shop-floor workers. *Everyone in a company must master appropriate communication skills and problem-solving tools and techniques to successfully perform his or her job, or team position, as redefined through the empowerment process.*

The knowledge and skills gained in training, along with positive attitudes, allow quality improvement to happen. Past movements, such as the zero defects programs of the 1960s and 1970s, failed in the United States primarily because they did not provide sufficient training nor inspire commitment (Crosby [19]). They mainly depended on flashy slogans and banners. In contrast, during this same period, the Japanese backed their TQC efforts with thorough and systematic education and on-the-job practice. Today, the successful results produced by the Japanese are evidenced by their international reputation for quality.

Quality education and training in Japan begins with top management and flows down through middle management and staff [20]. QC concepts, techniques, and statistical methods are the core subjects of learning at all levels. Figure 28.4 illustrates the Goal/QPC TQM wheel, which is based on JUSE (Union of Japanese Scientists and Engineers) quality education and training course content and structure. Figure 28.5 adds detail to the information summarized in the TQM wheel. By noting the level of exposure to each topic, we can see that the mastery levels fall heavily in engineering.

Many TQM-TQC training reference materials have been published and are widely available through JUSE [21, 22] and ASQC [23]. Other materials are developed and used internally by particular companies [24, 25, 26, 27, 28, 29]. TQM consultants often produce their own training material for distribution to their clients [30, 31, 32, 33].

**FIGURE 28.4**    Total quality management wheel.  Used with permission from GOAL/QPC, 13 Branch Street, Methuen, MA 01844-1953.  Tel: 508-685-3900.  Source: *TQM Wheel* © 1989–1994: GOAL/QPC.

\* Improvement oriented
\*\* Strategic planning/coordination oriented

## Total quality control terms

### TQC education in Japan

| Topic \ People | Top management | Middle mgmt./ staff | Engineers | Supervisors | Functional activities** | General workers |
|---|---|---|---|---|---|---|
| TQC concepts | ○ | ○ | ○ | ○ | ○% | ○ |
| QC techniques | ○ | ○ | ◎ | ◎ | ○ | ○ |
| Statistical methods | △ | ○ | ○ | ○ | ○ | ○ |
| Quality assurance | △ | △* | △ | ○* | △% | △ |
| Product development | △ | △* | ◎ |  | △% |  |
| Role in TQC | ◎ | ◎ | ◎ | ◎ | ◎ | ◎ |
| QC circle | △ | △* | △* | ○* | △ | △ |
| New product introduction | ○ | ○ | △* |  |  |  |
| Hoshin planning (policy deployment) | ◎ | ○ | △* | ◎ | △ | ○ |
| Company production system |  |  | ○ | ◎ |  |  |

Educated to:
△ = Understand
○ = Use
◎ = Master

\* No specific course listing.

\*\*Includes functions such as purchasing, sales, marketing, and so on.

*Sources:* JUSE, JSA, Deming Prize companies.

### TQC concepts

. Quality first
. PDCA cycle (plan, do, check, act)
. Fact control
. Process control
. Customer orientation
. Next step in process is our customer
. Hoshin planning
. QC techniques
. Quality assurance
. Promotion of TQC
. Quality circles

### QC techniques

. Seven QC tools:
  -Check sheet
  -Pareto diagram
  -Histogram
  -Cause and effect diagram
  -Stratification
  -Scatter diagram
  -Graph and control chart
. Seven management and planning tools for QC
  -Affinity diagram
  -Interrelationship diagram
  -Tree diagram
  -Matrix diagram
  -Prioritization matrix
  -Process decision
  -Program chart (PDPC)
. QC story: problem-solving procedure

### Statistical methods

. Statistical thinking
. Estimation
. Frequency distribution
. Sampling,
. Sampling inspection
. Simple analysis
. Multivariable analysis:
  Regression and principal component analysis
. Orthogonal polynomial
. Reliability engineering
  -Fault tree analysis (FTA)
  -Failure mode effect analysis (FMEA)
  -Weibull hazard
  -Cumulative hazard sheet

### Product development

. Quality deployment and quality table (QFD)
. FMEA - failure mode effect analysis
. Design review
. Process decision program chart (PDPC)

### Quality assurance

. PDCA (plan, do, check, act) of products, procedures, quality system
. Processing quality problems before and after sales
. Product planning
. Design / development
. Procurement
. Production
. Inspection
. Packaging
. Sales and distribution
. Installation and operation of products
. Documentation audits, quality material

*Quality-related education and training are long-term processes and should be systematically delivered in a top-down manner until everyone is equipped with the necessary philosophies, attitudes, and tools.* Then, continuing education and training at all levels must become a part of the corporate culture. Success in quality-related education and training demands careful selection of appropriate materials, an effective method of presentation, and opportunitunities to use the new knowledge and tools on the job. *Opportunity to practice what is learned in quality transformation is absolutely essential* to the empowerment process. *Quality transformation simply will not work without effective empowerment and practice.*

## 28.3 CREATIVITY

*Employee empowerment (aimed at improving quality and productivity) allows everyone the opportunity to use their powers of creativity, reasoning, and decision making.* Many foreign competitors, particularly the Japanese, recognize and actively encourage the creative contributions of their workers (see Ouchi [34]). Successful employee empowerment through work teams and participative management all presupposes the ability to tap these creative powers for the good of both the individual and the organization.

### Definitions and Concepts of Creativity

Creativity is a rather fuzzy concept. Shannon's definition of creativity appears below [35]:

> *Creativity . . . is basically the ability to produce new, interesting, and useful results . . .* the production and disclosure of a new fact, law, relationship, device, product, or material or process that is based generally on known and available knowledge but does not follow directly, easily, simply or even by logical process from the information at hand . . . a successful step across the borderline of knowledge, a defining of things previously unknown.

Shannon's definition covers most aspects of creativity in a technical and scientific sense. A definition of creativity oriented more toward fine arts might embrace the idea of the manipulation and combination of forms, colors, sounds, and movements in an original way so as to produce aesthetic enjoyment. While the exercise of creative thought may have vastly different sources of inspiration, knowledge bases, and final output or product, the creative process itself is, from a review of relevant literature, remarkably the same regardless of application.

*Any definition or concept of creativity involves imagination, newness, and originality,* whether the creative effort results in something totally new and previously unknown (e.g., Einstein's formulation of relativity) or original and novel combinations and extensions of what is already known (e.g., the integrated circuit, overnight delivery). In discussing and describing creative efforts, *a number of distinctions must be made which will assist us in understanding both individual and organization creativity.*

1 *Creative behavior. Creative behavior refers to the production of ideas that are both new and useful* (usefulness in itself is a relative concept) and implies translating such ideas into some form that others can understand.

2 *Creative ability. Creative ability is the capacity to produce ideas that are both new and useful.* This capacity can be more accurately described as a set of talents and abilities which enable the individual to make new combinations and associations. *There are two important aspects of creative ability:*

   a *It is a complex set of abilities and not a single, unidimensional capacity.* Being multidimensional, creativity does not lend itself easily to comprehensive, straightforward measurement.

   b *Creative ability involves coming up with new and novel associations between concepts and the ability to relate previously unrelated things or ideas.*

According to Reitz there are two types of creative ability [36]:

1 Divergent thinking. Divergent thinking is an aptitude for making new associations or combinations, a product of association between two or more already known elements (e.g., musical composition).

2 Bisociative activity. A bisociative act is an act of combining two previously unconnected cognitive matrices (e.g., Newton's formulation of the theory of gravity by relating the fall of an apple to the movements of heavenly bodies).

## Intelligence and Creativity

The relationship between intelligence and creativity is a complex one. Intelligence itself is not well understood, so any postulated relationship with creative ability or behavior is necessarily tentative. In general, creative ability is not parallel to intelligence. Most contemporary notions of intelligence involve the ability to learn quickly, reason abstractly, and retain information. Intelligence and creativity are both broad, loosely defined abilities which encompass a multiplicity of traits of which only some are interrelated. Some intellectual traits involve creativity, and some creative behavior requires an intellectual basis.

*According to Reitz* [37], *intelligence can explain only about 10 percent of the variation in creative ability* and largely reflects the intellectual components of creative behavior (certain abstract reasoning, memory, and learning components). *Hence, we conclude that intelligence and creativity are largely independent.*

Given that creative ability is widespread in the general population, *an organization may experience significant increases in the creative behavior of its employees by providing an environment conducive to creative thinking without relying on searches for exceptionally gifted individuals* [38, 39]. Creative talent, or the complex of abilities and talents which it comprises, has both genetic and environmental aspects. Hence, *to some degree creative behavior can be learned, but we must nonetheless cultivate natural creative abilities and furnish opportunities to use them.*

## Creative Thinking

In understanding creative thinking, one must recognize that there are processes and prerequisites associated with its occurrence.  Shannon identifies several *prerequisites for creative thinking* [40]:

1 *Knowledge* must be gained by passive experience (e.g., reading, listening) or by active experience (observing, investigating, tinkering, experimenting). The larger the knowledge base, the greater is the "raw material" for new ideas.
2 *Imagination* (through the manipulation of knowledge by combining and re-arranging facts and impressions into new patterns) requires knowledge to be productive.
3 *Evaluation* is the ability to develop embryonic ideas into usable ideas.  Shannon states, "creativity is more likely to occur when one lets the imagination soar and then engineers it back to earth."

Virtually all discussions of creative thinking and behavior include consideration of the creative process and its component phases.  *Reitz lists four phases of the creative process* [41].  *We have added an initial phase to this list:*

A *Perception of need.*  Perception of a need and the motivation (a condition that energizes and directs our actions) to take action triggers the creative process.
1 *Preparation.*  Preparation involves assembling materials or information passively (i.e., getting one's thoughts together) or actively.
2 *Incubation.*  Incubation is a period of apparent relaxation of conscious effort following intensive preparation and study.  It is thought that mental effort continues on an unconscious level even though conscious thought activity has been slowed or suspended.
3 *Insight.*  Insight is the first awareness of a new and valuable idea or association.  It may be the result of visible hard work or of seemingly accidental, serendipitous discovery.
4 *Verification.*  Verification consists of testing and confirmation of the new idea and communication to the destined audience.  Verification is a critical step and, in an organizational context, must be facilitated and expedited.

There are various techniques and training programs that have had success in developing individual and group potential for creative behavior and ability.  These techniques largely focus on the encouragement of the use of imagination and the suspension of judgmental thinking, along with an examination of the phases and workings of the creative process.  Brainstorming is one such technique in common use.

## Organizational Barriers to Creativity

*A number of organizational barriers to creative behavior exist.*  All eventually succeed in discouraging creative effort.  Some examples are the following:

1 *Preoccupation with immediate payoff.* The creative process by and large requires time in order to reach fruition. Any emphasis on obtaining immediate results will upset and sabotage the process. Organizations noted for encouraging creativity and regarding their employees as a strategic resource tend to think and plan for the long-term.

2 *Suspicion of creative people.* It is unfortunate that in many organizations creative talent is implicitly or even explicitly regarded as an aberration of some type. A creative department or group is sometimes perceived as a collection of "eccentrics" or "kooks" who are absent-minded, immature, and even child-like emotionally. Creative effort and behavior cannot be seen as somehow outside the normal scope of the organization and still be encouraged.

3 *Lack of long-range goals and support.* As mentioned above, creative effort and behavior require planning and support over an extended period of time. Chances are, if the organization does not allow for such efforts in goals, strategies, and plans, they will not be numerous or sustained.

4 *Organizational instability.* The literature notes that creative people (or people behaving creatively) tend to be very sensitive to the psychological climate of the organization. Changes in policies and wavering support of creative activities tend to inhibit creative efforts.

5 *Excessive organizational formality, rigidity, and autocracy.* The creative process requires fair amounts of both flexibility and autonomy on the part of the organization and its employees. While creative effort can be encouraged and nourished, it cannot be closely and rigidly programmed.

6 *Lack of recognition and reward.* Failure to recognize and reward outstanding and valuable efforts on the part of employees tends to discourage similar future efforts. Recognition needs tend to be very high for creative people and people behaving creatively, hence recognition is a necessity.

7 *Reluctance of management to take risks.* Encouraging creative efforts requires a degree of both trust and risk. It must be understood that the creative process does not always culminate in a fantastically successful idea or solution. A risk-adverse organization will focus on the consequences of potential failures rather than on the benefits of successful outcomes.

8 *Rigid programming of time.* Time programming is very much related to the organizational formality barrier cited above. During the incubation phase of the creative process it may appear as if nothing is happening. Rigid scheduling by management in order to eliminate employee "sloth" can interfere and disrupt the process. Deutsch & Shea state that one of the most fallacious, ingrained and harmful, culturally supported stereotypes (on the part of management) is that vigorous activity represents efficiency, whereas sitting and thinking are nothing but slothful idleness [42].

9 *Lack of effective communication* between technical and research functions and management. A lack of communication of goals, objectives, and progress inevitably leads to misunderstandings on both sides and disruption of the creative process.

It is apparent that there is a substantial degree of overlap between a number of the barriers listed above. A number of organizational characteristics that will stifle creativity tend to "cluster," or occur together, such as lack of planning, focus on the short-term (immediate payoffs), and lack of communication. The barriers may have a synergistic effect on each other, reinforcing an environment that thoroughly discourages creative behavior. The barriers to creative behavior are equally significant as obstacles to the implementation and success of quality transformation.

## Organization Support for Creativity

The literature on the subject of creativity, particularly that which reports on studies or surveys of creative individuals (see Deutsch & Shea [43]) suggests that there is a certain conflict between the needs and nature of an organization and the needs and nature of creative behavior. *In essence there are two cultures or value systems to contend with in the management of creative activities.*

1 *Business culture.* Business culture is based on authority derived from the organization—it is characterized by submission to the larger needs of the organization and is motivated by a desire to operate an organization and to identify with its success. The business culture looks to society and the business world as its reference point.

2 *Technical and scientific culture.* The technical-scientific culture derives its authority from knowledge about nature and is motivated by curiosity and the desire for mastery over and knowledge about technical and scientific problems. It looks to nature instead of business and society and regards scientific method and discipline as its reference points. It applies to both scientific researchers and empowered line workers.

The different values and motivations of the two cultures can make for a confusing, conflicting, and dynamic relationship. This relationship is one of the more important points to consider in the management of creative activities. These two value systems must be accommodated and integrated in order to elicit the creative behavior necessary for effective quality transformation.

A number of desirable characteristics of organizations and managers have been identified in studies of creative individuals and groups in research organizations. Those features include the following:

1 Upper management takes a visible and active role in setting the climate for creativity through positive attitudes, examples, encouragement, and support.
2 Support for research and creativity is designed into the long-range goals and plans of the organization.
3 Recognition of creative effort is provided and incentives spelled out.
4 Management is open-minded, receptive, and demonstrates a "let's go further" attitude.
5 Management allows flexibility for creators to freely choose their own areas of concentration (within reason) by indicating only broad areas of interest.
6 Individual personalities are considered in making assignments.
7 Open communications are maintained throughout the organization.

The description of an "ideal" manager stresses that he or she should not only be creative, but understand the creative process and its workings. Personality characteristics such as tact, compassion, understanding, and fairness rate very high. The manager as a defender of the freedoms of the group and as a leader with the ability to inspire and encourage (a coaching effect) is a valued asset. Physical arrangements that promote a creative atmosphere include private meeting rooms, up-to-date equipment and facilities, easy access to information, and adequate support activities and personnel (technical, secretarial, and so on).

The disagreements over the nature of rewards for creative efforts in organizations are apparent in the literature in the field. This controversy is also germane for participative management and teams. The consensus is that monetary rewards, especially sizable ones, may do more harm than good by triggering jealousies and rivalries. Meaningful, small, personalized awards such as gifts of nominal financial value are seen as more appropriate. Recognition in the form of written or verbal "pats on the back" by management is also viewed as appropriate and likely to be appreciated.

People accustomed to traditional organizations may expect monetary awards for creativity and innovation, due to past experience with employee suggestion reward programs; this also applies to management "perks" and bonuses. Many such programs have been resented and viewed with suspicion by workers due to perceived poor management, untimely and unfair handling of suggestions, and so on. Hence, all employees must clearly see the TQM paradigm transition "point" and accept a new recognition and reward structure as "fair." In short, effective quality transformations which foster creativity possess a consistent and rational recognition and reward structure which will not drive wedges between workers and managers.

## Automobile Rental — Creativity in Training and Analysis Case

A nationwide automobile rental company had noticed that their sales position had been declining for the last nine months, as measured in number of automobiles rented, sales dollars, and market share relative to their competitors. Over this same period more competitors than ever before had entered the local and regional markets. It was not clear whether the company's sales decline was a result of exemplary efforts by its competitors or its own lack of effort, or both. The company had about 3000 reservations agents, and most reservations were taken through telephone communications.

A number of discussions were held concerning this performance trend. Two general schools of thought emerged from the meetings: (1) hire a consulting firm to retrain all reservations agents, or (2) perform an internal, corrective-action team study of the reservation process with the goal of process improvement. The quoted cost of training for a three-day training program was estimated to be $500 per person. Although a full training program would thus cost $1.5 million, the net benefit of restoring sales to their previous level was estimated as $6 million per year. Hence, the expectation for a payback within three months seemed reasonable.

On the other hand, the cost of a thorough internal study performed by a corrective-action team was estimated to be on the order of $10,000 to $20,000. Benefits of the internal study were to restore sales to previous levels. Management opinion was split on the issue. However, personnel were available to form a corrective action team. Collectively, these individuals possessed a reasonable level of knowledge and analytical skill in basic process analysis and quality improvement.

Arguments were heard to support both the training and the analysis alternatives. On the one hand, the training advocates considered the problem to be of a strategic nature and its resolution beyond the capabilities of a corrective-action team. On the other hand, the analytical approach advocates pointed out that the company was committed to the continuous improvement philosophy and argued that, although a broad-spectrum training program might be successful, a solid cause-effect based solution developed in-house would add to employee confidence and moral, as well. In addition, they questioned the necessity of the three-day training program and new sales process in the belief that a certain amount of fine tuning in the current process might well solve the problem. Due to both commitment to quality transformation and confidence in employees, the decision was made to form a corrective-action team and empower it to address the sales gap problem.

The team activities are summarized and documented below.

*Goal*—Restore reservation sales to previous levels and then surpass those levels.

*Objective*—Identify the cause(s) of lagging sales. Develop an action plan to implement the goal. Test the action plan to verify its effectiveness. Pending effectiveness, install the action plan nationwide. Review the national results after six months.

*Procedure*—A sequence of process flowcharts was developed and check sheets devised to gather data pertaining to the activities of the telephone agents. Performance records of individuals were assessed, and three levels of sales performance (calculated as the ratio of firm reservations to total calls) were identified: (1) outstanding, (2) typical, and (3) poor.

Performance observations were made of workers in all three categories using volunteer operators; the agents being observed (video or audio recorded) had agreed to analyze the recordings as a team. The recorded calls were grouped into two categories: (1) rental demands from customers who knew what they wanted and (2) inquiries about rental availability and cost.

*Results*—A number of observations were made dealing with operator enthusiasm and concern for customers. However, one observation stood out from the others. In every case of the inquiry calls, the high sales performers asked directly for the customer's business to close the sale. Only sometimes did the average sales performer ask for the customer's business (about 35 percent of the time). The poor sales performers basically answered the inquirer's questions politely, but did not ask for the business.

*Action Plan*—The team made a number of recommendations but the most important one was to design and provide a reminder sticker for the agent's phone that read, "Most people who inquire need to rent a car. Therefore, remember—ALWAYS ASK FOR THE RESERVATION."

*Test*—The team members decided to give their new action plan a two-week trial. They used their previously developed check sheets to collect data. Results were very positive. Full-scale implementation of the action plan followed.

*Follow-up*—At the end of three months, lost sales were recaptured and, after six months, sales had increased by an additional 15 percent.

## REVIEW AND DISCOVERY EXERCISES

### Review

**28.1** Assume you enter a work environment and want to assess the level of employee empowerment, describe the clues or signs that would indicate
    **a** A high level of empowerment.
    **b** A low level of empowerment.

**28.2** Why is it difficult to transform a traditional workplace with a low level of empowerment to one with a high level?

**28.3** Discuss the possible advantages and possible disadvantages of empowerment
    **a** For employees (e.g., workers and managers).
    **b** For the organization (e.g., the stakeholders).

**28.4** Why is teamwork critical for successful employee empowerment?

**28.5** Compare the make-up and function of the work-unit team and the cross-functional team.

**28.6** Is the quality council team cross-functional by the very nature of its make-up? Explain.

**28.7** Which communication network or networks (Figure 28.2) seem most appropriate for a cross-functional team in a factory environment? Explain.

**28.8** Identify and explain a situation where the wheel network might yield high performance.

**28.9** Determine where we might find a chain communication network and describe its advantages and disadvantages.

**28.10** Explain why it is possible for both the employee and the company to win in an empowered-employee work environment.

**28.11** Where and how does employee empowerment impact Maslow's human needs hierarchy (see Figure 4.1)?

**28.12** Why is it important to develop limits on empowerment and communicate these bounds to employees?

**28.13** Explain and then support or refute Ishikawa's statement, "Quality control begins with education and ends in education."

**28.14** What is the difference between creative ability and creative behavior?

**28.15** Provide an example of
    **a** Divergent thinking.
    **b** A bisociative act.

**28.16** List and justify Shannon's three prerequisites for creative thinking.

**28.17** List and justify Reitz's four phases (plus the one we added) of the creative process.

**28.18** Using the list of nine organizational barriers to creativity, restate each in positive terms rather than negative terms (e.g., "Patience for long-term results" could replace "Preoccupation with immediate payoff").

**28.19**  Explain the difference between the business culture and the technical-scientific culture and the significance of the relationship between the two.

### Discovery

**28.20**  Many employees in the more traditional organization believe that they are employed to perform specific tasks and that ideas to improve the workplace processes and product should be purchased from them (e.g., through a suggestion system with a percent gainshare, 50–50). Is this belief legitimate? Is it consistent with TQ philosophy? Explain your answers.

**28.21**  Many managers take their titles literally and do not see any benefits to themselves in empowering employees (e.g., they see empowerment of employees as much more a burden than a benefit in terms of their needs). Assume you are appointed to facilitate an empowerment transformation. Does this attitude place the transformation in jeopardy? How would you respond to it?

**28.22**  Historically, we in the United States have witnessed the creative acts of individuals and hence associate creativity with individuals (Eli Whitney, Thomas Edison, Henry Ford, etc.). Now we see the empowerment of cross-functional and operations-unit work teams which are expected to act as a group in a creative manner. Specifically, how can leaders address recognition and incentives for team versus individual creative efforts in the same organization?

**28.23**  Critique the Covey empowerment model conditions. How realistic and reasonable are they? How could they be used as a guide to developing a higher level of empowerment in an organization? You may want to review the PCL discussion in Chapter 27 before addressing this question.

**28.24**  Refer back to the Automobile Rental Case. How was empowerment and creativity shown in this case? What result was obtained? Was the situation handled properly by the managers?

**28.25**  A steel pipe 24 in long with a sidewall thickness of 0.125 in is firmly embedded in a large slab of concrete with the open end about 18 in above the slab. The inside diameter of the pipe is 0.05 in greater than the diameter of a Ping-Pong ball. The ball has been dropped into the open end of the pipe. Develop as many workable, creative solutions as you can for getting the Ping-Pong ball out of the pipe, without damaging either.

## REFERENCES

1  S. R. Covey, *Principle Centered Leadership,* New York:  Simon and Schuster, Fireside, 1992.

2  R. S. Wellins, W. C. Byham, and J. M. Wilson, *Empowered Teams,* San Francisco: Jossey-Bass, 1991.

3  P. B. Crosby, *Quality is Free,* New York:  McGraw-Hill, 1979.

4  M. R. Weisbord, *Productive Workplaces, Organizing and Managing for Dignity, Meaning, and Community,* San Francisco: Jossey Bass, 1991.

5  S. A. Beebe and J. T. Masterson, *Communicating in Small Groups,* 3d ed., Glenview, IL: Harper Collins, 1989.

6  See reference 5.

7  See reference 5.

8  See reference 4.

9  J. K. Burgoon, D. B. Bullen, and W. G. Woodall, *Nonverbal Communication,* New York: Harper and Row, 1989.

10  See reference 9.

11  See reference 5.

12  See reference 9.

13  H. J. Reitz, *Behavior in Organizations,* Irwin Series in Management and the Behavioral Sciences, Richard D. Irwin, Inc., Homewood, IL, 1977.

14  B. H. Kantowitz and R. D. Sorkin, *Human Factors: Understanding People-System Relationships,* New York: John Wiley and Sons, Inc., 1983.

15  See reference 5.

16  *How to Operate QC Circle Activities,* Tokyo, Japan: QC Circle Headquarters, Union of Japanese Scientists and Engineers, JUSE, 1985.

17  K. Ishikawa, *What is Total Quality Control? The Japanese Way,* Englewood Cliffs, NJ: Prentice-Hall, 1985.

18  "Total Quality Control Education in Japan," Goal/QPC Research Committee, report no. 89–10–01, Methuen, MA: Goal/QPC Co., 1989.

19  See reference 3.

20  See reference 18.

21  See reference 16.

22  K. Ishikawa, *Guide to Quality Control,* 2d rev. ed., Asian Productivity Organization, Tokyo, Japan, 1982.

23  American Society for Quality Control (ASQC), 611 E. Wisconsin Ave., Milwaukee, WI, 53202.

24  "Advanced Quality System for Boeing Suppliers," D1-9000 Document, Seattle, WA: The Boeing Commercial Airplane Group, 1991.

25  J. Hoskins, B. Stuart, and J. Taylor, "Statistical Process Control," Phoenix: Motorola Inc., BR392/D.

26  "Process Control Capability and Improvement," Thornwood, NY: The Quality Institute (IBM), 1984.

27  "Using SPC to Be the Best," Detroit: Supplier Quality and Corporate Statistical Methods Office (Chrysler Motors), 1989.

28  "Continuing Process Control and Process Capability Improvement," Dearborn, MI: Statistical Methods Office (Ford Motor Company), 1984.

29  J. B. ReVelle, "The New Quality Technology," Hughes Aircraft, Los Angeles, CA, 1990.

30  C. Harwood, "Quality Improvement System," Cupertino, CA, Quality Improvement Company Training Manual, 1988.

31  M. Brassard, "The Memory Jogger Plus," Methuen, MA: GOAL/QPC, 1989.

32  B. King, "Better Designs in Half the Time," Methuen, MA: GOAL/QPC, 1989.

33  B. King, "Hoshin Planning the Developmental Approach," Methuen, MA: GOAL/QPC, 1989.

34  W. Ouchi, *Theory Z: How American Business Can Meet the Japanese Challenge,* Reading, MA: Addison-Wesley Publishing Company, 1981.

35  R. E. Shannon, *Engineering Management,* New York: John Wiley and Sons, 1980.

36  See reference 13.

37  See reference 13.

38  A. F. Osborne, *Applied Imagination: Principles and Procedures of Creative Problem Solving,* 3d rev. ed., New York: Charles Scribner's Sons, 1963.

**39**  F. Barron, *Creative Person and Creative Process,* New York:  Holt, Rinehart and Winston, 1969.

**40**  See reference 35.

**41**  See reference 13.

**42**  Deutsch & Shea, Inc.,  *Company Climate and Creativity:  105 Outstanding Authorities Present Their Views,* New York:  Industrial Relations News, 1959.

**43**  See reference 42.

# TOTAL QUALITY
# MANAGEMENT APPLICATION

## 29.0 INQUIRY

1 What does it take to implement TQM?
2 How does TQM encourage continuous improvement?
3 How do quality-award criteria relate to TQM?
4 How can TQM performance be measured?
5 What elements of TQM are common to all applications?

## 29.1 INTRODUCTION

In previous chapters we have addressed the creation of quality in primarily a generic fashion and used the term "quality transformation" to describe a movement from our present position of quality, customer satisfaction (external, internal, and stakeholders) to positions of improvement over the course of time. Now we will focus on a more specific course of action termed TQM (defined in Chapter 1). TQM applications vary widely with product lines, organization size, management philosophies and practices, and so on. Implementation of TQM requires a great deal of change in most organizations. Our present position must be assessed relative to our existing quality system and then TQM applications must be designed to link up at the present position. Then, action must be taken to move forward (e.g., toward the TQM paradigm) as rapidly as possible. The more we understand about the quality philosophies, leadership theories, the empowerment and creative processes, and the degree of commitment and persistance necessary to practice TQM, the more successful we will be.

## 29.2 TQM IMPLEMENTATION

*TQM is a long-term process which demands both strong leadership and sound management, in that order. Johnson* in his four-volume set on TQM implementation [1, 2, 3, 4] *cites five major differences between companies with outstanding TQM practice records and the rest*:

1 *Leadership.* Leadership begins at the top with these key people absolutely committed to TQM as a way of life. They will accept nothing less than quality and have proved to possess the energy to persevere over the long run. The Baldrige Award winners, and others who are recognized for global class quality, communicate that it requires approximately six years of hard work before the program is really cooking.

2 *Goals.* These companies have extremely ambitious goals that are not based on what they might be able to do, but on the achievements of other organizations that have been successful. By benchmarking others, they know the goals are attainable because other quality organizations already have achieved them.

3 *Action plan.* Each of these quality organizations have ongoing action plans that are closely monitored. Goals without solid plans to obtain them will be as successful as performance improvements gained by wishing instead of working. It just won't happen.

4 *Total organizational commitment.* Every person in every operation within the organization is attuned to quality. It is as important in marketing and finance as it is in design and engineering. Administration, human resources management, sales, and every other part of the organization must key to quality. Revolutionary changes in quality can happen only when the entire organization is tuned into excellence and working to improve performance in every area of the operation. Nothing less is acceptable.

5 *Training.* Each organization provides its people with ongoing training in every aspect of quality. Training is based on the strategic business plans which provide guidance to all aspects of the organization. Nothing is overlooked.

TQM initiatives in many forms have been designed and implemented over the past few years in the United States. Sometimes these efforts are initiated out of reflex responses to published TQM success stories, and sometimes they are initiated by a supplier because of pressure from a purchaser. Some TQM implementations are received with open minds by workers and managers, and some are not. In most organizations which have not been traditionally focused on quality and worker involvement, TQM implementation presents significant challenges.

*There have been many TQM failures and partial successes. Success rates are running at about 33 percent, Brown et al.* [5]. Sometimes initial failures can be reversed and success is gained by "in-flight" adjustments through trial-and-error methods. The trial-and-error method is typically slow and tends to create both people and technology problems that might be prevented through a more appropriate TQM design and implementation. Using experiences in the field, *Johnson provides five pragmatic guidelines, with comments, for initiating and implementing TQM in an organization* [6]:

1 *Determine the real current environment.* This important assessment requires management to determine how their people (everyone) perceive the environment (their current organization). Management beliefs and employee perceptions may be worlds apart.

One method begins with a checklist developed especially for this purpose. Individual managers could hold small group discussions and assess the answers to these and other TQM-related subjects. The results could then be combined to provide an organization overview in each area. This hands-on process supports the organization because employees see that the leadership is serious about understanding the situation which is a prerequisite for installing a TQM process.

A second method that suits larger organizations is to have an organization that specializes in such assessments come in and complete an assessment survey. This can be a stock survey or one tailored to exact organizational needs. Some organizations now are able to complete a benchmarking survey for you so that solid stretch goals can be developed.

2 *Determine what the environment should be.* Each work environment is and should be different. There are strengths and weaknesses in all of them. The organization must decide what their work environment should be, based on organizational vision, objectives, goals, and plans. Then training and work activities must be designed to develop this work environment.

3 *Outline the difference.* The difference between where the environment is and where it should be becomes the focal point for goals, training, improvement projects, and the rewards and recognition program.

4 *Develop the program.* The program must be based on need. Training for training's sake and rewards and recognition for the wrong things lead to nothing more than confusion. Key areas to review include the following:

*Employee focus.* Includes employee need, level of commitment, training requirements, rewards and recognition, skills, and the employee as a customer.

*Customer focus.* Includes all aspects of the customer's needs, wants, and desires and the organizational plan to meet these requirements.

*Cultural change.* The basic culture of the organization must be changed to support quality. Myths and perceptions that do not support TQM must be uncovered and eliminated. This could be a considerable undertaking if the organization has been an employee user and abuser. Positive employee-centered policies and programs that support the employee, the customer, and the organization must be developed and put in place. Employees must know their importance within the organization and the expectations the organizations has for them plus the rewards and recognition that can be expected for their efforts above and beyond normal expectations.

*Performance or quality teams.* Teamwork is essential to the TQM process. Everybody must be included in these teams, trained to participate as a team, and encouraged to be a real team player.

*Improvement projects.* Voluntary efforts have proven less than satisfactory. Every team must be trained to handle improvement projects effectively. The projects must be tracked and their efforts rewarded.

5 *The kickoff.* Getting started in a timely fashion is extremely important. Bold changes should be made at the outset. These should be followed by continuous change that exhibits the organization's seriousness about TQM. The chief executive should lead the charge. Battles are not won by armchair quarterbacks; they are won by field leaders who are out in front of the troops, leading through vision, solid goals, and example. In all too many cases, the TQM process never gets off the ground. A vision is developed, certain members of management are trained, and then the process slides into oblivion. It is more important to get started.

Reprinted, with permission, from R. S. Johnson, *TQM—Leadership for the Quality Transformation*, Milwaukee, WI: Quality Press, pages 4, 10–12, 1993.

## Radar Repair Process — TQM Implementation Teamwork Case [SR]

The government did not have a depot repair facility for one of its airborne radar systems. Instead it contracted the repair to a private company. The radar consisted of four separate units. When a failure occurred, the maintenance troops would troubleshoot the radar system, determine which of the four (different types of) units had failed, and replace it with a new one from their extra stock. They would then send the failed unit to the contractor's repair facility. The contractor would repair the unit and return it to the government.

There were not many of these extra units, so when one was broken, minimizing the repair turnaround time was very important in keeping all of the aircraft operating. When the program first started, the contractor's turnaround time was about 70 days. The firm had not put much effort into defining the repair processes to be carried out when a failed unit came to the factory.

In an effort to reduce the turnaround time and per unit cost, the contractor's program manager put together a cross-functional team to study the situation and to define and develop the proper repair processes for one type of failed unit. The team consisted of design engineers, assembly workers, manufacturing engineers, production-control dispatchers, managers, QC inspectors, and test technicians—the same people that had been doing the repair work without any clearly defined procedure, with respect to the entire system.

The team spent about a month observing the process and developing process flowcharts. These flowcharts included every step that a unit would go through from the time it entered the factory door to the moment it left the shop. A repair system was optimized, rather than its individual parts. The information was then synthesized, documented, and implemented.

The procedure was implemented and a turnaround time histogram chart was posted so everyone could see the results, if any, of their efforts. Within several months, the turnaround time dropped from 70 days to 35 days.

The team was then asked to repeat the operational analysis for the remaining three types of radar units. The developed procedure was followed for these units, with the same results. Costs were reduced, and the customer was delighted because the turnaround time had been cut in half.

## 29.3 TQM THEMES

*TQM initiatives are typically centered on a customer-service, a continuous-improvement, and/or a total-involvement theme.* TQM structures must be tailored to fit the current organizational environment and work toward the vision and goals set in the TQM strategic plan. Designing and fine tuning a TQM initiative to meet an organizational vision based on any one of the three themes can be challenging. *Two comprehensive criteria available to help structure a TQM initiative or assess TQM performance are the Baldrige Award guidelines document* [7] *and the ISO 9000 standards document* [8]. Each of these documents was briefly discussed in the context of quality systems in Chapter 3. We will expand our treatment of these documents and examine their potential to help in establishing effective TQM.

### Baldrige Award Criteria

The Malcolm Baldrige National Quality Award has drawn a great deal of attention in American manufacturing industries since its introduction in 1988. The Baldrige Award (named after Malcolm Baldrige, U.S. Secretary of Commerce, 1981–1987, who died in a tragic rodeo accident in 1987) is based on Public Law 100-107, which contains eight fundamental points [9]:

1 The leadership of the United States in product and process quality has been challenged strongly (and sometimes successfully) by foreign competition, and our nation's productivity growth has improved less than our competitors' over the last two decades.

2 American business and industry are beginning to understand that poor quality costs companies as much as 20 percent of sales revenues nationally and that improved quality of goods and services goes hand in hand with improved productivity, lower costs, and increased profitability.

3 Strategic planning for quality and quality improvement programs, through a commitment to excellence in manufacturing and services, are becoming more and more essential to the well-being of our nation's economy and our ability to compete effectively in the global marketplace.

4 Improved management understanding of the factory floor, worker involvement in quality, and greater emphasis on statistical process control can lead to dramatic improvements in the cost and quality of manufactured products.

5 The concept of quality improvement is directly applicable to small companies as well as large, to service industries as well as manufacturing, and to the public sector as well as private enterprise.

6 In order to be successful, quality improvement programs must be management-led and customer-oriented, and this may require fundamental changes in the way companies and agencies do business.

7 Several major industrial nations have successfully coupled rigorous private-sector quality audits with national awards giving special recognition to those enterprises the audits identify as the very best; and

8 A national quality award program of this kind in the United States would help improve quality and productivity by:

 a Helping to stimulate American companies to improve quality and productivity for the pride of recognition while obtaining a competitive edge through increased profits;

 b Recognizing the achievements of those companies that improve the quality of their goods and services and providing an example to others;

 c Establishing guidelines and criteria that can be used by business, industrial, governmental, and other organizations in evaluating their own quality improvement efforts; and

 d Providing specific guidance for other American organizations that wish to learn how to manage for high quality by making available detailed information on how winning organizations were able to change their cultures and achieve eminence.

The Baldrige Award has three basic recipient categories: manufacturing, service, and small business. Awards may or may not be given in each category each year. Table 29.1 lists the award winners by category from 1988 to 1993. Baldrige Award–winning companies have gained recognition nationwide for their quality

**TABLE 29.1    MALCOLM BALDRIGE NATIONAL QUALITY AWARD WINNERS 1988–1994**

|      | Manufacturing | Small Business | Service |
|------|---------------|----------------|---------|
| 1988 | Motorola, Inc.<br>Westinghouse Commercial<br>   Nuclear Fuel Div. | | |
| 1989 | Milliken & Co.<br>Xerox Business Products<br>   and Systems | Globe Metallurgical, Inc. | |
| 1990 | Cadillac Motor Car Co.<br>IBM Rochester | Wallace Co., Inc. | Federal Express |
| 1991 | Solectron Corp.<br>Zytec Corp. | Marlow Industries | |
| 1992 | AT&T Network Systems Group<br>   Transmission Systems Business Unit<br><br>Texas Instruments, Inc., Defense<br>   Systems & Electronics Group | AT&T Universal Card Services<br>The Ritz-Carlton Hotel Co. | Granite Rock Co. |
| 1993 | Eastman Chemical Co. | Ames Rubber Co. | |
| 1994 | | Wainwright Industries | AT&T Communication<br>   Services<br>GTE Directories |

achievements. In addition, they serve as role models for other manufacturing companies in their efforts toward total quality improvement.

There are seven examination categories involved in the Baldrige award [9]: (1) leadership, (2) information and analysis, (3) strategic planning, (4) human resource development and management, (5) process management, (6) business results, and (7) customer focus and satisfaction. (These seven categories and the award scoring weights and guidelines are shown in Table 3.4).

Figure 29.1 depicts the relationship between the seven categories. Here, we see a Baldrige model consisting of a driver, goal, system, and measurement components, all working together. *The Baldrige model serves as a very broad structural guide for quality transformations, and the assessment categories serve as a basis to evaluate the progress of implementation in any organization.*

**Leadership**    The Leadership Category examines senior executives' personal leadership and involvement in creating and sustaining a customer focus, clear values and expectations, and a leadership system that promotes performance excellence. Also examined is how the values and expectations are integrated into the company's management system, including how the company addresses its public responsibilities and corporate citizenship.

**Information and Analysis**    The Information and Analysis Category examines the management and effectiveness of use of data and information to support customer-driven performance excellence and marketplace success.

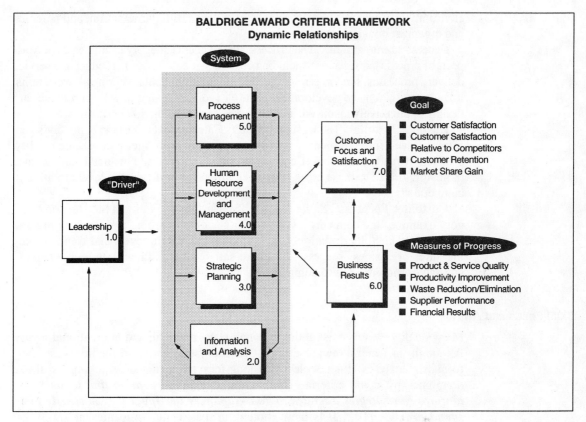

**Driver**

Senior executive leadership sets directions, creates values, goals, and systems, and guides the pursuit of customer value and company performance improvement.

**Goal**

The basic aims of the system are the delivery of ever-improving value to customers and success in the marketplace.

**System**

The System comprises the set of well-defined and well-designed processes for meeting the company's customer and performance requirements.

**Measures of Progress**

Measures of progress provide a results-oriented basis for channeling actions to delivering ever-improving customer value and company performance.

**FIGURE 29.1**   Baldrige Award criteria framework. Adapted from "Malcolm Baldrige National Quality Award—1995 Award Criteria," Washington, D.C.: U.S. Department of Commerce, 1994.

**Strategic Planning**   The Strategic Planning Category examines how the company sets strategic directions, and how it determines key plan requirements. Also examined is how the plan requirements are translated into an effective performance management system.

**Human Resource Development and Management**   The Human Resource Development and Management Category examines how the work force is enabled to develop and utilize its full potential, aligned with the company's performance objectives. Also examined are the company's efforts to build and maintain an environment conducive to performance excellence, full participation, and personal and organizational growth.

**Process Management** The Process Management Category examines the key aspects of process management, including customer-focused design, product and service delivery processes, support services and supply management involving all work units, including research and development. The category examines how key processes are designed, effectively managed, and improved to achieve higher performance.

**Business Results** The Business Results Category examines the company's performance and improvement in key business areas—product and service quality, productivity and operational effectiveness, supply quality, and financial performance indicators linked to these areas. Also examined are performance levels relative to competitors.

**Customer Focus and Satisfaction** The Customer Focus and Satisfaction Category examines the company's systems for customer learning and for building and maintaining customer relationships. Also examined are levels and trends in key measures of business success—customer satisfaction and retention, market share, and satisfaction relative to competitors.

## TQM Focus and Awards

Many quality awards exist at the regional, state, national, and international levels. Although, in general, award criteria provide sound short- to medium-term performance analyses, it is not clear that long-term strategies and thinking should be dominated by award criteria. *Bowles and Hammond caution that focusing on long-range strategies intended to "out-Japanese the Japanese" is clearly not a sound practice* [10]. It is their contention that many companies presently are placing too much emphasis on the pursuit of quality awards and ignoring such key elements of success as innovation, financial performance, and long-range planning.

The Bowles and Hammond perspective is compatible with our expressed philosophy and concept of proactive quality and the quality system triad (Figure 3.2). Clearly, the long-term issues involved in the product, technology, and service interfaces must be considered, and their overall impact on company survival and prosperity assessed. Simply stated, organizations survive on the basis of product, process, and customer service innovations, and profits or financial solvency, not awards.

TQM/TQC has, in one form or another, become commonplace in the United States and other industrialized nations—namely, Canada, Japan, and Germany. Programs in TQM/TQC vary widely because of diverse strategies and tactics both in the consulting community as well as within companies. In terms of its success or failure in obtaining significant results in product and process improvements, TQM effectiveness to date in the U.S. is reported to have about a 1 in 3 success rate in implementation, Brown et al. [11]. *A study based on a survey of 584 companies in the United States, Canada, Japan, and Germany suggests that many quality programs are too broadly focused to be viable* (Fuchsberg [12]). *The results of the survey indicate that the most productive quality programs are designed to focus on a small number of changes critical to product, process, and customer-service improvement.*

## ISO 9000 and TQM

*On the whole, the ISO 9000 series of standards constitutes a set of quality-system requirements which cover a broad range of topics* [13]. In the U.S., these standards are currently known as the ANSI/ASQC Q9000-9004-1994 quality standards [14]. (Table 3.2 summarizes the ISO 9001, 9002, and 9003 standards requirements.) These standards address purchaser-supplier relationships in a quality context (e.g., "Supplier shall establish and maintain . . ." terms). Hence, suppliers use these requirements to guide the development of their TQM structure.

The ISO 9004 (goods-oriented) and ISO 9004-2 (service-oriented) documents are written in a much more TQM-compatible manner than ISO 9001, 9002, and 9003. (Table 3.3 summarizes the ISO 9004 elements.) Here, we see both logical justification and guidance for the TQM structure (e.g., "The quality management system should . . .").

*If an organization is seeking ISO certification, it would be wise to consider the ISO standards and incorporate them into their TQM initiative.* Inevitably, some incompatibilities will arise when we design the people-oriented customer-service–continuous-improvement TQM initiative using a rather "lifeless" requirements document. But we should make every attempt to satisfy ISO 9000 where and when we can. If we are now seeking or later decide to seek ISO 9000 certification (or currently have certification), we do not want latent incompatibilities to emerge. For example, we do not want to see productive parts of our TQM practice outside ISO 9000 compliance. This situation will require us to make structural changes and adjustments that would not have been necessary had we done our homework earlier.

## 29.4  QUALITY TRANSFORMATION SUPPORT ACTIVITIES

A number of support activities are essential to keep quality transformations moving both within the organization and along the supplier network that surrounds the organization. We will discuss product and process certification and audits with regard to their functional relationship in quality transformation. We will also consider measurement strategies, with a focus on monitoring our progress and isolating areas requiring attention.

## Supplier Involvement and Certification

*Supplier (upstream) and customer (downstream) partnerships are essential for initiating modern production methods (e.g., focused factory, just-in-time, and so on) and effective quality improvement, and vital to the success of any business.* Effective supplier involvement requires a two-way communications channel and the firm belief that a win-win relationship is possible. This relationship's success depends on a long-term commitment of technical involvement where the supplier becomes a long-term partner. Usually, some form of supplier certification is used to consummate this relationship.

*Supplier certification programs contain multiple phases, focused on supplier involvement, training, and finally certification.* Each phase is outlined in Table 29.2. *A thorough certification program addresses both proactive and reactive quality control activities.* Both incoming materials and production processes are examined. As supplier development continues through the certification process, a multiyear contract is generally negotiated, and delivery requirements established. Suppliers are highly motivated to form effective partnerships, since many company (purchasing) strategic plans include a reduction in the supplier base (sometimes by as much as 50 percent).

Although the demands on suppliers for certification require a great deal of commitment, communication, coordination, and hard work in definition, design,

---

**TABLE 29.2    SUPPLIER CERTIFICATION PROGRAM OVERVIEW**

**I** Preliminary phase—supplier commitment
   **A** Assess supplier's receptivity to teamwork and process control —
      "becoming the best in their field."
   **B** Gather supplier performance history.
   **C** Understand product history.
   **D** Obtain formal commitment to quality improvement.

**II** Phase I—supplier program review and capability study (present operation)
   **A** Take facility tour—understand supplier's complete process (from order point, scheduling, manufacturing process, to shipping).
   **B** Conduct on-site survey—gather data to determine quality effectiveness.
   **C** Determine SPC comprehension and utilization.

**III** Phase II—quality control plan development (future operation)
   **A** Formulate proactive quality control process plan (relative to needs).
   **B** Formulate reactive, corrective action process and plan.
   **C** Monitor progress and establish review meetings with supplier until all assigned action items are completed.

**IV** Phase III—evaluate quality control.
   **A** Evaluate supplier quality control procedures.
   **B** Evaluate supplier quality control results.
   **C** Assure quality ownership (product and process) at the lowest level (e.g., that workers are source inspecting their own work and doing it right the first time).

**V** Phase IV—supplier certification (by part number)
   **A** Review results of company's inspection.
   **B** Review results of supplier's quality performance within company.
   **C** Review completeness of corrective action plan.
   **D** Review completeness of phases I, II, III, and IV.

**VI** Phase V—ongoing audit and maintenance
   **A** Audit supplier's materials.
   **B** Audit supplier's processes.
   **C** Disqualify or requalify supplier, as appropriate.

execution, and follow-through, the potential rewards are great.  The supplier gains a sole-source (or at least major) commitment from the company, which should lead to long-term profits.  The company gains a trustworthy source of supply and the assurance of component integrity and timeliness of delivery.

## Internal Process Certification

*Internal supply sources can be treated similar to external supply sources.*  In other words, internal processes are many times certified in order to develop and maintain component integrity and timely delivery to the next process (within our organization) or to the customer.  Each internal product line can be treated as a business unit (or focused factory) that is seeking certification.  However, internal certification is usually more informal.  A multiphase approach sometimes taken in manufacturing organizations is described below.

**Preliminary Phase**   The preliminary phase consists of building a cross-functional team, which will take an active role in the phases described below.  The team which includes managers, must be receptive to the teamwork process and be dedicated to preventing and eliminating defects and barriers to improvement.  The team should first gain a general understanding and appreciation of the product and associated processes.  A clear vision of long-term benefits is essential.  The purpose, goals, objectives, and targets for the certification should be clearly defined.

**Phase I—Review of Product Line**   The product line review requires a general assessment of the product line and its processes, backed by hard data.  Development of detailed process flow diagrams represents a first step.  Scrap rate and rework stations and the reasons for rework and scrap should be studied.  Current understanding and application of SPC techniques should be noted.  Consideration should be given to tooling and instrument calibration, as well as to other obvious considerations, such as materials, methods, and so on.

**Phase II—Identification and Evaluation of Critical Process Steps**   In phase II, critical process points are located and examined, so as to monitor, control, and improve the process in an effective manner.  Non-value-added activities should be removed.  This removal may require reorganization with respect to product and process engineering as well as on the production floor (e.g., direct production as well as production support services).  Mistake-proofing in processes should be considered and incorporated where justified.

**Phase III—Elimination of Inspection**   A major goal is to eliminate traditional sorting inspection.  It should be replaced (when possible) with source inspection and SPC.  For example, six-sigma processes need virtually no formal inspection, other than SPC monitoring to detect process shifts (e.g., the goal of six-sigma processes or ZD plans is to virtually eliminate sorting inspection).  The formulation of swift corrective action plans are also part of this phase.

**Phase IV—Process Certification, Audit, and Maintenance**    Process certification recognizes process integrity, a fully certified process should ideally result in a product which will meet requirements 100 percent of the time.  This concept of zero defects is functionally expressed through Motorola's Six-Sigma, 3.4-maximum-defective-parts-per-million concept.  The final phase also assures that ongoing periodic process audits and continuous process maintenance keep the process at the highest possible level of integrity.  Hence, continuous vigilance, in the spirit of constant improvement, is required.  Because of changes in technology, product lines, and so forth, quality presents a moving target.

## Quality Performance Audits

*Making high quality happen through customer satisfaction can be facilitated by conducting quality audits, which provide an ongoing examination of policy, strategy, planning, execution, and follow-through.*  Juran defines a quality audit as "an independent review of quality performance" [15].  Mizuno states that quality audits serve a diagnostic function regarding the broad spectrum of TQC [16].  Hence, we should expect constructive feedback aimed at continual quality improvement from our quality audits.

Before we examine the details of quality audits it is appropriate to consider their strategic importance.  A large commitment of time and resources is typically a prerequisite of quality audits.  We must make sure that the benefits of the audit outweigh the burdens associated with conducting it.  Otherwise, an audit bureaucracy may be established which serves no purpose other than perpetuating itself.

## Quality System Audit Structure

*The purpose of a quality audit is to view and assess a true picture of the "everyday" workings of the quality system.*  Juran poses six strategic questions that should be addressed with each audit [17]:

1  Are our quality policies and quality goals appropriate to our company's mission?
2  Does our quality provide product satisfaction to our clients?
3  Is our quality competitive with the moving target of the marketplace?
4  Are we making progress in reducing the cost of poor quality?
5  Is the collaboration among our functional departments adequate to assure optimal company performance?
6  Are we meeting our responsibilities to society?

*Juran, Ishikawa, and others tend to use the criteria set forth in prestigious quality awards such as the Japanese Deming Prize or the U.S. Baldrige Award as broad guidelines for quality audits.*  For example, Juran suggests using the Deming Prize guidelines (policy and objectives, organization and its operation, education and its dissemination, information flow and utilization, product and process quality, standardization, control and management, quality assurance functions, systems and methods, results, and future plans) to set up a comprehensive audit super-structure [18].

*Mizuno views quality audits as a means of assessing achievement or lack of achievement, education or lack of education, and understanding or lack of understanding through all levels of operations and management in TQC* [19]. He offers a generic quality-audit checklist (shown in Table 29.3). This checklist is tailored to the Japanese TQC philosophy and the Japanese system of registered quality circles. In lieu of QC circles, a U.S. based audit would substitute team-based structure and activities (e.g., quality councils, work teams, and corrective-action teams).

Mills provides a detailed treatment of quality audits, in which he notes a distinction between quality system and quality program audits [20]. He describes the organizational characteristics of quality audits with a focus on audit management. *Mills breaks the quality audit into a number of functional steps:* **(1)** *initiation,* **(2)** *planning,* **(3)** *implementation,* **(4)** *interpretation,* **(5)** *reporting,* **(6)** *corrective action requests,* **(7)** *audit effectiveness, and* **(8)** *follow-up with auditors.*

In the end, each organization must tailor its quality audits to its specific characteristics. In general, audits include critical parameters relative to products, processes, suppliers, and customers. Typically, audits are conducted on a 6- to 12-month basis, with advanced notification of areas to be audited.

### Quality-Business Metrics and Performance Evaluation

*The degree of success of quality transformation and improvement activities must be measured in order to gauge both internal and external customer satisfaction (and business success).* Effective performance measures, or metrics, provide an indication of effective actions and also calibrate their effectiveness. Hence, metrics help us to certify successes and/or provide support for corrective action. Furthermore, quality based metrics that can be converted or scaled to financial dimensions are extremely useful in justifying actions to skeptics.

There are literally an infinite number of combinations of potential quality metrics associated with customers and quality performance. Juran has developed four basic levels of metrics [21] and suggests a pyramid arrangement to represent the hierarchy (shown in Figure 29.2).

The hierarchy shown is quite general and forms the basis of a sound management information structure. In much of his work, Juran points out the similarities between sound financial planning and management and sound quality planning and management. He stops short of advocating an integration of the two critical, and related, parameters in a "master" information system. Crosby stresses the cost of nonconformance (quality) and its incorporation in ordinary accounting systems [22]. However, like Juran, he does not provide details on quality-metric design.

*Integrated quality-business metrics must be used throughout the product life cycle.* Critical quality metrics must be defined and developed for each stage. *In general, quality-business metric definition should consist of four basic steps:* **(1)** *determination of purpose or focus,* **(2)** *identification of the metric and units*

# TABLE 29.3   QUALITY SYSTEM AUDIT STRUCTURE

**I Scale of implementation (TQC)**
- A Whole company
- B Only in certain factories
- C Participating divisions
  - 1 Manufacturing
  - 2 Technology
  - 3 Clerical and administrative
  - 4 Sales
  - 5 Customer service
- D Only QC circles

**II Method of implementation (TQC)**
- A President has announced start of TQC
  - 1 No
  - 2 Last year
  - 3 Two to 3 years ago
  - 4 Four to 5 years ago
  - 5 More than 5 years ago
- B TQC promotion division (office, section, center) established
- C The purpose of introducing TQC
  - 1 Has been made clear
  - 2 Has not been made clear
- D TQC position plan
  - 1 Exists and all activities conform to it
  - 2 Does not exist or activities do not conform to it

**III QC education**
- A Is being carried out on a planned basis
- B Only those who request it are sent to outside quality control workshops
- C Percentages having received QC training (either inside or outside the company)
  - 1 Top executives (x%)
  - 2 Middle management (x%)
  - 3 Clerical and administrative staff (x%)
  - 4 QC circle leaders (x%)

**IV Policy control**
- A No policy has been defined
- B Policy is defined, but not managed
- C There is a long-term management plan according to which annual policy goals are set
- D Presidential policy has been made clear, middle management has developed this policy, and it is being carried out
- E Periodic checks are made to see how much progress has been made toward achieving policy goals, with the following year's policy goals established in line with the results of these checks
- F There is a clearly defined product quality policy

**V Quality assurance**
- A There is source control for new product and new technology
- B Critical quality problems have been recorded and are being analyzed

- C There are charts and diagrams of the whole quality assurance system and a list of the activities required at each step for organized quality assurance activities
- D There is a quality information network
- E Quality assurance is limited to reducing the number of defective products on the production line

**VI QC circle activities**
- A QC circle activities
  - 1 Are being carried out
  - 2 Were implemented one year ago
  - 3 Have been in force for 2–3 years
  - 4 Have been in force for 4–5 years
  - 5 Have been in force for more than 5 years
  - 6 Are not being carried out
- B QC circle activities cover
  - 1 Only part of the factory
  - 2 The whole factory
  - 3 The administrative, sales, and customer service divisions
- C Registration with QC circle headquarters (Japanese)
  - 1 All have been registered
  - 2 Some have been registered (x%)
- D Participation in regional meets
  - 1 Ongoing
  - 2 Have presented reports
  - 3 Have received award(s)
- E Effectiveness
  - 1 As expected
  - 2 Inadequate
  - 3 Extremely disappointing

**VII Control**
- A Daily activities and policy objectives are controlled using lists of control items
- B There is cross-functional management for
  - 1 New product development control
  - 2 Quality assurance
  - 3 Cost control
  - 4 Production volume control
- C Production line control
  - 1 QC process charts and tables
  - 2 Work standards

**VIII Effectiveness**
- A TQC has proved highly effective
  - 1 The nonconformance rate has been reduced
  - 2 Sales have gone up
  - 3 Profits have increased
- B TQC has been only partially effective
- C TQC has had very little effect

*Note:* Our readers may interpret TQC as TQM and should substitute teams for QC circles.

*Source:* Adapted, with permission, from S. Mizuno, *Company-Wide Total Quality Control*, Tokyo: Asian Productivity Organization, pages 278–280, 1988.

**FIGURE 29.2**  Juran's units-of-measure pyramid. Reprinted, with permission of The Free Press, a Division of Simon & Schuster, from J. M. Juran, *Juran on Leadership for Quality*, New York, NY: Free Press, page 259, 1989. Copyright 1989 by Juran Institute, Inc.

*of measure, (3) identification of the sensor or source of data, and (4) selection of the display method and linkage to the next higher hierarchical level.*

For example, the AT&T Power Systems "Golden Thread" quality-business measurement system [23] defines a hierarchy of measures dealing with customer satisfaction, employee satisfaction, and stakeholder satisfaction. This intricate (and integrated) network of metrics reflects business success in general and includes: customer satisfaction for product/offering acceptance, product performance in the short and long run, product delivery and support, employee satisfaction, stockholder profitability, market growth, and community involvement.

**Metric Focus**  *Every metric should make a critical contribution to quality control, quality improvement, or both, and ultimately to business success.* Hence, the benefits that result from a metric should justify the burdens associated with data collection, analysis, and storage. For example, a statistical control chart should be used only at critical process points where it contributes to detecting the presence of special causes. Its purpose is to help operators to produce better products at all times. The same chart or metric might also be useful to provide evidence, for customers, that the process is generating high-quality product and perhaps to eliminate downstream inspection operations.

**Potential Metrics and Units of Measure**  There will typically be a choice for metric design in terms of metric placement in the product-process hierarchy. *Metric choice should favor proactivity in order to prevent quality loss.* Hence, upstream metric placement should be considered for all critical production points. However, this recommendation is not meant to imply that endpoint metrics regarding stopgap measures and historic final product performance, customer satisfaction, and so forth, are unimportant.

The choice of units of measure is also very important. Creative thought processes will usually develop a number of alternatives. For example, *choices exist between broadly defined attributes metrics combining multiple observation points and narrowly defined variables metrics at single observation points.* Potential measurement-unit alternatives must be evaluated carefully. The selected unit should be justifiable in terms of its ability to contribute to quality at its level of origin in the hierarchy and its ability to convey meaningful and timely information to higher levels—ultimately to the top level, where it measures business success.

There is a tendency to favor highly quantitative metrics due to their compatibility with quantitative analysis techniques. This tendency favors historic perspectives and reactive quality assurance (unless the quantitative metrics are related to upstream process variables). Preventive means are driven by leading (in time) metrics and metrics which may be somewhat experience or judgment based. For example, once we see solid data that we are not meeting specifications downstream, we may have already produced a large amount of defective product upstream.

Typically, metrics are developed one at a time. However, single channels and single dimensions seldom are capable of providing more than a narrow view of a product or process. Hence, consideration must be given to metric definition both as individual measures and as multiple-channel or multiple-dimensional views. Broad vision is required to support broad decisions. Once again, AT&T's Golden Threads are offered as an excellent example of measuring the quality-business system performance as a whole, not as a collection of disjoint parts.

**Sensor or Data Source**     *Each measurement originates at a sensor.* Some sensors, such as thermocouples, are straightforward physical devices, whereas others may be a complex combination of physical devices and human observation and interpretation. The sensor may not pick up the true essence of the signal. Hence, inaccuracy or imprecision may pose a problem. *Dependable sensors and well-chosen sensor placement are critical in the chain of producing meaningful and timely data.*

To a large degree, experience and judgment remain critical factors in many quality related metrics. For example, measures of customer satisfaction relating to field performance or job satisfaction and so on, are difficult to determine and to translate into meaningful numbers. Each time data are translated from one form to another, something is lost.

Even though quality records (i.e., historic data) are useful in assessing performance, the validity of previously recorded data is always a legitimate concern. *The data should be relevant to the objective at hand. The integrity of the data (source, sensor, external conditions, and so on) must be established.* Ishikawa suggests that we regard any previously collected data as suspect, provided we cannot trace their source, conditions of collection, and so forth [24]. He contends that, in many cases, data are amassed (and sometimes selectively screened) to support only what the collector desires to show.

The matter of data integrity is critical. In principle, *each metric is carefully designed to support crucial decisions. Misinformation through the use of biased (inaccurate), imprecise (highly variable), or untimely data may easily lead to disastrous decisions.*

**Display Method and Hierarchical Linkage** *Data must be converted to information and made available to the right people at the right time in order to benefit the organization.* Typically, *metrics support the decision process in three ways:* **(1)** *to spot trends,* **(2)** *to compare alternatives, and/or* **(3)** *to predict performance.* In any case, metrics are designed to support decisions that impact the future (the past is beyond our influence). Hence, by definition, decision-relative projections, through trend analysis comparisons and numerical predictions, are the focus. The point is to produce timely, effective, and efficient communications, which is often best accomplished by presenting combinations of summary statistics, graphics, and descriptive statements, in the context of the issue at hand.

*Consideration of the pyramid of metrics, Figure 29.2, as well as the breadth of the quality system interfaces, Figure 3.2, are important in defining and developing metrics and building the quality information network.* It is clear that compression and concentration, as well as stratification, of data progress sequentially upward through the pyramid. The definition and nature of the higher-level ratios, indices, and summaries must be decided with care, as they are typically unique to each organization and depend on product lines and organizational formats.

## Harsh Environmental Conditions—Producer-Customer Cooperation for Robust Performance Case [TW]

I was involved in a project to demonstrate an infrared sight on an armored vehicle. I was asked to train the crew in the use of the sight and to direct the demonstration. I spent a week with the crew and vehicle and got an excellent opportunity to see their working environment and the field use and abuse our system might see. The tests were done at the Yuma Proving Grounds and included long drives through the desert and live firing with a 25-mm gun.

The most formidable design challenge was to compensate for the shock and vibration that was a standard part of the environment. The sight was to be used for firing and had to maintain accuracy in spite of significant gun shock. Our first experience was not successful; we discovered that no amount of hand tightening would prevent slippage. We had to pin our sight to the turret with a metal dowel. All internal mechanisms were made as simple as possible, using the smallest viable number of components. "Minimalism" was the fashion statement of the day.

Another problem resulted from the humidity and dustiness of the environment. The sight had to be hermetically sealed—not just to keep the moisture out, but also the dust. There's no way the crew could be expected to clean internal optics under field conditions. We were required to use chemical desiccants not because humidity was leaking in, but because the humidity that was sealed in at the factory would condense when the temperature fell.

The sight also had to be able to withstand being kicked and used as a step. The only way that the crew could get in and out of the vehicle was through the roof, and everything in the cabin was "climbed" on at various times. The support casting was made thick so it could easily support a person's weight.

A lot of ergonomic considerations had to be kept in mind. The sight had to operate at night, so all symbols on the control panel had to be large and clear enough to read under difficult lighting conditions. The toggle switches and knobs had to be large, robust, and well separated so that a crewman with large chemical suit gloves could operate them.

The automatic adjustment ability of the video display had to be maximized, since the crew would not have the time or opportunity to fine-tune the picture while bouncing along at 40 mph or while being fired upon.

A good deal of vital information was gained through first-hand field experience, inasmuch as the sight's intended, normal use here would constitute abuse under most other circumstances. In this situation, robust performance was clearly a requirement and had to be an integral part of the design from the very first.

## 29.5 QUALITY TRANSFORMATION DIVERSITY AND COMMONALITY

Total quality, TQ, which includes both TQC and TQM, is not a monolithic phenomena by any means, there is a great deal of commonality in intent, if not in actual mechanics. Here, we will address the general characteristics that most successful quality transformation applications have in common.

### Philosophy

*The most dominant feature in the modern philosophies of quality we have reviewed is the recognition that quality is a broad concept manifested in nearly everything that an organization does.* As stated by Hutchins [25], quality is everything that an organization does, in the eyes of its customers, that will encourage them to regard that organization as one of the best, if not the best, in its particular field of operation. In other words, quality is the overall measure of the achievement of customer satisfaction.

*Quality is not regarded as an absolute value, but is related to standards of performance that are ultimately determined by customer experience and satisfaction. The benchmarks are set by the best performer in a given area.* Quality improvement is seen as a dynamic process that can enable an organization to become more competitive across the entire spectrum of its activities and at all levels. There is also the complementary idea of continuous improvement. Rather than allowing the standards of quality (and productivity) to remain static, we must make a continuing effort to improve them.

The TQ literature regularly makes the point that, traditionally, quality has been too narrowly defined and has been regarded as incompatible with productivity, with one achieved or improved at the cost of the other. TQ philosophy regards this as a fallacy as the two are essentially inseparable in achieving an organization's goals. Deming's chain reaction [26] attests to the fact that quality and productivity are complementary, not competitive. Crosby points out that while 10 to 20 percent of sales dollars reflect the costs associated with traditional quality activities (inspection, scrap, repair), the costs need not be more than 4 or 5 percent [27]. Harrington points out that high-quality products yield 40 percent more return on investment than low-quality products [28].

*The TQ philosophy can be characterized as focusing on long-term goals, strategies, and activities.* Conventional practices of focusing solely on volume of production (accepting repair and scrap as cheaper in the short-term) are widely criticized in the TQ philosophy. *Overall, the TQ philosophies embrace long-range vision, competition-oriented innovation, and close relationships with customers and suppliers. Another universal feature of TQ philosophies is respect for employees and the role that they play in implementing and accomplishing improvements.* Indeed, the role of the employee can be characterized as crucial in quality transformation.

## Employee Empowerment

Most TQ approaches stress the development of a people-based philosophy and recognize that "If people don't care and aren't interested, then no amount of quality control will make a difference." Despite the emphasis on teamwork, TQ advocates treating and respecting employees as individuals. Management's primary role is to value the thoughts and desires of employees and to involve and empower workers in the quality process. *Development of employee potential through education and training and varying job assignments is a strong feature of TQ.*

The Deming cycle forms the basis for "hands-on" quality improvement [29]:

1 Planning, of all tasks to be performed and the sequence of events.
2 Doing, carrying out the tasks.
3 Checking, to ensure that tasks have been performed in the manner required.
4 Action, whatever may be required to improve the plan.

Hutchins presents an interesting paradigm, based on the four distinct Deming cycle phases, which applies to any form of cooperative effort on the production floor [30]. Hutchins's paradigm is compared to others in Figure 29.3. In the Taylor scientific-management scheme (see Figure 29.3a and b), the direct labor employee is charged only to "Do" and is excluded from the other three phases. Hutchins defines, or rather resurrects, the notion of a craftsmanship system in which the craftsperson is responsible for his or her own quality and essentially controls the day-to-day operation of the entire cycle (Figure 29.3c). Hutchins has updated the traditional concept by introducing craftsmanship elements into groups of workers rather than referring to individuals (see Figure 29.3d). In summary, the Deming PDCA cycle is portrayed in a Taylor-regime context, a scientific-management context, a traditional-craftsmanship context, and then in Hutchins's craft-group context. In the latter, the group is empowered to control the work cycle, and management and specialists act to assist the group (e.g., whole-process-based teamwork).

## Employee Responsibility and Recognition

The literature is unanimous in acknowledging that *the increased responsibility (and authority) taken on by workers must be rewarded with increased recognition from management.* However, there is considerable difference of opinion as to how recognition is to be manifested, particularly as it relates to financial rewards. The payment of bonuses and monetary rewards is sidestepped or minimized by many TQ authorities, such as Crosby and Deming. Rewards suggested for special accomplishments (e.g., for goals reached or jobs well done) by an individual or team consist of official management recognition, presentation of small symbolic gifts (such as "recognition" jewelry, certificates of performance, and so on), publication and promotion of the honor, and increased access to training, seminars, new responsibilities, and further job enrichment or promotion.

*Effective recognition and reward structures are critical for successful quality transformations.* Traditional individual performance reviews and piece-rate financial rewards tend to create problems, as noted earlier. Typically, *recognition and rewards must be associated with the team, group, or unit that achieved the accomplishment*; we must encourage the work unit. Otherwise, we send a signal to the team that our recognition and reward process is inconsistent with our new work structure.

## Management

The literature and field reports are unanimous in their opinion that *management must provide role models and continuously and visibly demonstrate (to the customers) the quality philosophy in everything that it does.* Upper management must first demonstrate its commitment and set the example. Crosby emphasizes the need for upper management to carefully formulate a quality policy for the company and introduce it in stages beginning with all management levels and proceeding down to the rank-and-file workforce.

The Crosby approach and other schools of thought on quality transformation recognize the need for a "waterfall effect"—a change in attitude that starts at the top and "washes each level of management clean of its old bad habits" (Harrington [31]). Harrington makes note of the tendency of many well-intentioned programs to bog down at the middle-management level. In this respect, securing support and involvement from middle- and lower-level management is also critical. Interestingly enough, it seems that most of the literature on TQ is aimed at upper management, goes into great detail as to how the rank-and-file workforce should be involved, but tends to side-step middle management. It is almost always assumed that middle- and lower-level managers will know what to do. Prescriptions for changing behavior at their level are not as detailed, although the principles are the same.

*TQ approaches, by and large, expect mature behavior from workers in the sense that a substantial degree of initiative and self-direction is expected.* Such behaviors

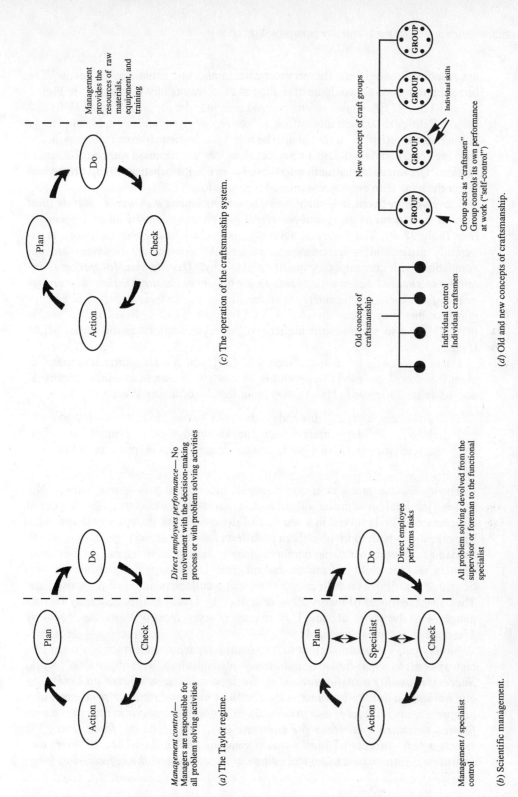

*Management control—*
Managers are responsible for
all problem solving activities

*(a) The Taylor regime.*

*Direct employees performance—* No
involvement with the decision-making
process or with problem solving activities

*(c) The operation of the craftsmanship system.*

New concept of craft groups

Individual skills

Group acts as "craftsmen"
Group controls its own performance
at work ("self-control")

Old concept of
craftsmanship

Individual control
Individual craftsmen

*(d) Old and new concepts of craftsmanship.*

*Management / specialist
control*

Direct employee
performs tasks

All problem solving devolved from the
supervisor or foreman to the functional
specialist

*(b) Scientific management.*

Management
provides the
resources of raw
materials,
equipment, and
training

**FIGURE 29.3** Craftsmanship and the PDCA cycle. Reproduced, with permission from Nichols Publishing Company, from D. Hutchins, *Quality Circles Handbook*, East Brunswick, NJ: Nichols Publishing Company, pages 18, 19, 22, and 24, 1985.

are associated with being able to meet esteem and self-actualization needs. The opportunities to do so would be classified as "motivators" by Herzberg (see Figure 4.1 and Table 27.6). It must be emphasized that, in order for the increased skills and responsibilities to become motivators, authority, opportunities, rewards, and recognition must be provided. It should also be noted that hygiene factors (salary, policies, working conditions) should not be neglected, as they are common sources of dissatisfaction. Organizational attention to these factors helps substantially to create and foster the idea of an employee-centered organization.

*The employee focus in quality transformations entails a theory Y attitude (that work is as natural as play) and pattern B behavioral approach to management, (see Table 27.5).* The theory X assumptions about human behavior (work is inherently distasteful) and consequent pattern A management behavior are not compatible with contemporary quality philosophy. *Quality transformation is intended to succeed based on appeals to higher-level (in the Maslow hierarchy) needs of workers.* Management, in addressing these needs, is expected to be supportive rather than strictly directive. The Covey PCL and its focus on the "whole person" [32] also addresses the higher level needs and reinforces the earlier MLM studies.

Ouchi describes what he calls "theory Z" [33], which is an approach to management based on Japanese practice and is, in many ways, an extension of theory Y, according to Dinsmore [34]. The two main lessons of theory Z are

1 Trust for employees. Trust and productivity go hand in hand in this approach.
2 Subtlety. Subtlety refers to specific knowledge of individuals and their characteristics and is thought to enhance assignments on projects and in work teams.

Theory Z management cultivates nonspecialized career paths through essentially lifelong job rotation to create a flexible and multiskilled workforce. Employees at all levels may be involved in a number of different work groups simultaneously. The purpose is both to involve them in different aspects of the organization and to acquaint them with increasing numbers of coworkers (seen as a means of nurturing subtlety and the ability to understand others). Decision making is heavily participative and draws on collective values and a collective sense of responsibility. The Japanese model of theory Z, as described by Ouchi [35] is generally incompatible with the styles of most U.S. managers, even most of those that could be described as theory Y–oriented.

Simply put, a *de facto* attempt to transplant a Japanese system (or any other foreign system) to a non-Japanese culture sounds appealing, but simply won't work. *Successful quality transformations in the United States are based on designing and building a quality-business system around our products, our processes, and our people, not on forcing our products, our processes, and our people into a preformed structure that works for someone else.* This type of *"force fitting"* is simply a fad. *Successful quality transformations are tailored to customers and constitute a systematic and logical course of action to meet the objectives of long-*

*term internal-external customer satisfaction, organizational prosperity and financial solvency, and benefits to our society in general.*

## 29.6 FACTORY ENGINEERING — TRANSITION CASE [HE]

### Background

This case study involves a factory facilities engineering (FE) department within a large company. The FE department is responsible for the definition and design of all rearrangements, installations, moves, and so forth, within the entire plant and consists of 20 employees. FE receives job requests from all product-process units and, in turn, works on the requests and provides an appropriate facilities design or retrofit plan.

The personnel within FE are classed as senior engineer, engineer, or engineering associate. Until recently, and for the past 15 years, one man held the manager's position. During that time, he ruled the department with an "iron fist." He communicated to the department only through the senior engineers. Engineers and associate engineers were left to execute the work under the direction of the senior engineers.

While other departments in the company were allowed overtime, the FE department was not. Even though the company stressed continuing education, both within the plant and outside the plant through bulletin board postings, very few FE employees were allowed time off from their duties to attend courses or seminars.

### The First Year (New Management)

Upon his announcement of retirement, the FE manager was honored with a small recognition ceremony organized by the plant manager to recognize the former's many years of service. At this point, a new FE manager was appointed. The new manager, an "insider," had watched the department closely and was well aware of the problems it had developed under the previous manager's direction. The new manager did away with strict senior engineer authority—the department members were all treated as equals. Jobs were assigned to each individual on a job-by-job basis. Creativity, morale, and motivation improved. FE employees were allowed to work overtime and to attend appropriate outside classes and seminars.

At this point in time, jobs were assigned to each of the 20 department members. Many of the jobs required more expertise than any one engineer possessed. All job requests came with a "drop-dead" date for scheduling purposes. Many times, the assigned employee would need to seek help from a colleague to address an area outside his or her own specialty. Some engineers and engineering associates held these job requests until the last minute, knowing a department peer needed to work on it—thus making the other person look bad.

On other occasions, engineers and engineering associates refused to respect the functional area of expertise of their peers. Thus, engineering work produced was not always as good as it should have been. Deficiencies were usually not apparent for some time (due to the nature of FE schedules and assignments). Still, they created problems that had to eventually be addressed in the facilities modifications.

Even though the department members felt good about their new level of empowerment, dissatisfaction was evident. Eventually, complaints from other departments (customers) regarding work performance (technical problems) and timeliness were received. FE also received complaints from the trade organization (those who actually "worked" the plans developed by FE). Customers were clearly dissatisfied with the service they were receiving (or not receiving). This situation grew steadily worse for one year.

Toward the end of the first year, the new FE manager was called into the plant manager's office and informed that the company intended to pursue a TQM program. Furthermore, he was told that he had been selected as a "guinea pig" to attend a seminar on TQM, try it in his area, and report back as to how well it worked.

### The Second Year

At the beginning of the second year, the manager called a meeting of the department in which he outlined the department's new strategy. The first item addressed was departmental goals and objectives (see Table 29.4). To accomplish the goals and objectives, the manager then proposed a team structure (see Table 29.5). He suggested that a cross-functional approach be taken in terms of team make-up. Then, team responsibilities were developed (see Table 29.6). In addition, the primary responsibilities within teams were defined, as well as team functions, accountability, and professional behavior (see Table 29.7).

Initially, resistance to this new structure and reorientation was encountered. But within one month, the team concept began to take hold. As a result, overall motivation, morale, and communication increased dramatically; customer satisfaction increased; compliments on work completed were received. The department averaged 25 new job requests per week. At the beginning of the year, the department had an open inventory of 180 jobs (that is, jobs waiting to be worked). By the end of the year, the open-inventory job list had been cut to 60 outstanding requests.

TABLE 29.4    DEPARTMENTAL GOALS AND OBJECTIVES

| Goals | Objectives |
| --- | --- |
| Improve job-request response time. | Clearly define quality for external and internal customers. |
| Emphasize cooperation and teamwork. | Realign authority and responsibilities in-line with job classifications. |
| Produce quality product. | Redefine roles, accountability, and performance-evaluation criteria. |
| Improve productivity. | |
| Enhance image with other departments. | |

TABLE 29.5    DEPARTMENTAL TEAM STRUCTURE

| Teams* | Team composition |
|---|---|
| Machine design team—focused on adapting new machinery to process needs | Mechanical engineer |
| High-bay–Mexico team—focused on high-bay plant area and the border plant in Mexico | Electrical engineer |
| Low-bay team—focused on low-bay production areas | Engineering associate |
| Office team—focused on office areas and needs | |

*Teams were focused to needs and areas so that they would be more responsive to and familiar with specific needs of the people operating in these areas.

### The Third Year

At the beginning of the third year, a substantial number of complaints were being received again.  The positive results of the second year seemed to disappear faster than they had appeared in the first place.  This quality-productivity erosion began immediately after the annual performance-rating review.

The open-job level climbed to 140 items within three months.  The morale and overall department attitude was deteriorating rapidly.  After six months, the department was at the 150 level of job inventory, and tensions were increasing.  Now, eight months into the third year, the manager has expressed his displeasure with TQM and the current FE performance level — no other action has been taken.

### Postscript

In Table 29.4, the final objective stated was to redefine "performance-evaluation criteria."  When the subject was addressed in the original meeting, the manager

TABLE 29.6    TEAM RESPONSIBILITIES

**Group responsibilities**
    Timely response to job requests
    Meet established schedules
    Support of other teams

**Individual responsibilities**
  **Engineers**
    Work order breakdown
    Conceptual design
    Detail design
    Technical buyout

  **Engineering associates**
    Detail design
    Drawing
    Specification writing

  **All personnel**
    Communication
    Coordination
    Cooperation

TABLE 29.7    TEAM FUNCTIONS, ACCOUNTABILITY, AND BEHAVIOR

**Team functions**

All job requests will be assigned to team engineers.
Team will decide how best to accomplish assignment.
Team will establish promise dates.
Team will meet weekly with manager to review promise dates and status.
Engineers will sign off all orders regardless of which team member wrote it.
Project activity within team areas of responsibility will be worked by the team.

**Team attitude and behavior**

Quality is judged by our customers.
We are first a service organization.
Goodwill is important to our organization.
We are all professionals and deserve to be treated as such.
When the team succeeds, I succeed.
Cooperation is essential to teamwork.

**Team member accountability**

Contribution to department effort.
Contribution to team effort.
Professional behavior.
Quality consciousness.

---

stated that this phase had not yet been finalized with top management. Nevertheless, a statement that team members would be rated on a team basis was made at the teamwork strategy meeting. Later, a careful study of the company operations manual indicated that the company performance guidelines did not include provisions for team evaluation. They provided only for performance ratings on an individual basis. The team evaluation issue resulted in a grid-lock with company policy and was never resolved. The manager had personally recognized teams for their good work. However, the company recognized only individual accomplishments. The manager did not know which individuals did what tasks and to what degree; nevertheless he did the best he could to fill out the individual job evaluations (for each employee) at the end of the second year.

---

## REVIEW AND DISCOVERY EXERCISES

### Review

**29.1**  Review Johnson's five major differences between companies with outstanding TQM and all others. Has Johnson missed anything? Explain.

**29.2**  In Johnson's guidelines on initiating TQM, he states that, "voluntary efforts have proven less than satisfactory" in improvement projects. What does this mean in terms of how we should go about improvement projects? Explain.

**29.3**  List the seven examination categories used in the Baldrige Quality Award evaluation. Can you think of any categories that should be added?

**29.4**  What motivation exists for the supplier and the customer to form a partnership?

**29.5**  How does supplier certification differ from internal process certification?

**29.6** Is the elimination of final inspection (sorting) a realistic goal in process certification? Explain.

**29.7** What is the purpose of a quality audit?

**29.8** Why is it common to see quality audits developed around quality award criteria?

**29.9** Provide an example set of hierarchical quality measures for an organization of your choice. Ensure that your response considers the four levels in the Juran units-of-measure pyramid (Figure 29.2) and the three quality system interfaces (Figure 3.2).

**29.10** The Crosby quality absolutes and steps (see Table 27.1) focus on the cost of nonconformance. Provide an example of a cost-of-nonconformance metric for
  **a** A food service organization.
  **b** A tool and die manufacturer.

**29.11** Explain how the AT&T Golden Thread policy represents a multidimensional view of quality.

**29.12** Provide an example of each of the ways quality-related metrics are used to support decisions for
  **a** Trends.
  **b** Comparisons.
  **c** Predictions.

**29.13** Regarding piece-rate-oriented production,
  **a** Why does the piece-rate method of employee compensation and recognition not tend to work well in TQ environments?
  **b** Would a group-oriented piece-rate system or gain sharing system be appropriate? Explain.

**29.14** Why is employee maturity (in work behavior) critical to quality transformation success?

## Discovery

**29.15** Reread Juran's sixth strategic quality-audit question dealing with responsibility to society. Then develop
  **a** A list of appropriate considerations.
  **b** An accompanying list of possible measures or evaluation points for the list in part *a*.

**29.16** After viewing the Golden Thread quality metrics, a visitor to the AT&T Power Systems plant asked, "Which metric is the most important?" "How can a manager run a plant without knowing which metric to focus on?" The host replied, "They are all important." "We expect good performance in all areas." Comment on the validity of the host's answer.

**29.17** Choose a product and provide an example of
  **a** A quality metric placed in a proactive position.
  **b** A quality metric placed in a reactive position.

**29.18** For each of the two quality metrics below, describe an appropriate sensor:
  **a** The "sharpness" of a cutting tool (on a lathe).
  **b** The "taste" of a soft drink.

**29.19** Piece-rate rewards have tended to not work well in TQM environments. Why do you think this is? What alternatives do we have?

**29.20** On the basis of Hutchins's depiction of craftsmanship, explain the advantages and disadvantages of the following model types:
  **a** Scientific management.
  **b** Individual craftsmanship.
  **c** Group craftsmanship.

*Questions 29.21 through 29.26 pertain to the Factory Engineering Case.*

**29.21** In the second year, what did the new manager do that was responsible for the turnaround in his department?  Explain.

**29.22** It appears that the new manager empowered his people in year 1, but did not obtain good results.  Explain why.

**29.23** Why did third-year performance results drop?  Explain.

**29.24** If you were the new manager, at eight months into the third year, what would you do (besides looking for a new job)?

**29.25** If you were the new manager's replacement in the ninth month of the third year, what would you do?

**29.26** What was (were) the root cause(s) for the "on-again, off-again" performance in the case study?  What went wrong?  Explain.

*Questions 29.27 and 29.28 pertain to Appendix 29A.*

**29.27** Why have the TQM/TQC movements received so much attention in the "real world" and so little attention in the "academic world?"

**29.28** Why has participative management received so much attention in the "academic world" and so little in the "real world?"

# APPENDIX 29A:  TQM/TQC AND PARTICIPATIVE MANAGEMENT PHILOSOPHIES

A number of specific conclusions can be drawn regarding the participative management movement and the total quality movement in the United States, according to Kolarik and Fox [36].

**1** Both philosophies seek to develop human resources through leadership and creativity.

**2** Both philosophies require high-level management commitment for effective results.

**3a** Total quality is customer directed, marketplace driven, and developed in response to competition for market share, productivity, and profits; it stresses respect for humanity.

**3b** Participative management is employee directed, social science driven, and developed in response to workplace productivity and the quality of worklife; it stresses fulfillment of social, esteem, and self-actualization needs.

**4a** The total quality movement embraces a contingency philosophy which includes respect for humanity and for workplace efficiency.  The results sought are internal and external customer satisfaction and greater market share.

**4b** Participative management philosophies formed one-half of the autocratic-participative philosophical dichotomy.  The result sought is the reconfiguration of the workplace environment to produce a higher quality of worklife (e.g., internal customer satisfaction).

## REFERENCES

**1** R. S. Johnson, *TQM—Leadership for the Quality Transformation,* Milwaukee, WI: ASQC Quality Press, 1993.

**2** R. S. Johnson, *TQM—Management Processes for Quality Operations,* Milwaukee, WI:  ASQC Quality Press, 1993.

**3** R. S. Johnson, *TQM—Quality Training Practices,* Milwaukee, WI:  ASQC Quality Press, 1993.

**4** R. S. Johnson, *TQM—The Mechanics of Quality Processes,* Milwaukee, WI: ASQC Quality Press, 1993.

**5** M. G. Brown, D. E. Hitchcock and M. L. Willard, *Why TQM Fails and What To Do About It,* New York, NY: Irwin, 1994.

**6** See reference 1.

**7** "Malcolm Baldrige National Quality Award — 1995 Award Criteria," Washington, DC: U. S. Department of Commerce, 1994.

**8** "ISO 9000," Geneva, Switzerland: International Organization for Standardization, 1992.

**9** See reference 7.

**10** J. Bowles and J. Hammond, "Being 'Baldrige-Eligible' Isn't Enough," *The New York Times,* Forum, September 22, 1991.

**11** See reference 5.

**12** G. Fuchsberg, "Quality Programs Show Shoddy Results," *Wall Street Journal,* May 14, 1992.

**13** See reference 8.

**14** "ANSI/ASQC Q9000-9004-1994," Milwaukee, WI: American Society for Quality Control, 1994.

**15** J. M. Juran, *Juran on Leadership for Quality,* New York: Free Press, 1989.

**16** S. Mizuno, *Company-Wide Total Quality Control,* Tokyo: Asian Productivity Organization, 1989.

**17** See reference 15.

**18** See reference 15.

**19** See reference 16.

**20** C. A. Mills, *The Quality Audit,* New York: McGraw-Hill, 1989.

**21** See reference 15.

**22** P. B. Crosby, *Quality is Free,* New York: McGraw Hill, 1979.

**23** "Quality Policy, Vision, and Mission," AT&T Power Systems, Mesquite, TX.

**24** K. Ishikawa, *What is Total Quality Control? The Japanese Way,* Englewood Cliffs, NJ: Prentice-Hall, 1985.

**25** D. Hutchins, *Quality Circles Handbook,* New York: Nichols Publishing Company, 1985.

**26** W. E. Deming, *Out of the Crisis,* Cambridge, MA: MIT Center for Advanced Engineering Studies, 1986.

**27** See reference 22.

**28** H. J. Harrington, *The Improvement Process: How America's Leading Companies Improve Quality,* New York: McGraw-Hill, 1987.

**29** See reference 26.

**30** See reference 25.

**31** See reference 28.

**32** S. R. Covey, *Principle Centered Leadership,* New York: Simon and Schuster, Fireside, 1992.

**33** W. Ouchi, *Theory Z: How American Business Can Meet the Japanese Challenge,* Reading, MA: Addison-Wesley Publishing Company, 1981.

**34** P. C. Dinsmore, *Human Factors in Product Management,* New York: American Management Association, 1984.

**35** See reference 33.

**36** W. J. Kolarik and R. R. Fox, "Total Quality Theory and Practice," *Proceedings of the 1992 Pacific Conference on Manufacturing,* pp. 966–973, 1992.

# STATISTICAL TABLES

## TABLE IX.1    TABLED RANDOM NUMBERS

| | | | | | | | | | | | | | |
|---|---|---|---|---|---|---|---|---|---|---|---|---|---|
| 16408 | 81899 | 04153 | 53381 | 79401 | 21438 | 83035 | 92350 | 36693 | 31238 | 59649 | 91754 | 72772 | 02338 |
| 18629 | 81953 | 05520 | 91962 | 04739 | 13092 | 97662 | 24822 | 94730 | 06496 | 35090 | 04822 | 86772 | 98289 |
| 73115 | 35101 | 47498 | 87637 | 99016 | 71060 | 88824 | 71013 | 18735 | 20286 | 23153 | 72924 | 35165 | 43040 |
| 57491 | 16703 | 23167 | 49323 | 45021 | 33132 | 12544 | 41035 | 80780 | 45393 | 44812 | 12515 | 98931 | 91202 |
| 30405 | 83946 | 23792 | 14422 | 15059 | 45799 | 22716 | 19792 | 09983 | 74353 | 68668 | 30429 | 70735 | 25499 |
| 16631 | 35006 | 85900 | 98275 | 32388 | 52390 | 16815 | 69298 | 82732 | 38480 | 73817 | 32523 | 41961 | 44437 |
| 96773 | 20206 | 42559 | 78985 | 05300 | 22164 | 24369 | 54224 | 35083 | 19687 | 11052 | 91491 | 60383 | 19746 |
| 38935 | 64202 | 14349 | 82674 | 66523 | 44133 | 00697 | 35552 | 35970 | 19124 | 63318 | 29686 | 03387 | 59846 |
| 31624 | 76384 | 17403 | 53363 | 44167 | 64486 | 64758 | 75366 | 76554 | 31601 | 12614 | 33072 | 60332 | 92325 |
| 78919 | 19474 | 23632 | 27889 | 47914 | 02584 | 37680 | 20801 | 72152 | 39339 | 34806 | 08930 | 85001 | 87820 |
| 03931 | 33309 | 57047 | 74211 | 63445 | 17361 | 62825 | 39908 | 05607 | 91284 | 68833 | 25570 | 38818 | 46920 |
| 74426 | 33278 | 43972 | 10119 | 89917 | 15665 | 52872 | 73823 | 73144 | 88662 | 88970 | 74492 | 51805 | 99378 |
| 09066 | 00903 | 20795 | 95452 | 92648 | 45454 | 09552 | 88815 | 16553 | 51125 | 79375 | 97596 | 16296 | 66092 |
| 42238 | 12426 | 87025 | 14267 | 20979 | 04508 | 64535 | 31355 | 86064 | 29472 | 47689 | 05974 | 52468 | 16834 |
| 16153 | 08002 | 26504 | 41744 | 81959 | 65642 | 74240 | 56302 | 00033 | 67107 | 77510 | 70625 | 28725 | 34191 |
| 21457 | 40742 | 29820 | 96783 | 29400 | 21840 | 15035 | 34537 | 33310 | 06116 | 95240 | 15957 | 16572 | 06004 |
| 21581 | 57802 | 02050 | 89728 | 17937 | 37621 | 47075 | 42080 | 97403 | 48626 | 68995 | 43805 | 33386 | 21597 |
| 55612 | 78095 | 83197 | 33732 | 05810 | 24813 | 86902 | 60397 | 16489 | 03264 | 88525 | 42786 | 05269 | 92532 |
| 44657 | 66999 | 99324 | 51281 | 84463 | 60563 | 79312 | 93454 | 68876 | 25471 | 93911 | 25650 | 12682 | 73572 |
| 91340 | 84979 | 46949 | 81973 | 37949 | 61023 | 43997 | 15263 | 80644 | 43942 | 89203 | 71795 | 99533 | 50501 |
| 91227 | 21199 | 31935 | 27022 | 84067 | 05462 | 35216 | 14486 | 29891 | 68607 | 41867 | 14951 | 91696 | 85065 |
| 50001 | 38140 | 66321 | 19924 | 72163 | 09538 | 12151 | 06878 | 91903 | 18749 | 34405 | 56087 | 82790 | 70925 |
| 65390 | 05224 | 72958 | 28609 | 81406 | 39147 | 25549 | 48542 | 42627 | 45233 | 57202 | 94617 | 23772 | 07896 |
| 27504 | 96131 | 83944 | 41575 | 10573 | 08619 | 64482 | 73923 | 36152 | 05184 | 94142 | 25299 | 84387 | 34925 |
| 37169 | 94851 | 39117 | 89632 | 00959 | 16487 | 65536 | 49071 | 39782 | 17095 | 02330 | 74301 | 00275 | 48280 |
| 11508 | 70225 | 51111 | 38351 | 19444 | 66499 | 71945 | 05422 | 13442 | 78675 | 84081 | 66938 | 93654 | 59894 |
| 37449 | 30362 | 06694 | 54690 | 04052 | 53115 | 62757 | 95348 | 78662 | 11163 | 81651 | 50245 | 34971 | 52924 |
| 46515 | 70331 | 85922 | 38329 | 57015 | 15765 | 97161 | 17869 | 45349 | 61796 | 66345 | 81073 | 49106 | 79860 |
| 30986 | 81223 | 42416 | 58353 | 21532 | 30502 | 32305 | 86482 | 05174 | 07901 | 54339 | 58861 | 74818 | 46942 |
| 63798 | 64995 | 46583 | 09765 | 44160 | 78128 | 83991 | 42865 | 92520 | 83531 | 80377 | 35909 | 81250 | 54238 |
| 82486 | 84846 | 99254 | 67632 | 43218 | 50076 | 21361 | 64816 | 51202 | 88124 | 41870 | 52689 | 51275 | 83556 |
| 21885 | 32906 | 92431 | 09060 | 64297 | 51674 | 64126 | 62570 | 26123 | 05155 | 59194 | 52799 | 28225 | 85762 |
| 60336 | 98782 | 07408 | 53458 | 13564 | 59089 | 26445 | 29789 | 85205 | 41001 | 12535 | 12133 | 14645 | 23541 |
| 43937 | 46891 | 24010 | 25560 | 86355 | 33941 | 25786 | 54990 | 71899 | 15475 | 95434 | 98227 | 21824 | 19585 |
| 97656 | 63175 | 89303 | 16275 | 07100 | 92063 | 21942 | 18611 | 47348 | 20203 | 18534 | 03862 | 78095 | 50136 |
| 03299 | 01221 | 05418 | 38982 | 55758 | 92237 | 26759 | 86367 | 21216 | 98442 | 08303 | 56613 | 91511 | 75928 |
| 79626 | 06486 | 03574 | 17668 | 07785 | 76020 | 79924 | 25651 | 83325 | 88428 | 85076 | 72811 | 22717 | 50585 |
| 85636 | 68335 | 47539 | 03129 | 65651 | 11977 | 02510 | 26113 | 99447 | 68645 | 34327 | 15152 | 55230 | 93448 |
| 18039 | 14367 | 61337 | 06177 | 12143 | 46609 | 32989 | 74014 | 64708 | 00533 | 35398 | 58408 | 13261 | 47908 |
| 08362 | 15656 | 60627 | 36478 | 65648 | 16764 | 53412 | 09013 | 07832 | 41574 | 17639 | 82163 | 60859 | 75567 |

*Source:* Reproduced with permission from *CRC Standard Mathematical Tables*, 28th ed., William H. Beyer (Ed.), Boca Raton, FL: CRC Press, p. 581, 1987.

**TABLE IX.2**    CUMULATIVE STANDARD NORMAL DISTRIBUTION TABLE

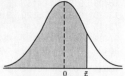

| Areas Under the Normal Curve [$\Phi(z)$] | | | | | | | | | |
| --- | --- | --- | --- | --- | --- | --- | --- | --- | --- |

| z | .00 | .01 | .02 | .03 | .04 | .05 | .06 | .07 | .08 | .09 |
| --- | --- | --- | --- | --- | --- | --- | --- | --- | --- | --- |
| −3.4 | .0003 | .0003 | .0003 | .0003 | .0003 | .0003 | .0003 | .0003 | .0003 | .0002 |
| −3.3 | .0005 | .0005 | .0005 | .0004 | .0004 | .0004 | .0004 | .0004 | .0004 | .0003 |
| −3.2 | .0007 | .0007 | .0006 | .0006 | .0006 | .0006 | .0006 | .0005 | .0005 | .0005 |
| −3.1 | .0010 | .0009 | .0009 | .0009 | .0008 | .0008 | .0008 | .0008 | .0007 | .0007 |
| −3.0 | .0013 | .0013 | .0013 | .0012 | .0012 | .0011 | .0011 | .0011 | .0010 | .0010 |
| | | | | | | | | | | |
| −2.9 | .0019 | .0018 | .0017 | .0017 | .0016 | .0016 | .0015 | .0015 | .0014 | .0014 |
| −2.8 | .0026 | .0025 | .0024 | .0023 | .0023 | .0022 | .0021 | .0021 | .0020 | .0019 |
| −2.7 | .0035 | .0034 | .0033 | .0032 | .0031 | .0030 | .0029 | .0028 | .0027 | .0026 |
| −2.6 | .0047 | .0045 | .0044 | .0043 | .0041 | .0040 | .0039 | .0038 | .0037 | .0036 |
| −2.5 | .0062 | .0060 | .0059 | .0057 | .0055 | .0054 | .0052 | .0051 | .0049 | .0048 |
| | | | | | | | | | | |
| −2.4 | .0082 | .0080 | .0078 | .0075 | .0073 | .0071 | .0069 | .0068 | .0066 | .0064 |
| −2.3 | .0107 | .0104 | .0102 | .0099 | .0096 | .0094 | .0091 | .0089 | .0087 | .0084 |
| −2.2 | .0139 | .0136 | .0132 | .0129 | .0125 | .0122 | .0119 | .0116 | .0113 | .0110 |
| −2.1 | .0179 | .0174 | .0170 | .0166 | .0162 | .0158 | .0154 | .0150 | .0146 | .0143 |
| −2.0 | .0228 | .0222 | .0217 | .0212 | .0207 | .0202 | .0197 | .0192 | .0188 | .0183 |
| | | | | | | | | | | |
| −1.9 | .0287 | .0281 | .0274 | .0268 | .0262 | .0256 | .0250 | .0244 | .0239 | .0233 |
| −1.8 | .0359 | .0352 | .0344 | .0336 | .0329 | .0322 | .0314 | .0307 | .0301 | .0294 |
| −1.7 | .0446 | .0436 | .0427 | .0418 | .0409 | .0401 | .0392 | .0384 | .0375 | .0367 |
| −1.6 | .0548 | .0537 | .0526 | .0516 | .0505 | .0495 | .0485 | .0475 | .0465 | .0455 |
| −1.5 | .0668 | .0655 | .0643 | .0630 | .0618 | .0606 | .0594 | .0582 | .0571 | .0559 |
| | | | | | | | | | | |
| −1.4 | .0808 | .0793 | .0778 | .0764 | .0749 | .0735 | .0722 | .0708 | .0694 | .0681 |
| −1.3 | .0968 | .0951 | .0934 | .0918 | .0901 | .0885 | .0869 | .0853 | .0838 | .0823 |
| −1.2 | .1151 | .1131 | .1112 | .1093 | .1075 | .1056 | .1038 | .1020 | .1003 | .0985 |
| −1.1 | .1357 | .1335 | .1314 | .1292 | .1271 | .1251 | .1230 | .1210 | .1190 | .1170 |
| −1.0 | .1587 | .1562 | .1539 | .1515 | .1492 | .1469 | .1446 | .1423 | .1401 | .1379 |
| | | | | | | | | | | |
| −0.9 | .1841 | .1814 | .1788 | .1762 | .1736 | .1711 | .1685 | .1660 | .1635 | .1611 |
| −0.8 | .2119 | .2090 | .2061 | .2033 | .2005 | .1977 | .1949 | .1922 | .1894 | .1867 |
| −0.7 | .2420 | .2389 | .2358 | .2327 | .2296 | .2266 | .2236 | .2206 | .2177 | .2148 |
| −0.6 | .2743 | .2709 | .2676 | .2643 | .2611 | .2578 | .2546 | .2514 | .2483 | .2451 |
| −0.5 | .3085 | .3050 | .3015 | .2981 | .2946 | .2912 | .2877 | .2843 | .2810 | .2776 |
| | | | | | | | | | | |
| −0.4 | .3446 | .3409 | .3372 | .3336 | .3300 | .3264 | .3228 | .3192 | .3156 | .3121 |
| −0.3 | .3821 | .3783 | .3745 | .3707 | .3669 | .3632 | .3594 | .3557 | .3520 | .3483 |
| −0.2 | .4207 | .4168 | .4129 | .4090 | .4052 | .4013 | .3974 | .3936 | .3897 | .3859 |
| −0.1 | .4602 | .4562 | .4522 | .4483 | .4443 | .4404 | .4364 | .4325 | .4286 | .4247 |
| −0.0 | .5000 | .4960 | .4920 | .4880 | .4840 | .4801 | .4761 | .4721 | .4681 | .4641 |

**TABLE IX.2**    CUMULATIVE STANDARD NORMAL DISTRIBUTION TABLE (*continued*)

Areas Under the Normal Curve [Φ(z)]

| z | .00 | .01 | .02 | .03 | .04 | .05 | .06 | .07 | .08 | .09 |
|---|---|---|---|---|---|---|---|---|---|---|
| 0.0 | .5000 | .5040 | .5080 | .5120 | .5160 | .5199 | .5239 | .5279 | .5319 | .5359 |
| 0.1 | .5398 | .5438 | .5478 | .5517 | .5557 | .5596 | .5636 | .5675 | .5714 | .5753 |
| 0.2 | .5793 | .5832 | .5871 | .5910 | .5948 | .5987 | .6026 | .6064 | .6103 | .6141 |
| 0.3 | .6179 | .6217 | .6255 | .6293 | .6331 | .6368 | .6406 | .6443 | .6480 | .6517 |
| 0.4 | .6554 | .6591 | .6628 | .6664 | .6700 | .6736 | .6772 | .6808 | .6844 | .6879 |
| 0.5 | .6915 | .6950 | .6985 | .7019 | .7054 | .7088 | .7123 | .7157 | .7190 | .7224 |
| 0.6 | .7257 | .7291 | .7324 | .7357 | .7389 | .7422 | .7454 | .7486 | .7517 | .7549 |
| 0.7 | .7580 | .7611 | .7642 | .7673 | .7704 | .7734 | .7764 | .7794 | .7823 | .7852 |
| 0.8 | .7881 | .7910 | .7939 | .7967 | .7995 | .8023 | .8051 | .8078 | .8106 | .8133 |
| 0.9 | .8159 | .8186 | .8212 | .8238 | .8264 | .8289 | .8315 | .8340 | .8365 | .8389 |
| 1.0 | .8413 | .8438 | .8461 | .8485 | .8508 | .8531 | .8554 | .8577 | .8599 | .8621 |
| 1.1 | .8643 | .8665 | .8686 | .8708 | .8729 | .8749 | .8770 | .8790 | .8810 | .8830 |
| 1.2 | .8849 | .8869 | .8888 | .8907 | .8925 | .8944 | .8962 | .8980 | .8997 | .9015 |
| 1.3 | .9032 | .9049 | .9066 | .9082 | .9099 | .9115 | .9131 | .9147 | .9162 | .9177 |
| 1.4 | .9192 | .9207 | .9222 | .9236 | .9251 | .9265 | .9278 | .9292 | .9306 | .9319 |
| 1.5 | .9332 | .9345 | .9357 | .9370 | .9382 | .9394 | .9406 | .9418 | .9429 | .9441 |
| 1.6 | .9452 | .9463 | .9474 | .9484 | .9495 | .9505 | .9515 | .9525 | .9535 | .9545 |
| 1.7 | .9554 | .9564 | .9573 | .9582 | .9591 | .9599 | .9608 | .9616 | .9625 | .9633 |
| 1.8 | .9641 | .9649 | .9656 | .9664 | .9671 | .9678 | .9686 | .9693 | .9699 | .9706 |
| 1.9 | .9713 | .9719 | .9726 | .9732 | .9738 | .9744 | .9750 | .9756 | .9761 | .9767 |
| 2.0 | .9772 | .9778 | .9783 | .9788 | .9793 | .9798 | .9803 | .9808 | .9812 | .9817 |
| 2.1 | .9821 | .9826 | .9830 | .9834 | .9838 | .9842 | .9846 | .9850 | .9854 | .9857 |
| 2.2 | .9861 | .9864 | .9868 | .9871 | .9875 | .9878 | .9881 | .9884 | .9887 | .9890 |
| 2.3 | .9893 | .9896 | .9898 | .9901 | .9904 | .9906 | .9909 | .9911 | .9913 | .9916 |
| 2.4 | .9918 | .9920 | .9922 | .9925 | .9927 | .9929 | .9931 | .9932 | .9934 | .9936 |
| 2.5 | .9938 | .9940 | .9941 | .9943 | .9945 | .9946 | .9948 | .9949 | .9951 | .9952 |
| 2.6 | .9953 | .9955 | .9956 | .9957 | .9959 | .9960 | .9961 | .9962 | .9963 | .9964 |
| 2.7 | .9965 | .9966 | .9967 | .9968 | .9969 | .9970 | .9971 | .9972 | .9973 | .9974 |
| 2.8 | .9974 | .9975 | .9976 | .9977 | .9977 | .9978 | .9979 | .9979 | .9980 | .9981 |
| 2.9 | .9981 | .9982 | .9982 | .9983 | .9984 | .9984 | .9985 | .9985 | .9986 | .9986 |
| 3.0 | .9987 | .9987 | .9987 | .9988 | .9988 | .9989 | .9989 | .9989 | .9990 | .9990 |
| 3.1 | .9990 | .9991 | .9991 | .9991 | .9992 | .9992 | .9992 | .9992 | .9993 | .9993 |
| 3.2 | .9993 | .9993 | .9994 | .9994 | .9994 | .9994 | .9994 | .9995 | .9995 | .9995 |
| 3.3 | .9995 | .9995 | .9995 | .9996 | .9996 | .9996 | .9996 | .9996 | .9996 | .9997 |
| 3.4 | .9997 | .9997 | .9997 | .9997 | .9997 | .9997 | .9997 | .9997 | .9997 | .9998 |
| 3.5 | .9997674 | | | | | | | | | |
| 4.0 | .9999683 | | | | | | | | | |
| 5.0 | .9999997133 | | | | | | | | | |
| 6.0 | .9999999990 | | | | | | | | | |

TABLE IX.3    t DISTRIBUTION TABLE—CRITICAL VALUES

| ν | 0.40 | 0.30 | 0.20 | 0.15 | 0.10 | 0.05 | 0.025 | 0.02 | 0.015 | 0.01 | 0.0075 | 0.005 | 0.0025 | 0.0005 |
|---|---|---|---|---|---|---|---|---|---|---|---|---|---|---|
| 1 | 0.325 | 0.727 | 1.376 | 1.963 | 3.078 | 6.314 | 12.706 | 15.895 | 21.205 | 31.821 | 42.434 | 63.657 | 127.322 | 636.590 |
| 2 | 0.289 | 0.617 | 1.061 | 1.386 | 1.886 | 2.920 | 4.303 | 4.849 | 5.643 | 6.965 | 8.073 | 9.925 | 14.089 | 31.598 |
| 3 | 0.277 | 0.584 | 0.978 | 1.250 | 1.638 | 2.353 | 3.182 | 3.482 | 3.896 | 4.541 | 5.047 | 5.841 | 7.453 | 12.924 |
| 4 | 0.271 | 0.569 | 0.941 | 1.190 | 1.533 | 2.132 | 2.776 | 2.999 | 3.298 | 3.747 | 4.088 | 4.604 | 5.598 | 8.610 |
| 5 | 0.267 | 0.559 | 0.920 | 1.156 | 1.476 | 2.015 | 2.571 | 2.757 | 3.003 | 3.365 | 3.634 | 4.032 | 4.773 | 6.869 |
| 6 | 0.265 | 0.553 | 0.906 | 1.134 | 1.440 | 1.943 | 2.447 | 2.612 | 2.829 | 3.143 | 3.372 | 3.707 | 4.317 | 5.959 |
| 7 | 0.263 | 0.549 | 0.896 | 1.119 | 1.415 | 1.895 | 2.365 | 2.517 | 2.715 | 2.998 | 3.203 | 3.499 | 4.029 | 5.408 |
| 8 | 0.262 | 0.546 | 0.889 | 1.108 | 1.397 | 1.860 | 2.306 | 2.449 | 2.634 | 2.896 | 3.085 | 3.355 | 3.833 | 5.041 |
| 9 | 0.261 | 0.543 | 0.883 | 1.100 | 1.383 | 1.833 | 2.262 | 2.398 | 2.574 | 2.821 | 2.998 | 3.250 | 3.690 | 4.781 |
| 10 | 0.260 | 0.542 | 0.879 | 1.093 | 1.372 | 1.812 | 2.228 | 2.359 | 2.527 | 2.764 | 2.932 | 3.169 | 3.581 | 4.587 |
| 11 | 0.260 | 0.540 | 0.876 | 1.088 | 1.363 | 1.796 | 2.201 | 2.328 | 2.491 | 2.718 | 2.879 | 3.106 | 3.497 | 4.437 |
| 12 | 0.259 | 0.539 | 0.873 | 1.083 | 1.356 | 1.782 | 2.179 | 2.303 | 2.461 | 2.681 | 2.836 | 3.055 | 3.428 | 4.318 |
| 13 | 0.259 | 0.537 | 0.870 | 1.079 | 1.350 | 1.771 | 2.160 | 2.282 | 2.436 | 2.650 | 2.801 | 3.012 | 3.372 | 4.221 |
| 14 | 0.258 | 0.537 | 0.868 | 1.076 | 1.345 | 1.761 | 2.145 | 2.264 | 2.415 | 2.624 | 2.771 | 2.977 | 3.326 | 4.140 |
| 15 | 0.258 | 0.536 | 0.866 | 1.074 | 1.341 | 1.753 | 2.131 | 2.249 | 2.397 | 2.602 | 2.746 | 2.947 | 3.286 | 4.073 |
| 16 | 0.258 | 0.535 | 0.865 | 1.071 | 1.337 | 1.746 | 2.120 | 2.235 | 2.382 | 2.583 | 2.724 | 2.921 | 3.252 | 4.015 |
| 17 | 0.257 | 0.534 | 0.863 | 1.069 | 1.333 | 1.740 | 2.110 | 2.224 | 2.368 | 2.567 | 2.706 | 2.898 | 3.222 | 3.965 |
| 18 | 0.257 | 0.534 | 0.862 | 1.067 | 1.330 | 1.734 | 2.101 | 2.214 | 2.356 | 2.552 | 2.689 | 2.878 | 3.197 | 3.922 |
| 19 | 0.257 | 0.533 | 0.861 | 1.066 | 1.328 | 1.729 | 2.093 | 2.205 | 2.346 | 2.539 | 2.674 | 2.861 | 3.174 | 3.883 |
| 20 | 0.257 | 0.533 | 0.860 | 1.064 | 1.325 | 1.725 | 2.086 | 2.197 | 2.336 | 2.528 | 2.661 | 2.845 | 3.153 | 3.849 |
| 21 | 0.257 | 0.532 | 0.859 | 1.063 | 1.323 | 1.721 | 2.080 | 2.189 | 2.328 | 2.518 | 2.649 | 2.831 | 3.135 | 3.819 |
| 22 | 0.256 | 0.532 | 0.858 | 1.061 | 1.321 | 1.717 | 2.074 | 2.183 | 2.320 | 2.508 | 2.639 | 2.819 | 3.119 | 3.792 |
| 23 | 0.256 | 0.532 | 0.858 | 1.060 | 1.319 | 1.714 | 2.069 | 2.177 | 2.313 | 2.500 | 2.629 | 2.807 | 3.104 | 3.768 |
| 24 | 0.256 | 0.531 | 0.857 | 1.059 | 1.318 | 1.711 | 2.064 | 2.172 | 2.307 | 2.492 | 2.620 | 2.797 | 3.091 | 3.745 |
| 25 | 0.256 | 0.531 | 0.856 | 1.058 | 1.316 | 1.708 | 2.060 | 2.167 | 2.301 | 2.485 | 2.612 | 2.787 | 3.078 | 3.725 |
| 26 | 0.256 | 0.531 | 0.856 | 1.058 | 1.315 | 1.706 | 2.056 | 2.162 | 2.296 | 2.479 | 2.605 | 2.779 | 3.067 | 3.707 |
| 27 | 0.256 | 0.531 | 0.855 | 1.057 | 1.314 | 1.703 | 2.052 | 2.158 | 2.291 | 2.473 | 2.598 | 2.771 | 3.057 | 3.690 |
| 28 | 0.256 | 0.530 | 0.855 | 1.056 | 1.313 | 1.701 | 2.048 | 2.154 | 2.286 | 2.467 | 2.592 | 2.763 | 3.047 | 3.674 |
| 29 | 0.256 | 0.530 | 0.854 | 1.055 | 1.311 | 1.699 | 2.045 | 2.150 | 2.282 | 2.462 | 2.586 | 2.756 | 3.038 | 3.659 |
| 30 | 0.256 | 0.530 | 0.854 | 1.055 | 1.310 | 1.697 | 2.042 | 2.147 | 2.278 | 2.457 | 2.581 | 2.750 | 3.030 | 3.646 |
| 40 | 0.255 | 0.529 | 0.851 | 1.050 | 1.303 | 1.684 | 2.021 | 2.125 | 2.250 | 2.423 | 2.542 | 2.704 | 2.971 | 3.551 |
| 60 | 0.254 | 0.527 | 0.848 | 1.045 | 1.296 | 1.671 | 2.000 | 2.099 | 2.223 | 2.390 | 2.504 | 2.660 | 2.915 | 3.460 |
| 120 | 0.254 | 0.526 | 0.845 | 1.041 | 1.289 | 1.658 | 1.980 | 2.076 | 2.196 | 2.358 | 2.468 | 2.617 | 2.860 | 3.373 |
| ∞ | 0.253 | 0.524 | 0.842 | 1.036 | 1.282 | 1.645 | 1.960 | 2.054 | 2.170 | 2.326 | 2.432 | 2.576 | 2.807 | 3.291 |

# TABLE IX.4  F DISTRIBUTION TABLES—CRITICAL VALUES

$F_{0.10, v_1, v_2}$;  $v_1$ = degrees of freedom numerator;  $v_2$ = degrees of freedom denominator

| $v_2$ \\ $v_1$ | 1 | 2 | 3 | 4 | 5 | 6 | 7 | 8 | 9 | 10 | 12 | 15 | 20 | 24 | 30 | 40 | 60 | 120 | ∞ |
|---|---|---|---|---|---|---|---|---|---|---|---|---|---|---|---|---|---|---|---|
| 1 | 39.86 | 49.50 | 53.59 | 55.83 | 57.24 | 58.20 | 58.91 | 59.44 | 59.86 | 60.19 | 60.71 | 61.22 | 61.74 | 62.00 | 62.26 | 62.53 | 62.79 | 63.06 | 63.33 |
| 2 | 8.53 | 9.00 | 9.16 | 9.24 | 9.29 | 9.33 | 9.35 | 9.37 | 9.38 | 9.39 | 9.41 | 9.42 | 9.44 | 9.45 | 9.46 | 9.47 | 9.47 | 9.48 | 9.49 |
| 3 | 5.54 | 5.46 | 5.39 | 5.34 | 5.31 | 5.28 | 5.27 | 5.25 | 5.24 | 5.23 | 5.22 | 5.20 | 5.18 | 5.18 | 5.17 | 5.16 | 5.15 | 5.14 | 5.13 |
| 4 | 4.54 | 4.32 | 4.19 | 4.11 | 4.05 | 4.01 | 3.98 | 3.95 | 3.94 | 3.92 | 3.90 | 3.87 | 3.84 | 3.83 | 3.82 | 3.80 | 3.79 | 3.78 | 3.76 |
| 5 | 4.06 | 3.78 | 3.62 | 3.52 | 3.45 | 3.40 | 3.37 | 3.34 | 3.32 | 3.30 | 3.27 | 3.24 | 3.21 | 3.19 | 3.17 | 3.16 | 3.14 | 3.12 | 3.10 |
| 6 | 3.78 | 3.46 | 3.29 | 3.18 | 3.11 | 3.05 | 3.01 | 2.98 | 2.96 | 2.94 | 2.90 | 2.87 | 2.84 | 2.82 | 2.80 | 2.78 | 2.76 | 2.74 | 2.72 |
| 7 | 3.59 | 3.26 | 3.07 | 2.96 | 2.88 | 2.83 | 2.78 | 2.75 | 2.72 | 2.70 | 2.67 | 2.63 | 2.59 | 2.58 | 2.56 | 2.54 | 2.51 | 2.49 | 2.47 |
| 8 | 3.46 | 3.11 | 2.92 | 2.81 | 2.73 | 2.67 | 2.62 | 2.59 | 2.56 | 2.54 | 2.50 | 2.46 | 2.42 | 2.40 | 2.38 | 2.36 | 2.34 | 2.32 | 2.29 |
| 9 | 3.36 | 3.01 | 2.81 | 2.69 | 2.61 | 2.55 | 2.51 | 2.47 | 2.44 | 2.42 | 2.38 | 2.34 | 2.30 | 2.28 | 2.25 | 2.23 | 2.21 | 2.18 | 2.16 |
| 10 | 3.29 | 2.92 | 2.73 | 2.61 | 2.52 | 2.46 | 2.41 | 2.38 | 2.35 | 2.32 | 2.28 | 2.24 | 2.20 | 2.18 | 2.16 | 2.13 | 2.11 | 2.08 | 2.06 |
| 11 | 3.23 | 2.86 | 2.66 | 2.54 | 2.45 | 2.39 | 2.34 | 2.30 | 2.27 | 2.25 | 2.21 | 2.17 | 2.12 | 2.10 | 2.08 | 2.05 | 2.03 | 2.00 | 1.97 |
| 12 | 3.18 | 2.81 | 2.61 | 2.48 | 2.39 | 2.33 | 2.28 | 2.24 | 2.21 | 2.19 | 2.15 | 2.10 | 2.06 | 2.04 | 2.01 | 1.99 | 1.96 | 1.93 | 1.90 |
| 13 | 3.14 | 2.76 | 2.56 | 2.43 | 2.35 | 2.28 | 2.23 | 2.20 | 2.16 | 2.14 | 2.10 | 2.05 | 2.01 | 1.98 | 1.96 | 1.93 | 1.90 | 1.88 | 1.85 |
| 14 | 3.10 | 2.73 | 2.52 | 2.39 | 2.31 | 2.24 | 2.19 | 2.15 | 2.12 | 2.10 | 2.05 | 2.01 | 1.96 | 1.94 | 1.91 | 1.89 | 1.86 | 1.83 | 1.80 |
| 15 | 3.07 | 2.70 | 2.49 | 2.36 | 2.27 | 2.21 | 2.16 | 2.12 | 2.09 | 2.06 | 2.02 | 1.97 | 1.92 | 1.90 | 1.87 | 1.85 | 1.82 | 1.79 | 1.76 |
| 16 | 3.05 | 2.67 | 2.46 | 2.33 | 2.24 | 2.18 | 2.13 | 2.09 | 2.06 | 2.03 | 1.99 | 1.94 | 1.89 | 1.87 | 1.84 | 1.81 | 1.78 | 1.75 | 1.72 |
| 17 | 3.03 | 2.64 | 2.44 | 2.31 | 2.22 | 2.15 | 2.10 | 2.06 | 2.03 | 2.00 | 1.96 | 1.91 | 1.86 | 1.84 | 1.81 | 1.78 | 1.75 | 1.72 | 1.69 |
| 18 | 3.01 | 2.62 | 2.42 | 2.29 | 2.20 | 2.13 | 2.08 | 2.04 | 2.00 | 1.98 | 1.93 | 1.89 | 1.84 | 1.81 | 1.78 | 1.75 | 1.72 | 1.69 | 1.66 |
| 19 | 2.99 | 2.61 | 2.40 | 2.27 | 2.18 | 2.11 | 2.06 | 2.02 | 1.98 | 1.96 | 1.91 | 1.86 | 1.81 | 1.79 | 1.76 | 1.73 | 1.70 | 1.67 | 1.63 |
| 20 | 2.97 | 2.59 | 2.38 | 2.25 | 2.16 | 2.09 | 2.04 | 2.00 | 1.96 | 1.94 | 1.89 | 1.84 | 1.79 | 1.77 | 1.74 | 1.71 | 1.68 | 1.64 | 1.61 |
| 21 | 2.96 | 2.57 | 2.36 | 2.23 | 2.14 | 2.08 | 2.02 | 1.98 | 1.95 | 1.92 | 1.87 | 1.83 | 1.78 | 1.75 | 1.72 | 1.69 | 1.66 | 1.62 | 1.59 |
| 22 | 2.95 | 2.56 | 2.35 | 2.22 | 2.13 | 2.06 | 2.01 | 1.97 | 1.93 | 1.90 | 1.86 | 1.81 | 1.76 | 1.73 | 1.70 | 1.67 | 1.64 | 1.60 | 1.57 |
| 23 | 2.94 | 2.55 | 2.34 | 2.21 | 2.11 | 2.05 | 1.99 | 1.95 | 1.92 | 1.89 | 1.84 | 1.80 | 1.74 | 1.72 | 1.69 | 1.66 | 1.62 | 1.59 | 1.55 |
| 24 | 2.93 | 2.54 | 2.33 | 2.19 | 2.10 | 2.04 | 1.98 | 1.94 | 1.91 | 1.88 | 1.83 | 1.78 | 1.73 | 1.70 | 1.67 | 1.64 | 1.61 | 1.57 | 1.53 |
| 25 | 2.92 | 2.53 | 2.32 | 2.18 | 2.09 | 2.02 | 1.97 | 1.93 | 1.89 | 1.87 | 1.82 | 1.77 | 1.72 | 1.69 | 1.66 | 1.63 | 1.59 | 1.56 | 1.52 |
| 26 | 2.91 | 2.52 | 2.31 | 2.17 | 2.08 | 2.01 | 1.96 | 1.92 | 1.88 | 1.86 | 1.81 | 1.76 | 1.71 | 1.68 | 1.65 | 1.61 | 1.58 | 1.54 | 1.50 |
| 27 | 2.90 | 2.51 | 2.30 | 2.17 | 2.07 | 2.00 | 1.95 | 1.91 | 1.87 | 1.85 | 1.80 | 1.75 | 1.70 | 1.67 | 1.64 | 1.60 | 1.57 | 1.53 | 1.49 |
| 28 | 2.89 | 2.50 | 2.29 | 2.16 | 2.06 | 2.00 | 1.94 | 1.90 | 1.87 | 1.84 | 1.79 | 1.74 | 1.69 | 1.66 | 1.63 | 1.59 | 1.56 | 1.52 | 1.48 |
| 29 | 2.89 | 2.50 | 2.28 | 2.15 | 2.06 | 1.99 | 1.93 | 1.89 | 1.86 | 1.83 | 1.78 | 1.73 | 1.68 | 1.65 | 1.62 | 1.58 | 1.55 | 1.51 | 1.47 |
| 30 | 2.88 | 2.49 | 2.28 | 2.14 | 2.05 | 1.98 | 1.93 | 1.88 | 1.85 | 1.82 | 1.77 | 1.72 | 1.67 | 1.64 | 1.61 | 1.57 | 1.54 | 1.50 | 1.46 |
| 40 | 2.84 | 2.44 | 2.23 | 2.09 | 2.00 | 1.93 | 1.87 | 1.83 | 1.79 | 1.76 | 1.71 | 1.66 | 1.61 | 1.57 | 1.54 | 1.51 | 1.47 | 1.42 | 1.38 |
| 60 | 2.79 | 2.39 | 2.18 | 2.04 | 1.95 | 1.87 | 1.82 | 1.77 | 1.74 | 1.71 | 1.66 | 1.60 | 1.54 | 1.51 | 1.48 | 1.44 | 1.40 | 1.35 | 1.29 |
| 120 | 2.75 | 2.35 | 2.13 | 1.99 | 1.90 | 1.82 | 1.77 | 1.72 | 1.68 | 1.65 | 1.60 | 1.55 | 1.48 | 1.45 | 1.41 | 1.37 | 1.32 | 1.26 | 1.19 |
| ∞ | 2.71 | 2.30 | 2.08 | 1.94 | 1.85 | 1.77 | 1.72 | 1.67 | 1.63 | 1.60 | 1.55 | 1.49 | 1.42 | 1.38 | 1.34 | 1.30 | 1.24 | 1.17 | 1.00 |

Source: Reproduced from Biometrika Tables for Statisticians, E. S. Pearson and H. O. Hartley (eds.), Cambridge: University Press, vol. 1, pp. 169–175, 1976. Reproduced with permission of the Biometrika Trustees.

**TABLE IX.4**  *F* DISTRIBUTION TABLES—CRITICAL VALUES *(continued)*

$F_{0.05,\,\nu_1,\,\nu_2}$;  $\nu_1$ = degrees of freedom numerator;  $\nu_2$ = degrees of freedom denominator

| $\nu_2$ \ $\nu_1$ | 1 | 2 | 3 | 4 | 5 | 6 | 7 | 8 | 9 | 10 | 12 | 15 | 20 | 24 | 30 | 40 | 60 | 120 | ∞ |
|---|---|---|---|---|---|---|---|---|---|---|---|---|---|---|---|---|---|---|---|
| 1 | 161.4 | 199.5 | 215.7 | 224.6 | 230.2 | 234.0 | 236.8 | 238.9 | 240.5 | 241.9 | 243.9 | 245.9 | 248.0 | 249.1 | 250.1 | 251.1 | 252.2 | 253.3 | 254.3 |
| 2 | 18.51 | 19.00 | 19.16 | 19.25 | 19.30 | 19.33 | 19.35 | 19.37 | 19.38 | 19.40 | 19.41 | 19.43 | 19.45 | 19.45 | 19.46 | 19.47 | 19.48 | 19.49 | 19.50 |
| 3 | 10.13 | 9.55 | 9.28 | 9.12 | 9.01 | 8.94 | 8.89 | 8.85 | 8.81 | 8.79 | 8.74 | 8.70 | 8.66 | 8.64 | 8.62 | 8.59 | 8.57 | 8.55 | 8.53 |
| 4 | 7.71 | 6.94 | 6.59 | 6.39 | 6.26 | 6.16 | 6.09 | 6.04 | 6.00 | 5.96 | 5.91 | 5.86 | 5.80 | 5.77 | 5.75 | 5.72 | 5.69 | 5.66 | 5.63 |
| 5 | 6.61 | 5.79 | 5.41 | 5.19 | 5.05 | 4.95 | 4.88 | 4.82 | 4.77 | 4.74 | 4.68 | 4.62 | 4.56 | 4.53 | 4.50 | 4.46 | 4.43 | 4.40 | 4.36 |
| 6 | 5.99 | 5.14 | 4.76 | 4.53 | 4.39 | 4.28 | 4.21 | 4.15 | 4.10 | 4.06 | 4.00 | 3.94 | 3.87 | 3.84 | 3.81 | 3.77 | 3.74 | 3.70 | 3.67 |
| 7 | 5.59 | 4.74 | 4.35 | 4.12 | 3.97 | 3.87 | 3.79 | 3.73 | 3.68 | 3.64 | 3.57 | 3.51 | 3.44 | 3.41 | 3.38 | 3.34 | 3.30 | 3.27 | 3.23 |
| 8 | 5.32 | 4.46 | 4.07 | 3.84 | 3.69 | 3.58 | 3.50 | 3.44 | 3.39 | 3.35 | 3.28 | 3.22 | 3.15 | 3.12 | 3.08 | 3.04 | 3.01 | 2.97 | 2.93 |
| 9 | 5.12 | 4.26 | 3.86 | 3.63 | 3.48 | 3.37 | 3.29 | 3.23 | 3.18 | 3.14 | 3.07 | 3.01 | 2.94 | 2.90 | 2.86 | 2.83 | 2.79 | 2.75 | 2.71 |
| 10 | 4.96 | 4.10 | 3.71 | 3.48 | 3.33 | 3.22 | 3.14 | 3.07 | 3.02 | 2.98 | 2.91 | 2.85 | 2.77 | 2.74 | 2.70 | 2.66 | 2.62 | 2.58 | 2.54 |
| 11 | 4.84 | 3.98 | 3.59 | 3.36 | 3.20 | 3.09 | 3.01 | 2.95 | 2.90 | 2.85 | 2.79 | 2.72 | 2.65 | 2.61 | 2.57 | 2.53 | 2.49 | 2.45 | 2.40 |
| 12 | 4.75 | 3.89 | 3.49 | 3.26 | 3.11 | 3.00 | 2.91 | 2.85 | 2.80 | 2.75 | 2.69 | 2.62 | 2.54 | 2.51 | 2.47 | 2.43 | 2.38 | 2.34 | 2.30 |
| 13 | 4.67 | 3.81 | 3.41 | 3.18 | 3.03 | 2.92 | 2.83 | 2.77 | 2.71 | 2.67 | 2.60 | 2.53 | 2.46 | 2.42 | 2.38 | 2.34 | 2.30 | 2.25 | 2.21 |
| 14 | 4.60 | 3.74 | 3.34 | 3.11 | 2.96 | 2.85 | 2.76 | 2.70 | 2.65 | 2.60 | 2.53 | 2.46 | 2.39 | 2.35 | 2.31 | 2.27 | 2.22 | 2.18 | 2.13 |
| 15 | 4.54 | 3.68 | 3.29 | 3.06 | 2.90 | 2.79 | 2.71 | 2.64 | 2.59 | 2.54 | 2.48 | 2.40 | 2.33 | 2.29 | 2.25 | 2.20 | 2.16 | 2.11 | 2.07 |
| 16 | 4.49 | 3.63 | 3.24 | 3.01 | 2.85 | 2.74 | 2.66 | 2.59 | 2.54 | 2.49 | 2.42 | 2.35 | 2.28 | 2.24 | 2.19 | 2.15 | 2.11 | 2.06 | 2.01 |
| 17 | 4.45 | 3.59 | 3.20 | 2.96 | 2.81 | 2.70 | 2.61 | 2.55 | 2.49 | 2.45 | 2.38 | 2.31 | 2.23 | 2.19 | 2.15 | 2.10 | 2.06 | 2.01 | 1.96 |
| 18 | 4.41 | 3.55 | 3.16 | 2.93 | 2.77 | 2.66 | 2.58 | 2.51 | 2.46 | 2.41 | 2.34 | 2.27 | 2.19 | 2.15 | 2.11 | 2.06 | 2.02 | 1.97 | 1.92 |
| 19 | 4.38 | 3.52 | 3.13 | 2.90 | 2.74 | 2.63 | 2.54 | 2.48 | 2.42 | 2.38 | 2.31 | 2.23 | 2.16 | 2.11 | 2.07 | 2.03 | 1.98 | 1.93 | 1.88 |
| 20 | 4.35 | 3.49 | 3.10 | 2.87 | 2.71 | 2.60 | 2.51 | 2.45 | 2.39 | 2.35 | 2.28 | 2.20 | 2.12 | 2.08 | 2.04 | 1.99 | 1.95 | 1.90 | 1.84 |
| 21 | 4.32 | 3.47 | 3.07 | 2.84 | 2.68 | 2.57 | 2.49 | 2.42 | 2.37 | 2.32 | 2.25 | 2.18 | 2.10 | 2.05 | 2.01 | 1.96 | 1.92 | 1.87 | 1.81 |
| 22 | 4.30 | 3.44 | 3.05 | 2.82 | 2.66 | 2.55 | 2.46 | 2.40 | 2.34 | 2.30 | 2.23 | 2.15 | 2.07 | 2.03 | 1.98 | 1.94 | 1.89 | 1.84 | 1.78 |
| 23 | 4.28 | 3.42 | 3.03 | 2.80 | 2.64 | 2.53 | 2.44 | 2.37 | 2.32 | 2.27 | 2.20 | 2.13 | 2.05 | 2.01 | 1.96 | 1.91 | 1.86 | 1.81 | 1.76 |
| 24 | 4.26 | 3.40 | 3.01 | 2.78 | 2.62 | 2.51 | 2.42 | 2.36 | 2.30 | 2.25 | 2.18 | 2.11 | 2.03 | 1.98 | 1.94 | 1.89 | 1.84 | 1.79 | 1.73 |
| 25 | 4.24 | 3.39 | 2.99 | 2.76 | 2.60 | 2.49 | 2.40 | 2.34 | 2.28 | 2.24 | 2.16 | 2.09 | 2.01 | 1.96 | 1.92 | 1.87 | 1.82 | 1.77 | 1.71 |
| 26 | 4.23 | 3.37 | 2.98 | 2.74 | 2.59 | 2.47 | 2.39 | 2.32 | 2.27 | 2.22 | 2.15 | 2.07 | 1.99 | 1.95 | 1.90 | 1.85 | 1.80 | 1.75 | 1.69 |
| 27 | 4.21 | 3.35 | 2.96 | 2.73 | 2.57 | 2.46 | 2.37 | 2.31 | 2.25 | 2.20 | 2.13 | 2.06 | 1.97 | 1.93 | 1.88 | 1.84 | 1.79 | 1.73 | 1.67 |
| 28 | 4.20 | 3.34 | 2.95 | 2.71 | 2.56 | 2.45 | 2.36 | 2.29 | 2.24 | 2.19 | 2.12 | 2.04 | 1.96 | 1.91 | 1.87 | 1.82 | 1.77 | 1.71 | 1.65 |
| 29 | 4.18 | 3.33 | 2.93 | 2.70 | 2.55 | 2.43 | 2.35 | 2.28 | 2.22 | 2.18 | 2.10 | 2.03 | 1.94 | 1.90 | 1.85 | 1.81 | 1.75 | 1.70 | 1.64 |
| 30 | 4.17 | 3.32 | 2.92 | 2.69 | 2.53 | 2.42 | 2.33 | 2.27 | 2.21 | 2.16 | 2.09 | 2.01 | 1.93 | 1.89 | 1.84 | 1.79 | 1.74 | 1.68 | 1.62 |
| 40 | 4.08 | 3.23 | 2.84 | 2.61 | 2.45 | 2.34 | 2.25 | 2.18 | 2.12 | 2.08 | 2.00 | 1.92 | 1.84 | 1.79 | 1.74 | 1.69 | 1.64 | 1.58 | 1.51 |
| 60 | 4.00 | 3.15 | 2.76 | 2.53 | 2.37 | 2.25 | 2.17 | 2.10 | 2.04 | 1.99 | 1.92 | 1.84 | 1.75 | 1.70 | 1.65 | 1.59 | 1.53 | 1.47 | 1.39 |
| 120 | 3.92 | 3.07 | 2.68 | 2.45 | 2.29 | 2.17 | 2.09 | 2.02 | 1.96 | 1.91 | 1.83 | 1.75 | 1.66 | 1.61 | 1.55 | 1.50 | 1.43 | 1.35 | 1.25 |
| ∞ | 3.84 | 3.00 | 2.60 | 2.37 | 2.21 | 2.10 | 2.01 | 1.94 | 1.88 | 1.83 | 1.75 | 1.67 | 1.57 | 1.52 | 1.46 | 1.39 | 1.32 | 1.22 | 1.00 |

TABLE IX.4   F DISTRIBUTION TABLES—CRITICAL VALUES (continued)

$F_{0.025, \, \nu_1, \, \nu_2}$;   $\nu_1$ = degrees of freedom numerator;   $\nu_2$ = degrees of freedom denominator

| $\nu_2 \backslash \nu_1$ | 1 | 2 | 3 | 4 | 5 | 6 | 7 | 8 | 9 | 10 | 12 | 15 | 20 | 24 | 30 | 40 | 60 | 120 | ∞ |
|---|---|---|---|---|---|---|---|---|---|---|---|---|---|---|---|---|---|---|---|
| 1 | 647.8 | 799.5 | 864.2 | 899.6 | 921.8 | 937.1 | 948.2 | 956.7 | 963.3 | 968.6 | 976.7 | 984.9 | 993.1 | 997.2 | 1001 | 1006 | 1010 | 1014 | 1018 |
| 2 | 38.51 | 39.00 | 39.17 | 39.25 | 39.30 | 39.33 | 39.36 | 39.37 | 39.39 | 39.40 | 39.41 | 39.43 | 39.45 | 39.46 | 39.46 | 39.47 | 39.48 | 39.49 | 39.50 |
| 3 | 17.44 | 16.04 | 15.44 | 15.10 | 14.88 | 14.73 | 14.62 | 14.54 | 14.47 | 14.42 | 14.34 | 14.25 | 14.17 | 14.12 | 14.08 | 14.04 | 13.99 | 13.95 | 13.90 |
| 4 | 12.22 | 10.65 | 9.98 | 9.60 | 9.36 | 9.20 | 9.07 | 8.98 | 8.90 | 8.84 | 8.75 | 8.66 | 8.56 | 8.51 | 8.46 | 8.41 | 8.36 | 8.31 | 8.26 |
| 5 | 10.01 | 8.43 | 7.76 | 7.39 | 7.15 | 6.98 | 6.85 | 6.76 | 6.68 | 6.62 | 6.52 | 6.43 | 6.33 | 6.28 | 6.23 | 6.18 | 6.12 | 6.07 | 6.02 |
| 6 | 8.81 | 7.26 | 6.60 | 6.23 | 5.99 | 5.82 | 5.70 | 5.60 | 5.52 | 5.46 | 5.37 | 5.27 | 5.17 | 5.12 | 5.07 | 5.01 | 4.96 | 4.90 | 4.85 |
| 7 | 8.07 | 6.54 | 5.89 | 5.52 | 5.29 | 5.12 | 4.99 | 4.90 | 4.82 | 4.76 | 4.67 | 4.57 | 4.47 | 4.42 | 4.36 | 4.31 | 4.25 | 4.20 | 4.14 |
| 8 | 7.57 | 6.06 | 5.42 | 5.05 | 4.82 | 4.65 | 4.53 | 4.43 | 4.36 | 4.30 | 4.20 | 4.10 | 4.00 | 3.95 | 3.89 | 3.84 | 3.78 | 3.73 | 3.67 |
| 9 | 7.21 | 5.71 | 5.08 | 4.72 | 4.48 | 4.32 | 4.20 | 4.10 | 4.03 | 3.96 | 3.87 | 3.77 | 3.67 | 3.61 | 3.56 | 3.51 | 3.45 | 3.39 | 3.33 |
| 10 | 6.94 | 5.46 | 4.83 | 4.47 | 4.24 | 4.07 | 3.95 | 3.85 | 3.78 | 3.72 | 3.62 | 3.52 | 3.42 | 3.37 | 3.31 | 3.26 | 3.20 | 3.14 | 3.08 |
| 11 | 6.72 | 5.26 | 4.63 | 4.28 | 4.04 | 3.88 | 3.76 | 3.66 | 3.59 | 3.53 | 3.43 | 3.33 | 3.23 | 3.17 | 3.12 | 3.06 | 3.00 | 2.94 | 2.88 |
| 12 | 6.55 | 5.10 | 4.47 | 4.12 | 3.89 | 3.73 | 3.61 | 3.51 | 3.44 | 3.37 | 3.28 | 3.18 | 3.07 | 3.02 | 2.96 | 2.91 | 2.85 | 2.79 | 2.72 |
| 13 | 6.41 | 4.97 | 4.35 | 4.00 | 3.77 | 3.60 | 3.48 | 3.39 | 3.31 | 3.25 | 3.15 | 3.05 | 2.95 | 2.89 | 2.84 | 2.78 | 2.72 | 2.66 | 2.60 |
| 14 | 6.30 | 4.86 | 4.24 | 3.89 | 3.66 | 3.50 | 3.38 | 3.29 | 3.21 | 3.15 | 3.05 | 2.95 | 2.84 | 2.79 | 2.73 | 2.67 | 2.61 | 2.55 | 2.49 |
| 15 | 6.20 | 4.77 | 4.15 | 3.80 | 3.58 | 3.41 | 3.29 | 3.20 | 3.12 | 3.06 | 2.96 | 2.86 | 2.76 | 2.70 | 2.64 | 2.59 | 2.52 | 2.46 | 2.40 |
| 16 | 6.12 | 4.69 | 4.08 | 3.73 | 3.50 | 3.34 | 3.22 | 3.12 | 3.05 | 2.99 | 2.89 | 2.79 | 2.68 | 2.63 | 2.57 | 2.51 | 2.45 | 2.38 | 2.32 |
| 17 | 6.04 | 4.62 | 4.01 | 3.66 | 3.44 | 3.28 | 3.16 | 3.06 | 2.98 | 2.92 | 2.82 | 2.72 | 2.62 | 2.56 | 2.50 | 2.44 | 2.38 | 2.32 | 2.25 |
| 18 | 5.98 | 4.56 | 3.95 | 3.61 | 3.38 | 3.22 | 3.10 | 3.01 | 2.93 | 2.87 | 2.77 | 2.67 | 2.56 | 2.50 | 2.44 | 2.38 | 2.32 | 2.26 | 2.19 |
| 19 | 5.92 | 4.51 | 3.90 | 3.56 | 3.33 | 3.17 | 3.05 | 2.96 | 2.88 | 2.82 | 2.72 | 2.62 | 2.51 | 2.45 | 2.39 | 2.33 | 2.27 | 2.20 | 2.13 |
| 20 | 5.87 | 4.46 | 3.86 | 3.51 | 3.29 | 3.13 | 3.01 | 2.91 | 2.84 | 2.77 | 2.68 | 2.57 | 2.46 | 2.41 | 2.35 | 2.29 | 2.22 | 2.16 | 2.09 |
| 21 | 5.83 | 4.42 | 3.82 | 3.48 | 3.25 | 3.09 | 2.97 | 2.87 | 2.80 | 2.73 | 2.64 | 2.53 | 2.42 | 2.37 | 2.31 | 2.25 | 2.18 | 2.11 | 2.04 |
| 22 | 5.79 | 4.38 | 3.78 | 3.44 | 3.22 | 3.05 | 2.93 | 2.84 | 2.76 | 2.70 | 2.60 | 2.50 | 2.39 | 2.33 | 2.27 | 2.21 | 2.14 | 2.08 | 2.00 |
| 23 | 5.75 | 4.35 | 3.75 | 3.41 | 3.18 | 3.02 | 2.90 | 2.81 | 2.73 | 2.67 | 2.57 | 2.47 | 2.36 | 2.30 | 2.24 | 2.18 | 2.11 | 2.04 | 1.97 |
| 24 | 5.72 | 4.32 | 3.72 | 3.38 | 3.15 | 2.99 | 2.87 | 2.78 | 2.70 | 2.64 | 2.54 | 2.44 | 2.33 | 2.27 | 2.21 | 2.15 | 2.08 | 2.01 | 1.94 |
| 25 | 5.69 | 4.29 | 3.69 | 3.35 | 3.13 | 2.97 | 2.85 | 2.75 | 2.68 | 2.61 | 2.51 | 2.41 | 2.30 | 2.24 | 2.18 | 2.12 | 2.05 | 1.98 | 1.91 |
| 26 | 5.66 | 4.27 | 3.67 | 3.33 | 3.10 | 2.94 | 2.82 | 2.73 | 2.65 | 2.59 | 2.49 | 2.39 | 2.28 | 2.22 | 2.16 | 2.09 | 2.03 | 1.95 | 1.88 |
| 27 | 5.63 | 4.24 | 3.65 | 3.31 | 3.08 | 2.92 | 2.80 | 2.71 | 2.63 | 2.57 | 2.47 | 2.36 | 2.25 | 2.19 | 2.13 | 2.07 | 2.00 | 1.93 | 1.85 |
| 28 | 5.61 | 4.22 | 3.63 | 3.29 | 3.06 | 2.90 | 2.78 | 2.69 | 2.61 | 2.55 | 2.45 | 2.34 | 2.23 | 2.17 | 2.11 | 2.05 | 1.98 | 1.91 | 1.83 |
| 29 | 5.59 | 4.20 | 3.61 | 3.27 | 3.04 | 2.88 | 2.76 | 2.67 | 2.59 | 2.53 | 2.43 | 2.32 | 2.21 | 2.15 | 2.09 | 2.03 | 1.96 | 1.89 | 1.81 |
| 30 | 5.57 | 4.18 | 3.59 | 3.25 | 3.03 | 2.87 | 2.75 | 2.65 | 2.57 | 2.51 | 2.41 | 2.31 | 2.20 | 2.14 | 2.07 | 2.01 | 1.94 | 1.87 | 1.79 |
| 40 | 5.42 | 4.05 | 3.46 | 3.13 | 2.90 | 2.74 | 2.62 | 2.53 | 2.45 | 2.39 | 2.29 | 2.18 | 2.07 | 2.01 | 1.94 | 1.88 | 1.80 | 1.72 | 1.64 |
| 60 | 5.29 | 3.93 | 3.34 | 3.01 | 2.79 | 2.63 | 2.51 | 2.41 | 2.33 | 2.27 | 2.17 | 2.06 | 1.94 | 1.88 | 1.82 | 1.74 | 1.67 | 1.58 | 1.48 |
| 120 | 5.15 | 3.80 | 3.23 | 2.89 | 2.67 | 2.52 | 2.39 | 2.30 | 2.22 | 2.16 | 2.05 | 1.94 | 1.82 | 1.76 | 1.69 | 1.61 | 1.53 | 1.43 | 1.31 |
| ∞ | 5.02 | 3.69 | 3.12 | 2.79 | 2.57 | 2.41 | 2.29 | 2.19 | 2.11 | 2.05 | 1.94 | 1.83 | 1.71 | 1.64 | 1.57 | 1.48 | 1.39 | 1.27 | 1.00 |

$f_{\alpha, \, \nu_1, \, \nu_2}$

TABLE IX.4    F DISTRIBUTION TABLES—CRITICAL VALUES (continued)

$F_{0.01, \nu_1, \nu_2}$; $\nu_1$ = degrees of freedom numerator;    $\nu_2$ = degrees of freedom denominator

| $\nu_2$ \ $\nu_1$ | 1 | 2 | 3 | 4 | 5 | 6 | 7 | 8 | 9 | 10 | 12 | 15 | 20 | 24 | 30 | 40 | 60 | 120 | ∞ |
|---|---|---|---|---|---|---|---|---|---|---|---|---|---|---|---|---|---|---|---|
| 1 | 4052 | 4999.5 | 5403 | 5625 | 5764 | 5859 | 5928 | 5981 | 6022 | 6056 | 6106 | 6157 | 6209 | 6235 | 6261 | 6287 | 6313 | 6339 | 6366 |
| 2 | 98.50 | 99.00 | 99.17 | 99.25 | 99.30 | 99.33 | 99.36 | 99.37 | 99.39 | 99.40 | 99.42 | 99.43 | 99.45 | 99.46 | 99.47 | 99.47 | 99.48 | 99.49 | 99.50 |
| 3 | 34.12 | 30.82 | 29.46 | 28.71 | 28.24 | 27.91 | 27.67 | 27.49 | 27.35 | 27.23 | 27.05 | 26.87 | 26.69 | 26.60 | 26.50 | 26.41 | 26.32 | 26.22 | 26.13 |
| 4 | 21.20 | 18.00 | 16.69 | 15.98 | 15.52 | 15.21 | 14.98 | 14.80 | 14.66 | 14.55 | 14.37 | 14.20 | 14.02 | 13.93 | 13.84 | 13.75 | 13.65 | 13.56 | 13.46 |
| 5 | 16.26 | 13.27 | 12.06 | 11.39 | 10.97 | 10.67 | 10.46 | 10.29 | 10.16 | 10.05 | 9.89 | 9.72 | 9.55 | 9.47 | 9.38 | 9.29 | 9.20 | 9.11 | 9.02 |
| 6 | 13.75 | 10.92 | 9.78 | 9.15 | 8.75 | 8.47 | 8.26 | 8.10 | 7.98 | 7.87 | 7.72 | 7.56 | 7.40 | 7.31 | 7.23 | 7.14 | 7.06 | 6.97 | 6.88 |
| 7 | 12.25 | 9.55 | 8.45 | 7.85 | 7.46 | 7.19 | 6.99 | 6.84 | 6.72 | 6.62 | 6.47 | 6.31 | 6.16 | 6.07 | 5.99 | 5.91 | 5.82 | 5.74 | 5.65 |
| 8 | 11.26 | 8.65 | 7.59 | 7.01 | 6.63 | 6.37 | 6.18 | 6.03 | 5.91 | 5.81 | 5.67 | 5.52 | 5.36 | 5.28 | 5.20 | 5.12 | 5.03 | 4.95 | 4.86 |
| 9 | 10.56 | 8.02 | 6.99 | 6.42 | 6.06 | 5.80 | 5.61 | 5.47 | 5.35 | 5.26 | 5.11 | 4.96 | 4.81 | 4.73 | 4.65 | 4.57 | 4.48 | 4.40 | 4.31 |
| 10 | 10.04 | 7.56 | 6.55 | 5.99 | 5.64 | 5.39 | 5.20 | 5.06 | 4.94 | 4.85 | 4.71 | 4.56 | 4.41 | 4.33 | 4.25 | 4.17 | 4.08 | 4.00 | 3.91 |
| 11 | 9.65 | 7.21 | 6.22 | 5.67 | 5.32 | 5.07 | 4.89 | 4.74 | 4.63 | 4.54 | 4.40 | 4.25 | 4.10 | 4.02 | 3.94 | 3.86 | 3.78 | 3.69 | 3.60 |
| 12 | 9.33 | 6.93 | 5.95 | 5.41 | 5.06 | 4.82 | 4.64 | 4.50 | 4.39 | 4.30 | 4.16 | 4.01 | 3.86 | 3.78 | 3.70 | 3.62 | 3.54 | 3.45 | 3.36 |
| 13 | 9.07 | 6.70 | 5.74 | 5.21 | 4.86 | 4.62 | 4.44 | 4.30 | 4.19 | 4.10 | 3.96 | 3.82 | 3.66 | 3.59 | 3.51 | 3.43 | 3.34 | 3.25 | 3.17 |
| 14 | 8.86 | 6.51 | 5.56 | 5.04 | 4.69 | 4.46 | 4.28 | 4.14 | 4.03 | 3.94 | 3.80 | 3.66 | 3.51 | 3.43 | 3.35 | 3.27 | 3.18 | 3.09 | 3.00 |
| 15 | 8.68 | 6.36 | 5.42 | 4.89 | 4.56 | 4.32 | 4.14 | 4.00 | 3.89 | 3.80 | 3.67 | 3.52 | 3.37 | 3.29 | 3.21 | 3.13 | 3.05 | 2.96 | 2.87 |
| 16 | 8.53 | 6.23 | 5.29 | 4.77 | 4.44 | 4.20 | 4.03 | 3.89 | 3.78 | 3.69 | 3.55 | 3.41 | 3.26 | 3.18 | 3.10 | 3.02 | 2.93 | 2.84 | 2.75 |
| 17 | 8.40 | 6.11 | 5.18 | 4.67 | 4.34 | 4.10 | 3.93 | 3.79 | 3.68 | 3.59 | 3.46 | 3.31 | 3.16 | 3.08 | 3.00 | 2.92 | 2.83 | 2.75 | 2.65 |
| 18 | 8.29 | 6.01 | 5.09 | 4.58 | 4.25 | 4.01 | 3.84 | 3.71 | 3.60 | 3.51 | 3.37 | 3.23 | 3.08 | 3.00 | 2.92 | 2.84 | 2.75 | 2.66 | 2.57 |
| 19 | 8.18 | 5.93 | 5.01 | 4.50 | 4.17 | 3.94 | 3.77 | 3.63 | 3.52 | 3.43 | 3.30 | 3.15 | 3.00 | 2.92 | 2.84 | 2.76 | 2.67 | 2.58 | 2.49 |
| 20 | 8.10 | 5.85 | 4.94 | 4.43 | 4.10 | 3.87 | 3.70 | 3.56 | 3.46 | 3.37 | 3.23 | 3.09 | 2.94 | 2.86 | 2.78 | 2.69 | 2.61 | 2.52 | 2.42 |
| 21 | 8.02 | 5.78 | 4.87 | 4.37 | 4.04 | 3.81 | 3.64 | 3.51 | 3.40 | 3.31 | 3.17 | 3.03 | 2.88 | 2.80 | 2.72 | 2.64 | 2.55 | 2.46 | 2.36 |
| 22 | 7.95 | 5.72 | 4.82 | 4.31 | 3.99 | 3.76 | 3.59 | 3.45 | 3.35 | 3.26 | 3.12 | 2.98 | 2.83 | 2.75 | 2.67 | 2.58 | 2.50 | 2.40 | 2.31 |
| 23 | 7.88 | 5.66 | 4.76 | 4.26 | 3.94 | 3.71 | 3.54 | 3.41 | 3.30 | 3.21 | 3.07 | 2.93 | 2.78 | 2.70 | 2.62 | 2.54 | 2.45 | 2.35 | 2.26 |
| 24 | 7.82 | 5.61 | 4.72 | 4.22 | 3.90 | 3.67 | 3.50 | 3.36 | 3.26 | 3.17 | 3.03 | 2.89 | 2.74 | 2.66 | 2.58 | 2.49 | 2.40 | 2.31 | 2.21 |
| 25 | 7.77 | 5.57 | 4.68 | 4.18 | 3.85 | 3.63 | 3.46 | 3.32 | 3.22 | 3.13 | 2.99 | 2.85 | 2.70 | 2.62 | 2.54 | 2.45 | 2.36 | 2.27 | 2.17 |
| 26 | 7.72 | 5.53 | 4.64 | 4.14 | 3.82 | 3.59 | 3.42 | 3.29 | 3.18 | 3.09 | 2.96 | 2.81 | 2.66 | 2.58 | 2.50 | 2.42 | 2.33 | 2.23 | 2.13 |
| 27 | 7.68 | 5.49 | 4.60 | 4.11 | 3.78 | 3.56 | 3.39 | 3.26 | 3.15 | 3.06 | 2.93 | 2.78 | 2.63 | 2.55 | 2.47 | 2.38 | 2.29 | 2.20 | 2.10 |
| 28 | 7.64 | 5.45 | 4.57 | 4.07 | 3.75 | 3.53 | 3.36 | 3.23 | 3.12 | 3.03 | 2.90 | 2.75 | 2.60 | 2.52 | 2.44 | 2.35 | 2.26 | 2.17 | 2.06 |
| 29 | 7.60 | 5.42 | 4.54 | 4.04 | 3.73 | 3.50 | 3.33 | 3.20 | 3.09 | 3.00 | 2.87 | 2.73 | 2.57 | 2.49 | 2.41 | 2.33 | 2.23 | 2.14 | 2.03 |
| 30 | 7.56 | 5.39 | 4.51 | 4.02 | 3.70 | 3.47 | 3.30 | 3.17 | 3.07 | 2.98 | 2.84 | 2.70 | 2.55 | 2.47 | 2.39 | 2.30 | 2.21 | 2.11 | 2.01 |
| 40 | 7.31 | 5.18 | 4.31 | 3.83 | 3.51 | 3.29 | 3.12 | 2.99 | 2.89 | 2.80 | 2.66 | 2.52 | 2.37 | 2.29 | 2.20 | 2.11 | 2.02 | 1.92 | 1.80 |
| 60 | 7.08 | 4.98 | 4.13 | 3.65 | 3.34 | 3.12 | 2.95 | 2.82 | 2.72 | 2.63 | 2.50 | 2.35 | 2.20 | 2.12 | 2.03 | 1.94 | 1.84 | 1.73 | 1.60 |
| 120 | 6.85 | 4.79 | 3.95 | 3.48 | 3.17 | 2.96 | 2.79 | 2.66 | 2.56 | 2.47 | 2.34 | 2.19 | 2.03 | 1.95 | 1.86 | 1.76 | 1.66 | 1.53 | 1.38 |
| ∞ | 6.63 | 4.61 | 3.78 | 3.32 | 3.02 | 2.80 | 2.64 | 2.51 | 2.41 | 2.32 | 2.18 | 2.04 | 1.88 | 1.79 | 1.70 | 1.59 | 1.47 | 1.32 | 1.00 |

## TABLE IX.5  CHI-SQUARED DISTRIBUTION TABLE—CRITICAL VALUES

υ = degrees of freedom

| υ | .995 | .99 | .975 | .95 | .90 | .80 | .70 | .50 | .30 | .20 | .10 | .05 | .025 | .01 | .005 |
|---|------|-----|------|-----|-----|-----|-----|-----|-----|-----|-----|-----|------|-----|------|
| 1 | $.0^4393$ | $.0^3157$ | $.0^3982$ | .00393 | .0158 | .0642 | .148 | .455 | 1.074 | 1.642 | 2.706 | 3.841 | 5.024 | 6.635 | 7.879 |
| 2 | .0100 | .0201 | .0506 | .103 | .211 | .446 | .713 | 1.386 | 2.408 | 3.219 | 4.605 | 5.991 | 7.378 | 9.210 | 10.597 |
| 3 | .0717 | .115 | .216 | .352 | .584 | 1.005 | 1.424 | 2.366 | 3.665 | 4.642 | 6.251 | 7.815 | 9.348 | 11.345 | 12.838 |
| 4 | .207 | .297 | .484 | .711 | 1.064 | 1.649 | 2.195 | 3.357 | 4.878 | 5.989 | 7.779 | 9.488 | 11.143 | 13.277 | 14.860 |
| 5 | .412 | .554 | .831 | 1.145 | 1.610 | 2.343 | 3.000 | 4.351 | 6.064 | 7.289 | 9.236 | 11.070 | 12.832 | 15.086 | 16.750 |
| 6 | .676 | .872 | 1.237 | 1.635 | 2.204 | 3.070 | 3.828 | 5.348 | 7.231 | 8.558 | 10.645 | 12.592 | 14.449 | 16.812 | 18.548 |
| 7 | .989 | 1.239 | 1.690 | 2.167 | 2.833 | 3.822 | 4.671 | 6.346 | 8.383 | 9.803 | 12.017 | 14.067 | 16.013 | 18.475 | 20.278 |
| 8 | 1.344 | 1.646 | 2.180 | 2.733 | 3.490 | 4.594 | 5.527 | 7.344 | 9.524 | 11.030 | 13.362 | 15.507 | 17.535 | 20.090 | 21.955 |
| 9 | 1.735 | 2.088 | 2.700 | 3.325 | 4.168 | 5.380 | 6.393 | 8.343 | 10.656 | 12.242 | 14.684 | 16.919 | 19.023 | 21.666 | 23.589 |
| 10 | 2.156 | 2.558 | 3.247 | 3.940 | 4.865 | 6.179 | 7.267 | 9.342 | 11.781 | 13.442 | 15.987 | 18.307 | 20.483 | 23.209 | 25.188 |
| 11 | 2.603 | 3.053 | 3.816 | 4.575 | 5.578 | 6.989 | 8.148 | 10.341 | 12.899 | 14.631 | 17.275 | 19.675 | 21.920 | 24.725 | 26.757 |
| 12 | 3.074 | 3.571 | 4.404 | 5.226 | 6.304 | 7.807 | 9.034 | 11.340 | 14.011 | 15.812 | 18.549 | 21.026 | 23.337 | 26.217 | 28.300 |
| 13 | 3.565 | 4.107 | 5.009 | 5.892 | 7.042 | 8.634 | 9.926 | 12.340 | 15.119 | 16.985 | 19.812 | 22.362 | 24.736 | 27.688 | 29.819 |
| 14 | 4.075 | 4.660 | 5.629 | 6.571 | 7.790 | 9.467 | 10.821 | 13.339 | 16.222 | 18.151 | 21.064 | 23.685 | 26.119 | 29.141 | 31.319 |
| 15 | 4.601 | 5.229 | 6.262 | 7.261 | 8.547 | 10.307 | 11.721 | 14.339 | 17.322 | 19.311 | 22.307 | 24.996 | 27.488 | 30.578 | 32.801 |
| 16 | 5.142 | 5.812 | 6.908 | 7.962 | 9.312 | 11.152 | 12.624 | 15.338 | 18.418 | 20.465 | 23.542 | 26.296 | 28.845 | 32.000 | 34.267 |
| 17 | 5.697 | 6.408 | 7.564 | 8.672 | 10.085 | 12.002 | 13.531 | 16.338 | 19.511 | 21.615 | 24.769 | 27.587 | 30.191 | 33.409 | 35.718 |
| 18 | 6.265 | 7.015 | 8.231 | 9.390 | 10.865 | 12.857 | 14.440 | 17.338 | 20.601 | 22.760 | 25.989 | 28.869 | 31.526 | 34.805 | 37.156 |
| 19 | 6.844 | 7.633 | 8.907 | 10.117 | 11.651 | 13.716 | 15.352 | 18.338 | 21.689 | 23.900 | 27.204 | 30.144 | 32.852 | 36.191 | 38.582 |
| 20 | 7.434 | 8.260 | 9.591 | 10.851 | 12.443 | 14.578 | 16.266 | 19.337 | 22.775 | 25.038 | 28.412 | 31.410 | 34.170 | 37.566 | 39.997 |
| 21 | 8.034 | 8.897 | 10.283 | 11.591 | 13.240 | 15.445 | 17.182 | 20.337 | 23.858 | 26.171 | 29.615 | 32.671 | 35.479 | 38.932 | 41.401 |
| 22 | 8.643 | 9.542 | 10.982 | 12.338 | 14.041 | 16.314 | 18.101 | 21.337 | 24.939 | 27.301 | 30.813 | 33.924 | 36.781 | 40.289 | 42.796 |
| 23 | 9.260 | 10.196 | 11.688 | 13.091 | 14.848 | 17.187 | 19.021 | 22.337 | 26.018 | 28.429 | 32.007 | 35.172 | 38.076 | 41.638 | 44.181 |
| 24 | 9.886 | 10.856 | 12.401 | 13.848 | 15.659 | 18.062 | 19.943 | 23.337 | 27.096 | 29.553 | 33.196 | 36.415 | 39.364 | 42.980 | 45.558 |
| 25 | 10.520 | 11.524 | 13.120 | 14.611 | 16.473 | 18.940 | 20.867 | 24.337 | 28.172 | 30.675 | 34.382 | 37.652 | 40.646 | 44.314 | 46.928 |
| 26 | 11.160 | 12.198 | 13.844 | 15.379 | 17.292 | 19.820 | 21.792 | 25.336 | 29.246 | 31.795 | 35.563 | 38.885 | 41.923 | 45.642 | 48.290 |
| 27 | 11.808 | 12.879 | 14.573 | 16.151 | 18.114 | 20.703 | 22.719 | 26.336 | 30.319 | 32.912 | 36.741 | 40.113 | 43.194 | 46.963 | 49.645 |
| 28 | 12.461 | 13.565 | 15.308 | 16.928 | 18.939 | 21.588 | 23.647 | 27.336 | 31.391 | 34.027 | 37.916 | 41.337 | 44.461 | 48.278 | 50.993 |
| 29 | 13.121 | 14.256 | 16.047 | 17.708 | 19.768 | 22.475 | 24.577 | 28.336 | 32.461 | 35.139 | 39.087 | 42.557 | 45.722 | 49.588 | 52.336 |
| 30 | 13.787 | 14.953 | 16.791 | 18.493 | 20.599 | 23.364 | 25.508 | 29.336 | 33.530 | 36.250 | 40.256 | 43.773 | 46.979 | 50.892 | 53.672 |

**TABLE IX.6   CUMULATIVE POISSON DISTRIBUTION TABLE**
(Probability of $c$ or fewer occurrences of event, having average number of occurrences equal to $\mu_c$ or $\mu_{np}$)

| $\mu_c$ or $\mu_{np}$ | c | | | | | | | | | |
|---|---|---|---|---|---|---|---|---|---|---|
| | 0 | 1 | 2 | 3 | 4 | 5 | 6 | 7 | 8 | 9 |
| 0.02 | .980 | 1.000 | | | | | | | | |
| 0.04 | .961 | .999 | 1.000 | | | | | | | |
| 0.06 | .942 | .998 | 1.000 | | | | | | | |
| 0.08 | .923 | .997 | 1.000 | | | | | | | |
| 0.10 | .905 | .995 | 1.000 | | | | | | | |
| 0.15 | .861 | .990 | .999 | 1.000 | | | | | | |
| 0.20 | .819 | .982 | .999 | 1.000 | | | | | | |
| 0.25 | .779 | .974 | .998 | 1.000 | | | | | | |
| 0.30 | .741 | .963 | .996 | 1.000 | | | | | | |
| 0.35 | .705 | .951 | .994 | 1.000 | | | | | | |
| 0.40 | .670 | .938 | .992 | .999 | 1.000 | | | | | |
| 0.45 | .638 | .925 | .989 | .999 | 1.000 | | | | | |
| 0.50 | .607 | .910 | .986 | .998 | 1.000 | | | | | |
| 0.55 | .577 | .894 | .982 | .998 | 1.000 | | | | | |
| 0.60 | .549 | .878 | .977 | .997 | 1.000 | | | | | |
| 0.65 | .522 | .861 | .972 | .996 | .999 | 1.000 | | | | |
| 0.70 | .497 | .844 | .966 | .994 | .999 | 1.000 | | | | |
| 0.75 | .472 | .827 | .959 | .993 | .999 | 1.000 | | | | |
| 0.80 | .449 | .809 | .953 | .991 | .999 | 1.000 | | | | |
| 0.85 | .427 | .791 | .945 | .989 | .998 | 1.000 | | | | |
| 0.90 | .407 | .772 | .937 | .987 | .998 | 1.000 | | | | |
| 0.95 | .387 | .754 | .929 | .984 | .997 | 1.000 | | | | |
| 1.00 | .368 | .736 | .920 | .981 | .996 | .999 | 1.000 | | | |
| 1.1 | .333 | .699 | .900 | .974 | .995 | .999 | 1.000 | | | |
| 1.2 | .301 | .663 | .879 | .966 | .992 | .998 | 1.000 | | | |
| 1.3 | .273 | .627 | .857 | .957 | .989 | .998 | 1.000 | | | |
| 1.4 | .246 | .592 | .833 | .946 | .986 | .997 | .999 | 1.000 | | |
| 1.5 | .223 | .558 | .809 | .934 | .981 | .996 | .999 | 1.000 | | |
| 1.6 | .202 | .525 | .783 | .921 | .976 | .994 | .999 | 1.000 | | |
| 1.7 | .183 | .493 | .757 | .907 | .970 | .992 | .998 | 1.000 | | |
| 1.8 | .165 | .463 | .731 | .891 | .964 | .990 | .997 | .999 | 1.000 | |
| 1.9 | .150 | .434 | .704 | .875 | .956 | .987 | .997 | .999 | 1.000 | |
| 2.0 | .135 | .406 | .677 | .857 | .947 | .983 | .995 | .999 | 1.000 | |

*Source:* Adapted with permission from E.L. Grant and R.S. Leavenworth, *Statistical Quality Control*, 6th ed., New York: McGraw-Hill, pp. 673–677, 1988.

**TABLE IX.6**    CUMULATIVE POISSON DISTRIBUTION TABLE (*continued*)

| $\mu_c$ or $\mu_{np}$ | c | | | | | | | | | |
|---|---|---|---|---|---|---|---|---|---|---|
| | 0 | 1 | 2 | 3 | 4 | 5 | 6 | 7 | 8 | 9 |
| 2.2 | .111 | .355 | .623 | .819 | .928 | .975 | .993 | .998 | 1.000 | |
| 2.4 | .091 | .308 | .570 | .779 | .904 | .964 | .988 | .997 | .999 | 1.000 |
| 2.6 | .074 | .267 | .518 | .736 | .877 | .951 | .983 | .995 | .999 | 1.000 |
| 2.8 | .061 | .231 | .469 | .692 | .848 | .935 | .976 | .992 | .998 | .999 |
| 3.0 | .050 | .199 | .423 | .647 | .815 | .916 | .966 | .988 | .996 | .999 |
| 3.2 | .041 | .171 | .380 | .603 | .781 | .895 | .955 | .983 | .994 | .998 |
| 3.4 | .033 | .147 | .340 | .558 | .744 | .871 | .942 | .977 | .992 | .997 |
| 3.6 | .027 | .126 | .303 | .515 | .706 | .844 | .927 | .969 | .988 | .996 |
| 3.8 | .022 | .107 | .269 | .473 | .668 | .816 | .909 | .960 | .984 | .994 |
| 4.0 | .018 | .092 | .238 | .433 | .629 | .785 | .889 | .949 | .979 | .992 |
| 4.2 | .015 | .078 | .210 | .395 | .590 | .753 | .867 | .936 | .972 | .989 |
| 4.4 | .012 | .066 | .185 | .359 | .551 | .720 | .844 | .921 | .964 | .985 |
| 4.6 | .010 | .056 | .163 | .326 | .513 | .686 | .818 | .905 | .955 | .980 |
| 4.8 | .008 | .048 | .143 | .294 | .476 | .651 | .791 | .887 | .944 | .975 |
| 5.0 | .007 | .040 | .125 | .265 | .440 | .616 | .762 | .867 | .932 | .968 |
| 5.2 | .006 | .034 | .109 | .238 | .406 | .581 | .732 | .845 | .918 | .960 |
| 5.4 | .005 | .029 | .095 | .213 | .373 | .546 | .702 | .822 | .903 | .951 |
| 5.6 | .004 | .024 | .082 | .191 | .342 | .512 | .670 | .797 | .886 | .941 |
| 5.8 | .003 | .021 | .072 | .170 | .313 | .478 | .638 | .771 | .867 | .929 |
| 6.0 | .002 | .017 | .062 | .151 | .285 | .446 | .606 | .744 | .847 | .916 |

| | 10 | 11 | 12 | 13 | 14 | 15 | 16 |
|---|---|---|---|---|---|---|---|
| 2.8 | 1.000 | | | | | | |
| 3.0 | 1.000 | | | | | | |
| 3.2 | 1.000 | | | | | | |
| 3.4 | .999 | 1.000 | | | | | |
| 3.6 | .999 | 1.000 | | | | | |
| 3.8 | .998 | .999 | 1.000 | | | | |
| 4.0 | .997 | .999 | 1.000 | | | | |
| 4.2 | .996 | .999 | 1.000 | | | | |
| 4.4 | .994 | .998 | .999 | 1.000 | | | |
| 4.6 | .992 | .997 | .999 | 1.000 | | | |
| 4.8 | .990 | .996 | .999 | 1.000 | | | |
| 5.0 | .986 | .995 | .998 | .999 | 1.000 | | |
| 5.2 | .982 | .993 | .997 | .999 | 1.000 | | |
| 5.4 | .977 | .990 | .996 | .999 | 1.000 | | |
| 5.6 | .972 | .988 | .995 | .998 | .999 | 1.000 | |
| 5.8 | .965 | .984 | .993 | .997 | .999 | 1.000 | |
| 6.0 | .957 | .980 | .991 | .996 | .999 | .999 | 1.000 |

**TABLE IX.6   CUMULATIVE POISSON DISTRIBUTION TABLE** (*continued*)

| $\mu_c$ or $\mu_{np}$ | c | | | | | | | | | |
|---|---|---|---|---|---|---|---|---|---|---|
| | 0 | 1 | 2 | 3 | 4 | 5 | 6 | 7 | 8 | 9 |
| 6.2 | .002 | .015 | .054 | .134 | .259 | .414 | .574 | .716 | .826 | .902 |
| 6.4 | .002 | .012 | .046 | .119 | .235 | .384 | .542 | .687 | .803 | .886 |
| 6.6 | .001 | .010 | .040 | .105 | .213 | .355 | .511 | .658 | .780 | .869 |
| 6.8 | .001 | .009 | .034 | .093 | .192 | .327 | .480 | .628 | .755 | .850 |
| 7.0 | .001 | .007 | .030 | .082 | .173 | .301 | .450 | .599 | .729 | .830 |
| 7.2 | .001 | .006 | .025 | .072 | .156 | .276 | .420 | .569 | .703 | .810 |
| 7.4 | .001 | .005 | .022 | .063 | .140 | .253 | .392 | .539 | .676 | .788 |
| 7.6 | .001 | .004 | .019 | .055 | .125 | .231 | .365 | .510 | .648 | .765 |
| 7.8 | .000 | .004 | .016 | .048 | .112 | .210 | .338 | .481 | .620 | .741 |
| 8.0 | .000 | .003 | .014 | .042 | .100 | .191 | .313 | .453 | .593 | .717 |
| 8.5 | .000 | .002 | .009 | .030 | .074 | .150 | .256 | .386 | .523 | .653 |
| 9.0 | .000 | .001 | .006 | .021 | .055 | .116 | .207 | .324 | .456 | .587 |
| 9.5 | .000 | .001 | .004 | .015 | .040 | .089 | .165 | .269 | .392 | .522 |
| 10.0 | .000 | .000 | .003 | .010 | .029 | .067 | .130 | .220 | .333 | .458 |

| | 10 | 11 | 12 | 13 | 14 | 15 | 16 | 17 | 18 | 19 |
|---|---|---|---|---|---|---|---|---|---|---|
| 6.2 | .949 | .975 | .989 | .995 | .998 | .999 | 1.000 | | | |
| 6.4 | .939 | .969 | .986 | .994 | .997 | .999 | 1.000 | | | |
| 6.6 | .927 | .963 | .982 | .992 | .997 | .999 | .999 | 1.000 | | |
| 6.8 | .915 | .955 | .978 | .990 | .996 | .998 | .999 | 1.000 | | |
| 7.0 | .901 | .947 | .973 | .987 | .994 | .998 | .999 | 1.000 | | |
| 7.2 | .887 | .937 | .967 | .984 | .993 | .997 | .999 | .999 | 1.000 | |
| 7.4 | .871 | .926 | .961 | .980 | .991 | .996 | .998 | .999 | 1.000 | |
| 7.6 | .854 | .915 | .954 | .976 | .989 | .995 | .998 | .999 | 1.000 | |
| 7.8 | .835 | .902 | .945 | .971 | .986 | .993 | .997 | .999 | 1.000 | |
| 8.0 | .816 | .888 | .936 | .966 | .983 | .992 | .996 | .998 | .999 | 1.000 |
| 8.5 | .763 | .849 | .909 | .949 | .973 | .986 | .993 | .997 | .999 | .999 |
| 9.0 | .706 | .803 | .876 | .926 | .959 | .978 | .989 | .995 | .998 | .999 |
| 9.5 | .645 | .752 | .836 | .898 | .940 | .967 | .982 | .991 | .996 | .998 |
| 10.0 | .583 | .697 | .792 | .864 | .917 | .951 | .973 | .986 | .993 | .997 |

| | 20 | 21 | 22 |
|---|---|---|---|
| 8.5 | 1.000 | | |
| 9.0 | 1.000 | | |
| 9.5 | .999 | 1.000 | |
| 10.0 | .998 | .999 | 1.000 |

**TABLE IX.6**   CUMULATIVE POISSON DISTRIBUTION TABLE (*continued*)

| $\mu_c$ or $\mu_{np}$ | c | | | | | | | | | |
|---|---|---|---|---|---|---|---|---|---|---|
| | **0** | **1** | **2** | **3** | **4** | **5** | **6** | **7** | **8** | **9** |
| 10.5 | .000 | .000 | .002 | .007 | .021 | .050 | .102 | .179 | .279 | .397 |
| 11.0 | .000 | .000 | .001 | .005 | .015 | .038 | .079 | .143 | .232 | .341 |
| 11.5 | .000 | .000 | .001 | .003 | .011 | .028 | .060 | .114 | .191 | .289 |
| 12.0 | .000 | .000 | .001 | .002 | .008 | .020 | .046 | .090 | .155 | .242 |
| 12.5 | .000 | .000 | .000 | .002 | .005 | .015 | .035 | .070 | .125 | .201 |
| 13.0 | .000 | .000 | .000 | .001 | .004 | .011 | .026 | .054 | .100 | .166 |
| 13.5 | .000 | .000 | .000 | .001 | .003 | .008 | .019 | .041 | .079 | .135 |
| 14.0 | .000 | .000 | .000 | .000 | .002 | .006 | .014 | .032 | .062 | .109 |
| 14.5 | .000 | .000 | .000 | .000 | .001 | .004 | .010 | .024 | .048 | .088 |
| 15.0 | .000 | .000 | .000 | .000 | .001 | .003 | .008 | .018 | .037 | .070 |
| | **10** | **11** | **12** | **13** | **14** | **15** | **16** | **17** | **18** | **19** |
| 10.5 | .521 | .639 | .742 | .825 | .888 | .932 | .960 | .978 | .988 | .994 |
| 11.0 | .460 | .579 | .689 | .781 | .854 | .907 | .944 | .968 | .982 | .991 |
| 11.5 | .402 | .520 | .633 | .733 | .815 | .878 | .924 | .954 | .974 | .986 |
| 12.0 | .347 | .462 | .576 | .682 | .772 | .844 | .899 | .937 | .963 | .979 |
| 12.5 | .297 | .406 | .519 | .628 | .725 | .806 | .869 | .916 | .948 | .969 |
| 13.0 | .252 | .353 | .463 | .573 | .675 | .764 | .835 | .890 | .930 | .957 |
| 13.5 | .211 | .304 | .409 | .518 | .623 | .718 | .798 | .861 | .908 | .942 |
| 14.0 | .176 | .260 | .358 | .464 | .570 | .669 | .756 | .827 | .883 | .923 |
| 14.5 | .145 | .220 | .311 | .413 | .518 | .619 | .711 | .790 | .853 | .901 |
| 15.0 | .118 | .185 | .268 | .363 | .466 | .568 | .664 | .749 | .819 | .875 |
| | **20** | **21** | **22** | **23** | **24** | **25** | **26** | **27** | **28** | **29** |
| 10.5 | .997 | .999 | .999 | 1.000 | | | | | | |
| 11.0 | .995 | .998 | .999 | 1.000 | | | | | | |
| 11.5 | .992 | .996 | .998 | .999 | 1.000 | | | | | |
| 12.0 | .988 | .994 | .997 | .999 | .999 | 1.000 | | | | |
| 12.5 | .983 | .991 | .995 | .998 | .999 | .999 | 1.000 | | | |
| 13.0 | .975 | .986 | .992 | .996 | .998 | .999 | 1.000 | | | |
| 13.5 | .965 | .980 | .989 | .994 | .997 | .998 | .999 | 1.000 | | |
| 14.0 | .952 | .971 | .983 | .991 | .995 | .997 | .999 | .999 | 1.000 | |
| 14.5 | .936 | .960 | .976 | .986 | .992 | .996 | .998 | .999 | .999 | 1.000 |
| 15.0 | .917 | .947 | .967 | .981 | .989 | .994 | .997 | .998 | .999 | 1.000 |

**TABLE IX.6**  CUMULATIVE POISSON DISTRIBUTION TABLE (*continued*)

| $\mu_c$ or $\mu_{np}$ | 4 | 5 | 6 | 7 | 8 | 9 | 10 | 11 | 12 | 13 |
|---|---|---|---|---|---|---|---|---|---|---|
| 16 | .000 | .001 | .004 | .010 | .022 | 043 | 077 | 127 | 193 | 275 |
| 17 | .000 | .001 | .002 | .005 | .013 | 026 | 049 | 085 | 135 | 201 |
| 18 | .000 | .000 | .001 | .003 | .007 | 015 | 030 | 055 | 092 | 143 |
| 19 | .000 | .000 | .001 | .002 | .004 | 009 | 018 | 035 | 061 | 098 |
| 20 | .000 | .000 | .000 | .001 | .002 | 005 | 011 | 021 | 039 | 066 |
| 21 | .000 | .000 | .000 | .000 | .001 | 003 | 006 | 013 | 025 | 043 |
| 22 | .000 | .000 | .000 | .000 | .001 | 002 | 004 | 008 | 015 | 028 |
| 23 | .000 | .000 | .000 | .000 | .000 | 001 | 002 | 004 | 009 | 017 |
| 24 | .000 | .000 | .000 | .000 | .000 | 000 | 001 | 003 | 005 | 011 |
| 25 | .000 | .000 | .000 | .000 | .000 | 000 | 001 | 001 | 003 | 006 |

| $\mu_c$ or $\mu_{np}$ | 14 | 15 | 16 | 17 | 18 | 19 | 20 | 21 | 22 | 23 |
|---|---|---|---|---|---|---|---|---|---|---|
| 16 | .368 | .467 | .566 | .659 | .742 | .812 | .868 | .911 | .942 | .963 |
| 17 | .281 | .371 | .468 | .564 | .655 | .736 | .805 | .861 | .905 | .937 |
| 18 | .208 | .287 | .375 | .469 | .562 | .651 | .731 | .799 | .855 | .899 |
| 19 | .150 | .215 | .292 | .378 | .469 | .561 | .647 | .725 | .793 | .849 |
| 20 | .105 | .157 | .221 | .297 | .381 | .470 | .559 | .644 | .721 | .787 |
| 21 | .072 | .111 | .163 | .227 | .302 | .384 | .471 | .558 | .640 | .716 |
| 22 | .048 | .077 | .117 | .169 | .232 | .306 | .387 | .472 | .556 | .637 |
| 23 | .031 | .052 | .082 | .123 | .175 | .238 | .310 | .389 | .472 | .555 |
| 24 | .020 | .034 | .056 | .087 | .128 | .180 | .243 | .314 | .392 | .473 |
| 25 | .012 | .022 | .038 | .060 | .092 | .134 | .185 | .247 | .318 | .394 |

| $\mu_c$ or $\mu_{np}$ | 24 | 25 | 26 | 27 | 28 | 29 | 30 | 31 | 32 | 33 |
|---|---|---|---|---|---|---|---|---|---|---|
| 16 | .978 | .987 | .993 | .996 | .998 | .999 | .999 | 1.000 | | |
| 17 | .959 | .975 | .985 | .991 | .995 | .997 | .999 | .999 | 1.000 | |
| 18 | .932 | .955 | .972 | .983 | .990 | .994 | .997 | .998 | .999 | 1.000 |
| 19 | .893 | .927 | .951 | .969 | .980 | .988 | .993 | .996 | .998 | .999 |
| 20 | .843 | .888 | .922 | .948 | .966 | .978 | .987 | .992 | .995 | .997 |
| 21 | .782 | .838 | .883 | .917 | .944 | .963 | .976 | .985 | .991 | .994 |
| 22 | .712 | .777 | .832 | .877 | .913 | .940 | .959 | .973 | .983 | .989 |
| 23 | .635 | .708 | .772 | .827 | .873 | .908 | .936 | .956 | .971 | .981 |
| 24 | .554 | .632 | .704 | .768 | .823 | .868 | .904 | .932 | .953 | .969 |
| 25 | .473 | .553 | .629 | .700 | .763 | .818 | .863 | .900 | .929 | .950 |

| $\mu_c$ or $\mu_{np}$ | 34 | 35 | 36 | 37 | 38 | 39 | 40 | 41 | 42 | 43 |
|---|---|---|---|---|---|---|---|---|---|---|
| 19 | .999 | 1.000 | | | | | | | | |
| 20 | .999 | .999 | 1.000 | | | | | | | |
| 21 | .997 | .998 | .999 | .999 | 1.000 | | | | | |
| 22 | .994 | .996 | .998 | .999 | .999 | 1.000 | | | | |
| 23 | .988 | .993 | .996 | .997 | .999 | .999 | 1.000 | | | |
| 24 | .979 | .987 | .992 | .995 | .997 | .998 | .999 | .999 | 1.000 | |
| 25 | .966 | .978 | .985 | .991 | .994 | .997 | .998 | .999 | .999 | 1.000 |

TABLE IX.7   STUDENTIZED RANGE—CRITICAL VALUES (SNK RANGE AND HSD TESTS)

Significant ranges $q_{0.05}$ ($p$, $df_{err}$)

| $df_{err}$ \ $p$ | 2 | 3 | 4 | 5 | 6 | 7 | 8 | 9 | 10 | 11 | 12 | 13 | 14 | 15 | 16 | 17 | 18 | 19 | 20 |
|---|---|---|---|---|---|---|---|---|---|---|---|---|---|---|---|---|---|---|---|
| 1 | 17.97 | 26.98 | 32.82 | 37.08 | 40.41 | 43.12 | 45.40 | 47.36 | 49.07 | 50.59 | 51.96 | 53.20 | 54.33 | 55.36 | 56.32 | 57.22 | 58.04 | 58.83 | 59.56 |
| 2 | 6.08 | 8.33 | 9.80 | 10.88 | 11.74 | 12.44 | 13.03 | 13.54 | 13.99 | 14.39 | 14.75 | 15.08 | 15.38 | 15.65 | 15.91 | 16.14 | 16.37 | 16.57 | 16.77 |
| 3 | 4.50 | 5.91 | 6.82 | 7.50 | 8.04 | 8.48 | 8.85 | 9.18 | 9.46 | 9.72 | 9.95 | 10.15 | 10.35 | 10.52 | 10.69 | 10.84 | 10.98 | 11.11 | 11.24 |
| 4 | 3.93 | 5.04 | 5.76 | 6.29 | 6.71 | 7.05 | 7.35 | 7.60 | 7.83 | 8.03 | 8.21 | 8.37 | 8.52 | 8.66 | 8.79 | 8.91 | 9.03 | 9.13 | 9.23 |
| 5 | 3.64 | 4.60 | 5.22 | 5.67 | 6.03 | 6.33 | 6.58 | 6.80 | 6.99 | 7.17 | 7.32 | 7.47 | 7.60 | 7.72 | 7.83 | 7.93 | 8.03 | 8.12 | 8.21 |
| 6 | 3.46 | 4.34 | 4.90 | 5.30 | 5.63 | 5.90 | 6.12 | 6.32 | 6.49 | 6.65 | 6.79 | 6.92 | 7.03 | 7.14 | 7.24 | 7.34 | 7.43 | 7.51 | 7.59 |
| 7 | 3.34 | 4.16 | 4.68 | 5.06 | 5.36 | 5.61 | 5.82 | 6.00 | 6.16 | 6.30 | 6.43 | 6.55 | 6.66 | 6.76 | 6.85 | 6.94 | 7.02 | 7.10 | 7.17 |
| 8 | 3.26 | 4.04 | 4.53 | 4.89 | 5.17 | 5.40 | 5.60 | 5.77 | 5.92 | 6.05 | 6.18 | 6.29 | 6.39 | 6.48 | 6.57 | 6.65 | 6.73 | 6.80 | 6.87 |
| 9 | 3.20 | 3.95 | 4.41 | 4.76 | 5.02 | 5.24 | 5.43 | 5.59 | 5.74 | 5.87 | 5.98 | 6.09 | 6.19 | 6.28 | 6.36 | 6.44 | 6.51 | 6.58 | 6.64 |
| 10 | 3.15 | 3.88 | 4.33 | 4.65 | 4.91 | 5.12 | 5.30 | 5.46 | 5.60 | 5.72 | 5.83 | 5.93 | 6.03 | 6.11 | 6.19 | 6.27 | 6.34 | 6.40 | 6.47 |
| 11 | 3.11 | 3.82 | 4.26 | 4.57 | 4.82 | 5.03 | 5.20 | 5.35 | 5.49 | 5.61 | 5.71 | 5.81 | 5.90 | 5.98 | 6.06 | 6.13 | 6.20 | 6.27 | 6.33 |
| 12 | 3.08 | 3.77 | 4.20 | 4.51 | 4.75 | 4.95 | 5.12 | 5.27 | 5.39 | 5.51 | 5.61 | 5.71 | 5.80 | 5.88 | 5.95 | 6.02 | 6.09 | 6.15 | 6.21 |
| 13 | 3.06 | 3.73 | 4.15 | 4.45 | 4.69 | 4.88 | 5.05 | 5.19 | 5.32 | 5.43 | 5.53 | 5.63 | 5.71 | 5.79 | 5.86 | 5.93 | 5.99 | 6.05 | 6.11 |
| 14 | 3.03 | 3.70 | 4.11 | 4.41 | 4.64 | 4.83 | 4.99 | 5.13 | 5.25 | 5.36 | 5.46 | 5.55 | 5.64 | 5.71 | 5.79 | 5.85 | 5.91 | 5.97 | 6.03 |
| 15 | 3.01 | 3.67 | 4.08 | 4.37 | 4.59 | 4.78 | 4.94 | 5.08 | 5.20 | 5.31 | 5.40 | 5.49 | 5.57 | 5.65 | 5.72 | 5.78 | 5.85 | 5.90 | 5.96 |
| 16 | 3.00 | 3.65 | 4.05 | 4.33 | 4.56 | 4.74 | 4.90 | 5.03 | 5.15 | 5.26 | 5.35 | 5.44 | 5.52 | 5.59 | 5.66 | 5.73 | 5.79 | 5.84 | 5.90 |
| 17 | 2.98 | 3.63 | 4.02 | 4.30 | 4.52 | 4.70 | 4.86 | 4.99 | 5.11 | 5.21 | 5.31 | 5.39 | 5.47 | 5.54 | 5.61 | 5.67 | 5.73 | 5.79 | 5.84 |
| 18 | 2.97 | 3.61 | 4.00 | 4.28 | 4.49 | 4.67 | 4.82 | 4.96 | 5.07 | 5.17 | 5.27 | 5.35 | 5.43 | 5.50 | 5.57 | 5.63 | 5.69 | 5.74 | 5.79 |
| 19 | 2.96 | 3.59 | 3.98 | 4.25 | 4.47 | 4.65 | 4.79 | 4.92 | 5.04 | 5.14 | 5.23 | 5.31 | 5.39 | 5.46 | 5.53 | 5.59 | 5.65 | 5.70 | 5.75 |
| 20 | 2.95 | 3.58 | 3.96 | 4.23 | 4.45 | 4.62 | 4.77 | 4.90 | 5.01 | 5.11 | 5.20 | 5.28 | 5.36 | 5.43 | 5.49 | 5.55 | 5.61 | 5.66 | 5.71 |
| 24 | 2.92 | 3.53 | 3.90 | 4.17 | 4.37 | 4.54 | 4.68 | 4.81 | 4.92 | 5.01 | 5.10 | 5.18 | 5.25 | 5.32 | 5.38 | 5.44 | 5.49 | 5.55 | 5.59 |
| 30 | 2.89 | 3.49 | 3.85 | 4.10 | 4.30 | 4.46 | 4.60 | 4.72 | 4.82 | 4.92 | 5.00 | 5.08 | 5.15 | 5.21 | 5.27 | 5.33 | 5.38 | 5.43 | 5.47 |
| 40 | 2.86 | 3.44 | 3.79 | 4.04 | 4.23 | 4.39 | 4.52 | 4.63 | 4.73 | 4.82 | 4.90 | 4.98 | 5.04 | 5.11 | 5.16 | 5.22 | 5.27 | 5.31 | 5.36 |
| 60 | 2.83 | 3.40 | 3.74 | 3.98 | 4.16 | 4.31 | 4.44 | 4.55 | 4.65 | 4.73 | 4.81 | 4.88 | 4.94 | 5.00 | 5.06 | 5.11 | 5.15 | 5.20 | 5.24 |
| 120 | 2.80 | 3.36 | 3.68 | 3.92 | 4.10 | 4.24 | 4.36 | 4.47 | 4.56 | 4.64 | 4.71 | 4.78 | 4.84 | 4.90 | 4.95 | 5.00 | 5.04 | 5.09 | 5.13 |
| ∞ | 2.77 | 3.31 | 3.63 | 3.86 | 4.03 | 4.17 | 4.29 | 4.39 | 4.47 | 4.55 | 4.62 | 4.68 | 4.74 | 4.80 | 4.85 | 4.89 | 4.93 | 4.97 | 5.01 |

*Source:* Reproduced from J. Pacharees, "Table of the Upper 10% Points of the Studentized Range," *Biometrika*, vol. 46, parts 3 and 4, pp. 465–466, March, 1959. Reproduced with permission of the *Biometrika* Trustees.

TABLE IX.7   STUDENTIZED RANGE—CRITICAL VALUES (SNK RANGE AND HSD TESTS) (continued)

Significant ranges $q_{0.01}$ ($p$, $df_{err}$)

| $df_{err}$ \ $p$ | 2 | 3 | 4 | 5 | 6 | 7 | 8 | 9 | 10 | 11 | 12 | 13 | 14 | 15 | 16 | 17 | 18 | 19 | 20 |
|---|---|---|---|---|---|---|---|---|---|---|---|---|---|---|---|---|---|---|---|
| 1 | 90.03 | 135.0 | 164.3 | 185.6 | 202.2 | 215.8 | 227.2 | 237.0 | 245.6 | 253.2 | 260.0 | 266.2 | 271.8 | 277.0 | 281.8 | 286.3 | 290.4 | 294.3 | 298.0 |
| 2 | 14.04 | 19.02 | 22.29 | 24.72 | 26.63 | 28.20 | 29.53 | 30.68 | 31.69 | 32.59 | 33.40 | 34.13 | 34.81 | 35.43 | 36.00 | 36.53 | 37.03 | 37.50 | 37.95 |
| 3 | 8.26 | 10.62 | 12.17 | 13.33 | 14.24 | 15.00 | 15.64 | 16.20 | 16.69 | 17.13 | 17.53 | 17.89 | 18.22 | 18.52 | 18.81 | 19.07 | 19.32 | 19.55 | 19.77 |
| 4 | 6.51 | 8.12 | 9.17 | 9.96 | 10.58 | 11.10 | 11.55 | 11.93 | 12.27 | 12.57 | 12.84 | 13.09 | 13.32 | 13.53 | 13.73 | 13.91 | 14.08 | 14.24 | 14.40 |
| 5 | 5.70 | 6.98 | 7.80 | 8.42 | 8.91 | 9.32 | 9.67 | 9.97 | 10.24 | 10.48 | 10.70 | 10.89 | 11.08 | 11.24 | 11.40 | 11.55 | 11.68 | 11.81 | 11.93 |
| 6 | 5.24 | 6.33 | 7.03 | 7.56 | 7.97 | 8.32 | 8.61 | 8.87 | 9.10 | 9.30 | 9.48 | 9.65 | 9.81 | 9.95 | 10.08 | 10.21 | 10.32 | 10.43 | 10.54 |
| 7 | 4.95 | 5.92 | 6.54 | 7.01 | 7.37 | 7.68 | 7.94 | 8.17 | 8.37 | 8.55 | 8.71 | 8.86 | 9.00 | 9.12 | 9.24 | 9.35 | 9.46 | 9.55 | 9.65 |
| 8 | 4.75 | 5.64 | 6.20 | 6.62 | 6.96 | 7.24 | 7.47 | 7.68 | 7.86 | 8.03 | 8.18 | 8.31 | 8.44 | 8.55 | 8.66 | 8.76 | 8.85 | 8.94 | 9.03 |
| 9 | 4.60 | 5.43 | 5.96 | 6.35 | 6.66 | 6.91 | 7.13 | 7.33 | 7.49 | 7.65 | 7.78 | 7.91 | 8.03 | 8.13 | 8.23 | 8.33 | 8.41 | 8.49 | 8.57 |
| 10 | 4.48 | 5.27 | 5.77 | 6.14 | 6.43 | 6.67 | 6.87 | 7.05 | 7.21 | 7.36 | 7.49 | 7.60 | 7.71 | 7.81 | 7.91 | 7.99 | 8.08 | 8.15 | 8.23 |
| 11 | 4.39 | 5.15 | 5.62 | 5.97 | 6.25 | 6.48 | 6.67 | 6.84 | 6.99 | 7.13 | 7.25 | 7.36 | 7.46 | 7.56 | 7.65 | 7.73 | 7.81 | 7.88 | 7.95 |
| 12 | 4.32 | 5.05 | 5.50 | 5.84 | 6.10 | 6.32 | 6.51 | 6.67 | 6.81 | 6.94 | 7.06 | 7.17 | 7.26 | 7.36 | 7.44 | 7.52 | 7.59 | 7.66 | 7.73 |
| 13 | 4.26 | 4.96 | 5.40 | 5.73 | 5.98 | 6.19 | 6.37 | 6.53 | 6.67 | 6.79 | 6.90 | 7.01 | 7.10 | 7.19 | 7.27 | 7.35 | 7.42 | 7.48 | 7.55 |
| 14 | 4.21 | 4.89 | 5.32 | 5.63 | 5.88 | 6.08 | 6.26 | 6.41 | 6.54 | 6.66 | 6.77 | 6.87 | 6.96 | 7.05 | 7.13 | 7.20 | 7.27 | 7.33 | 7.39 |
| 15 | 4.17 | 4.84 | 5.25 | 5.56 | 5.80 | 5.99 | 6.16 | 6.31 | 6.44 | 6.55 | 6.66 | 6.76 | 6.84 | 6.93 | 7.00 | 7.07 | 7.14 | 7.20 | 7.26 |
| 16 | 4.13 | 4.79 | 5.19 | 5.49 | 5.72 | 5.92 | 6.08 | 6.22 | 6.35 | 6.46 | 6.56 | 6.66 | 6.74 | 6.82 | 6.90 | 6.97 | 7.03 | 7.09 | 7.15 |
| 17 | 4.10 | 4.74 | 5.14 | 5.43 | 5.66 | 5.85 | 6.01 | 6.15 | 6.27 | 6.38 | 6.48 | 6.57 | 6.66 | 6.73 | 6.81 | 6.87 | 6.94 | 7.00 | 7.05 |
| 18 | 4.07 | 4.70 | 5.09 | 5.38 | 5.60 | 5.79 | 5.94 | 6.08 | 6.20 | 6.31 | 6.41 | 6.50 | 6.58 | 6.65 | 6.73 | 6.79 | 6.85 | 6.91 | 6.97 |
| 19 | 4.05 | 4.67 | 5.05 | 5.33 | 5.55 | 5.73 | 5.89 | 6.02 | 6.14 | 6.25 | 6.34 | 6.43 | 6.51 | 6.58 | 6.65 | 6.72 | 6.78 | 6.84 | 6.89 |
| 20 | 4.02 | 4.64 | 5.02 | 5.29 | 5.51 | 5.69 | 5.84 | 5.97 | 6.09 | 6.19 | 6.28 | 6.37 | 6.45 | 6.52 | 6.59 | 6.65 | 6.71 | 6.77 | 6.82 |
| 24 | 3.96 | 4.55 | 4.91 | 5.17 | 5.37 | 5.54 | 5.69 | 5.81 | 5.92 | 6.02 | 6.11 | 6.19 | 6.26 | 6.33 | 6.39 | 6.45 | 6.51 | 6.56 | 6.61 |
| 30 | 3.89 | 4.45 | 4.80 | 5.05 | 5.24 | 5.40 | 5.54 | 5.65 | 5.76 | 5.85 | 5.93 | 6.01 | 6.08 | 6.14 | 6.20 | 6.26 | 6.31 | 6.36 | 6.41 |
| 40 | 3.82 | 4.37 | 4.70 | 4.93 | 5.11 | 5.26 | 5.39 | 5.50 | 5.60 | 5.69 | 5.76 | 5.83 | 5.90 | 5.96 | 6.02 | 6.07 | 6.12 | 6.16 | 6.21 |
| 60 | 3.76 | 4.28 | 4.59 | 4.82 | 4.99 | 5.13 | 5.25 | 5.36 | 5.45 | 5.53 | 5.60 | 5.67 | 5.73 | 5.78 | 5.84 | 5.89 | 5.93 | 5.97 | 6.01 |
| 120 | 3.70 | 4.20 | 4.50 | 4.71 | 4.87 | 5.01 | 5.12 | 5.21 | 5.30 | 5.37 | 5.44 | 5.50 | 5.56 | 5.61 | 5.66 | 5.71 | 5.75 | 5.79 | 5.83 |
| ∞ | 3.64 | 4.12 | 4.40 | 4.60 | 4.76 | 4.88 | 4.99 | 5.08 | 5.16 | 5.23 | 5.29 | 5.35 | 5.40 | 5.45 | 5.49 | 5.54 | 5.57 | 5.61 | 5.65 |

**TABLE IX.8  DUNCAN'S MULTIPLE RANGE—CRITICAL VALUES (DM RANGE TEST)**

Significant Ranges $r_{0.05}$ ($p$, $df_{err}$)

| $df_{err}$ \ $p$ | 2 | 3 | 4 | 5 | 6 | 7 | 8 | 9 | 10 | 12 | 14 | 16 | 18 | 20 | 50 | 100 |
|---|---|---|---|---|---|---|---|---|---|---|---|---|---|---|---|---|
| 1 | 18.0 | 18.0 | 18.0 | 18.0 | 18.0 | 18.0 | 18.0 | 18.0 | 18.0 | 18.0 | 18.0 | 18.0 | 18.0 | 18.0 | 18.0 | 18.0 |
| 2 | 6.09 | 6.09 | 6.09 | 6.09 | 6.09 | 6.09 | 6.09 | 6.09 | 6.09 | 6.09 | 6.09 | 6.09 | 6.09 | 6.09 | 6.09 | 6.09 |
| 3 | 4.50 | 4.50 | 4.50 | 4.50 | 4.50 | 4.50 | 4.50 | 4.50 | 4.50 | 4.50 | 4.50 | 4.50 | 4.50 | 4.50 | 4.50 | 4.50 |
| 4 | 3.93 | 4.01 | 4.02 | 4.02 | 4.02 | 4.02 | 4.02 | 4.02 | 4.02 | 4.02 | 4.02 | 4.02 | 4.02 | 4.02 | 4.02 | 4.02 |
| 5 | 3.64 | 3.74 | 3.79 | 3.83 | 3.83 | 3.83 | 3.83 | 3.83 | 3.83 | 3.83 | 3.83 | 3.83 | 3.83 | 3.83 | 3.83 | 3.83 |
| 6 | 3.46 | 3.58 | 3.64 | 3.68 | 3.68 | 3.68 | 3.68 | 3.68 | 3.68 | 3.68 | 3.68 | 3.68 | 3.68 | 3.68 | 3.68 | 3.68 |
| 7 | 3.35 | 3.47 | 3.54 | 3.58 | 3.60 | 3.61 | 3.61 | 3.61 | 3.61 | 3.61 | 3.61 | 3.61 | 3.61 | 3.61 | 3.61 | 3.61 |
| 8 | 3.26 | 3.39 | 3.47 | 3.52 | 3.55 | 3.56 | 3.56 | 3.56 | 3.56 | 3.56 | 3.56 | 3.56 | 3.56 | 3.56 | 3.56 | 3.56 |
| 9 | 3.20 | 3.34 | 3.41 | 3.47 | 3.50 | 3.52 | 3.52 | 3.52 | 3.52 | 3.52 | 3.52 | 3.52 | 3.52 | 3.52 | 3.52 | 3.52 |
| 10 | 3.15 | 3.30 | 3.37 | 3.43 | 3.46 | 3.47 | 3.47 | 3.47 | 3.47 | 3.47 | 3.47 | 3.47 | 3.47 | 3.48 | 3.48 | 3.48 |
| 11 | 3.11 | 3.27 | 3.35 | 3.39 | 3.43 | 3.44 | 3.45 | 3.46 | 3.46 | 3.46 | 3.46 | 3.46 | 3.47 | 3.48 | 3.48 | 3.48 |
| 12 | 3.08 | 3.23 | 3.33 | 3.36 | 3.40 | 3.42 | 3.44 | 3.44 | 3.46 | 3.46 | 3.46 | 3.46 | 3.47 | 3.48 | 3.48 | 3.48 |
| 13 | 3.06 | 3.21 | 3.30 | 3.35 | 3.38 | 3.41 | 3.42 | 3.44 | 3.45 | 3.45 | 3.46 | 3.46 | 3.47 | 3.47 | 3.47 | 3.47 |
| 14 | 3.03 | 3.18 | 3.27 | 3.33 | 3.37 | 3.39 | 3.41 | 3.42 | 3.44 | 3.45 | 3.46 | 3.46 | 3.47 | 3.47 | 3.47 | 3.47 |
| 15 | 3.01 | 3.16 | 3.25 | 3.31 | 3.36 | 3.38 | 3.40 | 3.42 | 3.43 | 3.44 | 3.45 | 3.46 | 3.47 | 3.47 | 3.47 | 3.47 |
| 16 | 3.00 | 3.15 | 3.23 | 3.30 | 3.34 | 3.37 | 3.39 | 3.41 | 3.43 | 3.44 | 3.45 | 3.46 | 3.47 | 3.47 | 3.47 | 3.47 |
| 17 | 2.98 | 3.13 | 3.22 | 3.28 | 3.33 | 3.36 | 3.38 | 3.40 | 3.42 | 3.44 | 3.45 | 3.46 | 3.47 | 3.47 | 3.47 | 3.47 |
| 18 | 2.97 | 3.12 | 3.21 | 3.27 | 3.32 | 3.35 | 3.37 | 3.39 | 3.41 | 3.43 | 3.45 | 3.46 | 3.47 | 3.47 | 3.47 | 3.47 |
| 19 | 2.96 | 3.11 | 3.19 | 3.26 | 3.31 | 3.35 | 3.37 | 3.39 | 3.41 | 3.43 | 3.44 | 3.46 | 3.47 | 3.47 | 3.47 | 3.47 |
| 20 | 2.95 | 3.10 | 3.18 | 3.25 | 3.30 | 3.34 | 3.36 | 3.38 | 3.40 | 3.43 | 3.44 | 3.46 | 3.46 | 3.47 | 3.47 | 3.47 |
| 22 | 2.93 | 3.08 | 3.17 | 3.24 | 3.29 | 3.32 | 3.35 | 3.37 | 3.39 | 3.42 | 3.44 | 3.45 | 3.46 | 3.47 | 3.47 | 3.47 |
| 24 | 2.92 | 3.07 | 3.15 | 3.22 | 3.28 | 3.31 | 3.34 | 3.37 | 3.38 | 3.41 | 3.44 | 3.45 | 3.46 | 3.47 | 3.47 | 3.47 |
| 26 | 2.91 | 3.06 | 3.14 | 3.21 | 3.27 | 3.30 | 3.34 | 3.36 | 3.38 | 3.41 | 3.43 | 3.45 | 3.46 | 3.47 | 3.47 | 3.47 |
| 28 | 2.90 | 3.04 | 3.13 | 3.20 | 3.26 | 3.30 | 3.33 | 3.35 | 3.37 | 3.40 | 3.43 | 3.45 | 3.46 | 3.47 | 3.47 | 3.47 |
| 30 | 2.89 | 3.04 | 3.12 | 3.20 | 3.25 | 3.29 | 3.32 | 3.35 | 3.37 | 3.40 | 3.43 | 3.44 | 3.46 | 3.47 | 3.47 | 3.47 |
| 40 | 2.86 | 3.01 | 3.10 | 3.17 | 3.22 | 3.27 | 3.30 | 3.33 | 3.35 | 3.39 | 3.42 | 3.44 | 3.46 | 3.47 | 3.47 | 3.47 |
| 60 | 2.83 | 2.98 | 3.08 | 3.14 | 3.20 | 3.24 | 3.28 | 3.31 | 3.33 | 3.37 | 3.40 | 3.43 | 3.45 | 3.47 | 3.48 | 3.48 |
| 100 | 2.80 | 2.95 | 3.05 | 3.12 | 3.18 | 3.22 | 3.23 | 3.29 | 3.32 | 3.36 | 3.40 | 3.42 | 3.45 | 3.47 | 3.53 | 3.53 |
| ∞ | 2.77 | 2.92 | 3.02 | 3.09 | 3.15 | 3.19 | 3.23 | 3.26 | 3.29 | 3.34 | 3.38 | 3.41 | 3.44 | 3.47 | 3.61 | 3.67 |

*Source:* Reproduced with permission from D. B. Duncan, "Multiple Range and Multiple F Tests," *Biometrics,* vol. II, no. 1, pp. 3–4, March, 1955.

TABLE IX.8    DUNCAN'S MULTIPLE RANGE—CRITICAL VALUES (DM RANGE TEST) *(continued)*

Significant Ranges $r_{0.01}$ ($p$, $df_{err}$)

| $df_{err}$ \ $p$ | 2 | 3 | 4 | 5 | 6 | 7 | 8 | 9 | 10 | 12 | 14 | 16 | 18 | 20 | 50 | 100 |
|---|---|---|---|---|---|---|---|---|---|---|---|---|---|---|---|---|
| 1 | 90.0 | 90.0 | 90.0 | 90.0 | 90.0 | 90.0 | 90.0 | 90.0 | 90.0 | 90.0 | 90.0 | 90.0 | 90.0 | 90.0 | 90.0 | 90.0 |
| 2 | 14.0 | 14.0 | 14.0 | 14.0 | 14.0 | 14.0 | 14.0 | 14.0 | 14.0 | 14.0 | 14.0 | 14.0 | 14.0 | 14.0 | 14.0 | 14.0 |
| 3 | 8.26 | 8.5 | 8.6 | 8.7 | 8.8 | 8.9 | 8.9 | 9.0 | 9.0 | 9.0 | 9.1 | 9.2 | 9.3 | 9.3 | 9.3 | 9.3 |
| 4 | 6.51 | 6.8 | 6.9 | 7.0 | 7.1 | 7.1 | 7.2 | 7.2 | 7.3 | 7.3 | 7.4 | 7.4 | 7.5 | 7.5 | 7.5 | 7.5 |
| 5 | 5.70 | 5.96 | 6.11 | 6.18 | 6.26 | 6.33 | 6.40 | 6.44 | 6.5 | 6.6 | 6.6 | 6.7 | 6.7 | 6.8 | 6.8 | 6.8 |
| 6 | 5.24 | 5.51 | 5.65 | 5.73 | 5.81 | 5.88 | 5.95 | 6.00 | 6.0 | 6.1 | 6.2 | 6.2 | 6.3 | 6.3 | 6.3 | 6.3 |
| 7 | 4.95 | 5.22 | 5.37 | 5.45 | 5.53 | 5.61 | 5.69 | 5.73 | 5.8 | 5.8 | 5.9 | 5.9 | 6.0 | 6.0 | 6.0 | 6.0 |
| 8 | 4.74 | 5.00 | 5.14 | 5.23 | 5.32 | 5.40 | 5.47 | 5.51 | 5.5 | 5.6 | 5.7 | 5.7 | 5.8 | 5.8 | 5.8 | 5.8 |
| 9 | 4.60 | 4.86 | 4.99 | 5.08 | 5.17 | 5.25 | 5.32 | 5.36 | 5.4 | 5.5 | 5.5 | 5.6 | 5.7 | 5.7 | 5.7 | 5.7 |
| 10 | 4.48 | 4.73 | 4.88 | 4.96 | 5.06 | 5.13 | 5.20 | 5.24 | 5.28 | 5.36 | 5.42 | 5.48 | 5.54 | 5.55 | 5.55 | 5.55 |
| 11 | 4.39 | 4.63 | 4.77 | 4.86 | 4.94 | 5.01 | 5.06 | 5.12 | 5.15 | 5.24 | 5.28 | 5.34 | 5.38 | 5.39 | 5.39 | 5.39 |
| 12 | 4.32 | 4.55 | 4.68 | 4.76 | 4.84 | 4.92 | 4.96 | 5.02 | 5.07 | 5.13 | 5.17 | 5.22 | 5.24 | 5.26 | 5.26 | 5.26 |
| 13 | 4.26 | 4.48 | 4.62 | 4.69 | 4.74 | 4.84 | 4.88 | 4.94 | 4.98 | 5.04 | 5.08 | 5.13 | 5.14 | 5.15 | 5.15 | 5.15 |
| 14 | 4.21 | 4.42 | 4.55 | 4.63 | 4.70 | 4.78 | 4.83 | 4.87 | 4.91 | 4.96 | 5.00 | 5.04 | 5.06 | 5.07 | 5.07 | 5.07 |
| 15 | 4.17 | 4.37 | 4.50 | 4.58 | 4.64 | 4.72 | 4.77 | 4.81 | 4.84 | 4.90 | 4.94 | 4.97 | 4.99 | 5.00 | 5.00 | 5.00 |
| 16 | 4.13 | 4.34 | 4.45 | 4.54 | 4.60 | 4.67 | 4.72 | 4.76 | 4.79 | 4.84 | 4.88 | 4.91 | 4.93 | 4.94 | 4.94 | 4.94 |
| 17 | 4.10 | 4.30 | 4.41 | 4.50 | 4.56 | 4.63 | 4.68 | 4.72 | 4.75 | 4.80 | 4.83 | 4.86 | 4.88 | 4.89 | 4.89 | 4.89 |
| 18 | 4.07 | 4.27 | 4.38 | 4.46 | 4.53 | 4.59 | 4.64 | 4.68 | 4.71 | 4.76 | 4.79 | 4.82 | 4.84 | 4.85 | 4.85 | 4.85 |
| 19 | 4.05 | 4.24 | 4.35 | 4.43 | 4.50 | 4.56 | 4.61 | 4.64 | 4.67 | 4.72 | 4.76 | 4.79 | 4.81 | 4.82 | 4.82 | 4.82 |
| 20 | 4.02 | 4.22 | 4.33 | 4.40 | 4.47 | 4.53 | 4.58 | 4.61 | 4.65 | 4.69 | 4.73 | 4.76 | 4.78 | 4.79 | 4.79 | 4.79 |
| 22 | 3.99 | 4.17 | 4.28 | 4.36 | 4.42 | 4.48 | 4.53 | 4.57 | 4.60 | 4.65 | 4.68 | 4.71 | 4.74 | 4.75 | 4.75 | 4.75 |
| 24 | 3.96 | 4.14 | 4.24 | 4.33 | 4.39 | 4.44 | 4.49 | 4.53 | 4.57 | 4.62 | 4.64 | 4.67 | 4.70 | 4.72 | 4.74 | 4.74 |
| 26 | 3.93 | 4.11 | 4.21 | 4.30 | 4.36 | 4.41 | 4.46 | 4.50 | 4.53 | 4.58 | 4.62 | 4.65 | 4.67 | 4.69 | 4.73 | 4.73 |
| 28 | 3.91 | 4.08 | 4.18 | 4.28 | 4.34 | 4.39 | 4.43 | 4.47 | 4.51 | 4.56 | 4.60 | 4.62 | 4.65 | 4.67 | 4.72 | 4.72 |
| 30 | 3.89 | 4.06 | 4.16 | 4.22 | 4.32 | 4.36 | 4.41 | 4.45 | 4.48 | 4.54 | 4.58 | 4.61 | 4.63 | 4.65 | 4.71 | 4.71 |
| 40 | 3.82 | 3.99 | 4.10 | 4.17 | 4.24 | 4.30 | 4.34 | 4.37 | 4.41 | 4.46 | 4.51 | 4.54 | 4.57 | 4.59 | 4.69 | 4.69 |
| 60 | 3.76 | 3.92 | 4.03 | 4.12 | 4.17 | 4.23 | 4.27 | 4.31 | 4.34 | 4.39 | 4.44 | 4.47 | 4.50 | 4.53 | 4.66 | 4.66 |
| 100 | 3.71 | 3.86 | 3.98 | 4.06 | 4.11 | 4.17 | 4.21 | 4.23 | 4.29 | 4.33 | 4.38 | 4.42 | 4.45 | 4.48 | 4.64 | 4.65 |
| ∞ | 3.64 | 3.80 | 3.90 | 3.98 | 4.04 | 4.09 | 4.14 | 4.17 | 4.20 | 4.26 | 4.31 | 4.34 | 4.38 | 4.41 | 4.60 | 4.68 |

# TABLE IX.9 DUNNETT'S TEST—CRITICAL VALUES (ONE SIDED)

Critical values

## $d_{0.05}\ (a-1,\ df_{err})$

$(a-1)$ = Number of treatment means (excluding the control)

| $df_{err}$ | 1 | 2 | 3 | 4 | 5 | 6 | 7 | 8 | 9 |
|---|---|---|---|---|---|---|---|---|---|
| 5 | 2.02 | 2.44 | 2.68 | 2.85 | 2.98 | 3.08 | 3.16 | 3.24 | 3.30 |
| 6 | 1.94 | 2.34 | 2.56 | 2.71 | 2.83 | 2.92 | 3.00 | 3.07 | 3.12 |
| 7 | 1.89 | 2.27 | 2.48 | 2.62 | 2.73 | 2.82 | 2.89 | 2.95 | 3.01 |
| 8 | 1.86 | 2.22 | 2.42 | 2.55 | 2.66 | 2.74 | 2.81 | 2.87 | 2.92 |
| 9 | 1.83 | 2.18 | 2.37 | 2.50 | 2.60 | 2.68 | 2.75 | 2.81 | 2.86 |
| 10 | 1.81 | 2.15 | 2.34 | 2.47 | 2.56 | 2.64 | 2.70 | 2.76 | 2.81 |
| 11 | 1.80 | 2.13 | 2.31 | 2.44 | 2.53 | 2.60 | 2.67 | 2.72 | 2.77 |
| 12 | 1.78 | 2.11 | 2.29 | 2.41 | 2.50 | 2.58 | 2.64 | 2.69 | 2.74 |
| 13 | 1.77 | 2.09 | 2.27 | 2.39 | 2.48 | 2.55 | 2.61 | 2.66 | 2.71 |
| 14 | 1.76 | 2.08 | 2.25 | 2.37 | 2.46 | 2.53 | 2.59 | 2.64 | 2.69 |
| 15 | 1.75 | 2.07 | 2.24 | 2.36 | 2.44 | 2.51 | 2.57 | 2.62 | 2.67 |
| 16 | 1.75 | 2.06 | 2.23 | 2.34 | 2.43 | 2.50 | 2.56 | 2.61 | 2.65 |
| 17 | 1.74 | 2.05 | 2.22 | 2.33 | 2.42 | 2.49 | 2.54 | 2.59 | 2.64 |
| 18 | 1.73 | 2.04 | 2.21 | 2.32 | 2.41 | 2.48 | 2.53 | 2.58 | 2.62 |
| 19 | 1.73 | 2.03 | 2.20 | 2.31 | 2.40 | 2.47 | 2.52 | 2.57 | 2.61 |
| 20 | 1.72 | 2.03 | 2.19 | 2.30 | 2.39 | 2.46 | 2.51 | 2.56 | 2.60 |
| 24 | 1.71 | 2.01 | 2.17 | 2.28 | 2.36 | 2.43 | 2.48 | 2.53 | 2.57 |
| 30 | 1.70 | 1.99 | 2.15 | 2.25 | 2.33 | 2.40 | 2.45 | 2.50 | 2.54 |
| 40 | 1.68 | 1.97 | 2.13 | 2.23 | 2.31 | 2.37 | 2.42 | 2.47 | 2.51 |
| 60 | 1.67 | 1.95 | 2.10 | 2.21 | 2.28 | 2.35 | 2.39 | 2.44 | 2.48 |
| 120 | 1.66 | 1.93 | 2.08 | 2.18 | 2.26 | 2.32 | 2.37 | 2.41 | 2.45 |
| ∞ | 1.64 | 1.92 | 2.06 | 2.16 | 2.23 | 2.29 | 2.34 | 2.38 | 2.42 |

## $d_{0.01}\ (a-1,\ df_{err})$

$(a-1)$ = Number of treatment means (excluding the control)

| $df_{err}$ | 1 | 2 | 3 | 4 | 5 | 6 | 7 | 8 | 9 |
|---|---|---|---|---|---|---|---|---|---|
| 5 | 3.37 | 3.90 | 4.21 | 4.43 | 4.60 | 4.73 | 4.85 | 4.94 | 5.03 |
| 6 | 3.14 | 3.61 | 3.88 | 4.07 | 4.21 | 4.33 | 4.43 | 4.51 | 4.59 |
| 7 | 3.00 | 3.42 | 3.66 | 3.83 | 3.96 | 4.07 | 4.15 | 4.23 | 4.30 |
| 8 | 2.90 | 3.29 | 3.51 | 3.67 | 3.79 | 3.88 | 3.96 | 4.03 | 4.09 |
| 9 | 2.82 | 3.19 | 3.40 | 3.55 | 3.66 | 3.75 | 3.82 | 3.89 | 3.94 |
| 10 | 2.76 | 3.11 | 3.31 | 3.45 | 3.56 | 3.64 | 3.71 | 3.78 | 3.83 |
| 11 | 2.72 | 3.06 | 3.25 | 3.38 | 3.48 | 3.56 | 3.63 | 3.69 | 3.74 |
| 12 | 2.68 | 3.01 | 3.19 | 3.32 | 3.42 | 3.50 | 3.56 | 3.62 | 3.67 |
| 13 | 2.65 | 2.97 | 3.15 | 3.27 | 3.37 | 3.44 | 3.51 | 3.56 | 3.61 |
| 14 | 2.62 | 2.94 | 3.11 | 3.23 | 3.32 | 3.40 | 3.46 | 3.51 | 3.56 |
| 15 | 2.60 | 2.91 | 3.08 | 3.20 | 3.29 | 3.36 | 3.42 | 3.47 | 3.52 |
| 16 | 2.58 | 2.88 | 3.05 | 3.17 | 3.26 | 3.33 | 3.39 | 3.44 | 3.48 |
| 17 | 2.57 | 2.86 | 3.03 | 3.14 | 3.23 | 3.30 | 3.36 | 3.41 | 3.45 |
| 18 | 2.55 | 2.84 | 3.01 | 3.12 | 3.21 | 3.27 | 3.33 | 3.38 | 3.42 |
| 19 | 2.54 | 2.83 | 2.99 | 3.10 | 3.18 | 3.25 | 3.31 | 3.36 | 3.40 |
| 20 | 2.53 | 2.81 | 2.97 | 3.08 | 3.17 | 3.23 | 3.29 | 3.34 | 3.38 |
| 24 | 2.49 | 2.77 | 2.92 | 3.03 | 3.11 | 3.17 | 3.22 | 3.27 | 3.31 |
| 30 | 2.46 | 2.72 | 2.87 | 2.97 | 3.05 | 3.11 | 3.16 | 3.21 | 3.24 |
| 40 | 2.42 | 2.68 | 2.82 | 2.92 | 2.99 | 3.05 | 3.10 | 3.14 | 3.18 |
| 60 | 2.39 | 2.64 | 2.78 | 2.87 | 2.94 | 3.00 | 3.04 | 3.08 | 3.12 |
| 120 | 2.36 | 2.60 | 2.73 | 2.82 | 2.89 | 2.94 | 2.99 | 3.03 | 3.06 |
| ∞ | 2.33 | 2.56 | 2.68 | 2.77 | 2.84 | 2.89 | 2.93 | 2.97 | 3.00 |

Source: Reproduced with permission from C. W. Dunnett, "A Multiple Comparison Procedure for Comparing Several Treatments with a Control," *Journal of the American Statistical Association*, vol. 50, no. 272, pp. 1117–1118, December, 1955.

TABLE IX.9    DUNNETT'S TEST—CRITICAL VALUES (TWO SIDED) (continued)

Critical values

$d_{0.05}$ $(a - 1, df_{err})$

$(a - 1)$ = Number of treatment means (excluding the control)

| $df_{err}$ | 1 | 2 | 3 | 4 | 5 | 6 | 7 | 8 | 9 |
|---|---|---|---|---|---|---|---|---|---|
| 5 | 2.57 | 3.03 | 3.29 | 3.48 | 3.62 | 3.73 | 3.82 | 3.90 | 3.97 |
| 6 | 2.45 | 2.86 | 3.10 | 3.26 | 3.39 | 3.49 | 3.57 | 3.64 | 3.71 |
| 7 | 2.36 | 2.75 | 2.97 | 3.12 | 3.24 | 3.33 | 3.41 | 3.47 | 3.53 |
| 8 | 2.31 | 2.67 | 2.88 | 3.02 | 3.13 | 3.22 | 3.29 | 3.35 | 3.41 |
| 9 | 2.26 | 2.61 | 2.81 | 2.95 | 3.05 | 3.14 | 3.20 | 3.26 | 3.32 |
| 10 | 2.23 | 2.57 | 2.76 | 2.89 | 2.99 | 3.07 | 3.14 | 3.19 | 3.24 |
| 11 | 2.20 | 2.53 | 2.72 | 2.84 | 2.94 | 3.02 | 3.08 | 3.14 | 3.19 |
| 12 | 2.18 | 2.50 | 2.68 | 2.81 | 2.90 | 2.98 | 3.04 | 3.09 | 3.14 |
| 13 | 2.16 | 2.48 | 2.65 | 2.78 | 2.87 | 2.94 | 3.00 | 3.06 | 3.10 |
| 14 | 2.14 | 2.46 | 2.63 | 2.75 | 2.84 | 2.91 | 2.97 | 3.02 | 3.07 |
| 15 | 2.13 | 2.44 | 2.61 | 2.73 | 2.82 | 2.89 | 2.95 | 3.00 | 3.04 |
| 16 | 2.12 | 2.42 | 2.59 | 2.71 | 2.80 | 2.87 | 2.92 | 2.97 | 3.02 |
| 17 | 2.11 | 2.41 | 2.58 | 2.69 | 2.78 | 2.85 | 2.90 | 2.95 | 3.00 |
| 18 | 2.10 | 2.40 | 2.56 | 2.68 | 2.76 | 2.83 | 2.89 | 2.94 | 2.98 |
| 19 | 2.09 | 2.39 | 2.55 | 2.66 | 2.75 | 2.81 | 2.87 | 2.92 | 2.96 |
| 20 | 2.09 | 2.38 | 2.54 | 2.65 | 2.73 | 2.80 | 2.86 | 2.90 | 2.95 |
| 24 | 2.06 | 2.35 | 2.51 | 2.61 | 2.70 | 2.76 | 2.81 | 2.86 | 2.90 |
| 30 | 2.04 | 2.32 | 2.47 | 2.58 | 2.66 | 2.72 | 2.77 | 2.82 | 2.86 |
| 40 | 2.02 | 2.29 | 2.44 | 2.54 | 2.62 | 2.68 | 2.73 | 2.77 | 2.81 |
| 60 | 2.00 | 2.27 | 2.41 | 2.51 | 2.58 | 2.64 | 2.69 | 2.73 | 2.77 |
| 120 | 1.98 | 2.24 | 2.38 | 2.47 | 2.55 | 2.60 | 2.65 | 2.69 | 2.73 |
| ∞ | 1.96 | 2.21 | 2.35 | 2.44 | 2.51 | 2.57 | 2.61 | 2.65 | 2.69 |

$d_{0.01}$ $(a - 1, df_{err})$

$(a - 1)$ = Number of treatment means (excluding the control)

| $df_{err}$ | 1 | 2 | 3 | 4 | 5 | 6 | 7 | 8 | 9 |
|---|---|---|---|---|---|---|---|---|---|
| 5 | 4.03 | 4.63 | 4.98 | 5.22 | 5.41 | 5.56 | 5.69 | 5.80 | 5.89 |
| 6 | 3.71 | 4.21 | 4.51 | 4.71 | 4.87 | 5.00 | 5.10 | 5.20 | 5.28 |
| 7 | 3.50 | 3.95 | 4.21 | 4.39 | 4.53 | 4.64 | 4.74 | 4.82 | 4.89 |
| 8 | 3.36 | 3.77 | 4.00 | 4.17 | 4.29 | 4.40 | 4.48 | 4.56 | 4.62 |
| 9 | 3.25 | 3.63 | 3.85 | 4.01 | 4.12 | 4.22 | 4.30 | 4.37 | 4.43 |
| 10 | 3.17 | 3.53 | 3.74 | 3.88 | 3.99 | 4.08 | 4.16 | 4.22 | 4.28 |
| 11 | 3.11 | 3.45 | 3.65 | 3.79 | 3.89 | 3.98 | 4.05 | 4.11 | 4.16 |
| 12 | 3.05 | 3.39 | 3.58 | 3.71 | 3.81 | 3.89 | 3.96 | 4.02 | 4.07 |
| 13 | 3.01 | 3.33 | 3.52 | 3.65 | 3.74 | 3.82 | 3.89 | 3.94 | 3.99 |
| 14 | 2.98 | 3.29 | 3.47 | 3.59 | 3.69 | 3.76 | 3.83 | 3.88 | 3.93 |
| 15 | 2.95 | 3.25 | 3.43 | 3.55 | 3.64 | 3.71 | 3.78 | 3.83 | 3.88 |
| 16 | 2.92 | 3.22 | 3.39 | 3.51 | 3.60 | 3.67 | 3.73 | 3.78 | 3.83 |
| 17 | 2.90 | 3.19 | 3.36 | 3.47 | 3.56 | 3.63 | 3.69 | 3.74 | 3.79 |
| 18 | 2.88 | 3.17 | 3.33 | 3.44 | 3.53 | 3.60 | 3.66 | 3.71 | 3.75 |
| 19 | 2.86 | 3.15 | 3.31 | 3.42 | 3.50 | 3.57 | 3.63 | 3.68 | 3.72 |
| 20 | 2.85 | 3.13 | 3.29 | 3.40 | 3.48 | 3.55 | 3.60 | 3.65 | 3.69 |
| 24 | 2.80 | 3.07 | 3.22 | 3.32 | 3.40 | 3.47 | 3.52 | 3.57 | 3.61 |
| 30 | 2.75 | 3.01 | 3.15 | 3.25 | 3.33 | 3.39 | 3.44 | 3.49 | 3.52 |
| 40 | 2.70 | 2.95 | 3.09 | 3.19 | 3.26 | 3.32 | 3.37 | 3.41 | 3.44 |
| 60 | 2.66 | 2.90 | 3.03 | 3.12 | 3.19 | 3.25 | 3.29 | 3.33 | 3.37 |
| 120 | 2.62 | 2.85 | 2.97 | 3.06 | 3.12 | 3.18 | 3.22 | 3.26 | 3.29 |
| ∞ | 2.58 | 2.79 | 2.92 | 3.00 | 3.06 | 3.11 | 3.15 | 3.19 | 3.22 |

Source: Reproduced with permission from C. W. Dunnett, "New Tables for Multiple Comparisons with a Control," Biometrics, vol. 20, no. 3, pp. 488–489, September, 1964.

**TABLE IX.10**    *X*-BAR, *R*, AND *S* CONTROL CHART—3-SIGMA LIMIT CONSTANTS
(Based on Sampling from a Normal Distribution)

| Subgroup size $n$ | $d_2$ | $d_3$ | $c_4$ | $D_3$ | $D_4$ | $B_3$ | $B_4$ | $A$ | $A_2$ | $A_3$ |
|---|---|---|---|---|---|---|---|---|---|---|
| 2 | 1.128 | 0.8525 | 0.7979 | 0.00 | 3.27 | 0.00 | 3.27 | 2.12 | 1.88 | 2.66 |
| 3 | 1.693 | 0.8884 | 0.8862 | 0.00 | 2.57 | 0.00 | 2.57 | 1.73 | 1.02 | 1.95 |
| 4 | 2.059 | 0.8798 | 0.9213 | 0.00 | 2.28 | 0.00 | 2.27 | 1.50 | 0.73 | 1.63 |
| 5 | 2.326 | 0.8641 | 0.9400 | 0.00 | 2.11 | 0.00 | 2.09 | 1.34 | 0.58 | 1.43 |
| 6 | 2.534 | 0.8480 | 0.9515 | 0.00 | 2.00 | 0.03 | 1.97 | 1.22 | 0.48 | 1.29 |
| 7 | 2.704 | 0.8332 | 0.9594 | 0.08 | 1.92 | 0.12 | 1.88 | 1.13 | 0.42 | 1.18 |
| 8 | 2.847 | 0.8198 | 0.9650 | 0.14 | 1.86 | 0.19 | 1.81 | 1.06 | 0.37 | 1.10 |
| 9 | 2.970 | 0.8078 | 0.9693 | 0.18 | 1.82 | 0.24 | 1.76 | 1.00 | 0.34 | 1.03 |
| 10 | 3.078 | 0.7971 | 0.9727 | 0.22 | 1.78 | 0.28 | 1.72 | 0.95 | 0.31 | 0.98 |
| 11 | 3.173 | 0.7873 | 0.9754 | 0.26 | 1.74 | 0.32 | 1.68 | 0.90 | 0.29 | 0.93 |
| 12 | 3.258 | 0.7785 | 0.9776 | 0.28 | 1.72 | 0.35 | 1.65 | 0.87 | 0.27 | 0.89 |
| 13 | 3.336 | 0.7704 | 0.9794 | 0.31 | 1.69 | 0.38 | 1.62 | 0.83 | 0.25 | 0.85 |
| 14 | 3.407 | 0.7630 | 0.9810 | 0.33 | 1.67 | 0.41 | 1.59 | 0.80 | 0.24 | 0.82 |
| 15 | 3.472 | 0.7562 | 0.9823 | 0.35 | 1.65 | 0.43 | 1.57 | 0.77 | 0.22 | 0.79 |
| 16 | 3.532 | 0.7499 | 0.9835 | 0.36 | 1.64 | 0.45 | 1.55 | 0.75 | 0.21 | 0.76 |
| 17 | 3.588 | 0.7441 | 0.9845 | 0.38 | 1.62 | 0.47 | 1.53 | 0.73 | 0.20 | 0.74 |
| 18 | 3.640 | 0.7386 | 0.9854 | 0.39 | 1.61 | 0.48 | 1.52 | 0.71 | 0.19 | 0.72 |
| 19 | 3.689 | 0.7335 | 0.9862 | 0.40 | 1.60 | 0.50 | 1.50 | 0.69 | 0.19 | 0.70 |
| 20 | 3.735 | 0.7287 | 0.9869 | 0.41 | 1.59 | 0.51 | 1.49 | 0.67 | 0.18 | 0.68 |
| 30 | 4.086 | 0.6926 | 0.9914 | * | * | 0.60 | 1.40 | 0.55 | * | 0.55 |
| 40 | 4.322 | 0.6692 | 0.9936 | | | 0.66 | 1.34 | 0.47 | | 0.48 |
| 50 | 4.498 | 0.6521 | 0.9949 | | | 0.70 | 1.30 | 0.42 | | 0.43 |
| 60 | 4.639 | 0.6389 | 0.9958 | | | 0.72 | 1.28 | 0.39 | | 0.39 |
| 70 | 4.755 | 0.6283 | 0.9964 | | | 0.74 | 1.26 | 0.36 | | 0.36 |
| 80 | 4.854 | 0.6194 | 0.9968 | | | 0.76 | 1.24 | 0.34 | | 0.34 |
| 90 | 4.939 | 0.6118 | 0.9972 | | | 0.77 | 1.23 | 0.32 | | 0.32 |
| 100 | 5.015 | 0.6052 | 0.9975 | | | 0.79 | 1.21 | 0.30 | | 0.30 |

\* *R*-chart factors are given only up to $n = 20$; for larger subgroup sizes the *S* chart should be used.

*Source:* Adapted with permission from E. L. Grant and R. S. Leavenworth, *Statistical Quality Control*, 6th ed., New York: McGraw-Hill, pp 669–672, 1988.

**TABLE IX.11** X-BAR, R, AND S CONTROL CHART—PROBABILITY LIMIT CONSTANTS

**Probability based X-bar and R chart constants**

| Subgroup size n | $A_{2,0.001}$ | $A_{2,0.999}$ | $D_{0.001}$ | $D_{0.999}$ | $A_{2,0.005}$ | $A_{2,0.995}$ | $D_{0.005}$ | $D_{0.995}$ | $A_{2,0.025}$ | $A_{2,0.975}$ | $D_{0.025}$ | $D_{0.975}$ |
|---|---|---|---|---|---|---|---|---|---|---|---|---|
| 2 | 1.94 | 1.94 | 0.00 | 4.12 | 1.61 | 1.61 | 0.01 | 3.52 | 1.23 | 1.23 | 0.04 | 2.81 |
| 3 | 1.05 | 1.05 | 0.04 | 2.99 | 0.88 | 0.88 | 0.08 | 2.61 | 0.67 | 0.67 | 0.18 | 2.17 |
| 4 | 0.75 | 0.75 | 0.10 | 2.58 | 0.63 | 0.63 | 0.17 | 2.28 | 0.48 | 0.48 | 0.29 | 1.93 |
| 5 | 0.59 | 0.59 | 0.16 | 2.36 | 0.50 | 0.50 | 0.24 | 2.10 | 0.38 | 0.38 | 0.37 | 1.81 |
| 6 | 0.50 | 0.50 | 0.21 | 2.22 | 0.41 | 0.41 | 0.30 | 1.99 | 0.32 | 0.32 | 0.42 | 1.72 |
| 7 | 0.43 | 0.43 | 0.26 | 2.12 | 0.36 | 0.36 | 0.34 | 1.90 | 0.27 | 0.27 | 0.46 | 1.66 |
| 8 | 0.38 | 0.38 | 0.29 | 2.04 | 0.32 | 0.32 | 0.38 | 1.84 | 0.24 | 0.24 | 0.50 | 1.62 |
| 9 | 0.35 | 0.35 | 0.33 | 1.99 | 0.29 | 0.29 | 0.41 | 1.80 | 0.22 | 0.22 | 0.52 | 1.58 |
| 10 | 0.32 | 0.32 | 0.35 | 1.94 | 0.26 | 0.26 | 0.43 | 1.76 | 0.20 | 0.20 | 0.54 | 1.55 |

**Probability based X-bar and S chart constants**

| Subgroup size n | $A_{3,0.001}$ | $A_{3,0.999}$ | $B_{0.001}$ | $B_{0.999}$ | $A_{3,0.005}$ | $A_{3,0.995}$ | $B_{0.005}$ | $B_{0.995}$ | $A_{3,0.025}$ | $A_{3,0.975}$ | $B_{0.025}$ | $B_{0.975}$ |
|---|---|---|---|---|---|---|---|---|---|---|---|---|
| 2 | 2.74 | 2.74 | 0.00 | 2.92 | 2.28 | 2.28 | 0.00 | 2.49 | 1.74 | 1.74 | 0.03 | 1.99 |
| 3 | 2.01 | 2.01 | 0.03 | 2.43 | 1.68 | 1.68 | 0.07 | 2.12 | 1.28 | 1.28 | 0.15 | 1.77 |
| 4 | 1.68 | 1.68 | 0.09 | 2.19 | 1.40 | 1.40 | 0.14 | 1.94 | 1.06 | 1.06 | 0.25 | 1.66 |
| 5 | 1.47 | 1.47 | 0.14 | 2.04 | 1.23 | 1.23 | 0.21 | 1.83 | 0.93 | 0.93 | 0.33 | 1.59 |
| 6 | 1.33 | 1.33 | 0.20 | 1.94 | 1.10 | 1.10 | 0.27 | 1.76 | 0.84 | 0.84 | 0.39 | 1.53 |
| 7 | 1.22 | 1.22 | 0.24 | 1.87 | 1.01 | 1.01 | 0.32 | 1.70 | 0.77 | 0.77 | 0.44 | 1.50 |
| 8 | 1.13 | 1.13 | 0.28 | 1.80 | 0.94 | 0.94 | 0.36 | 1.65 | 0.72 | 0.72 | 0.48 | 1.47 |
| 9 | 1.06 | 1.06 | 0.32 | 1.75 | 0.89 | 0.89 | 0.40 | 1.61 | 0.67 | 0.67 | 0.51 | 1.44 |
| 10 | 1.00 | 1.00 | 0.35 | 1.72 | 0.84 | 0.84 | 0.43 | 1.58 | 0.64 | 0.64 | 0.53 | 1.42 |

*Source:* Adapted with permission from E. L. Grant and R. S. Leavenworth, *Statistical Quality Control*, 6th ed., New York: McGraw-Hill, p. 317, 1988.

TABLE IX.12    EWMA AND EWMD CONTROL CHART LIMIT
CONSTANTS

| Weighting factor | Equivalent sample size | Means | Standard deviations | | |
|---|---|---|---|---|---|
| $r$ | $n$ | $A^*$ | $D_1^*$ | $D_2^*$ | $d_2^*$ |
| 0.050 | 39 | 0.480 | 0.514 | 1.102 | 0.808 |
| 0.100 | 19 | 0.688 | 0.390 | 1.247 | 0.819 |
| 0.200 | 9 | 1.000 | 0.197 | 1.486 | 0.841 |
| 0.250 | 7 | 1.132 | 0.109 | 1.597 | 0.853 |
| 0.286 | 6 | 1.225 | 0.048 | 1.676 | 0.862 |
| 0.333 | 5 | 1.342 | 0.000 | 1.780 | 0.874 |
| 0.400 | 4 | 1.500 | 0.000 | 1.930 | 0.892 |
| 0.500 | 3 | 1.732 | 0.000 | 2.164 | 0.921 |
| 0.667 | 2 | 2.121 | 0.000 | 2.596 | 0.977 |
| 0.800 | | 2.449 | 0.000 | 2.990 | 1.030 |
| 0.900 | | 2.714 | 0.000 | 3.321 | 1.076 |
| 1.000 | 1 | 3.000 | | | |

*Source*: Reproduced with permission from A. L. Sweet, "Control Charts Using Coupled Exponentially Weighted Moving Averages," *Transactions of the IIE*, vol. 18, no. 1, pp. 26–33, 1986.

TABLE IX.13  MEDIAN, 5 PERCENT, AND 95 PERCENT RANK DISTRIBUTION TABLES

**Median Ranks**

| | Sample size | | | | | | | | | |
|---|---|---|---|---|---|---|---|---|---|---|
| j \ n | 1 | 2 | 3 | 4 | 5 | 6 | 7 | 8 | 9 | 10 |
| 1 | 50.000 | 29.289 | 20.630 | 15.910 | 12.945 | 10.910 | 9.428 | 8.300 | 7.412 | 6.697 |
| 2 | | 70.711 | 50.000 | 38.573 | 31.381 | 26.445 | 22.849 | 20.113 | 17.962 | 16.226 |
| 3 | | | 79.370 | 61.427 | 50.000 | 42.141 | 36.412 | 32.052 | 28.624 | 25.857 |
| 4 | | | | 84.090 | 68.619 | 57.859 | 50.000 | 44.015 | 39.308 | 35.510 |
| 5 | | | | | 87.055 | 73.555 | 63.588 | 55.984 | 50.000 | 45.169 |
| 6 | | | | | | 89.090 | 77.151 | 67.948 | 60.691 | 54.831 |
| 7 | | | | | | | 90.572 | 79.887 | 71.376 | 64.490 |
| 8 | | | | | | | | 91.700 | 82.038 | 74.142 |
| 9 | | | | | | | | | 92.587 | 83.774 |
| 10 | | | | | | | | | | 93.303 |

**Median Ranks**

| | Sample size | | | | | | | | | |
|---|---|---|---|---|---|---|---|---|---|---|
| j \ n | 11 | 12 | 13 | 14 | 15 | 16 | 17 | 18 | 19 | 20 |
| 1 | 6.107 | 5.613 | 5.192 | 4.830 | 4.516 | 4.240 | 3.995 | 3.778 | 3.582 | 3.406 |
| 2 | 14.796 | 13.598 | 12.579 | 11.702 | 10.940 | 10.270 | 9.678 | 9.151 | 8.677 | 8.251 |
| 3 | 23.578 | 21.669 | 20.045 | 18.647 | 17.432 | 16.365 | 15.422 | 14.581 | 13.827 | 13.147 |
| 4 | 32.380 | 29.758 | 27.528 | 25.608 | 23.939 | 22.474 | 21.178 | 20.024 | 18.988 | 18.055 |
| 5 | 41.189 | 37.853 | 35.016 | 32.575 | 30.452 | 28.589 | 26.940 | 25.471 | 24.154 | 22.967 |
| 6 | 50.000 | 45.951 | 42.508 | 39.544 | 36.967 | 34.705 | 32.704 | 30.921 | 29.322 | 27.880 |
| 7 | 58.811 | 54.049 | 50.000 | 46.515 | 43.483 | 40.823 | 38.469 | 36.371 | 34.491 | 32.795 |
| 8 | 67.620 | 62.147 | 57.492 | 53.485 | 50.000 | 46.941 | 44.234 | 41.823 | 39.660 | 37.710 |
| 9 | 76.421 | 70.242 | 64.984 | 60.456 | 56.517 | 53.059 | 50.000 | 47.274 | 44.830 | 42.626 |
| 10 | 85.204 | 78.331 | 72.472 | 67.425 | 63.033 | 59.177 | 55.766 | 52.726 | 50.000 | 47.542 |
| 11 | 93.893 | 86.402 | 79.955 | 74.392 | 69.548 | 65.295 | 61.531 | 58.177 | 55.170 | 52.458 |
| 12 | | 94.387 | 87.421 | 81.353 | 76.061 | 71.411 | 67.296 | 63.629 | 60.340 | 57.374 |
| 13 | | | 94.808 | 88.298 | 82.568 | 77.525 | 73.060 | 69.079 | 65.509 | 62.289 |
| 14 | | | | 95.169 | 89.060 | 83.635 | 78.821 | 74.529 | 70.678 | 67.205 |
| 15 | | | | | 95.484 | 89.730 | 84.578 | 79.976 | 75.846 | 72.119 |
| 16 | | | | | | 95.760 | 90.322 | 85.419 | 81.011 | 77.033 |
| 17 | | | | | | | 96.005 | 90.849 | 86.173 | 81.945 |
| 18 | | | | | | | | 96.222 | 91.322 | 86.853 |
| 19 | | | | | | | | | 96.418 | 91.749 |
| 20 | | | | | | | | | | 96.594 |

**Median Ranks**

| | Sample size | | | | | | | | | |
|---|---|---|---|---|---|---|---|---|---|---|
| j \ n | 21 | 22 | 23 | 24 | 25 | 26 | 27 | 28 | 29 | 30 |
| 1 | 3.247 | 3.101 | 2.969 | 2.847 | 2.734 | 2.631 | 2.534 | 2.445 | 2.362 | 2.284 |
| 2 | 7.864 | 7.512 | 7.191 | 6.895 | 6.623 | 6.372 | 6.139 | 5.922 | 5.720 | 5.532 |
| 3 | 12.531 | 11.970 | 11.458 | 10.987 | 10.553 | 10.153 | 9.781 | 9.436 | 9.114 | 8.814 |
| 4 | 17.209 | 16.439 | 15.734 | 15.088 | 14.492 | 13.942 | 13.432 | 12.958 | 12.517 | 12.104 |
| 5 | 21.890 | 20.911 | 20.015 | 19.192 | 18.435 | 17.735 | 17.086 | 16.483 | 15.922 | 15.397 |
| 6 | 26.574 | 25.384 | 24.297 | 23.299 | 22.379 | 21.529 | 20.742 | 20.010 | 19.328 | 18.691 |
| 7 | 31.258 | 29.859 | 28.580 | 27.406 | 26.324 | 25.325 | 24.398 | 23.537 | 22.735 | 21.986 |
| 8 | 35.943 | 34.334 | 32.863 | 31.513 | 30.269 | 29.120 | 28.055 | 27.065 | 26.143 | 25.281 |
| 9 | 40.629 | 38.810 | 37.147 | 35.621 | 34.215 | 32.916 | 31.712 | 30.593 | 29.550 | 28.576 |
| 10 | 45.314 | 43.286 | 41.431 | 39.729 | 38.161 | 36.712 | 35.370 | 34.121 | 32.958 | 31.872 |
| 11 | 50.000 | 47.762 | 45.716 | 43.837 | 42.107 | 40.509 | 39.027 | 37.650 | 36.367 | 35.168 |
| 12 | 54.686 | 52.238 | 50.000 | 47.946 | 46.054 | 44.305 | 42.685 | 41.178 | 39.775 | 38.464 |
| 13 | 59.371 | 56.714 | 54.284 | 52.054 | 50.000 | 48.102 | 46.342 | 44.707 | 43.183 | 41.760 |
| 14 | 64.057 | 61.190 | 58.568 | 56.162 | 53.946 | 51.898 | 50.000 | 48.236 | 46.592 | 45.056 |
| 15 | 68.742 | 65.665 | 62.853 | 60.271 | 57.892 | 55.695 | 53.658 | 51.764 | 50.000 | 48.352 |
| 16 | 73.426 | 70.141 | 67.137 | 64.379 | 61.839 | 59.491 | 57.315 | 55.293 | 53.408 | 51.648 |
| 17 | 78.109 | 74.616 | 71.420 | 68.487 | 65.785 | 63.287 | 60.973 | 58.821 | 56.817 | 54.944 |
| 18 | 82.791 | 79.089 | 75.703 | 72.594 | 69.730 | 67.084 | 64.630 | 62.350 | 60.225 | 58.240 |
| 19 | 87.469 | 83.561 | 79.985 | 76.701 | 73.676 | 70.880 | 68.288 | 65.878 | 63.633 | 61.536 |
| 20 | 92.136 | 88.030 | 84.266 | 80.808 | 77.621 | 74.675 | 71.945 | 69.407 | 67.041 | 64.852 |
| 21 | 96.753 | 92.488 | 88.542 | 84.912 | 81.565 | 78.471 | 75.602 | 72.935 | 70.450 | 68.128 |
| 22 | | 96.898 | 92.809 | 89.013 | 85.507 | 82.265 | 79.258 | 76.463 | 73.857 | 71.424 |
| 23 | | | 97.031 | 93.105 | 89.447 | 86.058 | 82.914 | 79.990 | 77.265 | 74.719 |
| 24 | | | | 97.153 | 93.377 | 89.847 | 86.568 | 83.517 | 80.672 | 78.014 |
| 25 | | | | | 97.265 | 93.628 | 90.219 | 87.042 | 84.078 | 81.309 |
| 26 | | | | | | 97.369 | 93.861 | 90.564 | 87.483 | 84.603 |
| 27 | | | | | | | 97.465 | 94.078 | 90.865 | 87.896 |
| 28 | | | | | | | | 97.555 | 94.280 | 91.186 |
| 29 | | | | | | | | | 97.638 | 94.468 |
| 30 | | | | | | | | | | 97.716 |

Source: Reproduced with permission from K.C. Kapur and L.R. Lamberson, Reliability in Engineering Design, New York: Wiley, pp. 486–497, 1977.

# TABLE IX.13  MEDIAN, 5 PERCENT, AND 95 PERCENT RANK DISTRIBUTION TABLES (continued)

## Median Ranks

| $j$ \ $n$ | 31 | 32 | 33 | 34 | 35 | 36 | 37 | 38 | 39 | 40 |
|---|---|---|---|---|---|---|---|---|---|---|
| 1 | 2.211 | 2.143 | 2.078 | 2.018 | 1.961 | 1.907 | 1.856 | 1.807 | 1.762 | 1.718 |
| 2 | 5.355 | 5.190 | 5.034 | 4.887 | 4.749 | 4.618 | 4.495 | 4.377 | 4.266 | 4.160 |
| 3 | 8.533 | 8.269 | 8.021 | 7.787 | 7.567 | 7.359 | 7.162 | 6.975 | 6.798 | 6.629 |
| 4 | 11.716 | 11.355 | 11.015 | 10.694 | 10.391 | 10.105 | 9.835 | 9.578 | 9.335 | 9.103 |
| 5 | 14.905 | 14.445 | 14.011 | 13.603 | 13.218 | 12.855 | 12.510 | 12.184 | 11.874 | 11.580 |
| 6 | 18.094 | 17.535 | 17.009 | 16.514 | 16.046 | 15.605 | 15.187 | 14.791 | 14.415 | 14.057 |
| 7 | 21.284 | 20.626 | 20.007 | 19.425 | 18.875 | 18.355 | 17.864 | 17.398 | 16.956 | 16.535 |
| 8 | 24.474 | 23.717 | 23.006 | 22.336 | 21.704 | 21.107 | 20.541 | 20.005 | 19.497 | 19.013 |
| 9 | 27.664 | 26.809 | 26.005 | 25.247 | 24.533 | 23.858 | 23.219 | 22.613 | 22.038 | 21.492 |
| 10 | 30.855 | 29.901 | 29.004 | 28.159 | 27.362 | 26.609 | 25.897 | 25.221 | 24.580 | 23.971 |
| 11 | 34.046 | 32.993 | 32.003 | 31.071 | 30.192 | 29.361 | 28.575 | 27.829 | 27.122 | 26.449 |
| 12 | 37.236 | 36.085 | 35.003 | 33.983 | 33.022 | 32.113 | 31.253 | 30.437 | 29.664 | 28.928 |
| 13 | 40.427 | 39.177 | 38.002 | 36.895 | 35.851 | 34.865 | 33.931 | 33.046 | 32.206 | 31.407 |
| 14 | 43.618 | 42.269 | 41.001 | 39.807 | 38.681 | 37.616 | 36.609 | 35.654 | 34.748 | 33.886 |
| 15 | 46.809 | 45.362 | 44.001 | 42.720 | 41.511 | 40.368 | 39.287 | 38.262 | 37.290 | 36.365 |
| 16 | 50.000 | 48.454 | 47.000 | 45.632 | 44.340 | 43.120 | 41.965 | 40.871 | 39.832 | 38.844 |
| 17 | 53.191 | 51.546 | 50.000 | 48.544 | 47.170 | 45.872 | 44.644 | 43.479 | 42.374 | 41.323 |
| 18 | 56.382 | 54.638 | 52.999 | 51.456 | 50.000 | 48.624 | 47.322 | 46.087 | 44.916 | 43.802 |
| 19 | 59.573 | 57.731 | 55.999 | 54.368 | 52.830 | 51.376 | 50.000 | 48.696 | 47.458 | 46.281 |
| 20 | 62.763 | 60.823 | 58.998 | 57.280 | 55.660 | 54.128 | 52.678 | 51.304 | 50.000 | 48.760 |
| 21 | 65.954 | 63.915 | 61.998 | 60.193 | 58.489 | 56.880 | 55.356 | 53.913 | 52.542 | 51.239 |
| 22 | 69.145 | 67.007 | 64.997 | 63.105 | 61.319 | 59.632 | 58.035 | 56.521 | 55.084 | 53.719 |
| 23 | 72.335 | 70.099 | 67.997 | 66.017 | 64.149 | 62.383 | 60.713 | 59.129 | 57.626 | 56.198 |
| 24 | 75.526 | 73.191 | 70.996 | 68.929 | 66.973 | 65.135 | 63.391 | 61.738 | 60.168 | 58.677 |
| 25 | 78.716 | 76.283 | 73.995 | 71.841 | 69.808 | 67.837 | 66.069 | 64.346 | 62.710 | 61.156 |
| 26 | 81.906 | 79.374 | 76.994 | 74.752 | 72.637 | 70.639 | 68.747 | 66.954 | 65.252 | 63.635 |
| 27 | 85.094 | 82.465 | 79.993 | 77.664 | 75.467 | 73.391 | 71.425 | 69.562 | 67.794 | 66.114 |
| 28 | 88.282 | 85.555 | 82.991 | 80.575 | 78.296 | 76.142 | 74.103 | 72.171 | 70.336 | 68.593 |
| 29 | 91.467 | 88.644 | 85.989 | 83.486 | 81.125 | 78.899 | 76.781 | 74.779 | 72.878 | 71.072 |
| 30 | 94.645 | 91.731 | 88.985 | 86.397 | 83.954 | 81.645 | 79.459 | 77.387 | 75.420 | 73.550 |
| 31 | 97.789 | 94.810 | 91.979 | 89.306 | 86.782 | 84.395 | 82.136 | 79.994 | 77.962 | 76.029 |
| 32 | | 97.857 | 94.966 | 92.213 | 89.608 | 87.145 | 84.813 | 82.602 | 80.503 | 78.508 |
| 33 | | | 97.921 | 95.113 | 92.433 | 89.894 | 87.490 | 85.209 | 83.044 | 80.986 |
| 34 | | | | 97.982 | 95.251 | 92.641 | 90.165 | 87.816 | 85.585 | 83.465 |
| 35 | | | | | 98.039 | 95.382 | 92.838 | 90.422 | 88.126 | 85.943 |
| 36 | | | | | | 98.093 | 95.505 | 93.025 | 90.665 | 88.420 |
| 37 | | | | | | | 98.144 | 95.622 | 93.202 | 90.897 |
| 38 | | | | | | | | 98.192 | 95.734 | 93.371 |
| 39 | | | | | | | | | 98.238 | 95.839 |
| 40 | | | | | | | | | | 98.282 |

## Median Ranks

| $j$ \ $n$ | 41 | 42 | 43 | 44 | 45 | 46 | 47 | 48 | 49 | 50 |
|---|---|---|---|---|---|---|---|---|---|---|
| 1 | 1.676 | 1.637 | 1.599 | 1.563 | 1.528 | 1.495 | 1.464 | 1.434 | 1.405 | 1.377 |
| 2 | 4.060 | 3.964 | 3.872 | 3.785 | 3.702 | 3.622 | 3.545 | 3.472 | 3.402 | 3.334 |
| 3 | 6.469 | 6.316 | 6.170 | 6.031 | 5.898 | 5.771 | 5.649 | 5.532 | 5.420 | 5.312 |
| 4 | 8.883 | 8.673 | 8.473 | 8.282 | 8.099 | 7.925 | 7.757 | 7.597 | 7.443 | 7.295 |
| 5 | 11.300 | 11.033 | 10.778 | 10.535 | 10.303 | 10.080 | 9.867 | 9.663 | 9.467 | 9.279 |
| 6 | 13.717 | 13.393 | 13.084 | 12.789 | 12.507 | 12.237 | 11.979 | 11.731 | 11.493 | 11.265 |
| 7 | 16.135 | 15.754 | 15.391 | 15.043 | 14.712 | 14.394 | 14.090 | 13.799 | 13.519 | 13.250 |
| 8 | 18.554 | 18.115 | 17.697 | 17.298 | 16.917 | 16.551 | 16.202 | 15.867 | 15.545 | 15.236 |
| 9 | 20.972 | 20.477 | 20.004 | 19.553 | 19.122 | 18.709 | 18.314 | 17.935 | 17.571 | 17.222 |
| 10 | 23.391 | 22.838 | 22.311 | 21.808 | 21.327 | 20.867 | 20.426 | 20.003 | 19.598 | 19.209 |
| 11 | 25.810 | 25.200 | 24.618 | 24.063 | 23.532 | 23.025 | 22.538 | 22.072 | 21.625 | 21.195 |
| 12 | 28.228 | 27.562 | 26.926 | 26.318 | 25.738 | 25.182 | 24.650 | 24.140 | 23.651 | 23.181 |
| 13 | 30.647 | 29.924 | 29.233 | 28.574 | 27.943 | 27.340 | 26.763 | 26.209 | 25.678 | 25.168 |
| 14 | 33.066 | 32.285 | 31.540 | 30.829 | 30.149 | 29.498 | 28.875 | 28.278 | 27.705 | 27.154 |
| 15 | 35.485 | 34.647 | 33.848 | 33.084 | 32.355 | 31.656 | 30.988 | 30.347 | 29.731 | 29.141 |
| 16 | 37.905 | 37.009 | 36.155 | 35.340 | 34.560 | 33.814 | 33.100 | 32.415 | 31.758 | 31.127 |
| 17 | 40.324 | 39.371 | 38.463 | 37.595 | 36.766 | 35.972 | 35.212 | 34.484 | 33.785 | 33.114 |
| 18 | 42.743 | 41.733 | 40.770 | 39.851 | 38.972 | 38.130 | 37.325 | 36.553 | 35.822 | 35.100 |
| 19 | 45.162 | 44.095 | 43.078 | 42.106 | 41.177 | 40.289 | 39.437 | 38.622 | 37.839 | 37.087 |
| 20 | 47.581 | 46.457 | 45.385 | 44.361 | 43.383 | 42.447 | 41.550 | 40.690 | 39.866 | 39.074 |
| 21 | 50.000 | 48.819 | 47.692 | 46.617 | 45.589 | 44.605 | 43.662 | 42.759 | 41.892 | 41.060 |
| 22 | 52.419 | 51.181 | 50.000 | 48.872 | 47.794 | 46.763 | 45.775 | 44.825 | 43.919 | 43.047 |
| 23 | 54.838 | 53.543 | 52.307 | 51.128 | 50.000 | 48.921 | 47.887 | 46.897 | 45.946 | 45.033 |
| 24 | 57.257 | 55.905 | 54.615 | 53.383 | 52.206 | 51.079 | 50.000 | 48.966 | 47.973 | 47.020 |
| 25 | 59.676 | 58.267 | 56.922 | 55.639 | 54.411 | 53.237 | 52.112 | 51.034 | 50.000 | 49.007 |
| 26 | 62.095 | 60.629 | 59.230 | 57.894 | 56.617 | 55.395 | 54.225 | 53.103 | 52.027 | 50.993 |
| 27 | 64.514 | 62.991 | 61.537 | 60.149 | 58.823 | 57.553 | 56.337 | 55.172 | 54.054 | 52.980 |
| 28 | 66.933 | 65.353 | 63.845 | 62.405 | 61.028 | 59.711 | 58.450 | 57.241 | 56.081 | 54.966 |
| 29 | 69.352 | 67.714 | 66.152 | 64.660 | 63.234 | 61.869 | 60.562 | 59.310 | 58.107 | 56.953 |
| 30 | 71.771 | 70.076 | 68.459 | 66.916 | 65.440 | 64.027 | 62.675 | 61.378 | 60.134 | 58.940 |
| 31 | 74.190 | 72.438 | 70.767 | 69.171 | 67.645 | 66.186 | 64.767 | 63.447 | 62.161 | 60.926 |
| 32 | 76.609 | 74.800 | 73.074 | 71.426 | 69.851 | 68.344 | 66.900 | 65.516 | 64.188 | 62.913 |
| 33 | 79.028 | 77.162 | 75.381 | 73.681 | 72.056 | 70.502 | 69.012 | 67.585 | 66.215 | 64.899 |
| 34 | 81.446 | 79.523 | 77.689 | 75.937 | 74.262 | 72.660 | 71.125 | 69.653 | 68.242 | 66.886 |
| 35 | 83.865 | 81.885 | 79.996 | 78.192 | 76.467 | 74.817 | 73.237 | 71.722 | 70.268 | 68.873 |
| 36 | 86.283 | 84.246 | 82.303 | 80.447 | 78.673 | 76.975 | 75.349 | 73.791 | 72.295 | 70.859 |
| 37 | 88.700 | 86.607 | 84.609 | 82.702 | 80.878 | 79.133 | 77.462 | 75.859 | 74.322 | 72.646 |
| 38 | 91.117 | 88.967 | 86.916 | 84.956 | 83.083 | 81.291 | 79.574 | 77.928 | 76.349 | 74.832 |
| 39 | 93.531 | 91.327 | 89.222 | 87.211 | 85.283 | 83.448 | 81.686 | 79.997 | 78.375 | 76.819 |
| 40 | 95.940 | 93.684 | 91.527 | 89.465 | 87.493 | 85.606 | 83.798 | 82.065 | 80.402 | 78.805 |
| 41 | 98.324 | 96.036 | 93.830 | 91.718 | 89.697 | 87.763 | 85.910 | 84.133 | 82.428 | 80.791 |
| 42 | | 98.363 | 96.127 | 93.969 | 91.900 | 89.920 | 88.021 | 86.201 | 84.455 | 82.778 |
| 43 | | | 98.401 | 96.215 | 94.102 | 92.075 | 90.132 | 88.269 | 86.481 | 84.764 |
| 44 | | | | 98.437 | 96.298 | 94.229 | 92.243 | 90.337 | 88.507 | 86.750 |
| 45 | | | | | 98.471 | 96.378 | 94.351 | 92.403 | 90.532 | 88.735 |
| 46 | | | | | | 98.504 | 96.455 | 94.468 | 92.557 | 90.721 |
| 47 | | | | | | | 98.536 | 96.528 | 94.580 | 92.705 |
| 48 | | | | | | | | 98.566 | 96.598 | 94.688 |
| 49 | | | | | | | | | 98.595 | 96.666 |
| 50 | | | | | | | | | | 98.623 |

TABLE IX.13   MEDIAN, 5 PERCENT, AND 95 PERCENT RANK DISTRIBUTION TABLES (continued)

## 5 Percent Ranks

Sample size

| j \\ n | 21 | 22 | 23 | 24 | 25 | 26 | 27 | 28 | 29 | 30 |
|---|---|---|---|---|---|---|---|---|---|---|
| 1 | 0.244 | 0.233 | 0.223 | 0.213 | 0.205 | 0.197 | 0.190 | 0.183 | 0.177 | 0.171 |
| 2 | 1.719 | 1.640 | 1.567 | 1.501 | 1.440 | 1.384 | 1.332 | 1.284 | 1.239 | 1.198 |
| 3 | 4.010 | 3.822 | 3.651 | 3.495 | 3.352 | 3.220 | 3.098 | 2.985 | 2.879 | 2.781 |
| 4 | 6.781 | 6.460 | 6.167 | 5.901 | 5.656 | 5.431 | 5.223 | 5.031 | 4.852 | 4.685 |
| 5 | 9.884 | 9.411 | 8.981 | 8.588 | 8.229 | 7.899 | 7.594 | 7.311 | 7.049 | 6.806 |
| 6 | 13.245 | 12.603 | 12.021 | 11.491 | 11.006 | 10.560 | 10.148 | 9.768 | 9.415 | 9.087 |
| 7 | 16.818 | 15.994 | 15.248 | 14.569 | 13.947 | 13.377 | 12.852 | 12.367 | 11.917 | 11.499 |
| 8 | 20.575 | 19.556 | 18.634 | 17.796 | 17.030 | 16.328 | 15.682 | 15.085 | 14.532 | 14.018 |
| 9 | 24.499 | 23.272 | 22.164 | 21.157 | 20.238 | 19.396 | 18.622 | 17.908 | 17.246 | 16.633 |
| 10 | 28.580 | 27.131 | 25.824 | 24.639 | 23.559 | 22.570 | 21.662 | 20.824 | 20.050 | 19.331 |
| 11 | 32.811 | 31.126 | 29.609 | 28.236 | 26.985 | 25.842 | 24.793 | 23.827 | 22.934 | 22.106 |
| 12 | 37.190 | 35.254 | 33.515 | 31.942 | 30.513 | 29.508 | 28.012 | 26.911 | 25.894 | 24.953 |
| 13 | 41.720 | 39.516 | 37.539 | 35.756 | 34.139 | 32.664 | 31.314 | 30.072 | 28.927 | 27.867 |
| 14 | 46.406 | 43.913 | 41.684 | 39.678 | 37.862 | 36.209 | 34.697 | 33.309 | 32.030 | 30.846 |
| 15 | 51.261 | 48.454 | 45.954 | 43.711 | 41.684 | 39.842 | 38.161 | 36.620 | 35.200 | 33.889 |
| 16 | 56.302 | 53.151 | 50.356 | 47.858 | 45.607 | 43.566 | 41.707 | 40.004 | 38.439 | 36.995 |
| 17 | 61.559 | 58.020 | 54.902 | 52.127 | 49.636 | 47.384 | 45.336 | 43.464 | 41.746 | 40.163 |
| 18 | 67.079 | 63.091 | 59.610 | 56.531 | 53.779 | 51.300 | 49.052 | 47.002 | 45.123 | 43.394 |
| 19 | 72.945 | 68.409 | 64.507 | 61.086 | 58.048 | 55.323 | 52.861 | 50.621 | 48.573 | 46.691 |
| 20 | 79.327 | 74.053 | 69.636 | 65.819 | 62.459 | 59.465 | 56.770 | 54.327 | 52.099 | 50.056 |
| 21 | 86.705 | 80.188 | 75.075 | 70.773 | 67.039 | 63.740 | 60.790 | 58.127 | 55.706 | 53.493 |
| 22 | | 87.269 | 80.980 | 76.020 | 71.828 | 68.176 | 64.936 | 62.033 | 59.403 | 57.007 |
| 23 | | | 87.788 | 81.711 | 76.896 | 72.810 | 69.237 | 66.060 | 63.200 | 60.605 |
| 24 | | | | 88.265 | 82.388 | 77.711 | 73.726 | 70.231 | 67.113 | 64.299 |
| 25 | | | | | 88.707 | 83.017 | 78.470 | 74.583 | 71.168 | 68.103 |
| 26 | | | | | | 89.117 | 83.603 | 79.179 | 75.386 | 72.038 |
| 27 | | | | | | | 89.498 | 84.149 | 79.844 | 76.140 |
| 28 | | | | | | | | 89.853 | 84.661 | 80.467 |
| 29 | | | | | | | | | 90.185 | 85.140 |
| 30 | | | | | | | | | | 90.497 |

## 5 Percent Ranks

Sample size

| j \\ n | 1 | 2 | 3 | 4 | 5 | 6 | 7 | 8 | 9 | 10 |
|---|---|---|---|---|---|---|---|---|---|---|
| 1 | 5.000 | 2.532 | 1.695 | 1.274 | 1.021 | 0.851 | 0.730 | 0.639 | 0.568 | 0.512 |
| 2 | | 22.361 | 13.535 | 9.761 | 7.644 | 6.285 | 5.337 | 4.639 | 4.102 | 3.677 |
| 3 | | | 36.840 | 24.860 | 18.925 | 15.316 | 12.876 | 11.111 | 9.775 | 8.726 |
| 4 | | | | 47.237 | 34.259 | 27.134 | 22.532 | 19.290 | 16.875 | 15.003 |
| 5 | | | | | 54.928 | 41.820 | 34.126 | 28.924 | 25.137 | 22.244 |
| 6 | | | | | | 60.696 | 47.930 | 40.031 | 34.494 | 30.354 |
| 7 | | | | | | | 65.184 | 52.932 | 45.036 | 39.338 |
| 8 | | | | | | | | 68.766 | 57.086 | 49.310 |
| 9 | | | | | | | | | 71.687 | 60.584 |
| 10 | | | | | | | | | | 74.113 |

## 5 Percent Ranks

Sample size

| j \\ n | 11 | 12 | 13 | 14 | 15 | 16 | 17 | 18 | 19 | 20 |
|---|---|---|---|---|---|---|---|---|---|---|
| 1 | 0.465 | 0.426 | 0.394 | 0.366 | 0.341 | 0.320 | 0.301 | 0.285 | 0.270 | 0.256 |
| 2 | 3.332 | 3.046 | 2.805 | 2.600 | 2.423 | 2.268 | 2.132 | 2.011 | 1.903 | 1.806 |
| 3 | 7.882 | 7.187 | 6.605 | 6.110 | 5.685 | 5.315 | 4.990 | 4.702 | 4.446 | 4.217 |
| 4 | 13.507 | 12.285 | 11.267 | 10.405 | 9.666 | 9.025 | 8.464 | 7.969 | 7.529 | 7.135 |
| 5 | 19.958 | 18.102 | 16.566 | 15.272 | 14.166 | 13.211 | 12.377 | 11.643 | 10.991 | 10.408 |
| 6 | 27.125 | 24.530 | 22.395 | 20.607 | 19.086 | 17.777 | 16.636 | 15.634 | 14.747 | 13.955 |
| 7 | 34.981 | 31.524 | 28.705 | 26.358 | 24.373 | 22.669 | 21.191 | 19.895 | 18.750 | 17.731 |
| 8 | 43.563 | 39.086 | 35.480 | 32.503 | 29.999 | 27.860 | 26.011 | 24.396 | 22.972 | 21.707 |
| 9 | 52.991 | 47.267 | 42.738 | 39.041 | 35.956 | 33.337 | 31.083 | 29.101 | 27.395 | 25.865 |
| 10 | 63.564 | 56.189 | 50.535 | 45.999 | 42.256 | 39.101 | 36.401 | 34.060 | 32.009 | 30.195 |
| 11 | 76.160 | 66.132 | 58.990 | 53.434 | 48.925 | 45.165 | 41.970 | 39.215 | 36.811 | 34.693 |
| 12 | | 77.908 | 68.366 | 61.461 | 56.022 | 51.560 | 47.808 | 44.595 | 41.806 | 39.358 |
| 13 | | | 79.418 | 70.327 | 63.656 | 58.343 | 53.945 | 50.217 | 47.003 | 44.197 |
| 14 | | | | 80.736 | 72.060 | 65.617 | 60.436 | 56.112 | 52.420 | 49.218 |
| 15 | | | | | 81.896 | 73.604 | 67.381 | 62.332 | 58.088 | 54.442 |
| 16 | | | | | | 82.925 | 74.988 | 68.974 | 64.057 | 59.897 |
| 17 | | | | | | | 83.843 | 76.234 | 70.420 | 65.634 |
| 18 | | | | | | | | 84.668 | 77.363 | 71.738 |
| 19 | | | | | | | | | 85.413 | 78.389 |
| 20 | | | | | | | | | | 86.089 |

## TABLE IX.13 MEDIAN, 5 PERCENT, AND 95 PERCENT RANK DISTRIBUTION TABLES (continued)

**5 Percent Ranks**

| $j$ \ $n$ | Sample size | | | | | | | | | |
|---|---|---|---|---|---|---|---|---|---|---|
| | 31 | 32 | 33 | 34 | 35 | 36 | 37 | 38 | 39 | 40 |
| 1 | 0.165 | 0.160 | 0.155 | 0.151 | 0.146 | 0.142 | 0.138 | 0.135 | 0.131 | 0.128 |
| 2 | 1.158 | 1.122 | 1.086 | 1.055 | 1.025 | 0.996 | 0.969 | 0.943 | 0.919 | 0.896 |
| 3 | 2.690 | 2.604 | 2.524 | 2.448 | 2.377 | 2.310 | 2.246 | 2.186 | 2.129 | 2.075 |
| 4 | 4.530 | 4.384 | 4.246 | 4.120 | 3.999 | 3.885 | 3.778 | 3.676 | 3.580 | 3.488 |
| 5 | 6.578 | 6.365 | 6.166 | 5.978 | 5.802 | 5.636 | 5.479 | 5.331 | 5.190 | 5.057 |
| 6 | 8.781 | 8.495 | 8.227 | 7.976 | 7.739 | 7.516 | 7.306 | 7.107 | 6.919 | 6.740 |
| 7 | 11.109 | 10.745 | 10.404 | 10.084 | 9.783 | 9.499 | 9.232 | 8.979 | 8.740 | 8.513 |
| 8 | 13.540 | 13.093 | 12.675 | 12.283 | 11.914 | 11.567 | 11.240 | 10.931 | 10.638 | 10.361 |
| 9 | 16.061 | 15.528 | 15.029 | 14.561 | 14.122 | 13.708 | 13.318 | 12.950 | 12.601 | 12.271 |
| 10 | 18.662 | 18.038 | 17.455 | 16.909 | 16.396 | 15.913 | 15.458 | 15.028 | 14.622 | 14.237 |
| 11 | 21.336 | 20.618 | 19.948 | 19.319 | 18.730 | 18.175 | 17.653 | 17.160 | 16.694 | 16.252 |
| 12 | 24.077 | 23.262 | 22.501 | 21.788 | 21.119 | 20.491 | 19.898 | 19.340 | 18.812 | 18.312 |
| 13 | 26.883 | 25.966 | 25.111 | 24.310 | 23.560 | 22.855 | 22.191 | 21.565 | 20.973 | 20.413 |
| 14 | 29.749 | 28.727 | 27.775 | 26.884 | 26.049 | 25.265 | 24.527 | 23.832 | 23.175 | 22.553 |
| 15 | 32.674 | 31.544 | 30.491 | 29.507 | 28.585 | 27.719 | 26.905 | 26.138 | 25.414 | 24.729 |
| 16 | 35.657 | 34.415 | 33.258 | 32.177 | 31.165 | 30.216 | 29.324 | 28.483 | 27.690 | 26.940 |
| 17 | 38.698 | 37.339 | 36.074 | 34.894 | 33.789 | 32.754 | 31.781 | 30.865 | 30.001 | 29.185 |
| 18 | 41.797 | 40.317 | 38.940 | 37.657 | 36.457 | 35.332 | 34.276 | 33.283 | 32.346 | 31.461 |
| 19 | 44.956 | 43.349 | 41.656 | 40.466 | 39.167 | 37.951 | 36.809 | 35.736 | 34.725 | 33.770 |
| 20 | 48.175 | 46.436 | 44.823 | 43.321 | 41.920 | 40.609 | 39.380 | 38.224 | 37.136 | 36.109 |
| 21 | 51.458 | 49.581 | 47.841 | 46.225 | 44.717 | 43.309 | 41.988 | 40.748 | 39.581 | 38.480 |
| 22 | 54.810 | 52.786 | 50.914 | 49.177 | 47.560 | 46.049 | 44.634 | 43.307 | 42.058 | 40.881 |
| 23 | 58.234 | 56.055 | 54.344 | 52.181 | 50.448 | 48.832 | 47.320 | 45.902 | 44.569 | 43.314 |
| 24 | 61.739 | 59.314 | 57.235 | 55.239 | 53.385 | 51.658 | 50.045 | 48.534 | 47.114 | 45.778 |
| 25 | 65.336 | 62.810 | 60.493 | 58.355 | 56.374 | 54.532 | 52.812 | 51.204 | 49.694 | 48.275 |
| 26 | 69.036 | 66.313 | 63.824 | 61.534 | 59.416 | 57.454 | 55.624 | 53.914 | 52.311 | 50.805 |
| 27 | 72.563 | 69.916 | 67.237 | 64.754 | 62.523 | 60.429 | 58.483 | 56.666 | 54.966 | 53.370 |
| 28 | 76.650 | 73.640 | 70.748 | 68.113 | 65.695 | 63.483 | 61.392 | 59.463 | 57.661 | 55.972 |
| 29 | 81.054 | 77.518 | 74.375 | 71.535 | 68.944 | 66.561 | 64.357 | 62.309 | 60.399 | 58.612 |
| 30 | 85.591 | 81.606 | 78.150 | 75.069 | 72.282 | 69.732 | 67.384 | 65.209 | 63.185 | 61.294 |
| 31 | 90.789 | 86.015 | 82.127 | 78.747 | 75.728 | 72.990 | 70.482 | 68.168 | 66.021 | 64.021 |
| 32 | | 91.063 | 86.415 | 82.619 | 79.312 | 76.352 | 73.663 | 71.196 | 68.916 | 66.797 |
| 33 | | | 91.322 | 86.793 | 83.085 | 79.848 | 76.946 | 74.304 | 71.876 | 69.629 |
| 34 | | | | 91.566 | 87.150 | 83.526 | 80.357 | 77.510 | 74.915 | 72.525 |
| 35 | | | | | 91.797 | 87.488 | 83.946 | 80.841 | 78.048 | 75.497 |
| 36 | | | | | | 92.015 | 87.809 | 84.344 | 81.302 | 78.560 |
| 37 | | | | | | | 92.222 | 88.115 | 84.723 | 81.741 |
| 38 | | | | | | | | 92.419 | 88.405 | 85.085 |
| 39 | | | | | | | | | 92.606 | 88.681 |
| 40 | | | | | | | | | | 92.784 |

**5 Percent Ranks**

| $j$ \ $n$ | Sample size | | | | | | | | | |
|---|---|---|---|---|---|---|---|---|---|---|
| | 41 | 42 | 43 | 44 | 45 | 46 | 47 | 48 | 49 | 50 |
| 1 | 0.125 | 0.122 | 0.119 | 0.116 | 0.114 | 0.111 | 0.109 | 0.107 | 0.105 | 0.102 |
| 2 | 0.874 | 0.853 | 0.833 | 0.814 | 0.795 | 0.778 | 0.761 | 0.745 | 0.730 | 0.715 |
| 3 | 2.024 | 1.975 | 1.928 | 1.884 | 1.842 | 1.801 | 1.762 | 1.725 | 1.689 | 1.655 |
| 4 | 3.402 | 3.319 | 3.240 | 3.165 | 3.093 | 3.025 | 2.959 | 2.897 | 2.836 | 2.779 |
| 5 | 4.930 | 4.810 | 4.695 | 4.586 | 4.481 | 4.382 | 4.286 | 4.195 | 4.108 | 4.024 |
| 6 | 6.570 | 6.409 | 6.256 | 6.109 | 5.969 | 5.836 | 5.708 | 5.586 | 5.469 | 5.357 |
| 7 | 8.298 | 8.093 | 7.898 | 7.713 | 7.536 | 7.366 | 7.205 | 7.050 | 6.902 | 6.760 |
| 8 | 10.097 | 9.847 | 9.609 | 9.382 | 9.166 | 8.959 | 8.762 | 8.573 | 8.392 | 8.218 |
| 9 | 11.958 | 11.660 | 11.377 | 11.107 | 10.850 | 10.605 | 10.370 | 10.146 | 9.931 | 9.725 |
| 10 | 13.872 | 13.525 | 13.195 | 12.881 | 12.582 | 12.296 | 12.023 | 11.762 | 11.512 | 11.272 |
| 11 | 15.833 | 15.436 | 15.058 | 14.698 | 14.355 | 14.028 | 13.715 | 13.416 | 13.130 | 12.856 |
| 12 | 17.838 | 17.389 | 16.961 | 16.554 | 16.166 | 15.796 | 15.443 | 15.105 | 14.782 | 14.472 |
| 13 | 19.883 | 19.379 | 18.901 | 18.445 | 18.012 | 17.598 | 17.203 | 16.825 | 16.464 | 16.117 |
| 14 | 21.964 | 21.406 | 20.875 | 20.370 | 19.889 | 19.430 | 18.993 | 18.574 | 18.174 | 17.790 |
| 15 | 24.081 | 23.466 | 22.881 | 22.326 | 21.796 | 21.292 | 20.810 | 20.350 | 19.910 | 19.488 |
| 16 | 26.230 | 25.557 | 24.918 | 24.311 | 23.732 | 23.180 | 22.654 | 22.151 | 21.671 | 21.210 |
| 17 | 28.412 | 27.679 | 26.984 | 26.323 | 25.694 | 25.095 | 24.523 | 23.977 | 23.455 | 22.955 |
| 18 | 30.624 | 29.831 | 29.078 | 28.363 | 27.683 | 27.034 | 26.416 | 25.825 | 25.261 | 24.721 |
| 19 | 32.867 | 32.011 | 31.200 | 30.429 | 29.696 | 28.997 | 28.331 | 27.696 | 27.088 | 26.507 |
| 20 | 35.138 | 34.219 | 33.348 | 32.520 | 31.733 | 30.984 | 30.269 | 29.588 | 28.936 | 28.313 |
| 21 | 37.440 | 36.455 | 35.522 | 34.636 | 33.794 | 32.993 | 32.229 | 31.500 | 30.804 | 30.138 |
| 22 | 39.770 | 38.719 | 37.722 | 36.777 | 35.879 | 35.025 | 34.210 | 33.434 | 32.692 | 31.980 |
| 23 | 42.129 | 41.009 | 39.949 | 38.943 | 37.987 | 37.078 | 36.212 | 35.387 | 34.599 | 33.845 |
| 24 | 44.518 | 43.328 | 42.201 | 41.133 | 40.118 | 39.154 | 38.235 | 37.360 | 36.524 | 35.726 |
| 25 | 46.937 | 45.674 | 44.480 | 43.347 | 42.273 | 41.251 | 40.279 | 39.353 | 38.469 | 37.625 |
| 26 | 49.388 | 48.050 | 46.785 | 45.587 | 44.451 | 43.371 | 42.344 | 41.366 | 40.432 | 39.541 |
| 27 | 51.869 | 50.454 | 49.117 | 47.852 | 46.652 | 45.513 | 44.430 | 43.398 | 42.415 | 41.476 |
| 28 | 54.385 | 52.889 | 51.478 | 50.143 | 48.878 | 47.678 | 46.537 | 45.451 | 44.416 | 43.428 |
| 29 | 56.935 | 55.356 | 53.868 | 52.461 | 51.129 | 49.866 | 48.666 | 47.524 | 46.436 | 45.399 |
| 30 | 59.522 | 57.857 | 56.288 | 54.807 | 53.406 | 52.078 | 50.817 | 49.618 | 48.477 | 47.388 |
| 31 | 62.149 | 60.393 | 58.741 | 57.183 | 55.710 | 54.315 | 52.991 | 51.734 | 50.537 | 49.396 |
| 32 | 64.820 | 62.968 | 61.228 | 59.590 | 58.042 | 56.578 | 55.190 | 53.871 | 52.617 | 51.423 |
| 33 | 67.539 | 65.585 | 63.753 | 62.029 | 60.404 | 58.868 | 57.413 | 56.032 | 54.720 | 53.470 |
| 34 | 70.311 | 68.248 | 66.318 | 64.505 | 62.798 | 61.187 | 59.662 | 58.217 | 56.844 | 55.538 |
| 35 | 73.146 | 70.963 | 68.927 | 67.020 | 65.227 | 63.537 | 61.940 | 60.427 | 58.991 | 57.627 |
| 36 | 76.053 | 73.738 | 71.587 | 69.578 | 67.694 | 65.921 | 64.247 | 62.664 | 61.164 | 59.738 |
| 37 | 79.049 | 76.584 | 74.306 | 72.185 | 70.203 | 68.341 | 66.587 | 64.931 | 63.362 | 61.874 |
| 38 | 82.160 | 79.517 | 77.093 | 74.849 | 72.759 | 70.805 | 68.963 | 67.228 | 65.589 | 64.034 |
| 39 | 85.429 | 82.561 | 79.964 | 77.580 | 75.370 | 73.309 | 71.378 | 69.561 | 67.846 | 66.222 |
| 40 | 88.945 | 85.759 | 82.944 | 80.392 | 78.046 | 75.870 | 73.838 | 71.932 | 70.136 | 68.440 |
| 41 | 92.954 | 89.196 | 86.073 | 83.310 | 80.802 | 78.494 | 76.350 | 74.347 | 72.465 | 70.691 |
| 42 | | 93.116 | 89.437 | 86.374 | 83.661 | 81.196 | 78.924 | 76.812 | 74.836 | 72.978 |
| 43 | | | 93.270 | 89.666 | 86.662 | 83.998 | 81.573 | 79.337 | 77.256 | 75.306 |
| 44 | | | | 93.418 | 89.887 | 86.939 | 84.321 | 81.936 | 79.734 | 77.683 |
| 45 | | | | | 93.560 | 90.098 | 87.204 | 84.631 | 82.285 | 80.117 |
| 46 | | | | | | 93.695 | 90.300 | 87.459 | 84.929 | 82.621 |
| 47 | | | | | | | 93.825 | 90.494 | 87.703 | 85.216 |
| 48 | | | | | | | | 93.950 | 90.681 | 87.939 |
| 49 | | | | | | | | | 94.069 | 90.860 |
| 50 | | | | | | | | | | 94.184 |

TABLE IX.13   MEDIAN, 5 PERCENT, AND 95 PERCENT RANK DISTRIBUTION TABLES (continued)

## 95 Percent Ranks

| j \ n | 1 | 2 | 3 | 4 | 5 | 6 | 7 | 8 | 9 | 10 |
|---|---|---|---|---|---|---|---|---|---|---|
| | | | | | Sample size | | | | | |
| 1 | 95.000 | 77.639 | 63.160 | 52.713 | 45.072 | 39.304 | 34.816 | 31.234 | 28.313 | 25.887 |
| 2 | | 97.468 | 86.465 | 75.139 | 65.741 | 58.180 | 52.070 | 47.068 | 42.914 | 39.416 |
| 3 | | | 98.305 | 90.239 | 81.075 | 72.866 | 65.874 | 59.969 | 54.964 | 50.690 |
| 4 | | | | 98.726 | 92.356 | 84.684 | 77.468 | 71.076 | 65.506 | 60.662 |
| 5 | | | | | 98.979 | 93.715 | 87.124 | 80.710 | 74.863 | 69.646 |
| 6 | | | | | | 99.149 | 94.662 | 88.889 | 83.125 | 77.756 |
| 7 | | | | | | | 99.270 | 95.361 | 90.225 | 84.997 |
| 8 | | | | | | | | 99.361 | 95.898 | 91.274 |
| 9 | | | | | | | | | 99.432 | 96.323 |
| 10 | | | | | | | | | | 99.488 |

## 95 Percent Ranks

| j \ n | 11 | 12 | 13 | 14 | 15 | 16 | 17 | 18 | 19 | 20 |
|---|---|---|---|---|---|---|---|---|---|---|
| | | | | | Sample size | | | | | |
| 1 | 23.840 | 22.092 | 20.582 | 19.264 | 18.104 | 17.075 | 16.157 | 15.332 | 14.587 | 13.911 |
| 2 | 36.436 | 33.868 | 31.634 | 29.673 | 27.940 | 26.396 | 25.012 | 23.766 | 22.637 | 21.611 |
| 3 | 47.009 | 43.811 | 41.010 | 38.539 | 36.344 | 34.383 | 32.619 | 31.026 | 29.580 | 28.262 |
| 4 | 56.437 | 52.733 | 49.465 | 46.566 | 43.978 | 41.657 | 39.564 | 37.668 | 35.943 | 34.366 |
| 5 | 65.019 | 60.914 | 57.262 | 54.000 | 51.075 | 48.440 | 46.055 | 43.888 | 41.912 | 40.103 |
| 6 | 72.875 | 68.476 | 64.520 | 60.928 | 57.744 | 54.835 | 52.192 | 49.783 | 47.580 | 45.558 |
| 7 | 80.042 | 75.470 | 71.295 | 67.497 | 64.043 | 60.899 | 58.029 | 55.404 | 52.997 | 50.782 |
| 8 | 86.492 | 81.898 | 77.604 | 73.641 | 70.001 | 66.663 | 63.599 | 60.784 | 58.194 | 55.803 |
| 9 | 92.118 | 87.715 | 83.434 | 79.393 | 75.627 | 72.140 | 68.917 | 65.940 | 63.188 | 60.641 |
| 10 | 96.668 | 92.813 | 88.733 | 84.728 | 80.913 | 77.331 | 73.989 | 70.880 | 67.991 | 65.307 |
| 11 | 99.535 | 96.954 | 93.395 | 89.595 | 85.834 | 82.223 | 78.809 | 75.604 | 72.605 | 69.805 |
| 12 | | 99.573 | 96.914 | 93.890 | 90.334 | 86.789 | 83.364 | 80.105 | 77.028 | 74.135 |
| 13 | | | 99.606 | 97.195 | 94.315 | 90.975 | 87.623 | 84.366 | 81.250 | 78.293 |
| 14 | | | | 99.634 | 97.577 | 94.685 | 91.535 | 88.357 | 85.253 | 82.269 |
| 15 | | | | | 99.659 | 97.732 | 95.010 | 92.030 | 89.009 | 86.045 |
| 16 | | | | | | 99.680 | 97.868 | 95.297 | 92.471 | 89.592 |
| 17 | | | | | | | 99.699 | 97.989 | 95.553 | 92.865 |
| 18 | | | | | | | | 99.715 | 98.097 | 95.783 |
| 19 | | | | | | | | | 99.730 | 98.193 |
| 20 | | | | | | | | | | 99.744 |

## 95 Percent Ranks

| j \ n | 21 | 22 | 23 | 24 | 25 | 26 | 27 | 28 | 29 | 30 |
|---|---|---|---|---|---|---|---|---|---|---|
| | | | | | Sample size | | | | | |
| 1 | 13.295 | 12.731 | 12.212 | 11.735 | 11.293 | 10.883 | 10.502 | 10.147 | 9.814 | 9.503 |
| 2 | 20.673 | 19.812 | 19.020 | 18.289 | 17.612 | 16.983 | 16.397 | 15.851 | 15.339 | 14.860 |
| 3 | 27.055 | 25.947 | 24.925 | 23.980 | 23.104 | 22.289 | 21.530 | 20.821 | 20.156 | 19.533 |
| 4 | 32.921 | 31.591 | 30.364 | 29.227 | 28.172 | 27.190 | 26.274 | 25.417 | 24.614 | 23.860 |
| 5 | 38.441 | 36.909 | 35.193 | 34.181 | 32.961 | 31.824 | 30.763 | 29.769 | 28.837 | 27.962 |
| 6 | 43.698 | 41.980 | 40.390 | 38.914 | 37.541 | 36.260 | 35.062 | 33.940 | 32.887 | 31.897 |
| 7 | 48.739 | 46.849 | 45.097 | 43.469 | 41.952 | 40.535 | 39.210 | 37.967 | 36.800 | 35.701 |
| 8 | 53.594 | 51.546 | 49.643 | 47.873 | 46.221 | 44.677 | 43.230 | 41.873 | 40.597 | 39.395 |
| 9 | 58.280 | 56.087 | 54.046 | 52.142 | 50.364 | 48.700 | 47.139 | 45.673 | 44.294 | 42.993 |
| 10 | 62.810 | 60.484 | 58.315 | 56.289 | 54.393 | 52.616 | 50.948 | 49.379 | 47.901 | 46.507 |
| 11 | 67.189 | 64.746 | 62.461 | 60.321 | 58.316 | 56.434 | 54.664 | 52.998 | 51.427 | 49.944 |
| 12 | 71.420 | 68.874 | 66.485 | 64.244 | 62.138 | 60.158 | 58.293 | 56.536 | 54.877 | 53.309 |
| 13 | 75.501 | 72.869 | 70.391 | 68.058 | 65.861 | 63.791 | 61.839 | 59.996 | 58.254 | 56.605 |
| 14 | 79.425 | 76.728 | 74.176 | 71.764 | 69.487 | 67.336 | 65.303 | 63.380 | 61.561 | 59.837 |
| 15 | 83.182 | 80.444 | 77.836 | 75.361 | 73.015 | 70.792 | 68.686 | 66.691 | 64.799 | 63.005 |
| 16 | 86.755 | 84.006 | 81.366 | 78.843 | 76.441 | 74.158 | 71.988 | 69.927 | 67.970 | 66.111 |
| 17 | 90.116 | 87.397 | 84.752 | 82.204 | 79.762 | 77.430 | 75.207 | 73.089 | 71.073 | 69.154 |
| 18 | 93.219 | 90.589 | 87.978 | 85.431 | 82.970 | 80.604 | 78.338 | 76.173 | 74.106 | 72.133 |
| 19 | 95.990 | 93.540 | 91.019 | 88.509 | 86.052 | 83.672 | 81.378 | 79.176 | 77.066 | 75.047 |
| 20 | 98.281 | 96.178 | 93.832 | 91.411 | 88.994 | 86.623 | 84.318 | 82.092 | 79.950 | 77.894 |
| 21 | 99.756 | 98.360 | 96.348 | 94.099 | 91.771 | 89.440 | 87.148 | 84.915 | 82.753 | 80.669 |
| 22 | | 99.767 | 98.433 | 96.505 | 94.344 | 92.101 | 89.851 | 87.633 | 85.468 | 83.367 |
| 23 | | | 99.777 | 98.499 | 96.648 | 94.569 | 92.406 | 90.232 | 88.083 | 85.981 |
| 24 | | | | 99.786 | 98.560 | 96.780 | 94.777 | 92.689 | 90.584 | 88.501 |
| 25 | | | | | 99.795 | 98.616 | 96.902 | 94.969 | 92.951 | 90.913 |
| 26 | | | | | | 99.803 | 98.668 | 97.015 | 95.148 | 93.194 |
| 27 | | | | | | | 99.810 | 98.716 | 97.120 | 95.314 |
| 28 | | | | | | | | 99.817 | 98.761 | 97.218 |
| 29 | | | | | | | | | 99.823 | 98.802 |
| 30 | | | | | | | | | | 99.829 |

**95 Percent Ranks**

Sample size

| j | 41 | 42 | 43 | 44 | 45 | 46 | 47 | 48 | 49 | 50 |
|---|----|----|----|----|----|----|----|----|----|----|
| 1 | 7.046 | 6.884 | 6.730 | 6.582 | 6.440 | 6.305 | 6.175 | 6.050 | 5.931 | 5.816 |
| 2 | 11.055 | 10.804 | 10.563 | 10.334 | 10.113 | 9.902 | 9.700 | 9.506 | 9.319 | 9.140 |
| 3 | 14.571 | 14.241 | 13.927 | 13.626 | 13.338 | 13.061 | 12.796 | 12.541 | 12.297 | 12.061 |
| 4 | 17.840 | 17.439 | 17.056 | 16.690 | 16.339 | 16.002 | 15.679 | 15.369 | 15.071 | 14.784 |
| 5 | 20.951 | 20.483 | 20.036 | 19.608 | 19.198 | 18.804 | 18.427 | 18.064 | 17.715 | 17.379 |
| 6 | 23.947 | 23.416 | 22.907 | 22.420 | 21.954 | 21.506 | 21.076 | 20.663 | 20.266 | 19.883 |
| 7 | 26.854 | 26.262 | 25.694 | 25.151 | 24.630 | 24.130 | 23.650 | 23.188 | 22.744 | 22.317 |
| 8 | 29.689 | 29.037 | 28.413 | 27.814 | 27.241 | 26.691 | 26.162 | 25.623 | 25.164 | 24.694 |
| 9 | 32.461 | 31.752 | 31.073 | 30.422 | 29.797 | 29.198 | 28.622 | 28.068 | 27.535 | 27.022 |
| 10 | 35.180 | 34.415 | 33.682 | 32.980 | 32.306 | 31.659 | 31.037 | 30.439 | 29.864 | 29.309 |
| 11 | 37.851 | 37.032 | 36.247 | 35.495 | 34.773 | 34.079 | 33.413 | 32.772 | 32.154 | 31.560 |
| 12 | 40.478 | 39.607 | 38.772 | 37.971 | 37.202 | 36.463 | 35.753 | 35.069 | 34.411 | 33.778 |
| 13 | 43.065 | 42.143 | 41.259 | 40.410 | 39.596 | 38.813 | 38.060 | 37.336 | 36.698 | 35.933 |
| 14 | 45.615 | 44.644 | 43.712 | 42.817 | 41.958 | 41.132 | 40.338 | 39.573 | 38.836 | 38.126 |
| 15 | 48.131 | 47.110 | 46.132 | 45.193 | 44.290 | 43.422 | 42.587 | 41.783 | 41.008 | 40.262 |
| 16 | 50.612 | 49.546 | 48.522 | 47.539 | 46.594 | 45.665 | 44.810 | 43.968 | 43.156 | 42.373 |
| 17 | 53.062 | 51.950 | 50.883 | 49.857 | 48.871 | 47.922 | 47.009 | 46.129 | 45.280 | 44.462 |
| 18 | 55.482 | 54.326 | 53.215 | 52.148 | 51.122 | 50.134 | 49.183 | 48.266 | 47.382 | 46.530 |
| 19 | 57.871 | 56.672 | 55.520 | 54.413 | 53.348 | 52.322 | 51.334 | 50.382 | 49.463 | 48.577 |
| 20 | 60.230 | 58.991 | 57.799 | 56.653 | 55.549 | 54.487 | 53.463 | 52.476 | 51.523 | 50.604 |
| 21 | 62.560 | 61.281 | 60.051 | 58.867 | 57.727 | 56.629 | 55.570 | 54.549 | 53.563 | 52.612 |
| 22 | 64.861 | 63.545 | 62.273 | 61.057 | 59.882 | 58.749 | 57.556 | 56.602 | 55.584 | 54.801 |
| 23 | 67.133 | 65.781 | 64.478 | 63.223 | 62.013 | 60.846 | 59.721 | 58.634 | 57.585 | 56.572 |
| 24 | 69.376 | 67.989 | 66.652 | 65.363 | 64.121 | 62.922 | 61.765 | 60.647 | 59.568 | 58.524 |
| 25 | 71.588 | 70.169 | 68.800 | 67.480 | 66.205 | 64.975 | 63.787 | 62.640 | 61.531 | 60.459 |
| 26 | 73.769 | 72.320 | 70.922 | 69.571 | 68.267 | 67.007 | 65.790 | 64.613 | 63.476 | 62.375 |
| 27 | 75.919 | 74.443 | 73.016 | 71.637 | 70.304 | 69.016 | 67.771 | 66.566 | 65.401 | 64.274 |
| 28 | 78.035 | 76.534 | 75.082 | 73.677 | 72.317 | 71.002 | 69.730 | 68.500 | 67.308 | 66.155 |
| 29 | 80.117 | 78.594 | 77.119 | 75.689 | 74.306 | 72.966 | 71.668 | 70.412 | 69.196 | 68.017 |
| 30 | 82.162 | 80.621 | 79.125 | 77.674 | 76.268 | 74.905 | 73.584 | 72.304 | 71.064 | 69.862 |
| 31 | 84.166 | 82.611 | 81.099 | 79.630 | 78.203 | 76.819 | 75.477 | 74.175 | 72.912 | 71.687 |
| 32 | 86.128 | 84.564 | 83.039 | 81.554 | 80.111 | 78.708 | 77.346 | 76.023 | 74.739 | 73.493 |
| 33 | 88.042 | 86.475 | 84.942 | 83.446 | 81.988 | 80.569 | 79.190 | 77.848 | 76.545 | 75.279 |
| 34 | 89.903 | 88.340 | 86.805 | 85.302 | 83.834 | 82.402 | 81.007 | 79.650 | 78.329 | 77.045 |
| 35 | 91.702 | 90.153 | 88.623 | 87.119 | 85.645 | 84.204 | 82.797 | 81.426 | 80.090 | 78.790 |
| 36 | 93.430 | 91.907 | 90.391 | 88.892 | 87.418 | 85.972 | 84.557 | 83.175 | 81.826 | 80.511 |
| 37 | 95.070 | 93.591 | 92.102 | 90.618 | 89.150 | 87.704 | 86.285 | 84.895 | 83.536 | 82.210 |
| 38 | 96.598 | 95.190 | 93.744 | 92.287 | 90.834 | 89.395 | 87.977 | 86.584 | 85.218 | 83.882 |
| 39 | 97.976 | 96.681 | 95.305 | 93.891 | 92.464 | 91.041 | 89.630 | 88.238 | 86.870 | 85.528 |
| 40 | 99.126 | 98.025 | 96.760 | 95.414 | 94.030 | 92.633 | 91.238 | 89.854 | 88.488 | 87.144 |
| 41 | 99.875 | 99.147 | 98.071 | 96.835 | 95.518 | 94.164 | 92.795 | 91.427 | 90.069 | 88.728 |
| 42 |  | 99.878 | 99.167 | 98.116 | 96.907 | 95.618 | 94.291 | 92.950 | 91.608 | 90.275 |
| 43 |  |  | 99.881 | 99.186 | 98.158 | 96.975 | 95.714 | 94.414 | 93.098 | 91.781 |
| 44 |  |  |  | 99.883 | 99.205 | 98.199 | 97.041 | 95.805 | 94.531 | 93.240 |
| 45 |  |  |  |  | 99.886 | 99.222 | 98.238 | 97.103 | 95.892 | 94.643 |
| 46 |  |  |  |  |  | 99.889 | 99.239 | 98.275 | 97.163 | 95.976 |
| 47 |  |  |  |  |  |  | 99.891 | 99.255 | 98.311 | 97.221 |
| 48 |  |  |  |  |  |  |  | 99.893 | 99.270 | 98.345 |
| 49 |  |  |  |  |  |  |  |  | 99.895 | 99.285 |
| 50 |  |  |  |  |  |  |  |  |  | 99.897 |

**95 Percent Ranks**

Sample size

| j | 31 | 32 | 33 | 34 | 35 | 36 | 37 | 38 | 39 | 40 |
|---|----|----|----|----|----|----|----|----|----|----|
| 1 | 9.211 | 8.937 | 8.678 | 8.434 | 8.203 | 7.985 | 7.778 | 7.581 | 7.394 | 7.216 |
| 2 | 14.409 | 13.985 | 13.585 | 13.207 | 12.850 | 12.512 | 12.191 | 11.885 | 11.595 | 11.319 |
| 3 | 18.946 | 18.394 | 17.873 | 17.381 | 16.915 | 16.474 | 16.054 | 15.656 | 15.277 | 14.915 |
| 4 | 23.150 | 22.482 | 21.850 | 21.253 | 20.688 | 20.152 | 19.643 | 19.159 | 18.698 | 18.259 |
| 5 | 27.137 | 26.360 | 25.625 | 24.931 | 24.272 | 23.648 | 23.054 | 22.490 | 21.952 | 21.440 |
| 6 | 30.964 | 30.084 | 29.252 | 28.465 | 27.718 | 27.010 | 26.337 | 25.696 | 25.085 | 24.503 |
| 7 | 34.665 | 33.687 | 32.763 | 31.887 | 31.056 | 30.268 | 29.518 | 28.804 | 28.124 | 27.475 |
| 8 | 38.261 | 37.190 | 36.176 | 35.216 | 34.305 | 33.439 | 32.616 | 31.832 | 31.084 | 30.371 |
| 9 | 41.766 | 40.606 | 39.507 | 38.466 | 37.477 | 36.537 | 35.643 | 34.791 | 33.979 | 33.203 |
| 10 | 45.190 | 43.945 | 42.765 | 41.645 | 40.582 | 39.571 | 38.608 | 37.691 | 36.815 | 35.979 |
| 11 | 48.542 | 47.214 | 45.956 | 44.761 | 43.626 | 42.546 | 41.517 | 40.537 | 39.601 | 38.706 |
| 12 | 51.825 | 50.419 | 49.086 | 47.819 | 46.615 | 45.468 | 44.376 | 43.334 | 42.339 | 41.388 |
| 13 | 55.044 | 53.564 | 52.159 | 50.823 | 49.552 | 48.341 | 47.187 | 46.086 | 45.034 | 44.028 |
| 14 | 58.203 | 56.651 | 55.177 | 53.775 | 52.440 | 51.168 | 49.955 | 48.796 | 47.689 | 46.630 |
| 15 | 61.302 | 59.683 | 58.144 | 56.678 | 55.282 | 53.951 | 52.680 | 51.466 | 50.305 | 49.195 |
| 16 | 64.343 | 62.661 | 61.060 | 59.534 | 58.080 | 56.691 | 55.366 | 54.098 | 52.886 | 51.725 |
| 17 | 67.326 | 65.585 | 63.926 | 62.343 | 60.833 | 59.391 | 58.012 | 56.693 | 55.431 | 54.222 |
| 18 | 70.251 | 68.456 | 66.742 | 65.106 | 63.543 | 62.049 | 60.620 | 59.252 | 57.942 | 56.686 |
| 19 | 73.117 | 71.272 | 69.509 | 67.823 | 66.210 | 64.668 | 63.190 | 61.776 | 60.419 | 59.119 |
| 20 | 75.922 | 74.034 | 72.225 | 70.493 | 68.835 | 67.246 | 65.723 | 64.264 | 62.864 | 61.520 |
| 21 | 78.664 | 76.738 | 74.889 | 73.116 | 71.415 | 69.784 | 68.219 | 66.717 | 65.275 | 63.891 |
| 22 | 81.338 | 79.382 | 77.499 | 75.689 | 73.951 | 72.280 | 70.676 | 69.135 | 67.654 | 66.230 |
| 23 | 83.939 | 81.961 | 80.052 | 78.212 | 76.440 | 74.735 | 73.094 | 71.517 | 69.999 | 68.539 |
| 24 | 86.460 | 84.472 | 82.545 | 80.680 | 78.881 | 77.145 | 75.473 | 73.862 | 72.310 | 70.815 |
| 25 | 88.891 | 86.907 | 84.971 | 83.091 | 81.270 | 79.509 | 77.809 | 76.168 | 74.586 | 73.060 |
| 26 | 91.219 | 89.255 | 87.325 | 85.439 | 83.604 | 81.825 | 80.101 | 78.435 | 76.825 | 75.270 |
| 27 | 93.422 | 91.505 | 89.596 | 87.717 | 85.878 | 84.087 | 82.347 | 80.660 | 79.027 | 77.447 |
| 28 | 95.470 | 93.635 | 91.772 | 89.916 | 88.086 | 86.292 | 84.542 | 82.840 | 81.188 | 79.587 |
| 29 | 97.310 | 95.615 | 93.834 | 92.024 | 90.217 | 88.433 | 86.682 | 84.972 | 83.306 | 81.688 |
| 30 | 98.841 | 97.396 | 95.752 | 94.021 | 92.261 | 90.501 | 88.760 | 87.050 | 85.378 | 83.746 |
| 31 | 99.835 | 98.878 | 97.476 | 95.880 | 94.198 | 92.483 | 90.768 | 89.069 | 87.399 | 85.763 |
| 32 |  | 99.840 | 98.912 | 97.552 | 96.001 | 94.364 | 92.694 | 91.021 | 89.362 | 87.729 |
| 33 |  |  | 99.845 | 98.845 | 97.623 | 96.114 | 94.521 | 92.893 | 91.260 | 89.639 |
| 34 |  |  |  | 99.849 | 98.975 | 97.690 | 96.222 | 94.669 | 93.081 | 91.487 |
| 35 |  |  |  |  | 99.854 | 99.004 | 97.754 | 96.324 | 94.810 | 93.260 |
| 36 |  |  |  |  |  | 99.858 | 99.031 | 97.814 | 96.420 | 94.943 |
| 37 |  |  |  |  |  |  | 99.861 | 99.057 | 97.871 | 96.511 |
| 38 |  |  |  |  |  |  |  | 99.865 | 99.081 | 97.925 |
| 39 |  |  |  |  |  |  |  |  | 99.869 | 99.104 |
| 40 |  |  |  |  |  |  |  |  |  | 99.872 |

**TABLE IX.14**    NORMAL PROBABILITY PLOTTING PAPER

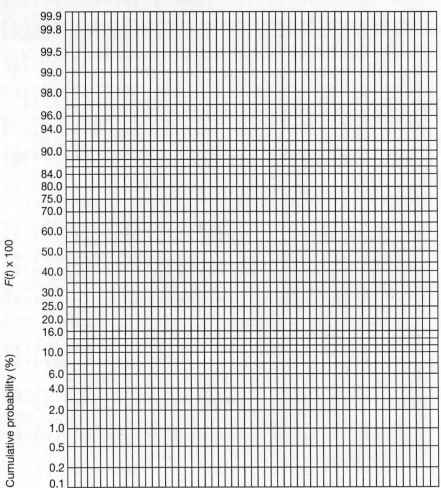

**TABLE IX.15**  EXPONENTIAL PROBABILITY PLOTTING PAPER

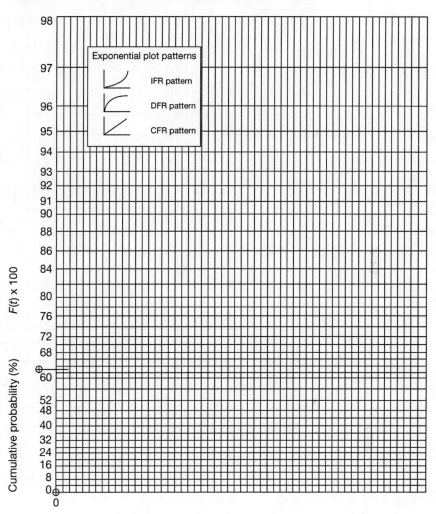

**TABLE IX.16** WEIBULL PROBABILITY PLOTTING PAPER

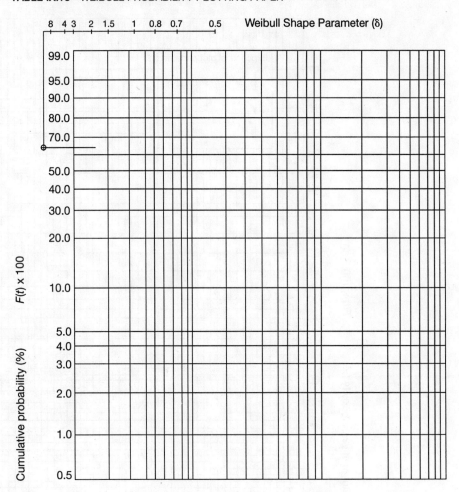

TABLE IX.17   LOGNORMAL PROBABILITY PLOTTING PAPER

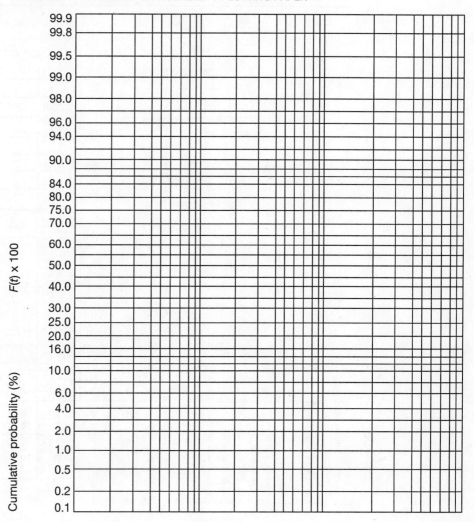

## TABLE IX.18   SELECTED ORTHOGONAL ARRAY TABLES

$$L_4(2^3)$$

Orthogonal array

| Expt. No. | Column 1 | 2 | 3 |
|---|---|---|---|
| 1 | 1 | 1 | 1 |
| 2 | 1 | 2 | 2 |
| 3 | 2 | 1 | 2 |
| 4 | 2 | 2 | 1 |

Linear graph for $L_4$

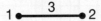

$$L_8(2^7)$$

Orthogonal array

| Expt. No. | Column 1 | 2 | 3 | 4 | 5 | 6 | 7 |
|---|---|---|---|---|---|---|---|
| 1 | 1 | 1 | 1 | 1 | 1 | 1 | 1 |
| 2 | 1 | 1 | 1 | 2 | 2 | 2 | 2 |
| 3 | 1 | 2 | 2 | 1 | 1 | 2 | 2 |
| 4 | 1 | 2 | 2 | 2 | 2 | 1 | 1 |
| 5 | 2 | 1 | 2 | 1 | 2 | 1 | 2 |
| 6 | 2 | 1 | 2 | 2 | 1 | 2 | 1 |
| 7 | 2 | 2 | 1 | 1 | 2 | 2 | 1 |
| 8 | 2 | 2 | 1 | 2 | 1 | 1 | 2 |

Interaction table

| Expt. No. | Column 1 | 2 | 3 | 4 | 5 | 6 | 7 |
|---|---|---|---|---|---|---|---|
| 1 | (1) | 3 | 2 | 5 | 4 | 7 | 6 |
| 2 | | (2) | 1 | 6 | 7 | 4 | 5 |
| 3 | | | (3) | 7 | 6 | 5 | 4 |
| 4 | | | | (4) | 1 | 2 | 3 |
| 5 | | | | | (5) | 3 | 2 |
| 6 | | | | | | (6) | 1 |
| 7 | | | | | | | (7) |

Linear graph for $L_8$

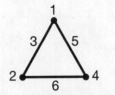
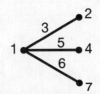

Reprinted from G. Taguchi, *Systems of Experimental Design: Engineering Methods to Optimize Quality and Minimize Cost*, New York: UNIPUB/Kraus International Publications, pp. 1128–1139 and 1153–1155, 1987. Copyright 1987 by UNIPUB/Kraus International Publications. Reprinted with permission of the Kraus Organization and Quality Resources, White Plains, NY 10601.

**TABLE IX.18    SELECTED ORTHOGONAL ARRAY TABLES** (*continued*)

$$\boxed{L_9(3^4)}$$

Orthogonal Array

| Expt. No. | Column | | | |
|:---:|:---:|:---:|:---:|:---:|
| | 1 | 2 | 3 | 4 |
| 1 | 1 | 1 | 1 | 1 |
| 2 | 1 | 2 | 2 | 2 |
| 3 | 1 | 3 | 3 | 3 |
| 4 | 2 | 1 | 2 | 3 |
| 5 | 2 | 2 | 3 | 1 |
| 6 | 2 | 3 | 1 | 2 |
| 7 | 3 | 1 | 3 | 2 |
| 8 | 3 | 2 | 1 | 3 |
| 9 | 3 | 3 | 2 | 1 |

Linear Graphs for $L_9$

3,4

1 ●━━━━━● 2

**TABLE IX.18    SELECTED ORTHOGONAL ARRAY TABLES** (*continued*)

$$L_{16}(2^{15})$$

**Orthogonal Array**

| Expt. No. | Column | | | | | | | | | | | | | | |
|---|---|---|---|---|---|---|---|---|---|---|---|---|---|---|---|
| | 1 | 2 | 3 | 4 | 5 | 6 | 7 | 8 | 9 | 10 | 11 | 12 | 13 | 14 | 15 |
| 1 | 1 | 1 | 1 | 1 | 1 | 1 | 1 | 1 | 1 | 1 | 1 | 1 | 1 | 1 | 1 |
| 2 | 1 | 1 | 1 | 1 | 1 | 1 | 1 | 2 | 2 | 2 | 2 | 2 | 2 | 2 | 2 |
| 3 | 1 | 1 | 1 | 2 | 2 | 2 | 2 | 1 | 1 | 1 | 1 | 2 | 2 | 2 | 2 |
| 4 | 1 | 1 | 1 | 2 | 2 | 2 | 2 | 2 | 2 | 2 | 2 | 1 | 1 | 1 | 1 |
| 5 | 1 | 2 | 2 | 1 | 1 | 2 | 2 | 1 | 1 | 2 | 2 | 1 | 1 | 2 | 2 |
| 6 | 1 | 2 | 2 | 1 | 1 | 2 | 2 | 2 | 2 | 1 | 1 | 2 | 2 | 1 | 1 |
| 7 | 1 | 2 | 2 | 2 | 2 | 1 | 1 | 1 | 1 | 2 | 2 | 2 | 2 | 1 | 1 |
| 8 | 1 | 2 | 2 | 2 | 2 | 1 | 1 | 2 | 2 | 1 | 1 | 1 | 1 | 2 | 2 |
| 9 | 2 | 1 | 2 | 1 | 2 | 1 | 2 | 1 | 2 | 1 | 2 | 1 | 2 | 1 | 2 |
| 10 | 2 | 1 | 2 | 1 | 2 | 1 | 2 | 2 | 1 | 2 | 1 | 2 | 1 | 2 | 1 |
| 11 | 2 | 1 | 2 | 2 | 1 | 2 | 1 | 1 | 2 | 1 | 2 | 2 | 1 | 2 | 1 |
| 12 | 2 | 1 | 2 | 2 | 1 | 2 | 1 | 2 | 1 | 2 | 1 | 1 | 2 | 1 | 2 |
| 13 | 2 | 2 | 1 | 1 | 2 | 2 | 1 | 1 | 2 | 2 | 1 | 1 | 2 | 2 | 1 |
| 14 | 2 | 2 | 1 | 1 | 2 | 2 | 1 | 2 | 1 | 1 | 2 | 2 | 1 | 1 | 2 |
| 15 | 2 | 2 | 1 | 2 | 1 | 1 | 2 | 1 | 2 | 2 | 1 | 2 | 1 | 1 | 2 |
| 16 | 2 | 2 | 1 | 2 | 1 | 1 | 2 | 2 | 1 | 1 | 2 | 1 | 2 | 2 | 1 |

**Interaction Table**

| Column | Column | | | | | | | | | | | | | | |
|---|---|---|---|---|---|---|---|---|---|---|---|---|---|---|---|
| | 1 | 2 | 3 | 4 | 5 | 6 | 7 | 8 | 9 | 10 | 11 | 12 | 13 | 14 | 15 |
| 1 | (1) | 3 | 2 | 5 | 4 | 7 | 6 | 9 | 8 | 11 | 10 | 13 | 12 | 15 | 14 |
| 2 | | (2) | 1 | 6 | 7 | 4 | 5 | 10 | 11 | 8 | 9 | 14 | 15 | 12 | 13 |
| 3 | | | (3) | 7 | 6 | 5 | 4 | 11 | 10 | 9 | 8 | 15 | 14 | 13 | 12 |
| 4 | | | | (4) | 1 | 2 | 3 | 12 | 13 | 14 | 15 | 8 | 9 | 10 | 11 |
| 5 | | | | | (5) | 3 | 2 | 13 | 12 | 15 | 14 | 9 | 8 | 11 | 10 |
| 6 | | | | | | (6) | 1 | 14 | 15 | 12 | 13 | 10 | 11 | 8 | 9 |
| 7 | | | | | | | (7) | 15 | 14 | 13 | 12 | 11 | 10 | 9 | 8 |
| 8 | | | | | | | | (8) | 1 | 2 | 3 | 4 | 5 | 6 | 7 |
| 9 | | | | | | | | | (9) | 3 | 2 | 5 | 4 | 7 | 6 |
| 10 | | | | | | | | | | (10) | 1 | 6 | 7 | 4 | 5 |
| 11 | | | | | | | | | | | (11) | 7 | 6 | 5 | 4 |
| 12 | | | | | | | | | | | | (12) | 1 | 2 | 3 |
| 13 | | | | | | | | | | | | | (13) | 3 | 2 |
| 14 | | | | | | | | | | | | | | (14) | 1 |
| 15 | | | | | | | | | | | | | | | (15) |

**TABLE IX.18    SELECTED ORTHOGONAL ARRAY TABLES** (*continued*)

Linear graphs for $L_{16}$

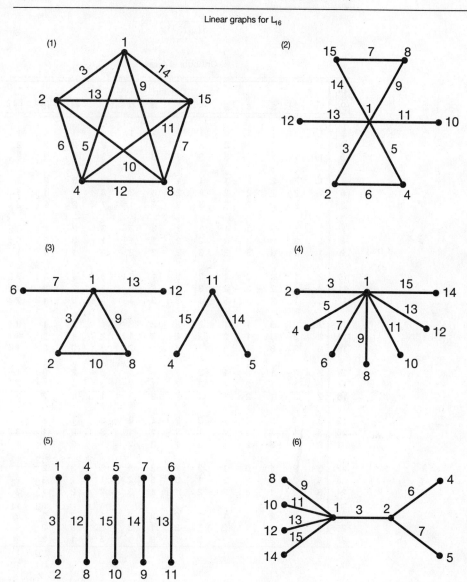

**TABLE IX.18    SELECTED ORTHOGONAL ARRAY TABLES** (*continued*)

$$L_{27}(3^{13})$$

Orthogonal Array

| Expt. No. | Column | | | | | | | | | | | | |
|---|---|---|---|---|---|---|---|---|---|---|---|---|---|
| | 1 | 2 | 3 | 4 | 5 | 6 | 7 | 8 | 9 | 10 | 11 | 12 | 13 |
| 1 | 1 | 1 | 1 | 1 | 1 | 1 | 1 | 1 | 1 | 1 | 1 | 1 | 1 |
| 2 | 1 | 1 | 1 | 1 | 2 | 2 | 2 | 2 | 2 | 2 | 2 | 2 | 2 |
| 3 | 1 | 1 | 1 | 1 | 3 | 3 | 3 | 3 | 3 | 3 | 3 | 3 | 3 |
| 4 | 1 | 2 | 2 | 2 | 1 | 1 | 1 | 2 | 2 | 2 | 3 | 3 | 3 |
| 5 | 1 | 2 | 2 | 2 | 2 | 2 | 2 | 3 | 3 | 3 | 1 | 1 | 1 |
| 6 | 1 | 2 | 2 | 2 | 3 | 3 | 3 | 1 | 1 | 1 | 2 | 2 | 2 |
| 7 | 1 | 3 | 3 | 3 | 1 | 1 | 1 | 3 | 3 | 3 | 2 | 2 | 2 |
| 8 | 1 | 3 | 3 | 3 | 2 | 2 | 2 | 1 | 1 | 1 | 3 | 3 | 3 |
| 9 | 1 | 3 | 3 | 3 | 3 | 3 | 3 | 2 | 2 | 2 | 1 | 1 | 1 |
| 10 | 2 | 1 | 2 | 3 | 1 | 2 | 3 | 1 | 2 | 3 | 1 | 2 | 3 |
| 11 | 2 | 1 | 2 | 3 | 2 | 3 | 1 | 2 | 3 | 1 | 2 | 3 | 1 |
| 12 | 2 | 1 | 2 | 3 | 3 | 1 | 2 | 3 | 1 | 2 | 3 | 1 | 2 |
| 13 | 2 | 2 | 3 | 1 | 1 | 2 | 3 | 2 | 3 | 1 | 3 | 1 | 2 |
| 14 | 2 | 2 | 3 | 1 | 2 | 3 | 1 | 3 | 1 | 2 | 1 | 2 | 3 |
| 15 | 2 | 2 | 3 | 1 | 3 | 1 | 2 | 1 | 2 | 3 | 2 | 3 | 1 |
| 16 | 2 | 3 | 1 | 2 | 1 | 2 | 3 | 3 | 1 | 2 | 2 | 3 | 1 |
| 17 | 2 | 3 | 1 | 2 | 2 | 3 | 1 | 1 | 2 | 3 | 3 | 1 | 2 |
| 18 | 2 | 3 | 1 | 2 | 3 | 1 | 2 | 2 | 3 | 1 | 1 | 2 | 3 |
| 19 | 3 | 1 | 3 | 2 | 1 | 3 | 2 | 1 | 3 | 2 | 1 | 3 | 2 |
| 20 | 3 | 1 | 3 | 2 | 2 | 1 | 3 | 2 | 1 | 3 | 2 | 1 | 3 |
| 21 | 3 | 1 | 3 | 2 | 3 | 2 | 1 | 3 | 2 | 1 | 3 | 2 | 1 |
| 22 | 3 | 2 | 1 | 3 | 1 | 3 | 2 | 2 | 1 | 3 | 3 | 2 | 1 |
| 23 | 3 | 2 | 1 | 3 | 2 | 1 | 3 | 3 | 2 | 1 | 1 | 3 | 2 |
| 24 | 3 | 2 | 1 | 3 | 3 | 2 | 1 | 1 | 3 | 2 | 2 | 1 | 3 |
| 25 | 3 | 3 | 2 | 1 | 1 | 3 | 2 | 3 | 2 | 1 | 2 | 1 | 3 |
| 26 | 3 | 3 | 2 | 1 | 2 | 1 | 3 | 1 | 3 | 2 | 3 | 2 | 1 |
| 27 | 3 | 3 | 2 | 1 | 3 | 2 | 1 | 2 | 1 | 3 | 1 | 3 | 2 |

**TABLE IX.18**   SELECTED ORTHOGONAL ARRAY TABLES (*continued*)

| Columns | 1 | 2 | 3 | 4 | 5 | 6 | 7 | 8 | 9 | 10 | 11 | 12 | 13 |
|---|---|---|---|---|---|---|---|---|---|---|---|---|---|
| | | | | | | Interaction table | | | | | | | |
| 1 | (1) | 3 4 | 2 4 | 2 3 | 6 7 | 5 7 | 5 6 | 9 10 | 8 10 | 8 9 | 12 13 | 11 13 | 11 12 |
| 2 | | (2) | 1 4 | 1 3 | 8 11 | 9 12 | 10 13 | 5 11 | 6 12 | 7 13 | 5 8 | 6 9 | 7 10 |
| 3 | | | (3) | 1 2 | 9 13 | 10 11 | 8 12 | 7 12 | 5 13 | 6 11 | 6 10 | 7 8 | 5 9 |
| 4 | | | | (4) | 10 12 | 8 13 | 9 11 | 6 13 | 7 11 | 5 12 | 7 9 | 5 10 | 6 8 |
| 5 | | | | | (5) | 1 7 | 1 6 | 2 11 | 3 13 | 4 12 | 2 8 | 4 10 | 3 9 |
| 6 | | | | | | (6) | 1 5 | 4 13 | 2 12 | 3 11 | 3 10 | 2 9 | 4 8 |
| 7 | | | | | | | (7) | 3 12 | 4 11 | 2 13 | 4 9 | 3 8 | 2 10 |
| 8 | | | | | | | | (8) | 1 10 | 1 9 | 2 5 | 3 7 | 4 6 |
| 9 | | | | | | | | | (9) | 1 8 | 4 7 | 2 6 | 3 5 |
| 10 | | | | | | | | | | (10) | 3 6 | 4 5 | 2 7 |
| 11 | | | | | | | | | | | (11) | 1 13 | 1 12 |
| 12 | | | | | | | | | | | | (12) | 1 11 |
| 13 | | | | | | | | | | | | | (13) |

Linear graph for $L_{27}$

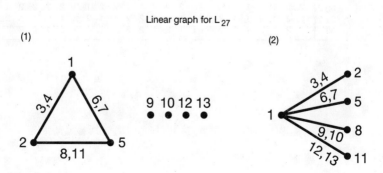

(1)

(2)

**TABLE IX.18**    SELECTED ORTHOGONAL ARRAY TABLES (*continued*)

## Orthogonal Array

| Expt. No. | 1 | 2 | 3 | 4 | 5 | 6 | 7 | 8 | 9 | 10 | 11 | 12 | 13 | 14 | 15 | 16 | 17 | 18 | 19 | 20 | 21 | 22 | 23 | 24 | 25 | 26 | 27 | 28 | 29 | 30 | 31 |
|---|---|---|---|---|---|---|---|---|---|---|---|---|---|---|---|---|---|---|---|---|---|---|---|---|---|---|---|---|---|---|---|
| 1 | 1 | 1 | 1 | 1 | 1 | 1 | 1 | 1 | 1 | 1 | 1 | 1 | 1 | 1 | 1 | 1 | 1 | 1 | 1 | 1 | 1 | 1 | 1 | 1 | 1 | 1 | 1 | 1 | 1 | 1 | 1 |
| 2 | 1 | 1 | 1 | 1 | 1 | 1 | 1 | 1 | 1 | 1 | 1 | 1 | 1 | 1 | 1 | 2 | 2 | 2 | 2 | 2 | 2 | 2 | 2 | 2 | 2 | 2 | 2 | 2 | 2 | 2 | 2 |
| 3 | 1 | 1 | 1 | 1 | 1 | 1 | 1 | 2 | 2 | 2 | 2 | 2 | 2 | 2 | 2 | 1 | 1 | 1 | 1 | 1 | 1 | 1 | 1 | 2 | 2 | 2 | 2 | 2 | 2 | 2 | 2 |
| 4 | 1 | 1 | 1 | 1 | 1 | 1 | 1 | 2 | 2 | 2 | 2 | 2 | 2 | 2 | 2 | 2 | 2 | 2 | 2 | 2 | 2 | 2 | 2 | 1 | 1 | 1 | 1 | 1 | 1 | 1 | 1 |
| 5 | 1 | 1 | 1 | 2 | 2 | 2 | 2 | 1 | 1 | 1 | 1 | 2 | 2 | 2 | 2 | 1 | 1 | 1 | 1 | 2 | 2 | 2 | 2 | 1 | 1 | 1 | 1 | 2 | 2 | 2 | 2 |
| 6 | 1 | 1 | 1 | 2 | 2 | 2 | 2 | 1 | 1 | 1 | 1 | 2 | 2 | 2 | 2 | 2 | 2 | 2 | 2 | 1 | 1 | 1 | 1 | 2 | 2 | 2 | 2 | 1 | 1 | 1 | 1 |
| 7 | 1 | 1 | 1 | 2 | 2 | 2 | 2 | 2 | 2 | 2 | 2 | 1 | 1 | 1 | 1 | 1 | 1 | 1 | 1 | 2 | 2 | 2 | 2 | 2 | 2 | 2 | 2 | 1 | 1 | 1 | 1 |
| 8 | 1 | 1 | 1 | 2 | 2 | 2 | 2 | 2 | 2 | 2 | 2 | 1 | 1 | 1 | 1 | 2 | 2 | 2 | 2 | 1 | 1 | 1 | 1 | 1 | 1 | 1 | 1 | 2 | 2 | 2 | 2 |
| 9 | 1 | 2 | 2 | 1 | 1 | 2 | 2 | 1 | 1 | 2 | 2 | 1 | 1 | 2 | 2 | 1 | 1 | 2 | 2 | 1 | 1 | 2 | 2 | 1 | 1 | 2 | 2 | 1 | 1 | 2 | 2 |
| 10 | 1 | 2 | 2 | 1 | 1 | 2 | 2 | 1 | 1 | 2 | 2 | 1 | 1 | 2 | 2 | 2 | 2 | 1 | 1 | 2 | 2 | 1 | 1 | 2 | 2 | 1 | 1 | 2 | 2 | 1 | 1 |
| 11 | 1 | 2 | 2 | 1 | 1 | 2 | 2 | 2 | 2 | 1 | 1 | 2 | 2 | 1 | 1 | 1 | 1 | 2 | 2 | 1 | 1 | 2 | 2 | 2 | 2 | 1 | 1 | 2 | 2 | 1 | 1 |
| 12 | 1 | 2 | 2 | 1 | 1 | 2 | 2 | 2 | 2 | 1 | 1 | 2 | 2 | 1 | 1 | 2 | 2 | 1 | 1 | 2 | 2 | 1 | 1 | 1 | 1 | 2 | 2 | 1 | 1 | 2 | 2 |
| 13 | 1 | 2 | 2 | 2 | 2 | 1 | 1 | 1 | 1 | 2 | 2 | 2 | 2 | 1 | 1 | 1 | 1 | 2 | 2 | 2 | 2 | 1 | 1 | 1 | 1 | 2 | 2 | 2 | 2 | 1 | 1 |
| 14 | 1 | 2 | 2 | 2 | 2 | 1 | 1 | 1 | 1 | 2 | 2 | 2 | 2 | 1 | 1 | 2 | 2 | 1 | 1 | 1 | 1 | 2 | 2 | 2 | 2 | 1 | 1 | 1 | 1 | 2 | 2 |
| 15 | 1 | 2 | 2 | 2 | 2 | 1 | 1 | 2 | 2 | 1 | 1 | 1 | 1 | 2 | 2 | 1 | 1 | 2 | 2 | 2 | 2 | 1 | 1 | 2 | 2 | 1 | 1 | 1 | 1 | 2 | 2 |
| 16 | 1 | 2 | 2 | 2 | 2 | 1 | 1 | 2 | 2 | 1 | 1 | 1 | 1 | 2 | 2 | 2 | 2 | 1 | 1 | 1 | 1 | 2 | 2 | 1 | 1 | 2 | 2 | 2 | 2 | 1 | 1 |
| 17 | 2 | 1 | 2 | 1 | 2 | 1 | 2 | 1 | 2 | 1 | 2 | 1 | 2 | 1 | 2 | 1 | 2 | 1 | 2 | 1 | 2 | 1 | 2 | 1 | 2 | 1 | 2 | 1 | 2 | 1 | 2 |
| 18 | 2 | 1 | 2 | 1 | 2 | 1 | 2 | 1 | 2 | 1 | 2 | 1 | 2 | 1 | 2 | 2 | 1 | 2 | 1 | 2 | 1 | 2 | 1 | 2 | 1 | 2 | 1 | 2 | 1 | 2 | 1 |
| 19 | 2 | 1 | 2 | 1 | 2 | 1 | 2 | 2 | 1 | 2 | 1 | 2 | 1 | 2 | 1 | 1 | 2 | 1 | 2 | 1 | 2 | 1 | 2 | 2 | 1 | 2 | 1 | 2 | 1 | 2 | 1 |
| 20 | 2 | 1 | 2 | 1 | 2 | 1 | 2 | 2 | 1 | 2 | 1 | 2 | 1 | 2 | 1 | 2 | 1 | 2 | 1 | 2 | 1 | 2 | 1 | 1 | 2 | 1 | 2 | 1 | 2 | 1 | 2 |
| 21 | 2 | 1 | 2 | 2 | 1 | 2 | 1 | 1 | 2 | 1 | 2 | 2 | 1 | 2 | 1 | 1 | 2 | 1 | 2 | 2 | 1 | 2 | 1 | 1 | 2 | 1 | 2 | 2 | 1 | 2 | 1 |
| 22 | 2 | 1 | 2 | 2 | 1 | 2 | 1 | 1 | 2 | 1 | 2 | 2 | 1 | 2 | 1 | 2 | 1 | 2 | 1 | 1 | 2 | 1 | 2 | 2 | 1 | 2 | 1 | 1 | 2 | 1 | 2 |
| 23 | 2 | 1 | 2 | 2 | 1 | 2 | 1 | 2 | 1 | 2 | 1 | 1 | 2 | 1 | 2 | 1 | 2 | 1 | 2 | 2 | 1 | 2 | 1 | 2 | 1 | 2 | 1 | 1 | 2 | 1 | 2 |
| 24 | 2 | 1 | 2 | 2 | 1 | 2 | 1 | 2 | 1 | 2 | 1 | 1 | 2 | 1 | 2 | 2 | 1 | 2 | 1 | 1 | 2 | 1 | 2 | 1 | 2 | 1 | 2 | 2 | 1 | 2 | 1 |
| 25 | 2 | 2 | 1 | 1 | 2 | 2 | 1 | 1 | 2 | 2 | 1 | 1 | 2 | 2 | 1 | 1 | 2 | 2 | 1 | 1 | 2 | 2 | 1 | 1 | 2 | 2 | 1 | 1 | 2 | 2 | 1 |
| 26 | 2 | 2 | 1 | 1 | 2 | 2 | 1 | 1 | 2 | 2 | 1 | 1 | 2 | 2 | 1 | 2 | 1 | 1 | 2 | 2 | 1 | 1 | 2 | 2 | 1 | 1 | 2 | 2 | 1 | 1 | 2 |
| 27 | 2 | 2 | 1 | 1 | 2 | 2 | 1 | 2 | 1 | 1 | 2 | 2 | 1 | 1 | 2 | 1 | 2 | 2 | 1 | 1 | 2 | 2 | 1 | 2 | 1 | 1 | 2 | 2 | 1 | 1 | 2 |
| 28 | 2 | 2 | 1 | 1 | 2 | 2 | 1 | 2 | 1 | 1 | 2 | 2 | 1 | 1 | 2 | 2 | 1 | 1 | 2 | 2 | 1 | 1 | 2 | 1 | 2 | 2 | 1 | 1 | 2 | 2 | 1 |
| 29 | 2 | 2 | 1 | 2 | 1 | 1 | 2 | 1 | 2 | 2 | 1 | 2 | 1 | 1 | 2 | 1 | 2 | 2 | 1 | 2 | 1 | 1 | 2 | 1 | 2 | 2 | 1 | 2 | 1 | 1 | 2 |
| 30 | 2 | 2 | 1 | 2 | 1 | 1 | 2 | 1 | 2 | 2 | 1 | 2 | 1 | 1 | 2 | 2 | 1 | 1 | 2 | 1 | 2 | 2 | 1 | 2 | 1 | 1 | 2 | 1 | 2 | 2 | 1 |
| 31 | 2 | 2 | 1 | 2 | 1 | 1 | 2 | 2 | 1 | 1 | 2 | 1 | 2 | 2 | 1 | 1 | 2 | 2 | 1 | 2 | 1 | 1 | 2 | 2 | 1 | 1 | 2 | 1 | 2 | 2 | 1 |
| 32 | 2 | 2 | 1 | 2 | 1 | 1 | 2 | 2 | 1 | 1 | 2 | 1 | 2 | 2 | 1 | 2 | 1 | 1 | 2 | 1 | 2 | 2 | 1 | 1 | 2 | 2 | 1 | 2 | 1 | 1 | 2 |

**TABLE IX.18** SELECTED ORTHOGONAL ARRAY TABLES (continued)

Interaction Table

Column

| Column | 1 | 2 | 3 | 4 | 5 | 6 | 7 | 8 | 9 | 10 | 11 | 12 | 13 | 14 | 15 | 16 | 17 | 18 | 19 | 20 | 21 | 22 | 23 | 24 | 25 | 26 | 27 | 28 | 29 | 30 | 31 |
|---|---|---|---|---|---|---|---|---|---|---|---|---|---|---|---|---|---|---|---|---|---|---|---|---|---|---|---|---|---|---|---|
| 1 | (1) | 3 | 2 | 5 | 4 | 7 | 6 | 9 | 8 | 11 | 10 | 13 | 12 | 15 | 14 | 17 | 16 | 19 | 18 | 21 | 20 | 23 | 22 | 25 | 24 | 27 | 26 | 29 | 28 | 31 | 30 |
| 2 | | (2) | 1 | 6 | 7 | 4 | 5 | 10 | 11 | 8 | 9 | 14 | 15 | 12 | 13 | 18 | 19 | 16 | 17 | 22 | 23 | 20 | 21 | 26 | 27 | 24 | 25 | 30 | 31 | 28 | 29 |
| 3 | | | (3) | 7 | 6 | 5 | 4 | 11 | 10 | 9 | 8 | 15 | 14 | 13 | 12 | 19 | 18 | 17 | 16 | 23 | 22 | 21 | 20 | 27 | 26 | 25 | 24 | 31 | 30 | 29 | 28 |
| 4 | | | | (4) | 1 | 2 | 3 | 12 | 13 | 14 | 15 | 8 | 9 | 10 | 11 | 20 | 21 | 22 | 23 | 16 | 17 | 18 | 19 | 28 | 29 | 30 | 31 | 24 | 25 | 26 | 27 |
| 5 | | | | | (5) | 3 | 2 | 13 | 12 | 15 | 14 | 9 | 8 | 11 | 10 | 21 | 20 | 23 | 22 | 17 | 16 | 19 | 18 | 29 | 28 | 31 | 30 | 25 | 24 | 27 | 26 |
| 6 | | | | | | (6) | 1 | 14 | 15 | 12 | 13 | 10 | 11 | 8 | 9 | 22 | 23 | 20 | 21 | 18 | 19 | 16 | 17 | 30 | 31 | 28 | 29 | 26 | 27 | 24 | 25 |
| 7 | | | | | | | (7) | 15 | 14 | 13 | 12 | 11 | 10 | 9 | 8 | 23 | 22 | 21 | 20 | 19 | 18 | 17 | 16 | 31 | 30 | 29 | 28 | 27 | 26 | 25 | 24 |
| 8 | | | | | | | | (8) | 1 | 2 | 3 | 4 | 5 | 6 | 7 | 24 | 25 | 26 | 27 | 28 | 29 | 30 | 31 | 16 | 17 | 18 | 19 | 20 | 21 | 22 | 23 |
| 9 | | | | | | | | | (9) | 3 | 2 | 5 | 4 | 7 | 6 | 25 | 24 | 27 | 26 | 29 | 28 | 31 | 30 | 17 | 16 | 19 | 18 | 21 | 20 | 23 | 22 |
| 10 | | | | | | | | | | (10) | 1 | 6 | 7 | 4 | 5 | 26 | 27 | 24 | 25 | 30 | 31 | 28 | 29 | 18 | 19 | 16 | 17 | 22 | 23 | 20 | 21 |
| 11 | | | | | | | | | | | (11) | 7 | 6 | 5 | 4 | 27 | 26 | 25 | 24 | 31 | 30 | 29 | 28 | 19 | 18 | 17 | 16 | 23 | 22 | 21 | 20 |
| 12 | | | | | | | | | | | | (12) | 1 | 2 | 3 | 28 | 29 | 30 | 31 | 24 | 25 | 26 | 27 | 20 | 21 | 22 | 23 | 16 | 17 | 18 | 19 |
| 13 | | | | | | | | | | | | | (13) | 3 | 2 | 29 | 28 | 31 | 30 | 25 | 24 | 27 | 26 | 21 | 20 | 23 | 22 | 17 | 16 | 19 | 18 |
| 14 | | | | | | | | | | | | | | (14) | 1 | 30 | 31 | 28 | 29 | 26 | 27 | 24 | 25 | 22 | 23 | 20 | 21 | 18 | 19 | 16 | 17 |
| 15 | | | | | | | | | | | | | | | (15) | 31 | 30 | 29 | 28 | 27 | 26 | 25 | 24 | 23 | 22 | 21 | 20 | 19 | 18 | 17 | 16 |
| 16 | | | | | | | | | | | | | | | | (16) | 1 | 2 | 3 | 4 | 5 | 6 | 7 | 8 | 9 | 10 | 11 | 12 | 13 | 14 | 15 |
| 17 | | | | | | | | | | | | | | | | | (17) | 3 | 2 | 5 | 4 | 7 | 6 | 9 | 8 | 11 | 10 | 13 | 12 | 15 | 14 |
| 18 | | | | | | | | | | | | | | | | | | (18) | 1 | 6 | 7 | 4 | 5 | 10 | 11 | 8 | 9 | 14 | 15 | 12 | 13 |
| 19 | | | | | | | | | | | | | | | | | | | (19) | 7 | 6 | 5 | 4 | 11 | 10 | 9 | 8 | 15 | 14 | 13 | 12 |
| 20 | | | | | | | | | | | | | | | | | | | | (20) | 1 | 2 | 3 | 12 | 13 | 14 | 15 | 8 | 9 | 10 | 11 |
| 21 | | | | | | | | | | | | | | | | | | | | | (21) | 3 | 2 | 13 | 12 | 15 | 14 | 9 | 8 | 11 | 10 |
| 22 | | | | | | | | | | | | | | | | | | | | | | (22) | 1 | 14 | 15 | 12 | 13 | 10 | 11 | 8 | 9 |
| 23 | | | | | | | | | | | | | | | | | | | | | | | (23) | 15 | 14 | 13 | 12 | 11 | 10 | 9 | 8 |
| 24 | | | | | | | | | | | | | | | | | | | | | | | | (24) | 1 | 2 | 3 | 4 | 5 | 6 | 7 |
| 25 | | | | | | | | | | | | | | | | | | | | | | | | | (25) | 3 | 2 | 5 | 4 | 7 | 6 |
| 26 | | | | | | | | | | | | | | | | | | | | | | | | | | (26) | 1 | 6 | 7 | 4 | 5 |
| 27 | | | | | | | | | | | | | | | | | | | | | | | | | | | (27) | 7 | 6 | 5 | 4 |
| 28 | | | | | | | | | | | | | | | | | | | | | | | | | | | | (28) | 1 | 2 | 3 |
| 29 | | | | | | | | | | | | | | | | | | | | | | | | | | | | | (29) | 3 | 2 |
| 30 | | | | | | | | | | | | | | | | | | | | | | | | | | | | | | (30) | 1 |
| 31 | | | | | | | | | | | | | | | | | | | | | | | | | | | | | | | (31) |

**TABLE IX.18    SELECTED ORTHOGONAL ARRAY TABLES** (*continued*)

## Linear Graphs

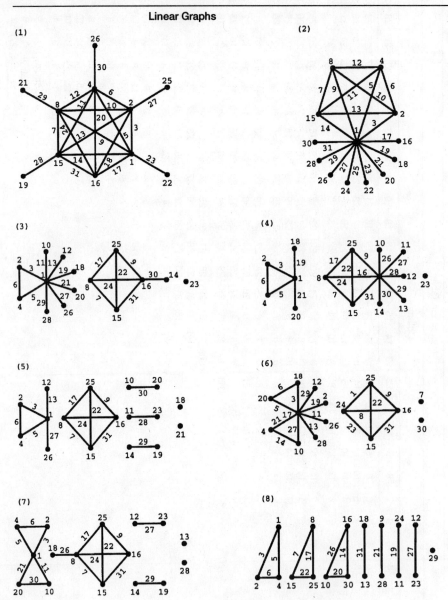

TABLE IX.18    SELECTED ORTHOGONAL ARRAY TABLES (*continued*)

**Linear Graphs**

(9)

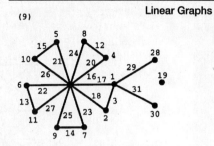

(10)

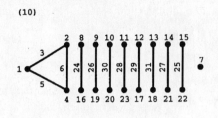

(11)

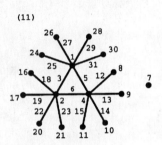

(12)

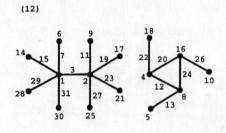

(13)

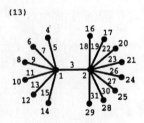

# INDEXES

# AUTHOR INDEX

# SUBJECT INDEX